ELECTRICITY AND MAGNETISM IN BIOLOGY AND MEDICINE

Martin Blank, Ph.D.

Editor

Review and research papers presented at
The First World Congress for
Electricity and Magnetism in Biology and Medicine
Orlando, Florida, 1992

San Francisco Press, Inc.
Box 426800, San Francisco, CA 94142-6800, USA

Printed in the U.S.A.

ISBN 0-911302-67-0

Library of Congress Catalog Card Number 92-060738

EDITOR'S PREFACE

We have come a long way since our intial discussions at the Biochemistry Gordon Conference in June 1986. At that time, representatives of the Bioelectromagnetics Society (BEMS), Biological Repair and Growth Society (BRAGS), and Bioelectrochemical Society (BES)—the European Bioelectromagnetics Association (EBEA) was not yet in existence, but some of its current members were present—formalized cooperative arrangments that eventually culminated in the First World Congress for Electricity and Magnetism in Biology and Medicine. The Congress was organized by representatives of these four societies, with the participation of the EMB and PE Societies of the Institute of Electrical and Electronics Engineers, and Commission K of URSI; S. R. Pollack served as President, B. Greenbaum as Vice President, K. C. Hoffman as Secretary-Treasurer, the undersigned as Board member, and W. G. Wisecup as Executive Director of the Congress Governing Board. The Congress took place during 14-19 June 1992 in Orlando, Fla., with a full program of scientific, social, and Society activities.

The scientific activities of the Congress are summarized in this volume, which inculdes both invited and contributed papers. The arrangement is approximately in the order set by the Program Committee, starting with the five plenary symposia (organized on scientific, medical, environmental, and biotechnological issues), followed by three special symposia: Electroporation and Electrofusion (organized by J. Weaver), Molecular Electronics (organized by F. Hong), and EMF issues for Electrified Transportation Systems (organized by A. Brecher, I. Gyuk, and M. Nakagawa). The major part of the volume is devoted to the contributed lectures and poster presentations grouped according to the program topics.

To accommodate the large number of presentations in a single volume, individual papers have been kept short. The result is a birds-eye view of research in 1992 around the world on various aspects of our rapidly developing field. The authors have provided references to more detailed treatments.

Both the Congress and this volume were the result of the efforts of many individuals, to whom we express our thanks. For this volume, I should like to thank especially the members of the editorial board—C. A. L. Bassett, R. Cadossi, O. P. Gandhi, B. F. Sisken, and B. Veyret—for their help in reviewing the manuscripts; and to W. G. Wisecup and the staff at W/L Associates, and to my secretary, K. Winkleman, for their cooperation.

Martin Blank, Ph.D.
Editor

TABLE OF CONTENTS

PLENARY SYMPOSIA

Electromagnetic Fields in Modern Society

Page

Beneficial Effects of Electromagnetic Fields: Medical Devices and Technology

Biological Effects of EMF: Basic Sciences

Environmental Issues

Recent Advances in Science and Techology

SYMPOSIA

Electroporation and Electrofusion

(Organized by J. Weaver)

Molecular Electronics

(Organized by F. Hong)

EMF Issues for Electrified Transportation System

(Organized by A. Brecher, I. Gyuk, and M. Nakagawa)

SCIENCE POLICY

CALCIUM AND SIGNAL TRANSDUCTION

EMBRYOLOGY AND DEVELOPMENT

CANCER-RELATED STUDIES

EPIDEMIOLOGY

MEMBRANE TRANSPORT

GENE EXPRESSION AND MUTAGENESIS

THEORY AND MODELS

HIGH-FREQUENCY FIELDS

CELL GROWTH AND PROLIFERATION

IMMUNE SYSTEMS

HIGH-POWER MICROWAVES

ELECTROMECHANICAL TRANSDUCTION

NERVE REGENERATION

SOFT TISSUE AND ISCHEMIA

MAGNETIC AND ELECTRIC NEUROSTIMULATION AND IMAGING

IN VIVO STUDIES

[*See also* pp. 925, 927, 930]

EXPOSURE ASSESSMENT

INSTRUMENTATION, EXPOSURE SYSTEMS

IN VITRO STUDIES

ADDITIONAL PAPERS

Martin Blank, Ed., *Electricity and Magnetism in Biology and Medicine*

CELLULAR AND MOLECULAR PATHWAYS OF EXTREMELY LOW FREQUENCY ELECTROMAGNETIC FIELD INTERACTIONS WITH LIVING SYSTEMS

T.S. Tenforde*

Pacific Northwest Laboratories, Richland, Wash.

There is growing evidence that environmental electric and magnetic fields in the extremely low frequency (ELF) band below 300 Hz can influence biological functions by mechanisms that are only poorly understood at the present time. The primary objectives of this paper are to review the physical properties of ELF fields; their interactions with living systems at the tissue, cellular, and subcellular levels; and the key role of cell membranes in the transduction of signals from imposed ELF fields. Topics of discussion include signal-to-noise ratios for single cells and cell aggregates, resonance phenomena involving a combination of static and ELF magnetic fields, and the possible influence of ELF fields on molecular signaling pathways that involve membrane receptors and cytoplasmic second messengers.

The continuing controversy over the possible health effects of environmental levels of ELF fields has underscored the need to gain a fundamental understanding of the pathways by which these fields interact with living systems. The weak 50/60Hz electric and magnetic fields to which humans are commonly exposed in the home or workplace, which typically range from 10 to 50 V/m and 0.1 to 0.3 μT, induce electric fields within the body of the order of 5 μV/m or less. These field levels are several orders of magnitude below the field intensities that generally result in reproducible biological effects in a laboratory setting. A major challenge in both laboratory research and biophysical modeling is to elucidate the mechanisms by which living cells and organized tissues could exhibit a response to weak ELF fields at signal levels comparable to or below the intrinsic physical and biological electric noise present in living systems.

PHYSICAL PROPTERTIES OF ELF FIELDS

ELF fields in tissue have a long wavelength (~ 1000 m at 60 Hz) and skin depth (~ 150 m at 60 Hz), as a result of which these fields behave as though they were composed of independent, quasistatic electric and magnetic field components.[1] As a consequence, the radiating properties of ELF fields can be neglected in their interactions with tissue. Another important property of ELF fields is their extremely small energy. For example, the energy of a 60Hz photon is 2.5×10^{-13} eV, which is smaller by 11 orders of magnitude than the Boltzmann thermal energy kT ($=2.7\times10^{-2}$ eV at 310 K), and by 14 orders of magnitude than the energy required to break a chemical bond. Substantial laboratory evidence supports the physical expectation that ELF fields do not disrupt the chemical bonds in DNA, proteins, or other biological molecules. Direct genotoxic effects of ELF fields leading to cell death, gene mutation, or neoplastic transformation would therefore not be expected, and have, in fact, not been observed.

A third important feature of ELF fields applied to living organisms through air is the nonthermal nature of their interactions. The highest field in tissue that can be induced by an ELF field applied through air is about 1 V/m, which leads to a specific energy absorption rate of 10^{-4} W/kg. This rate of energy deposition is smaller by four orders of magnitude than the body's basal metabolic rate and produces a negligible rate of temperature rise (about 3×10^{-8} C/s). The interactions of ELF fields applied to the body through air are therefore of a nonthermal nature.

*Correspondence to: T. S. Tenforde, Life Sciences Ctr (K1-50), Pacific Northwest Laboratories, Richland WA 99352.

BIOLOGICAL EFFECTS AND MEMBRANE INTERACTIONS

Despite the minimal perturbations of molecular structure by environmental levels of ELF fields, there is abundant evidence for responses to these fields at the tissue and cellular levels.[2-5] For example, small functional changes in excitable tissues and neuroendocrine alterations have been reported in response to ELF fields that induce tissue voltage gradients ≤10 mV/m. These alterations include changes in evoked brain potentials,[6] heart rate,[6] and nocturnal synthesis of pineal melatonin.[7,8] In addition, a large number of cellular phenomena, including alterations in growth rate, gene expression, and macromolecular synthesis, have been reported to occur in response to fields of moderate to weak intensity.[5] These effects include alterations in biosynthesis of specific messenger RNA and proteins at field levels below 1 mV/m.[9] Finally, there is a rapidly growing body of information that implicates the cell membrane as a primary site of ELF field interactions.[2,5,10,11] A wide variety of cell-membrane structural and functional properties have been reported to be altered in response to ELF fields, including changes in Ca^{++} binding to anionic fixed charges at the cell surface (e.g., sialic acid residues of membrane glycoproteins), changes in the transport of ions such as Ca^{++} and the secretion of small solutes such as insulin, and alterations in ligand-receptor interactions that trigger changes in the biosynthetic and functional states of cells. The threshold tissue field levels that lead to such effects appear to vary widely depending on the end point studied, but in nearly all cases are smaller than 0.1 V/m. In the specific case of field-induced Ca^{++} desorption from fixed-charge sites on the membrane surface, the effective field level in tissue has been reported to be as low as 1-10 μV/m.[12,13]

The phospholipid bilayer that forms the structural matrix in membranes of living cells is an electrical insulator, and the membrane electrical conductivity is smaller by about 5 orders of magnitude than that of the extracellular medium or the cytoplasm. As a result, the membrane of a living cell forms an excellent electrical barrier, as well as a superb chemical barrier, that mediates cellular interactions with the external environment. For this reason, it is generally believed that cellular responses to weak ELF fields are initiated by membrane interactions that serve as the primary mechanism of field transduction. Under typical exposure conditions with induced ELF fields in the extracellular medium below 1 V/m, the "leakage" field in the cell cytoplasm is less than 1 μV/m. This conclusion also holds for the circulating electric field induced directly in the cell cytoplasm by magnetic induction. For example, a sinusoidal 60Hz, 0.1mT field induces a maximum electric field of 0.2 μV/m in the cytoplasm of a cell with a 10μm radius. These considerations reinforce the importance of the cell membrane in ELF signal reception and transduction.

Another point to be made regarding the role of cell membranes in ELF signal transduction is the amplification of the extracellular field that occurs across the membrane. By solving Maxwell's equations for the specific case of a dielectric shell (membrane) surrounding a spherical conductor (cytoplasm), one can easily show that the electric field across the membrane is greater than that in the extracellular medium by a factor of $1.5\,R/d$, where R is the cell radius and d is the membrane thickness. For a spherical cell with a radius of 10 μm and a membrane thickness of 5 nm, the field across the membrane is therefore predicted to be 3000 times greater than that in the extracellular medium. A weak environmental field that induces a voltage gradient of 5 μV/m in the extracellular fluid thus produces a field of about 15 mV/m across the cell membrane.

ELECTRICAL NOISE IN BIOMEMBRANES

Several physical and biological sources of electrical noise within the cell membrane may impose a lower limit on the strength of an ELF field that can be recognized as a coherent signal.[14] The four major sources of electrical noise in biological membranes are: (1) Johnson-Nyquist thermally generated electrical noise, which produces a 3μV transmembrane voltage shift at physiological temperatures; (2) $1/f$ noise associated with ion current flows through membrane channels, which typically produces a 10μV transmembrane voltage shift; (3) "shot" noise, which results from the discrete nature of ionic charge carriers and is a minor source of membrane electrical noise; and (4) endogenous biological background fields produced by electrically active organs such as the heart, muscles, and the nervous system, which can exceed the contribution of physical noise sources by an order of magnitude or more.

In the presence of these various sources of membrane noise, which produce time-averaged fields greater than 100 V/m across the cell membrane, it is therefore of interest to explore the minimum strength of induced electric fields in tissue that can achieve a signal-to-noise ratio greater than 1 within the cell membrane. For example, Weaver and Astumian arrived at an estimate of approximately 0.1 V/m as the smallest applied electric field that exceeds the Johnson-Nyquist noise signal in the membrane of a single cell with an elongated cylindrical geometry (such as a fibroblast, neuron, or muscle cell).[15] Larger threshold field levels, greater than 1 V/m, were predicted for small spherical cells such as lymphocytes. Further increases in the threshold field level are expected if other sources of electrical noise within the cell membrane are taken into account.

These calculations of signal-to-noise ratio consider only the transmembrane electric potential shift introduced by an extracellular field. It has been argued rather convincingly that the field and noise sources that are most relevant to biological signal transduction are those that reside within the highly charged electrical double layer that exists at the cell surface. This double layer, frequently referred to as the Helmholtz-Stern layer, is comprised of fixed anionic charges on the outer membrane surface and diffusible cations in the surrounding medium. As described originally by Debye and Hückel, the average thickness of the diffuse electrical double layer at cell surfaces is approximately 0.8 nm in a physiological medium.[16] By considering Johnson-Nyquist noise and other sources of electrical noise within the double layer, one can conclude that the minimum electric field required in the extracellular medium to achieve a signal-to-noise ratio greater than 1 is about 10 mV/m.

Another important factor to be considered in calculating the threshold field level that exceeds intrinsic electrical noise in biological membranes is the junctional coupling that occurs between cells in organized tissues. For an aggregate of N cells with electrically coupled membranes, the threshold field level to achieve a signal-to-noise ratio of 1 is lower than the threshold for single cells by approximately $N^{-5/6}$ as a result of two factors: (1) field amplification due to the larger size of the aggregate, and (2) reduction of membrane voltage noise as a result of the larger capacitance of the aggregate.[17] For example, if an aggregate of 10^6 electrically coupled cells is considered, the threshold level to achieve a signal-to-noise ratio of one is predicted to be lower by 100000 than the threshold for a single cell. Because of the junctional coupling exhibited by most biological tissues, the threshold field to achieve membrane and cellular responses may therefore be of the order of 1 μV/m, as contrasted with the value of about 0.1 V/m or higher predicted for single cells. A similar conclusion has been reached by Pilla from an electrical network model of cells that communicate electrically via junctional coupling.[18]

BIOLOGICAL SIGNAL TRANSDUCTION PATHWAYS

The most singly important issue in understanding the pathways by which weak ELF signals could influence membrane and cellular functions is the elucidation of mechanisms by which these fields are transduced within the cell membrane. One approach that has been taken in addressing this question is to consider the intricate biochemical pathways that have evolved as a mechanism by which a living cell communicates with its extracellular environment. As illustrated in Fig. 1, the binding of a single molecule of a mitogenic substance to a specific receptor within the membrane triggers a cascade of events that involve conformational shifts in membrane-associated proteins. These events in turn lead to signal transduction and amplification via the production of cytoplasmic second messengers and internal effectors such as free Ca^{++} and protein phosphorylases (kinases) that regulate DNA transcription and protein biosynthesis.[19,20] The end result of a single mitogen binding event at the membrane surface is thus a cytoplasmic signal that is amplified to a level that can produce robust effects on macromolecular synthesis and cellular responses involving significant changes in functional and proliferative states.

The interaction of ELF fields with biological mechanisms could in principle lead to alterations in each component of this elegant signaling process that occurs in living cells. A useful working hypothesis, illustrated in Fig. 2, is that the pericellular fields and currents induced by an applied ELF field initiate electrochemical events within the cell membrane that are important elements of the primary signal transduction and amplification process. These biochemically mediated events then produce cytoplasmic second messenger responses that trigger changes in the biosynthesis of macromolecules and alterations in cellular growth, differentiation, and functional properties.

During the past decade, a growing body of experimental evidence has been acquired that supports this general picture of the sequence of events leading to ELF signal transduction and amplification at the cellular level. For example, it has been demonstrated that pulsed electromagnetic fields (PEMF) with ELF repetition frequencies inhibit the production of cAMP by bone cells in response to the binding of parathyroid hormone (PTH) to surface receptors.[21] Further studies have shown that the PEMF action leads to inability of the PTH-receptor complex to activate the alpha subunit of G protein, thereby interfering with the sequence of events that lends to activation of adenylate cyclase at the cytoplasmic membrane interface.[22] Other studies using human lymphocytes have shown that exposure to microwave fields with amplitude modulation at ELF frequencies leads to the inhibition of non-cAMP-dependent histone kinases.[23]

1

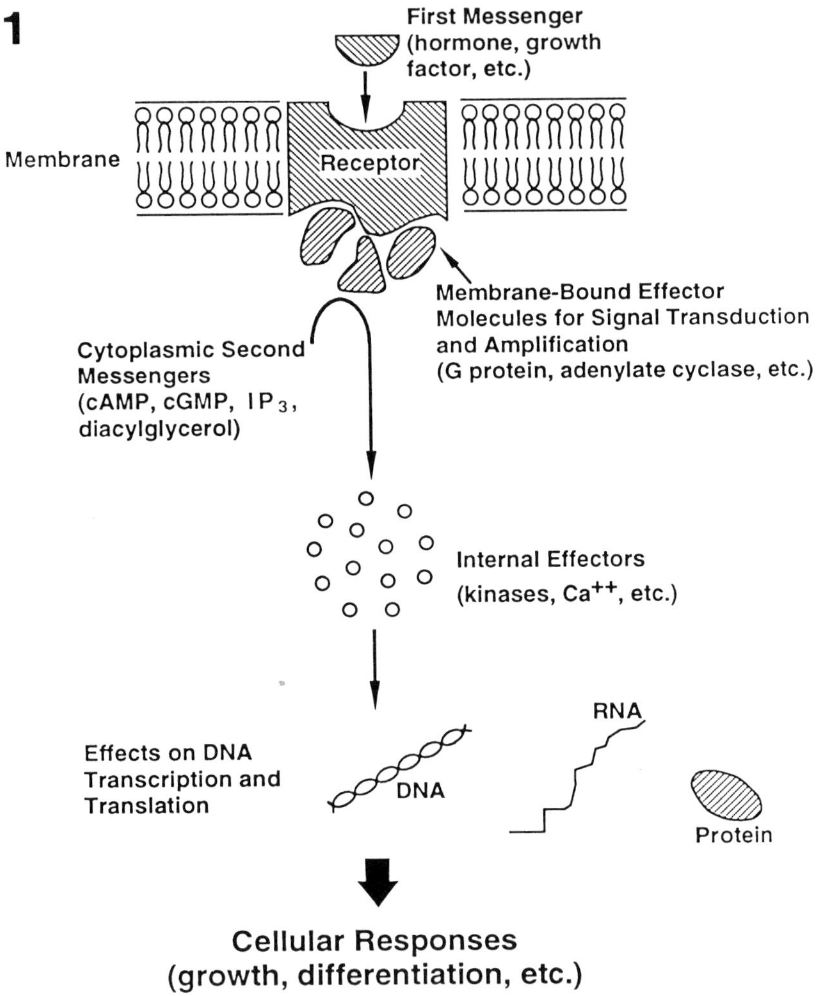

2

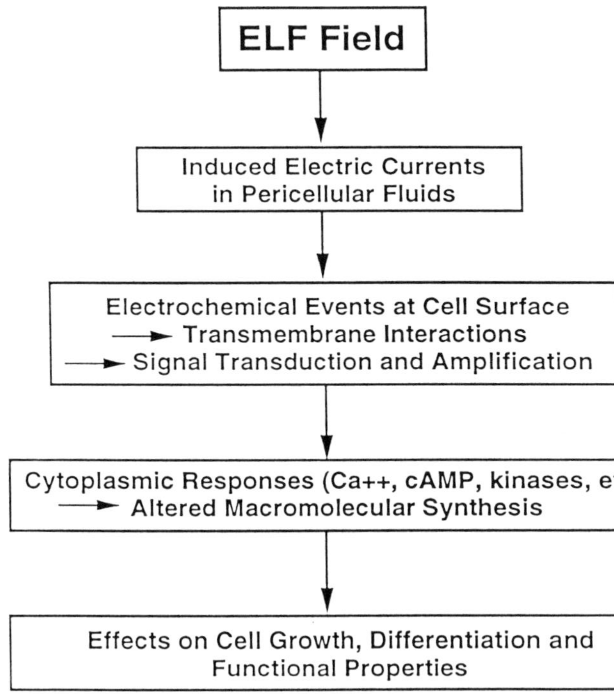

FIG. 1.—Biological signal transduction pathways involving first- and second-messenger systems. Abbreviations—cAMP: cyclic adenosine monophosphate; cGMP: cyclic guanosine monophosphate; IP_3: triphosphoinositol (which releases Ca^{++} from intracellular stores).

FIG. 2.—Hypothesized sequence of membrane-mediated events leading to cellular and tissue responses to applied ELF fields.

The possible effects of ELF fields on the kinase-C signaling pathway are of particular interest because this pathway is known to be activated by the binding of tumor promoters such as phorbol esters. Activation of this pathway by the binding of a first messenger leads to a cascade of events that produce activated kinase-C and free cytosolic Ca^{++} ions. Recent studies have demonstrated a significant 70% elevation of kinase-C activity in human HL-60 cells exposed to a 50Hz pulsed magnetic field.[24] The field effect was considerably damped by the addition of EGTA (a Ca^{++} chelator) to the medium. This observation is consistent with previous findings that kinase-C activation relies on the mobilization of Ca^{++} ions. Another relevant study is the recent finding that stimulation of Ca^{++} uptake into rat thymocytes by the plant lectin Concanavalin A (Con-A) is significantly augmented by exposure to a sinusoidal 60Hz magnetic field.[25] In lymphoid cells the binding of Con-A to surface receptors triggers cytoplasmic signaling events involved in the kinase-C pathway, including the activation of kinase-C and an elevation of the cytosolic Ca^{++} concentration. Enhancement of the Con-A effect by a 60Hz magnetic field was shown to depend on the strength of the electric field induced in the extracellular medium.

An in vivo study of particular interest in the context of possible ELF field effects on the kinase-C signaling pathway is the recent finding by Stuchly et al. that a 60Hz, 2mT magnetic field exerts a co-promoting effect on mouse skin carcinogenesis.[26] In these experiments, skin tumors were initiated by the topical application of 10 nanomole of the carcinogen 7,12-dimethylbenzanthracene (DMBA), and were then promoted by the application of 4.9 nanomole of the phorbol ester 12-O-tetradecanoylphorbol-13-acetate (TPA), once per week for a total of 23 weeks. In the group of mice that received DMBA and TPA plus exposure to the 60Hz field for 6h/day, 5 days/week throughout the tumor promotion phase, both the percentage of mice with tumors and the number of tumors per mouse increased more rapidly with time than in the group of mice that received DMBA plus TPA alone. For example, at week 18 the percentage of mice with tumors in these two groups were, respectively, 25% and 8% and the mean number of tumors per mouse were 1.90 ± 0.69 (SEM) and 0.65 ± 0.46. By week 23 the differences between the mice exposed to the magnetic field and the nonexposed group were no longer statistically significant. Other studies by the same group of investigators have shown that a 60Hz, 2mT magnetic field acting alone on DMBA-initiated skin cells has no tumor-promoting effect.[27] It would be of considerable interest to conduct further experiments in which comparative measurements are made of kinase-C activity in the affected skin cells throughout the tumor promotion phase in the field-exposed and nonexposed groups of mice. Such experiments would provide a test of the hypothesis that ELF fields may exert a co-promoting effect on tumor development through an influence on the kinase-C signaling pathway activated by phorbol esters.

Recent studies of rat mammary tumor development following initiation with the carcinogen nitrosomethylurea (NMU) suggest that ELF fields may also exert direct tumor-promoting activity.[28] In these experiments rats were injected intravenously with a 50mg/kg dose of NMU that produced mammary tumors in 54% of the rats. This percentage increased to 86% in a group of rats initiated with NMU and exposed to a 50Hz, 0.2mT magnetic field for 3 h/day for 5 weeks. The latency time to tumor development was also found to decrease significantly in the rats exposed to both NMU and the magnetic field relative to rats that received only NMU. These results are consistent with a tumor-promoting effect of the field, but could also be related to a field-induced suppression of pineal melatonin synthesis and a resultant elevation in breast cancer risk.[29,30] Further research on the effects of ELF fields on chemically induced mammary tumors in rodents may provide useful insights into the pathways by which these fields could exert carcinogenic effects.

PHYSICAL MECHANISMS OF ELF FIELD INTERACTIONS

In the search for mechanisms by which extremely weak ELF fields could exert significant biological effects, a number of innovative models have been proposed over the course of the last two decades. One class of models that has received particular attention during the past several years are resonance models that involve the combined action of an ELF field and the static geomagnetic field. For example, ion cyclotron resonance (ICR) was proposed by Liboff in 1985 as a possible mechanism that could facilitate the transport of ions such as Ca^{++} through membrane channels in the presence of the geomagnetic field and a weak ELF field tuned to the ICR frequency.[31] Although some data on Ca^{++} uptake by lymphocytes[32] and diatoms[33] have been cited as lending support to this model, there are also negative experimental findings.[34,35] In addition, a number of physical arguments can be raised against the ICR model,[5] the most serious of which is the collisional damping of resonant ion motion that is expected to occur in a condensed phase.[36]

Two other resonance models that have been proposed recently are the "quantum beats" model of Lednev,[37] in which combined static and ELF fields affect vibrational energy levels and transition probabilities of bound ions (e.g., Ca^{++} ions bound to calmodulin); and the model of Zhadin and Fesenko,[38] in which the rotational energy levels of bound ions are pumped by combined static and ELF fields. Shuvalova et al. have presented data on the effect of combined fields on the rate of calmodulin-dependent phosphorylation of myosin that appear to support the predictions of the Lednev model.[39] However, a recent study in which an optical technique was used to study Ca^{++} binding to calmodulin and to metallochromic dyes failed to find any effects of combined static and time-varying fields under the resonance conditions predicted by Lednev's model.[40] Adair has pointed out that the Lednev model of resonant field effects is improbable because of the long lifetime of the excited vibrational states (about 8 s), during which de-excitation would occur as a result of collisional damping.[41] Similar arguments can be raised against the type of resonant field effects envisioned in the Zhadin and Fesenko model. In addition, the rate of transition of a bound ion to an excited state predicted by this model is so low that a transition would occur only once in several months at typical environmental magnetic field levels.

A large number of nonequilibrium models have been proposed in which field-induced structural and functional perturbations result from membrane interactions that exploit the existence of unstable or metastable states. Examples of such interactions are dissipative instabilities involving cooperative transitions of allosteric membrane proteins in the presence of an applied field. Critical-state instabilities in membrane physical properties near the phase transition temperature can also be amplified by an applied electromagnetic field. A host of other models involve coherent field interactions that lead to the excitation of oscillating dipolar modes in membrane proteins or the production of nonlinear oscillations (solitons) that facilitate vibrational energy transfer in macromolecular structures. Several reviews discussing the physical basis of these nonequilibrium models have been published.[2,10,42-44]

The discovery of biogenic magnetite particles in the tissues of a large number of organisms, including several mammalian species,[45,46] has led to speculation that oscillatory magnetomechanical forces and torques on these particles could provide a mechanism for the transduction of signals from weak ELF magnetic fields. Of particular interest is the recent demonstration of magnetite crystals in various anatomic locations within the human brain.[47] Kirschvink et al. have proposed a model in which oscillatory magnetic forces on magnetite particles at ELF frequencies are visualized as producing the opening and closing of pressure-sensitive ion channels in membranes.[48] The 60Hz field level required to overcome the effects of Brownian motion is predicted from this theoretical model to be of the order of 0.1 mT, which is within the range of local fields at locations close to the surfaces of several types of household appliances and machine tools.

One difficulty with this model is the scarcity of magnetite crystals relative to the number of cells in the brain tissue. For example, human brain tissue is reported to contain a few million magnetite crystals per gram, distributed in $5\text{-}10 \times 10^5$ discrete clusters.[47] The number of cells in brain tissue thus exceeds the number of magnetite crystals by a factor of about 100. It is therefore difficult to envision how oscillating magnetomechanical interactions of an ELF field with magnetite crystals could affect a significant number of pressure-sensitive ion channels in the brain. However, the effects of such interactions on neural signaling in localized brain regions could possibly result in a biological response, although there is no evidence at present to support this hypothesis. Further studies are clearly needed to reveal the biological role of magnetite and the possible mechanisms through which this mineral could play a role in ELF signal transduction.

CONCLUDING REMARKS

Evidence is mounting for a central role of cell membranes in the reception, transduction, and amplification of signals imposed by ELF fields. A major challenge for the future will be the elucidation of specific molecular pathways through which these fields can influence transmembrane signaling events and affect the functional and proliferative states of cells and organized tissues. Further research is also needed to gain an understanding of the ELF signal parameters above which predictable biological responses occur. Recent laboratory studies have provided a number of clues on the pathways through which ELF fields may operate at the cellular and subcellular levels. However, a great deal of research lies ahead in order to characterize fully the molecular substrates of ELF field interactions and the resultant cascade of electrical and biochemical signals that lead to cellular and tissue responses.[49]

REFERENCES

1. T. S. Tenforde, "Biological interactions of extremely-low-frequency electric and magnetic fields," *Bioelectrochem. Bioenerg.* 25: 1, 1991.

2. W. R. Adey, "Nonlinear electrodynamics in cell membrane transductive coupling," in R. C. Aloia, C. C. Curtain, and L. M. Gordon, Eds., *Membrane Transport and Information Storage,* vol. 4, New York: Wiley-Liss, 1990, 1.

3. L. E. Anderson, "Exposure levels, bioeffects, and epidemiology," *Health Phys.* 61: 41, 1991.

4. C. Polk, "Biological effects of low-level, low-frequency electric and magnetic fields," *IEEE Trans.* ED-34: 243, 1991.

5. T. S. Tenforde, "Biological interactions and potential health effects of extremely-low-frequency magnetic fields from power lines and other common sources," *Ann. Rev. Publ. Health* 13: 173, 1992.

6. C. Graham, M. R. Cook, and H. D. Cohen, *Immunological and Biochemical Effects of 60-Hz Electric and Magnetic Fields in Humans,* Midwest Research Institute Final Report Contract No. DE-FC01-84-CE-76246 (Order No. DE90006671), Oak Ridge, Tenn.: U.S. Department of Energy Office of Science and Technology Information, 1990.

7. B. W. Wilson, L. E. Anderson, D. I. Hilton, and R. D. Phillips, "Chronic exposure to 60-Hz electric fields: Effects on pineal function in the rat," *Bioelectromagnetics* 2: 371, 1981 [erratum: 4: 293, 1983].

8. A. Lerchl, K. O. Nonaka, K. A. Stokkan, and R. J. Reiter, "Marked rapid alterations in nocturnal pineal serotonin metabolism in mice and rats exposed to weak intermittent magnetic fields," *Biochem. Biophys. Res. Commun.* 169: 102, 1990.

9. R. Goodman and A. S. Henderson, " Transcription and translation in cells exposed to extremely-low-frequency electromagnetic fields," *Bioelectrochem. Bioenerg.* 25: 335, 1991.

10. T. S. Tenforde and W. T. Kaune, "Interaction of extremely-low-frequency electric and magnetic fields with humans," *Health Phys.* 53: 585, 1987.

11. W. R. Adey, "Electromagnetic fields, cell membrane amplification, and cancer promotion," in B. W. Wilson, R. G. Stevens, and L. E. Anderson, Eds., *Extremely-Low-Frequency Electromagnetic Fields: The Question of Cancer,* Columbus, Ohio: Battelle Press, 1990, 211.

12. S. M. Bawin and W. R. Adey, "Sensitivity of calcium binding in cerebral tissue to weak environmental electrical fields oscillating at low frequency," *Proc. Natl. Acad. Sci. USA* 73: 199, 1976.

13. C. F. Blackman, S. G. Benane, L. S. Kinney, D. E. House, and W. J. Joines, "Effects of ELF fields on calcium-ion efflux from brain tissue in vitro," *Bioelectromagnetics* 6: 1, 1985.

14. H. M. Fishman and H. R. Leuchtag, "Electrical noise in physics and biology," *Curr. Top. Membr. Transport* 37: 3, 1990.

15. J. C. Weaver and R. D. Astumian, "The response of living cells to very weak electric fields: The thermal noise limit," *Science* 247: 459, 1990.

16. T. Tenforde, "Microelectrophoretic studies on the surface chemistry of erythrocytes," *Adv. Biol. Med. Phys.* 13: 43, 1970.

17. J. C. Weaver and R. D. Astumian, "ELF biological threshold estimates: An approach for electric fields based on noise and mechanism models," *Bioelectromagnetics* Suppl. 1: 119, 1992.

18. A. A. Pilla, P. R. Nasser, and J. J. Kaufman, "On the sensitivity of cells and tissues to therapeutic and environmental electromagnetic fields," *Bioelectrochem. Bioenerg.* 30: 161, 1993.

19. D. L. Alkon and H. Rasmussen, "A spatial-temporal model of cell activation," *Science* 239: 998, 1988.

20. R. A. Luben, "Effects of low-energy electromagnetic fields (pulsed and DC) on membrane signal transduction processes in biological systems," *Health Phys.* 61: 15, 1991.

21. R. A. Luben, C. D. Cain, M. C. Y. Chen, D. M. Rosen, and W. R. Adey, "Inhibition of parathyroid hormone actions on bone cells in culture by induced low energy electromagnetic fields," *Proc. Natl. Acad. Sci. USA* 79: 4180, 1982.

22. C. D. Cain, W. R. Adey, and R. A. Luben, "Evidence that pulsed electromagnetic fields inhibit coupling of adenylate cyclase by parathyroid hormone in bone cells," *J. Bone Min. Res.* 2: 437, 1987.

23. C. V. Byus, R. L. Lundak, R. M. Fletcher, and W. R. Adey, "Alterations in protein kinase activity following exposure of cultured human lymphocytes to modulated microwave fields," *Bioelectromagnetics* 5: 341, 1984.

24. M. G. Monti, L. Pernecco, M. S. Moruzzi, R. Battini, P. Zaniol, and B. Barbiroli, "Effects of ELF pulsed electromagnetic fields on protein kinase C activation process in HL-60 leukemia cells," *J. Bioelectr.* 10: 119, 1991.

25. R. P. Liburdy, "Calcium signaling in lymphocytes and ELF fields," *FEBS Lett.* 301: 53, 1992.

26. M. A. Stuchly, J. R. N. McLean, R. Burnett, M. Goddard, D. W. Lecuyer, and R. E. J. Mitchel, "Modification of tumor promotion in the mouse skin by exposure to an alternating magnetic field," *Cancer*

Lett. 65: 1, 1992.

27. J. R. N. McLean, M. A. Stuchly, R. E. J. Mitchel, D. Wilkinson, H. Yang, M. Goddard, D. W. Lecuyer, M. Schunk, E. Callary, and D. Morrison, "Cancer promotion in a mouse-skin model by a 60-Hz magnetic field: II. Tumor development and immune response," *Bioelectromagnetics* 12: 273, 1991.

28. D. Sh. Beniashvili, V. G. Bilanishvili, and M. Z. Menabde, "Low-frequency electromagnetic radiation enhances the induction of rat mammary tumors by nitrosomethylurea," *Cancer Lett.* 61: 75, 1991.

29. L. Tamarkin, M. Cohen, D. Roselle, C. Reichert, M. Lippman, and B. Chabner, "Melatonin inhibition and pinealectomy enhancement of 7,12-dimethylbenz(a)anthracene-induced mammary tumors in the rat," *Cancer Res.* 41: 4432, 1981.

30. R. G. Stevens, S. Davis, D. B. Thomas, L. E. Anderson, and B. W. Wilson, "Electric power, pineal function, and the risk of breast cancer," *FASEB J.* 6: 853, 1992.

31. A. R. Liboff, "Geomagnetic cyclotron resonance in living cells," *J. Biol. Phys.* 13: 99, 1985.

32. A. R. Liboff, R. J. Rozek, M. L. Sherman, B. R. McLeod, and S. D. Smith, "$^{45}Ca^{++}$ cyclotron resonance in human lymphocytes," *J. Bioelectr.* 6: 13, 1987.

33. S. D. Smith, B. R. McLeod, A. R. Liboff, and K. Cooksey, "Calcium cyclotron resonance and diatom mobility," *Bioelectromagnetics* 8: 215, 1987.

34. W. C. Parkinson and C. T. Hanks, "Search for cyclotron resonance in cells in vitro," *Bioelectromagnetics* 10: 129, 1989.

35. A. R. Liboff and W. C. Parkinson, "Search for ion-cyclotron resonance in a Na^+-transport system," *Bioelectromagnetics* 12: 77, 1991.

36. B. Halle, "On the cyclotron resonance mechanism for magnetic field effects on transmembrane ion conductivity," *Bioelectromagnetics* 9: 381, 1988.

37. V. V. Lednev, "Possible mechanism for the influence of weak magnetic fields on biological systems," *Bioelectromagnetics* 12: 71, 1991.

38. M. N. Zhadin and E. E. Fesenko, "Ionic cyclotron resonance in biomolecules," *Biomed. Sci.* 1: 245, 1990.

39. L. A. Shuvalova, M. V. Ostrovskaya, V. A. Sosunov, and V. V. Lednev, "Influence of a weak magnetic field under conditions of paramagnetic resonance on the rate of calmodulin-dependent phosphorylation of myosin in solution," *Proc. Natl. Acad. Sci. USSR* 317: 227, 1991.

40. C. Bruckner-Lea, C. H. Durney, J. Janata, C. Rappaport, and M. Kaminski, "Calcium binding to metallochromic dyes and calmodulin in the presence of combined AC-DC magnetic fields," *Bioelectromagnetics* 13: 147, 1992.

41. R. K. Adair, "Criticism of Lednev's mechanism for the influence of weak magnetic fields on biological systems," *Bioelectromagnetics* 13: 231, 1992.

42. W. R. Adey, "Tissue interactions with nonionizing electromagnetic fields," *Physiol. Rev.* 61: 435, 1981.

43. L. S. Taylor, "The mechanisms of athermal microwave biological effects," *Bioelectromagnetics* 2: 259, 1981.

44. E. Postow and M. L. Swicord, "Modulated fields and 'window' effects," in E. Postow and C. Polk, Eds., *Handbook of Biological Effects of Electromagnetic Fields,* Boca Raton, Fla.: CRC Press, 1986, 425.

45. J. L. Kirschvink, D. S. Jones, and B. J. MacFadden, Eds., *Magnetite Biomineralization and Magnetoreception in Animals: A New Biomagnetism,* New York: Plenum Press, 1985.

46. J. L. Kirschvink, "Biogenic magnetite and magnetoreception," *Bioelectromagnetics* 10: 239, 1989.

47. J. L. Kirschvink, A. Kobayashi-Kirschvink, and F. B. Woodford, "Magnetite biomineralization in the human brain," *Proc. Natl. Acad. Sci. USA* 89: 7683, 1992.

48. J. L. Kirschvink, A. Kobayashi-Kirschvink, J. Diaz-Ricci, and S. J. Kirschvink, "Magnetite in human tissues: A mechanism for the biological effects of weak ELF magnetic fields," *Bioelectromagnetics* Suppl. 1: 101, 1992.

49. Research support is received from the U.S. Department of Energy under Contract DE-AC06-76RLO 1830 with the Pacific Northwest Laboratory. The Pacific Northwest Laboratory is operated for the U.S. Department of Energy by the Battelle Memorial Institute.

Martin Blank, Ed., *Electricity and Magnetism in Biology and Medicine*

MEDICAL BENEFITS OF ELECTRIC AND MAGNETIC FIELDS: CHALLENGES FOR BROADER USE

C. A. L. Bassett*

Columbia University, New York

Clinical success and safety of *selected*, time-varying electric and magnetic fields for ununited fractures have been well documented for 20 years. Cellular and subcellular mechanisms of action are effective for this disorder and others. Modification of abnormal cell function, with increasingly precise field parameters, will facilitate new therapeutic indications. However, progress is hampered by skepticism engendered by poor experimental reproducibility, preoccupation with hazards and "the" physical mechanism of coupling, medical bias toward biochemistry, conservatism, cost, and regulatory impediments. Bioelectromagnetics must develop a consensus and an effective educational outreach for physicians, biologists, and administrators before principles will enter the medical curriculum.

During the past two decades, selected, time-varying electric and magnetic fields have played an increasingly successful role in the care of several challenging medical problems,[1] mainly fractures that have failed to heal in children and adults and chronic skin wounds. Other clinical conditions also appear to benefit from this technology: osteonecrosis, osteoporosis, osteochondritis dessicans, chronic refractory tendinitis, and adult-onset diabetes.[1,2] Laboratory studies also suggest a rational basis for treating severed nerves to speed regeneration, to control skin flap and graft necrosis, and to limit effects of acute myocardial ischemia.[1,2]

Many medical studies, including double-blind trials and experimental data on pertinent bioeffects, attest to the efficacy and safety of certain types of weak low-frequency magnetic fields. More than a quarter million patients with nonunions, osteonecrosis, and chronic wounds have been treated with pulsed electromagnetic fields (PEMFs). PEMFs have field parameters that induce voltage waveforms in tissues similar to those arising from electromechanical transduction during cyclic deformation of connective tissue.[1]

With this background of safety and efficacy, why do only 7% of the 579 Congress papers and posters deal with medical uses and benefits of electric and magnetic fields? Why is the medical community so very sparsely represented? As members of the world's leading scientific bodies, we must address root causes for this situation and shape strategies to correct it. If successful, a revolution in management of disease will occur in the coming decades as bioelectromagnetics assumes an ever expanding therapeutic role in the service of medicine, based on precise physical control of abnormal cell function.

CAUSES FOR LIMITED MEDICAL ACCEPTANCE AND USE OF BIOELECTROMAGNETICS

Skepticism

Reproducibility of Bioeffects. Bioelectromagnetics has a rigorous requirement for interdisciplinary research. Getting the physics side of the equation right and the biological or medical side wrong has obscured understanding of the myriad of factors that influence bioresponses to electromagnetic fields.[3] Just as often, the reverse has been true. One laboratory fails to confirm a response reported by another. All too often, failures to control critical elements of exposure or the biotarget are responsible. Many reports do not contain enough experimental details to permit duplicate studies. Model systems and field parameters vary widely and few investigations focus on comparative designs in which a single variable is tested. Field strength may be described precisely, while frequency content, waveform details, orientation of the field/target, and critically important features of the biosystem are ignored. Some clinical skepticism doubtless stems from conflicting experimental data.

Fortunately, this situation is improving, as understanding of the need for precision in experimental design and reporting has become more universal among physicists, engineers, geneticists, biologists, and physicians.

---

*Correspondence to: C. A. L. Bassett (Bioelectric Res. Ctr), Columbia University, 2600 Netherland Ave., New York NY 10463.

Improvements reflect increasing awareness that *specificity* requires matching of field parameters with the bioprocess being studied. "*Window*" and *threshold* requirements must be coupled with proper *exposure* conditions (e.g., timing and orientation) and a *receptive functional state in the biotarget* for progress to occur.

Physics and "Chaos." Controversy surrounding "the" mechanism (singular) has contributed greatly to the slow pace at which bioelectromagnetics is penetrating medical therapeutics. In fact, there are probably several mechanisms in the physical realm (e.g., resonance, quantum effects, etc.) and several more in the biological realm (e.g., at the organism, cellular, and subcellular levels). At present, many mechanisms in the biorealm have been defined and, for pathologic states currently treated with PEMFs, known cell responses are appropriate to correct or modify the abnormality.[1,2] Unfortunately, these facts are ignored too frequently while the focus is on what is happening at the physical level!

Lack of substantive physical answers has resulted in enough sniping by some in the physics community to have an adverse effect on both the clinic and on governmental agencies. Bioeffects of weak fields have even been denigrated as "pathological science" and the "Emperor's clothes" by those who assume thermal noise over-rides any effect of an imposed weak field. Lost in these assumptions are the marvelous complexities of biology (e.g., multiprocessed cell arrays linked by junctions, nonlinear responses, and signal averaging; among a host of other pertinent issues). Investigations of ion cyclotron resonance, parametric resonance, "coulombic" models (with Zeeman or Stark effects) and, more recently, quantum effects on singlet-triplet states represent important probing for mechanistic answers.

Preoccupation with Biochemistry. From their earliest days in medical school, students live in a world surrounded by biochemistry. They may have learned earlier that chemistry would not exist but for specific charge-change interactions. A hundred years ago Solvay wrote: "The phenomena of life can and should be explained by the action of only physical forces which govern the Universe and that, among these forces, electricity plays a predominant role." Until principles of physics that impact cell function (beyond action potentials) are better known, tomorrow's physicians will be as ill prepared as today's to understand and practice therapeutics that involve electric and magnetic fields.

The Hazard Issue. One of the most difficult topics for patients and their doctors in the decision to use PEMFs is a *perception* that "electromagnetic fields may be hazardous for health." Bombarded by lay press sensationalism on *epidemiologic association* of ills with power lines, VDTs, etc., many avoid PEMF treatment, despite a 20-year record of safety. Few dealing with hazards have yet focused on the issue of specificity of bioresponses to selected field parameters. Yet the level of specificity to various frequencies, amplitudes, and exposure conditions can exceed that of drugs (e.g., transcription and translation[4]). To the uninformed, electromagnetic fields are lumped regardless of type (e.g., they are sinusoidal or square and pulsed or continuous, or containing high or low frequencies).

Conservatism and Dogma

Physicians are properly conservative by training. Those at the "cutting edge" of any new discipline must be prepared to persist until principles and practice are established and accepted. Thus, it is with bioelectromagnetics. Conservatism arises not only from the Hippocratic Oath ("Do no harm") but also as self-protection against "ignorance." After physicians have mastered chemistry and pharmacology, it is a major challenge for them to start over with the basics of resonance, solid-state properties, Maxwell's equations, and the other fundamentals required before time-varying fields can be understood and put to rational therapeutic use. The obvious answer is to learn these topics at the beginning of a medical career. The medical curriculum needs upgrading, so that those steeped in the dogma of nerve and muscle physiology have a basis of reference for observations that fields as weak as 1 μV/cm can exert biological effects.

Regulatory Impediments

In the USA, the Food and Drug Administration (FDA) assures that drugs and devices are safe and effective. The FDA often approves new drugs on the basis of their mechanism of action, without limiting use to one specific medical condition. For example, if a new antibiotic is effective against streptococci, it is left to the doctor to decide whether it is also useful for an infection of the throat or an abscess of the thigh. PEMFs, on the other hand, are approved on an indication-by-indication basis, because authors fail to cite *bio*mechanisms that are proven. Report after report appears with the tenet that "the" mechanisms of action is unknown, which serves to provide the FDA with a rationale for its position. If faced with a patient with a debilitating fragment of dead bone, I choose field parameters known to stimulate growth of new vessels. This is an appropriate matching of a proven biomechanism to therapeutic need. In the end, the patient does not care from what physical parameter success derived. If the same mechanistic yardstick applied by the FDA to PEMFs were applied *at the physics level* to aspirin, that drug would be pulled from our therapeutic armamentarium.

SUMMARY

Bioelectromagnetics holds significant promise for treating disease by modifying the pathologic behavior of local cell populations with great precision. Scientific and clinical principles are already well established for certain musculoskeletal, neural, and epidermal conditions, many of which resist standard medical and surgical regimens. This foundation of success and knowledge serves as a base upon which therapeutic expansion will occur. As new understanding adds mechanistic rationale for matching specific field characteristics with selective cell responses and, ultimately, with individual disease states, a wide range of human ills (even cancer) will come under better control.

The medical promise of bioelectromagnetics can be realized only through rigorous interdisciplinary research efforts and coordinated, educational outreach programs. Medicine must be made more aware of the advances that will impact the practice of many of its members. Correcting this situation is our collective responsibility. Unfortunately, most of our research is reported in journals with editorial and audience profiles matched to the topic. The net result, insofar as the average medical student, physician, or dean is concerned, is to bury the knowledge. At present, we speak to ourselves, for ourselves—a situation that must change for educational progress to occur. Communication skills that reduce this most complex of disciplines to reasonable dimensions for the student and physician must be developed. Our societies must promote clearly written, authoritative, state-of-the-art reviews for a wide range of medical and biological journals—even the *New England Journal of Medicine*. Articulate, knowledgeable members should be identified for symposia and panels and sponsored at major national and international specialty meetings. A comprehensive program will eventually change the medical curriculum. We must be prepared to aid in designing, promoting, and supporting these changes. There is much to be done; together, we can do it!

REFERENCES

1. C. A. L. Bassett, "Fundamental and practical aspects of therapeutic uses of pulsed electromagnetic fields (PEMFs)," *CRC Crit. Rev. Biomed. Eng.* 17: 451-529, 1989.

2. C. A. L. Bassett and D. O. Carpenter, "Therapeutic uses of electric and magnetic fields in orthopaedics," in D. O. Carpenter, Ed., *Biologic Effects of Electric and Magnetic Fields,* Academic Press (in press).

3. C. A. L. Bassett, "Physical and biological principles affecting weak ELF electromagnetic bioresponses," in C. T. Brighton and S. R. Pollack, Eds., *Electromagnetics in Biology and Medicine,* San Francisco: San Francisco Press, 1991, pp. 1-13.

4. R. Goodman and A. S. Henderson, "Transcription and translation in cells exposed to extremely low frequency electromagnetic fields," *Bioelectrochem. Bioenerget.* 25: 335-355, 1991.

Martin Blank, Ed., *Electricity and Magnetism in Biology and Medicine*

POSSIBLE HEALTH RISKS FROM POWER-FREQUENCY FIELDS: ISSUES IN RISK PERCEPTION, COMMUNICATION, ASSESSMENT, AND MANAGEMENT*

M. Granger Morgan†

Carnegie Mellon University, Pittsburgh, Pa.

Many concerned citizens and government officials are not willing to suspend all risk management activities until laboratory and epidemiological research provides a definitive answer to the simple question, "does exposure to power-frequency fields pose significant risks to public health?" For this reason, issues of risk perception, communication, assessment, and management are of great and growing importance. This paper reports results from empirical studies of the factors that contribute to public understanding and perceptions of this issue. It discusses how communication about this topic to nontechnical decision makers and the public can be improved. It explores how and to what extent the incomplete scientific evidence that is now available can be usefully used in risk assessment. It concludes by describing the range of options available for risk management.

Over the past decade a great deal of evidence has accumulated that power-frequency electric and magnetic fields can produce changes in biological systems.[1] Epidemiological evidence has pointed to a possible role in cancer promotion.[2] Whether exposure to fields does in fact pose health risks is still an open question, but many concerned citizens, managers, and regulatory officials are not willing to suspend all risk management activities until laboratory and epidemiological research provides a definitive answer to the simple question, "does exposure to power-frequency fields pose significant risks to public health?" For this reason, issues of risk perception, communication, assessment, and management are of great and growing importance.

PUBLIC UNDERSTANDING AND RISK COMMUNICATIONS

What the public and nontechnical government officials believe about power-frequency fields will play a deciding role in the way this issue unfolds over the next several years. For this reason our group at Carnegie Mellon has conducted a number of studies of lay understanding and preferences. Perhaps the most basic question is, what do people know about fields? We have explored this question with three questionnaires. The first involved a set of true-false statements about the nature and properties of fields. The second involved a more complex pictorial design, which explores people's knowledge of the relative strengths of fields present in different settings and some of the spatial and temporal variations associated with these fields. The third involved a comparison of power-frequency fields with four other physical agents. Details were reported previously.[3]

Lay respondents correctly knew only a few of the simplest facts but displayed significant awareness of the limits of their knowledge and little clearly wrong knowledge. On average these respondents correctly rank-order several common field-exposure situations by field strength but could not differentiate between electric and magnetic fields and believed that the field strengths associated with various appliances are all about the same, whereas in fact they vary widely. Most respondents dramatically underestimated the dynamic range of actual field strengths. Many respondents understood that field strength decreases with distance from sources, but as shown in Fig. 1, most dramatically underestimate the rate of this decrease. This misunderstanding can lead laypeople to conclude incorrectly that any transmission line they can see is exposing them. Although lay respondents display a variety of incomplete and confused ideas, we concluded that, in contrast to some other risks we have studied (e.g., radon[4]), most of these misunderstandings do not pose significant obstacles to learning a correct simple mental model.

---

*Portions of this paper draw heavily upon the author's earlier writings.

†Correspondence to: M. G. Morgan (Engg. and Public Policy), Carnegie Mellon University, Pittsburgh PA 15213.

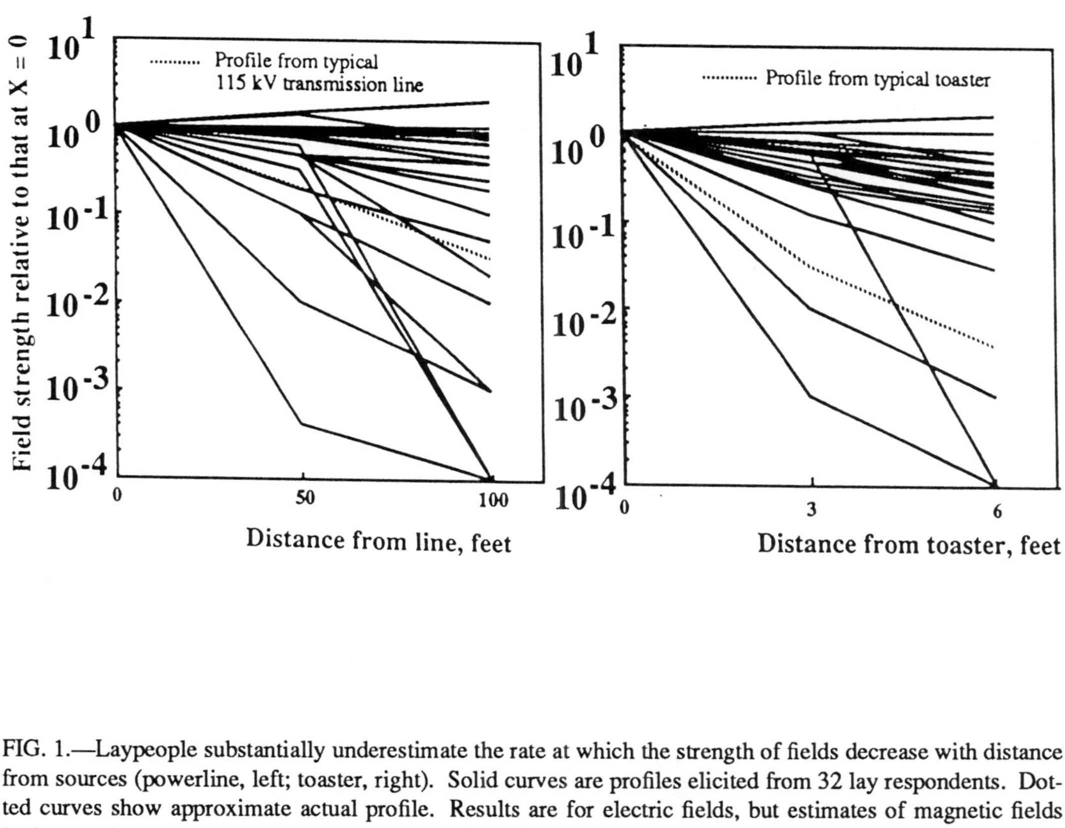

FIG. 1.—Laypeople substantially underestimate the rate at which the strength of fields decrease with distance from sources (powerline, left; toaster, right). Solid curves are profiles elicited from 32 lay respondents. Dotted curves show approximate actual profile. Results are for electric fields, but estimates of magnetic fields look essentially identical. Similar results have been obtained with other groups, e.g., Wharton School MBAs. (Figure adapted from Ref. 3.)

We have recently completed a study of the preferences of lay opinion leaders for alternative possible strategies for exposure management. Respondents were presented with 40 plausible options to reduce or eliminate exposure to power-frequency fields ranging from voluntary design, measurement, and information guidelines to banning sources of higher than average fields. A majority of subjects favored implementing twelve of the options proposed. These twelve options are characterized by restrictions on new sources of high fields, including overhead distribution and transmission lines and wiring in new buildings, and providing field information on new appliances. A surprising finding was that when respondents were asked to estimate costs, no effect was observed on preferences for exposure control options, although the cost estimates that the subjects provided tended to be reasonable. Insensitivity to the fact that field strength decreases with distance was approximately equally prevalent in this study as in our previous study of lay perceptions of fields. Although most subjects were moderate in their beliefs about the risks posed by fields, stronger belief in health effects correlated positively with preferences for implementing more exposure-reduction strategies. Overall, subjects appeared to favor field limitation measures that could entail significant investments, especially for new sources. A manuscript report on the findings has been prepared.[5]

The author and Gordon Hestor have studied the issue of involving the public in transmission-line siting.[6] Given the proper context and support, the public can provide very reasonable and appropriate advice. We are currently exploring how well the public understands the design of scientific experiments. Initial results indicate that many laypeople are able to recognize straightforward weaknesses in experimental design (e.g., failure to have persons categorizing disease outcomes in test animals be "blind" with respect to exposures), but may be less able to identify more subtle weaknesses (e.g., covariation between exposures of interest and other possible risk factors in epidemiologic studies). There is also some evidence of a tendency toward being uncritical of positive studies (i.e., those that show an apparent increase in risk with exposure). Another striking result is that whereas laypeople differ widely in their opinions about what assumptions (i.e., null hypotheses) scientists *should* adopt prior to doing a health effects study, a majority apparently think the assumptions they believe scientists *do* make are inappropriate. A similar question about public health officials' assumptions yielded similar but even more pronounced results. We conclude that communication with the public about the implications of specific scientific studies can be effective only if accompanied by a substantial education effort.

One of the principal motivations for the studies described in this section has been to improve risk communication. We began our work in risk communication with a 45-page brochure titled *Electric and Magnetic Fields from 60 Hertz Electric Power: What Do We Know About Possible Health Risks?*.[7] Roughly 100000 copies are now in circulation in the USA and the brochure has been translated and reproduced in several other countries. New communications materials, including a two-part brochure for use with in-home measurement programs and a revised version of our general brochure, are under development, based on insights from the studies reported above.

PROBLEMS IN RISK ASSESSMENT

The primary obstacle to doing risk assessment on power-frequency fields is that we do not know how to define dose. Our early work illustrated techniques in parametric risk analysis and bounding analysis but results were of little use to public policy.[8]

In the spring of 1991 the writer and his colleague Indira Nair ran a workshop for the Department of Health Services of the State of California in which we asked a group of leading biologists for their judgments about the plausibility of seven possible relations between exposure and biological effects.[9] Seven "effects functions" were considered: (1) effect proportional to time-average of field strength; (2) effect proportional to cumulative field strength or some function of field strength; (3) effect is binary in field presence (i.e., depends only on presence or absence of field); (4) effect depends on "switching;" (5) effect occurs only within field strength window(s); (6) effect occurs only within frequency windows; and, in combination with one or more of the preceding, (7) effect requires the presence of additional external agent(s).

Structure was provided for the workshop with a 19-page questionnaire booklet that all participants completed before arrival. Discussions at the workshop worked through the evidence for and against each effects functions in some detail. Substantial agreement was found about several effects functions in a few specific contexts such as calcium efflux and cell signaling and biosynthesis pathways. However, there was no significant agreement in many other contexts. Sometimes disagreement resulted from alternative readings of the literature. In other cases, such as in immune response, it apparently resulted from the very limited available evidence. When pressed to make arguments that one or more of the effects functions that we considered could be ruled out across the board, participants reached a strong consensus that *none* could be excluded at this time. It is noteworthy that time-average field strength, which is widely used by epidemiologists, engineers, risk analysts and regulators, did not emerge from the ranking exercise as clearly the most likely, or most favored, candidate. Results are now in press.[9]

Building on this work the author and his colleagues Indira Nair and Jun Zhang have written computer simulations that permit the application of several of the effects functions discussed at the Berkeley workshop to samples of EMDEX personal monitor data.[10] Relatively low correlations were observed between most of the effects function outpouts and average field strengths. While many of the low correlation coefficients we observed result from a lack of association between our effects function output and average field strength, some result from a noisy nonlinear relationship between the two measures. Simple nonlinear curves were fitted and a series of arguments constructed to sort effects functions by degree of plausibility under the assumption that available epidemiological findings on childhood leukemia are not artifactual. The results of such simulations can then be compared with available epidemiological evidence, and in this way it may be possible to rank effects functions loosely in terms of their plausibility. Once an effects function has been identified as particularly plausible, a histogram of relative risk can be generated by applying it to results from an epidemiological study. This work was presented at the 1991 DOE Contractors' meeting;[10] a detailed manuscript has been prepared for publication.

Given a set of plausible alternative effects functions one can perform parametric analysis of the cost effectiveness of alternative strategies for exposure management. Others may prove to be cost effective across a much broader set. Jack Adams, Jun Zhang, Indira Nair, and the author are currently estimating such exposure reduction supply curves for 345kV transmission lines. Preliminary findings were presented in the oral version of this paper and will appear in a separate technical paper.[11]

EXPOSURE MANAGEMENT OPTIONS

Whatever one's views about the possible risks posed by human exposure to power-frequency fields may be, individuals, managers, and regulators face three broad policy options with respect to managing exposure: (1) a minimum response involving no action on exposure; (2) a response based on "prudent avoidance," which only takes actions that can avoid exposing people to specific field environments at modest cost; or, (3) a dramatic response, which entails major action on modifying or eliminating exposure. For reasons elaborated elsewhere,[12] I believe that *prudent avoidance* is the most viable current option. I choose these two words with care. *Prudent* because the idea is to search for steps that could get people out of fields, especially new fields, at *low cost* and with modest inconvenience. *Avoidance* because, given the ambiguity about dose, the idea is to try to get people *out* of certain exposure situations (e.g., by widening rights-of-way or not using electric blankets), not just to lower field strength a little bit. Like such widely used legal phrases as "beyond a shadow of a doubt" or "due diligence," prudent avoidance is a concept that only has meaning after regulators, courts, and others have worked carefully to give it meaning. Tom Watson and Mark Warnquist,[13] two lawyers with the firm of Cromwell and Moring who are active in the litigation of powerline cases in the USA, report that the process of giving meaning to the phrase "prudent avoidance" is now well along in the USA and that they are "getting fairly consistent and reasonable answers." We exercise prudence all the time in our private lives, yet in the public domain our society has fallen into the habit of trying to categorize everything as either *safe* or *hazardous*. Companies fear that there is a risk that if they take some modest low-cost steps toward exposure avoidance in the name of prudence, they may soon be forced by public hysteria to pull out all the stops and do a lot of other things that are much more expensive and go well beyond what people would consider prudent. As they give meaning to the words "prudent avoidance," regulators and legislators must recognize this risk of going overboard and build in appropriate protections.

Beyond thinking about current exposure, utilities, regulators, research organizations, and others would be wise to think about the question, "what comes after a smoking gun?" and make prudent preparations now. In short, while we wait for better scientific understanding, wise managers and regulators should be doing three things:

- working to avoid exposing many more people to fields when it can be done at modest costs and with little inconvenience;
- working to preserve future options and avoid setting precedents that will cause serious problems in the future if fields do (or do not) turn out to present significant risks to public health; and
- working to keep a sense of perspective and help the public keep a sense of perspective about this issue. Even if fields do turn out to pose significant health risks, the risks are likely to be modest for most ex-

posed individuals, and small compared with many other risks in modern life. We should be prudent, not paranoid.[14]

REFERENCES

1. I. Nair, M. G. Morgan, and H. K. Florig, *Biological Effects of Power-frequency Electric and Magnetic Fields: Background Report,* Office of Technology Assessment, U.S. Congress, OTA-BP-E-53, Washington: U.S. Government Printing Office, 1989.

2. Electric Power Research Institute, *Extremely Low Frequency Electric and Magnetic Fields and Cancer: A Literature Review,* Final Report prepared by W. A. Creasey and R. B. Goldberg, Information Ventures, Inc., EN-6674 Research Project 2965-12, December 1989.

3. M. G. Morgan, H. K. Florig, I. Nair, C. Cortés, K. Marsh, and K. Pavlosky, "Lay understanding of power-frequency fields," *Bioelectromagnetics* 11: 313, 1990.

4. A. Bostrom, B. Fischhoff, and M. G. Morgan, "Eliciting mental models of hazardous processes: A methodology and an application to radon," *J. Social Issues* (in press).

5. A. Bostrom, M. G. Morgan, J. Adams, and I. Nair, "Lay preferences for exposure control of power-frequency fields," (manuscript in review).

6. G. Hester, M. G. Morgan, I. Nair, and H. K. Florig, "Small-group studies of regulatory decision-making for power frequency electric and magnetic fields," *Risk Analysis* 10: 213, 1990.

7. M. G. Morgan, *Electric and Magnetic Fields from 60 Hertz Electric Power: What Do We Know About Possible Health Risks?* Pittsburgh, Pa.: Carnegie Mellon University, 1989, 45 pp.

8. M. G. Morgan, H. K. Florig, I. Nair, and D. R. Lincoln, "Some first steps toward assessing possible health risks from exposure to 50/60 Hz electromagnetic fields," in L. E. Anderson, B. J. Kelman, and R. J. Weigel, Eds., *Interaction of Biological Systems with Static and ELF Electric and Magnetic Fields,* Proc. 23rd Hanford Life Sciences Symposium, Battelle Pacific Northwest Laboratories, Richland, Wash., CONF-84, 1984.

9. M. G. Morgan and I. Nair, "Alternative functional relationships between ELF field exposure and possible health effects: Report of an expert workshop," *Bioelectromagnetics* (in press).

10. M. G. Morgan, I. Nair, and J. Zhang, "Studies of the plausibility of alternative effects functions for 60 Hz fields: A presentation at the 1991 Annual Review of Research on Biological Effects of 50/60 Hz Electric and Magnetic Fields Contractors Review," Milwaukee, Wisc., 1991.

11. J. Zhang, J. Adams, M. G. Morgan, and I. Nair, "Estimates of exposure reduction supply curves for several different effects functions," this volume.

12. M. G. Morgan, "Prudent avoidance," *Public Utilities Fortnightly* 26: March 15, 1992.

13. T. Watson and M. Warnquist, "EMF: Liability, prudent avoidance and communication," *Electricity J.* 4: April 1991.

14. The very considerable assistance of the author's colleagues Indira Nair, Keith Florig, Gordon Hester, Ann Bostrom, Baruch Fischhoff, Jun Zhang, Jack Adams, and several others is gratefully acknowledged. The work reported here was supported by the Electric Power Research Institute, the Southern California Edison Company in cooperation with the State of California, the U.S. Department of Energy, and the Congressional Office of Technology Assessment. The views expressed are the author's own.

Martin Blank, Ed., *Electricity and Magnetism in Biology and Medicine*

STATE OF THE ART IN ELECTROMAGNETIC THERAPEUTICS

A. A. Pilla*

Mount Sinai School of Medicine, New York

Noninvasive electromagnetic fields (EMF) are now a common part of the orthopaedist's armamentarium for the care of delayed and nonunion fractures. Emerging clinical orthopaedic applications of EMF include avascular necrosis, spinal fusion, tendenitis, and osteoporosis. EMF has been shown to modulate soft tissue and tendon healing in animals and in clinical trials. There is active research in the use of EMF for cardiac muscle protection and regeneration, peripheral nerve regeneration, and immune system activation. The physical mechanism of electromagnetic bioeffects is not yet completely understood. Nevertheless, promising hypotheses have led to the creation of several waveforms with demonstrated biological effect. In nearly all approaches, ionic or ligand processes at the cell membrane are considered the most likely targets. Electrical dosimetry considerations based on such models have led to the creation of therapeutic devices for which the spectral density of the in situ electric field satisfies the dielectric properties of a cell-array tissue model. The EMF sensitivity analysis presented in this study shows that all of the currently known EMF signals in use for bone repair appear to satisfy signal-to-noise thresholds for a membrane target. The cell-array model predicts a more effective range of EMF waveform parameters for enhanced EMF clinical effects. This model also addresses the sensitivity of skeletal tissue to endogenous electric fields (streaming potentials).

The use of electrical signals to modulate the rate of in vitro and in vivo biological growth and repair processes has been widespread since the first clinical report of the use of weak DC currents for bone repair.[1] The most common application of electrical signals at present is via pulsed electromagnetic fields (EMF). This modality was originally created to provide a noninvasive, no-touch means of inducing an electrical waveform at the cell/tissue level.[2] The origin of the EMF waveform in current clinical use began with the early work of Becker, Yasuda, Brighton, and Bassett,[3-6] who considered that an electrical pathway may be the means through which bone adaptively responds to mechanical input. Subsequently, a linear physiochemical approach was taken by Pilla,[7-10] who used an electrochemical model of the cell membrane to predict a range of EMF waveform parameters for which bioeffects might be expected. This approach was based on the assumption that the cell membrane was the most likely EMF target. A range of waveform parameters for which the induced electric field could couple to electrochemical (i.e., voltage-dependent) kinetics at the cell's surfaces was thereby defined. Extensions of this linear model involved Lorentz force considerations,[11,12] which led to the suggestion that the magnetic field alone could be a dominant stimulus. From this suggestion evolved ion-resonance and quantum theories that accounted for combined AC and DC magnetic field effects.[13,14]

In order for an EMF *electrical* bioeffect to be possible, the electromagnetic signal should not only satisfy the dielectric properties of the target, but also induce sufficient voltage to be detectable above thermal noise. For many of the EMF signals discussed in this study, that may be possible only if cell-cell communication exists via electrical pathways (gap junctions) between cells in tissues. The EMF sensitivity of such cell arrays can be larger by orders of magnitude than that for isolated cells and/or tissues with no (or nonfunctional) gap junctions.[15,16] Important characteristics to assure adequate electric dosimetry are the geometry *and* size of the target, its dielectric properties *and* the time rate of change dB/dt of the magnetic field for electromagnetically induced E.[17] Many dosimetry errors have been made by not taking both the dielectric properties of the target and its geometry into account. This study considers the electric dosimetry for the various EMF signals in use (or being readied for use) for bone repair. Special attention is given to the effects of target size and dielectric properties based on the cell-array model. It is shown that this model can readily explain observations of remarkable tissue and organism sensitivity to both endogenous (streaming potentials) and exogenous (EMF) electric fields.

---

*Correspondence to: A. A. Pilla (Bioelectrochemistry), Dept. of Orthopaedics, Mt. Sinai School of Medicine, New York NY 10029.

COMMON METHODS OF THERAPEUTIC EMF STIMULATION

All electromagnetic methods in use or under study for therapeutic applications can be categorized according to the relative values of the electric and magnetic fields. The oldest technique for orthopaedic applications uses semi-invasive DC electrodes.[1] Dosimetry in this approach depends not only on current density in the vicinity of the electrode, but also upon faradaic electrochemical processes that change the chemical environment in the treatment site.[7] The biological response to DC currents from implanted electrodes is localized to the electrode vicinity and is always accompanied by some degree of tissue inflammation or even necrosis. This approach has been largely outdated by noninvasive techniques that are significantly more useful and effective for clinical applications.

The majority of electromagnetic clinical devices utilize a time-varying magnetic field $B(t)$ that induces a time-varying electric field $E(t)$ at the treatment site. $E(t)$ is proportional to the rate of change of current dI/dt in a conductor. The evaluation of this quantity for a given driving voltage function results in a description of the shape of $E(t)$ in vacuum, air, and all nonmagnetic homogeneous conducting media in which the resulting current flow is not high enough to produce significant back electromotive force (the case for body fluids).[18] A simple working relationship to assess the induced electric field $E(t)$ can be obtained for a target with cylindrical geometry as

$$E(t) = -1/2r\ dB/dt \tag{1}$$

which states that the amplitude of the electric field within the area of the target penetrated by the magnetic field is proportional to dB/dt (i.e., dI/dt) *and* the radius of the target (i.e., to target position or size). The waveform of $E(t)$ also depends on dB/dt, which in turn determines the amplitude spectrum and permits optimal frequency matching for a given target impedance.

Typical pulse-type waveforms for $E(t)$ in common clinical use are shown in Fig. 1. Peak E from all waveforms, including two sinusoidal signals (not shown), ranges from 10^{-3} to 10 mV/cm (Eq. 1) for a target 2 cm in radius. The lowest value was for the 76.6Hz sinusoidal signal developed according to ion-resonance theory, which also requires a parallel weak static magnetic field.[13] Clinical EMF signals for bone repair are inductively coupled, except for one capacitively coupled 60kHz sinusoidal signal, for which E dosimetry, estimated from the geometry and dielectric properties of the target,[9] is in the mV/cm range.

THE QUESTION OF SIGNAL-TO-THERMAL NOISE RATIO IN REAL TISSUE

That therapeutic EMF can affect the behavior of living cells and tissues remains controversial.[20] The physical argument against the possibility of an EMF bioeffect is usually based on the ratio of induced transmembrane voltage signal to the root mean square (RMS) thermal noise voltage. This signal-to-noise ratio (SNR) is usually calculated on the assumption that the EMF target is a spherical cell 10 μm in radius. These calculations rightly predict that SNR $\ll 1$ for clinical EMF signals. However, this is a grossly oversimplified model, having little basis for real tissue structure. Most cells are anisotropic in shape and function and could be oriented to capture (detect) significantly more of the applied field than the corresponding isotropic shape.[21] In addition, gap junctions provide pathways for ionic and molecular intercellular communication.[22]

The role of cooperative organization in the EMF sensitivity of biological systems has been considered qualitatively.[23] Gap junctions allow ionic coupling and metabolic cooperation, without which disorders in growth control and tissue repair, as well as neoplastic transformations, could occur.[24] Functional modification of gap junctions by modulated microwave fields has been reported.[25] It has been suggested that the thermal threshold of cell sensitivity to external electric fields can be as low as 10^{-3} mV/cm only if a time-averaging process for the biochemical sensor is postulated.[26]

Bone tissue may be represented by an array of cells in gap junction contact. For an array of length L, a useful model is a distributed parameter linear electrical analog that permits a description of impedance per unit length for cells electrically connected in series (transmission line). Such a model allows the induced transmembrane voltage V_m to be evaluated as a function of frequency ω and position x along the array.[15,16]

The propagation of V_M in a cell array is given by:

$$V_M(x,\omega) = -\frac{E(\omega)}{\gamma}\frac{\sinh(\gamma x)}{\cosh(\gamma L)} \tag{2}$$

where $\gamma = [(R_e + R_i)Y_m(\omega)]^{1/2}$, $Y_M(\omega) = 1/Z_M(\omega)$, and R_e and R_i are respectively the extra- and intracellu-

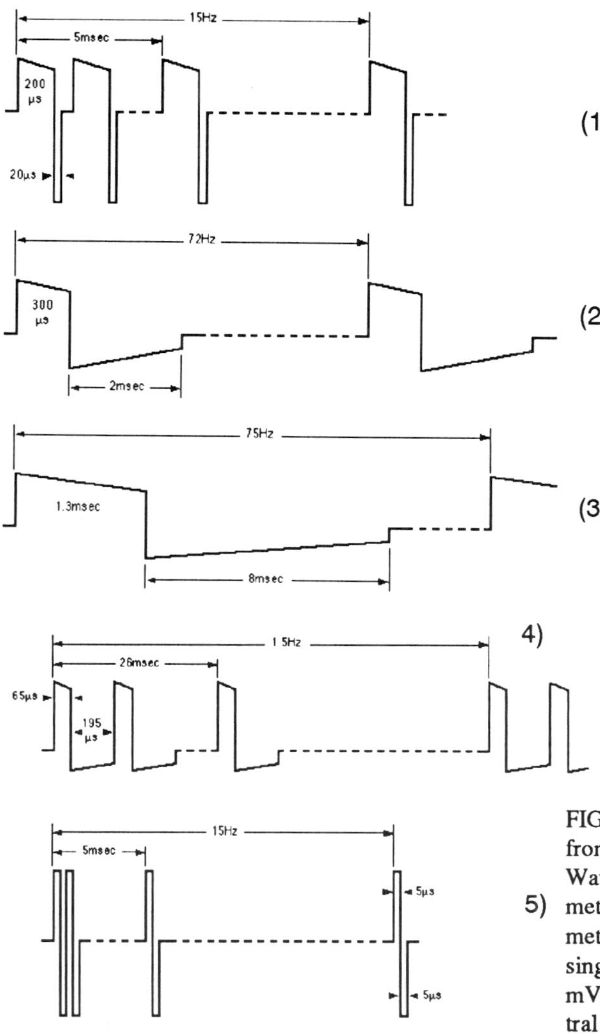

FIG. 1.—Induced electric field $E(t)$ in tissue from EMF devices for clinical applications. Waveforms 1 and 4 consist of bursts of asymmetrical pulses; 5 is a burst of very narrow symmetrical pulses; 2 and 3 are wide asymmetrical single pulses. For all signals peak E is 1-10 mV/cm. All have sufficient low-frequency spectral density to be detectable by cell arrays. Biological effects have been reported for all signals.

lar electrolyte resistances, Z_M is the membrane impedance (all per unit array length), and $E(\omega)$ is the applied (extracellular) EMF. The exact form of Z_M depends upon the model assumed. The simplest and most classic case is the membrane capacitance C_d in parallel with an ionic leak pathway R_M. The impedance of the cell array $Z(x, \omega)$ is

$$Z(x, \omega) = \frac{(R_e + R_i)}{\gamma} \tanh(\gamma x) \tag{3}$$

Spatial amplification (V_M/EL) may be evaluated by use of Eq. (1) for an array varying from $L = 10$ µm (single cell) to a physiologically relevant $L = 1\text{-}10$ mm.[15,16] The results show that V_M/EL, for $L = 1$ mm, is larger by orders of magnitude than that for $L = 10$ µm. However, maximum gain occurs only at low frequencies (<100 Hz) because of the dielectric properties of the cell array. The power spectral density $S_n(\omega)$ of thermal noise is given by DeFelice[27] as

$$S_n(\omega) = 4k\,\mathrm{TRe}[Z(x, \omega)] \tag{4}$$

where Re is the real part of $Z(x, \omega)$.

An evaluation of signal-to-noise ratio over the entire frequency range by use of the ratio of spectral densities ($V_M/S_n^{1/2}$) makes no assumptions about noise bandwidth, as shown in Fig. 2 for three EMF signals (waveforms 1, 4, and 5 in Fig. 1). It may be seen that $V_M(L, \omega)/[S_n(\omega)]^{1/2}$ increases with decreasing ω for all signals, which is characteristic of array response. All signals appear to have adequate (i.e., within an order of magnitude of thermal noise amplitude) SNR to be detectable, in the relatively low-frequency range, by the membrane target, despite widely different pulse characteristics. Fourier spectral analysis of these waveforms[28] shows that the two most important signal parameters to match the low-pass high-gain characteristics of the cell array are pulse width and burst count (see Fig. 1). It is for this reason that a narrow pulse waveform with high burst count (5/5 μs, 500) has *low-frequency* behavior similar to a wider pulse with lower burst count (200/20 μs, 21), both at identical peak E and repetition rate. The very wide (ms range) single-pulse signals in Fig. 1 also result in adequate SNR for a 1mm cell array.

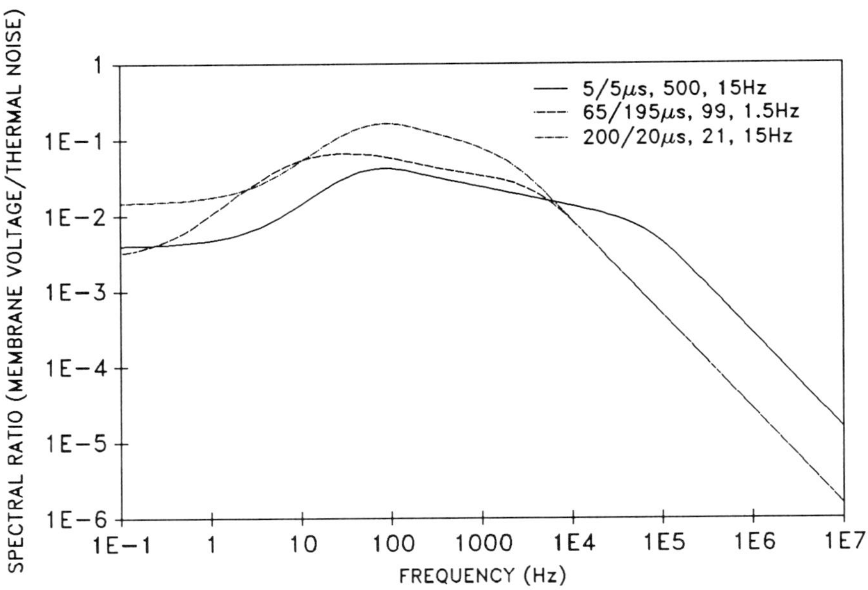

FIG. 2.—Comparison of the spectral ratio of induced transmembrane voltage to thermal noise for three common pulse waveform EMF signals in clinical use or trial for modulating bone repair. Fourier analysis reveals that pulse width and burst count are important to achieve adequate SNR at the low frequencies required for a cell array (here $L = 1$ *mm*).

It is of interest to examine SNR for the 76.6Hz sinusoidal waveform currently under clinical trials for bone repair. This frequency was originally chosen on the basis of the ion resonance theory[13] representing harmonics of Ca^{2+} and Mg^{2+} at a parallel DC magnetic field of 20 μT. Peak signal amplitude for induced E is 10^{-3} mV/cm, significantly below that of all the pulse waveforms considered here. As shown in Fig. 3, the spectral ratio of 76.6 Hz actually *exceeds* that for a common pulsating EMF signal.

DISCUSSION AND CONCLUSIONS

It is apparent that all waveforms evaluated in this study, including the extremely weak 76.6Hz ion-resonance sinusoidal signal, induce sufficient transmembrane voltage at the ends of a 1mm cell array for adequate SNR. Clearly, for a given pulse width (≤200 μs) and peak E (1-10 mV/cm), burst count is one of the most important waveform parameters to satisfy the frequency and sensitivity requirements of the cell array model. Waveform 4 (Fig. 1) has the largest burst count for the widest pulses and is the only waveform, to the author's knowledge, that has accelerated the overall biomechanical healing curve for bone repair in an animal model.[29] This finding is further supported by results from a similar animal model,[30] in which the authors concluded that 20μs symmetrical pulses with 250 burst count, vs 200μs waveform with approximately 20 burst count, was most effective in causing an increase of three-point bending stiffness vs control. It is evident that all the EMF waveforms analyzed here according to the cell-array model are at the threshold of effectiveness. This result may explain why most signals approved at present require several months to achieve a satisfactory clinical result for delayed and nonunion fractures.

A remaining question is whether the B or the E field is the effective stimulus. The sensitivity analysis presented here strongly suggests that $E(t)$ from all clinically relevant EMF waveforms has the spectral density and peak amplitude to be detectable in thermal noise. Experimental support comes from studies of calcium

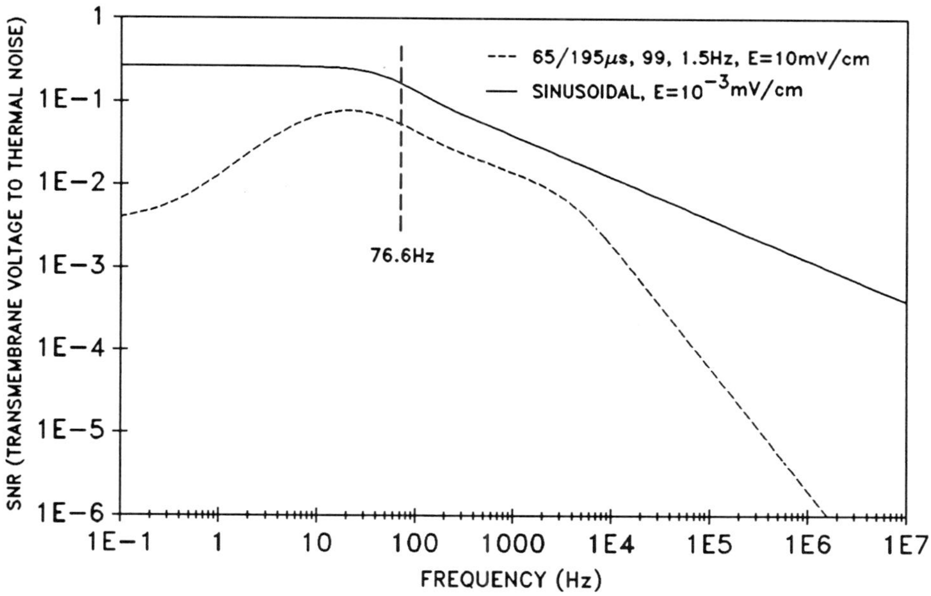

FIG. 3.—Evaluation of SNR (ratio of transmembrane voltage to thermal noise spectra) for the 76.6Hz ion resonance signal under trial for bone repair. SNR is higher for sinusoidal signal ($E_{peak} = 10^{-3}$ mV/cm) than for a pulsating EMF signal ($E_{peak} = 10$ mV/cm) in clinical use. Both signals induce sufficient transmembrane voltage for adequate SNR at frequencies between 10 and 100 Hz. As for Fig. 2, $L = 1$ mm.

transport in mitogen-stimulated thymic lymphocytes that scale with the induced electric field, not with the applied magnetic field.[31] The basis of the experiment was control of the level of the induced E field, at constant dB/dt, by the size of the current loop in specially constructed annular culture dishes. As shown in Eq. (1), induced electric field amplitude depends on loop radius for cylindrical geometry. The in vivo experiments consisted of exposure of fibular osteotomies in the rabbit to a clinically effective pulse burst EMF applied via an external coil or implanted electrodes.[32] The results showed that the observed 1.4 times biomechanical acceleration of bone repair depended only on the in situ electric field, not on the large magnetic component present in the external coil signal.

The cell-array model may offer a relatively simple approach to explaining the possible role of streaming potentials in skeletal tissue growth, repair, and maintenance.[33-36] Streaming potentials are an endogenous source of $E(t)$ which can be detected and spatially amplified by the cell array as a function of its length, just as for exogenous E. The problem has always been that $E(t)$ from streaming potentials is most often a slowly time-varying signal, i.e., one having negligible amplitude at frequencies above 100 Hz. The frequency response of ion or ligand binding at a single cell membrane is too high for it to detect $E(t)$ from streaming, which has made it difficult to accept streaming potentials as the primary stimulus. On the other hand, the much higher EMF sensitivity and much lower frequency response of the cell-array tissue model means that the overall response time of the same membrane process is within the frequency range of $E(t)$ from streaming potentials. It thus appears that this tissue model may offer a relatively simple explanation for the response of hard tissue to mechanical input via electrochemical information transfer, providing further evidence for the elusive mechano-electric hard-tissue regulatory pathway.

As the EMF sensitivity analysis presented here continues to gain experimental support, it becomes relatively straightforward to predict the waveform parameters around which clinical signals should be designed. The objective would be to satisfy the frequency and amplitude response of physiologically relevant cell arrays (real tissue) such that SNR could be comfortably above unity. Certainly all the clinically relevant signals studied here have marginally effective SNR. Studies utilizing signals constructed for more optimal SNR would surely advance the state of the art in electromagnetic therapeutics, and shed significant light upon the physical mechanism of action of both therapeutic and environmental EMF.[37]

REFERENCES

1. Z. B. Friedenberg, M. C. Harlow, and C. T. Brighton, *J. Trauma* 11: 8831, 1971.
2. C. A. L. Bassett, A. A. Pilla, and R. Pawluk, *Clin. Orthop.* 124: 117-131, 1977.
3. R. O. Becker, *J. Bone Joint Surg.* 43A: 643, 1961.
4. I. J. Yasuda, *Japanese Orthop. Surg. Soc.* 28: 267, 1954.
5. C. A. L. Bassett, *Calc. Tiss. Res.* 1: 252, 1968.
6. C. T. Brighton, *J. Bone Joint Surg.* 63A: 847, 1981.
7. A. A. Pilla, *Proc. 7th IECEC,* Washington, D.C.: American Chemical Society, 1972, 761-764.
8. A. A. Pilla, *Ann. N.Y. Acad. Sci.* 238: 149, 1974.
9. A. A. Pilla, R. E. Schmukler, J. J. Kaufman, and G. Rein, in A. Chiabrera, C. Nicolini, and H. P. Schwan, Eds., *Interactions Between Electromagnetic Fields and Cells,* New York: Plenum Press, 1985, 423-436.
10. A. A. Pilla, J. J. Kaufman, and J. T. Ryaby, in M. Blank and E. Findl, Eds., *Mechanistic Approaches to Interactions of Electric and Electromagnetic Fields with Living Systems,* New York: Plenum Press, 1987, 39-62.
11. A. Chiabrera, M. Grattarola, R. Viviani, and C. Braccini, *Stud. Biophys.* 90: 77-81, 1982.
12. A. Chiabrera, M. Grattarola, and R. Viviani, *Bioelectromagnetics* 5: 173-178, 1984.
13. B. R. McLeod and A. R. Liboff, *Bioelectromagnetics* 7: 177-189, 1986.
14. V. V. Lednev, *Bioelectromagnetics* 12: 71-75, 1991.
15. A. A. Pilla, P. R. Nasser, and J. J. Kaufman, in M. J. Allen, S. F. Cleary, A. E. Sowers, and D. D. Shillady, Eds., *Charge and Field Effects in Biosystems—3,* Boston: Birkhauser, 1992, 231-241.
16. A. A. Pilla, P. R. Nasser, and J. J. Kaufman, *Bioelectrochemistry and Bioenergetics* 30, 1992.
17. A. A. Pilla, P. Sechaud, and B. R. McLeod, *J. Biol. Phys.* 11: 51, 1983.
18. W. H. Hayt, *Engineering Electromagnetics,* New York: McGraw-Hill, 1981, 346-358.
19. C. T. Brighton and S. R. Pollack, *J. Bone Joint Surg.* 67A: 577-585, 1985.
20. R. K. Adair, *Phys. Rev. A* 43: 1038-1049, 1991.
21. K. J. McLeod, R. C. Lee, and H. P. Ehrlich, *Science* 236: 1465-1468, 1987.
22. J. D. Sheridan and M. M. Atkinson, *Ann. Rev. Physiol.* 47: 337-353, 1985.
23. W. R. Adey, *Neurochem. Res.* 13: 671-692, 1988.
24. W. R. Loewenstein, *Physiol. Rev.* 61: 829-841, 1981.
25. W. H. Fletcher, W. W. Shiu, D. A. Haviland, C. F. Ware, and W. R. Adey, *Proc. BEMS 8th Annual Mtg.,* Madison, Wis., 1986, p. 12.
26. J. C. Weaver and R. D. Astumian, *Science* 247: 459-462, 1990.
27. L. J. DeFelice, *Introduction to Membrane Noise,* New York: Plenum Press, 1981, 243-245.
28. A. A. Pilla and J. J. Kaufman, *J. Bioelectricity* 3: 3-18, 1984.
29. A. A. Pilla, M. Figueiredo, P. R. Nasser, J. J. Kaufman, and R. S. Siffert, *Trans. Orthoped. Res. Soc.* 17: 529, 1992.
30. D. Pienkowski, S. R. Pollack, C. T. Brighton, and N. J. Griffith, *J. Orthopaedic Res.* 10: 247-255, 1992.
31. R. P. Liburdy, *FEBS Letters* 301: 53-59, 1992.
32. A. A. Pilla, M. Figueiredo, P. R. Nasser, J. J. Kaufman, and R. S. Siffert, this volume.
33. C. A. L. Bassett, *Calcif. Tiss. Res.* 1: 252-272, 1968.
34. S. R. Pollack, *Orthopaedic Clinics of North America* 15: 3-14, 1984.
35. M. Otter, V. R. Palmieri, D. D. Wu, K. G. Seiz, L. A. MacGinitie, and G. V. B. Cochran, *J. Orthopaedic Res.* 10: 710-719, 1992.
36. W. R. Walsh and N. Guzelsu, *J. Orthopaedic Res.* (in press).

37. The author sincerely thanks Richard Dugot (Biolectron), John Erickson (AME), Neil Griffith (IMD), Ewa Herbst (EBI), and James Ryaby (Orthologic) for graciously providing technical and clinical information concerning their commercially available EMF devices for bone repair.

Martin Blank, Ed., *Electricity and Magnetism in Biology and Medicine*

NERVE REGENERATION: IMPLICATIONS FOR CLINICAL APPLICATIONS OF ELECTRICAL STIMULATION

B. F. Sisken* and J. Walker

University of Kentucky, Lexington

Injuries involving the extremities frequently cause trauma to the peripheral nerves innervating these structures, resulting in chronic functional impairment. This secondarily affects restoration of associated tissues, and may contribute to the onset of muscle atrophy. It is apparent that the faster the nerve regenerates, the sooner restorative processes and recovery begin. Following a crush or transection injury, the nerve fibers distal to the level of injury degenerate; new fibers must regrow from the proximal fibers to innervate the end organs and restore function. Two major problems may occur: regenerating fibers grow into the distal stump incorrectly due to misalignment at the transection site, and distal end organs undergo atrophy in the interim as a consequence of this misalignment and subsequent slow growth rate of nerve fibers. The importance of enhancing nerve regeneration cannot be overestimated since it is known that healing of almost all soft and hard tissues depends crucially on a functional nerve supply. Therefore, any procedure that hastens the regenerative process simultaneously reduces the incidence of muscle atrophy and accelerates return of normal function.

An important aim of many clinical and basic investigations in neurobiology is to develop new, improved methods to treat damaged nerves. Studies performed in our laboratory and elsewhere have documented enhancement of neurite outgrowth when specific electric or electromagnetic fields are applied to primary neurons in vitro[1] or to lesioned peripheral nerves in animal models.[1-10] Therapeutic use of electromagnetic fields for human peripheral nerve injuries has great potential and could make a significant contribution, since few noninvasive technologies are available for such problems.

CLINICAL PROBLEM: NERVE INJURY

Sunderland[11] has classified peripheral nerve injuries into five categories:

1. Axonal continuity without degeneration, fully reversible local changes, recovery occurs in weeks. No neuronal loss.
2. Axonal injury is acute so that axonal degeneration occurs below the injury. Recovery occurs in months; the higher (closer to the spinal cord) the injury, the longer the time to recovery. No neuronal loss.
3. Axonal degeneration, neuronal loss, scar tissue formation, misdirected growth, muscle atrophy with some loss of function.
4. Acute injury so that neuroma may form; requires excision of damaged segment and surgical repair of nerve.
5. Injury so severe that little function is restored.

In injuries involving transection of axonal fibers (numbers 3-5 in Sunderland's categories) have the least probability of functional return, since the nerve cell body is isolated from its distal components. Axons distal to the injured site undergo degeneration and many of the nerve cell bodies do not survive. Reapproximation of the proximal and distal portions often results in morphological and functional misalignment. Furthermore, the rate of degeneration is 1-3 mm/day, so that the target tissue (primarily skin and muscle) is essentially denervated for long periods and undergoes atrophy. After a crush lesion (injury number 2 above), regeneration occurs faster and more precisely; very few nerve cell bodies degenerate.

*Correspondence to: B. F. Sisken (Anatomy and Neurobiology), University of Kentucky, Lexington KY 40506.

RECOVERY: ASSESSMENT OF NERVE FUNCTION

In considering this topic readers should consult the following article and texts: Berger and Schaumberg,[12,13] and texts by Lundborg,[14] Terzis and Smith,[15] and Sunderland.[11]

Clinically, sensory recovery is measured by response to touch (as cotton wisps), pain (pinprick), and temperature. Touch and pain are more reliably tested by use of a pin scratch, which produces a series of temporal and spatial stimuli. A well-performed two-point discrimination test can be used to determine tactile sense accurately. Assessment of sensory return is measured in animal models by the ''pinch'' test,[3,7] which assesses the earliest growth cones of regenerating nerve fibers. Other, more technology-sophisticated tests involve determination of evoked sensory potentials[13] and use of the vibrating probe.[4] Motor recovery can be measured by nerve conduction velocity and surface-recorded compound muscle action potential. The continued increase in recovery of motor potential amplitude after injury indicates axonal regeneration.

The following sequence for recovery of motor function has been presented by Sunderland:[11]

1. regeneration of motor axons
2. reinnervation of muscle fibers as revealed by electrical activity
3. maturation of nerve fibers to end organs in muscle
4. voluntary muscle contractions and movement
5. restoration of co-ordinated motor functions
6. complete recovery

Assessment of recovery of motor function in animal models frequently employs compound action-potential determinations, as well as histologic and morphometric procedures. Both have deficiencies: the former reflects activity only in the largest-diameter axons, the latter (whether counting large or small-diameter axons) may not be relevant to functional return. According to Berger and Schaumberg,[12,13] a more clinically useful technique is to measure functional performance (number of stairs climbed; ability to open bottles, button shirts, walk on toes and heels, etc.) which employs both sensory and motor components as well as central nervous system integration. By noting distances covered or time required to complete the task, one can gain an appreciation and quantitation of functional improvement as a function of time after injury. For the same reasons, many basic neuroscientists are employing a gait analysis[16,17] to assess time-dependent recovery of function after peripheral nerve injury. Data from such studies have been reported in animals treated with electrical stimulation.[4,5,9,10]

ELECTROMAGNETIC FIELDS

The application of electromagnetic fields from Helmholtz coils to induce healing appears to be one of the most popular of the noninvasive techniques. This technology is fairly easy to use, does not cause undue heating or abrasion of the skin, and is now available in portable units. Most of the clinical uses of these fields have addressed hard-tissue repair.[18] More recently, this technology has been extended to treat soft-tissue lesions such as ulcers, and eventually could be a treatment of choice for injuries involving skin, nerve, muscle, tendon, etc.

Most of the basic studies testing the efficacy of this technology on the progress of regeneration of injured peripheral nerves have used small mammals (rodents, cats) as models. These models have been accepted as being appropriate, since many of the sequences of degeneration and regeneration as a function of time, type of injury, and distance from the injury site are similar to those found in humans.

Tests by various electric and electromagnetic fields have used central and peripheral nervous tissue in animal models and in culture preparations. The data indicate that fields of widely different characteristics (DC to high frequencies) are effective.

Some of the early research on electromagnetic field stimulation of nerve tissue used Diapulse equipment (a 27MHz RF signal, 5-120 mW/cm^2).[6] These studies indicated some early enhancement of regeneration. Pulsed electromagnetic fields (PEMF, repetition rate 72 Hz, ElectroBiology, Inc.) at much lower energy were used by Ito and Bassett in the rat after transection of the sciatic nerve,[2] and 15 Hz PEMF (ElectroBiology, Inc.) were used in the cat after transection of the common peroneal nerve.[5] In the former study, recovery time for plantar flexion was shortened with PEMF. In the latter study, preservation of neuron populations in the spinal cord was maintained after nerve transection, whereas sham control exhibited significant neuronal loss. In another study, a 0.3mT PEMF signal repeating a 2 Hz (Bietic Research, Inc., Lyndhurst, N.J.) was tested on a transected rat sciatic nerve model.[10] Based on the sciatic-function walking test, this PEMF signal demonstrated significant improvement in gait at 140 days and continued to increase relative to other groups up to the end of the experiment (165 days).

Our studies with a 0.3mT PEMF signal at a repetition rate of 2 Hz (Bietic Research, N.J.) has been tested after a crush lesion of the rat sciatic nerve and found to stimulate regeneration significantly. The enhancement in regeneration rate (22%) of peripheral nerve is comparable to that found after conditioning lesions[19d] application of hormones (testosterone)[20] or forskolin.[21,22] When animals were pretreated with PEMF before crush injury, with no further treatment, significant increases in the nerve regeneration rate were found.[3] This "pretreatment" with PEMF strongly resembles the heightened regenerative response obtained after a "conditioning" (pre-injury wound) lesion.

Similar increases in regeneration rate (21%) have been reported in experiments with 50Hz, 0.4mT sinusoidal fields.[7] It is of interest that no pretreatment effects were obtained with these sinusoidal fields.

CONCLUSION

Although electrical stimulation devices have yet to be used in the clinic for healing for nerve injuries, there is little doubt that such "healing aids" will be common in the future. For this to become a reality, more experimentation on transection lesions on rodent and larger mammalian models should be encouraged. In addition, development of more specific functional testing is needed so that one can assess total integration of sensory and motor components.[23]

REFERENCES

1. B. F. Sisken, "Electric and pulsed electromagnetic field effects on nerve tissue regeneration," in C. T. Brighton and S. R. Pollack, Eds., *Electromagnetics in Medicine and Biology,* San Francisco: San Francisco Press, 1991, 259-274.

2. H. Ito and C. A. L. Bassett, "Effect of weak pulsing electromagnetic fields on neural regeneration in the rat," *Clinical Orthop. and Related Research* 181: 283-290, 1983.

3. M. Kanje, A. Rusovan, B. F. Sisken, and G. Lundborg, "Pretreatment of rats with pulsed electromagnetic fields enhance regeneration of the rat sciatic nerve," *Bioelectromagnetics* (in press).

4. J. M. Kerns, A. J. Fakhouri, H. P. Weinrib, and J. A. Freeman, "Electrical stimulation of nerve regeneration in the rat," *Neuroscience* 40: 93-107, 1991.

5. M. G. Orgel, W. J. O'Brien, and H. M. Murray, "Pulsing electromagnetic field therapy in nerve regeneration: An experimental study in the cat," *Plast. Reconstr. Surg.* 73: 173-183, 1984.

6. A. R. M. Raji and R. E. M. Bowden, "Effects of high-peak pulsed electromagnetic field on the degeneration and regeneration of the common peroneal nerve in rats," *J. Bone and Joint Surg.* 65: 478-492, 1983.

7. A. Rusovan and M. Kanje, "Stimulation of regeneration of the rat sciatic nerve by 50 Hz sinusoidal magnetic fields," *Exp. Neurol.* 112: 312-316, 1991.

8. B. F. Sisken, M. Kanje, G. Lundborg, E. Herbst, and W. Kurtz, "Stimulation of rat sciatic nerve regeneration with pulsed electromagnetic fields," *Brain Res.* 485: 309-316, 1989.

9. J. Walker, S. Guarnieri, P. Meade, and B. F. Sisken, "Effect of pulsed electromagnetic fields on functional recovery after rat sciatic nerve crush as assessed by video footprint analysis: Preliminary results," this volume.

10. R. J. Zienowicz, B. A. Thomas, W. H. Kurtz, and M. G. Orgel, "A multivariate approach to the treatment of peripheral nerve transection injury: The role of electromagnetic field therapy," *Plast. Reconstr. Surg.* 87: 122-129, 1991.

11. S. Sunderland, *Nerve Injuries and Their Repair: A Critical Appraisal,* Paris: Churchill Livingstone, 1991.

12. A. R. Berger and H. H. Schaumberg, "Rehabilitation of Peripheral Neuropathies (review)," *J. NeuroRehab.* 2: 25-36, 1988.

13. A. R. Berger and H. H. Schaumberg, "Rehabilitation of Focal Nerve Injuries (review)," *J. NeuroRehab.* 2: 65-91, 1988.

14. G. Lundborg, *Nerve Injury and Repair,* Paris: Churchill Livingstone, 1989.

15. J. Terzis, *The Peripheral Nerve: Structure, Function and Reconstruction,* New York: Raven Press, 1990.

16. C. J. Brown, S. E. MacKinnon, P. J. Evans, J. R. Bain, A. P. Makino, R. T. Hunter, and G. M. T. Hare, "Self-evaluation of walking-track measurement using a sciatic function index," *Microsurgery* 10:

226-235, 1989.

17. L. de Medinaceli, W. J. Freed, and R. J. Wyatt, "An index of the functional condition of rat sciatic nerve based on measurements made from walking tracks," *Exp. Neurol.* 77: 634, 1982.

18. C. A. L. Bassett, "Fundamental and practical aspects of therapeutic uses of pulsed electromagnetic fields (PEMFs)," *Crit. Rev. in Biomed. Engineering* 17: 451-529, 1989.

19. I. G. McQuarrie, "The effect of a conditioning lesion on the regeneration of motor axons," *Brain Res.* 152: 597-602, 1978.

20. K. A. Kujawa, E. Emeric, and K. J. Jones, "Testosterone differentially regulates the regenerative properties of injured hamster facial motoneurons," *J. Neurosci.* 11: 3898-3906, 1991.

21. S. L. Kilmer and R. C. Carlsen, "Forskolin activation of adenylate cyclase in vivo stimulates nerve regeneration," *Nature* 307: 455-457, 1984.

22. H. W. Klein, S. Kilmer, and R. C. Carlsen, "Enhancement of peripheral nerve regeneration by pharmacological activation of the cyclic AMP second messenger system," *Microsurgery* 10: 122-126, 1989.

23. This work was supported by Office of Naval Research contract N00014-86-K0221 and NIH grant NS 29621-01, and by the Orthopedic Research Education Foundation.

Martin Blank, Ed., *Electricity and Magnetism in Biology and Medicine*

CLINICAL HYPERTHERMIA: TECHNOLOGICAL ADVANCES WILL REALIZE THE BIOLOGICAL PROMISE

T. C. Cetas*

University of Arizona, Tucson

The biological rationale for hyperthermia is that temperatures a few degrees above normal kill cells, damage vascular structures, inhibit radiation repair, and enhance the effects of some chemotherapeutic agents. Clinical benefits include improved tumor reduction and extended duration of tumor response. Heating systems must provide detailed control of the heating pattern. Regional electromagnetic systems are less ineffective when the patient's temperature is maintained near 37 C. They are effective for systemic heating with chemotherapeutic agents and for prewarming for a local hyperthermia boost. Ultrasonic systems with controlled power distribution appear to have applicability to the widest range of localized tumors. Interstitial systems have been developed for combination with radioisotopes. Microwave arrays are promising for widespread superficial lesions such as chest-wall recurrences, solid breast tumors, and many neck tumors. They may be useful for brain and lung.

Hyperthermia has gone through the common cycle of early promising investigations through wild expectations to disillusionment. Now, a more realistic understanding is leading to the introduction of heating systems and methods that can effectively heat some tumors. We also appreciate how difficult that can be in the general case. The Sixth International Congress on Hyperthermic Oncology (6th-ICHO) was held in Tucson in April 1992. With over 450 proffered papers and nearly 100 invited presentations, the field is still scientifically strong. This brief review is based mostly on that meeting.

BIOLOGICAL RATIONALE REVISITED

The biological rationale for hyperthermia follows from a number of observations. First, cells heated above about 42 C for a time will die. The degree of cell kill depends on the time at final temperature, the rate of heating, the initial temperature, the cell type, the cellular environment, and the thermal history. A recent advance in biological understanding is the identification of heat shock proteins with the induction of thermal tolerance, which unites two fields of basic biology. A second rationale is that hyperthermia sensitizes cells to radiation and to drugs. With radiation therapy, hyperthermia inhibits radiation repair through denaturization of proteins. With chemotherapy, increased uptake of drugs and increased activation rates are positive factors for some drugs. With others, the higher temperatures may increase the rate of expulsion of the cytotoxic agent from the cell and result in thermal protection.

Physiological effects are important both for direct cytotoxicity and for the physical ability to heat tumors. Poor blood flow leads to hypoxia, which protects cells from radiation damage. It also inhibits chemotherapies from reaching the target tumors. On the other hand, poorly perfused tumors are easier to heat owing to reduced convective cooling.

CLINICAL OBSERVATIONS

Clinical studies confirm the biological expectations. When hyperthermia is used in conjunction with radiation, the clinical response in terms of "complete disappearance of the tumor" (CR) is approximately doubled and the duration of that response is increased. At the 6th-ICHO, Gonzalez Gonzalez reported preliminary results of one of the first randomized, multi-institutional studies in hyperthermia with adequate quality control.[1] In 124 malignant melanoma tumors from 11 institutions, the CR rate for combined hyperthermia and radiation was 60% compared with 33% for radiation alone. The long-term control at 5 years is 46% and

*Correspondence to: T. C. Cetas (Radiation Oncology), University of Arizona, Tucson AZ 85724.

29%, respectively. (This trial is now completed; the numbers are final.)

The Radiation Therapy Oncology Group (RTOG) completed a randomized trial (81-04) in 1987 of the combined treatment vs radiation alone for superficial tumors (300 tumors entered).[2] If all tumors are considered, no difference exists (30% CR for radiation alone vs 32% for the combined therapy). For tumors smaller than 3 cm, the respective rates were 39% vs 52% (statistically insignificant; $p = 0.51$). However, the probability of the tumors for which a complete response was achieved remaining a CR after 12 months was 12% for radiation alone and 82% for the combined therapy. Furthermore, the long-term control for the subset (139) of patients who completed a radiation dose of 29 Gy or more was 28% for the radiation alone vs 53% for the combined therapy. Waveguide apertures operating at 915 MHz were used to treat these patients and the 3cm limitation becomes obvious once the field patterns for such applicators are understood. Unfortunately, until very recently, hyperthermia equipment has been developed, distributed, and used in ways that cannot reliably treat tumors of clinical importance. Now, in select academic centers, new systems are being tested that can do so.[3,4]

REGIONAL EM HEATING SYSTEMS

Regional electromagnetic heating systems have been under investigation for about a decade but the clinical results are not exciting. The usual procedure is to attempt to keep the patient's core temperature close to normal while depositing energy from relatively long-wavelength EM fields (ca 60-110 MHz) regionally into the pelvis or abdomen. It is hoped that restricted blood flow in these large tumors will result in high tumor temperatures. Generally, few temperatures measured in tumors reach 42 C. Raising the patient's temperature systematically to 39 or 40 C would result in a much larger portion of the tumor exceeding 42 C. If regional EM heating systems are combined with systemic insulation, then systemic temperatures will quickly rise to 41.8 C within 20 to 30 min,[5] thus reducing the likelihood of inducing thermal tolerance. EM regional systems are also useful when used for producing whole-body hyperthermia in conjunction with chemotherapeutic agents.[1]

ULTRASONIC HEATING SYSTEMS

The primary physical modality for noninvasive localized power deposition in tissues is focused ultrasound. The wavelength, absorption, propagation, and focusing properties are more favorable than electromagnetic techniques. The combination of these properties permits control of the power deposition spatially and temporally. The primary difficulty associated with ultrasound arises from its extremely high absorption in bone and its inability to penetrate air.

Lele[6] initially developed this approach, but the most advanced clinical system development at present is that of Hynynen et al.[7] They have mounted up to six heating transducers and an imaging transducer on a mechanically scanned gantry situated in a waterbed on which the patient lies.

Table 1 lists the tumor sites that have been treated along with the temperature parameters they use to describe the effectiveness of the heating.[8] Problems associated with the presence of bone can be overcome in many circumstances. For example, in a Phase I trial of heat and radiation, ultrasound was used for brain tumors by surgical removal of a skull flap to provide a window for the ultrasound. Now this method is being used in conjunction with carbo-platin for recurrent brain tumors.

TABLE 1.—Scanned focused ultrasound treatment temperature analysis.

Site	No. of patients	Mean T_{max}/T_{min} (C)	% sensors $T \geq 42.5$ C
Pelvis	22	43.6/40.2	33
Chest wall/breast	14	45.1/40.1	43
Head and neck	8	43.4/39.0	22
Groin, perineum	8	44.8/40.6	58
Axilla	7	46.6/40.3	64
Extremity	5	45.6/40.1	37

Prostate heating is difficult with external ultrasound since pelvic bones frequently shield critical portions of the tumor or hyperplasia. Transrectal ultrasonic transducers coupled with imaging transducers have been proved by Lewis et al. to be quite successful.[1] The temperatures achieved are comparable with those listed in Table 1 for the other ultrasound sites.

A final example is provided by Anhalt et al. for scanned focused ultrasound in treating superficial chest-wall tumors.[1] Plane-wave ultrasound is frequently limited by bone pain from the ultrasound hitting the ribs. However, if a focused high-frequency (ca 3MHz) beam is used, the energy sharply diverges beyond the focus and the intensity reaching the ribs is reduced. By using two sets of transducers of differing frequencies simultaneously, one can make adjustments for the varying thickness of the flesh over the entire field. The scanning and variable gain give control in two dimensions and the variable frequency mix gives control in the third. A warm-water bolus must be used to help smooth out the thermal fields and reduce heat loss. Another improvement comes from the use of a patient-controlled pain button. When the ultrasonic beam causes pain at a location as it scans, the patient presses a button that causes the computer to reduce power whenever the beam passes that point. The power can be thus maintained at a high level for the rest of the scanned volume (Table 2).

TABLE 2.—Superficial scanned ultrasound temperature summary.

No. of treatments	Mean T_{max}/T_{min} (C)	% sensors $T \geq 42.5$ C
First series with 4MHz and 1MHz transducers		
56 total	44.1/39.8	36
20 best	44.5/40.0	43
Second series with 4MHz and 1MHz transducers, pain button, and power control		
12 total	45.9/41.1	70
5 best	46.4/41.2	74

Ultrasound is also being investigated for applications using high-temperature pulses. Here the thermal dose is the same, but the treatment is delivered in a short pulse to a small volume. Perfusion effects can be ignored and the treatments can be thus guided by SAR determinations alone. An exciting application under investigation by Hynynen et al. is to place the ultrasound transducer on subjects while they are in an MRI scanner.[1] The production of lesions can be observed in real time with MRI providing a means of noninvasive surgery.

ELECTROMAGNETIC ARRAYS

The primary principle for effective hyperthermia treatments is multiple-parameter control with fine resolution both spatially and temporally. Such control is less precise with electromagnetic means. Nevertheless, the presence of bone and air, and occasionally the limitations of mechanical scanning mechanisms, prevent the ultrasound from reaching some tumors. There are many important sites where electromagnetic methods may be more appropriate. Microwave heating of brain tumors would eliminate the need for a craniectomy, as required for ultrasound. The head has very complicated tissue heterogeneities that are not easily amenable to ultrasound. Lung tumors cannot be treated effectively by any means of hyperthermia at present. Finally, microwave arrays already have demonstrated effectiveness in treating widespread superficial diseases (e.g., Kapp et al.,[1] Fessenden et al.,[1] Hand et al.[1]).

The group at Stanford (Kapp et al. and Fessenden et al.) first developed scanning spiral applicators and achieved good success. They then proceeded to develop arrays of large numbers of small spiral applicators, which also have been well tested. Table 3 shows their reported thermometry results for both types of applicator systems. Recently, a clinically robust current sheet applicator has been used in small arrays. In contrast to the Stanford spiral applicators, the CSAs are linearly polarized and have less sensitivity in coupling to the presence of air bubbles and to tissue variations. CSAs have been used in arrays of up to four applicators. Thermometry results of Hand et al.[1] with these CSAs are also given in Table 3. Both the Stanford group and Hand point out that incoherent arrays are preferred over coherent arrays for heating widespread superficial disease. An advantage of linearly polarized applicators is that they can be operated coherently in order to achieve heating that would otherwise be impossible with microwave sources. Effective heating to 5 cm of a phantom model and of a real tumor in an otherwise intact breast has been demonstrated with these CSAs. This phenomenon is well understood and has been discussed previously by Nilsson et al. in reference to waveguides.[9]

TABLE 3.—Time-averaged temperatures for microwave array systems.

Stanford (California) experience	Mean T_{max}/T_{min} (C)	% sensors $T \geq 42.5$ C (interpolated)
Scanning spiral 10	45.2/39.9	48.5
Scanning spiral 15	44.9/40.3	51.0
Array of 6	45.2/39.5	42.5
Array of 12	44.6/39.5	44.5
Array of 25	44.6/39.5	35.0
MRC Hammersmith (London) experience		
Array of 4 CSAs	44.0/39.8	40.5
Array of 2 CSAs	43.5/40.9	47.5

Richards has considered the theoretical problem of determining, under the best of circumstances, which tumors can be treated by electromagnetic fields with a fully optimized aperture field on the surface of the body in question.[1] He uses a "mini-max" optimization technique for which the SAR in the target is fixed and the algorithm searches for the maximum, undesired hot spot, and then minimizes it subject to the constraints imposed. The routine continues until the optimal aperture field is determined. Thus far, he has shown that it is theoretically possible to heat tumors in the brain (layered-sphere model) noninvasively with microwave frequencies between 300 and 900 MHz. What is perhaps surprising is that the intuitive approach of setting the phase of all sources to add at the target does not even approximate the optimal configuration. The same approach can be used for lung and for other head and neck tumors.

CONCLUDING THOUGHTS

The biological promise of hyperthermia as an adjunct to cancer therapy for both radiation and drugs remains. New systems are being developed that will make this promise a clinical reality. But the new systems require sophisticated engineering to develop and proper scientific understanding in their application. The most important mindset for the engineers is to provide multiple-parameter control over the heating in spatial and temporal detail. The clinical benefits are great, but so are the risks to patients from systems that can do harm.[10]

REFERENCES

1. E. W. Gerner and T. C. Cetas, Eds., *Hyperthermic Oncology 1992* (in press).

2. C. A. Perez, T. Pajak, B. Emami, N. B. Hornback, L. Tupchong, and P. Rubin, "Randomized Phase III study comparing irradiation and hyperthermia with irradiation alone in superficial measurable tumors," *Am. J. Clin. Oncol. (CCT)* 14: 133-141, 1991.

3. M. D. Sapozink, G. Joszef, M. A. Astrahan, F. A. Gibbs, Z. Petrovich, and J. R. Stewart, "Adjuvant pelvic hyperthermia in advanced cervical carcinoma: I. Feasibility, thermometry and device comparison," *Int. J. Hyperthermia* 6: 985-996, 1990.

4. R. J. Myerson, L. Leybovich, B. Emami, P. W. Grigsby, W. Straube, and D. von Gerichten, "Phantom studies and preliminary clinical experience with the BSD 2000," *Int. J. Hyperthermia* 7: 937-951, 1990.

5. M. K. Croghan, D. S. Shimm, K. H. Hynynen, D. P. Anhalt, S. S. Dean, A. M. Fletcher, S. H. Dyet, and T. C. Cetas, "Effect of surface warming on patients receiving regional-systemic hyperthermia," *Int. J. Radiat. Oncol. Biol. Phys.* 19: 51, 202, 1990.

6. P. P. Lele, "Physical aspects and clinical studies with ultrasound hyperthermia," in F. C. Storm, Ed., *Hyperthermia in Cancer Therapy*, Boston: Hall Medical Publishers.

7. K. Hynynen, R. Roemer, D. Anhalt, C. Johnson, Z. X. Xu, W. Swindell, and T. C. Cetas, "A scanned, focused, multiple transducer ultrasonic system for localized hyperthermia treatments," *Int. J. Hyperthermia* 3: 21-35, 1987.

8. P. M. Harari, K. Hynynen, R. B. Roemer, D. P. Anhalt, D. S. Shimm, B. Stea, and J. R. Cassady, "Development of scanned focussed ultrasound hyperthermia: Clinical response evaluation," *Int. J. Radiat. Oncol. Biol. Phys.* 21: 831-840, 1991.

9. P. Nilsson, T. Larsson, and B. Persson, "Absorbed power distributions from two tilted waveguide applicators," *Int. J. Hyperthermia* 1: 29-43, 1985.

10. This work was supported in part by grants CA 29653 and CA 39468 from the National Cancer Institute.

Martin Blank, Ed., *Electricity and Magnetism in Biology and Medicine*

ION TRANSPORT THROUGH BIOMEMBRANES, SKIN, AND OTHER TISSUES: APPLICATIONS TO DRUG DELIVERY

R. O. Potts*

Cygnus Therapeutic Systems, Redwood City, Calif.

Drug delivery is often impeded by barrier membranes that occur throughout biological systems. Recent evidence suggests that these barriers are primarily lipid in nature, although the specific composition and morphology of each may vary. The skin is a particularly important site of potential drug administration where delivery is impeded by a multilamellar lipid domain, the stratum corneum. The use of electric fields to permeabilize barrier membranes (electroporation) reversibly has been widely demonstrated in other lipid systems. Arguments are presented here that suggest that a similar reversible alteration of stratum corneum lipids may result in enhanced drug delivery through the skin.

The therapeutic uses of electromagnetic fields are widely reported throughout biology and medicine. Existing clinical applications such as enhanced bone and wound repair are extensively discussed in the present volume. Moreover, a fundamental understanding of bioelectrical phenomena such as signal transduction, nerve conduction, and electrofusion is crucial to the development of new drug therapies. In addition, since molecular migration can be controlled by an electric-field gradient, and since these fields can reversibly alter the very barrier properties that impede transport, novel means of drug delivery are now possible. This paper deals with the use of electric fields to enhance and control drug delivery through the skin and related tissues.

Mass transport in biological systems occurs primarily via passive diffusion in a chemical potential gradient or via facilitated transport involving energy-mediated, biological pumps.[1] The former involves exceptionally slow rates ($D_{passive} \leq 10^{-5}$ cm^2/s); the latter processes are highly specific (e.g., Na^+, K^+, or glucose transport). Thus, biological barriers such as cell membranes, skin, and other epithelia are very effective at preventing the uptake of xenobiotic materials. Recent evidence suggests that these biological barriers are primarily lipid in nature, consisting of single lamella or multilamellar composites, leading to the possibility of a common mechanism to enhance transport.[2] An impermeable and/or selective barrier is of clear advantage when an organism is exposed to infections or toxic substances, but the delivery of therapeutically beneficial drugs is limited.

Phamaceutical compounds are often delivered via oral administration in order to build systemic levels, which in turn equilibrate with local tissue to achieve the therapeutic concentration needed at the affected site. As a consequence, systemic toxicity is often a requisite risk associated with the therapeutic benefit of drug therapy. Furthermore, once ingested, the rate of drug delivery is controlled by passive diffusion and biological processes, leading to variation in systemic drug concentration among individuals, as well as imprecise control over the kinetics of drug delivery. Finally, orally administered drugs are often extensively metabolized in the liver before they are further distributed within the body (the "first-pass" effect), leading to potential liver toxicity and reduced drug bioavailability in systemic circulation.

A number of nonoral drug-delivery strategies have been developed to control the input rate, to reduce "first-pass" metabolism and/or to achieve a therapeutic concentration solely at the target site. For example, temporal control of drug delivery is often achieved by encapsulation of the compound in a polymer matrix, where release characteristics are determined by drug diffusion out of the matrix.[3] Many of the site-specific approaches are biological in origin, such as the covalent attachment of drug to a ligand that is recognized by the target organ.[4] Electric fields have also been widely used to affect enhanced temporal and/or localized drug delivery, especially in superficial tissue. One of the best examples is that of iontophoresis, the use of DC electric fields to enhance and control the rate of drug delivery through the skin. Recent evaluations have provided considerable insight into the mechanism of iontophoretic transport through the skin.[5-7]

---

*Correspondence to: R. O. Potts, Cygnus Therapeutic Systems, 400 Penobscot Dr., Redwood City CA 94063.

Iontophoretic transport involves both diffusion in a chemical (activity) potential gradient ($d\mu/dx$), and ion migration in an electric field gradient (dV/dx), and can be described by the Nernst-Planck equation

$$J_i = u_i c_i [(d\mu_i/dx) + z_i F(dV/dx)] \tag{1}$$

In Eq. (1), the flux of the ith ion (J_i) is related to its mobility u_i, charge z_i, and concentration c_i; Faraday's constant is denoted as F. In addition, the skin has a net negative charge and is therefore permselective with preferred transport of positive ions. As a consequence of this permselectivity, net transport of uncharged molecules occurs in the direction of positive ion flow, a process called electroosmosis. Thus, the net transport through the skin is described by

$$J_i = u_i c_i [(d\mu_i/dx) + z_i F(dV/dx)] + c_i v \tag{2}$$

where v is the permeant's velocity due to convective (electroosmotic) transport. It can be shown that the electroosmotic term $c_i v$ in Eq. (2) is proportional to the applied current.[5]

Since iontophoretic transport is proportional to the potential gradient across the skin, the extent and rate of delivery can be precisely maintained by the applied electric field. Several important potential drug delivery applications rely heavily on this concept, including enhanced delivery of therapeutic drugs and proteins.[5-7] Of equal importance, the temporal aspects of drug delivery through the skin can be controlled, which leads to a number of applications involving "pulsitile" drug delivery.[6] Electrically enhanced delivery holds particular promise for the administration of compounds requiring complex kinetics. Reverse iontophoresis (extraction of compounds through the skin) also provides the opportunity for noninvasive diagnostic assays. For example, glucose extraction through the skin could replace existing invasive blood assays required of diabetics. In vitro results obtained with human and mouse skin[8] showed that glucose (an unionized compound) was preferentially extracted through the skin at the cathode (due to the permselective nature of the skin), in direct proportion to the glucose concentration beneath the skin and the applied potential (or current in these experiments, in which the resistance remained constant). When coupled with a glucose biosensor, an iontophoretic extraction device could provide the potential for continuous noninvasive evaluation of serum glucose levels.

The electrical and permeability resistance of mammalian skin resides primarily in the thin outer layer called the stratum corneum, a complex composite of cellular remnants embedded in a three-dimensional lipid matrix (Fig. 1). The stratum corneum lipids form multilamellar domains and comprise the only continuous phase in this tissue. In spite of profound compositional differences between the lipids of the stratum corneum and more conventional phospholipid bilayers, the mechanisms of diffusional transport in each appear remarkably similar involving free-volume fluctuations in the alkyl chains.[9]

The mechanism of ion transport through the stratum corneum and cell membrane lipids may also be similar. We have measured the complex electrical impedance of skin[10,11] and found that the electrical properties of the stratum corneum can be modeled by a 50kΩ-cm^2 resistor and a 30nf/cm^2 capacitor in parallel (Fig. 2). The high electrical impedance of the skin is primarily attributed to the lipids of the stratum corneum. Several lines of evidence support the assignment of electrical impedance to the stratum corneum lipids. First, due to their extracellular localization, stratum corneum lipids can be extracted, which results in a 100-fold decrease in impedance.[12] Second, molecules (such as oleic acid) that are known to perturb the lipid structure of the stratum corneum[13] cause a dramatic decrease in the electrical impedance (Fig. 2). In addition, we have measured the impedance as a function of temperature[11] and found that an increase in temperature to 60 C resulted in a large increase in C and decrease in R. Nearly identical results have been obtained with human skin.[12] Since stratum corneum lipids undergo a phase transition at 60 C,[14] these results suggest strongly that the impedance to ion transport through the stratum corneum is associated with extracellular lipids. A similar temperature-induced change in C for erythrocyte membranes[15] was attributed to an increased dielectric constant due to water uptake into fluctuating free-volume voids within the lipid hydrocarbon chains of the erythrocyte membrane. Similar lipid free-volume fluctuations have been demonstrated in the stratum corneum.[9] Thus, the electrical properties of the skin can be associated with the stratum corneum lipids in a manner similar to results obtained with phospholipid-based bilayer cell membranes.

The surface of mammalian skin contains hair follicles, sweat glands, and other appendages (Fig. 3). Due to the high resistance of the surrounding stratum corneum, iontophoretic drug delivery occurs primarily through these shunts.[5] However, they comprise less than 0.1% of the total skin surface area, so that high current densities occur near these structures. Tissue damage that can result at iontophoretic potentials above about 10 V is most likely due to this high localized current density.[5] One possible means to overcome such

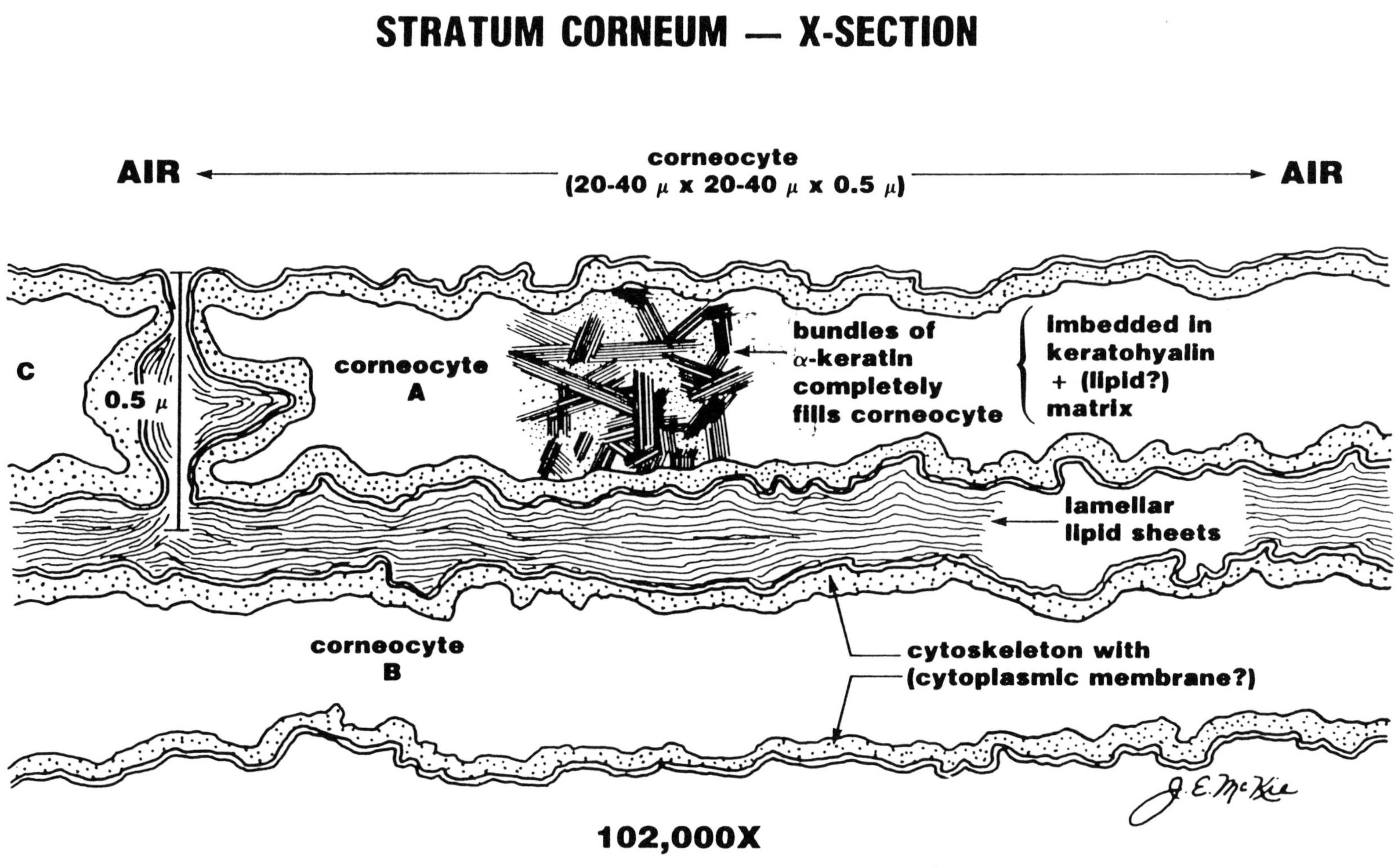

FIG. 1.—Schematic representation of outer surface of stratum corneum. (Courtesy of Dr. J. E. McKie.)

tissue damage and at the same time increase delivery is to permeabilize the surrounding stratum corneum reversibly. This goal has been achieved via chemical means by the use of compounds that reversibly alter the stratum corneum lipid structure;[13] or transient electric fields may alter skin permeability. Such tissue permeabilization has not been widely investigated, yet the lipoidal nature of the stratum corneum, coupled with results showing reversible electroporation of cell membranes and other lipid lamellar structures, suggests that similar effects could occur in skin. Tissue electroporation of frog skin has been demonstrated by Powell et al.[16] Moreover, these investigators noted that the voltage-time characteristics of frog-skin electroporation were similar to those seen in simpler bilayer systems. Subsequent work by this group (reported in the present volume) suggests that impressive and reversible enhancement of transdermal delivery can be achieved in human skin by use of transient electric fields of millisecond duration and 100-1000V amplitudes. Although in vivo results are yet to be obtained, these results point to the potential for enhanced drug delivery through the skin by electroporative devices.

Finally, although the use of electroporation to enhance transdermal drug delivery is a logical first step, there are clearly other applications. Electrically enhanced drug delivery to peripheral tumors was recently demonstrated by Mir and co-workers.[16,17] Bleomycin, a potent cytotoxic anticancer drug, was used in conjunction with a pulsed electric field to treat both spontaneous and transplanted tumors in mice. In treated animals, bleomycin was administered systemically (via injection) at a site distal to the tumor. After a 30min delay to allow the drug to be distributed uniformly throughout the body, a square-wave electric pulse (1500 V/cm, 100 μs) was applied to the skin above the tumor site. There was significantly greater tumor regression in animals treated by the drug and a pulse than in animals that received either treatment alone or no treatment at all. The importance of these results can best be demonstrated by the fact that 5 μg of bleomycin followed by an electrical pulse to the tumor site achieved regression equivalent to more conventional therapy in which 1 mg of drug per day was administered via injection for 5 days. Given the extreme systemic toxicity of bleomycin, electrochemotherapy can lead to greater drug toleration, yet produce equivalent clinical efficacy.

In conclusion, site-specific and temporally controlled drug delivery is impeded by the formidable barrier membranes predominant throughout nature. These barriers appear primarily lipoidal in nature, and it is therefore reasonable to suggest that reversible lipid membrane permeabilization (i.e., electroporation) is a viable means for enhancing drug delivery in tissues throughout the body.

REFERENCES

1. W. D. Stein, *Transport and Diffusion Across Cell Membranes,* New York: Academic Press, 1986.

2. R. O. Potts, and R. H. Guy, "Do lamellar lipid structures share a common transport mechanism?" *Proc. 19th Int. Symp. Control. Rel. Bioact. materials,* 1992.

3. H. A. Nash, "Controlled release systems for contraception," in R. S. Langer and D. L. Wise, Eds., *Medicinal Application of Controlled Release,* Boca Raton, Fla.: CRC Press, 1984.

4. J. S. Friedberg, R. G. Tompkins, S. L. Rakestraw, S. W. Warren, A. J. Fischman, and M. L. Yarmush, "Antibody-targeted photolysis," in W. J. M. Hrushesky, R. Langer, and F. Theewes, *Temporal Control of Drug Delivery, Ann. N.Y. Acad. Sci.,* 1991, vol. 618, p. 383.

5. R. R. Burnette, "Iontophoresis," in J. Hadgraft and R. H. Guy, Eds., *Transdermal Drug Delivery, Developmental Issues and Research Initiatives,* New York: Marcel Dekker, 1989, 247.

6. G. A. Lattin, R. V. Padmanabhan, and J. B. Phipps, "Electronic control of iontophoretic drug delivery," in *Temporal Control of Drug Delivery,* in Ref. 4.

7. V. Srinivasan, S. M. Sims, W. I. Higuchi, C. R. Behl, A. W. Malick, and S. Pons, "Iontophoretic transport of drugs: A constant voltage approach," in J. Kost, Ed., *Pulsed and Self-regulated Drug Delivery,* Boca Raton, Fla.: CRC Press, 1990, 65-88.

8. P. Glikfeld, R. S. Hinz, and R. H. Guy, "Noninvasive sampling of biological fluids by iontophoresis," *Pharm. Res.* 6: 988, 1989.

9. R. O. Potts and M. L. Francoeur, "Lipid biophysics of water loss through the skin," *Proc. Natl. Acad. Sci. USA* 87: 3871-3873, 1990.

10. R. O. Potts, R. H. Guy, and M. L. Francoeur, "Routes of ionic permeability through mammalian skin," *Solid State Ionics* 53-56: 165-169, 1992.

11. S. Y. Oh, L. Leung, D. Bommannan, R. H. Guy, and R. O. Potts, "Effect of current, ionic strength, and temperature on the electrical properties of skin," *J. Controlled Release* (in press).

12. A. C. Allenby, J. Fletcher, C. Schock, T. F. S. Tees, "The effect of heat, pH, and organic solvents on the electrical impedance and permeability of human skin," *Brit. J. Dermatol.* 81: 31, 1969.

13. M. L. Francoeur, G. M. Golden, and R. O. Potts, "Oleic acid: Its effects on stratum corneum in relation to (trans)dermal drug delivery," *Pharm. Res.* 7: 621, 1990.

14. G. M. Golden, D. B. Guzek, A. H. Kennedy, J. E. McKie, and R. O. Potts, "Stratum corneum lipid phase transitions and water barrier properties," *Biochemistry* 26: 2382, 1987.

15. J-Z. Bao, C. C. Davis, and R. E. Schmukler, "Frequency domain impedance measurements of erythrocytes," *Biophys. J.* 61: 1427, 1992.

16. K. T. Powell, A. W. Morgenthaller, and J. C. Weaver, "Tissue electroporation: Observation of reversible electrical breakdown in viable frog skin," *Biophys. J.* 56: 1163, 1986.

17. L. M. Mir, S. Orlowski, J. Bělehrádek, C. Paolietti, "Electrochemotherapy potentiation of antitumour effect of bleomycin by local electric pulses," *Eur. J. Cancer* 27: 68, 1991.

18. J. Bělehrádek, J. Orlowski, S. Peddevin, C. Paoletti, and L. M. Mir, "Electrochemotherapy of spontaneous mammary tumours in mice," *Eur. J. Cancer* 27: 73, 1991.

19. R. R. Burnette and T. M. Bagniefsky, "Influence of constant current iontophoresis on the impedance and passive Na^+ permeability of excised nude mouse skin," *J. Pharm. Sci.* 77: 492, 1988.

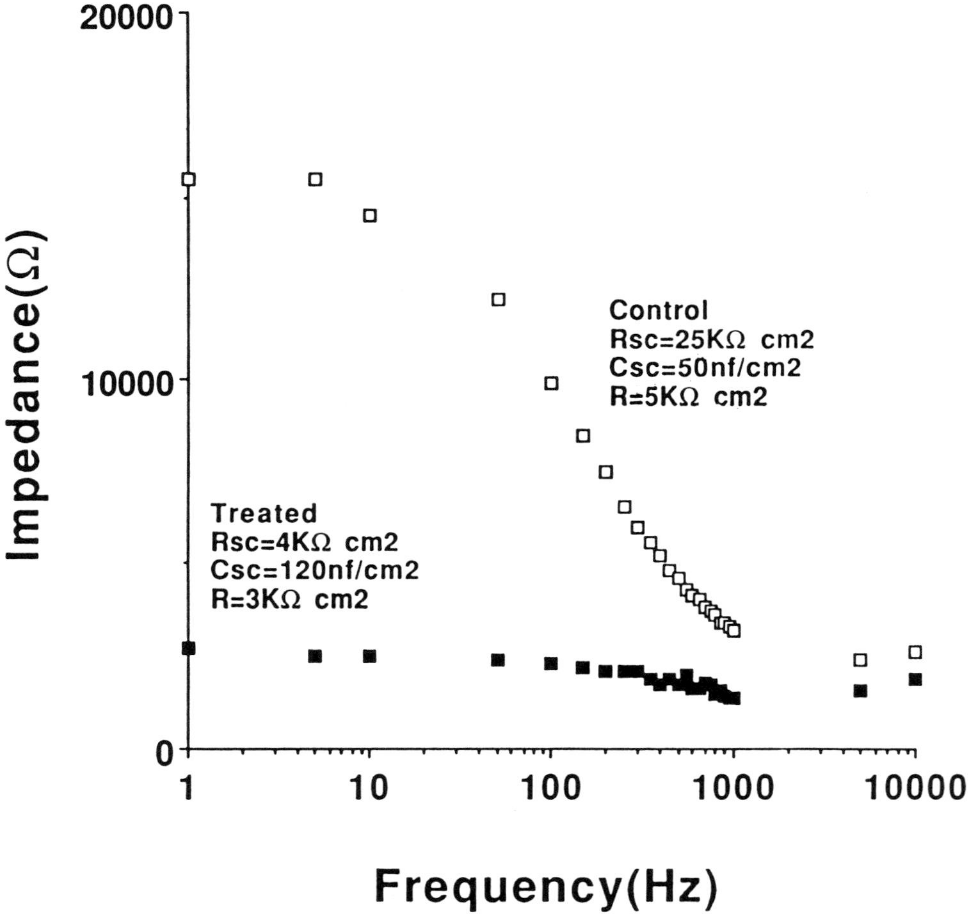

FIG. 2.—Electrical impedance of hairless mouse skin as a function of frequency. A four-electrode arrangement was used to apply sinusoidal current of 1 μA across full-thickness skin mounted between the chambers of a side-by-side diffusion apparatus. Each side of skin was in contact with HEPES buffer. The stratum corneum side of the treated sample was in contact with a solution of 0.25% oleic acid in ethanol-water (40-60 w/w/) for 2 h prior to impedance measurements. Oleic acid was removed and chamber was flushed with several changes of buffer immediately prior to the electrical measurements. Data were analyzed by conventional means on the assumption of a parallel RC circuit.[19]

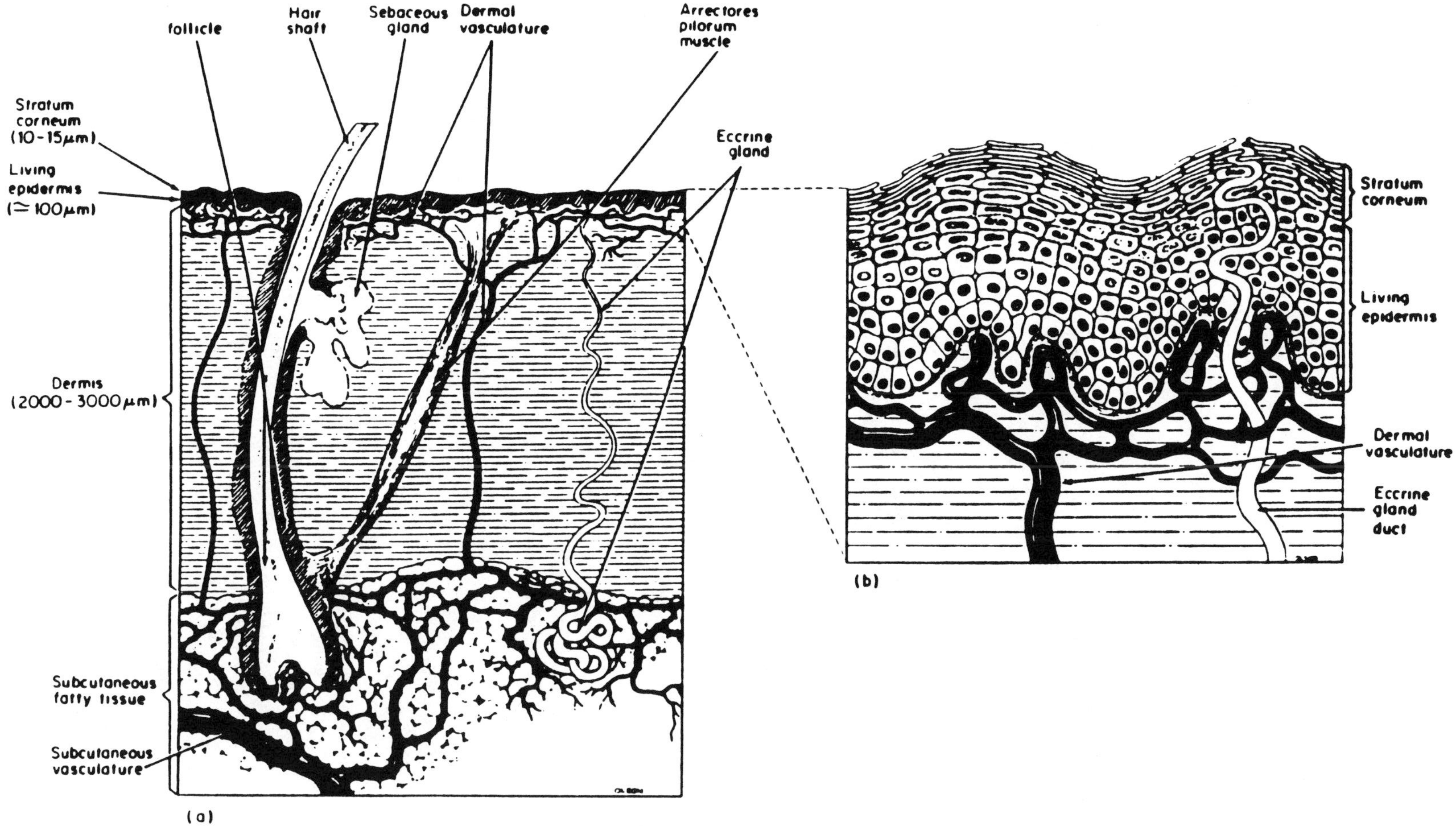

FIG. 3.—Major structures of skin.

Martin Blank, Ed., *Electricity and Magnetism in Biology and Medicine*

ANALYSIS OF GENE REGULATION

J. D. Saffer*

Pacific Northwest Laboratories, Richland, Wash.

Understanding the interaction of electromagnetic fields with biological systems requires a description of events at the cellular and molecular levels. Much current research is utilizing modern biological techniques to address these issues. Several experimental approaches are discussed and their advantages and pitfalls are highlighted. The techniques to be covered encompass methods for analyzing gene expression, including transcriptional assays. The information provided should help researchers critically evaluate and plan cellular and molecular studies.

Several new techniques in molecular biology have made it possible to address many facets of gene regulation and signal transduction. Some of these techniques may be considered relatively old in molecular biology terms (i.e., 10 to 15 years), but their application to problems in bioelectromagnetics is still very recent. We provide information needed to choose the appropriate methods and understand their limitations. Experimental details are not given, but can be found in the cited references.

The correct technique to investigate any specific problem depends on the biology of the processes in question. For biological effects of electromagnetic fields, work is being done or evaluated by scientists of diverse training and expertise. Accordingly, some basic biological concepts along with the discussion of techniques are included, and the need to understand the biochemistry as well as the method is demonstrated.

OVERVIEW OF THE REGULATION OF RNA TRANSCRIPT LEVELS

Stimulated by the work of Goodman and Henderson[1-5] considerable interest has developed in electromagnetic-field effects on RNA transcript levels. Before describing the methods available to examine messenger RNA levels, let us review how production of RNA is regulated.

A typical mammalian gene (Fig. 1) for a protein product consists of exons (blocks of coding information), introns (divider regions that do not appear in the final mRNA), and regulatory information that directs when and to what extent RNA copies (transcripts) of the gene are made. The primary transcript, covering all the exons and introns, is processed to remove the intron sequences, and a length of adenosine bases, the poly(A) tail, is added; not all mammalian genes contain introns and not all RNAs have poly(A) tails. The final processed transcript is transported to the cytoplasm where it is stored, translated into protein, or degraded. All the processes from initiation of transcription to the accumulation of mRNA in the cytoplasm are highly regulated.

The first level of control involves the initiation of transcription. Regulatory information is contained in several regions in and around the gene (Fig. 1). The promoter (**P**) contains the basic signals that identify the beginning of a transcription unit. As such, it is always located just upstream (i.e., 5′) of the first exon (1). The promoter binds RNA polymerase II through interaction with several other proteins that recognize DNA sequences in the promoter. The establishment of this transcription-capable complex can be regulated by DNA sequences called modulators (**M**). Modulators, including positive-acting enhancers and negative-acting reducers, are often located at a distance (along the DNA strand, not necessarily in space); these regulatory elements can be located upstream of the gene (5′), downstream (3′), or within the transcription unit. Promoters and modulators are under complex control including the pathways of signal transduction and second messenger signaling, and together act to regulate the initiation of transcription.

In some cases, the rate of initiation of transcription is the limiting step in RNA production and is directly reflected in the abundance of mRNA. However, there are many other levels of control. The production of the primary transcript can be regulated following the beginning of transcription. Regulatory sequences called

*Correspondence to: J. D. Saffer (Biology & Chemistry), Pacific Northwest Laboratories, Richland WA 99352.

attenuators (A) within the gene can cause the transcription machinery to pause providing control over elongation. The processing of the primary transcript to a final mRNA and the transport of the message to the cytoplasm also provide the cell with additional control options, although these steps are usually very rapid and efficient. Finally, the rate of degradation of the mRNA in the cytoplasm contributes to its availability for translation.

METHODS FOR ISOLATING RNA

Isolation of RNA is the first step in most analysis procedures. The major concerns are to extract the RNA quantitatively and to avoid degradation. The various locations of RNA in the compartmentalized cell affect the ability to purify all species equally. The quantitative extraction of RNA is also affected by the number of steps in the procedure; in general, more manipulations lead to less quantitative isolations. Because ribonucleases (RNases) are abundant and stable, most procedures attempt to get the RNA quickly into conditions in which RNases are inactive.

Isolation of "total RNA" (i.e., all RNA in the cell) addresses both major concerns.[6] Agents such as guanidinium thiocyanate can be used to lyse the cell rapidly and denature all proteins including the RNases. Because the entire cell is solubilized, extraction is efficient with a minimal number of steps and intact RNA can be obtained. On the negative side, care must be taken to avoid DNA contamination, which in some cases can affect subsequent analysis. If information is needed about cellular localization, this method is inappropriate.

Although the nuclear RNA contributes very little to the total amount of RNA, in some cases it is important to analyze specifically the RNA in the cytoplasm. Cells can be lysed with nonionic detergents that leave the nuclear membrane intact.[7] The nuclei with the DNA can then be quickly removed by centrifugation. This simple and fast method has the disadvantage that RNases may be active during the initial lysis steps, so that intact RNA is more difficult to isolate. RNase inhibitors added to the lysis buffer may improve this somewhat.

Messenger RNA usually represents only about 2% of the total RNA in a cell. When analyzing very low abundance transcripts, the mRNA may need to be separated from the total or cytoplasmic RNA. The usual approach is to select for RNA containing poly(A) tails by hybridization to immobilized polymers of thymidine.[8] Most, but not all, mRNAs have poly(A) tails, which permits substantial enrichment of this population. Although greater sensitivity is an advantage, isolation of poly(A)-containing RNA is much less quantitative and problems with RNase activity become critical.

More or less mRNA as detected by hybridization does not necessarily mean that the amount being used for protein production changes accordingly. Therefore, it is sometimes necessary to isolate the RNA that is specifically associated with polysomes, the translation apparatus.[9] Polysomes are separated from cytoplasmic extracts by gradient centrifugation and the associated RNA is purified. Because of the additional steps, quantitative extraction of intact RNA is problematic.

ANALYSIS OF RNA

The steady-state level of any mRNA species in the cell can be determined by several methods, each with particular advantages and pitfalls. The method of choice is determined by what processes are being examined and what other information is needed. In some cases, the data from one method are complementary to those from another, so that multiple methods can be combined to strengthen conclusions. For example, the steady-state level of any mRNA species depends on the rates of synthesis and degradation; measurements of the steady-state level, the synthesis rate, and the degradation rate should provide consistent results. Combining approaches is particularly important when quantitation is difficult (e.g., when small changes are observed).

In each method described below, substantial controls are required for obtaining quantitative results. It is difficult to isolate RNA with precisely the same efficiency for each preparation, and dispensing equal amounts of RNA can be equally problematic. Internal controls must be utilized; unfortunately the best controls are not always easy to identify. RNAs for structural proteins (e.g., actin, tubulin, or histone) have often been used, but the amounts of these mRNAs also vary considerably with cell changes. In addition, the integrity of RNA preparations and the linearity of hybridization conditions must be defined for each experiment. The parameters of the detection method also need to be defined; for example, autoradiography is not necessarily linear.

"Northern blot" analysis[10] is perhaps the most common method for analyzing RNA. Briefly, the RNA is fractionated by gel electrophoresis, transferred to a nylon membrane, and specific sequences are detected by hybridization to a labeled probe. This procedure not only defines how much RNA of a specific gene is present, but also its size. For example, transcription of the c-*myc* gene (Fig. 2) originates from two major promoters, **P1** and **P2**, 167 base pairs apart. High-resolution electrophoresis can resolve the resulting mRNAs so that two closely spaced bands can be observed. When the Northern blots are used for quantitation, this size information helps distinguish specific hybridization from nonspecific background. As with many genes, transcription of c-*myc* is known to occur simultaneously from the opposite (noncoding) strand. Additional hybridization from this antisense transcript can be eliminated by use of a single-stranded probe specific for the message. As low as 2-fold changes in specific RNA concentration can be observed, but quantitation of this change from bands on an autoradiograph can be difficult. In addition to the general controls noted above, the amount of RNA loaded in each gel lane to be compared and the amount transferred to the membrane should be verified. Both desiderata can be readily achieved by blot shadowing.[11]

When size information is not needed, "slot blots" can allow quantitation of changes in mRNA levels more easily than Northern blots. As the name implies, RNA is fixed to the membrane in a well-defined slot (rectangle about 1×5 mm) or a dot of 3-5 mm in diameter (dot blot). As with Northern blots, specific mRNAs are then detected by hybridization to a labeled probe. The two major advantages of slot blots are that the fixed shape and location of the slot or dot make it possible to count radioactivity directly or to determine density on a film; and slight degradation of the RNA does not affect the results because all the RNA is fixed in one location. This method also has some drawbacks. Accurate quantitation requires no cross-hybridization, since all signal shows up in the slot; adequate controls must establish this criterion. Because of variability, the slot blot internal controls should be done with additional hybridization on the same filter; care must be taken to remove all signal from the previous detection. For the c-*myc* example, the use of single-strand probes is absolutely essential, since the sense and antisense transcripts do not vary coordinately.

When only minute amounts of material are available for analysis, the "polymerase chain reaction" can provide an excellent alternative. This method is not only the most sensitive available, but is very fast and does not require radiolabeled probes. For the amplification of RNA,[12] a cDNA copy of the RNA, made by use of reverse transcriptase, is amplified in a standard PCR reaction; the resulting product(s) are then analyzed by gel electrophoresis. Because of the possible nonlinearity in the amplification reaction, quantitation is accomplished either by inclusion of various concentrations of a competitor or by careful establishment of the linear range.[13] With appropriate controls the method can be accurate to less than 2-fold differences. For genes such as c-*myc,* the primers could be chosen to allow detection of all sense-strand transcripts, or **P**1-specific messages alone.

When more information is needed about the size and structure of mRNA, "protection assays" can be very powerful. These methods, including S1 nuclease analysis[7] and RNase protection assays[14] can precisely (to the nucleotide) define the ends of the mRNA and the splice junctions. A short probe (homologous to sequences near the endpoint or junction to be examined) is hybridized in solution to the isolated RNA; then single-stranded sequences are removed with the appropriate nuclease. The remaining probe, which was protected by being duplexed with the specific RNA, is analyzed by gel electrophoresis. Because the RNA for these assays only needs to cover the region from the short probe to the end or junction in question, slight endonucleolytic degradation can be tolerated. The solution hybridizations also have the advantage of being highly quantitative so that 2-fold differences can be observed. Depending on the method used, there is trade-off between sensitivity and background problems. Protection assays can be used for quantitating c-*myc* expression since the probes are strand-specific.

"Primer extension" is another procedure capable of defining endpoints of RNA molecules and is particularly useful for determining the sequences at splice junctions. A short probe is hybridized to RNA and extended to the end of the transcript by use of reverse transcripterase.[15] Depending on the gene, 3-fold changes can be observed. However, it is difficult to reverse transcribe some sequences and quantitation is less reliable.

The preceding methods are designed to measure the amount of mRNA in the cell. For some questions, the rate of synthesis of the transcript is more relevant. "Nuclear run-off analysis" can be used to measure the transcription rate directly.[16] Previously initiated transcripts are extended in isolated nuclei under conditions in which no new initiation can occur. Labeled precursors are used during the extension, and the resulting RNA is hybridized to an excess of DNA fixed on a membrane, usually in a slot blot format. Because the RNA can be labeled to high specific activity, this method is very sensitive and can provide excellent quantitation. However, the nuclei are susceptible to damage, which affects RNA synthesis and hence the quantitation. Another drawback is that this method assumes the overall rate of RNA synthesis is the same in the control and experimental groups, an assumption that must be confirmed for each set of experiments. Nuclear runoffs can also disclose the presence of an attenuator. For example, the attenuator in c-*myc* was found after observation of a signal that differed according to which region of the gene was immobilized on the filter; fewer transcripts hybridized to the end of the first exon than the beginning.[17] In general, several regions of the gene should be probed not only to locate attenuators, but also because hybridization to a 3′ probe most accurately represents production of full-length mRNA.

To complement the determination of synthesis rate, "message stability assays" can be used to measure the decay rate of mRNAs.[18] Cells are treated with an inhibitor of RNA synthesis (e.g., actinomycin). Analysis of the RNA abundance (by any of the above methods) as a function of time then defines the decay rate. In general, the quantitative abilities of this method reside in the actual method used for assaying RNA abundance. If the treatment of the cells with the RNA synthesis inhibitor itself affects the decay rate, the results can be erroneous; but in general this is a useful method.

TESTING REGULATION WITH "DESIGNER GENES"

The results from the RNA analyses described provide substantial information, but each method has limitations that must be considered in interpreting data. To corroborate data from these assays, specific hypotheses can be tested by means of "designer genes." It is possible to create nearly any artificial gene imaginable by genetic-engineering techniques. These artificial genes or "constructs" can be transfected into mammalian cells or used to create transgenic animals.

To illustrate how this approach can be used, consider the following hypothetical situation. Northern blot analysis has indicated that the amount of mRNA from the *emf* locus increases on exposure to pulsed magnetic fields. This increase was reflected in synthesis rate as measured by nuclear run-off assays. To verify a transcriptional activation mechanism, a carefully designed gene could be constructed using the promoter and modulator sequences from *emf* coupled to a marker gene (typically an easily assayed nonmammalian gene like chloramphenicol acetyltransferase or luciferase). If magnetic-field exposure causes a change in transcription initiation rate, expression from this construct should also respond and not depend on the sequences in the marker gene.

A complementary construct could also be made using the *emf* exons and introns coupled to the promoter from a gene known to be unresponsive to the stimulus in question. In our working model, there should be no effect of the magnetic field on this construct. On the other hand, if the mRNA analyses have shown that the response was due to a change in mRNA stability, the results with the two constructs would be reversed.

This scenario could be extended to define the underlying mechanism for the magnetic field effect. Specific regulatory elements from the responsive gene can be removed from the first designer gene above or added to other constructs. This procedure lets the researcher "map" the field-responsive element. Standard means can then be applied to identify the protein that interacts with the responsive element and determine the signal transduction pathways that alter its activity. Although the construction of the test genes and the definition of transduction pathways are complex and time-consuming, the approach has been used often and is entirely realistic.

This hypothetical exercise covers only some of the areas that can be approached with engineered genes. The major advantage is that specific models can be tested, not only in cells in vitro, but also in whole animals. Furthermore, in appropriately designed experiments, the major points can be made with qualitative observations rather than reliance on more difficult, and sometimes unconvincing, quantitations. The approach has drawbacks as well, the most important of which is that artificial genes do not always correctly represent their natural counterparts. Overall, however, no other existing methods are more likely to give access to the mechanisms of electromagnetic field effects.

REFERENCES

1. R. Goodman and A. Henderson, "Some biological effects of electromagnetic fields," *Bioelectrochem. Bioenergetics* 15: 39, 1986.

2. R. Goodman and A. Henderson, "Stimulation of RNA synthesis in the salivary gland cells of *Sciara coprophila* by an electromagnetic signal used for treatment of skeletal problems in horses," *J. Bioelectricity* 6: 37, 1987.

3. R. Goodman, J. Abbott, and A. Henderson, "Transcriptional patterns in the X chromosome of *Sciara coprophila* following exposure to magnetic fields," *Bioelectromagnetics* 8: 1, 1987.

4. R. Goodman, L.-X. Wei, J.-C. Xu, and A. Henderson, "Exposure of human cells to low-frequency electromagnetic fields results in quantitative changes in transcripts," *Biochim. Biophys. Acta* 1009: 216, 1989.

5. L.-X. Wei, R. Goodman, and A. Henderson, "Changes in levels of c-*myc* and histone H2b following exposure of cells to low-frequency sinusoidal electromagnetic fields: Evidence for a window effect," *Bioelectromagnetics* 11: 269, 1990.

6. P. Chomczynski and N. Sacchi, "Single-step method of RNA isolation by acid guanidinium thiocyanate-phenol-chloroform extraction," *Anal. Biochem.* 162: 156, 1987.

7. A. J. Berk and P. A. Sharp, "Sizing and mapping of early adenovirus mRNAs by gel electrophoresis of S1 endonuclease-digested hybrids," *Cell* 12: 721, 1977.

8. H. Aviv and P. Leder, "Purification of biologically active globin messenger RNA by chromotography on oligothymidylate acid-cellulose," *Proc. Natl. Acad. Sci. USA* 69: 1408, 1972.

9. E. W. Müllner, B. Neupert, and L. C. Kühn, "A specific mRNA binding factor regulates the iron-dependent stability of cytoplasmic transferrin receptor mRNA," *Cell* 58: 373, 1989.

10. P. S. Thomas, "Hybridization of denatured RNA and small DNA fragments transferred to nitrocellulose," *Proc. Natl. Acad. Sci. USA* 77: 5201, 1980.

11. S. J. Thurston and J. D. Saffer, "Ultraviolet shadowing nucleic acids on nylon membranes," *Anal. Biochem.* 178: 41, 1989.

12. M. A. Frohman, M. K. Dush, and G. R. Martin, "Rapid production of full length cDNAs from rare transcripts: Amplification using a single gene-specific oligonucleotide primer," *Proc. Natl. Acad. Sci. USA* 85: 8998, 1988.

13. T. Horikoshi, K. D. Danenberg, T. H. W. Stadlbauer, M. Volkenandt, L. C. C. Shea, K. Aigner, B. Gustavsson, L. Leichman, R. Frösing, M. Ray, et al., "Quantitation of thymidylate synthase, dihydrofolate reductase, and DT-diaphorase gene expression in human tumors using the polymerase chain reaction," *Cancer Res.* 52: 108, 1992.

14. D. A. Melton, P. A. Kreig, M. R. Rebagliati, T. Maniatis, K. Zinn, and M. R. Green, "Efficient in vitro synthesis of biologically active RNA and RNA hybridization probes from plasmids containing a bacteriophage SP6 promoter," *Nucleic Acids Res.* 12: 7035, 1984.

15. S. L. McKnight and R. Kingsbury, "Transcription control signals of a eukaryotic protein-coding gene," *Science* 217: 316, 1982.

16. M. E. Greenberg and E. B. Ziff, "Stimulation of 3T3 cells induces transcription of the c-*fos* proto-oncogene," *Nature* 311: 433, 1984.

17. S. Wright and J. M. Bishop, "DNA sequences that mediate attenuation of transcription from the mouse protooncogene *myc*," *Proc. Natl. Acad. Sci. USA* 86: 505, 1989.

18. R. D. Press, D. Samols, and D. A. Goldthwait, "Expression and stability of c-*sis* mRNA in human glioblastoma cells," *Biochemistry* 27: 5736, 1988.

Additional References

F. M. Ausubel et al., Eds., *Current Protocols in Molecular Biology,* New York: Greene Publishing Associates and Wiley, 1991.

M. Kriegler, *Gene Transfer and Expression,* New York: Stockton Press, 1990.

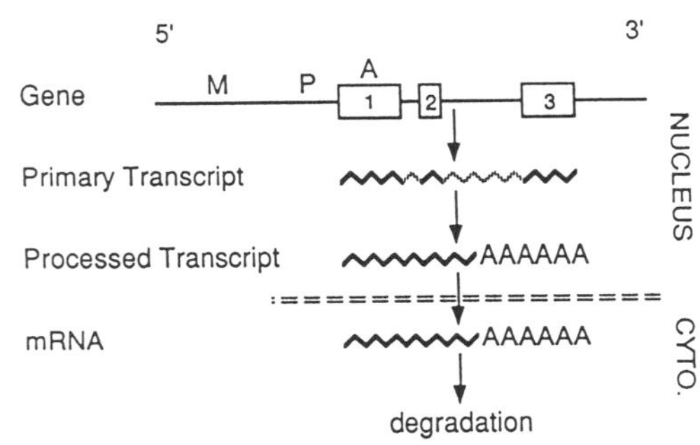

FIG. 1.—Production of mRNA.

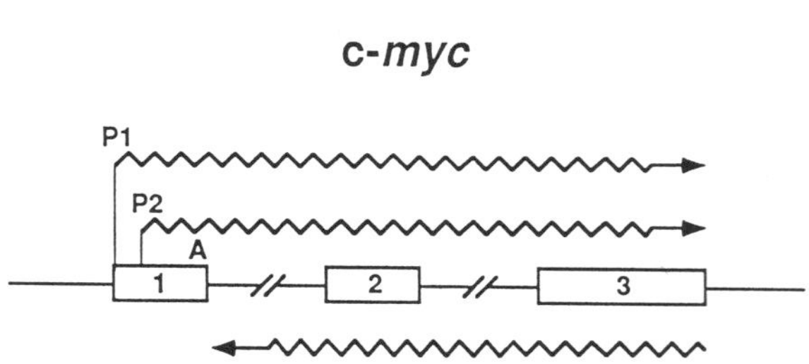

FIG. 2.—RNA transcription of c-*myc*.

Martin Blank, Ed., *Electricity and Magnetism in Biology and Medicine*

BIOPHYSICAL MECHANISMS

A. Chiabrera* and B. Bianco

University of Genoa, Italy

G. D'Inzeo

University of Rome "La Sapienza"

The assessment of the biological effectiveness of low-intensity electromagnetic exposure is an open problem. There is growing experimental evidence for such an effectiveness, but experiments are hardly reproducible. Moreover, the theoretical approaches are still inadequate to handle satisfactorily such complicated phenomena, though there are promising indications. Major points of the subject are briefly stressed, and a list of key references is given.

Understanding the mechanisms of action of the electromagnetic fields on biosystems implies investigations at various physical scales (i.e., organization levels) of the biological media. The macroscale deals with a biological system exposed to EMFs generated by external sources (exogenous fields). The mesoscale corresponds to the tissue level, and the microscale is the realm of cellular and molecular phenomena.

From Macroscale to Mesoscale

The first step, from macroscale to mesoscale, is necessary for the evaluation of the intensities of electric and magnetic field components or the electromagnetic power density inside a tissue or a composite medium, as a function of the external sources. This approach implies knowledge of the electromagnetic characteristics (e.g., effective complex electric permittivity and magnetic permeability, at low exposure levels) of the materials and the solution of Maxwell's equations with the appropriate boundary conditions. The experimental and theoretical techniques needed have long been well established, but have been used only for solving a limited number of practical cases. To produce a correct interpretation of the induced effects, this approach should be applied to each experimental situation. That is not always possible, because of the limited availability in the various laboratories of expertise, of measured data on specific biomaterials or tissues, and of sufficient computer power.

From Mesoscale to Microscale

The second step, from mesoscale to microscale, is the most important from a biological point of view and deserves long-term interdisciplinary efforts of the scientific community. A better description and understanding of this problem can be achieved if we consider, first, the exogenous field interaction with the cell components (field interactions); and second, the consequent modification of cellular activity. Among the available models of field interactions a large body of literature exists, corroborated by experimental techniques and data, concerning dielectrophoretic cell drift and rotation, membrane permeability alteration, and protein conformational changes. This knowledge has already led to applications such as cell fusion, electroporation, and electrochemioterapy. These models furnish a theoretical support for the understanding of high- and moderate-intensity field effects. At low field intensities (electric field lower than 10 V/m and magnetic induction lower than 1 mT), the support offered by the proposed models is far from satisfactory. Two classes of models have been studied so far. The first is based on the classical Langevin equation applied to charge messenger (e.g., calcium) movements near the membrane surface. The interaction is described by means of the Lorentz force, so that the action of the exogenous electric field and/or magnetic field, both steady and time varying, can be taken into account. The available theoretical results show the possibility of cyclotron resonances, which parallel some experimental findings. In the second class, the awareness of quantum

---

*Correspondence to: A. Chiabrera (Biophys. & Electronic Engg.), University of Genoa, Via Opera Pia 11a, I-16145 Genoa, Italy.

effects, which may occur in specific membrane sites, has been examined by means of quantum models based on the weak Zeeman and Stark effects.

The inability of the available models to provide a comprehensive explanation of several experimental data points up the need for their refinement, including background noise, or for new proposals.

The next step is the evaluation of the changes in cellular activity. A gap exists between the previous models and this step. The authors simply postulate the existence of the interaction, and they describe it by means of phenomenological parameters that enter one or more biological processes under study. The available models address known membrane processes to explain large biochemical effects induced by low-intensity EMFs. They can be grouped into two classes. Some models are based on the built-in membrane nonlinearities (e.g., Hodgkin and Huxley type of equation); others rely on dynamical field-induced conformation change of membrane proteins.

The lack of models suitable for the variety and the complexity of the above effects show the necessity for a coordinated scientific effort and for direct interaction between theoreticians and experimentalists before a full comprehension of the bioelectromagnetic phenomena can be achieved.

In the Reference Section (which is by no means exhaustive) we report selected books and papers in such a way as to cover the various abovementioned points of interest.

REFERENCE CLASSIFICATION

Exposure Evaluation

Berntsen and Bach Andersen, 1991; Bolomey et al., 1991; Borup and Gandhi, 1985; Broquetas et al., 1991; Caorsi et al., 1990, 1991; Foster and Schepps, 1981; Foster and Schwan, 1986; Fricke, 1953; Gabriel et al., 1987; Grant et al., 1985; Grodzinsky, 1983; Hagmann and Levin, 1990; Kosterich et al., 1983, 1984; Paulsen et al., 1988; Pilla et al., 1983; Schwan, 1984, 1985; Stuchly and Stuchly, 1985; Su, 1989; Takashima et al., 1984; Takashima, 1989; Tsai et al., 1986.

Direct Force Effects

Frankel and Blackmore, 1984; Frankel, 1984; Kirshvink et al., 1992; Molinari and Viviani, 1978; Pastushenko et al., 1985; Pethig, 1979; Sauer, 1985; Sauer and Schloegl, 1985; Schwan, 1989; Semm, 1992.

Electrokinetic Transduction

Basset and Pawluck, 1972; Grodzinsky et al., 1978; Gross and Williams, 1982; Lotke et al., 1974; McGinitie et al., 1987; Pienkowski and Pollack, 1983.

Electrochemical Information Transfer

Asami et al., 1990; Pilla, 1974; Pilla and Margules, 1977.

Interaction with Phospholipids

Bond and Wyeth, 1986, 1987; Cevc and Marsh, 1987; Liburdy and Penn, 1984; Liburdy, 1986; Papahadjopoulos et al., 1973; Scherer and Seelig, 1987; Seelig et al., 1987; Shepherd and Büldt, 1978.

Interaction with Channel Proteins

Alekseyev et al., 1980; Allis and Sinha, 1981, 1982; Sandblom et al., 1985; Arber, 1976; Arber and Lin, 1983, 1984, 1985a&b; Bernardi et al., 1988; Caddemi et al., 1986; Cain, 1980, 1981; Cito et al., 1992; D'Inzeo et al., 1988; Galvin and McRee, 1986; Hille, 1984; Lai et al., 1988; Liburdy and Penn, 1984; Lords et al., 1973; Philippova et al., 1988; Pickard and Rosenbaum, 1978; Seaman and Wachtel, 1978; Seaman et al., 1982; Sheridan et al., 1979; Tinney et al., 1976; Wachtel et al., 1975.

The Classical Langevin-Lorentz Model

Adey et al., 1982; Adey, 1984; Athey, 1981; Bawin and Adey, 1976, 1977; Bawin et al., 1975, 1978; Blackman et al., 1979, 1980, 1985; Chiabrera et al., 1984a, 1985, 1986, 1989; Chiabrera and Bianco, 1987; Durney et al., 1988; Dutta et al., 1984; Halle, 1988; Kaczmarek and Adey, 1974; Liboff, 1985; Liboff and McLeod, 1988; McLaughlin, 1981; McLeod and Liboff, 1986; Merrit et al., 1982; Smith et al., 1987; Thomas et al., 1987.

The Zeeman and Stark Quantum Model

Bianco and Chiabrera, 1992; Chiabrera et al., 1991a, 1991b; Hart, 1990; Lednev, 1991.

Interaction with Transport Processes Across the Membrane

Albanese and Bell, 1984; Blank and Britten, 1978; Blank and Kavanaugh, 1982; Blank, 1983, 1984, 1987a&b.

Nonlinear Ionic Fluxes in Exposed Cells

Borgens et al., 1980; Marconi et al., 1985; McCann et al., 1981; Ortoleva, 1981.

Catalyzed Reactions under EM Exposure

Chiabrera and Rodan, 1984b; Robertson and Astumian, 1990, 1991.

REFERENCES

Adey, W. R., Bawin, S. M., and Lawrence, A. F., "Effects of weak amplitude-modulated microwave fields on calcium efflux from awake cat cerebral cortex," *Bioelectromag.* 3: 295, 1982.

Adey, W. R., "Nonlinear, nonequilibrium aspects of electromagnetic field interactions at cell membranes," in W. R. Adey and A. F. Lawrence, Eds., *Nonlinear Electrodynamics in Biological Systems,* New York: Plenum Press, 1984.

Albanese, R. A. and Bell, E. L., "Radiofrequency radiation and chemical reaction dynamics," in W. R. Adey and A. F. Lawrence, Eds., *Nonlinear Electrodynamics in Biological Systems,* New York: Plenum Press, 1984.

Alekseyev, S. I., Tyazhelov, V. V., Grigor'ev, P. A., and Siden, G. I., "Some aspects of microwave effect on the bilayer membranes modified with gramicidin," *Biophysics* 25: 750, 1980.

Allis, J. W. and Sinha, B. L., "Fluorescence depolarization studies of red cell membrane fluidity: The effect of exposure to 1.0 GHz microwave radiation," *Bioelectromag.* 2: 13, 1981.

Allis, J. W. and Sinha, B. L., "Fluorescence depolarization studies of the phase transition in multilamellar phospholipid vesicles exposed to 1.0 GHz microwave radiation," *Bioelectromag.* 3: 323, 1982.

Arber, S. L., "Effect of microwaves on resting potential of giant neurons of mollusk, *Helix pomatia,*" *Elektron. Obrab. Mater.* 6: 78, 1976.

Arber, S. L. and Lin, J. C., "Microwave enhancement of membrane conductance in snail neurons: Role of temperature," *Physiol. Chem. Phys.* 15: 259, 1983.

Arber, S. L. and Lin, J. C., "Microwave enhancement of membrane conductance in snail neurons: Effects of EDTA, caffeine and tetracaine," *Physiol. Chem. Phys.* 16: 469, 1984.

Arber, S. L. and Lin, J. C., "Microwave-induced changes in nerve cells: Effects of modulation and temperature," *Bioelectromag.* 6: 257, 1985a.

Arber, S. L. and Lin, J. C., "Extracellular calcium and microwave enhancement of membrane conductance in snail neuron," *Radiat. Environ.* 24: 149, 1985b.

Asami, K., Takahashi, Y., and Takashima, S., "Frequency domain analysis of membrane capacitance of culture cells (HeLa and myeloma) using the micropipette technique," *Biophys. J.* 58: 143, 1990.

Athey, T. W., "Comparison of RF-induced calcium efflux from chick brain at different frequencies: Do the scaled power density windows align?," *Bioelectromag.* 2: 407, 1981.

Bassett, C. A. L. and Pawluk, R. J., "Electrical behavior of cartilage during loading," *Science* 178: 982, 1972.

Bawin, S. M., Kaczmarek, K. L., and Adey, W. R., "Effects of modulated VHF fields on the central nervous system," *Ann. N.Y. Acad. Sci.* 247: 74, 1975.

Bawin, S. M. and Adey, W. R., "Sensitivity of calcium binding in cerebral tissue to weak environmental electric fields oscillating at low frequencies," *Proc. Natl. Acad. Sci. USA* 73: 1999, 1976.

Bawin, S. M. and Adey, W. R., "Calcium binding in cerebral tissues," in D. G. Hazzard, Ed., *Biological Effects and Measurement of Radiofrequency Microwaves,* HEW Publ., FDA, 1977.

Bawin, S. M., Sheppard, A. R., and Adey, W. R., "Possible mechanisms of weak electromagnetic field coupling in brain tissue," *Bioelectrochem. Bioenerg.* 5: 67, 1978.

Bernardi, P., D'Inzeo, G., Eusebi, F., Grassi, F., and Tamburello, C., "Microwave-induced desensitization of acetylcholine receptor channels in cultured quail myotubes," *Proc. 10th Annual Meeting of Bioelectromagnetic Society,* Stamford, Connecticut, 1988, abstract.

Berntsen, S. and Bach Andersen, J., "A general formulation of applied potential tomography," *Radio Sci.* 26: 535, 1991.

Bianco, B. and Chiabrera, A., "From the Langevin-Lorentz equation to the Zeeman model of electromagnetic effects on ligand-receptor binding," *Bioelectrochem. Bioenerg.* 28: 355, 1992.

Blackman, C. F., Elder, J. A., Weil, C. M., Benane, S. G., Eichinger, D. C., and House, D. E., "Induction of calcium ion efflux from brain tissue by radiofrequency radiation: Effects of modulation, frequency and field strength," *Radio Sci.* 14(6S): 93, 1979.

Blackman, C. F., Benane, S. G., Elder, J. A., House, D. E., Lampe, J. A., and Faulk, J. M., "Induction of calcium ion efflux from brain tissue by radiofrequency radiation: Effect of sample number and modulation frequency on the power density window," *Bioelectromag.* 1: 35, 1980.

Blackman, C. F., Benane, S. G., Rabinowitz, J. R., House, D. E., and Joines, W. T., "A role for the magnetic field in the radiation-induced efflux of calcium ions from brain tissue in vitro," *Bioelectromag.* 6: 327, 1985.

Blank, M. and Britten, J. S., "The surface compartment model of the steady state excitable membrane," *Bioelectrochem. Bioenerg.* 5: 528, 1978.

Blank, M. and Kavanaugh, W. P., "The surface compartment model (SCM) during transients," *Bioelectrochem. Bioenerg.* 9: 427, 1982.

Blank, M., "The surface compartment model (SCM) with a voltage sensitive channel," *Bioelectrochem. Bioenerg.* 10: 451, 1983.

Blank, M., "Properties of ion channels inferred from the surface compartment model (SCM)," *Bioelectrochem. Bioenerg.* 13: 93, 1984.

Blank, M., "Ionic processes at membrane surfaces: The role of electrical double layers in electrically stimulated ion transport," in M. Blank and E. Findl, Eds., *Mechanistic Approaches to Interaction of Electromagnetic Fields with Living Systems*, New York: Plenum Press, 1987a.

Blank, M., "The influence of surface charge on oligomeric reactions as a basis for channel dynamics," in M. Blank and E. Findl, Eds., *Mechanistic Approaches to Interaction of Electromagnetic Fields with Living Systems*, New York: Plenum Press, 1987b.

Bolomey, J. C., Pichot, C., and Garboriaud, G., "Planar microwave imaging camera for biomedical applications: Critical and prospective analysis of reconstruction algorithms," *Radio Sci.* 26: 51, 1991.

Bond, J. D. and Wyeth, N. C., "Are membrane microwave effects related to a critical phase transition?," *J. Chem. Phys.* 85(12): 7377, 1986.

Bond, J. D. and Wyeth, N. C., "Membrane, electromagnetic fields, and critical phenomena," in M. Blank and E. Findl, Eds., *Mechanistic Approaches to Interaction of Electromagnetic Fields with Living Systems*, New York: Plenum Press, 1987.

Borgens, R. B., Jaffe, L. F., and Cohend, M. J., "Large and persistent electrical currents enter the transected lamprey spinal cord," *Proc. Natl. Acad. Sci. USA* 77: 1209, 1980.

Borup, D. T. and Gandhi, O. P., "Calculation of high resolution SAR distributions in biological bodies using the FFT algorithm and conjugate gradient method," *IEEE Trans.* MTT-33: 417, 1985.

Broquetas, A., Romeu, J., Rius, J. M., Elias-Fuste, A. R., Cardama, A., and Jofre, L., "Cylindrical geometry: A further step in active microwave tomography," *IEEE Trans.* MTT-39: 844, 1991.

Caddemi, A., Tamburello, C., Zanforlin, L., and Torregrossa, V., "Microwave effects on isolated chick embryo hearts, *Bioelectromag.* 7: 359, 1986.

Cain, C. A., "A theoretical basis for microwave and RF field effects on excitable cellular membranes," *IEEE Trans.* MTT-28: 142, 1980.

Cain, C. A., "Biological effects of oscillating electric fields: Role of voltage sensitive ion channels," *Bioelectromag.* 2: 23, 1981.

Caorsi, S. Gragnani, G. L., and Pastorino, M., "Overconstrained version of moment method for prediction of electromagnetic energy deposition inside dielectric objects," *El. Letters* 26: 1997, 1990.

Caorsi, S. Gragnani, G. L., and Pastorino, M., "A multiview microwave imaging system for two-dimensional penetrable objects," *IEEE Trans.* MTT-39: 845, 1991.

Cevc, G., and Marsh, D., *Phospholipid bilayers: Physical principles and models*, New York: Wiley, 1987.

Chiabrera, A., Grattarola, M., and Viviani, R., "Interaction between electromagnetic fields and cells: Microelectrophoretic effect on ligands and surface receptors," *Bioelectromag.* 5: 173, 1984a.

Chiabrera, A. and Rodan, G. A., "The effect of electromagnetic fields on receptor-ligand interaction: A theoretical analysis," *J. Bioelectricity* 3: 509, 1984b.

Chiabrera, A., Bianco, B., Caratozzolo, F., Giannetti, G., Grattarola, M., and Viviani, R., "Electric and magnetic field effects on ligand binding to the cell membrane," in A. Chiabrera, C. Nicolini, and H. P. Schwan, Eds., *Interaction Between Electromagnetic Fields and Cells*, New York: Plenum Press, 1985.

Chiabrera, A., Gyebyi, K., Kaufman, J. J., Ryaby, J., Smith Sonneborn, J., and Pilla, A. A., "Lorentz magnetic force effect on biological systems: Application to paramecium," *Proc. 6th Annual Meeting of BRAGS*, Utrecht, The Netherlands, 1986, abstract.

Chiabrera, A. and Bianco, B., "The role of the magnetic field in the EM interaction with ligand binding," in M. Blank and E. Findl, Eds., *Mechanistic Approaches to Interaction of Electromagnetic Fields with Living Systems*, New York: Plenum Press, 1987.

Chiabrera, A., Morro, A., and Parodi, M., "Water concentration and dielectric permittivity in molecular crevices," *Il Nuovo Cimento*, 11D: 7, 1989.

Chiabrera, A., Bianco, B., Kaufman, J. J., and Pilla, A. A., "Quantum dynamics of ions in molecular crevices under electromagnetic exposure," in C. T. Brighton and S. R. Pollack, Eds., *Electromagnetics in Medicine and Biology*, San Francisco: San Francisco Press, 1991a.

Chiabrera, A., Bianco, B., Kaufman, J. J., and Pilla, A. A., "Quantum analysis of ion binding kinetics in electromagnetic bioeffects," in C. T. Brighton and S. R. Pollack, Eds., *Electromagnetics in Medicine and Biology,* San Francisco: San Francisco Press, 1991b.

Cito, C., D'Inzeo, G., and Tarricone, L., "Microwave fields modify Ach-channel's gating: Theoretical study by means of a stochastic approach," P-3, *First World Congress for Electricity and Magnetism in Biology and Medicine,* Lake Buena Vista, Fla., 1992, abstract.

D'Inzeo, G., Bernardi, P., Eusebi, F., Grassi, F., Tamburello, C., and Zani, B. M., "Effects of microwaves on the acetylcholine-induced channels in cultured chick myotubes," *Bioelectromag.* 9: 363, 1988.

Durney, C. H., Rushforth, G. K., and Anderson, A. A., "Resonant AC-DC magnetic fields: Calculated response, *Bioelectromag.* 9: 315, 1988.

Dutta, S. K., Subramoniam, A., Ghosh, B., and Parshad, R., "Microwave radiation-induced calcium ion efflux from human neuroblastoma cells in culture," *Bioelectromag.* 5: 71, 1984.

Foster, K. R. and Schwan, H. P., "Dielectric properties of tissues," in C. Polk and E. Postow, Eds., *CRC Handbook of Biological Effects of Electromagnetic Fields,* Boca Raton, Fla.: CRC Press, 1986, 27-96.

Foster, K. R. and Schepps, J. L., "Dielectric properties of tumor and normal tissues at radio through microwave frequencies," *J. Microwave Power* 16: 107, 1981.

Frankel, R. B., "Magnetic guidance of organisms," *Biophys. Bioeng.* 13: 85, 1984.

Frankel, R. B. and Blakemore, R. P., "Precipitation of Fe_3O_4 in magnetotactic bacteria," *Phil. Trans. R. Soc. Lond.* B-304: 567, 1984.

Fricke, H., "Relation of the permittivity of biological cell suspensions to fractional cell volume," *Nature* 172: 731, 1953.

Gabriel, C., Grant, E. H., Tata, R., Braun, P. R., Gestblom, B., and Noreland, E., "Microwave absorption in aqueous solution of DNA," *Nature* 328: 145, 1987.

Galvin, M. J. and McRee, D. I., "Cardiovascular, hematologic and biochemical effects of acute ventral exposure of conscious rats to 2450-MHz (CW) microwave radiation," *Bioelectromag.* 7: 223, 1986.

Grant, E. H., McLean, V. E. R., Nightingale, N. R. V., and Gabriel, C., "Dielectric properties of water in biological solutions," in A. Chiabrera, C. Nicolini, and H. P. Schwan, Eds., *Interactions Between Electromagnetic Fields and Cells,* New York: Plenum Press, 1985.

Grodzinsky, A. J., Lipshitz, H., and Glimcher, M. J., "Electromechanical properties of articular cartilage during compression and stress relaxation," *Nature* 275: 448, 1978.

Grodzinsky, A. J., "Electromechanical and physiochemical properties of connective tissue," *CRC Critical Reviews in Biomedical Engineering* 9: 133, 1983.

Gross, D. and Williams, W. S., "Streaming potential and the electromechanical response of physiologically moist bone," *J. Biomech.* 15: 227, 1982.

Hagman, M. J. and Levin, R. L., "Procedures for noninvasive electromagnetic property and dosimetry measurements," *IEEE Trans.* AP-38: 99, 1990.

Halle, B., "On the cyclotron resonance mechanism for magnetic field effects on transmembrane ion conductivity," *Bioelectromag.* 9: 381, 1988.

Hart, F. X., "A quantum mechanical model for bioelectromagnetic resonance phenomena," *J. Bioelec.* 9: 1, 1990.

Hille, B., *Ionic Channels of Excitable Membranes,* Sunderland: Sinauer Associates Inc., 1984.

Kaczmarek, L. K. and Adey, W. R., "Weak electric gradients change ionic and transmitter fluxes in cortex," *Brain Res.* 66: 537, 1974.

Kirschvink, J. L. and Kobayashi-Kirschvink, A., "Magnetite biomineralization in human tissues: A solution to the thermal noise problem of ELF bioeffects," *First World Congress for Electricity and Magnetism in Biology and Medicine,* B-8, Lake Buena Vista, Fla., 1992, abstract.

Kosterich, J. D., Foster, K. R., and Pollack, S. R., "Dielectric permittivity and electrical conductivity of fluid saturated bone," *IEEE Trans.* BME-30: 81, 1983.

Kosterich, J. D., Foster, K. R., and Pollack, S. R., "Dielectric properties of fluid-saturated bone: The effect of variation in conductivity of immersion fluid," *IEEE Trans.* BME-31: 369, 1984.

Lai, H., Horita, A., and Guy, A. W., "Acute low-level microwave exposure and central cholinergic activity: Studies on irradiation parameters," *Bioelectromag.* 9: 355, 1988.

Lednev, V. V., "Possible mechanism for influence of weak magnetic fields on biological system," *Bioelectromag.* 12: 71, 1991.

Liboff, A. R., "Cyclotron resonance in membrane transport," in A. Chiabrera, C. Nicolini, and H. P. Schwan, Eds., *Interacton Between Electromagnetic Fields and Cells,* New York: Plenum Press, 1985.

Liboff, A. R. and McLeod, B. R., "Kinetics of channelized membrane ions in magnetic fields," *Bioelectromag.* 9: 39, 1988.

Liburdy, R. P. and Penn, A., "Microwave bioeffects in the erythrocyte are temperature and pO dependent: Cation permeability in protein shedding occur at the membrane phase transition," *Bioelectromag.* 5: 283, 1984.

Liburdy, R. P., "Microwave associated protein shedding in human erythrocite: Quantitative HPLC analysis," *8th Annual Meeting of Bioelectromag. Soc.,* Madison, Wisc., 1986, abstract.

Lords, J. L., Durney, C. H., Borg, A. M., and Tinney, C. E., "Rate effects in isolated hearts induced by microwave irradiation," *IEEE Trans.* MTT-21: 834, 1973.

Lotke, P. A., Black, J., and Richardson, S. L., "Electromechanical properties in human articular cartilage," *J. Bone and Joint Surg.* 56A: 1040, 1974.

Marconi, L., Morgavi, G., Ridella, S., and Rolando, C., "Nonlinear ionic fluxes in EMF exposed cells," in A. Chiabrera, C. Nicolini, and H. P. Schwan, Eds., *Interaction Between Electromagnetic Fields and Cells,* New York: Plenum Press, 1985.

McCann, F. V., Pettengill, O. S., Cole, J. J., Russell, J. A. G., and Sorenson, G. D., "Calcium spike electrogenesis and other electrical activity in continuously cultured small cell carcinoma of the lung," *Science* 212: 1155, 1981.

McGinitie, L. A., Grodzinsky, A. J., Frank, E. M., and Gluzband, Y. A., "Frequency and amplitude dependence of electric field interactions: Electrokinetics and biosynthesis," in M. Blank and E. Findl, Eds., *Mechanistic Approaches to Interactions of Electric and Electromagnetic Fields with Living Systems,* New York: Plenum Press, 1987.

McLaughlin, S., "The role of electro-osmosis in the electric-field-induced movement of charged macromolecules on the surfaces of cells," *Biophys. J.* 34: 85, 1981.

McLeod, B. R. and Liboff, A. R., "Dynamic characteristics of membrane ions in multifield configurations of low frequency electromagnetic radiation," *Bioelectromag.* 7: 177, 1986.

Merritt, J. G., Shelton, W. S., and Chamnes, A. F., "Attempt to alter $^{45}Ca^{2+}$ binding to brain tissue with pulse-modulated microwave energy," *Bioelectromag.* 3: 475,. 1982.

Molinari, G., and Viviani, A., "Analytical evaluation of the electro-dielectrophoretic forces acting on spherical impurity particles, in dielectric fluids," *J. Electrostatics* 5: 343, 1978.

Ortoleva, P. J., "Biological effects of non-ionizing radiation" in K. H. Illinger, Ed., *ACM Symposium Series,* Washington, D.C., 1981.

Papahadjopoulos, D., Jacobson, K., Nir, S., and Isac, T., "Phase transitions in phospholipid vesicles: Fluorescence polarization and permeability measurements concerning the effect of temperature and cholesterol," *Biochim. Biophys. Acta* 311: 330, 1973.

Pastushenko, V. Ph., Kuzmin, P. I., and Chizmadzhev, Yu. A., "Dielectrophoresis and electrorotation: A unified theory of spherically symmetrical cells," *Stud. Biophys.* 110: 47, 1985.

Paulsen, K. D., Lynch, D. R., and Strohbehn, J. W., "3-D finite, boundary and hybrid elements solutions of the Maxwell equations for lossy media," *IEEE Trans.* MTT-36: 687, 1988.

Pethig, R., *Dielectric and Electronic Properties of Biological Materials,* Chichester, England: Wiley, 1979.

Philippova, T. M., Novoselov, V. I., Bystrova, M. F., and Alekseev, S. I., "Microwave effects on champhor binding to rat olfactory epithelium," *Bioelectromag.* 9: 347, 1988.

Pickard, W. F. and Rosenbaum, F. J., "Biological effects of microwaves at the membrane level: Two possible athermal electrophysiological mechanisms and a proposed experimental test," *Math. Biosci.* 39: 235, 1978.

Pienkowski, D. and Pollack, S. R., "The origin of stress-generated potentials in fluid-saturated bone," *J. Orth. Res.* 1: 30, 1983.

Pilla, A. A., "Mechanisms of electrochemical phenomena in tissue repair and growth," *Bioelectrochem. Bioenerg.* 1: 227, 1974.

Pilla, A. A. and Margules, G. S., "Dynamic interfacial electrochemical phenomena of living cells: Application to the toad urinary bladder membrane system," *Electrochem. Soc.* 124: 1697, 1977.

Pilla, A. A., Sechaud, P., and McLeod, B. R., "Electrochemical and electrical aspects of low-frequency electromagnetic current induction in biological systems," *J. Biological Physics* 11: 51, 1983.

Robertson, B. and Astumian, R. D., "Kinetics of a multistate enzyme in a large oscillating field," *Biophys. J.* 57: 689, 1990.

Robertson, B. and Astumian, R. D., "Frequency dependence of catalyzed reactions in a weak oscillating field," *J. Chem. Phys.* 94: 7414, 1991.

Sandblom, J. Teander, S., and Baltzer, P., "The effect of microwave radiation on the stability and formation of gramicidin A channels in lipid bilayer membranes," *Proc. XIV ICMBE,* Espoo, Finland, 1985.

Sauer, F. A., "Interaction forces between microscopic particles in an external electromagnetic field," in A. Chiabrera, C. Nicolini, and H. P. Schwan, Eds., *Interactions Between Electromagnetic Fields and Cells,* New York: Plenum Press, 1985.

Sauer, F. A. and Schloegl, R. W., "Torques exerted on cylindres and spheres by external electromagnetic fields: A contribution to the theory of field induced cell rotation," in A. Chiabrera, C. Nicolini, and H. P. Schwan, Eds., *Interactions Between Electromagnetic Fields and Cells,* New York: Plenum Press, 1985.

Scherer, P. G. and Seelig, J., "Reorientation of lipid head groups at membrane-water interface," *EMBO J.* 6: 2915, 1987.

Schwan, H. P., "Electrical and acoustic properties of biological materials and biomedical applications," *IEEE Trans.* BME-31: 872, 1984.

Schwan, H. P., "Dielectric properties of cells and tissues," in A. Chiabrera, C. Nicolini, and H. P. Schwan, Eds., *Interactions Between Electromagnetic Fields and Cells,* New York: Plenum Press, 1985.

Schwan, H. P., "Dielectrophoresis and rotation cells," in E. Neumann, A. E. Sowers, and C. A. Jordan, Eds., *Electrophoretion and Electrofusion in Cell Biology,* New York: Plenum Press, 1989.

Seaman, R., and Wachtel, H., "Slow and rapid response to CW and pulsed microwave radiation by individual Aplysia pacemakers," *J. Microwave Power* 13: 77, 1978.

Seaman, R., Ajer, R. K., and DeHaan, R. L., "Changes in cardiac-cell membrane noise during microwave exposure," *Proc. IEEE-MTT Symp.* 1: 436, 1982.

Seelig, J., Macdonald, P. M., and Scherer, P. G., "Phospholipid head groups as sensors of electric charge in membranes," *Biochem.* 24: 7535, 1987.

Semm, P., "Three different magnetic systems in the vertebrate brain," *First Congress EBEA*, Brussels, Belgium, January 23-25, 1992.

Shepherd, J. C. W. and Büldt, G., "Dielectric measurements on reorientation rates in lipid head groups," *Biochim. Biophys. Acta* 514: 83, 1978.

Sheridan, J. P., Gaber, B. P., Cavatorta, F., and Schoen, P. E., "Molecular level effects of microwaves on natural and model membranes: A Raman spectroscopic investigation," in L. S. Taylor and A. Y. Cheung, Eds., *Proc. National Radio Science Meeting and Bioelectromagnetics Symp.*, Seattle: University of Washington, 1979.

Smith, S. D., McLeod, B. R., Liboff, A. R., and Cooksey, K., "Calcium cyclotron resonance and diatom mobility," *Bioelectromag.* 8: 218, 1987.

Stuchly, M. A. and Stuchly, S. S., "A six-octave experimental system for in-vivo and in-vitro dielectric measurements," in A. Chiabrera, C. Nicolini, and H. P. Schwan, Eds., *Interactions Between Electromagnetic Fields and Cells*, New York: Plenum Press, 1985.

Su, C. C., "Electromagnetic scattering by a dielectric body with arbitrary inhomogeneity and anisotropy," *IEEE Trans.* AP-37: 384, 1989.

Takashima, S., Gabriel, C., Sheppard, R. J., and Grant, E. H., "Dielectric behavior of DNA solution at radio and microwave frequencies (at 20°C)," *Biophys. J.* 46: 29, 1984.

Takashima, S., *Electrical properties of biopolymers and membranes*, Bristol and Philadelphia: Hilger, 1989.

Thomas, J. R., Schrot, J., and Liboff, A. R., "Low-intensity magnetic fields alter operant behavior in rats," *Bioelectromag.* 7: 349, 1987.

Tinney, C. E., Lords, J. L., and Durney, C. H., "Rate effects in isolated turtle hearts induced by microwave irradiation," *IEEE Trans.* MTT-24: 18, 1976.

Tsai, C. T., Massoudi, H., Durney, C. H., and Iskander, M. F., "A procedure for calculating fields inside arbitrary shaped, inhomogeneous dielectric bodies using linear basis functions with the moment method," *IEEE Trans.* MTT-34: 1131, 1986.

Wachtel, H., Seaman, R., and Joines, W., "Effects of low-intensity microwaves on isolated neurons," *Ann. N.Y. Acad. Sci.* 247: 46, 1975.

Martin Blank, Ed., *Electricity and Magnetism in Biology and Medicine*

MECHANISMS OF ELECTROTRANSFORMATION

Yu. A. Chizmadzhev,* S. I. Sukharev, V. A. Klenchin, V. F. Pastushenko, and L. V. Chernomordik

Frumkin Institute of Electrochemistry, Moscow

S. M. Serov

Englehardt Institute of Molecular Biology, Moscow

The task of our investigation was to elucidate the mechanism of DNA electroinjection into eukaryotic cells. Using transient expression of β-galactosidase gene in Cos-1 cells, we show that DNA uptake is a fast process in which DNA electrophoresis plays the key role. The assay of electrically induced uptake of fluorescent dextrans (FD) by cells in conjunction with the two-pulse technique (first pulse 6 kV/cm, 10 μs; second pulse 200 V/cm, 10 ms) shows that the presence of DNA during electroporation leads to a sharp increase of membrane permeability for FDs of M.W. ≤ 20 kDa (pore radius about 4 nm). The effect can be seen even if FDs are added to the cell suspension a few minutes after the pulse application (pore lifetime about 10-100 s). The higher the DNA fragment, the greater the increase in permeability. The use of a two-pulse technique allows us to separate electroporation and DNA electrophoresis. Developed theory has shown that DNA in electric field exerts high pressure on the membrane, which can induce various structural rearrangements. In conclusion, electroporesis is the driving force of DNA uptake which not only brings it to the cell surface but also drags DNA through the membrane, at first using preexisting pores, which are then enlarged and deformed in the course of translocation. The theory predicts the kind of DNA percolation proposed by E. Neumann at the Frumkin Symposium in Suzdal (USSR) in 1987.

Electrotransfection of cells, i.e., transfection induced by electric field pulses, has become increasingly popular as an effective technique for introduction of foreign DNA into cells of any origin.[1] Various kinds of mammalian cells,[2] insect,[3] protozoan[4] and plant cells,[5] intact bacteria,[6] and yeasts [7] have been successfully transformed by means of this technique. The underlying mechanism of DNA electrotranslocation through the cell plasma membrane is not yet understood, and experimental parameters are varied empirically. One may expect that the results obtained in many cases are far from being potentially achievable. Understanding the mechanism is important for improving efficiency and expanding applications for the electrotransfection technique. For example, increase in transfection efficiency should improve expression cloning in mammalian cells,[8] cell transformation in vivo for gene therapy,[9] and obtaining transgenic cereals.[10]

The first hypothesis for DNA electrotranslocation into cells was proposed immediately after the technique was invented by Neumann and co-workers.[11] They assumed that DNA diffuses freely through membrane pores created by a high electric field. Later this hypothesis was extended by additional stages, such as DNA binding to the membrane and subsequent lateral diffusion through the pore.[12] It was also proposed that DNA can be transferred into cells with the flow of water during colloid-osmotic cell swelling upon electropermeabilization,[13] or with the electro-osmotic water flux that takes place near a charged cell surface in a high electric field.[14] The electrically induced DNA uptake by liposomes was shown to be mediated by a formation of membrane invaginations "swallowing" the DNA,[15] which allowed the authors to assume the same process in cells. The possibility of electrophoretic transfer of DNA as a polyanion was also considered,[16,17] but only indirect evidence was provided for this hypothesis. The task of our investigation was to elucidate the mechanism of DNA electroinjection into eucaryotic cells, using transient expression of β-galactosidase gene in Cos-1 cells. We attempted to answer two questions: which force pushes DNA into the cell, and what kind of rearrangement in the structure of the plasma membrane mediates DNA translocation?

---

*Correspondence to: Yu. A. Chizmadzhev, Frumkin Institute of Electrochemistry, Leninsky pr. 31, Moscow, Russia.

RESULTS

The main features of experimental protocol used in the first part of the study are illustrated in Table 1; details can be found elsewhere.[18-21] First of all we tried to determine how fast the process of penetration of DNA into the cell is. The plasmid was introduced into the cell suspension at various times before and after pulse application. The efficiency of penetration of DNA added after the pulse was two orders of magnitude lower than for DNA added before the pulse. Under the same conditions, the permeability of cells for Lucifer Yellow decreased twofold within 20 min. At the same time, the transfection efficiency of the postpulse addition of plasmid exceeded by 2-3-fold the value without electrical treatment and within 30 min did not depend noticeably on the time of addition. We estimated the characteristic time of plasmid penetration into the cells using plasmid accessibility for DNAase. Cell suspension with DNA was supplemented with DNAase at various times after the pulse application. When DNAase was introduced 2 s before the pulse, the transfection efficiency decreased to the background level, whereas addition of DNAase 2 s after the pulse did not decrease transfection efficiency as compared with the control. This result indicates that all transfecting DNA becomes inaccessible for DNAase, i.e., translocated into the cells in 3 s or less after the electric treatment.

TABLE 1.—Experimental system.

Cells: Cos-1 (amplify DNA, containing SV40 ori)

Plasmid: pCH110 (contains β-gal gene and SV40 ori)

Standard electrotransfection conditions:

Cells in PBS without Ca^{2+} and Mg^{2+}	(5×10^5/55 μl)
Plasmid, supercoiled form	(1-3 μg/55 μl)
One rectangular pulse	3.5 kV/cm 100 μs
Room temperature	

Cells + DNA → 10 min → pulsation → 10 min → plating →

→ 20 min → 37 C, 48 h → measurement of transfection efficiency (TE).

TE: β-galactosidase activity per viable cell

$$TE = \frac{10^6 (A_{420} - 1.1\, A_{550})}{NT}$$

where A = light absorbance
N = number of viable cells
T = time of incubation

To elucidate the role of electric field direction we used a monolayer of cells on a porous film for transfection. In this system mainly the upper surface of the cells growing in a monolayer is accessible to DNA. The electric field normal to the monolayer plane induced DNA electrophoresis either toward or away from the cells, depending on the polarity of the pulse. The data indicate that when DNA moves toward the cells, transfection efficiency is an order of magnitude higher than with a reverse polarity. Permeability of the cells for propidium iodide, added 1 min after the pulse, was the same at both pulse polarities. This result indicates that the degree of membrane electroporation does not depend on the direction of electric field.

In two further experiments we investigated the effects of factors that decrease the electrophoretic mobility of DNA in the pulsation medium (Ficoll 400, Mg^{2+}). It was shown that electrophoresis is an important stage of electrotransformation. But two question still remained unanswered: first, is electrophoresis responsible only for transport of DNA to cell surface or does it drive DNA through the pores; and second, is transport through the pores or endocytosis-like entrapment the real mechanism of DNA uptake?

To elucidate these problems we studied the electropermeabilization of cells in the presence of DNA using fluorescent dyes of various sizes (see Table 2 and Ref. 22 for details). To separate two effects provided by a pulsed electric field (electroporation and DNA electrophoresis) we introduced a two-pulse technique. The first pulse (6 kV/cm, 10 μs) creates pores efficiently and transfection efficiency (TE) is low. The second pulse, of much lower amplitude but substantially longer (0.2 kV/cm, 10 ms) does not cause poration and transfection by itself, but enhances TE by about one order of magnitude.

TABLE 2.—DNA-induced increase of electropermeabilization of cells.

DNAs: Linear form of plasmid pCH110 (7.2 Kb); linear form of λ phage DNA (48 Kb); calf thymus DNA fragments (20-100 Kb).

Dyes: Lucifer Yellow (LY, radius 0.75 nm); FITC-dextrans, M.W.: 4000 (FD-4, 1.5 nm); 20000 (FD-20, 3.2 nm); 40000 (FD-40, 4.4 nm); 70000 (FD-70, 5.8 nm).

Cells ± dye ± DNA → pulsation → ± dye ± → 10 min →

→ 1 h resealing in growth medium → 4 times washes with medium → fluorescence measurement

The Effect of DNA on Electrically Induced Uptake of FITC—Dextrans by Cells

The presence of DNA in the cell suspension during electric pulse treatment markedly increases the membrane permeability to molecules as large as dextrans. The dependence of FD4 uptake by cells on thymus DNA concentration in the pulsation medium, obtained at constant pulse parameters (1.5 kV/cm, 1 ms), is shown in Fig. 1. A reliable increase in cell fluorescence intensity is observed beginning at 1 μg/ml DNA. The midpoint of the sigmoid curve is at about 10 μg/ml and saturation takes place at approximately 100 μg/ml DNA. The DNA-dependent increase in permeability is a long-lasting membrane alteration. The effect of DNA can also be seen if FD is added not prior to, but following pulse application. The amounts of FD20 taken up by cells when added at various time points after pulse treatment is shown in Fig. 2. Curve 1, obtained in the presence of 100 μg/ml of thymus DNA, lies above the control (curve 2) over the entire range of time intervals studied, without reaching the control level even 100 s after the pulse. The relative amounts of fluorescent label entrapped by cells in the presence of DNA and in the control for FDs of four different sizes are plotted in Fig. 3. The effect of DNA-dependent increase in plasma membrane permeability is observed only for the "light" dextrans FD4 and FD20. DNA does not influence the entrapment of the larger dextrans FD40 and FD70. The rate of membrane permeability increase was found to depend not only on the size of FD, but also on the size of DNA present in the pulsation medium. The highest increase of permeability is observed for light dextran FD4 in the presence of λ-DNA, the larger DNA used.

Electrotransfection in a Two-pulse Mode

The TE level as a function of electric charge passed through the chamber during the second pulse is shown in Fig. 4. In this experiment we varied the duration of the second pulse. The TE dependence rises monotonously, but not linearly. In the second series of experiments we studied TE as a function of time delay between the two pulses (Δt). As the delay increases, TE declines, reaching half its initial value at Δt = 10 s (Fig. 5). Even at an interval of 100 s between pulses, the TE level is twice as high as that obtained by application of the first pulse alone. It is important to note that introduction of DNA into the cell suspension between two pulses gave a negligibly low level of TE. This two-pulse technique also allowed us to determine whether the DNA uptake by cells correlates to the entrapment of fluorescent dye. We observed a more than 3-fold rise in FD4 uptake upon the second (low-voltage) pulse application in the presence of DNA. At the same time, the amount of FD4 taken up by cells declines with increasing interval between pulses, similar to TE.

DISCUSSION

We showed that DNA translocation into cells using a high-voltage pulse takes less than 3 s, and that DNA electrophoresis plays an important role in this process. When the DNA crosses the membrane, it interacts with the membrane such that membrane permeability to indifferent water-soluble dyes becomes markedly higher. If the membrane is subjected to a short electroporative pulse in the presence of DNA, subsequent DNA translocation through this membrane can be driven by a relatively low electric field, which is unable to form electropores by itself. These regularities have new implications for the mechanism of the electrically induced DNA uptake by cells.

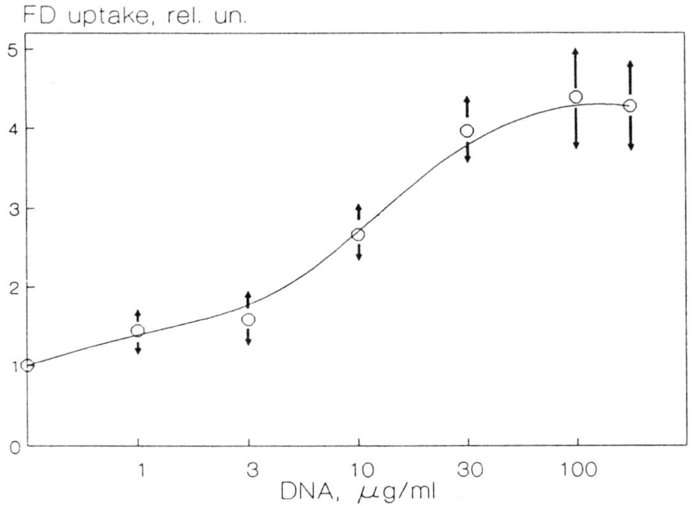

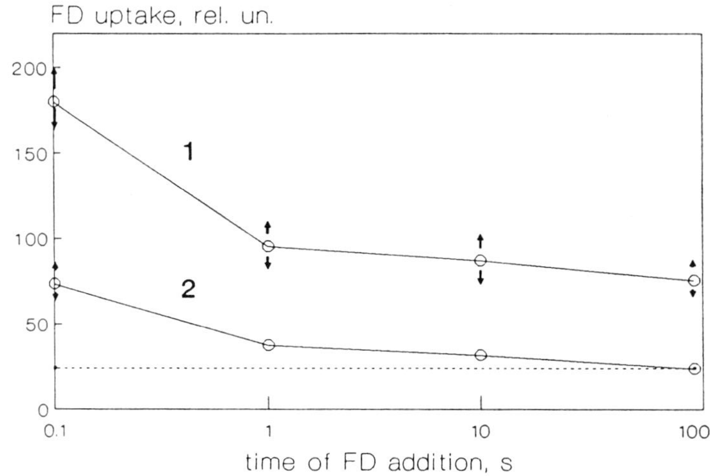

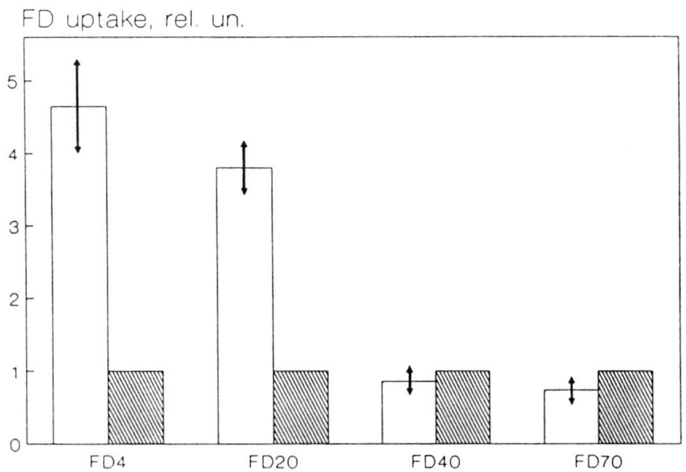

FIG. 1.—Dependence of FD4 uptake by cells on concentration of calf thymus DNA in pulsation medium. FD4 and DNA were added to cell suspension a few minutes prior to application of a single pulse of 1.5 kV/cm, 1 ms.

FIG. 2.—Cellular uptake of FD20 as a function of time of the dextran introduction after pulse application. Cells were subjected to single pulse of 1.5 kV/cm, 1 ms in the presence of 0.1 mg/ml of calf thymus DNA (curve 1) and without DNA (curve 2). The values at $t = 0$ were obtained with FD20 added to cell suspension prior to pulse application. Level of fluorescence of cells in control (no pulse applied) is shown by dashed line.

FIG. 3.—Normalized amounts of four different FDs taken up by cells upon electric treatment by a single pulse of 1.5 kV, 1 ms in the presence of 0.1 mg/ml of calf thymus DNA. Control values assayed in absence of DNA (shaded bars) are taken as unity.

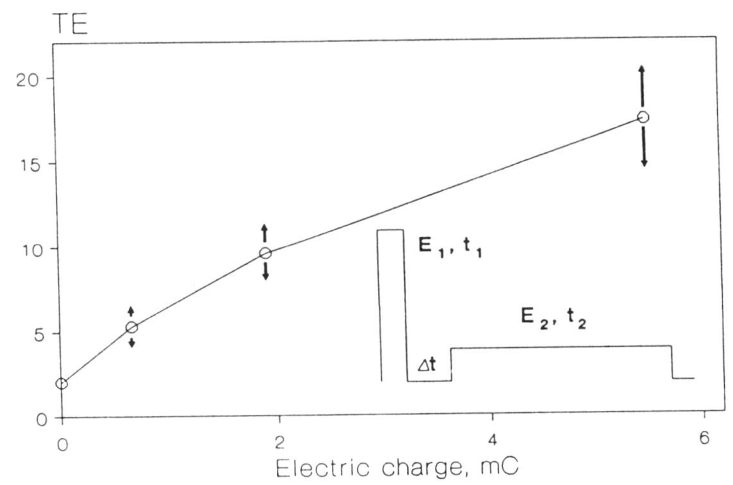

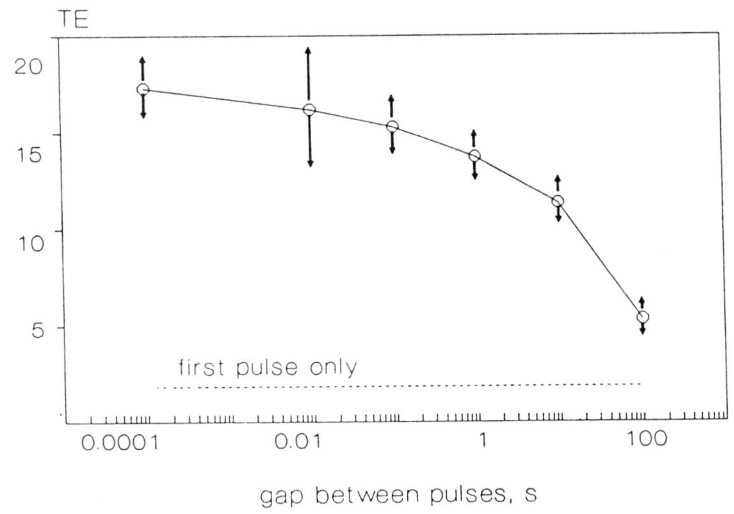

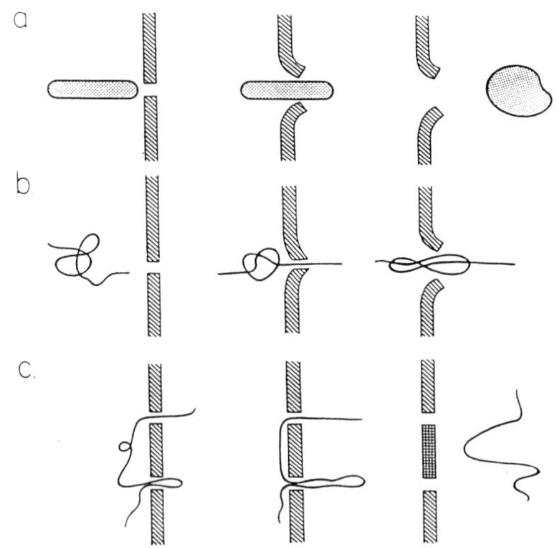

FIG. 4.—Transfection efficiency (TE) as a function of electric charge passed through chamber during second (low-voltage) pulse. Duration of second pulse was varied in this experiment. Inset: Sequence of two unequal pulses which was used for separation of two different effects of electric field in electrotransformation. First pulse of $E_1 = 6$ kV/cm and $t_1 = 10$ μs, and second pulse of $E_2 = 0.2$ kV/cm and t_2 varying from 0 to 10 ms, were applied with time gap Δt of 100 μs.

FIG. 5.—Dependence of transfection efficiency (TE) on time delay between two pulses. Parameters of pulses are as noted in legend to Fig. 4. Cells were treated in a low-conductance medium. Level of TE obtained with first pulse only is shown by dashed line.

FIG. 6.—Possible ways of DNA interaction with electroporated membrane in a high electric field. (a) DNA coil is aligned in an electric field and at the appropriate pulse polarity it moves toward the porated membrane. DNA may interact with a single pore: the pore becomes wider upon DNA passage by the action of electrophoretic force. (b) Passage of DNA coil can be initiated by penetration of one end of thread, which then leads whole molecule through one pore. (c) DNA molecule can be involved in two pores (or more) and moving with electrophoretic force it cuts the membrane between these pores as a sharp thread can do (cut in membrane is hatched).

The Motive Force of DNA Electrotranslocation

Transfection of cells by sequences of two unequal pulses enables us to distinguish two different effects of the electric field: membrane electroporation and electrophoresis of DNA. Since the low-voltage pulse does not influence membrane permeability in the absence of DNA, even if it is applied after the high-voltage pulse, the sharp increase in TE in two-pulse experiments (Figs. 4 and 5) should be caused exclusively by DNA electrophoresis during the second pulse. The increase in TE by the second pulse is very similar to the observations made by Andreason and Evans,[17] in which they reported a significant rise in transfection yield when cells were treated by a train of low-voltage pulses following a short electroporative pulse of high amplitude. The TE dependence on the electric charge Q passed in the circuit during the second pulse (Fig. 4) rises monotonously, consistent with the hypothesis of electrophoretic DNA transfer. Since a relatively low (0.2 kV/cm) electric field provides an efficient DNA transfer through the electropermeabilized membrane, one may suggest that application of rectangular or exponentially decaying pulses of significantly higher amplitude (0.8-4 kV/cm, the range widely used for eukaryotic cells), in addition to forming pores, causes DNA translocation into cells by the same electrophoretic mechanism. Note that use of exponential pulses appears to be especially suitable, since these pulses have a high peak providing for poration, and a long tail of relatively low amplitude, which is sufficient for effective DNA injection into the cell. However, the existence of electropores in the membrane and simultaneous application of low electric field is insufficient for DNA entry into cells. Addition of DNA into the cell suspension between the high "electroporative" and low "injecting" pulses gave extremely low TE.

This result strongly indicates that DNA influences electroporation and should be present during the first pulse as well. Indeed, we observed increases in cell permeability upon electric treatment in the presence of DNA. More exactly, this effect is due not only to the presence of DNA in the medium; as seen from the correlation between TE and the amount of FD taken up by cells in two-pulse experiments, permeability rises due to DNA transfer into cells. In other words, DNA translocation leads to an alteration of electropores in the membrane. This effect can be easily explained in terms of a mechanical interaction between DNA and membrane due to the electrophoretic force, whereas it cannot be explained by free DNA diffusion through the electropores. It appears that the long-lived alterations of membrane permeability, and the increase in TE upon second (low-intensity) pulse application only when DNA was added prior to the first (high-voltage) pulse, make direct DNA interaction with electropore(s) a more likely explanation than endosome formation.

Summarizing the data presented above, we conclude that DNA electrotranslocation through the plasma membrane of mammalian cells is not a simple penetration via "ready-to-use" electropores (Fig. 6). This process includes mechanical interaction between the pores and the DNA driven by electrophoretic force. The proposed mechanism[23] is similar to some kind of DNA percolation discussed by E. Neumann at the Frumkin Symposium in Suzdal (USSR) in 1987.

REFERENCES

1. H. Potter, "Electroporation in biology: Methods, applications, and instrumentation," *Anal. Biochem.* 174: 361-373, 1988.

2. G. Chu, H. Hayakawa, and P. Berg, "Electroporation for the efficient expression of mammalian cells with DNA," *Nucl. Acids Res.* 15: 1311-1326, 1987.

3. S. G. Mann and L. A. King, "Efficient transfection of insect cells with baculovirus DNA using electroporation," *J. Gen. Virol.* 70: 3501-3505, 1989.

4. A. Laban and D. Wirth, "Transfection of *Leishmania enriettii* and expression of chloramfenicol acetyltransferase gene," *Proc. Natl. Acad. Sci. USA* 86: 9119-9123, 1989.

5. R. A. Dekeyser, B. Claes, R. M. U. De Rycke, M. E. Habets, M. C. Van Montagu, and A. B. Caplan, "Transient gene expression in intact and organized rice tissues," *Plant Cell* 2: 591-602.

6. J. B. Luchansky, P. M. Muriana, and T. R. Klaenhammer, "Application of electroporation for transfer of plasmid DNA to Lactobacillus, Lactococcus, Leuconostoc, Listeria, Pediococcus, Bacillus, Staphylococcus, Enterococcus and Propionibacterium," *Mol. Microbiol.* 2: 637-646, 1988.

7. E. Meilhoc, J.-M. Masson, and J. Teissié, "High efficiency transformation of intact yeast cells by electric field pulses," *Bio/Technology* 8: 223-227, 1990.

8. K. Sasaki, Y. Yamano, S. Bardhan, N. Iwai, J. J. Murray, M. Hasegawa, Y. Matsuda, and T. Inagami, "Cloning and expression of a complementary DNA encoding a bovine adrenal angiotensin II type-1 receptor," *Nature* 351: 230-232, 1991.

9. A. V. Titomirov, S. Sukharev, and E. Kistanova, "In vivo electroporation and stable transformation of skin cells of newborn mice by plasmid DNA," *Biochim. Biophys. Acta* 1088: 131-134, 1991.

10. Y. Tada, M. Sakamoto, M. Matsuoka, and T. Fujimura, "Expression of a monocot LHCP promoter in transgenic rice," *EMBO J* 10: 1803-1808, 1991.

11. E. Neumann, M. Schaefer-Rider, Y. Wang, and P. H. Hofschneider, "Gene transfer into mouse lyoma cells by electroporation in high electric fields," *EMBO J.* 1: 841-845, 1982.

12. T.-D. Xie, L. Sun, and T. Y. Tsong, "Study of mechanisms of electric field-induced DNA transfection: I. DNA entry by surface binding and diffusion through membrane pores," *Biophys. J.* 58: 13-19, 1990.

13. H. Stopper, H. Jones, and U. Zimmermann, "Large-scale transfection of mouse L-cells by electropermeabilization," *Biochim. Biophys. Acta* 900: 38-44, 1987.

14. D. S. Dimitrov and A. E. Sowers, "Membrane electroporation: Fast molecular exchange by electroosmosis," *Biochim. Biophys. Acta* 1022: 381-392, 1990.

15. L. V. Chernomordik, A. V. Sokolov, and V. G. Budker, "Electro-stimulated uptake of DNA by liposomes," *Biochim. Biophys. Acta* 1024: 179-183, 1990.

16. D. J. Winterbourne, S. Thomas, J. Hermon-Taylor, I. Hussain, and A. P. Johnstone, "Electric shock-mediated transfection of cells," *Biochem. J.* 251: 427-434, 1988.

17. G. L. Andreason and G. A. Evans, "Optimization of electroporation for transfection of mammalian cells," *Anal. Biochem.* 180: 269-275, 1989.

18. V. A. Klenchin, S. I. Sukharev, S. M. Serov, L. V. Chernomordik, and Yu. A. Chizmadzhev, "Electrically induced DNA uptake by cells is a fast process involving DNA electrophoresis," *Biophys. J.* 60: 804-811, 1991.

19. Y. Gluzman, "SV40-transformed simian cells support the replication of early SV40 mutants," *Cell* 23: 175-182, 1981.

20. M. Hattori and Y. Sakaki, "Dideoxy sequencing method using denatured plasmid templates," *Anal. Biochem.* 152: 232-238, 1986.

21. G. An, K. Hidaka, and L. Siminovitch, "Expression of bacterial β-galactosidase in animal cells," *Mol. Cell. Biol.* 2: 1628-1632, 1982.

22. S. I. Sukharev, V. A. Klenchin, S. M. Serov, L. V. Chernomordik, and Yu. A. Chizmadzhev, "Electroporation and electrophoretic DNA transfer into cells: The effect of DNA interaction with electropores," *Biophys. J.* (in press).

23. V. Ph. Pastushenko and Yu. A. Chizmadzhev, "Energetic estimates of deformations of DNA and cell membranes during electrotransformation," *Biol. Membrany.* (in Russian; in press).

Martin Blank, Ed., *Electricity and Magnetism in Biology and Medicine*

EFFECTS OF LOW-ENERGY ELECTROMAGNETIC FIELDS (EMF) ON SIGNAL TRANSDUCTION BY G PROTEIN-LINKED RECEPTORS

R. A. Luben*

University of California, Riverside

Previous studies in our laboratory showed that EMF decreased signal transduction by the parathyroid hormone (PTH) receptor in bone cells. We theorize that other members of the G protein-linked receptor family also might be susceptible to EMF. To test this theory, we examined effects of β-adrenergic receptors in cultured cells. Using pulsed fields (~ 10 G, 15-72 Hz), we demonstrated significant inhibition on ligand binding and adenylyl cyclase activation by β-adrenergic receptors in bone cells and pineal cells, but not in fibroblasts or hepatoma cells. Similar findings were observed with 1G 60Hz sine-wave fields in preliminary experiments. The results are consistent with a model in which EMF causes tissue-specific changes in conformation of the membrane-resident portion of G protein-linked receptors. This model may allow us to develop a more detailed theory of how EMF interacts with other cell membrane molecules and with biological systems in general.

Low-energy electrical and magnetic fields have been widely used in orthopedics to promote the healing of bone fractures and pseudoarthroses.[1] Our laboratory has studied the effects of EMF on bone cell function in vitro for several years.[2-4] Our work has shown that the pulsed EMF used in clinical fracture-treatment devices produces activation of mouse osteoblasts in vitro by means of a strong inhibition of parathyroid hormone (PTH) responsiveness in the cells,[2] leading to increased synthesis of collagen,[5] decreased levels of bone resorption,[3] and accelerated differentiation of osteoblasts from stem cells.[4] These are the same changes in osteoblast activity that are observed during stimulation of fracture healing in vivo.[1] The effects of EMF on PTH responses were shown to be consistent with decreased coupling of receptors to adenylyl cyclase via the stimulatory G protein.[4] Forskolin and cholera toxin had no effects on the changes induced by EMF,[4,5] which suggests that the hormone receptor, rather than any of the other components of the membrane signaling pathway, was the probable locus of the effects.

Associated studies using flow cytometry and immunohistochemistry indicated that exposure to EMF produces a change in the accessibility of some PTH receptor epitopes to monoclonal antibodies.[6,7] which in turn suggests that the fields induce a relatively persistent conformational change in the medium-accessible domains of the receptor and a corresponding decrease in signal transduction capacity (i.e., desensitization). We have proposed that some or all of the fracture healing effects of these devices in vivo may be due to localized desensitization of osteoblasts to endogenous PTH at the site of application of the electromagnetic energy.[8] In the present work, we have examined the possibility that electromagnetic energy may have a more generalized effect on signal transduction by G protein-linked membrane receptors. We chose to examine the β-adrenergic receptor because it is widespread in tissues, because reagents for its study are readily available, and because some of the effects of LEMF on behavior and diurnal rhythms have been proposed as consistent with modulation of β-adrenergic receptor activity.[9]

MATERIALS AND METHODS

Primary osteoblast-enriched bone cell cultures were isolated by collagenase digestion of cranial bones from 5-day-old mice, as described previously.[2] These cell preparations are approximately 75% osteoblasts as evaluated by alkaline phosphate staining. Cells were cultured in Eagle's MEM containing 5% fetal bovine serum (FBS) at 37 C in a humidified atmosphere of 5% CO_2 in air. Pineal cells were prepared by trypsinization of pineal glands excised from 10-day-old mice, with a modification of procedures described by Buda and Klein.[10] These cells were grown in BGJb culture medium supplemented with 10% FBS, 0.2mM glutamine,

---

*Correspondence to: R. A. Luben (Biomedical Sciences), University of California, Riverside CA 92521-0121.

and 0.1mM ascorbic acid. Each cell type was grown in the presence or absence of pulsed EMF at a repetition rate of either 15 Hz (PTP field) or 72 Hz (SPP field), produced by a clinical bone fracture treatment device (ElectroBiology Inc., Fairfield, N.J.). These fields have been described in our previous work.[2-4] The peak magnetic field during each pulse is approximately 10 G; it induces a peak electric field of approximately 0.06 V/m in the medium surrounding the cells.[2] Control cells were grown in the same device but with the current switched off. Some experiments were done with a 1 G sine-wave field produced by a Helmholtz coil device described previously.[11] The activity of β-adrenergic receptors was determined by measurement of adenylyl cyclase activation by the β-adrenergic agonist isoproterenol.[12,13] The highly specific β-adrenergic binding ligand ^{125}I-cyanopindolol was used to more specifically characterize the receptor. Saturation binding curves were done as previously described.[14]

RESULTS AND DISCUSSION

Figure 1 shows a comparison of the effects of exposure to pulsed magnetic fields on responses of the PTH receptor and the β-adrenergic receptor in bone cells. Figure 1A shows that the half-maximal response to PTH in terms of cAMP accumulation was increased approximately eight-fold by exposure to the field. Thus, at submaximal doses of PTH, the cells were considerably desensitized. However, at higher doses of PTH, the magnitude of the *maximum* response was not changed by exposure to the field. In previous studies we showed that this desensitization also extended to effects of PTH on collagen synthesis, alkaline phosphatase activity and bone resorption, all responses that depend on activation of the PTH receptor.[2-4,6,8] It is of interest that under the same exposure conditions, the binding of the biologically active peptide ^{125}I-PTH(1-34) to the cells was not changed by EMF exposure (Fig. 1B). We have interpreted these data (which replicate our findings over the past 10 years) to indicate that the hormone binding domain(s) of the PTH receptor are not being modified by exposure of the field, but the signal transduction functions of the receptor are being decreased, possibly by a conformational change induced by interaction of the field with the membrane. These data are consistent with a structural model of the receptors for PTH, TSH, LH and other peptide hormones,[15] which postulates a large amino-terminal hormone binding domain that extends from the surface of the cell into the extracellular fluid. This extracellular domain apparently binds hormone relatively independently of the signal-transduction portion of the receptor, which is localized to the seven hydrophobic helices grouped in the membrane. EMFs, which very probably interact with the charged ionic layer within about 0.1 nm of the membrane,[16] could plausibly affect the charge structure and conformation of the membrane-resident domain of the receptor without having major effects on the extracellular hormone binding domain of this class of receptors.

The probability that the effect of EMF on PTH responsiveness is specific to the membrane is increased by our previous observations[5] that the effects of 1,25-dihydroxyvitamin D_3 and dexamethasone on collagen synthesis by bone cells were not modified by EMF exposure; both of these agents act primarily by means of cytosolic-nuclear receptors rather than membrane receptors. Moreover, the effect appears to be receptor-specific, since EMF exposure did not change the responses of the cells to PGE_2 or insulin.[4,6] However, the exposure of bone cells to EMF did modify the responsiveness of the β-adrenergic receptor (Figs. 1C and D). In Fig. 1(C) we see that a field used for clinical fracture treatment inhibits the cAMP response of bone cells to isoproterenol, an agonist that primarily stimulates β-adrenergic receptors. As we observed for PTH, the inhibition appears to be primarily a matter of shifting the dose response curve to the right, i.e., desensitizing the receptor without decreasing the maximum cAMP response levels at high doses of agonist. Unlike the case of the PTH receptor, however, the field did inhibit the binding of the β-adrenergic receptor specific ligand ^{125}I-cyanopindolol (Fig. 1D). Since, unlike the peptide hormones, the binding of adrenergic ligands to their receptors is inside the membrane pocket formed by the seven transmembrane protein helices of the G protein-linked receptors[15] (i.e., in the same portion of the receptor responsible for signal transduction), it would be expected that any conformational change in the membrane-resident portion of the receptor would very likely change not only signal transduction, but also ligand binding. Thus these data are consistent with the hypothesis that EMF may act at the membrane surface to cause a change in conformation of the G protein-linked receptor.

In order to test further the hypothesis that EMF exposure could modulate the β-adrenergic receptor, we tested the effect of the pulsed-field clinical EMF fracture-treatment devices on the response of β-adrenergic receptors in mouse fibroblasts, liver hepatoma cells, and dispersed primary cultures of pineal cells. The results of these experiments are summarized in Fig. 2 and Table 1. No effect of EMF was noted in the fibroblasts or the hepatoma cells, but a significant inhibition of the cAMP response to isoproterenol was found in

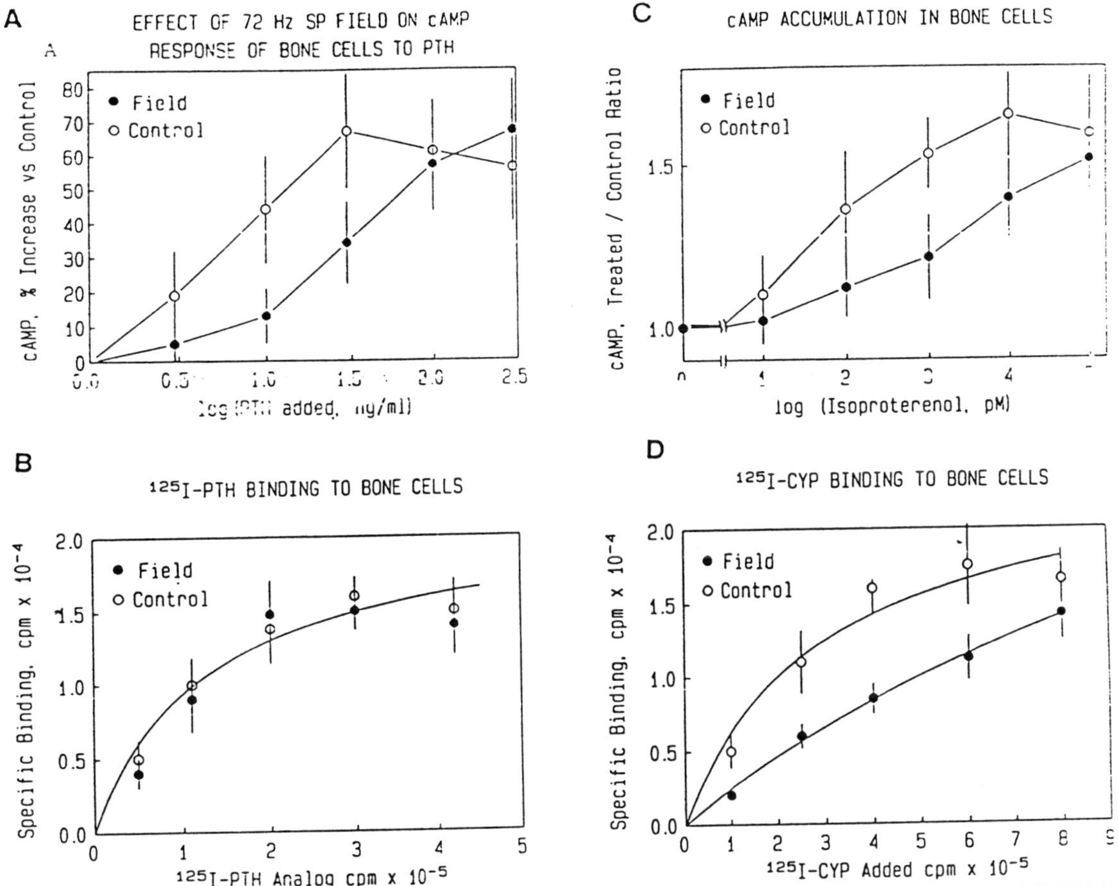

FIG. 1.—Inhibition of PTH receptor and β-adrenergic receptor responses by pulsed EMF in mouse bone cell cultures. (A) Inhibition of cAMP accumulation stimulated by PTH. Cells were incubated for 6 h with Bi-osteogen clinical fracture healing device, which produces pulsed 10G field. Last hour of incubation included 0.1 mM isobutylmethylxanthine. Cells were then treated with indicated dose of PTH for 5 min and cAMP was determined by RIA. (B) Lack of effect of pulsed EMF on binding of ^{125}I-PTH analog. Cells were incubated with same field as in (A); then PTH binding was determnined at 15 C for 2 h. (C) Effect of pulsed EMF on β-adrenergic receptor activity (cAMP accumulation) stimulated by isoproterenol in mouse bone cells. (Same protocol as in A.) (D) Inhibition of binding of β-adrenergic receptor-specific ligand ^{125}I-cyanopindolol in bone cells treated as described in (C).

the pineal cells (Table 1). This finding is of particular interest, since the pineal gland has been hypothesized to be a site of EMF actions on diurnal rhythms and possibly on carcinogenesis.[17] Moreover, the major regulatory neurochemical input to the pineal gland is β-adrenergic.[9] Closer examination of this phenomenon indicated (Fig. 2) that EMF exposure caused a desensitization of the pineal β-adrenergic receptor to isoproterenol, similar to that seen in mouse bone cells. Other preliminary experiments (Table 1) indicate that the binding of ^{125}I-cyanopindolol to pineal cells is inhibited similarly to the inhibition shown for bone cells in Fig. 1.

TABLE 1.—Effects of β-adrenergic receptors in various cell types: field-treated/untreated ratios.

Cell line tested	cAMP Accumulation at 10^{-6} M isoproterenol	^{125}I-CYP binding at half-maximal activity
Bone cells	0.64 ± 0.13*	0.52 ± 0.09*
HEP hepatoma	1.06 ± 0.11	0.93 ± 0.12
Skin fibroblasts	0.96 ± 0.13	1.06 ± 0.11
Pineal cells	0.59 ± 0.14*	0.71 ± 0.15*

*Significant inhibition by exposure to field, $p < 0.05$, Dunnett's multiple range test. Cells were cultured and treated as described in legend to Fig. 1. Values represent the mean ±SEM of quadruplicate cultures.

The fact that not all cells that possess β-adrenergic receptors are affected by EMF suggests that factors other than simple expression of the receptor are involved. Such factors might include the chemical and physical status of the membrane, different subtypes of the receptor, and/or differential glycosylation of the receptor in different cell types.[15]

In more recent experiments we have begun to extend the findings obtained with clinical field generators both to lower field strengths and to the 60Hz sine-wave alternating field most commonly encountered in environmental and occupational settings. We have done a series of experiments similar to those in Figs. 1 and 2 using a 60Hz, 1.0G sine-wave field.[11] This field, although still stronger than constant exposures normally encountered in the environment, is much closer to the levels at which health hazards have been suggested by epidemiological studies.[18] It is entirely possible for an ordinary individual to encounter transient 60Hz fields of 1 G or more during the day, for example while using appliances such as hair dryers, electric shavers, and power saws.[19] Moreover, determination of whether fields of 1 G can significantly modulate biological processes is an important test of theoretical calculations such as those of Adair,[20] who asserts that fields lower than tens of gauss cannot possibly contribute to membrane information because of "the thermal noise limit."

Figure 3 shows that the cyclic AMP responses of mouse bone cells to isoproterenol were inhibited by a 1G 60Hz sine-wave field, although not as strongly as with the 10G pulsed fields of Figs. 1 and 2. Nevertheless, comparison of the treated vs control dose response curves by ANOVA indicated that at submaximal doses of agonist, it would take from 30 to 50% more hormone or drug to produce an equivalent response in EMF-exposed cells than in corresponding controls. Preliminary experiments with pineal cells suggests that a similar inhibition by the 1G field occurs in these cells (data not shown).

To summarize, in preliminary experiments we have demonstrated significant inhibition of the activation of cAMP accumulation by β-adrenergic agonists in both osteoblasts and pineal cell cultures; and we have observed changes in the binding kinetics of the β-adrenergic ligand ^{125}I-cyanopindolol consistent with a field-induced change in receptor conformation. When combined with our previous observations of the effects of EMF on PTH receptor function, the results support the hypothesis that EMF may produce cellular responses by specifically modulating the conformation of G protein-linked membrane receptors, thus perturbing associated signal transduction processes. These findings have led us to hypothesize[8] that G protein-linked receptors may be a particular target for the interaction of EMF with cell membranes in responsive cells. In addition to possible effects on the pineal,[9] neurological and behavior effects of EMF[21] have been linked to neurotransmitter receptors, many of which are G protein linked. Changes in the functions of G protein-linked receptors also can be a factor in cell transformation. Overexpression of the serotonin receptor, for example, has been shown to cause neoplastic transformation of cells.[22] G protein-linked receptors can be influential in controlling cell growth and development,[23] e.g., by controlling gene expression via transcriptional regulators (*fos*, CREB) that are phosphorylated by PKC, PKA, or both.[24] It is entirely possible that small changes in the surface charge distribution of membranes, comparable to the increased negative surface charge observed by Smith et al.[25] in cells exposed to fields very similar to those used in our studies, could lead to significant changes in the signal transduction properties of G protein-linked receptors in those membranes. If substantiated, this hypothesis would ameliorate objections that low energy fields are too weak to have any biological effects, since the energy needed to produce small (but functionally significant) conformational changes in surface domains of membrane proteins may be very low indeed.[26]

REFERENCES

1. C. A. L. Bassett, "Pulsing EMF field treatment in ununited fractures and failed arthrodeses," *JAMA* 247: 623, 1982.

2. R. A. Luben, C. D. Cain, M. C.-Y. Chen, D. M. Rosen, and W. R. Adey, "Effects of electromagnetic stimuli on bone and bone cells in vitro: Inhibition of responses to parathyroid hormone by low-energy low-frequency fields," *Proc. Natl. Acad. Sci.* 79: 4180, 1982.

3. C. D. Cain and R. A. Luben, "Pulsed electromagnetic field effects on PTH-stimulated cAMP accumulation and bone resorption in mouse calvariae," in L. E. Anderson, B. J. Kelman, and R. J. Wiegel, Eds., *Interaction of Biological Systems with Static and ELF Electric and Magnetic Fields: Proceedings of the 23rd Hanford Life Sciences Symposium,* Columbus, Ohio: Batelle Press, 24, 1987, 269.

4. C. D. Cain, W. R. Adey, and R. A. Luben, "Evidence that pulsed electromagnetic fields inhibit coupling of adenylate cyclase by parathyroid hormone in bone cells," *J. Bone & Min. Res.* 2: 437, 1987.

5. D. M. Rosen and R. A. Luben, "Multiple hormonal mechanisms for the control of collagen synthesis in an osteoblast-like cell line, MMB-1," *Endocrinology* 112: 992, 1983.

6. R. A. Luben and C. D. Cain, "Modulation by low energy electromagnetic fields of bone cell membrane activities, bone resorption and expression of parathyroid hormone (PTH) receptor-related antigens," *J. Cell Biol.* 103: 457a, 1986.

7. R. A. Luben, "Comparison of electromagnetic effects on parathyroid hormone receptors and beta-adrenergic receptors in bone cells," *J. Cell Biol.* 109: 172a, 1989.

8. R. A. Luben, "Effects of low energy electromagnetic fields on membrane signal transduction processes in biological systems," *Health Physics* 61: 15, 1991.

9. R. J. Reiter, "Pineal melatonin: Cell biology of its synthesis and of its physiological interactions," *Endocrine Reviews,* 12: 151, 1991.

10. M. Buda and D. C. Klein, "A suspension culture of pinealocytes: Regulation of N-acetyl transferase activity," *Endocrinology* 103: 1482, 1978.

11. J. L. Phillips, W. Haggren, W. J. Thomas, et al., "Magnetic field-induced changes in specific gene transcription," *Biochim. Biophys. Acta* (in press).

12. A. M. Spiegel, "Signal transduction by guanine nucleotide binding proteins," *Ml. Cell. Endocrinol.* 49: 1, 1987.

13. R. J. Lefkowitz and M. G. Caron, "Molecular and regulatory properties of adrenergic receptors," *Rec. Prog. Hormone Res.* 43: 469, 1987.

14. B. K. Kobilka et al., "Chimeric α2-, β2-adrenergic receptors: Delineation of domains involved in effector coupling and ligand binding specificity," *Science* 240: 1310, 1988.

15. H. G. Dohlman, J. Thorner, M. G. Caron, and R. J. Lefkowitz, "Model systems for the study of seven-transmembrane-segment receptors," *Ann. Rev. Biochem.* 60: 653, 1991.

16. W. R. Adey, "Cell membranes: The electromagnetic environment and cancer," *Neurochem. Res.* 13: 671, 1988.

17. B. W. Wilson, R. G. Stevens, and L. E. Anderson, "Neuroendocrine mediated effects of electromagnetic-field exposure: Possible role of the pineal gland," *Life Sci.* 45: 1319, 1989.

18. D. A. Savitz, N. E. Pearce, and C. Poole, "Methodological issues on epidemiology of electromagnetic fields and cancer," *Epidemiological Reviews* 11: 59, 1989.

19. A. Leonard, R. Neutra, M. Yost, and G. Lee, "Electric and magnetic fields: Measurements and possible effects of human health," *Health Physics Newsletter* 9(No. 10): 13, 1992.

20. R. N. Adair, "Constraints on biological effects of weak extremely-low-frequency electromagnetic fields," *Physics Reviews* A43: 1039, 1991.

21. R. F. Seegal, J. R. Wolpaw, and R. Dowman, "Chronic exposure of primates to 60 Hz electric and magnetic fields: II. Neurochemical effects," *Bioelectromagnetics* 10: 289, 1989.

22. D. Julius, T. J. Livelli, T. M. Jessell, and R. Axel, "Ectopic expression of the serotonin 1c receptor and the triggering of malignant transformation," *Science* 244: 1057, 1989.

23. S. K. Gupta, C. Gallego, and G. L. Johnson, "Mitogenic pathways regulated by G protein oncogenes," *Molecular Biol. of the Cell* 3: 123, 1992.

24. M. Kaleko et al., "Overexpression of the human IGF-1 receptor promotes ligand-dependent neoplastic transformation," *Mol. Cell. Biol.* 10: 464, 1990.

25. O. M. Smith, E. M. Goodman, B. Greenebaum, and P. Tipnis, "An increase in the negative surface charge of U937 cells exposed to a pulsed magnetic field," *Bioelectromagnetics* 12: 197, 1991.

26. J. C. Weaver and R. D. Astumian, "The response of living cells to very weak electric fields: The thermal noise limit," *Science* 247: 459, 1990.

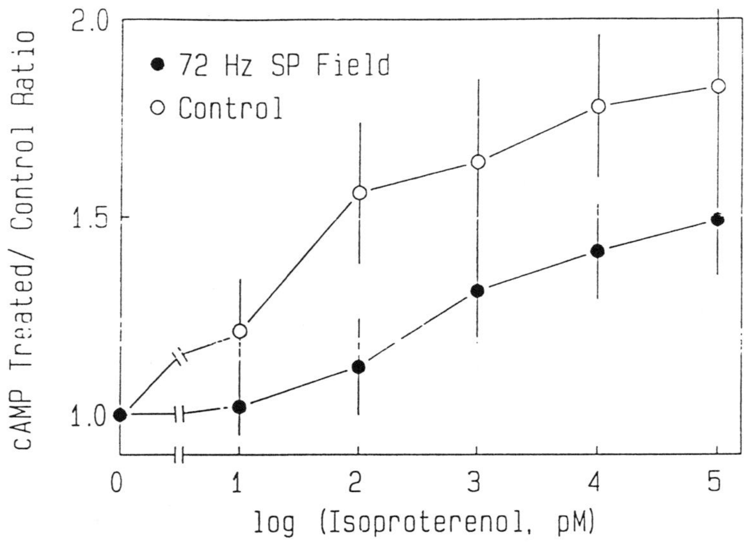

FIG. 2.—Inhibition of isoproterenol-stimulated cAMP accumulation in mouse pineal cell cultures by 72Hz pulsed EMF. Mouse pineal cultures were prepared as described in text and exposure to the same field under the same protocol as described in Fig. 1(A).

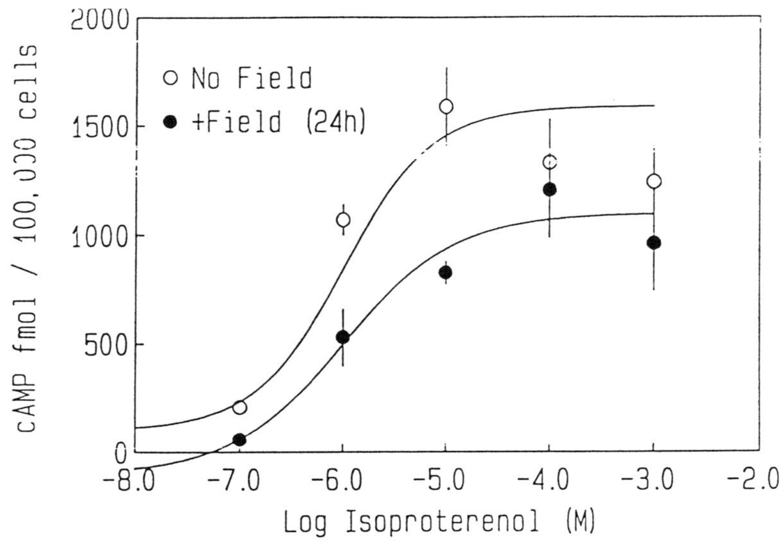

FIG. 3.—Magnetic field (1 G) inhibition of cAMP responses to isoproterenol in bone cells. Cells were treated as described in Fig. 1, except that time of exposure was 24 h and field was a 60Hz sine wave, 1.0G field.

Martin Blank, Ed., *Electricity and Magnetism in Biology and Medicine*

MOLECULAR BIOLOGY IN BIOELECTROMAGNETICS: THE ROAD TO THE FUTURE

J. L. Phillips*

Pettis Memorial VA Medical Center, Loma Linda, Calif.

Research in bioelectromagnetics is taking on a new look. There is a decreased emphasis on simply cataloging EMF-induced bioeffects and an increased emphasis on understanding how EMFs produce these various effects, on characterizing that complex elusive entity called "mechanism." Using an observed bioeffect as a starting point, investigators are beginning to "move backward" toward what is the initial event in the interaction of EMFs with a living system. That is why researchers in bioelectromagnetics are turning to the vast literature and valuable techniques of molecular biology in their quest to understand "mechanism."

This review concentrates on two areas. First, some basic concepts in molecular biology are presented. Second, representative studies that have appeared in the peer-reviewed literature are summarized and evaluated.

MOLECULAR BIOLOGY

It has been known since the turn of the century that genetic information is carried by chromosomes found in the cell's nucleus; and it has been known virtually from their discovery that chromosomes consist of both nucleic acid and protein. The nucleic acid is DNA, the substance of which genes are made, and, with the exception of certain viruses, it is the molecule of heredity in nearly all organisms. All the processes in which living systems participate can be viewed in terms of the functioning of DNA molecules. It is the ultimate function of DNA to give rise to the diverse complement of proteins that characterize the very nature of a biological system at any given time. Hence, changes in a cell's functional requirements necessitate changes in the nature and quantity of cellular proteins, thus altering cell-specific patterns of gene expression.

The molecular events involved in producing protein from the code contained within the sequence of bases (A, deoxyadenine; G, deoxyguanine; C, deoxycytosine; T, deoxythymidine) that constitute a gene include the basic processes of transcription and translation. Transcription is the production of a specific molecule of messenger RNA (mRNA) from a given sequence of DNA; translation is the production of a specific amino acid sequence (protein) by interaction of mRNA, transfer RNAs to which specific amino acids are attached, ribosomes (including ribosomal RNA), and a number of requisite protein factors. In this review, emphasis is placed on the process of transcription.

The synthesis of a molecule of mRNA from a specific gene is catalyzed by an enzyme known as DNA-dependent RNA polymerase II or, more commonly, RNA polymerase II or pol II. The key to this synthesis is production of nucleic acid, the mRNA, whose base sequence is complementary to the reaction template, the DNA. The transcription is initiated at a particular region of the DNA molecule called the promoter, which may be thought of as a group of transcriptional control modules or "response elements" clustered around the initiation site for pol II. The response elements are then defined as DNA sequences that reversibly enhance or repress transcription in response to molecular signals. These signals, such as induced changes in second-messenger level or kinase activity, alter the binding of protein transcription factors, which recognize specific base sequences contained within the response elements. It is this protein-DNA interaction that is in part responsible for determining the pattern of gene transcription that occurs in a cell at any given time.

Also relevant to gene transcription is our current view of DNA organization. Genes have been shown to be composed of coding sequences, called exons, separated by noncoding intervening sequences, called introns. Thus, when DNA is transcribed in the nucleus, the product, heterogeneous nuclear RNA or hnRNA, must be processed to remove sequences corresponding to introns before an mRNA is produced that can direct

*Correspondence to: J. L. Phillips (Research Serv. 151), Pettis Mem. VA Med. Ctr., Loma Linda CA 92357.

protein synthesis in the cytoplasm. However, even though much of the genome constitutes noncoding sequences, these regions contain many regulatory sequences that are, for instance, involved in the regulation of gene transcription and in determining mRNA half-life. Indeed, because of the complex nature of eukaryotic gene transcription, there are multiple points at which the process can be controlled.

MOLECULAR BIOLOGY IN BIOELECTROMAGNETICS

In 1983 Goodman and colleagues reported that two different electromagnetic signals, a 72Hz single pulse and a 15Hz pulse train, each increased gene transcription in dipteran salivary gland cells, compared to unexposed control cells, as assessed by transcription autoradiography and cytological nick translation.[1] These two techniques permit a qualitative assessment of gene transcription based on the incorporation of a radioactive RNA precursor into nascent RNA chains attached to chromosome regions. This group also separated RNAs by size class via sucrose density centrifugation. Their results indicated increased precursor incorporation into RNAs of a size class consistent with that expected for mRNAs. Furthermore, the two signals produced different kinetics of precursor incorporation. Similar results were obtained in a study by Phillips and McChesney,[2] who reported increased incorporation of ^{3}H-uridine into both total cellular RNA and mRNA in T-lymphoblastoid cells exposed to a 72Hz pulsed magnetic field. These studies, while pointing to general changes in gene transcription in EMF-exposed cells, do not address the important issue of which *specific* genes were affected. In addition, other key questions were raised. Do different signals affect the same or different genes? Are there genes in EMF-exposed cells whose transcription is decreased? Which signal characteristics are important determinants of cell responsiveness? By what mechanism does EMF exposure alter gene transcription?

Goodman and colleagues have attempted to answer some of these questions in subsequent studies. For instance, they reported that a 72Hz sine wave produced general changes in gene transcription in dipteran salivary gland cells similar to those observed with the 72Hz pulsed signal, as judged by transcription autoradiography.[3] They also report that increasing the sine-generated frequency to 222 or 4400 Hz produced the same pattern of induced transcription, although it was not as dramatic as observed when cells were exposed to a 72Hz sine wave field. However, it was not clear whether the observed changes in gene transcription were related to the signal frequency, signal intensity, or both.

Goodman and Henderson have attempted to determine the relative effects of signal waveform and frequency on gene transcription in dipteran salivary-gland cells.[4] Frequencies tested were 1.5, 2, 4, 15, 72, 222, and 4400 Hz. Repetitive single pulses, repetitive pulse burst, and sine waveforms were used at various intensities. (Reference ? gives specifics about waveform characteristics.) Again using transcription autoradiography, these investigators found general increases in gene transcription with 13 of the 15 signals tested, although the greatest effect was observed with the 72Hz signals. A similar study with 72Hz signals was reported by this group in 1987.[5]

Although the amount of gene transcript, or mRNA, in a cell is not necessarily a measure of gene transcription, several investigators have reported changes in specific mRNA levels in EMF-exposed cells. Before proceeding, I must emphasize one point: we have heard all too often at meetings that changes in transcript or mRNA levels equate with changes in gene transcription, which is *not* necessarily so. The "life cycle" of a gene transcript begins with its transcription and processing in the cell's nucleus. The processed mRNA is then transported to the cytoplasm, where it is ultimately degraded. Variations in cytoplasmic steady state mRNA levels can result from any change along this pathway, including changes in the nuclear stability of a (pre-)mRNA. To determine with certainty that altered transcript levels are the result of altered gene transcription, additional studies with appropriate techniques (e.g., nuclear run-off assay) must be performed.

Cadossi et al. exposed mice to a pulsed EMF after total body x-ray irradiation.[6] They extracted RNA from the spleens at day 25 after irradiation. They then determined the levels of histone H3 and nuclear protein p53 mRNA. Their results indicated a slight increase in cycling cells among spleens of mice exposed to the pulsed signal. Fitzsimmons et al. exposed TE-85 human osteosarcoma cells to a low-amplitude (10^{-7} V/cm), low-frequency (10-16 Hz), capacitively coupled electric field in order to study the mechanism through which an electric field may increase bone-cell proliferation.[7] They found that electric-field stimulation was associated with increased accumulation of mRNA for the insulin-like growth factor II (IGF-II). They acknowledge that accumulation of a transcript can occur from either transcriptional or post-transcriptional mediated processes, and are at present determining the mechanism underlying their observed effect. In the studies of Cadossi et al.[6] and Fitzsimmons et al.,[7] Northern blot analysis was employed to measure transcript levels. The Northern blot permits more confidence in specific transcript measurement, since the size of the transcript hybridizing with the gene-specific probe is determined and can be compared with the size expected for that transcript based on available scientific literature.

Goodman and colleagues have also reported transcript-level measurements in HL-60 human myelogenous leukemia cells exposed to a variety of electromagnetic signals.[8,9] However, their studies employed a dot-blot assay, rather than the Northern blot, to assess transcript levels. In this technique, total cellular RNA is immobilized on an inert support without prior separation of the RNAs by class size, and is then hybridized to a gene-specific probe. These investigators have reported results that on the surface appear to be very interesting. For example, they reported that transcripts with homology to β-actin, histone H2B, and the *my* proto-oncogene increased in response to 4 signals as follows: 60Hz sine wave ≥ 72Hz single pulse = 15Hz pulse train > 1.5Hz pulse train > control levels.[8] In a subsequent paper,[9] they extend their studies to, among other parameters, the effect of signal amplitude on transcript level, this time reporting effects of c-*myc*, β-actin, histone H2B, β-tubulin, and v-*src* transcripts. They report that (a) response may occur as early as in 3-4 min; (b) the patterns for each transcript measured were identical; and (c) the level of each transcript peaked when cells were exposed to a setting of 5 μV for 20 min. Herein lie the reasons these data must be viewed with caution. Indeed, the five genes of interest to this group do not normally respond to a various stimuli with the same time course. In HL-60 cells, c-*myc* mRNA levels generally peak about 1 h after cell treatment;[10] *src* and *src*-family mRNA levels at 3-6 h;[11] and histone H2B expression is tightly coupled to DNA synthesis.[12] It is thus unreasonable to expect all genes studied to respond with the same pattern and with the same time course. The significance of Goodman and Henderson's data in this area remains to be determined.

Phillips and colleagues have addressed the question of EMF effects on specific gene transcription using the nuclear run-off assay.[13] This technique lets mRNA strands that are already initiated in vivo to be completed in vitro, thus providing an accurate assessment of gene transcription at the time of cell lysis. This group reported that exposure of a T-lymphoblastoid cell line to a 1G sinusoidal magnetic field at 60 Hz increased the transcription of the c-fos, c-jun, c-myc, and protein kinase C genes in a time-dependent manner consistent with the scientific literature. The mechanism underlying these specific effects is under investigation.

In summary, the potential of the knowledge and techniques of molecular biology to contribute to our understanding of EMF-induced effects is tremendous. We look forward to the results of studies detailing EMF-induced effects on gene promoters, transcription factors, and signal-transduction pathways. In providing answers to key questions, we are limited only by our own imaginations and creativity.[14]

REFERENCES

1. R. Goodman, C. A. L. Bassett, and A. S. Henderson, "Pulsing electromagnetic fields induce cellular transcription," *Science* 220: 128, 1983.

2. J. L. Phillips and L. McChesney, "Effect of 72 Hz pulsed magnetic field exposure on macromolecular synthesis in CCRF-CEM cells," *Cancer Biochem. Biophys.* 12: 1, 1991.

3. R. Goodman and A. S. Henderson, "Sine waves enhance cellular transcription," *Bioelectromagnetics* 7: 23, 1986.

4. R. Goodman and A. S. Henderson, "Some biological effects of electromagnetic fields," *Bioelectrochem. Bioenerg.* 15: 39, 1986.

5. R. Goodman, J. Abbott, and A. S. Henderson, "Transcriptional patterns in the X chromosomes of *Sciara coprophilia* following exposure to magnetic fields," *Bioelectromagnetics* 8: 1, 1987.

6. R. Cadossi, V. R. Hentz, J. Kipp, R. Eiverson, G. Ceccherelli, P. Zucchini, G. Emilia, G. Torelli, C. Franchesi, and A. Cossarizza, "Effect of low frequency low energy pulsing electromagnetic field (PEMF) on x-ray-irradiated mice," *Exp. Hematol.* 17: 88, 1989.

7. R. J. Fitzsimmons, D. D. Strong, S. Mohan, and D. J. Baylink, "Low-amplitude, low-frequency electric field-stimulated bone cell proliferation may in part be mediated by increased IGF-II release," *J. Cell. Physiol.* 150: 84, 1992.

8. R. Goodman, L.-X. Wei, J.-C. Xu, and A. Henderson, "Exposure of human cells to low-frequency electromagnetic fields results in quantitative changes in transcripts," *Biochim. Biophys. Acta* 1009: 216, 1989.

9. R. Goodman and A. S. Henderson, "Transcription and translation in cells exposed to extremely low frequency electromagnetic fields," *Bioelectrochem. Bioenerg.* 25: 335, 1991.

10. Z. Salehi and J. E. Niedel, "Multiple calcium-mediated mechanisms regulate c-myc expression in HL60 cells," *J. Immunol.* 145: 276, 1990.

11. M. Furue, S. I. Katz, Y. Kawakami, and T. Kawakami, "Coordinate expression of *src* family protooncogenes in T cell activation and its modulation of cyclosporine," *J. Immunol.* 144: 736, 1990.

12. A. R. Shakoori, A. M. Oberdort, T. A. Owen, L. A. Weber, E. Hickey, J. L. Stein, J. B. Lian, and G. S. Stein, "Expression of heat shock genes during differentiation of mammalian osteoblasts and promyelocytic leukemia cells," *J. Cell. Biochem.* 48: 277, 1992.

13. J. L. Phillips, W. Haggren, W. J. Thomas, T. Ishida-Jones, and W. R. Adey, "Magnetic field-induced changes in specific gene transcription," *Biochim. Biophys. Acta* 1132: 140, 1992.

14. This work was supported by U.S. Department of Energy contract DE-AI01-90CE35035.

Martin Blank, Ed., *Electricity and Magnetism in Biology and Medicine*

PARAMETRIC AND MECHANISTIC STUDIES OF PERTURBATIONS OF THE CIRCADIAN MELATONIN RHYTHM BY MAGNETIC-FIELD EXPOSURE

R. J. Reiter* and K. Yaga

University of Texas Health Science Center, San Antonio

The pineal gland, an endocrine organ in the brain, produces the hormone melatonin only during the night. The nocturnal synthesis and secretion of melatonin by the pineal gland is normally controlled by the light:dark cycle, with light inhibitory to the production of the hormone. Besides light, the exposure of mammals to inverted geomagnetic fields at night suppresses melatonin synthesis, as well as the activity of the enzymes required for melatonin formation and the intracellular second messenger cAMP. The pineal may respond to magnetic (B) fields directly or indirectly via the retinas. In the retinas, changes in dopamine and 11-*cis*-retinal have been implicated as possibly being involved in the inhibitory effects of magnetic fields on the pineal gland.

The pineal gland, a hormone-producing organ in the brain of vertebrates, is of interest to scientists working on bioeffects of electromagnetic radiation by virtue of its ability to respond to both visible and nonvisible portions of the electromagnetic spectrum. The major hormone of the pineal gland is N-acetyl-5-methoxytryptamine, commonly called melatonin. In mammals, including man, the pineal is an end organ of the visual system, with the production of melatonin being controlled by the prevailing light:dark cycle. Normally, visible light detected by the retinas prevents the formation and secretion of melatonin; whereas in darkness, its production increases markedly.[1] Since melatonin is quickly released from the pineal gland, blood levels of the hormone are high at night and low during the day. Information about the light:dark cycle is transferred from the retinas to the pineal gland over a series of neurons that include ganglion cells of the retinas, the suprachiasmatic nuclei (SCN), the paraventricular nuclei, and pre- and postganglionic sympathetic neurons. Besides 400-720nm light, wavelengths in the ultraviolet (UV) range as well as extremely low frequency wavelengths and pulsed static magnetic fields alter the ability of the gland to synthesize its chief hormonal product.[2]

THE CIRCADIAN PRODUCTION OF MELATONIN BY THE PINEAL GLAND

The mechanisms governing the nocturnal rise in pineal melatonin production are well understood. In the pineal gland, the process is initiated by the release of the catecholaminergic neurotransmitter norepinephrine (NE) from the postganglionic sympathetic nerves that end in the gland (Fig. 1). NE interacts primarily with β- and to a lesser degree with α-adrenergic receptors to promote the production of 3′,5′-cyclic adenosine monophosphate (cAMP), the key intracellular messenger which eventually causes the expression of the rate limiting enzyme in melatonin production *N*-acetyltransferase (NAT). NAT *N*-acetylates serotonin (5HT) to *N*-acetylserotonin, which is subsequently converted to melatonin by the enzyme hydroxyindole-*O*-methyltransferase (HIOMT). Once produced, melatonin is quickly released from the pineal into the blood vascular system; thus, as melatonin synthesis increases at night, blood levels of the hormone rise soon after.

5HT, a product of the amino acid tryptophan, also has other metabolic fates in the pineal gland. One of these pathways involves its oxidative deamination and eventual conversion to 5-hydroxyindole acetic acid (5HIAA) (Fig. 1).

The primary regulatory factor determining the 24h rhythm in melatonin synthesis and release is the light:dark cycle. During the day, NE release from the intrapineal neurons is low and the critical intracellular second messenger cAMP is not produced in significant amounts; thus, NAT activity and melatonin synthesis

---

*Correspondence to: R. J. Reiter (Cellular & Structural Biol.), University of Texas Health Science Ctr., 7703 Floyd Curl Dr., San Antonio TX 78284-7762.

remain low. At night, NE release increases and leads to a rise in cAMP, an activation of NAT, and a marked increase in melatonin synthesis and release.[1] If animals and/or man are exposed to light at night, even for a short time, pineal melatonin production ceases and blood levels fall. Besides visible light, near UV light (290-400 nm), which is not perceptible by the mammalian retinas, inhibits melatonin production in the few animals where it has been tested.

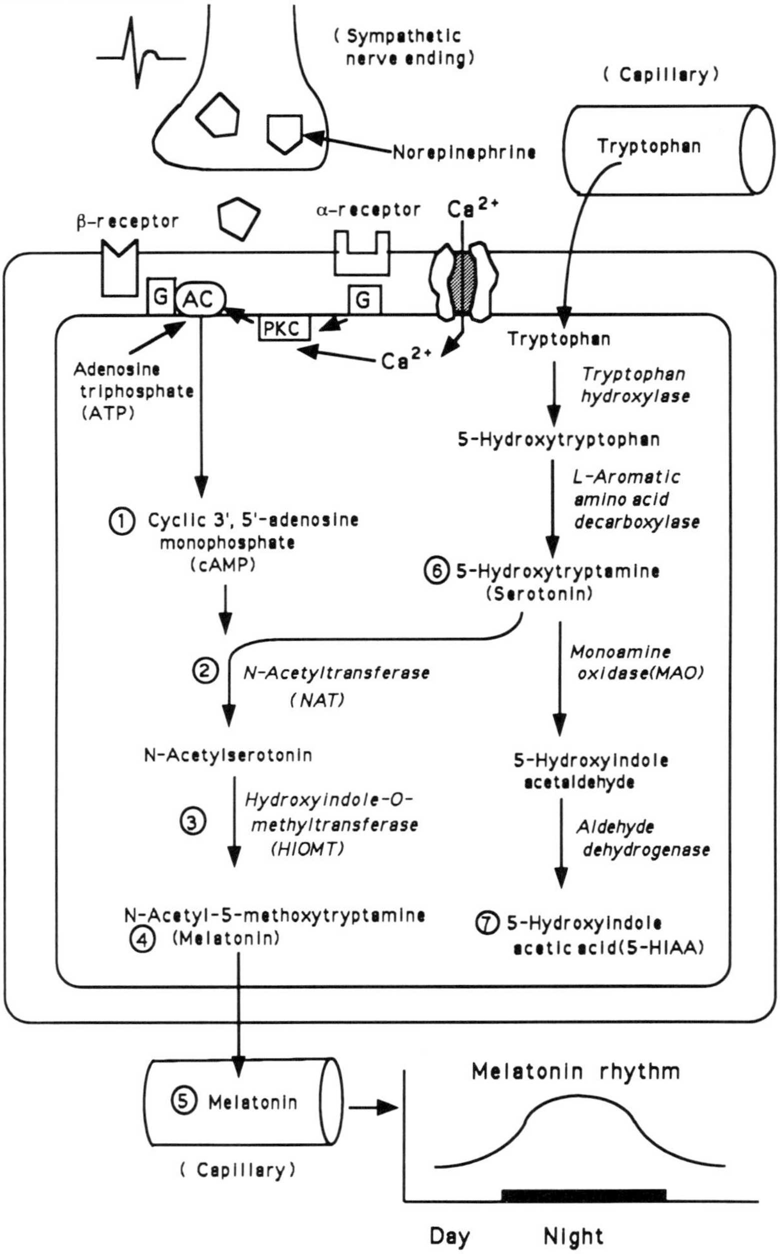

FIG. 1.—Schematic representation of melatonin synthesis and secretion in mammalian pineal gland. Nocturnal rise in pineal melatonin production is regulated by increased release of norepinephrine (NE) from sympathetic neurons that end in the pineal gland. NE interacts with receptors, which leads to intracellular increase in cAMP and expression of NAT, the rate limiting enzyme in melatonin production. Numbers 1-7 indicate constituents that have been shown to be changed by B-field exposure. Constituents 1-5 have been reported to decrease, which results in a rise in serotonin and 5HIAA (6 and 7, respectively). Melatonin rhythm is depicted at lower right.

MAGNETIC FIELD EFFECTS ON THE MELATONIN RHYTHM

Both sinusoidal and static (time-invariant) magnetic field (B-field) exposures have been reported to change the circadian production of melatonin, but the current report centers on the effects of static B fields, since their interactions with the melatonin cycle have been more thoroughly defined. Over the past decade, several investigators have reported that static B-field exposure decreases the nocturnal metabolism of 5HT to melatonin, which results in changes in one or more constituents in this synthetic pathway.[2,3] Most frequently, the level of pineal melatonin was the index measured, and in all studies where a change was observed, a decrease occurred. These observations were most often made in birds and rodents (usually rats). The experimental design usually consisted of the exposure of the animals to an inverted geomagnetic field during the dark portion of the light:dark cycle. In a few cases the field was inverted (pulsed) frequently during the exposure period. In most cases the exposure periods were short (up to 1 h) and the observed effects typically were measured within 30-60 min after exposure onset.

As mentioned, pineal melatonin levels were the usual endpoint in these studies; other parameters that were estimated include cAMP levels, the activities of NAT and HIOMT, 5HT and 5HIAA concentrations, and melatonin levels in the blood (items 1-7 in Fig. 1).

The measurements that have been made in the pineal and blood of animals exposed to static B fields suggest that the fields inhibit melatonin production at a site proximal to the intracellular second messenger cAMP. The resulting reduction in cAMP causes a drop in NAT activity that, along with a lowered activity of the melatonin forming enzyme HIOMT, leads to a decreased level of newly synthesized melatonin; as a consequence, blood levels of melatonin drop accordingly. Conversely, since the activity of NAT that metabolizes 5HT is diminished, 5HT accumulates and is thereafter converted to 5HIAA. Thus, whereas cAMP, NAT, HIOMT, and pineal and blood melatonin levels may decrease, rises in 5HT and 5HIAA would be expected and have been reported (Fig. 1).[2,3]

The magnitude of the pineal changes induced by the exposure of animals to static B fields can generally be classified as moderate; a total suppression of nocturnal melatonin production has rarely been observed. However, the consistency of the observations (made by a variety of workers using various exposure paradigms) provides strong support for the notion that the pineal gland is one organ whose activity is rather routinely influenced by B-field exposure. It is possible that the less-than-total inhibition of melatonin production could relate to the exposure parameters used in these studies; thus, use of other exposure parameters (i.e., variation of the field strength, frequency of pulsing, etc.) may yet identify situations in which the circadian melatonin rhythm is even more severely perturbed.

IMPORTANT PARAMETERS OF EXPOSURE AND POSSIBLE BIOPHYSICAL MECHANISMS

The geomagnetic field strength, depending on the location on the earth's surface, usually varies between 0.2 and 0.7 G (20-70 μT). Within these limits, the partial rotation or total inversion of the field suppresses the synthetic activity of the pineal gland. Whereas perturbations of the geomagnetic field have proven effective in inducing changes in the pineal gland, static B-field strengths outside the geomagnetic range have not been tested. Furthermore, no careful field strength-response studies have been performed, so that it is not known whether there is a relationship between the perturbed field strength and the magnitude of change in pineal melatonin synthesis.

The ability of a pulsed static B field to inhibit the pineal gland may depend on the time of the day the exposure occurs. Since melatonin synthesis is minimal during the day, the ability of pulsed fields to suppress this synthetic pathway further has not been thoroughly tested. During the night, the gland exhibits an increasing sensitivity to perturbed B fields as darkness continues; thus, the gland appears to respond more robustly to the fields later in the night than shortly after darkness onset.[4] One potential explanation for this finding is that larger quantities of NE are released onto the pinealocytes in the early portions of the dark phase, which makes it more difficult to suppress the activation of the melatonin-forming pathway. Conversely, later in the dark phase, NE release drops, making the system more sensitive to inhibition.

In early studies, it was common merely to place animals between Helmholtz coils and invert the geomagnetic field only once during the exposure period. More recently it has become common to invert (pulse) the field at either 1 or 5min intervals during the exposure period.[5] There is little evidence to suggest that the frequency of field inversion is related to the magnitude of change in melatonin production.

Yet according to Lerchl et al.[5] the rapidity of inversion of the static B field may be critical to its ability to suppress pineal melatonin synthesis. In their study, rats at night were exposed for 1 h to a geomagnetic field of 0.44 G repeatedly inverted at 1min intervals. One group of rats was exposed to a field "instantaneously" inverted by means of a manual relay switch, and another group was exposed to a field "slowly" inverted by means of an integrating potentiometer. A depression in pineal NAT activity and melatonin levels was observed only in the case of the "instantaneously" inverted fields; i.e., the change seemed to depend on dB/dt.[5] The implication is that the melatonin-synthesis changes were due to the resulting induced electric currents (eddy currents), i.e., the Faraday effect. By use of a sinewave oscillator and an audio power amplifier to drive the Helmholtz coils it would be possible to vary dB/dt virtually continuously, thereby sustaining the electric currents. This experiment may help to clarify further the role of the induced electric currents in causing melatonin synthesis suppression.

The site and biophysical mechanisms of B-field detection remains unknown. Several authors have come to the conclusion that the retinas serve as the detectors for the magnetic fields.[3] This conclusion is based on at least two observations. First, surgical removal of the eyes reportedly prevents the inhibitory effects of the geomagnetic field on pineal melatonin synthesis in mammals.[3] Second, the concentration of the catecholaminergic neurotransmitter dopamine (DA) in the retina reportedly changes substantially (either decreases or increases in a species specific manner) when mammals are exposed to altered geomagnetic fields (Fig. 2). The DA changes were interpreted to mean that retinas respond to the fields, with the information being transferred to the pineal gland over the neural pathways described in the introductory remarks.[3] However, since DA is not a neurotransmitter in the primary visual pathway in the retinas, it is difficult to envisage how a change in this constituent in the retina would alter neural transmission to the pineal gland.

Other evidence implicates the retinas as the site at which B fields interact to regulate pineal melatonin synthesis. According to Olcese,[3] low-level background red light (which itself does not suppress melatonin production) must be present before B fields are effective in curtailing nocturnal pineal melatonin synthesis. If this finding is confirmed, it would certainly argue for an involvement of the retinas in B-field detection.

Rather than DA being involved in the mechanism by which B fields are detected, we feel an alternate possibility exists. It is well documented that light exposure at night rapidly depresses pineal melatonin. Light has this effect presumably because of its ability to cause the isomerization of 11-*cis*-retinal to all-*trans*-retinal.[6] Retinal is a component of rhodopsin, the rod photoreceptor visual pigment. The isomerization of 11-*cis*-retinal is the initial step in the photoactivation of the rod cell and the energy of a single photon of visible electromagnetic radiation can activate this process (Fig. 3). B fields, directly or indirectly via eddy currents, might induce the isomerization of 11-*cis*-retinal leading to the photoactivation of rhodopsin. This response in the retina would be interpreted as light (a mini-magnetophosphene) by the pineal gland and reduction of melatonin synthesis would follow.

The above-mentioned hypotheses are based on the assumption that the retinas are the sites of magnetoreception. However, the possibility remains that the pineal gland is directly magnetosensitive.[2] For example, the gland may be directly coupled to B fields because of its content of the ferromagnetic mineral magnetite.[7]

REFERENCES

1. R. J. Reiter, "Pineal melatonin: Cell biology of its synthesis and of its physiological interactions," *Endocrine Rev.* 12: 151, 1991.

2. R. J. Reiter and B. A. Richardson, "Magnetic field effects on pineal indoleamine metabolism and possible biological consequences," *FASEB J.* 6: 2283, 1992.

3. J. M. Olcese, "The neurobiology of magnetic field detection in rodents," *Prog. Neurobiol.* 35: 325, 1990.

4. K. Yaga, R. J. Reiter, L. C. Manchester, H. Nieves, J.-H. Sun, and L.-D. Chen, "Pineal sensitivity to pulsed static magnetic fields changes during the photoperiod," *Brain. Res. Bull.* 30: 153, 1953.

5. A. Lerchl, K. O. Nonaka, and R. J. Reiter, "Pineal gland magnetosensitivity to static magnetic fields is a consequence of induced electric currents (eddy currents)," *J. Pineal Res.* 10: 109, 1991.

6. P. A. Hargrave and J. H. McDowell, "Rhodopsin and phototransduction: A model system for G protein-linked receptors," *FASEB J.* 6: 2323, 1992.

7. J. L. Kirshvink, A. Kobayashi-Kirshvink, and B. J. Woodford, "Magnetic biomineralization in the human brain," *Proc. Natl. Acad. Sci. USA* (in press).

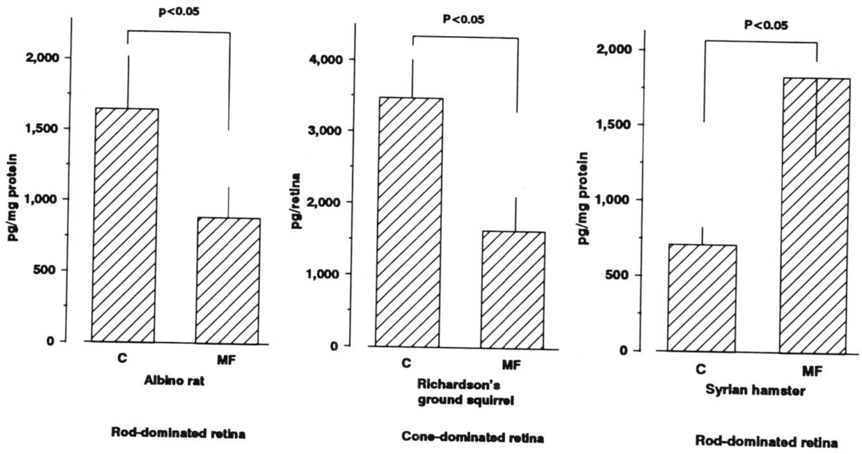

FIG. 2.—Daytime retinal dopaminergic changes in 3 species of rodents after 30min exposure to perturbed static B field. In retinas of one nocturnal rodent (rat) and one diurnal rodent (ground squirrel), retinal DA levels dropped; in Syrian hamster, a nocturnal rodent, retinal levels of DA increased.

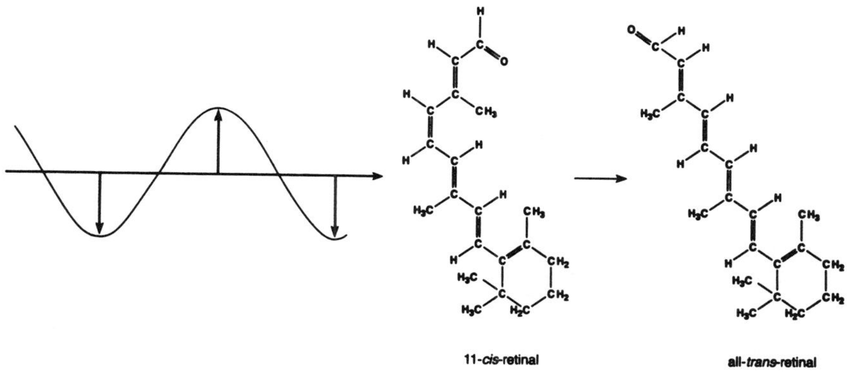

FIG. 3.—Isomerization of 11-*cis*-retinal to all-*trans*-retinal in the retina. Retinal is a component of the rod photopigment rhodopsin and its isomerization is usually induced by light; as little as 1 photon of light can achieve this change. It is theoretically possible that pulsed B fields also induce the isomerization of 11-*cis*-retinal.

Martin Blank, Ed., *Electricity and Magnetism in Biology and Medicine*

BIOLOGICAL EFFECTS IN ANIMALS EXPOSED TO EXTREMELY LOW FREQUENCY (ELF) MAGNETIC AND ELECTRIC FIELDS

L. E. Anderson*

Pacific Northwest Laboratories, Richland, Wash.

There is now convincing evidence from a large number of laboratories that exposure to extremely low frequency (ELF) magnetic and electric fields produces biological responses in animals. Many of the observed effects appear to be directly or indirectly associated with the neural or neuroendocrine systems. Such effects include increased neuronal excitability, chemical and hormonal changes in the nervous system, altered behavioral responses (some of which are related to sensing the presence of the field), and changes in endogenous biological rhythms. Additional indices of general physiological status appear relatively unaffected by exposure, although effects have been described in bone growth and fracture repair, reproduction and development, and immune system function. Three major challenges exist for ongoing research: (1) knowledge about the mechanisms underlying observed bioeffects is incomplete, (2) we do not yet understand what physical aspects of exposure produce biological responses, and (3) health consequences resulting from ELF exposure are unknown. Although no animal studies clearly demonstrate deleterious effects of ELF fields, several are suggestive of potential health impacts. Biological responses to ELF magnetic and/or electric field exposures are discussed from the perspective of laboratory animal studies.

Over the past several decades, the likelihood for humans and animals to be exposed to ELF magnetic (H) and electric (E) fields has increased substantially. Furthermore, the level of exposure has increased by orders of magnitude over the natural background of generally very low E and H fields. Such elevated fields span all frequency ranges; however, specific interest is currently focused on the potential biological impact of increased power frequency (50 and 60Hz) fields. Detailed reviews on the bioeffects of ELF fields are available.[1-3]

In the past 25 years, research programs throughout the world have made significant progress in describing interactions between H and E fields found in the environment and in living organisms. Also addressed have been questions regarding biological effects from such fields, both real and potential, and whether such effects are permanent or transient, detrimental, or beneficial. In H and E field studies that have been reported, exposure levels have been utilized from 0.1 to 30 mT and from a few volts per meter (V/m) to more than 100 kV/m. Similarly, a broad range of biological endpoints have been examined for evidence of possible ELF field-exposure effects.

Most indices of general physiological status appear to be relatively unaffected by exposure to ELF fields. It is also generally recognized that exposure to such fields does produce responses in specific biological systems; however, health implications for humans and animals have yet to be determined. Where experimental effects are demonstrated, the mechanisms of interaction between the field and the organisms remain largely unknown. For instance, it is not known whether observed effects result from fields acting at the surface of the body, from E fields induced in the interior of the body, or from H fields penetrating the body.

Even though the effects of H and E fields in humans are of primary importance, many areas of biological interest are more effectively investigated by use of other animal species. Results from experimental laboratory research in rats, mice, birds, swine, and nonhuman primates form the basis for this paper, with emphasis on areas that appear to be most sensitive to ELF fields and that may provide information of most relevance to human ELF exposure.

---

*Correspondence to: L. E. Anderson (Bioelectromagnetics), Pacific Northwest Laboratories, Richland WA 99352.

NERVOUS SYSTEM

Many of the biological effects that have been reported in animals or humans exposed to ELF H or E fields appear to be associated with the nervous system. Such system responsiveness to fields might be anticipated, since the nervous system plays a basic role in the interaction of animals with their environment. Indeed, other biological systems may be influenced indirectly by ELF exposure through neural/neuroendocrine (hormonal) functions. Reported nervous-system effects from ELF exposures include changes in behavioral and activity response, chemical changes in nerve cells, changes in the excitability of nerves, altered neurotransmitter and neurohormone levels, and disruption of biological rhythms.

Behavior. Among the most sensitive measures of perturbation in a biological system are tests that determine modifications in the behavioral patterns of animals. Behavioral studies in several species provide evidence of E field perception, as well as indications that behavior can be altered by exposure. The threshold of detection of 60Hz E fields has been reported to be between 4 and 10 kV/m in rats.[4] E-field thresholds in other animal species, including mice,[5] pigs,[6] and birds,[7] have been reported in the 25-35kV/m range. Perception of H fields has not been observed in the submillitesla range,[8] but visual perception in the form of phosphenes has been demonstrated with H fields above 20 mT in humans.[9,10]

Avoidance behavior of animals has been investigated at several E-field strengths. With some exceptions, animals generally avoid exposure to E fields over 50-75 kV/m.[11] Below those levels, changes in physical activity respones have been reported, although the changes are usually transitory.[12] Effects of H fields on behavior are curious in that many of the investigations performed at low field intensities showed behavioral alterations (primarily activity changes).[13] In contrast, studies conducted at higher H-field intensities demonstrated no evidence of effects on animal behavior.

Neurochemistry. A number of studies have investigated the central nervous system (CNS) for changes in various chemicals in the brain upon exposure to ELF fields. In general, these studies report small albeit highly variable changes in certain neurotransmitters.[14] The data provide limited evidence that exposure to E fields in the ELF range may cause slight changes in nervous system function. The number of experiments is not large, and there are significant questions about the strength and validity of several of the studies.

Neurophysiology. In neurophysiology a confusing array of studies claims both effects and noneffects of ELF-field exposure. A case in point is the commonly used measure of general CNS activity, the electroencephalogram (EEG), where significant alterations have been reported in some studies but not in others.[14] In an assessment of more specific electrical "fingerprints" of the brain, evoked responses, no effects caused by exposure were observed in visual evoked response;[15] however, changes have been demonstrated in somatosensory evoked response.[16]

Biological Rhythms. A number of investigations have been conducted to examine the effects of ELF fields on natural biological rhythms. The phase and duration of activity and rhythms of oxidative metabolism have been shown to be shifted in male mice by exposure to ELF fields.[17] Wilson and coworkers[18] measured the changing levels of indolamines and enzymes in the pineal glands of rats exposed to E fields. A significant reduction in the normal night-time rise of melatonin and its associated synthetic enzymes was observed in rats exposed to either 1.5 or 40 kV/m. In more recent studies, nocturnal pineal components in mice and rats were shown to be sensitive to rotated H fields.[19]

It is difficult to interpret the possible health consequences from the work thus far published on circadian or biological rhythm effects from ELF exposures. However, it is evident that E and H fields may alter the biological timing mechanisms in mammals. It is possible that such changes could mediate significant changes in other areas in which effects have been observed (e.g., behavior, reproduction, and development). In addition, plausible mechanisms involving ELF-induced hormonal changes have been proposed as a possible basis for increased risk of adverse health outcomes.[20]

REPRODUCTION AND DEVELOPMENT

It is generally assumed that developing organisms, including pre- and postnatal mammals, are more sensitive to physical or chemical agents than are adult animals. This greater sensitivity, when it occurs, is thought to originate in subtle effects on the processes and controls that guide the developing cellular interactions. A number of studies have been conducted to examine the effects of ELF exposure on reproduction and development of both mammalian and nonmammalian species. Most of the nonmammalian studies have been performed in birds, either chickens or pigeons. E-field exposure of chicks at several fields strengths, both before and after hatching, did not produce any significant effects in viability, morphology, behavior, or growth.[21] There have been some studies that reported deleterious effects of E fields on postnatal growth and survival in prenatal mammals; however, these studies are countered by others in which rats, rabbits, or mice were exposed to broad ranges of field strengths with no evident effects on reproduction, survival, or growth.[21,22]

Few studies have been performed to examine the effects of ELF H fields on growth and development. In the most comprehensive study published to date, no reproductive or developmental effects of 60Hz H fields were observed.[23] A great deal of interest has been generated by reports from Delgado's laboratory that marked increases in malformation rates were observed in chick eggs exposed to low levels of pulsed magnetic fields.[24] Subsequent efforts have partially confirmed these results.[25]

BONE GROWTH AND REPAIR

In animals exposed to 60Hz E fields, it appears that bone growth in rats and mice, per se, was not affected by exposure to 100 kV/m; however, bone-fracture repair was retarded.[26] Pulsed H-field exposure of bone produces quite different responses, with enhanced repair.[27]

IMMUNOLOGY

Exposure of animals to electric fields does not appear to affect the immune system. In a comprehensive investigation of the immune system, no effects of exposure at very low field strengths (150-250 V/m) in mice or rats were observed.[28] In contrast to the apparent lack of E-field influence in vivo on the immune system, magnetic fields are reported to affect system responses to mitogens and antigens strongly.[29]

CARCINOGENESIS AND MUTAGENESIS

Because of several recent epidemiological reports of positive correlations between ELF fields and cancer, considerable research interest has been generated concerning a possible connection between H-fields and cancer. To date few published laboratory animal studies bear directly on this question. One study has reported increased skin tumors in mice initiated with a chemical carcinogen followed by exposure to ELF plus a chemical promoter.[30]

SUMMARY AND CONCLUSIONS

Numerous studies have been initiated to determine the nature of the physical mechanisms involved in ELF field-induced effects and to establish to what extent an electrical environment containing H and E fields poses a health hazard to living organisms. The biological effects reported in many of the experiments have not confirmed pathological effects, even after prolonged exposures to high-intensity (10mT) and high-strength E fields (100kV/m) fields. However, the question of whether an occurrence of a "biological effect" from field exposure constitutes a health hazard has yet to be answered. Areas in which effects have been demonstrated appear to be associated primarily with the nervous system. In addition, in several instances, unconfirmed or controversial data exist where observed effects may be due to the fields (e.g., changes in brain chemistry and morphology, alterations in reproduction and development). It is not yet known whether confirmed or putative effects are due to a direct interaction of the electric field with tissue or to an indirect interaction, e.g., a physiological response due to detection and/or sensory stimulation by the field. Future research thrusts, at the minimum, should address the following topics: (1) an investigation of what aspects of field exposure constitutes dose to living systems; (2) a determination of the interaction mechanisms between ELF fields and animals/humans; (3) an exploration of the implications of observed effects on nervous system function; (4) an evaluation of the potential effects of ELF fields on reproduction and early development; and (5) a determination of whether cancer promotion and/or progression is influenced by ELF fields.

Research being undertaken at present in several countries is designed to provide data to help set maximum levels of ELF field intensities that are biologically acceptable for both workers and the general public. This research also addresses the broader issues of increased electromagnetic fields in the environment from a variety of sources, and the potential health implications for humans.[31]

REFERENCES

1. L. E. Anderson, "Biological effects of extremely low frequency electromagnetic fields: In vivo studies," *Proc. Sci. Workshop on Health Effects of Electric and Magnetic Fields on Workers," National Institute for Occupational Safety and Health, 1991, 45-89.*

2. L. E. Anderson and W. T. Kaune, "Electric and magnetic fields at extremely low frequencies," in M. J. Suess, Ed., *Nonionizing Radiation Protection,* Geneva: World Health Organization, 1989, 175-243, 2nd. ed.

3. A. R. Sheppard and M. Eisenbud, *Biological Effects of Electric and Magnetic Fields on Extremely Low Frequency,* New York: New York University Press, 1977.

4. S. Stern, V. G. Laties, C. V. Stancampiano, C. Cox, and J. O. de Lorge, "Behavioral detection of 60-Hz electric fields by rats," *Bioelectromagnetics* 4: 214-247, 1983.

5. R. Sander, J. Brinkmann, and B. Kuhne, "Laboratory studies on animals and human beings exposed to 50-Hz electric and magnetic fields," *International Conference on Large High Voltage Electrical Systems,* Paris, 1982 (Abstr. 36-01).

6. W. T. Kaune, R. D. Phillips, D. L. Hjeresen, R. L. Richardson, and J. L. Beamer, "A method for the exposure of miniature swine to vertical 60-Hz electric fields," *IEEE Trans.* BME-25: 276-283, 1978.

7. H. B. Graves, J. H. Carter, D. Kellmel, L. Cooper, D. T. Poznaniak, and J. W. Bankoske, "Perceptibility and electrophysiological response of small birds to intense 60-Hz electric fields," *IEEE Trans.* PAS-97: 1070-1073, 1978.

8. R. H. Lovely, J. A. Creim, W. T. Kaune, M. C. Miller, R. D. Phillips, and L. E. Anderson, "Rats are not aversive when exposed to 60-Hz magnetic fields at 3.03 mT, *Bioelectromagnetics* 13: 351-362, 1992.

9. H. B. Barlow, H. I. Kohn, and E. G. Walsh, "Visual sensations aroused by magnetic fields," *Am. J. Physiol.* 148: 376, 1947.

10. P. Lovsund, P. A. Oberg, S. E. G. Nilsson, and T. Reuter, "Magnetophosphenes: A quantitative analysis of thresholds," *Med. Biol. Eng. Comput.* 18: 326, 1980.

11. D. L. Hjeresen, W. T. Kaune, J. R. Decker, and R. D. Phillips, "Effects of 60-Hz electric fields on avoidance behavior and activity of rats," *Bioelectromagnetics* 1: 299-312, 1980.

12. R. H. Lovely, "Recent studies in the behavioral toxicology of ELF electric and magnetic fields," in M. E. O'Connor and R. H. Lovely, Eds., *Electromagnetic Fields and Neurobehavioral Function," New York: Liss, 1988, 327-348.*

13. T. S. Tenforde, "Interaction of ELF magnetic fields with living matter," in C. Polk and E. Postow, Eds., *Handbook of Biological Effects of Electromagnetic Fields,* Boca Raton, Fla.: CRC Press, 1986, 197-225.

14. L. E. Anderson, "Interaction of ELF electric and magnetic fields with neural and neuroendocrine systems," in B. W. Wilson, R. G. Stevens, and L. E. Anderson, Eds., *Extremely Low Frequency Electromagnetic Fields: The Question of Cancer,* Columbus, Ohio: Battelle Press, 1990, 139-158.

15. R. A. Jaffee, C. A. Lopresti, D. B. Carr, and R. D. Phillips, "Perinatal exposure to 60-Hz electric fields: Effects on the development of the visual-evoked response in the rat," *Bioelectromagnetics* 4: 327-339, 1983.

16. J. R. Wolpaw, R. F. Seegal, R. I. Dowman, and S. Satya-Murti, *Chronic Effects of 60 Hz Electric and Magnetic Fields on Primate Central Nervous System Function,* Final Report to the New York State Power Lines Project, New York, 1987.

17. C. F. Ehret, R. S. Rosenberg, G. A. Sacher, P. H. Duffy, K. R. Groh, and J. J. Russell, *Biomedical Effects Associated with Energy Transmission Systems: Effects of 60-Hz Electric Fields and Circadian and Ultradian Physiological and Behavioral Functions in Small Rodents,* Washington, D.C.: Annual Report to U.S. Department of Energy, 1980.

18. B. W. Wilson, L. E. Anderson, D. I. Hilton, and R. D. Phillips, "Chronic exposure to 60-Hz electric fields: Effects on pineal function in the rat," *Bioelectromagnetics* 2: 371-380, 1981.

19. A. Lerchl, K. O. Nonaka, K. A. Stokkan, and R. J. Reiter, "Marked rapid alterations in nocturnal pineal serotonin metabolism in mice and rats exposed to weak intermittent magnetic fields," *Biochem. Biophys. Res. Commun.* 169: 102-108, 1990.

20. R. G. Stevens, S. Davis, D. B. Thomas, L. E. Anderson, and B. W. Wilson, "Electric power, pineal function, and the risk of breast cancer," *FASEB* 6: 853-860, 1992.

21. N. Chernoff, "Reproductive and developmental effects in mammalian and avian species from exposure to ELF fields," in Ref. 13, pp. 227-240.

22. D. N. Rommereim, R. L. Rommereim, R. L. Buschbom, M. R. Sikov, and L. E. Anderson, "Reproduction, growth and development of rats during chronic exposure to multiple field strengths of 60-Hz electric fields," *Fund. Appl. Toxicol.* 14: 608-621, 1990.

23. D. N. Rommereim, R. L. Rommereim, D. L. Miller, R. L. Buschbom, and L. E. Anderson, "Developmental toxicology evaluation of 60-Hz horizontal magnetic fields in rats," *Curr. Top. Occup. Health* (in press).

24. J. M. R. Delgado, J. Leal, J. L. Monteagudo, and M. G. Garcia, "Embryological changes induced by weak, extremely low frequency electromagnetic fields," *J. Anat.* 134: 533, 1982.

25. E. Berman, L. Chacon, D. House, B. A. Koch, W. E. Koch, J. Leal, S. Lovtrup, E. Mantiply, A. H. Martin, G. I. Martucci, K. H. Mild, J. C. Monahan, M. Sandstrom, K. Shamsaifar, R. Tell, M. A. Trillo, A. Ubeda, and P. Wagner, "Development of chicken embryos in a pulsed magnetic field," *Bioelectromagnetics* 11: 169-187, 1990.

26. B. J. McClanahan and R. D. Phillips, "The influence of electric field exposure on bone growth and fracture repair in rats," *Bioelectromagnetics* 4: 11-20, 1983.

27. C. A. L. Bassett, S. N. Mitchell, and S. R. Gaston, "Pulsing electromagnetic field treatment in ununited fractures and failed arthrodeses," *JAMA* 247: 623-628, 1982.

28. J. E. Morris and R. D. Phillips, "Effects of 60-Hz electric fields on specific humoral and cellular components of the immune system," *Bioelectromagnetics* 3: 341-348, 1982.

29. P. Conti, G. E. Gigante, M. G. Cifone, E. Alesse, G. Ianni, M. Reale, and P. U. Angeletti, "Reduced mitogenic stimulation of human lymphocytes by extremely low frequency electromagnetic fields," *FEBS Lett.* 162: 156-160, 1983.

30. J. R. N. McLean, M. A. Stuchly, R. E. J. Mitchell, D. Wilkinson, H. Yang, M. Goddard, D. W. Lecuyer, M. Schunk, E. Callary, and D. Morrison, "Cancer promotion in a mouse-skin model by a 60-Hz magnetic field: II. Tumor development and immune response," *Bioelectromagnetics* 12: 273-288, 1991.

31. This work was funded in part by the U.S. Department of Energy under contract DE-AC06-76RLO 1830.

Martin Blank, Ed., *Electricity and Magnetism in Biology and Medicine*

RISK ASSESSMENT AND IRPA/INIRC GUIDELINES ON PROTECTION AGAINST ELECTROMAGNETIC FIELDS AND WAVES

M. Grandolfo*

National Institute of Health, Rome

M. H. Repacholi

Australian Radiation Laboratory, Yallambie, Victoria

In 1977 the International Radiation Protection Association (IRPA) formed the International Non-Ionizing Radiation Committee (INIRC). The IRPA/INIRC reviews the literature on nonionizing radiation and makes assessment of the health risks of human exposure to such radiation. IRPA/INIRC has recently proposed new criteria and standards for limiting the exposure of workers and the general population to 50/60Hz electromagnetic fields, and to radiofrequency and microwave radiation in the frequency range from 100 kHz to 300 GHz. This paper presents the main conclusions of the IRPA/INIRC documents.

The primary purpose of the International Radiation Protection Association (IRPA) is to serve as a medium for international communication and cooperation in radiation protection with the goal of advancing sound and effective radiation protection in all parts of the world. During the 4th International Congress of IRPA in Paris in 1977, the IRPA General Assembly decided to modify its constitution to include explicitly nonionizing radiation in its field of activities, and created the International Non-Ionizing Radiation Committee (INIRC). During the 15 years that have elapsed since it was created, IRPA/INIRC has produced several reports either alone or in cooperation with other international organizations. These reports then became the scientific database for the development of the IRPA/INIRC guidelines on limits of exposure of various types of nonionizing radiation.[1] During the 8th International Congress of IRPA in Montreal in 1992, IRPA/INIRC became the International Commission on Non-Ionizing Radiation Protection (ICNIRP). The relationship with IRPA will be based to some extent on that between the International Commission on Radiological Protection (ICRP) and the International Society for Radiology.

With reference to nonionizing electromagnetic fields, IRPA/INIRC issued interim guidelines on limits of exposure to 50/60Hz electric and magnetic fields, and guidelines on limits of exposure to radiofrequency (RF) fields and microwaves (MW) in the frequency range from 100 kHz to 300 GHz.

IRPA/INIRC GUIDELINES FOR 50/60Hz FIELDS

IRPA/INIRC published limits of exposure to 50/60Hz electric and magnetic fields[2] are given in Table 1; the rationale for limits on electrical and magnetic field exposures is stated in the footnotes.

IRPA/INIRC proposed as a criterion that the internal electric and magnetic fields should not induce a current density greater than 10 mA/m^2 in the body, since this value is within the range of the endogenous current densities. Knowledge about the possible effects of long-term exposures to fields inducing currents near the criterion value is still limited and most evidence is based on short-term observations. In view of these reservations, the continuous occupational exposure was limited to 10 kV/m, inducing a current density of 4 mA/m^2 on the average. There is substantial workplace experience (in addition to controlled laboratory studies on volunteers) which indicates that short-term exposures to fields up to 30 kV/m produce proportionally increasing discomfort and stress and should be limited in duration accordingly. A practical approach to limiting the duration of exposure to fields between 10 and 30 kV/m is to use the formula $t \leq 80/E$ over the whole working day, where t is the duration of exposure in hours to a field strength E in kV/m. A further

*Correspondence to: M. Grandolfo (Physics Lab.), National Institute of Health, Viale Regina Elena 299, I-00161 Rome, Italy.

safety factor was incorporated for exposure of the general public: a safety factor of 5 with respect to the criterion of 10 mA/m^2, which corresponds to an electric field strength of 5 kV/m.

TABLE 1.—IRPA/INIRC limits of exposure to 50/60Hz electric and magnetic fields.

Exposure characteristics	Electric field strength kV/m (rms)	Magnetic flux density mT (rms)
Occupational		
Whole working day	10	0.5
Short term	30[a]	5[b]
For limbs	--	25
General Public		
Up to 24 h/d[c]	5	0.1
Few hours per day[d]	10	1

(a) The duration of exposure to fields between 10 and 30 kV/m may be calculated from the formula $t \leq 80/E$, where t is the duration in hours per work day and E is the electric field strength in kV/m.

(b) Maximum exposure duration is 2 h per work day.

(c) This restriction applies to open spaces in which members of the general public might reasonably be expected to spend a substantial part of the day, such as recreational areas, meeting grounds, and the like.

(d) These values can be exceeded for a few minutes per day provided precautions are taken to prevent indirect coupling effects.

The limit of 5 kV/m for continuous exposure of the general public provides substantial protection from annoyance caused by steady-state contact currents or transient discharges. However, this limit cannot completely eliminate perception of the electric field effects, since the perception threshold for some people is below 5 kV/m. In such cases additional technical measures (e.g., grounding) may be instituted to avoid indirect coupling effects arising from touching charged ungrounded objects. Continuous exposures of the general public outdoors rarely exceed 1-2 kV/m.

Some epidemiological reports present data indicative of an increase in the incidence of cancer among children, adults, and occupational groups. The studies suggest an association with exposure to weak 50 or 60Hz magnetic fields. These associations cannot be satisfactorily explained by the available theoretical basis for the interaction of 50/60Hz electromagnetic fields with living systems. The magnetic flux densities in some epidemiological studies suggesting an increased cancer incidence are at values near 0.1 mT. This magnetic flux density would induce a current density that is well below the levels normally occurring in the body. The epidemiological studies are not conclusive. Although these epidemiological data cannot be dismissed, there must be additional studies before they can serve as a basis for health-hazard assessment. Furthermore, scant laboratory evidence is available to support the hypothesis that there is an association between 50/60Hz fields and increased cancer risk.

The total number of direct observations of the effect of magnetic flux densities in humans is limited. Controlled laboratory studies on human volunteers exposed for 4 to 6 h/day for several days to magnetic flux densities up to 5 mT (together with electric fields up to 20 kV/m) did not demonstrate significant effects. Therefore, the short-term occupational exposure should not exceed 5 mT (inducing current densities of 10 mA/m^2, the criterion value) and 25 mT for the extremities. Because of the sparseness of data on long-term exposures to magnetic fields, the magnetic flux density for continuous exposure in the occupational environment is limited to 0.5 mT.

The limit for continuous exposure of the general public was set at 0.1 mT, a factor of 5 below the limit for continuous occupational exposure; the short-term exposure limit was set at 1 mT.

IRPA/INIRC GUIDELINES FOR RADIOFREQUENCY FIELDS

The IRPA/INIRC guidelines on radiofrequency and microwaves[3] were developed on the basis that an SAR of 4 W/kg was considered to be an RF energy deposition rate above which there was an increasing likelihood that adverse health consequences could occur. Below this level, no adverse health effects have been established from acute exposures. Incorporating a safety factor of 10 to allow for possible consequences of long-term exposures, 0.4 W/kg was used as the basic limit for deriving exposure limits for occupational exposure. A further safety factor of 5 was incorporated to derive general-public limits.

According to the IRPA/INIRC guidelines, occupational exposure to radiofrequency fields at frequencies below and up to 10 MHz should not exceed the levels of unperturbed RMS electric and magnetic field strengths given in Table 2, when the squares of the electric and magnetic field strengths are averaged over any 6min period during the working day, provided that the body-to-ground current does not exceed 200 mA and that any hazards of radiofrequency burns are eliminated. For pulsed fields, a conservative approach consists of limiting pulsed electric and magnetic field strengths as averaged over the pulse width to 32 times the values given in Table 2.

TABLE 2.—IRPA/INIRC occupational exposure limits to radiofrequency electromagnetic fields.

Frequency f(MHz)	Unperturbed rms field strength		Equivalent plane wave power density	
	Electric E(V/m)	Magnetic H (A/m)	P_{eq}(W/m^2)	P_{eq}(mW/cm^2)
0.1-1	614	$1.6/f$	--	--
>1-10	$614/f$	$1.6/f$	--	--
>10-400	61	0.16	10	1
>400-2000	$3f^{1/2}$	$0.008f^{1/2}$	$f/40$	$f/400$
>2000-300000	137	0.36	50	5

Occupational exposure to frequencies above 10 MHz should not exceed an SAR of 0.4 W/kg when averaged over any 6min period and over the whole body, provided that in the extremities (hands, wrists, feet, and ankles) 2 W per 0.1 kg shall not be exceeded and that 1 W per 0.1 kg shall not be exceeded in any other part of the body.

The limits of occupational exposure given in Table 2 for the frequencies between 10 MHz and 300 GHz are the working limits derived from the SAR value of 0.4 W/kg. They represent a practical approximation of the incident plane wave power density needed to produce a whole-body average specific absorption rate of 0.4 W/kg. These limits apply to whole body exposure from either continuous or modulated electromagnetic fields from one or more sources, averaged over any 6min period during the working day.

Exposures of the general public to frequencies above 10 MHz should not exceed an SAR of 0.08 W/kg when averaged over the whole body and over any 6min period.

The limits of radiofrequency exposure to the general public given in Table 3 for the frequencies between 10 MHz and 300 GHz are derived from the SAR value of 0.08 W/kg. They represent a practical approximation of the incident plane wave power density needed to produce a whole-body average SAR of 0.08 W/kg.

TABLE 3.—IRPA/INIRC general public exposure limits to radiofrequency electromagnetic fields.

Frequency f(MHz)	Unperturbed rms field strength		Equivalent plane wave power density	
	Electric E(V/m)	Magnetic H (A/m)	P_{eq}(W/m^2)	P_{eq}(mW/cm^2)
0.1-1	87	$0.23/f^{1/2}$	--	--
>1-10	$87/f^{1/2}$	$0.23/f^{1/2}$	--	--
>10-400	27.5	0.073	2	0.2
>400-2000	$1.37f^{1/2}$	$0.0037\,f^{1/2}$	$f/200$	$f/2000$
>2000-300000	61	0.16	10	1

The limits given in Table 3 apply to whole body exposure from either continuous or modulated electromagnetic fields from one or more sources, averaged over any 6min period during the 24h day.

CONCLUSION

At present there are insufficient biological and epidemiological data to make a health risk assessment or even to determine whether there is potential hazard with regard to carcinogenic effects of power frequency fields. Epidemiological studies do not provide a firm causal link with electric- or magnetic-field exposures. A number of epidemiological studies scheduled to be completed in the coming years should add significantly to our knowledge of the effects of human exposure to ELF fields. Laboratory studies still have to prove to be reproducible for low-level fields; none of the hypothetical mechanisms proposed in the literature can be regarded as proven by the experiment.

At first glance "prudent avoidance" of exposure, whenever possible, seems to be a good strategy for people to follow, but in a subtle way this policy undermines the role of scientific knowledge as the best basis for determining appropriate action at a political level. It shifts responsibility from the political authority to the single private citizen; in this scenario, the citizen will become the decisionmaker on what to use and/or when and how to reduce personal exposure to the physical agents.

A problem strictly correlated to this displacement of responsibility is that individual behavior may be based on a perceived risk level that differs significantly from the objective level of risk. Psychological research indicates the degree to which levels of risk may be grossly misjudged by the public. Low-probability events present additional difficulties for individual perception. In conclusion, the strategy of risk communication is also an important approach in trying to smooth this situation by correcting public misperceptions on the risk levels associated with various activities or sources.[4]

REFERENCES

1. A. S. Duchêne, J. R. A. Lakey, and M. H. Repacholi, Eds., *The IRPA Guidelines on Protection Against Non-Ionizing Radiation,* New York: Pergamon Press, 1991.

2. International Radiation Protection Association/International Non-Ionizing Committee, "Interim guidelines on limits of exposure to 50/60 Hz electric and magnetic fields," *Health Physics* 58: 113, 1990.

3. International Radiation Protection Association/International Non-Ionizing Committee, "Interim guidelines on limits of exposure to radiofrequency electromagnetic fields in the frequency range from 100 kHz to 300 GHz," *Health Physics* 54: 115, 1988.

4. The authors are very grateful to Ms. Alma Paoluzi for her skilled technical assistance.

Martin Blank, Ed., *Electricity and Magnetism in Biology and Medicine*

SCIENTIFIC UNCERTAINTY, RISK ASSESSMENT, AND STANDARD-SETTING

I. Nair*

Carnegie Mellon University, Pittsburgh, Pa.

RISK ASSESSMENT AND THE STANDARD-SETTING PROCESS

The ideal approach to setting guidelines or standards would begin with an objective assessment of the science. However, the problem is never that simple. Standard setting has been fraught with controversy in many cases of "known" agents such as ionizing radiation, particularly in the early days of the science. This happens because the stage of standard setting usually comes after we use a technology and there is suggestive evidence that exposure to an environmental agent from that technology has potentially harmful health effects. Ideally, again, we would estimate to what extent people are exposed, estimate the effects of the exposure, assess the risk of the relevant health effect, then based on a set of value judgments such as cost-benefit analysis, set exposure limits, and hence standards.[1]

But because of the historical sequence in which standard setting occurs, it is often fraught with controversy. The technology has societal and commercial value; it may even be indispensable as in the case of electric power. The science is usually uncertain at this stage. Indeed, at this point, the science has just begun—it is often uncertain because it is incomplete. The science may indicate that the agent may expose populations to health risk, but is far from being able to provide an understanding of the health risk. The potential cost of complying with the standards often conflicts with economic interests. This is a period in which uncertain and incomplete scientific evidence may be interpreted differently by the various interested parties.

Thus the process of arriving at standards usually takes a long time. Even when we have standards, people continue to argue especially about effects at low levels. For example, in the case of ionizing radiation, 0.1 roentgen per day was the standard for workers from 1920 to 1954 because the science was just evolving and there was controversy about health effects of low levels of ionizing radiation. In fact, there were no standards for public exposure till 1956.[2]

In this paper, I explore briefly some current issues in standard setting in the case of ELF fields, focusing on the underlying philosophy, and the sociology of the science and of risk assessment. Before we discuss ELF fields, let me note that standard setting requires a knowledge of the dose metric, the exposure patterns—residential and workplace—and a dose-response relation for the effect or effects of concern.

SPECIAL FEATURES OF THE ELF FIELD PROBLEM

What about ELF fields and standards? The problem of ELF fields has several features that muddle the picture.

Field-cell Interaction

ELF fields are not analogous to ionizing radiation, where energy transfer is itself the primary basic interaction leading to ionization. The energy transfer in the case of ELF fields is followed by a set of cell responses that apparently amplify the signal and cause effects far greater than the energy transfer alone would predict. Yet because fields are part of the electromagnetic spectrum, we started the process of thinking about the biological effects of ELF fields with two erroneous assumptions:

1. "Energy" as represented by the field strength is the measure, or at least a good surrogate for the dose. A lot of epidemiology and the recent attempts at reducing fields of VDT displays reflect this thinking.

*Correspondence to: I. Nair, CMU Dept. Engg and Public Policy, Pittsburgh PA 15213.

2. The response of the cell to the field is modeled as though the cell is a uniform conducting medium with no time-dependent variations in structure and function. Yet much of the biological work tells us that the cell response is highly selective, depending on time and period of exposure, frequency, and intensity.

Yet for the most part, in the community that tries to assess risk, one still sees the tendency to think of the interaction as the action of an 'active' physical agent on a 'passive' target system. Standards and exposure are mainly in the realm of those of us who are physicists and engineers and still think of the field rather than the cell as the primary actor in the story.

Exposure Metric

Because of the above "mind-fix," and the lack of knowledge of a mechanism of interaction, we have not found (or cannot see) a clear exposure metric. Time, or (equivalently) the phase or state of both the signal and of the cellular system are probably important. Not being clear how time enters the picture, we continue to use field strength measures for does. This assumption may invalidate all current epidemiological data for use when we eventually find the metric.

Meanwhile, is field strength a good enough surrogate on which to set exposure control measures? Most people who talk of standards are assuming it is. For instance, it is the basis of the guidelines set for VDT fields by the Swedish Standards Committee.

The important point is that biology is central in defining the metric; it will not come from physics alone.

Reproducibility

In view of the above, and of the limited funding for the research on the basic biology, reproducibility has to be defined carefully and precisely. Different experiments that lead to the same conclusions about phenomenology or mechanism are perhaps the best way to build basic understanding, rather than insistence on repetition of identical observations, although such observations would certainly be welcome. If key parameters of the field exposure and biological system conditions are clearly specified, the scientific community can judge which experiments represent "replications."[3] For instance, we can say that the effect of fields on calcium movement across cell membranes has been demonstrated "reproducibly," even though under different conditions the efflux may be increased or decreased.

THE SCIENTIST AS RISK ASSESSOR

When the science is known, the scientist may be the legitimate advisor in standard setting. But is that true when the science is very poorly known, as in the case of ELF fields? Risk assessment is as much (if not more) a societal activity as a scientific activity. In the presence of high scientific uncertainty, is a scientist still the best person to lead the structuring of the thinking on risk assessment or standard setting? What should the guiding principles be in that case?

In order to distil scientific "truth," the scientists' strategy in the face of uncertainty is to avoid "Type I errors" (or, false positives) at all costs. Will this strategy assure the best outcome for the public that are exposed to a potential risk? Some argue not.[4,5] Some even argue that it is better to lean in favor of commiting a "Type I error" rather than a "Type II" (false negatives) from the points of view of public interest and of the principle of "do least harm."

COMPLETE SCIENCE VS INCOMPLETE SCIENCE IN STANDARD SETTING[6]

As the science of a health risk emerges, it stands to reason that the acute, high-dose effects are the first to be discovered and verified. "Safety standards" involving these phenomena are usually introduced first. In the case of ionizing radiation, erythema (reddening of the skin) was used to set the early standards.

At the stage in which we are with respect to ELF fields now, there is a realm of such complete science. Such is the case of electric shock and ventricular defibrillation effects in the case of ELF fields. The standards of electrical safety were set on the basis of the complete science of acute effects.

That is not the realm of the present concern. The potential effects of chronic exposure from low-intensity fields are the focus of the standard setting now required. The public demand for standards revolves around these potential risks—of carcinogenicity, developmental, and neurobehavioral effects. These risks involve the incomplete (and hence uncertain) understanding of the science, but that does not mean that we simply declare that standards set for the completely understood effects are the only valid standards. The preference of the scientist for avoiding false positives usually induces the scientific standard-setting community to take this stance. We have to continue to struggle with the question of how the uncertain science is to be factored in, not declare the *uncertain* science as invalid.

ARE CURRENT PRINCIPLES OF HELP IN STANDARD SETTING?

One possible way for standard setting is to look for precedents or guiding principles that may point to a level of exposure that may be reasonable at the present level of knowledge.

Is prudent avoidance a reasonable guideline? Prudent avoidance, originally defined by Granger Morgan in this context, is the *avoidance* (not reduction, because we do not know what reduction of exposure means) of any field we can avoid without significant cost to the quality of life. In the case of the consumer of electricity, the practice of prudent avoidance is left to the individual. Prudent avoidance on the part of the organizations such as power companies is more complex. The ethics of prudent avoidance here is to avoid exposing consumers to a new field condition that is not of their choice and has no balancing benefit. In this case, prudent avoidance may become a difficult guideline to follow, since the idea of prudence may be interpreted differently by different people.

ALARA (as low as reasonably achievable), a standard maxim in radiation protection, is essentially meaningless here, because we do not know what "low" is in the context of exposure to ELF fields. Using field strengths alone to define and implement an ALARA policy would again lead to a state of complacency. Established endpoints such as heating or electric shock are meaningless in standard setting for public exposure to 60Hz fields.

CONCLUSION: KEEP THE QUESTIONS OPEN

Given all the above, how does one approach standard setting? I suggest that at this stage, we should not. Standards set now would be based not on scientific knowledge, but rather on the lack of it. At this point

- allocation of scarce research funds to good basic science is more important for ultimate risk assessment and standard setting than direct risk assessment type of activities or epidemiological studies;
- large measurement programs that measure average fields strengths may prove to have no validity when the science is uncovered;
- setting standards that may be hard to change and that may lead to a state of societal complacency that we have done something about the problem may be more unethical than not setting a standard.

In summary, this is not a time for quick answers, standard setting, and feeling comfortable. This is a time for keeping the questions alive, to strive to answer the questions meaningfully rather than merely quickly, to make sure that the public is kept fully aware of the scientific uncertainty and its ramifications. This is a time to remember that ethical decisionmaking requires decisions made in the light of knowledge, not decisions made merely to allay public concern, and to allow the decisionmakers to go on to conduct business as usual.[7]

..
Do not now seek the answers,
which cannot be given
...
Live the questions now.
Perhaps you will then
....
Live along some distant day
into the answers.

Rainer Maria Rilke

REFERENCES

1. M. G. Morgan, "Probing the question of technology-induced risk," *IEEE Spectrum* 58-64, November 1981; "Choosing and managing technology-induced risk," *IEEE Spectrum* 53-60, December 1981.
2. M. Eisenbud, *Environmental Radioactivity*, New York: Academic Press, 1987, 3rd ed.
3. J. L. Phillips, personal communication.

4. M. G. Morgan, "Prudent avoidance," *Public Utilities Fortnightly* 15 March 1992, 26-29.

5. K. S. Shrader-Frechette, *Risk and Rationality,* Berkeley: University of California Press, 1991.

6. C. F. Blackman, personal communication.

7. The author is deeply indebted to Dr. Jerry Phillips and Dr. Carl Blackman for discussing several aspects of this article with her and clarifying many of her thoughts.

Martin Blank, Ed., *Electricity and Magnetism in Biology and Medicine*

THE NEED FOR SCIENCE-BASED SAFE EXPOSURE STANDARDS FOR ELECTROMAGNETIC FIELDS

J. M. Osepchuk*

Raytheon Research Division, Lexington, Mass.

Development throughout the world of limits for safe exposure to radiofrequency energy has been based on a scientific consensus on the thresholds for adverse effects. Now there is a special need for development of safe exposure limits below 3 kHz. Proposals for "prudent avoidance" as an alternative to safe exposure standards must be rejected in order to permit technological progress in society for the well-being of humanity.

Health physicists used to apply the term "radiation hazards" only to ionizing radiation and ignored the frequencies below infrared because of the negligible quantum energies. Still fears of electromagnetic fields or nonionizing radiation arose. In World War II some soldiers exhibited fears of the cathode-ray tube (CRT), and soon "microwaves" had their own mythology because of their demonstrated focusing properties—hence the death-ray concept.

Therefore, when work on safe-exposure standards first began in the 1950s, the focus was on the "microwave frequencies." The first ANSI C95 standard was a single number, 10 mW/cm^2, for the whole frequency range of 10 MHz to 100 GHz (as averaged over any 6 min of exposure). In the USA both federal and state agencies have selected the C95 standards as models for regulations of all types. This step has led to the largest transnational broadbased consensus process on standards for safe exposure to electromagnetic energy under IEEE SCC 28.

Figure 1 shows the normal standards-setting process; thresholds for effects and hazards are universally assumed.[1,2] The standards process involves much more than scientific deliberations (including literature review). It also involves the development of rules in technical terms, measuring techniques and dosimetry, and the art of writing these rules in unambiguous but readable language, all of which may take 5 to 8 years. The result is protection against known hazards with some safety factor not influenced by public fears and unknown speculative hazards. In a sense, every time a new standard revision is issued, the "jury" has made a decision.

Like the actions of courts, standards must be continually updated, extended to larger scope, and revised to eliminate errors and to improve the correspondence of the standard to existing science. Thus the standards become more complicated.

Figures 2 and 3 show existing long-term exposure limits for several countries. We show the limits at low frequency for B fields only because of today's special interest in magnetic-field exposure at ELF frequencies. These standards tend to extrapolate to around roughly 1 G at 60 Hz. The more conservative standards in the former USSR are characterized by lack of detail, rationale, and clarity. In a sense they are only primitive, rough rules. Definitions of important parameters such as time-averaging, spatial averaging, etc., are absent or ambiguous.

Figure 4 is a capsule guide for the new U.S. standard, IEEE C95.1-1991. This standard includes important improvements such as rules to prevent RF shock and burns at low frequency, and lowered averaging time at millimeter-wave frequencies so that burns are prevented. Averaging times are frequency dependent and permit short-term exposure levels higher than long-term limits.[3]

In Figure 5 we show what all standards should show: dose (specific absorption) limits at short exposure duration and dose rate (specific absorption rate) limits at long exposure times. The characteristic time T (averaging time) should reflect something meaningful like thermal time constant, although this is not required per se. Constant SA (energy) is consistent with the meaning of dose in chemical and ionizing radiation toxicology.

---

*Correspondence to: J. M. Osepchuk, Raytheon Res. Div., 131 Spring St., Lexington MA 02173.

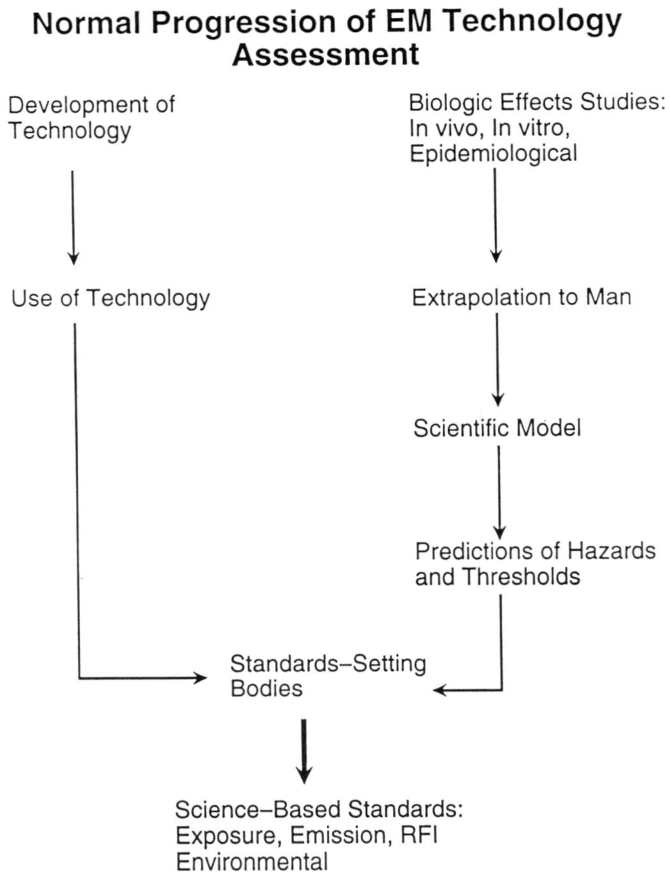

FIG. 1.—Process of standards-setting.

Several tasks are already apparent for revising the new C95 standard.

1. Develop rules for avoiding contact discharges.
2. Develop rules on relaxed partial-body exposure consistent with SAR rules.
3. Delineate more explicit different requirements in the different frequency regimes—quasi-static, resonance, and quasi-optical.
4. Review averaging-time values to make them more compatible with electro-stimulation at low frequency.
5. Simplify the rules on pulses and extend them to more complicated waveforms.
6. Explain time-averaging better.
7. With a more transnational membership, work toward worldwide harmonization of standards.

Attacks on the standards-setting process have been occurring for the past 20 years; the proposal for "prudent avoidance" is the most recent. Typical of these previous attacks is the antiscience stance taken by the science historian Steneck, who has said, "The scientific bias of all of ANSI's RF standards is clear. . . . C95 members could have spent their time discussing how low standards could be set without infringing on economics development instead of figuring out how standards could be set to avoid injury. . . ."[8]

The arguments against the "prudent avoidance" approach include:

1. It ignores existing science.
2. There is confusion about what should be reduced.
3. Weak-field effects are implausible.
4. Reduction in one field may result in an increase in another field of perhaps greater potential hazard.

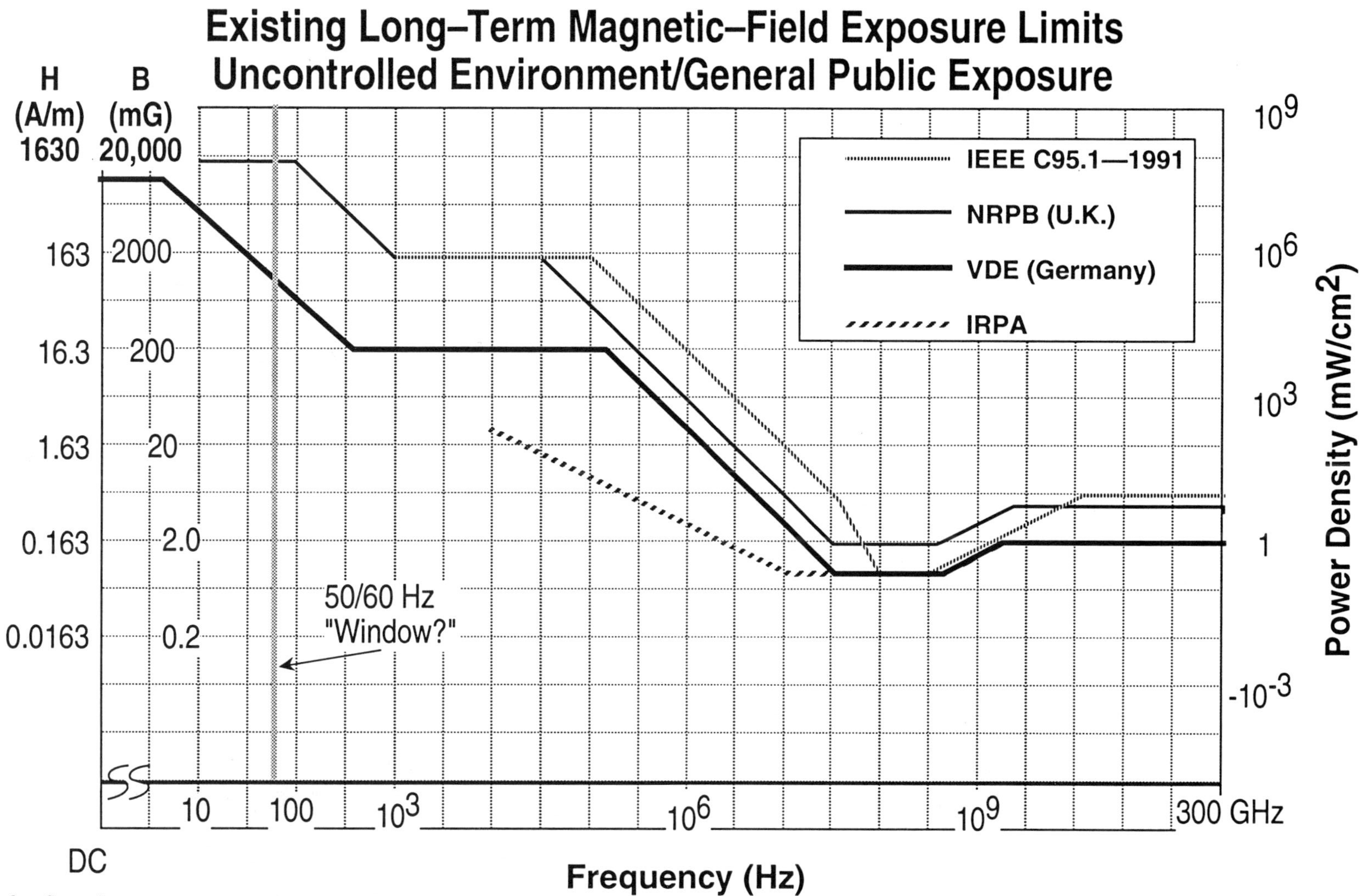

FIG. 2.—Some long-term exposure standards: General population.

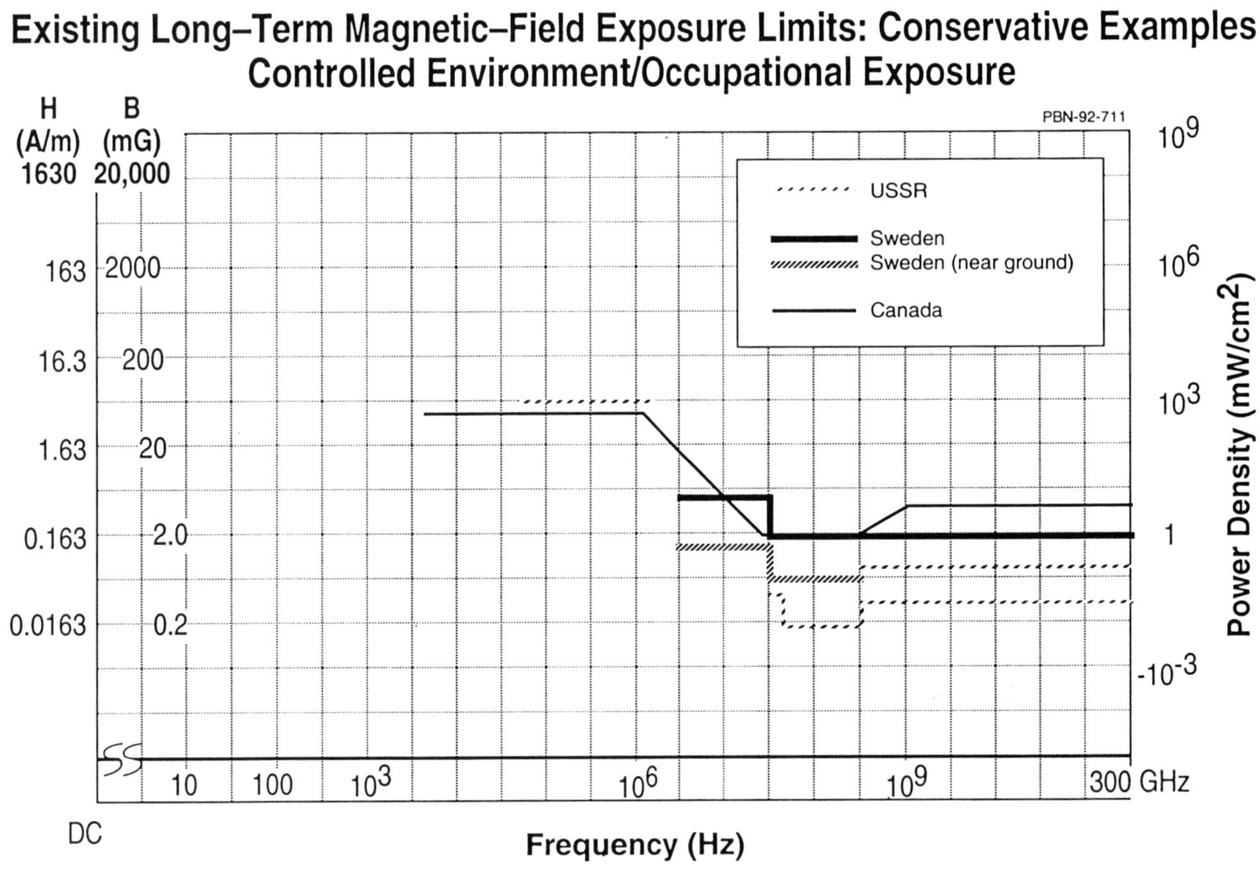

FIG. 3.—Some long-term exposure standards: Occupational

A Capsule Guide to the Final Draft Revision; IEEE C95.1-1991

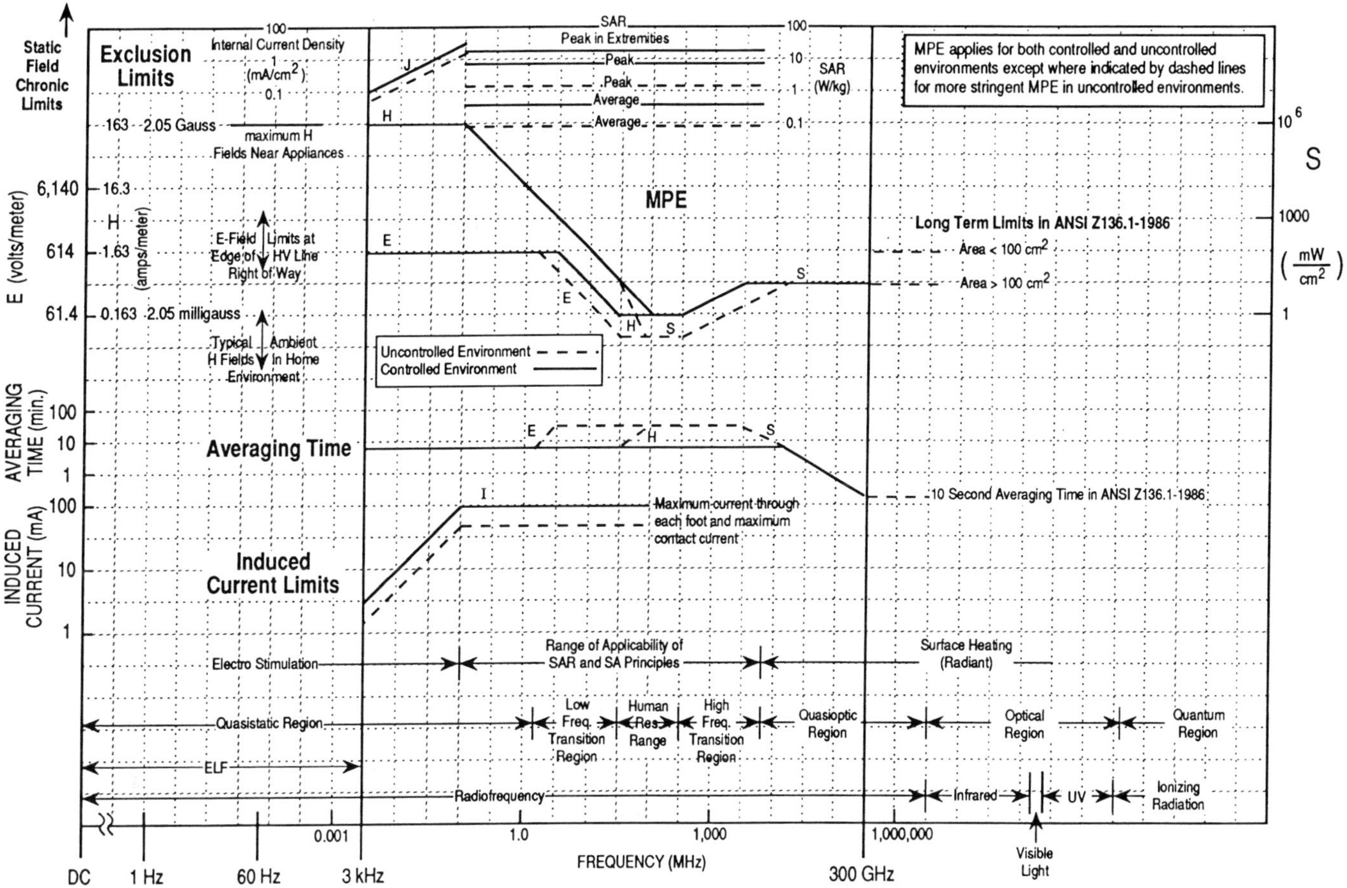

FIG. 4.—Capsule guide for IEEE C95.1-1991.

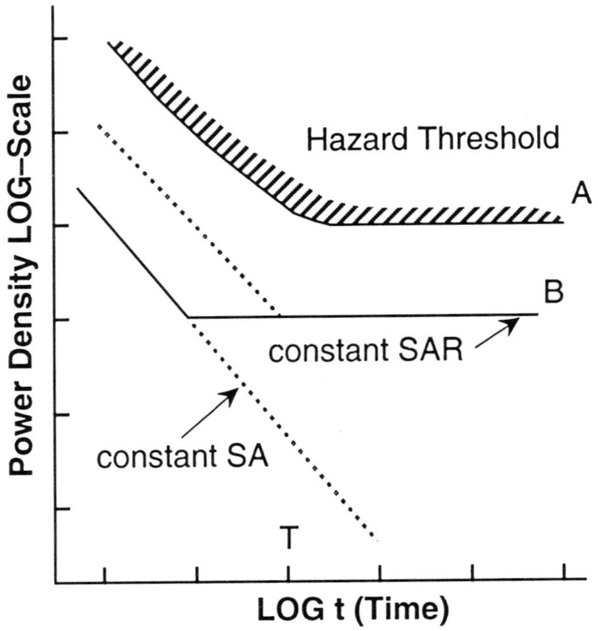

FIG. 5.—Exposure diagram.

5. "Prudent avoidance" leads to electrophobia and those who exploit it, notably litigators.
6. It delays development of beneficial effects of EM fields.

Since the probability of weak-field hazards is quite small, we should spend as much time preparing for a judgment of no hazard as time spent anticipating a hazard.

We all agree that

1. Ill-based litigation should be decried.
2. High-quality research is necessary.
3. Public education is desirable.
4. Standards setting is the ultimate objective.

REFERENCES

1. S. Barański and P. Czerski, *Biological Effects of Microwaves,* Stroudsburg, Pa.: Dowden, Hutchinson, and Ross, 1976, 11-15.

2. B. A. Minin, *Microwaves and Human Safety,* translation from U.S. Joint Publications Research Service, JPRS 66506-1, U.S. Department of Commerce, 1965.

3. J. M. Osepchuk (Moderator), "Panel discussion on standards," in J. C. Lin, Ed., *Electromagnetic Interaction with Biological Systems,* New York: Plenum, 1989, 281-289.

4. M. G. Morgan et al., *Electric and Magnetic Fields from 60 Hz Electric Power: What Do We Know About Possible Health Risks?* Pittsburgh, Pa.: Carnegie-Mellon University, 1989.

5. V. Delpizzo, *Bioelectromagnetics* 11: 139-147, 1990.

6. G. Hester, "Electric and magnetic fields: Managing an uncertain risk," *Environment* 34: 139-147, 1990.

7. I. Nair, M. G. Morgan, and K. Florig, *Biological Effects of Power-frequency Electric and Magnetic Fields,* OTA Report (OTA-BP-E-53), Washington, D.C.: Office of Technology Assessment, U.S. Congress, 1989.

8. N. Steneck, *The Microwave Debate,* Cambridge, Mass.: MIT Press, 1984, 234-235.

Martin Blank, Ed., *Electricity and Magnetism in Biology and Medicine*

REDOX REACTIONS IN BIOLOGICAL SYSTEMS

H. Metzner*

Institut für Chemische Pflanzenphysiologie, Tübingen, Germany

Most biochemists are used to characterizing charge transfer reactions even in intracellular organelles by redox potential changes. However, that is inappropriate in many biochemical systems owing to the very small volume of the relevant compartments. The author recommends consideration of charge-exchange processes by the application of energy term schemes. Living systems may not be compared to liquids. The charge-exchanging units are bound to membranes. Photosynthesis model experiments have demonstrated that irradiated semiconductors may act as electron donors that can reduce carboxyl groups to the oxidation state of carbonyl and/or alcoholic hydroxyl groups.

As part of the Symposium on Recent Advances in Science and Technology, this contribution is not a purely theoretical introduction into more or less trivial electrochemical events. It discusses redox reactions in the context of possible technical applications.

If we start from the very general equation

$$\mathrm{Acc} + \mathrm{Don} = \mathrm{Acc}^- + \mathrm{Don}^+$$

we can—at least theoretically—regard it as the summary of two consecutive steps,

$$\mathrm{Don} \quad - \mathrm{Don}^+ + e^-$$

$$\mathrm{Acc} + e^- \ - \mathrm{Acc}^-$$

Using the mass action law, we obtain

$$[e^-] = k_1 \frac{[\mathrm{Acc}^-]}{[\mathrm{Acc}]} = k_2 \frac{[\mathrm{Don}]}{[\mathrm{Don}^+]}$$

which means that

$$[e^-] = \mathrm{const} \cdot \frac{[\mathrm{Red}]}{[\mathrm{Ox}]}$$

If we insert this relation into the well-known equation for the redox potential, we obtain

$$E = E_0 + \frac{RT}{n\mathcal{F}} \ln \frac{[\mathrm{Ox}]}{[\mathrm{Red}]} = E_0 + \frac{RT}{n\mathcal{F}} \cdot \mathrm{const} \cdot \frac{1}{[e^-]}$$

In this way we link a hypothetical electron concentration with the "classical" redox potential. Of course, nobody expects to find a measurable concentration of free electrons within a liquid. The situation is very similar to that of free protons. The chemist does not hesitate to calculate in terms of so-called hydrogen ions. If we symbolize them with H^+ we should keep in mind that H^+ means a naked proton; like electrons it does not exist in a solution. Any free proton is immediately caught by a neutral molecule, e.g., water. Thus we obtain

$$H_2O + p \ - H_3O^+$$

The biochemist is often interested in knowing whether a charge exchange between two redox couples seems feasible or not. In these cases he looks it up in the electrochemical tables, in which the redox potentials of the relevant compounds are listed. For example, in the case of plant photosynthesis, biochemists have

---

*Correspondence to: H. Metzner, Institut f. Chemische Pflanzenphysiologie, University of Tübingen, D-7400 Tübingen, Germany.

identified several redox couples that form a characteristic chain at the surface of the thylakoid membrane, i.e., the membrane of chlorophyll-containing vesicles inside the chloroplast. Avoiding any particular chemical formula, we may call these redox systems A/A^+, B/B^+, C/C^+, and D/D^+. Now an electrochemical table may show us that the redox potential of A/A^+ is the most negative, whereas that of D/D^+ may even have a positive value. If these four systems are coupled, we may predict that the electrons will reduce part of the D^+ to D, while A is oxidized to A^+. Of course both systems retain a certain percentage of the oxidized and the reduced form, respectively, but the biochemist will be inclined to calculate the degree of charge accumulation at the level of the D/D^+ system.

Physical chemists might object that the redox potential is a macrophysical term, but that every thylakoid contains just a few molecules of the relevant redox system. As biochemists we know that the thylakoid is a vesicle with a volume of only 10^{-4}–10^{-3} μm^3.[1] If we assume a neutral reaction, the "number" of H_3O^+ ions is about 0.03. Se we can merely state a probability of finding a proton or the H_3O^+ cation inside a thylakoid vesicle. This example shows very clearly how problematic it is to talk about proton gradients. Reacting to the critical remark of the physical chemist, some electrochemists might be inclined to replace the concentration values by probabilities and rewrite the redox potential equation as

$$E = E_0 + \frac{RT}{n\mathcal{F}} \cdot \ln \frac{p(\text{Ox})}{p(\text{Red})}$$

However, that would be of doubtful value. If we consider very small systems such as we face inside cell organelles, we should avoid macrophysical terms. This caution is valid even for the factor T. The temperature is defined as the average speed of particles. For a single particle T makes no sense.

After decades of photosynthesis research, we know that the electron transport across a given redox chain is driven by chlorophyll. We could use electrochemical data sets to find the redox potential of this sensitizer. We would then realize that its value is even more positive than that of the system D/D^+. That might explain why all chlorophyll molecules are in the reduced state after dark periods.

Actually, chlorophyll does not work in its ground state; it affects the electron distribution in the redox chain only after its excitation by absorbed photons. How can we explain that? Let us put electrochemical considerations aside and turn to microphysics. As we see from the term scheme (Fig. 1), each chlorophyll molecule may be energetically characterized by the position of three relevant energy values. In its ground state (S_0) the chlorophyll molecules have several energy levels, each populated by two electrons. Between the highest occupied level and the ionization limit there is an energy difference that we call the ionization energy. On the other hand there is a definitely smaller distance between the ionization limit and the energy level of the lowest unoccupied or incompletely occupied energy level; we call it the electron affinity of the molecule.

What happens if we irradiate the chlorophyll? One of its electrons will be lifted to a higher term and leave behind an electron "hole." Comparing the term scheme of Chl and Chl^* we realize that the ionization energy is now decreased while the electron affinity has been increased. In other words, the excited molecule is both a better electron donor and a more efficient electron acceptor.[2]

In any case, the electron affinity of the redox system A/A^+ is higher than the ionization energy of Chl^*. The excited chlorophyll is thus able to inject one of its electrons into the redox chain ($A \rightarrow D$). On the other hand, the electron affinity of Chl^* is higher than the ionization energy of D. So chlorophyll can refill the electron hole with one of the electrons of D. Without discussing the details of the photosynthetic electron transfer, we merely note that there are two different chlorophylls, one acting in contact with the system D/D^+ (Chl_I) and the other with the system A/A^+ (Chl_{II}).

Here we face a critical question. After a short lifetime of the excited state—of the order of nanoseconds—both chlorophylls (Chl_I and Chl_{II}) return to their ground state. This process may be accompanied by fluorescence emission, or it may happen in connection with electron transfer to an acceptor molecule. If such an oxidation step occurs, the result is that the chlorophyll is missing one of its electrons. From what source does Chl_{II}^+ regain this missing charge? This turned out to be one of the essential and still unsolved problems of photosynthesis research. Since irradiated chloroplasts release a constant stream of oxygen, the prevailing opinion is that the chloroplast has the power to split H_2O molecules. Yet a comparison of the redox potentials would demonstrate that the redox potential of Chl_{II}/Chl_{II}^+ is less positive than that of the system O/O^{2-}.

This is not the place to discuss the present state of photosynthesis research. Instead, we should ask whether the given consideration of the primary events inside the chloroplast is really correct. The photosynthetic charge transport is not performed in solution; electron exchange reactions inside thylakoids occur at surfaces. The membranes to which the redox couples (A/A^+, B/B^+, C/C^+, D/D^+) and the two chlorophylls are fixed to have quite distinct properties that distinguish them decisively from a liquid. In any solution we may presuppose that both the solvent molecules and the dissolved charged or uncharged species are in a permanent statistical movement within a given volume. That is not the case at membranes. There may be a two-dimensional shift of constituents and absorbed species, but in any case the charge exchange takes place within a plane. In other words, biological membranes behave much more like a crystal surface than like a liquid. We should keep that in mind when we discuss electron-transfer steps in biological systems.

My co-workers and I have over many years tried to simulate the process of plant photosynthesis in inorganic models. To do so we were obliged to study the biophysical essentials of charge transfer steps. To give the essential ideals of the experiments, let us first consider a very special kind of electron donor, the surface of an illuminated semiconductor. As we know, irradiated alkali metals emit electrons. If (as is the case for silicon) the energy of the absorbed photons is not sufficient for the release of electrons, the negative charges can be nevertheless transferred into the semiconductor surface. In contact with a suitable acceptor molecule, the electron may then switch from the charged crystal to the attached acceptor.

We can readily use this effect for biotechnological studies. If we irradiate an aqueous suspension of small rare-earth oxide particles, we observe a charge transfer leading to strongly reducing oxide surfaces. In contact with organic acids, irradiated lanthanum oxide crystals convert carboxyl groups first into carbonyl and then into hydroxyl groups.[3] For example, if we bind formic acid (H-COOH) to the semiconductor surface, this compound is efficiently reduced to methanol. We may say that in this case a kind of "bound CO_2" acts as an electron acceptor.

Keeping this possibility in mind, we have to ask ourselves how photoreactions inside the cell compartments really work. Is the comparison of a cell compartment with an inorganic semiconductor justified or not? Over 30 years ago Arnold and Sherwood applied a thin layer of freshly isolated chloroplasts to the surface of an insulator; they chose a sapphire crystal. After drying the plastid layer they found that the chloroplast film acted as a photosemiconductor.[4]

That need not mean that the transfer of energy inside a chloroplast works in accordance with this principle; but this example shows at any rate that redox reactions in living systems should not be treated with the theoretical apparatus of liquid-state electrochemistry.

In recent years our understanding of biological systems has reached a state in which we can use some of these complicated structures as models for technical achievements. Photosynthesis models might well turn out to be a good example of how we can learn from biophysical processes that Nature had optimized over billions of years.

REFERENCES

1. H. Metzner, "Reflections on the proper description of photobiological charge exchange processes," *Physiol. Vég.* 21: 601, 1983.

2. M. Almgren, B. Holmström, and L. Tegnér, "Solar energy and photoinduced redox processes: General principles," in S. Claesson and L. Engström, Eds., *Solar Energy: Photochemical Conversion and Storage,* Stockholm: National Swedish Board for Energy Source Development, 1977.

3. H. Metzner, *Verfahren und Vorrichtung zur Reduktion, insbesondere Methanisierung von Kohlendioxid,* Europ. Patent No. 111870.

4. W. Arnold and H. K. Sherwood, "Are chloroplasts semiconductors? *Proc. Natl. Acad. Sci. USA* 43: 105, 1957.

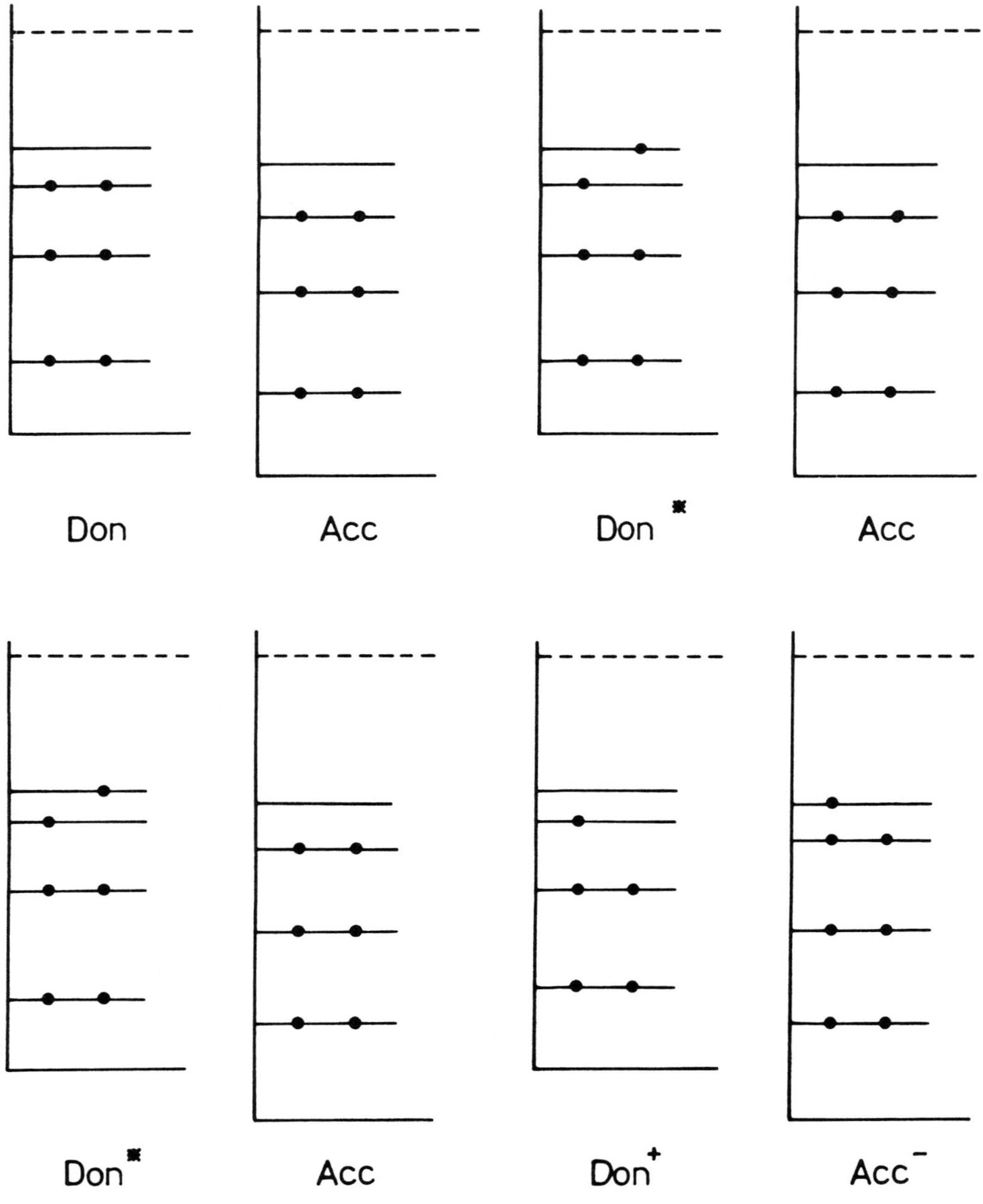

FIG. 1.—Term schemes of a donor-acceptor couple. Upper left: Don and Acc in S_0 state; upper right: Don in S_1 (excited) state, Acc in S_0 state; lower left: same as upper right; lower right: donor-acceptor couple after charge transfer.

Martin Blank, Ed., *Electricity and Magnetism in Biology and Medicine*

ELECTROPORATION: A DRAMATIC NONTHERMAL ELECTRIC FIELD PHENOMENON

J. C. Weaver*

Harvard-MIT Division of Health Sciences and Technology, Cambridge, Mass.

Electroporation is of growing interest because it is a fascinating membrane phenomenon for which major applications have been identified. As understood at present, electroporation occurs in cell and artificial planar bilayer membranes if the transmembrane voltage $U(t)$ reaches 0.5 to 1.5 V for short times. In the case of cells, brief electric field pulses of 10^3 to 10^4 V/cm are used to cause reversible electrical breakdown (REB), which is accompanied by a tremendous increase in molecular transport across the membrane. REB involves a rapid membrane discharge, so that U returns to low values within a few microseconds of the pulse. Membrane recovery can be orders of magnitude longer. Electroporation is widely used for DNA introduction, as well as for delivering drugs, fluorescent probes, enzymes, and antibodies into isolated cells. More recently, tissue electroporation has become of interest with the prospect of gene therapy of subcutaneous tissue, enhanced chemotherapy of tumors, and electrically controlled transdermal drug delivery. Basic phenomena, understanding of mechanism, and existing and potential are applications are briefly reviewed.

Electroporation is widely used to refer to the striking phenomena that occur because of large transmembrane voltages.[1-3] Other terms such as electropermeabilization and electroinjection are also used, but suggest a more specific mechanism such as permeation or a specific end result such as introduction of molecules. A common idea is that the natural barrier function of bilayer and cell membranes is tremendously reduced, so that ordinarily blocked water-soluble molecules can cross the membrane.[4-7] However, the microscopic mechanisms by which molecular transport occurs are not yet established. With REB, a tremendous increase in electrical conduction occurs. In addition, rupture (electrical breakdown) of planar membranes is a destructive mechanical event.[8] Moreover, the ability to insert membrane proteins stably into the membrane and the occurrence of electrofusion are companion phenomena that go beyond membrane permeability changes and introduction of molecules into cells. For this reason, an attractive working hypothesis is that some type of aqueous pathways ("pores") appear because of the increased value of U. Finally, physical models based on transient aqueous pores appear capable of explaining many essential features of the mechanical, electrical, and molecular transport behavior. For these reasons, the term "electroporation" is increasingly widely accepted.

DISCUSSION

Nature and Origin of Pores

Electroporation is thought to involve the stochastic creation of microscopic pores through the combined contributions of thermal fluctuations ("kT energy") and a large U ("electrical energy"). These "primary" pores are believed to be transient, with fluctuating size and shape. They are also believed to have a minimum radius of about 1 nm, and a statistical distribution of larger sizes. The primary pores probably cannot be visualized by any known form of microscopy, because of their small size, transient nature, and a lack of contrast between water and the membrane for photons, electrons, and phonons. Various types of atomic microscopy may have an adequate combination of spatial resolution and contrast, but a nearby microscopic probe will itself greatly perturb the electric field. If a probe is used to cause electroporation, the likely result is pore formation immediately proximate to the probe; although that may be a way to study pores, it will not really correspond to imaging. Thus, our understanding of the structure of primary electropores is likely to be inferred from nonimaging optical measurements[9] and from ionic and molecular transport properties measurements. Finally, the very large pores observed by electron microscopy in erythrocyte membranes are believed

---

*Correspondence to: J. C. Weaver, Harvard-MIT Division of Health Sciences and Technology, Cambridge MA 02139.

to be secondary pores,[10] since they appear long after membranes are known to discharge.

Bilayer membranes are microscopic and experience a variety of conformational fluctuations.[11] Those with the most relevance to electroporation are the membrane conformational fluctuations that involve entry of water and water-soluble molecules into the membrane (Fig. 1). According to the transient aqueous pore hypothesis, the energy needed to form an aqueous pore is

$$\Delta E(r, U) = 2\pi\gamma r - \pi\Gamma r^2 - \frac{\pi(\varepsilon_w - \varepsilon_l)U^2}{h}\int_{r_{min}}^{r} \alpha^2 r\, dr \tag{1}$$

where α is a function of the pore size and pore internal conductivity,[12,13] Γ is the surface energy density of the membrane, ε_w and ε_l are the relative dielectric constants of water and lipid, and γ is the pore edge energy. The stochastic entry and departure of water governs pore evolution. Hydrophobic pores are assumed to form first, possibly from pre-pore conformational fluctuations of the membrane. Transitions from hydrophobic pores to hydrophilic pores occurs at larger r because γ for hydrophobic pores is much larger than for hydrophilic pores. The hydrophilic pores are believed to be metastable over short time scales. Composite pores involving membrane proteins may also form, with smaller effective values of γ and longer metastable lifetimes. Entry of transported or of cell-attached intracellular macromolecules into a pore should prevent the pore from shrinking because of coulombic repulsion and to persistent pores. Finally, very large secondary pores may evolve from the primary transient aqueous pores because of pressure-driven flows.[10]

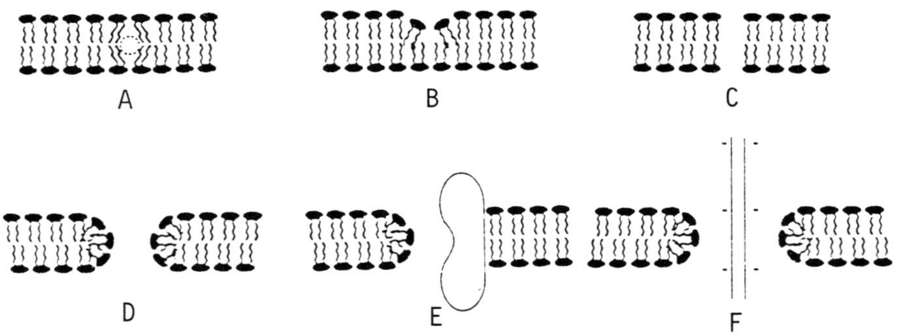

FIG. 1.—Hypothetical structures of transient and metastable membrane conformations related to electroporation. A: free volume fluctuation;[34] B: aqueous protrusion or "dimple";[35] C: hydrophobic pore;[8] D: hydrophilic pore;[8] E: composite pore with one or more proteins at inner edge;[36] F: composite pore with "foot-in-the-door" charged macromolecule inserted into a hydrophilic pore. The transient aqueous pore model assumes that transitions A $\rightarrow$ B $\rightarrow$ C $\rightarrow$ D occur with increasing frequency as U is increased. A similar transition series at the sites of membrane proteins may lead to E. Persistent pores of type F are hypothesized to occur by entry of a charged, linear molecule into a hydrophilic pore, and may exist long after U has decayed because of pore conduction. Protruding cytoplasmic molecules, components of extracellular matrix, and "in transit" transported molecules are examples of pore-entering molecules that may lead to persistent pores. Indeed, evidence for "in transit" participation by DNA has recently been reported.[37]

The electrical conditions that have typically been found to cause electroporation can be simply stated in approximate form: elevation of U to about $0.5 \leq U \leq 1.5$ V for about $10^{-6} \leq t_{exposed} \leq 10^{-3}$ s. A great many observations together suggest that this form is essentially universal and is consistent with the idea that electroporation is predominantly a physical phenomenon, with the chemical properties of various membranes overwhelmed through only slight changes in $U(t)$. The electrical and mechanical behavior of artificial planar bilayer membranes has been studied, with mechanical destruction (rupture = irreversible electrical breakdown) typically occurring for $U = 200$ to 500 mV.[8] Paradoxically, for short pulses some membranes (e.g., oxidized cholesterol) allow a much larger U to be reached, rupture does not occur, and instead the membrane rapidly discharges through "reversible electrical breakdown" (REB).[14]

Vesicles and cells are separated from the external electrolyte by a closed membrane. In this case, interfacial polarization produces large changes in the transmembrane voltage $\Delta U(t)$ by external electric field changes $\Delta E(t)$:[5]

$$\Delta U(t) = 1.5\, E_e\, R_{cell} \cos\theta \tag{2}$$

where R_{cell} is the cell radius and θ is the angle between the applied electric field $\vec{E}_e$ and the site on the cell membrane at which U is measured. Equation (2) provides a general coupling mechanism, one with considerable amplification, which relates an external electric field and the transmembrane electric field. The case of an isolated spherical membrane for *small* electric fields has long been known to be described by Eq. (2). However, once pores appear and significantly change the membrane conductivity in some regions of the membrane, Eq. (2) is *invalid,* because the membrane conductivity is neither small nor constant with respect to space and time.[16]

In the case of isolated cells, mammalian cells experience electroporation starting at electric fields of $E \approx 1$ kV/cm for short pulses. Bacteria, which are significantly smaller, require a much larger E. In most work to date the electric field exposure has been very brief. The degree of electroporation occurring at lower fields and longer times has received only slight attention.[17] Finally, even though electroporation is universal, individual cells can experience significantly different degrees of electroporation because of cell-to-cell variations in $\Delta U(t)$, which in turn arise from differences in cell size, cell shape, cell orientation, and cell-to-cell proximity for the same applied uniform electric field.[18] Consistent with this expected variability, a wide distribution in the amount of molecular uptake by individual cells within a large cell population has been observed.[19-21] Use of nonuniform electric fields (e.g., nonparallel plane electrodes, adherent cells cultured on insulating glass or plastic surfaces) can cause even greater variations.

Electroporation Is a Nonthermal Phenomenon

Electroporation should be regarded as a nonthermal phenomenon because the dramatic membrane rearrangement associated with pore formation occurs extremely rapidly, *before* any significant temperature rise of the medium. In physiologic media the ohmic heating is of order $dT/dt \approx 10^4$ to 10^5 C sec^{-1}, but because of the short exposure time, the temperature rise is only about 1 C per pulse. Moreover, electroporation can also be demonstrated in special low-conductivity media for which the extracellular heating is even smaller. Electroporation also appears to occur usually under mild biochemical conditions. With the exception of the large electric fields used for electroporation of small microorganisms,[22] direct, damaging interaction of the field with molecules in solution appears to be negligible.

Prompt Membrane Destruction (Rupture) Due to Electroporation

Many experiments show that planar and cell membranes can be significantly damaged by some pulses. Early nonpore theories[23] could not explain the magnitude of the critical transmembrane voltage or the stochastic nature of rupture (irreversible electrical breakdown), but transient aqueous-pore models correctly predict both.[8,13]

Electrical Behavior for a Short Pulse

The behavior of the transmembrane voltage $U(t)$ during and after application of a pulse is intimately connected with the number and size of the pores in the heterogeneous pore population which emerges. For a planar membrane, the success of a transient aqueous pore model in providing a reasonable, quantitative description of $U(t)$ under these conditions provides confidence that electroporation is a valid concept. Very small pulses simply charge the membrane, with a time constant (e.g., 10^{-6} s) associated with the membrane capacitance and the resistance of the external charging pathway. Somewhat larger pulses bring the membrane to several hundred millivolts and lead to rupture. Progressively larger pulses lead to a larger membrane conductance $G(t)$ and increasingly faster decay of $U(t)$. This is the origin of reversible behavior, as the membrane discharges for large pulses before a single large pore can evolve and lead to rupture. In theoretical models, REB is found to be caused by the nonlinear appearance of pores of many sizes, such that the membrane conductance becomes progressively lower for larger pulses. According to both experimental evidence[14] and theoretical models,[24] 0.1% or less of the membrane area becomes aqueous during REB.

Molecular Transport Due to Electroporation

Conditions that cause REB also cause a large increase in molecular transport across cell membranes. In the largest number of applications to date, DNA is introduced into cells and the degree of transformation is scored. Large numbers of smaller molecules can also be introduced into cells, which suggests widespread potential applications (Table 1). The detailed mechanism by which molecular transport occurs during electroporation is not fully understood, but candidates include electrical drift, electro-osmosis, diffusion, and endocytosis.[18,25] The first three mechanisms can clearly lead to increased molecular transport for water-soluble molecules if pores are involved, but the relationship between endocytosis and electroporation is less clear.

TABLE 1.—Existing and likely electroporation applications.

Cell killing	DNA introduction
Loading drugs into cells	Loading dyes and tracers into cells
In situ enzymology (load reagents)	Intracellular immunoassays
Electro-insertion of proteins into membranes	Release of intracellular compounds
Tumor tissue drug delivery	Transdermal drug delivery
Isolated cell fusion*	Cell/tissue fusion*

*The many possible applications of cell fusion are worthy of a separate discussion.

The molecular size, shape, and charge can all be expected to be important. Simple exclusion by geometric size ("sieving") can be considered, but distortions of the pore by a nearby or entering molecule may also occur. Similarly, the shape of a molecule, particularly if a molecule is long, should be important (e.g., Fig. 1F). Exclusion because of a Born-energy repulsion associated with the lower dielectric constant of the membrane interior regions, and interactions with charged head groups within hydrophilic pores and any composite pores should also be significant. The few quantitative determinations of molecular transport at the individual cell level by experiment show a large cell-to-cell variation.[20,21]

Electro-insertion of Macromolecules into the Membrane

In contrast to transmembrane transport, membrane proteins can also be stably inserted *into the membrane* by electroporation.[26] As suggested in Fig. 1F, if a molecule with a hydrophobic region separated by two hydrophilic regions enters a large hydrophilic pore, that pore can be expected to shrink down to the less repulsive hydrophobic region of the molecule, trapping the molecule. As is well known qualitatively, and also quantitatively by consideration of the electrostatic energy of the system,[27] a membrane protein within a collapsed pore should often be stable. The gentleness of electroporation is supported by the observation that red blood cells with electro-inserted molecules have long circulating halflives in laboratory animals.

Membrane Recovery

In contrast to the onset of electroporation, the rate and variation in pore shrinkage and disappearance appear to depend on the membrane properties and the external environment. Some artificial planar bilayer membranes may rupture more readily than others due to differences in Γ and γ. For cells, postpulse behavior can vary significantly. For example, some primary pores may expand into very large pores by secondary processes particular to the cell type and the experimental conditions. As another example, some experiments show a long-term (seconds to minutes) persistence of transport in a subpopulation of cells. Finally, for the same conditions, one cell type may mostly survive while another may be mostly killed. The processes that occur after U has decayed to low levels is poorly understood.

Cell Stress and Death Due to Electroporation

A general trend is found in the use of electroporation to introduce molecules into cells. For a given pulse shape, small pulses cause no effect, but starting at about 1 kV/cm (mammalian cells, short pulses), some cells experience significant uptake. As larger fields are used, the percentage of participating cells increases, but the percentage of surviving cells simultaneously decreases. Eventually, for very large fields, essentially no cells survive. Why? There are at least two hypotheses: (1) a prompt membrane rupture occurs in some portions of the cell membrane, leading to a large, permanent hole in the membrane; and (2) lethal chemical imbalances occur, due to the influx and efflux through both transient and metastable pores. Although important to almost all applications, a good understanding of cell stress and resultant cell death does not yet exist.

Generally Attractive Features of Applications

The study of electroporation is compelling for two general reasons: (1) fundamental aspects of membrane structure and behavior is involved, and (2) significant applications in research, biotechnology, and medicine are likely (Table 1). It is significant that electroporation achieves a chemical and biological result by a purely physical means. Unlike chemical manipulations of biological systems, there is no chemical residue. Further, as noted above, electroporation appears to be universal at the membrane level. Finally, the fact that electrical phenomena are involved has the important technical consequence that electronic systems can be used both to cause electroporation and then to measure its electrical consequences. For example, in the emerging application of tissue electroporation, rapid electrical measurements permit the occurrence and aftereffects of electroporation to be monitored.[28] As suggested in Table 1, the use of electroporation with cultured cells is increasingly used to introduce a variety of molecules from the extracellular solution. Already the number of published studies is large, as can be seen by considering some recent reviews and their references.[1-3] Although the potential applications of tissue electroporation are large, only a few studies have been reported. Some initial work has emphasized the ability to make electrical measurements,[28] or combined molecular transport and electrical measurements;[29] others have focused on the biological consequences of molecular transport.[30-32] Finally, electrofusion of cells to tissue is being investigated.[33] Much more can be expected in the future.

SUMMARY AND FUTURE PROSPECTS

Compared to many scientific and engineering topics, electroporation is still at an early stage. Many basic phenomena have been discovered but remain incompletely understood, even as many new applications have been identified. The apparent universality of electroporation, and its predominantly electrical nature, suggest that electroporation will receive increasing attention and grow into an important technology.

REFERENCES

1. E. Neumann, A. Sowers, and C. Jordan, Eds., *Electroporation and Electrofusion in Cell Biology,* New York: Plenum, 1989.

2. T. Y. Tsong, "Electroporation of cell membranes," *Biophys. J.* 60: 297-306, 1991.

3. D. C. Chang, B. M. Chassy, J. A. Saunders, and A. E. Sowers, Eds., *Guide to Electroporation and Electrofusion,* New York: Academic Press, 1992.

4. E. Neumann and K. Rosenheck, "Permeability changes induced by electric impulses in vesicular membranes," *J. Membrane Biol.* 10: 279-290, 1972.

5. U. Zimmermann, G. Pilwat, and F. Riemann, "Preparation of erythrocyte ghosts by dielectric breakdown of the cell membrane," *Biochim. Biophys. Acta* 375: 209-219, 1975.

6. K. Kinosita Jr. and T. Y. Tsong, "Formation and resealing of pores of controlled sizes in human erythrocyte membrane," *Nature* 268: 438-441, 1977.

7. A. E. Sowers and M. R. Lieber, "Electropore diameters, lifetimes, numbers, and locations in individual erythrocyte ghosts," *FEBS Lett.* 205: 179-184, 1986.

8. I. G. Abidor, V. B. Arakelyan, L. V. Chernomordik, Yu. A. Chizmadzhev, V. F. Pastushenko, and M. R. Tarasevich, "Electric breakdown of bilayer membranes: I. The main experimental facts and their qualitative discussion," *Bioelectrochem. Bioenerget.* 6: 37-52, 1979.

9. E. Neumann, A. Sprafke, E. Boldt, and H. Wolf, "Biophysical digression on membrane electroporation," in Ref. 3, 77-90.

10. D. C. Chang, "Strucutre and dynamics of electric field-induced membrane pores as revealed by rapid-freezing electron microscopy," in Ref. 3, 9-27.

11. R. Lipowsky, "The conformation of membranes," *Nature* 349: 475-481, 1991.

12. V. F. Pastushenko and Yu. A. Chizmadzhev, "Stabilization of conducting pores in BLM by electric current," *Gen. Physiol. Biophys.* 1: 43-52, 1982.

13. A. Barnett and J. C. Weaver, "Electroporation: A unified, quantitative theory of reversible electrical breakdown and rupture," *Bioelectrochem. and Bioenerg.* 25: 163-182, 1991.

14. R. Benz, F. Beckers, and U. Zimmermann, "Reversible electrical breakdown of lipid bilayer membranes: A charge-pulse relaxation study," *J. Membrane Biol.* 48: 181-204, 1979.

15. H. P. Schwan, "Dielectrophoresis and rotation of cells," in Ref. 1, 3-21.

16. K. Kinosita Jr., M. Hibio, H. Itoh, M. Shigemori, K. Hirano, and T. Hayakawa, "Electroporation visualized on the time scale from a submicrosecond to seconds," in Ref. 3, 29-47.

17. T.-D. Xie and T. Y. Tsong, "Study of mechanisms of electric field-induced DNA transfection: II. Transfection of low amplitude, low frequency alternating electric fields," *Biophysical J.* 58: 897-903, 1990.

18. J. C. Weaver and A. Barnett, "Progress towards a theoretical model of electroporation mechanism: Membrane electrical behavior and molecular transport," in Ref. 3, 91-117.

19. J. C. Weaver, G. I. Harrison, J. G. Bliss, J. R. Mourant, and K. T. Powell, "Electroporation: High frequency of occurrence of the transient high permeability state in red blood cells and intact yeast," *FEBS Lett.* 229: 30-34, 1988.

20. D. C. Bartoletti, G. I. Harrison, and J. C. Weaver, "The number of molecules taken up by electroporated cells: Quantitative determination," *FEBS Lett.* 256: 4-10, 1989.

21. M. R. Prausnitz, B. S. Lau, C. D. Milano, S. Conner, R. S. Langer, and J. C. Weaver, "A quantitative study of electroporation showing a plateau in molecular transport," (in press).

22. A. G. Sabelnikov and E. S. Cymbalyuk, "Clotting of donor DNA during electroporation of prokaryotes," *Bioelectrochem. Bioenerget.* 24: 313-321, 1990.

23. J. M. Crowley, "Electrical breakdown of bimolecular lipid membranes as an electromechanical instability," *Biophys. J.* 13: 711-724, 1973.

24. M. Wang, S. Freeman, V. G. Bose, S. Dyer, and J. C. Weaver, "Theoretical modelling of electroporation: Electrical behavior and molecular transport," this volume.

25. U. Zimmermann, R. Schnettler, G. Klöck, and H. Watzka, "Mechanisms of electrostimulated uptake of macromolecules into living cells," *Naturwissenschaft* 77: 543-545, 1990.

26. Y. Mouneimne, P.-F. Tosi, Y. Gazitt, and C. Nicolau, "Electro-insertion of Xeno-glycophorin into the red blood cell membrane," *Biochem. Biophys. Res. Comm.* 159: 34-40, 1989.

27. D. Mycue, J. Zahn, M. Zahn, and J. C. Weaver, "Interaction mechanism of electric fields with cells: Alteration of binding site access of membrane proteins," this volume.

28. K. T. Powell, A. W. Morgenthaler, and J. C. Weaver, "Tissue electroporation: Observation of reversible electrical breakdown in viable frog skin," *Biophys. J.* 56: 1163-1171, 1989.

29. M. R. Prausnitz, V. G. Bose, R. S. Langer, and J. C. Weaver, "Transtissue molecular transport due to electroporation of skin," this volume.

30. R. C. Lee and M. S. Kolodney, "Electrical injury mechanisms: Electrical breakdown of cell membranes," *Plast. Reconstr. Surg.* 80: 672-679, 1987.

31. A. V. Titomirov, S. Sukharev, and E. Kistoanova, "In vivo electroporation and stable transformation of skin cells of newborn mice by plasmid DNA," *Biochim. Biophys. Acta* 980: 9-14, 1989.

32. L. M. Mir, S. Orlowski, J. Belehradek Jr., and C. Paoletti, "In vivo potentiation of the bleomycin cytotoxicity by local electric pulses," *Eur. J. Cancer* 27: 68-72, 1991.

33. R. J. Grasso, R. Heller, J. C. Cooley, and E. M. Haller, "Electrofusion of individual animal cells directly to interaction corneal epithelial tissue," *Biochim. Biophys. Acta* 980: 9-14, 1989.

34. R. O. Potts and M. L. Francoeur, "Lipid biophysics of water loss through the skin," *PNAS* 87: 3871-3873, 1990.

35. Yu. A. Chizmadzhev, V. B. Arakelyan, and V. F. Pastushenko, "Electric breakdown of bilayer membranes: III. Analysis of possible mechanisms of defect origin," *Bioelectrochem. Bioenerget.* 6: 63-70, 1979.

36. J. Teissié and T. Y. Tsong, "Evidence of voltage-induced channel opening in Na/K ATPase of human erythrocyte membrane," *J. Membrane Biol.* 55: 133-140, 1980.

37. V. A. Klenchin, S. I. Sukharev, S. M. Serov, L. V. Chernomordik, and Yu. A. Chizmadzhev, "Electrically induced DNA uptake by cells is a fast process involving DNA electrophoresis," *Biophysical J.* 60: 804-811, 1991.

Martin Blank, Ed., *Electricity and Magnetism in Biology and Medicine*

ELECTROMAGNETIC TECHNOLOGY AND THE FUTURE OF BIOELECTROMAGNETICS

W. R. Adey*

University School of Medicine, Loma Linda, Calif.

We have traveled far in recent days on a journey of the mind and spirit, witnessing the full panorama of a newly explored realm of science. We are aware that, for the first time, acolytes of an unprecedented gamut of scientific disciplines have met together to explore and identify major areas of common interest in bioelectromagnetics. It has been a new and splendid vista, for much has been accomplished in recent years to achieve our present platform of biological and biophysical understanding.

Our objective here is to examine contributions of technologies to past research with which, in many instances, they have evolved; and thus, to speculate on ways in which future research may be equivalently linked to, or limited by, availability of needed new technologies. The topic is obviously important, if past history is to mirror future opportunity. To this time, there have been few areas of biological science so closely linked to research at the cutting edge of a key area of the physical sciences; in this case, to research in the physics of matter.

Indeed, biological observations may well become benchmarks and pacesetters in the future of this field of physics.[1,2] Observations on biological coherence and cooperativity now challenge physical scientists to seek models and to design experiments in the realm of biology. For the future, progress in elucidating these sensitivities and subtleties of electromagnetic field interactions with biomolecular systems will be jointly the province of physicists and biologists. At every turn, there will be an increasing demand for new technologies.

PRINCIPAL AREAS OF RECENT RESEARCH IN BIOELECTROMAGNETICS

The main endeavors of the world's small community of scholars in this field have been focused in limited but widely separated areas of biology and medicine. In many respects, they form an hierarchical sequence:

(a) Coupling mechanisms between fields and tissues at the cellular level.[3-7]
(b) Field effects on embryonic and fetal development.[8]
(c) Modulation of central nervous and neuroendocrine functions.[9-11]
(d) Modified immune functions.[12-14]
(e) Regulation of cell growth, and field action as promoter in tumor formation.[15-25]
(f) Diagnostic and therapeutic applications.

THE TECHNOLOGICAL FOCUS: KEY AREAS OF FUTURE RESEARCH THAT MAY DEPEND ON NEW TECHNOLOGIES

With an acute awareness of accountability in future years, we may embark on a brief speculative journey. Of necessity, the selection of topics is eclectic, focusing on but a few areas that may nonetheless pass into the byways of science; yet it is an opportunity for a synthesis of current research that bears a stamp of future opportunity and new options.

Possible Role of Free Radicals in Cell Transduction of Low-level Electromagnetic Fields

As evidence has mounted confirming occurrence of bioeffects of electromagnetic fields that are not only dwarfed by much larger intrinsic bioelectric processes but may also be substantially below the level of tissue thermal noise, there is a mainstream of theoretical and experimental studies seeking an understanding of the first transductive steps. McLauchlan[26] has proposed a role for chemically very reactive free radicals.

*Correspondence to: W. R. Adey, Research Service (151), VA Medical Center, Loma Linda CA 92357.

In chemical reactions, bonds break and re-form. Most bonds consist of paired electrons with opposite spins, with one electron derived from each partner in the union. In a chemical reaction, the bond breaks; then each partner reclaims its electron from the bond and moves away to encounter a new partner. It is now an unattached, highly charged free radical. Reforming a bond requires a meeting between two radicals with opposite electron spins, the union producing a *singlet pair*. No union results from encounters between radicals with the same spins. Unsuccessful pairings are more likely than successful, because a singlet pair can form in only one way but there are three ways in which electrons with similar spins can unite. The latter are therefore known as *triplet pairs*.

In McLauchlan's model, very low static magnetic fields cause triplet pairs to break and form singlets. But as the field is increased, typically to a level of about 8 mT, two of the three triplet states become entirely decoupled from the singlet state. Thus, at this field level two-thirds of the radical pairs may not react as they would in a weaker field, "an enormous effect of a small magnetic field on a chemical reaction, *and the effect begins at the lowest applied field strength*." The effect is quite general and does not depend on any specific chemical identity of the radicals.

Addition of an oscillating field to this system introduces intensity windows that will cause triplets to return to singlet states and react with one another. "The all-important interaction has an energy very much less than the thermal energy of the system, and is effective exclusively through its influence on the kinetics; this is counter-intuitive to most scientists." The major effect of the field is to remove degeneracies of sublevels of the triplet pair, whose energies are equal in zero field, but differ progressively in an increasing field as a Zeeman splitting effect.

As McLauchlan points out, it remains for future research to establish the significance of radical reactions in biological systems. At this time, many of these investigations will remain technology-limited.

Determination of the Role of Iron-containing Molecules in Sensitivities to Electromagnetic Fields

There is a wealth of evidence that iron-containing molecules are widespread in the chemistry of cellular transductive mechanisms. They include transferrins with specific receptors on cell surfaces, hem-containing G proteins that couple receptors to enzymes at cell membranes, and cytochrome P-450 enzymes that are key elements in cellular respiration. Transferrin mechanisms are sensitive to 60Hz electromagnetic fields.[27] These iron atoms are not arranged in ferromagnetic configurations and are too scattered to form magnetic dipoles. Nonetheless, future research may focus on their interactions with biomolecules that exhibit paramagnetic properties.[28,29]

Neuromelanins are polymeric pigments found in the brain and are suspected of a role in Parkinson's disease.[30] Neuromelanin exhibits paramagnetism due to the presence of large numbers of unpaired electrons, which qualify this molecule as a stable free radical.

These paramagnetic molecules are well known to influence proton relaxation times in nuclear magnetic resonance studies, including responses in malignant melanomas in magnetic-resonance imaging (MRI) studies.[31] Although the short T1 relaxation times in these studies were attributed to paramagnetism of melanin-free radicals, it has now been shown that this shortening requires interactions of paramagnetic ferric iron (FE^{3+}) with melanin;[32] synthetic L-DOPA melanin alone in agar phantoms was without effect on either T1 or T2 relaxation times.[33] Future developments of these techniques may allow MRI studies of nigrostriatal regions of the brain in Parkinson's disease, based on the compromised status of neuromelanin and ferric iron.

Possible Role of Highly Cooperative Electron Transfer in Proteins as a Basis for Electromagnetic-Field Interactions

In 1966 pioneering studies by Chance[34,35] disclosed the physical nature of first steps in activation of a cytochrome *c*-bacteriochlorophyll complex by millisecond electron transfer. Of great importance was the finding that the transfer was temperature-independent from 120 K to 4 K. They deduced that the reaction proceeded through a quantum-mechanical tunneling mechanism, perhaps taking place over distances as large as 30 Å. They suggested that this long-range electron transfer may be essential in the energetics of mitochondria. These studies coincided with Mitchell's development of his Nobel Prize-winning chemiosmotic model, which proposed that key electron-transfer steps of respiration operated across the full 35 Å width of the cell membrane.

Extension of these early concepts by Moser et al.[36] has opened doors on exciting new vistas, with options for much further research on detection and coupling of electromagnetic fields at cell membranes. Using powerful first-order analysis of electron-transfer measurements in biological and chemical systems, they found that a variation of 20 Å in the distance between donors and acceptors in protein changed the electron-transfer rate by 10^{12}-fold. In the time domain, there is also a strong dependence on distance. Thus, in considering electron transfer across the full thickness of the cell membrane, with edge-to-edge distances between 25 and 35 Å, the optimal electron transfer rate is pushed into seconds and days time scales, respectively.

In these studies, protein behaves like an organic glass, presenting a uniform electronic barrier to electron tunneling and a uniform nuclear characteristic frequency. Future technological developments, with appropriate spatial and temporal resolution, may make this powerful technique feasible for studies in biological membranes and intact cells. Selection of distance, free energy, and reorganization energy would be sufficient to define rate and directional specificity of biological electron transfer, thus meeting physiological requirements in a wide range of cytochrome systems.

Experimental Evaluation of Biophysical Models of Transmembrane Signaling: Some Disparate, Some Complementary

In electron-spin resonance studies with labeled tails of phospholipid molecules in an artificial phospholipid bilayer, McConnell[37] noted that intrusion of a protein strand into this bilayer induces coherent states between charges on the tails of adjoining phospholipid molecules for considerable distances from the protein strand. At the same time, motions of the phospholipid tails are constrained and they behave more rigidly. These interactions suggest establishment of energetic domains determined by joint states of intramembranous proteins and surrounding phospholipid molecules. Moreover, states of *dielectric strain* associated with these lipoprotein interactions are likely to determine optical properties within their cooperative domains. As in fiber-optic systems, dependence of these optical properties on membrane excitation states may determine stability of *dark soliton* propagation as a means of transmembrane signaling.

Dark solitons and their analogies with electro-optical properties of cell membranes. A model with possible relevance to organization of cell membranes is suggested by studies of *dark solitons* in single-mode optic fibers.[38] Due to nonlinear interactions between light and the transmitting medium in these fibers (the refractive index varies with light intensity), the normal tendency of light pulses to disperse can be negated. As a result, optical solitons can propagate without distortion. Dark pulses, short breaks in an otherwise continuous laser beam, can be made to propagate stably for 4000 km. In an analogy with water waves, wave breaking occurs because the group velocity of the wave is greater at the peak than at the edges; the wave overtakes itself. The nonlinear Schrödinger equation also predicts wave-breaking effects in solitons, and determines a wide range of frequencies within the pulse (chirping).

These dark solitons suggest analogies with the sharp changes in optical properties of living vertebrate and invertebrate axons accompanying polarizing currents,[39] and with the highly cooperative movements of about 18 Å reported with laser interferometry at the axon surface within 1 ms of excitation.[40] These electro-optical properties of cell membranes suggest an important avenue for future research in exploring the cooperative domains that exist between protein strands and their mantle of phospholipids within the plasma membrane.

Inward signaling and energy transfer along cell membrane receptor proteins. In their simplest forms, membrane receptor proteins appear to cross the plasma membrane only once. In a second group, the strand crosses the membrane seven times. The most striking feature of these proteins is in the structure of the extremely short chain of 23 amino acids that lies within the cell membrane at each consecutive membrane crossing. These short segments are composed of *hydrophobic* amino acids (and therefore nonconducting), in contrast to the hundreds of *hydrophilic* amino acids forming the rest of the strand inside and outside the membrane.

Is there evidence that this transmembrane movement of ions is mediated by these receptor protein segments, despite their hydrophobicity and their location in a highly hydrophobic environment? Ullrich et al.[41] conclude that a hydrophobic segment as short as 23 amino acids is probably too short to be involved in conformation changes; and that its hydrophobic character makes unlikely its participation in either ionic or proton movement by coulombic forces.

Direct measurement of receptor-mediated inward calcium currents in epidermal cells[7] and mast cells[5] indicates that it is not voltage-activated and would thus be consistent with ion translocation by molecular vibrational modes. It shows a characteristic inward rectification. Future research may determine the validity of the proposal by Hoth and Penner that this may be the mechanism by which electrically nonexcitable cells maintain raised intracellular Ca^{2+} concentrations and replenish their diminished Ca^{2+} stores after receptor stimulation.

Fluorescence Spectrometry and Microscopy of Intracellular Calcium Fluxes Induced by Electromagnetic Fields

Intracellular calcium is a ubiquitous second messenger. Information is encoded in the magnitude, frequency, and spatial organization of changes in concentration of cytosolic free Ca^{2+}. Ion-sensitive fluophores offer valuable techniques in displaying the spatiotemporal distribution of Ca^{2+} in single cells and in domains of cultured cells. Ca^{2+} can be detected with substances requiring excitation at either single or dual wavelengths.

Fluorescence microscopy indicates that, in confluent cell cultures, the aggregate levels of intracellular calcium may rise over large domains of the culture while decreasing in other domains.[42] As yet, there is no understanding of the signaling mechanism by which such a highly cooperative behavior can occur over many hundreds of neighboring cells. It is likely to be a manifestation of an important modulating behavior within a tissue. An important goal of future research will be testing the possible role of electromagnetic fields in mediating such a pervasive influence.

Initial applications of fluorescence microscopy have measured calcium uptake as an aggregate behavior in cultures of lymphocytes and other cells as a function of electromagnetic field frequency and intensity.[43] Future research may examine relations between calcium oscillations in individual cells of a domain in confluent cultures, based on chaotic models and experimentally testing sensitivity to frequency and intensity of electromagnetic fields. Using fluorescence techniques and digital image analysis, one can study distribution of calcium signals in space and time within a single cell.[44] Calcium signals may be uniform, or they may arise in discrete regions and propagate as waves. For example, in bovine adrenal cells, opening of voltage-operated channels gives a uniform calcium signal, whereas activation of muscarinic receptors causes a highly focal rise in calcium. In fertilization, the calcium signal begins at the point of sperm-egg fusion from where it spreads to the opposite pole.[45] These waves propagate in many different cell types in a nondecremental fashion at velocities in the range 10-200 μm/s.

Intracellular levels of calcium often oscillate, especially during stimulation of the phosphoinositide pathway in cell membranes. Oscillatory patterns vary, usually taking the form of calcium spikes. The oscillatory mechanism is based on the periodic release of internal calcium (cytosolic oscillators), with periods ranging from 5 to 60 s.[46]

Regenerative spiral waves of release of free Ca^{2+} occur in eggs of the toad *Xenopus laevis*.[44] The spiral activity has no apparent structural basis, such as an unusual formation of organelles. A prominent feature of these Ca^{2+} release patterns is mutual annihilation of colliding wavefronts. Annihilation is independent of the type of wave propagation and occurs between planar, circular, and spiral waves. Absence of Ca^{2+} release in the wake of colliding waves may imply that the Ca^{2+} stores are refractory, and that the Ca^{2+} mobilizing mechanism behaves as an excitable medium.

Evaluation of Reported Occurrence of Ferromagnetic Materials in Mammalian Tissues

The presence of even trace amounts of ferromagnetic material in human tissues would have significant implications in clinical procedures such as MRI, involving static magnetic fields in excess of 1.5 T. At the other extreme, if tissue transduction of static or oscillating magnetic fields occurs through equilibrium thermodynamic mechanisms (i.e., by heating), it would be difficult for weak magnetic fields to produce effects at the cellular level in competition with background thermal energy. With these permanent magnets, however, a variety of energetic effects would be possible.[47]

In past studies of iron storage, it has been assumed that no ferromagnetic materials are present in human tissues.[48,49] However, Kirschvink (1992, in press) claims to have detected minute amounts in human brain tissue, using a SQUID system shielded by 6 tons of steel. Ferromagnetic materials interact very strongly with magnetic fields. Unlike diamagnetic and paramagnetic substances, quantum-mechanical interactions acting on unpaired electrons within these materials force the electron magnetic moments *(Bohr magnetons)* to form long-range alignments. Magnetic moments for each Bohr magneton within such crystals add vectorially, and particles only a few tens of nanometers in size will have magnetic interaction energies with the geomagnetic field that are in excess of kT.

There remain critical questions about possible mechanisms for its biomolecular transductive coupling, the determination of its cellular loci, and possible structural and functional links to pathways subserving electromagnetic-field transduction.

THE FUTURE OF BIOELECTROMAGNETICS IN DEVELOPING DIAGNOSTIC AND THERAPEUTIC APPLICATIONS

With little doubt about their increasing utilization and broadening applications, we may set aside for the moment impressive therapeutic results from use of electromagnetic fields in tissue healing, regeneration, and growth, and turn to possibilities for a medical imaging system based on microwave scattering.

Structural and Functional Imaging by Microwave Scattering

The goal of this development would be a combined structural and functional image, specifically for the brain and other central nervous structures. Despite spectacular success of separate developments in structural imaging with CT and MRI scanning, and metabolic imaging with positron emission tomography (PET), attempts to combine these techniques in a single structural and functional image have been only moderately successful, lacking either spatial or temporal resolution, or both. There is a further option in the use of microwave scattering techniques.

Larsen and Jacobi[50] have successfully imaged the isolated dog kidney, using a 3.9GHz field. A resolution of about 5 mm can be achieved with a coupling medium with a high dielectric constant that results in wavelength contraction from 7.5 cm in air to about 8.5 mm in fluid and brain tissue. In their feasibility studies so long ago, it was possible to identify major subdivisions of kidney substance, including cortical zones containing proximal and distal convoluted tubules, deeper cortical layers containing glomeruli, and an outer medullary zone containing loops of Henle and the collecting system.

Evidence for existence of functional information attributable to physiological dielectric scattering in brain tissue comes from animal studies by Kelley et al.[51] Using carrier frequencies between 100 and 900 MHz, they applied CW fields to the rat's head and examined the scattered field for dynamic variations in amplitude that might correspond to concurrent electrophysiological waveforms. They detected amplitude modulation strongly correlated with simultaneous electrical evoked potentials elicited by light flashes.

We may speculate that, if future developments combine the essence of both techniques, we may construct a functional image, giving substance at last to Sherrington's notion of the brain's myriad functions as "an enchanted loom."

Tailored Drugs with Tailored EM Fields in the Future of Pharmacotherapy

Twenty years ago in Warsaw, the late Professor P. Czerski[52] predicted that microwave fields might serve to enhance the effectiveness of therapeutic substances, reducing required does of efficacious drugs and permitting focal action at cell and tissue sites.

Studies in France by Mir et al. (in the present volume) show the feasibility of joint drug/EM field action. In patients with head and neck cancers, action of the cancer chemotherapy drug bleomycin was enhanced by as much as 650000 times when pulsed electric fields were applied a few minutes after intravenous administration of the drug. This field effect may result from electroporation. For the future, this technique of *electrochemotherapy* may be the forerunner of vastly improved techniques of drug administration, meeting criteria for specificity of drug action and diminished toxicity scarcely conceived at this time.

SUMMARY

There is a reasonable prospect that bioelectromagnetics may attain the status of a separate biological discipline, having developed unique tools and experimental philosophies in a search for essential order in living systems. Future research on submolecular transductive coupling will be diversified and increasingly dependent on new technologies, such as high-resolution magnetic resonance spectroscopy and electro-optical techniques. New physical techniques will open avenues to such challenging problems as structural modifications during receptor-ligand binding, vibration modes in cell membrane lipoprotein domains during excitation, and possible coherent millimeter-wave emissions accompanying enzyme action.

Our journey in realms of biology has been swift. In little more than a century, our vista has moved from organs to tissues, to cells, and most recently to the molecules that are the exquisite fabric of living systems. Now we stand at a new frontier. It may be more difficult to understand, but it is of vastly greater significance. For it is at the atomic level that physical processes, rather than chemical reactions in the fabric of molecules, appear to shape the transfer of energy and the flow of energy in living systems. Trullinger has summarized our present position:

Once the uninitiated can bring himself to revolt from his "linear" upbringing (via the harmonic oscillator and the Schrödinger equation), and can accept the fact that large-amplitude localized objects can exist which *do not* spread in time, then he stands on the threshold of developing that wonderfully simple and beautiful pattern of thought that emphasizes the remarkable stability of such objects, their very natural occurrence in the ground state, their coexistence in many cases with extended linearized solutions, their striking particle-like behavior in the presence of perturbations, and perhaps most important of all, their essential role in explaining so many physical phenomena that *cannot* be explained in any other way.[53]

REFERENCES

1. H. Fröhlich, "Coherent excitation in coherent biological systems," in F. Gutmann and H. Keyzer, Eds., *Modern Bioelectrochemistry,* New York: Plenum, 1986, 241.

2. H. Fröhlich, "Theoretical physics and biology," in H. Fröhlich, Ed., *Biological Coherence and Response to External Stimuli,* New York-Berlin-Heidelberg: Springer-Verlag, 1988, 1.

3. W. R. Adey, "Electromagnetic fields and the essence of living systems," in J. Bach Andersen, Ed., *Modern Radio Science,* Oxford: Oxford University Press, 1990, 1.

4. W. R. Adey, "Collective properties of cell membranes," in C. Ramel and B. Norden, Eds., *Resonance and Other Interactions of Electromagnetic Fields with Living Systems,* Royal Swedish Academy of Sciences, Stockholm, Oxford University Press, 1992, 47.

5. M. Hoth and R. Penner, "Depletion of intracellular calcium stores activates a calcium current in mast cells," *Nature* 355: 353-355, 1992.

6. R. A. Luben, "Effects of low-energy electromagnetic fields (pulsed and DC) on membrane signal transduction processes in biological systems," *Health Physics* 61: 15-28, 1991.

7. W.H. Moolenaar, W. J. Aerts, L. G. J. Tertoolen, and S. W. Delast, "The epidermal growth-factor induced calcium signal in A431 cells," *J. Biol. Chem.* 261: 279-285, 1986.

8. R. F. McGivern, R. Z. Sokol, and W. R. Adey, "Prenatal exposure to a low-frequency electromagnetic field demasculinizes adult scent marking behavior and increases accessory sex organ weight in rats," *Teratology* 41: 1-8, 1990.

9. A. Lerchl, K. O. Nonaka, and R. J. Reiter, "Pineal gland 'magnetosensitivity' is a consequence of induced electric currents (eddy currents)," *J. Pineal Res.* 10: 109-116, 1990.

10. R. J. Reiter, "Changes in circadian melatonin synthesis in the pineal gland of animals exposed to extremely low frequency electromagnetic radiation: A summary of observations and speculation on their implications," in M. C. Moore-Ede, S. S. Campbell, and R. J. Reiter, Eds., *Electromagnetic Fields and Circadian Rhythmicity,* Boston: Birkhauser, 1992, 13.

11. B. W. Wilson and L. E. Anderson, "Electromagnetic-field effects on the pineal gland," in B. W. Wilson, R. G. Stevens, and L. E. Anderson, Eds., *Extremely Low Frequency Electromagnetic Fields: The Question of Cancer,* Columbus, Ohio: Battelle Press, 1990, 159.

12. C. V. Byus, R. L. Lundak, R. M. Fletcher, and W. R. Adey, "Alterations in protein kinase activity following exposure of cultured lymphocytes to modulated microwave fields," *Bioelectromagnetics* 5: 34-51, 1984.

13. D. B. Lyle, P. Schechter, W. R. Adey, and R. L. Lundak, "Suppression of T lymphocyte cytotoxicity following exposure to sinusoidally amplitude-modulated fields," *Bioelectromagnetics* 4: 281-292, 1983.

14. D. B. Lyle, R. D. Ayotte, A. R. Sheppard, and W. R. Adey, "Suppression of T lymphocyte cytotoxicity following exposure to 60 Hz sinusoidal electric fields," *Bioelectromagnetics* 9: 303-313, 1988.

15. H. C. Pitot and Y. P. Dragan, "Facts and theories concerning the mechanisms of carcinogenesis," *FASEB J.* 5: 2280-2286, 1991.

16. I. B. Weinstin, "The origins of human cancer: Molecular mechanisms of carcinogenesis and their implications for cancer prevention and treatment," *Cancer Res.* 48: 4135-4143, 1988.

17. W. R. Adey, "Joint actions of environmental nonionizing electromagnetic fields and chemical pollution in cancer promotion," *Environ. Health Perspectives* 86: 297-305, 1990.

18. W. R. Adey, "Electromagnetic fields, cell membrane amplification, and cancer promotion," in B. W. Wilson, R. G. Stevens, and L. E. Anderson, Eds., *Extremely Low Frequency Electromagnetic Fields: The Question of Cancer,* Columbus, Ohio: Battelle Press, 1990, 211.

19. W. R. Adey, "Signal functions of brain electrical rhythms and their modulation by external electromagnetic fields," in E. Basar and T. H. Bullock, Eds., *Induced Rhythms of the Brain,* Boston: Birkhauser, 1991, 323.

20. C. V. Byus, S. E. Pieper, and W. R. Adey, "The effects of low-energy 60 Hz environmental electromagnetic fields upon the growth-related enzyme ornithine decarboxylase," *Carcinogenesis* 8: 1385-1389,

1987.

21. C. V. Byus, K. Kartun, S. Pieper, and W. R. Adey, "Increased ornithine decarboxylase activity in cultured cells exposed to low energy modulated microwave fields and phorbol ester tumor promoters," *Cancer Res.* 48: 4222-4226, 1988.

22. Y. Hiraki, N. Endo, M. Takigawa, A. Asada, H. Takahashi, and F. Suzuki, "Enhanced responsiveness to parathyroid hormone and induction of functional differentiation of cultured rabbit costal chondrocytes by a pulsed electromagnetic field," *Biochim. Biophys. Acta* 931: 94-100, 1987.

23. R. A. Luben, M.-Y. Chen, D. M. Rosen, and W. R. Adey, "Effects of electromagnetic stimuli on bone and bone cells, in vitro: Inhibition of responses to parathyroid hormone by low-energy, low-frequency fields," *Proc. Natl. Acad. Sci. USA* 79: 4180-4183, 1982.

24. R. Goodman and A. S. Henderson, "Exposure of salivary gland cells to low-frequency electromagnetic fields alters polypeptide synthesis," *Proc. Natl. Acad. Sci. USA* 8: 3928-3932, 1988.

25. J. L. Phillips, W. Haggren, W. J. Thomas, T. Ishida-Jones, and W. R. Adey, "Magnetic field-induced changes in specific gene transcription," *Biochim. Biophys. Acta* (in press).

26. K. McLauchlan, "Are environmental magnetic fields dangerous?" *Physics World* January: 41-45, 1992.

27. J. L. Phillips, "Transferrin receptors and natural killer cell lysis: A study using Colo 205 cells exposed to 60 Hz electromagnetic fields," *Immunology Letters* 13: 295-299, 1986.

28. S. S. Chio, J. S. Hyde, and R. C. Sealy, "Temperature-dependent paramagnetism in melanin polymers," *Arch. Biochem. Biophys.* 199: 13?-139, 1980.

29. R. C. Sealy, J. S. Hyde, C. C. Felix, I. A. Menon, G. Prota, H. M. Swartz, S. Perad, and H. F. Haberman, "Novel free radicals in synthetic and natural pheomelanins: Distinction between DOPA melanins and cysteinyldopa melanins by ESR spectroscopy," *Proc. Nat. Acad. Sci. USA* 79: 2885-2889, 1982.

30. M. B. Youdim, D. Ben-Shahar, and S. Yehuda, "Putative biochemical mechanism of the effect of iron deficiency on brain biochemistry and behavior," *Amer. J. Clin. Nut.* 50(Suppl. 3): 607-615.

31. R. B. Chambers, F. H. Davidorf, J. F. McAdoo, and D. W. Chakeres, "Magnetic resonance imaging of uveal melanomas," *Arch. Ophthalmol.* 105: 917-921, 1987.

32. J. M. Tosk, B. A. Holshouser, R. C. Aloia, D. B. Hinshaw, A. N. Hasso, J. P. MacMurray, A. D. Will, and L. P. Bozzetti, "Effects of the interaction between ferric iron and L-Dopa melanin on T1 and T2 relaxation times determined by magnetic resonance imaging," *Magnetic Resonance in Medicine* 26: 40-45, 1992.

33. W. S. Enochs, W. B. Hyslop, H. F. Bennett, R. D. Brown, S. H. Koenig, and H. M. Swartz, "Sources of the increased longitudinal relaxation rates observed in melanotic melanomas: An *in vitro* study of synthetic melanins," *Investig. Radiol.* 24: 794-804, 1989.

34. B. Chance, D. Devault, and W. W. Hildreth, "Early chemical events in photosynthesis: Kinetics of oxidation of cytochromes of types *c* or *f* in cells, chloroplasts and chromatophores," *Brookhaven Symp. Biol.* 19: 115-131, 1966.

35. D. Devault and B. Chance, "Studies of photosynthesis using a pulsed laser: I. Temperature dependence of cytochrome oxidation rate in chromatium: Evidence for tunneling," *Biophys. J.* 6: 825-847, 1966.

36. C. C. Moser, J. M. Keske, K. Warncke, R. S. Farid, and P. L. Dutton, "Nature of biological electron transfer," *Nature* 355: 796-82, 1992.

37. H. M. McConnell, "Coupling between lateral and perpendicular motion in biological membranes," in F. O. Schmitt, D. M. Schneider, and D. M. Crothers, Eds., *Functional Linkage in Biomolecular Systems,* New York: Raven Press, 1975, 123.

38. P. L. Christiansen, "Shocking optical solitons," *Nature* 339: 17-20, 1989.

39. J. M. Tobias and S. Solomon, "Opacity and diameter changes in polarized nerve," *J. Cell. Comp. Physiol.* 35: 25-34, 1950.

40. B. C. Hill, E. D. Schubert, M. A. Nokes, and R. P. Michelson, "Laser interferometer measurements of changes in crayfish axon diameter concurrent with axon potential," *Science* 196: 426-428, 1977.

41. A. Ullrich, L. Coussens, J. S. Hayflick, T. J. Dull, A. Gray, A. W. Tam, J. Lee, Y. Yarden, T. A. Libermann, J. Schlessinger, E. L. V. Mayes, N. Whittle, M. D. Waterfield, and P. H. Seburg, "Human epidermal growth factor receptor cDNA sequence and aberrant expression of the amplified gene in A431 epidermoid carcinoma cells," *Nature* 309: 428-431, 1985.

42. R. Y. Tsien, "New tetracarboxylate chelators for fluorescence measurement and photochemical manipulation of cytosolic free calcium concentration," *Soc. Gen. Physiol. Ser.* 40: 327-345, 1986.

43. J. Walleczek, ''Electromagnetic field effects on cells of the immune system: The role of calcium signaling,'' *FASEB J.* (in press).

44. J. Lechleiter, S. Girard, E. Peralta, and D. Clapham, ''Spiral calcium wave propagation and annihilation in *Xenopus laevis* oocytes,'' *Science* 252: 123-127, 1991.

45. W. B. Busa, J. E. Ferguson, S. K. Joseph, J. R. Williamson, and R. J. Nuccitelli, ''Activation of frog (*Xenopus laevis*) eggs by inositol phosphate: I. Characterization of Ca^{2+} release from intracellular stores,'' *J. Cell. Biol.* 101: 677-682, 1985.

46. M. J. Berridge, P. H. Cobbold, and K. S. R. Cuthbertson, ''Spatial and temporal aspects of cell signalling,'' *Phil. Trans. Roy. Soc.* 320: 325-343, 1988.

47. J. L. Kirschvink, D. S. Jones, and B. J. McFadden, Eds., *Magnetite Biomineralization and Magnetoreception in Organisms,* New York: Plenum Press, 1985, 682 pp.

48. G. M. Brittenham, D. E. Farrell, and J. W. Harris, ''Magnetic susceptibility measurements of human iron stores,'' *New England J. Med.* 307: 1671-1675, 1982.

49. J. M. Gomori, R. I. Grossman, H. I. Goldberg, R. A. Zimmerman, and L. T. Bilaniuk, ''Intracranial hematomas: Imaging by high-field MR,'' *Radiology* 157: 87-93, 1985.

50. L. E. Larsen and J. H. Jacobi, ''Microwave scattering parameter imagery of an isolated canine kidney,'' *Med. Phys.* 5: 394-403, 1979.

51. J. T. Kelley, R. Everett, E. L. Reilly, and G. S. Colton, ''The relationship between flash evoked potentials and evoked amplitude modulation patterns of an applied UHF electromagnetic field in the rat,'' *Bioelectromagnetics* 5: 365-375, 1984.

52. P. Czerski, Ed., *Biologic Effects and Health Hazards of Microwave Radiation,* Warsaw: Polish Medical Publishers, 1973, 350 pp.

53. S. E. Trullinger, ''Where do we go from here?'' in A. R. Bishop and T. Schneider, Eds., *Solitons and Condensed Matter Physics,* Berlin: Springer-Verlag, 1978, 338-340.

Martin Blank, Ed., *Electricity and Magnetism in Biology and Medicine*

MEMBRANE ELECTROPORATION: TOWARD A MOLECULAR MECHANISM

E. Neumann,* E. Werner, A. Sprafke, and K. Krüger

University of Bielefeld, Germany

Plasmid DNA exhibits a large and rapid counterion dissociation field effect (0.1-1 μs) followed by a slower conductance decrease (5 μs) concomitant with a partial uncoiling. Turbidity dichroism and linear dichroism of the membrane probe DPH, in combination with conductivity relaxations of NaCl (0.2 M) filled lipid vesicles ($\phi \approx 320$ nm) in 0.2 mM NaCl iso-osmolar sucrose solution indicate that membrane electroporation is two-phasic. The data are consistent with a rapid transition ($\tau \approx 0.5$ μs) between hydrophobic and hydrophilic, inverted pores followed by a slower rearrangement ($\tau \approx 5$ μs) of the pore-wall lipids to a crater-like electropore of minimum mean radius $\langle r \rangle = 0.4 \pm 0.04$ nm at $E = 25$ kV cm^{-1}.

In recent years, membrane electroporation methods have gained increasing importance in cell biology, biotechnology, and medicine; for reviews see Sowers (1987) and Neumann et al. (1989). Specifically, the electroporation techniques are used to manipulate biological cells and organelles, cell aggregates, and tissue. The various applications include the direct transfer of genes or other nucleic acids and of proteins into the cell interior, as well as the electro-release of cellular components into the medium. Further applications are electrofusion to produce hybridoma cells or insertion of foreign glycoproteins into the membrane of blood organelles. The applied field strengths are $E = 2\text{-}4$ kV cm^{-1} and the pulse durations are $\Delta t = 1$ μs to 50 ms.

Evidence suggests that it is probably the lipid part of the biological membrane which is transiently permeabilized by an electroporation pulse (Teissié and Tsong, 1981; Chernomordik et al., 1987). Therefore, lipid bilayer vesicles may be used as a model system not only to study the molecular mechanisms of lipid rearrangements during membrane electroporation, but also to analyze the electrically mediated transport of DNA through cell membranes and to elucidate the molecular processes of membrane electrofusion and of electroinsertion phenomena. Indeed, recent electro-optic and conductometric studies on plasmid DNA and on bilayer membrane vesicles of charged lipids show that both polyelectrolyte systems display major structural changes under electroporation conditions. The lipid-vesicle data are consistent with electrically induced transitions from hydrophobic (HO) aqueous pores to hydrophilic, inverted (HI) electropores that can form crater-like pore walls. The optical and conductometric properties of pDNA are found to be very similar to short linear DNA fragments, for instance 262 bp (Müller et al., 1989). At very low ionic strengths, the pDNA Na^+ salt dissolved in distilled water shows extremely large relative conductance changes (up to 500%), involving a pronounced negative Wien effect accompanied by base-pair unstacking, partial pDNA uncoiling (Werner, 1990), and finally, at pulse durations $\Delta t > 10$ μs, aggregate formation (Sabelnikov and Cymbalyuk, 1990) that may reduce the efficiency of gene transfer by electroporation.

TABLE 1.—Electroporation phenomena.

I. Primary processes
   1. Interfacial ionic polarization of cells (Maxwell-Wagner)
   2. Electric pore formation, phase transition-like lipid rearrangements:
      Hydrophobic pore → hydrophilic pore → crater-like electropore

II. Secondary processes
   1. Transmembrane material transport (electro-release, electro-uptake)
   2. Fusion of electroporated membranes (electrofusion)
   3. Insertion of (glyco)proteins into electroporated cell membranes (electroinsertion)

---

*Correspondence to: E. Neumann (Chemistry), University of Bielefeld, P.O. Box 10 01 31, D-4800 Bielefeld, Germany.

STRUCTURAL CHANGES IN PLASMID DNA

Similar to linear DNA fragments (Müller et al., 1989), the circular supercoiled plasmid DNA (pDNA, 2000 bp) also exhibits large conductance changes in electric fields. In the low field strength range ($E \leq 20\,\text{kV cm}^{-1}$) there is only one rapid conductance relaxation, which indicates counterion dissociation of pDNA. In the same time range the optical signals show negative absorbance dichroism: $\Delta A / A_0 < 0$, indicating rotational movements of parts of the pDNA superhelix into the direction of the applied field. Since there is no rapid isotropic absorbance mode, ΔA^{iso} (260 nm) = 0 in the low-field range, no base-pair unstacking or other conformation changes affecting the chromophores have occurred.

The analysis of the Wien effects causing a conductance increase shows that the slow phase (II), which apparently saturates at $E = 20\,\text{kV cm}^{-1}$ for $[Na^+]$ = 0.2 mM, is in the same time range as the negative dichroism $\Delta A(I)/A_0$ and thus contains conductance contributions due to rotational movements of pDNA. The first, rapid phase (I) exhibits a linear field-strength dependence and may be analyzed in terms of the Onsager field dissociation effect, which yields an effective charge of the polyionic pDNA system of $z \approx -6$; for details see Müller et al. (1989).

Again, similar to the 262 bp rod, all amplitudes of the various conductometric and optical relaxations decrease with increasing amounts of added NaCl up to 1 mM. Therefore, the pDNA superhelix may be also considered as a flexible overall rod-like system which is oriented by counterion flow polarization like the small linear DNA fragments.

ELECTROPORE FORMATION IN LIPID VESICLES

The electro-optic data of the lipid bilayer vesicles ($\phi \approx 320$ nm), filled with 0.2 M NaCl and suspended in 0.2 mM NaCl solution, indicate that at sufficiently large field strengths ($E > E_c(\text{II}) \approx 25\,\text{kV cm}^{-1}$) there are two relaxation modes that are related to structural changes of the membrane phase. In the low-field range $[E < E_c(\text{II})]$ there is only one optical relaxation mode (I). The anisotropic turbidity mode $\Delta T(\text{I})$, the DPH absorbance dichroism mode $\Delta A(\text{I})$, as well as the isotropic modes $\Delta T^{iso}(\text{I})$ and $\Delta A^{iso}(\text{I})$, all have a sigmoidal onset and relax with similar time constants. The sigmoidal onset is consistent with the initial displacement current phase P, which reflects the ionic interfacial charging of the vesicles in the direction of E; see, for instance, Neumann et al. (1989). Since the charging by a rectangular electric field builds up exponentially, the membrane processes immediately coupled to the charging have a sigmoidal onset.

In the high-field range $[E \geq E_c(\text{II})]$, the amplitudes of the OD-dichroism phases (II) are opposite in sign to those of the phases (I). The isotropic modes $\Delta(\text{OD})^{iso}$ are also two-phasic. It is remarkable that the conductivity relaxation phase (II) becomes visible at the same field strength $E_c(\text{II})$ as the optical phases (II). The dependence of the critical field strength $E_c(\text{II})$ on the concentration of NaCl inside the vesicles suggests that the conductance mode (II) reflects the release of internal salt through the electroporated pole caps of the vesicles.

The time constants of the optical changes are the same within the margin of error over the entire field-strength range. Furthermore, if the actual amplitudes are referred to their saturation values at high fields, all data points fall on the same curve for mode (I) and mode (II), respectively. It thus appears that the different optical signal changes are due to the same two membrane processes. The reaction scheme therefore comprises at least two steps and three lipid states:

$$L_1 \overset{(\text{I})}{\rightleftharpoons} L_2 \overset{(\text{II})}{\rightleftharpoons} L_3 \tag{1}$$

The signs of the DPH absorbance dichroism $\Delta A(\text{I}) < 0$ and of the turbidity dichroism of the vesicles $\Delta T(\text{I}) > 0$ are consistent with rotational movements of DPH molecules taken along with the lipid molecules. The finite values of $\Delta A^{iso}(\text{I})$ and $\Delta T^{iso}(\text{I})$ indicate that the structural change has changed the immediate environment of the DPH in the membrane and of the scattering centers. A likely candidate for the isotropic contributions is the entrance of more water molecules into the membrane phase.

Taken together, the conductometric and electro-optic data of the phase (I) are consistent with the Litster-Chizmadzhev model of a lipid phase transition from hydrophobic (HO) pores to hydrophilic (HI) pores. Thus the step $L_1 \rightleftharpoons L_2$ in scheme (1) is specified to HO $\rightleftharpoons$ HI. The data of the field-strength dependence of $\Delta A^{iso}(\text{I})$ cannot at present differentiate whether the HO $\rightleftharpoons$ HI transition is smooth or whether there is a threshold value $E_c(\text{I})$ for the onset of the HI electropore formation.

If the threshold model is applied, we obtain $E_c(\mathrm{I}) = 8.3$ kV cm^{-1}. The respective transmembrane voltage can be calculated from the transmembrane potential difference $\Delta\psi_c$, caused by interfacial polarization:

$$\Delta\psi_c = -1.5 E_c(a + l_D) \qquad (2)$$

where $a = \phi/2 \approx 160$ nm is the mean radius of the vesicles and $l_D \approx 20$ nm is the Debye length at 0.2 mM NaCl. Therefore $\Delta\psi_c(\mathrm{I}) \approx -0.23$ V.

The analysis of the $\Delta A^{iso}(\mathrm{I})$ data yields a mean pore radius $\langle r \rangle = 0.4 \pm 0.04$ nm for the range $E \leq 25$ kV cm^{-1}. At $E_c(\mathrm{II}) \approx 25$ kV cm^{-1} there is the onset of the optical relaxation phases (II) and the onset of the drastic conductivity increase. The step (II) in scheme (1) can be modeled by a transition from a normal HI-electropore to a crater-like pore wall. According to Eq. (2), $E_c(\mathrm{II})$ yields the critical transmembrane potential difference for the onset of the crater formation: $\Delta\psi_c(\mathrm{II}) = -0.68$ V. Both values $\Delta\psi_c(\mathrm{I})$ and $\Delta\psi_c(\mathrm{II})$ are independent of the vesicle size and of l_D, they are thus constants of the system. The mean pore radius of lipid vesicles is comparable with that of planar lipid bilayers (0.3-0.5 nm; see, for instance, Glaser et al., 1988). But unlike planar bilayers, the vesicle electropores do not appreciably conduct Na^+ and Cl^- ions until the normal HI pore forms a crater-like pore wall.

ELECTRODIFFUSION OF DNA AND OF CELLS

Cell surface structures such as a cell wall and the outer side of the plasma membrane usually carry a net negative charge and are therefore surrounded by a net positively charged counterion atmosphere. Rodlike cells such as bacteria are similarly flow-polarized as DNA and orient themselves with their long axis in the direction of the external field, provided the pulse duration is long enough compared to the rotational time constant.

As a whole, cell surfaces at pH 7 are normally polyanionic. Elongated cells such as bacteria and DNA not only orient their long axis in the direction of the external field; as polyanions, they move electrophoretically in a drift direction opposite to the direction of the external field vector. It is known that the mobility of DNA, $u(\mathrm{D}) \approx 1.5 (\pm\ 0.5) \times 10^{-4}$ cm^2 V^{-1} s^{-1}, is independent of the chain length. Generally, the electric mobility tensor u_i of (geometrically anisotropic) particles depends on the surface charge density and on particle size (as well as on the direction of the longest particle axis relative to E). Cell surface structures are usually less charged than DNA and cells are larger than DNA; therefore $u(\mathrm{D}) > u(\mathrm{C})$. As a consequence, in the presence of a field E, DNA moves faster than larger cells. Hence it is conceivable that DNA is electrophoretically inserted in, and drawn through, the electroporated membrane cap facing the cathodic electrode of the measuring chamber. Electroporative bacterium transformation requires large pulse durations. The reason is most likely the network structure of the porous cell wall. Externally applied DNA must first (electro)diffusively migrate through the wall structures (equivalent to a higher solution viscosity) before entering the plasma membrane.

In conclusion, the combination of electro-optic and conductometric relaxation spectrometries is a powerful technique for the study of molecular details of macromolecules and membrane systems in particular and of electroporation phenomena in general.

ACKNOWLEDGMENT

We gratefully acknowledge grants Ne 227/6-1 and Ne 227/9-1 from the Deutsche Forschungsgemeinschaft to E.N.

REFERENCES are cited in:

E. Neumann, E. Werner, A. Sprafke, and K. Krüger, in B. R. Jennings and S. P. Stoylov, Eds., *1992 Colloid and Molecular Electrooptics,* Bristol: IOP Publishers, 197-206.

E. Neumann, *Bioelectrochem. Bioenerg.* 28: 247-267, 1992.

Martin Blank, Ed., *Electricity and Magnetism in Biology and Medicine*

ELECTROINSERTION OF PROTEINS INTO MEMBRANES: A NOVEL APPROACH TO THE STUDY OF MEMBRANE RECEPTORS

Y. Mouneimne,* P.-F. Tosi, and C. Nicolau

Harvard University Medical School, Boston, Mass.

Electroinsertion is the implantation of receptors in the cell's plasma membrane by use of electrical pulses. This technique was used in the insertion of the CD4 receptor and the human glycophorin A in erythrocytes, various types of nucleated cells, and large liposome membranes. Human red blood cells bearing CD4 on their membranes were able to internalize HIV-1 in vitro and compete with T4 cells for HIV attachment. In circulation, CD4 exposed by RBC subjected to electroinsertion has the RBC's normal life span in various species. RBC-CD4 may thus provide a useful therapeutic agent against AIDS as suggested by in vitro results and phase I clinical trial observations on AIDS patients.

Insertion of membrane proteins into red blood cells (RBC), nucleated cells, or liposomes membrane was achieved by use of a pulsed electric field. Electroinsertion occurs at field strength lower than the critical field E_c for electroporation,[1-5] with pulse duration up to milliseconds.[6] The field intensity is mostly determined by the size of the cells or liposomes.[7] The amount of inserted proteins per cell depends on (among other factors) the protein concentration. The method has been described.[8] Briefly, after washing the cells in 10 mM phosphate buffered saline (ph 8.8), the proteins and cells are mixed and exposed to four pulses at 37 C at 15min intervals between pulses. Cells are incubated for 1 h at 37 C and then washed with 5 mM phosphate buffered saline (pH 7.4). The field intensity E was 1.3 kV/cm for human, 1.5 kV/cm for piglet, 1.6 kV/cm for rabbit, and 2.1 kV/cm for mouse RBC. For nucleated cells, E was 700 V/cm for CEM-CM3 cells, 500 V/cm for Hela S3 cells, and 670 V/cm for $CD8^+$ T cells. For liposomes with an average radius of 2 μm, E was 1.6 kV/cm.

ELECTROINSERTION OF PROTEINS INTO THE RED BLOOD CELLS MEMBRANE

Insertion of proteins into erythrocytes membrane provides long-lived circulating proteins. Electroinserted human glycophorin bound two monoclonal antibodies, 10F7 and D2 10, specific for two different epitopes on the molecule.[9] Up to 10000 human glycophorin molecules were inserted per mouse RBC, as detected by cytofluorimetry. Full-length CD4 electroinserted in RBC membranes reacted specifically with Leu 3a, OKT4A, OKT4C, OKT4D, and BL4/10T4;[10] these monoclonal antibodies are specific of different epitopes on the CD4 molecule.[11] Up to 5000 CD4 molecule per human RBC were detected cytofluorimetrically, after electroinsertion, by means of stained monoclonal antibodies. This finding indicates that the immunological integrity of the inserted proteins was preserved. Human erythrocytes bearing inserted CD4 were able to internalize HIV-1 in vitro, as shown by a fluorescence dequenching technique. The pre-incubation of HIV-1 with RBC-CD4 reduced the infection of target cells by HIV-1 by more than 90%, as detected by viral reverse transcriptase and the amount of P24 antigen produced by the target cells.[12]

Red blood cells subjected to electroinsertion retained normal hematological parameters (Table 1). Mouse RBC inserted with human glycophorin had a normal half-life span of 12 days.[9] Pig RBC-CD4 had a half-life span of 14 days; human RBC-CD4, of 50 days (manuscript in preparation). The normal life span of the erythrocytes suggests that no major damage was induced by the electroinsertion technique.

Neither inserted human glycophorin in mouse erythrocyte[9] nor inserted CD4 in pig erythrocyte induced an immune response in vivo, but the same amount of proteins injected as a solution triggered a humoral immune response.

---

*Correspondence to: Y. Mouneimne, Center for Blood Research Labs., Harvard University Medical School, Boston MA 02115.

TABLE 1.—Hematological parameters of human RBC after electroinsertion of CD4.

	RBC (control)	RBC-CD4
Mean corpuscular Hb pg (MCH)	25.3	25.3
Mean corpuscular volume fl (MCV)	78.2	79.2
Mean corpuscular hemoglobin Concentration g/dl (MCHC)	32.3	31.9
P_{50} (Hb half saturation pressure, Torr)	25	25.5

ELECTROINSERTION OF PROTEINS IN THE PLASMA MEMBRANE OF NUCLEATED CELLS

The electroinsertion of foreign membrane proteins into the plasma membranes of nucleated cells yielded immunologically active epitopes inserted into viable cells.[13] This technique may be used to study the protein mobility and fate in natural membranes.

The survival rate of the cells subjected to electroinsertion was between 55 and 70%, depending on the cell type. Up to 1.4×10^5 glycophorin molecules were inserted per CEM CM3 cell, 2.4×10^5 per Hela S3 cell, and 2.2×10^4 molecule per bovine $CD8^+$ T cell.[13] The cells cultured after electroinsertion lost electroinserted glycophorin through a temperature and cell-type dependent mechanism with two different rates. For 2 h after electroinsertion, at 37 C, the loss of the inserted glycophorin was 12.5%/h for the $CD8^+$ T cells, 6.5%/h for CEM CM3 cells, and only 0.8%/h for the Hela S3 cells. After 2 h the rate increased substantially, up to 41.7%/h for the $CD8^+$ T cells, 13.5%/h for CEM CM3 cells, and 8.9%/h for the Hela S3 cells. Cytocholasin D, a partial inhibitor of endocytosis,[14] efficiently inhibited the disappearance of electroinserted glycophorin during the first 2 h after electroinsertion. Cells maintained at 4 C after the electroinsertion lost little of the inserted glycophorin, but viability dropped to 50% or below after 8 h.

These preliminary results show the feasibility of inserting proteins into the plasma membrane of nucleated cells as a basis for further studies.

ELECTROINSERTION OF PROTEINS INTO LIPOSOMES

Reconstitution of membrane proteins into liposomes has been achieved by various methods, such as reverse-phase evaporation[15] or detergent removal.[16] Considerable loss of proteins due to encapsulation inside the liposome during these reconstitutions can occur and the liposome size may influence the proportion of "correctly" oriented proteins in the liposome membrane. Electroinsertion yields 70% of the inserted proteins in the "correct" orientation;[8] almost no loss of protein due to encapsulation inside the liposomes is observed. Electroinsertion of proteins into liposomes might be used for the study of lipid-protein interaction, or to associate antigens with powerful adjuvants (manuscript in preparation). We observed that the insertion of human glycophorin into liposomes depends on the lipid composition. We tried various lipid composition and observed good yields of insertion in 1:0.63 phosphatidylcholine-cholesterol liposomes. Up to 6500 human glycophorin molecules were detected per liposome (4 μm average diameter).[17]

REFERENCES

1. K. Kinosita Jr. and T. Y. Tsong, *Biochim. Biophys. Acta* 47: 227-242, 1977.

2. R. W. Glaser, S. L. Leiken, L. V. Chernomordik, V. F. Pastushenko, and A. I. Sokirko, *Biochim. Biophys. Acta* 940: 275-287, 1988.

3. R. Benz, F. Beckers, and U. Zimmermann, *J. Membrane Biol.* 48: 181-204, 1979.

4. E. Neumann, in E. Neumann, A. E. Sowers, and C. A. Jordan, Eds., *Electroporation and Electrofusion in Cell Biology,* New York: Plenum Press, 1989, 61-82.

5. J. C. Weaver and A. Barnett, in D. C. Chang, B. M. Chassy, J. A. Sanders, and A. E. Sowers, *Guide to Electroporation and Electrofusion,* New York: Academic Press, 1992, 91-117.

6. Y. Mouneimne, P. F. Tosi, Y. Gazitt, and C. Nicolau, *Biochem. Biophys. Res. Commun.* 9: 34-40, 1989.

7. H. J. Fricke, *Appl. Phys.* 24: 644-646, 1953.

8. Y. Mouneimne, P. F. Tosi, R. Barhoumi, and C. Nicolau, in Ref. 5, 327-346.
9. Y. Mouneimne, P. F. Tosi, R. Barhoumi, and C. Nicolau, *Biochim. Biophys. Acta* 1066: 83-89, 1991.
10. Y. Mouneimne, P. F. Tosi, R. Barhoumi, and C. Nicolau, *Biochim. Biophys. Acta* 1027: 53-58, 1990.
11. P. F. Tosi, H. Benoist, and C. Nicolau, *J. Fluorescence* 1: 141-148, 1991.
12. M. Zeira, P. F. Tosi, Y. Mouneimne, J. Lazarte, L. Sneed, D. J. Volsky, and C. Nicolau, *Proc. Natl. Acad. Sci. USA* 88: 4409-4413, 1991.
13. Y. Mouneimne, W. C. Brown, C. Nicolau, and P. F. Tosi, 1992 (submitted for publication).
14. K. Sandrig and B. Von Deuss, *J. Biol. Chem.* 205: 6382-6388.
15. J. L. Rigaud, A. Bluzat, and M. Buschleen, *Biochem. Biophys. Res. Commun.* 111: 373-382, 1983.
16. J. Philippot, S. Mutaftschiev, and J. P. Liautard, *Biochim. Biophys. Acta* 734: 137-143, 1983.
17. We thank Dr. C. V. Black for preparing the liposomes.

Martin Blank, Ed., *Electricity and Magnetism in Biology and Medicine*

INCORPORATION OF INDIVIDUAL CELLS INTO INTACT TISSUE BY ELECTROFUSION

R. Heller*

University of South Florida, Tampa, Fla.

Model systems for the study of species-specific diseases can be created by the transfer of membrane components from susceptible cells to intact tissue of naturally resistant animals. Cell-tissue electrofusion (CTE) protocols have been used to incorporate individual human cells into intact rabbit corneal epithelial tissue both in vitro and in vivo. In order to develop additional biomedical applications, it is important to examine the characteristics of the fusion process. Thus, this study was initiated to establish the time course of the cellular incorporation and to examine some of the parameters that control the success or failure of the fusion process.

Cell fusion can be accomplished by exposure of juxtaposed cell membranes to electrical fields.[1,2] Senda et al. were the first to describe the fusion of two cells by electrofusion.[3] Subsequently, other investigators reported similar results with animal cells[4,5] and yeast cells.[6,7] Electrofusion has been used for many applications, including formation of hybridomas,[8-11] production of monoclonal antibodies,[12,13] study of membrane fusion mechanisms,[14-16] and examination of membrane interaction between two cells or within a single cell.[17-19] Cell-tissue electrofusion (CTE) represents an area of electrofusion that can be utilized to incorporate individual cells into intact tissue.[20] CTE has been employed both in vitro and in vivo for the inter-species transfer of surface membrane components.[21]

CTE has already been demonstrated to be a valuable tool in the creation of model systems for the study of species-specific diseases. It is believed that CTE will prove useful in several other biomedical applications, including site specific delivery systems, wound healing, skin grafts, and various surgical procedures. The initial step in the development of these and other applications is the ability to characterize the fusion process and to begin to examine the relationship between the various fusion parameters. This paper summarizes the work initiated to obtain this goal.

RESULTS AND DISCUSSION

The ability to increase the fusion yield in CTE and to obtain a reproducible number of cells fused depends on maintaining contact between the individual cells and the tissue. For this purpose Zeta-Probe (ZP) disks (electrostatically charged membranes) can be used to place the cells. One can layer a reproducible number of cells onto these disks by centrifuging for 5 min at 180 G.[22] CTE experiments were performed by placing the disk, cell side down, onto excised rabbit corneas and administering one to three 20μs, 0.6 kV/cm square-wave pulses. The corneas were washed to remove unfused cells, fixed in 2.5% glutaraldehyde, and examined by scanning electron microscopy (SEM). This procedure was used successfully to increase the number of cells fused (Fig. 1) when compared to previous yields.[20] This procedure allows control over the number of cells placed in contact and maintains consistency between samples.

A spectrofluorometric assay for the quantitation of electrofusion (FAQE) was developed and then used to demonstrate the reproducibility of the fusion process.[23] The mean number of cells fused, in five experiments, was determined to be approximately 5000. The variation was minimal and all the experimental corneas had the same number of fused cells ±10%. These results demonstrate two main points: (1) this assay could now be used to analyze quantitatively various fusion parameters; and (2) several corneas could be prepared, each with approximately the same number of cells fused to their surface.

---

*Correspondence to: R. Heller (Surgery, MDC Box 16), University of South Florida College of Medicine, 12901 Downs Blvd., Tampa FL 33612-4799.

CTE procedures use mechanical force to juxtapose the cells prior to initiation of the fusion process. FAQE was used to examine the effect of various amounts of mechanical force on the number of cells fused. We measured the amount of force applied to the system in grams/electrode diameter by placing the fusion chamber on a torsion balance. ZP disks containing HL60 cells were placed cell side down onto rabbit corneas. By use of a 2mm-diameter electrode, between 25 and 400 g of force was applied to the cornea through the ZP disk. A single 20μs, 0.6kV/cm square-wave pulse was administered. The corneas were washed, fixed in 2.5% glutaraldehyde, and examined by SEM. The results of these experiments indicate that a threshold is reached at 100 g (Table 1). The best results of fused cells was obtained at this level. Although a higher yield was obtained at 400 g, there was also a large amount of tissue damage. These results established a set of standard parameters to obtain reproducible CTE as follows: cells on a ZP disk are placed on the tissue and a single 20μs, 0.6kV/cm square-wave pulse is administered through a flat, 2mm-diameter circular electrode while 100 g of force is applied.[24]

TABLE 1.—Effect of mechanical pressure on cell-tissue electrofusion.

Mechanical pressure (g)	# Cells fused (s.d.)
25	877 (547)
100	4693 (250)
200	5009 (220)
400	6649* (1138)

Experiments were performed in triplicate.
*damage to corneal tissue.

Several corneas were prepared at the standard CTE parameters in order to examine the time frame in which the cells were incorporated into the tissue. The corneas were then fixed with 2.5% glutaraldehyde at various time intervals ranging from 3 to 20 min. Examination by SEM revealed that the cells were gradually incorporated into the tissue (Fig. 2). The entire fusion process was completed within 20 min after the initiation of the fusion process.

SUMMARY

Currently, experimental procedures are being worked out to use FAQE to examine other important fusion parameters. Electromechanical parameters of interest include pulse width and amplitude, number of pulses, and number of cells placed in contact with the tissue. The interaction of these and other fusion parameters are being investigated to determine the set of fusion conditions that will yield an optimal number of fused cells.

REFERENCES

1. G. Bates, J. Saunders, and A. E. Sowers, in A. E. Sowers, Ed., *Cell Fusion* R, New York: Plenum Press, 1987, 367-395.

2. U. Zimmermann and J. Vienken, *J. Membrane Biol.* 67: 165-182, 1982.

3. M. Senda, J. Takeda, S. Abe, and T. Nakamura, *Plant Cell Physiology* 20: 1441-1443, 1979.

4. E. Neumann, G. Gerisch, and K. Opatz, *Naturwissenschaften 67: 414-415, 1980.*

5. H.-P. Richter, P. Scheurich, and U. Zimmermann, *Dev. Growth Diff.* 23: 479-486, 1981.

6. H. Weber, W. Forster, H. Berg, and H.-E. Jacob, *Curr. Genet.* 4: 165-166, 1981.

7. H. Weber, W. Forster, H.-E. Jacob, and H. Berg, *Allg. Mikrobiol.* 21: 555-562, 1981.

8. M. Glassy, *Nature* 333: 579-580, 1988.

9. M. C. Glassy and M. Pratt, in E. Neumann, A. E. Sowers, and C. A. Jordan, Eds., *Electroporation and Electrofusion in Cell Biology*, New York: Plenum Press, 1989, 271-282.

10. D. R. Hewish and J. A. Werkmeister, *J. Immunol. Meth.* 120: 285-289, 1989.

11. J. White, M. Blackman, J. Bill, J. Kappler, P. Marrack, D. P. Gold, and W. Born, *J. Immunol.* 143: 1822-1825, 1989.

12. S. H. K. Foung and S. Perkins, *J. Immunol. Meth.* 116: 117-122, 1989.

13. M. M. S. Lo, T. Y. Tsong, M. K. Conrad, S. M. Strittmatter, L. D. Hester, and S. H. Snyder, *Nature* 310: 792-794, 1984.

14. A. E. Sowers, *Biophys. J.* 52: 1015-1020, 1987.

15. A. E. Sowers, *ibid.*, 54: 619-626, 1988.

16. A. E. Sowers, *Biochim. Biophys. Acta* 985: 339-342, 1989.
17. P. M. Frederik, M. C. A. Stuart, and A. J. Verkleij, *ibid.*, 979: 275-278, 1989.
18. H. Oberleithner, *NIPS* 6: 181-184, 1991.
19. A. E. Sowers, *Biochim. Biophys. Acta* 1025: 247-251, 1990.
20. R. J. Grasso, R. Heller, J. C. Cooley, and E. M. Haller, *ibid.*, 980: 914, 1989.
21. R. Heller and R. J. Grasso, *ibid.*, 1024: 185-188, 1989.
22. R. Heller and R. J. Grasso, *J. Tiss. Cult. Meth.* 13: 25-30, 1991.
23. R. Heller, *Analytical Biochemistry* 202: 286-292, 1992.
24. R. Heller and R. Gilbert, in D. C. Chang, B. M. Chassy, J. A. Saunders, and A. E. Sowers, Eds., *Guide to Electroporation and Electrofusion*, New York: Academic Press, 1992, 393-410.

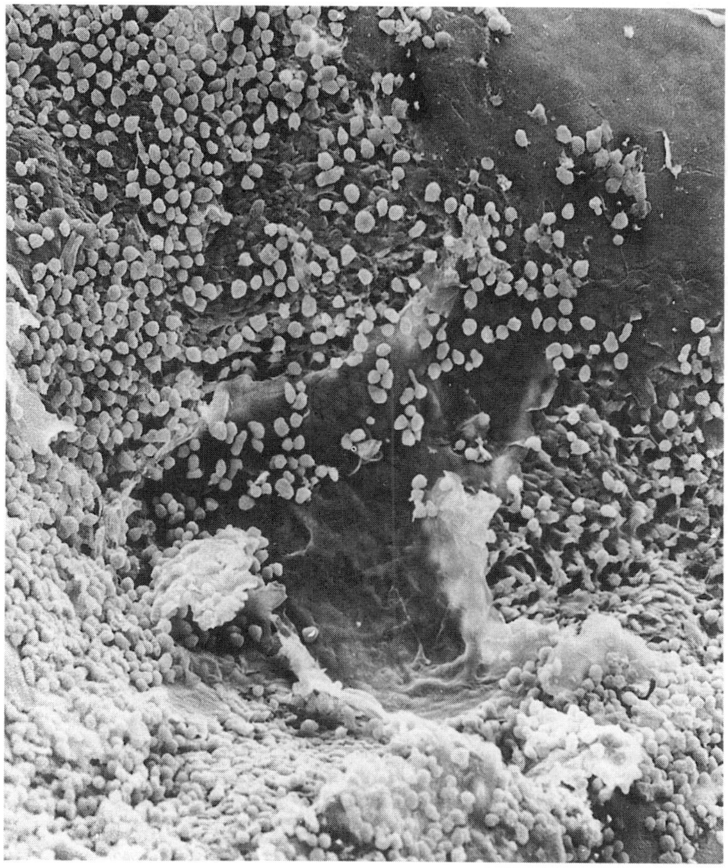

FIG. 1.—Scanning electron photomicrograph illustrating increased number of cells electrofused to corneal surface: electrofusion of human HL 60 lymphoma cells directly to tissue-immobilized rabbit cells within superficial corneal epithelium. Corneal surface shows thousands of HL60 cells fused to superficial layer of rabbit corneal epithelium (200×).

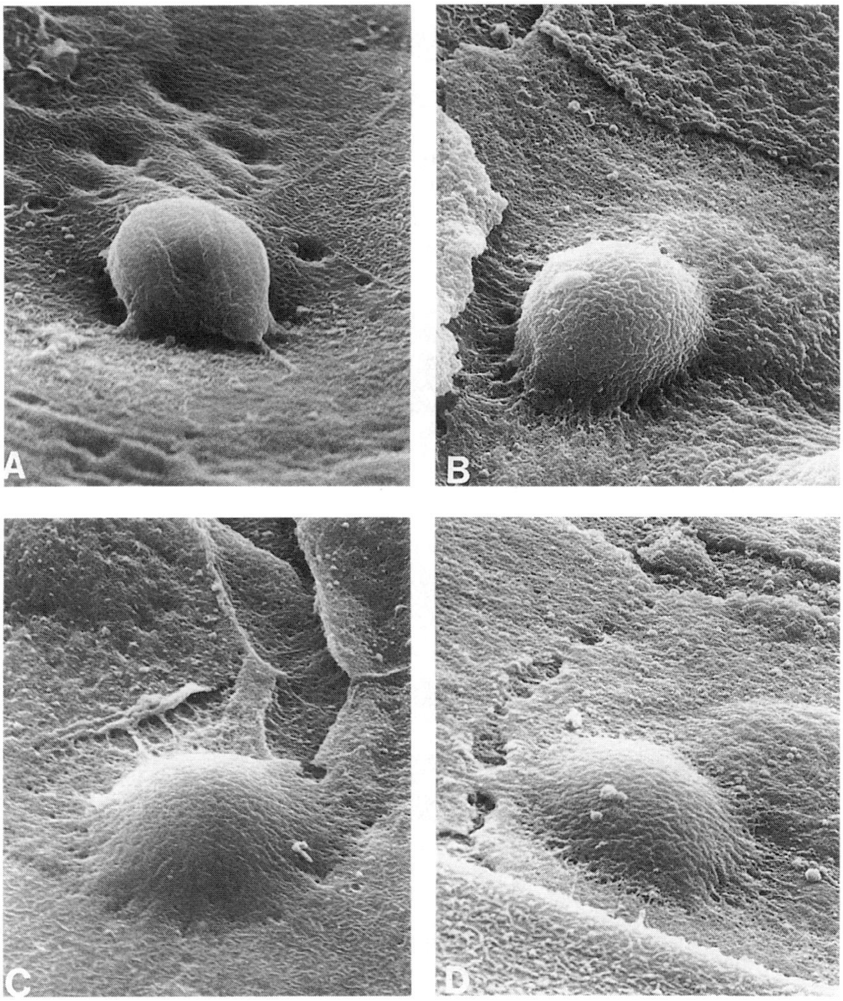

FIG. 2.—Scanning electron photomicrograph showing time course of incorporation of individual cells into intact tissue, illustrating various stages of incorporation of individual HL60 cells into intact rabbit corneal epithelium. A: 3 min postfusion (4400×); B: 6 min (4800×); C: 9 min (4800×); D: 12 min (4800×).

Martin Blank, Ed., *Electricity and Magnetism in Biology and Medicine*

ELECTROCHEMOTHERAPY: A NEW ANTITUMOR TREATMENT USING ELECTRIC PULSES

L. M. Mir,* S. Orlowski, J. Belehradek Jr., B. Poddevin, G. Pron, C. Domenge, M. Belehradek, B. Luboinski, and C. Paoletti

Institut Gustave-Roussy, Villejuif, France

Electrochemotherapy (ECT) is a new antitumor treatment that consists in delivering appropriate electric pulses (EP) to the tumor a few minutes after an intravenous injection of bleomycin (BLM). It achieves a targeting of BLM cytotoxic effects to the tumor exposed to EP. The basic finding that led us to ECT was the huge increase of BLM cytotoxicity after in vitro cell electropermeabilization (EPN). The bases of ECT efficacy both at the cell and the organism levels are reported and discussed. The first clinical trial shows that ECT is well tolerated by the patients. Net antitumor effects were observed.

Since 1986-87 we have been interested in the pharmacological applications of the electropermeabilization (EPN) of living cells by 3 to 8 square-wave electric pulses (EP) at 1000-1700 V/cm, 100 μs long, and pulse repetition frequency of 1 Hz. The possibilities of such an approach have been explored by use of the uptake of a highly fluorescent, nonpermeant, low-molecular-weight compound, Lucifer Yellow (LY; 457 Daltons); nonpermeant means that the molecule is unable to cross the plasma membrane. The qualitative analysis of this uptake provided us with a rapid assay for determination of the degree of EPN and viability of the pulsed cell population. The quantitative measurement of the uptake revealed that the internalized concentration can equilibrate with the external concentration of LY at the time of EPN.[1]

We have also introduced into the cells molecules of pharmacological or biological interest, such as oligonucleotides,[2] vegetal toxins,[3] antibodies,[4], or DNA.[5,6]

We also electropermeabilized the cells in the presence of various anticancer drugs and analyzed the increase in cytotoxicity due to the facilitated transmembrane crossing. No increase in cytotoxicity was obtained with drugs actively transported across the membrane, and only a slight increase with lipophilic drugs. On the contrary, with bleomycin (BLM), a large (1500 Daltons) hydrophilic drug currently used for cancer treatment in humans, toxicity increased considerably: up to 650000 times at the level required to kill 90% of the treated cells.[3] We prepared a radiolabelled 57-cobalt BLM possessing a very high specific radioactivity, taking advantage of the irreversibility of the chelation complex that BLM forms with the cobalt.[7] We measured the BLM association to electropermeabilized and to nonpermeabilized cells in drug-cells contact conditions identical to those used for the BLM cytotoxicity measurements. It was possible to calculate the BLM electroloading in cells in culture. Indeed, 57-CoBLM association to the electropermeabilized cells is the sum of the 57-CoBLM association to the nonpermeabilized cells and of a second term whose extent is strictly proportional to the extracellular 57-CoBLM concentration used. If we assume a homogeneous distribution of the electroloaded BLM in the cellular volume, the electroloaded internalized CoBLM concentration represents 35-40% of the external BLM concentration. This proportionality allowed us to determine the exact amounts of BLM introduced directly into the cytosol of pulsed cells no matter what the external BLM concentration was. Our results show (1) that as few as some hundreds of BLM molecules internalized in the cell cytosol are sufficient to kill the cell, and (2) that BLM cannot diffuse through the unperturbed plasma membrane. Thus, BLM is an antitumor drug possessing a very high intrinsic cytotoxicity restricted by the plasma membrane of the cell unless permeabilized.[8]

Our current activity is focused on applications of the potentiation of the BLM cytotoxicity after EPN. Indeed, the electric pulses (EP) used in vitro also potentiate BLM cytoxicity in vivo. Subcutaneous transplanted[9] or spontaneously occurring[10] tumors (mean tumor diameter: 6-7 mm) were successfully treated in mice by the combination of BLM and EP; the EP were delivered at the tumor site by external electrodes 30

---

*Correspondence to: L. M. Mir (Biochimie URA147-PRII), Institut Gustave-Roussy, 39 rue C. Desmoulins, F-94805 Villejuif Cédex, France.

min after intramuscular injection of 500 or 50 μg of BLM. Tumor growth arrest was immediately observed, almost always followed by partial regressions and even complete regressions of the treated tumors. Cures were obtained in all experimental models used (B16 melanoma and LPB sarcoma in C57B1/6 mice, KB human carcinoma in nude mice, and spontaneous mammary carcinoma of C3H/Bi mice). Under the same conditions, EP or BLM alone did not alter tumor growth. We now inject BLM intravenously only 3 min before the EP, and with a single combined treatment we cure 57% of the mice bearing LPB sarcomas. We termed this new anticancer approach *Electrochemotherapy* (ECT).[9-11]

We investigated the basis of ECT efficacy. A field intensity above a threshold value was found to be necessary to obtain the ECT benefits; it may correspond to the electric field necessary for the in vivo tissue EPN. Moreover, in order to study the electroloading of BLM in vivo, we injected 57-CoBLM into mice bearing a tumor at their flank. Three minutes later, half the mice received the EP at the tumor site, as in the case of usual ECT, and a blood sample was taken from each mouse to determine its individual 57-CoBLM blood level at that moment. Since 57-CoBLM is noncytotoxic, tumor development was similar in the two groups of mice. Three days later, after the renal and fecal elimination of most of the 57-CoBLM injected, the mice were sacrificed and several organs (including tumors) were sampled. We measured the total radioactivity of each organ as well as its weight, and normalized the radioactivity per mg with respect to the individual initial blood level of 57-CoBLM. No difference in radioactivity retention were detected between the mice receiving and those not receiving the EP for any of the organs, except the tumors. EP-treated tumors retained 4 times more 57-CoBLM (report in preparation). This increase is similar to that obtained for cells in culture exposed to similar 57-CoBLM external concentrations. Thus, the initial massive cell lysis of the tumors treated by ECT seems to be due to the same principle as that demonstrated in vitro on cells in culture.

However, we never permeabilized 100% of the cells in vitro. Furthermore, the treatment of the same murine tumor in nude mice and in immunocompetent syngeneic mice did not result in the same rate of cures achieved: in both cases complete regressions were obtained, but the nude mice all showed recurrencies a few days later. We investigated the possible effect of the host's immune response in the elimination of the residual cells that were not permeabilized and consequently not killed by the BLM. We transiently immunodepressed immunocompetent mice the day before the ECT of their tumors, which dramatically decreased the rate of cures achieved. On the contrary, systemic intraperitoneal injections of interleukine-2 after ECT increased the number of cures.[12] These results demonstrate that the host's immune system participates in the destruction of the residual tumor cells. The optimization of this response is under current investigation.

ECT has already been applied to patients with permeation nodules of head and neck squamous-cell carcinomas. The treated nodules were located in the anterior cervical region and the upper part of the thorax. In the first 10 clinical ECT sessions (8 patients, two of them receiving ECT twice 2-3 weeks apart), only few nodules per patient and only nodules of relatively small size were treated. Patients received only a slight anxiolytic premedication and the treatment was well tolerated. Treatment did not present any danger or toxic effect. In all cases the growth of the treated nodules was at least slowed down compared with that of neighboring nodules not submitted to EP but exposed to the same dose of BLM. Moreover, in 53% of the cases (18 nodules among 34), ECT resulted in the complete disappearance of the treated nodules. This clinical trial is at present prolonged by the treatment of much larger permeation nodules of head and neck squamous-cell carcinomas. Extension of the trial to other tumor localizations will be undertaken in the near future.

In conclusion, we have defined a new antitumor approach that uses electric pulses. The efficacy of Electrochemotherapy is based on (1) that under usual conditions BLM is almost totally unable to enter the cells; and (2) that by means of EP achieving EPN of cells either in culture or in situ, BLM enters the cells and its cytotoxicity is largely potentiated. In mice, cures are regularly obtained; and we have demonstrated that, after an initial massive cell lysis due to the BLM *plus* EP treatment, the host's immune response participates in achieving a cure. The first clinical trial demonstrates the tolerance and the feasibility of this new treatment in human patients.

REFERENCES

1. L. M. Mir, H. Banoun, and C. Paoletti, "Introduction of definite amounts of nonpermeant molecules into living cells after electropermeabilization: Direct access to the cytosol," *Exp. Cell Res.* 175: 15-25, 1988.

2. D. Bazile, L. M. Mir, and C. Paoletti, "Voltage dependent introduction of a d(alpha)octothymidylate into electropermeabilized cells," *Biochem. Biophys. Res. Commun.* 159: 633-639, 1989.

3. S. Orlowski, J. Belehradek Jr., C. Paoletti, and L. M. Mir, "Transient electropermeabilization of cells in culture: Increase of the cytotoxicity of anticancer drugs," *Biochem. Pharmacol.* 37: 4727-4733, 1988.

4. M. R. Casabianca-Pignède, L. M. Mir, J. B. Le Pecq, and A. Jacquemin-Sablon, "Protection against ricin conferred by the introduction of antiricin antibodies into Chinese hamster cells by electropermeabilization," *J. Cell. Pharmacol.* 2: 54-60, 1991.

5. F. Presse, A. Quillet, L. M. Mir, C. Marchiol-Fournigault, J. Feunteun, and D. Fradelizi, "An improved electrotransfection method using square shaped electric impulsions," *Biochem. Biophys. Res. Commun.* 151: 982-990, 1988.

6. S. Emami, L. M. Mir, C. Gespach, and G. Rosselin, "Transfection of fetal rat intestinal epithelial cells by viral oncogenes: Establishment and characterization of the E1A-immortalized SLC-11 cell line," *Proc. Natl. Acad. Sci. USA* 86: 3194-3198, 1989.

7. B. Poddevin, J. Belehradek Jr., and L. M. Mir, "Stable [^{57}Co]-bleomycin complex with a very high specific radioactivity for use at very low concentrations," *Biochem. Biophys. Res. Commun.* 173: 259-264, 1990.

8. B. Poddevin, S. Orlowski, J. Belehradek, Jr., and L. M. Mir, "Very high cytotoxicity of bleomycin introduced into the cytosol of cells in culture," *Biochem. Pharmacol.* 42(S): 67-75, 1991.

9. L. M. Mir, S. Orlowski, J. Belehradek Jr., and C. Paoletti, "Electrochemotherapy: Potentiation of antitumour effect of bleomycin by local electric pulses," *Eur. J. Cancer* 27: 68-72, 1991.

10. J. Belehradek Jr., S. Orlowski, B. Poddevin, C. Paoletti, and L. M. Mir, "Electrochemotherapy of spontaneous mammary tumours in mice," *Eur. J. Cancer* 27: 73-76, 1991.

11. L. M. Mir, M. Belehradek, C. Domenge, S. Orlowski, B. Poddevin, J. Belehradek Jr., G. Schwaab, B. Luboinski, and C. Paoletti, "L'électrochimiothérapie, un nouveau traitement antitumoral: Premier essai clinique," *C. R. Acad. Sci. (Paris)* Sér. III, 313: 613-618, 1991.

12. L. M. Mir, S. Orlowski, B. Poddevin, and J. Belehradek Jr., "Electrochemotherapy tumor treatment is improved by interleukin-2 stimulation of the host's defenses," *Eur. Cytokine Netw.* 3: 331-334, 1992.

Martin Blank, Ed., *Electricity and Magnetism in Biology and Medicine*

TRANSTISSUE MOLECULAR TRANSPORT DUE TO ELECTROPORATION OF SKIN

M. R. Prausnitz, V. G. Bose, R. S. Langer,* and J. C. Weaver

Massachusetts Institute of Technology, Cambridge, Mass.

The creation of transient pores in lipid bilayers by electroporation is a well-recognized phenomenon in artificial and cellular systems. The present study demonstrates, for the first time, what appears to be electroporation of human cadaver skin. We suggest that pores are being created in the intercellular lipids of the stratum corneum. Dramatic increases, of up to four orders of magnitude, in transdermal flux due to electroporation are reported.

Electroporation is believed to be the creation of transient pores universally in lipid bilayer membranes caused by a high transmembrane voltage, typically by the application of short electric pulses. Pores have been created in artificial lipid bilayers as well as in many types of cell membranes of different compositions.[1] Given the proper conditions, these pores remain open after the electric pulse, but eventually close, with recovery found to be a complicated process. Electroporation of intact tissues, as opposed to single cells, has received little attention. We propose and demonstrate that electroporation of lipid bilayers in the skin can be performed and causes greatly increased transdermal transport.

Electroporation of the skin may have important applications to transdermal drug delivery; such delivery offers many advantages over conventional routes of administration (e.g., oral, parenteral).[2] Currently only relatively small and lipophilic drugs, which are effective at low concentrations, can be delivered transdermally at therapeutic levels. Approaches to flux enhancement such as the use of chemical penetration enhancers and iontophoresis have met with limited success. Electroporation could therefore significantly increase the number of drugs capable of being delivered by the transdermal route.

EXPERIMENTAL METHODS

Epidermis is heat separated (60 C, 2 min) from human cadaver skin; loaded into permeation chambers (Fig. 1); exposed to isotonic phosphate-buffered saline (PBS), pH 7.4; and allowed to hydrate fully (12-18 h at 4 C). The temperature is then raised to 37 C and a 10^{-3} M solution of calcein in PBS is added to the donor compartment, while fresh PBS is added to the receptor compartment. Calcein is a highly charged ($z = -4$) fluorescent molecule of molecular weight 623 Da. Once a steady state flux through pre-existing pathways is established (within a few hours), electric pulsing protocols are applied. The receptor compartment is sampled periodically by a complete emptying of its contents and replacement with fresh PBS. Calibrated analysis by spectrofluorimetry allows measurement of receptor compartment calcein concentrations, and thereby calculation of transdermal calcein fluxes.

RESULTS AND DISCUSSION

Representative results of transdermal fluxes before, during, and after pulsing are shown in Fig. 2. Curves A, B, and C represent samples that received one 1 ms pulse every 5 s for 1 h. Curve D shows the flux due to 0.2 V DC applied across an unpulsed skin specimen for 1 h. Curve E is a control, where the skin was not exposed to electric fields at all.

Fluxes before pulsing and for curve E (no-pulsing control) are below the detection limit of 6×10^{-4} µg/cm²h. Fluxes during pulsing (curves A, B, and C) are orders of magnitude higher than fluxes without pulsing; pulsed-sample fluxes return completely or partially to pre-pulsing values within a few hours. Although complete recovery is not always observed, fluxes decrease from enhanced levels by at least two

---

*Correspondence to: R. S. Langer, Harvard-MIT Division of Health Sciences and Technology, MIT, Cambridge MA 02139.

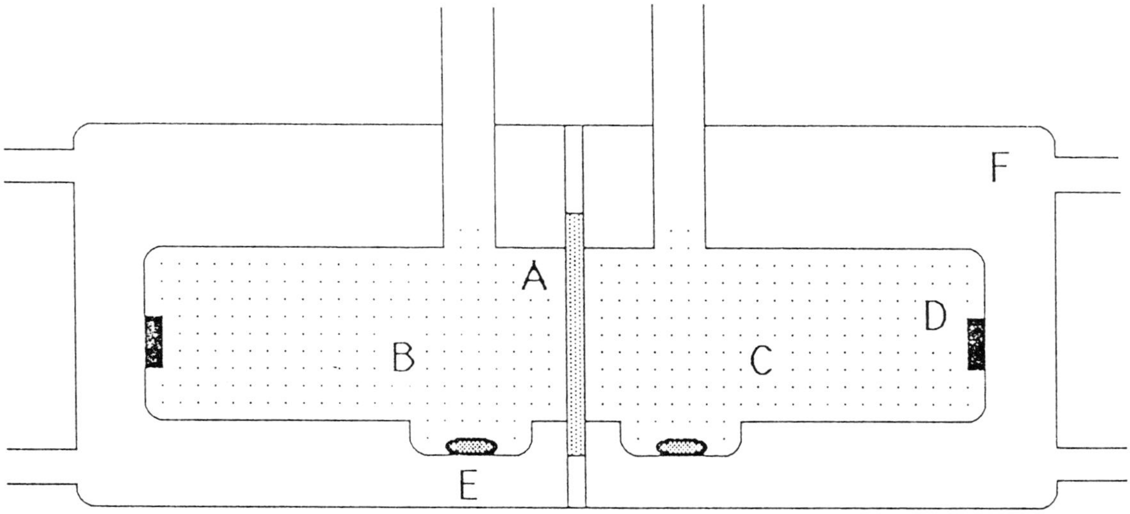

FIG. 1.—Permeation chamber used to measure transdermal fluxes due to electroporation. Human cadaver epidermis (A) separates the donor (B, outer/stratum corneum side) and receptor (C, inner/viable epidermis side) compartments, each filled with PBS. Donor compartment also contains 10^{-3} M calcein. For application of electric pulses, Ag/AgCl electrodes (D) are placed approximately 3 cm from the skin on each side. The chamber contains stir bars (E) to assure good mixing and is water jacketed (F) for temperature control at 37 C.

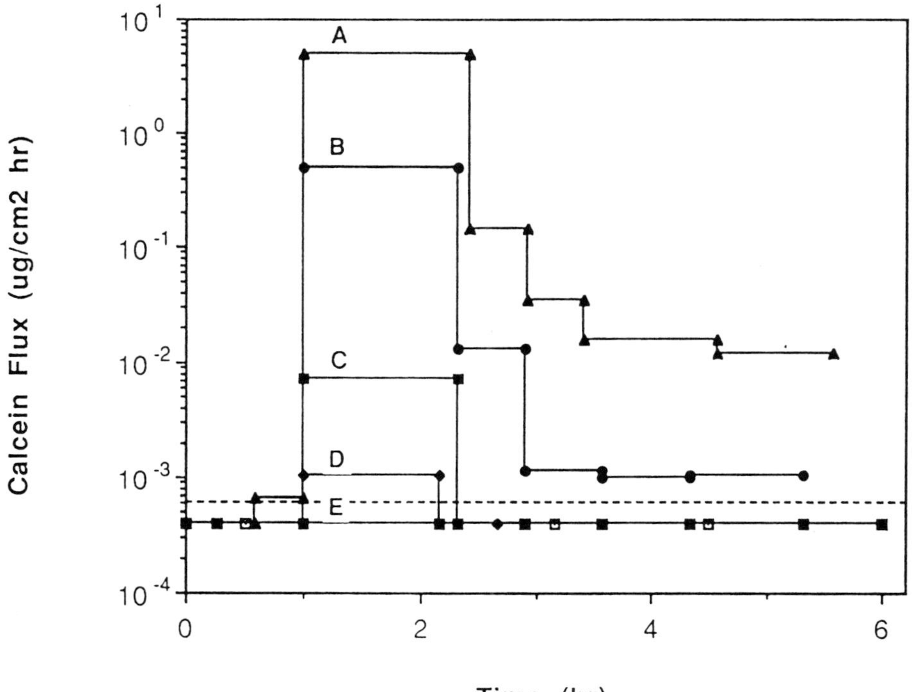

FIG. 2.—Transdermal fluxes of calcein due to exposure to electric fields. Curves A (Δ), B (•), and C (■) represent samples pulsed at 1000, 250, and 150 V, respectively, for 1 h starting at time equal to 1 h. Pulses (exponential-decay pulses with decay time constants of 1.0-1.3 ms) were applied once every 5 s. Curve D (♦) represents a sample exposed to 0.2 V DC for 1 h. Curve E (⊡) is a control, not exposed to electric fields. Voltages represent potential drops between electrodes; drops across skin are 30% of drops between electrodes. Calcein detection limit of 6×10^{-4} μg/cm²h is indicated by dashed line.

orders of magnitude within 2 h, which indicates 99% reversibility. In contrast to this recovery from multiple pulses over hours, electrical measurements on electroporated frog skin (a much simpler system) have been reported and show return within a few minutes after the application of a single pulse.[3] However, reversibility over a time scale of hours is consistent with electroporation behavior often found in cells.[1]

Direct current, at 0.2 V, applied constantly for 1 h is electrically "equivalent" to pulsing at 1000 V. If the skin is unaltered (i.e., if its resistance remains the same), then constant application of 0.2 V would transfer the same amount of charge across the tissue as 1000 V applied at a 1/5000 duty cycle (1 pulse for 1 ms every 5 s). However, if changes in the skin occurred due to the 1000V pulses that allowed more charge across the skin, transdermal calcein transport would probably increase. We have observed difference in flux of well over three orders of magnitude between "equivalent" pulsing and DC conditions. Therefore, electrophoresis alone cannot account for the large fluxes observed due to pulsing.

The stratum corneum, the outermost layer of the skin, has been shown to be the primary barrier to transport across the skin.[2] Most molecules cross the stratum corneum via intercellular routes, through the intercellular lipids. Many domains of these lipids are believed to be organized in bilayer structures,[4] which could be electroporated when subjected to appropriate electrical pulses. We therefore suggest that the dramatic increases in transdermal calcein flux shown in Fig. 2 are due to electroporation of the skin; in particular, electroporation of lipid bilayers located in the intercellular region of the stratum corneum.

In conclusion, we have demonstrated that flux of a small, highly charged molecule across human skin can be increased by at least four orders of magnitude by the application of electrical pulses. These increases cannot be explained by electrophoresis alone. We propose that dramatic and largely reversible changes in the skin are occurring due to electroporation, probably electroporation of lipid bilayers in the stratum corneum. This is the first determination of increased transdermal flux due to electroporation. That it occurs in human skin is potentially important for medical applications such as transdermal drug delivery.[5]

REFERENCES

1. D. C. Chang, B. M. Chassy, J. A. Saunders, and A. E. Sowers, *Guide to Electroporation and Electrofusion,* New York: Academic Press, 1992.
2. C. Cullander and R. H. Guy, *Adv. Drug Deliv. Rev.* 8: 291-329, 1992.
3. K. T. Powell, A. W. Morgenthaler, and J. C. Weaver, *Biophys. J.* 56: 1163-1171, 1989.
4. J. A. Bouwstra et al., *J. Controlled Release* 15: 209-220, 1991.

5. We thank C. S. Lee for help in the laboratory and R. O. Potts and I. H. Blank for helpful discussions. This work was supported in part by a fellowship to M. R. Prausnitz provided by Cygnus Therapeutic Systems and in part by U.S. Army Research Office DAAL03-90-G-0218 and National Institutes of Health grant GM34077 to J. C. Weaver.

Martin Blank, Ed., *Electricity and Magnetism in Biology and Medicine*

THEORY OF ELECTROINJECTION OF DNA INTO CELLS

V. F. Pastushenko* and Yu. A. Chizmadzhev

A. N. Frumkin Institute of Electrochemistry, Russian Academy of Sciences, Moscow

In recent years the possibility of electrophoretic permeation of DNA into the cell in electrotransformation experiments has been discussed.[1-3] In the first approximation, considering DNA as a charged particle, the electrophoretic injection of DNA into the cell can be analyzed by a study of the electric potential distribution near the electropermeabilized cell. We did these calculations, using the method presented by Zeks et al.[4] for arbitrary intra- and extracellular conductivities, on the assumption that the resistance of the permeabilized portion of the membrane is negligible.[5] The electric field strength is increased near the permeabilized portion of the membrane, exceeding the external field strength essentially for small enough permeabilized regions of the cell membrane.

Assuming that the optimal pulse duration should be limited by the condition that the molecule should not traverse the cell, we concluded that the optimal intracellular conductivity should be somewhat higher than the extracellular one. The trajectories of the DNA molecules (field lines) were calculated. The trajectories entering the cell form the reaction zone, which is approximately a tube whose cross section decreases with increasing extracellular conductivity. For similar conductivities, this area is close to the area of the permeabilized portion of the membrane.

Due to inhomogeneity of the electric field near the cell, the molecule of DNA approaching the cell becomes partially oriented. The conditions of the electrophoretic permeation of the molecule through the pore can be considered taking into account the bending elasticity modulus of the molecule. It is shown that the bending rigidity of the molecule is almost negligible for pores several nanometers across at a typical electric-field strength of about 10^5 V/m. Thus the molecule is deformed as a nonelastic fiber, so that the radius of the pore cannot be increased by the elastic forces in the molecule.

If different parts of the molecule enter different pores, the competition between the possible pathways creates a resultant force acting on the membrane.[6,7] These forces are essentially higher than those capable of creating the membrane invagination. Thus, energetically, electroinduced membrane invagination is shown to be possible. However, an alternative membrane deformation in the form of the membrane cut between the competing pores is in principle also possible. For the field strength of the order of 10^5 V/m and a linear tension coefficient of the pores of about 10^{-11}N, such membrane cuts become possible if the minimal length of the molecular part that has penetrated into the cell has reached at least 1 nm. According to energy estimates, the probability of the membrane cut increases in comparison with the probability of the membrane invagination with the decrease of the linear tension coefficient and the size of the permeabilized portion of the membrane. For typical values of these parameters, both types of the membrane deformation are equally probable for a permeabilized membrane size of the order of 0.1 nm. Thus, the mechanism of electrophoretic translocation of DNA may be thought of as follows. DNA freely permeates into the cell through the single pore, much like a nonelastic fiber. In cases of competition between different pores to serve as electrophoretic pathways, the electrically induced membrane invagination or membrane cut may serve as intermediate membrane readjustments accompanying the DNA translocation.

*Correspondence to: V. F. Pastushenko, A. N. Frumkin Institute of Electrochemistry, Leninski prosp. 31, 117071 Moscow, Russia.

REFERENCES

1. E. Neumann, M. Schaefer-Ridder, Y. Wang, and P. H. Hofscheider, *EMBO J.* 1: 841-845, 1982;
2. D. J. Winterbourne, S. Thomas, J. Hermon-Taylor, J. Hussain, and A. P. Johnstone, *Biochem. J.* 251: 427-434, 1988.
3. V. A. Klenchim, S. M. Serov, S. I. Sukharev, L. V. Chernomordik, and Yu. A. Chizmadzhev, *Biological Membranes* 7: 446-448, 1990.
4. B. Zeks, S. Svetina, and V. F. Pastushenko, *Biological Membranes* 8: 430-438, 1991.
5. V. F. Pastushenko, *ibid.,* 1110-1119.
6. V. F. Pastushenko, *Appl. Theoret. Electrophoresis* 1: 313-316, 1991.
7. V. F. Pastushenko, *ibid.,* 317-321.

Martin Blank, Ed., *Electricity and Magnetism in Biology and Medicine*

ELECTROLOADING OF LEUCOCYTES WITH DRUGS

S. Sixou and J. Teissié*

CNRS Centre de Recherche de Biochimie et de Génétique Cellulaires, Toulouse, France

Targeted drug delivery can be obtained by use of electroloaded phagocytes. Specific electropermeabilization and associated loading by hydrophilic molecules of neutrophils in large volumes of whole blood was operated in a flow system by taking advantage of the physical definition of cell electropulsation. A slow release of the entrapped molecule was observed. The functionality of phagocytes was preserved after electropermeabilization, as shown in vitro by the oxydative burst after stimulation. Targeting of electropulsed neutrophils was obtained to a local inflammation in vivo on rats.

Targeting of specific sites is a powerful way to increase the efficiency of drug delivery. The use of circulating cells would provide a convenient vehicle for such purposes. The first step in drug loading is easily obtained by electropulsation.[1] Electroloaded erythrocytes, lymphocytes, or platelets have been tested as drug vehicles previously,[2-4] but no specific targeting to infectious areas can be obtained. From that point of view, circulating phagocytic cells (neutrophils and monocytes) are of particular interest. Phagocytes can kill invading microorganisms , after chemotactism to the infected area, by phagocytosis and oxygenating agent generation. In the present communication, we checked that phagocytes can be specifically electroloaded; that the oxydative metabolism of the drug carrying leukocytes is preserved; and that, in vivo, a normal accumulation of the electropulsed cells occurs in an inflamed area.

MATERIALS AND METHODS

Blood Cells

For permeabilization experiments erythrocytes were obtained by centrifuging blood samples (2000 g, 5 min, room temperature) and removing supernatant. Red blood cells permeabilization was assayed through the associated hemolysis. Leucocytes were used as "leucocytes concentrates": after elimination of most of the erythrocytes by centrifugation of blood samples (1100 g, 30 min, 4° C), a leucocytes and erythrocytes mixture in a 1/100 ratio was obtained. For chemiluminescence experiments, peripheral blood was harvested from adult healthy donors into heparinized tubes and was stored at room temperature for less than 8 h after drawing.

Rat neutrophils were obtained 3 h after intraperitoneal injection of 0.55 mg of carrageenan to 250g male Sprague-Dawley rats. Cells were washed twice by centrifugation at 4° C with Krebs-Ringer buffer without calcium or magnesium (KR buffer), resuspended at a concentration of 5×10^7 cells/ml in an 80μM propidium iodide solution or in a trypan blue solution in the KR buffer, and pulsed. After 20 min incubation at room temperature, the suspension was diluted with 10 volumes of ice-cold KR buffer and kept at 4° C for 0-21 h. Permeabilization was evaluated by dye penetration. Mortality was calculated by spontaneous labeling.

Reagent Solutions

- A 3.2 mM stock solution of Phorbol 12-myristate 13-acetate (PMA, Sigma, USA) was prepared in dimethylsulfoxide (DMSO, Fluka, Switzerland) and stored at −20° C. This stock solution was diluted in Krebs-Ringer phosphate buffer, pH 7.4, containing 11.7 mM glucose (KR buffer), to a 16.2μM concentration, just before the assay.
- To obtain opsonized zymosan (OZ), 20 mg of Zymosan A (Sigma, USA) were washed twice in phosphate-buffered saline medium (PBS) by centrifugation (350 g, 10 min at room temperature). Then, 2 ml of fresh autologous serum were added and the tube was incubated for 30 min at 37° C. The opsonized

---

*Correspondence to: J. Teissié, CNRS Centre de Recherche de Biochimie et de Génétique Cellulaires, 118 route de Narbonne, F-31062 Toulouse Cédex, France.

zymosan was then washed as above and resuspended in 2ml PBS. This suspension was diluted 10-fold in KR buffer, just before the assay.

- A 20mM stock solution of luminol (Sigma, USA) was prepared in DMSO and stored at 4° C. This stock solution was diluted in KR buffer to a 1.6 mM concentration, just before the assay.

Electropermeabilization

Experiments were performed at room temperature directly on intact blood or on rat neutrophiles. Electropermeabilization was carried out with a CNRS electropulser (Jouan, France) that can deliver square-wave pulses whose parameters (voltage E, pulse duration T, number N, and pulse frequency) are all independently adjustable.

The gap between the electrodes was then filled with whole blood samples (200 μl) and repetitive pulses were applied. Control cells were treated in a similar way except that no pulse was applied. Large volume (several milliliters) were treated by a flow pulsing chamber.[5]

Chemiluminescence Measurements

Chemiluminescence was measured in a luminometer (Luminometer 1250, LKB Wallac, Sweden). The reaction was started by mixing 200 μl of luminol and 100 μl of pulsed blood (directly or diluted in KR buffer). Phagocytes were activated (time = 0) by addition of 200 μl of the diluted solution of stimulating agent (PMA or OZ). After the tubes were gently vortexed, samples were incubated at 37° C. Kinetics were then determined by repeated measurements of light emission, every 2 min after stirring.

Radioactive Labeling of Neutrophils

Cells were washed as described previously and resuspended at a 10^8 cell/ml concentration in KR buffer. Electropulsation was then performed as described above. After 10 min of incubation at 30° C, both control and pulsed cells were labeled in the same way with extemporaneously formed ^{99m}Tc-HMPAO by 10min incubation at room temperature. After washing, labeled cells were resuspended at a concentration of 10^8 cells/ml in KR buffer again; 5×10^7 control or pulsed cells were then injected into the tail vein of each rat. Inflammation was induced immediately after intravenous administration by injection of 10 mg of zymosan in the right hind paw of each rat. Scintigraphic imaging was then performed on whole animal bodies.

RESULTS

Specific Electropermeabilization of a Subpopulation in a Mixture

Cell electropermeabilization is size dependent as predicted by the physical definition of the field-induced potential difference

$$\Delta U_p = FrE_{p,r} \tag{1}$$

where ΔU_p is the critical membrane potential difference which triggers permeabilization, F is a constant, and $E_{p,r}$ is the critical field intensity for a cell of radius r.

When working with a mixed cell population, under the hypothesis that ΔU_p and F are constant, one can specifically permeabilize large cells while leaving the smaller ones unaffected by using a field intensity E larger than the critical $E_{p,r}$ of the larger ones, but smaller than the one of the smaller population.[6] Such a property would open the opportunity for specific loading of larger cells. The flow technique was checked to be capable of such applications.

The target cells are leucocytes because of their functional specificity. The procedure must be such as to leave the smaller red blood cells unaffected. Cells were directly pulsed in plasma. Leucocyte viability decreased only above 1.6 kV/cm. At this field value, 70% of the leucocytes (polymorphs and monocytes) are permeabilized without alteration of viability. No hemolysis of erythrocytes was observed.

Stable Electroloading of Neutrophils

This critical parameter was tested by observation of the release of an electroloaded hydrophilic molecule in vitro. Neutrophils were electroloaded with propidium iodide, a molecule with a strong fluorescence emission. We used electrical conditions that led to the permeabilization (i.e., the fluorescence labeling) of 65% of rat peritoneal neutrophils: two square-wave pulses of 1.3 kV/cm lasting 1 ms. The post pulse leakage (the fluorescence decrease) for cells kept at 4° C shows in vitro that propidium iodide is always partly trapped in cells, even 21 h after electroloading. A parallel experiment with PI electroloaded CHO cells (Chinese hamster ovary cells) incubated at 37° C showed that 50% of propidium iodide is still trapped in cells 4 h after cell electroloading. With the hydrophilic character of PI taken into account, these observations indicate that a long-lived delivery of hydrophilic drugs should be obtained from electroloaded neutrophils.

Functionality Preservation In Vitro: PMA and OZ Stimulation

Oxygenation activity of stimulated phagocytes leads to oxidation of luminol, which gives an emission of light that can be detected with a luminometer. This activity was quantitated by two parameters: A_{max}, the maximum amplitude of chemiluminescence measurement obtained with luminol excess (640 μM); and the time interval T_{max} from addition of stimulation agent to A_{max}[7] that represents the reaction speed.[8,9]

Almost no chemiluminescence was detected on unstimulated samples, neither control nor pulsed ones. For the short pulse condition (10 pulses of 1.6 kV/cm lasting 100 μs), T_{max} decreased significantly by 21 ± 9% (n = 6) and A_{max} increased by 24% with a sharp increase in the error bar from 10 to 26% (Table 1). This change is apparently due to blood physiological variations from one donor to another. For the longer pulse conditions, the results of Table 1 show that no significant changes of A_{max} were observed and only a statistically insignificant decrease of T_{max}. These second pulsing conditions do not impair the oxygenation activities. Preservation of chemiluminescence production was maintained even 7 h after pulse application. Similar effects were obtained after OZ stimulation.

TABLE 1.—Effect of electrical pulses on relative values of chemiluminescence parameters from phagocytic leukocytes stimulated with PMA and OZ. Results are expressed as a percentage of the control sample and are the mean ± standard deviation of 3 to 6 experiments (i.e., 3 to 6 different donors).

Electrical pulses	A_{max}	T_{max}	A_{max}	T_{max}
	PMA		OZ	
None	100±9	100±10	100±10	100±14
10×100 μs 1.6 kV/cm	124±26	79±9	104±6	74±2
2×1 ms 1.2 kV/cm	104±8	87±4	104±2	90±9

A local permeabilization of cell membrane appears as soon as the electric field intensity E becomes higher than a threshold value $E_{p,r}$ but the permeability efficiency depends only on the pulse number N and duration T.[10] Flow Φ is given by

$$\Phi = kX(N,T)[1-(E_{p,r}/E)] \tag{2}$$

where k is a coefficient dependent on both the test molecule and the cell membrane properties.

According to Eq. (2), application of long-duration and low-intensity electric fields (2×1 ms ×1.2 kV/cm compared with 10×100μs×1.6 kV/cm) results in a decrease of the permeabilized membrane area (from 31% to 25%) and in a simultaneous increase of the local permeability in this area. Consequently, alteration of oxygenating activities (after stimulation) is related to the perturbation of a large area of the phagocyte membrane. On the contrary, a strong local perturbation allows preservation of this activity.

Chemotaxis of Electroloaded Neutrophils

To be targeted drug vectors to inflamed or infected areas, electroloaded neutrophils must keep their specific properties of migration and chemotaxis. We checked these functions in vivo by labeling control and pulsed rat neutrophils with ^{m99}Tc-HMPAO and by injecting them back into rats. At the same time, an experimental edema was induced in one paw of each animal.[11] With this technique, edema was maximal in 30 min. Scintigraphic imaging was then performed 30 min and 20 h after cell injection.

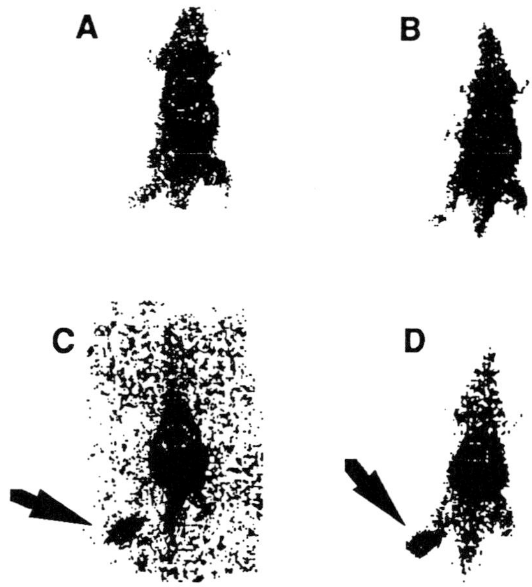

FIG. 1.—Time dependence of mapping of ^{99m}Tc-HMPAO labeled neutrophils intravenously administered to rats with experimental paw edema. Images A and C are for control cells and images B and D for pulsed cells for various periods of time (30 min for images A and B and 20 h for images C and D). Local accumulation of radioactive labeled neutrophils is easily seen in the right paw.

Images at 30 min (A and B) and 2 h (C and D) of animals injected with either control (A and C) or pulsed (B and D) cells are presented in Fig. 1. After 20 h most of the radioactivity is observed in the liver, the spleen, and the gut, but significant accumulation of neutrophils is revealed in the inflamed area (image C). The comparison with pulsed cells in images B and D shows similar cell distribution, first in the heart and then in the liver. In the same way, the edema is easily spotted after 20 h. In both cases, the accumulation ratio was ten times larger than in the left control paw. These images testify to the ability of pulsed neutrophils to be attracted to an inflamed area and to accumulate there.

DISCUSSION

Electropulsed phagocytes are still functional several hours after pulse application. The generation of oxygenating agents is preserved even 7 h after electropulsing, whereas the permeabilization state disappeared some minutes after pulse application, at room temperature.[1] The stability of this phenomenon means that phagocytic cells electroloaded with a drug would keep their functionality after drug incorporation, if the drug is chosen so as to avoid a specific drug toxicity on cell functions.

The results on radioactivity accumulation at the level of inflammation demonstrate that neutrophils electroloaded with hydrophilic molecules are efficient targeted drug vehicles. The fast local accumulation of electroloaded phagocytes and the slow release of the drug across the resealed membrane, as observed in the present work by use of an hydrophilic molecule, will induce its continuous delivery precisely in the inflamed (or infected) areas. Drug-targeted delivery will then be obtained.[12]

REFERENCES

1. E. Neumann, A. E. Sowers, and C. Jordan, Eds., *Electroporation and Electrofusion in Cell Biology,* Plenum Press, 1989.

2. K. Kinosita Jr. and T. Y. Tsong, *Nature* 272: 258-260, 1978.

3. K. Hughes and N. Crawford, *Biochim. Biophys. Acta* 981: 277-287, 1989.

4. U. Zimmermann, P. Scheurich, G. Pilwat, and R. Benz, *Angew. Chem. Int. Ed. Engl.* 20: 325-344, 1981.

5. J. Teissié, S. Sixou, and M. P. Rols, in M. J. Allen, S. F. Cleary, A. E. Sowers, and D. D. Shillady, Eds., *Charge and Field Effects in Biosystems—3,* Birkhäuser, 1992, pp. 449-466.

6. S. Sixou and Teissié, *Biochim. Biophys. Acta* 1028: 154-160, 1990.

7. M. Markert, P. C. Andrews, and B. Babior, in V. Ginsburg, Ed., *Methods in Enzymology,* San Diego: Academic Press, 1984, vol. 103, pp. 358-364.

8. R. C. Allen, R. L. Stjernholm, and R. H. Steele, *Biochem. Biophys. Res. Comm.* 47: 679-684, 1972.

9. R. C. Allen and L. D. Loose, *Biochem. Biophys. Res. Comm.* 69: 245-252, 1976.

10. M. P. Rols and J. Teissié, *Biophys. J.* 58: 1089-1098, 1990.

11. J. P. Tarayre, A. Delhon, M. Aliaga, M. Barbara, E. Bruniquel, V. Caillol, L. Puech, N. Consul, and J. Tisné-Versailles, *Pharmacol. Res.* 21: 375-384, 1989.

12. This work was partly supported by research grants from AFLM (Association Française de Lutte contre la Mucoviscidose) to S. Sixou and to Dr. J. Teissié. Technical facilities were provided by Drs. Tarbes (CTS Toulouse), Favre (CRAC Toulouse), and Puget (LPTF CNRS Toulouse).

Martin Blank, Ed., *Electricity and Magnetism in Biology and Medicine*

CHARACTERIAZATION OF LONG-LIVING ELECTROPORES IN GIANT LIPID VESICLES BY AN ELECTROMECHANICAL MICROPIPET TECHNIQUE

D. V. Zhelev

Central Laboratory of Biophysics, Bulgarian Academy of Sciences, Sofia

D. Needham*

Duke University, Durham, N.C.

Giant liposomes (25-56 μm in diameter) were held by a micropipet and a small membrane tension was created by control of the pipet suction pressure. The liposomes were then subjected to single square electric pulses until a stable pore (with a lifetime of seconds) was formed. The pores could be kept in a quasi-equilibrium state before they closed spontaneously in milliseconds. It was clear from the experiments that only one pore remained open after the pulse. Both the pore radius and the balancing membrane tension were determined for each liposome, and from these two parameters the line tension was found. The line tension for two different lipid compositions was: $(0.92 \pm 0.07) \times 10^{-11} N$ for SOPC and $(3.05 \pm 0.12) \times 10^{-11} N$ for SOPC:CHOL (1:1).

The application of strong electric fields across a cell or lipid bilayer results in the formation and expansion of membrane pores. This process is a useful tool in cellular and molecular biology and gene technology because it affords access to cellular contents for relatively short times. One of the basic parameters that characterizes membrane pores is the line tension (the energy per unit length of the membrane contour at the pore edge). Values for this basic parameter have been found to be in the range 10^{-12} to $10^{-11} N$.[1-4]

In previous electromechanical micropipet measurements[5] we found that after a permeabilizing pulse at low membrane tensions, the pores resealed. We assumed that membrane breakdown was caused by the synergistic action of the line tension, the membrane tension, and the electric field.[5,6] The objectives of the present study were (1) to determine the pore radius, (2) to determine the membrane tension that just opposed pore closure, and (3) to derive a relationship between these measured parameters in order to give an estimate of the pore line tension.

METHODS

The experimental procedures have been described elsewhere.[7] We used liposomes made from stearoyloleoylphosphatidylcholine (SOPC) with and without 50 mol% cholesterol (CHOL). The liposomes were held with a micropipet and were positioned between two square platinum electrodes 1 mm^2 in area, separated by 250 μm. Rectangular pulses 150 μs long were applied, of amplitude above the critical voltage for membrane breakdown, while the suction pressure was increased in increments until the liposome started to flow into the pipet. After a second (or seconds) the flow ceased (Fig. 1).

RESULTS AND DISCUSSION

By controlling the suction pressure and measuring the vesicle volume changes we determined the pore radius,

$$R_{pr}^3 = 3 Q \mu_i / \Delta P$$

where Q is the outgoing volume through the pore, μ_i is the viscosity of the inside fluid, and ΔP is the pressure difference across the pore.

---

*Correspondence to: D. Needham (ME and Materials Science), Duke University, Durham NC 27706.

Before electroporation, the liposome membrane was subjected to static tension.[8] After permeabilization, the liposome started to flow, and the measured pressure difference was distributed among (1) pressure drop along the pipet, (2) pressure drop for the viscous drag of the liposome membrane inside the pipet, and (3) pressure drop creating the tension in the outside liposome membrane.

The tension in the outside membrane portion of the liposome T_{osp} was[7]

$$T_{osp} = \frac{\dfrac{\Delta P_m}{2} - \dfrac{\mu_0 V_{rt}}{R_p}\left[\dfrac{4(L_{eff} - L_p)}{R_p} + \dfrac{L_p}{h}\right]}{\left[\dfrac{1}{R_p} - \dfrac{1}{R_0}\right]}$$

where μ_0 is the viscosity of the outside fluid, V_{rt} is the measured entry velocity, L_p is the projection length of the liposome membrane in the pipet, L_{eff} is the effective pipet length, and h is the thickness of the liquid film between pipet wall and liposome membrane.

We used a thermodynamic approach to define the line tension. When the pore forms, the free energy of the liposome membrane changes. The porated membrane can be divided into two regions:[7] (1) a pore region and (2) an unperturbed membrane region having the same structure and properties as before pore formation.

In this simplistic view, the change of the pore free energy W_p at a given membrane tension is defined by the difference between the free energy of the membrane material in the pore region and the free energy stored in an equivalent part of the planar membrane having the same area as the pore,

$$W_p = 2\pi R_p \Gamma - R_p^2 T$$

where Γ is the line tension and T is the membrane tension. The pore free energy has a maximum at unstationary equilibrium when $R = \Gamma / T$. Thus, the line tension can be found from the dependence of R_p on $1/T$.

Figure 2 shows the determined pore radii versus corresponding reciprocal tensions for SOPC and SOPC:CHOL (1:1) liposomes. For each of the two lipid compositions, the bigger slopes are twice as large as the smaller ones, a result of testing both single- and double-walled liposomes. The line tension for a single SOPC bilayer is $(0.92 \pm 0.07) \times 10^{-11}\,N$ and for a SOPC:CHOL (1:1) bilayer, $(3.05 \pm 0.12) \times 10^{-11}\,N$.

The time evolution of the pore free energy is characterized by changes of the pore radius. The dependence of the pore free energy on the pore radius for one of our experiments is shown in Fig. 3. When the critical voltage is applied, the pore free energy is negative and the pore radius increases (curve 1). After the electric field ceases, the pore free energy becomes positive and fluctuates around its maximum because, in our experiments, the line tension term is balanced by the membrane tension (curve 2). This is an unstationary equilibrium and the pore will either reclose (curve 3) or increase in size (curve 4), which results in liposome membrane breakdown. The electrostatic term of the pore free energy for curve 1 was calculated following Pastushenko and Chizmadzhev,[9] and curve 5 represents the same term calculated following Winterhalter and Helfrich.[10]

In summary, a new electromechanical method is presented for the determination of the membrane line tension of lipid vesicles; the line tension for two lipid compositions is found. Our results predict that when the membrane tension is negligible, the evolution of pores in electroporation is a cyclic process.[11]

REFERENCES

1. J. D. Lister, *Phys. Lett.* 53A: 193, 1975.
2. C. Taupin, M. Dvolaitzky, and C. Sauterey, *Biochemistry* 14: 4771, 1975.
3. W. Harbich and W. Helfrich, *Z. Naturforsch.* 34A: 1063, 1979.
4. L. V. Chernomordik, M. M. Kozlov, G. B. Melikyan, I. G. Abidor, V. S. Markin, and Y. A. Chizmadzhev, *Biochim. Biophys. Acta* 812: 643, 1985.
5. D. Needham and R. M. Hochmuth, *Biophys. J.* 55: 1001, 1989.
6. D. Zhelev, D. Dimitrov, and I. Tsoneva, *Bioelectrochem. Bioenerg.* 19: 217, 1988.
7. D. V. Zhelev and D. Needham, *Biochim. Biophys. Acta* (in press).
8. E. A. Evans and R. Skalak, *Mechanics and Thermodynamics of Biomembranes,* Boca Raton, Fla.: CRC Press, 1980.
9. V. F. Pastushenko and Y. A. Chizmadzhev, *Biofizika* (in Russian) 28: 1036, 1983.
10 M. Winterhalter and W. Helfrich, *Phys. Rev.* A-36: 5874, 1987.

11. This work was supported by grant GM 40162 from the National Institutes of Health.

1

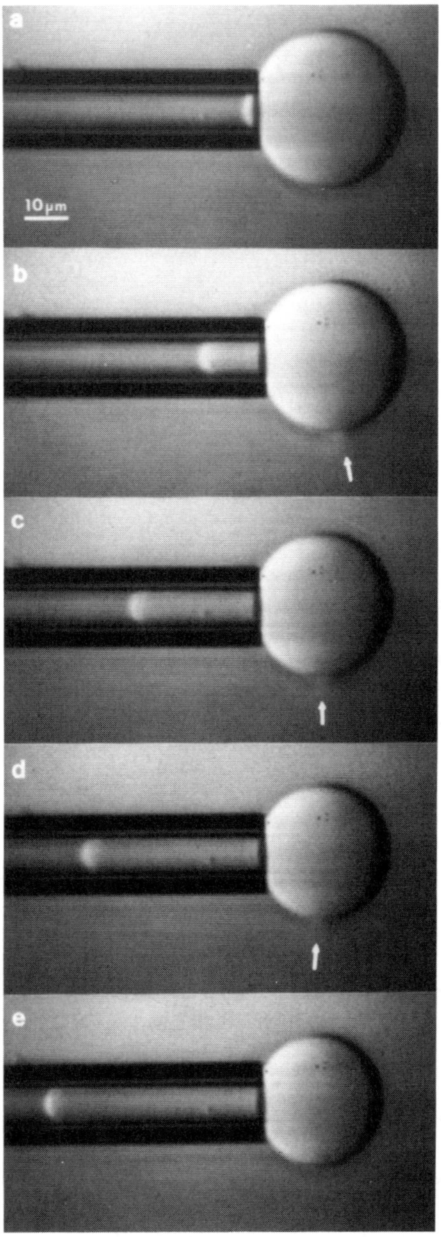

2

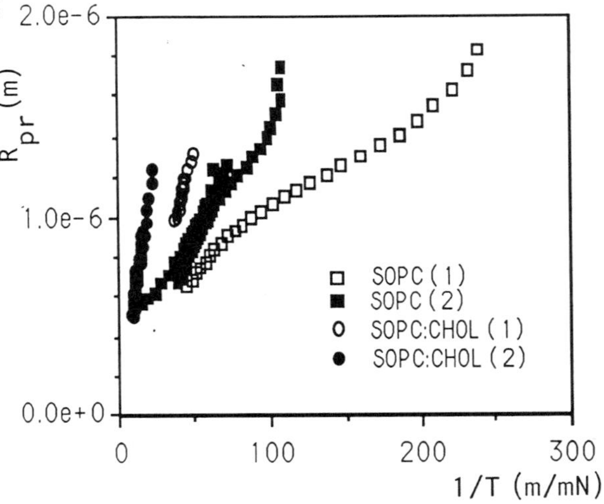

3

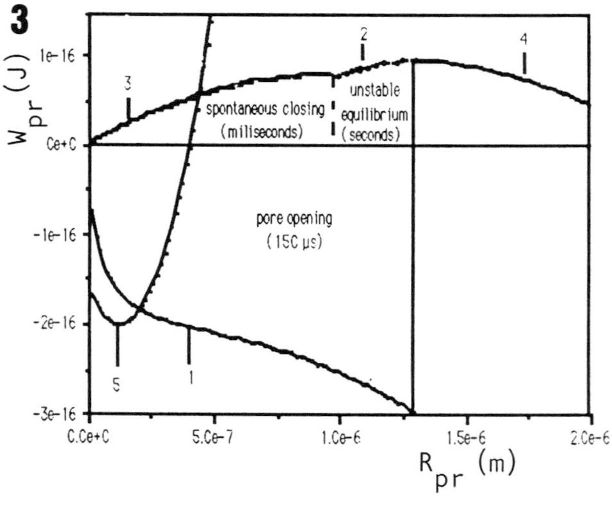

FIG. 1.—Liposome electroporation experiment: (a) $t = 0$ s, liposome in the suction pipet before pulse application; (b) $t = 0.5$ s, after application of pulse the pore forms and liposome membrane starts to flow into pipet; (c) $t = 1.0$ s and (d) $t = 1.5$ s, jet of solution is indicated by arrow; (e) $t = 2.0$ s, pore spontaneously closes and membrane flow ceases.

FIG. 2.—Pore radius R_p vs reciprocal far-field tension T for SOPC and SOPC:CHOL (1:1) liposomes; numbers in brackets denote number of bilayers.

FIG. 3.—Pore free energy changes during pore opening and closure.

Martin Blank, Ed., *Electricity and Magnetism in Biology and Medicine*

GRADUAL CHANGES OF MEMBRANE PROPERTIES AT HIGH TRANSMEMBRANE ELECTRIC POTENTIAL SIMULATE A BREAKDOWN THRESHOLD

R. W. Glaser* and J. Gimsa

Humboldt University, Berlin

Pore formation, expansion, and resealing are time-dependent processes that depend strongly on transmembrane voltage, but have no threshold voltage. The electric properties of a system of membranes and electrolyte solutions undergo qualitative changes in a narrow range of voltage between low voltages at which transmembrane voltage is determined by the external field and time constant of membrane charging, and high voltages at which the conductance of the membrane limits the transmembrane voltage. This transition is described in the literature as reversible electrical breakdown. During pulses of higher voltage, the current through the membrane is of the same order of magnitude as the current that would flow in the system without a membrane. Voltage across the membrane is independent of external voltage. The size and number of pores that are necessary to downregulate transmembrane voltage depend on pore resistance, surrounding solutions, and the geometry of the system. This phenomenon is a key to an understanding of cell damage and other secondary effects following electrical pulsation.

Electroporation and electrofusion of cells have been applied in biotechnology, medicine, and other fields. These techniques are based on a reversible increase of membrane permeability and conductivity at electric potential differences across the membrane (TMP) of the order of 1 V. Despite numerous studies of the mechanism of electroporation, the interpretation of many experimental results is still controversial.

Most studies on cells find the onset of measurable effects above a threshold or breakdown voltage.[1] Gradual changes of membrane properties as a function of voltage and time have been found by other methods.[2] The latter are consistently explained by theories of pore formation.[3]

Most electroporation studies tend to neglect the effect of the increasing current that flows through the membrane during the pulse and reduces the TMP, thus restricting the further development of the pores. However, it is precisely this mechanism that provides the link between the observation of an apparent threshold voltage of reversible electrical breakdown and gradual changes of membrane properties with increasing TMP.[1,2] Experimental and theoretical investigations concerning the mechanism of electroporation are summarized and new experimental results concerning the asymmetric effects of electric pulses are presented.

Pore Formation

Pore formation has been studied in uranyl-modified azolectin membranes. Membrane conductivity increases at a nearly constant rate during the pulse. The rate of pore formation was determined as a function of voltage.[3] Theory predicts an exponential increase of the rate of pore formation with the square of TMP.[4] The experimental data were consistent with the theoretical prediction in the range between 0.7 ;and 1.45 V, covering nearly three orders of magnitude in dI/dt. It was calculated from these data that an energy barrier of 40-50 kT at a radius between 0.3 and 0.6 nm had to be overcome for hydrophilic pore formation.[3]

Pore Conductance

Ions crossing a narrow pore in a membrane undergo an interaction with the edges of the membrane and have to overcome an energy barrier.[5] Voltage across the membrane lowers this barrier. Thus conductivity increases with increasing voltage. The effect is strongest with small pores. Charged head groups in or around the pore produce additional interactions and selectivity for charged molecules. The radius of pores in uranyl-modified membranes was estimated from the voltage dependence of conductivity to be of the order of

---

*Correspondence to: R. W. Glaser, Inst. Med. Immunologie, Schumannstr. 20/21, D-1040 Berlin, Germany.

1 nm.[3]

The permeability of the erythrocyte membrane for Na and Cl ions after pulsation in NaCl solution of physiological ionic strength was six orders of magnitude lower than necessary to account for its conductivity during the pulse. Apart from the nonlinear conductivity, this phenomenon is probably due to a reduction of pore radius within less than 1 s.

Limitation of Pore Formation

As pores arise and enlarge in a membrane during the pulse, ions cross the membrane and current begins to flow. The resistance of the solution causes a potential difference to be built up around the membrane that reduces the potential difference across the membrane (TMP) and slows down the formation of new pores. As a result, at the end of the pulse TMP reaches a certain value that depends on the length of the pulse but not on external voltage.[1] This relationship was directly confirmed by Hibino et al., who measured TMP in sea-urchin eggs during pulsation using a voltage-sensitive fluorescent dye.[2] Consequently, the exponential dependence of current on external voltage turns into a linear one. This behavior is generally observed in cell membranes and artificial membranes that exhibit reversible electric breakdown.

This mechanism is the most important link between (a) theories of pore formation (and experiments at low voltage) that find an exponential dependence of pore formation on voltage (e.g., the increase in membrane conductance of L cells as a linear function of time and the approximately exponential dependence of the rate of increase on voltage);[6] and (b) the observation of a threshold voltage for reversible electrical breakdown of membrane properties, which was invariably observed when cells in the orifice of a Coulter counter were exposed to increasing voltage.[7]

Asymmetric Breakdown

Under various experimental conditions, electrical pulses have strongly asymmetric effects on cells.[8,9] There is an obvious parallel to the higher sensitivity of cells to electric pulses at lower ionic strength.

We observed the influx of propidium iodide (PI) into CB-F7 cells after various electric pulses. The electric field was produced between two platinum electrodes by a computer with a home-made D/A converter and amplifier. Pulses were controlled with a storage oscilloscope. (CB-F7 is a heteromyeloma cell line that has been used for electrofusion.)[10]

We obtained the following results:

1. Pulses that lead to staining of only the positive hemisphere modify the membrane in the negative hemisphere in such a way that subcritical pulses of opposite orientation lead to PI staining of the other hemisphere. This behavior was found in isotonic sucrose solution with or without 10 mM phosphate buffer and in the low-ionic-strength MCM solution used for electrofusion.[10]
2. Eventual mechanical disintegration after repeated pulses started in the hemisphere that was stained first.
3. In solutions of physiological ionic strength (150 mM KCl or RPMI 1640 cell culture solution), the same pulses do not induce PI staining. Longer pulses lead to symmetrical staining.

In the literature, discussion has focused on superposition of the resting membrane potential and the externally applied potential,[8,11] and electro-osmosis through the pores.[12] The first interpretation can be disproved by the experiment described above, by other experiments,[13] and from theoretical considerations. Electro-osmosis can explain only some of the observations.

Obviously, several mechanisms produce asymmetric effects after pulsation. We suggest consideration of the following mechanisms.

1. Directed transport of charged molecules by the electric field during the pulse.
2. Different pore conductivity in the two hemispheres due to anion/cation selectivity of the pores. The current through the membrane is the same on both sides of the cell. Thus, more or bigger pores are formed in the hemisphere in which the individual pores have a lower conductance.
3. Different nonlinear properties or different resealing of the pores. Asymmetry of the cell membrane and surrounding media may lead to small pores in one hemisphere and a few big pores in the other one. Thus the permeability for large molecules may be totally different.

In addition, it would be interesting to investigate possible contributions of different stress of the membrane as a result of hydrodynamic forces of electrophoresis, asymmetric membrane proteins, osmotic equilibration after asymmetric concentration changes of ions due to anion- or cation selective pores, or electrophoretic redistribution of membrane components.[14]

REFERENCES

1. R. W. Glaser, (a) *Studia Biophysica* 116: 77-86, 1986; (b) Dissertation (A), Humboldt University, Berlin, 1986; (c) (with A. Wagner and E. Donath), *Bioelectrochem. Bioenerg.* 16: 455-467, 1986.
2. M. Hibino, M. Shigemore, H. Itoh, K. Nagayama, and K. Kinosita Jr., *Biophys. J.* 59: 209-220, 1991.
3. R. W. Glaser, S. L. Leikin, L. V. Chernomordik, V. F. Pastushenko, and A. I. Sokirko, *Biochim. Biophys. Acta* 940: 275-287, 1988.
4. S. L. Leikin, R. W. Glaser, and L. V. Chernomordik, *Biol. Membr.* 3: 944-951, 1986.
5. V. A. Parsegian, *Ann. N. Y. Acad. Sci.* 264: 161-174, 1975.
6. U. Zimmermann and H. B. Urnovitz, *Methods Enzymol.* 151: 194-221, 1987.
7. G. Pilwat and U. Zimmermann, *Biochim. Biophys. Acta* 820: 305-314, 1985.
8. W. Mehrle, R. Hampp, and U. Zimmermann, *Biochim. Biophys. Acta* 978: 267-275, 1989.
9. A. E. Sowers and M. R. Lieber, *FEBS Lett.* 205: 179-184, 1986.
10. R. W. Glaser, H.-D. Volk, C. Liebenthal, S. Jahn, and R. Grunow, *Human Antibodies and Hybridomas* 1: 111-114, 1990.
11. E. Tekle, R. D. Astumian, and P. B. Chock, *Biochem. Biophys. Res. Commun.* 172: 282-287, 1990.
12. D. S. Dimitrov and A. E. Sowers, *Biochim. Biophys. Acta* 1022: 381-392, 1990.
13. A. E. Sowers, *Biophys. J.* 54: 619-626, 1988.

14. The work on artificial membranes was the result of fruitful cooperation with Sergei L. Leikin, Leonid V. Chernomordik, Vasili F. Pastushenko, and Artjom I. Sokirko in the laboratory of Yuri A. Chizmadzhev. We thank Robert Arndt for his help with the construction of the D/A converter and amplifier. We acknowledge stimulating discussions with Edwin Donath.

Martin Blank, Ed., *Electricity and Magnetism in Biology and Medicine*

THEORETICAL MODELING OF ELECTROPORATION: ELECTRICAL BEHAVIOR AND MOLECULAR TRANSPORT

M. Wang,* S. Freeman, V. Bose, S. Dyer, and J. C. Weaver

Harvard-MIT Division of Health Sciences and Technology, Cambridge, Mass.

Quantitative predictions of the transmembrane voltage $U(t)$, molecular transport of a small, charged molecule, and the aqueous fractional area $F_w(t)$ of an electroporated membrane are presented. The present model is based on the hypothesis of transient aqueous pores created by thermal fluctuations and local electric-field energy. The model's predictions are in reasonable agreement with experimental observations.

The fascinating membrane phenomena that occur at large transmembrane voltages $U(t)$ are generally thought to involve the electrically assisted formation of pores, which is in turn believed to cause striking mechanical behavior (e.g., membrane rupture), electrical behavior (e.g., reversible electrical breakdown or REB), and tremendously enhanced molecular transport (e.g., DNA uptake).[1,2] Basic understanding of electroporation involves an interplay between experimental measurements and theoretical predictions. Here we describe recent progress with a version of a transient aqueous pore model of electroporation.

METHODS

As described previously in detail,[3-7] a physical model for electroporation can be constructed by use of a relatively simple description of a planar membrane, pore geometry, pore formation energy, pore interior conductivity, and the localized potential drops just outside the pore entrances. The membrane itself is represented by the parallel combination of an almost constant capacitance C and a significantly changing conductance $G(t)$; the latter is due to ionic conduction through the pore population. The membrane is assumed to charge in response to an externally applied pulse, with the charging pathway characterized by the external pathway resistance R_0 = constant and C. Entry of ions into a pore with nearby low dielectric constant material is repelled because of the Born energy. In the present version, we use a Born energy estimate obtained by Jordan[8] and have rewritten the computer code to allow treatment of both square and exponential pulses. In summary form, the main equations of the model are

$$\Delta E(r,U) = 2\pi\gamma r - \pi\Gamma r^2 - \frac{\pi(\varepsilon_w - \varepsilon_l)U^2}{h} \int_{r_{\min}}^{r} \alpha^2 r \, dr \tag{1}$$

which is the pore formation energy as a function of pore radius r and of U. The function α depends on the pore radius r and the pore's internal conductivity.[5,6] Evolution of the pore population is described by[3]

$$\frac{\partial n}{\partial t} = D_p \left[\frac{\partial^2 n}{\partial r^2} + \frac{\partial}{\partial r} \left[\frac{n}{kT} \frac{\partial \Delta E}{\partial r} \right] \right] \tag{2}$$

where D_p is the effective diffusion constant for the pore radius and $n(r,t)$ is a pore probability density function that gives the probability of having a pore with radius r and $r + dr$ as $n(r,t)\,dr$. This reflects the fundamentally stochastic nature of pore formation, and leads to a heterogeneous pore population (a range of pore sizes is expected) which changes rapidly with time. The interaction of a planar bilayer membrane with its electrical environment is described by[6]

$$C\frac{dU}{dt} = \begin{cases} \dfrac{I_p R_N}{R_E + R_N} - U\left[G(t) + \dfrac{1}{R_E + R_N}\right] & \text{if } t < t_{\text{pulse}} \\ -U/R & \text{if } t > t_{\text{pulse}} \end{cases} \tag{3}$$

---

*Correspondence to: M. Wang, Harvard-MIT Division of Health Sciences and Technology, Cambridge MA 02139.

which is the differential equation that describes the equivalent circuit of the membrane and the external charging pathway. ($R = 1/G$ is the membrane resistance.) In short, the model is based on a modular approach, in which the pore properties are represented by simple physical descriptions and are then used to solve self-consistently the two differential Eqs. (2) and (3).

RESULTS AND DISCUSSION

Predicted electrical behavior for a very short (0.4μs) square pulse has been presented elsewhere[6,7] and agrees reasonably, but not completely, with experimental measurements.[9] We have since extended the numerical simulation of electroporation to predict the fraction of the membrane area that becomes aqueous, i.e., is occupied by pores (Fig. 1).[10] This extension also allows prediction of the fractional change in the membrane capacitance, $\Delta C/C$, which is found to be small (less than 4%), in agreement with experimental observation.[11]

The predicted behavior of an exponential pulse has also been explored as shown in Fig. 1 for several different pulse magnitudes V_0. It is interesting that a temporary plateau in transmembrane voltage is predicted for the larger pulses. This behavior is not intuitively obvious, and is being studied further. Predictions of the transport of a small charged molecule (propidium iodide, $z = +2$, 660 Dalton) for the same exponential pulse have also been made; they show first a sharp rise above an approximate threshold and then a gradual decline as the pulse magnitude V_0 is increased (Fig. 2). In this case, a planar membrane much smaller in area was used, to approximate the participating portion of a cell membrane. The molecular transport dependence on V_0 is reminiscent of the plateau observed for calcein ($z = -4$, 623 Dalton) uptake by red blood cell ghosts,[12] but the model's prediction is not a true plateau.[12]

REFERENCES

1. T. Y. Tsong, "Electroporation of cell membranes," *Biophys. J.* 60: 297-306, 1991.

2. D. C. Chang, B. M. Chassy, J. A. Saunders, and A. E. Sowers, Eds., *Guide to Electroporation and Electrofusion,* New York: Academic Press, 1992.

3. I. G. Abidor, V. B. Arakelyan, L V. Chernomordik, Yu. A. Chizmadzhev, V. F. Pastushenko, and M. R. Tarasevich, "Electric breakdown of bilayer membranes: I. The main experimental facts and their qualitative discussion," *Bioelectrochem. Bioenerg.* 6: 37-52, 1979.

4. J. C. Weaver and R. A. Mintzer, "Decreased bilayer stability due to transmembrane potentials," *Phys. Lett.* 86A: 57-59, 1981.

5. V. F. Pastushenko and Yu. A. Chizmadzhev, "Stabilization of conducting pores in BLM by electric current," *Gen. Physiol. Biophys.* 1: 43-52, 1982.

6. A. Barnett and J. C. Weaver, "Electroporation: A unified, quantitative theory of reversible electrical breakdown and rupture," *Bioelectrochem. Bioenerg.* 25: 163-182, 1991.

7. J. C. Weaver and A. Barnett, "Progress towards a theoretical model of electroporation mechanism: Membrane electrical behavior and molecular transport," in D.C. Chang, B. M. Chassy, J. A. Saunders, and A. E. Sowers, Eds., *Guide to Electroporation and Electrofusion,* New York: Academic Press, 1992, 91-117.

8. P. C. Jordan, "Effect of pore structure on energy barriers and applied voltage profiles: I. Symmetrical channels," *Biophys. J.* 45: 1091-1100, 1984.

9. R. Benz, F. Beckers, and U. Zimmermann, "Reversible electrical breakdown of lipid bilayer membranes: A charge-pulse relaxation study," *J. Membrane Bio.* 48: 181-204, 1979.

10. L. V. Chernomordik, S. I. Sukharev, I. G. Abidor, and Yu. A. Chizmadzhev, "The study of the BLM reversible electrical breakdown mechanism in the presence of UO_2^{2+}," *Bioelectrochem. Bioenerg.* 9: 149-155, 1982.

11. M. R. Prausnitz, B. S. Lau, C. D. Milano, S. Conner, R. S. Langer, and J. C. Weaver, "A quantitative study of electroporation showing a plateau in net molecular transport" (in preparation).

12. This work was supported by Army Research Office Grant No. DAAL03-90-G-0218, by NIH grant ES06010, and by a computer equipment grant from Stadtwerke Düsseldorf, Düsseldorf, Germany.

1

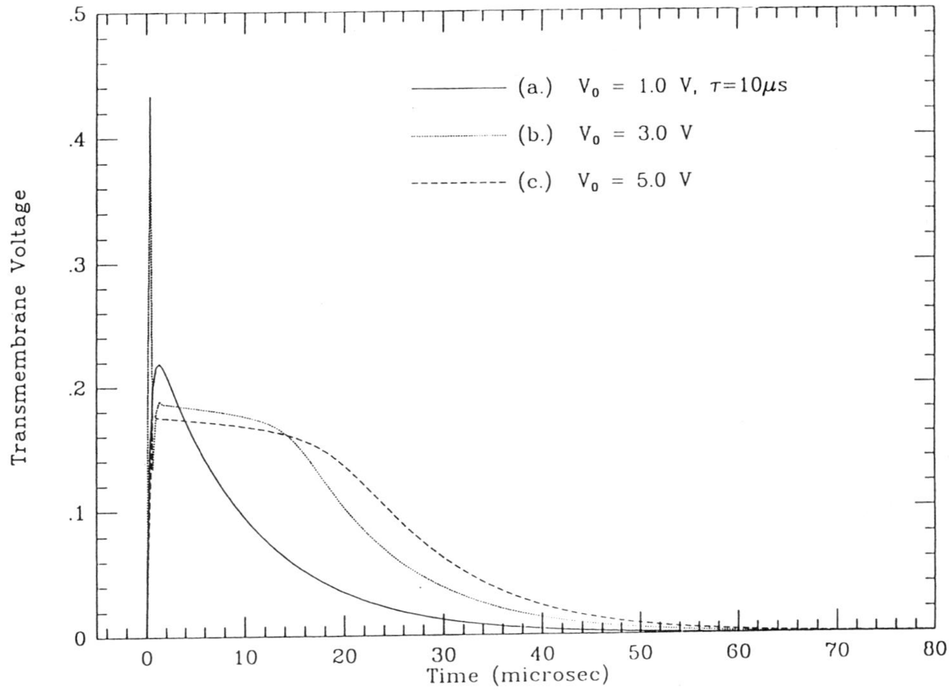

2

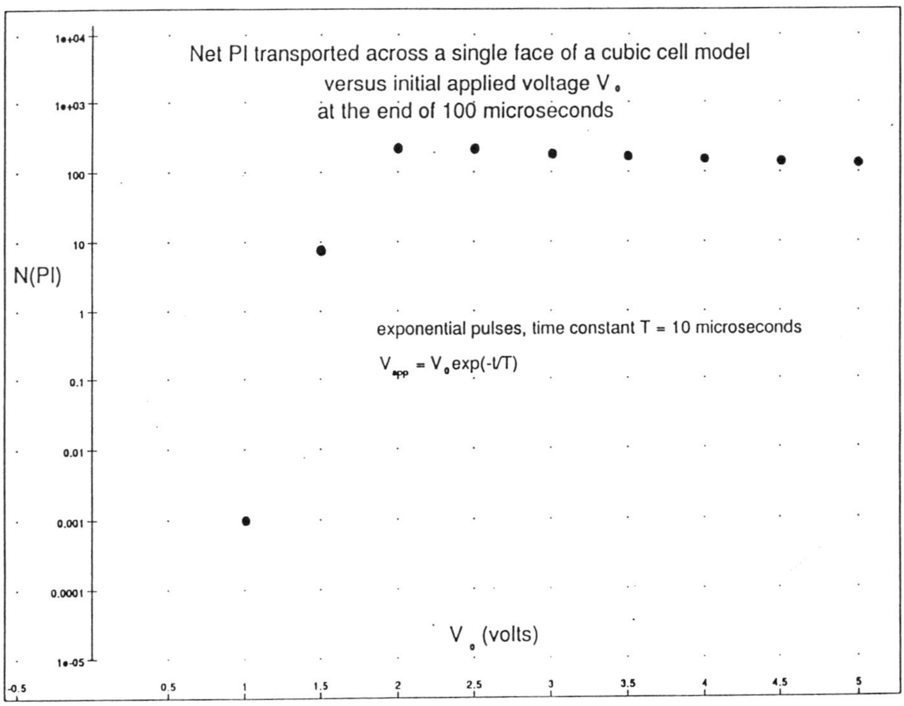

FIG. 1.—Predicted $U(t)$ due to an exponential pulse of the form $V = V_0 \exp(-t/\tau)$ with $\tau = 10$ μs, based on cubic cell model in which participating membrane was assumed to be a 10×10μm square.

FIG. 2.—Predicted cumulative transport of propidium iodide (•) due solely to electrical drift through the transient aqueous pores for the same exponential pulse as in Fig. 1. Extracellular concentration is 80 M.

Martin Blank, Ed., *Electricity and Magnetism in Biology and Medicine*

QUANTITATIVE DETERMINATION OF MOLECULAR TRANSPORT ACROSS ERYTHROCYTE GHOST MEMBRANES BY ELECTROPORATION

B. S. Lau,* C. D. Milano, M. R. Prausnitz, R. S. Langer, and J. C. Weaver

Massachusetts Institute of Technology, Cambridge, Mass.

Preliminary results of a systematic, quantitative study of net calcein transport due to electroporation of erythrocyte ghosts are presented. Calcein ($z = -4$, 623 Da) is a fluorescent molecule which ordinarily does not cross cell membranes. In this study, uptake increases with field strength E at lower E above a threshold of less than 1 kV/cm, but reaches a plateau at higher E. However, the molecular concentration inside the ghosts is significantly lower than the concentration outside the ghosts, suggesting a nonequilibrium, or altered equilibrium, condition. Increases in the external molecule concentration and in the number of pulses both increase uptake throughout the ranges investigated. At higher concentrations and after multiple pulses, uptake still reaches a plateau at high voltages. However, these plateaus occur at higher uptake values.

The creation of a transient high-permeability state in lipid bilayer membranes, known as electroporation, has received attention as an intriguing biophysical phenomenon, as well as a tool for applications in biotechnology and medicine.[1] Of primary interest for applications of electroporation is the associated transmembrane transport of macromolecules otherwise unable to cross cell membranes. Most electroporation studies focus on transfection, an event involving many other factors in addition to transmembrane transport; others emphasize transport, but simply note whether it occurs, without quantifying various amounts of transport. The following study presents a systematic and quantitative analysis of the transmembrane transport caused by electroporation of erythrocyte ghosts, including absolute numbers of molecules transported.

EXPERIMENTAL METHODS

The experimental methods are described in detail elsewhere.[2] Briefly, human red blood cells are washed, converted into ghosts by hypotonic lysis, and stored at 4 C. Within 3 days, these ghosts are suspended in a solution containing a known concentration of a fluorescent molecule and are given one or more electric pulses. Pulses are exponential decay pulses with a decay time constant between 1.0 and 1.5 ms. Five minutes after pulsing, ghosts are washed to remove any fluorescent molecules not entrapped inside ghosts. Solutions are kept at 4 C before and after pulsing.

Analysis is performed by flow cytometry, where light scatter and fluorescence data are collected from 20000 individual ghosts for each experimental condition. By use of quantitative calibration beads, ghost fluorescences are used to determine the number of molecules inside each ghost. Each condition has been replicated by at least five samples from 3 or more different days.

RESULTS AND DISCUSSION

Above a threshold of less than 1 kV/cm, transport of calcein into ghosts increases with increasing field strength E up to approximately 2 kV/cm, where calcein transport reaches a plateau (Fig. 1). A plateau would be expected if the molecule concentration inside the ghost reached the concentration outside the ghost, as there would then be no further net transport. However, if the number of molecules inside each ghost is divided by the average 150μm^3 ghost volume,[3] these concentrations are far from equal (Table 1). This suggests that a nonequilibrium, or altered equilibrium, condition exists. Evidence of a transient uptake mechanism has been previously found in yeast.[4]

---

*Correspondence to: B. S. Lau (Chemical Engg), MIT, Cambridge MA 02139.

TABLE 1.—Summary of calcein uptake experiments under various electroporation conditions. Pulses were separated by 5 s for multiple pulses. "Concentration" refers to external concentration of the molecule. "Plateau uptake" refers to average number of molecules per ghost in field-strength plateau region. "% External concentration" is a measure of average concentration inside each ghost in field-strength plateau region relative to external concentration.

Concentration	# of pulses	Plateau uptake	% External concentration
10^{-5} M	1	1.4×10^4	1.5
3×10^{-5} M	1	3.0×10^4	1.1
10^{-4} M	1	5.0×10^4	0.6
3×10^{-4} M	1	9.3×10^4	0.3
10^{-3} M	1	2.2×10^5	0.2
10^{-4} M	1	6.1×10^4	0.7
10^{-4} M	2	8.9×10^4	1.0
10^{-4} M	3	1.5×10^5	1.7

To learn whether the plateau concentration achieved at high E can be exceeded by changes in other parameters, we varied the external concentration over two orders of magnitude and used multiple pulses (range = 1 to 3). In each case, a plateau in E is still observed, but the corresponding uptake per ghost increases (Table 1). Not surprisingly, increasing the external calcein concentration increases the amount of uptake. However, the departure from "equilibrium" actually decreases with increasing external concentration. Also, increasing the number of pulses increases plateau uptake to within 1.7% of the external concentration.

These results demonstrate that there are at least two domains of transport: a low-E region in which transport increases with increasing E, and a high-E region in which transport is independent of E. However, unlike with E, the effects of external concentration and of multiple pulses do not saturate over the ranges investigated. This finding has important implications for understanding and modeling mechanisms of transport by electroporation: any mechanism must explain why uptake reaches a plateau at high E, but there is no plateau when other parameters are increased. There are also important implications for those interested in applications such as gene transfection, or any other protocol in which molecular transport is to be maximized: increasing E may be less effective at increasing uptake than providing a greater external concentration, applying multiple pulses, or perhaps changing other parameters.[5]

REFERENCES

1. D. C. Chang, B. M. Chassy, J. A. Saunders, and A. E. Sowers, *Guide to Electroporation and Electrofusion,* New York: Academic Press, 1992.
2. M. R. Prausnitz et al., *Biophys. J.* (in press).
3. A. E. Sowers and M. R. Lieber, *FEBS Lett.* 205: 179-184, 1986.
4. D. C. Bartoletti, G. I. Harrison, and J. C. Weaver, *FEBS Lett.* 256: 4-10, 1989.
5. We thank J. H. Kim, S. Conner, and G. A. Paradis for help in the laboratory, and A. E. Sowers for helpful discussions. This work was supported in part by a fellowship to M. R. Prausnitz provided by Cygnus Therapeutic Systems, and in part by ARO grant DAAL03-90-G-0218 and NIH grant GM34077 to J. C. Weaver.

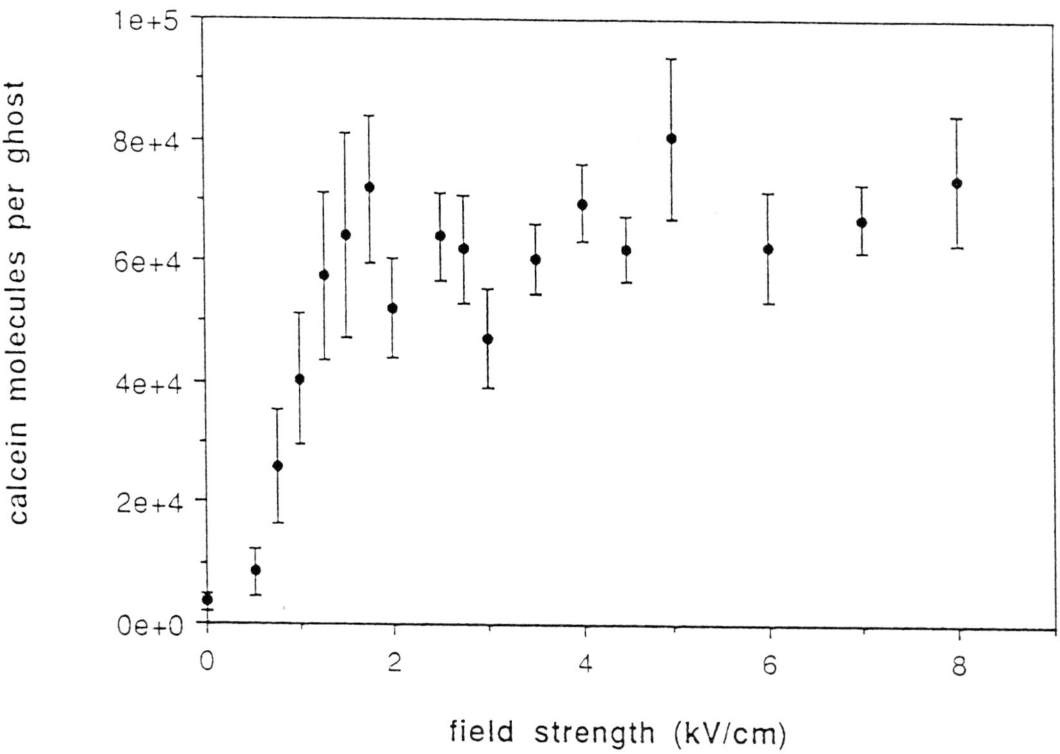

FIG. 1.—Uptake of calcein as a function of field strength. Average number of calcein molecules taken up by each ghost following a single pulse at various field strengths is shown. This graph includes data from more than 10^6 individual ghosts from five or more replicate experiments. Standard error bars are shown.

Martin Blank, Ed., *Electricity and Magnetism in Biology and Medicine*

ELECTROPORATION OF CELL MEMBRANES BY ROTATING ELECTRIC FIELDS

J. Gimsa*

Humboldt University, Berlin

U. Löwe

Federal Health Office, Berlin

P. Marszalek and T. Y. Tsong

University of Minnesota, St. Paul

We monitored the entry of the fluorescence dye, propidium iodide, into murine myeloma cells after poration of the cytoplasmic membrane by application of a rotating electric field for 100ms. The dye becomes strongly fluorescent only by binding to DNA after entering the cell. The poration field was generated between four electrodes driven by four 90°-phase-shifted square-topped signals.[1] Field strength and frequency were about 77 kV/m and 2.5 kHz, respectively. Various results for the fluorescence increase of the cytoplasm after the pulse were obtained. Spherical cells showed either a ring-shaped fluorescence increase or an increase starting at one or more points. For ellipsoidal cells, the increase started from membrane points close to their long axis. These experiments confirm the advantages of rotating fields for biotechnological applications. Gentler field conditions can be used because these fields scan the cell surface. Scanning increases the probability of a poration of randomly oriented nonspherical cells or of the weakest membrane points of electrically inhomogeneous cells.

Electroporation is an important method in immunochemistry and in medical and agricultural research.[2] For some applications, AC fields have been shown to be preferable to DC fields. Chang et al. reported a higher survival rate of fibroblast cells.[3] It has been shown that DC fields primarily induce membrane breakdown on one side of the cell.[4,5] This phenomenon was explained by the assumption of a superposition of external field and physiological transmembrane potential, which causes different stresses for the cell membrane areas facing the positive and negative electrode, respectively. For AC fields these stresses would be equal, causing poration at both poles and thus a higher poration efficiency.[2] The transmembrane potential generated by a rotating electric field, like the one generated by AC, decreases with increasing field frequency.[6,7] In the low-frequency range (below 10 kHz), the polarization process of the cell is dominated by charges induced outside the low-conducting cell membrane and the balancing countercharges in the cytoplasm, i.e., induced positive and negative charges outside the cell facing the positively and negatively charged electrode, respectively. Consequently, the cell interior remains almost field free and the voltage over the cell drops mainly across the membrane. For higher frequencies, the induced transmembrane voltage decreases due to a capacitive bridging of the membrane. In AC fields there are two points for which the transmembrane potential is at a maximum. These points face the electrodes. If the external field frequency is of the order of the time constant of the membrane, a phase shift between external field and induced membrane potential occurs.[6]

MATERIALS AND METHODS

Murine myeloma cells (line Tib9) were chosen because of their spherical shape in suspension. The cells were obtained from American Type Culture Collections. Cells were grown at 37 C in RPMI 1640 medium

---

*Correspondence to: J. Gimsa (Biology), Institute of Biophysics, Humboldt University, Invalidenstr. 42, D(O)-1040, Germany.

(Sigma Chemical Co.) supplemented by 10% fetal bovine serum, 2% L-glutamine, and 0.4% v/v gentamicin (10 mg/ml), in 5% CO_2. For measurements they were collected by centrifugation at 800 g for 5 min, washed once under the same conditions in 0.3M sucrose, and suspended in a small amount of 0.3M sucrose. For poration experiments, a solution of 0.3M sucrose containing 10μM propidium iodide was used. A small amount of the original growth medium was added to adjust the external conductivity to about 5 mS/m. All experiments were carried out in a four-electrode chamber at room temperature (23 C). The distance between two opposite electrodes was about 200 μm. Four 90° phase shifted square-topped signals with a duty cycle of 50% and a signal voltage of 10 V peak-to-peak were applied to the electrodes for 100 ms. Within this time the poration field rotates 250 times for a frequency of 2.5 kHz. Photographs were taken on a 3200 ASA Kodak black-and-white film.

EXPERIMENTS AND RESULTS

The application of the rotating poration field produced a torque acting on the cells.[6] Depending on their adhesion to the microscopic slide, some of them rotated during the pulse, but much slower than the external field (up to four revolutions per pulse). Usually the fluorescence increase took between 5 and 60 s. In general, four different cases must be considered in the analysis of the shape of fluorescence patterns: spherical and nonspherical cells that either rotated or did not rotate during pulse application.

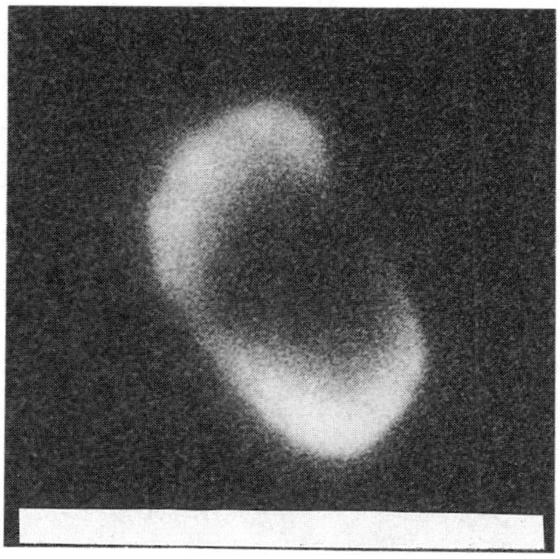

FIG. 1.—Nonspherical cell which rotated during field application 30 s after the pulse. The bar represents 20 μm.

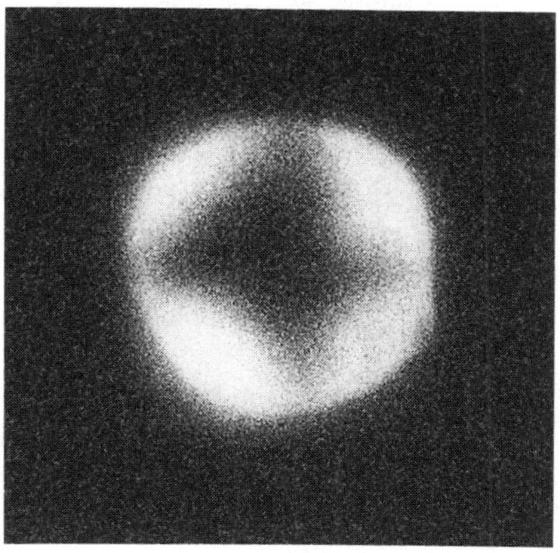

FIG. 2.—Fluorescence pattern of a spherical cell which did not rotate during pulse application 6 s after the pulse. There are four preferential areas of the membrane for the entrance of the fluorescence dye.

DISCUSSION

Field in the Chamber. The square-topped driving signals used for the poration experiments generate a field that does not rotate in a continuous way. In the center of the chamber the field vector jumps four times by 90° during one cycle of the external field. After every jump it points into the center between the two temporarily negative electrodes.

Frequency of the Poration Field. It can be estimated from the time constant of the membrane charging process that for a field frequency of 2.5 kHz almost the entire voltage drop over a cell drops over its membrane, causing dielectric breakdown. On the other hand, the frequency is high enough to insure a sufficient number of field cycles during the pulse time and to avoid electrode polarization problems.

Fluorescence Increase. Fluorescence probe techniques are well established.[4,5,7,8] In poration experiments with DC or pulsed fields, an asymmetrical uptake of the dye has been found,[4,5] which was explained by a superposition of the field-induced transmembrane and the resting potentials. In rotating as well as in AC fields, the induced transmembrane potential at a certain point of the membrane changes its sign and has a positive and an equally high negative maximum during every cycle of the external field. This means that in rotating fields the fluorescence patterns cannot be explained by the polarity of the field: whether a poration at a certain point occurs depends on the absolute value of the induced transmembrane potential and on the quality of the membrane at this point. Figure 1 suggests that the poration of a nonspherical cell occurs preferably at membrane parts close to its long axis. Because this cell did rotate during pulse application, it must be assumed that field strength differences acting on various cell parts due to the electrode arrangement were equilibrated.

The latter was not the case for the cell in Fig. 2. This cell is spherical, so that the transmembrane potential induced by a continuously rotating field should have been constant in the equatorial plane of the cell. The cell did not rotate during pulse application, so that the nonuniform field could cause different stresses for the membrane. With respect to the cell, the electrodes were arranged vertically and horizontally. It can be seen that the fluorescence increase is weaker at membrane parts that face an electrode. The four brighter spots reflect the membrane areas facing the interelectrode spaces that were exposed to a higher field stress.

Special Features of Rotating Fields. In AC fields, only the two points facing the electrodes are permanently under maximum AC-field stress. With regard to a fixed point in continuously rotating fields on the cell membrane, a sinusoidal amplitude change is established by the continuous change of the external field direction. The rotating field scans the cell equator (with respect to the plane of field rotation). Thus stress conditions of the two points in case of normal AC apply to a horizontal ring over the membrane laying in the plane of the rotating field. The nonstressed parts of the membrane become smaller. These considerations suggest that the poration efficiency must be higher. In addition, the probability of a poration of randomly oriented nonspherical cells or of the weakest membrane points of electrically inhomogeneous cells is increased.

REFERENCES

1. J. Gimsa, R. Glaser, and G. Fuhr, "Remarks on the field distribution in four electrode chambers for electrorotational measurements," *Stud. Biophys.* 125: 71-76, 1988.

2. E. Neumann, A. E. Sowers, and C. A. Jordan, *Electroporation and Electrofusion in Cell Biology*, New York: Plenum Press, 1989.

3. D. C. Chang, "Cell poration and cell fusion using an oscillating electric field," *Biophys. J.* 56: 641-652, 1989.

4. W. Mehrle, U. Zimmermann, and R. Hampp, "Evidence for asymmetrical uptake of fluorescence dyes through electropermealized membranes of Avena mesophyll protoplasts," *FEBS Lett.* 185: 89-94, 1985.

5. E. Tekle, R. D. Astumian, and P. B. Chock, "Electropermeabilization of cell membranes: Effect of the resting membrane potential," *Biochem. Biophys. Res. Comm.* 172: 282-287, 1990.

6. G. Fuhr, R. Hagedorn, R. Glaser, J. Gimsa, and T. Müller, "Membrane potentials induced by external rotating electrical fields," *J. Bioelectricity* 6: 49-69, 1987.

7. P. Marszalek, D.-S. Liu, and T. Y. Tsong, "Schwan equation and transmembrane potential induced by alternating electric field," *Biophys. J.* 58: 1053-1058, 1990.

8. C.-J. G. Yeh, B.-L. Hsi, and W. P. Faulk, "Propidium iodide as a nuclear marker in immunofluorescence: II. Use with cellular identification and viability studies," *J. Immunol. Methods* 43: 269-275, 1981.

Martin Blank, Ed., *Electricity and Magnetism in Biology and Medicine*

CELL SURVIVAL FOLLOWING ELECTROPORATION: QUANTITATIVE ASSESSMENT USING LARGE NUMBERS OF MICROCOLONIES

E. Gift,* G. I. Harrison, D. Cho, and J. C. Weaver

Massachusetts Institute of Technology, Cambridge, Mass.

Most electroporation applications are based on greatly increased transport of molecules across the cell membrane and on cell survival. We are applying a new method, Microdrop Technology, to the basic problem of determining which subpopulations of cells experience electroporation and then survive. Microdrop Technology is based on microscopic gel drops (GMDs), which are formed so as to have a high probability of containing zero or one initial cell. Here GMDs were used to isolate cells by placement within physically manipulable microenvironments. The cells were then exposed to electroporating electric field pulses, transferred to a growth medium, and incubated. The resulting microcolonies (viable cells after pulsing) or single cells (dead cells after pulsing) were then measured by flow cytometry.

Electroporation is increasingly used to introduce molecules into living cells. A common observation is that although molecular uptake increases with the magnitude of the electric field pulse used to cause electroporation, cell death also increases.[1] However, relatively little is known about the desired subpopulation: cells that take up significant numbers of molecules and also survive. Short-term viability assays based on "dye exclusion" cannot be relied on, because electroporation fundamentally results in dramatic alteration of dye transport for unknown times after electric-field pulsing.[2]

Clonal growth assays are the most stringent test of cell viability, because a large number of biochemical pathways must participate. Traditional "viable enumeration" of cells is based on clonal growth, but is tedious and very slow. Microdrop Technology provides clonal growth determinations on individual cells within one or two generation's time, and is compatible with exposing cells to electric-field pulses. The method is based on the formation of small volume drops, usually gel microdrops (GMDs), which can have a high probability of initially containing zero or one cell, and allow their clonal growth to be measured (Fig. 1). Following incubation, the relative amount of biomass of individual cells and microcolonies is measured for several thousand cell-containing GMDs, e.g., by flow cytometry.[3] This procedure makes it possible to look for subpopulations of killed and surviving cells based on the stringent test of clonal growth.

METHODS

In summary form, our experimental protocol involves preparing cells in GMDs and exposing a portion of the GMD preparation to a series of different electric-field pulses. Following an incubation to allow for cell recovery and growth, the amount of clonal growth within GMDs is determined.[4] Essential details of the protocol are given below.

Cell and GMD Preparation

Yeast (*Saccharomyces cerevisiae;* ATCC 2341) was cultured in YPD medium at 27 C. Agarose GMDs 37-53 μm in diameter were made from type IX agarose. This material was inoculated with the cell suspension and dispersed into mineral oil, which produced agarose GMDs with the following occupation statistics: 89% unoccupied, 10% occupied by one cell, and 1% occupied by more than one cell.[3]

---

*Correspondence to: E. Gift, Harvard-MIT Div. of Health Sciences and Technology, MIT, Cambridge MA 02139.

Experimental Protocol

Two series of experiments were performed. Both series measured clonal growth in GMDs after electroporation. The first series measured this growth by FITC staining and flow cytometry. The second series measured clonal growth by light scatter in flow cytometry. In addition, electroporative uptake was measured quantitatively by a fluorescent marker (calcein). In both experiment series, GMDs inoculated with yeast were prepared and suspended in YPD. The GMDs were then centrifuged (600-750 g, 3 min) and resuspended in phosphate buffer (first series) or 1 mM calcein (green fluorescence, GF; 623 Da; Molecular Probes C-481) in phosphate buffer (second series). For each electric-field strength, an aliquot (0.8 ml) was placed in a parallel-plate (aluminum electrodes, 0.4cm gap) electroporation cuvette at room temperature (20 ± 5 C). A single exponential pulse ($\tau = 0.6$ ms) was applied to the electrodes. The GMD suspension was then held for 5 min, and in the case of the second series, centrifuged, and washed twice with phosphate buffer. Finally, the GMDs were suspended in YPD and placed in a 27 C incubator for 3-4 h.

Flow Cytometry

In the first series of experiments, the samples were stained with FITC prior to analysis, which gave a fluorescent signal proportional to the size of the microcolony in each GMD. After staining, the samples were washed and resuspended in phosphate buffer. Analysis was done with an Ortho IIs and an argon laser 488 nm line at 35 mW. In the second series of experiments, the samples were resuspended in phosphate buffer and analyzed directly. Analysis was done with a FACStar Plus and an argon laser 488 nm line at 100mW excitation. In both series of experiments, forward light scatter (FS) was used as the trigger event. FS and GF were measured for each of at least 10^4 GMDs per sample. In the second series, GMDs were sorted to microscope slides and visually inspected under a fluorescence microscope to verify our interpretation of the data. Quantitative interpretation of the fluorescence data from bin number to number of molecules was done by use of FITC calibration beads.[5]

RESULTS AND DISCUSSION

Data were obtained for several different electric field pulses, with measurements made on 10^4 agarose GMDs that contained either individual cells (dead; no growth) or microcolonies of about eight cells of the yeast *S. cerevisiae*. Empty (sterile) GMDs and occupied GMDs could be readily distinguished (Fig. 2), as could the difference in stainable biomass between microcolonies, individual cells, and cells that appear to have lost biomass.

Furthermore, the uptake of a small, highly charged molecule (Calcein; $z = -4$, 623 Dalton) can be measured for single yeast cells that are entrapped in GMDs. In this case, the average number of molecules per cell was found to be negligible until an approximate threshold of about $E = 2.5$ kV/cm, and then to rise to a maximum of about 2×10^6 molecules/cell. This preliminary result represents on-going studies. The intent is to extend the measurements to provide simultaneous determinations of electroporative uptake and clonal growth, by use of the uptake of a nondegraded fluorescent molecule (e.g., calcein; FITC-dextran) as a quantitative indicator of electroporation occurrence.

REFERENCES

1. D. C. Chang, B. M. Chassy, J. A. Saunders, and A. E. Sowers, Eds., *Guide to Electroporation and Electrofusion,* New York: Academic Press, 1992.

2. J. C. Weaver, G. I. Harrison, J. G. Bliss, J. R. Mourant, and K. T. Powell, "Electroporation: High frequency of occurrence of the transient high permeability state in red blood cells and intact yeast," *FEBS Lett.* 229: 30-34, 1988.

3. J. C. Weaver, J. G. Bliss, K. T. Powell, G. I. Harrison, and G. B. Williams, "Rapid clonal growth measurements at the single-cell level: Gel microdroplets and flow cytometry," *Bio/Technology* 9: 873-877, 1991.

4. G. I. Harrison and J. C. Weaver, "Effect of an electroporating electric field pulse on cell survival: Determination of clonal growth in gel microdrops" (in preparation).

5. D. C. Bartoletti, G. I. Harrison, and J. C. Weaver, "The number of molecules taken up by electroporated cells: Quantitative determination," *FEBS Lett.* 256: 4-10, 1989.

6. This work was supported by NIH Grant GM34077, U.S. Army Research Office grant DAAL03-90-G-0218, and by the DoD-University Research Instrumentation Program (grant DAAG29-84-G-0066).

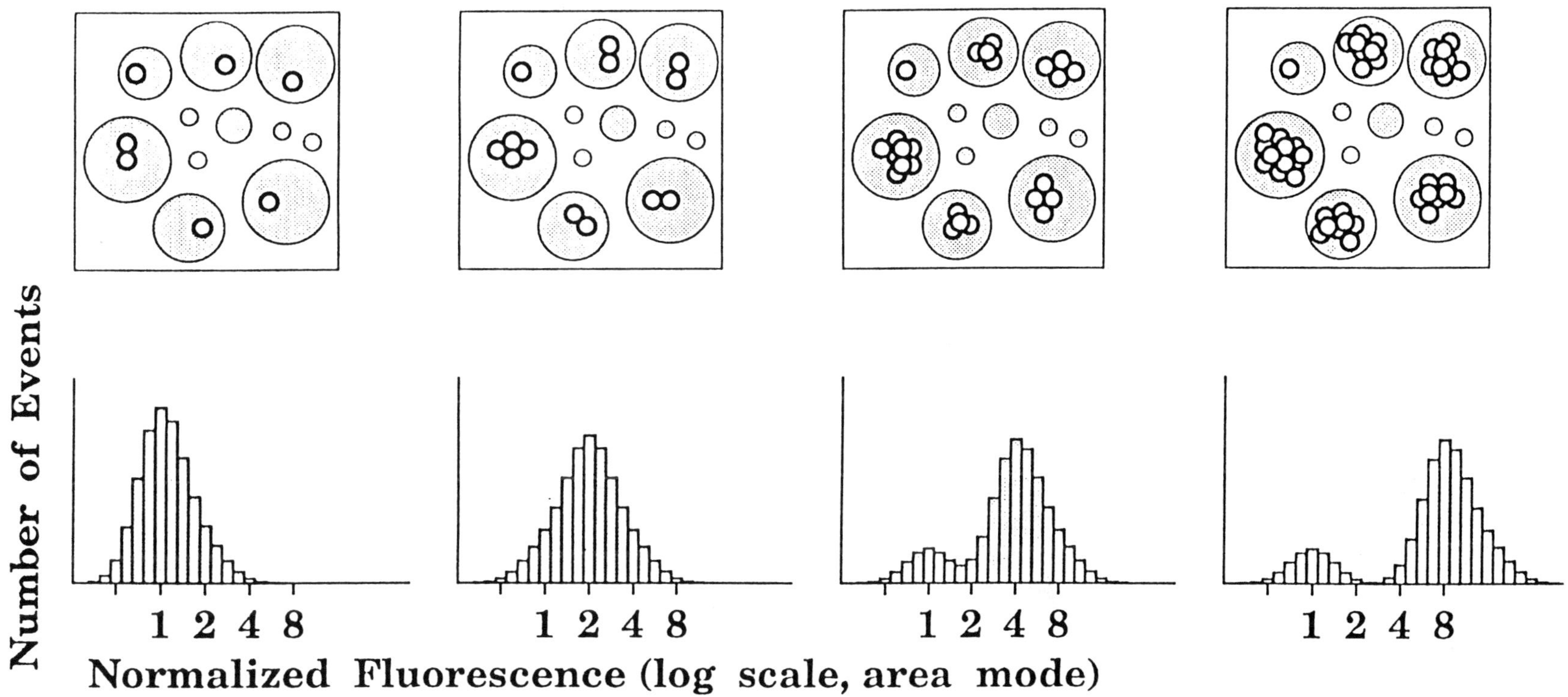

FIG. 1.—Basic idea of using gel microdrops (GMDs) to determine cell survival by clonal growth. Statistical inoculation of microdrops is governed by Poisson statistics. Large permeability of microdrops allows microcolony formation to occur (top sequence) while GMDs are suspended by gentle stirring, and subsequently allows conventional fluorescence stains for biomass to be used with flow cytometry. Flow cytometric measurements are typically one-dimensional histograms (bottom sequence), or two-dimensional cytograms (Fig. 2). In both cases, comparison of one or more biomass signals to the nonincubated case provides a determination of growth.

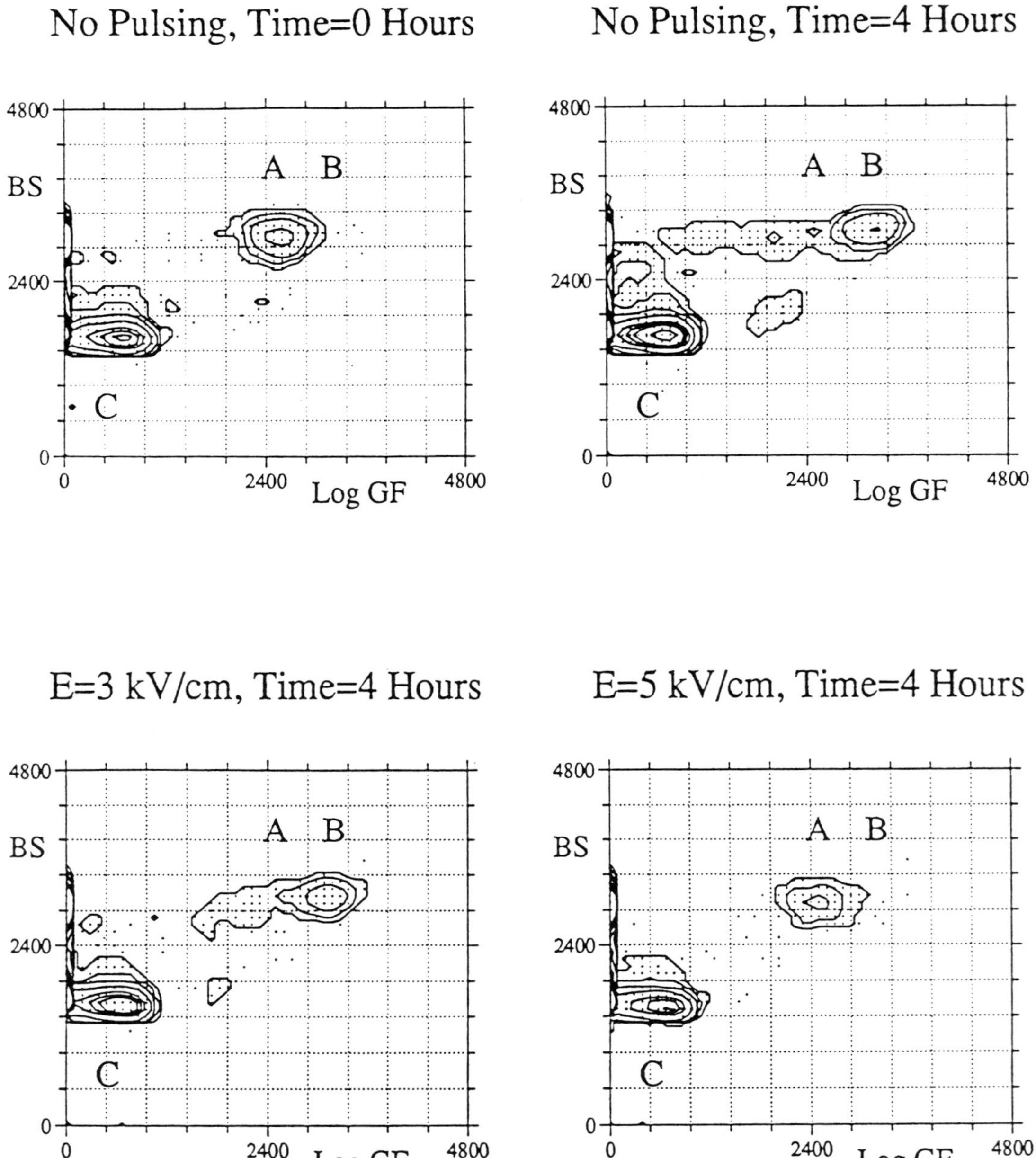

FIG. 2.—Pair of cytograms (BS = "blue scatter" vs GF = "green fluorescence") for 10^4 individual GMDs that contained either individual cells or microcolonies of the yeast *S. cerevisiae*. Populations below "A" have the same GF as double control group and are interpreted as individual cells. Populations below "B" have significantly more GF than the control, and are interpreted as microcolonies. Populations above "C" are empty (unoccupied) GMDs.

Martin Blank, Ed., *Electricity and Magnetism in Biology and Medicine*

TIME COURSE OF ELECTROPERMEABILIZATION

M.-P. Rols* and J. Teissié

CNRS-LPTF, Toulouse, France

Application of electric-field pulses leads to the reversible permeabilization of the plasma membrane of cells. The phenomenon is under the control of the electric-field intensity, the number of pulses N, and their duration T. If the electric-field intensity is the driving force leading to membrane permeabilization, the extent of this phenomenon, i.e., the flow of exchanged molecules between cell and external medium, is controlled by the pulse number and duration. The effect of these two parameters appears not to be the same: at a given cumulated pulse duration (NT = constant) a higher permeabilization efficiency is observed at higher pulse number.

Electropulsation, i.e., submitting cells to strong shortlived electric field pulses, is a new method for the experimental manipulation of the cytoplasm. Its consequences (electropermeabilization, electrofusion, and electrotransformation) are very potent tools for bioengineering.[1]

When applied to a complex system such as a cell, the description of the effects induced by the field pulse can only be phenomenological. It is at the present time impossible to provide information at the molecular level. The most important primary consequence of the field is a position-dependent modulation of the membrane potential difference, which is superimposed on the resting potential difference. An asymmetry in the distribution of the transmembrane field is then present during the pulse.[2] Field effects are then nonuniform on the cell surface. When the resulting membrane potential difference is brought locally above a critical value E_p, a dramatic alteration of the membrane organization state is induced wherever this electric condition is fulfilled.[3] Locally the membrane is then permeable and fusogenic.[4] This process occurs by a two-step kinetic: creation of defects (in a few nanoseconds) followed by an expansion during the field pulse (μs to ms). A very fast stabilization step is present at the end of the pulse (ms). The permeable state of the membrane is local. The cell membrane is then topologically heterogeneous, with no apparent mixing. The part of the cell surface affected by the field is in a metastable organization and returns slowly to the normal organization with a selective permeability (s to min). At the molecular level, a change in the average orientation of the phospholipid polar headgroups is present.[5] This resealing is under the control of the cytoskeleton.[6]

Only few experimental results have been reported concerning the role of the electric-field parameters other than the field strength, i.e., the pulse number and duration. In most cases, this is due to the use of an exponential-decay electric field (capacitor discharge) where the pulse duration is related to τ (i.e., RC) and where only one pulse is applied.

We present here a study concerning the respective role of the pulse number and duration on the electropermeabilization phenomenon of mammalian cells. A square-wave electric-field pulse generator was used because of its advantage in applying constant electric pulses intensity over a time range from μs to ms. This experiment was performed by applying constant total pulse duration (pulse number N times pulse duration T) and following the effect of changing the pulse number and concomitantly of the pulse duration on the permeabilization and resealing of Chinese hamster ovary (CHO) cells.

MATERIALS AND METHODS

Cell Culture. CHO cells of the wild-type cell line (WTT clone) were grown plated in a monolayer (generation time 16-18 h) but adapted for suspension culture in spinner flask (generation time 18-20 h). Cells grown in suspension can be replated very easily. They were cultured according to Ref. 7.

---

*Correspondence to: M.-P. Rols, CNRS Laboratoire de Pharmacologie et Toxicologie Fondamentales, 118 route de Narbonne, F-31062 Toulouse Cédex, France.

Electropermeabilization. The cell permeabilization protocol for cells in monolayer or in suspension has been described elsewhere.[7] Briefly, the culture medium was discarded and replaced by a low-ionic-content saline buffer, the "pulsing buffer" (10mM phosphate buffer, 250mM sucrose, 1mM $MgCl_2$, pH 7.2). Two thin flat stainless-steel parallel electrodes were used. They were connected to a voltage generator of square-wave electric pulses (CNRS electropulser; Jouan, France). The pulses applied to the cells were monitored with an oscilloscope incorporated in the cell electropulser.

Determination of the Permeabilization. Electropermeabilization of CHO cells was quantified by penetration of unpermeant dyes. Cells pulsed in the presence of trypan blue (4 mg/ml, Sigma), were incubated for 5 min at room temperature and then observed under an inverted light microscope (Leitz, Germany). The percentage of permeabilized CHO cells was 100 times the ratio of the number of blue-stained cells to the total number of cells. Cells pulsed in the presence of propidium iodide (100 μM, Sigma) were analyzed by flow cytometry (Becton-Dickinson FACScan).

Determination of Reversibility. Reversibility was assayed by the trypan blue test. Cells were pulsed in the pulsing buffer and suspended in the dye containing buffer after the indicated time span. Cells were incubated 5 min thereafter. Dye penetration was used to monitor the induced permeabilization of the membrane, not to assay the cell viability.

RESULTS

The relative importance of the pulse number and duration parameters was studied by variation of pulse number and duration values, with their product kept constant. Cells were pulsed either plated on a Petri dish or in suspension. A preliminary study showed that the extent of permeabilization at a given (E, N, T) did not dependent on the delay between the pulse (from 0.1 ms to 10 s).

Effect of the Electric Field Parameters on the Permeabilization of CHO Cells Growing in Monolayer

Cells were directly pulsed on the dish in medium containing trypan blue to detect the permeabilization. The relative effect of pulse intensity, number, and duration on permeabilization was studied both on the creation and the resealing phenomena. Permeabilization was detected only for electric-field intensities greater than a threshold value E_p close to 0.3 kV/cm (Fig. 1A). Above this value, the permeabilization was increased to 100%. At a given electric-field strength, the permeabilization efficiency appears to depend on pulse number and duration. Moreover, for a given total cumulated pulse duration, the permeabilization was increased by an increase in the number of pulses: the threshold value E_p was decreased and the slope of the curve at 50% permeabilization increased. As an example, 10 pulses of 0.3 ms had a greater efficiency on permeabilization than 2 pulses of 1.5 ms. A similar dependence of the resealing process was observed with the lifetime of the permeable state enhanced at a high number of pulses (Fig. 1B).

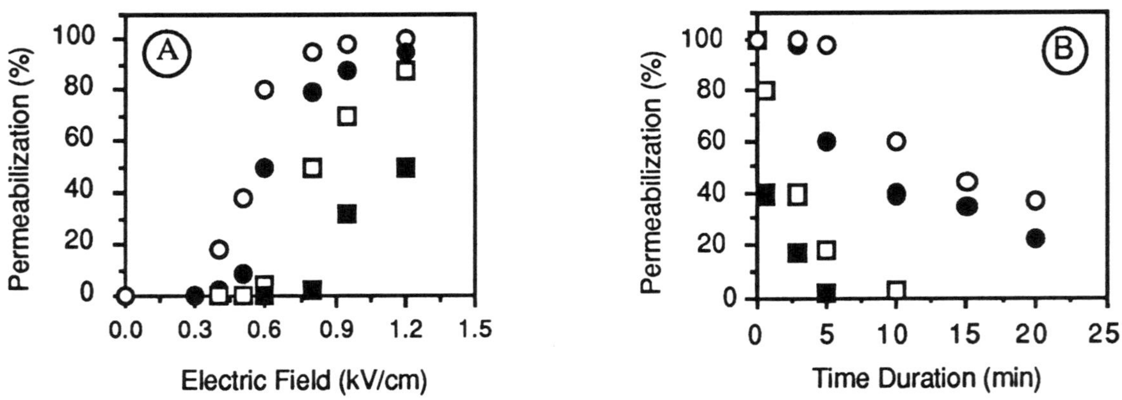

FIG. 1.—Effect of electric-field parameters on monolayer-grown CHO cell permeabilization and resealing. Cells were pulsed in the presence of trypan blue with (A) electric field of increasing intensity, or (B) at constant electric field (1.2 kV/cm). ○ 10 pulses, 0.3ms duration; ● 2 pulses, 1.5ms duration; □ 10 pulses, 0.03ms duration; □ 1 pulse, 0.3ms duration.

Effect of the Electric Field Parameters on the Permeabilization of CHO Cells Growing in Suspension

Cells were pulsed in a medium containing propidium iodide to detect and quantify the permeabilization level by flow cytometry. As for cells growing plated, the percentage of permeabilized cells in suspension was dependent on pulse intensity, number, and duration, and was only detected for electric-field strengths larger than about 0.45 kV/cm. At a given electric-field intensity, permeabilization efficiency was also increased when a high number of pulses was used. This result was observed at two levels. If the percentage of fluorescent cells was only slightly increased (Fig. 2A), the amount of dye accumulated into the pulsed cells (Fig. 2B) was increased by 70 or 100% at constant pulse duration (5 and 0.5 ms, respectively) for electric-field intensities between 0.8 and 1.4 kV/cm. We also observed that the level of fluorescence into pulsed cells was increased by an increase in the pulse duration up to a maximum value, indicating a saturation of the process (data not shown).

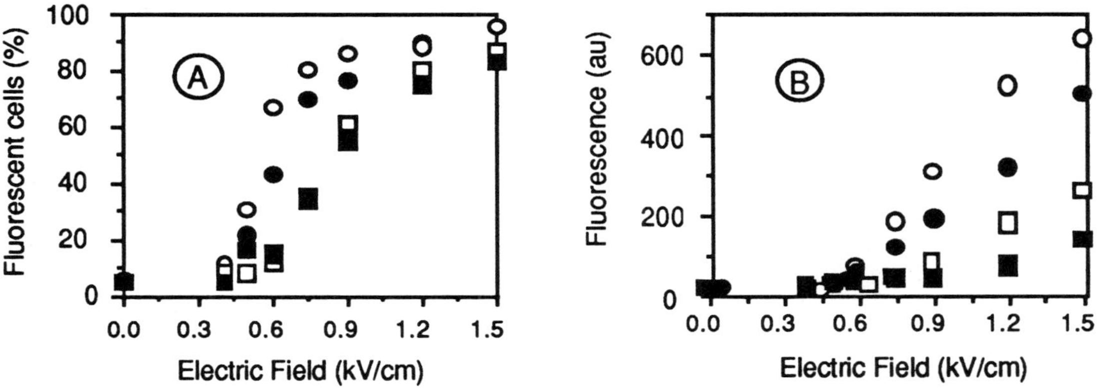

FIG. 2.—Effect of electric-field parameters on suspension-grown CHO cell permeabilization. Cells were pulsed in the presence of propidium iodide with electric field of increasing intensity. ○ 10 pulses, 0.5ms duration; ● 2 pulses, 2.5ms duration; □ 10 pulses, 0.05ms duration; □ 1 pulse, 0.5ms duration.

DISCUSSION

The flow ϕ of a molecule M across the electropermeabilized cell membrane is given by[8]

$$\phi(M,t) = P_M A\, \Delta M(t) X(T,N)[1-(E_p/E)\exp-K(T,N,M)t]$$

where t is the delay after the pulse, P_M is the permeability coefficient when the membrane is permeabilized, and $\Delta M(t)$ is the concentration difference. $X(T,N)$ is the probability that permeabilization indeed occurs at a given point in the critical cap of the cell surface; it depends on the pulse number N and pulse duration T, the critical field intensity E_p, and the field intensity E. Permeabilization of cells is detected for electric-field intensities greater than a threshold ($E > E_p$) and it is increased with the number of pulses and their duration according to an X law. The electric-field strength controls the extent of the critical cap prone to permeabilization and the pulse number and duration act as modulating factors affecting the extent of the flow as well as the lifetime of the phenomenon, i.e, the value of K (dK/dT and $dK/dN < 0$).

Electropermeabilization of cells could be associated with the existence of natural structural defects in the membrane. Under the action of an electric field, such defects could develop and lead to the formation of permeated transient structures or (PTS).[9] When the density of PTS is large enough, coalescence may occur with PTS fusion and could then be responsible for ultrastructural alterations observed in the electropulsed cell membrane.[10,11]

We propose that the fusion of PTS could lead to the formation of a membrane area with a higher flow exchange. The effect of pulse duration T and number N is to increase this flow, but not with the same efficiency. This state is associated with a lower flow of exchanged molecules, as when two pulses with half the duration are applied. The pulse duration T controls the local extent of membrane defects and gives rise to the high permeability of the membrane. New defects are created when new pulses are applied while previous PTS are expanded. Our results show that the kinetics of the local expansion is not uniform but decreases with pulse duration, which could explain why a large number of short pulses gives a higher exchange of molecules across the membrane than one pulse with a long duration (at a same cumulative duration).[12]

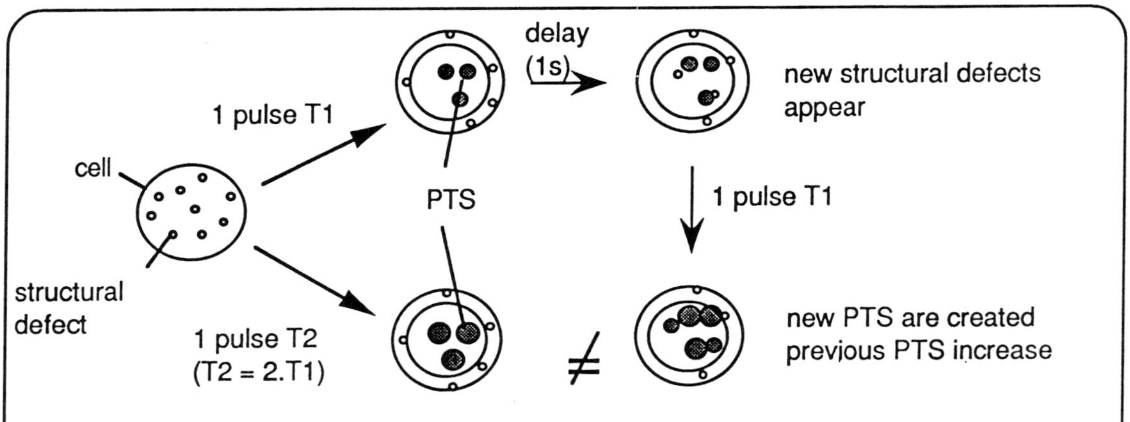

scheme of the effect of pulse number and duration on permeabilization efficiency

the circles into the cell represent the membrane area permeabilizable (where E > Ep)

REFERENCES

1. D. C. Chang, B. M. Chassy, J. A. Sawnders, and A. E. Sowers, Eds., *Guide to Electroporation and Electrofusion*, San Diego: Academic Press, 1992.
2. K. Kinosita, I. Ashikawa, N. Saita, M. Yoshimura, H. Itoh, K. Nagayama, and A. Ikegami, *Biophys. J.* 53: 1015-1019, 1988.
3. J. Teissié and T. Y. Tsong, *Biochemistry* 20: 1548-1554, 1981.
4. A. E. Sowers, *J. Cell Biol.* 102: 1358-1362, 1986.
5. A. Lopez, M. P. Rols, and J. Teissié, *Biochemistry* 27: 1222-1227, 1988.
6. M. P. Rols and J. Teissié, in C. Troyanowsky, Ed., *The Living Cell in Its Four Dimensions*, 1991, pp. 251-260.
7. J. Teissié and M. P. Rols, in N. Latruffe, Y. Gaudemer, P. Vignais, and A. Azzi, Eds., *Dynamics of Membrane Proteins and Cellular Energetics*, 1988.
8. M. P. Rols and J. Teissié, *Biophys. J.* 58: 1089-1098, 1990.
9. I. G. Abidor, K. B. Arakelyan, L. V. Chernomordik, Yu. A. Chizmadzhev, V. F. Pastushenko, and M. R. Rarasevich, *Bioelectrochem. Bioenerg.* 6: 37-52, 1979.
10. D. A. Stenger and S. W. Hui, *J. Membrane Biol.* 93: 43-53, 1986.
11. I. P. Sugar, W. Forster, and E. Neumann, *Biophys. Chem.* 26: 321-335, 1987.
12. This work was partly supported by a grant by the French association Vaincre les Maladies Lysosomiales.

Martin Blank, Ed., *Electricity and Magnetism in Biology and Medicine*

PROTEIN ELECTROLOADING INTO MAMMALIAN CELLS BY "PULSE-FIRST" ELECTROHETEROFUSION WITH ERYTHROCYTE GHOSTS

M.-P. Rols,* F. Dahhou, and J. Teissié

CNRS-LPTF, Toulouse, France

Exposing cells to short high-voltage electric-field pulses leads to the reversible permeabilization of their plasma membranes. Electropermeabilized cells have been shown to fuse by creating contact after electropulsation. Experiments using this fusion method were performed to introduce (under viability conditions) macromolecules loaded in erythrocytes. Compared with other methods developed with this aim (chemicals or virus-mediated cell fusion), such a fusion method is of widest interest and leads to a quantitative and qualitative effective introduction of macromolecules.

Application of electric-field pulses to cells can lead to the transient permeabilization of their plasma membrane.[1] This electro-induced permeated state of the cell membrane is reversible and can lead to cell fusion when cells are put into contact after application of the pulses:[2,3] the electropermeabilized state is fusogenic. Electropulsation is now widely used in cell biology to introduce macromolecules into cells.

Modification of a cell cytoplasmic content can be obtained experimentally by two different approaches: (a) direct injection and (b) fusion with a specifically loaded vesicle. The two procedures may be mediated by electropulsation. Since 1982, plasmid transfer can be operated by electrotransformation.[4] By a similar procedure, macromolecules (such as dextran or antibodies) can be transferred.[5] But it was recently shown that a large fraction of exogenous molecules remain embedded in the membrane. This is a serious artifact of the procedure. Fusion is routinely obtained by electropulsation (electrofusion). The contact-first (CF) procedure has a major drawback when cells of widely differing sizes are to be fused. This is a strong limitation to the use of electrofusion with fusion.

In the present report, we demonstrate that these two problems associated with electropulsation can be avoided by use of a pulse-first (PF) approach, in which the two partners are permeabilized separately.

MATERIALS AND METHODS

CHO Cells

Chines hamster ovary cells (CHO cells, clone WTT[6]) were used. They grow at 37 C in an air/CO_2 (95:5) incubator (Jouan) in Petri dishes (35 mm in diameter, Nunc, Denmark), but have been adapted for suspension culture under gentle agitation (100 rpm). Cells are maintained in exponential growth phase (4×10^5 to 10^6 cells/ml) by daily dilution of the suspension. Cells grown in suspension could be easily replated.

Ghosts Preparation[5]

After washing, packed RBCs were hemolyzed by addition of 4 ml of distilled water. The resulting lysate was centrifuged at 11 000 G for 10 min and the supernatant was discarded. This procedure was repeated five times.

Loading RBCs Ghosts

By a press-well procedure, macromolecules (FITC-dextran 70 kDa, β galactosidase 540 kDA) can be enclosed in RBCs ghosts and remain trapped within them after restoration of impermeability.[6,7]

*Correspondence to: M. P. Rols, CNRS Laboratoire de Pharmacologie et Toxicologie Fondamentales, 118 route de Narbonne, F-31062 Toulouse Cédex, France.

Electropermeabilization

Electropulsation is operated by use of a CNRS cell electropulsator (Jouan, France).[3] It gives square-wave uniform electric-field pulses. Penetration of trypan blue (Sigma T0887, 4 mg/ml in the pulsing buffer) is used to quantify permeabilization of CHO cells. FD70 leakage from electropulsed RBC ghosts is used to monitor permeabilization.

Electrofusion

Electrofusion is obtained as previously described[2,3] by the creation of closed contact between the two cell populations of electropermeabilized cells: CHO cells and loaded RBCs. Electropermeabilized loaded RBCs are pelleted by centrifugation (5 min at 3000 G); then electropermeabilized CHO cells are brought into contact with RBC pellets by gentle centrifugation (5 min at 120 G).

Extension of this technique can be obtained for CHO cells growing on Petri dishes: after electromobilization of plated CHO cells, pulsed ghost suspension is added. Contact between the two cell populations is obtained by centrifugation of the Petri dish (5 min at 35 G).

β Galactosidase Activity Determination

The activity of β galactosidase enzyme is determined 48 h after electrofusion by Xgal metabolization. Blue-stained cells are observed under an inverted light microscope (Leitz, Germany).

RESULTS

1. *Electropermeabilization of CHO Cells.* As shown in Fig. 1(a), permeabilization of the cells is observed for electric-field intensity larger than a threshold value of 0.6 kV/cm (for 10 pulses of 100μs duration). Under the same conditions (number of pulses and duration), cells growing in Petri dishes were permeabilized at field intensities larger than 0.3 kV/cm. Under these conditions (10 pulses of 100μs duration), more than 90% of CHO cells were permeabilized if the field intensity was larger than 1.4 kV/cm. Cell viability was not altered as long as electric-field intensity remained smaller than 2 kV/cm. When the cells were incubated at room temperature, electric-field-induced permeability disappeared spontaneously. All cells were still permeable 5 min after electropulsing, but 90% of them recover plasma membrane impermeability after 20 min post pulse. When electropermeabilized cells were kept on ice, permeability remained present for rather long periods; this property was utilized to fuse the cells with electropermeabilized erythrocytes ghosts loaded with macromolecules.

2. *Electropermeabilization of Loaded Erythrocytes Ghosts.* Erythrocytes ghosts loaded with FITC-dextran 70 kDa or β galactosidase were electropermeabilized as described for CHO cells. Loaded RBCs began to be permeabilized at electric-field intensities larger than 2.5 kV/cm with 10 pulses of 100μs duration, and more of 90% of this cell population were permeabilized for 4 kV/cm field intensity (Fig. 1b). Conditions for permeabilization of RBCs were evidently rather different from those for permeabilization of CHO cells, due to the size difference between the two cell populations.

3. *Fusion is Obtained by Creating Contact Between Electropermeabilized CHO Cells and Loaded Erythrocytes Ghosts.* As noted above, the size difference in CHO and RBC cells means that they have different optimal conditions for permeabilization. Electropermeabilization of loaded ghosts was obtained only under field conditions (4 kV/cm) where CHO cells had lost their viability. Fusion can nevertheless be obtained when a derivative of the pulse-first procedure (PF) is used. The two cell populations were electropermeabilized separately (step A) and contact between them was established by gentle centrifugation (5 min at 35 G, step B). The fusogenic state was preserved by keeping the pulsed cells on ice. Heterofusion was performed by incubation for 30 min at 37 C, to allow CHO cells to reseal.

As shown in Table 1, we could obtain penetration of FITC-dextran 70 kDa (A) and β galactosidase (B) in CHO cells without affecting their viability. No fusion was observed if close contact was not established between the two cell populations, or if one of them was not previously electropermeabilized. The labeling of 10% of the CHO cell population when brought into contact with ghosts that are not permeabilized may be explained by the metastable organization of the ghost plasma membranes induced by their preparation.

TABLE 1.

Electric Field Intensity (kV/cm) (10 pulses of 100 μs were applied)			
CHO cells	Loaded Ghosts	Fluorescent CHO cells (%)	CHO cells Viability (%)
1.4	4	90-95	100
1.4	1.4	10	100
4	4	<0.5	<10
1.4	0	10	100
0	4	5	100

DISCUSSION

The present study provided direct evidence that the pulse-first procedure (PF) electroheterofusion occurred as a consequence of electropermeabilization, two events that do not occur simultaneously. The ability to fuse is conferred on the cells by the creation of a new long-lived metastable state of their plasma membrane that we called TPS (transient permeated structures).[3] In the present example, electroheterofusion is obtained between mammalian cells and vector vesicles (erythrocyte ghosts) loaded with the macromolecules to be transferred.

With regard to other methods developed in this field, the choice of loaded erythrocyte ghosts appears to be of interest. Not only can RBCs be obtained in large quantities, absence of a nucleus and prior hemolysis prevent contamination by their own material during heterofusion. Under the most favorable conditions, it is possible to obtain penetration of large quantities of FITC-dextran 70 kDa or β galactosidase enzyme into 90-95% of the CHO cell population without affecting viability. Another advantage of this approach is that no field-mediated insertion occurs. That was described to be the case for the electroinjection method, in which a mixture of macromolecules and cells was pulsed. A large fraction of the macromolecules associated with the cells by the pulse was not in the cytoplasm but remained in the membrane, as shown by their protease sensitivity.[10] Such an artifact is not present in the PF heteroelectrofusion.[11]

REFERENCES

1. E. Neumann and J. Rosenheck, *J. Membrane Biol.* 10: 279-290, 1972.
2. A. E. Sowers, *J. Cell Biol.* 102: 1358-1362, 1986.
3. J. Teissié and M. P. Rols, *Biochem. Biophys. Res. Commun.* 140: 258-266, 1986.
4. E. Neumann, M. Schaefer-Ridder, Y. Wang, and P. H. Hofschneider, *Embo J.* 1: 841-845, 1982.
5. H. Lambert, R. Pankov, J. Gauthier, and R. Hancock, *Biochem. Cell Biol.* 68: 729-734, 1990.
6. M. M. Gottesman, *Molecular Cell Genetics,* New York: Wiley, 1985.
7. M. Furusawa, T. Nishimura, M. Yamaizumi, and Y. Okada, *Nature* 249: 449, 1974.
8. M. Rechteiner, *Exp. Cell Res.* 93: 487-492, 1975.
9. R. A. Schlegel and L. McEvoy, *Methods Enzymol.* 149: 293-300, 1986.
10. M. Glogauer and A. G. McCulloch, *Experimental Cell Res.* 200: 227-234, 1992.

11. Thanks are due to Mrs. Zalta for her help with the cell culture. Dr. Rols was recipient of a Fondation pour la Recherche Médicale fellowship.

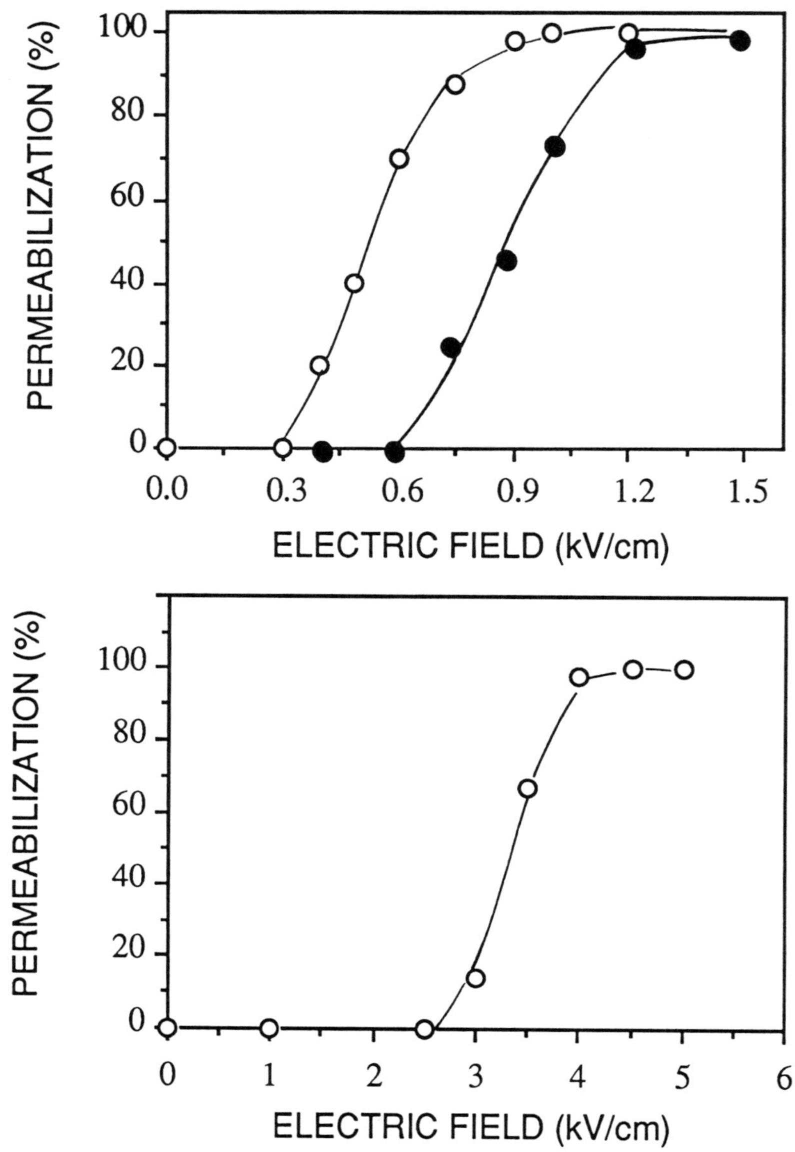

FIG. 1.—Cell permeabilization is a function of field strength and species. (a) CHO cells are pulsed (10 times for 0.1 ms) in presence of trypan blue; ○ cells growing plated on Petri dishes, ● cells growing in suspension. (b) RBC ghosts are first loaded with fluorescent FD 70, then pulsed (10 times for 0.1 ms); leakage of fluorescent molecules is measured by spectrofluorimeter.

Martin Blank, Ed., *Electricity and Magnetism in Biology and Medicine*

ARE DISPERSIONS IN DIELECTROPHORETIC AND ELECTROROTATIONAL CELL SPECTRA IN THE 10kHZ REGION RELATED TO PROTEIN PROPERTIES?

J. Gimsa* and E. Donath

Humboldt University, Berlin

Electrical methods offer the possibility of very sensitive measurements of ionic currents. DC measurements of electric properties of proteins have attained high sensitivity (patch clamp), but AC measurements are very rare. Dielectric properties of single cells can be measured either by dielectrophoresis (DP) or electrorotation (ER) over a wide frequency range. When DP and ER data of different cell species were compared on the basis of the single-shell model, discrepancies in membrane properties derived from DP and ER data were found. This finding may be explained by the introduction of an additional relaxation in the 10kHz region that may be based on two different mechanisms: relaxation of an additional internal membrane charging process, and relaxation of a protein-mediated ion transport process. Model calculations were done to consider the effect of the fast anion-transporter of human red cells (band 3) on the first critical frequency of DP and the first characteristic frequency of ER, respectively, over the external conductivity.

DP and ER can be described by a unified theory.[1] A single-shell model describing the cell as a homogeneous sphere covered by a membrane is the simplest model used to obtain dielectric cell parameters from measured spectra. If this model were applicable to the description of cells, DP and ER data should yield identical cell parameters. Both DP and ER spectra of this model are well defined by certain characteristic points (Fig. 1),[2] which can be easily determined experimentally. Experiments were carried out to investigate the dependence of the first critical frequency f_{ct1} of DP, and the first characteristic frequency f_{c1} of ER, respectively, on the external conductivity. If the frequency dependence of the dielectric membrane properties could be attributed to the relaxation of a protein-mediated ion transport process, an interesting possibility for the study of ion transport via membrane proteins would be offered.

MATERIALS AND METHODS

Materials and methods have been described in detail previously.[2] In all experiments, uniformly spherical cells (discocytes in the case of erythrocytes) were chosen. The isotonic measuring solutions were adjusted to the desired external conductivity. DP and ER measurements were performed in a two- (needle and plate) and four-electrode chamber (4 needles), respectively. For ER measurements, four 90° phase-shifted signals of 2.4 V were applied to the chamber electrodes, and measurements were made at the center of the chamber. For DP measurements, two 180° phase-shifted driving signals were used. For myeloma, square-topped signals with a duty cycle of 50% were used.

Theory

It has been shown that the real part of the complex dipole moment of a particle determines the magnitude of the DP force experienced by the particle in a nonuniform electric field, and the imaginary part determines the torque in a rotating field.[1] The frequency-dependent part of the induced dipole moment of a single-shell sphere can be expressed by a dimensionless parameter $\vec{m}'$ (Fig. 1).

---

*Correspondence to: J. Gimsa (Biology), Institute of Biophysics, Humboldt University, Invalidenstr. 42, D(O)-1040, Germany.

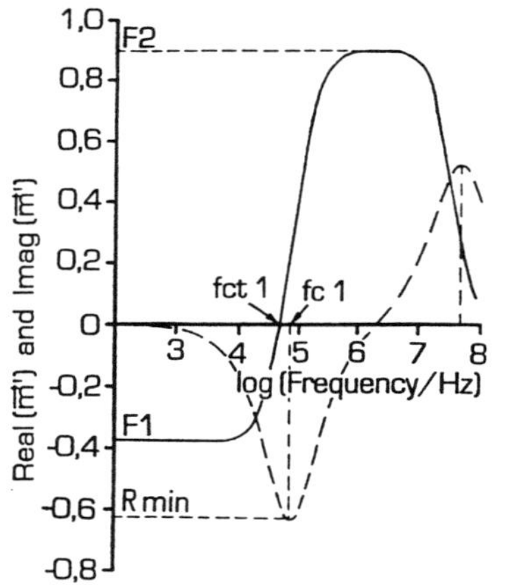

FIG. 1.—Theoretical DP (—) and ER (---) spectra of single-shell model.

Both spectra can be described by certain characteristics. In DP, the cell experiences no net force at a certain frequency of the external field if the real part of the induced dipole moment is zero. In the low-frequency range of the DP spectrum this is true for the critical frequency f_{ct1}. Related to this dispersion is a peak of the imaginary part of the induced dipole moment. This peak is characterized by the first characteristic frequency f_{c1}. At f_{c1} the cell spins at maximum speed against the rotation direction of the field (R_{min}). All characteristic points of Fig. 1 (F_1, F_2, and f_{ct1} for DP and R_{min} and f_{c1} for ER, respectively) can be expressed by cell radius, membrane capacitance, area specific membrane conductance, and external and internal conductivity, respectively.[2]

RESULTS

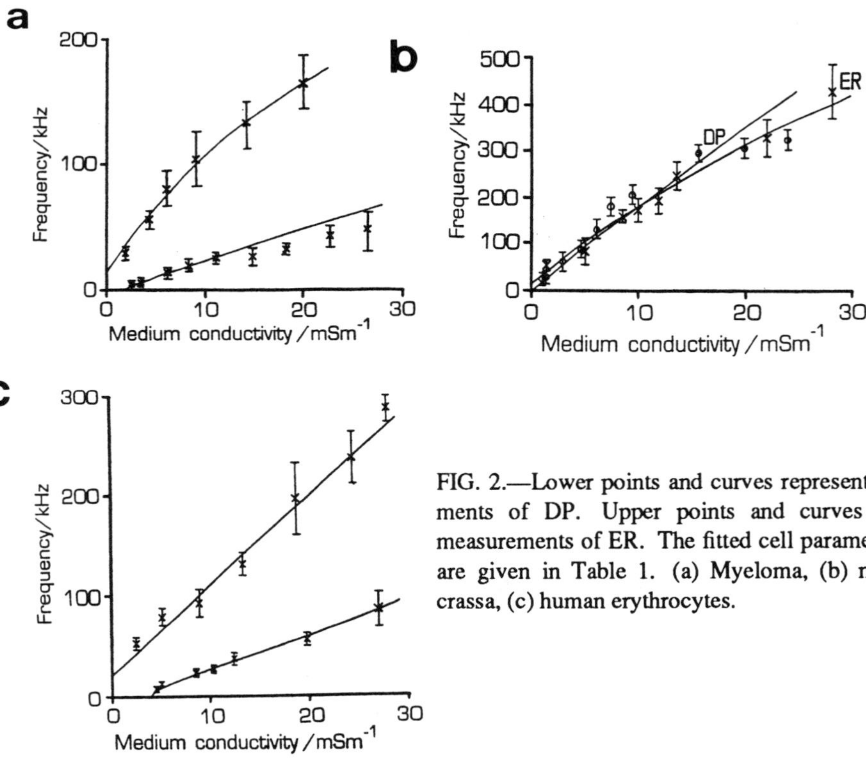

FIG. 2.—Lower points and curves represent measurements of DP. Upper points and curves represent measurements of ER. The fitted cell parameter values are given in Table 1. (a) Myeloma, (b) neurospora crassa, (c) human erythrocytes.

Measured data were fitted with a nonlinear regression program based on the equations given in Ref. 2. Parameters found are given in Table 1.

TABLE 1.—Parameters used to calculate f_{ct1} and f_{c1} curves.

Parameter	Neurospora f_{ct1}	Neurospora f_{c1}	Myeloma f_{ct1}	Myeloma f_{c1}	Erythrocyte f_{ct1}	Erythrocyte f_{c1}
Radius (μm)	8.8	8.8	5.81	5.81	3.0	3.0
Membrane capacitance (10^{-2}F/m^2)	0.81	0.40	1.55	0.45	0.41	0.52
Membrane conductance (S/m^2)	500	500	380	380	44	480
Internal conductivity (S/m)	0.061	>1.0	>1.0	0.068	0.1	0.154

DISCUSSION

Three-shell Model. For neurospora and myeloma cells, attempts to fit f_{ct1} and f_{c1} values with the same parameter set failed. The membrane capacitances found from the analysis of f_{ctl} plots are much higher than those found from f_{c1} analysis. Thus, the behavior of both cell species have large nuclei, the spectra may be influenced by the nucleus membrane. This additional internal membrane system can be considered by means of the three-shell model. It has been shown that the three-shell model can explain the contradictions found.[2,3]

Protein Dispersion Model. For erythrocytes higher membrane capacitances and internal conductivities are usually reported than the values given in Table 1. For mouse erythrocytes, Asami et al.[3] measured 0.72×10^{-2} F/m^2. The three-shell model explanation for this discrepancy is not plausible for erythrocytes, since no internal membrane systems exist. In human erythrocytes the band 3 protein is responsible for the fast anion exchange. There are of the order of 10^6 copies per cell working[4] at transfer rates of up to 50000 ions, which would correspond to AC membrane conductivities of the order of 1000 S/m^2. It is known that band 3 does not contribute to the DC conductivity, because the sequential transport mechanism only allows one ion at a time to be transported across the membrane. The analysis of this transport system yielded a frequency-dependent, area-specific membrane conductance and capacitance, respectively.[4] The contribution of band 3 to the low-frequency capacitance can be calculated from the high-frequency conductance and can be expected to be between 10^{-2} F/m^2 and 2.4×10^{-3} F/m^2. Although the contribution of the band 3 to membrane capacitance is of the order of that of the lipid membrane in the f_{ct1} and f_{c1} plots, only deviations that would be hard to measure could be found (see Fig. 2). Theoretical considerations showed that the low-frequency parts of the curves are particularly affected. Above the dispersion frequency of protein transport, the curves approach the frequency-independent parameter curves. Inhibition of the band 3 system by the specific inhibitor DIDS did not significantly change f_{ct1} over the external conductivity plot. From $F1$ (as well as from R_{min} and f_{c1} changes) it could be concluded that inhibition of band 3 significantly decreases the membrane conductance.[5]

REFERENCES

1. H. P. Schwan, "Dielectrophoresis and rotation of cells," in E. Neumann, A. E. Sowers, and C. A. Jordan, Eds., *Electrorotation and Electrofusion in Cell Biology*, New York: Plenum, 1989.

2. J. Gimsa, P. Marszalek, U. Löwe, and T. Y. Tsong, *Biophys. J.* 60: 5-14, 1991.

3. K. Asami, Y. Takahashi, and S. Takashima, *Biochim. Biophys. Acta* 1010: 49-55, 1989.

4. E. Donath, M. Egger, and V. Ph. Pastushenko, *Bioelectrochem. Bioenerg.* 23: 337-360, 1990.

5. This work was supported by grant I/66158 from the Volkswagen-Stiftung. The aid of J. Lochmann for the DP measurements on red cells is gratefully acknowledged.

Martin Blank, Ed., *Electricity and Magnetism in Biology and Medicine*

PULSE IMPEDANCE MEASUREMENTS: THEIR APPLICATION IN MEDICINE AND RESEARCH

U. Pliquett* and F. Pliquett

University of Leipzig, Germany

A biological object such as tissue, cell suspensions, or individual cells can be described by suitable models which treat physical, chemical, or biological behavior. One point of view uses the object's passive electrical properties, which can be obtained by application of an electrical signal to the object and measurements of the response of the system. A special method, discussed here, is the pulse impedance measurement.

METHOD

The simplest case of passive electrical behavior is ohmic dissipation in a resistor. But a biological tissue is a complex composition of resistive and capacitive components. These components are mainly the resistance inside and outside the cells. All of the cells' membranes (cellular membrane, membranes of intracellular structures) lead to a capacitive component. Because the membranes contain lipids and dynamic pores, the membrane capacitor has a variable loss factor. Another component, not significantly dependent on the object, is electrode polarization. It is approximately comprised of a resistive and a capacitive component.

We have made use of special equipment based on two electrodes. The application of a rectangular pulse wave leads to a pulse deformation, approximately expressed by

$$I = c(1 - a e^{-t/\tau_1} - b e^{-t/\tau_2})$$

where c is a constant that depends on the interface, a and b are coefficients of the exponential functions, τ_1 and τ_2 are time constants, I is the measured current, and t is the time after the pulse.

The two exponential functions have very different time constants, differing by a factor of more than 100. It is thus possible to calculate four independent parameters (two coefficients and two time constants) from the deformed pulse. The advantage of using an averaging technique in the measurement is a lower noise level. From these four parameters we calculate an electrical equivalent circuit containing four elements. Two elements describe the current path inside the cell: a resistor for the ohmic path and a capacitor for the sum of the membrane capacitances. The outside cell path is modeled by a resistor and a capacitor, whose capacitance depends mainly on the electrode polarization. In this model there are two approximations: (1) all the elements are linear and time independent; and (2) the loss factor of the membranes is negligible. However, there are also advantages: (1) all elements can be calculated from the deformed pulse wave; and (2) each element corresponds primarily to a biophysical parameter.

In this paper we describe two applications in medicine and research.

TUMOR DIAGNOSTICS BY PULSE IMPEDANCE MEASUREMENT

The biophysical differences between normal tissue and a tumor leads to a difference in their passive electrical behavior. We determined that a parameter h_β is significantly different for normal and for tumor tissue. This parameter, obtained from the deformed pulse, is the difference between the height of the pulse after the rise and the height after 125 μs. From the biophysical point of view this parameter corresponds to the volume contained within intact cell membranes, relative to the whole volume of the tissue. A special device we developed uses h_β to differentiate between normal tissue and tumor. Because each individual tissue has its characteristic behavior, one must in each case normalize tumor tissue measurements to normal tissue in the same organ. The device works at a pulse frequency of 4 kHz and a pulse voltage of 1.2 V. The

*Correspondence to: Uwe Pliquett, Universität Inst. f. Biophysik, Liebigstr. 27, D(O)-7010 Leipzig, Germany.

electrodes may be of various types, invasive or noninvasive. Both kinds are round, with a radius of 1 mm and a separation of 3 mm. As material we use stainless steel, clamped in Teflon. Results obtained by this device are scaled between 1 and 99, with normal tissue having the higher values. In cooperation with Drs. Kühndel and Stafri, we carried out measurements on a large number of patients at the site of the *portio uteri*. The results are shown in Fig. 1.

We used three kinds of invasive electrodes, 1, 2, and 3 mm deep. The figure shows the number of patients. We found a significant decrease from the high values of normal tissue to the unknown tissue. We found a very high degree of correlation between our measurements and the tissue type (normal or malignant), with malignancy determined by histology.

DYNAMIC OF PASSIVE ELECTRICAL BEHAVIOR AFTER ELECTRICAL BREAKDOWN

To observe a change in behavior after an electrical breakdown one must use a very rapid method,[2] because it is a very transient event. At a pulse frequency of 4 kHz, a time resolution of 250 μs is possible. The disadvantage in this case is the high level of noise. Using average values from about 40 periods one obtains a time resolution of 10 ms and an acceptable level of noise. For this application we developed special equipment, containing a switch between the pulse measuring device and the high-voltage pulse. The time regime at the measuring object is shown in Fig. 2.

The voltage and duration of the breakdown impulse is variable. We use a two-electrode interface so that the same electrodes are used for the breakdown impulse and for the measurement. From the measurement we obtain four parameters: two for the exponential function belong to the fast part, and two to the slow one. They are usually averaged, for example for 10 ms. Based on these values and additional information about the interface, we calculate the four elements of the previously described equivalent electrical circuit. We normalize the calculated values of the parameters to the values obtained before breakdown. In a special series of experiments we used erythrocytes, which were washed, suspended in phosphate buffer, and centrifuged to yield a high concentration. In half the cases we applied a high-voltage pulse (1500 V/cm, 24 μs), which led to a reversible breakdown. In all other cases a larger pulse (9000 V/cm, 24 μs) led to a nonreversible breakdown. In each case we observed a large change in the passive electrical properties. Specifically, we found a large decrease in the intracellular resistance, and after about 1 s, an increase of the extracellular resistance. The fractional change in capacitance associated with the membranes changes only slightly. In the case of the nonreversible breakdown, it increased slightly. If the breakdown is reversible, the values of the elements return within 30 min. In comparison, the elements after a nonreversible pulse had significantly different values. For example, a very large extracellular resistor and a small intracellular resistance were found, possibly as a result of a cell swelling.

CONCLUSIONS

The pulse impedance measurement may be a suitable method to differentiate between different kinds of biological tissue. It may be of use to confirm a cancer diagnosis or to diagnose a cancer at an early stage. For this application there now exists easy-to-use equipment, which provides immediate measurements. The high time resolution, compared with other methods (e.g., measurement in the frequency domain), can be used for the observation of fast events such as the dynamic changes that occur after electrical breakdown. The disadvantage may be the lower accuracy than in the frequency domain, but it is sufficient for most applications.

REFERENCES

1. F. Pliquett and U. Pliquett, "Impulsimpedanzmessgerät UP-3 und UP-4 zur medizinischen Diagnostik," *Neurervorschlag* KMU IX/6410/87, 1987.

2. U. Pliquett, *Elektrische Netzwerkanalyse biologischer Gewebe,* Dissertation, Mittweida College of Engineering, 1991.

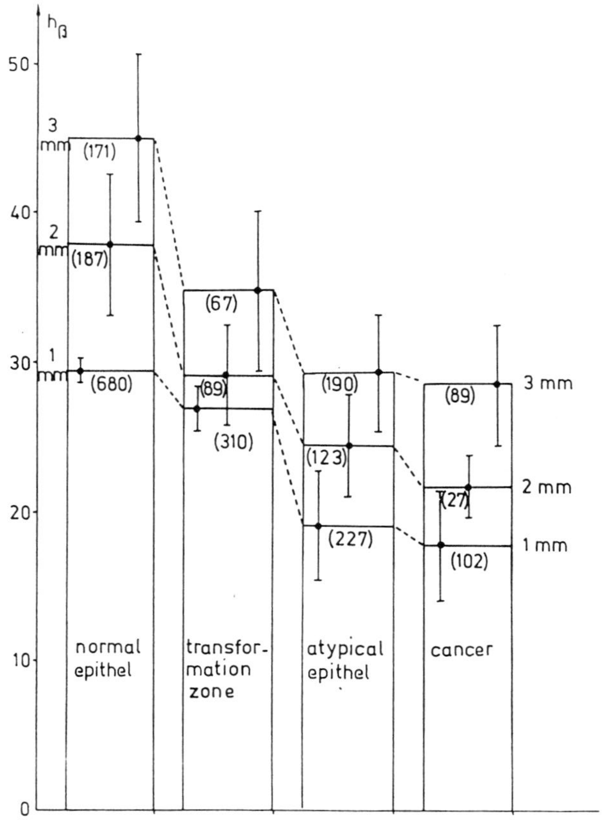

FIG. 1.—h_β from normal and cancer *portio uteri* epithelium tissue measured in vivo at a depth of 1, 2, and 3 mm (in cooperation with Drs. Kühndel and Stafri).

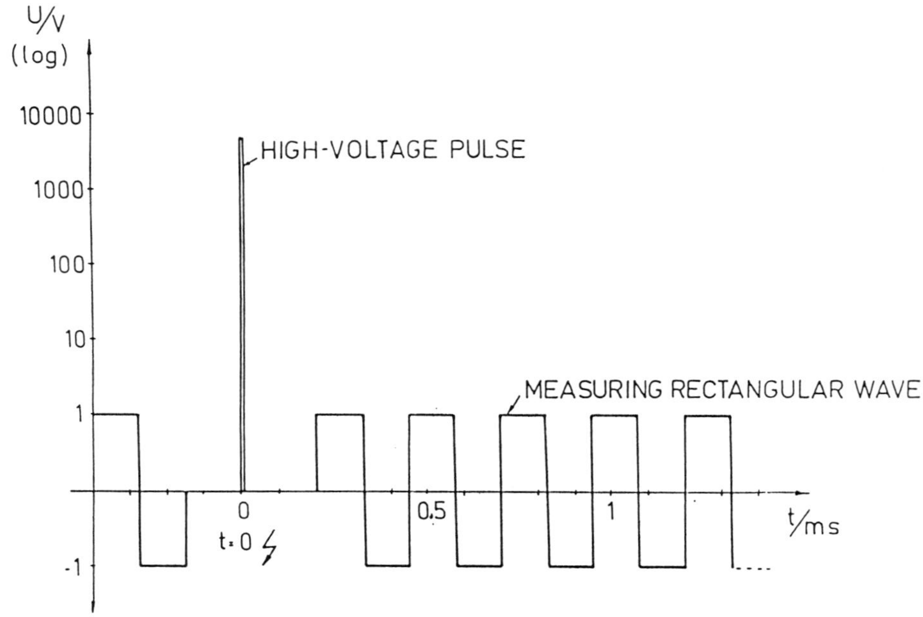

FIG. 2.—Breakdown and measuring pulses.

Martin Blank, Ed., *Electricity and Magnetism in Biology and Medicine*

FAST KINETIC STUDES OF *ESCHERICHIA COLI* ELECTROTRANSFORMATION

N. Eynard and J. Teissié*

Laboratoire de Pharmacologie et Toxicologie Fondamentales, Toulouse, France

Direct gene transfer is obtained in *E. coli* by use of a square-wave electric pulse. The field pulse is observed by video monitoring to induce an orientation of the bacteria parallel to the field lines. Fast kinetic turbidity changes indicate that this process is very fast. As such, under pulsing conditions likely to induce transformation, only the ends ("caps") are affected by the field. A strong cytoplasmic ion leakage occurs during the pulse and affects the interfacial ionic concentration. The contact between the plasmid and the bacteria can be very short before the pulse, but the plasmid must be present during the pulse. The plasmid remains accessible to externally added DNAses up to 5 s after the pulse, which suggests that the transfer step is slow. Electric-field-mediated transfer takes place in two steps: an anchoring process during the pulse followed by crossing of the membrane.

Several processes are induced at the cell level upon electropulsation. The well-known Joule heating phenomenon cannot be neglected under electrical conditions likely to induce electrotransformation. Dipoles are induced by the electric field and are oriented by a field effect. A reorientation of the pulsed vesicles may occur. But the most dramatic consequence is a size-dependent modulation of the cell membrane potential difference.[1] When this effect is larger than a critical threshold, a membrane structural reorganization takes place and induces a reversible permeable state (so called *electropermeabilization*)[1]. In 1982 gene transfer was mediated in eukaryotic cells by electropulsation (*electrotransformation*).[2] Electrotransformation was successfully applied to many systems of mammalian cells and plant protoplasts.[3] More recently, gene transfer was obtained with walled systems such as bacteria[4] or yeasts.[5] Electrotransformation is in fact now routinely used in many laboratories of bacterial genetics. Yields as high as several percent have been reported when strains known to be easily transformed by the $CaCl_2$ method are used.[6,7] Mechanisms of electrotransformation of bacteria remain unexplained, however, just as they are in other systems. Nevertheless, we have demonstrated a correlation between electropermeabilization and electrotransformation in the case of *E. coli* and *S. thyphimurium*.[8]

In the present work, we investigate the processes by square-wave pulses, in a pulsing buffer with a low ionic content as routinely used (i.e., 10% glycerol). The model studied is *E. coli* CB0129 and pBR 322. As long pulse durations (several milliseconds) are needed to obtain high yields in transformation,[8] the bacteria are always suspended in an almost ion-free solution, which tends to limit Joule heating during the pulse. A direct investigation of the processes affecting the bacteria during the pulse was performed by various approaches with very fast kinetic resolutions: video monitoring, turbidity, and conductance. Transfer of plasmid DNA was analyzed by rapid mixing techniques in which a time resolution of the order of 1-2 s was obtained.

RESULTS

Orientation Changes of the Bacteria During the Pulse

Since they are rods, *E. coli* are affected by the field through dipole induction.[9] The field polarizes bacteria, which are thus submitted to a torque. This phenomenon, observed previously but not quantified, was observed here by two methods.

The direct video observation shows that the random distribution of the bacterial population is changed to a well-ordered distribution just after the pulse; the time resolution is 40 ms, much longer than the pulse duration, but several orders of magnitude shorter than the relaxation process induced by the Brownian motion.[9] This well-ordered distribution was detected even under very mild pulsing conditions, when neither

---

*Correspondence to: J. Teissié, LPTF-CNRS, 118 route de Narbonne, F-31062 Toulouse Cédex, France.

transformation nor permeabilization were observed (low field intensity).

Studies of fast change in transmitted light confirm this conclusion indirectly. Light transmitted by a bacterial suspension is under the control of the orientation distribution. We know than an electric field applied to a suspension of nonspherical cells creats an optical anisotrophy that can be measured by changes in birefringence[9] or light transmission.[10] A decrease of the transmitted light during the pulse (Fig. 1) is in agreement with the alignment of the long axis of the bacteria's rod along the field lines (perpendicular to the light direction in our experiments). But the most important information provided by the transmitted-light experiments is the rise time of the orientation change. Under field conditions likely to induce a high yield in transformation (E ˜>˜ 2 kV/cm), the orientation process occurs with a half life shorter than 1 ms; i.e., in a minute fraction of the pulse duration. The conclusion is that for a pulse of several milliseconds, the field is applied to a bacterial population in which all the rods are parallel to the field lines.

Leakage of Cytoplasmic Content

We have shown previously that electropulsation of intact bacteria induces a leakage of ATP.[8] If ATP can leak out, one would expect that other cytoplasmic molecules and ions should be able to cross the membranes. As the experiments were done in a medium with a very low conductivity in order to minimize Joule heating, the outflow of cytoplasmic ions should cause an increase of the conduct of the suspension (a decrease of resistance), as had been observed in the case of red blood cells.[11] The comparison of the kinetics in transmitted light and conductance changes demonstrates that the contribution of the orientation process to the change is negligible.

Kinetics of the Gene Transfer

The time the plasmids are added to the bacteria plays a critical role in the yield of transformants (Fig. 2). If the plasmid is added before pulsing, the yield remains the same whatever the time of contact (up to 1 h, on ice), as reported previously.[12] But when the DNA is not added until after the pulse, no transformant is detected, not even when the delay between pulsing and plasmid addition is less than 1 s.

DNAse treatment of the plasmid-bacteria mixture affects the yield in transformants. If DNAse is present before pulsing, no transformant is obtained, which merely shows that the enzyme is active enough to inactivate all the plasmids we add. The viability of cells is not modified by the presence of DNAse. But if DNAse is added after pulsing, a recovery in the yield in transformants is obtained; a lag in the resistance to DNAse action is clearly present. We checked to make sure that this outcome was not due to a diffusion of DNAse inside the cells, and found that DNAse does not affect the viability of the pulsed bacteria nor the resistance of pulsed bacteria containing a plasmid. One should then conclude that plasmids remain subject to cleaving by DNAses up to 3 s after the pulse.

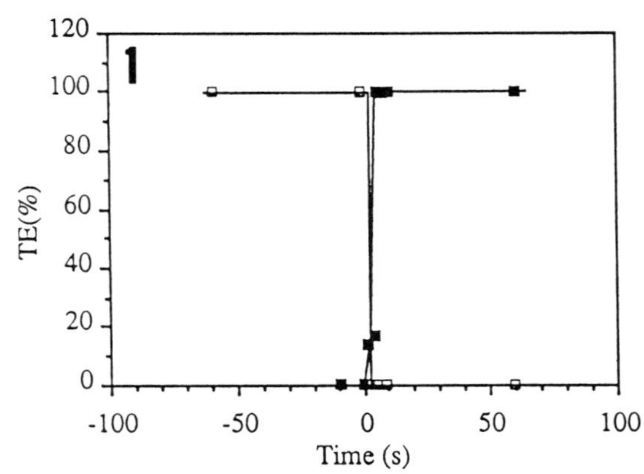

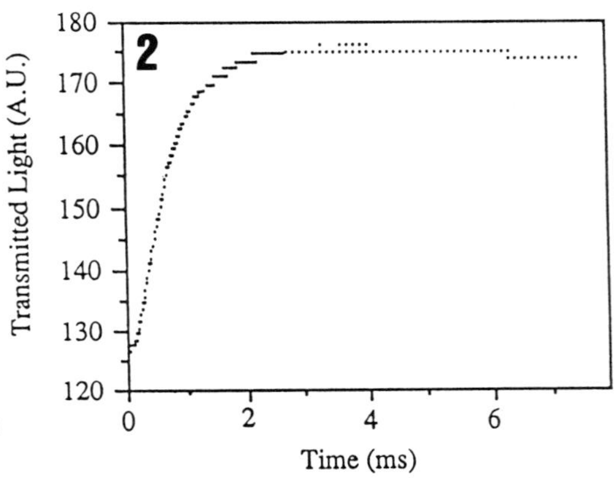

FIG. 1.—Changes in light transmitted through bacterial suspension during pulse: light path 2 cm, one square-wave uniform-field pulse of 2.7 kV/cm applied during 7 ms.

FIG. 2.—Effect of pre- and post-pulse DNA addition on transformation efficiency (□) and time dependence of effect of DNAse addition on transformation efficiency (■). Field pulse applied at time zero; negative time values indicate that plasmids or DNAse is added during pulsing. Field parameters are 1 pulse of 5 kV/cm lasting 7 ms. A relative scale in transformants is used; 100% corresponds to 3×10^4 transformants per 2×10^8 pulsed cells per microgram of DNA.

By comparing the results on addition times for plasmids and treatment times by DNAses, one must conclude that under our conditions the plasmid must be present during the pulse, but nevertheless remains transiently accessible after the pulse to the outer buffer. The gene transfer should be considered to be triggered by the field pulse, but apparently occurs rather slowly (on a time scale of seconds, compared to the millisecond time scale of the pulse).

DISCUSSION

It is well known that when a cell is immersed in an electric field, the most affected parts of the cell are the areas that face the electrodes.[1] One of the major consequences of the very fast orientation of the bacteria during the pulse is that under transforming conditions (i.e., field strength about 5 kV/cm, pulse duration longer than 5 ms),[1] the longer axis of the cell is parallel to the field lines during most of the duration of the pulse. Taking into account the shape of *E. coli*, a rod with two hemispherical ends (caps), permeabilization and the associated gene transfer would then occur only in this highly curved fraction of the bacteria, with the normal to the caps surface perpendicular to the field lines. The cylindrical central part would not be affected except during the orientation process, i.e., a fraction of millisecond. The fraction of the cell surface which is prone to gene transfer is then very small (10%).

The observation of ion leakage during the pulse is of importance to the understanding of the electrotransformation processes. It has been suggested that ions at low concentration may have a specific effect on the efficiency of plasmid transfer.[12] Our observation indicates that under optimized conditions, the local interfacial ionic concentration is mainly due to the contribution of the cytoplasmic outflow. The local K^+ concentration should then be very high.

A critical step occurs when the field is applied because the plasmid must be present during the pulse. It is well known that in eukaryotic cells, a dramatic alteration of the membrane occurs during the permeabilizing pulse, but relaxes in a few microseconds. We can then suggest that an anchoring step takes place during the pulse that causes insertion of the plasmid in the membrane after a partial disruption of the membrane's integrity.

The observation that the yield of transformation is affected by the addition of DNAses after the pulse (Fig. 2) shows that one of the steps of gene transfer takes place after the pulse. Such a process is not driven by the electrophoresis forces associated with the external field, but is mediated by the cell itself.

Our model is that, in an ion-free buffer, electrotransformation is a two-step process. During the pulse, the field plays two roles: (a) induction of a local competent organization of membranes, and (b) strong adsorption of DNA to the cell surface. This adsorption is facilitated by the high interfacial ionic concentration due to the cytoplasm leakage or by electrophoretic forces. As bacteria are oriented in less than 1 ms, caps would be a more potent site for anchorage. The structure of the transient alterations of the cell wall and membranes, which support DNA anchoring and transfer, remains to be elucidated.

After the pulse, a transmembrane diffusion step occurs that is mediated by the cell. This second process is slow (time scale of seconds) and strongly temperature dependent, as are many other membrane phenomena.

REFERENCES

1. E. Neumann, in E. Neumann, A. E. Sowers, and C. A. Jordan, Eds., *Electroporation and Electrofusion in Cell Biology,* New York: Plenum, 1989, chap. 4, pp. 61-82.

2. E. Neumann, M. Schaefer-Ridder, Y. Wang, and P. H. Hofschneider, *EMBO J.* 1: 841-845, 1982.

3. W. Förster and E. Neumann, Ref. 1, chap. 19, pp. 299-318.

4. B. Chassy and J. Flickinger, *FEMS Microbiol. Letters,* 1988, 173-177.

5. E. Meilhoc, J. M. Masson, and J. Teissié, *Biotechnology* 8: 223-227, 1990.

6. W. J. Dower, J. F. Miller, and C. W. Ragsdale, *Nucl. Acids Res.* 16: 6127-6145, 1988.

7. B. M. Chassy, A. Mercenier, and J. Flickinger, *Tibtech* 6: 303-309, 1988.

8. S. Sixou, N. Eynard, J. M. Escoubas, E. Werner, and J. Teissié, *Biochim. Biophys. Acta* 1088: 135-138, 1991.

9. C. T. O'Konski and B. H. Zimm, *Science* 111: 113-116, 1950.

10. C. M. Cerda, R. T. Foister, and S. G. Mason, *J. Colloid Interface Sci.* 82: 577-579, 1981.

11. K. Kinosita and T. Y. Tsong, *Biochim. Biophys. Acta* 554: 479-497, 1979.

12. T. D. Xie, L. San, and T. Y. Tsong, *Biophys. J.* 58: 13-19, 1990.

13. D. S. Dimitrov and A. E. Sowers, *Biochim. Biophys. Acta* 1022: 381-392, 1990.

14. K. Kinosita, I. Ashikawa, N. Saita, H. Yoshimura, H. Itoh, K. Nagayama, and A. Ikegami, *Biophys. J.* 53: 1015-1019, 1988.

Martin Blank, Ed., *Electricity and Magnetism in Biology and Medicine*

BACTERIORHODOPSIN AS AN ADVANCED BIOELECTRONIC MATERIAL: AN ELECTROCHEMICAL APPROACH TO MOLECULAR DEVICES

Felix T. Hong*

Wayne State University, Detroit, Mich.

The mechanism of the pulsed light-induced electric signal in bacteriorhodopsin (BR) thin films is analyzed by a combined electrochemical and electrophysiological approach that permits identification of the signal components with molecular processes occurring at various domains of BR. Applications of the photoelectric effect of BR thin films to molecular electronics are discussed.

Bacteriorhodopsin is a membrane-bound protein found in the purple membrane of *Halobacterium halobium*. It is a single-chain polypeptide with 248 amino acid residues, which form seven transmembrane α-helices, with the C-terminus located intracellularly and the N-terminus extracellularly. Under illumination with visible light, protons are transported from the intracellular space to the extracellular space. Concurrently, a net (DC) photocurrent can be detected flowing in the outward direction. If a brief light pulse is used to stimulate the membrane, a photochemical reaction cycle is initiated during which BR undergoes several conformation changes with continuous shifts of its absorption maximum; i.e., there is a photochromic effect. Concurrently, a transient photoelectric current can be detected with several exponential decay components. The latter photocurrent is due to light-induced rapid charge separation and is therefore capacitative in nature (AC photoelectric effect). It is also known as the fast photocurrent or displacement photocurrent. Both the photochromic effect and the AC photoelectric effect have an ultrafast risetime, 0.2 and <5 ps, respectively.

Its photochromic and photoelectric properties, together with its remarkable chemical stability, make BR suitable for use in the construction of prototype molecular devices. Prototype devices based on its photochromic effect have been reported by Vsevolodov et al., Hampp, and Birge et al.[1-3] Prototype devices based on its AC photoelectric properties have been reported by Haronian and Lewis[4] and Miyasaka et al.[5] In the present paper, the AC photoelectric effect of BR will be analyzed from an electrochemical point of view, with its applications for device construction kept in mind.

ELECTROCHEMICAL AND EQUIVALENT-CIRCUIT ANALYSIS

Interpretation of AC photoelectric signals is not straightforward. Most investigators decompose the transient signal into several exponential components and attempt to correlate them with similar components of the photocycle event. However, there is a lack of agreement on these relaxation time constants among various laboratories, mainly because a hidden parameter was not under control. This hidden parameter is the access impedance, which varies from zero under short-circuit measurement conditions to infinity under open circuit conditions. This peculiar effect, which causes the discrepancies, can be understood in terms of membrane electrochemistry.

We started by considering two simple molecular mechanisms of light-induced charge separation. Since BR is asymmetrically oriented in the membrane, and since BR carries a number of charged amino acid side chains, it is conceivable that light action could cause these charged groups to move. This process results in the formation of a transient array of electric dipoles, which subsequently vanishes upon charge recombination. This process manifests itself macroscopically as a displacement photocurrent (oriented dipole mechanism). Alternatively, a displacement photocurrent can also be generated by light-induced interfacial proton transfer at the membrane-water interface, which constitutes another kind of charge separation (interfacial proton transfer mechanism). The subsequent reverse interfacial proton transfer constitutes the charge

---

*Correspondence to: F. T. Hong (Physiology), Wayne State University, Detroit MI 48201.

recombination. By applying the Gouy-Chapman diffuse double-layer theory to these two mechanisms, we can derive two slightly different equivalent circuits. These two circuits can be further reduced to a common irreducible equivalent circuit. The most salient feature of this circuit is the inclusion of a capacitance that is connected in series with the photo emf (photocurrent source). We have demonstrated that this capacitance is physically distinct from the ordinary membrane capacitance. A complete equivalent circuit that includes both DC and AC photoelectric events as well as inert supporting structures (RC network formed by the ordinary membrane capacitance and resistance) correctly predicts the time course of experimentally measured photoelectric signals, provided that the access impedance is included in the analysis. The time course of the observed signal is distorted by membrane RC relaxation except under strictly short-circuit conditions. However, the intrinsic photochemical relaxation time is unaltered by these changes and can always be recovered by deconvolution of the measured photosignal.

Using this electrochemical approach, along with three reconstitution methods to construct oriented BR thin films on a Teflon substrate or in lipid bilayer membranes, we have developed a protocol to decompose the photoelectric signals according to the underlying physiochemical process. Three separate components are identified: a fast component (B1) is associated with the initial rapid intramolecular charge separation and is temperature insensitive and pH independent; and two slower components (B2, B2′) are associated with proton uptake at the cytoplasmic surface and with proton release at the extracellular surface, respectively. Both B2 and B2′ are pH sensitive but in opposite directions.

FUNCTIONAL SIGNIFICANCE OF AC PHOTOELECTRIC EFFECT

The B1 component is most likely associated with photoisomerization of the chromophore. During that time, light energy is stored as charge separation and is subsequently used to drive conformational changes. In BR these conformational changes are important for efficient unidirectional (vectorial) proton transport by retarding reverse proton transfer (charge recombination). In fact, bacteriorhodopsin is an intelligent material in the sense that BR adjusts and enhances its proton pumping capability when it faces the "back pressures" of a proton gradient that appears as a result of the pumping action.

The visual pigment rhodopsin also exhibits similar displacement current components, R1 and R2. The functional significance of R1 is similar to B1, but the significance of R2 is probably quite different from that of B2 and B2′. Whereas B2 and B2′ are a manifestation of an obligatory process in proton pumping, R2 is not, because rhodopsin does not pump protons. We suspect R2 and the associated surface potential change may function as the mechanistic trigger to initiate the visual transduction that starts with binding and activation of transducin by photoactivated rhodopsin.

IMPLICATIONS FOR MOLECULAR DEVICE CONSTRUCTION

One can modulate the waveform of AC photoelectric signals by altering the loading condition of the photogenerator and/or by modifying the inert supporting structure. These features are also shared by conventional microelectronic devices. What has set molecular devices apart from conventional microelectronic devices is the possibility of modulating the signal waveform by altering the underlying chemical processes. Bacteriorhodopsin can be considered a bifunctional sensor molecule. It is a photosensor that can be modulated by proton concentration at the membrane surface. It can also be exploited as an ion-sensitive field-effect transistor (ISFET) that can be activated by light.

The potential of BR for photoelectric-device construction can be further expanded to include similar retinal proteins such as halorhodopsin and sensory rhodopsins, as well as mutant retinal proteins obtained by site-directed mutagenesis. Our preliminary results indicate that a single-point mutation of BR could lead to dramatic changes in its photoelectric effect. This possibility gives investigators additional latitude to custom-tailor BR for their specific applications.[6]

REFERENCES

1. N. N. Vsevolodov, A. B. Druzhko, and T. V. Djukova, in F. T. Hong, Ed., *Molecular Electronics: Biosensors and Biocomputers,* New York: Plenum Press, 1989, 381-384.
2. N. Hampp, C. Bräuchle, and D. Oesterhelt, *Biophys. J.* 58: 83-93, 1990.
3. R. R. Birge et al., *Trans. IEEE* Computer-25(11): 56-57, 1992.
4. D. Haronian and a. Lewis, *Appl. Optics* 30: 597-608, 1991.

5. T. Miyasaka, K. Koyama, and I. Itoh, *Science* 255: 342-344, 1992.
6. This research was supported by contracts from the Office of Naval Research (N00014-87-K-0047) and the Naval Surface Warfare Center (N60921-91-M-G761).

SELECTED BIBLIOGRAPHY OF AUTHOR'S WORKS

Photochem. Photobiol. 24: 155-189, 1976.

Bioelectrochem. Bioenerg. 5: 425-455, 1978.

In M. Blank, Ed., *Bioelectrochemistry: Ions, Surfaces, Membranes* (ACS Advances in Chemistry Series No. 188), Washington, D.C.: American Chemical Society, 1980, 211-237.

In M. Blank, Ed., *Electrical Double Layers in Biology,* New York: Plenum Press, 1986, 129-147 (with T. L. Okajima).

BioSystems 19: 223-236, 1986.

J. Electrochem. Soc. 134: 3044-3052, 1987.

J. Molec. Electronics 5: 163-185, 1989.

In P. I. Lazarev, Ed., *Molecular Electronics,* Dordrecht: Kluwer, 1991, 291-310.

Nanobiology, 1: 39-60, 1992.

In M. Blank and I. Vodyanoy, Eds., *Biomembrane Electrochemistry,* Washington, D.C.: American Chemical Society (in press).

In A. Aviram, Ed., *Molecular Electronics: Science and Technology,* New York: American Institute of Physics (with F. H. Hong, R. B. Needleman, B. Ni, and M. Chang), 1992, 204-217.

Martin Blank, Ed., *Electricity and Magnetism in Biology and Medicine*

PHOTOCHROMIC MEMORIES BASED ON 1,4-DIHYDROPYRIDINES: A NEW APPROACH TO THE ANALYSIS OF MOLECULAR MOVEMENTS

S. Nešpůrek*

Institute of Macromolecular Chemistry, Czechoslovak Academy of Sciences, Prague

L. Pecen

Institute of Informatics and Computer Science, Czechoslovak Academy of Sciences, Prague

The bleaching process of the photochromic reaction of 1-methyl-2,4,4,6-tetraphenyl-1,4-dihydropyridine (DHP) in chloroform solution (a) without and (b) with the presence of poly(methylmethacrylate) (PMMA) may be described by (a) the monomolecular reaction and (b) the dispersive first-order-reaction, respectively. The application of the analysis of a chaotic signal makes it possible to write a differential equation describing the bleaching process and to determine characteristic frequencies and the molecular dynamics.

The photochromic phenomenon may be schematically characterized as a photoreversible transformation of colorless species A to colored ones B. The behavior of photochromic materials depends on the photochemical mechanism[1] and can involve (among others) cis-trans isomerism,[1-3] homolytic or heterolytic bond cleavage,[1,4-7] and tautomeric migration of hydrogen or phenyl.[1,8,9] The fact that photochromic molecules can switch between two alternative forms may provide a phenomenological entrée into the desired novel technology of molecular memories and switches.

PROBLEM

The behavior of photochromic materials is strongly influenced by the interactions and dynamics of molecules surrounding the photochromic center. The molecular motion can provoke a special electrical or optical noise and affect the signal-to-noise ratio. Thus, the noise of molecular electronic elements represents a real problem that has not been discussed so far. In this communication, we present a new approach to the analysis of the electronic signal based on the measurement of its amplitude and fluctuations. This technique allows us to distinguish between the deterministic and stochastic parts of the signal, to write a differential equation characterizing the physical process, and to determine the correlation function and power spectral density of both the deterministic and stochastic part of the signal.

MATERIAL AND MEASUREMENTS

The details of the synthesis of DHP have been described elsewhere.[8,10] Samples were irradiated with 15ns flashes of 266nm light produced by an Nd-YAG laser equipped with two frequency doublers. Transient optical absorption in the perpendicular direction was measured by a conventional technique consisting of a pulsed xenon lamp, monochromator, photomultiplier, and a digitizing oscilloscope (Hewlett Packard 54510A, 250 MHz).

EXPERIMENTAL RESULTS

The absorption spectrum of DHP shows two absorption bands with the maxima at 235 and 288 nm. Irradiation in solution (acetonitrile, chloroform) led to the appearance of the new transient absorption with the maxima located at 330 and 520 nm. The photochromic species are formed very rapidly (a rise time of about

---

*Correspondence to: L. Nešpůrek, CSAV Institute of Macromolecular Chemistry, Heyrovský Sq. 2, Prague 6 Czech Republic.

20 ns) and directly (no photochemical intermediates were detected). From the dependence of the logarithm of the normalized reactant concentration on time, the lifetime of the colored species in an oxygen-free chloroform solution was found to be 0.02 ms at room temperature. The bleaching process could be described by the monomolecular reaction (~ $\exp[-\beta t]$). An addition of PMMA has changed this reaction to a nonexponential time dependence that could be analyzed in terms of a dispersive first-order reaction (~ $\exp[-\beta_1 t^{\alpha}]$), where $\alpha = 0.8$ (Fig. 1).

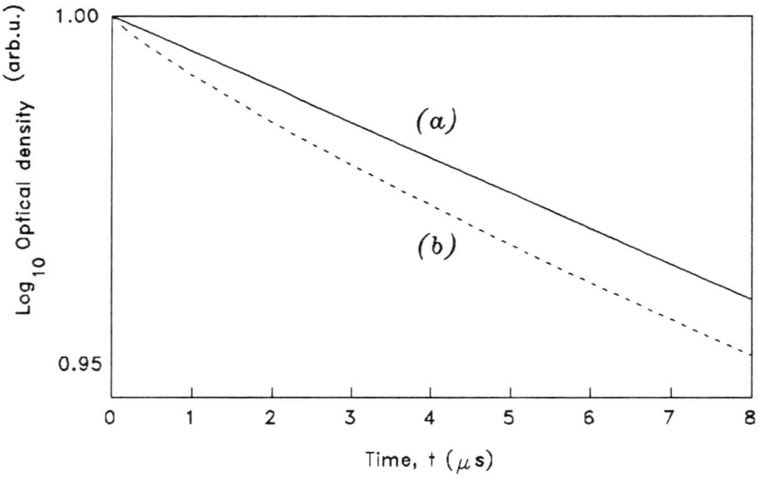

FIG. 1.—Experimental decay curves for thermal bleaching of (a) DHP in chloroform, (b) DHP in chloroform in the presence of PMMA. Curve (b) can be linearized in semilogarithmic plot of the logarithm of the normalized rectant concentration $[M(t)]/[M(0)]$ vs t^{α}.

EXPERIMENTAL SIGNAL ANALYSIS

Until recently, power spectra and correlation functions were used to characterize the states of fluctuating signals, but by these methods one could hardly distinguish between very complicated time series. We propose a procedure for finding a difference equation that has only one solution as a deterministic part of the signal. The difference equation can be transformed into an ordinary differential equation.

Before using the classical algorithm,[11] we decide whether any deterministic part of the signal exists. If we find the deterministic part of the signal as a first step, we must estimate the order of the related differential equation. Therefore a new method,[12] based on the calculation of multivariable mutual information among k components of Takens's vectors[13,14] with increasing k is proposed. Each difference equation of order n can be written in the following form:

$$I_n\,[X(t), X(t+\delta), \cdots, X(t+n\delta), t, \delta] = 0 \tag{1}$$

for all t for which the arguments are defined. The function I_n is not identically equal to zero. For a real signal containing some part of random noise, condition (1) cannot be satisfied exactly. Therefore we are looking for a function I_n (satisfying some normalization condition), for which the integral has the minimal value

$$\int I_n^2\, dt \tag{2}$$

We presuppose that I_n can be expressed in the form of Taylor series, and for practical calculation we cut off the series at a power m. (Accuracy increases with increasing m, Fig. 2.) The equations are obtained by differentiation of (2) over Taylor coefficients; and by solving these equations under the normalization condition we obtain a difference equation. Now we can use a standard transformation between a difference equation and a related ordinary differential equation.

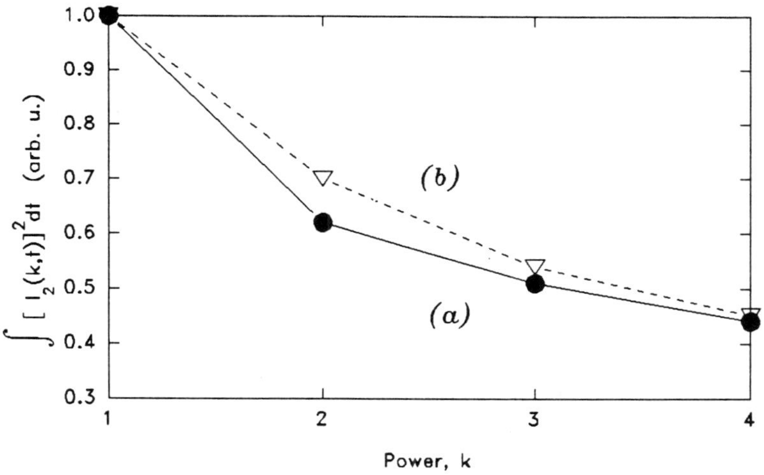

FIG. 2.—Plot of the integral $\int I_n\,(t,k)^2 dt$ vs power k.

The differential equations for both cases, i.e., (a) DHP/chloroform and (b) DHP/PMMA/chloroform, were found to be of the form of

$$Atyy'' + Byy' - Ct(y')^2 - Dt^2y'' + Ety' - Ft = 0 \tag{3}$$

with the majority coefficients

$$A = 1,\ B = 0.063,\ C = 1.121,\ D = 0.011,\ E = 0.012,\ F = 0.009$$

for case (a) and

$$A = 1,\ B = 0.225,\ C = 0.989,\ D = 0.013,\ E = 0.016,\ F = 0.011$$

for case (b). The equations differ mainly in the magnitudes of the coefficients of the second term (yy'), which characterize the dispersion of the bleaching process. (The process is more dispersive when a polymer is present.) The calculated autocorrelation function (AF) of the overall experimental signal for samples (b) decreases approximately 2-3 times faster than that for samples (a). The polymer shortens the memory of the bleaching process. Two characteristic maxima in AF were found at 46 and 11 ns. The corresponding frequencies were found in the power spectral density (Sp). The frequency 22 MHz (~ 46 ns) together with its harmonics is the characteristic one of the deterministic part of the signal and is related to molecular dynamics. For samples (b) these peaks are more dispersive and new peaks shifted by about 7 MHz lower were detected.

CONCLUSION

The presence of the polymer increases the dispersivity of the bleaching reaction of 1-methyl-2,4,4,6-tetraphenyl-1,4-dihydropyridine. Whereas the process in chloroform solution is very close to the monomolecular reaction, the addition of the polymer changes this reaction to a nonexponential time dependence, which can be analyzed in terms of a dispersive first-order reaction. The mathematical approach based on the analysis of a chaotic signal represents a powerful tool for the analysis of the photochromic bleaching processes. This technique can be used generally for the noise and fluctuations studies in electronic molecular devices.

REFERENCES

1. G. H. Brown, *Protochromism*, London: Focal Press, 1971.
2. J. Labský, I. Koropecký, S. Nešpůrek, and J. Kálal, *Eur. Polym. J.* 17: 309, 1981.
3. L. Matějka, S. Nešpůrek, M. Kucharski, and K. Dušek, *Polym. Bull.* 1: 167, 1978.
4. S. Nešpůrek and M. Šorm, *Czech. J. Phys.* B25: 1051, 1975.

5. S. Nešpůrek and M. Šorm, *Collect. Czech. Chem. Commun.* 42: 811, 1977.
6. S. Nešpůrek, J. Hašek, M. Šorm, K. Haml, J. Obrda, J. Lipinski, and H. Chojnacki, *J. Mol. Struct.* 82: 95, 1982.
7. S. Nešpůrek, J. Böhm, and J. Kuthan, *J. Mol. Struct.* (Theochem) 136: 261, 1986.
8. S. Nešpůrek, M. Schwartz, S. Böhm, and J. Kuthan, *J. Photochem. Photobiol.* (A: Chem.) 60: 345, 1991.
9. S. Nešpůrek and W. Schnabel, *ibid.*, 62: 151, 1991.
10. A. Kurfürst, J. Zelený, M. Schwartz, and J. Kuthan, *Chem. Pap.* 41: 623, 1984.
11. P. Grassberger and I. Procaccia, *Physica* 13D: 34, 1984.
12. L. Pecen, Ph.D. Thesis, Charles University, Prague, 1992.
13. T. S. Parker and L. O. Chua, *Practical Numerical Algorithms for Chaotic Systems*, New York: Springer-Verlag, 1989.
14. H. G. Schuster, *Deterministic Chaos: An Introduction*, Weinheim: Physik-Verlag, 1984.

Martin Blank, Ed., *Electricity and Magnetism in Biology and Medicine*

THE LONG ROAD TOWARD ORGANIC UNIMOLECULAR RECTIFIERS

R. M. Metzger*

University of Alabama, Tuscaloosa

The progress toward assembly and testing of an organic unimolecular rectifier of the type D-σ-A is reviewed. Langmuir-Blodgett (LB) films of D-σ-A molecules (D = one-electron donor, σ = carbamate, A = one-electron acceptor) have been prepared. Electrical currents have been measured through monolayers, by macroscopic means and by scanning tunneling microscopy. Schottky barriers have been found, but the 1974 Aviram-Ratner Ansatz of unimolecular rectification[1] has not yet been verified.

Let us define a unimolecular, or zero-dimensional device as an electronic (or other) device, whose function depends on the electronic structure or conformation of a single organic molecule or of a very small cluster of molecules. With a *reductio ad absurdum* size (5 to 100 nm^3), and a fast operating speed (<1 ns), such a device should have tremendous technological potential; this is the realm of "molecular electronics," narrowly defined.[2]

In 1973 Ansatz by Aviram, Ratner, and co-workers[1] for a unimolecular rectifier of the type D-σ-A based on organic donor (D) and acceptor (A) molecules connected by a covalent sigma bridge (σ) was a seminal idea in molecular electronics. This proposed zero-dimensional device would rectify electrical current between two metal electrodes, M_1 and M_2, because the D end is a good organic one-electron donor (but poor acceptor) with a low ionization potential I_D, such as tetrathiafulvalene (TTF, **1** in Fig. 1a); A is a good organic one-electron acceptor (but poor donor) with high electron affinity A_A, such as 7,7,8,8-tetracyanoquinodimethane (TCNQ, **2** in Fig. 1a); and σ is a covalent bridge, which separates the molecular orbitals of donor and acceptor, but permits through-bond electron tunneling. The energy of the neutral ground state D-σ-A of **3** would be closer, by several eV, to the zwitterionic excited state D^+-σ-A^- than to the state D^--σ-A^+ (Fig. 1b): electrons would flow from M_2 through D-σ-A to M_1 (Fig. 1c).

A synthetic collaboration with C. A. Panetta[3-8] has afforded several candidate D-σ-A molecules. This brief review summarizes synthetic and assembly criteria for unimolecular rectification, chronicles the synthesis and characterization of several D-σ-A molecules, discusses "rectification" measurements, and proposes ideas for future progress. More details are given elsewhere.[5,7]

CRITERIA FOR AN AVIRAM-RATNER D-σ-A DEVICE

Except for the scanning tunneling microscope (STM), one cannot today connect a single molecule to an external circuit. Therefore, rectification was initially sought for one-molecule-thick monolayers of molecules transferred to metallic substrates by the Langmuir-Blodgett (LB) technique. The LB monolayer domain under study should be defect-free; in reality, it has many microscopic defects.

Some criteria for the rational synthesis of D-σ-A systems are:

(a) I_D for the donor D must be small (typical range 6.3-8 eV) and approach the work function of M_1 (range 3.5-5.5 eV);

(b) A_A for the acceptor A must be large (range 2-3.3 eV) and approach the work function of M_2 (range 3.5-5.5 eV);

(c) the bridge should keep the D and A properties apart: a saturated σ bridge is best, but an unsaturated unconjugated π bridge is also acceptable;

(d) the syntheses should be relatively easy;

---

*Correspondence to: R. M. Metzger (Chemistry), University of Alabama, Tuscaloosa AL 35487-0336.

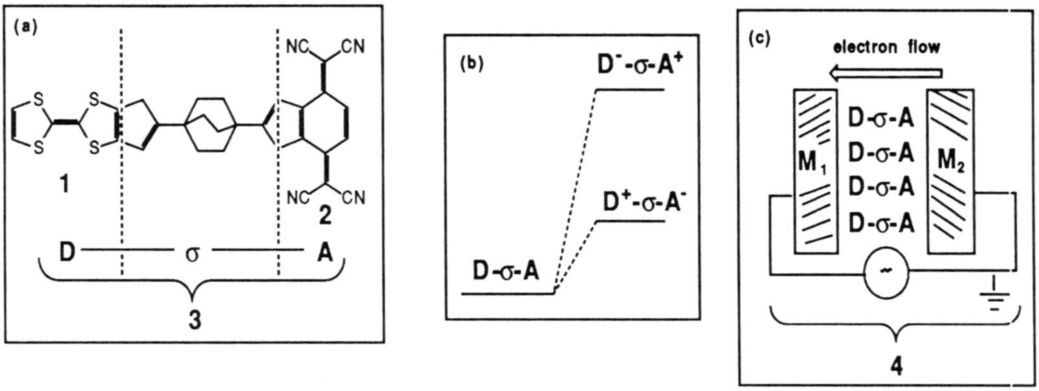

FIG. 1.—(a) D-σ-A *Gedankenmolekül* **3** (Ref. 1); (b) ordering of zwitterionic states; (c) proposed Aviram-Ratner device **4.**

(e) the molecules must pack efficiently (often with the help of "greasy" hydrophobic tails) and transfer well onto solid substrates, or alternately, the D-σ-A molecule could be anchored covalently to a metal surface; and

(f) the "through-bond" electron transfer must be fast.

RESULTS

Figure 2 shows the D-σ-A molecules **(5-23)** prepared by C. A. Panetta and co-workers, and one D^{+}-π-A^{-} molecule **(24)** prepared by G. J. Ashwell, which self-assemble as "Pockels-Langmuir" monolayers at the air-water interface, and which transfer well onto solid supports as LB monolayers. Table 1 summarizes their electronic and monolayer properties.

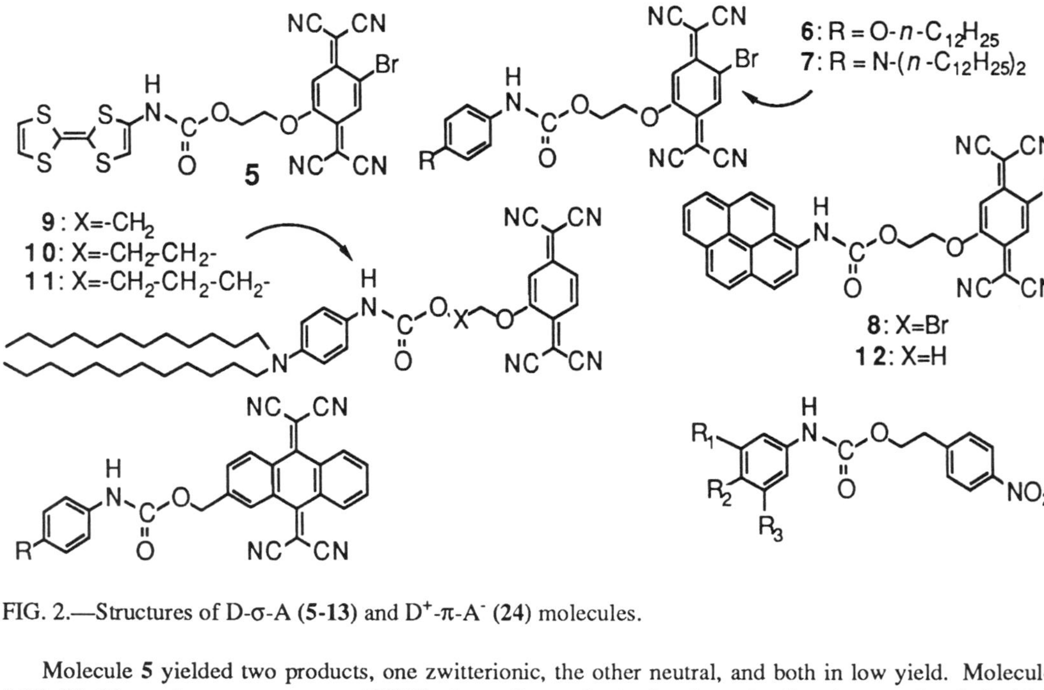

FIG. 2.—Structures of D-σ-A **(5-13)** and D^{+}-π-A^{-} **(24)** molecules.

Molecule **5** yielded two products, one zwitterionic, the other neutral, and both in low yield. Molecules **5-12, 19, 20** use the strong acceptor TCNQ; the easily synthesized anthraquinodimethane molecule in **13,14** is a nonplanar weak two-electron acceptor. Molecules **15-20** are model systems for the incorporation of D-σ-A systems into LB-film-forming polymerizable diacetylenes (like **21-23**), as potential nonlinear optical

TABLE 1.—Data for D-σ-A Pockels-Langmuir films at temperature T; Π_c , A_c are pressure and molecular area at film collapse; D = donor, A = acceptor is S = strong (low I_D or high A_A), M = medium, or W = weak.[5,7]

Molecule	No.	D	A	T/K	Π_c/mN/m	A_c/Å^2
TTF-C-BHTCNQ	**5**	S	S	292	12.7	134±50
DDOP-C-BHTCNQ	**6**	W	S	292	20.2	50±1
BDDAP-C-BHTCNQ	**7**	M	S	293	47.3	57±1
Py-C-BHTCNQ	**8**	M	A	283	28.2	53±1
BDDAP-C-HETCNQ	**9**	M	S	293	40.0	44±1
BDDAP-C-HPTCNQ	**10**	M	S	293	45.1	44.4±1
BDDAP-C-HBTCNQ	**11**	M	S	293	46.4	26.7±3?
Py-C-HETCNQ	**12**	M	S	283	46	--
BDDAP-C-HMTCAQ	**13**	M	W	293	22.3	58±1
BHAP-C-HMTCAQ	**14**	M	W	293	35.8	42±1
DDOP-C-ENP	**15**	W	W	278	23.7	38±1
TDDOP-C-ENP	**16**	W	W	278	34.0	76±1
3,5-BHDOAP-C-ENP	**17**	W	W	293	49.6	39±2
3,4-BHDOAP-C-ENP	**18**	W	W	293	54.5	35.8±0.5
TDDOP-C-HETCNQ	**19**	W	S	283	47.5	54±1
3,4-BHDOAP-C-HETCNQ	**20**	W	S	293	55.2	51.0±1
MTDAP-C-ENP	**21**	W	W	278	16.5	63±1
3,5-BTDYOAP-C-ENP	**22**	W	W	293	18.7	58±3
3,4-BTDYOAP-C-ENP	**23**	W	W	293	49.4	50±1

devices; **15, 16, 19, 21** form Z-type multilayers on a glass substrate, but LB multilayers or **19** did not show second-harmonic generation; **21** forms an orange polymer; **23** forms a blue polymer.

Cyclic Voltametry. The oxidation (reduction) potentials of the D(A) ends are preserved in D-σ-A, as expected.

Crystal Structures were solved for D and A molecules, and for 2 non-film-forming D-σ-A: the carbamate linkage is extended.

RECTIFICATION MEASUREMENTS

Rectification has been observed in organic LB *multi*layers[9,10] (for more than 6 monolayers). An LB monolayer *photo*diode has also been achieved.[11] We review here several attempts, some in our own laboratory, to detect *unimolecular rectification* in D-σ-A molecules.

(1) We used a Hg droplet to probe the conductivity across a single monolayer LB film of **6** and **8** (on Pt, or on ITO glass). Only the background ohmic conductivity of the solid support was detected (pinholes).

(2) To avoid domain-edge defects, we coated 75 Al strips with an LB monolayer of **7**, and then with new Al strips on top, with low Al-Al overlap: of the 75 junctions (area 0.5×0.5 mm^2), many were open circuits; the rest were ohmic.

(3) A monolayer of **6** (weak D end), sandwiched between Pt and Mg|Ag electrodes, behaved as a rectifier,[12] but probably due to a Schottky barrier at the Mg|TCNQ interface.[13] Samples did eliminate the film defects.

(4) An LB film of **24**, sandwiched between Pt and Mg|Ag, showed rectification or, more likely, some Schottky barrier effect.[13,14]

STM seems to be the ideal tool to measure unimolecular rectification. For instance, STM was used to detect negative resistance at certain B-free sites of a B-doped Si tunnel diode.[15]

(5) A molecule designed for intramolecular H atom transfer was studied in a modified STM; the asymmetric I-V plot, first ascribed to molecular rectification, was later deemed an artifact.[16]

(6) Pomerantz studied, with his modified STM, an LB monolayer of **7** deposited on an Au|Ag|mica substrate. ''Rectification'' currents were found, but were again ascribed to an instrumental artifact.

(7) Pomerantz recently found that anchoring Cu tetraazaphthalocya-nine to a base-coated graphite yielded an asymmetric STM I-V plot.[17]

(8) We studied monolayers of **8-11** and **24,** transferred onto highly oriented pyrolytic graphite (Union Carbide, ZYA) by the horizontal Langmuir-Schaefer lifting technique in air at room temperature, using a Digital Instruments Nanoscope II STM, and Pt-Ir Nanotips.[18] The best resolution was found for **24,** where the bright regions could be interpreted as being due to the cyano nitrogen and carbon atoms of the malononitrile end of **24,** and the dark regions as due to the quinolinium ring hidden below. However, the results for **24** could not be repeated, because of poor adhesion of **24** to graphite. Recently, good STM images of LB multilayers of **10** and **11** were obtained: the molecules **8, 10** and **11** arrange themselves differently on the graphite as a function of carbamate bridge length between D and A. The usual STM direct tunneling I-V plot is a symmetrical S-shaped curve. Many asymmetric I-V plots were seen for **8, 9, 10, 11,** and **24,** which may indicate rectification, but these curves were not reproducible.[18]

PROSPECTS FOR FUTURE WORK

STM may be the best technique for verifying the Aviram-Ratner Ansatz. Covalent attachment to atomically flat substrates (e.g., using S on Au) may be preferable to physisorption of LB films on substrates: LB films transfer poorly to graphite. For fast intramolecular electron transfer, short greasy tails (e.g., C_6 in **14**) or short anchoring groups are preferred.

I_D should be as small, and A_A as large, as possible; low I_D is attainable (in metalloporphyrins or tetramethyl-*para*-phenylenediamine), but high A_A is found only in reactive cyanocarbons or bulky perfuorinated species, or fullerenes. Low I_D and high A_A can cause high chemical reactivity, even an unstable neutral state.[19]

REFERENCES

1. A. Aviram and M. A. Ratner, *Chem. Phys. Lett.* 29: 277, 1974.
2. R. M. Metzger, in R. M. Metzger, P. Day, and G. C. Papavassiliou, Eds., *Lower-Dimensional Systems and Molecular Electronics, New York: Plenum Press, 1991, NATO ASI Series B248, 659.*
3. R. M. Metzger and C. A. Panetta, in F. L. Carter, Ed., *Molecular Electronic Devices,* New York: Dekker, vol. 2, 1987, 1.
4. R. M. Metzger and C. A. Panetta, Ref. 1, 611.
5. R. M. Metzger and C. A. Panetta, *New J. Chem.* 15: 209, 1991.
6. R. M. Metzger and C. A. Panetta, *Synth. Metals* 41-43: 1407, 1991.
7. R. M. Metzger, in R. R. Birge, Ed., *ACS Symposium Series* (in press).
8. C. A. Panetta, N. E. Heimer, C. L. Hussey, and R. M. Metzger, *Synlett* 301, 1991.
9. E. E. Polymeropoulos, D. Möbius, and H. Kuhn, *Thin Solid Films* 68: 173, 1980.
10. M. Sugi, K. Sakai, M. Saito, Y. Kawabata, and S. Iizima, *Thin Solid Films* 132: 69, 1985.
11. M. Fujihira, K. Nishiyama, and H. Yamada, *Thin Solid Films* 132: 77, 1985.
12. N. J. Geddes, J. R. Sambles, D. J. Jarvis, W. G. Parker, and D. J. Sandman, *Appl. Phys. Lett.* 56: 1916, 1990.
13. D. J. Sandman, J. Laski, N. J. Geddes, J. R. Sambles, D. J. Jarvis, and W. G. Parker, *Synth. Metals* 41-43: 1415, 1991.
14. G. J. Ashwell, J. R. Sambles, A. S. Martin, W. G. Parker, and M. Szablewski, *J. Chem. Soc. Chem. Commun.* 1374, 1990.
15. P. Bedrossian, D. M. Chen, K. Mortensen, and J. A. Golovchenko, *Nature* 342: 258, 1989.
16. A. Aviram, C. Joachim, and M. Pomerantz, *Chem. Phys. Lett.* 146: 490, 1988; *Chem. Phys. Lett.* 162: 416, 1989.
17. M. Pomerantz, to be published.
18. X.-L. Wu, P. Wang, R. M. Metzger, and G. J. Ashwell (unpublished results).
19. Thanks are due to C. A. Panetta, N. E. Heimer, D. L. Mattern (Univ. of Mississippi), X.-L. Wu and P. Wang (Univ. of Alabama), and G. J. Ashwell (Cranfield) for their collaboration; and to the NSF (DMR-880124) and to JVC America for their support.

Martin Blank, Ed., *Electricity and Magnetism in Biology and Medicine*

ON THE CONTROL OF ELECTRONIC AND OPTICAL PROPERTIES FOR MOLECULAR ELECTRONIC APPLICATION

S. Roth*

Max-Planck-Institut für Festkörperforschung, Stuttgart, Germany

There are two motivations for molecular electronics. One is to make electronic devices smaller and smaller until they are finally as small as molecules. The second is to start from molecular materials and ask what they might be good for. Molecular materials are materials whose properties can be tailor-made by molecular engineering; e.g., by attachment of appropriate functional groups. Polyenes combine both motivations. The concept of functional groups on polyenes is demonstrated in Fig. 1.

A particular case is attaching donor and acceptor groups on the chain ends (Fig. 2).[1] If such a molecule is excited, for example by absorption of light, the donor pushes electrons into the chain and the acceptor pulls electrons out, thus creating a substantial dipole moment. If a third group is incorporated into the molecule (Fig. 3), this third group could control the communication between the chain ends; e.g., it might act as a tunnel barrier. If the transparency of the barrier can be modulated from the outside (e.g., again by light absorption), the molecule can carry out logic functions. Molecules like those in Figs. 2 and 3 could be the building blocks of future molecular electronics. Of course, the molecules that eventually might be used will be much more complicated. But these molecules allow us to study some of the fundamental problems related to molecular electronics, both experimentally and in theory.

In an "arm-chair-model" one could imagine periodic chains of donor-acceptor substituted polyenes. Then each polyene would form the cell of a cellular automaton (Fig. 4).[2] The automaton will be hit by intensive pulses of laser light (which serve as clock pulses) and the cells will change their state by absorption of light, whereby the switching probability of a particular cell will depend on the state of the neighbor cells (because of the detuning of the resonance by the electrostatic field of the dipoles).

Molecules with the functional groups could be the cells of a molecular shift register.[3,4] Absorption of light would open the barrier and an electron would move from one side of the cell to the other. From there it would tunnel to the near end of the next cell and the next light pulse would move it one step further. Such molecular shift registers placed on top of present-day microelectronic memories could largely extend the memory capacity by use of the third dimension.

Molecules as drawn in Figs. 2 and 3 have been synthesized by various groups.[5] They are now being studied mainly by optical methods to evaluate such properties as switching efficiency and the relevant time constants. The results of this evaluation will help in the design of more appropriate molecules, and so on. For certain investigations, particularly for measurements of nonlinear optical properties, the molecules have to sit in a well-ordered way in homogeneous thin films. Wagner et al. report on the preparation of Langmuir-Blodgett films and their characterization.[6] Useful information can be obtained from time resolved spectroscopy[7] and transient photoconductivity.[8] To stabilize the excited state it will be necessary to consider more complicated molecules in which electronic excitation couple strongly to the conformational degrees of freedom (twists, puckering, etc.). Bianthrones[9] are a typical example.[10]

REFERENCES

1. S. Roth, *Z. Physik* B64: 25, 1986.
2. S. Roth, G. Mahler, Y. Q. Shen, and F. Coter, *Synthetic Metals* 28: C815, 1989.
3. J. J. Hopfield, J. N. Onuchic, and D. N. Beratan, *Science* 241: 817, 1988; *J. Phys. Chem.* 93: 6350, 1989.
4. P. E. Burrows and E. G. Wilson, *J. Mol. Electronics* 6: 209, 1990.

*Correspondence to: S. Roth, MPI-Festkörperforschung, Heisenbergstr. 1, D-7000 Stuttgart 80, Germany.

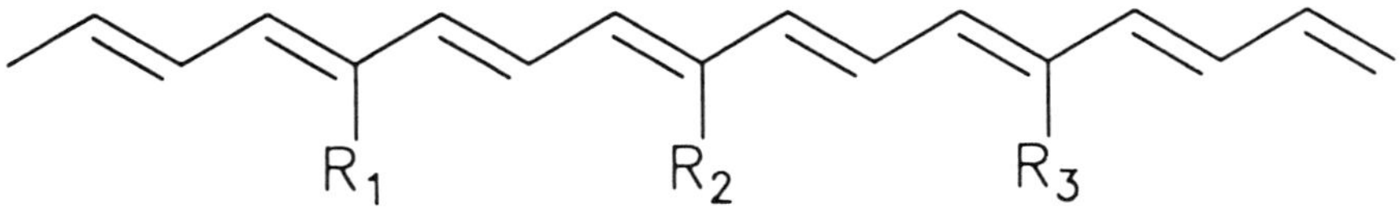

FIG. 1.—Chain of a polymer with conjugated double bonds (polyene) with attached functional groups $R_1, R_2, \cdots$. The functional groups are envisaged as communicating via π-electron system along polymer backbone.

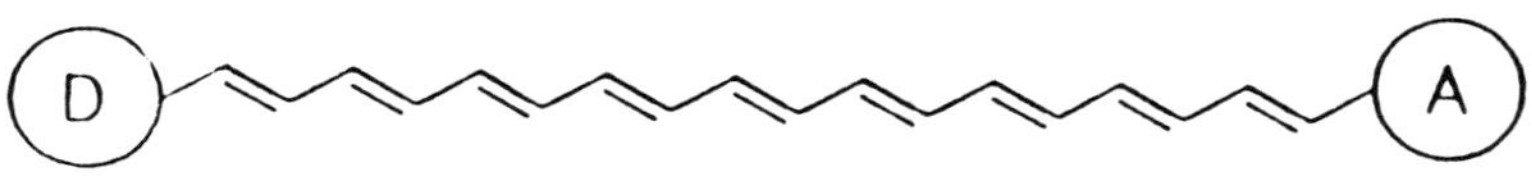

FIG. 2.—Donor and acceptor groups at the ends of polyenes.

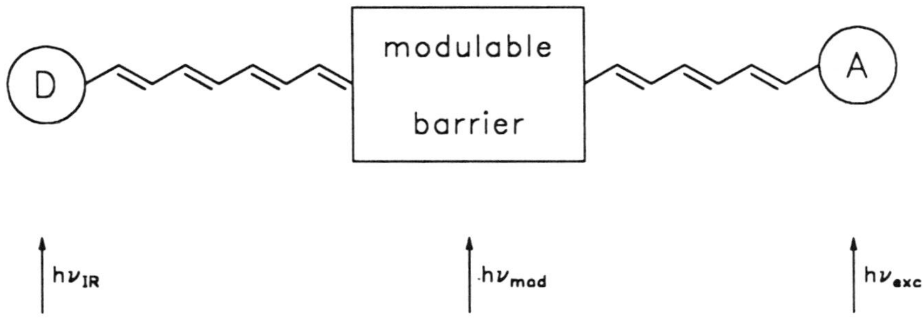

FIG. 3.—Donor-acceptor substituted polyene with a third functional group incorporated, the "barrier."

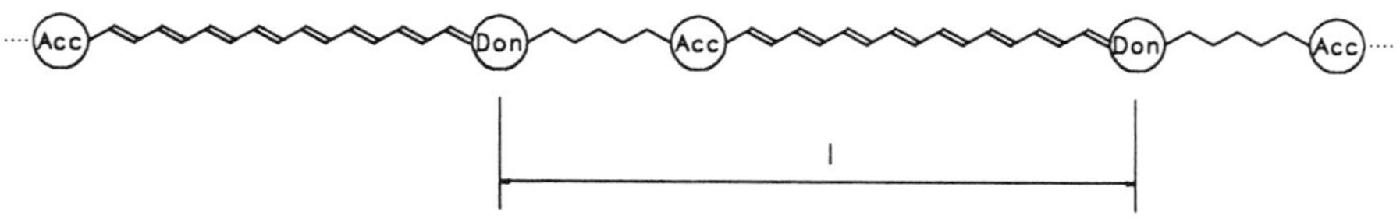

FIG. 4.—Donor-acceptor situated polyenes as cells of a linear cellular automaton.

5. W. Göhring, S. Roth, and M. Hanack, *Synthetic Metals* 17: 259, 1987; F. Effenberger, H. Schlosser, P. Bäuerle, S. Maier, H. Port, H. C. Wolf, *Angew. Chem.* 100: 274, 1988; S. Hagen, S. Roth, and M. Hanack, *Synthetic Metals* 41-43: 1557, 1991; S. Hagen, Ph.D. Thesis, Tübingen, Germany, 1991.

6. T. Wagner, S. Hagen, H. Schier, and S. Roth, *Synthetic Metals* 41-43: 1519, 1991.

7. J. Anders, H. J. Byrne, J. Poplawski, S. Roth, P. Sommer-Larsen, and K. Schaumburg, presented at ECMC '92, Padova, Italy, August 1992.

8. J. Reichenbach, M. Kaiser, J. Anders, H. J. Byrne, and S. Roth, *Europhys. Letters* 18: 251, 1992; J. Reichenbach, M. Kaiser, J. Anders, H. J. Byrne, and S. Roth, *Synthetic Metals* (in press).

9. K. Schaumburg, J.-M. Lehn, V. Goulle, S. Roth, H. J. Byrne, S. Hagen, J. Poplawski, K. Brunfeldt, K. Bechgaard, T. Bjornholm, P. Frederiksen, M. Jorgensen, K. Lerstrup, P. Sommer-Larsen, O. Goscinski, J.-L. Calais, and L. Eriksson, Workshop on Nanostructures Based on Molecular Materials, Monrepos, 1991, by VCH Weinheim.

10. The work outlined in this paper was supported by DFG, Sonderforschungsbereich Molekularelektronik and by the Esprit Basic Research Actions OLDS and MOL SWITCH. The author cooperates with the Esprit Network NEOME.

Martin Blank, Ed., *Electricity and Magnetism in Biology and Medicine*

HARNESSING QUANTUM MECHANICAL PARALLELISM FOR THE CONTROL OF MEMBRANE ELECTROCHEMISTRY

M. Conrad*

Wayne State University, Detroit, Mich.

The specificity of macromolecules allows the parallelism inherent in the quantum-mechanical wave function (the superposition of electronic states) to increase information processing power at the cellular and multicellular levels. The effect is illustrated through a model of computing and control based on macromolecular self-assembly. The model is paradigmatic of the contribution that structural features of biochemical dynamics can make to nerve impulse activity, and through it to brain function as a whole.

Macromolecular self-assembly provides a powerful mode of structural pattern formation inside cells. The same mechanisms can also serve as a basis of cellular pattern recognition. The resulting computational capabilities vastly exceed those obtainable with conventional models of computation since many more interactions are exploited, including nonclassical interactions connected with the quantum-mechanical superposition principle.

This note briefly outlines the general scheme.[1-3] The mathematical treatment is presented elsewhere.[4]

MACROSCOPIC-MICROSCOPIC INTERFACE

The basic flow of information is: signal pattern impinging on membrane → released macromolecules → self-assembled mosaic → readout adaptor → membrane electric activity. Each input signal line triggers the release of a specifically shaped macromolecule (or the appearance of a specific conformational change). The signal pattern is thus represented as a set of conformations, which self-assemble into a mosaic, somewhat in the fashion of jigsaw puzzle pieces. The mosaic will have numerous shape features, each associated with common features of different input patterns. Adaptor enzymes connect these shape features to effector proteins that control membrane electric activity or other cellular activities. The number of different groupings of input signal patterns that the cell can recognize depends on the number of different types of adaptor enzymes.

The self-assembly mode of computing converts a symbolic pattern-recognition problem into a free-energy minimization process. The cell can be viewed as a unit that transforms spatiotemporal input patterns into output actions (e.g., patterns of firing in the case of a neuron). It is possible to knit together simple pattern recognizers (such as threshold elements) to form a universal computer. *A fortiori* it is possible to knit together processors that operate in the self-assembly mode of computing to form a universal system—since they are more powerful as processors, in particular more powerful with respect to their real-time pattern processing capabilities. The reason is that all the "clever physics" of self-assembly speeds up the search of the energy landscape that maps the pattern recognition problem.

QUANTUM PARALLELISM AND THE SPEEDUP EFFECT

Self-assembly necessarily involves major changes in the nuclear coordinates of the interacting macromolecules, and consequently electronic-conformational interactions play an important role.[5] The effects of the electronic motions on the nuclear coordinates cannot be ignored (the Born-Oppenheimer approximation is inapplicable). Under these conditions the superposition of electronic states can facilitate the search for the free-energy minimum.[4] Roughly, the picture is that the electronic system explores for a lower-energy electronic minimum, then drags the nuclear system to a corresponding conformational minimum, and is in turn

---

*Correspondence to: M. Conrad (CS Dept.), Wayne State University, Detroit MI 48202.

dragged into a minimum more consistent with the conformational minimum.

Hartree-Fock type self-consistent field theory provides a useful formal framework for describing the self-assembly process in a manner that elicits the role of the superposition principle. The idea is to use Hartree-Fock to represent the self-organization of the complex rather than as a calculation scheme for approximating the wave function of a static molecular form. The interacting molecules plus heat bath are described by some true time-independent potential that governs both the electronic and nuclear dynamics. This potential is unknown, and if it were known it would not be very useful since it would hide relevant features in information about the heat bath. Some information must be discarded to bring out the essential features (as is usual in passing from a fully detailed microscopic description to a macroscopic description). As an idealization the electronic and nuclear wave functions are separated; this is the extreme and unrealistic case in which the potential of the two systems do not share coordinate dependencies except at updating points. If the electronic-nuclear interaction is significant under these conditions then it should be even more significant under more realistic conditions.

The electronic wave function under the above idealization determines the potential that governs the time development of the nuclear system; similarly, the nuclear wave function determines the potential that governs the time development of the electronic system. The two wave functions (and hence the two potentials) must be self-consistent, and the self-organization of the complex corresponds to the evolution of self-consistency. The electronic and nuclear potentials are time-dependent approximations to the true potential. As the electronic wave function changes (for example, due to perturbation by the changing nuclear coordinates or in response to perturbation from the radiation field) the potential governing the nuclear motions will change (since the electronic charge distribution will change), leading to a change in the nuclear wave function. The potential governing the electronic system will then change, leading to a further change in the electronic wave function and therefore to further change in the potential governing the nuclear system.

The process could converge to self-consistency in a local minimum, in which case the whole system will remain in a metastable state until fluctuation carries it to another minimum; or it could diverge, due to a self-amplifying (positive feedback) interaction between the electronic and nuclear systems. The nuclei will undergo a bobbing motion due to the interference of electronic states of different energy; the nuclear motions will further perturb the electronic wave functions, which will further perturb the electronic system, and so on, until the nuclei jump out of the local minimum. Because of the self-amplifying effect, the speed with which the interacting molecules discover the global minimum (or, alternatively, dissociate) will increase. The rate of complex formation (and, in the case of enzyme catalysis, the equally critical rate of dissociation) is in this way sped up by the parallelism inherent in the electronic wave function.

Proteins and other biological macromolecules are colorless, apart from chromophores. The electronic wave function can nevertheless spread out (become a superposition of eigenfunctions of different energies) as a consequence of energy perturbations much smaller than ultraviolet energies, because the system is non-Born-Oppenheimer. Some of the spectra ordinarily attributed to the infrared will be due to accelerations of electrons relative to the nuclear coordinates, which corresponds to the obvious fact that the electronic spectra will undergo slight shifts in the course of self-assembly.

As noted above, the hydrogen bonds are intermediate between the electrons and the more massive nuclei. Tunneling motions of hydrogen bonds can further contribute to the amplification effect, especially as hydrogen bonds provide preferred pathways for tunneling motions of electrons. This contribution is important, since it is favorable to the occurrence of degeneracies and therefore to significant shifts in the structure of the wave function.

It might initially be thought that thermal effects would dominate the quantum effect. However, that is not so because the potential surface is time dependent. The complete system (electrons, nuclei, heat bath) has an initial kinetic and potential energy that *in toto* is conserved. The potential and kinetic energies can individually increase or decrease in the course of time due to energy exchanges between the electronic and nuclear coordinates (think of a bouncing ball). The lowering of potential barriers by enzymes, and also the self-assembly process, are better conceived in terms of dynamic surfaces than in terms of a static construct. The true picture is that the time dependence of the surface is strongly influenced by the superposition of electronic states.

CONCLUDING REMARKS

The quantum mechanical speedup effect will occur whenever features of the milieu (acting as signals) influence the conformations of molecules that enter into a complex whose conformational features exert a controlling effect on some cellular activity. This is a ubiquitous feature of cellular biochemistry. A number of lines of evidence suggest that second messengers, in particular cyclic nucleotides, link nerve impulse activity to molecular interactions at the subcellular level.[6] Structural alterations in the cytoskeleton appear to play an important role.[7-9] The implication is that the parallelism inherent in the quantum mechanical wave function vastly enhances the capacity of cells, including neurons, to recognize input patterns and to transform these to functionally useful outputs.[10]

REFERENCES

1. M. Conrad, "Microscopic-macroscopic interface in biological information processing," *BioSystems* 16: 345-363, 1984.
2. M. Conrad, "Molecular computing," in M. C. Yovits, Ed., *Advances in Computers,* San Diego: Academic Press, 31: 234-235, 1990.
3. M. Conrad, "Quantum mechanics and cellular information processing: The self-assembly paradigm," *Biomedica Biochimica Acta* 49: 743-755, 1990.
4. M. Conrad, "Quantum molecular computing: The self-assembly model," *Int. J. Quant. Chemistry: Quantum Biology Symposium* 19: 125-143, 1992.
5. M. V. Volkenstein, *Physics and Biology,* New York: Academic Press, 1982.
6. E. A. Liberman, S. V. Minina, N. E. Shklovsky-Kordy, and M. Conrad, "Microinjection of cyclic nucleotides provides evidence for a diffusional mechanism of intraneuronal control," *BioSystems* 15: 127-132, 1982.
7. E. A. Liberman, S. V. Minina, O. L. Mjakotina, N. E. Shklovsky-Kordy, and M. Conrad, "Neuron generator potentials evoked by intracellular injection of cyclic nucleotides and mechanical distension," *Brain Res.* 338: 33-44, 1985.
8. S. R. Hameroff, *Ultimate Computing,* Amsterdam: North-Holland, 1987.
9. G. Matsumoto, S. Tsukita, and T. Arai, "Organization of the axonal cytoskeleton: Differentiation of the microtubule and actin filament arrays," in *Cell Movement (vol. 2: Kinesin, Dynein, and Microtubule Dynamics),* New York: Alan R. Liss, Inc., 1989, 335-356.

10. This work was supported in part by U.S. National Science Foundation grant ECS-9109860.

Martin Blank, Ed., *Electricity and Magnetism in Biology and Medicine*

A MODEL FOR MOLECULAR COMPUTATION VIA BOOLEAN OPERATIONS IN CYTOSKELETAL MICROTUBULES AND MICROTUBULE ASSOCIATED PROTEINS

R. Lahoz-Beltra

University of Arizona College of Medicine, Tucson and Complutense University of Madrid

S. R. Hameroff*

University of Arizona College of Medicine, Tucson

J. E. Dayhoff

University of Maryland, College Park

Dynamic activities within living cells are organized by cytoskeletal microtubules (MT) and cross-bridging MT-associated proteins (MAPs), shown essential for learning and synaptic plasticity in neurons. Automata models and some evidence predict propagating acousto-conformational signals and patterns in MT. Propagation is predicted to occur via a Fröhlich pumping mechanism involving GTP hydrolysis of MT subunits and phosphorylation of MAPs. Propagating signals interact through MT and MAPs with molecular logic, of which Boolean algebra is one example. We consider MAP connections among MT as OR, AND, XOR, and BUS operators modifying MT conformational automata patterns. "Output" may result in restructuring of intracellular functional architectures such as neuronal synapses. Computer simulations demonstrate capabilities for computation and numerical calculation.

Living cells are dynamically organized by networks of protein polymers called the cytoskeleton, which includes microtubules (MT), actin, intermediate filaments, and microtubule associated proteins (MAPs). MAPs attach at specific MT subunit sites and interconnect MT, other cytoskeletal elements, membrane proteins and organelles to form a dynamic intracellular scaffolding, whose control of cellular movement, growth, and adaptive behavior (e.g., synaptic formation and modification in neurons) has suggested some type of intelligent orchestration.[1] MT are hollow cylinders, 25 nm in diameter, whose walls are assemblies of 13 longitudinal protofilaments that are each a series of subunit polar dimer proteins known as tubulin (Fig. 1). Previous papers have modeled information processing in MT and MT/MAP networks by cooperative changes in tubulin conformational states in the nanosecond to 10ps time scale based on Fröhlich's predicted coherent excitations.[2] For example, cellular automata models[3-5] predict propagating conformational signals and patterns in a velocity range of 8-800 m/s in MT and MT/MAP networks. We present a theoretical model for computing via such propagating signals using arithmetic operations by Boolean algebra in which MAPs act as Boolean operators (OR, AND, NOT, XOR) and the computing structures are MT/MAP networks.[6]

We consider signals along MT which are "read" by interconnected MAPs, presumed capable of Boolean operations (via allosteric conformational effects) or may act as "bus" MAPs which transfer signals unchanged between MT. We propose that the arrangement of MT and MAPs in a specific architecture might result in a network capable of arithmetic operations such as addition, subtraction, and multiplication, as well as Boolean logic operations.

To represent bits and perform computations we use the BCD binary-code system employed by most calculator circuit designs. We assume that each tubulin subunit dimer within an MT can be in one of two possible conformational states (Fig. 1) that have been labeled as either alpha state (bit state 1) or beta state (bit state 0). Since BCD code requires a 4-bit word per single decimal digit, 4 parallel adjacent protofilaments are necessary to codify a single decimal digit. Thus a bit-line can be defined as an automata glider that is

*Correspondence to: S. R. Hameroff (Anesthesiology), University of Arizona Med. Ctr., Tucson AZ 85724.

transported along a protofilament. For a multidigit number (e.g., 628), each single digit would be expressed by a 4-bit word (0110-0010-1000) and occupy 12 parallel protofilaments. Our choice of this system is arbitrary and serves merely to demonstrate the possibility of MT-based computation.

In the proposed model each one of 4 protofilaments used to codify a BCD single digit on one microtubule (MT1) is connected by one or more MAPs with protofilaments located on a second microtubule (MT2). MAPs are considered either as bit-lines (like protofilaments) where a glider signal can be transported unchanged between MT ("BUS MAP"), or as bit-lines where a Boolean operation is performed in one of the two attachments between a MAP and the parallel MT ("LOGIC-MAP"). Figure 2 shows a schematic for addition of 3 + 2; Fig. 3, a computer simulation for addition of 5 + 4. MT are modeled as two-dimensional lattices; each tubulin has two conformational states (alpha, 1, black; beta, 0, white), and MAPs are simple connections attached between MT. During the simulation, alpha states move one tubulin subunit per time unit from the right MT end to the left MT end. Information is processed from the top (upper MT) to the bottom (lower MT) of the network and from the right side (right end on MT) to the left one (left end on MT). This model demonstrates one possible mode of intracellular information processing that may be implemented in future molecular computing devices.

REFERENCES

1. S. R. Hameroff, *Ultimate Computing: Biomolecular Consciousness and Nanotechnology,* Amsterdam: Elsevier-North Holland, 1987.

2. H. Fröhlich, "The extraordinary dielectric properties of biological materials and the action of enzymes," *Proc. Natl. Acad. Sci. USA* 72: 4211-4215, 1975.

3. S. R. Hameroff, S. Rasmussen, and B. Mansson, "Molecular automata in microtubules," in C. Langton, Ed., *Artificial Life,* Reading, Mass.: Addison-Wesley, 1989.

4. S. Rasmussen, H. Karampurwala, R. Vaidyanath, K. S. Jensen, and S. Hameroff, "Computational connectionism within neurons: A model of cytoskeletal automata subserving neural networks," *Physica* D42: 428-449, 1990.

5. J. E. Dayhoff, S. R. Hameroff, C. Swenberg, and R. Lahoz-Beltra, *Biological Plausibility of Back-error Propagation in Microtubules,* University of Maryland SRC Technical Report, TR 92-127, 1992.

6. R. Lahoz-Beltra, S. Hameroff, and J. E. Dayhoff, "Cytoskeletal logic: A model for molecular computation via Boolean operations in microtubules and microtubule-associated proteins," *BioSystems* (in press).

7. Rafael Lahoz-Beltra is supported by Ministerio de Educacion y Ciencia (Spain) under Fulbright/MEC (1989-1991) and FPI/MEC (1991-1992) grants. J. E. Dayhoff is supported by the Naval Surface Warfare Center, Dahlgren, Va., and the Systems Research Center, University of Maryland; and S. R. Hameroff is supported in part by NSF grant DMS-9114503.

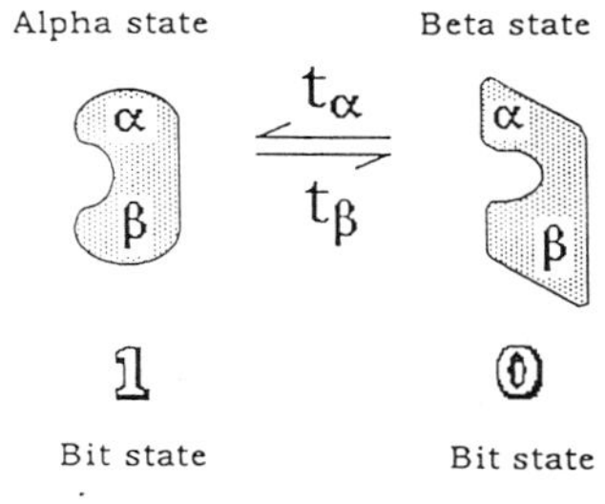

BCD Code and Tubulin States

0	0000	ββββ	5	0101	βαβα
1	0001	βββα	6	0110	βααβ
2	0010	ββαβ	7	0111	βααα
3	0011	ββαα	8	1000	αβββ
4	0100	βαββ	9	1001	αββα

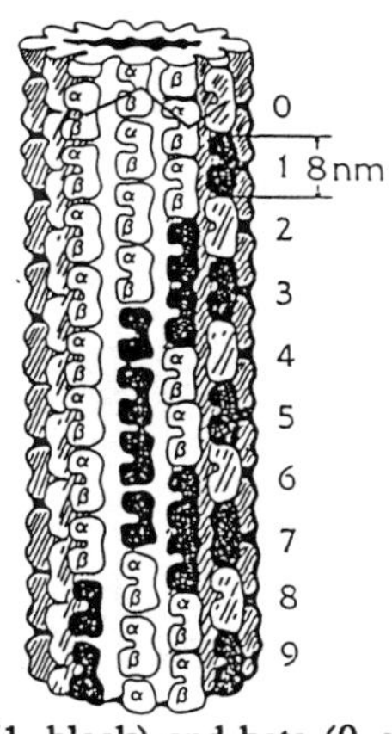

FIG. 1.—Left, top: Conformational switching between alpha (1, black) and beta (0, white) conformational states in tubulin subunits. Left, bottom: Table showing conversion from single decimal digits (from 0 to 9) to BCD (4-bit word) and equivalence between BCD and tubulin conformational states (4-states word). Right: MT showing single decimal digits which have been codified as 4-conformational states words; alpha and beta states are represented in black (1) and white (0), respectively. Zig-zag line drawn on MT indicates how to read binary number codified on MT lattice.

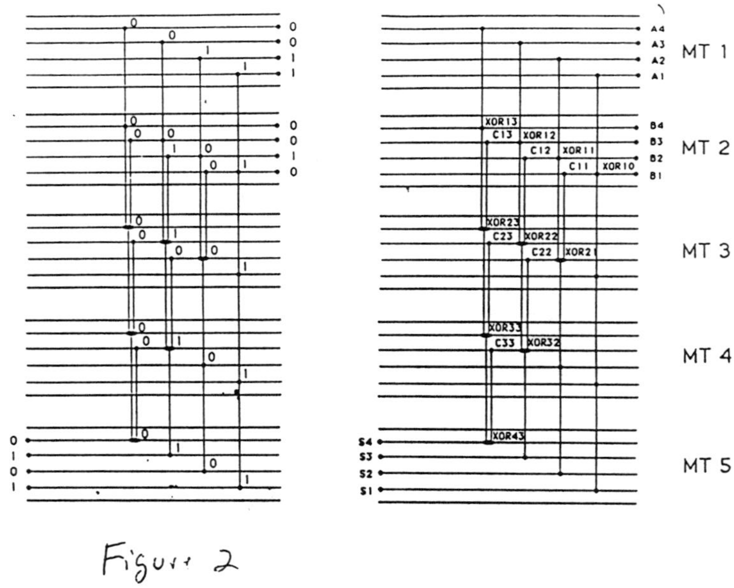

FIG. 2.—Schematic of artificial MT-MAP adder/subtracter network for single BCD digits. Right: Addition/subtraction performed by Boolean logic equations (C: carry term). Left: Two numbers, 0011 (3) and 0010 (2), are summed showing carry terms, intermediate values, and final output.[3]

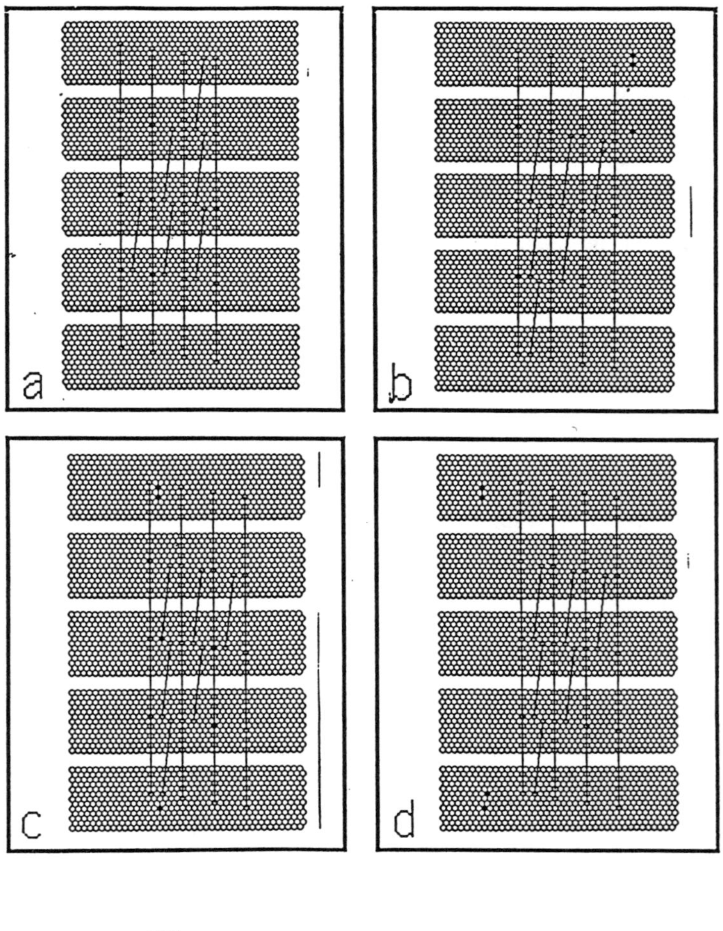

FIG. 3.—Computer simulation of MT-MAP ADDER network. (a) Network without input data (resting state). (b) Input values are 0101 (5 in decimal code) on MT1 and 0100 (4 in decimal code) on MT2. Only alpha state (digit 1 or black subunit) is visible, since beta state (digit 0 or white subunit) is the same as background. (c) A carry term equal to 1 (black subunit) is located over MAP bond site at third MT. (d) Output value 1001 (9 in decimal code) is shown at left-end on MT4.

Martin Blank, Ed., *Electricity and Magnetism in Biology and Medicine*

TWO-DIMENSIONAL CRYSTALLIZATION OF PROTEINS AND FINE PARTICLES

K. Nagayama*

ERATO, JRDC, Tsukuba, Japan

A unique new surface process has been recently discovered in our project, involving the use of polystyrene latex particles ranging in size from 50 nm to 2 μm. What is remarkable in the fine-particle assembly is the sudden and rapid growth of crystal domains and the well-ordered crystal arrays. The new assembly process can only be explained by a consideration of the synergetic combination of many surface effects and can explain the long-term puzzle associated with the protein 2D crystal on mercury. The 2D crystallization of proteins and latex particles could yield an insight on how to engineer the patterning of proteins and fine particles on two-dimensional surfaces in general.

Protein Arrays

A protein array is a kind of protein assembly two-dimensionally developed on the surface of a liquid or solid; i.e., two-dimensional (2D) crystals of proteins or complex patterns made of several protein species.[1] In contrast to conventional bioelectronics practice, in which the individual nature of proteins as a simple entity is utilized, we try to use proteins in their integral form in our protein arrays. To prepare a protein array with a desired 2D pattern, the properties of the original interprotein interactions can be deliberately modified by an appropriate method such as chemical modification, genetical mutation of the amino acids on the surface, or reshuffling of domain structures by the use of genetic engineering. The pattern formation by proteins is clearly analogous to recent silicon technology in the large-scale integration (LSI) industries. Densely packed units have their own advantages in integration, even when the functioning of the units themselves has not been improved. Protein arrays merely use the original functions of proteins and organize them to a higher degree.

A reduction in dimensions always simplifies our thinking, but in the real 3D world this argument is sometimes deceptive; for example, it may lead to new complications in the interaction of proteins with the substance. We have to define the physico-chemical and morphological properties of the substrate surface precisely in order to match the sophisticated structure of the protein surface satisfactorily. The choice of the appropriate supporting surface is then very important for the formation of highly ordered protein monolayers. A protein-membrane system including hexagonally packed bacteriorhodopsin, whose optoelectronic function is attracting the interest of many molecular-electronics engineers, is one example. To replace the bilayer membrane by artificial substrates, various kinds of interfaces such as air/water, air/metal, water/graphite, and air/lipid/water have been considered. Our recent innovation in the 2D crystallization of proteins has adopted 2D surfaces of a liquid metal, mercury, as a likely alternative to the water surface as the working field for 2D crystal or array formation.

To date more than a dozen different proteins and protein complexes have been crystallized by this method. One-third of the trials have resulted in satisfactory ordering in arrays, including water-soluble proteins; ferritin;[2] the water-soluble part of H^+-ATPase[3] and a chaperonin[4] of thermophilic bacteria; and a DNA-protein complex, bacteriophage.[5]

Particle Arrays

The process of array formation is governed by various physico-chemical factors. They range from those relevant in the early stage of lattice formation to those in the successive stages of molecular orientation. The governing factors of lattice formation are usually long-range attractive forces as observed in nonspecific interprotein interactions (colloidal interactions), protein/water interaction, nonspecific protein/substrate

---

*Correspondence to: K. Nagayama, ERATO/JRDC, Higashiarai 18-1, Tsukuba 305 Japan.

interaction, and water capillary forces appearing at the air/water interface. The governing factors of molecular orientation are mainly short-range attractive or repulsive forces of the sort observed in specific interprotein interactions and specific protein/substrate interactions. The formation of a protein array requires the combination of well-defined short-range and long-range forces. When proteins are separated by distances large enough compared with their own sizes, they are simply considered as classical colloidal particles that express only nonspecific short- and long-range interactions. This is one reason we are also interested in the condensation reaction of colloidal particles. In this respect the 2D lattice formation of polystyrene latexes at the water/air interface will yield basic knowledge for the formation process of the protein array.[6] By use of variously sized polystyrene latex particles ranging from 50 nm to 2 μm, a unique new surface process has recently been discovered in our project.[7] What is remarkable in the fine-particle assembly is the sudden and rapid growth of crystal domains and the well-ordered crystal arrays. This feature differs significantly from the usual 2D or 3D crystals observed in solution, whose growth is relatively slow due to the rate-limiting step of diffusion. One can explain the new assembly process only by taking into account the combination of many surface effects such as convection-flow driven by water evaporation and a long-range capillary force acting in the lateral direction,[8] which could also explain the old puzzle related to the protein 2D crystal on mercury[3]—a growth thousands times faster than that observed in the crystal at the lipid/water interface.[9] The new phenomenon discovered in the initial stage of protein assembly in the 2D crystallization latex particles could provide an insight on how to engineer the assembly and alignment of proteins and fine particles in general.

Component Techniques in Array Engineering

The protein and particle array engineering is based on the following scientific and technological elements.

Surface Processing. Substrate surfaces well defined in their clearness, roughness, and physico-chemical properties should be prepared.

Characterization of Substrate Surface. There are two kinds of substrates in array engineering: the primary substrate, and the secondary substrate onto which the arrays are transferred from the primary substrate. The most promising primary substrates we found are a liquid metal for proteins and glass or mica for particles. The flatness and roughness of the solid surface should be well characterized by means of various surface techniques, such as scanning electron microscopy.

Array Formation. The protein array was processed with the rapid spreading of solutions on clean mercury based on their high wettability under 1 atm. oxygen gas. The particle array was prepared on a solid surface with the flow toward assembling using solvent convection driven by evaporation. Control of the thickness of the liquid film for immersing the particles is crucial.

Characterization of Array Structure. The crystal form, the degree of crystallization, 2D sequential order of several protein species, and protein molecular orientation should be precisely characterized by morphological and spectroscopical methods. When atomic resolution was called for, a transmission electron microscope was used. For the direct observation of protein and particle array on the substrate, scanning electron microscopy and scanning probe microscopy were applied.

Array Design. Array structure can be well simulated by computational chemistry on the basis of interprotein or interparticle interactions analogous to the simulation of protein structures based on interresidue interactions of constituent amino acids.

Protein Synthesis. Protein molecules well defined in their surface for interprotein interactions should be expressed in cloning systems and purified with chromatographic equipment. Genetic engineering is our nuts-and-bolts approach to modifying the surface structure of proteins.

Thermodynamics of Protein Array. Phenomena such as protein adsorption to the substrate surface, the lattice formation process, and the molecular reorientation process should be carefully studied from a thermodynamical point of view. In this respect the conventional techniques used in colloid and surface science are important. Contact-angle measurement of the aqueous solution of proteins, surface-tension measurement of protein solutions, and several kinds of light-scattering measurement of protein solutions may give some insight.[10]

REFERENCES

1. K. Nagayama, *Nanobiology* 1: 25-37, 1992.

2. H. Yoshimura, M. Matsumoto, S. Endo, and K. Nagayama, *Ultramicroscopy* 32: 265-274, 1990.

3. H. Yoshimura, S. Endo, M. Matsumoto, K. Nagayama, and Y. Kagawa, *J. Biochemistry* 106: 958-960, 1989.

4. N. Ishii, H. Taguchi, K. Yoshida, H. Yoshimura, and K. Nagayama, *J. Biochemistry* 110: 905-908, 1991.

5. H. Yoshimura, S. Endo, K. Nagayama, and M. Matsumoto, *JEOL News* 29E: 2-5, 1991.

6. P. Pieranski, *Phys. Rev. Lett.* 45: 569-572, 1980.

7. N. D. Denkov, D. B. Velev, P. A. Kralchevsky, I. B. Ivanov, H. Yoshimura, and K. Nagayama, *Nature* (in press).

8. P. A. Kralchevsky, V. N. Paunov, I. B. Ivanov, and K. Nagayama, *J. Coll. Interface Sci.* 151: 74-94, 1992.

9. P. Fromherz, *Nature* 231: 267-268, 1971.

10. Our sincere thanks go to Drs. S. Ebina, H. Yoshimura, S. Endo, C. Dushkin, P. Kralchevsky, I. Ivanov, N. Denkov, and M. Matsumoto for their collaboration in this study.

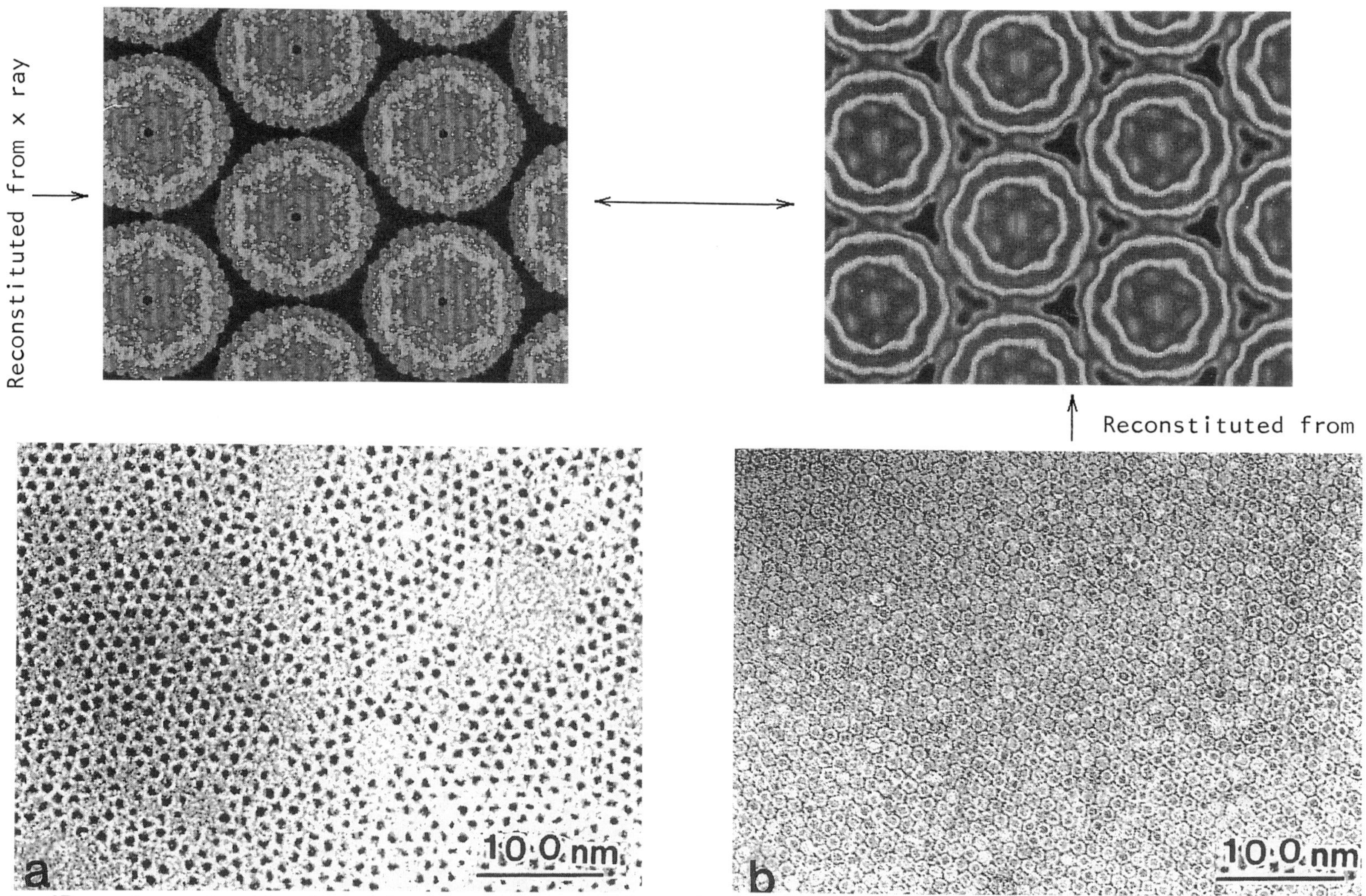

FIG. 1.—2D monolayers of proteins observed by electron microscope: micrographs of (a) amorphous monolayer of ferritin molecules developed on a water/lipid interface, (b) crystalline monolayer of ferritin molecules developed on a mercury surface with water.

Martin Blank, Ed., *Electricity and Magnetism in Biology and Medicine*

SITE-DIRECTED MUTAGENESIS OF CYTOCHROME b_{562} FOR STABILIZATION TO FORM TWO-DIMENSIONAL REGULAR ARRAY

K. Kobayashi,* K.-H. Kim, K. Saito, T. Nagamune, J. Otomo, Y. Kobayashi, T. Furuno, and H. Sasabe

The Institute of Physical and Chemical Research (RIKEN), Saitama, Japan

For the development of bioelectronic devices and/or the structural analyses of proteins, two-dimensional crystallization of proteins is required. Cytochrome b_{562}, which is one of the heme proteins, has been chosen as a model protein for the fabrication of two-dimensional crystals. Heme proteins are a particularly interesting class due to an electron transfer characteristic associated with the porphyrin-based prosthetic group. Two approaches in making two-dimensional crystals of cytochrome b_{562} are presented here. One is the formation of a monomolecular layer of cytochrome b_{562} reconstituted with alkylated heme. The other is increasing stability of proteins by protein engineering.

For the development of bioelectronic devices, the critical step will be the development of general methods for controlling the spatial arrangement and orientation of proteins to form highly ordered macroscopic materials. In our investigations, cytochrome b_{562} has been chosen as a model protein for the fabrication of two-dimensional crystals. Cytochrome b_{562} is a small (M_r 12 000) water-soluble protein found in *Escherichia coli,* consisting of 106 amino acid residues.[1,2] It is classified as an electron transfer protein. The structure of the *E. coli* ferricytochrome b_{562} has been defined by x-ray crystallography to a resolution of 2.5 Å.[3] The heme, protoporphyrin IX, is noncovalently bound and is ligated to the polypeptide chain through methionine 7 on the N-terminal helix and histidine 102 on the C-terminal helix.

Two approaches to making two-dimensional crystals of cytochrome b_{562} are presented in this report. One is the formation of a monomolecular layer of cytochrome b_{562} reconstituted with alkylated heme. When crystalizing molecules in two dimensions, the general method utilized is the Langmuir-Blodgett method. If the protein molecules are amphiphilic at the air/water interface, the protein solution can be spread directly onto the water surface. As mentioned, the heme prosthetic group of cytochrome b_{562} is not covalently bound to the protein, and can therefore be removed and replaced in vitro by other prosthetic groups such as modified porphyrin. As a model case, we synthesized protoheme IX with the long alkyl chain, and reconstituted apocytochrome b_{562} with the modified protoheme IX. This process would give cytochrome b_{562} an amphiphilic property. The other approach is to increase the stability of proteins by protein engineering. When proteins are crystallized at interfaces, most proteins denature due to surface-pressure effects. There are several approaches to overcoming this problem. One is to enhance the structural stability of the protein by site-directed mutagenesis to prevent denaturation. An effective method of making a thermally stable mutant protein is to introduce a disulfide bridge formed by two cysteine residues.[4,5] Thus, we made a mutant protein having both glycine 22 and asparagine 82 changed to cysteines. These positions were calculated to be suitable for disulfide bridge formation by means of computer simulation. Another effective method would be to introduce an increased hydrophobic interaction between the α-helices of the protein and/or to replace a poor helix-forming residue within an α-helix with a good helix former. Thus, we made another mutant protein having glycine 64 changed to an alanine.

FORMATION OF A MONOMOLECULAR LAYER OF ALKYLATED CYTOCHROME b_{562}

Sample Preparation

Cytochrome b_{562} was obtained from *E. coli* strain TB-1 harboring pNS 207 plasmid grown in LB medium and purified as previously described[6] with some modifications. The apoprotein was prepared at 4 C by the butanone extraction method.[7] The apoprotein solution was dialyzed against pure water. The modified

*Correspondence to: K. Kobayashi, FRP (RIKEN), 2-1 Hirosawa, Wako, Saitama 351-01 Japan.

protoheme IX (stearylated protoheme IX in this study) was obtained from the reaction of protoheme IX with stearylamine in pyridine at 4 C, in the presence of dicyclohexylcarbodiimide.

Reconstitution of Apocytochrome b_{562} with Alkylated Heme

The reconstitution of the apocytochrome b_{562} with protoheme IX can be achieved by a reaction in alkaline pH buffer,[8] but since stearylated protoheme is insoluble in water, the mixture of dimethylsulfoxide and water was used as the solvent for reconstitution reaction. The effect of solvent composition of dimethylsulfoxide and water on the reconstitution reaction was examined. As a result, the reconstitution reaction was successful when the dimethylsulfoxide content was 55-65%. The effect of the temperature on the reconstitution reaction was also examined and it was found that the lower the temperature of reconstitution reaction, the higher the fraction of the reconstitution. From the time-course experiment of the reconstitution reaction monitored by the characteristic (Soret) band at 418 nm, it was found that the reconstitution reaction takes 30-40 min to complete, which is a rather slow reaction.

Surface Property of Modified Cytochrome b_{562}

Figure 1 shows the π-A isotherms of stearylated holocytochrome b_{562} spread on 0.25 mM $CdCl_2$, pH 7.7, at 21 C. Stearylated holocytochrome b_{562} was prepared with reference to the above results, namely, dimethylsulfoxide content 63%, temperature 10 C, reconstitution time 40 min. Generally, it is difficult to spread the protein solution directly onto a buffer surface. Therefore, we utilized dimethylformamide or isopropyl alcohol to aid the procedure. In Fig. 1, curve A shows the isotherm of native b_{562} and curve B shows that of stearylated holocytochrome b_{562}. It is clear that curve B shows a surface-active property. Similar results were obtained from curves C and D with the addition of dimethylformamide and isopropyl alcohol, respectively. Thus, we conclude that the presence of organic solvent is not necessary in spreading stearylated holocytochrome b_{562} onto a buffer surface. The surface pressure of ca. 26 dyn cm^{-1} represented in Fig. 1 is thought to be the value inherent to stearylated holocytochrome b_{562}. After the compression, the surface pressure decreased gradually and showed a stable value of ca. 20 dyn cm^{-1} (data not shown), which indicates that the stearylated holocytochrome b_{562} forms a stable monomolecular layer on the buffer surface.

COMPARISON OF THERMAL STABILITIES BETWEEN NATIVE AND MUTANT PROTEINS

A site-directed mutagenesis was carried out to produce mutants, G64A (glycine 64 → alanine) and A22C/G82C (alanine 22 → cysteine, glycine 82 → cysteine). The thermal stabilities of these mutants with oxidized heme were examined and were compared with that of the wild-type protein with oxidized heme. Thermally induced denaturation of cytochrome b_{562} was monitored by the UV-Vis (Soret band at 418 nm) and CD (at 222 nm) spectral changes, and the differential scanning microcalorimetry (DSC) measurement. Each can follow the dissociation of the heme moiety, the transition of α-helix to random coil, and the thermal transition of heat capacity, respectively. The results of UV-Vis, CD, and DSC measurements indicate that the dissociation of the heme moiety initiates the denaturation of α-helices. It is clear from Fig. 2 that the denaturation temperature of G64A mutant is almost 78° and is 2° higher than that of wild type, whereas that of A22C/G82C mutant is about 10° lower. It is suggested that the unexpected instability in A22C/G82C mutant is due to the weak binding of the heme moiety to the peptide chain which was induced by structural distortion of apo-mutant protein.

CONCLUSION

The electron-transferring protein cytochrome b_{562} was chosen as the target material for bioelectronics and was modified genetically and chemically in order to increase stability and orientational control in two-dimensional crystallization by the LB method. These modifications of the protein were nearly successful and the purpose was partially accomplished. More efforts are thus needed to develop techniques for two-dimensional crystallization and the protein design for constructing the frontier of future electronic devices.[9]

REFERENCES

1. E. Itagaki and L. P. Hager, *J. Biol. Chem.* 241: 3687, 1966.
2. F. Lederer, A. Glatigny, P. H. Bethge, H. D. Bellany, and F. S. Mathews, *J. Mol. Biol.* 148: 427, 1981.
3. F. S. Mathews, P H. Bethge, and E. W. Czerwinski, *J. Biol. Chem.* 254: 1699, 1979.

4. C. Mitchinson and J. A. Wells, *Biochemistry* 28: 4807, 1989.
5. L. J. Perry and R. Wetzel, *Science* 226: 555, 1984.
6. F. W. J. Teale, *Biochem. Biophys. Acta* 35: 543, 1959.
7. H. Nikkila, R. B. Gennis, and S. G. Sligar, *Eur. J. Biochem.* 202: 309, 1991.
8. E. Itagaki, G. Palmer, and L. P. Hager, *J. Biol. Chem.* 242: 2272, 1967.
9. The authors would like to thank Prof. S. G. Sligar (University of Illinois) for providing the over-expressed *E. coli* strain, and Drs. H. Uedaira and H. Morii (Research Institute for Polymers and Textiles) for carrying out CD and DSC analyses.

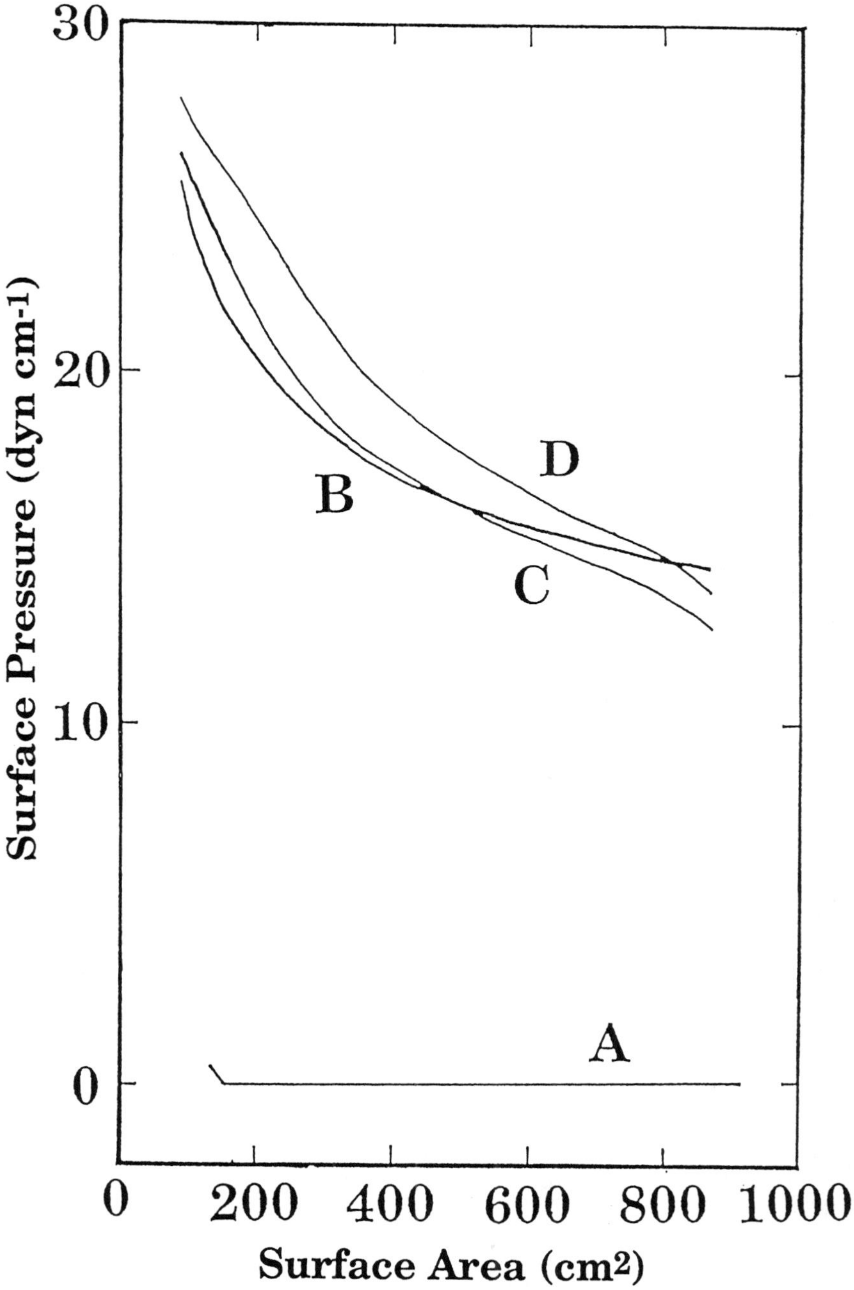

FIG. 1.—π-A isotherms on 0.25 mM $CdCl_2$, pH 7.7. A: native b_{562}, B: st-cytochrome b_{562} in pure water, C: st-cytochrome b_{562} in pure water and DMF, D: st-cytochrome b_{562} in pure water and isopropyl alcohol.

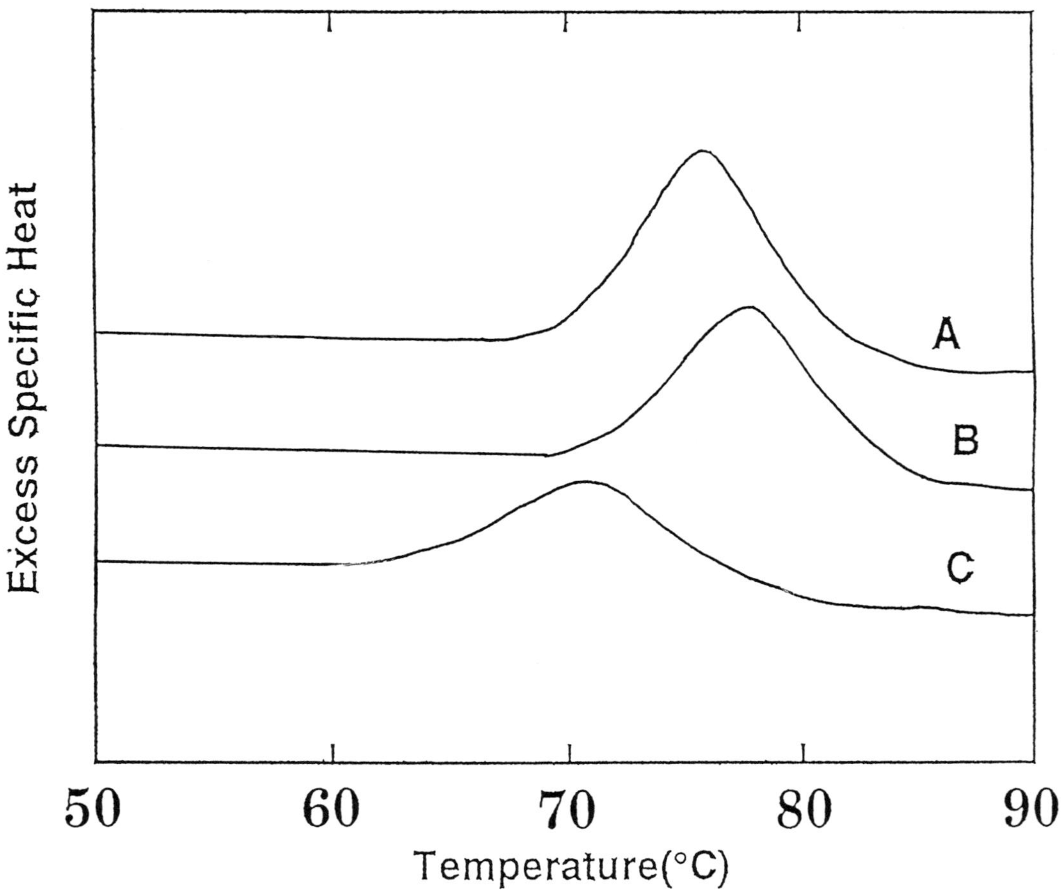

FIG. 2.—DSC data for native cytochrome b_{562}, G64A, and 22C/82C. These results were obtained in 10 mM phosphate buffer. The excess specific heat as a function of temperature was measured with a DASM-1 microcalorimeter. The temperature was increased at a rate of 0.5°C/min from 5 to 90°C. A: native cytochrome b_{562}, B: G64A, C: 22C/82C.

Martin Blank, Ed., *Electricity and Magnetism in Biology and Medicine*

MOLECULAR NANOTECHNOLOGY WITH TWO-DIMENSIONAL PROTEIN CRYSTALS

U. B. Sleytr,* D. Pum, M. Sára, and P. Messner

Universität für Bodenkultur, Vienna

Biological self-assembly systems represent particularly appealing model systems for the study of structural principles for miniaturized functional units and devices in molecular nanotechnology. Supramolecular structures such as two-dimensional crystalline bacterial cell surface layers (S-layers) have been optimized for specific functions in the course of billions of years of evolution.[1-3]

S-layer lattices formed of assemblies of a single protein or glycoprotein species (Mw 40000 to 200000), can exhibit oblique (p2), square (p4), or hexagonal (p6) symmetry. Generally the center-to-center spacing of the morphological units in the 2D-crystalline arrays can range from 5 to 32 nm.[3] Isolated S-layer subunits have shown the inherent ability to self-assemble into 2D arrays, either in suspension or on surfaces suitable for adhesion.[4,5] These self-assembly studies have shown that all the information required for generating the 2D-crystalline arrays is entirely contained in the individual protomeric unit.

High-resolution electron microscopy, in combination with digital image reconstruction procedures and chemical modification procedures, revealed that S-layers are anisotropic structures. They exhibit considerable asymmetry in the topography and charge distribution of the outer and inner surface.[6,7] Furthermore, electron microscopial and permeability studies showed that S-layers function as selective isoporous molecular sieves in the ultrafiltration range.[8,9]

A remarkable feature of many bacteria is their ability to produce glycosilated S-layer proteins.[3,10] High-resolution electron microscopy in combination with chemical modification and labeling experiments have shown that the carbohydrate chains are exposed on the cell surface.[11] Since S-layers are composed of identical subunits arranged according to a crystal lattice, functional groups such as amino-, carboxyl-, or hydroxyl-groups occur on each constituent subunit in an identical position and orientation.

Based on the data obtained by fundamental studies on S-layers, a considerable potential in biotechnological and nonbiological applications became evident.[12-14]

ISOPOROUS ULTRAFILTRATION MEMBRANES

S-layer ultrafiltration membranes (SUMs) were produced by deposition of S-layer fragments as a coherent layer on microfiltration membranes. The mechanical and chemical stability of this composite structure is obtained by inter- and intramolecular crosslinking. The chemical and thermal resistance of these membranes was demonstrated to be comparable to polyamide membranes.[8] By immobilizing molecules of various size in the pore areas, SUMs with defined shifts in the molecular sieving properties and the cut off values were obtained.[15]

MATRIX FOR IMMOBILIZATION OF MACROMOLECULES

Since S-layers possess identical physicochemical properties on each molecular unit area they represent unique structures for a defined binding of macromolecules.[12,16-18] S-layers composed of glycosilated protein subunits have an even greater potential for controlled immobilization. Since their carbohydrate residues are located on the surface of the protein network at well-defined positions, binding of molecules is possible in the third dimension. By high-resolution electron microscopy, it was shown that molecules immobilized on both the protein and the carbohydrate chains are frequently very regularly arranged, reflecting the periodicity of the supporting S-layer lattice. Molecules can be bound to the S-layer through electrostatic, hydrophobic, or

*Correspondence to: U. B. Sleytr (Center for Ultrastructure Research), Universität für Bodenkultur, A-1180 Vienna, Austria.

covalent bonds.

S-layer lattices reconstituted on flat solid supports have shown a great potential as immobilization matrix for studying molecules by scanning-probe microscopy.[12] Mono- and multi-enzyme biosensors were the first applications involving S-layers as a structural principle and immobilization matrices.[13]

S-LAYERS AS SUPPORTS FOR LANGMUIR-BLODGETT FILMS

S-layers and S-layer ultrafiltration membranes can be used as carriers for immobilizing mono- and multilayers of surfactants and biological membranes.[14] Since the surface properties of S-layers can be adjusted by use of a broad spectrum of chemical modification reactions, the more hydrophilic of hydrophobic Langmuir-Blodgett films can bind with either their hydrophilic or hydrophobic side. The combination of S-layers with bimolecular lipid membranes where functional molecules (e.g., carriers, ion channels, ionophores, proton pumps, receptor molecules) are incorporated, opens a broad spectrum of applications.[13]

REFERENCES

1. U. B. Sleytr and P. Messner, "Crystalline surface layers in procaryotes," *J. Bacteriol.* 170: 2891-2897, 1988.

2. U. B. Sleytr and P. Messner, "Crystalline surface layers on bacteria," *Ann. Rev. Microbiol.* 37: 311-319, 1983.

3. P. Messner and U. B. Sleytr, "Crystalline bacterial cell-surface layers," in A. H. Rose, Ed., *Advances in Microbial Physiology,* vol. 3, London: Academic Press, 1992, 213-275.

4. U. B. Sleytr and P. Messner, "Self-assembly of crystalline bacterial cell surface layers (S-layers)," in H. Plattner, Ed., *Electron Microcopy of Subcellular Dynamics,* Boca Raton, Fla.: CRC Press, 1989, 13-31.

5. D. Pum, M. Sára, and U. B. Sleytr, "Use of two-dimensional protein crystals from bacteria for non-biological applications," *J. Vac. Sci. Technol.* B-7: 1391-1397, 1990.

6. M. Sára and U. B. Sleytr, "Charge distribution on the S-layer of *Bacillus stearothermophilus* NRS 1536/3c and the importance of charged groups for morphogenesis and function," *J. Bacteriol.* 169: 2804-2809, 1987.

7. D. Pum, M. Sára, and U. B. Sleytr, "Structure, surface charge and self-assembly of the S-layer lattice from *Bacillus coagulans* E38-66," *J. Bacteriol.* 171: 5296-5303, 1989.

8. M. Sára and U. B. Sleytr, "Membrane biotechnology: Two-dimensional protein crystals for ultrafiltration purposes," in H.-J. Rehm, Ed., *Biotechnology 6b,* Weinheim VCH, 1988, 615-636.

9. M. Sára, S. Küpcü, and U. B. Sleytr, "Crystalline bacterial cell surface layers used as ultrafiltration membranes and immobilization matrix," *Gen. Eng. Biotechnol.* 10: 10-13, 1990.

10. P. Messner and U. B. Sleytr, "Bacterial surface layer glycoproteins," *Glycobiology* 1: 545-551, 1991.

11. M. Sára, S. Küpcü, and U. B. Sleytr, "Localization of the carbohydrate residue of the S-layer glycoprotein from *Clostridium thermohydrosulfuricum* L111-69," *Arch. Microbiol.* 151: 416-420, 1989.

12. U. B. Sleytr, D. Pum, M. Sára, and P. Messner, "Two-dimensional protein crystals as patterning elements in molecular nanotechnology," in A. Aviram, Ed., *Molecular Electronics—Science and Technology,* AIP Conference Proceedings 262, New York: American Institute of Physics, 1992, 167-177.

13. D. Pum, M. Sára, and U. B. Sleytr, "Two-dimensional (glyco)protein crystals as patterning elements and immobilisation matrices for the development of biosensors," in U. B. Sleytr, P. Messner, D. Pum, and M. Sára, Eds., *Immobilised Macromolecules: Application Potentials,* London: Springer-Verlag, 1993, 141-160.

14. U. B. Sleytr, M. Sára, and D. Pum, "Application potentials of two dimensional protein crystals," *Microelectronic Engineering* 9: 13-20, 1989.

15. S. Küpcü, M. Sára, and U. B. Sleytr, "Chemical modification of crystalline ultrafiltration membranes and immobilization of macromolecules," *J. Membr. Sci.* 61: 167-175, 1991.

16. M. Sára and U. B. Sleytr, "Use of regularly structured bacterial cell envelope layers as matrix for the immobilization of macromolecules," *Appl. Microbiol. Biotechnol.* 30: 184-189, 1989.

17. M. Sára, S. Küpcü, C. Weiner, S. Weigert, and U. B. Sleytr, "Crystalline protein layers as isoporous molecular-sieves and immobilisation and affinity matrices," in Sleyter et al. (Ref. 13), 71-86.

18. D. Pum, M. Sára, P. Messner, and U. B. Sleytr, "Two-dimensional (glyco)protein crystals as patterning elements for the controlled immobilisation of functional molecules," *Nanotechnology,* 2: 196-202, 1991.

Martin Blank, Ed., *Electricity and Magnetism in Biology and Medicine*

MOLECULAR ELECTRONICS: NATURAL AND ARTIFICIAL MICROTUBULES BASED ON FULLERENE C_{60}

D. Koruga* and S. Hameroff

University of Arizona, Tucson

J. Simic-Krstic, S. Jankovic, and M. Trifunovic

University of Belgrade, Yugoslavia

J. Withers and R. Loutfy

MER Corp., Tucson, Ariz.

We propose to synthesize C_{60} as a helical artificial microtubule with properties similar to natural microtubules (MT), which participate in bioinformation processes such as memory and learning. We have already used scanning tunneling microscopy (STM) to characterize tubulin in MT with submolecular resolution and C_{60} with atomic resolution. Devices based on a dimer of C_{60} will be similar to biological MT in terms of structure, energy, and information. Devices based on C_{116} can be useful both as an interface between biological and electronic devices on a cell scale and as a new kind of intelligent hardware in future molecular information machines.

Carbon is the primary atomic component of biomolecular structures, including proteins. From a semiconductor standpoint, nitrogen, hydrogen, oxygen, and sulfur may be viewed as "impurity atoms"; however, proteins differ greatly from man-made semiconductor materials. Proteins are polymers whose primary structures are linear molecules built up from simple repeating units, the amino acids. Sneddon et al. classified amino acids relevant to electronic transfer.[1] This classification ranks side chains based on their doublet acceptor energy levels, i.e., increasing ability to accept an extra electron. Thus the amino acids fall into four groups termed *repulsive, insulating, semiconducting,* and *attractive*. Amino acids which possess negative charges (e.g., *aspartic* and *glutamic* acids) may be viewed as "n-type" semiconductors; those which possess positive charges (e.g., *lysine, arginine,* and *histidine*) may be viewed as "p-type" semiconductors.

NATURAL MICROTUBULES

Major components of the cytoskeleton, microtubules (MT) are hollow cylinders composed of subunit proteins known as tubulin. Tubulin is a globular protein 50 kD in weight consisting of about 450 amino acids (Fig. 1a). Two monomer types, α and β tubulin, form α-β heterodimers approximately $4\times4\times8$ nm in size. MT are primarily involved in cell growth, movement, differentiation, adaptation, and formation and modulation of synaptic connections in neurons (Fig. 1b). Experimental evidence suggests they are involved in learning and cognition.[2,3] Previous investigations have considered information-processing capabilities of MT based on their symmetry and semiconductor properties, conformational dynamics, and capabilities for automata behavior.[4-6]

ARTIFICAL MICROTUBULES

The recently discovered giant carbon molecule fullerene C_{60} has a unique structure and a totally different shape from other known forms of inorganic carbon: graphite and diamond.[7,8] Osawa has been involved in research on C_{60} longer than anyone else, having predicted its stability more than 20 years ago.[9] A recent

*Correspondence to: D. Koruga (Anesthesiology), University of Arizona Health Sciences Center, Tucson AZ 85724.

paper states: "Carbon does exist in several allotropic forms out of which C_{60} and C_{70} are most recent representatives. C_{60} had been predicted by Osawa already in 1970. His paper, written in Japanese, attracted little attention due to language barriers and also due to the role of computational chemistry in those days; a quantitative and reliable confirmation of a reasonable suggestion by computational methods was intractable in 1970 (at least for molecules of that size)."[10]

However, the structure of fullerene is similar to that of the biological clathrins. Reminiscent of Buckminster Fuller's geodesic domes, spherical buckminsterfullerene (C_{60}) has its atoms arrayed in a regular collection of 12 pentagon and 20 hexagon rings of carbon to form a hollow cage that resembles a soccer ball. C_{60} and other fullerenes may exhibit superconductivity, and can trap ions, molecules, and atoms (as well as electrons) within their cage structure. Superconductivity and the ability to sequester and protect molecules open a host of possibilities for scientific and technological innovation. Iijima has reported the preparation of a new type of finite carbon structure consisting of needle-like tubes ranging from 4 to 30 nm in diameter and up to 1 μm in length.[11]

On each tube the carbon atom hexagons are arranged in a helical fashion about the needle axis (Fig. 2). Fewer hexagons are drawn than would form a real tube, but the essential needle configuration correctly represents an experimental diffraction pattern. A cylindrical tube can be formed by rolling up the hexagonal sheet about the filament axis (heavy line) so as to superimpose hexagons labeled A and B at the top of A′ and B′ at the bottom. Hexagons are aligned perfectly to make a helix around the needle axis. The helical arrangement of hexagons is responsible for the mirror symmetry. The orientation of a hexagon sheet can be determined uniquely by reference to the needle axis.

The goal is to make an artificial dimer based on C_{60} as C_{116} (Fig. 3a) that will possess biophysical-like properties useful for "smoothly" interacting with natural (biological) microtubules. The inner diameter of artificial microtubules will be about 30 nm. This value is the same as the outer diameter of natural microtubules in the cell. The results of this research will provide unique opportunities for using C_{116} in practical applications such as an interface between cells and electronic devices. We expect that it will be possible to take a signal from the neuron directly to an artificial neural chip (an adaptive learning device based on an artificial neural network in VLSI) that will be the controller of such electromechanical devices as artificial hands.

Artificial microtubules (Fig. 3b) based on C_{116} (AMT/C_{116}) will be capable of providing the correct signal coming from the brain (when a person wants to move a finger and/or fingers and/or a hand) to the chip (electronic device). When a person desires to move an artificial finger, someone else will move the finger for that person, sending electrical signals to the chip. Signals from the brain (through a motor neuron and AMT/C_{116}) and artificial hand will couple and, after many repetitions (100-1000), the adaptive learning chip will make a link between signals from the brain and signals from the hand (or fingers) and will recognize which signal corresponds to each finger and/or hand. Similar applications could be created for artificial legs and eyes.

CONCLUSIONS

We propose to synthesize C_{116} as a dimer of C_{60}. The dimer will be used to synthesize artificial microtubules with properties similar to biological ones. Devices based on C_{116} can be realized as an interface between biological and electronic devices and as intelligent molecular machines.[12]

REFERENCES

1. S. F. Sneddon, "A new classification of amino acid side chains based on doublet acceptor energy levels," *Biophys. J.* 53: 83-89, 1988.

2. R. Mileusnic, S. P. Rose, and P. Tilson, "Learning and chick brain tubulin," *J. Neurochem.* 34: 4, 1980.

3. G. Bensimon and R. Chermat, "Microtubule disruption and cognitive defects: Effect of colchicine on learning behavior in rats," *Pharmacol. Biochem. Behavior* 38: 141-145, 1991.

4. D. Koruga, "Microtubular screw symmetry: Packing of spheres as a latent bioinformation code," *Ann. N.Y. Acad. Sci.* 166: 953-955, 1986.

5. S. Hameroff, *Ultimate Computing: Biomolecular Consciousness and Nanotechnology*, Amsterdam: North-Holland, 1987.

6. S. Rasmussen, H. Karampurwala, R. Vaidyanath, K. S. Jensen, and S. Hameroff, "Computational connectionism within neurons: A model of cytoskeletal automata subserving neural networks," *Physica D*

42: 428-449, 1990.

7. H. W. Kroto, J. R. Heath, C. D. O'Brien, R. F. Curl, and R. E. Smalley, "C_{60}: Buckminsterfullerene," *Nature* 318: 162-163, 1985.

8. W. Krätschmer, D. Lamb, K. Fostiropoulos, and R. Huffman, "Solid C_{60}: A new form of carbon," *Nature* 347: 354-358, 1990.

9. E. Osawa, *Kagaku* (Kyoto) 25: 85, 1970.

10. H. R. Karfunkel and T. Dressler, *J. Am. Chem. Soc.* 114: 2285, 1992.

11. S. Iijima, "Helical microtubules of graphic carbon," *Nature* 354: 56-58, 1991.

12. This research is supported by the Serbian Research Council.

1a

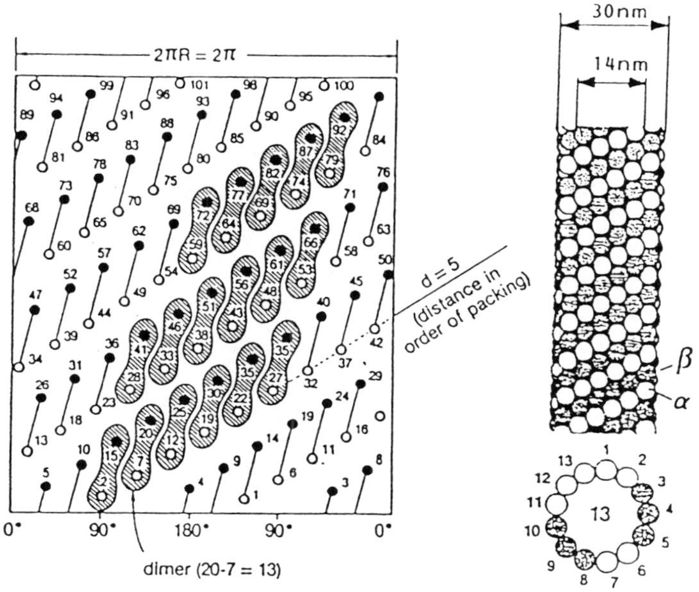

1b

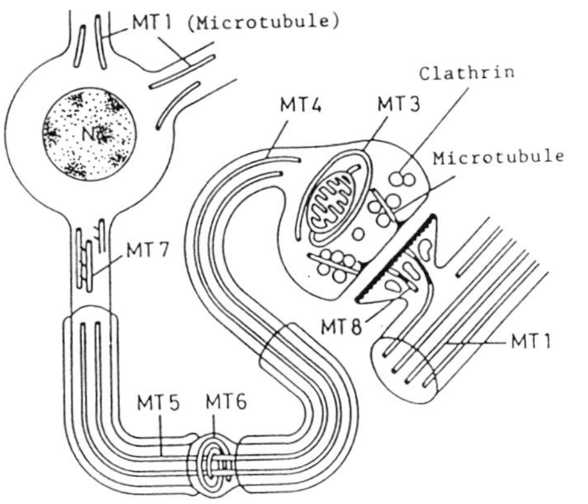

2

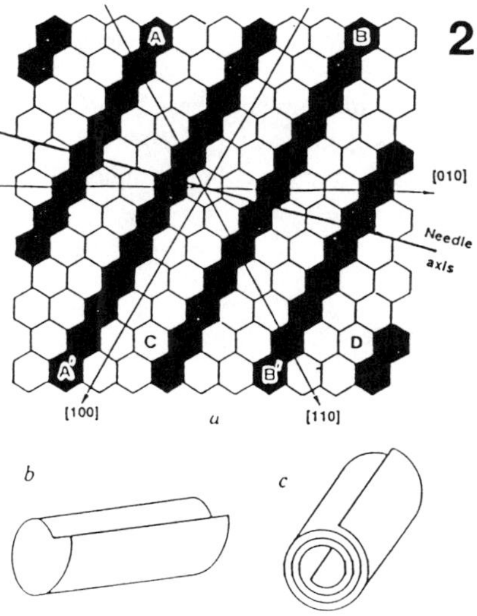

FIG. 1 (after Ref. 11).—Microtubules with 13 protofilaments in neurons. (a) Microtubules based on tubulin dimers with same value in order of packing as number of protofilaments in microtubule. (b) Five clathrins (same symmetry properties as C_{60}) joined with microtubules at synapse.

FIG. 2.—Artificial microtubules produced from C_{60} with hexagonal packing of carbon atoms on surface of a cylinder.

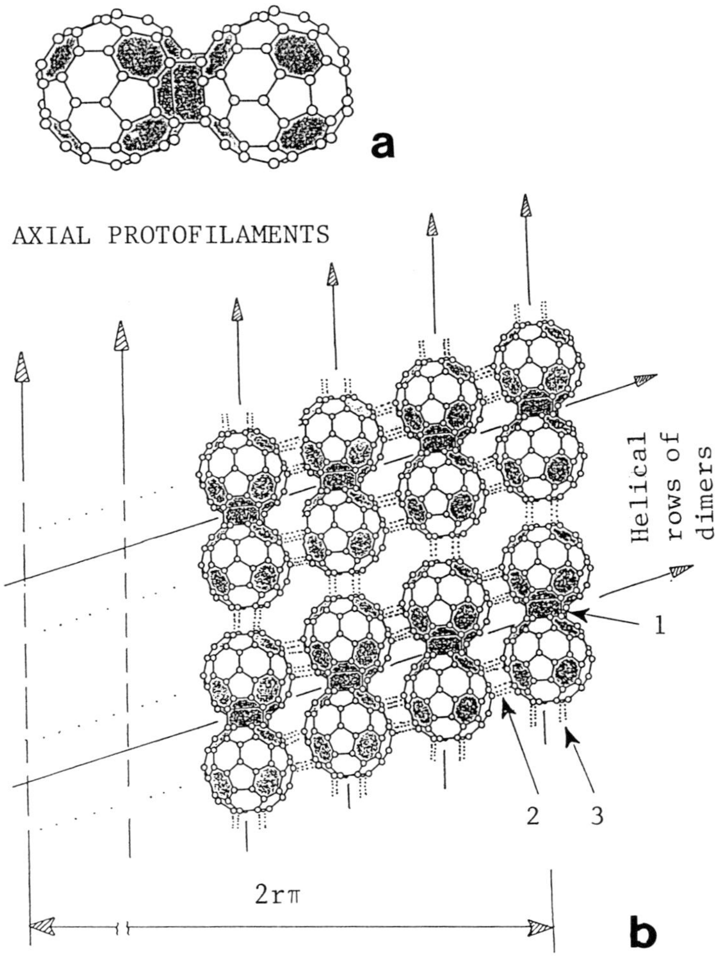

FIG. 3.—Artificial microtubules based on C_{116} as a dimer of C_{60}. (a) C_{116} as a dimer of C_{60}. Two C_{60} molecules joined together as C_{58}. (b) Tube composed of C_{116} molecules with three links: link 1 between monomers; link 2 between two dimers in helical direction; link 3 between two dimers in axial direction.

Martin Blank, Ed., *Electricity and Magnetism in Biology and Medicine*

PROTOTYPE OF PHOTOCHEMICAL SWITCHING IN LANGMUIR-BLODGETT FILMS

H. Tachibana,* R. Azumi, T. Nakamura, E. Manda, M. Matsumoto, and Y. Kawabata

National Chemical Laboratory for Industry, Ibaraki, Japan

New types of photochemical switching phenomena in LB films were observed in amphiphilic molecules consisting of switching, transmission, and working units. The conductivity of the LB film is controlled by the irradiation of UV or visible light through photoisomerization of the switching unit. Three types of photochemical switching were observed. Two of them showed reversible processes; the third exhibited a behavior which is compared to the learning process based on the plasticity of synaptic transmission. Multiple switching devices can be fabricated by successive deposition of two LB films consisting of molecules with different switching units.

Langmuir-Blodgett (LB) films have attracted considerable interest in terms of the fabrication of molecular electronic and bioelectronic devices.[1-3] Recently LB films have been developed that transform photons into the other functions.[4,5]

We have demonstrated a novel type of photochemical switching device in which absorbed photons induce a change in the conductivity of LB films of multifunctional molecules having azobenzene, alkyl chain, and a charge transfer complex as a switching, a transmission, and a working unit, respectively.[6-8]

In this paper, we show that one can fabricate different types of photochemical switching devices by simply changing the length of the transmission unit. Furthermore, the fabrication of the multiple switching device is described.

EXPERIMENTAL

The molecules of APT(8-n) were synthesized as described previously.[7] Monolayer measurements and the photochemical switching experiments were done as described previously.[7]

RESULTS AND DISCUSSION

Reversible photoisomerization of azobenzene was observed in the LB films of APT(8-n). These LB films exhibited an intense absorption at ca. 360 nm, corresponding to a monomeric state of the trans-azobenzene. On irradiation with UV light (365 nm), the absorption due to the trans-isomer decreased and the absorption around 450 nm due to the cis-isomer appeared. The trans-isomers were restored by the irradiation with VIS light (436 nm).

Figure 1 shows the change in the absorbance at 356 nm due to the absorption of trans-azobenzene and the conductivity of the LB films on alternate irradiation with UV and VIS light. The value of A_t/A_0 in Fig. 1(a) represents a fraction of trans-azobenzene, where A_t is the absorbance at time t, and A_0 at $t = 0$. The fraction of trans-azobenzene changed similarly for the three samples except for the magnitude.

Conductivities of the LB films of APT(8-6), APT(8-12), and APT(8-14) changed differently on photoirradiation. The LB films of APT(8-12) and APT(8-14) showed a reversible change in conductivity on alternate irradiation with UV and VIS light, but each in the opposite direction.

The conductivity of APT(8-6) increased with trans-to-cis isomerization by ca. 6% at the photostationary state. But the conductivity was unchanged on irradiation with visible light, though photoisomerization of azobenzene was observed. Another irradiation with UV light caused the conductivity to show another increase. However, the increment per one cycle decreased with increasing number of cycles and the conductivity tended to become constant at a cycle of more than thirty times. This behavior is compared to the learning process based on the plasticity of synaptic transmission. In other words, electrons can pass through LB

---

*Correspondence to: H. Tachibana, National Chemical Laboratory for Chemistry, Tsukuba, Ibaraki 305 Japan.

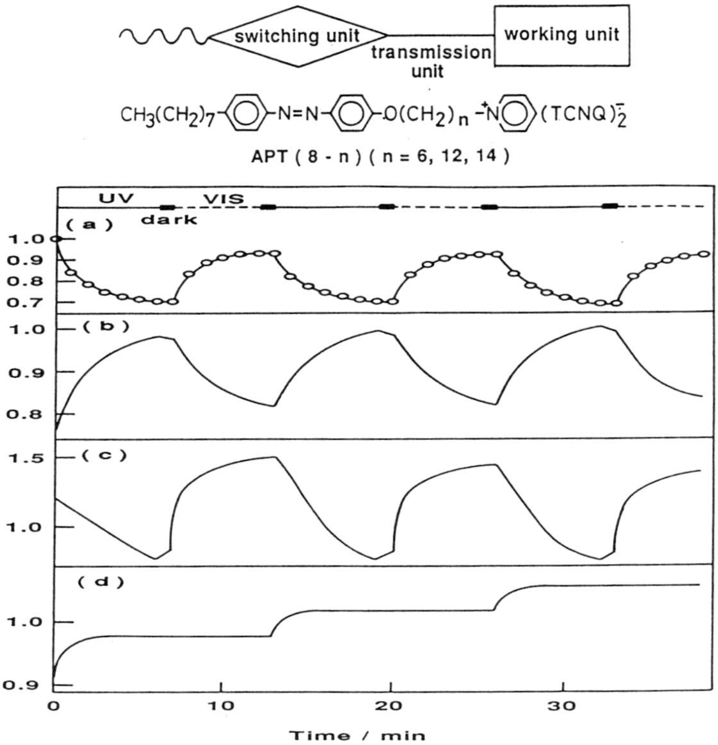

FIG. 1.—Change in fraction of trans-azobenzene for (a) APT(8-12) and change in conductivity of the LB films of (b) APT(8-12), (c) APT(8-14), and (d) APT(8-6) on alternate irradiation with UV and VIS light.

films more easily as the stimulation (photoirradiation) is repeated.

Multiple switching devices can be fabricated by a transfer of two types of photochemical switching LB films successively; the LB films of molecules have different switching units. This arrangement allows use of four types of control light compared with only two types of control light available in the above case.[9]

CONCLUSIONS

We have demonstrated prototypes of photochemical switching devices based on LB films. Furthermore, multiple switching devices have been fabricated by successive transfer of two LB films of molecules having different switching units. By a further examination of the effect of the chemical modification on the switching phenomena and optimization of the structure of amphiphilic molecules, it will be possible to fabricate molecular electronic or bioelectronic devices possessing a variety of functions.

REFERENCES

1. R. M. Metzger and C. A. Panetta, *New J. Chem.* 15: 209, 1991.
2. K. Sakai, H. Matsuda, H. Kawada, K. Eguchi, and T. Nakagiri, *Appl. Phys. Lett.* 53: 1274, 1988.
3. Z. F. Liu, K. Hashimoto, and A. Fujishima, *Nature* 347: 658, 1990.
4. T. Seki, T. Tamaki, Y. Suzuki, Y. Kawanishi, and K. Ichimura, *J. Am. Chem. Soc.* 22: 3505, 1989.
5. M. Iwamoto, Y. Majima, N. Naruse, T. Noguchi, and H. Fuwa, *Nature* 353: 645, 1991.
6. H. Tachibana, T. Nakamura, M. Matsumoto, H. Komizu, E. Manda, H. Niino, A. Yabe, and Y. Kawabata, *J. Am. Chem. Soc.* 111: 3080, 1989.
7. H. Tachibana, A. Gota, T. Nakamura, M. Matsumoto, E. Manda, H. Niino, A. Yabe, and Y. Kawabata, *Thin Solid Films* 179: 207, 1989.
8. H. Tachibana, R. Azumi, T. Nakamura, M. Matsumoto, and Y. Kawabata, *Chem. Lett.* 173, 1992.
9. H. Tachibana, E. Manda, R. Azumi, T. Nakamura, M. Matsumoto, and Y. Kawabata, *Appl. Phys. Letters* 61: 2420, 1992.

Martin Blank, Ed., *Electricity and Magnetism in Biology and Medicine*

A NOVEL GAS-SENSING METHOD BASED ON THE DYNAMIC RESPONSES of AN SnO_2 SEMICONDUCTOR

S. Nakata, Y. Kaneda, and S. Akakabe

Nara University of Education, Japan

K. Yoshikawa*

Nagoya University, Japan

We propose an intelligent gas-sensing system based on the utilization of nonlinear dynamics. A sinusoidal voltage is applied to a heater equipped with a SnO_2 semiconductor gas sensor and the resulting output resistance of the sensor is analyzed by the fast Fourier transformation (FFT). The higher harmonics of FFT imply the nonlinear property in the response. It has been found that the intensities of the higher harmonics exhibit characteristic change depending on the chemical structure of gases. The reproducibility of this method is very good. The physical and chemical meaning of the higher harmonics is theoretically discussed.

In the past, many investigators have tried to modify the electrode surface with functional chemical compounds in order to improve the selectivity on the response as a linear function with respect to the external stimuli. However, if one wishes to detect thousands of chemical compounds, identification based on the approximation of linear response is not sufficient.[1-4] The difference between the classical chemical sensor and the present method can be schematically represented as follows.

Classical Chemical Sensor (linear response)

Chemical compound → [transducer] → DC voltage or current (one-dimensional)

Present Method (nonlinear response)

Chemical compound → [transducer] → $\omega_0, 2\omega_0, 3\omega_0, \ldots$ (multidimensional)

Recently, we have invented a novel sensing method with quantitative characterization of electrochemical nonlinearity in aqueous samples by the analysis of the higher harmonics in FFT of output current.[1] Using this method, one can not only detect but also quantify various chemical species contained in a sample solution. We have reported that a chemical sensor based on the information of nonlinear dynamics serves as a good model of taste and olfactory reception.[1] As an extension of this study, we have carried out a study on the dynamic response of a ceramic gas sensor by the application of a sinusoidal input voltage to a heater situated by the sensor.[2-4] It was found that the waveform of the resistance changes characteristically, depending on the chemical structure of the gas. The waveform has been quantitatively analyzed by using FFT. In the present paper, we report the new result on this sensing system.

Experiments were performed as in the previous studies.[2-4] The AC voltage and the resistance of the gas sensor were successively stored in a computer and then Fourier-transformed to the frequency domain. A SnO_2 sensor, TGS813 manufactured by Figaro Engineering Inc. (Osaka), was used.

---

*Correspondence to: K. Yoshikawa, Graduate School of Human Informatics, Nagoya University, Nagoya 464-01, Japan.

Figure 1 shows the conductance G vs heater voltage of the gas sensor for clean air, ethylene, and propane in steady states with fixed temperatures. Here, the DC heater voltage was applied to the ceramic body and measurement was carried out after the equilibration of the temperature. It was found that the conductance of the gas sensor changed with the gas species in various ways. Such a characteristic change in the conductance is attributable to the temperature dependence on the barrier potential of the semiconductor and the equilibration on adsorption of gas molecules onto the ceramic surface.

As a next step, we have measured the dynamic response for a mixed gas sample. Figure 2(A) shows the phase portrait of the sensor resistance R vs heater voltage for the mixture gases composed of ethylene and propane. It was found that these phase portraits show characteristic nonlinear response depending on the relative concentration of ethylene and propane. Figure 2(B) shows the relationship between the intensities of higher harmonics and the ratio of ethylene and propane. It is apparent that one can distinguish the two different gases based on the information of higher harmonics by use of a single sensor. The importance of the nonlinearity in the gas sensor has also been indicated by several other research groups.[5,6] We report an advanced study on quantitative analysis of the nonlinearity based on the theoretical consideration on the kinetics of adsorption, chemical reaction, and desorption of gas on the ceramic surface.[2-4] We believe that utilization of the nonlinear dynamical response is quite promising for the development of an "intelligent" gas sensor.

REFERENCES

1. K. Yoshikawa, S. Maeda, and H. Kawakami, *Ferroelectrics* 86: 117, 1989.
2. S. Nakata, K. Yoshikawa, M. Shoji, H. Kawakami, and T. Ishii, *Biophys. Chem.* 34: 201, 1989.
3. S. Nakata and K. Yoshikawa, *J. Molec. Elec.* 7: 101, 1991.
4. S. Nakata, H. Nakamura, and K. Yoshikawa, *Sensors and Actuators* B-8: 187, 1992.
5. S. Wlodek, K. Colbow, and F. Consdori, *ibid.* B-3: 63, 1991.
6. J. W. Gardner, *ibid.* B-1: 166, 1991.

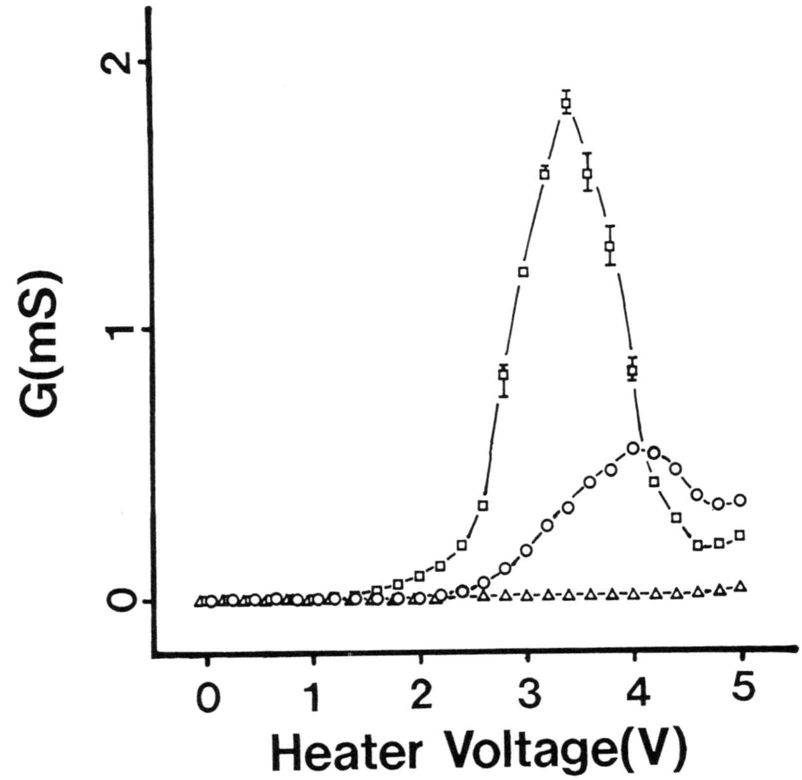

FIG. 1.—Conductance (G) vs heater voltage curve in SnO_2 gas sensor with exposure to air as a control Δ, 0.2 mol/m^3 propane gas O, and 0.2 mol/m^3 ethylene gas □.

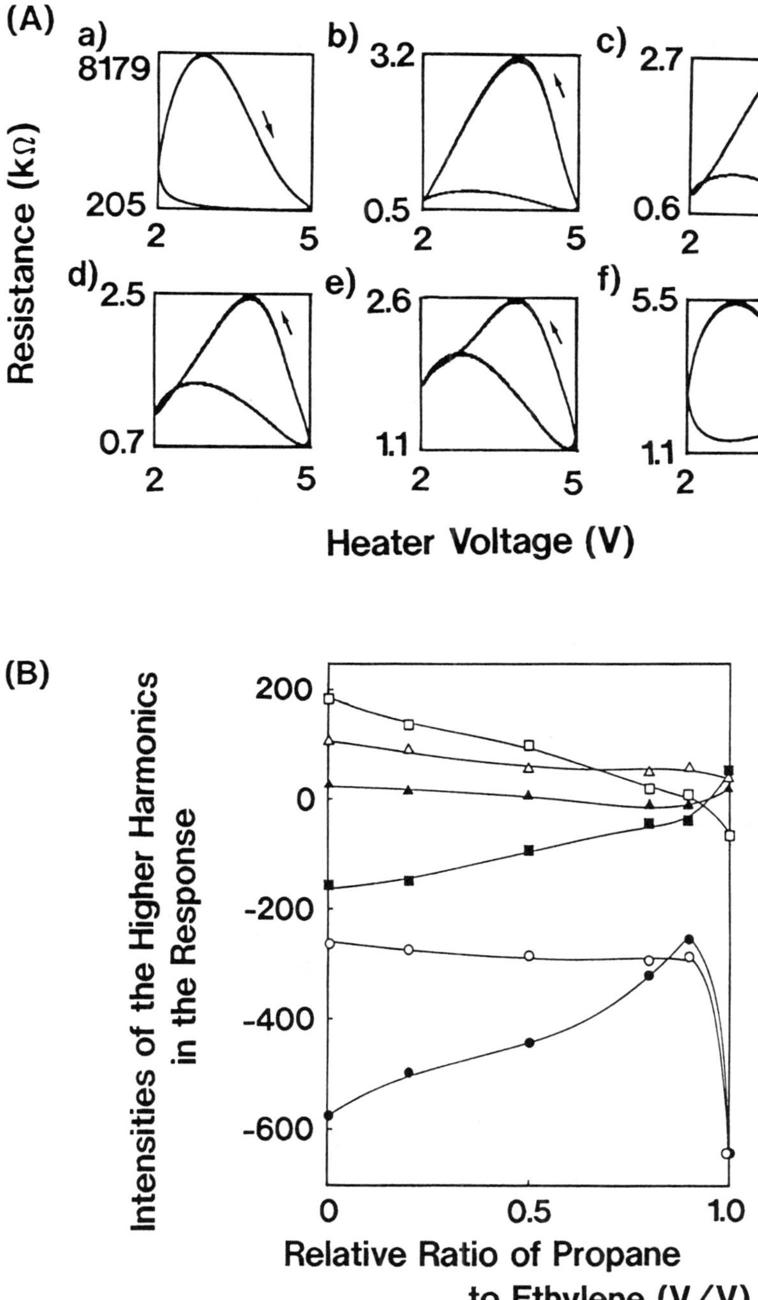

FIG. 2.—(A) Dynamic responses for mixed gases composed of ethylene and propane: (a) air as a control, (b) 0.2 mol/m^3 ethylene, (c) 0.16 mol/m^3 ethylene and 0.04 mol/m^3 propane, (d) 0.1 mol/m^3 ethylene and 0.1 mol/m^3 propane, (e) 0.04 mol/m^3 ethylene and 0.16 mol/m^3 propane, and (f) 0.2 mol/m^3 propane; heater voltage , $3.5+1.5\cos(2\pi ft)/V$, $f = 0.04/\mathrm{Hz}$. (B) Relationship between relative concentration of ethylene-propane mixed gas vs intensity of higher harmonics in FFT: ● R_2, ■ R_3, ▲ R_4, ○ I_2, □ I_3, Δ I_4.

Martin Blank, Ed., *Electricity and Magnetism in Biology and Medicine*

RECTIFICATION OF ELECTRIC CURRENT BY MOLECULES: MOLECULAR-SCALE OBSERVATIONS WITH A SCANNING TUNNELING MICROSCOPE

M. Pomerantz,* A. Aviram, R. A. McCorkle, L. Li, and A. G. Schrott

IBM Watson Research Center, Yorktown Heights, N.Y.

Experiments are described in which a useful electronic function is caused by a few molecules observed with a scanning tunneling microscope. The methods of attaching the molecules, characterizing them on the sample surface, and the observation of electrical characteristics are described. A particular molecule, an aza-substituted Cu phthalocyanine, was found to make the I-V characteristics of the STM much more asymmetrical, or rectifying. Other molecules did not produce this effect. Evidence is given that the electronic energy levels are responsible for the differences of behavior.

It has been proposed that individual molecules may perform useful electrical functions, such as rectification of current.[1] We present experiments in which the electrical effects of molecules are measured on the molecular scale. The only instrument capable of observing the electrical characteristic of individual molecules at present is the scanning tunneling microscope (STM). The sharp tip of the STM is used as an electrode, positioned within 1 nm of the molecules. Because of the strong distance dependence of the electron tunneling process, only a small number of molecules, perhaps only one, can carry current between the tip and surface.

It is now widely appreciated that the tip of the STM exerts a force on the surface when their separation is small enough to allow tunneling to be observable.[2] This effect introduces one of the major uncertainties in the use of STM: it distorts the surface it is measuring, or can move molecules on the surface.[3] Nevertheless, topographs showing atomic resolution and recognizable molecules have been recorded in some cases.[4] To inhibit molecular motion, we sought chemical reactions by which molecules of interest might be strongly bound to the surface. The substrate of choice was highly oriented pyrolytic graphite (HOPG), because it is not reactive in air, is stable against damage by the tunneling current except at high currents,[5] and easily yields atomic resolution with flat surfaces. It proved to be possible to bond basic molecules to surfaces of HOPG that had been acidified.

ATTACHMENT AND CHARACTERISTICS OF MOLECULES ON GRAPHITE

Our bonding procedure was inspired by the reported use of chromic anydride to generate carboxylic acid groups on the surface of HOPG.[6] We therefore tried molecules with some basic functionalities, which would react with the acid groups and attach by ionic bonds. This attempt was successful for several molecules in which N served as the base. They included the amine thiol ($H\text{-}S\text{-}(CH_2)_2\text{-}NH_2$), and copper tetra aza phthalocyanine, Cu 4, 4′, 4″, 4‴-tetraaza-29H, 31H-phthalocyanine (Fig. 1b, inset). We refer to the latter as CuTAP. Purchased from the Aldrich Chemical Co. (#31,020-4), it was used without further purification. The N substitutions on its outer rings are bases. Auger spectroscopy verified[7] that the treatment of the surface with a chromic oxide solution resulted in at least a seven-fold increase in the number of molecules on the surface, compared with merely immersing the HOPG in a solution of CuTAP. At maximum, the fraction of the surface covered by the molecules is estimated from these data to be about 1/2. Angle-dependent XPS showed that the molecules were not flat on the surface. Infrared absorption showed that the molecular structure of the attached molecules corresponded to the initial material, as did the ultraviolet photoemission spectrum (UPS). These findings are evidence that the structure of the molecules is not changed by the attachment process.[8]

---

*Correspondence to: M. Pomerantz, Cedar Lane Apts C-1, Ossining NY 10562.

ELECTRICAL CHARACTERISTICS WITH THE STM

The STM we used was a Nanoscope I (Digital Instruments Corp., Santa Barbara, Calif.). Its electronic controls were modified to allow digital control and recording of topographs. A sample-and-hold circuit was installed so that arbitrary pulse shapes could be applied while the tip position was fixed, in order to produce current-voltage characteristics. The sample was in air at room temperature. The tips were freshly cut Pt/Ir wire, tested on virgin HOPG to be sure they gave atomic resolution before each experiment.

In Fig. 1(a), the upper trace is the voltage varying from +1.2 to −1.2 V, and the resulting current for a pure HOPG surface. The current is symmetric within a factor of 2. With the attachment of CuTAP the current becomes much larger at about −0.7 V to the surface (Fig. 1b). On a larger scale (Fig. 1c), the asymmetry is seen to be a factor of at least 40, which is observed over about 10% of the surface; asymmetry of 10 is noted over about 50% of the area, which is close to the measured coverage of molecules.

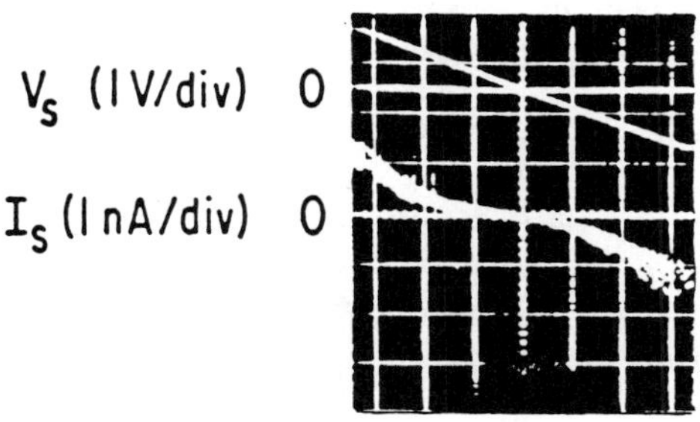

FIG. 1.—Storage-scope traces of the voltages and currents by the STM. (a) Pure graphite sample; V_s is voltage between the sample and tip, scanned in 38 ms; I_s is resulting tunnel current. (b) Sample is acidified graphite after reaction with CuTAP, with the same scales as in (a); current has gone off scale at $V_s = -1.2$; the inset shows the planar structure of CuTAP. (c) Same as (b) but I_s scale is larger by 20-fold to show entire asymmetric current. Feedback current = 0.4 nA at + 0.7 V in all the above.

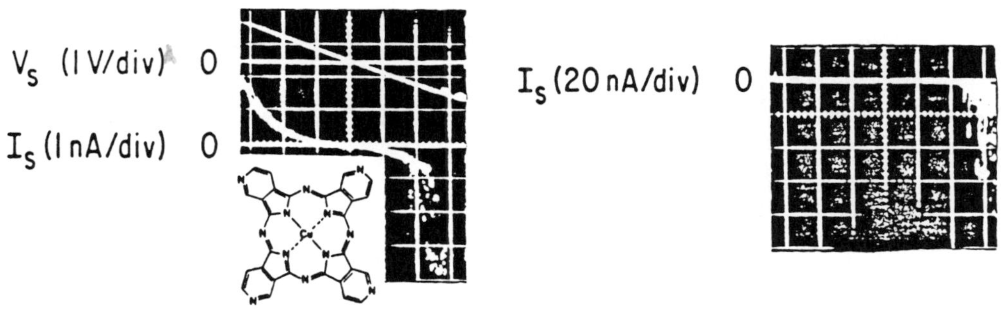

From UPS measurements[8] we found that the highest occupied electronic states (HOMO) of the CuTAP were centered at about 1.3 eV below the Fermi energy of the HOPG. This led us to postulate that the mechanism of rectification was resonant tunneling through the HOMO, which would begin at about the observed bias of −0.7 V applied to the surface. The physical asymmetry needed to produce the electrical asymmetry is presumed to be the differences in the separations of the molecule between the surface and the tip. (Differences in the barrier thickness are known to cause asymmetry in resonant tunneling.[9]) Other molecules that we attached to the surface, such as the amine thiol, gave the same electrical characteristics as pure HOPG. They lack the CuTAP energy levels accessible by biasing to 1 V.

CONCLUSIONS

This demonstration of rectification resulting from the presence of molecules is a step toward utilization of very small numbers of molecules for electrical purposes. It shows that a knowledge of the electronic and chemical properties is required to achieve desired characteristics and physical structures. The considerable task of wiring to groups of molecules remains before us, but the advantages of using ready-made molecules and their energy levels are now manifest.[10]

REFERENCES

1. A. Aviram and M. A. Ratner, *Chem. Phys. Lett.* 29: 277, 1974.
2. G. Binnig, C. F. Quate, and C. Gerber, *Phys. Rev. Lett.* 56: 930, 1986.
3. Z. Nawaz, T. R. Cataldi, J. Knall, R. Somekh, and J. B. Pethica, *Surf. Sci.* 265: 139-155, 1992.
4. L. L. Soethout, H. Van Kempen, and G. F. Van De Walle, in P. W. Hawkes, Ed., *Advances in Electronics and Electron Physics,* New York: Academic Press, 1990, vol. 79, 153.
5. T. R. Albrecht, M. M. Dovek, M. D. Kirk, C. A. Lang, C. F. Quate, and D. P. Smith, *Appl. Phys. Lett.* 55: 1727, 1989.
6. W. M. Heckl, K. M. Kallury, M. Thompson, C. Gerber, H. J. Hörber, and G. Binnig, *Langmuir* 5: 1433, 1989.
7. M. Pomerantz, A. Aviram, R. A. McCorkle, P. A. Roland, and A. G. Schrott, in A. Aviram, Ed., *Proc. 2nd Intern. Conf. on Molecular Electronics,* New York: AIP, 1992.
8. M. Pomerantz, A. Aviram, R. A. McCorkle, L. Li, and A. G. Schrott, *Science* 255: 1115-1118, 1992.
9. W. D. Goodhue, T. C. Sollner, H. Q. Le, E. R. Brown, and B. A. Vojak, *Appl. Phys. Lett.* 49: 1086-1088, 1986.
10. R. Kaufman and C. Joachim contributed to the design and implementation of the Nanoscope modifications, R. G. Schad and P. A. Roland made the Auger measurements, D. W. Abraham lent us a Nanoscope base, and A. C. Callegari, J. E. Demuth, and D. P. E. Smith made helpful suggestions, for all of which we are grateful.

Martin Blank, Ed., *Electricity and Magnetism in Biology and Medicine*

SIMULATION OF SCANNING TUNNELING MICROSCOPY SYSTEMS

A. Cambiaso,* E. Di Zitti, B. Bianco, and D. Ricci

University of Genoa, Italy

This paper presents a computer simulation of a scanning tunneling microscope (STM), developed as a tool to study the effect of the equipment parameters that determine the STM image formation. The model of the STM system is developed in a general way starting from the representation of the STM as a set of interacting modules. The simulation of the system is performed by computation of the dynamical evolution of the state of each module in the time domain. Examples of simulated images of graphite, obtained under various set-up parameters, are also presented.

An unresolved problem in scanning tunneling microscopy (STM)[1] is the correct interpretation of images. A first step toward its solution is the development of specific tools for studying the role of the acquisition parameters concerning the STM equipment in the image formation. Following this guideline, we present a simulation program that emulates the image acquisition by scanning a graphite surface via electron tunneling with a sharp tip. The simulation regards the operating modes of the STM as an instrument designed to control a given current during the scanning by means of piezoelectric ceramic drivers used as electromechanical transducers. Attention is focused on the machine parts that constitute the automatic control loop, in order to distinguish and evaluate the contributions to the imaging process. This environment has been modeled in a general way (constant current and quasi-constant-height operating modes). The specific characterization of such modules concerns the Nanotec STM 100 (HUND, Germany).

MODELING

A general STM equipment[2] is constituted by a large variety of modules (transducers, analog and digital electronic modules, digital-software blocks) that interact with the specimen under investigation by the tunnel current flow between a sharp conductive tip and the conductive specimen. Figure 1 shows the block description of the STM system considered, as implemented in the computer program; in the following a brief description of the blocks and variables of Fig. 1 is provided.

Scanning generator—generation of the piezo voltages (Vx , Vy) for scanning the sample surface (xy plane) and the synchronization of the overall system, according to the scan parameters selected by the user (window position and width, scan speed, number of picture samples).

Piezo scanner—piezoelectric transducer for tip positioning at X, Y, Z coordinates.

Tip/sample interaction—representation of the interaction between the tip (biased at a voltage Vt) and the specimen, giving rise to the tunnel current It.

I/V converter—generation of the voltage signal, proportional to the tunnel current.

Amplifier—linear or logarithmic (function selected by the user) voltage amplifier.

I/V_{ref} converter—generation of a voltage proportional to the reference value Ir of the tunnel current to be controlled.

Σ—difference amplifier.

Feedback amplifier—Integral/proportional amplifier for generating the voltage (Vz) to be applied to the z-piezoelectric driver.

AD1, AD2—Analog-to-digital converters for current (quasi-constant height) and topographic (constant current) signal acquisition.

---

*Correspondence to: A. Cambiaso, DIBE, University of Genoa, Via Opera Pia 11a, I-16145 Genoa Italy.

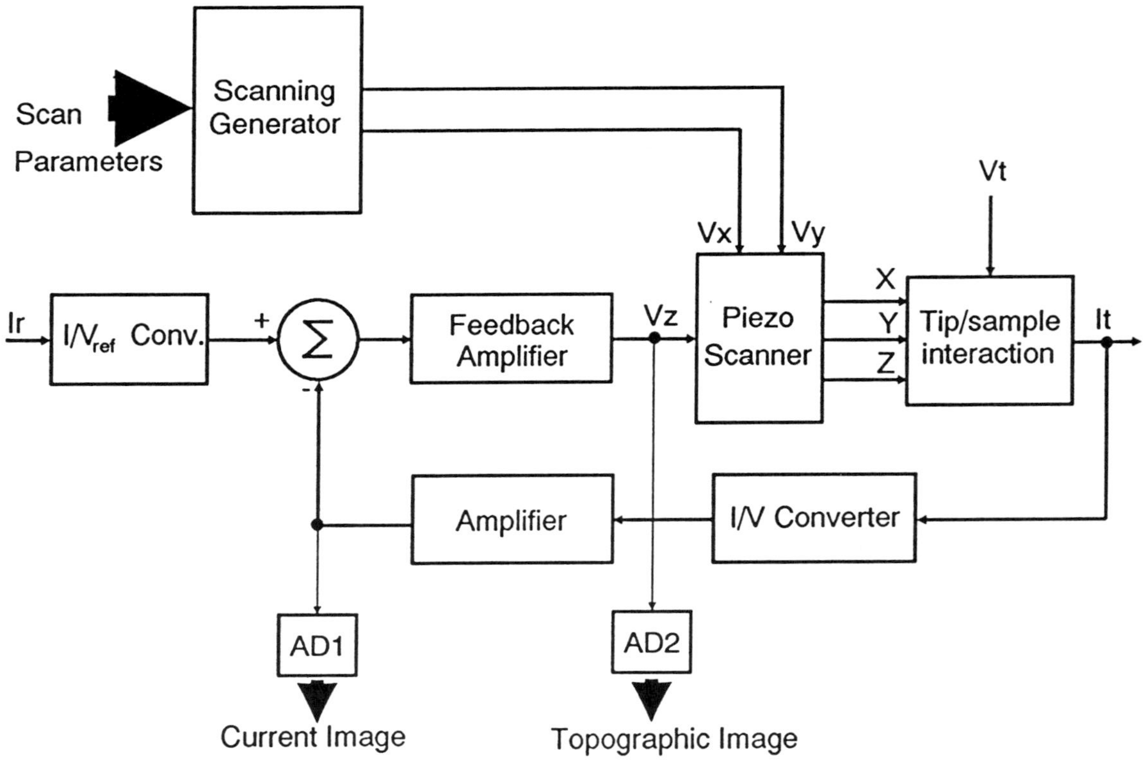

FIG. 1.—Block diagram of the simulated STM system.

Let us consider some detail on specific blocks. The sample-tip interaction is considered in a quasi-static situation for a graphite multilayer[3] and the current calculation is performed according to the Tersoff and Hamann approach:[4] the tunnel current is approximately proportional both to the bias voltage and to the sample electronic density in the position of the center of symmetry of the probe potential. (The latter should be identified with the center of the atom forming the probe apex.)

The I/V converter amplifies the tunnel current by 10^8 times and converts it into a voltage signal. Starting from the actual electrical scheme this stage has been analytically studied, and a two-pole transfer function has been derived. The numerical values of its coefficients have been successfully checked by numerical simulations with the SPICE program.

A geometrical model of the piezoelectric transducers has been developed to simulate systematic image defects due to the scanning. The scanner is constituted by four equal cylindrical sectors of known material. The piezo voltages are applied to the piezo sectors in the following mode: Vx is applied to a couple of opposite elements (in the x direction) using opposite polarity in order to stretch one element and to shrink the other; Vy is applied to the other pair of piezo elements (in the y direction) in the same way as Vx; Vz is added to all piezo elements in order to stretch/shrink the whole scanner in the z direction. This driving mode can produce systematic distortions in the images by introducing a second-order term in the linear relationship between piezo voltages and displacements.

The feedback amplifier consists of a pure integrator and a proportional module operating in parallel. The quasi-constant-height operating mode of the Nanotec STM can be performed by a setting of the time constant of the integrator to the maximum value allowed by the equipment.

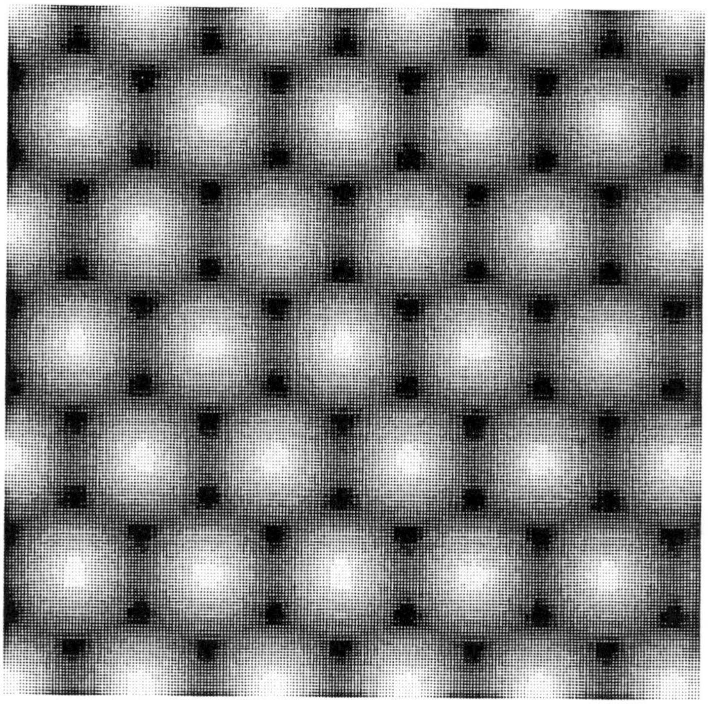

FIG. 2.—Image of a simulated STM scan of a graphite surface (1.28 nm×1.28 nm in size) in constant-current mode.

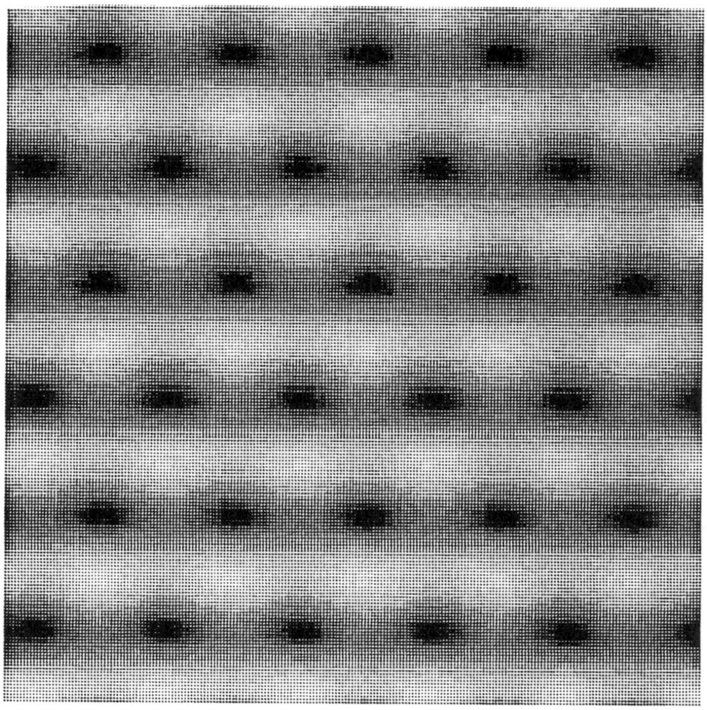

FIG. 3.—Image of a simulated STM scan, obtained using the same parameters of Fig. 2, except the time constant of the integrator of the feedback amplifier, taken 5 times higher.

SIMULATION TECHNIQUES

The simulation of the imaging process is performed by computation of the time evolution of the state of the overall STM system through numerical integration of the input/output equations of each module. A backward-Euler method has been selected for this purpose.[5] The simulation occurs in two steps. In a first phase, the stationary tunnel condition is determined when the state of the STM loop reaches the equilibrium condition. In the second phase, the dynamical evolution of the state during the scan is calculated.

The software package has been developed in FORTRAN-77 language on a Hewlett-Packard HP360 graphic workstation. The results of the simulation are digital images stored in files for a further display or processing. Two significant examples are shown in Figs. 2 and 3.

CONCLUSIONS

A flexible model for the simulation of the STM imaging process has been described, which allows us to specify the behavior of the several parts of the system to the desired degree of accuracy. The developed computer simulation can be employed to point out the role of the acquisition parameters on the STM image formation process. This approach can also constitute a useful tool both to optimize the acquisition parameters and to study the mechanisms of artifact formation on the images. Moreover, different STM machines and/or materials can be easily simulated by substitution of only few modules in the developed software package.[6]

REFERENCES

1. R. J. Behm, N. Garcia, and H. Rohrer, Eds., *Scanning Tunneling Microscopy and Related Methods,* NATO-ASI Series, Series E: Applied Sciences, vol. 184, Dordrecht: Kluwer Academic Publishing, 1989.

2. Y. Kuk and P. J. Silverman, *Rev. Sci. Instrum.* 60: 165-179, 1989.

3. H. Mizes, *Interpretation of Scanning Tunneling Microscopy Images of Graphite,* Ph.D. Dissertation, Stanford University, USA, 1988.

4. J. Tersoff and D. R. Hamann, *Phys. Rev.* B-31: 805-813, 1985.

5. A. L. Sangiovanni-Vincentelli, in *Computer Aids for VLSI Circuits,* NATO-ASI Series E (48) 19-112, 1981.

6. This work was supported by ESPRIT BRA 3200 *OLDS* and MURST 40% *Elettronica Molecolare.*

Martin Blank, Ed., *Electricity and Magnetism in Biology and Medicine*

SCANNING TUNNELING MICROSCOPY IMAGING OF ORGANIC MOLECULES: FROM ALKANES TO POLYMERS

A. Bonfiglio,* S. Cincotti, E. DiZitti, R. Paradiso, D. Ricci, and L. Elementi

University of Genoa, Italy

A. Bolognesi and W. Porzio

Istituto di Chimica delle Macromolecole, CNR, Milan, Italy

This paper presents scanning tunneling microscopy (STM) investigations on dotriacontane, stearic acid, and tetracosanoic acid dissolved in octylbenzene at the liquid/graphite interface. An analogous technique has been applied in order to obtain preliminary results on octadecylmercaptane and poly(3-decylthiophene). The images of alkanes and alkyl derivatives presented show that the tunnel current produced by the graphite atoms is modulated by the carbon atoms of the alkyl chains of the compounds deposited, giving rise to characteristic structures superimposed on graphite. Possible explanations of these structures are illustrated.

Scanning tunneling microscopy (STM) has revealed its potential in a number of fields.[1,2] A particularly interesting application is the observation of organic molecules.[3] These compounds have recently attracted attention for their characteristic properties, very promising for molecular electronics purposes. STM investigations on ultrathin organic films formed with such molecules provide information on their electron tunneling properties and structural organization. This knowledge provides a better insight into the fundamental mechanisms on which their utilization for information processing relies.

High-resolution voltage bias dependent images were acquired over a range of tunneling conditions. The STM employed was a Nanotec STM 100 (Hund), with cut Pt/Ir tips (Goodfellow). Dotriacontane, stearic acid, and tetracosanoic acid (Fluka Chem.) were dissolved near the saturation value in octylbenzene (Fluka Chem.) and immediately deposited on highly oriented pyrolytic graphite (HOPG—Union Carbide). Maximum contrast was observed for images taken in the quasi-constant height mode, with bias voltage of about 1 V and currents of about 1 nA. Similar steps, developed through a preliminary training on conventional alkyl derivatives, have also been followed for the observation of octadecylmercaptane and poly(3-decylthiophene), a polymer which, opportunely doped, shows conductive properties.[4,5] For octadecylmercaptane (melting point about 27 C) it was necessary to melt crystals directly on the heated graphite. Poly(3-decylthiophene) was synthesized and showed the following characteristics: molecular weight $\approx$ 6000, average repetitive unit number $\approx$ 26.

RESULTS AND DISCUSSION

All images of alkane and alkyl derivatives highlight the typical two-dimensional organization of the long organic molecules in lamellae. In every lamella, only half the expected number of carbon atoms of the chain can be counted, since what we observe is a modulation of the underlying graphite by the molecules and not the real molecular structure. Another typical characteristic that appears in each image is the shift of each chain within a lamella of one inter-chain distance with respect to the next one.[3,6] All the geometrical and dimensional information available from the images is consistent with the results derived from other techniques. In Figs. 1(a) and (b) it is possible to observe dotriacontane and tetracosanoic acid, respectively. Neither presents any kind of modulation along the lamellae. Stearic acid, on the other hand, presents a modulation along the lamella. In Figs. 1(c) and (d), we present two different images of stearic acid at two distinct solution concentrations, the first at 13 and the second at 15 mg/ml.

---

*Correspondence to: A. Bonfiglio (Engg Biophys. & Electronics), University of Genoa, Via Opera Pia 11a, I-16145 Genoa Italy.

The interpretation of the experimental observations of alkanes and alkyl derivatives on graphite requires a model of the two-dimensional arrangement of the alkyl chains on graphite. We must address two principal questions: to what extent are the alkyl chain carbon atoms in registry with the substrate atoms; and which are the possible orientations of the chain with respect to graphite? Calorimetric studies, performed years before the advent of the STM technique,[7] showed that graphite adsorbs hydrocarbon in preference to water and, preferentially, long alkyl chain hydrocarbons from low-molecular-weight solvents. The heat of adsorption of n-paraffin vapors increases of 6.28 kJ/mol for each methylenic group added. This fact suggests that the longer the chain, the more contacts the carbon atoms can form with the graphite atoms below. Hence the chains should lie flat on the substrate and the hydrogen atoms joined to the hydrocarbon chain should fit on some specific sites of the substrate, namely on the center of the hexagons formed by the atoms of graphite. This theory has been revived[6] in order to explain dotriacontane layers disposition; in this hypothesis, only a perfect registry between alkyl chain carbons and hexagon centers is possible. Even if supported by energetic arguments, this proposition does not succeed in explaining a contrast modulation along the molecule chain which is clearly evident in images (Fig. 1a). An alternative model has been proposed,[3] for which the contrast modulation perpendicular to the lamellae boundary is attributed to the fact that the carbon-carbon bond lengths in graphite and in the alkyl chains are not perfectly identical, giving rise to a mismatch of a few picometers for every molecule that, in propagating, causes a substantial change in contrast along the chain. Finally, the presence of a superstructure along the lamella, as in the case of stearic acid (Fig. 1c), indicates that the lateral spacing between parallel chains is not perfectly commensurate with the graphite interatomic distance, probably due to the presence of a head group that requires more space. Our results support this second explanation.

Octadecylmercaptane shows a behavior similar to alkanes and alkyl derivatives. If deposited on graphite, it forms lamellar structures clearly detectable (Fig. 2a) in which it is more difficult to obtain atomic resolution. Finally, images of poly(3-decylthiophene), observed both in constant-current and in quasi-constant height mode, show periodical structures (Fig. 2b) that can be related to the repetitive polimeric units in the chain.

CONCLUSIONS

Alkanes and fatty acids form on graphite ordered structures. The images show a modulation of the tunnel current of the graphite atoms by the alkyl chains that causes the appearance of the characteristic lamella-like structures. Various models have been analyzed in order to explain the possible dispositions of the molecules on graphite. Finally, preliminary results on octadecylmercaptane and poly(3-decylthiophene) have been presented.[8]

REFERENCES

1. G. Binnig and H. Rohrer, *IBM J. Res. Develop.* 30: 355-369, 1986.
2. P. K. Hansma and J. Tersoff, *J. Appl. Phys.* 61: R1, 1987.
3. S. Buchholz and J. P. Rabe, *J. Vac. Sci. Technol.* B9: 1126-1128, 1991.
4. R. D. McCullough and R. D. Lowe, *J. Chem. Soc. Chem. Commun.* 70: 1992.
5. I. Paloheimo, H. Stubb, and P. Y. Lahti, "Thin solid films," *Proc. LB5 Conference,* Paris, August 1991 (in press).
6. G. L. McGonigal, R. H. Bernhardt, and D. J. Thompson, *Appl. Phys. Lett.* 57: 28-30, 1990.
7. A. J. Groszek, *Proc. Roy. Lond.* A314: 473-498, 1970.
8. This work was supported by ESPRIT BRA 3200 OLDS and MURST (40%) "Elettronica Molecolare."

FIG. 1.—STM images of (a) dotriacontane, (b) tetracosanoic acid, (c) and (d) stearic acid on HOPG. Concentrations, voltages, currents, and viewed areas: (a) 18 mg/ml, 0.1 V, 0.2 nA, 9.8×9.8 nm; (b) 19 mg/ml, 0.9 V, 0.7 nA, 9.8×9.8 nm; (c) 13 mg/ml, 1.0 V, 0.3 nA, 12×12 nm; (d) 15 mg/ml, 1.0 V, 0.3 nA, 12×12 nm.

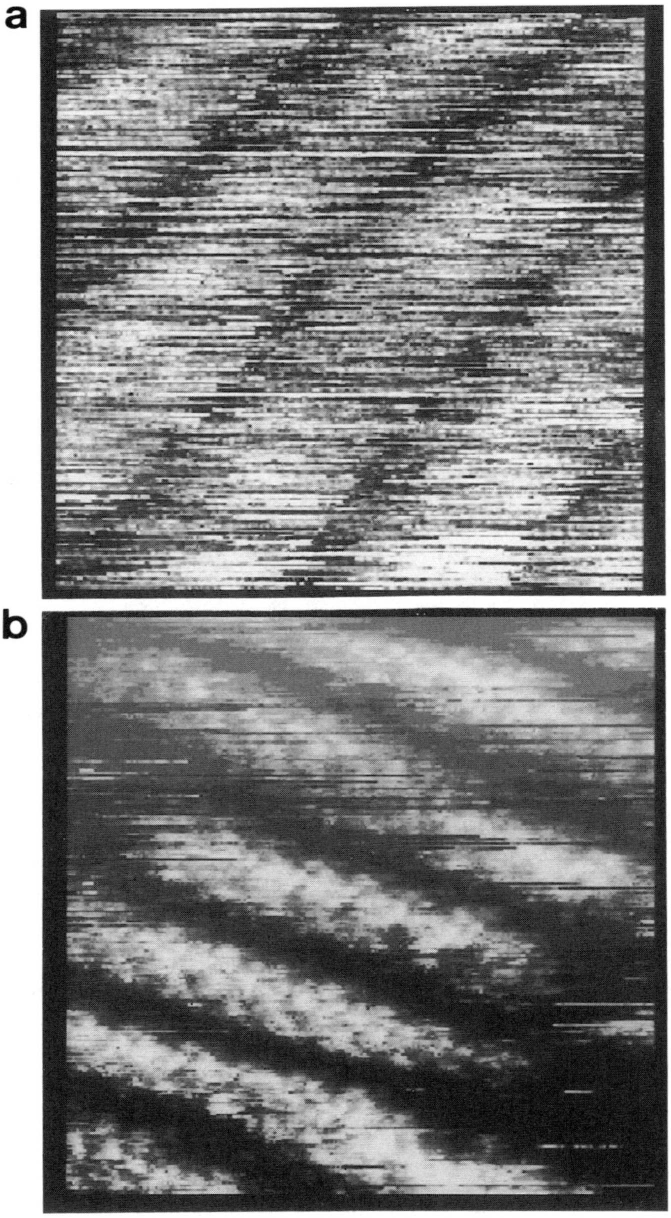

FIG. 2.—(a) Octadecylmercaptane, directly melted on heated HOPG, 1.4 V, 1.0 nA, 14.7×14.7 nm; (b) Poly(3-decylthiophene) on graphite, 1.6 V, 0.3 nA, 9.8×9.8 nm.

Martin Blank, Ed., *Electricity and Magnetism in Biology and Medicine*

LOCAL SURFACE MODIFICATIONS OF CONDUCTIVE SUBSTRATES FOR ORGANIC FILM DEPOSITION BY STM VOLTAGE PULSES

D. Ricci* and E. Di Zitti

University of Genoa, Italy

Scanning Tunneling Microscopy (STM) investigations (performed by the application of voltage pulses between tip and sample in order to produce local modifications) are presented. Holes were created on graphite and mounds on gold evaporated on mica. Experimental conditions for modifications are described.

The characteristic of scanning tunneling microscopy (STM)[1] to approach a conductive surface at nanometer scale by a sharp tip makes this technique powerful for many applications. A very promising one is the use of the tip as a probe to supply voltage pulses to the underlying surface to produce controlled local modifications.[2,3] In molecular electronics, this potentiality is under study with regard to the interaction with molecular films deposited on solid substrates. Some specific requirements are preliminary for this application: (i) the choice of appropriate conductive substrates suitable for film adhesion, (ii) the experimental set-up for this STM operating model, and (iii) the development of an experimental technique capable of controlled and reproducible modification.

This paper concerns such investigations as performed on highly oriented pyrolitic graphite (HOPG) and gold substrates.

EXPERIMENTAL SET-UP FOR VOLTAGE STM PULSES

To obtain voltage pulses between tip and sample, a programmable pulse generator (HP 8116A) was interfaced with the STM (Nanotec STM 100, HUND) by an RC highpass connected to the sample holder. The sample holder was modified to insulate the sample from the system ground. To this end, a teflon disk was interposed between the top and bottom parts of the sample holder, where contact pins were provided. The experimental set-up is sketched in Fig. 1. The voltage pulse range available is from −8 V to 8 V; the minimum pulse width is 8 ns.

Substrates used for tests were highly oriented pyrolitic graphite (HOPG, Union Carbide, grade Z), and gold evaporated on mica. Tips used were cut from Pt/Ir wires and electrochemically etched tungsten tips.

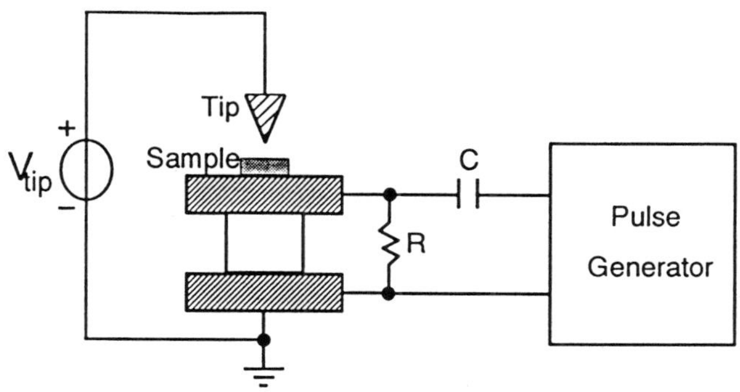

FIG. 1.—Experimental set-up to provide pulse operating mode to Nanotec STM 100.

---

*Correspondence to: D. Ricci (Biophysical and Electronic Engg), University of Genoa, Via Opera Pia 11A, I-16145 Genoa, Italy.

RESULTS

The effect of the external circuit was tested by atomic-resolution imaging on HOPG with the generator connected but disabled. Atomic resolution was still achievable, and the image quality was comparable to that obtained during normal operation.

Results were obtained by imaging of a given area before and after a programmed pulse was superimposed to the bias voltage in a selected point of the scanning area. While the surface was being imaged, the tunneling current and bias voltage (tip-substrate) were in the ranges 0.1-1.2 nA and 0.05-0.3 V, respectively. Pulse amplitudes were increased above 0 V by steps of 0.2 V: positive pulse means positive at the substrate. Pulse lengths between 0.01 and 10 μs were used.

We have found that there is a threshold value for the voltage pulse under which no modification was performed. Relationships between static tunneling current, threshold value, pulse length, and modification dimensions were found.

In the case of HOPG a minimum threshold value of 3 V for pulse amplitude that gives rise to modifications was found.

The lower the static tunneling current the higher the threshold value: amplitudes vary from 3 to 5 V. With higher bias current the modifications generally have larger dimensions. An example is shown in Fig. 2.

Short pulse lengths tend to give more controlled and localized modifications. Multiple hole creation is more common with long pulse lengths. Typical modified areas have a diameter of 10-30 nm. Negative pulses in the same range of amplitudes did not produce modifications.

In the case of gold samples, a minimum threshold value of 5 V for pulse amplitudes that gives rise to modifications was found. Values used throughout experiments ranged from 5 to 8 V. The typical modifications produced were mounds of approximately 30 nm in diameter. An example is shown in Fig. 3.[4]

REFERENCES

1. G. Binnig, and H. Rohrer, *IBM J. Res. Develop.* 30: 355-369, 1986.

2. J. A. Miller and R. J. Hocken, *J. Appl. Phys.* 68(2): 905-907, 1990.

3. J. Schneir, H. H. Harary, J. A. Dagata, P. K. Hansma, and R. Sonnenfeld, *Scanning Microscopy* 3: 719-724, 1989.

4. This work was supported by C.E.C. Project ESPRIT II BRA 3200 OLDS and the Italian National Project MURST 40% Elettronica Molecolare.

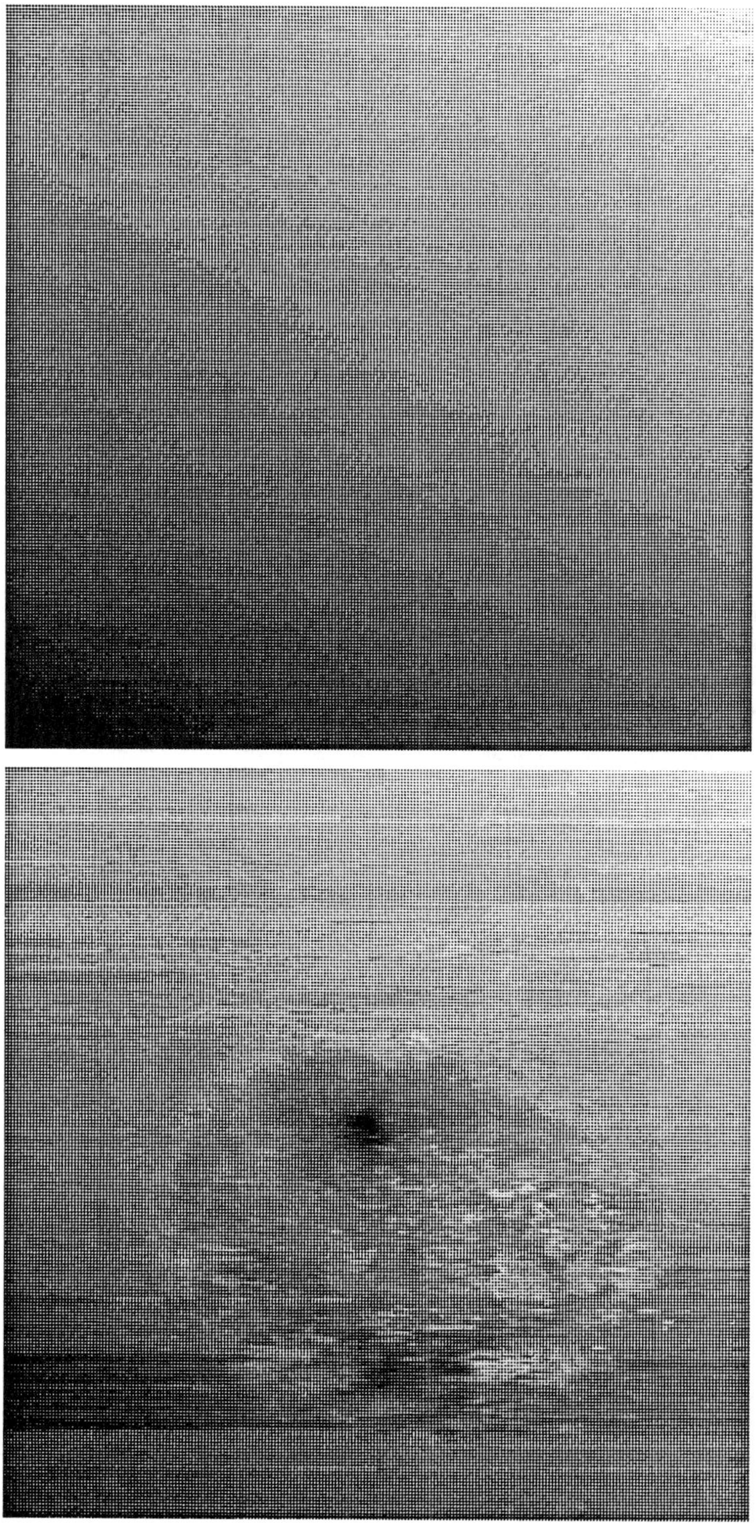

FIG. 2.—Topview images of flat region of graphite before (top) and after (bottom) a −3.2V, 0.1μs pulse, supplied at center of scanning area.

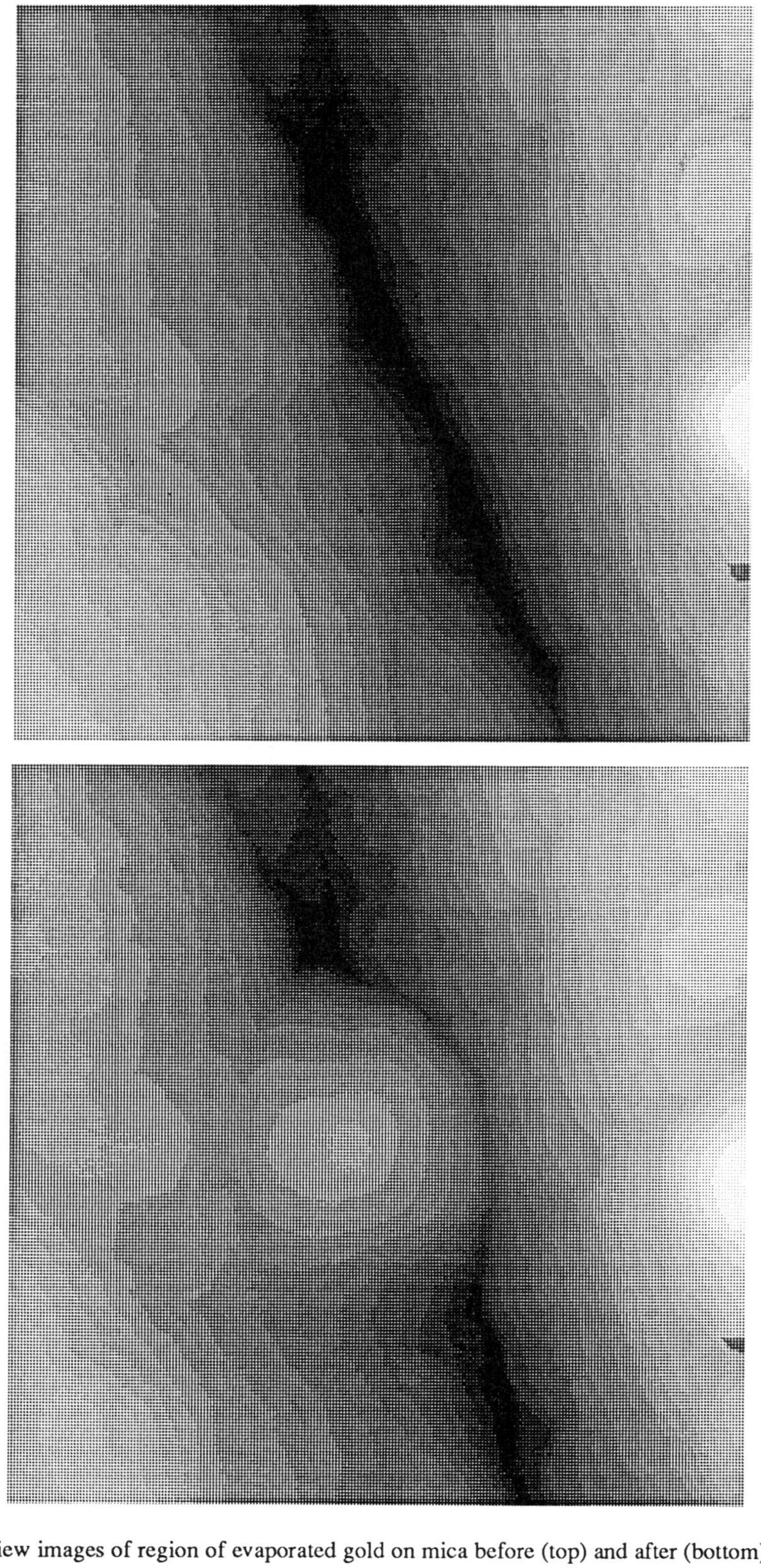

FIG. 3.—Topview images of region of evaporated gold on mica before (top) and after (bottom) a -6V, 0.1μs pulse, supplied at center of scanning area.

Martin Blank, Ed., *Electricity and Magnetism in Biology and Medicine*

APPLICATIONS OF BACTERIORHODOPSIN VARIANTS IN OPTICAL INFORMATION PROCESSING

N. Hampp* and C. Bräuchle

University of Munich, Germany

D. Oesterhelt

Max Planck Institute for Biochemistry, Martinsried, Germany

The biological photochrome bacteriorhodopsin (BR) is contained in the purple membranes (PM) of *Halobacterium halobium*. Under illumination, BR passes reversibly through a cyclic sequence of intermediates. Two dominant states are observed that have spectrally well-separated absorption maxima at 570 nm and 410 nm. This large photochromic shift, the high resolution (>5000 lines/mm), the excellent reversibility, and the fast time scale of the rise and decay of holograms in BR-films make them promising media for dynamic holographic applications, such as pattern recognition. In this paper the use of BR-films containing wildtype BR and mutated bacteriorhodopsins in holographic real-time pattern recognition is demonstrated.

Structure and Function of Bacteriorhodopsin

Bacteriorhodopsin (BR) from *Halobacterium halobium* is embedded as a two-dimensional crystalline lattice of BR trimers in the lipid bilayer of the cell membrane.[1] A retinylidene molecule bound via a Schiff base to lysine-216 and an inner shell of amino acids form the chromophoric group of BR. Proton transport through the cell membrane is intimately connected to the photocycle of BR (Fig. 1). Under illumination with light at about 570 nm the photochemical reaction B $\rightarrow$ J is induced. From the J-state, BR passes through a couple of short-lived intermediates to the M-state in about 50 μs. The M-intermediate has the longest lifetime, in the millisecond range. From the M-state BR can relax to the B-state either thermally or photochemically. The photochemical transition from M $\rightarrow$ B is initiated by the absorption of a photon of, for example, 413 nm by the M-intermediate. Due to the extreme stability of BR toward photochemical degradation, the state of absorption can be changed reversibly at least several hundred thousand times. The high specific absorption of the initial B-state of 63000 $l\,mol^{-1}cm^{-1}$ and the high quantum efficiency[2] of at least $\phi = 0.64$ of the primary photoreaction B $\rightarrow$ J result in a strong intensity dependence of the light-induced changes of the absorption and of the refractive index in the PM film. For hologram formation and erasure both conversions B $\rightarrow$ M ("B-type hologram") and M $\rightarrow$ B ("M-type hologram") can be used.[3-5]

Holographic Properties of BR Films

Dried films of BR or BR embedded into inert matrices such as polyvinylalcohol or polyacrylamide can be used as holographic media. The amino-acid sequence of BR can be changed by gene-technological methods and thereby the photophysical properties can be modified. A detailed discussion of the generation and characterization of such materials is found in Ref. 6. BR films containing the variant BR_{D96N} that was selected in the experiments have a recording sensitivity about 50% higher and a two-fold higher diffraction efficiency compared to wildtype films, due to the prolonged retardation of the thermal relaxation M $\rightarrow$ B. Systematic experimental investigations about the optimization of the holographic properties were done in Refs. 7 and 8.

---

*Correspondence to: N. Hampp, Institut f. Physikalische Chemie, Sophienstr. 11, D-8000 Munich 2, Germany.

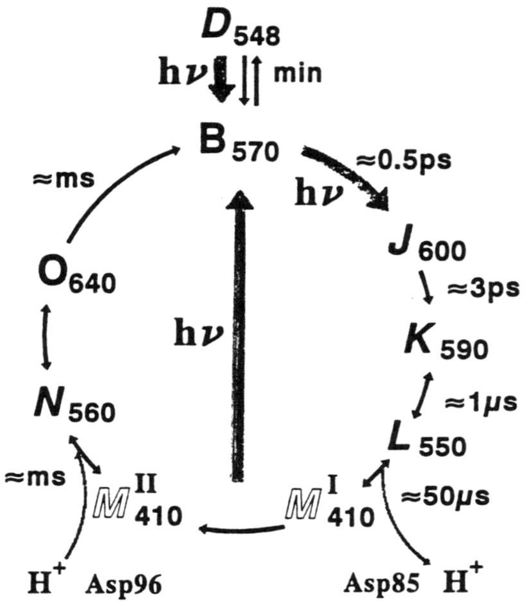

FIG. 1.—Photocycle of bacteriorhodopsin: a scheme of its photochemical and thermal conversions.

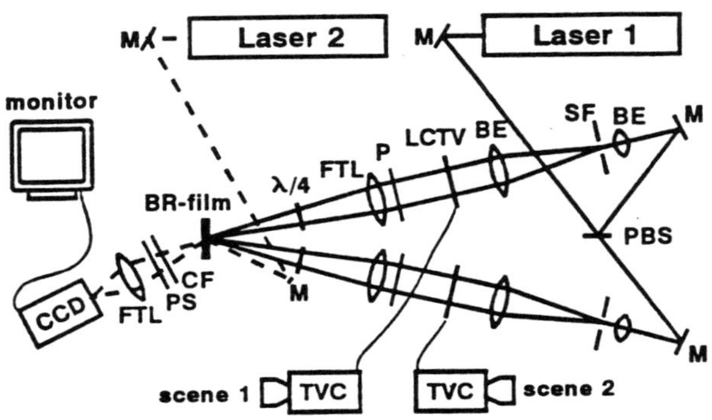

FIG. 2.—Setup for dynamic holographic pattern recognition. The two input patterns are displayed on LCTVs; the correlation signal is detected by a CCD sensor.

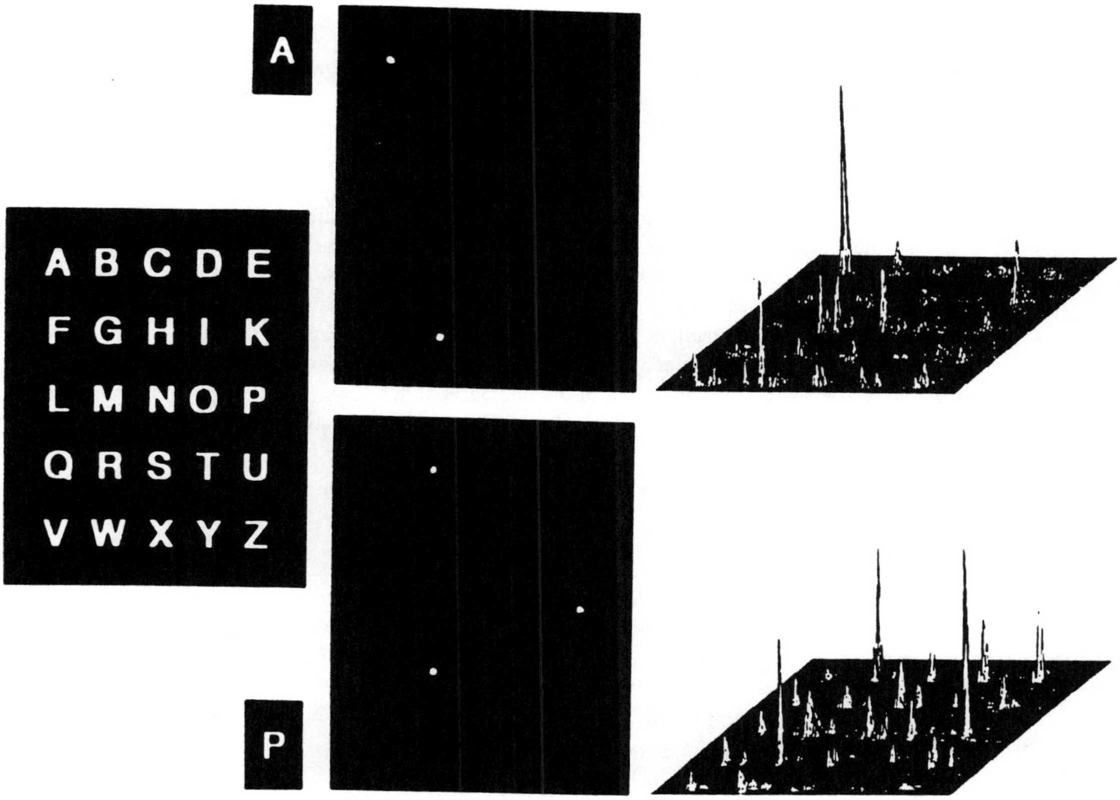

FIG. 3.—Correlation of a 25-letter pattern with two single letters and the recorded intensity distribution.

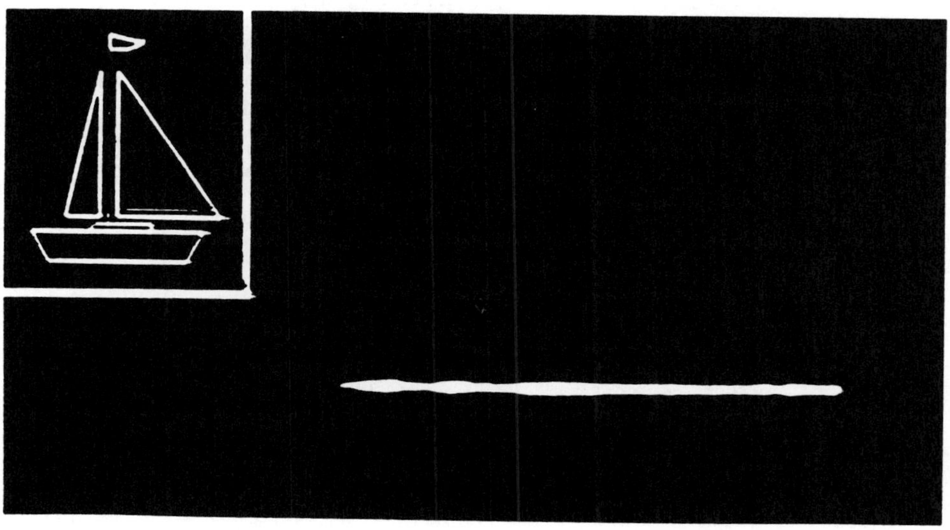

FIG. 4.—The sailboat shown in the upper left corner was tracked during its passage through the viewing range of the TV camera.

Application of BR Films in Holographic Pattern Recognition

A dual-axis joint-Fourier-transform correlator[9] was chosen for holographic pattern recognition with BR films. In this correlator type, two Fourier-transformed input patterns are superimposed (Fig. 2). In the plane where the coherent Fourier spectra interfere, a BR-film is placed that records the resulting holographic gratings. In our experiments, M-type holograms with a read/pump-beam of 530 nm wavelength and recording beams of 413 nm were used. The diffracted waves were again Fourier transformed and projected onto a CCD detector.[10] Intensity distributions like those in Fig. 3 were monitored without any nonoptical postprocessing. Figure 3 shows the ''master'' pattern (25 letters), the two single letters used for the correlating, and pseudo 3D-plots of the correlation signals obtained. For real-time applications, two TV cameras are used as the input devices of the patterns on the LCTVs. From the first one a constant picture of a sailing boat is transmitted to one of the SLMs. Another TV camera monitors a moving boat and displays it on the second SLM. The correlation signal was recorded on video tape. Figure 4 is the superposition of the visible correlation spots. Stopping of the boat leads to a brighter spot in the photograph, due to the longer exposure time. The thickness of the line is thus related to the speed of the boat. The response dynamic of the BR films determines the data throughput rate of the system.

SUMMARY

The use of BR films containing genetical modified BR variants in optical pattern recognition was demonstrated. The fast photoresponse of BR films and their high reversibility permit an easy adjustment of the correlator system. The throughput rate of such a real-time correlator system is limited by the input and output devices (e.g., spatial light modulators) and the processing speed of the particular recording medium, which depends on its hologram rise times and decay times and is in the range of milliseconds for M-type holograms. The attractive properties of films based on the described biological material should stimulate further investigations in this field.[11]

REFERENCES

1. D. Oesterhelt and W. Stoeckenius, *Nature* 233: 149, 1971.
2. J. Tittor and D. Oesterhelt, *FEBS Lett.* 263: 269, 1990.
3. N. Hampp, C. Bräuchle, and D. Oesterhelt, *Biophys. J.* 58: 83, 1990.
4. D. Oesterhelt, C. Bräuchle, and N. Hampp, *Quart. Rev. Biophys.* 24: 425, 1991.
5. C. Bräuchle, N. Hampp, and D. Oesterhelt, *Adv. Mater.* 3: 420, 1991.
6. J. Soppa, J. Otomo, J. Straub, J. Tittor, S. Meessen, and D. Oesterhelt, *J. Biol. Chem.* 264: 13049, 1989.
7. N. Hampp, A. Popp, C. Bräuchle, and D. Oesterhelt, *J. Phys. Chem.* 96: 4679, 1992.
8. D. Zeisel and N. Hampp, *J. Phys. Chem.* 96: 7778, 1992.
9. T. C. Lee, J. Rebholz, and P. Tamuva, *Opt. Lett.* 4: 121, 1979.
10. R. Thoma and N. Hampp, *Opt. Lett.* 17: 1158, 1992.
11. This work was supported by the Bundesministerium für Forschung und Technologie (FKZ 0319231 B).

Martin Blank, Ed., *Electricity and Magnetism in Biology and Medicine*

MOLECULAR MEMORY DEVICES WITH LANGMUIR-BLODGETT FILMS

K. Takimoto,* H. Kawade, K. Hatanaka, K. Eguchi, and T. Nakagiri

Canon Inc., Kanagawa, Japan

The current-voltage characteristics have been measured for a probe/Langmuir-Blodgett (LB) film/metal structure with the scanning tunneling microscope (STM). Rapid increase of current and substantial increase in conductance have been found when the positive voltage was applied to the probe. A bright spot in the STM image has been observed at the position where the increase in the conductance occurred. The changes in the STM images are attributed to the change in the electronic state of LB films, rather than the surface topography, and may be associated with the switching phenomena in the LB films.

The Langmuir-Blodgett (LB) technique has been recognized as a method for organizing molecules and forming ultrathin films with precisely controlled thickness. Recently, a novel electrical switching and memory phenomenon was reported for a metal/LB film/metal (M/LB/M) sandwich structure in which the LB film consisted of molecules having a number of conjugated bonds.[1] This switching and memory phenomenon was not associated with the growth and rupture of metallic filaments through the LB film.[1] Furthermore, it has been shown that the top electrodes could be fabricated by a standard lithography technique on polyimide LB films.[2]

On the other hand, the proposal has been made to replace the top electrode with a scanning tunneling microscope (STM) probe for miniaturizing the M/LB/M device beyond the limits of lithography. The atomic resolution of the STM leads to a possibility of establishing a high-density and huge-capacity memory device.[2,3] In this paper, we report the results of current-voltage (I-V) characteristics observed in the probe/LB film/metal (probe/LB/M) structure with the STM, and of the changes in the STM images accompanied with the I-V characteristics.

This work was performed with polyimide LB films deposited on a gold electrode that was epitaxially grown on mica. We used a homemade STM system. As the probe, a mechanically cut Pt/Rh(20% Rh) wire was used. The I-V characteristics were measured by application of a triangular voltage pulse, 0.5 s wide, with the feedback loop disabled. All measurements were carried out in air and at room temperature.

Figure 1 shows the I-V characteristics for 6 monolayers of polyimide with the probe positive. The current gradually increased with increasing voltage and showed a rapid increase at 3.3 V. The current exceeded 100 nA above 3.3 V, and showed a constant value due to the saturation of the current-to-voltage converter of the STM. The current decreased with decreasing voltage, but the conductance showed a higher value than that before the rapid increase of the current was observed. For low peak values of the triangular voltage pulse (less than 3.0 V), the rapid increase of the current and the change in the conductance was not observed.

Figure 2 shows the changes in the STM images due to the applications of voltage pulses. Figure 2(a) is the STM image after application of a 3.6V pulse. A bright spot was observed at the position where the pulse was applied. The spot was about 10 nm in diameter and 2 nm in apparent height. Furthermore, the probe was positioned above a point B (Fig. 2a) 25 nm from the point A, and another 3.6V pulse was applied. Figure 2(b) in the STM image after the application of voltage. Another bright spot similar to that in Fig. 2(a) was observed at the point B. For both cases, the rapid increase of the current and the substantial increase in the conductance were shown in the I-V characteristics. The spots were not found when the conductance did not show the increase. The distance between the probe and the sample surface can be set to various values by a change in the bias voltage. In this work, the STM was operated with a probe bias between ±0.1 and ±1.0 V. Similar I-V characteristics and bright spots in the STM images were observed under various conditions.

---

*Correspondence to: K. Takimoto, Canon Research Center, 5-1 Morisato Wakamiya, Kanagawa 243-01 Japan.

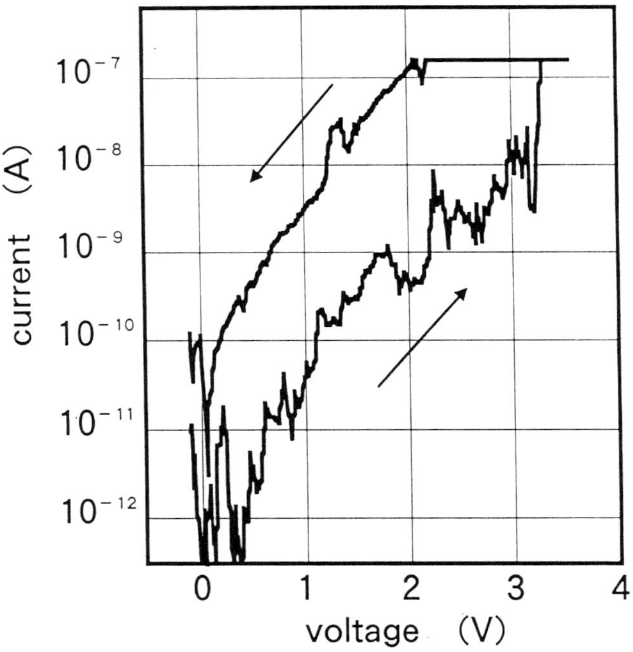

FIG. 1.—Current-voltage characteristics for 6 monolayers of polyimide.

These results show that the changes in the STM images due to the application of voltage are consistent with the corresponding I-V characteristics. However, two explanations are possible for the origin of the bright spots in the STM image. One is a real topographical modification of the sample surface; the other is the modification of the electronic state of the sample surface. Generally speaking, it is difficult to distinguish them with the STM. The observation of the topographical image with the atomic force microscope (AFM) is one of the methods to distinguish these cases. The simultaneous observation of the topographical and current images with the combined AFM/STM is in progress.[4] The preliminary results show that the high-contrast spot in the current image was observed after the application of a voltage pulse, but that a topographical change in the AFM image was not observed. Probably, our results can be attributed to the change in an electronic state of the LB film rather than the surface topography.

We have measured the I-V characteristics for the probe/LB/M structures under the STM, and found the rapid increase of current and the substantial increase of conductance caused by the application of voltage. Subsequently, a bright spot was observed at the position where the increase in the conductance occurred. The changes in the STM images are attributed to the change in the electronic states of LB films rather than to the surface topography, and may be associated with the switching phenomena in the LB films.

The application of the STM is rapidly growing in the field of nanofabrication,[5] in which the STM is used as a tool for the manipulation of an atom or a molecule or their clusters, i.e., transferring them from and/or to a surface. However, it is also expected that the STM will also be used to modify the electronic states of local areas in nanoscale dimensions even of individual molecules without material transfer. Our results suggest this possibility. It should be emphasized that the combination of STM and LB technology is very important if molecular electronics is to come true.

REFERENCES

1. K. Sakai, H. Matsuda, H. Kawada, K. Eguchi, and T. Nakagiri, *Appl. Phys. Lett.* 53: 1274, 1988.
2. K. Sakai, H. Kawada, O. Takamatsu, H. Matsuda, K. Eguchi, and T. Nakagiri, *Thin Solid Films* 179: 137, 1989.
3. O. Albrecht, K. Sakai, K. Takimoto, H. Matsuda, K. Eguchi, and T. Nakagiri, "Biomolecular Electronics," in R. R. Birge, Ed., *Advances in Chemistry Series* (to be published by American Chemical Society).
4. Y. Sugawara, T. Ishizaka, S. Morita, S. Imai, and N. Mikoshiba, *Jpn. J. Appl. Phys.* 29: L157, 1990.
5. C. F. Quate, "Scanning tunneling microscopy and related methods," in R. J. Behm, N. Garcia, and H. Rohrer, Eds., *NATO ASI* Ser. E: Applied Sciences (Kluwer Academic, Dordrecht), 184: 281-297, 1990.

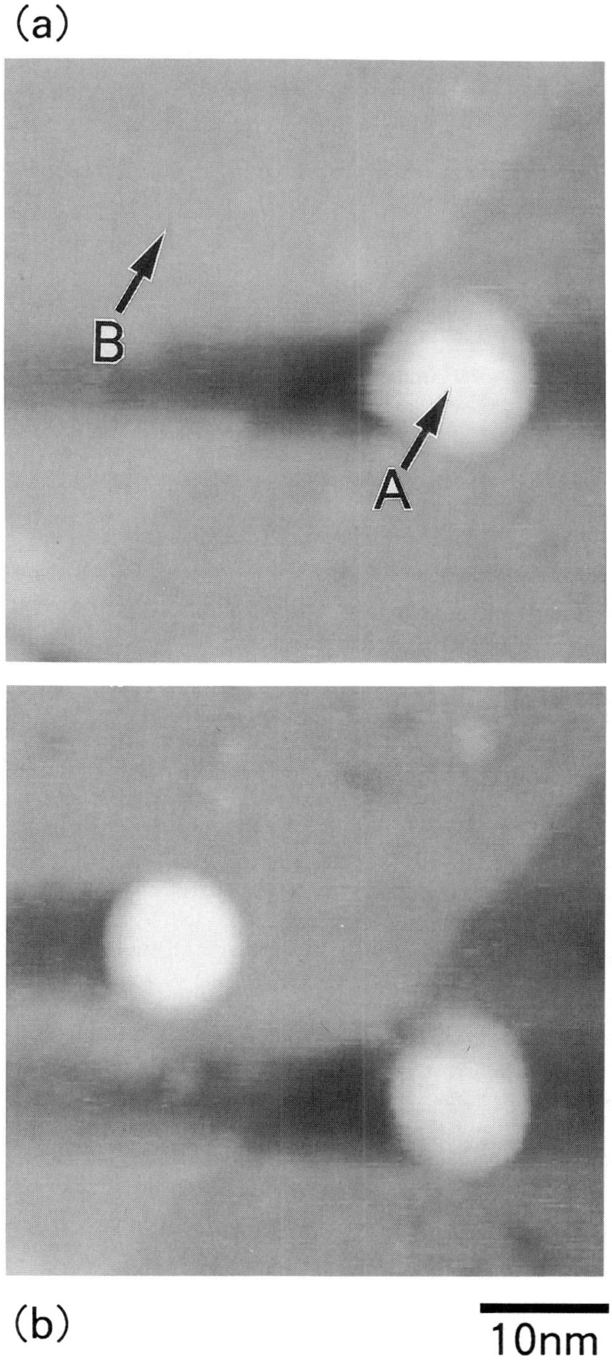

FIG. 2.—(a) Change in STM image for 6 monolayers of polyimide on Au after application of a voltage pulse. (b) Same area after application of a pulse at point B indicated in (a). All images were observed under a constant current of 10 pA and a probe bias of -0.1 V. Scanning area of these images was 46×46 nm.

Martin Blank, Ed., *Electricity and Magnetism in Biology and Medicine*

MEMBRANE CHANNEL ENERGETICS: SURFACE CHARGE IN PROTEIN INTERACTIONS

M. Blank*

Columbia University, New York, N.Y.

Ion channels are molecular switches in biological membranes, and insight into their properties is useful in the development of molecular electronics technology. The opening and closing of these cylindrical assemblies of proteins in response to electrical and chemical stimuli can be modeled as aggregation/disaggregation reactions of protein assemblies that respond to changes in surface charge. This model describes the disaggregation of hemoglobin tetramers, the function of voltage-gated channels during excitation, and the cooling/heating observed in excitable membranes.

Living systems have evolved highly specialized protein structures with unusual advantages (e.g., small size, speed, sensitivity, selectivity, and stability) that can be used in molecular electronic devices for sensory and information storage/retrieval applications.[1,2] For progress in this aspect of biotechnology, it is important to understand how these biological structures achieve their unusual properties. In this paper, we shall review an electrochemical model that describes the opening/closing reactions and the energetics of voltage-gated ion channels in terms of the aggregation/disaggregation reactions of proteins.

SURFACE FREE ENERGY MODEL OF PROTEIN AGGREGATION

Oligomeric proteins do not generally change their internal structure as a result of aggregation/disaggregation reactions, so the total free energy change in a reaction can be approximated by the change in surface free energy. The surface free energy is determined by the surface charge, so disaggregation varies with the molecular charge (or bulk pH); see Fig. 1. The quantitative predictions of this model for the disaggregation of hemoglobin tetramers[3] agree with the measured disaggregation as a function of pH.[4] The surface free energy model also explains the physical meaning of the commonly used measure of cooperativity, the Hill coefficient, since it can be shown to be a classic surface excess quantity.[5] The very high viscosities observed in concentrated hemoglobin solutions can be calculated on the assumption of aggregation.[6]

Protein aggregation reactions are generally described as "entropy driven" because they proceed spontaneously with the evolution of heat. The reverse reaction, disaggregation, is accompanied by cooling. Both forward and reverse reaction heats are consequences of changes in the protein interfacial area in contact with the aqueous solution, and when similar reactions occur in channels, opening is accompanied by cooling and closing by heating.

VOLTAGE-GATED ION CHANNELS

In excitable membranes, the gating current transfers charge across the membrane between channel surfaces, and changes the energetics to favor the opening of a constriction or a "gate" in the channel.[7] The gating current is different for sodium and potassium channels, and we calculate the different ionic currents for the two types of channel (Fig. 2). The fast gating sodium channels show an early inward ionic current, and the slow gating potassium channels show a delayed outward current.[8] Gating charge kinetics in channels determine ion selectivity and inactivation.[8,9]

This model accounts for the unusually large thermal changes associated with channel opening and closing. In protein reactions, aggregation is exothermic and disaggregation endothermic, so channel opening should lead to cooling. The magnitude of the initial cooling suggests that only about 3% of the protein

---

*Correspondence to: M. Blank (Physiology), Columbia University, 630 W 168 St., New York NY 10032.

surface is involved in gating. This model of channels also explains the changes in activity due to heating of nerves by microwaves.

Ion Pumping

AC fields change the ion concentrations at membrane surfaces that normally control the activity of the Na,K-ATPase, the enzyme that "pumps" Na^+ and K^+ ions across cell membranes. The increases calculated with the model are in the range of frequencies at which biological effects of AC have been demonstrated.[10]

During the operation of the Na,K-ATPase, an ion binding surface alternates contact with the solutions on the two sides of the membrane. The channel model suggests a mechanism. Assume a total fixed charge of 60 per 100 nm^2 divided between the inner and outer surfaces at 25 and 35, respectively. According to Fig. 1, one protein face of the channel would be open and the other closed. If the ATPase shifts 10 charges at constant total charge, the probabilities reverse. A change in total charge, via a change in pH, would uncouple the probabilities and make such a mechanism less efficient, in line with the pH optimum of the Na,K-ATPase.

The shift of charge that causes a channel to open can also alter the ion gradients through changes in the binding of counter ions. (Figure 2 shows how the direction of an ion flux depends on the rate of charge shift.) The two changes would cause a transient reversed local concentration gradient for driving ions through the opened channel, at the same time as there is a diminished electrical gradient across the channel. The ATPase could function as an ion pump through charge transfer between the two enzyme surfaces by virtue of its catalytic function. The splitting of ATP releases redox energy that is comparable to electron shifts; the reverse reaction, the formation of ATP from ADP and P, is driven by redox processes in the inner mitochondrial membrane. The correlation between the nonspecific inhibitory ability of a cation and its redox potential,[11] as well as parallels with the mitochondrial ATPase, suggest a role for redox processes and rapid charge conduction within the Na,K-ATPase.[12]

REFERENCES

1. M. Blank, "Biological switches," *Chemtech* 6: 434-438, 1988.

2. M. Blank, "Electrochemical processes in membrane channels and biosynthetic structures," in F. T. Hong, Ed., *Molecular Electronics*, New York: Plenum Press, 1987, 77-81.

3. M. Blank, "A surface free energy model for protein structure in solution: Hemoglobin equilibria," *Colloids and Surfaces* 1: 139-149, 1980.

4. M. Blank and L. Soo, "Surface free energy as the potential in oligomeric equilibria: Prediction of hemoglobin disaggregation constant," *Bioelectrochem. Bioenerg*. 17: 349-360, 1987.

5. M. Blank, "Hemoglobin reactions as interfacial phenomena," *J. Electrochem. Soc*. 123: 1653-1656, 1976.

6. M. Blank, "Molecular association and the viscosity of hemoglobin solutions," *J. Theoret. Biol*. 108: 55-64, 1984.

7. M. Blank, "The surface compartment model (SCM): Role of surface charge in membrane permeability changes," *Bioelectrochem. Bioenerg*. 9: 615-624, 1982.

8. M. Blank, "The surface compartment model: A theory of ion transport focused on ionic processes in the electrical double layers at membrane protein surfaces," *Biochem. Biophys. Acta* 906: 277-294, 1987.

9. M. Blank, "Theory of frequency-dependent ion concentration changes in oscillating electric fields," *J. Electrochem. Soc*. 134: 1112-1117, 1987.

10. M. Blank, "Na,K-ATPase function in alternating electric fields," *FASEB J*. 6: 2434-2438, 1992.

11. J. S. Britten and M. Blank, "Effects of cations on biologically active surfaces: Specific binding sites in the Na,K-ATPase," *J. Colloid and Interface Science* 43: 564-570, 1973.

12. We thank ONR and EPRI for their support.

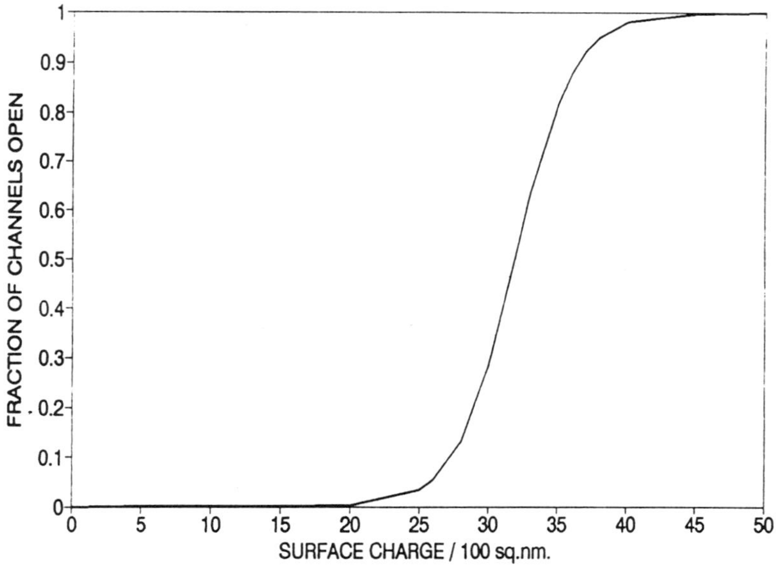

FIG. 1.—Fraction of oligomeric channels that are open as a function of charge density on inner surface of membrane in charges per 100 nm^2. Dissociation equilibrium constant for proteins in oligomeric channel is assumed to equal value for hemoglobin disaggregation at the point of zero charge. (Reproduced with permission from *Bioelectrochem. Bioenerg.*)

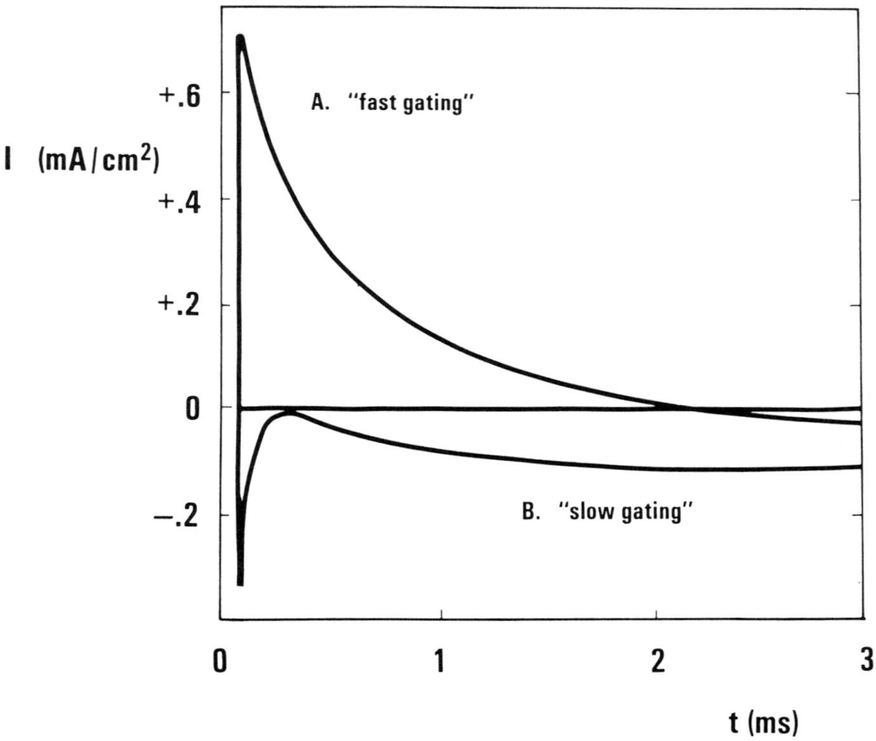

FIG. 2.—Ionic current I (in mA/cm^2) vs time t (in ms) for two different gating rates (A is 15 times faster than B). Parameters of curves A and B are identical except for the one that controls gating current. (Reproduced with permission from *Bioelectrochem. Bioenerg.*)

Martin Blank, Ed., *Electricity and Magnetism in Biology and Medicine*

EFFECT OF RADIO-FREQUENCY RADIATION (RFR) AND DIAZOLUMINOMELANIN (DALM) ON THE GROWTH POTENTIAL OF BACILLI

J. G. Bruno and J. L. Kiel*

Armstrong Laboratory, Brooks AFB, Tex.

DALM, which is thought to generate increased levels of free radicals in response to RFR, was synthesized in situ by *Bacillus anthracis*. Bacilli grown under DALM-producing conditions showed significantly lower plate counts in response to circularly polarized 2450MHz RFR (100 W/kg) with exposures as short as 30 min. Bacilli grown on media for 3 days prior to irradiation demonstrated lower colony counts than 1-day-old cultures under DALM-producing and nonproducing conditions. No significant RFR effects were observed for endospore preparations of bacilli grown in the absence of DALM substrates. Thus, DALM appears to function as a microwave-sensitive antibiotic that may operate through a free radical-mediated mechanism. In addition, DALM-laden bacteria have demonstrated temperature-dependent "slow fluorescence," which supports the hypothesis that DALM toxicity is mediated by free radicals and suggests that DALM may be of utility in molecular electronics.

Free radicals are known mediators of ionizing radiation bioeffects. At low levels, nonionizing radiation imparts insufficient energy to generate free radicals directly. Yet several investigators[1,2] have proposed or examined free-radical-driven biochemical processes to ascertain whether natural free-radical levels might be enhanced by RFR. Results indicated little evidence for microwave-induced cellular damage by free radicals secreted to the extracellular environment.[2] Hence, we reasoned that coupling RFR to a robust free-radical generator such as DALM, which is probably synthesized intracellularly or in the periplasmic space,[3] might amplify suspected bioeffects.

We initiated studies of *Bacillus anthracis* (nonpathogenic Sterne strain) under conditions thought to generate elevated levels of free radicals (i.e., DALM loaded) to determine whether nonthermal RFR-driven free-radical formation could kill bacteria. Bacilli were chosen for two reasons. First, one may examine RFR effects on purified, quiescent endospores (vaccine) as well as metabolically active vegetative cells. Second, when fed appropriate substrates (i.e., 3-amino-tyrosine, nitrate, and luminol), the bacteria synthesize a chemiluminescent polymer known as DALM.[3-5] DALM is thought to generate increased levels of free radicals in response to RFR, because its free radical-driven chemiluminescence increases with microwave heating.[5]

METHODS AND RESULTS

Figures 1(a) and (b) summarize results obtained from colony counts of bacteria cultivated for 1 day on TSA (trypticase soy agar) only, or TSA plus DALM substrates (320 mg/l 3-amino-L-tyrosine, 12 g/l potassium nitrate), and 100 mg/l 5-amino-2, 3-dihydro-1,4-phthalazinedione (luminol) with (solid line) or without (dashed line) RFR exposure. Similar results were obtained for bacteria cultured 3 days on TSA only ($n = 4, p = 0.0627$) and TSA plus DALM substrates ($n = 3, p = 0.0001$). Figure 1(c) illustrates that RFR had no appreciable effect on the growth potential of a live endospore vaccine (Thraxol-2, Mobay Corp., Shawnee, Kans.). These data were collected from samples irradiated with 2450MHz CW RFR for 30 min at 37 C in buffer consisting of sterile phosphate buffered saline (PBS), 0.75% $NaHCO_3$, and 0.03% H_2O_2, followed by serial 10-fold dilution in PBS and plating in liquid TSA cooled to 45 C. Controls (-RFR) were kept in a 37 C water bath to mimic temperature conditions during RFR exposure. All cells were initially diluted to an absorbance of 2.0 at 600 nm. Temperature was maintained at 37 C by means of a computer-controlled air cooling-heating system (Questech, San Antonio, Tex.). Colony counts greater than 300 were scored as too numerous to count, but assigned a value of 300 for plotting and statistical analysis. All p values were

*Correspondence to: J. L. Kiel, Armstrong Laboratory (AFSC), Brooks AFB TX 78235-5000.

derived from 2-way analysis of variance (ANOVA). Subsequent data (not shown) suggest extinction of DALM-loaded bacteria somewhere between 4 and 8 h of RFR exposure.

SUMMARY AND CONCLUSIONS

This work demonstrates the microwave-augmented antibiotic effect of DALM. In addition, observations of reduced colony counts for DALM-laden bacilli and the lack of an RFR effect on quiescent endospores suggests that RFR-induced cellular damage is mediated by free radicals. The identity of free radicals generated by DALM is still unknown; however, we suspect the formate radical is a primary damaging species. Other radicals, such as the semiquinone, may also be involved, but probably to a much lesser extent. Global assessment of macromolecules from exposed and unexposed bacteria by use of silver-stained SDS polyacrylamide gels has revealed no obvious evidence of free-radical damage (data not shown). Only age-related and buffer-related changes (i.e., the carbonate/bicarbonate concentration was varied) in banding patterns are evident.

Although it is difficult to produce tangible evidence of free-radical damage by the methods employed here, three lines of evidence point to a free-radical mechanism (data not shown). First, only DALM-laden bacteria die or are growth inhibited in response to microwave exposures. Second, DALM-laden bacterial chemiluminesce strongly in an RF field, even at 37 C. This chemiluminescence is believed to originate from free-radical interaction with luminol moieties in DALM. Third, such bacteria exhibit the little-known phenomenon of "slow fluorescence" (unpublished observation).[6,7]

Slow fluorescence is a process distinct from both fluorescence and phosphorescence in that the emission is relatively long lived, but proceeds from an excited triplet (in this case possibly a semiquinone radical).[7] We have observed both red and green thermally dependent fluorescence emissions of a relatively long-lived nature that are quickly regenerated in DALM-loaded bacilli upon subsequent excitation. The observation of slow fluorescence strongly suggests the possibility of free-radical generation by DALM.

Thus, these experiments suggest that DALM, biosynthesized in situ, in conjunction with microwave energy may be used as a potentially effective antibiotic and photonic switching device (with the microwave field turning on the emission). Future work will focus on establishing the postulated free radical-mediated mechanism by addition of various free radical scavengers to abolish or abrogate the differences in colony counts reported here.[8]

REFERENCES

1. D. N. Erwin, *Aviation Space Environ. Med.* 59(11, Suppl.): A21-31, 1988.
2. J. L. Kiel and D. N. Erwin, *Int. J. Hyperthermia* 2: 201-212, 1986.
3. J. L. Kiel, J. E. Parker, J. L. Alls, and R. A. Weber, *Proc. Int. Conf. of IEEE-MBS* 13: 1605-1606, 1991.
4. J. L. Kiel, G. J. O'Brien, J. Dillon, and J. R. Wright, *Free Rad. Res. Comm.* 8: 115-121, 1990.
5. J. L. Kiel, C. Gabriel, D. M. Simmons, D. N. Erwin, and E. H. Grant, in J. L. Kiel, J. E. Parker, J. L. Alls, and R. A. Weber, *Proc. Int. Conf. of IEEE-MBS* 12: 1689-1690, 1990.
6. J. March, *Advanced Organic Chemistry*, New York: McGraw-Hill, 1977, p. 218.
7. C. C. Felix, J. S. Hyde, and R. C. Sealy, *Biochem. Biophys. Res. Comm.* 88: 456-461, 1979.

8. This work was supported by the Air Force Office of Scientific Research. The work was performed while Dr. Bruno held a National Research Council-AFSC Research Associateship.

FIG. 1

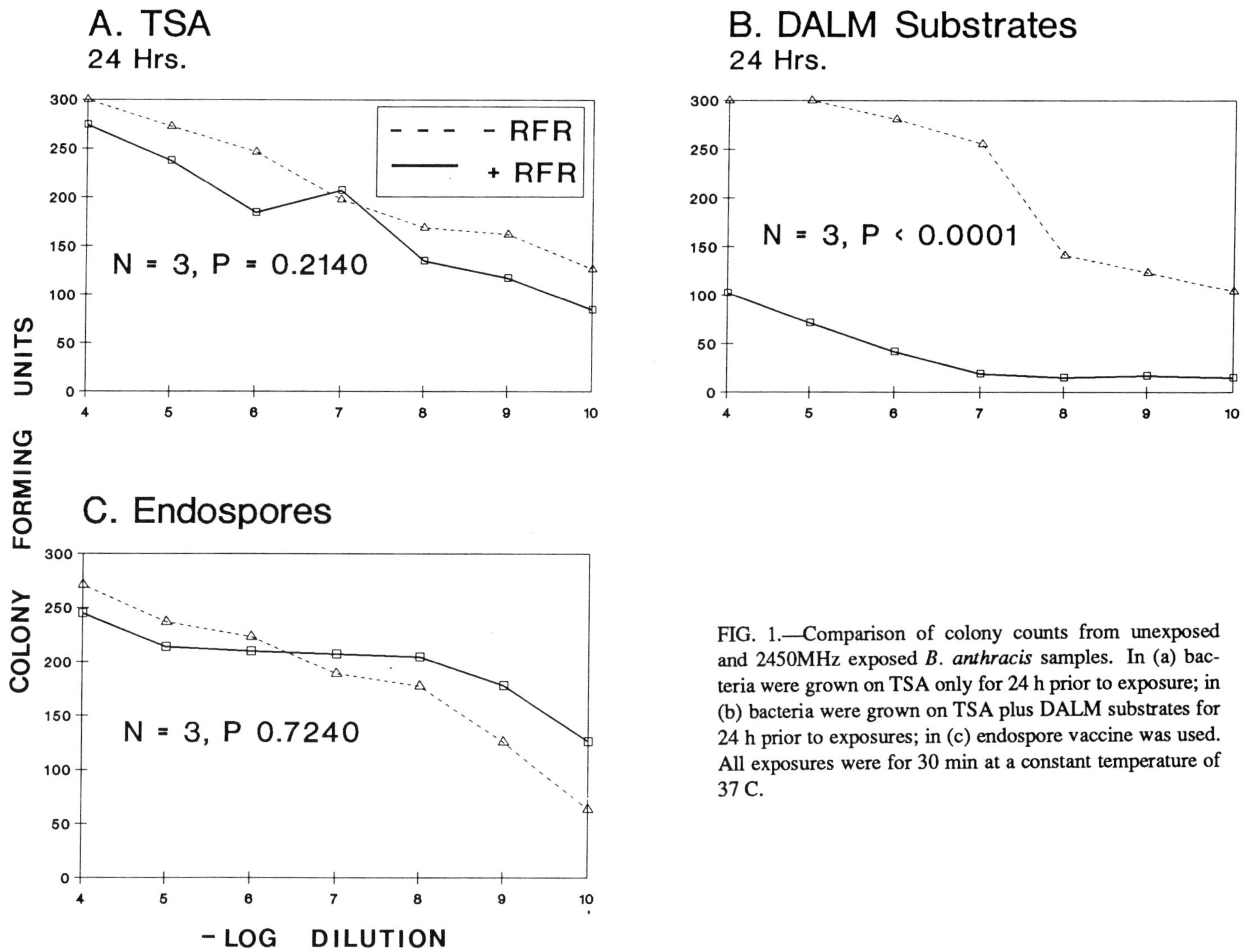

FIG. 1.—Comparison of colony counts from unexposed and 2450MHz exposed *B. anthracis* samples. In (a) bacteria were grown on TSA only for 24 h prior to exposure; in (b) bacteria were grown on TSA plus DALM substrates for 24 h prior to exposures; in (c) endospore vaccine was used. All exposures were for 30 min at a constant temperature of 37 C.

Martin Blank, Ed., *Electricity and Magnetism in Biology and Medicine*

GLUCOSE OXIDASE ELECTRODES WITH POLYURETHANE MATRIX

S. Alva,* R. S. Phadke, and G. Govil

Tata Institute of Fundamental Research, Bombay

Glucose oxidase has been immobilized in a polyurethane (PU-6) matrix by the method of physical entrapment and has been used for the estimation of glucose under in vitro conditions. The effects of mediators (potassium ferricyanide) in the electrolyte and co-immobilized relays (ferrocene) in the immobilization matrix on the performance of the electrode has been studied.

The performance of a biosensor depends largely on appropriate interfacing of the biological component with a transducer.[1] Entrapment in natural or synthetic polymer gels is the most suitable method of immobilization of the biological component since it can be used for various systems, such as enzymes or cells.[2] The isocyanate functional groups at the terminal ends of the polyurethane prepolymer [a polymer of poly(ethylene) glycol and poly(propylene) glycol] react in the presence of an aqueous solution of the enzyme to form crosslinked polymeric structure by urea linkage, which entraps the enzyme. The hydrophilic/hydrophobic properties of the prepolymer can be controlled by variation of the ratio of the poly(ethylene glycol) part and the poly(propylene glycol) part in the polyether diol moiety of the prepolymer. Polyurethane prepolymer (PU-6) has a high content of poly(ethylene glycol) (91%) and hence displays a hydrophilic character.[3] Ferrocene or its derivatives are sometimes used to attain improved electrical communication between the enzyme and the transducer.[4,5] Redox couples such as potassium ferricyanide and quinhydrone have been reported to serve as electron mediators for enzymes that have flavine adenine dinucleotide (FAD) as a coenzyme.[6]

Here, we report a method of preparation of the enzyme electrode by use of polyurethane PU-6 as the immobilizing matrix and performance of such an electrode for in vitro estimation of glucose in the presence of mediators and relays. We also report similar experiments done on a carbon plate as substrate instead of expensive platinum strips.

MATERIALS AND METHODS

Glucose oxidase (EC 1.1.3.4, type VII from aspirgillus niger) and horse raddish peroxidase (POD, EC1.1.11.7) were purchased from Sigma Chemicals, USA. Polyurethane prepolymer PU-6, a noncommercial product, was a gift from Profs. A. Tanaka and K. Sonomoto, Kyoto University, Yoshida, Koyoto, Japan. Blood serum and lyophilized blood plasma were obtained from Haffkine Biopharmaceuticals Ltd., India. All reagents used were of analytical grade or better.

Preparation of the Enzyme Electrode

Aqueous solution of the enzyme (~ 1000 units) was prepared in 200 μl of 0.1 M sodium phosphate buffer. A platinum strip of surface area 2.2 cm^2 was annealed by heating to red hot before use. It was washed first with dilute nitric acid and then thoroughly rinsed with double-distilled water. 100 μl of the enzyme solution was spread on one side of the platinum strip. Along 5 mg of the prepolymer PU-6 was mixed with the enzyme solution and allowed to polymerize. Polymerization was complete after about 5 min. The process was repeated for the other side of the platinum strip. The enzyme electrode thus formed was kept in dark overnight. The electrode was washed repeatedly with phosphate buffer to remove free enzyme and/or prepolymer present on the surface. The electrode was stored dry in dark. The co-immobilization of ferrocene was done by mixing it thoroughly in aqueous solution of the enzyme prior to adding PU-6, which resulted in uniform distribution of ferrocene throughout the film. The average thickness of the sample under

---

*Correspondence to: S. Alva (Chemical Physics), Tata Institute of Fundamental Research, Homi Bhaba Rd., Bombay 400 005 India.

dry conditions was found to be 0.12 ± 0.02 mm.

The activity of GOD in aqueous solution was measured by standard methods, with o-dianisidine as the coloring agent.[7] For amperometric experiments, a three-electrode setup was used with the enzyme electrode as the working, saturated calomel electrode as the reference and platinum mesh as the counter electrode.

RESULTS AND DISCUSSION

It has been observed that the activity of immobilized GOD is 45% of that of free solution. Addition of ferrocene to the immobilizing matrix shows no further deterioration of the enzyme activity.

Potentiometric Studies

The response of the electrode has been recorded at various intervals of time. It increases initially linearly and attains a constant value after 100 s. However, all the measurements have been done after 120 s of the introduction of the sample to insure the equilibrium conditions. Figure 1 shows the change in the potential as a function of logarithm of the concentration of glucose in the solution under various experimental conditions. Curve (a) depicts the performance in the presence of the mediator; curves (b) and (c) are for electrodes with relay (ferrocene) in the matrix in the absence and the presence of mediators, respectively. One notices that the combined use of mediators and the relays gives the best performance.

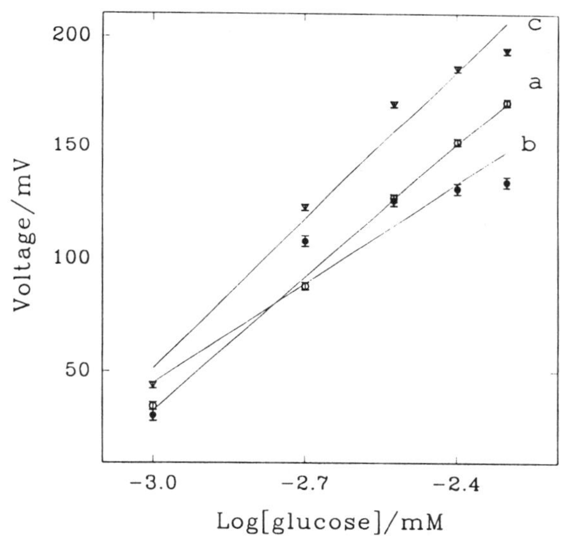

FIG. 1.—Change in potential as a function of logarithm of glucose concentration for electrode with GOD: (a) with mediator (0.62mM of potassium ferricyanide), (b) with relay (ferrocene), (c) with relay and mediator. Mediator (ferricyanide) concentration was optimized to 0.62 mM after extensive experimentations.

Amperometric Studies

Figure 2 shows the change in the current measured between the working electrode and the counter electrode when a potential (−0.36 V with the mediator, −0.8 V without the mediator) is applied to the working electrode with respect to reference electrode. Curve (a) represents the changes in the current as a function of glucose concentration in the presence of the mediator. Curves (b) and (c) are for electrodes with relay (ferrocene) in the matrix in the absence and presence of mediators, respectively. One notices that simultaneous use of mediators and relays is advantageous.

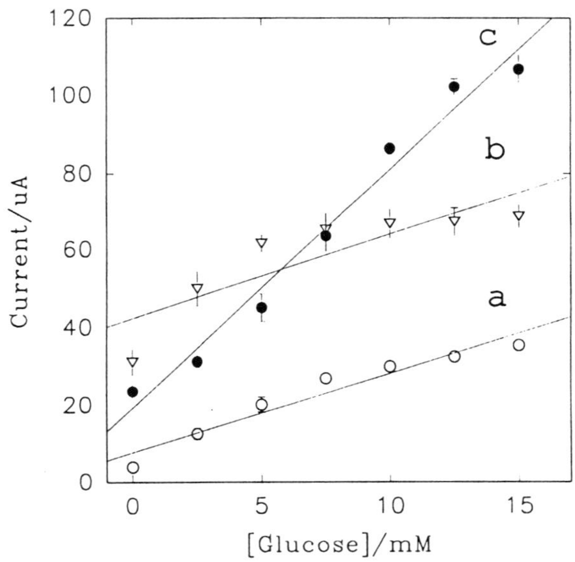

FIG. 2.—Current vs glucose concentration for glucose oxidase immobilized in polyurethane matrix on platinum (a) with mediator, (b) with relay (ferrocene), (c) with mediator and relay. Supporting electrolyte 0.1M sodium phosphate buffer, pH 6.8. Mediator (ferricyanide) concentration = 0.62 mM.

In a separate series of experiments, conducting vitreous carbon has been used to substitute for expensive platinum strip. It has been observed that such an electrode exhibits equally good performance as the platinum strip based electrodes. Thus, our investigations indicate that combined use of mediators and relays is advantageous and use of carbon does not adversely affect the performance of the electrode.

REFERENCES

1. A. E. G. Cass, G. Davis, G. D. Francis, H. A. O. Hill, W. J. Aston, I. J. Higgins, E. V. Plotkin, L. D. L. Scott, and A. P. F. Turner, *Anal. Chem.* 56: 667-671, 1984.

2. G. G. Guilbault and G. O. Neto, in *Immobilized Cells and Enzymes: A Practical Approach*, Oxford: IRL Press, 1985, pp. 58-74.

3. S. Fukui and A. Tanaka, *Adv. Bioeng. Biotech.* 29: 1-31, 1984.

4. Y. Degani and A. Heller, *J. Am. Chem. Soc.* 110: 2615-2620, 1988.

5. B. A. Gregg and A. Heller, *Anal. Chem.* 62: 258-263, 1990.

6. D. L. Williams, A. R. Diog Jr., and A. Korosi, *Anal. Chem.* 42: 118-121, 1970.

7. R. Bentley, in P. D. Boyer, Ed., *The Enzymes*, New York: Academic Press, 7: 567-586, 1963.

Martin Blank, Ed., *Electricity and Magnetism in Biology and Medicine*

MODIFICATION, CHARACTERIZATION, AND UTILITY OF FAD

S. Alva,* R. S. Phadke, and G. Govil

Tata Institute of Fundamental Research, Bombay

Flavine adenine dinucleotide (FAD) has been made amphiphilic by the attachment of palmitic acid to the adenine amino group. It has been characterized chemically and biochemically. The redox property has been verified by cyclic voltametric methods.

Flavine adenine dinucleotide can exist in two stable oxidation states, oxidized (FAD) and reduced ($FADH_2$). Switching of the molecule from one state to the other can be attained chemically or electrochemically.[1] The flavine undergoes redox reactions reversibly by two one-electron steps or one two-electron step, which changes the chemistry of enediamine subfunction of the isoalloxazine moiety.[2] The adenine moiety is not involved in the catalysis, but helps in anchoring the coenzyme at the active site.

The substance is highly hydrophilic and forms a coenzyme to a number of enzymes.[1] We have changed the hydrophilicity of FAD by chemically attaching hydrocarbon chains to the noncatalytic adenine moiety, and verified that the modified coenzyme retains its electrochemical, optical, and biochemical properties.

MATERIALS AND METHODS

Flavine adenine dinucleotide, glucose oxidase (EC1.1.3.4, type VII from *A. niger*), and peroxidase (EC1.1.11.7) were purchased from Sigma Chemicals, USA. All other chemicals were of analytical grade or better. Absorbance studies were done with a CARY 17 spectrophotometer. Fluorescence measurements were done with a Shimadzu spectrophotofluorimeter. Cyclic voltametric experiments were done with a Potentiostat/Galvanostat EG&G Princeton Applied Research Model 273 in a three-electrode setup, with argon-saturated 0.1M sodium phosphate buffer as the electrolyte and a scan rate of 50 mV/s.

RESULTS AND DISCUSSION

Synthesis of Fatty Acid Derivatives of FAD

FAD is a hydrophilic molecule containing polar groups such as phosphate and amine. Our strategy to impart amphiphilicity to FAD is based on covalent attachment of a long alkyl chain to the amino group of the adenine ring. It is known that the adenine ring does not directly participate in redox processes. Hence the attachment of hydrocarbon chain to the adenine ring is expected to leave electrochemical and biochemical properties of FAD unaltered.

Palmitic acid was converted to acyl chloride with the help of thionyl chloride. Appropriate amounts of fatty acid and thionyl chloride were refluxed in a round-bottomed flask fitted with a water-cooled condenser maintained at 340 K for 1 h. Unreacted excess thionyl chloride was distilled off from the reaction mixture. The acyl chlorides and FAD were heated in a round-bottomed flask fitted with an air condenser at 340 K for 2 h. The reaction mixture was extracted with ether to separate acyl chloride and fatty-acid derivatives of FAD [(Flavin (6-palmitoyl amido) adenine diphosphate—PFAD] from the mixture. The solution was washed with water to remove unreacted FAD. The process also helps to convert the unreacted acyl chloride back to fatty acid. Ether can then be evaporated to give a solid mass, which is a mixture of free fatty acid and a fatty acid derivative of FAD (PFAD). This sample was subjected to a variety of characterization and analysis techniques.

---

*Correspondence to: S. Alva (Chemical Physics), Tata Institute of Fundamental Research, Homi Bhaba Rd., Bombay 400 005 India.

Characterization

The nature of the structure of PFAD (Fig. 1) suggests that this compound is likely to be less soluble in water than the native compound. We have found that the compound is readily soluble in ether and chloroform and sparingly soluble in the aqueous medium. The absorption spectrum of PFAD in chloroform exhibits peaks at 447 nm and 401 nm, which correspond to the characteristic absorption maxima of flavine ring. (FAD in aqueous solution has absorption maxima at 445 nm and 380 nm, respectively.) The observed red shift of the peak at 380 nm could be attributed to a change in the overall structure of the molecule, as well as to a change in the polarity of the solvent.

The emission spectrum of PFAD in chloroform with excitation at 445 nm exhibits a band with a maximum around 520 nm. We may recall that FAD in aqueous solution shows a broad emission spectrum with an emission maximum around 520 nm on excitation at 445 nm. Thus, one can infer from these observations that various steps in the chemical reactions have not adversely affected the characteristic optical properties of the FAD molecule.

A thin film of modified FAD was coated onto a glass surface by the solvent-evaporation method. Identical glass plate was used as the reference. The absorption peaks appear at the same positions as in the case of solution spectrum. The emission band of the film also appears at a characteristic position (520 nm) of modified FAD in solution. These results indicate that the optical properties of modified FAD in the solid phase are identical with those in the solution phase.

It is well known that the amide bond has characteristic absorption maxima at 190 nm that can be used as a monitor for the detection of bond formation. We observed that PFAD shows an intense peak around 190 nm. This finding clearly establishes the formation of a fatty acid-FAD complex. Additional support for the formation of a complex was obtained by monitoring of the breaking of the amide bond. Hydrolysis of the amide bond in the complex can be effected by the addition of dilute hydrochloric acid to the ethanolic solutions of the compound. Breaking of the bond is indicated by a decrease in the intensity of the absorption peak. We have observed that the intensity of the peak decreased slowly with time and reached the noise level 1 h after the addition of hydrochloric acid. Thus, amide bond formation between FAD-NH_2 and COOH of fatty acid is unequivocally established.

Biological Activity

FAD forms a coenzyme to a number of redox enzymes. Glucose oxidase is one of the flavine enzymes that are known to be stable and easily available. The biological activity of the modified FAD was assayed by a standard procedure.[3] The apoenzyme was prepared after the method described by Swoboda.[4] The activity of the reconstituted enzyme was measured by use of o-dianisidine as the coloring agent and was observed to be 2100 U/g. These results thus indicate that the incorporation of modified coenzyme in GOD retains the catalytic capability of the enzyme.

Electrochemical Activity

Redox properties can be studied conveniently by cyclic voltametry. A thin film of PFAD was formed on a platinum strip by the solvent-evaporation technique. A three-electrode setup consisting of a coenzyme-coated electrode as the working electrode, a saturated calomel electrode as the reference, and a platinum mesh as the counter electrode was used. The characteristic redox peaks appear at -0.57 and -0.63 V, respectively. These values are identical to those of native FAD.[5] This finding shows that the redox properties of FAD do not alter on modification.

We have thus succeeded in synthesizing an amphiphilic derivative of FAD that exhibits characteristic properties of the native molecule.

REFERENCES

1. S. Ghisla and V. Massey, *Eur. J. Biochem.* 181: 1, 1991.
2. H. Shinohara, M. Gratzel, N. Vlachopoulor, and M. Aizawa, *Bioelectrochemistry & Bioenerg.* 26: 307, 1991.
3. R. Bently, in P. D. Boyer, Ed., *The Enzymes*, New York: Academic Press, 7: 567, 1963.
4. B. E. P. Swoboda and V. Massey, *J. Bio. Chem.* 240: 2209, 1965.
5. H. M. Sonawat, R. S. Phadke, and G. Govil, *Biotechnol. and Bioeng.* 26: 1066, 1984.

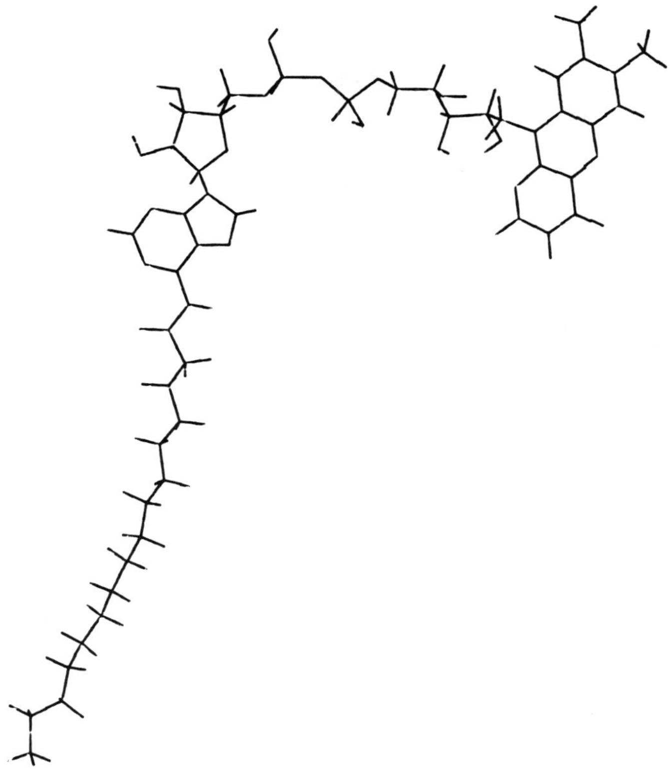

FIG. 1.—Structure of flavine (6-palmitoyl amido) adenine dinucleotide.

Martin Blank, Ed., *Electricity and Magnetism in Biology and Medicine*

DIRECT OBSERVATION OF THE CONFORMATIONAL DYNAMICS OF LARGE DNA MOLECULES BY FLUORESCENCE MICROSCOPY

K. Minagawa, Y. Matsuzawa, K.Yoshikawa,* M. Matsumoto, and M. Doi

Nagoya University, Japan

Individual DNA molecules were visualized by use of fluorescence microscopy. The observation was carried out (i) in the presence of cationic polypeptides, and (ii) in gel electrophoresis. Higher-order structures and dynamic behaviors of DNA have been quantitatively analyzed. It has become clear that the conformation of DNA changes in a cooperative manner in the presence of polyarginine. Moreover, microscopic observation enables us to analyze promptly both the electrophoretic mobility and conformation of DNA.

Conformational dynamics of DNA molecules has been an important subject from both the biological and physical point of view. We have been studying the dynamics of higher-order structures of DNA molecules using fluorescence microscopy.[1,2] In the present paper, we report our recent results on the dynamical conformational change of DNA molecules observed under two conditions: (i) in a buffer solution containing a cationic polypeptide as a model for nucleoproteins, and (ii) in an agarose gel with external electric field.

EXPERIMENTAL

T4DNA (166kbp, full length: 55 μm) was purchased from Nippon Gene. Calf thymus histone and poly-L-arginine (DP120) were purchased from SIGMA. The samples were diluted with TBE buffer solution (45 mM Tris, 45 mM borate, 1 mM EDTA) containing a fluorescent dye, DAPI (4′,6-diamidino-2-phenylindole), and 2-mercaptoethanol as an antioxidant. A Nikon TMD microscope equipped with a 100× oil-immersed objective and a highly sensitive SIT camera were used. The time-dependent conformational data were analyzed from video images by use of a personal computer. Gel electrophoresis was carried out by means of a small gel placed on the microscope stage. The electric-field conditions were controlled by a wave generator.

RESULTS AND DISCUSSION

Biphasic Conformational Change Induced by Cationic Polypeptides

Figures 1(a)-(c) show the fluorescence images of T4 DNA exhibiting large fluctuating motion in an aqueous solution. Calf thymus histone was added to the T4 DNA solution and the resulting complex of DNA and histone was observed. It was found that DNA molecules exhibit shrunk spherical shapes in the presence of histone. The complex showed the Brownian fluctuational movement and kept the spherical shape during the observation. This observation suggests that the stretching motion is restrained by the binding of histone due to the effect of cross linking between the phosphate groups outside the double-helix.

In order to shed light on the interaction, poly-L-arginine (p-Arg) was used as a simple model for the basic protein. When p-Arg was added to the T4 DNA solution, DNA exhibited spherical shapes similar to those observed in the presence of histone, as shown in Figs. 1(d)-(f). The conformation of DNA was observed under various concentrations of p-Arg. It was found that the binding of p-Arg decreases the persistent length[3] of the DNA molecule and that the DNA molecules exhibit two distinguishable conformations: the totally complexed (spherical) and the "naked" (elongated) forms at the intermediate concentration of p-Arg; whereas at the higher concentrations of p-Arg, all DNA molecules exhibit shrunk shape. This result implies that the p-Arg molecules bind to DNA in a cooperative manner.

---

*Correspondence to: K. Yoshikawa, Grad. School of Human Informatics, Nagoya University, Nagoya 464-01 Japan.

It is expected that this sort of cooperative interaction between DNA and basic proteins concerns the mechanism of the control of gene expression in living cells.

Observation of DNA Undergoing Gel Electrophoresis

Microscopic observations of the DNA molecules were performed under steady and biased sinusoidal[4] electric fields. Under the steady field, DNA molecules are often trapped by gel fibers and exhibit characteristic V-conformation, causing a decrease in velocity. Figure 2 shows a typical conformational change of T4DNA. It was found that the trapped molecule keeps the elongated conformation for several seconds, and is then released from the gel fiber accompanied by rapid contraction.

The velocity of center of mass was determined from the video image. The electrophoretic mobility μ was calculated from $\mu = v/E$, where v and E are the velocity of center of mass and the electric field strength, respectively. The electrophoretic mobility of DNA was found to increase with the increase in the field strength.

For the biased sinusoidal field, it was found that the mobility depends both on the strength of the field and on the frequency of the sinusoidal field. In some frequency conditions, higher mobility was observed compared to that under steady field with the same field strength.[5]

REFERENCES

1. K. Minagawa, Y. Matsuzawa, K. Yoshikawa, M. Matsumoto, and M. Doi, *FEBS Lett.* 295: 65-67, 1991.

2. M. Matsumoto, T. Sakaguchi, H. Kimura, M. Doi, K. Minagawa, Y. Matsuzawa, and K. Yoshikawa, *J. Polym. Sci. B* 30: 779-783, 1992.

3. M. Doi and S. F. Edwards, *The Theory of Polymer Dynamics*, Oxford: Clarendon, 1986.

4. T. Shikata and T. Kotaka, *Biopolymers* 31: 253-254, 1991.

5. This work was partly supported by a Grant-in-Aid for Scientific Research from the Ministry of Education, Science and Culture, Japan.

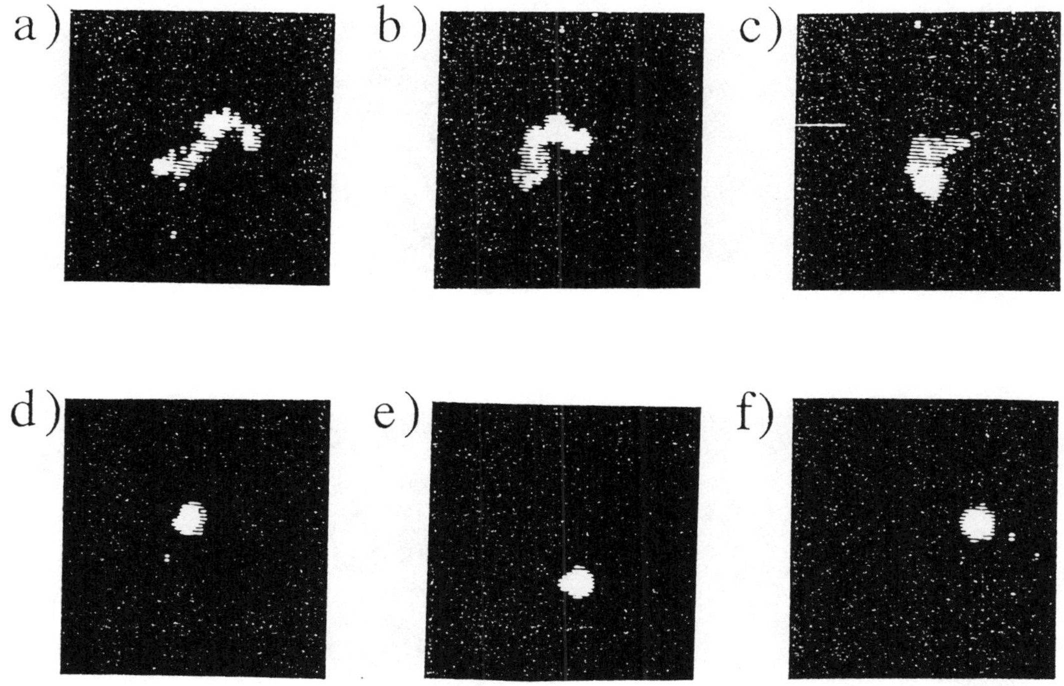

FIG. 1.—Micrographs of T4 DNA molecules showing Brownian motion in aqueous solutions: (a)-(c) T4 DNA in TBE buffer solution, (d)-(f) T4 DNA complexed with p-Arg, [Arg]/[DNA] = 1.0 (mole ratio of monomer units).[1]

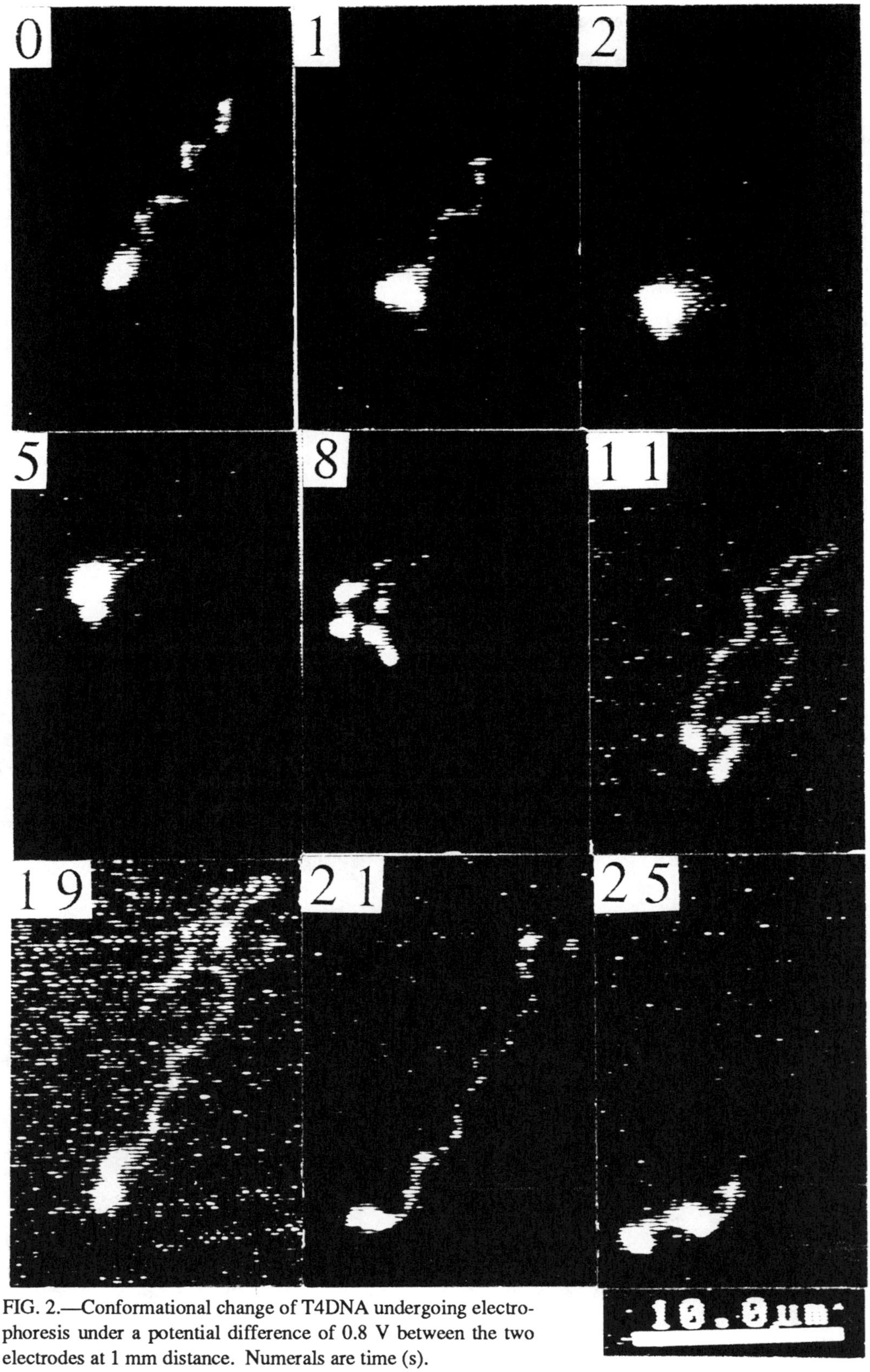

FIG. 2.—Conformational change of T4DNA undergoing electrophoresis under a potential difference of 0.8 V between the two electrodes at 1 mm distance. Numerals are time (s).

Martin Blank, Ed., *Electricity and Magnetism in Biology and Medicine*

MODULATION OF THE PHOTOELECTRIC SIGNAL IN BACTERIORHODOPSIN THIN FILMS BY MEANS OF SITE-DIRECTED MUTAGENESIS

Filbert H. Hong,* B. Ni, M. Chang, R. Needleman, and Felix T. Hong

Wayne State University, Detroit, Mich.

The effect of point mutation of bacteriorhodopsin (BR) primary amino acid sequence on its fast photoelectric effect has been investigated. We developed a new system that permits the expression of mutant BRs in *Halobacterium halobium* instead of *E. coli*. We then used an electrochemical method to evaluate their photoelectric responses. Contrary to the conventional method of exponential analysis, this method permits unequivocal assignment of the signal components to specific underlying molecular processes. Our preliminary results suggest that it is possible to use site-directed mutagenesis to custom-tailor the electrical characteristics of mutant BRs for applications in molecular electronics.

Bacteriorhodopsin (BR), a purple membrane-bound protein from *Halobacterium halobium* has been widely used for the construction of prototype molecular devices. Successful photonic applications based on the use of a genetic mutant of bacteriorhodopsin[1] or a bacteriorhodopsin with a substituted synthetic chromophore[2] have been reported. Prototype molecular devices based on the fast photoelectric effect of wild type BR have also been reported.[3,4] In this paper, we examine the effect of point mutations of BR on its fast photoelectric effect. Such a study has been previously reported by Holz et al.[5] Our approach differs in two ways. First, mutant BRs are expressed in *H. halobium* rather than *E. coli*. Second, we use an electrochemical method that permits the unequivocal assignment of signal components to the specific underlying molecular processes, whereas the previous approach decomposed the signal into several exponential terms.

The fast photoelectric effect is caused by the pulsed light-induced transient charge movement within the BR molecule or at the membrane-water interface. It has been previously shown that the photocurrent signal contains three components, called B1, B2, and B2′.[6] The B1 component is due to the light-induced charge separation within the BR molecule. The B2 and B2′ components are due to proton uptake and release at the cytoplasmic and extracellular surfaces, respectively.

We used two different methods to reconstitute BR membranes on a Teflon substrate. The first method allows us to observe the B1 and B2 components; the second method exhibits only the B1 component. The B2′ component is eliminated, since the extracellular aqueous phase is shielded by the Teflon film. We previously found that the B1 component is pH-independent, but the B2 component exhibits a progressive change with respect to varying pH. Four mutant BRs, each having a single point mutation (aspartic acid → asparagine) at residues 85, 96, 115, and 212, were examined.

EXPERIMENTAL

The mutant BR's studied were expressed in *H. halobium* strain L33, a strain in which the gene encoding BR was disrupted by the integration of the insertion element ISH2 (i.e., *bop*::ISH2). The expression plasmids used were derivatives of plasmid p319 described by Ni et al.,[7] but lacking the *brp* gene. Transformation was also performed according to Ni et al.,[7] except that the transformants were plated on solid plates containing sucrose and mevinolin rather than on soft overlay agar plates, and the concentration of mevinolin was decreased to 5 μg/ml. Transformants expressing a mutant BR were grown in the absence of mevinolin for the preparation of membranes by a linear sucrose gradient; yields were comparable to those seen in wild type strains.

The bacteriorhodopsin was reconstituted onto a thin (6.35μm) Teflon film in one of two ways. In the first method, an aqueous suspension of BR membrane fragments was placed on a Teflon film previously coated

*Correspondence to: F. H. Hong (Physiology), Wayne State University, Detroit MI 48201.

with a layer of hexadecane/heptane solution, according to the method of Trissl and Montal.[8] In the second method, an electrolyte suspension of BR membrane fragments was placed on a Teflon film previously coated with an octadecylamine/azolectin mixture, and left to dry for about 4 days, according to the method of Okajima and Hong.[9] A single layer of bacteriorhodopsin molecules was placed on the Teflon film by the first method; multiple layers were deposited by the second method. In both types of thin films, the BR molecules were oriented with the extracellular side toward the Teflon. The prepared film was mounted in a Lucite chamber with the Teflon divided into two separate compartments, representing the extracellular and cytoplasmic phases. A platinum electrode was immersed in each aqueous phase. The electrodes were connected to a current amplifier, as previously described.[9] The light source was a microsecond light pulse from a flash lamp-pumped dye laser (590 nm).

RESULTS AND DISCUSSION

The photocurrent signal of the Trissl-Montal film reconstituted with the D212N mutant showed no changes as the pH varied from 5 to 11, which suggests the absence of a B2 component. These signals were found to superimpose the signal of the corresponding multilayer film after normalization. We conclude that in this pH range, the B2 is absent from the D212N mutant.

At pH levels below 4, the signal did exhibit pH dependence. This dependence was more pronounced in 4M NaCl solution than in 0.1M NaCl solution. We do not believe that this enhancement effect was due to ionic strength, because the effect was absent in chloride-free solution (such as Na_2SO_4). In such a solution, there was no B2 component at pH 1.8, but addition of minute amounts of sodium chloride reproduced the pH-dependent signals. The pH-dependent portion of the photocurrent at low pH is due probably to interfacial chloride ion transfer.

The photocurrent signals of films reconstituted with the mutants D85N, D96N, and D115N showed distinct patterns of pH dependence. These patterns were more complex; further investigation is needed.

CONCLUSIONS

Our D212N BR mutant data provide further experimental support for the idea that the decomposition of the photosignal into pH-dependent and pH-independent components is not arbitrary. Instead, the B1 and B2 components are most likely natural entities corresponding to two separate molecular processes. Our preliminary results with the mutant BRs suggest that it is possible to custom-tailor the photoelectric properties of bacteriorhodopsin for use in the construction of molecular electronic devices by site-directed mutagenesis.[10]

REFERENCES

1. N. Hampp, C. Bräuchle, and D. Oesterhelt, *Biophys. J.* 58: 83-93, 1990.
2. N. N. Vsevodolov, A. B. Druzhko, and T. V. Djukova, in F. T. Hong, Ed., *Molecular Electronics: Biosensors and Biocomputers,* New York: Plenum Press, 1989, 381-384.
3. D. Haronian and A. Lewis, *Appl. Optics* 30: 597-608, 1991.
4. T. Miyasaka, K. Koyama, and I. Itoh, *Science* 255: 342-344, 1992.
5. M. Holz, L. A. Drachev, T. Mogi, H. Otto, A. D. Kaulen, M. P. Heyn, V. P. Skulachev, and H. G. Khorana, *Proc. Natl. Acad. Sci. USA* 86: 2167-2171, 1989.
6. F. T. Hong, in M. Blank and I. Vodyanoy, Eds., *Biomembrane Electrochemistry,* Washington: American Chemical Society (in press).
7. B. Ni, M. Chang, A. Duschl, J. Lanyi, and R. Needleman, *Gene* 90: 169-172, 1990.
8. H.-W. Trissl and M. Montal, *Nature* 266: 655-657, 1977.
9. T. L. Okajima and F. T. Hong, *Biophys. J.* 50: 901-912, 1986.

10. Partial funding of this project was provided by ONR contract N00014-87-K-0047 to FTH, NSF grant DMB-9007365 to RBN, and Naval Surface Warfare Center contract N60921-91-M-G761 to FTH and RBN.

Martin Blank, Ed., *Electricity and Magnetism in Biology and Medicine*

EMF EFFECTS AND TRANSPORTATION: RESEARCH DESIDERATA

I. Gyuk*

U.S. Department of Energy, Washington, D.C.

Biological Effects: The Background

In recent years it has become increasingly clear that electric and magnetic fields (EMF) of low frequency (<150 Hz) and fairly low intensity (<1 Gauss) can have biological effects, usually under carefully controlled laboratory conditions. Correspondingly, there has also been increasing concern that these biological effects might result in health effects.

Although the energy involved in electric and magnetic fields of extremely low frequency is quite small, there is a broad consensus that there are indeed biological effects due to these fields. The locus of cell-field interaction is apparently the cell membrane. In particular, it is the flux of calcium ions and other biologically important messengers that appears to be altered by the presence of EMF. Following changes at the cell surface, the subsequent cascade of intracellular processes, such as production of ornithine decarboxylase, can be altered as well.[1] Although there is not enough energy for direct interaction with DNA, gene expression and protein synthesis are apparently affected.[2] Melatonin is an important hormone linked to diurnal rhythms of the body and regulating most other hormone systems. Even short exposure of rodents to EMF may affect the nightly rise of production of this hormone.[3] Sheep, baboons, and humans may be more resistant. There is also some indication that EMF may affect the immune system. Although experimentation with humans must proceed with extreme caution, it has been shown that human heart rate is slightly lowered by mixed EMF.[4]

Unlike ionizing radiation and most chemical factors, the effect of EMF at low frequencies appears to be highly nonlinear. In fact, there seem to be "windows" in intensity range and possibly even resonance-like phenomena. Intermittency, transients, and possibly the geomagnetic field may play a role. The dose-effect relationship is little known thus far.

Observed effects are to be considered as biological effects only; they do not constitute actual health effects. Nonetheless, concern is focused by these studies on potential risks of abnormal cell growth, reproductive issues, and neurophysiology. Epidemiological studies have not yielded conclusive evidence of health hazards. Instead, there appears to be a linkage of diseases such as childhood leukemia with exposure surrogates such as wiring codes. Risk factors appear to have values of about 2.0, barely out of the noise level.

Interest and attention of potential health effects of EMF have largely been focused on power transmission lines. However, more recently, electrical transportation modes have also come under scrutiny, particularly novel systems such as magnetically levitated (maglev) trains. The question of potential health effects of these systems deserves serious consideration. It therefore becomes important to formulate a research agenda specifically addressing electrical transportation modes.

Measurements and Exposure Assessment

Since we do not really know which parameters of the fields are biologically effective, and what range of values is relevant, evaluation of the electric and magnetic field environment becomes rather difficult. Nonetheless, by use of simple metrics, extensive exposure measurements for utility workers are being undertaken.

Thorough measurements of EMF characteristics of a variety of transport modes needs to be performed, and must include intensity ranges, frequency distribution, and timing features, both under standard operating conditions as well as during special conditions such as acceleration, braking, and idling. Preliminary work has revealed that most transport modes have two characteristics that differ considerably from those of transmission and distribution lines: fields are highly intermittent and show a very broad range of contributing frequencies. Detailed measurements on the German maglev system and a number of other rail systems are

*Correspondence to: I. Gyuk, U.S. Department of Energy, CE-141, Washington DC 20585.

under way.[5] Unexpected results emerge at times, such as a measured 270Hz pulsed DC field with an average 150 G at floor level on the Washington Metro.[6]

In addition to source characterization, it is also important to determine the actual exposure dose of transport workers when realistic operations are considered.

Replication of Key Biological Experiments

Although we are far from understanding the biology of cell-field interaction, a number of results have emerged reliably showing the effect of field exposure at low frequencies. In many cases exposure is in the vicinity of 60 Hz, since this is the dominant power frequency. Most of this work should be repeated under exposure conditions similar to various transportation modes (e.g., a broad frequency spectrum, characteristic transients, and intermittency). Among prime candidates are cellular phenomena, gene transcription experiments, melatonin studies, and human cardiac function. If replication of key experiments under transportation EMF exposure conditions proves to be negative, the likelihood of health effects becomes small.

Epidemiology of Electric Transport Workers

Some 50 occupational studies on workers in "electrical" professions have been performed or are currently under way. Unfortunately, most of these studies rely on simple job classification rather than detailed measurements of the field exposure. Elevated risks of diseases such as leukemia, brain tumor, and male breast cancer have been identified in a variety of professions. Values of risk factors are often 2 to 3 times higher than for residential studies.

Few studies have been performed thus far for electrical transport workers. There is a Norwegian study that revealed increased risk of male breast cancer;[7] and a Swiss study, which showed an excess of leukemia and lymphoma.[8] But the Norwegian study dealt with very small numbers, and the Swiss results depend on the group of nontransport workers used for comparison. Special epidemiological studies should be performed on electrical transport workers ranging from locomotive engineers to airline pilots.

Evaluation of Shielding Options

Strictly speaking, mitigation of field effects should only be undertaken after definite health effects have actually been proven. Furthermore, the dose-effect relationship should be determined, so that one knows what to mitigate against. However, there is already sufficient public concern, so that various mitigation options should be considered and implemented where it can be done with a modicum of expense. Excessive field gradients and transients pose a special problem here.

First among the design options is configuration. Layout of trains and other transportation vehicles should be arranged to minimize exposure of human occupants. Passive shielding is effective, although it carries a heavy weight penalty. Bucking coils can be employed in a number of situations. Finally, whole new classes of materials, based on thin film technologies, await development. Such materials would be extremely valuable, not only for biological reasons, but also for standard electromagnetic interference problems.

Conclusions

Whether one is dealing with trains, planes, or automobiles, electric and magnetic fields are ubiquitous in transportation. The question of potential health effects deserves serious consideration. This is particularly the case when one is dealing with novel systems such as maglev, electronic guideways, electric cars, or magnetohydrodynamically propelled ships.

REFERENCES

1. C. V. Byus, S. E. Pieper, and W. R. Adey, "The effect of low energy 60Hz environmental electromagnetic fields upon the growth-related enzyme ornithine decarboxylase," *Carcinogenesis* 8: 1385-1389, 1987.

2. R. Goodman and A. Henderson, "Transcription in cells exposed to extremely low frequency electromagnetic fields: A review," *Bioenergetics* 25: 335-355, 1991.

3. B. W. Wilson, R. G. Stevenson, and L. E. Anderson, Eds., *Extremely Low Frequency Electromagnetic Fields: The Question of Cancer,* Columbus, Ohio: Battelle Press, 1990.

4. M. R. Cook, C. Graham, H. D. Cohen, and M. M. Gerkovitch, "A replication study of human exposure to 60 Hz fields: Effects on neurobehavioral measures," *Bioelectromagnetics* (in press).

5. F. Dietrich and W. Feero, *Magnetic Field Testing of the TR07 Maglev Vehicle and System,* Final Report, DOT-VNTSC-FRA-92-6.1, 1992.

6. E. Mantiply, private communication (1992).
7. T. Tynes and A. Anderson, ''Electromagnetic fields and male breast cancer,'' *Lancet* 1596, 1990.
8. M. Balli-Antunes, D. H. Pfluger, and C. E. Minder, ''The mortality from malignancies of the haematopoietic and lymphatic systems among railway engine drivers,'' *Environmetrics* 1: 121-130, 1990.

Martin Blank, Ed., *Electricity and Magnetism in Biology and Medicine*

REVIEW OF EMF HEALTH AND SAFETY ISSUES AND FINDINGS ASSOCIATED WITH ELECTRIFIED TRANSPORTATION SYSTEMS

R. B. Goldberg* and W. A. Creasey

Information Ventures, Inc., Philadelphia

K. R. Foster

University of Pennsylvania, Philadelphia

The TR07 magnetic levitation (maglev) produces AC magnetic fields similar in intensity to common electrical devices (<100 mG), but with more AC components from 2 to 45 Hz and a complex variation in frequency and intensity as a function of train speed. We reviewed the EMF bioeffects research literature relevant to maglev systems using the Information Ventures *EMF Database*. The limited results available indicate that magnetic fields comparable in strength to maglev exposures can produce a range of biological effects, but their relevance to potential human health hazards remains speculative.

The introduction of new large-scale electronic technology such as maglev rail transport raises concerns over possible human health effects resulting from electromagnetic-field (EMF) exposures of workers and the general public. We have been reviewing the scientific literature on EMF bioeffects for the U.S. Department of Transportation's Volpe National Transportation Systems Center (VNTSC) to identify studies that might be relevant to maglev fields and provide some basis for making preliminary recommendations of the possible human health impact of this technology. We outline our approach to the problem and some preliminary conclusions.

THE PROBLEM OF ELF EMF RISK ASSESSMENT

Traditional risk assessment applied to maglev EMF exposures would involve several steps: (1) defining a hazard, (2) establishing a dose-response relationship for that endpoint, (3) establishing the exposure from the new technology, and (4) projecting the potential health impact. Well-defined EMF hazards do exist (such as electric shock, pacemaker interference, and burns), but they require fields of much higher intensity than personnel are likely to encounter with maglev. However, there is considerable concern over possible hazards from chronic exposure to relatively weak extremely low frequency (ELF) magnetic fields. Endpoints of concern include several types of cancer, reproductive problems, and behavioral effects. The epidemiologic studies that have raised these concerns are inconclusive and difficult to interpret for reasons that have been widely discussed;[1,2] there have been few large-scale animal screening studies. There is thus little basis for risk assessment. Nevertheless, it is possible to compare magnetic-field exposure of maglev with exposure to other systems. In addition, we have searched the literature for other relevant studies, including epidemiologic studies of workers in existing rail systems (who are also exposed to considerable ELF magnetic fields) and basic biological studies that might indicate possible mechanisms for hazard.

EXPOSURE ASSESSMENT OF THE TR07 MAGLEV TRAIN (GERMANY)

The only detailed measurements of a maglev system available to us were made by Electric Research & Management, Inc. (ERM), at the Emsland Transrapid Test Facility, described in part in the present volume.[3,4] AC magnetic-field emissions from the TR07 maglev system occur primarily in the ELF range (0-300 Hz) with more intermittent (transient) and intense exposures in the 2-45Hz range than is typical of power

*Correspondence to: R. B. Goldberg, Information Ventures, Inc., 1500 Locust St. (Suite 3216), Philadelphia PA 19102-4321.

frequency (50 or 60Hz) exposures. Total ELF EMF field intensities are usually less than 100 mG at floor level and under 20 mG at head level for passengers, with occupational exposures slightly higher in intensity and richer in higher-frequency components. The frequency components vary with vehicle speed. The DC magnetic field is within 400 mG of geomagnetic-field levels at the maglev train floor, and within about 100 mG of ambient levels at head height. These levels fall well within existing international exposures guidelines in terms of time-averaged field intensity, the basis for current exposure standards.

REVIEW OF THE RELEVANT EMF RESEARCH LITERATURE

Our review considers three general classes of studies relevant to concerns raised by maglev: (1) epidemiologic and occupational-health studies involving railway and other workers with exposures analogous to anticipated maglev exposures, (2) animal screening studies, and (3) basic studies of ELF EMF bioeffects that suggest a potential interaction with maglev fields.

Information Ventures, Inc., staff members have prepared analytical summary abstracts of the diverse interdisciplinary EMF research literature that consistently include details of experimental design and results not commonly found in the author's abstract. We have used this material to establish a unique computer database employing retrieval software that makes it possible for any term used in the abstract text or citation to serve as the basis for retrieval. This *EMF Database*, which at present contains more than 12000 analytical abstracts, is a particularly effective tool for asking new questions of old literature. For example, using the truncated search term "rail" we rapidly identified a number of epidemiologic studies that included railroad workers within study groups.[5-9] Some of these studies subsume rail workers among other "electrical workers," and the very limited data suggest that they have a modest increase in cancer risk (approximately doubled) as reported by several authors for other occupational groups with presumed ELF EMF exposure. Interpretation of these reports is limited by small sample sizes, unreliable exposure estimates, and the possibility of confounding factors.

Few large-scale animal toxicology studies of ELF EMF fields have been completed that include a magnetic-field component. Several animal studies are still in progress, including some that test for any copromotional effects of ELF EMF.[10] Behavioral studies with baboons suggest possible stress and disruption of learning and task performance with chronic moderate ELF electric-field exposure, and modification of the response by simultaneous exposure to ELF magnetic-field components.[11]

ELF EMF bioeffects studies are much more diverse and harder to interpret in terms of potential health hazards than toxicology studies. Some work is just getting under way to evaluate maglev-type fields directly; the results of one in vitro study are reported in the present volume.[12] Low-intensity 16Hz magnetic fields have been reported to cause altered calcium flux in brain tissue.[13] There are reports of altered mobility of diatoms,[14,15] but other groups were unable to replicate these results.[16] Disruption of the nocturnal rise in melatonin levels by a sudden inversion of the DC magnetic fields in rats[17] and by exposure to 60Hz AC magnetic fields in hamsters[18] have been reported. Time-varying magnetic fields have been reported to produce teratological effects in early chicken embryos (reviewed by Berman[19]), although the response is not always replicated and the effects do not show a clear pattern related to dose. Finally, cells in vitro have shown transitory alterations in patterns of mRNA synthesis in response to AC magnetic fields at 45 and 60 Hz.[20] The implications of these effects are not clear and any extrapolation to risk is unwarranted.

CONCLUSION

The ELF magnetic-field exposure from the TR07 maglev system is comparable in amplitude to other sources of ELF EMF exposure, but with a different frequency spectrum. More dosimetry data are needed, for the proposed American maglev system as well as from existing electrified rail systems. There is no clear evidence of hazard with the fields from this system, but many of the concerns that have been raised about power-line fields can be extended to maglev fields as well.[21]

REFERENCES

1. W. A. Creasey and R. B. Goldberg, Electric Power Research Inst. Research Project 2965-12, Final Report, EPRI Publication EN-6674, 1989.

2. K. R. Foster, *Health Physics* 62: 429-435, 1992.

3. F. M. Dietrich, W. E. Feero, D. C. Robertson, and G. A. Steiner, *Annual Review of Research on Biological Effects of 50 and 60 Hz Electric and Magnetic Fields*, Milwaukee: DOE, 1991, P-23.

4. F. M. Dietrich, D. C. Steiner, D. C. Robertson, and W. E. Feero, "Review of EMF health and safety issues and findings associated with electrified transportation systems," this volume.

5. P. Baroncelli et al., *Am. J. Ind. Med.* 10: 45-55, 1986.

6. J. R. Wilkins III and V. D. Hundley, *Am. J. Epidemiol.* 131: 995-1008, 1990.

7. M. R. Spitz and C. C. Johnson, *Am. J. Epidemiol.* 121: 924-929, 1985.

8. R. S. Lin, P. C. Dischinger, J. Conde, and K. P. Farrell, *J. Occup. Med.* 27: 413-419, 1985.

9. T. Tynes and A. Andersen, *Lancet* 336(8730): 1596, 1990.

10. J. R. N. McLean et al., *Bioelectromagnetics* 12: 273-287, 1991.

11. A. M. Coelho Jr., S. P. Easley, and W. R. Rogers, in Ref. 3, A-32.

12. K. R. Groh, C. B. Chubb, F. R. Collart, and E. Huberman, this volume.

13. C. F. Blackman, B. W. Wilson, R. G. Stevens, and L. E. Anderson, Eds., *Extremely Low Frequency Electromagnetic Fields: The Question of Cancer*, Columbus, Ohio: Battelle Press, 1989, 187-208.

14. J. A. Reese, M. E. Frazier, J. E. Morris, R. L. Buschbom, and D. L. Miller, *Bioelectromagnetics* 12: 21-25, 1991.

15. A. R. Liboff, B. R. McLeod, and S. D. Smith, in Ref. 13, 251-289.

16. W. C. Parkinson and G. L. Sulik, *Annual Review of Research on Biological Effects of 50 and 60 Hz Electric and Magnetic Fields*, Denver, Colo.: DOE, 1990, P-5.

17. R. J. Reiter et al., *ibid.*, A-30.

18. S. M. Yellon, in Ref. 3, A-25.

19. E. Berman, *Reproductive Toxicology* 4: 45-49, 1990.

20. R. Goodman, L.-X. Wei, and A. Henderson, in Ref. 3, A-18.

21. The work discussed in this paper was supported in part by VNTSC under Order DTRS57-91-P-81743.

Martin Blank, Ed., *Electricity and Magnetism in Biology and Medicine*

REVIEW OF EMF BIOEFFECTS LITERATURE RELATIVE TO MAGLEV-GENERATED MAGNETIC FIELDS

B. W. Wilson,* R. J. Reiter, and A. A. Pilla

Battelle Pacific Northwest Laboratory, Richland, Wash.

Introduction of magnetically levitated (maglev) trains into the U.S. transportation sector is being planned at a time when uncertainty exists regarding the possibility of adverse health effects from exposure to power-frequency (and its associated broadband) electromagnetic fields (EMF). To assess the state of knowledge relative to anticipated EMF exposures from electrically operated transportation systems, including light rail and maglev, we have reviewed laboratory and clinical studies concerning biological effects of EMF exposure. Epidemiologic studies, including several on electrical transport workers, were also considered. A primary objective of the review was to consider, based on present knowledge, the potential health implications for maglev-associated electric and magnetic fields. Special attention was given to the broadband-frequency characteristics of magnetic fields and their possible importance in the biological effects observed from EMF exposure. This short paper highlights what we believe are the main points contained in a full reported entitled *Broadband Magnetic Fields: Their Possible Role in EMF-Associated Bioeffects*.

Although work remains to be carried out in characterizing magnetic fields produced by existing maglev systems, it is clear that maglev operation will produce magnetic fields with main component frequencies other than 60 Hz along the guideways and in the vehicle itself. Measurement data from one demonstration vehicle shows that its magnetic fields will vary in frequency from 0 to 250 Hz, with the main components at less than 15 Hz. In measurements taken to data on this system (the German Trans-Rapid Vehicle TR-07), maximum levels in the passenger compartments were 300-500 mG, measured near the floor.[1,2] The report by Dietrich et al. elsewhere in this volume[1] also provides information on the power distribution of the magnetic fields by frequency.

Table 1 lists maximum measured or estimated magnetic fields that have been reported for the passenger compartments or train crew areas for several existing electrically powered rail systems. Also listed are the rail systems for which epidemiologic or bioeffects data have been reported in the literature. As can be seen from the table, among those systems for which magnetic field data are reported, the maglev vehicle TR-07 has the lowest maximum magnetic field flux density in the passenger compartments or crew areas.

TABLE 1.—Listing of electrically operated mass transit systems for which magnetic-field flux densities, epidemiologic data, or bioeffects laboratory results have been reported. Note that the maglev measurement data indicate that passenger and train crew compartment levels are lower than for other systems measured.

System	Magnetic fields	Bioeffects/epidemiology	Reference
British Rail	Measured, 25 G 0.5 m above floor	None reported	3
German Maglev TR-07	Mostly below 0.5 G[1]	Effects on transformed cells at 2 G (preliminary)	4
Swiss National Rail System	Calculated >10 G at 16.7 Hz	Excess leukemia/lymphoma	5
Norwegian Municipal Tram	No measurements available	Excess male breast cancer	6
American Light Rail (Washington, D.C., Metro)	Measured > 10 G in some cars near floor	None	7

*Correspondence to: B. W. Wilson (P8-01), Battelle Pacific Northwest Laboratory, Box 999, Richland WA 99352.

Figure 1 illustrates approximate relationships between imposed magnetic field flux density, induced electric field, and induced currents for a human-sized object. Flux densities or induced current for biological effects over which there is no controversy (visual phosphenes and electrically induced convulsions) are indicated in this figure. Field strengths at which other effects have been reported are shown, along with the approximate corresponding magnetic flux densities represented by a number of magnetic-field sources, including those reported for the TR-07 maglev vehicle.

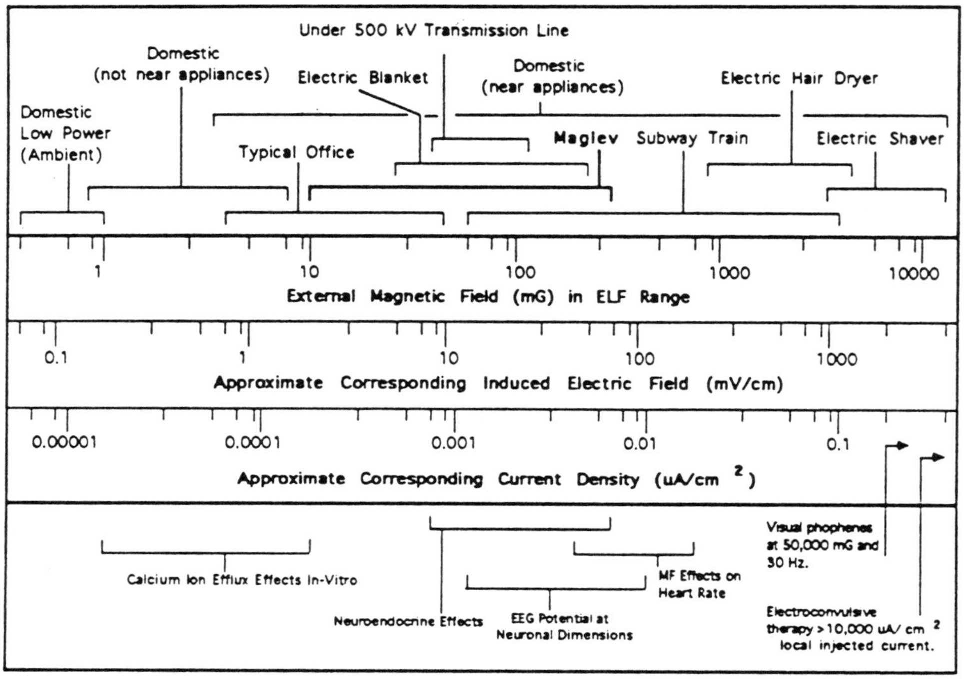

FIG. 1.—Range of magnetic-field flux densities measured for a number of commonly encountered sources are shown together with their approximate corresponding induced electric fields and resulting induced currents, as calculated for human-sized model. Also indicated are approximate ranges for reported biological effects. (This illustration is not intended to imply that biological effects listed will necessarily occur at the field levels indicated. Other field parameters, including frequency, are important in determining whether many of these effects can be observed.)

Animal studies have been carried out to address the areas of interest as raised by results from epidemiologic studies. These studies have focused attention on possible increased risk for cancer, miscarriage, and depression.[8] Well-done large-population studies in rodents have shown no effects of electric or magnetic field exposure on reproductive outcome. There is some evidence from animal studies that EMF exposure may affect short-term memory[9] and there is limited evidence from laboratory studies with humans that is consistent with effects on central nervous system (CNS) function.[10] A few recent studies have suggested that EMF exposure may enhance the development of cancer in animals that have been treated with chemical carcinogens.[11,12]

On the cellular level, there is a growing consensus that the cell membrane is the site of action for the fields. It appears that signal transduction occurs at the cell membrane, and that many of the effects observed in studies with cultured cells may be a consequence of changes in membrane-bound ion channels or receptors.[13] Important in detection are organized biological structures that are electrically connected to facilitate cell-to-cell communication. It is in consideration of these larger, more complex biological structures (compared with individual isolated cells) that the evidence and theoretical underpinnings for a biological response to EMF become strongest.

At the subcellular level there is interest in EMF effects on gene regulation. Although there are no reliable reports of direct chemical damage to DNA from magnetic field exposure, several laboratories have reported alterations in the gene transcription (e.g., Goodman et al.[14]). EMF-induced changes in the regulation of genes, while not constituting a mutation or heritable change, could play a role in the expression of oncogenes or other genes that code for key channels or receptor proteins and thereby affect cancer risk.

There is growing recognition that the pineal gland and its principal hormone, melatonin, are likely to play an important role in communication among the neurological, endocrine, and immune systems. Alterations in the melatonin circadian rhythm have been linked, through animal or human studies, with each of the broad areas of interest arising from EMF epidemiologic studies: cancer, reproduction, and mood disorders.[15] EMF effects on neuroendocrine function are an example of how an observed biological effect (e.g., reduction in pineal melatonin synthesis) may be related to a health effect (e.g., increased breast cancer risk in males), as suggested by the epidemiologic studies.

Based on available data, we suggest that any demonstrated beneficial or possible adverse biological effects are most likely to be determined by the broadband characteristics of electromagnetic fields. Fields exhibiting high time rates of change, including pulses and other transient phenomena, may be classified as broad-spectrum or broadband because of their high-frequency components. Likewise, intermittent field exposure is included in the broadband definition because of low-frequency components, including those below 1 Hz.

It appears that the currents induced by time-varying magnetic fields are associated more strongly with beneficial effects in humans, as observed in clinical studies. Neither various magnetic-resonance models, nor those postulating changes in ion motion or binding as a direct consequence of interaction with the magnetic component of the field appear as powerful as the induced-current models in explaining EMF effects in humans as observed in clinical studies or as postulated from review of the EMF epidemiological data.

Important in determining biological effects of EMF exposure is the reactivity of the biological system to the exogenous field. Many results to date, from both clinical and in vitro studies, suggest that biological systems are less reactive or unreactive to weak EMF unless shifted from homeostasis by chemical stimuli, trauma, pathology, etc. Among examples of this apparent characteristic are the need for mitogen stimulation in cellular responses involving calcium uptake and the fact that pulsed EMF stimulates bone growth only in cells at or near the fracture interface. Also consistent with this view is the relatively stronger evidence for a copromotional effect than for a promotion-only effect in the development of cancers in chemically initiated laboratory animals.

Literature suggests that the following physical characteristics determine the amplitude and spectral density of the induced electric field at the target: (1) maximum time rate of change of the magnetic field (dB/dt) determines the peak E field at the target; (2) pulse shape determines the field's spectral or frequency content which may allow the dielectric properties of the target to be matched; and (3) the relative configurations of the EMF (e.g., coil position, polarization, etc., in bone-growth stimulation applications) and of the target determine which part of the target receives the peak induced electric field.

Characteristics of the EMF signal on which the reactivity of biological system may depend are: (1) the periodicity or repetition rate which accommodates refractory times in the responding biochemical pathway(s), (2) the total exposure time in one 24h period, relating to such processes as cell cycle, clock time or circadian phase, and (3) the maximum size of the induced current loop, which determines the peak induced electric field.

Characteristics of biological systems that affect their sensitivity to imposed EMF include: (1) presence and function of gap junctions (the cell array tissue model clearly shows that cell-cell communication via electrically conducting gap junctions increases EMF sensitivity by several orders of magnitude vs a single cell exposed to the same EMF source), and (2) the biochemical and physiological state of the target (the detected EMF signal is likely to result in a bioeffect if the target is not in a resting or maintenance state).

One objective of this review was to consider whether, based on present knowledge, exposure to maglev-associated magnetic fields would warrant any special consideration with regard to possible adverse health effects. Based on currently available occupational epidemiologic data for electrical-transport workers, available clinical and laboratory data on EMF biological effects, and available information on maglev-generated magnetic fields as compared with those reported for existing systems, we conclude that EMF-associated risks from maglev, if any, are not likely to be greater than those from electrical-transport systems already in use.

REFERENCES

1. F. M. Dietrich, G. A. Steiner, D. C. Robertson, and W. E. Feero, this volume.

2. ERM, *Final Report on Magnetic Field Testing of TR07 Maglev Vehicle and System,* prepared by Electric Research and Management for the Federal Railroad Administration under Contract DTFR53-91-C-00047, 1992. (Available through NTIS.)

3. P. J. Chadwick and F. I. Lowes, "Magnetic fields from transport systems in the UK" (Abstract), presented at First World Congress on Electricity and Magnetism in Biology and Medicine, Lake Buena Vista, Fla., 1992.

4. K. R. Groh, F. R. Chub, F. R. Collart, and E. Huberman, this volume.

5. M. Balli-Antunes, D. H. Pfluger, and C. E. Minder, "The mortality from malignancies of the haematopoietic and lymphatic systems (MHLS) among railway engine drivers," *Environmetrics* 1: 121-130, 1990.

6. T. Tynes and A. Anderson, "Electromagnetic fields and male breast cancer," *Lancet* 1596, 1990.

7. E. Mantiply et al., 1992.

8. E. F. Walborg, "Extremely low frequency electromagnetic fields and cancer: Focus on tumor initiation, promotion, and progression," Report prepared for National Electric Manufacturers Association (NEMA), 1991.

9. R. H. Lovely, D. L. Miller, and L. E. Anderson, "Assessment of rats' behavior in a radial arm maze during exposure to magnetic fields" (Abstract), *13th Ann. Mtg. BEMS*, Salt Lake City, 1991.

10. G. B. Bell, A. A. Marino, A. L. Chesson, and F. A. Struve, "Human sensitivity to weak magnetic fileds," *Lancet* 338: 251, 1991.

11. J. R. N. McLean, M. A. Stuchly, R. E. J. Mitchel, M. Goddard, and D. W. Lecuyer, "Cancer copromotion in the mouse skin model by 60-Hz magnetic fields: Tumor development and immune response" (Abstract), *13th Ann. Mtg. BEMS* Salt Lake City, 1991.

12. D. Sh. Beniashvili, V. G. Bilanishvili, and M. Z. Menabde, "Low frequency electromagnetic radiation enhances the induction of rat mammary tumors by nitrosomethyl urea," *Cancer Letters* 61: 75-79, 1991.

13. W. R. Adey, "Physiological signaling across cell membranes and cooperative influences of low-frequency electromagnetic fields," in H. Fröhlich, Ed., *Biological Coherence and Response to External Stimuli*, Heidelberg: Springer-Verlag, 1988, 148-170.

14. R. Goodman, D. Weisbrot, U. Alun, and A. Henderson, "Transcription in *Drosophila melangaster* salivary gland cells is altered following exposure to low frequency electromagnetic fields: Analysis of chromosome 3R," *Bioelectromagnetics* 13: 111-118, 1992.

15. B. W. Wilson, R. G. Stevens, and L. E. Anderson, "Neuroendocrine mediated effects of electromagnetic-field exposure: Possible role of the pineal gland," *Life Sciences* 45: 1319-1332, 1989.

Martin Blank, Ed., *Electricity and Magnetism in Biology and Medicine*

EFFECTS OF MAGLEV-SPECTRUM MAGNETIC FIELD EXPOSURE ON CEM T-LYMPHOBLASTOID HUMAN CELL GROWTH AND DIFFERENTIATION

K. R. Groh,* C. B. Chubb, F. R. Collart, and E. Huberman

Argonne National Laboratory, Ill.

Exposure to magnetic fields similar to those produced by maglev vehicles (combined ac and dc components) was studied for the ability to alter cell and culture growth and chemically induced cellular differentiation processes in human CEM T-lymphoblastoid leukemia cells. A series of continuous and intermittent magnetic field (MF) exposures for varying lengths of time were tested at intensities up to 7-fold greater than that produced by the German TR07 maglev vehicle. Phorbol 12-myristate 13-acetate (PMA) or mycophenolic acid (MPA) were used to induce cell differentiation. Changes in cell number, morphology, and fluorescence expression of antigenic markers of differentiation were monitored. The results indicated that maglev-spectrum magnetic field exposure had little effect on culture growth characteristics, final culture cell titer, or chemically induced cellular differentiation under the conditions tested when exposed to maglev-spectrum magnetic fields up to 2 G, compared to chemically-treated but MF-unexposed cultures.

Many studies over the past 15 years have shown that extremely low frequency (ELF) electromagnetic (EM) fields can produce biological and physiological changes in biological systems.[1,2] Magnetic levitation (maglev) transportation systems offer many advantages over conventional systems in reliability, economics, energy savings, and reduced noise and environmental pollution, but the potential health impact on passengers and support personnel from exposure to the EM fields produced in the levitation and guidance of maglev vehicles must be examined. Since maglev EM field exposures are quite different in frequency range and duration from those from 60Hz EM fields in our environment, their effect on biological systems must be examined to determine how they compare to other EM field exposure effects prior to being installed for public use. In this study, a sensitive cell differentiation assay system was used to test the effects of maglev EM field exposure on a CEM T-lymphoblastoid cell line, which has demonstrated reduced immunological function due to 60Hz EM field exposure.[3]

MATERIALS AND METHODS

Exponentially growing CEM T-lymphoblastoid human leukemia cells were inoculated into 60mm Petri dishes at a concentration of 2×10^4 cells/ml in 10 ml of medium (RPMI 1640, GIBCO) containing 15% fetal calf serum, penicillin (100 units/ml), and streptomycin (10 μg/ml), and grown under standard conditions for 24 h (37 C, 85% RH, 8% CO_2). Duplicate plates were then chemically induced to differentiate by addition of either mycophenolic acid (MPA) or phorbol 12-myristate 13-acetate (PMA) at appropriate concentrations, or left untreated. The cultures were placed either in the control (no MF exposure) or in the nonferrous MF-exposure incubator (36 cm w × 55 cm h × 32 cm d). Previous environmental measurements and culture-growth experiments had demonstrated equivalent conditions in the two incubators. Duplicate plates were removed daily and counted by a hemocytometer. Horizontal continuous or intermittent MF coils (300 turns, 0.5 m radius) and a bipolar power supply (Kepco, BOP 50-2M). The maglev ac and dc component fields were simulated from actual Transrapid TR07 MF measurements made by Electric Research and Management, Inc.,[4] using a PC software program (LabWindows, National Instruments, Austin, Texas). At the end of exponential growth, plates were examined for morphological changes and fixed, and specific surface antigen immunofluorescent antibody (OKT3) reactivity was measured to determine the extent of culture differentiation under the various culture and MF-exposure conditions.[5]

---

*Correspondence to: K. R. Groh (Biological & Medical Res. Div.), Argonne National Laboratory, Argonne IL 60439.

RESULTS

Growth of CEM T-lymphoblastoid cells showed the same standard growth kinetics in both the control and the MF-exposure incubators, with no significant difference in growth rate or final cell titer (Fig. 1). Mycophenolic acid (MPA) at a concentration of 0.1 μM had no significant effect on growth rate or cellular differentiation in the presence or absence of intermittent or continuous MF exposure for 4 days.

PMA at a concentration of 3.0 μM inhibited normal exponential culture growth more effectively than MPA (Fig. 2). After 3 days in the culture medium, PMA had effectively inhibited further culture growth. Differentiation of cells in both the control and MF-exposure incubators increased from 0 to 20% after 4 days of incubation with PMA, which suggests that maglev MF exposure did not influence differentiation.

DISCUSSION

Electromagnetic field exposure has been linked in an associative manner with carcinogenesis, especially to the promotional step of carcinogenesis, but no cause and effect relationship has yet been established.[1] Carcinogenesis can be described as a process in which uncontrolled cellular replication and de-differentiation replaces normal growth and differentiation,[6] and ELF-EM fields have been shown to alter cellular replication.[7] Lyle et al. have shown that T-lymphocytic cells are metabolically changed and that their immune function is decreased by exposure to ELF-EM fields.[3,8]

The results of the present study suggest that maglev EM-field exposure up to 7 times the specific frequency and intensity produced by the Transrapid TR07 has no deleterious effect on the growth or differentiation of CEM T-lymphoblastoid cells, which have been shown to respond to EM-field exposure. Other MF exposures cannot be ruled out until additional exposure conditions are examined.[9]

REFERENCES

1. I. Nair, M. G. Morgan, and W. K. Florig, *Biological Effects of Power Frequency Electric and Magnetic Fields—Background Paper,* OTA-BP-E-53, Washington, D. C.: U.S. Government Printing Office, 1989, 104 pp.

2. "Magnetic Fields," *Environmental Health Criteria 69,* Geneva: World Health Organization, 1987, 197 pp.

3. D. B. Lyle, R. D. Ayotte, A. R. Sheppard, and W. R. Adey, "Suppression of T-lymphocyte cytotoxicity following exposure to 60-Hz sinusoidal electric fields," *Bioelectromag.* 9: 303-313, 1988.

4. *Final Report on Magnetic Field Testing of TR07 Maglev Vehicle and System,* prepared for the Federal Railway Administration by Electric Research and Management, Inc., P. O. Box 165, State College, PA, 1992, 2 volumes.

5. B. Ryffel, C. D. Henning, and E. Huberman, "Differentiation of human T-lymphoid leukemia cells into cells that have a suppressor phenotype is induced by phorbol 12-myristate 13-acetate," *Proc. Natl. Acad. Sci. USA* 79: 7336-7340, 1982.

6. E. Huberman, "The potential role of cell differentiation in carcinogenesis," in R. Langenbach et al., Eds., *Tumor Promoters: Biological Approaches for Mechanistic Studies and Assay Systems,* New York: Raven Press, 1988, 247-258.

7. S. M. Ross, "Combined DC and ELF magnetic fields can alter cell proliferation," *Bioelectromag.* 11: 27-36, 1990.

8. F. B. Lyle, X. Wang, R. D. Ayotte, A. R. Sheppard, and W. R. Adey, "Calcium uptake by leukemic and normal T-lymphocytes exposed to low frequency magnetic fields," *Bioelectromag.* 12: 145-156, 1991.

9. We wish to thank Dr. P. Klingensmith and C. Miller for technical support. This work was supported by an interagency agreement between the U.S. Departments of Transportation and of Energy.

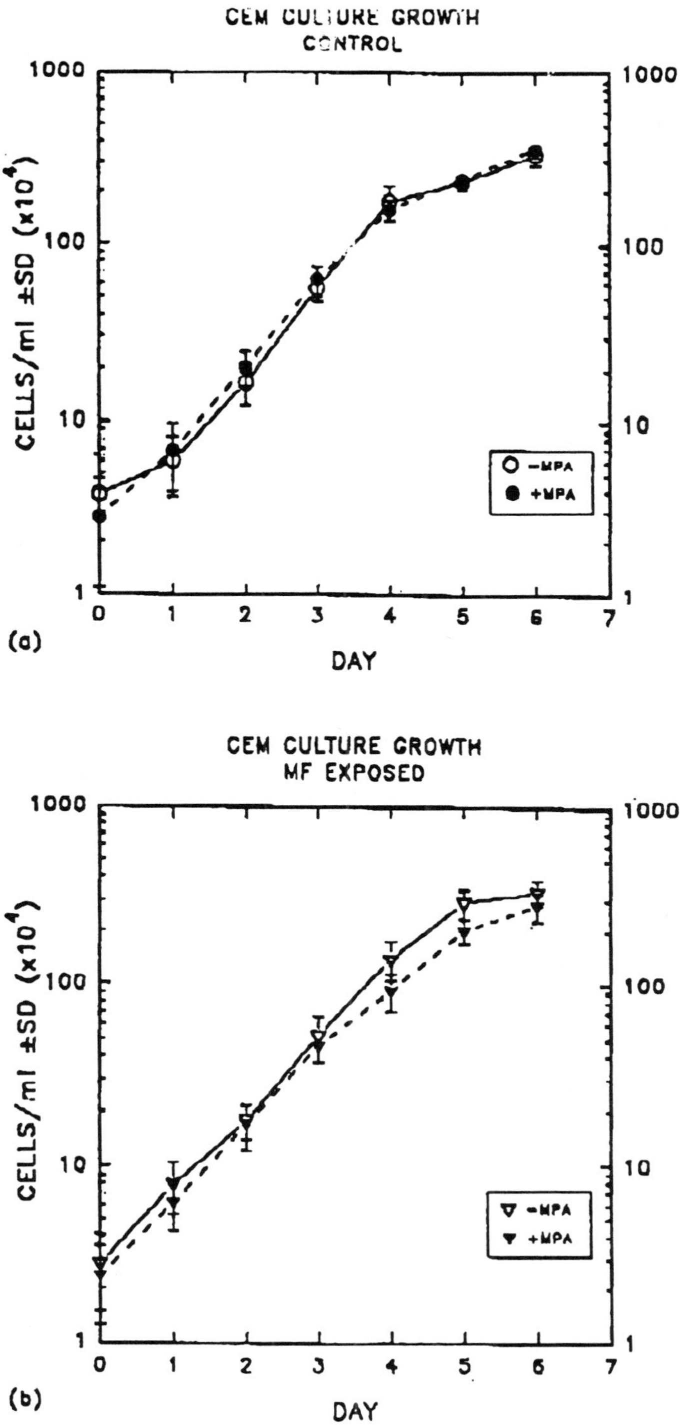

FIG. 1.—On day 0, MPA (final concentration, 0.1 μM) was added to 50% of the plates and incubated for 4 days under (a) control, (b) maglev MF-exposure. Culture growth was determined by daily hemocytometer counts, in duplicate. On day 4, samples were assayed for percent differentiation. ● MP-treated, O untreated.

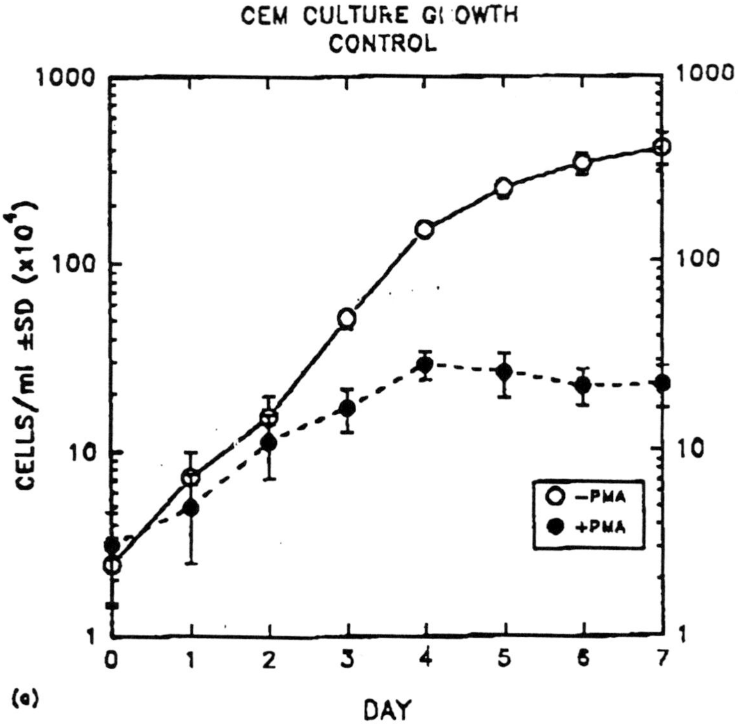

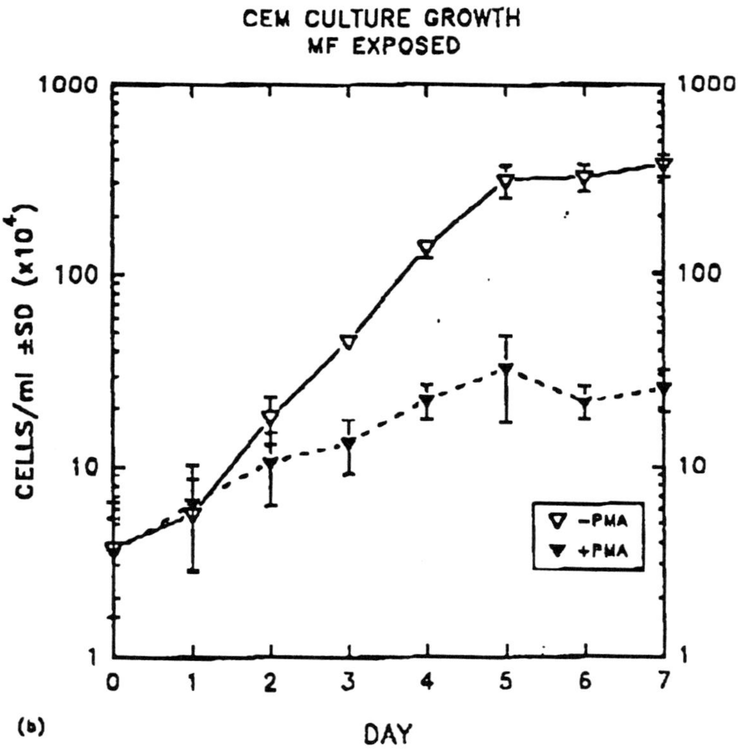

FIG. 2.—On day 0, PMA (final concentration, 3.0 μM) was added to 50% of the plates and incubated for 4 days under (a) control, (b) maglev MF-exposure. Culture growth was determined by daily hemocytometer counts, in duplicate. On day 4, samples were assayed for percent differentiation. ● PMA-treated, O untreated.

Martin Blank, Ed., *Electricity and Magnetism in Biology and Medicine*

EFFORTS OF FRENCH RAILROADS TO REDUCE TRACTION ELECTROMAGNETIC FIELDS

C. G. Gourdon*

French National Railroads (SNCF), Paris

This paper provides a railroad engineer's advice about reduction of the traction electromagnetic fields at the source, considered to be reasonable from the engineering viewpoint only by balancing of the power-supply installation. A description is given of the arrangement applied in France, in particular for high-speed (TGV) trains, the so-called "2×25kV" system, which is based on a power distribution at 50 kV between the traction wires and a return feeder, with autotransformers connected to them on the average every 10 miles, and giving the usual 25kV traction voltage. Engine filtering is also briefly discussed.

The French National Railroads (SNCF) have had no experience with the direct effects of low-frequency fields on health. However, they were faced with so many problems of coping with the health effects of adjacent telecommunications networks associated with railroads that its engineers have acquired a substantial experience in field reduction "at the source." A description of the remedies that can be used on railroads in case those fields are considered to be too high may be of interest. The following is a railroad engineer's point of view, to draw attention to the complexity of some solutions that come to mind but whose cost would be out of proportion to the environmental gain.

THE FRENCH TRACTION SYSTEM

In the 1950s the French National Railroads experimented with the "Aerotrain," wich was based on the hovercraft principle, with an air pressure sustained over a concrete inverted T-shaped track. They decided to stick with conventional trains, for reasons that are beyond the scope of this paper. The requisite increase in speed was obtained by the design of lighter trains and more powerful locomotives, and by the construction of new lines with "softer" curves (minimum radius 4000 m instead of 1000 m). The traction system basically conforms to Fig. 1. Each track is equipped with a catenary system comprising poles, a messenger wire, hangers, and a contact wire under which the locomotives collect the current by pantographs (Fig. 2). The catenary system is fed at 25 kV, 50 Hz by fixed substations, which are in turn supplied by the National Electricity Grid (usually at 90, 220, or 480 kV at 50 Hz). The current is returned to the substation via the rails and ground, which leads to voltages between ground and nearby metallic parts that may be dangerous either by electrostatic effects or by magnetic induction.

WHAT CAN BE ALTERED WITHOUT ENGINEERING NONSENSE

This system proved to be very satisfactory from the technical and economical viewpoints and is now used all over the world, including the USA (at the standardized frequency of 60 Hz). It seems to be out of the question to demand a basic change for environmental reasons. In fact, the voltage used is high enough to give lower currents (and hence more economic wiring) wires than the previous traction systems operating at 1.5, 3, or 15 kV, and low enough to keep the insulators, etc., simple. As for the fundamental frequency, adopting a value different from that of the National Grid would lead to an economic disaster for the railroads, by necessitating a dedicated primary grid instead of the present very simple voltage transformation from the National Grid. To impose such a system on the entire French electrified railroad network (8500 miles) would cost an estimated $4 billion, compared with $2.5 billion for the present system. No railroad could now afford such a surcharge.

---

*Correspondence to: C. G. Gourdon, SNCF, 162 rue de Saussure, F-75840 Paris Cédex 18, France.

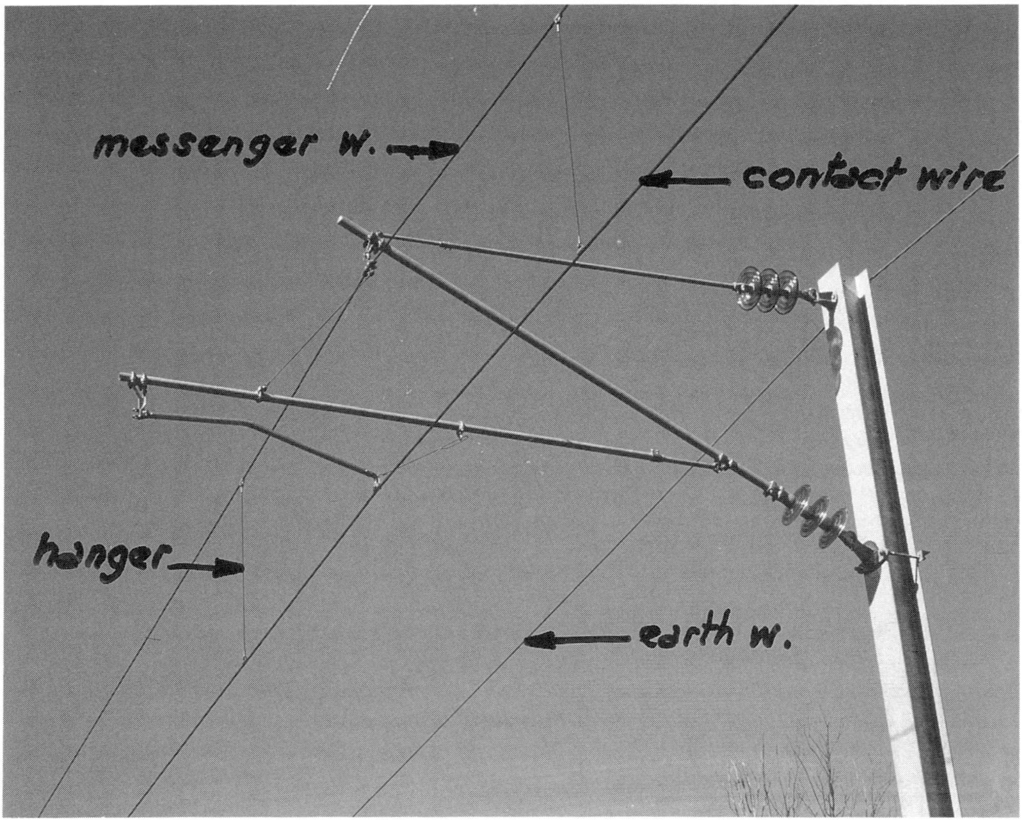

FIG. 1.—Typical French 25kV 50Hz catenary equipment.
FIG. 2.—Atlantic TGV set.

The French National Railroads chose to solve the difficult electromagnetic-compatibility problems simply by locomotive filtering (see below) and by a balanced mounting of the overhead system. This mounting, the "2×25kV" or "autotransformer" (AT) system is described below, following a short historical account of the circumstances that led to its adoption.

ELECTROMAGNETIC COMPATIBILITY (EMC) HISTORY AND STANDARDS IN FRANCE

EMC in France has always been a matter of (a) hazards to persons simultaneously touching the ground a a metallic wire or structure in the vicinity of the track, and (b) interference with telecommunications circuits (telephone noise, abnormal calls, data errors, etc.). The French Government has adopted the standards established by the International Telecommunications Union (ITU), as follows:

Electrostatic effects: 10 mA maximum through the human body; interference minor.

Magnetic induction (limits for electro-motive force (emf) between ground and wire): open wires, 60 V (normal train running) to 430 V (short circuit); protected wires (to be touched only by skilled staff), up to 60% of the dielectric strength of the wires (usually 1200 V); interference, 1 mV psophometric emf (a combination of emfs at the various frequency components, weighted according to the sensitivity of the human ear).

From the beginning of the 25kV AC traction standard in 1956 until 1976, the problem was handled totally separately by the National Railroads and its main perturbed neighbor, the Telecom Administration. The former applied the simplest (but most perturbing) mounting (Fig. 1) and paid for the countermeasures taken by the latter. Matters came to a head in 1976 with the simultaneous arrival of very heavy Telecom charges and of plans for the first high-speed train (TGV South-East to Paris to Lyons), whose requirements are much higher than those of a conventional train (600 A vs 200 A). Interest shifted to a reduction at the source, first applied in the form of the 2×25kV system on the part of the TGV South-East line (commissioned in 1981).

From that time on, all new high-speed lines have been fully equipped with the 2×25kV system; and a new Railroad/Telecom procedure has been initiated for the electrification of conventional lines with the setting up of a comprehensive economical balance between the 2×25kV and the 1×25kV systems (Fig. 1).

THE 2×25kV SYSTEM

All systems of reduction at source are based on the symmetrization (balancing) of the overhead power-supply facilities with respect to ground, by use of a return feeder for the catenary current (Fig. 3). Ideally, at a point *W* at some distance from the track, the induced emf is +U by the catenary system and −U by the return feeder, which results in zero induction. (Actually, it is only a reduction, since no balancing can be perfect.) A first system using transformers in series on the catenary to "extract" the return current from the ground (booster transformer or "BT" system) was employed by a few European railroads. The Japanese Railways also used it on their Shin Kansen (new high-speed lines), but they had so many traction problems with it that they switched over to a new system based on autotransformers (AT) in 1972. Actually, this approach had been first used in 1910 by the Newark Railroad, and in fact for telephone immunization purposes as well, but had been forgotten by the engineers. The French system is similar.

The AT system makes use of 50kV substations and autotransformers (connected between catenary and return feeder) whose midpoint is put at the rail potential, which results in the usual 25kV traction voltage between catenary and rail (Fig. 4). The return feeder is therefore at the same voltage but with an opposite phase angle, hence the designation "2×25kV" system; the previous system (Fig. 1) then becomes "1×25 kV." At 10 yards from the track, maximum 50Hz fields of 10 A/m (125 mG) are encountered with the 1×25kV system, whereas the 2×25kV system usually reduces this value to 60 mG, for an average AT spacing of 10 miles. In addition, the interference is of much shorter duration, and the reduction is even greater for points farther from the track. The above are only average values, as the distribution of the currents is rather complicated and the actual reduction factor depends on the frequency, the ground resistivity, the autotransformer spacing, the placement of the perturbed point considered, the instantaneous locations and power demands of the trains, and other variables. More accurate forecasts about a contemplated arrangement call for a statistical computer study.

On the average, the 2×25kV/1×25kV surcharge amounts to about 20%. This is not at all negligible and justifies the recommendation to keep the environmental requirements within reasonable bounds.

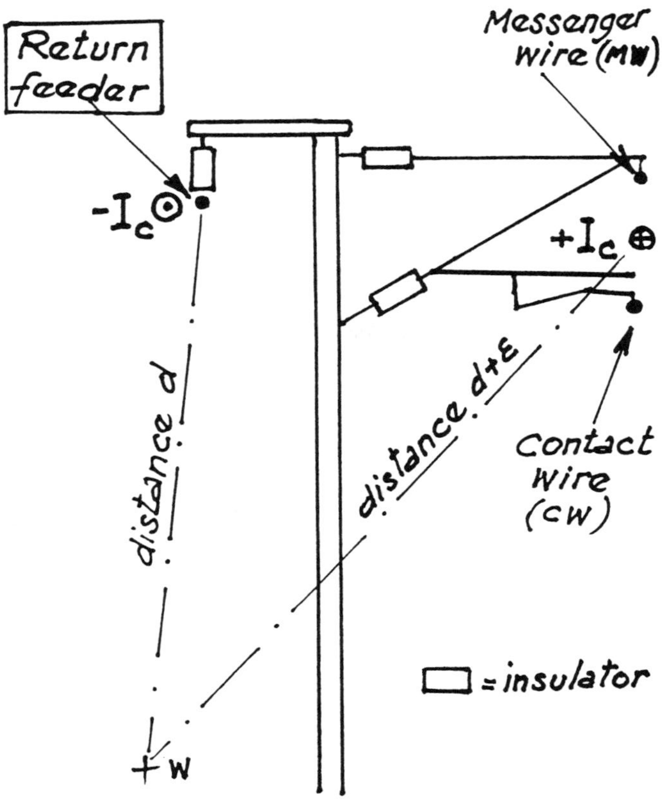

FIG. 3.—Principle of systems with reduction at source.

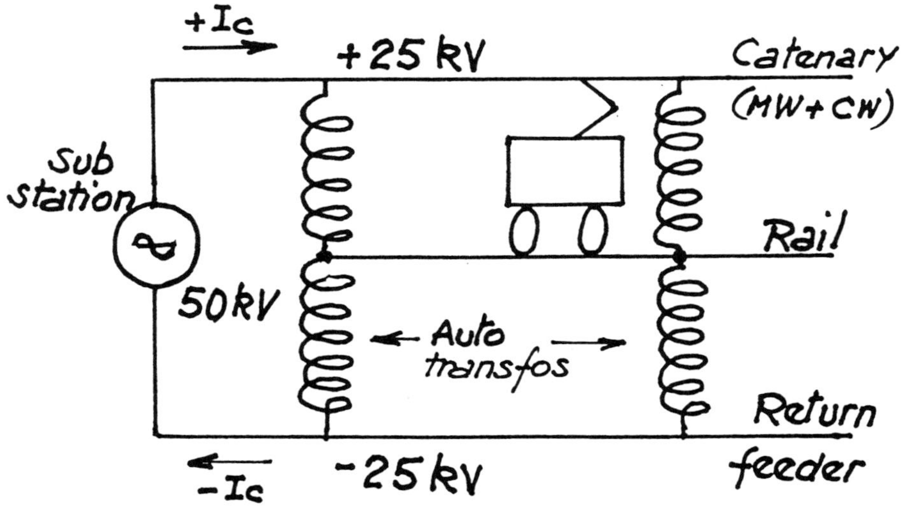

FIG. 4.—Principle of the 2 × 25kV power supply system.

LOCOMOTIVE FILTERING

Reduction at the source obviously relates to field generation by the engines as well. For a very long time, mechanical engineers were reluctant to filter the output of the locomotive engine, as the weight and volume of high-power filters are of the order of a ton and a cubic yard, which was particularly not feasible for TGV sets.

However, it turned out that filters would be of interest not only to the signaling and telecom engineers, by reducing audio frequency interference, but would also improve the "power factor" of the engines, thus decreasing the current drawn from the National Electricity Grid and the corresponding charges. Figure 5 shows part of a TGV electric diagram with filters.

However, a filter only changes the power distribution in its frequency range and may worsen interference outside it. As an example, here are the variations in power of typical frequencies between the first-generation TGV set (South-East) and the second generation (Atlantic), for 50Hz currents assumed to be identical: filter set at 150 Hz, −40%; at 250 Hz, +50%; at 350 Hz, +80%. For that reason, one must bear in mind that a modification for environmental reasons may be not only economically unjustified, but also that a reduction at a given frequency may worsen the situation in another, equally critical respect.

CONCLUSION

The 2×25kV system has proved very satisfactory in France for all parties (Telecom, mechanical engineers, and signal engineers of the railroad) in solving EMC telecom problems, including those of the high-power TGV lines. It is doubtless also helpful from the health-protection standpoint if minimization of the magnetic field is desirable for that purpose.

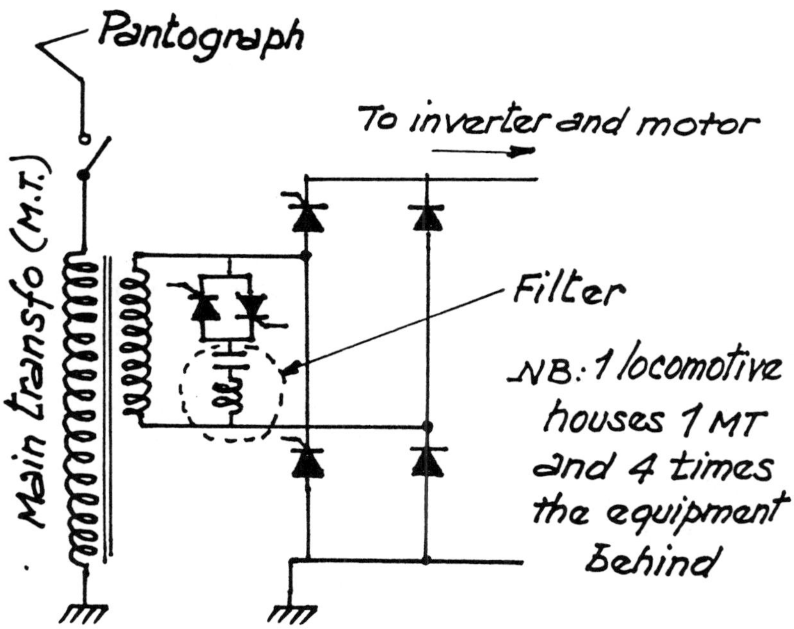

FIG. 5.—Principle of filtering on board a TGV locomotive.

Martin Blank, Ed., *Electricity and Magnetism in Biology and Medicine*

EMF ISSUES WITH MAGLEV IN JAPAN

M. Nakagawa* and T. Koana

Railway Technical Research Institute, Tokyo

Concerns of the residents near the Yamanashi maglev experimental track, especially about electromagnetic fields (EMFs) presumed to be generated by the system, are increasing. The consensus seems to be spreading that static magnetic fields of relatively moderate intensity have no specific bioeffects, but those caused by time-varying magnetic fields are not yet determined. We measured AC and DC magnetic fields in electric railways and compared the results with those of the electrodynamic suspension (EDS) maglev system. EMFs in the maglev coach are rather smaller than those in electric railways. We evaluated the bioeffects of EMFs by the standardized mortality ratio (SMR) obtained from a Japanese National Railways (JNR) cohort. SMRs of electric train crews, who are relatively close to EMFs, are smaller than those of the clerical workers taken as control. Low-level extremely low frequency (ELF) EMFs in MDS maglev appear to be too weak to raise health problems.

ELF MEASUREMENT ON ELECTRIC RAILWAYS AND EMF PRESUMPTION OF EDS MAGLEV SYSTEM

In August 1989 the Japan Ministry of Transport authorized the construction of an experimental track for maglev trains in Yamanashi (adjoining Tokyo), soon to go into practical operation. Construction was started in autumn 1991, and the interest of local residents in economic and environmental problems with EMFs, noise, and vibration has been increasing.

As to the EMF issue, engineers in the Japanese Railways Group know that intensities in electric transportation systems are higher than those of EDS maglev. Since none of the materials needed to evaluate EMFs of the electric railway train was available to us, we could not evaluate the presumed generation of maglev EMFs. We did measure some AC and DC magnetic fields on the electric railways transport system, and found the values on routine paths in the facilities or in the coaches at coasting speed to be as follows: power plant 0.1-50 μT; substation 0.1-3 μT; railway station 0.2-10 μT; DC train 0.5-5 μT (static MF, 50-200 μT); AC train 0.2-150 μT (static, 100-4000 μT); AC/DC train 0.5-75 μT (static, 200-1000 μT); and AC/DC locomotive less than 4 μT (static, 50 μT). Maximum values around facilities and power-running vehicles turned out to be several times greater than these values. Table 1 shows the expected or calculated EMFs of an EDS maglev system. Superconductive magnets will be relocated to the connecting portion between coaches in commercial-type vehicles used at Yamanashi (Fig. 1). A higher intensity of the static magnetic field is found at the connecting swiveling undercarriage (bogie); that of the AC field is to be found only when two maglev trains pass each other or one passes a waiting train. Accordingly, the static and time-varying magnetic fields in the maglev coach proper are both lower than those of the usual electrical railroad vehicles.

EVALUATION ON THE BIOEFFECTS OF EMFs WITH MAGLEV BY SMRs

Some epidemiological reports have appeared claiming positive evidence of weak ELF EMF effects. Evidence of weak EMF effects under experimental conditions is difficult to obtain. We attempted an epidemiological study to determine whether health problems could be found among electrical-railroad workers.

Nonexecutive JNR workers ($n = 23594$) over 40 years old in a metropolitan area in 1968 were divided into 6 groups. Group I consists of clerical workers and station workers, who serve as controls; group II are drivers of electric locomotives and train crews; group III are workers in electric power facilities. Men in groups II and III were workers relatively close to ELF EMFs. Men in the remaining groups were in yards (group IV), railroad and vehicle maintenance (group V), and motorbus (group VI), the conditions of which

*Correspondence to: M. Nakagawa, Railway Technical Research Institute, Hikaricho, Kokubunji, Tokyo 185 Japan.

were complicated; 2325 men had died by the end of 1985.

TABLE 1.—Presumed electromagnetic fields with EDS maglev system in operation type.

Situation	AC magnetic fields (μT)		DC magnetic fields (mT)		Main frequency
	Maximum point	Seat or routine path	Maximum point	Seat or routine path	
In the train					
Connecting portion	<1000[a]	<100[a]	20		
Coach	<100	<10	1	<0.1	≈ 150 Hz
Out of the train					
Resident	<200[a]	--	≈ 0.2[a]		≈ 50 Hz
Maintenance	--	--	<1000	<100	

[a]Only at passing or passing each other.

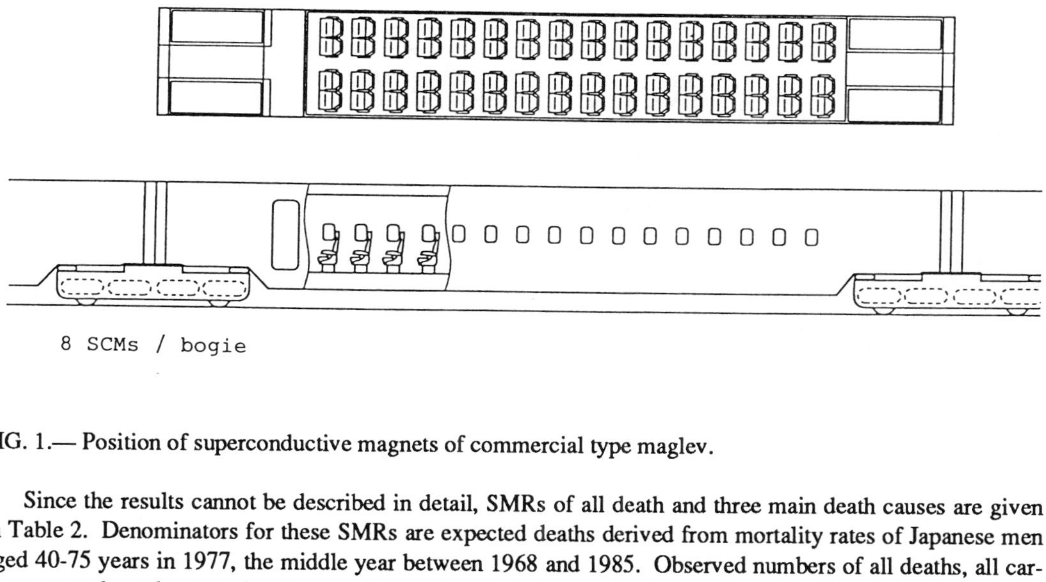

FIG. 1.— Position of superconductive magnets of commercial type maglev.

Since the results cannot be described in detail, SMRs of all death and three main death causes are given in Table 2. Denominators for these SMRs are expected deaths derived from mortality rates of Japanese men aged 40-75 years in 1977, the middle year between 1968 and 1985. Observed numbers of all deaths, all carcinomas and cerebrovascular diseases are much less than expected numerics, which illustrate the so-called "healthy worker effect." Ninety-five percent confidence intervals (CI) were given in parentheses after SMRs; a result was regarded as statistically significant if the 95% CI did not exceed 100. Table 3 indicates the ratio of SMRs from groups II-VI with group I taken as control, which serves as a table of comparison among JNR workers. In groups II and III (the workers close to EMFs) the all-deaths ratio was lower in group II and death ratios for all carcinomas were lower than for the controls both in groups II and III, without significant difference. Health impediments among the workers near EMFs is considered similar to (or rather better than) controls. Because EMFs in the coach of maglev are lower than those of usual electric railroads, we think that the issue of the maglev effect on passengers and workers' health is nothing fundamental.

TABLE 2.—SMRs of main death causes (95% confidence intervals; SMR denominators are expected deaths calculated from Japanese men aged 40-75 in 1977).

Group	All death	All carcinoma	Ischemic heart diseases	Cerebrovascular diseases
I Deskworker, station worker	61 (56-67)	69 (60-81)	71 (51-98)	52 (42-64)
II Train crew	55 (46-67)	60 (43-84)	71 (35-145)	55 (35-85)
III Electric power facilities	70 (55-89)	54 (33-86)	85 (34-210)	53 (29-96)
IV Yard worker	67 (59-76)	85 (69-104)	89 (57-140)	52 (38-71)
V Railroad/vehicle maintenance	63 (57-70)	69 (57-83)	46 (29-74)	60 (47-77)
VI Motorbus	80 (67-96)	80 (58-112)	81 (39-168)	63 (40-99)
Total	64 (61-68)	71 (65-78)	69 (60-85)	55 (49-62)

TABLE 3.—Ratios of SMR for group I as control (95% confidence intervals).

Group	All death	All carcinoma	Ischemic heart diseases	Cerebrovascular diseases
II Train crew	0.90 (0.73-1.11)	0.87 (0.61-1.25)	1.00 (0.47-2.17)	1.05 (0.63-1.75)
III Electric power facilities	1.15 (0.88-1.51)	0.77 (0.46-1.28)	1.20 (0.44-3.25)	1.00 (0.50-2.02)
IV Yard worker	1.12 (0.97-1.28)	1.22 (0.97-1.54)	1.25 (0.76-2.06)	1.00 (0.70-1.42)
V Railroad/vehicle maintenance	1.07 (0.94-1.21)	1.00 (0.81-1.22)	0.65 (0.39-1.07)	1.15 (0.86-1.53)
VI Motorbus	1.33 (1.08-1.65)	1.16 (0.80-1.66)	1.14 (0.52-2.55)	1.20 (0.71-2.04)

Martin Blank, Ed., *Electricity and Magnetism in Biology and Medicine*

MAGNETIC FIELD TESTING OF TR07 MAGLEV SYSTEM

F. M. Dietrich,* G. A. Steiner, D. C. Robertson, and W. E. Feero

Electric Research and Management, Inc., State College, Penna.

A sophisticated magnetic field measurement system, the *MultiWave*^{TM} System,† was used to record the temporal, spatial, and frequency distributions of the stray static and ELF magnetic fields in areas onboard or near the Transrapid TR07 facility, where workers or the general public could be exposed to stray magnetic fields from the Transrapid Maglev System. The measurement system captured data in the 0-2kHZ portion of the magnetic field present at each of its ten sensor locations recording the field waveforms. The magnetic-field magnitudes were well within the range of our normal electrical environments. The salient characteristic of the magnetic fields is that the frequencies at which these magnitudes were measured are not primarily at the power frequencies.

This paper summarizes the magnetic field measurements conducted onboard the Transrapid TR07 Maglev Vehicle and in the vicinity of the vehicle, guideway, passenger station, control center, and power-supply facilities at the Transrapid Test Facility in Emsland, Germany, in August 1990.[1] At the outset of the TR07 test it was recognized that the simple magnetic-field survey instruments used for transmission and distribution-line surveys would be inappropriate for quantifying the fields associated with the TR07. Many of the laboratory and clinical studies have reported effects that appear to result from exposure to fields over a wide range of magnitudes and frequencies. Most of the existing field-survey meters were sensitive only to a relatively narrow band of frequencies, centered on the power frequencies of 50 and 60 Hz Only recently has it been possible to record and store all the discrete segments of the frequency bands that biological reports suggest may be important.

The magnetic-field data collected from TR07 were significantly more complex than those found in the vicinity of powerlines.

Multiple Sources. The magnetic fields onboard or in the vicinity of the TR07 Maglev System arise from multiple sources, hence spatial distributions cannot be expressed as simple attenuation curves that are temporally stable.

Continuous Frequency Distribution. The time-varying component of the guidance and levitation magnetic fields have a distributed rather than discrete frequency distribution. Consequently, one must address field intensities in various bands of finite width rather than fields at specific frequencies.

Variable Frequency Components. The magnetic field from the longstator of the linear synchronous motor (or from the power equipment driving that field) is a discrete frequency component that varies over time. Consequently, simple averaging of repetitive frequency spectra will obliterate this component of the field as a discrete component.

ORGANIZING ELECTROMAGNETIC DATA

To measure the magnetic-field environment of the TR07, a new monitoring system developed under EPRI sponsorship, was used.[2] The system, marketed under the brand name *MultiWave*^{TM}, can record and store 3-axis magnetic fields in up to 40 different physical locations nearly simultaneously. The 3-axis feature, combined with waveforms captured in each axis (vertical, transverse, and longitudinal) preserves the direction, magnitude, polarization, and frequency spectrum of the magnetic field at each point in space measured. To gain temporal information on the nature of the magnetic field, the field is sampled repetitively at intervals

---

*Correspondence to: F. M. Dietrich, Electric Research and Management, Inc., Box 165, State College PA 26804.

†This product incorporates technology developed for the electric power industry under the sponsorship of the Electric Power Research Institute (EPRI).

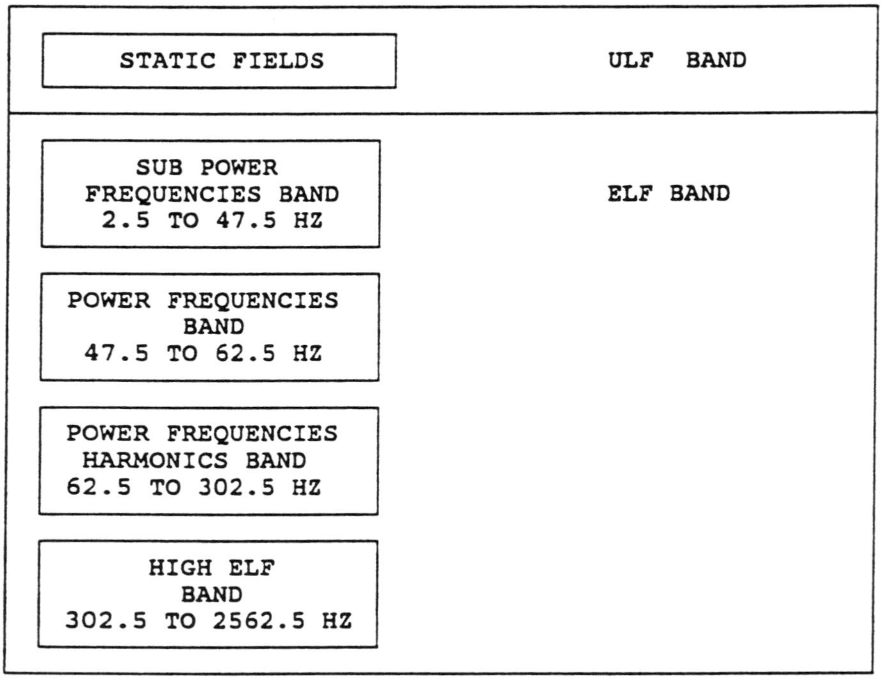

FIG. 1.—Magnetic-field flux densities grouped by frequency partitions within ELF and ULF bands.

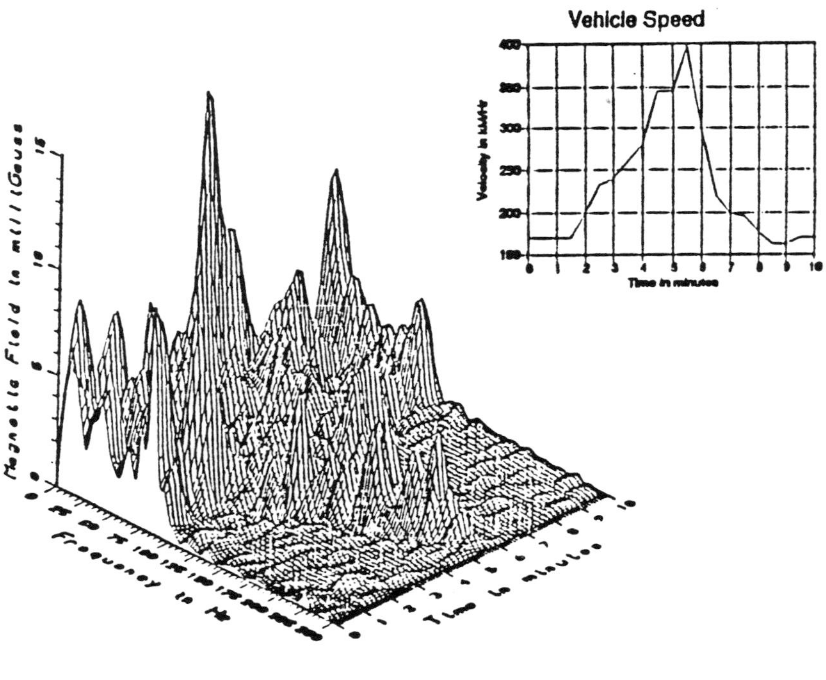

FIG. 2.—Plot of typical magnetic-flux density vs time and frequency showing dynamics of field characteristics within TR07 vehicle in response to various vehicle operating conditions.

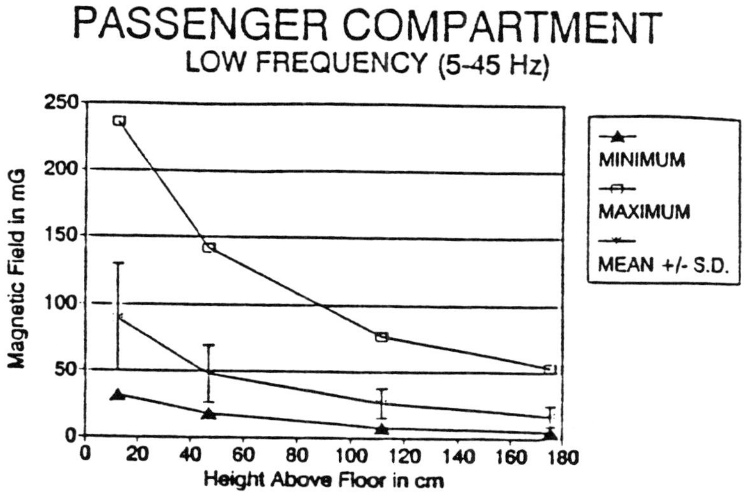

FIG. 3.—Minimum, average, and maximum-time varying magnetic field in low-frequency (5-45Hz) band measured in passenger compartments of TR07 vehicle for various heights above floor.

set by the test requirements. Use of this system generates excellent but extensive data. To evaluate the magnetic field characteristics of the TR07, it was necessary to develop subsets of the data which partitioned them into frequency bands within the ULF and ELF bands that would facilitate understanding its significance relative to previous research.[3] The subsets used in this study are depicted in Fig. 1. The data were decomposed into narrow frequency bands (5 Hz). Frequency spectra from approximately 2.5 to 2562.5 Hz were evaluated by Fourier transform. The data can be shown graphically over this wide bandwidth (Fig. 2), but for the data to relate to reported health concerns, it seemed appropriate to segregate the data into groupings.

FINDINGS

A general finding is that the magnetic fields at all of these locations have complex frequency spectra that are highly variable over time. Specifically, we found the following.

Passenger Compartments. The time-varying magnetic fields in the passenger compartments of the TR07 vehicle extended from 2.5 Hz to over 2 kHz and were highly variable over time. The average field levels ranged from approximately 100 mG near the floor of the vehicle to approximately 20 mG at standing head level. Except for the strong height dependence, the magnetic fields did not depend strongly on location within the passenger compartment. The standard deviation of the magnetic field intensity over time was typically 40% of the mean value, but maximum values at any height above the floor were often 2.5 times the average values. Typically, 70-80% of the time-varying magnetic fields were at frequencies below 47.5 Hz. The static magnetic field level within the passenger compartments was dominated by the geomagnetic fields at standing head level. Higher static magnetic field levels near the floor were 835 mG (compared with the earth's magnetic field in that vicinity of 500 mG) and reduced to approximately 500 mG at standing level. Static fields were more spatially variable but slightly more stable over time than the alternating or time-varying fields. Figure 3 demonstrates the type of data collected at various heights above the floor.

Near the Guideway. Elevated magnetic field levels only exist near the guideway for very brief periods of time when the guideway is energized and a vehicle is passing. The maximum time-varying magnetic field levels within 3 m of the centerline of the guideway are comparable with average field levels onboard the vehicle, 65-95 mG, but occur only for a brief period of time. At 10 m from the centerline, the maximum time-varying magnetic field levels are approximately 20 mG.

Passenger Station. Magnetic-field levels at the passenger station are similar to those along the guideway. Within the waiting area at the station the time-varying magnetic fields produced by the passing vehicle on the energized guideway is approximately 20 mG.

COMPARISON OF TR07 FIELDS WITH OTHER ENVIRONMENTAL MAGNETIC FIELDS

Public acceptance of magnetic-field exposures is at present more on an equitable and comparable basis to other exposures than it is to quantifiable characteristics of the field itself. It is in this light that the comparison and contrast of magnetic fields produced by the TR07 to other environments is presented.

Static Fields. The static fields encountered onboard the TR07 vehicle (mean values 500-1000 mG) are generally within the range of static fields encountered daily in the perturbed geomagnetic fields (250-1000 mG). The magnitude of the static field onboard the TR07 vehicle has substantial temporal variability. Sizable changes in static-field intensity occur more frequently and more rapidly than most changes in the geomagnetic field that result from walking past a perturbing object such as a steel column of beam. Static-field conditions at the station, along the guideway, or near power-supply equipment are essentially all geomagnetic fields, with no significant contribution from active operation of the TR07 system.

Amplitude Characteristics of ELF Fields. Notwithstanding the aforementioned need to compare the temporal and frequency characteristics for various low-frequency magnetic fields, this subsection compares the measured magnetic-field levels onboard and near the TR07 facility with reported magnetic-field levels near various power-frequency sources. The lack of comparable environmental field data in several of the frequency bands makes any other approach unworkable. It must be realized that the comparison of the total magnetic-field bandwidth between 2.5 and 2562.5 Hz of the TR07 with the predominantly power-frequency environmental field levels concurrently reported in the literature is an apples-and-oranges comparison. With that important caveat, generalizations are offered.

The magnetic fields of magnitudes found in the passenger compartment of the TR07 can be shown to fall well within the intensity levels that characterize magnetic fields found near appliances and at the edge of transmission-line rights-of-way. The magnetic-field intensities found near the guideways exhibit a highly temporal characteristic. When a train passes by, the magnetic-field magnitudes close to the guideway are typical of those found at the edge of transmission line rights-of-way and near household appliances. The magnetic field levels remote from the guideway are similar to fields found near distribution-line rights-of-way.[4]

REFERENCES

1. *Safety of High Speed Magnetic Levitation Transportation Systems: Final Report on Magnetic Field Testing of TR07 Maglev Vehicle,* U.S. Department of Transportation, Federal Railroad Administration, April 1992; vol. I, Analysis (DOT/FRA/ORD-92/09.I, DOT-VNTSC-FRA-92-6.1); vol. II, Appendices (DOT/FRA/ORD-92/09.II, DOT-VNTSC-FRA-92-6.2).

2. *Measurement of Power System Magnetic Fields by Waveform Capture,* EPRI TR-100061, Project RP2942-1, Final Report, February 1992.

3. *IEEE Standard Dictionary of Electrical and Electronic Terms,* ANSI/IEEE Std 100-1988, New York, NY: IEEE, 4th ed.

4. This work was sponsored by the U.S. Department of Energy and the U.S. Department of Transportation (Federal Rail Administration).

Martin Blank, Ed., *Electricity and Magnetism in Biology and Medicine*

THE PRESENT STATE OF EMF RESEARCH: WHY MORE IS NEEDED

R. R. Neutra*

California Department of Health Services, Emeryville, Calif.

The scientific evidence regarding the mode of action of electric and magnetic fields differs in a number of ways from that of the more familiar chemical agents that society regulates. One difference is that EMFs are not a single agent. Although there is a unifying theory that describes EMFs, exposures from the electrical utility system represent a complex mixture of frequencies, polarizations, waveforms, directions, and relations to the earth's static magnetic field. In addition, exposures vary in intensity and duration.

Another difference from the usually regulated chemical carcinogens is that some of the epidemiological evidence relates directly to general environmental exposures. Four of the seven major epidemiological studies conducted to date suggest that people who live in homes located near high-current distribution wires may have higher rates of childhood leukemia and/or brain cancer than people who live near underground or lower-current overhead configuration wires. It is not clear what aspect of the EMF mixture, if any, is responsible for these statistical associations.

Such studies have focused public attention on the EMF issue.

It is rare to have epidemiological evidence that suggests an effect of an agent at the low levels found in the general environment. (Residential lead exposure is one salient exception.) Usually, relevant epidemiology has to do with higher *occupational* exposures. When we must rely on occupational epidemiology there is often disagreement as to whether a lower *environmental* exposure would (1) also cause an untoward effect, (2) be detectable by epidemiologists, or, (3) be of sufficient magnitude to generate regulatory concern. Some of the epidemiology seems to suggest the possibility of detectable risks of cancer from general environmental exposures, which if true would be of regulatory concern. Thus, if these studies were to be confirmed, and if laboratory studies were supportive, environmental EMF regulation would have a rather stronger mix of supporting science to guide it.

EMFs also have some relevant, if not conclusive, epidemiological evidence from occupational data What aspect of the EMF mixture, if any, is responsible for these occupational findings is not clear, nor do we known the intensity or duration of exposure that might produce such results.

The evidence relevant to EMFs is also unusual in that we have no solid evidence one way or the other from whole-animal bioassays. Studies of the possible reproductive and carcinogenic effects of magnetic fields are only just getting under way. Earlier studies of electric fields tended to show inconclusive results. Usually whole-animal bioassays are the mainstay of environmental regulation. Experimental production of cancer or birth defects with high doses of an agent can show that it is a hazard, at least at high doses. This type of animal evidence, together with theoretical downward extrapolation to lower doses (with or without epidemiological evidence of risk), has in the past provided a rationale for regulation. Although animal studies now under way are exploring the effects of a few of the components of the total EMF mixture, they will probably not generate results until the mid-1990s, if then. Scientists are reluctant to rely completely on nonexperimental epidemiological results when there is controversy about the mechanisms by which the EMF mixture might have an effect.

That brings us to another unusual aspect of the scientific evidence regarding EMFs: the degree of argumentation about the hypothesized mechanisms of action. There is controversy as to whether the 0-5mG alternating magnetic fields characteristic of residential and even most occupational exposures would be predicted, under current physical theory, to produce any effect in cells or in humans. The energy these low-level fields are thought to induce in cells is many orders of magnitude less than the energy in random thermal "noise" in such tissues. Such physicists argue that this finding suggests that the epidemiological results must be due to

*Correspondence to: R. R. Neutra (Environmental Health), California Dept. of Health Services, 5900 Hollis St. (Suite E), Emeryville CA 94608.

some non-EMF confounding factor, and that the laboratory studies mentioned below are simple laboratory errors. Others point to the possibility of direct coupling of the magnetic field to molecules with a magnetic moment, or of cooperative phenomena occurring with membrane-bound molecules. Thus, EMF has no generally agreed-upon physical mechanism by which these low levels of exposure can produce alterations in normal functioning.

Arguments about EMF mechanisms are unlike those about chemical carcinogens, which tend to focus on disagreements about which of the classes of already identified mechanisms is in operation, whereas arguments about EMFs involve the postulation of new classes of mechanisms; if these mechanisms are proven true, this discovery would radically change our views on the role of electricity and magnetism in biology, which would introduce a new paradigm. Scientists have historically resisted new paradigms until the evidence for dropping the old one is very compelling.

Another difference between EMFs and chemical carcinogens may relate to the slope of the dose-response curves. There are cellular experiments with magnetic fields in the hundreds to thousands of milligauss that produce alterations in calcium metabolism, RNA transcription, and growth enzymes. Some of these experiments suggest that the effects may not become more pronounced as intensity increases but instead may show intensity and frequency "windows," i.e., zones of intensity or frequency that produce effects even though levels just above and below these zones produce no effect. If these results are borne out, they suggest that the strategy of simply lowering exposure levels, which diminishes the effect of *chemical* carcinogen mixtures, may not work in this case. It also suggests that the standard practice of testing an agent at very high doses may not increase the sensitivity of the experiment and that strong effects may be seen at low exposures. But these results are preliminary, so there is much work to be done in "dose exploration," a term we use to refer to experiments that include descriptions of how these biological effects (at the cellular and whole-animal level of organization) vary as elements of the EMF mixture are applied at different intensities and durations.

It is generally believed that EMFs do not seem to affect the cellular DNA, although they may affect RNA transcription and other phenomena related to cell growth. There is a lack of understanding of the pathway(s), if any, leading from the documented changes in normal cellular functioning to fullblown carcinogenic change. Thus there is uncertainty about *biological mechanism*.

Although some have advocated aggressive regulation at this time, most scientists and regulators feel that all the above mentioned differences between the kind of evidence we have about EMFs and the kind of evidence we have about agents that we regulate make standard setting for EMFs premature. To summarize, there is a lack of solid experimental carcinogenesis and reproductive effects in animals, an uncertainty about epidemiological studies, a lack of a demonstrated physical mechanism or biological mechanism, and a lack of understanding of what elements of the EMF mixture at what intensity and duration, if any, would cause an effect. However, most concur that further research is needed.

Martin Blank, Ed., *Electricity and Magnetism in Biology and Medicine*

A PUBLIC-HEALTH FRAMEWORK FOR ADDRESSING A LAYPERSON'S PERCEPTION OF EMF HEALTH RISKS

P. A. Valberg*

Harvard School of Public Health, Boston, Mass. and Gradient Corporation, Cambridge, Mass.

The electric power that comes into homes and workplaces is a source of electric and magnetic fields (EMF) that pervade working and living areas. Some people have become alarmed about possible injury to health arising from exposure to EMF associated with electric-power generation, distribution, and use. Anxiety about the health effects of powerline EMF have derived mainly from epidemiological studies showing an association of childhood cancer with utility-wire configuration outside the child's home. Due to the ubiquitous presence of electric power, the public-health implications of this association are potentially very significant; however, evidence for a causal relationship between exposure to EMF and any health effect has so far not been found. This uncertainty, plus the suggestive research findings, have fueled public anxiety and media attention. When discussing the case for EMF health effects with a layperson, one should do so within a context of causes of morbidity and mortality in the U.S. population. A lay audience cannot gauge the relative importance of EMF health risks without understanding baseline public health information. Any discussion of steps to avoid potential EMF health-effects needs a perspective on other hazards to life and health.

A public-health orientation differs from a scientific or a medical orientation, although the goals of these disciplines are intertwined. Scientific research aims to *understand* disease causation. Medicine aims to *cure* disease or ameliorate symptoms in the afflicted individual. Public health aims to *prevent* or minimize disease occurrence in the population. The following steps outline how environmental problems are addressed:

- Evaluate the nature and extent of the exposure. Is it new, different from the norm, unusual in population distribution, distinguishable from background, or otherwise quantifiable?
- Assess what is known about biological responses or biochemical alterations that result from exposure. What are the mechanisms by which cell function may be compromised?
- Analyze available risk data to determine whether the results are statistically significant and demonstrate a dose-response relationship.
- Compare the potential health risks in relation to established health risks so that the importance of taking action can be appreciated.
- Take action, but be wary of solutions that allocate resources to a minor risk at the expense of neglecting major tasks.

EMF EXPOSURE AND MECHANISM

Atmospheric electric fields and geomagnetic fields of substantial magnitude have existed throughout human history. Movement in these "static" fields results in exposure to time-varying E- and B-fields. Electric currents generated naturally in body tissues can exceed by a considerable margin the electric currents induced in body tissues by environmental sources of E- and B-fields. Thus, exposure from powerline sources is difficult to distinguish, aside from their pure frequency content, from natural EMFs with which humans have always existed.

The 60Hz EMFs are at the lowest end of the electromagnetic radiation frequency spectrum containing x rays, visible light, radiowaves, etc. The energy of electromagnetic radiation decreases with frequency. X-ray photons can ionize and fragment biological molecules; light photons can bend (isomerize) biological molecules; microwave photons can vibrate biological molecules; but the energy contained in photons of 60Hz electromagnetic radiation is too feeble to alter biological molecules directly.[1,2] The 60Hz EMF can induce weak currents and voltages in the body, but scientists have been unable to identify the mechanism by which

*Correspondence to: P. A. Valberg, Gradient Corp., 44 Brattle St., Cambridge MA 02138.

these weak currents and voltages could make their presence felt in the midst of the much greater electrical activity that naturally occurs in living organisms.[3-6] Prior to the EMF epidemiological results summarized below, most scientists considered the EMF produced by powerlines and appliances far too weak to have any important biological effects. Moreover, per capita consumption of electric power has increased dramatically in this century, going from nearly zero to the present per capita consumption rate of about 1210 W continuous. Even since 1950 per capita consumption has increased sixfold (Fig. 1).

EMF HEALTH CONCERNS RAISED BY EPIDEMIOLOGY

Interest in the health effects of electric-power EMF was spurred by publication in 1979 of an epidemiological study showing an association between childhood cancer and certain "wire codes" (utility-wire configurations outside the home).[7] The wire configuration rankings were defined in an attempt to rank the probable contribution of outside distribution lines to magnetic fields within the home and were employed as surrogates for EMF exposure; an increased childhood cancer risk was reported as associated with living near high-current-configuration wire codes. Subsequent childhood cancer epidemiological studies have reported both "positive" and "no association" between EMF surrogates and childhood brain cancer or childhood leukemia.[8-10] The association of cancer risk with wiring configuration can be expressed as an "odds ratio" (OR), which conveys the strength of the association. The original study reported a leukemia risk OR of 3.0,[7] but in more recent studies the OR has dropped to 2.8 and to 2.2.[9,10] In these studies, correlation of leukemia risk with *measured* EMF failed to reach statistical significance (OR = 1.5 to 1.9, ns). Studies of EMF and adult cancer have not uncovered any clear association.[11-13]

Historical Trends in Public Health

From the early 1900s to the present time, per capita electric power consumption in the U.S. has increased over tenfold. Over this same period of time U.S. health statistics have reflected dramatic changes in national health:

- Mortality has been decreasing for all age groups, with the steepest decline (over 10-fold) in the 5-14 year-old age group. Life expectancy increased from 47 in 1900 to 74 in 1990.
- The death rate due to most cancers has been decreasing. Stomach cancer mortality has dropped sixfold; liver and uterine cancer have also shown a dramatic drop. For cancers with increasing death rates, etiologic factors such as tobacco consumption have been identified.
- Childhood leukemia incidence has not increased from 1950 to the present in the USA (Fig. 1) nor in Denmark (Fig. 2).[14,15]
- Health hazards can be mirrored in national statistics. Historical increases in consumption of tobacco shows a close correlation with lung-cancer rates 20 years later (Fig. 3). For an exposure as pervasive in the population as powerline EMF, health impacts might be expected to be reflected in national statistics.

A comparison of leukemia and lung cancer case-control studies is also instructive. The initial case-control study of lung cancer involved 1385 men, 605 hospital patients with lung cancer, and 780 control hospital patients.[16] Both groups were interrogated as to smoking, occupational, family, and medical history. The odds ratio for lung-cancer risk between the highest and lowest smoking categories was in excess of 28, more than ten times the reported association between EMF and childhood leukemia.

Childhood leukemia incidence within the U.S. shows large ethnic and geographical variations. For example, Los Angeles Hispanics have a leukemia rate of 4.4, whereas Detroit Blacks have a rate of 1.2 (annual incidence per 100000).[15] Boys have a higher rate than girls in every locale except Hawaii. Worldwide variations show that leukemia is high in Costa Rica (4.9) and low in Kuwait (1.2). These dramatic differences are seemingly unrelated to EMF.

Information on Competing Health Risks

Of the U.S. population of about 254 million, approximately 2.1 million people die annually. Of annual U.S. deaths, heart and vascular disease are responsible for about on half, and cancer is responsible for about one quarter (Fig. 4). Of the remaining deaths, the two largest components are lung disease on the one hand, and accidents, suicide, and homicide together on the other hand. Childhood cancer deaths are less than 0.1% of all annual deaths.

Annual deaths for children ages 1-14 in the U.S. are 16200. Nearly half of these childhood deaths are due to accidents in the home, automobiles, pedestrians, drowning, firearms, choking, and poisoning (Fig. 5).

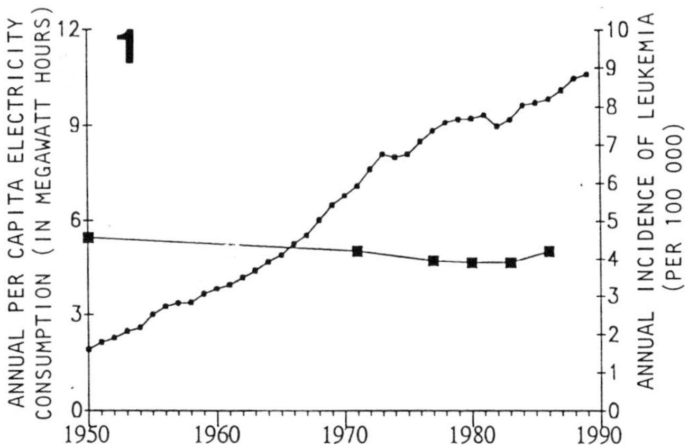

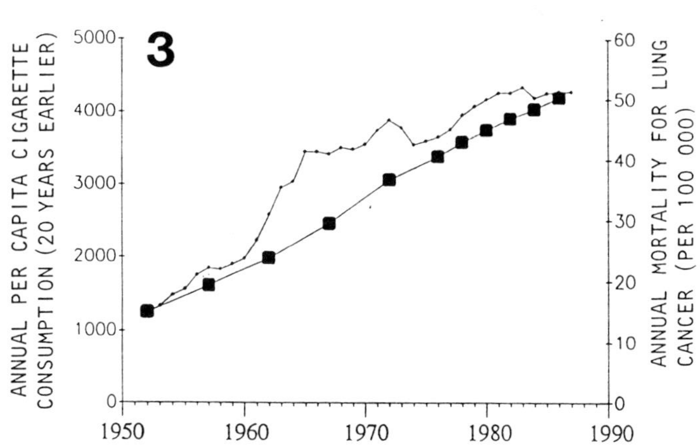

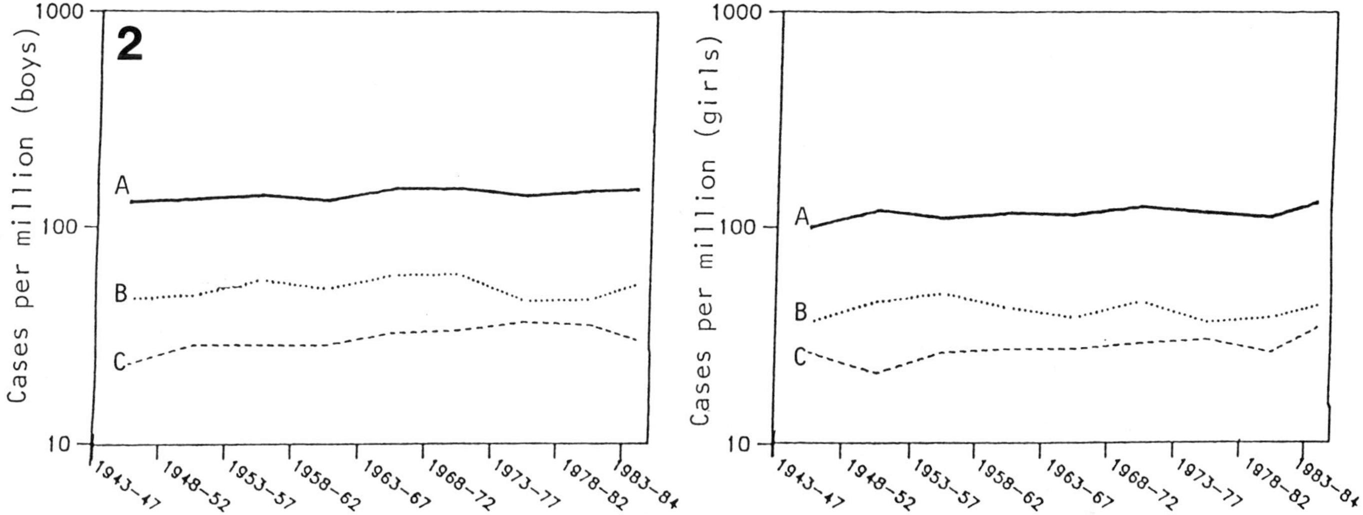

FIG. 1.—Per capita U.S. electric power consumption (dots) and leukemia incidence in children aged 1-14 years (squares).

FIG. 2.—Childhood cancer incidence in Danish boys (left) and girls (right): A, all cancers; B, leukemia; C, CNS neoplasms.

FIG. 3.—U.S. per capita cigarette consumption, 1930-1970 (dots), and lung cancer mortality per 100000, 1950-1990 (squares). Unlike childhood leukemia, lung cancer has long latency period; therefore time frame for cigarette consumption has been shifted 20 years prior to lung cancer statistics.

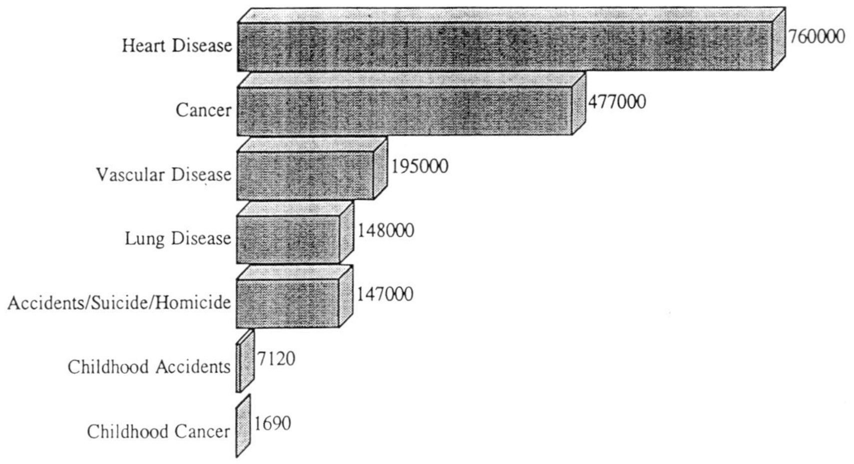

FIG. 4.—Annual U.S. deaths by specific causes.

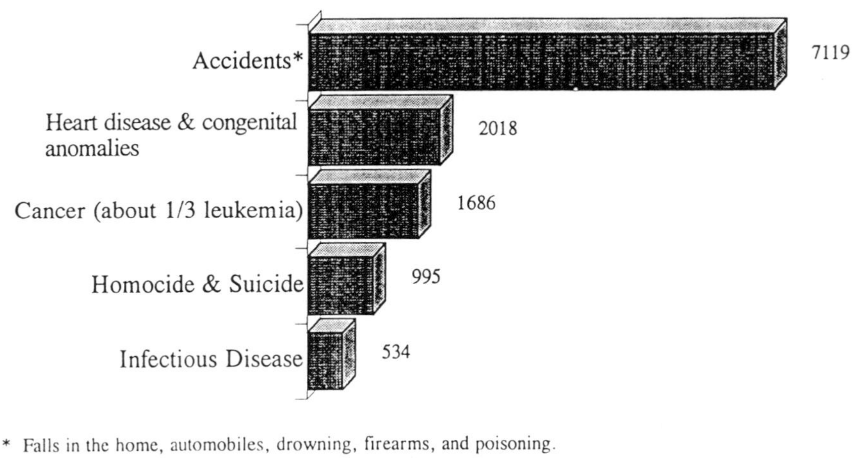

FIG. 5.—Annual U.S. deaths for children aged 1-14 years.

The number of injuries caused by accidents is many times larger than the number of deaths. Total leukemia incidence in the USA is less than 3% of all cancer incidence, and leukemia deaths are 4% of all childhood deaths (ages 1-14). For example, the frequency of leukemia death is less than the frequency of death from any one of the following accidental causes: inhalation of a foreign object, home fire, drowning, bicycle accident. Automobile accidents are a major cause of childhood injury and death. Even deaths from homicide and suicide in childhood exceed leukemia deaths.

A recent article that discusses primary prevention of cancer lists the major cancers (lung, colon and rectum, bladder, etc.) where, in addition to genetic factors, an environmental factor is suspected.[17] Examples of environmental factors given are tobacco, animal fat, alcohol, salt, UV sunburn, viral infection, overeating, x rays. The article concludes that "the majority of the causes of cancer . . . are associated with personal choices and not with the environment in general."[17]

All health risks that derive from preventable causes ought to be minimized. However, the costs per health benefit vary widely among different risks. Haphazard regulations can be a hazard to public health. Moreover, cancer mortality rates correlate inversely with socioeconomic status, so that poorer is riskier. Finally, if regulations eliminate jobs, that also has a health cost. The riskiest occupation is no occupation.

CONCLUSION

We should not rest assured of the safety of EMF simply on the basis of an absence of effects in national health statistics. Yet it is important to draw attention to the overall health picture in discussions of EMF health risks with a group of people without a public-health background The widespread public perception that pollutants, EMF, or chemicals in the general environment are a major cancer hazard is incorrect. In the context of communicating information on potential EMF health risks, one can avert needless anxiety by emphasizing that our current knowledge indicates that lifestyle, diet, and genetic factors have a far more significant effect on overall human health than even our worst fears of EMF health effects. Furthermore, communicating information on historical patterns of EMF exposure and the lack of any coincident pattern of cancer incidence provides reassurance that the magnitude of any EMF effect is unlikely to be very large. Such knowledge is essential to allocating wisely society's limited resources for prevention of ill health.

REFERENCES

1. R. K. Adair, "Tissue interactions with nonionizing electromagnetic fields," *Physiol. Rev.* 61: 435-514, 1981.
2. American Institute of Biological Sciences (AIBS), *Biological and Human Health Effects of Extremely Low Frequency Electromagnetic Fields,* Technical Report prepared by the Committee on Biological and Human Health Effects of Extremely Low Frequency Electromagnetic Fields, Report AD/A152731 (NTIS, Springfield, VA 22161), 1985.
3. R. K. Adair, "Biological effects on the cellular level of electric field pulses," *Health Physics* 61: 395-399, 1991.
4. R. K. Adair, "Constraints on biological effects of extremely-low-frequency electromagnetic fields," *Phys. Rev.* A-43: 1039-1048, 1991.
5. R. K. Adair, "Criticism of Lednev's mechanism for the influence of weak magnetic fields on biological systems," *Bioelectromagnetics* 13: 231-235, 1992.
6. J. C. Weaver and R. D. Astumian, "The response of living cells to very weak electric fields: The thermal noise limit," *Science* 247: 459-462, 1990.
7. N. Wertheimer and E. Leeper, "Electrical wiring configurations and childhood cancer," *Am. J. Epidemiology* 109: 273-284, 1979.
8. A. Ahlbom, "A review of the epidemiologic literature on magnetic fields and cancer," *Scand. J. Work Environ. Health* 14: 337-343, 1988.
9. S. J. London, D. C. Thomas, J. D. Bowman, E. Sobel, T.-C. Cheng, and J. M. Peters, "Exposure to residential electric and magnetic fields and risk of childhood leukemia," *Am. J. Epidemiol.* 134: 923-937, 1991.
10. D. A. Savitz, H. Wachtel, F. A. Barnes, E. M. John, and J. G. Tvrdik, "Case-control study of childhood cancer and exposure to 60-Hz magnetic fields," *Am. J. Epidemiol.* 128: 21-38, 1988.
11. M. Coleman and V. Beral, "A review of epidemiological studies of the health effects of living near or working with electricity generation and transmission equipment," *Int. J. Epidemiol.* 17: 1-13, 1988.
12. S. Preston-Martin and J. M. Peters, "Prior employment as a welder associated with the development of chronic myeloid leukemia," *Brit. J. Cancer* 58: 105-108, 1988.
13. R. K. Severson, R. G. Stevens, W. T. Kaune, D. B. Thomas, L. Heuser, S. Davis, and L. E. Sever, "Acute nonlymphocytic leukemia and residential exposure to power frequency magnetic fields," *Am. J. Epidemiol.* 128: 10-20, 1988.
14. P. N. Brown, H. Ertz, J. H. Olsen, M. Yssing, E. Scheibel, and O. M. Jensen, "Incidence of childhood cancer in Denmark 1943-1984," *Int. J. Epidemiol.* 18: 546-555, 1989.
15. M. S. Linet and S. S. Devesa, "Descriptive epidemiology of childhood leukemia," *Brit. J. Cancer* 63: 424-429, 1991.
16. E. L. Wynder and E. A. Graham, "Tobacco smoking as a possible etiologic factor in bronchiogenic carcinoma," *JAMA* 143: 329-336, 1950.
17. B. E. Henderson, R. K. Ross, and M. C. Pike, "Toward the primary prevention of cancer," *Science* 254: 1131-1138, 1991.

Martin Blank, Ed., *Electricity and Magnetism in Biology and Medicine*

IMPACT OF CONCISE INFORMATION ON THE LAYPERSON'S OPINION OF THE RISKS OF CHRONIC EXPOSURE TO ELF MAGNETIC FIELDS

V. Delpizzo* and G. Elliot

Australian Radiation Laboratory, Yallambie

We report the results of a mail survey conducted in Melbourne, Australia, aimed at ascertaining to what extent a concise information brochure could affect the perception of the possible long-term health hazards of extremely low frequency (ELF) magnetic-field exposure. We found that in at least two important respects, perception did not change significantly: (1) the public's concern is still focused disproportionately on high-voltage transmission lines, rather than on the much more ubiquitous distribution lines; and (2) a significant fraction of the population sampled maintained the belief that the most appropriate course of action at this stage was to introduce some form of regulation of exposure.

The Australian Radiation Laboratory (ARL) has among its functions that of providing advice to the public on matters of radiation health and safety. To this purpose it produces a series of "Information Bulletins"—brief, plain-language leaflets on topics of public concern that are widely circulated among the community. Over the past several years the Laboratory has received numerous requests for advice from members of the public concerned about the possible health hazards of chronic exposure to ELF magnetic fields. Although much of the required information is given verbally, it is in the interest of both parties to provide some written material on the subject. At this stage we have not yet decided to produce an Information Bulletin on this topic and have relied instead on the booklet written by Morgan.[1] However, several parties have expressed a preference for a more concise document. One of us (VD) had doubts about the capability of a typical ARL information bulletin to cover adequately the complexities of an issue still far from being well understood even in the scientific community. We therefore decided to survey a random sample of adult Melbourne residents to determine to what extent a concise document could convey accurate information on this topic and influence public perception.

We wrote a 500-word Information Sheet that, in the best judgment of the authors, contained an accurate statement of the current scientific understanding of the issue. The points we tried to emphasize were: (1) there is evidence of possible adverse health effects of ELF magnetic fields that, though far from being convincing, cannot be dismissed; (2) there is little if any evidence that chronic exposure to electric fields is hazardous; and (3) there are no data to indicate whether a specific magnetic-field exposure level is safe or hazardous. Although not explicitly stated, the conclusions we hoped a reader would reach were: (1) there is a rather pressing need for further research; (2) voluntary avoidance of ELF magnetic fields is a possible course of action that should be given consideration, although not necessarily an overriding priority; (3) there is no basis on which to establish regulations, even interim; and (4) the emphasis on high-voltage transmission lines is misplaced.

RESULTS

One thousand four hundred questionnaires were mailed to 700 randomly selected addresses. We asked that if more than two adults lived at that address, the two closest to 40 years of age should each complete a questionnaire. By means of multiple-choice questions, the respondents were asked to state their opinion on the hypothesis that ELF magnetic fields could be responsible for an increase in cancer incidence, and on the course of action that should be adopted at this stage. They were also asked to assess to what extent their home could be regarded as "at risk" and why. They were then asked to open and read the sealed Information Sheet. The respondents were then again asked to state their understanding of the problem. The results presented here are based on the analysis of the 439 questionnaires returned by 30 April 1992.

---

*Correspondence to: V. Delpizzo, Australian Radiation Laboratory, Lower Plenty Rd., Yallambie VIC 3085, Australia.

As in all mail surveys, the possibility of nonresponse bias exists. We tried to minimize it by carefully phrasing the introductory letter so that persons of different points of view felt equally motivated to respond. However, it is certainly possible that the opinions reported reflect prevalently the views of individuals who feel more strongly about the issue. It is also possible that persons who are more interested in this topic (such as those who contact our Laboratory for advice) may read more carefully and learn more from the Information Sheet than unconcerned respondents. Therefore the Information Sheet may prove more effective in practice than the results of this survey suggest.

We found that almost all respondents (93%) were aware of the hypothesized cancer risk, mainly through media reports. A large majority (79%) believed that this risk is either possible (60%) or likely (19%) and that further research is warranted. Approximately 43% thought that voluntary avoidance of the fields should be practiced, and about 27% advocated interim guidelines. Most respondents (93%) believed that both children and adults were potentially at risk, and the majority (70%) identified only high-voltage transmission lines as a source of potentially hazardous fields.

The overall impact of the Information Sheet was to increase marginally the level of concern. The fraction of the sample that, after reading the Information Sheet, believed the risk to be possible or likely rose to 62% and 22% respectively, and the percentage of those who regarded it as unlikely or absurd decreased from 14% to 9%. Almost 35% of respondents believed that residents of their home are or may be at risk because of transmission or distribution lines or both, up from 21% before reading the Information Sheet. (The number of those who believed the exposure of their co-residents to be due to home appliances remained almost unchanged at about 12%.)

The Information Sheet proved to be rather ineffectual in correcting existing views. The statement in the Sheet that present knowledge "does not allow us to determine whether there is a specific magnetic field level above which chronic exposure is dangerous or compromises human health" did not deter many of the respondents from requesting "provisional" (30%) or even "strong" regulations (6%).

In the first version of the Information Sheet, we explained that electric currents generated magnetic fields and that distribution lines may carry weak or strong currents, depending on the number of houses they supply. In a pilot study conducted to test the questionnaire, the percentage of respondents that identified distribution lines as a possible source of significant magnetic fields rose, after reading the Information Sheet, from only 22% to 54%. This result prompted us to modify the Information Sheet to state that "distribution lines can generate a weak or strong magnetic field depending on the number of houses they supply." In the main study, the awareness of distribution lines as a potential source of magnetic fields rose from 29% to 75%, possibly as a result of the simplified information.

The ultimate purpose of the Information Sheet should be to make individuals capable of assessing whether they are at risk. That is of course extremely difficult, because very little is known about the hypothesized risk at this stage. In the case of residential exposure, the best we can do is to determine whether a house meets criteria that, in an epidemiological study, would result in a classification of "exposed." Our ongoing exposure assessment program suggests that no more than half of Melbourne homes could be confidently regarded as "low exposure." (The median front-door measurement in Melbourne is 1 mG.) However, of the houses that could be regarded as "high" or "medium" exposure, only about 5% should be regarded as predominantly affected by transmission lines, and in (possibly) another 5% of cases, an elevated magnetic field may be the result of either relatively distant transmission lines or close distribution lines. In most cases, only distribution lines are the cause of the elevated magnetic field.

However, our survey indicates that even after reading the Information Sheet, the respondents' focus remained on high-voltage transmission lines. Less than 10% of respondents believe or suspect that residents of their own home may be at risk because of distribution lines alone, and about 24% believe or suspect that transmission lines might pose a hazard within their own home, either alone (7%) or together with distribution lines (17%).

Although this lack of concern about distribution lines may be welcomed by some, it has two negative connotations: it implies that the risk is perceived to be very unequally distributed among the community, and it raises the possibility that important decisions (including the purchase and, in extreme cases, even the hasty sale of the family home) may be taken in the light of a mistaken assessment of the potential risk.

CONCLUSION

The main effect of the Information Sheet appears to be that information from what is perceived to be an authoritative source lends some credibility to media reports. This effect is measured by a slight shift of belief from ''absurd'' or ''unlikely'' toward ''possible'' or ''likely'' and by a noticeable increase of the fraction of respondents that believe or suspect themselves and their co-residents to be personally at risk. However, the Information Sheet appears to be ineffective in imparting information that leads to a real understanding of the issue. Until more quantitative scientific evidence (including a clearer definition of ''exposure'') becomes available, it is unlikely that this situation can be changed.

According to our results, the main topics on which a public education program should focus are the impossibility of introducing regulations and the redirection of the layperson's attention from transmission lines to more common sources of ELF magnetic fields. It will require great sensitivity to achieve this goal without creating still greater public anxiety.

REFERENCES

1. M. G. Morgan. *Electric and Magnetic Fields from 60 Hertz Electric Power: What Do We Know About Possible Health Risks?* Department of Engineering and Public Policy, Carnegie Mellon University, Pittsburgh, PA 15213, 1989.

Martin Blank, Ed., *Electricity and Magnetism in Biology and Medicine*

ESTIMATES OF EXPOSURE REDUCTION SUPPLY CURVES FOR SEVERAL EFFECTS FUNCTIONS

J. Zhang,* J. Adams, M. G. Morgan, and I. Nair

Carnegie Mellon University, Pittsburgh, Penn.

How much reduction in public exposure to power-frequency fields can be obtained for various levels of investment? The answer depends on the exposure context and the "effects function" that is assumed to describe the relationship between field exposure and biological effect. A variety of technologies might be used for reducing exposure in any particular context (e.g., from power transmission, power distribution, building wiring, appliances and other end products). In addition there are various effects functions that might describe the relationship between field and possible health risk (e.g., average field, *dB/dt*, resonant "windows," etc.). In this paper we describe a strategy by which exposure reduction supply curves (i.e., the cost of purchasing various amounts of exposure reduction) can be estimated parametrically for various exposure conditions and effects functions, and we display preliminary results.

In this paper we describe and demonstrate a procedure for estimating the cost effectiveness of various strategies for reducing exposures to magnetic fields. Suppose we know what attribute of the field is of importance. Suppose further that we have a set of mitigation technologies A, B, C, D. For each of them we can ask, how much exposure reduction can we get for a given level of investment? For simplicity, assume the technologies are mutually exclusive. Then if we plot the cost curve for each, the lower envelope of the resulting plots provides us with an estimate of the "exposure reduction supply curve," as illustrated below.

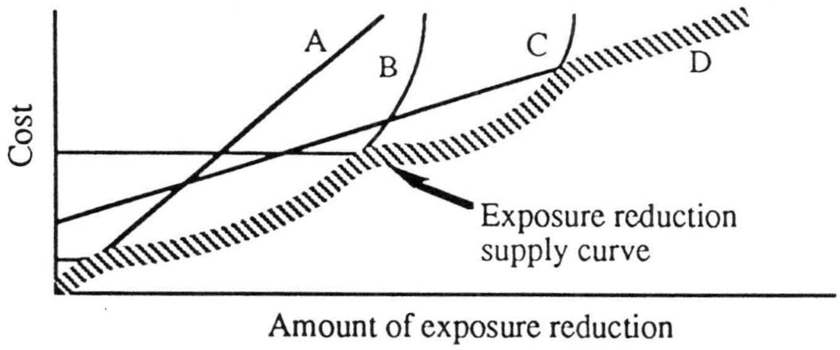

It is unclear what aspect (if any) of field exposure is important. We shall term the functional relations between the aspects of field that matters and the resulting effect an "effects function." Several possible effects functions have been postulated based on current literature: average field, cumulative field, binary in field presence, switching, field-strength window, frequency window, presence of some other agent such as DC field.[1] In other contributions, we have described how one can write a computer program to apply a number of different effects functions to personal monitor time series data.[2]

For example, suppose one wants to implement an effects function based on an intensity window. A sampled personal monitor time series is obtained, and a computer program then searches the time series for instances in which the field strength lies inside the range of the window, as shown below.

In computing an exposure reduction supply curve, one can perform such a simulation for the base-case exposure and then perform it for each of the various control technologies of interest. In this way, it becomes possible to produce a different exposure-reduction supply curve for each of the effects functions of interest. With these results one can ask, to what extent does a particular exposure management technology remain cost-effective across different assumptions about effects function? Such information should prove useful in future decision making about exposure management.

---

*Correspondence to: Jun Zhang, CMU Dept. of Engg and Public Policy, Pittsburgh PA 15213.

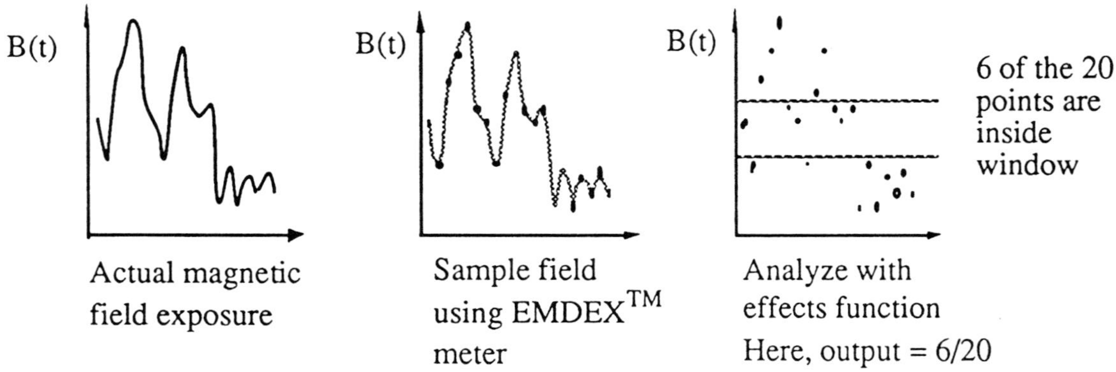

A Transmission-line Example

We demonstrate the procedure for the specific case of a "typical" 345kV transmission line. We take the current as roughly 500 A per phase and for simplicity assume no variation over time. The base case line configuration is flat with a 27ft phase-to-phase spacing. The line is assumed to pass through urban, suburban, and rural areas, with the line length distributed in proportion to the distribution of the U.S. population (26% urban, 52% suburban, 22% rural).

Population distributions around the line are based on a linearization of results from Florig and Morgan[3] and on 1990 U.S. census data.[4]

Emdex™ data were collected with a 1.5s sampling interval for a sample of 24 Pittsburgh residents. These data are taken as the base case exposure in the absence of a line. Thanks to superposition we can synthesize exposure time series for those living near the line, taking into consideration possible phase difference between the two and possible motion of the person within the home, as illustrated below.

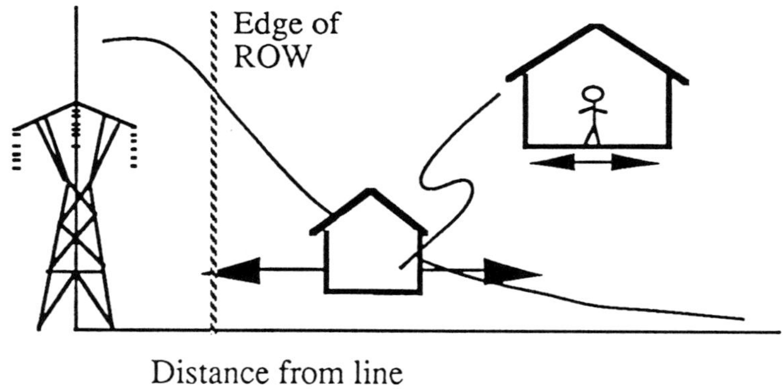

Once exposure has been estimated for a single individual, we can sum over all individuals in order to estimate the total exposure for the line, as shown in the diagram below.

We use three effects functions in this illustration: average field, threshold ($\geq$ 3 mG), and an amplitude window (1.5 mG $\leq B \leq$ 2.5 mG). Exposure control technologies considered were: convert from flat to delta configuration; compact the phases; convert to six conductors (same load); expand RoW to 120 or 160 m, and reroute (assume we can move half to next lowest population density by doubling length). To get the percentage exposure reduction for each migration strategy, we take the difference between the effects function output for the mitigate line and effects function output for the horizontal base case.

Estimates of costs of the various control options were obtained by use of data from Eseerco,[5] Zaffanella,[6] and various interviews with utility experts. Switching from flat to delta configuration was assumed to be costless. Compaction was taken to increase project costs by 10%. Switching to 6 conductors was estimated to increase costs by 33%. Right-of-way costs were estimated to be 16% of total project costs, and expanded right-of-way was taken to cost the same per unit area as original right-of-way. Rerouting was assumed to double the length of the line segment. All these values are likely to vary substantially from project to project.

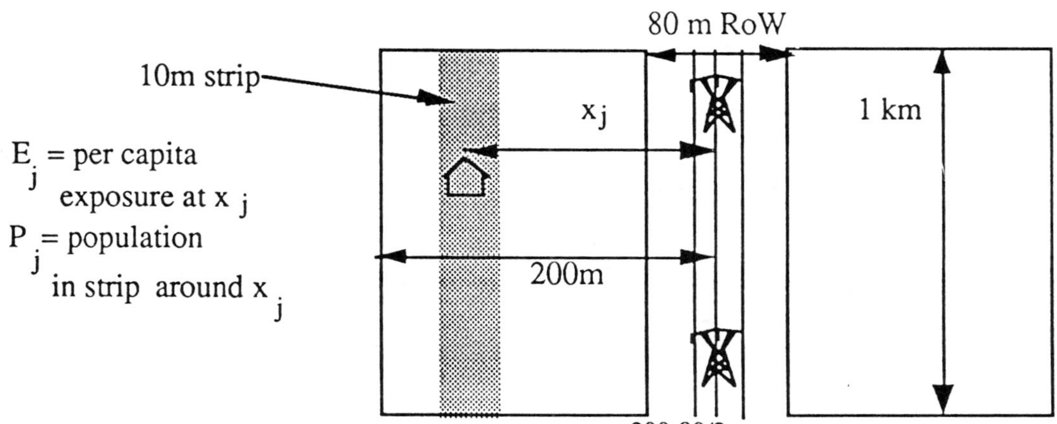

$$\text{Change in exposure per km of line} = 2 \sum_{j=1}^{\frac{200-80/2}{10}} \Delta (E_j)\, P_j$$

RESULTS

Assuming first that each control option is mutually exclusive, we estimate cost-effectiveness curves for the three effects functions as follows.

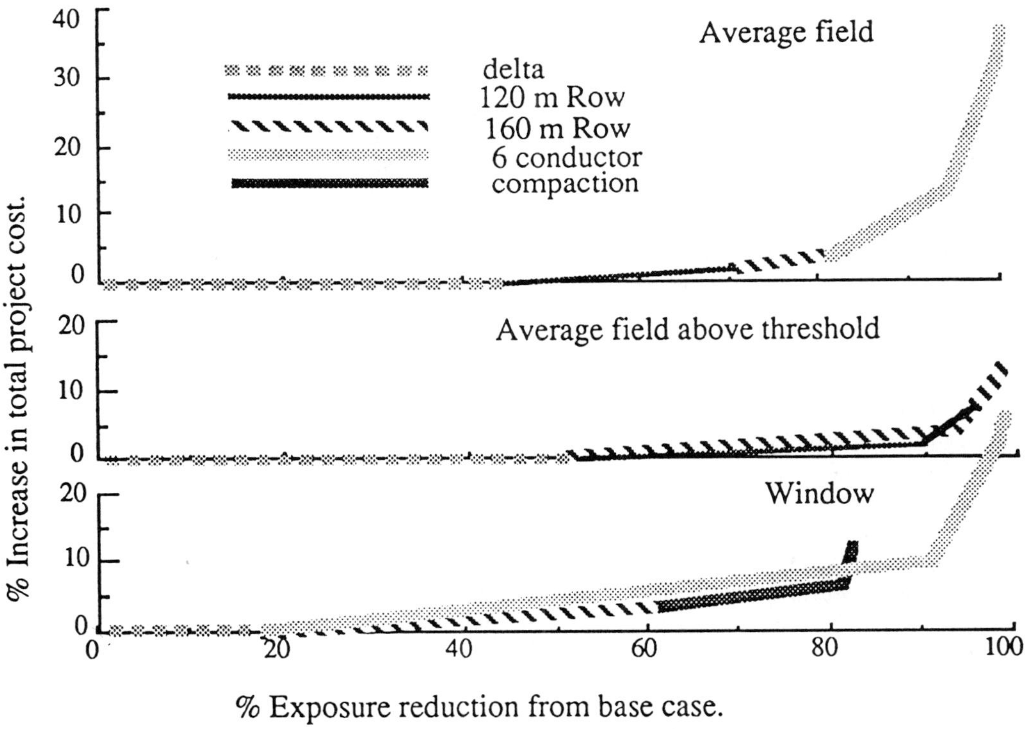

In fact, after we have chosen the first most cost-effective control option, we can go on to rerun the analysis to find the next most cost-effective option and so on, until we have traced out the entire supply curve. When we do that we obtain the following result.

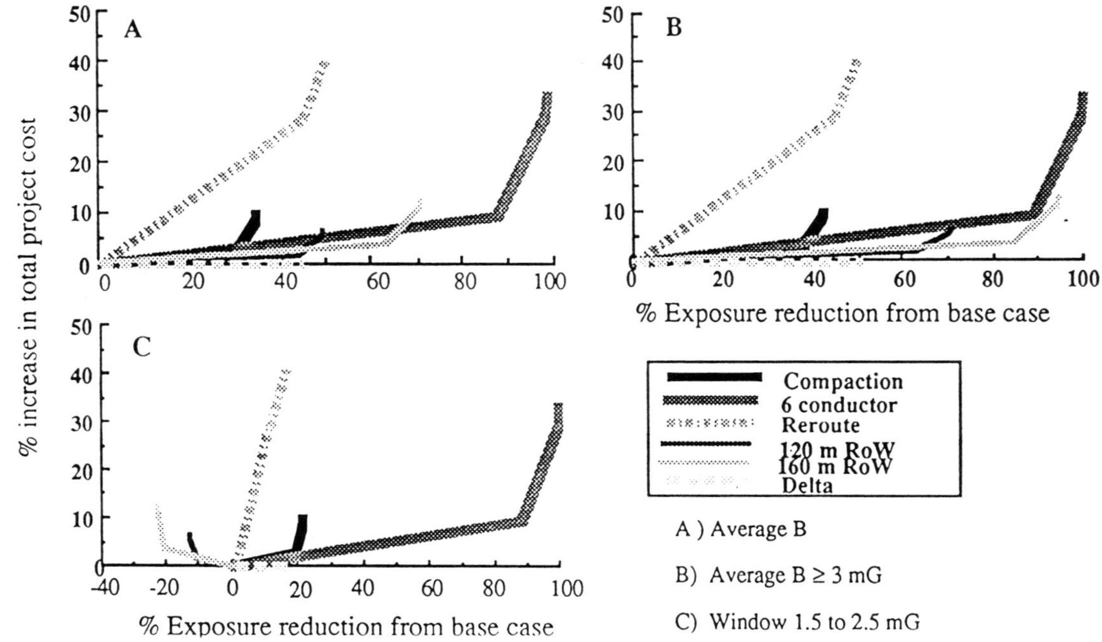

A) Average B

B) Average B ≥ 3 mG

C) Window 1.5 to 2.5 mG

CONCLUSION

The sample calculations performed in this paper clearly demonstrate that the best mitigation strategy may depend on which effects function is used. We plan to perform calculations similar to those demonstrated here with more complete data for transmission, distribution, and residential and occupational exposure situations. The results may provide a basis for guiding exposure management choices.

The method we have demonstrated will become more useful as the biological and epidemiological evidence grows more complete, since better knowledge will lead to more refined definition and use of effects functions.[7]

REFERENCES

1. M. G. Morgan and I. Nair, "Alternative functional relationships between ELF field exposure and possible health effects: Report of an expert workshop," *Bioelectromagnetics* (in press).

2. M. G. Morgan, I. Nair, and J. Zhang, "A method for assessing alternative effects functions that uses simulation with EMDEX^{TM} data," *Bioelectromagnetics* (in press).

3. K. Florig and M. G. Morgan, "Measurements of housing density along transmission lines," *Bioelectromagnetics* 9: 87, 1988.

4. *Statistical Abstract of the United States,* 1991, 11th ed., 21-35.

5. Empire State Electric Energy Research Corporation, *Magnetic Field Levels Associated with 345kV Transmission Circuits in New York State,* June 1989.

6. L. Zaffanella, "Overhead power lines," *EMF Science & Communication Seminar,* San Jose, Calif., 1991.

7. We thank J. Greenlea and her colleagues and J. Sandford for advice and data, and P. Steranchak for assistance in preparing the manuscript. The work was supported by Southern California Edison and by EPRI. The authors are solely responsible for the contents.

Martin Blank, Ed., *Electricity and Magnetism in Biology and Medicine*

ASPECTS OF PRESENT-DAY RADIATION PROTECTION FOR THE FREQUENCY REGION 30 kHz-300 GHz

R. Matthes*

Institute of Radiation Hygiene, Neuherberg, Germany

Today in many countries the national standards for radiation protection on nonionizing radiation are under revision or have been recently revised. In the USA, the revised American National Standard (ANSI C95.1-1991) recommending safety levels with respect to human exposure to radiofrequency electromagnetic fields from 3 kHz to 300 GHz was developed in 1991. In Germany a general revision of the corresponding national standard (DIN 0848, Part 2, draft) for the frequency range from 30 kHz to 300 GHz had also been published in 1991. A new standard in the United Kingdom is expected. The Commission of the European Communities is drafting a directive which will, after approval, become a legal EC standard, at least for workplaces. A complete technical standardization at the European level in the same sphere has begun. The reason for all these activities is the continuing scientific progress concerning the interaction of nonionizing radiation with people, as well as the increasing public concern about possible health hazards from new widespread technical applications of nonionizing radiation.

The scope of all standards in this field is the protection of people from possibly harmful effects of electromagnetic-field exposure. In the revised German standard (DIN 0848, Part 2) a distinction is made between direct exposure effects in the body and indirect exposure effects arising from touching electrically conductive parts in the electromagnetic field or from electromagnetic incompatibility of medical implants such as cardiac pacemakers (Fig. 1). The indirect exposure hazards are not considered in the electromagnetic field limits but a separate limitation of the resulting body currents is recommended (Fig. 2).

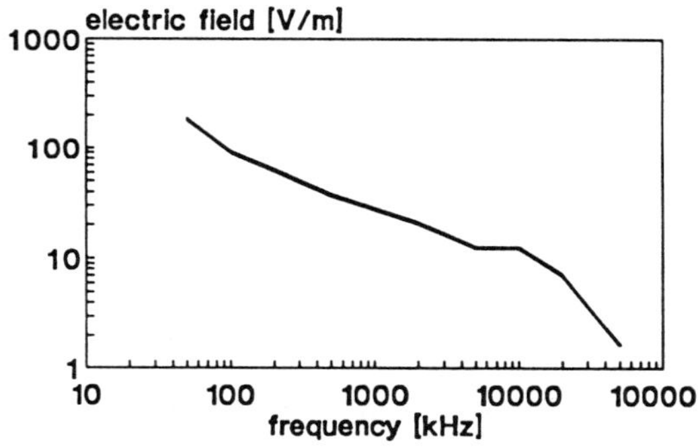

FIG. 1.—Limit for interference of electric fields with cardiac pacemakers.

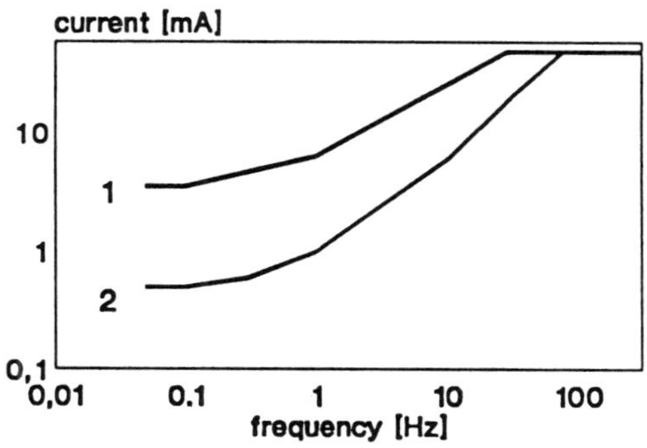

FIG. 2.—Limits for body current resulting from touching conductive objects. (Numerals correspond to Table 1 areas.)

Another distinction is made concerning various exposure environments or situations with respect to various parts of the population for which precautionary aspects should be considered (Table 1). In addition to the safety levels, a second lower value is necessary in environments where people have no knowledge or control of their exposure, where chronic exposure is possible, and where prevention of nonhazardous but uncomfortable exposure effects is needed, e.g., in living quarters.

*Correspondence to: R. Matthes, Bundesamt f. Strahlenschutz, Institut f. Strahlenhygiene, Ingolstädter Landstr. 1, D-8042 Neuherberg, Germany.

TABLE 1.—Distinction for two tiers for exposure limits.

exposure area 1	exposure area 2
controlled areas, for example: - operating areas - areas which can be controlled by the operator - generally accessible areas where it is ensured, because of the operation mode of the installations or the duration of stay, that the exposure is only of limited duration (no longer than 6 hours per day)	areas where long term exposure is possible such as for example: - areas with residential and social buildings - separate residential estates - sports, leisure time and holiday facilities - workplaces where the generation of fields is not expected

Like most of the international standards for high-frequency electromagnetic fields, the German standard is based on the limitation of induced body currents and on the concept of specific absorption rate (SAR) (Table 2). These quantities are defined to be basic. They must always be observed in every exposure situation.

TABLE 2.—Basic limits for electric current density, specific absorption, and specific absorption rate resulting from exposure to high-frequency electromagnetic fields.

frequency range		exposure area 1	exposure area 2
30 kHz to 100 kHz	spatial	f/12 mA/(m²Hz)	f/30 mA/(m²Hz)
0.1 MHz to 1 MHz	spatial whole body	f/12 mA/(m²Hz) 0.4 W/kg	f/30 mA/(m²Hz) 0.08 W/kg
> 1 MHz	whole body spatial hand, wrist, ankle	0.4 W/kg 10 W/kg 20 W/kg	0.08 W/kg 2 W/kg 4 W/kg
short pulses		10 mJ/kg	2 mJ/kg
current densities averaged over 1 cm² and 1 s all SAR-values averaged over 6 min spatial peak values averaged over 10 g tissue mass			

The limit values of current density were derived from a classification (UNEP/WHO/IRPA, 1987) of biological effects related to body current density in the extremely low frequency range between 3 Hz and 300 Hz (Fig. 3). For higher frequencies a relaxation proportional to the increasing frequency was assumed. This concept leads to conservative current density limits compared to other standards; for example, ANSI C95.1-1991. This current density limit (Fig. 3) is the only base limit for frequencies up to 100 kHz. Above 100 kHz this current density leads to an energy absorption higher than the SAR limit of 0.4 W/kg. Therefore the current density limit is not valid for whole-body exposure above this frequency. For spatial exposure this current density limit is valid up to approximately 1 MHz. Above 1 MHz only the SAR concept is valid. In other standards the current density limits are derived from the research on sensory effects and radiofrequency burns. In the German standard this is considered separately as an indirect field effect (Fig. 2) when touching conducting objects in electromagnetic fields.

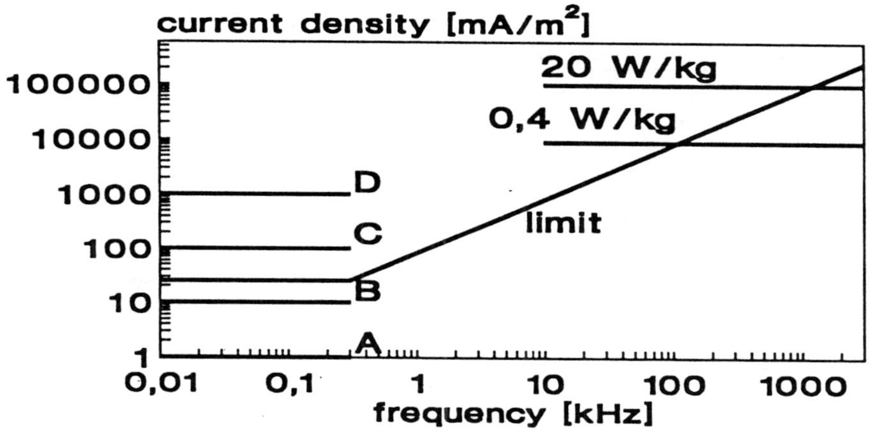

FIG. 3.—Derivation of current-density limit for low-frequency effects. A-D are threshold levels (UNEP 87) for (A) minor biological effects, (B) well-established effects, (C) possible health hazards, (D) acute health hazards. The current density limit is only valid up to frequencies at which energy absorption becomes dominant (0.4 W/kg whole body ≈ 100 kHz, 20 W/kg spatial ≈ 1 MHz).

The basic SAR limits can be derived from the current scientific status of biological effects of high-frequency electromagnetic exposition. The values shown in Table 1 are in good agreement with the corresponding values of other standards or guidelines. At the high-frequency end, above several gigahertz the penetration depth of electromagnetic fields in the body decreases and most of the energy is absorbed in a thin surface layer. In some standards the SAR concept is therefore not valid in this frequency region; it is replaced by the power flux density. In most cases the limit of the power flux density is adapted to existing infrared standards. In the German standard, the SAR concept is kept up all the way to the highest frequency. The decrease of the absorption layer must be considered in the SAR limits for spatial exposure. In this case, proper selection of the shape of the averaging mass is essential. It is not recommended to use a fixed cube for spatial averaging at all exposure conditions. However, the volume must be selected so as to gather the main absorbing parts of the tissue. With regard to very sensitive organs (e.g., the eyes) and to excessive absorption peaks within small portions of tissue, the averaging mass for spatial SAR of 10 g was chosen, which is between the two extremes of 1 g and 100 g used in other standards.

There is no exclusion clause in this standard, especially not for low-power devices. The absorption in close vicinity to such devices shows a very complex dependency from power, frequency, distance, and type of antenna used. Therefore, such devices should be designed in compliance with the basic SAR limits. Beside the standard, there are recommendations in Germany for mobile telecommunications. This recommendation provides guidance for the user (Table 3). The given safety distances for mobile telephones are not meant to be a standard for the design of such equipment.

TABLE 3.—Minimum safety distances (cm) between mobile telephones and the body. (Derived from basic spatial SAR value of 2 W/kg.)

	0.5 W	1 W	2 W	4 W	5 W	8 W	20 W
450 MHz analog	0	4	--	--	20	--	40
900 MHz analog	0	5	--	--	25	--	50
900 MHz digital	--	--	0	3	--	5	8
1.8 GHz digital	--	0	3	--	--	7	12

For practical purposes, limits for electric and magnetic field strength and for the power flux density are derived. They consider worst-case exposure conditions and body resonances.

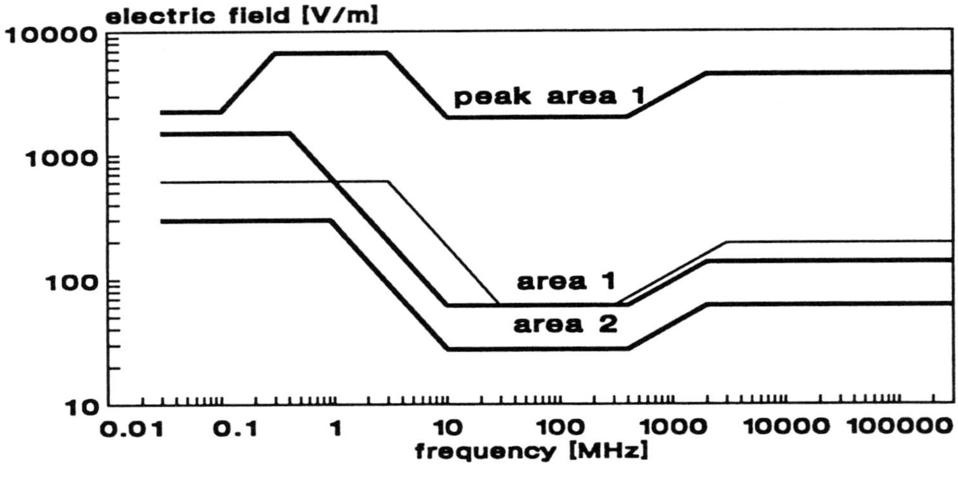

FIG. 4.—Limits of electric-field strength for direct exposure effects in exposure areas 1 and 2. Exposure duration longer than 6 min. Absolute peak value limits for exposure area 1 are also shown. Thin line shows corresponding ANSI limit.

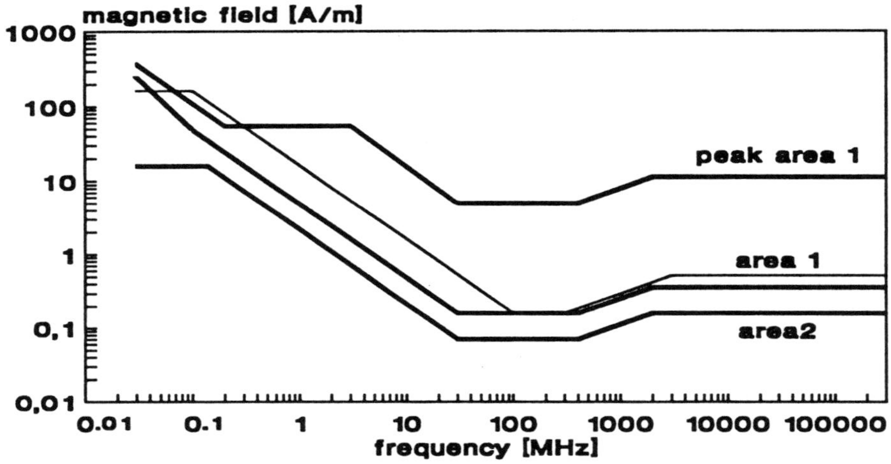

FIG. 5.—Limits of magnetic-field strength for direct exposure effects in exposure areas 1 and 2. Exposure duration longer than 6 min. Absolute peak value limits for exposure area 1 are also shown. Thin line shows corresponding ANSI limit.

They are only valid for far-field conditions. In Figs. 4 and 5 the electric and the magnetic field limits including the absolute peak value for extreme short pulses are shown. For the frequency range above 30 MHz, the power flux density limit is given as calculated from far-field relations.

BIBLIOGRAPHY

American National Standard (ANSI C95.1-1991), *Safety Levels with Respect to Human Exposure to Radio Frequency Electromagnetic Fields, 3 kHz to 300 GHz.*

DIN VDE 0848, Part 2, 1991 (draft), *Safety in Electromagnetic Fields: Protection of Persons in the Frequency Range 30 kHz to 300 GHz.*

S. G. Allen et al., "Proposals for basic restrictions for protection against occupational exposure to electromagnetic non-ionizing radiations: Recommendations of an international working group set up under the auspices of the Commission of the European Communities," *Physica Media* 7(2): 1991.

United Nations Environment Programme/World Health Organisation/International Radiation Protection Association, *Environmental Health Criteria 69: Magnetic Fields,* Geneva: World Health Organisation, 1987.

Martin Blank, Ed., *Electricity and Magnetism in Biology and Medicine*

30 YEARS OF BIOELECTRICAL REPAIR AND GROWTH: A BIBLIOGRAPHIC ANALYSIS

J. A. Spadaro*

Health Science Center, State University of New York, Syracuse

An analysis was made of the "demographics" of the published literature on bioelectrical research relevant to the growth and repair of skeletal tissues. It was based on a laboratory database of 892 publications, collected during 1962-1992 (excluding BRAGS abstracts). The tool for the analysis was the 'Endnote Plus' bibliographic database program. Results to data show the rapid expansion of interest in the mid 1970s, peaking in the mid 1980s. The distribution also shows a shift of interest from DC methods to inductively coupled (PEMF) and capacitive applications after 1980.

It is of interest to track the development of bioelectrical research applied to the growth and repair of skeletal tissues since the early 1960s, when the present era began. One approach is to analyze the "demographics" of the published literature. We attempted this approach by using a laboratory database of 892 publications, a significant portion of the worldwide literature on the subject (1962-1992). Such an analysis reflects past and present changes in amount of activity, methodology, funding, commercialization, and other aspects.

METHOD

The bibliographic database ($n = 892$) used for this study has been collected over 20 years. It represents over 90% of published reports concerning bioelectrical research and practice applied to bone, cartilage, and other skeletal structures. Even so, some selection biases exist in such a collection owing to problems such as foreign languages and availability of journals. No selection was made in the analysis on the basis of quality or uniqueness of the work reported. Citations from the meetings of the Bioelectrical Repair and Growth Society (BRAGS) are *not* included in this database, unless they were also published separately.

For maintenance and analysis the bibliographic database program "Endnote Plus" (Niles & Assoc., Emeryville, Calif.) was used in the Apple Macintosh computer environment. The subject classifications were done by referring, in most cases, to the journal texts. In the accompanying charts, the term "direct current" is used here to include continuous DC and slowly (≤ 1 Hz) interrupted DC; "alternating current" denotes strictly sinusoidal as well as electrode-delivered pulsating waveforms; "PEMF" (pulsating electromagnetic fields) denotes inductively coupled pulses of various waveforms; "capacitive" denotes capacitive electric fields or skin-coupled stimuli; "endogenous SGP" denotes studies of natural and stress-generated electrical activity. Reports that discuss more than one type of stimulus are included with each type.

RESULTS AND CONCLUSIONS

The figures show an expansion of activity from 1974, peaking in the mid 1980s, possibly associated with the start of clinical application to ununited fracture and commercialization (Fig. 1). The shift in annual publication rate from DC and AC to noninvasive PEMF and capacitive types after 1980 is seen in Fig. 2 and 3. The decline in clinical reporting (Fig. 3, bottom) is open to interpretation. It is interesting that only 17 of the hundreds of authors appeared in 10 or more articles (Fig. 4). Of the approximately 400 first authors, 17% are currently members of the Bioelectrical Repair and Growth Society.

ACKNOWLEDGMENT

This project was supported by the SUNY-HSC-Syracuse Department of Orthopedic Surgery. The author thanks all who have helped with this task, particularly Mary Spadaro, Sharon Chase, Norma Henderson, Eleni Mantis, and Analise Brause. For more information on the database, please contact the author.

---

*Correspondence to: J. A. Spadaro (Orthopedic Surgery), SUNY-HSC Rm 4400, 750 E Adams St., Syracuse NY 13210.

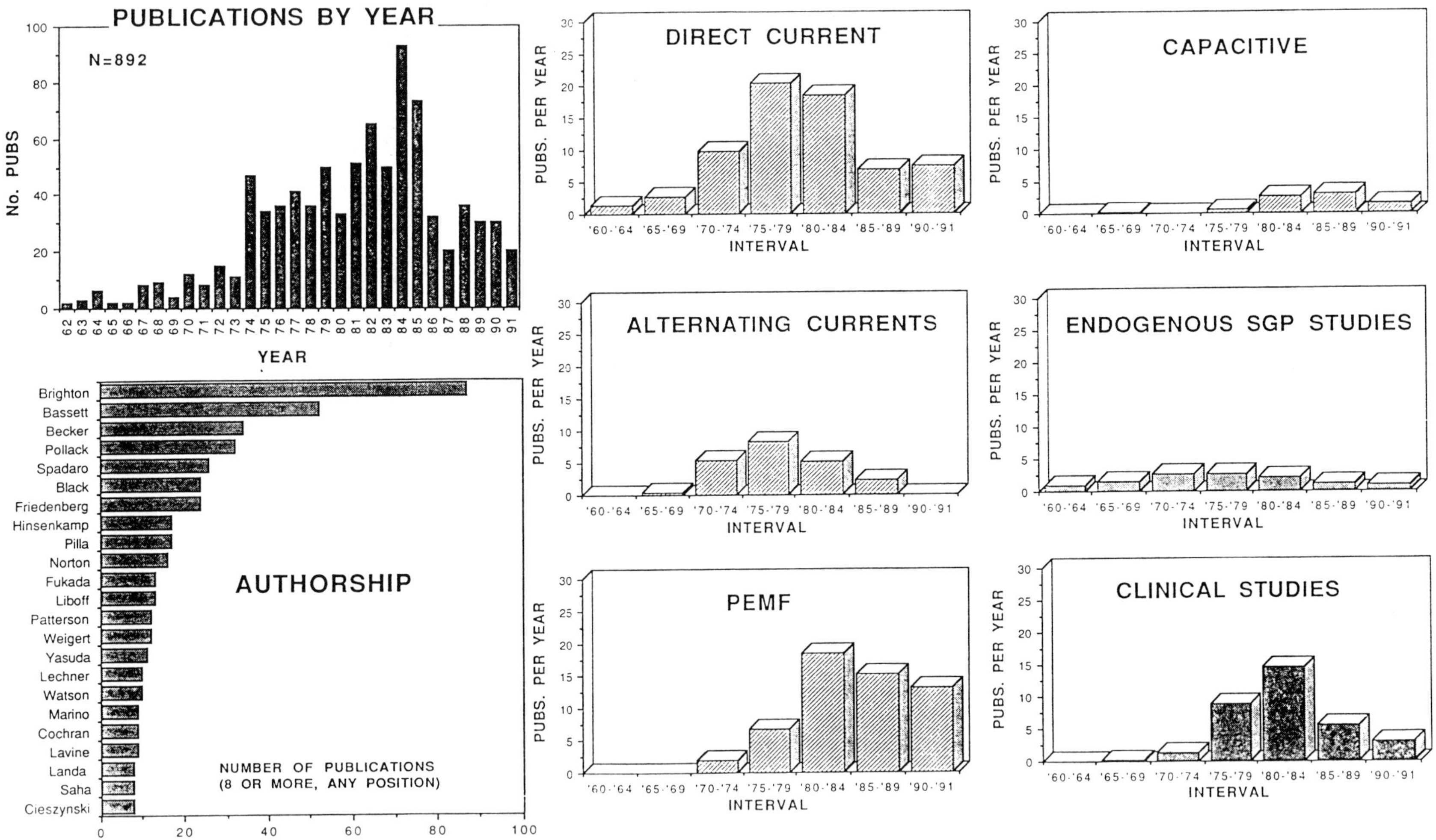
PUBLICATIONS BY YEAR
N=892
No. PUBS
YEAR
62 63 64 65 66 67 68 69 70 71 72 73 74 75 76 77 78 79 80 81 82 83 84 85 86 87 88 89 90 91
AUTHORSHIP
Brighton
Bassett
Becker
Pollack
Spadaro
Black
Friedenberg
Hinsenkamp
Pilla
Norton
Fukada
Liboff
Patterson
Weigert
Yasuda
Lechner
Watson
Marino
Cochran
Lavine
Landa
Saha
Cieszynski
NUMBER OF PUBLICATIONS
(8 OR MORE, ANY POSITION)
DIRECT CURRENT
ALTERNATING CURRENTS
PEMF
CAPACITIVE
ENDOGENOUS SGP STUDIES
CLINICAL STUDIES
PUBS. PER YEAR
'60-'64 '65-'69 '70-'74 '75-'79 '80-'84 '85-'89 '90-'91
INTERVAL

Martin Blank, Ed., *Electricity and Magnetism in Biology and Medicine*

CELL DENSITY EFFECTS ON THE INTERACTION OF ELECTRIC FIELDS WITH CELLS IN VITRO

K. J. McLeod,* H. J. Donahue, M.-A. Fontaine, and C. T. Rubin

State University of New York, Stony Brook

The effect of cell size and shape on the interaction of extremely low frequency (ELF) electric fields with cells was investigated by manipulation of bone cell morphology through changes in cell density. When osteoblast-like cells were exposed to 30Hz electric fields of 6 μV/cm rms, the alkaline phosphatase activity of the cells, as normalized to either cell number or total protein, increased significantly. This increased enzyme activity depended critically on the cell density, with the optimal density for cell sensitivity in the range of 200×10^3 cells/cm^2. Analysis of these results indicates that cell shape is a critical parameter defining the sensitivity of osteoblasts to electric fields.

Independent of the biological state of a cell, there exists specific physical characteristics of the cell that may significantly affect the extent to which an ELF electric field can perturb the cell. The initial interaction between ELF electric fields and the cell must occur through electric charges located at the extracellular surface of cells (either in the cell glycocalyx or on the plasma membrane). Therefore, changes in the physical characteristics of a cell population that alter either the spatial distribution of the electric fields around the cells or the amount of exposed charge on the cells should substantially affect field-cell interactions. Indeed, we have previously shown that for cells of similar size and shape, the cellular orientation with respect to the electric field alters the magnitude of the cellular response to the electric field.[1] Perhaps of more physiologic relevance are the roles of cell size and cell shape in mediating electric-field interactions. These two morphologic characteristics affect field cell interactions in an interdependent manner; therefore, we have investigated the extent to which both cell size and cell shape can affect cellular interactions with ELF electric fields.

We have investigated the role of cell size and shape using osteoblast-like cells grown in monolayer culture. Bone cells are uniquely suited for studies on the interaction of electric fields with cells because endogenous electric fields of substantial intensity normally occur within bone,[2,3] and exogenously applied fields are known to stimulate bone formation in vivo.[4] We have utilized a clonal osteoblastic rat osteosarcoma cell line (ROS 17/2.8) and have obtained variations in cell size and shape by controlling maximum cell densities in the culture wells. In addition, to insure that a sufficient number of experimental samples could experience simultaneous and identical exposure conditions, and to provide a credible sham exposure, we have developed a new electric-field exposure apparatus. We have utilized this apparatus to examine the effect of electric fields on cell proliferation and alkaline phosphatase activity as a function of cell density.

MATERIALS AND METHODS

Cells and Growth Conditions

ROS 17/2.8 cells were cultured in Ham's F-12 medium supplemented with 10% fetal bovine serum (FBS), 2 mM L-glutamine and 1 mM sodium pyruvate, and transferred to eight-well Tissue-tek culture plates (Nunc) at nominal plating densities of 6, 12, 24, and 48×10^3 cells/cm^2 and covered with 0.5 ml of supplemented medium. Cells were maintained for 60 h prior to use.

Exposure Apparatus

An exposure system was developed for this study which insures a relatively uniform exposure for the cells within each exposed sample and provides a rigorous sham exposure. This system incorporates two critical elements. First, exposure to the low-frequency electric fields is accomplished by bifilar wound solenoidal magnetic coils to induce fields into the monolayer cultures. Second, the cells are grown in monolayer culture

*Correspondence to: K. J. McLeod, SUNY Dept. of Orthopaedics, Stony Brook NY 11794-8181.

in square well culture dishes, which forces the induced electric field to be relatively uniform across the monolayer surface.

The individual solenoid coils used for both the sham and treatment exposures consisted of 4cm diameter, 25cm long hollow cores with bifilar windings. In the treatment configuration, the two windings on the solenoid coil were attached in series so as to produce additive magnetic flux densities summing to 1.8mT for a 30mA root mean square (rms) current. The calculated induced electric field in the eight-well plates at 30 Hz, with 0.5 ml medium (for a medium height of 0.7 cm and well width of 0.9 cm) was 6 μV/cm rms for a 1.8mT flux. In the sham exposure configuration, the two windings of the solenoid coil were connected so as to produce opposing magnetic flux densities. In this configuration, the measured magnetic flux in the interior of the coil was less than 0.01 mT for a 30mA rms current.

Assays

Following the 12h exposure period, cell number in the eight-well Tissue-tek plates was determined by counting, with a hemacytometer, the cells in two of eight wells at each plated cell density. Projected area of the cells in monolayer culture was calculated from perimeter tracings of photographic images of the cultures obtained at densities of approximately 10, 50, 150, and 450×10^3 cells/cm^2. Total protein per well was determined using the method of Bradford,[5] and alkaline phosphatase activity was determined as the rate of conversion of p-nitrophenyl phosphate to p-nitrophenol.

RESULTS

Increases in cell density were associated with a substantial decrease in cell area (Fig. 1). Average projected cell area decreased from approximately 4400 μm^2 for cell densities near 10×10^3 per cm^2; to 2400 μm^2 at 60×10^3 per cm^2; to 630 μm^2 at 150×10^3 per cm^2; and to 180 μm^2 at 450×10^3 per cm^2. Total cell-associated protein also decreased over this same density range, from 420 μg/10^6 cells at a density of 10×10^3 cells/cm^2 to less than 90 μg/10^6 cells at a density of 550×10^3 cells per cm^2. Alkaline phosphatase activity was found to be relatively stable as a function of cell density when normalized to either cell number or total protein. When normalized to cell number, alkaline phosphatase activity averaged 165 ± 13 nanomoles p-nitrophenol/min/10^6 cells, and when normalized to total protein, average alkaline phosphatase activity was found to be 670 nanomoles p-nitrophenol/min/mg protein.

Following the 12h electric-field exposure at 30 Hz, 6μV/cm rms, alkaline phosphatase activity was seen to increase, as normalized to either cell number (Fig. 2) or protein. This stimulation occurred in a density-dependent manner. Scatterplot curve fitting identified an optimal density near 200×10^3 cells/cm^2. Regression analyses of the data above and below this cutoff point confirmed a significant difference in the alkaline phosphatase activity as normalized to either protein or cell number.

DISCUSSION

The field-exposure apparatus designed for this study has allowed us to demonstrate a response of ROS 17/2.8 cell alkaline phosphatase activity to electric fields induced into the media. Moreover, the apparent sensitivity of the cells to this ELF field appears to be strongly dependent on the density of the growing cells.

An important question is whether the observed density-dependent response is due to differences in the biological state of the cells at different cell densities, or to density-dependent physical differences in the cells. Because the growth rates of the cultures were uniform at all densities, and because the control cultures showed no substantial differences in alkaline phosphatase activity with density, we believe the differential responses to the induced electric fields reflect true differences in sensitivity due to physical differences in the cells at different densities.

We investigated the possibility that sensitivity may depend on the shape of the individual cells in the cell population by calculating a size-independent index for the cells as a function of density, based on the surface area measurements of the cells. As cell area is a function of both cell shape and cell size (volume), we have utilized a shape index (area$^{1.5}$/volume, using protein/cell as a measure of cell volume, and protein area as a measure of cell height) to characterize the three-dimensional shape changes in the cell, independent of the size of the cell (Fig. 3).

Although the cells progressively decrease in area and volume with increasing cell density, the sensitivity of the cells to the extracellular electric fields first increases and subsequently decreases with increasing cell density, which suggests that neither cell area nor cell volume controls the cell response to the electric fields in

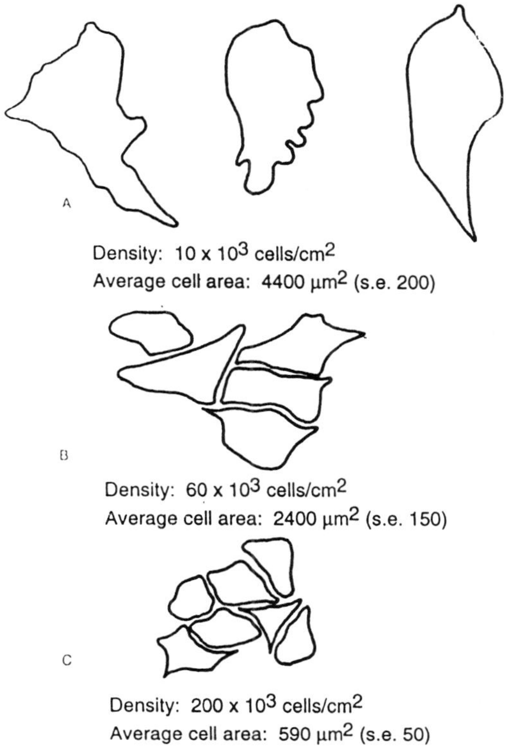

FIG. 1.—Effect of cell density on projected areas of cells. Projected areas of cells are shown at approximate cell densities of (a) 10, (b) 60, (c) 200×10^3 cells/cm^2. Average cell areas at these densities were (a) 4400 ± 200, (b) 2400 ± 150, (c) 590 ± 120 μm^2.

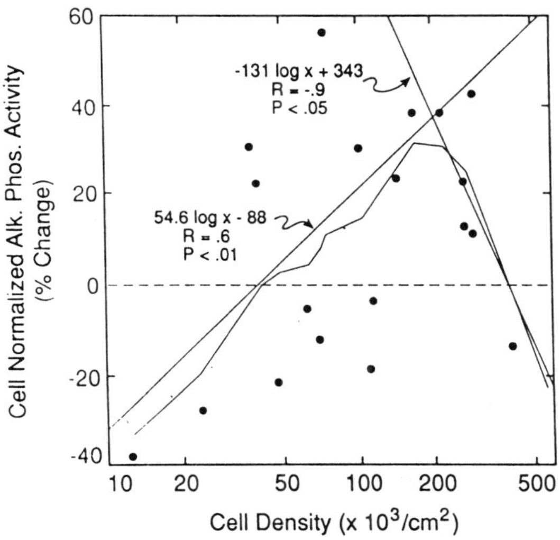

FIG. 2.—Cell density dependent effect of electric fields on alkaline phosphatase activity normalized to cell number. Statistical values within the figure reflect confidence that regression lines differ significantly from zero slope. Scatterplot data have been fitted to a curve constructed according to algorithm of Cleveland;[6] each data point is average of four samples.

its environment. However, the responsiveness of the cells does demonstrate an inverse correlation with the shape index. This analysis identifies cell shape, or perhaps more accurately the effective height of the cell, as a critical physical factor mediating the interaction of the ELF field with cells. From 10×10^3 to 60×10^3 cells/cm^2 the cell height or thickness is increasing, consistent with the increase in cellular sensitivity to the electric fields. Above 60×10^3 cells/cm^2 there is additional increase in cell thickness, but the increase in density results in a dense, uniform cell sheet in which the individual cells do not appear to rise above the cell layer. From the perspective of an electric-field interaction, the effective height of the cell appears to return to a small value.

SUMMARY

A dominant parameter for defining cellular sensitivity to electric fields for monolayer cultures appears to be cell shape, or alternately the height of the cells above the substrate. The critical factor in establishing a field-cell interaction may therefore be the degree to which the extracellular electric field is distorted around the cell. If it is the magnitude of the spatial gradients induced in the electric field that is critical to effect a cellular response, the results can be extended to three-dimensional tissue structures in order to identify the tissues that may be most susceptible to extracellular electric field induced responses. In addition, the identification of the role of cell shape in the regulation of electric field-cell interactions may help to define the physical mechanism by which electric fields initially interact with living cells.

REFERENCES

1. K. J. McLeod, R. C. Lee, and M. P. Ehrlich, "Frequency dependence of electric field modulation of protein synthesis in fibroblasts," *Science* 236: 1465-1469, 1987.
2. I. Yasuda, "Piezoelectric activity of bone," *J. Japanese Orthop. Surg. Soc.* 28: 267, 1954.
3. C. A. L. Bassett and R. O. Becker, "Generation of electric potentials by bone in response to mechanical stress," *Science* 137: 1063-1064, 1962.
4. C. T. Rubin, K. J. McLeod, and L. E. Lanyon, "Prevention of osteoporosis by pulsed electromagnetic fields," *J. Bone & Joint Surg.* 71-A(3): 411-417, 1989.
5. M. Bradford, "A rapid and sensitive method for the quantification of microgram quantities of proteins using the principle of protein-age binding," *Anal-Biochem.* 72: 248-254, 1976.
6. W. S. Cleveland, "Robust locally weighted regression and smoothing scatterplots," *J. Amer. Stat. Assoc.* 74: 829-836, 1979.

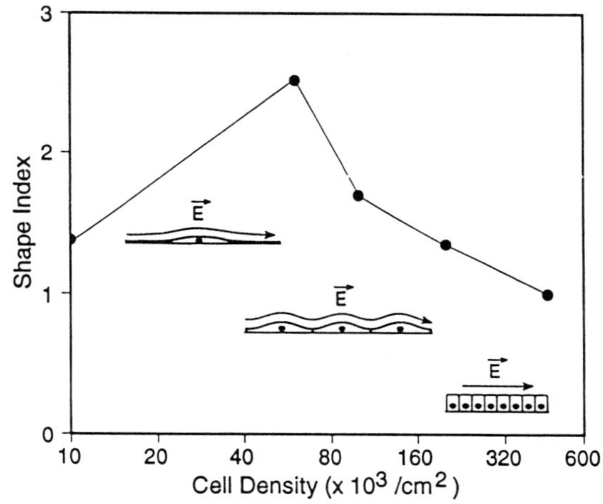

FIG. 3.—Shape index and cell profile representations as a function of cell density. Shape index is calculated as cell area$^{1.5}$/cell volume, with cell protein used as estimator of cell density. Insets of cell profiles show schematic representation of one interpretation of ELF electric field with cells growing in monolayer. At low densities cells appear as flat structures, which minimally perturb extracellular field. At higher densities, cells become more rounded, producing substantial gradients in extracellular fields. At confluency, cells approach a cuboidal shape, but the effect of neighboring cells is to make the sheet of cells appear flat, which again precludes any interaction with electric field.

Martin Blank, Ed., *Electricity and Magnetism in Biology and Medicine*

MEASUREMENT OF CYTOSOLIC CALCIUM CONCENTRATION OF PRIMARY BONE CELL CULTURE UNDER CALCIUM ION RESONANCE CONDITIONS

K. A. Reinbold* and S. R. Pollack

University of Pennsylvania, Philadelphia

O. L. da Silva

University of Sao Paulo, Brazil

Significant biological effects are seen in living organisms and cells when exposed to parallel AC and DC electromagnetic fields with frequency set according to the resonance condition, $\omega_c = qB_0/m$. Primary rat calvarial bone cells were exposed to these fields and cytosolic calcium was measured by fura-2 fluorescence technology. Bradykinin added directly after field exposure was the agonist control. No significant calcium concentration increase occurred during 20 min of exposure. However, a marked alteration in the bradykinin response occurred after the field was turned off. We believe that the resonance condition may operate through a modulation of an agonist-receptor system.

Specific combinations of low-frequency, low-amplitude AC and DC electromagnetic (EM) fields have been shown experimentally to produce significant biological effects in living organisms and cells.[1-5] These effects have a striking resonance character when the applied EM field is given by

$$B = B_0 + B_1 \cos(\omega_c t) \tag{1}$$

such that $\omega_c = qB_0/m$, where ω_c is the resonance frequency, q and m are the charge and mass of ions such as Ca^{2+} and Na^+, and B_0 and B_1 are the magnitudes of the parallel DC and AC EM fields, respectively.

Several mechanisms have been proposed that offer the possibility of EM field interactions according to the relationship $\omega_c = qB_0/m$. This relationship suggests interactions between the applied field and various ionic species. Liboff proposed the ion cyclotron resonance (ICR) theory,[6] and McLeod proposed trajectory specific gating of ion channels.[7] Other hypothesized mechanisms involve modulation of the binding energies of Ca^{2+}-binding proteins such as calmodulin or troponin C, and Larmor precession of orbits.[8-11]

Since Ca^{2+} is ubiquitous to many cellular biochemical pathways, specifically cell signaling, many experiments were designed to measure the effects of EM fields at frequencies tuned to this ion. Liboff et al.[2] and Rozek et al.[4] measured increased $^{45}Ca^{2+}$ incorporation in human lymphocytes as measured after field exposure, which suggests a transmembrane ionic influx mechanism is involved, The ICR effects on diatoms[1] as well as the fact that nifedipine blocks $^{45}Ca^{2+}$ uptake[4] seemed to point to a channel-like site for signal transduction, which further strengthened the influx argument.

On the other hand, Parkinson et al. attempted but did not observe real-time changes of cytosolic Ca^{2+} concentration using fura-2, a Ca^{2+}-selective fluorescent probe, and spectrofluorometry with various cell lines when exposed to these fields.[12] This result may be due to the fact that cells from these assumedly asynchronous cells may have been in varying phases of the cell cycle and may respond to field exposure to varying degrees.[13,14] If only a small portion of the cell population had responded to the field exposure, it may be that the signal-to-noise ratio under the conditions of their experiment was too low for a response to be observed. Single-cell experiments can obviate this question; that was the technology used in this study to determine whether rapid Ca^{2+} influx occurs in primary-culture rat calvarial bone cells when subjected to EM fields that satisfy the ICR conditions.

---

*Correspondence to: K. A. Reinbold (Bioengineering), 120 Hayden Hall, 240 S. 33rd St., Philadelphia PA 19104-6392.

MATERIALS AND METHODS

Calvaria were dissected from 1 to 3 day old Sprague-Dawley rats (Charles River) and serially digested in 0.2% collagenase.[15] Cells of the first two digestions were discarded; those of the last two were pooled in tissue culture medium (37.5% MEM, 36.5% Earle's B.S.S., 10% NCTC-135, and 10% newborn calf serum and antibiotics). The cell-separation method used in this study indeed yields bone cells, as shown by the expression of markers of the osteoblast phenotype.[16-18] These cells express alkaline phosphatase activity, synthesize osteocalcin and type I collagen, and exhibit a 15- to 30-fold increase in cAMP when exposed to parathyroid hormone. The cells were then plated onto modified 35mm glass bottom microwell dishes (MatTek Corp., Ashland, Mass.) at a density of 20000 cells/cm^2 and subjected to incubation at 37 C in 21% oxygen, 5% carbon dioxide, and maximal humidity with the tissue culture medium (TCM) being changed on days 1 and 4. The cells were used 2-5 days after plating before the culture had reached confluence.

Next, 20 µl of an 8% (w/v) pluronic:DMSO solution was added to 50 µg of Fura-2 AM (Molecular Probes, Inc., Eugene, OR) and allowed to sit in the dark for 5 min. This solution was then added to 20 ml Liebovitz (L-15) medium (Sigma Chem. Co., St. Louis, MO), vortexed for 2 min and sonicated for 5 min, which resulted in a 2.5 µM Fura 2-AM:L-15 cell-loading solution. The TCM was then aspirated from a cell culture dish and the dish was rinsed three times with L-15; 1 ml of the cell-loading solution was then added to the dish and floated in a covered 37 C water bath for 45 min. The cells were then rinsed three times with a modified Hanks' solution (137 mM NaCl, 5.4 mM KCl, 0.44 mM KH_2PO_4, 0.34 mM Na_2HPO_4, 1.25 mM $CaCl_2$, 0.8 mM $MgSO_4$, 5.6 mM D-glucose, 20 mM HEPES, and 4.2 mM $NaHCO_3$) that was adjusted to a pH of 7.4. The dishes each contained 3 ml Hanks' prior to experimentation.

The culture dishes were placed in a glass water-jacketed incubator (37±0.5 C) on a lucite microscope stage that was mounted on an Olympus IMT-2 inverted microscope (Tokyo, Japan). Epifluorescent excitation at wavelengths of 340 and 360 nm yielded 510nm emission images of the field of view of the 20x objective that were captured by a SIT camera with associated image processing hardware and stored on CD-ROM. Pixel-to-pixel ratios of the emitted image intensities were computed and compared to calibration curves to yield values for the cytosolic Ca^{2+} concentration on a single-cell basis.

Vertically and horizontally oriented Helmholtz coil pairs were designed to fit around the microscope stage and objective and produced the required AC and DC EM fields as well as those required to cancel out the earth's field. Four different controls were developed to compare to the resonance condition experimental field. The controls were: (1) ambient earth's field present (with the other three controls having the earth's field cancellation) in addition to (2) a DC field of 20 µT, and (3) an AC field of 40 µT or (4) zero field (no additional field added). The resonance-condition experimental field was 40 µT peak-peak AC in parallel with 20 µT DC. All fields other than those required to cancel out the earth's field were directed vertically, i.e., perpendicular to the culture dish. The horizontal DC field amplitude was cancelled to within 2.0 µT. Field amplitude was measured by a FGM-3D1 fluxgate magnetometer (Walker Scientific Inc., Worcester, Mass.). The applied field frequency was scanned about the calculated resonance frequency of 15.3 Hz, from 10.3 to 20.3 Hz, in both the AC controls and the experimentals.

Temperature equilibration was achieved when dish temperature was held between 36 and 38 C for at least 5 min. After a 2min baseline, the field was turned on for 20 min and then turned off. To verify cellular integrity, 0.75 ml of 0.5 µM bradykinin (BK), a known Ca^{2+} agonist, was then added to the 3 ml of solution already in the dish, to yield a final concentration of 0.1 µM BK, and images were recorded for another 7.5 min. The fluorescence intensity images from the 340 and 360nm excitations were recorded every 30 s over the total of 29.5 min for a total of 60 measurements.

RESULTS

Figure 1 shows a typical Ca^{2+} concentration profile for a control run with ambient earth's field present, where the ordinate is the calibrated cytosolic Ca^{2+} concentration of a single cell and the abscissa represents a time of 29.5 min. Note the low background prior to BK addition to the medium and the nature of the kinetic response of the cell to BK. This type of response, typified by a rapid increase followed by an exponential decrease, will be called "normal." Figure 2 shows a concentration profile for the case of exposure to a resonant EM field of 20 µT DC and 40 µT p-p AC at 15.3 Hz with the earth's field cancelled. The BK was introduced at 22 min at the same time the EM fields were turned off. Note the change in the kinetics and amplitude (with respect to baseline) of the BK response. This response will be called "abnormal." Note also that there is no significant jump in baseline during the 20 min field exposure, other than a slight drift that is sometimes present even in controls.

Table 1 shows the results of these studies at conditions of resonance, at off-resonance frequencies, and for all controls grouped together. Only cells responsive to BK, 193 out of the 232 total cell studied (83%) were counted in the analysis. Sometimes spontaneous Ca^{2+} spikes or oscillations occurred both in the controls and the experimentals.

TABLE 1.—Results of resonance studies. B_z = 20 μT DC + 40 μT p-p AC (parallel), earth's field cancelled, 37 C. (*Not including 15.3 Hz data, ** not including 15.3 Hz nor 15.3±2.5Hz data.)

Exposure variable	Number of cells studied	Response to 0.1 μM BK (normals/abnormals)	Number of cells w/ spikes or baseline shifts	Total number of spikes
All controls	87	79/8	12	19
15.3 Hz	25	7/18	2	5
15.3 Hz* ± 2.5 Hz	66	47/19	2	2
15.3 Hz** ± 5.0 Hz	15	8/7	0	0

DISCUSSION

It is evident for the conditions of this experiment that no significant cytosolic Ca^{2+} concentration increase occurs within the first 20 min of stimulation as would be measured by sustained baseline shifts or an increase in spiking frequency or amplitude. What is noted is a marked change in the BK response where the normal response of controls (Fig. 1) is altered to an abnormal response (Fig. 2), as listed in Table 1. In controls, 91% of the BK responses were normal, whereas only 28% of the cells studied under exposure to the 15.3Hz resonant field had normal responses. Exposure to the resonant field at frequencies on either side of 15.3 Hz shows an intermediate effect.

Given these results, we believe that the resonance condition may operate through a modulation of an agonist-receptor system, rather than primary Ca^{2+} channel gating or other ion translocation mechanism. Studies with other agonist-receptor systems are being planned. Investigation of cell plating density and synchronized cell cultures will continue in the search for a more robust response.

REFERENCES

1. S. D. Smith, B. R. McLeod, A. R. Liboff, and K. Cooksey, "Calcium cyclotron resonance and diatom mobility," *Bioelectromagnetics* 8: 215-227, 1987.

2. A. R. Liboff, R. J. Rozek, M. L. Sherman, B. R. McLeod, and S. D. Smith, "Ca^{2+}-45 cyclotron resonance in human lymphocytes," *J. Bioelectricity* 1987: 13-22, 1987.

3. A. R. Liboff, S. D. Smith, and B. R. McLeod, "Experimental evidence for ion cyclotron resonance mediation of membrane transport," in M. Blank and E. Findl, Eds., *Mechanistic Approaches to Interaction of Electric and Electromagnetic Fields with Living Systems,* London: Plenum Press, 1987.

4. R.J. Rozek, M. L. Sherman, A. R. Liboff, B. R. McLeod, and S. D. Smith, "Nifedipine is an antagonist to cyclotron resonance enhancement of ^{45}Ca incorporation in human lymphocytes," *Cell Calcium* 8: 413-427, 1987.

5. R. J. Fitzsimmons, J. T. Ryaby, F. P. Magee, and D. J. Baylink, "IGF-II receptor number is increased in TE-85 cells by low-amplitude, low-frequency electromagnetic field (EMF) exposure," (in preparation).

6. A. R. Liboff, "Cyclotron resonance in membrane transport," in A. Chiabrera, C. Nicolini, and H. P. Schwan, Eds., *Interactions Between Electromagnetic Fields and Cells,* London: Plenum Press, 1985, 281-296.

7. B. R. McLeod, A. Liboff, and D. Smith, "Theoretical model that predicts frequency, amplitude, and harmonic resonances in cell channels," *Trans. BEMS* 11(abst): 33, 1989.

8. V. V. Lednev, "Possible mechanism for the influence of weak magnetic fields on biological systems," *Bioelectromagnetics* 12: 71-75, 1991.

9. A. Chiabrera, B. Bianco, J. J. Kaufman, and A. A. Pilla, "Quantum analysis of ion binding kinetics in electromagnetic bioeffects," in C. T. Brighton and S. R. Pollack, Eds., *Electromagnetics in Biology and Medicine,* San Francisco: San Francisco Press, 1991, 27-31.

10. A. Chiabrera, B. Bianco, J. J. Kaufman, and A. A. Pilla, "Quantum dynamics of ions in molecular crevices under electromagnetic exposure," Ref. 9, pp. 21-26.

11. D. T. Edmonds, "Larmor precession as a mechanism for the detection of static and alternating magnetic fields," (this volume).

12. W. C. Parkinson and C. T. Hanks, "Search for cyclotron resonance in cells in vitro," *Bioelectromagnetics* 10: 129-145, 1989.

13. H. Ozawa, E. Abe, Y. Shibasaki, T. Fukuhara, and T. Suda, "Electric fields stimulate DNA synthesis of mouse osteoblast-like cells (MC3T3-E1) by a mechanism involving calcium ions," *J. Cellular Physiology* 138: 477-483, 1989.

14. K. Hanley, L. A. Norton, and G. A. Rodan, "Electromagnetic field effects on periosteal and osteoblast-like cells," *J. Dental Res.* 60A(abst. 365): 402, 1981.

15. J. S. Brand and T. J. Hefley, "Collagenase and the isolation of cells from bone," in T. G. Pretlow II and T. P. Pretlow, Eds., *Cell Separation: Methods and Applications,* New York: Academic Press, 1984, 265-283.

16. C. T. Brighton and W. P. McCluskey, "Response of cultured bone cells to a capacitively coupled electric field: Inhibition of cAMP response to parathyroid hormone," *J. Orthopaedic Res.* 6: 567-571, 1987.

17. J. L. Schaffer, C. T. Brighton, C. C. Clark, and F. J. Brinley, "In vitro growth of rat calvarial bone cells at various oxygen tensions," *Trans. Orthopaedic Res. Soc.* 12: 176, 1987.

18. K. P. Spindler, D. B. Shapiro, J. B. Gross, C. T. Brighton, and C. C. Clark, "The effect of ascorbic acid on the metabolism of rat calvarial bone cells in vitro," *J. Orthopaedic Res.* 7: 696-701, 1989.

19. This work was supported by National Science Foundation grant BCS89-21492.

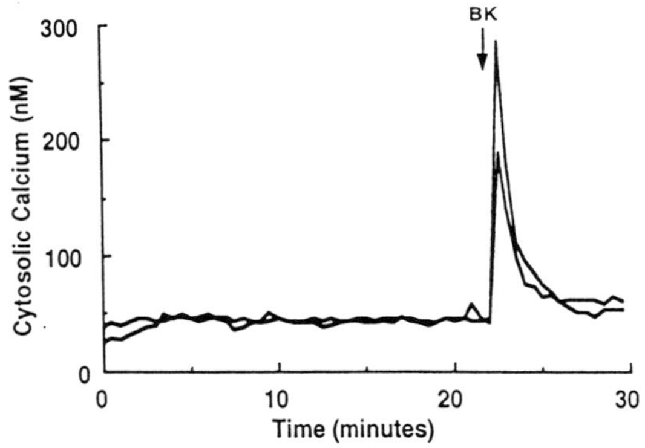

FIG. 1.—Calcium concentration profiles for two different control cells with ambient earth's field present (37 C).

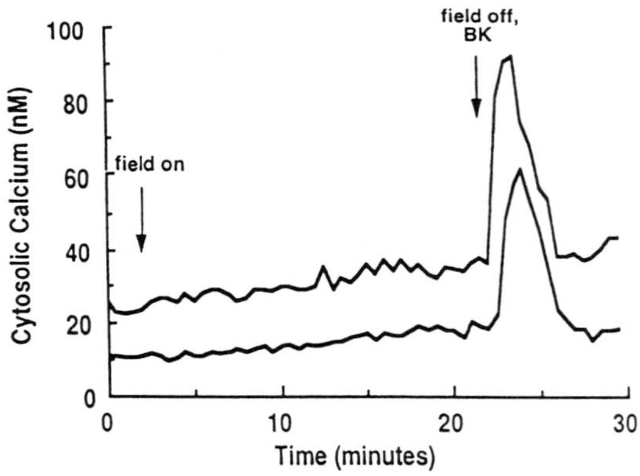

FIG. 2.—Calcium concentration profiles for two different cells under resonance conditions (B_z = 20 μT DC + 40 μT p-p AC, f = 15.3 Hz, earth's field cancelled, 37 C).

Martin Blank, Ed., *Electricity and Magnetism in Biology and Medicine*

CHANGES IN THE CYTOSOLIC CALCIUM CONCENTRATION OF PRIMARY BONE CELL CULTURES DUE TO ELECTRIC FIELDS AT 0.1 V/m FROM 6 to 600 kHz

S. R. Pollack* and K. A. Reinbold

University of Pennsylvania, Philadelphia

O. L. da Silva

University of Sao Pauli, Brazil

Electric fields have been shown to produce significant changes in proliferation and protein synthesis in several bone cell lines. We have investigated fields of constant magnitude 0.1 V/m in tissue culture from 6 to 600 kHz in primary rat calvaria cells using Fura-2 fluorescence technology. Pronounced cytosolic calcium oscillations (spiking) were observed due to the presence of the electric field from 20 to 300 kHz. Bradykinin was used as a positive agonist control during these experiments. The calcium response due to bradykinin was significantly altered by the field even though the field was turned off prior to the insertion of the bradykinin.

Considerable progress has been achieved in determining dose-response relationships in the electric-field amplitude effect on bone cells. Field amplitudes of the order of 0.1 to 1 V/m have been computed for 60 kHz devices used in clinical fracture nonunion treatment and for in vitro studies in bone cells.[1] This field amplitude has been shown to produce statistically significant changes in protein synthesis in primary cultures of rat calvarial cells.[2,3] In addition, the same field amplitude produced increased fracture strength in the healing osteotomy in the rabbit fibula model at the midtime (14 days) point during the time course of healing.[4] In this project we have asked whether a real-time Ca^{2+} concentration change in the cytosol of primary bone cell cultures is attributable to an electric field 0.1 V/m in amplitude and, if so, how the frequency in the range of 6 to 600 kHz influences this response.

McLeod et al.[5] studied real-time Ca^{2+} concentration changes in ensembles of ROS cells for fields of 0.1 to 1 V/m and frequencies from 6 to 600 Hz as functions of extracellular Ca^{2+} concentration. They showed increased cytosolic Ca^{2+} concentration and Ca^{2+} spiking with only 6 min of electric-field stimulation when the extracellular Ca^{2+} concentration was 0.5 mM. For higher frequencies (200 to 600 Hz) they noted no change in Ca^{2+} fluctuations. Accordingly, we investigated high-frequency electric-field transduction mechanisms by measuring the real-time electric-field effect of cytosolic Ca^{2+} concentrations using Fura-2 in a system capable of measurement on a cell-by-cell basis.

MATERIALS AND METHODS

Calvaria were dissected from 1 to 3 day old Sprague-Dawley rats (Charles River) and serially digested in 0.2% collagenase.[6] Cells of the first two digestions were discarded, those of the last two were pooled in tissue culture medium (37.5% MEM, 36.5% Earle's B.S.S., 10% NCTC-135, 10% newborn calf serum and antibiotics). The cells were then plated onto modified 35mm glass bottom microwell dishes (MatTek Corp. Ashland, Mass.) at a density of 20000 cells/cm^2 and subjected to incubation at 37 C in 21% oxygen, 5% carbon dioxide, and maximal humidity with the tissue culture medium (TCM) being changed on days 1 and 3. The cells were used 2-5 days after plating, before the culture had reached confluence. These cells exhibit the osteoblast phenotype. We examined 143 cells from the isolations of 86 rat puppies from 8 mothers.

We added 20 μl of an 8% (w/v) pluronic: DMSO solution to 50 μg of Fura-2 AM (Molecular Probes, Inc., Eugene, Ore.) and let them sit in the dark for 5 min. This solution was then added to 20 ml Liebovitz

*Correspondence to: S. R. Pollack (Bioengineering), University of Pennsylvania, Philadelphia PA 19104-6392.

(L-15) medium (Sigma Chem. Col, St. Louis, Mo.) vortexed for 2 min, and sonicated for 5 min, which resulted in a 2.5 μM Fura 2-AM:L-15 cell loading solution. The TCM was then aspirated from a cell culture dish and the dish was rinsed three times with L-15. One ml of the cell-loading solution was then added to the dish and floated in a covered 37 C water bath for 45 min. The cells were then rinsed three times with a modified Hanks' solution (137 mM NaCl, 5.4 mM KCl, 0.44 mM KH_2PO_4, 1.25 mM $CaCl_2$, 0.8 mM $MgSO_4$, 5.6 mM D-glucose, 20 mM HEPES, and 4.2 mM $NaHCO_3$) adjusted to pH 7.4. The dishes each contained 2 ml Hanks' prior to experimentation. The cultures were studied at room temperature or 23 ± 2 C.

The culture dishes were placed on the microscope stage of an Olympus IMT-2 inverted microscope (Tokyo). Epifluorescent excitation at wavelengths of 340 and 360 nm yielded 510nm emission images of the field of view of the 20x objective that were captured by a SIT camera with associated image processing hardware and stored on CD-ROM. Pixel-by-pixel ratios of the emitted image intensities were computed and compared to calibration curves to yield values for the cytosolic Ca^{2+} concentration on a single-cell basis.

Dipole electrodes of 0.25mm-diameter platinum wire spaced 1.0 cm apart were immersed in the Hanks' solution over the glass coverslip cemented to the base of the culture dish. A sinusoidal voltage of 0.01 V p-p was applied at frequencies varying from 6 to 600 kHz. The cells of interest for the study lay along the line joining the two electrodes, where the field was 0.1 V/m.

After a 2min baseline, the field was turned on for 20 min and then turned off. To verify cellular integrity, 0.5 ml of 0.5 μM bradykinin (BK), a known Ca^{2+} agonist, was then added to the 2 ml of solution already in the dish, to yield a final concentration of 0.1 μM BK, and images were recorded for another 7.5 min. The fluorescence intensity images from the 340 and 360 nm excitations were recorded every 30 s for a total time of 29.5 min.

RESULTS

Figure 1 shows a typical response of a room temperature control experiment. The ordinate, in nM Ca^{2+}, is the ratio of the 510nm fluorescence emission images acquired by excitation at either 340 or 360 nm. Conversion to Ca^{2+} concentration is achieved by comparison to calibration curves. The abscissa is in units of time with a total of 29.5 min shown. Note the constant baseline and the BK response used here as the positive agonist control, as well as the spontaneous oscillation at an amplitude of 218% over baseline. The characteristic shape of the BK response—a rapid rise followed by a nearly exponential decay to the baseline—is called the "normal" response.

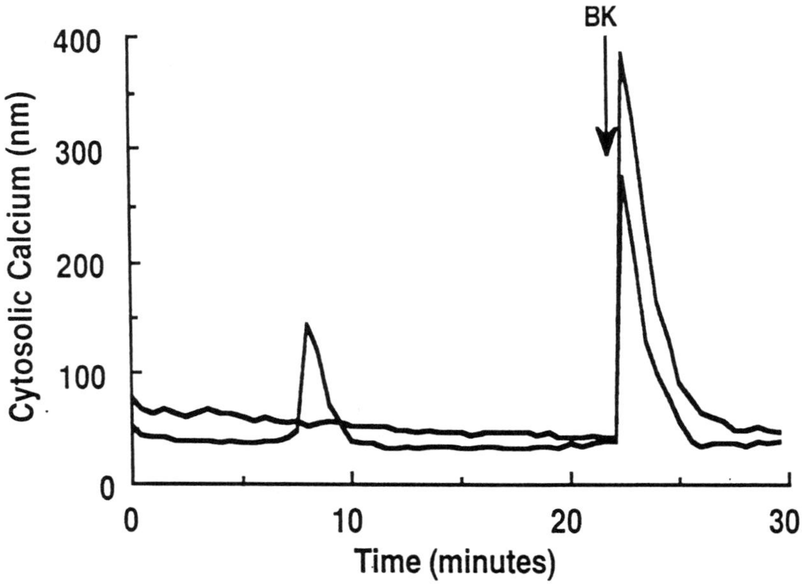

FIG. 1.—Calcium concentration profiles for two different E-field control cells (23 C).

Figure 2 shows a similar experiment, except that after 2 min an electric field of 0.1 V/m p-p at 60 kHz was turned on for 20 min. Immediately following the cessation of the applied field, BK was added as the positive agonist control. In this figure one notes the occurrence of pronounced spiking (oscillation) with variable latency and a retarded BK response, which we label as "abnormal" in contrast to Fig. 1.

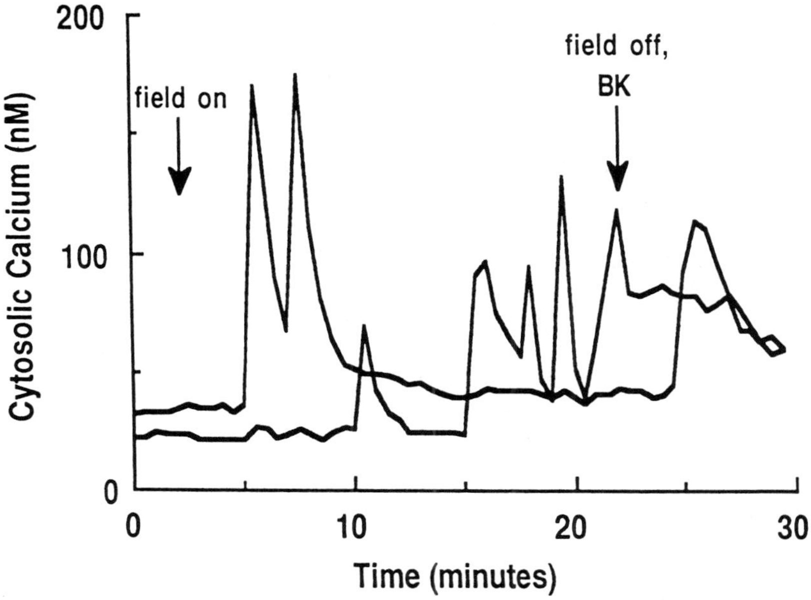

FIG. 2.—Calcium concentration profiles for two different cells during electric-field exposure: 60 kHz, $E = 0.1$ V/m, 23 C.

Table 1 shows the total results to date for each frequency studied at $E = 0.1$ V/m. The total number of cells studies is smaller than the amount stated previously because some cells (13% of total) were unresponsive to BK stimulation and could not be counted in the total analysis.

TABLE 1.—Results of the electric-field study.

Frequency (kHz)	Number of cells	Bradykinin nrm/abnrm	Number of cells w/ spikes	Number of spikes	Spike amplitude (mean of % BL)
Control	32	25/7	2	3	121
$6 \le f \le 10$	15	11/4	0	0	0
$20 \le f \le 100$	68	46/22	18	28	308
$300 \le f \le 600$	9	4/5	3	6	296

DISCUSSION

A total of 27% of all BK-responsive cells studied between 20 and 600 kHz displayed Ca^{2+} spiking. In all cases where significant spiking was observed, the spike amplitudes were of the order of the BK Ca^{2+} peak, 150 to 300 nM, with an average of 306% over baseline in amplitude. Latency ranged from less than 1 min to as much as just under 20 min. The mean time between spikes is 2.8 min; however, the range is from 1 to 8.5 min. Spontaneous spiking was observed in controls in only 6% of the cell; however, these spikes were only 121% over baseline in amplitude. Considerable variability was observed with regard to the BK response following E-field stimulation. Abnormal responses as a percent of total was 17% for room temperature controls, 33% between 20 and 600 kHz, and 27% between 6 and 10 kHz. Abnormal responses are characterized by marked latency and significant kinetic alterations. The inositol phosphate pathway and voltage gated calcium channels will be investigated to explore this finding further.[7]

REFERENCES

1. E. J. Vreslovic, S. R. Pollack, and C. T. Brighton, "Comparison of theoretically determined electric fields in a rabbit fibular fracture and a human tibial fracture during capacitive coupling," *Trans. BRAGS* 5: 22, 1985.

2. E. Okereke, C. T. Brighton, and S. R. Pollack, "Proliferative response of bone cells in vitro to capacitively coupled electrical stimulation," *Trans. BRAGS* 9: 56, 1989.

3. J. R. Corsetti, S. E. Levine, S. R. Pollack, and C. T. Brighton, "Enhanced proliferation of cultured bone cells by a capacitively coupled electric field is inhibited by verapamil and W7," *Trans. BRAGS* 11: 36, 1991.

4. C. T. Brighton, W. J. Hozack, M. D. Brager, R. E. Windsor, S. R. Pollack, E. J. Vreslovic, and J. E. Kotwick, "Fracture healing in the rabbit fibula when subjected to various capacitively coupled electrical fields," *J. Orthopaedic Res.* 3: 331-340, 1985.

5. K. J. McLeod, H. J. Donahue, P. E. Levin, and C. T. Rubin, "Low-frequency sinusoidal electric fields alter calcium fluctuations in osteoblast-like cells," in C. T. Brighton and S. R. Pollack, Eds., *Electromagnetics in Biology and Medicine,* San Francisco: San Francisco Press, 1991, 111-115.

6. J. S. Brand and T. J. Hefley, "Collagenase and the isolation of cells from bone," in T. G. Pretlow II and T. P. Pretlow, Eds., *Cell Separation: Methods and Applications,* New York: Academic Press, 1984, 265-283.

7. This work was supported by NSF grant BCS89-21492.

Martin Blank, Ed., *Electricity and Magnetism in Biology and Medicine*

SIMULTANEOUS DUAL-SAMPLE FLUORIMETRIC DETECTION OF REAL-TIME EFFECTS OF ELF ELECTROMAGNETIC FIELDS ON CYTOSOLIC FREE CALCIUM AND DIVALENT CATION FLUX IN HUMAN LEUKEMIC T-CELLS (JURKAT)

J. Walleczek,* P. L. Miller, and W. R. Adey

Pettis Memorial Veterans' Hospital, Loma Linda, Calif.

A new fluorimetric technique, dual-sample (or differential) fluorescence spectroscopy (DFS), was developed that enables the real-time observation of Ca^{2+}-dependent fluorescent changes in two cuvette samples (i.e., exposure and control) simultaneously. With DFS it is possible for the first time to monitor real-time changes in the cytosolic free calcium concentration $[Ca^{2+}]_i$ in the absence *and* presence of electromagnetic fields (EMF) at the same time. We report preliminary results on 50/60Hz sinusoidal electric or magnetic influences on $[Ca^{2+}]_i$, and Ca^{2+} transmembrane influx in cells of the human leukemic T-cell line JURKAT. Our results support the possibility that 50/60Hz sinusoidal EMFs may *nonthermally* affect Ca^{2+}-mediated signal transduction pathways and subsequently alter human T-lymphocyte function.

Membrane-mediated signal transduction pathways are thought to be an early target of EMF-interactions with cellular systems, including cells of the immune system.[1] In particular, the early modulation of Ca^{2+} signals by EMFs is a plausible candidate for inducing later-stage EMF effects on lymphocyte function.[2] The challenge of reliable monitoring of potential EMF-induced real-time Ca^{2+} changes in lymphocytes was the primary motivation for constructing a DFS instrument. Its design was based on the assumption that a reliable and efficient approach for probing bioelectromagnetic phenomena would be to monitor real-time fluorescent changes in an EMF-exposed cell sample and in an identical isothermal control sample simultaneously. Commercially available fluorimetry systems (one-sample systems) are limited to measuring either the experimental *or* the control sample and thus cannot be used in this manner. Here, we report the first results with this new fluorescence technique.

MATERIALS AND METHODS

Dual-sample Fluorescence Spectroscopy (DFS)

The general principles of the DFS technique are outlined above. Figure 1 gives a schematic representation of the key components of the DFS magnetic field exposure system. The cells were prevented from sedimenting to the bottom of the cuvette during the experiments by a *non*-magnetic cell mixing system. (Magnetic fields generated by conventional magnetic cell stirrers are of the order of 0.1 to 1.0 mT and therefore preclude the conduct of a meaningful exposure experiment.)

Magnetic and Electric Field Exposures

Cells were exposed to the 60Hz sinusoidal magnetic field for 7 min. The exposure was started 2 min before the addition of the T-cell activator. Magnetic field uniformity over the entire cell sample was 100%. Each experimental series (exposures at B_{rms} = 0, 0.1, 1, 5, and 10 mT) was carried out on a separate day. At the beginning and at the end of an experimental series, sham exposures—with the experimental sample in the nonenergized exposure system—were carried out to verify the reliability of the DFS method.

Sinusoidal electric fields of 50 mV/cm at 50 Hz were generated inside the cuvette by passage of an appropriate current signal (I_{rms} = 2.4 mA) through the cell suspension. Two platinum foil sheets, of dimensions to fit inside the cuvette, were used as electrodes (covered with agarose). An opening was cut into one of the platinum sheets to permit passage of excitation light. The electric and magnetic field experiments were done at 25.0 ± 0.1 and 37.0 ± 0.1 C, respectively.

*Correspondence to: J. Walleczek, Pettis Memorial Veterans' Hospital, Loma Linda CA 92357.

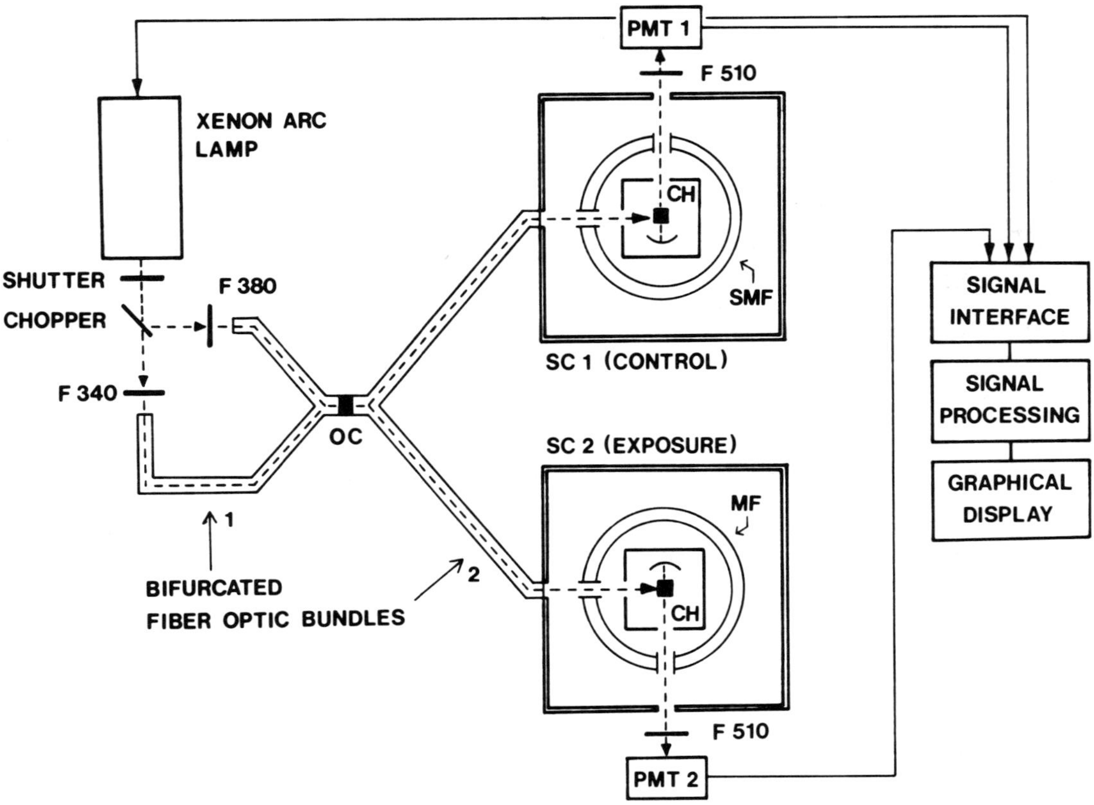

FIG. 1.—Dual-sample FURA-2 fluorescence spectroscopy (DFS) magnetic field-exposure system. Abbreviations: CH, nonmetallic thermostatted cuvette holder; F, interference filter (340, 380, or 510 nm bandpass filters); MF, water-cooled bifilar Lee-Whiting magnetic-field-exposure coil; OC, optical coupler; PMT, photomultiplier tube; SC, nonmetallic sample compartment; SMF, sham magnetic field exposure coil (same as MF). Solid lines represent electrical pathways; dashed lines represent excitation and emission light paths.

JURKAT T-cell FURA-2 Experiments

The human leukemic T-cell line JURKAT (clone E6-1) was maintained in RPMI-1640 medium supplemented by 5% heat-inactivated fetal calf serum and 2 mM L-glutamine at 37 C in a humidified 3% CO_2 atmosphere. Cells were loaded with FURA-2/AM as described previously.[3] All experiments described here were carried out with FURA-2-loaded cells resuspended in buffer (in mM; NaCl, 145; KCl, 5; Na_2HPO_4, 1; $CaCl_2$, 1; $MgSO_4$, 0.5; glucose, 5; Na-HEPES, 10; pH 7.4 at 37 C) at a concentration of 2×10^6 cells/ml. Whenever appropriate, the results have been expressed as the mean ± S.E.M. Statistical evaluation was done by comparison of variances (ANOVA).

RESULTS

50Hz Electric Field Effect on Ca^{2+} (Mn^{2+}) Influx

We attempted the first real-time measurements of potential nonthermal effects of sinusoidal electric fields (calculated E_{rms} = 50 mV/cm, f = 50 Hz) on Ca^{2+} flux across the cell membrane. For this purpose, a method based on Mn^{2+}-induced quenching of intracellular FURA-2 emission after single-wavelength excitation (at 360 nm) was employed. The detection of Mn^{2+} fluxes is regularly used in real-time membrane transport studies as a method for examining cellular Ca^{2+} fluxes.[3] Figure 2 shows an original dual-sample recording from one such DFS experiment carried out with resting JURKAT T-cells. These results provide *direct* evidence for nonthermogenic field-stimulated cellular Ca^{2+} (Mn^{2+}) membrane fluxes.

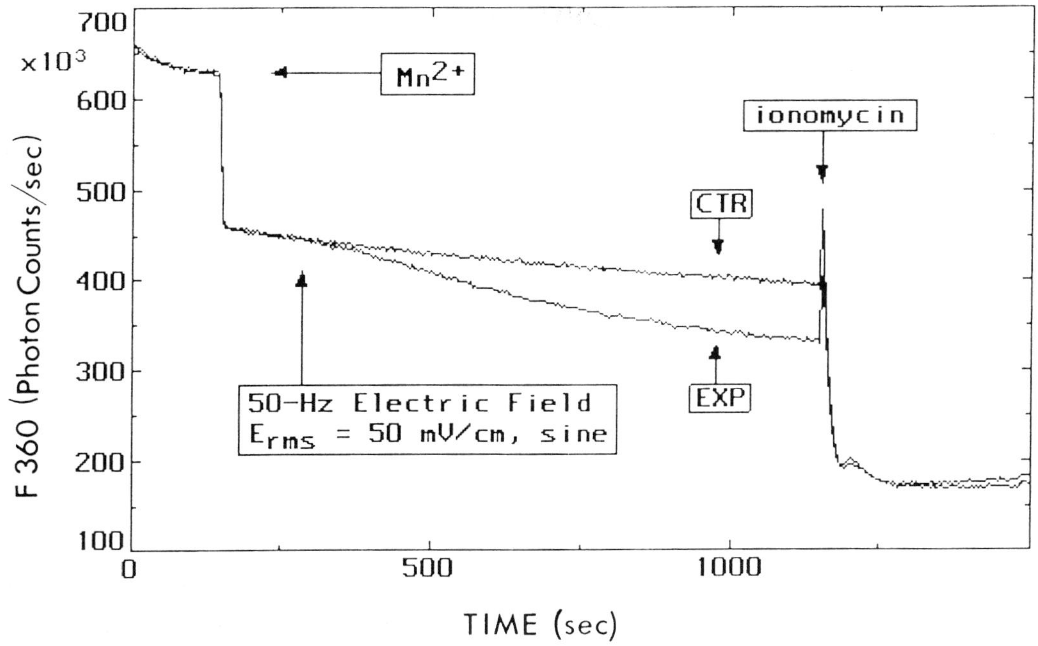

FIG. 2.—Original dual-sample recording of electric field effect on Ca^{2+} transmembrane influx in FURA-2-loaded JURKAT T-cells. At t = 150 s: simultaneous addition of 100 μM Mn^{2+} to control and experimental cuvettes quenches photon emission signal derived from *extracellular* FURA-2; t = 270 s: initiation of field exposure in the experimental sample only, Mn^{2+} influx rate into exposed cells appears to be enhanced after only 60 to 120 s; t = 1180 s: as a calibration step, 5 μM ionomycin (Mn^{2+} ionophore) was simultaneously added to both cuvettes. This procedure induces a rapid Mn^{2+} influx that completely eliminates photon emission from intracellular FURA-2. EXP = exposure, CTR = control.

60Hz Magnetic Field Effect on $[Ca^{2+}]_i$: Intensity Dependence

The DFS method was also applied to the real-time study of the 60Hz sinusoidal magnetic field interaction with T-cell receptor (TCR) induced Ca^{2+} signaling in JURKAT T-lymphocytes. The results from three independent sets of field intensity-response experiments are summarized in Fig. 3. The addition of the anti-CD3 mAb specific for the CD3-complex of the TCR induced a 2.36-fold increase in $[Ca^{2+}]_i$, reaching a plateau at 286.7 ± 11.4 nM after 5 min from a resting level of 121.2 ± 8.2 nM (N = 6). In contrast, in the presence of the 1mT (10G) magnetic field, the TCR-induced $[Ca^{2+}]_i$ rise reached only 252.3 ± 10.4 nM (N = 3). A similar effect was observed when the 0.1 mT (1 G) field was applied. However, the higher field intensities (5 and 10 mT) appeared to have no detectable effect (Fig. 3).

DISCUSSION

Our initial findings with the dual-sample fluorescence technique demonstrate for the first time that electric-field influences on cellular Ca^{2+} (Mn^{2+}) ion influx can be monitored in real time. However, so far we have only observed this effect at a relatively high (albeit nonthermal) electric-field intensity (50 mV/cm). Similar experiments at E_{rms} = 1 mV/cm proved unsuccessful (not shown).

The above report also summarizes the first evidence for real-time magnetic field effects on cellular Ca^{2+} signaling. However, in the interpretation of the 60Hz magnetic-field results it has to be borne in mind that the observed field-induced Ca^{2+} changes are rather small. For example, the largest of the observed effects (at 1 mT) is only about 12%. The biological significance of a field-induced change of this magnitude in TCR-triggered $[Ca^{2+}]_i$ is at present unknown. Nevertheless, if confirmed, our results would suggest that weak sinusoidal magnetic fields (of the order of a few Gauss) can, at least to a small extent, modulate Ca^{2+} signaling in human T-cells.[4]

REFERENCES

1. W. R. Adey, "Cell membranes: The electromagnetic environment and cancer promotion," *Neurochem. Res.* 13: 671-677, 1988.

2. J. Walleczek, "Electromagnetic field effects on cells of the immune system: The role of calcium signaling," *FASEB J.* 6: 3177-3185, 1992.

3. J. E. Merrit, R. Jacob, and T. J. Hallam, "Use of manganese to discriminate between calcium influx and mobilization from intracellular stores in stimulated human neutrophils," *J. Biol. Chem.* 264: 1522-1527, 1989.

4. This work is supported by the Fetzer Institute and by the U.S. Department of Energy.

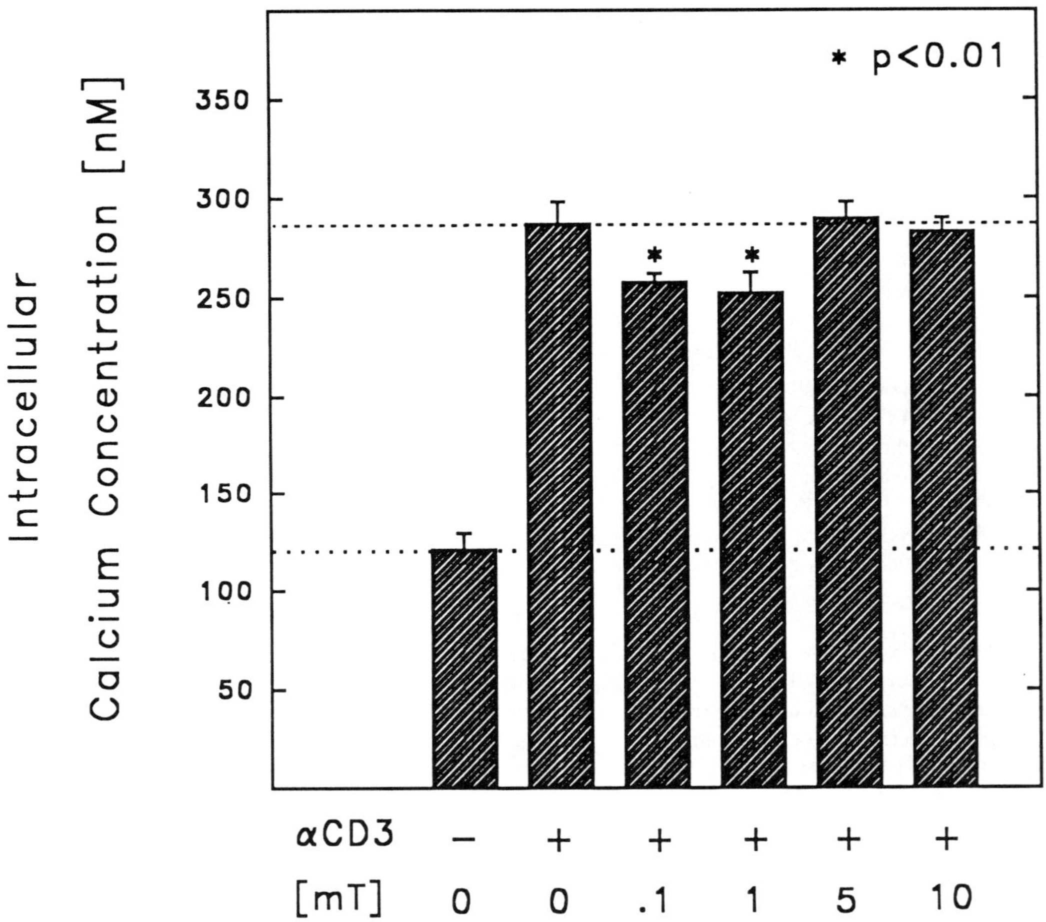

FIG. 3.—Intensity-dependent 60Hz magnetic field effect on T-cell receptor-induced $[Ca^{2+}]_i$ rise in JURKAT T-Lymphocytes. Bars represent $[Ca^{2+}]_i$ in nM (N = 3 to 6). In stimulated cells, $[Ca^{2+}]_i$ was determined 5 min after addition of 0.5 μg/ml anti-CD3 mAb in the presence or absence of the 60Hz field.

Martin Blank, Ed., *Electricity and Magnetism in Biology and Medicine*

INTRACELLULAR CALCIUM SIGNALING IN JURKAT E6-1 CELLS EXPOSED TO AN INDUCED 1 mV/cm, 60Hz SINUSOIDAL ELECTRIC FIELD

D. B. Lyle,* J. Doshi, T. A. Fuchs, J. P. Casamento, Y. Sei, P. K. Arora, and M. L. Swicord

Food and Drug Administration, Rockville, Md., and National Institutes of Health, Bethesda, Md.

Intracellular calcium signaling was monitored in Jurkat E6-1 cells following or during exposure to a 1mV/cm 60Hz electric field induced by a 238G magnetic field from a large solenoid. Cells were exposed in protein-containing culture medium in the outer annular ring of a Falcon 3037 organ tissue culture dish. Fluo-3-loaded cells were stimulated with anti-CD3 monoclonal antibody at desired time points, and increases in fluorescence (indicating an increased intracellular calcium concentration) were monitored with a flow cytometer. Cells were exposed at room temperature and at 37 C. In all cases, neither the calcium signal nor concomitant cell function from field-exposed cells differed significantly from controls.

Several laboratories have reported alterations in the association of $^{45}Ca^{2+}$ with cultured tumor cells, normal lymphocytes, tissue preparations, or in vivo intact cat brains during or following exposure to various extremely low frequency (ELF) electromagnetic fields (for review see Lyle et al.[1]). This finding has suggested the possibility that ELF fields might alter transmembrane calcium signaling. Calcium signaling is a key pathway involved in the activation, differentiation, and proliferation of many cell types.[2] Furthermore, it is associated with aspects of tumor promotion.[3] However, changes in $^{45}Ca^{2+}$ association could be due to differences in binding at the cell surface and/or movement into the cell's interior. Jurkat is a human helper T-lymphocyte cell line that has been extensively used for studying calcium signaling pathways.[4] We monitored an intracellular calcium signal stimulated by a monoclonal antibody against CD3 (a complex of cell-surface molecules involved with transmembrane signaling) in Jurkat E6-1 following and during ELF field exposure. The cells were exposed to an induced 1mV/cm 60Hz electric field in a solenoidal exposure regime that has been reported to alter $^{45}Ca^{2+}$ association with mitogen-stimulated rat thymocytes.[5]

MATERIALS AND METHODS

The Jurkat clone E6-1 was grown in RPMI 1640 medium containing 10% FCS, 2 mM glutamine, 100 U/ml penicillin, and 100 µg/ml streptomycin ("complete medium"). For experiments conducted in air, 10 mM HEPES was added to the complete medium to control pH. For monitoring of intracellular calcium, cells were loaded with 1 µM fluo-3 (Molecular Probes) via a 30min incubation in the dark with the acetoxy-methyl ester. Cells were then washed three times in HEPES-buffered complete medium, resuspended at $1\text{-}2\times10^6$/ml in HEPES-buffered complete medium, and placed 4 ml/dish into two annular-ring Petri dishes (Falcon 3037 organ tissue culture dishes). The dishes were placed on supports sitting in water priorly stabilized at the desired temperature, in a central reservoir within a water-jacketed closed container. Water at the appropriate temperature was constantly pumped from a circulating water bath through the walls of the incubation chambers. Temperatures were routinely checked before and after all field and control exposures with a Fluhe 80T-150U digital-readout universal temperature probe.

One water-jacketed incubator was situated within the bore of a water-cooled solenoid; the other at a distance of 9ft from the solenoid. For a solenoid horizontal field of 60 Hz, 238 G, the horizontal 60Hz value at the control site was 13.0 mG. Temperature variation between exposed and control dishes was 0.1 C or less. The vertical DC component, parallel to the AC, was 380 mG in the solenoid and 480 mG at the control site. The solenoid (designed and built by Dr. Robert Liburdy at Lawrence Berkeley Laboratory: $N = 400$, 2.3 Ω, 20.1 mH) was energized by a 60Hz current. Heat generated by the solenoid was effectively removed by its own water-jacketed, cooling housing (6 L/min). The 238_{rms} G, 60Hz field used in these experiments induced an average 1mV/cm 60Hz electric field in the medium of the annular ring of the organ-tissue culture dishes.

*Correspondence to: D. B. Lyle (HFZ-114), FDA-CDRH, Rockville MD 20857.

The associated current density was 16 μA/cm^2; conductivity of the medium, 1.7 S/m.

Petri dishes were allowed to equilibrate in place on their respective water-bath holders for at least 30 min, with the temperature set to 24 or 37 C. Then, at a 1min stagger, fluo-3-loaded cells were placed into medium at the desired temperature and immediately poured into control and exposure sites. For 30min pre-exposure experiments, maintaining the 1min stagger, the Petri dishes were removed; cells were quickly resuspended, placed to test tubes, and stimulated with a murine monoclonal antibody against human CD3. For simultaneous exposure during calcium signaling, cells were stimulated with anti-CD3 at the same time they were placed into the field and control sites, and 0.3ml samples were taken directly from the Petri dishes for analysis at time points up to 1 h. Samples were analyzed on a Becton Dickenson FACScan flow cytometer for the number of cells out of 10000 examined showing fluorescence intensity above that of unstimulated fluo-3 loaded cells.[6] Significance of changes between field and control samples was determined within a single experiment by Kolmogorov-Smirnov statistics, or between groups of experiments via 2-tailed Student's *t*-tests or ANOVA.

RESULTS AND DISCUSSION

Figure 1 shows the anti-CD3 stimulated calcium signal time course over a 10min period after cells were exposed to an induced 1mV/cm 60Hz electric field for 30 min. Results are the means of four experiments normalized to the 2min peak medium value. Compared with controls, no significant changes within individual experiments or among the groups of experiments occurred. Prior experiments based on $^{45}Ca^{2+}$ endpoint noted a rapid effect by a field close to calcium ion resonance conditions that was present during exposure but not detected following a 30min pre-exposure.[1] Consequently, we then conducted the same experiment by taking samples during field exposure. In addition, two phases of calcium signaling are being examined in our system. There is a rapid release of intracellular calcium stores, which accounts for most of the initial peak, followed by a longer plateau phase that is maintained by movement of calcium across the plasma membrane (data not shown). It is conceivable that a field effect might not occur during the initial peak, yet be noted during the plateau phase. We therefore extended the assay period during the field exposure out to 1 h. For a normalized group of four experiments, no significant changes occurred over a 1h time course during constant field exposure (Fig. 2). Another possible contributor to a lack of an effect could be that the assays were being performed at room temperature. We then conducted the same experiment at 37 C (Fig. 3). For this experiment, no significant changes were noted between field and control. However, a rapid loss of the dye was observed due to the 37 C incubation, which has been attributed in some cells to an anion transport system that actively excretes the dye.[7] This, then, was a way to look concomitantly at cellular function during the 1h period of calcium-signaling monitoring. Figure 4 shows the percentage of cells over time which lost fluorescence below the fluo-3 loaded cell baseline. Again, no significant changes were observed between field and control samples.

Since individual cell sensitivity to electric field stimulation has been reported,[8] it may be that the rapidly proliferating, fully transformed cell line Jurkat is insensitive to modulation by the field used in this study, whereas partially activated normal cells such as thymocytes are sensitive.[5] Also, calcium modulation at the surface of the cell might be affected in Jurkat, whereas intracellular signaling processes might not be. Experiments are proceeding with Jurkat and also mitogen-activated thymocytes in order to address these further questions.

REFERENCES

1. D. B. Lyle, X. H. Wang, R. D. Ayotte, A. R. Sheppard, and W. R. Adey, "Calcium uptake by leukemic and normal T-lymphocytes exposed to ELF magnetic fields," *Bioelectromagnetics* 12: 145-156, 1991.

2. M. J. Berridge and R. F. Irvine, "Inositol phosphates and cell signalling," *Nature* 341: 197-205, 1989.

3. J. F. Whitfield, *Calcium, Cell Cycles, and Cancer," Boca Raton, Fla.: CRC Press, 1990.*

4. A. Weiss, J. Imboden, D. Shoback, and J. Stobo, "Role of T3 surface molecules in human T-cell activation: T3-dependent activation results in an increase in cytoplasmic free calcium," *Proc. Natl. Acad. Sci.* 81: 4169-4173, 1984.

5. J. Walleczek and R. P. Liburdy, "Nonthermal 60 Hz sinusoidal magnetic-field exposure enhances $^{45}Ca^{2+}$ uptake in rat thymocytes: Dependence on mitogen activation," *FEBS* 271: 157-160, 1990.

6. Y. Sei and P. K. Arora, "Quantitative analysis of calcium (Ca^{2+}) mobilization after stimulation with mitogens or anti-CD3 antibodies: Simultaneous fluo-3 and immunofluorescence flow cytometry," *J. Immunol. Meth.* 137: 237-244, 1991.

7. F. Di Virgilio, T. H. Steinberg, and S. C. Silverstein, "Inhibition of fura-2 sequestration and secretion with organic anion transport blockers," *Cell Calcium* 11: 57-62, 1990.

8. I. Binderman, D. Somjen, Z. Shimshoni, J. Levy, H. Fischler, and R. Korenstein, "Stimulation of skeletal-derived cell cultures by different electric field intensities is cell-specific," *Biochim. Biophys. Acta* 844: 273-279, 1985.

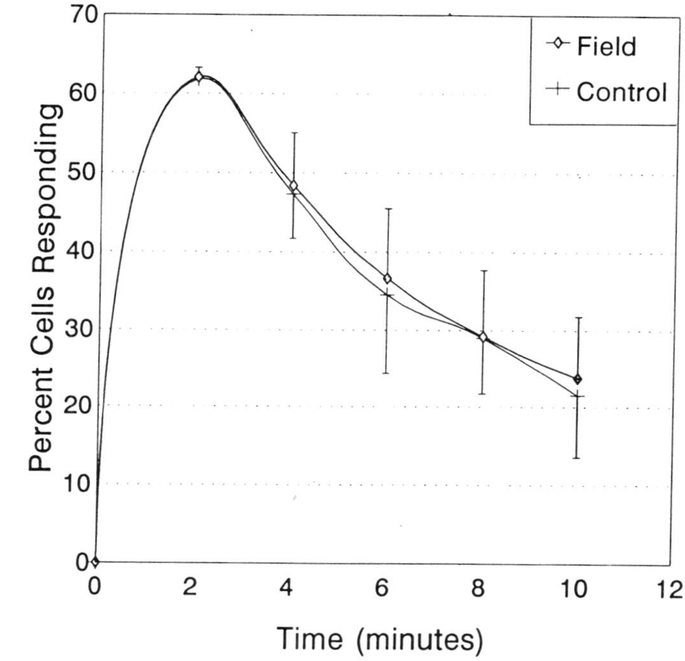

FIG. 1.—Time course of calcium signal in Jurkat E6-1 cells after 30min at 24 C exposure to 238G sinusoidal 60Hz field inducing 1mV/cm electric field in culture medium. Results are normalized means of 4 exp ± 1 S.D. (n = 10 000 cells/exp).

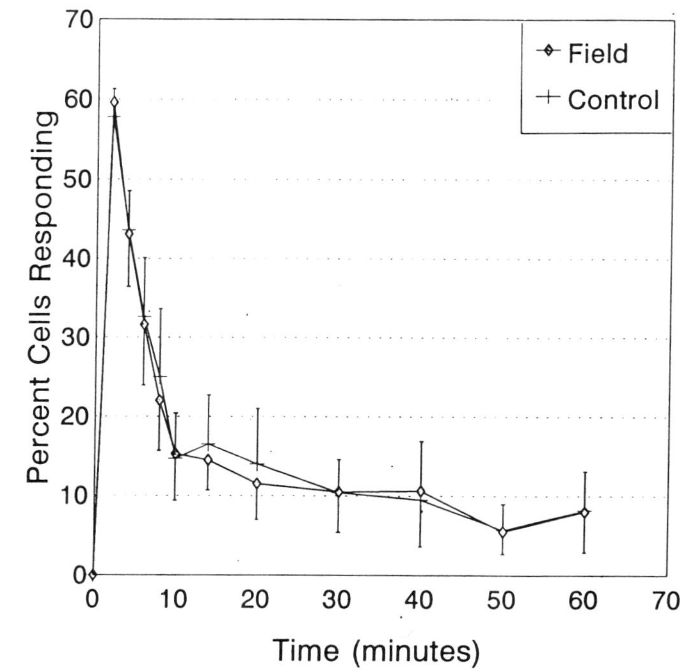

FIG. 2.—Time course of calcium signal in Jurkat E6-1 cells during exposure at 24 C to 238G sinusoidal 60Hz field inducing 1mV/cm electric field in culture medium. Results are normalized means of 4 exp ± 1 S.D. (n = 10 000 cells/exp).

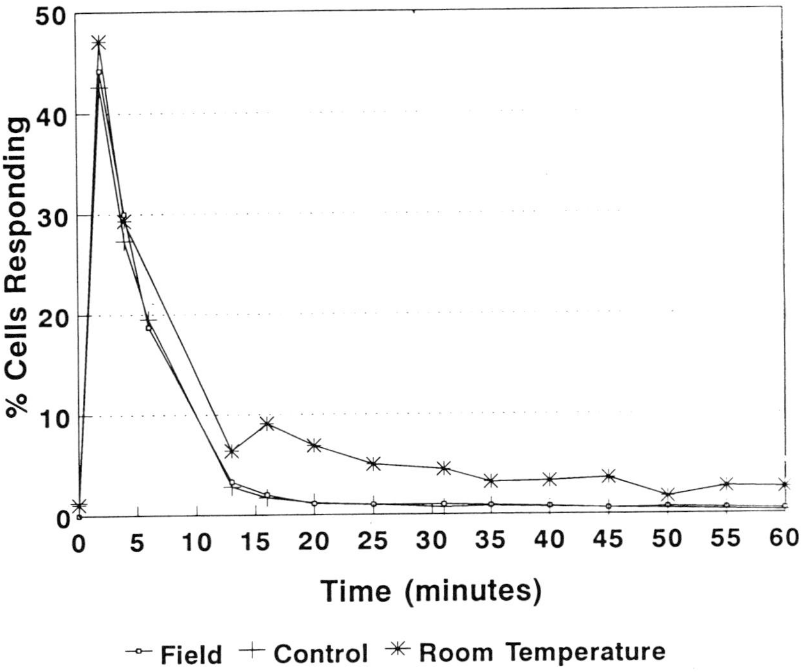

FIG. 3.—Time course of calcium signal in Jurkat E6-1 during exposure to 238G sinusoidal 60Hz field inducing 1mV/cm electric field in culture medium, conducted at 37 C (field and control) and 24 C (room temperature). Each point is derived from the analysis of 10000 cells.

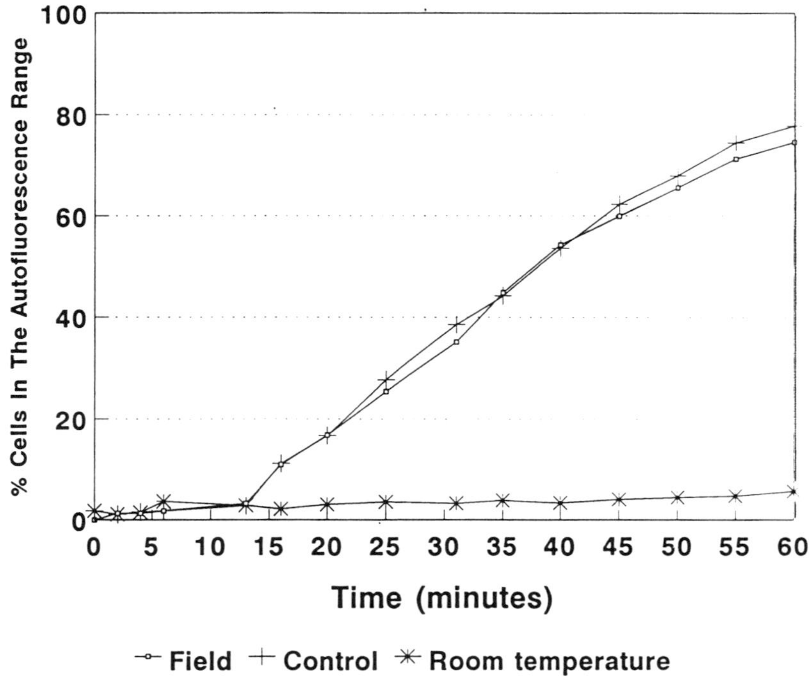

FIG. 4.—Time course of lose of baseline fluorescence in fluo-3-loaded Jurkat E6-1 cells exposed to 238G sinusoidal 60Hz field inducing 1mV/cm electric field in culture medium, conducted at 37 C.

Martin Blank, Ed., *Electricity and Magnetism in Biology and Medicine*

INTRACELLULAR CALCIUM, CALCIUM TRANSPORT, AND c-MYC mRNA INDUCTION IN LYMPHOCYTES EXPOSED TO 60Hz MAGNETIC FIELDS: THE CELL MEMBRANE AND THE SIGNAL TRANSDUCTION PATHWAY

R. P. Liburdy,* D. E. Callahan, T. R. Sloma, and P. Yaswen

Lawrence Berkeley Laboratory, Calif.

The first real-time measurements of $[Ca^{2+}]_i$ during exposure to a 60Hz electric field (1.7 mV/cm) reveal an increase in $[Ca^{2+}]_i$ detected 100 s after mitogen activation.[1] This implicates the plateau phase and calcium influx across the cell membrane, rather than the release of calcium from intracellular stores during the early phase in the interaction. This hypothesis was tested by separation of these two phases of Ca^{2+} signaling in time; 60Hz fields only exerted influence on the calcium influx during the plateau phase of calcium signaling. We hypothesized that such alterations in Ca^{2+} signaling will be propagated along the signal-transduction (ST) pathway. This hypothesis was tested by performing the first simultaneous measurements of Ca^{2+} influx (early ST marker) *and* mRNA transcripts for the proto-oncogene c-MYC (mid-state ST marker) in lymphocytes exposed to a 60Hz field (1.7 mV/cm). Both ST markers are enhanced, which indicates linkage between a field effect acting at the cell membrane involving calcium and subsequent ST pathway processes relevant to gene activation.

We tested two hypotheses in these studies. First, we asked whether the cell membrane, and specifically the calcium ion channel, is directly involved in ELF field interactions. This question is important because receptor sites and ion channels are located in the cell membrane and are the first structures involved in the signal transduction process. Second, we asked whether alterations in calcium ion flux are propagated along the signal-transduction cascade to alter events such as gene activation. This question follows from our interaction model, in which ELF fields alter calcium ion flux and thereby influence subsequent cellular events in the signal-transduction cascade, such as gene activation.[1-3] Results presented here indicate that (1) the cell membrane is a site of ELF field interaction, (2) the early signal transduction markers of calcium ion influx and intracellular calcium are altered by ELF fields, and, most important, (3) these changes in calcium ion second messenger are associated with induction of c-MYC mRNA, a mid-stage signal-transduction marker. Our observation of increased mRNA transcripts confirms the pioneering work of Goodman and Henderson.[4] They employed transformed HL60 leukemia cells, whereas we used mitogen-activated normal lymphocytes. Taken together, our calcium ion transport, $[Ca^{2+}]_i$, and c-MYC mRNA findings provide the first evidence supporting a field interaction in which changes in intracellular calcium are propagated along the signal-transduction cascade to alter gene activation events.

Preparation of rat thymic lymphocytes, the $^{45}Ca^{2+}$ influx assay, the temperature-regulated, solenoidal exposure system and dosimetry procedures, details of FURA-2 loading, and real-time measurements of intracellular calcium during field exposures were recently described.[1-3] Northern analyses were as described.[5]

To evaluate early calcium responses to 60Hz electric fields, the first real-time measurements of $[Ca^{2+}]_i$ during field exposure were performed. Figure 1 illustrates a typical time course for $[Ca^{2+}]_i$ during Con-A activation in the absence or presence of a 1.7 mV/cm 60Hz electric field. When Con-A (1 μg/ml) was added at 200 s, $[Ca^{2+}]_i$ increased during the early phase of calcium signaling to a value of approximately 300 nM at 800 s. In the presence of the 60Hz field, when Con-A was added at 200 s, an increase in $[Ca^{2+}]_i$ was observed that was identical to that shown for unexposed cells over the first 100 s. However, $[Ca^{2+}]_i$ increased at a greater rate than for cells in the absence of the field. By 800 s the plateau phase of $[Ca^{2+}]_i$ reached approximately 380 nM. Alteration of the plateau phase of calcium signaling implicates the calcium ion channel of the outer plasma membrane as a site of field interaction, in contrast to structures associated with calcium release from intracellular stores.

---

*Correspondence to: R. P. Liburdy, LBL Bioelectromagnetics Research Facility (Bldg 74), University of California, Berkeley CA 94720.

To test the above hypothesis we separated the early and plateau phase of calcium signaling in time by placing cells in calcium-free media and then adding calcium to the buffer. Figure 2 shows that at approximately 200 s Con-A was added and $[Ca^{2+}]_i$ increased to approximately 200 nM at 600 s, corresponding to the early phase with release of calcium from intracellular stores. When extracellular calcium was added to the cells at approximately 670 s, the plateau phase of $[Ca^{2+}]_i$ reached 390 nM at 800 s. In the absence of Con-A, when the 60Hz electric field was applied no change in $[Ca^{2+}]_i$ was detected, a result that is consistent with the observations presented in Fig. 1. In Fig. 2, when Con-A was added to control cells in calcium-free buffer, an increase in $[Ca^{2+}]_i$ followed and an essentially identical increase was observed for Con-A treated cells exposed to the 60Hz electric field. Thus, the electric field does not affect the early phase of calcium signaling. When extracellular calcium was added to thymocytes during field exposure, an increase in $[Ca^{2+}]_i$ was observed to levels of approximately 490 nM at 800 s. These findings demonstrate that the ELF field influences calcium movement across the outer cell membrane, which implicates the cell membrane at a site of field interaction.

We hypothesize that such alterations in early signal transduction will affect subsequent "downstream" cellular events in the signal transduction pathway, such as gene activation and cell proliferation.[1-3] To test this hypothesis, we performed the first experiments to quantitate simultaneously Ca^{2+} influx (early ST marker) *and* c-MYC mRNA (mid-stage ST marker) in the same cell preparation exposed to 60Hz ELF fields (1.7 mV/cm, 37 C, 60 min).

Figure 3 shows three separate Northern analyses of the same sample that were imaged by a cooled-CCD camera. There are three bands for each Northern blot, which correspond to −Con-A, +Con-A (1 μg/ml), and +Con-A plus 60 Hz (left to right). Panel A (top row) shows the flat-fielded image of the Northern blots, Panel B (second row) shows a binary image indicating regions of interest (ROI), Panel C (third row) shows a pixel-by-pixel subtraction of the ROIs minus a threshold value, and Panel D (bottom row) depicts a 3D-contour plot representing the quantitated intensity of each band. We performed a multivariate analysis of variance for mean pixel intensity (mean ± SD) of each band across the three Northern analyses: −Con-A = 332 ± 128; +Con-A = 398 ± 48; +Con-A plus 60 Hz = 864 ± 181. In the presence of Con-A we detect a significant 2-fold enhancement of c-MYC mRNA during exposure to the 60Hz field ($p = 0.0114$). This same sample was assessed for $^{45}Ca^{2+}$ influx: (CMP/10^7 cells), −Con-A = 9878 ± 749; +Con-A = 10014 ± 1046; +Con-A plus 60 Hz = 15713 ± 1403. A statistically significant 1.6-fold enhancement was detected at the $p = 0.029$ level for 60Hz field treatment plus Con-A vs Con-A alone. These calcium results are consistent with mRNA findings presented in Fig. 3. Taken together they provide the first experimental support for an ELF interaction mechanism involving the cell membrane and Ca^{2+} signaling triggering "downstream" ST pathway processes[1-3] such as gene activation.[6]

REFERENCES

1. R. P. Liburdy, "Calcium signaling in lymphocytes and ELF fields: Evidence for an electric fields metric and a site of interaction involving the calcium ion channel," *FEBS Letters* 301: 53-59, 1992.

2. R. P. Liburdy, "ELF fields and the immune system: Signal transduction, calcium metabolism, and mitogenesis in lymphocytes with relevance to carcinogenesis," in C. Ramel and B. Norden, Eds., *Interaction Mechanisms of Low-level Electromagnetic Fields in Living Systems,* London: Oxford University Press (in press).

3. R. P. Liburdy, "Biological interactions of cellular systems with time-varying magnetic fields," in R. L. Magin, R. P. Liburdy, and B. Persson, Eds., *Biological Effects and Safety Aspects of Nuclear Magnetic Resonance Imaging and Spectroscopy, Annals N.Y. Acad. Sci.* 649: 74-95, 1992.

4. R. Goodman, "Transcription and translation in cells exposed to extremely low frequency electromagnetic fields," *Bioelectrochem. & Bioenerg.* 25: 335-355, 1991.

5. P. Yaswen, A. Smoll, D. Peehl, D. Trask, R. Sager, and M. Stampfer, "Down regulation of a calmodulin-related gene during transformation of human mammary epithelial cells," *Proc. Natl. Acad. Sci. USA* 87: 7360-7364, 1990.

6. Support provided in part by the U.S. Department of Energy under contract DE-AC-3-76SF00098, and NCI under grant CA53711.

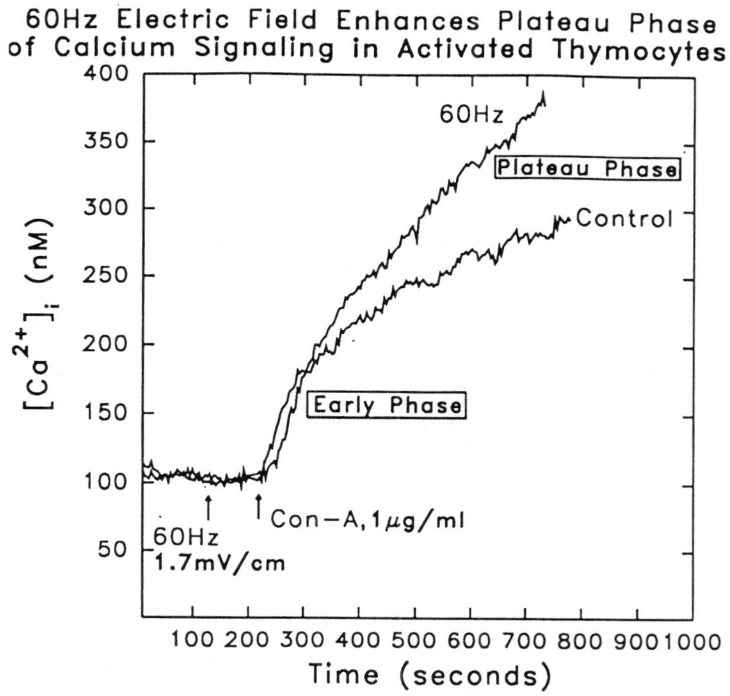

FIG. 1.—Real-time measurements of intracellular calcium in mitogen-activated thymic lymphocytes during 60Hz field exposure.[1]

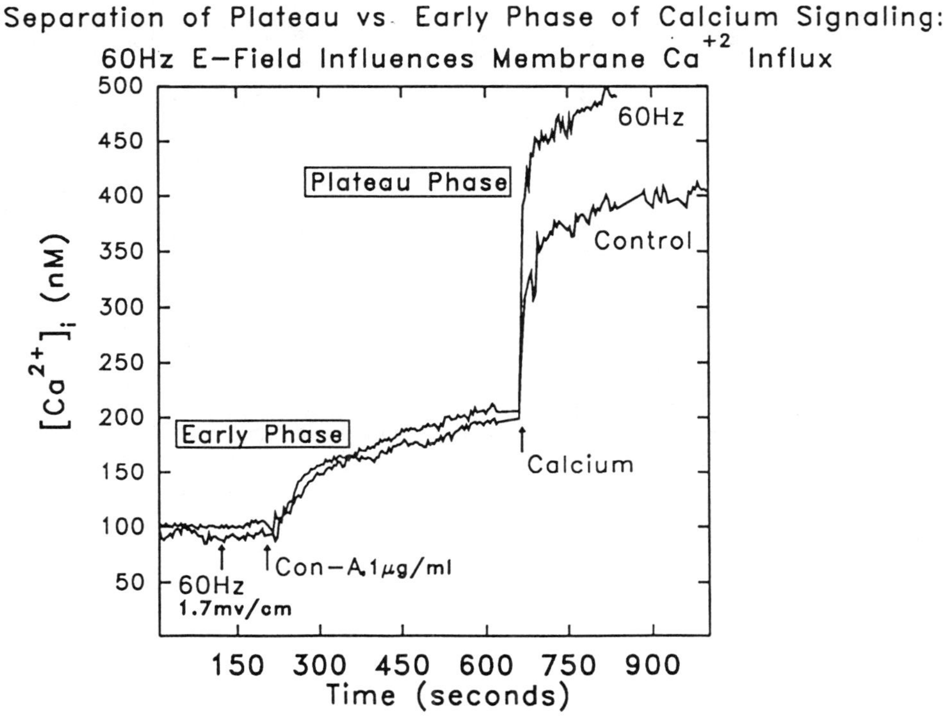

FIG. 2.—Real-time measurements of intracellular calcium in mitogen-activated thymic lymphocytes during 60Hz field exposure: Separation of early and plateau phase of calcium signaling.[1]

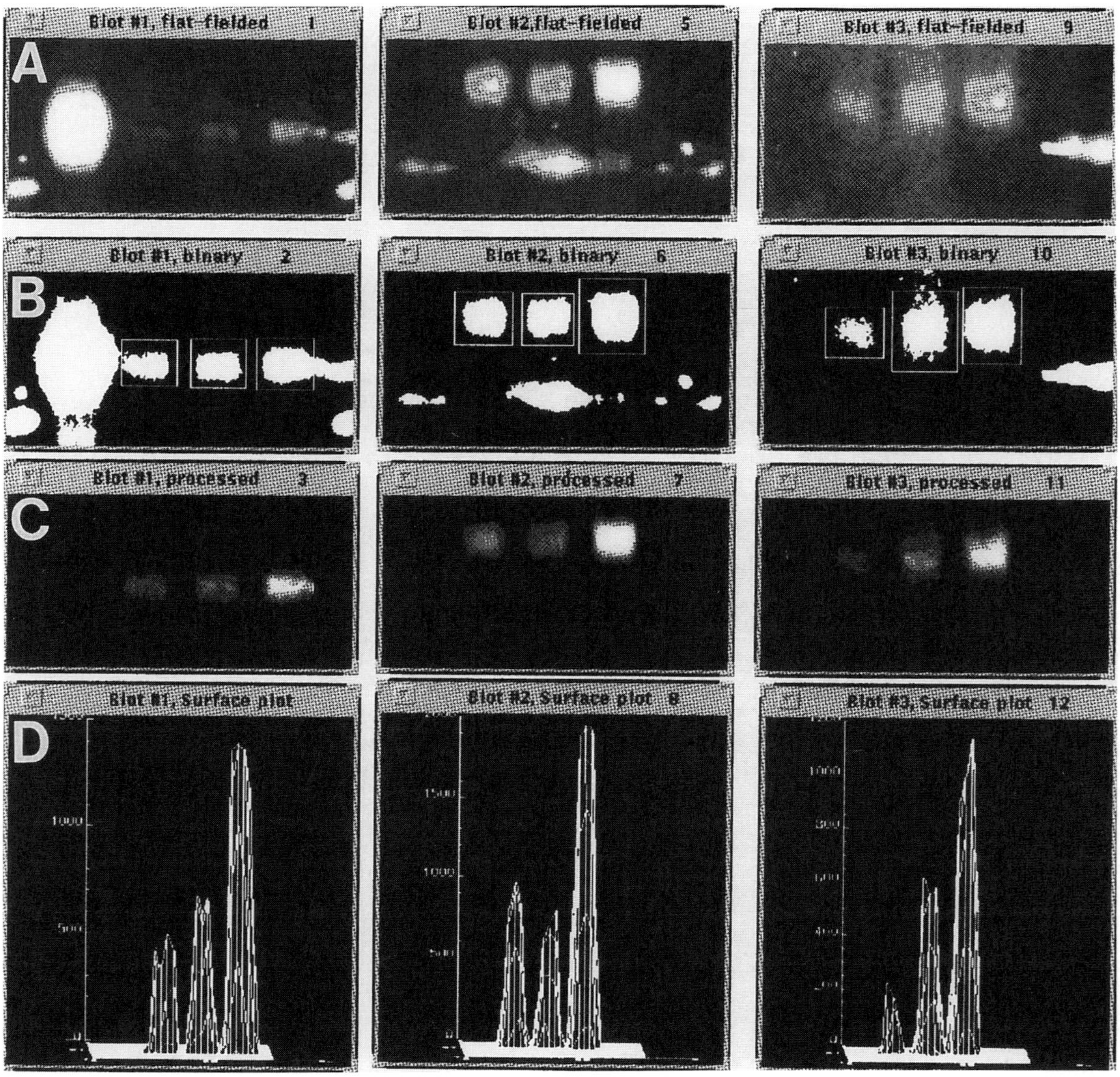

FIG. 3.—Northern analyses of c-MYC mRNA: Quantitation by digital image analysis with a cooled CCD-camera. Samples analyzed on three blots: −Con-A, +Con-A, and +Con-A + 60 Hz.

Martin Blank, Ed., *Electricity and Magnetism in Biology and Medicine*

MEASUREMENT OF CYTOSOLIC FREE CALCIUM IN HL-60 CELLS DURING EXPOSURE TO LOW-FREQUENCY MAGNETIC FIELDS

J. J. L. Carson,* F. S. Prato, D. J. Drost, and S. J. Dixon

University of Western Ontario, London, Ont., Canada

We have previously shown that magnetic fields associated with magnetic resonance imaging (MRI) increase ($[Ca^{2+}]_i$) in HL-60 cells. In addition, we have shown that this 30nM increase is due to the extremely low frequency component (ELF) of MRI). To investigate this phenomenon further, we have build a system that can measure $[Ca^{2+}]_i$ during simultaneous ELF exposure. Preliminary experiments on HL-60 cells at 25 ± 0.1 C indicate a small increase in $[Ca^{2+}]_i$ within 2 min of ELF exposure onset. Furthermore, this increase in $[Ca^{2+}]_i$ was observed only when cells were exposed to both an ELF magnetic field (2.5 mT, rectangular pulses, rise time 0.25 ms, 47 ms repetition rate) and a collinear static magnetic field (0.14 mT). A downward drift in baseline was often encountered and has been attributed to cell settling within the sample cuvette. We have adopted a new mixing technique that eliminates this artifact and maintains a constant fluorescence signal during the course of an experiment.

Cytosolic free calcium concentration ($[Ca^{2+}]_i$) is known to play a role in regulating several cellular functions including membrane transport, secretion, proliferation, muscle contraction, and motility. A number of in vitro studies have implicated $[Ca^{2+}]_i$ as a possible target for extremely low frequency (ELF) electromagnetic fields. Several studies have found changes in $^{45}Ca^{2+}$ uptake or release by or from biological preparations;[1-3] these results suggest but do not indicate changes in $[Ca^{2+}]_i$. Using $^{45}Ca^{2+}$ uptake and release assays, investigators have shown that ELF-mediated effects depend on temperature,[4] degree of cell activation,[5-7] strength of the local static magnetic field,[8,9] and the frequency and amplitude of the ELF component.[10] Stronger evidence for a $[Ca^{2+}]_i$-related ELF electromagnetic field effect has been demonstrated in the ELF-related potentiation of DNA synthesis, neurotransmitter release and motility in mammalian cells.[11-13]

Some studies have attempted to measure $[Ca^{2+}]_i$ with fluorescence calcium probes during or after exposure to ELF electromagnetic fields. Parkinson and Hanks used the fluorescent calcium probe fura-2 to measure $[Ca^{2+}]_i$ in a variety of cell types during ELF electromagnetic field exposure.[14] Although no measurable changes in $[Ca^{2+}]_i$ coincident with ELF field exposure were observed, only limited frequency and amplitude ranges of ELF fields were tested. Using similar fluorescence techniques and the calcium-sensitive probe indo-1, our group has observed a small but reproducible increase in $[Ca^{2+}]_i$ in HL-60 cells after a short (23min) exposure to the three fields of magnetic resonance imaging.[15] We have further demonstrated that samples exposed to the ELF gradient magnetic field (approximately 0.1 mT, 0-100 Hz) alone elicited a response, but samples exposed to the static (0.15 T) or radiofrequency (6.25 MHz, AM-modulated) fields did not. Also we have shown that the effect could not be explained by sample heating, sound, mechanical vibration, or cell death. With the intent to extend our initial work, we have built an apparatus that can monitor $[Ca^{2+}]_i$ during on-line exposure to a wide range of ELF and gradient electromagnetic fields.

SYSTEM FOR MEASUREMENT OF $[Ca^{2+}]_i$ DURING ELF EXPOSURE

An overview of our current exposure/measurement system appears in Fig. 1. The low-frequency magnetic-field exposure systems consists of five coils, all centered on a quartz cuvette (contains cell suspension) in the sample chamber (Fig. 1, SC). Three square Helmholtz-like coils (diameters—0.8, 0.9, and 1.0 m) are arranged so that a constant magnetic field of up to 0.1 mT can be generated in any direction (B_x, B_y, B_z) when used in conjunction with their constant-current power studies. Two Helmholtz-like coils (each 0.21 m in diameter) generate an ELF electromagnetic field. This coil is also capable of generating a switched

---

*Correspondence to: J. J. L. Carson (Nuclear Medicine), St. Joseph's Health Centre, London ON Canada N6A 4V2.

gradient magnetic field similar to that found in clinical magnetic resonance imagers. With a power-supply amplifier, the coil system can generate magnetic fields up to 4.0 mT with or without a linear gradient of up to 12 mT/m and a time rate of change as high as 1 T/s. ELF or gradient waveforms are generated with 1 in 4095 resolution by use of a microcomputer equipped with a digital-to-analog converter board.

The quartz cuvette is optically coupled to a conventional fluorescence spectrophotometer (Fig. 1, FS) by a system of lenses and mirrors (Fig. 1, OC). This device isolates the cuvette by approximately 1.3 m from any materials or devices that would perturb the magnetic field and is capable of measuring changes in fluorescence during on-line electromagnetic-field exposure. To maintain the sample temperature to within 0.1 C, the cuvette is in direct contact with water from a temperature-controlled circulating water bath. The sample compartment (Fig. 1, SC) allows light to enter and leave the sample cuvette on all four sides through watertight UV-transmitting windows.

RESULTS

Originally, cells undergoing measurement within this system were maintained in suspension with a nonmagnetic pipetting mechanism (Fig. 2). By this technique, indo-1-loaded HL-60 cell samples (maintained at 25 ± 0.1 C and prepared as described in Ref. 15) displayed large decreases in fluorescence on account of cells settling out of the light path toward the bottom of the cuvette. The current technique for maintaining cells in suspension involves movement of a disk (7 mm in diameter, normal to surface of disk in the *z* direction) vertically (*z* direction), and repeatedly through the entire sample. Computer control of mixing and data collection permits rejection of scattered light when the disk is in the light path. Mixing performance is shown in Fig. 3. Stable fluorescence readings from freshly isolated indo-1-loaded human peripheral blood lymphocytes (maintained at 37 ± 0.1 C and prepared as described in Ref. 16) were obtained when the disk moved to the bottom of the cuvette in less than 0.22 s. The upward movement of the disk was slower by a factor of 3 than the downward movement; the time between mixing movements was approximately 2 s in all cases.

These results (Fig. 2 and 3) demonstrate that the exposure/measurement system can provide high-resolution stable readings of $[Ca^{2+}]_i$ from indo-1-loaded HL-60 and human peripheral blood lymphocytes. Furthermore, these measurements can be made during on-line ELF or gradient electromagnetic-field exposure.

REFERENCES

1. C. F. Blackman, S. G. Benane, J. R. Rabinowitz, D. E. House, and W. T. Joines, "A role for the magnetic field in the radiation-induced efflux of calcium ions from brain tissue in vitro," *Bioelectromagnetics* 6: 327-337, 1985 and 7: 347, 1986.

2. R. J. Rozek, M. L. Sherman, A. R. Liboff, B. R. McLeod, and S. D. Smith, "Nifedipine is an antagonist to cyclotron resonance enhancement of $^{45}Ca^{2+}$ incorporation in human lymphocytes," *Cell Calcium* 8: 413-427, 1987.

3. D. B. Lyle, X. H. Wang, R. D. Ayotte, A. R. Sheppard, and W. R. Adey, "Calcium uptake by leukemic and normal T-lymphocytes exposed to low frequency magnetic fields," *Bioelectromagnetics* 12: 145-156, 1991.

4. C. F. Blackman, S. G. Benane, and D. E. House, "The influence of temperature during electric- and magnetic-field-induced alteration of calcium-ion release from in vitro brain tissue," *ibid.*, 173-182, 1991.

5. J. Walleczek and R. P. Liburdy, "Nonthermal 60 Hz sinusoidal magnetic-field exposure enhances $^{45}Ca^{2+}$ uptake in rat thymocytes: Dependence on mitogen activation," *FEBS Lett.* 271: 157-160, 1990.

6. G. Emilia, G. Torelli G. Ceccherelli, A. Donelli, S. Ferrari, P. Zucchini, and R. Cadossi, "Effect of low-frequency low-energy pulsing electromagnetic fields on the response to lectin stimulation of human normal and chronic lymphocytic leukemia lymphocytes," *J. Bioelectricity* 4: 145-161, 1985.

7. P. Conti, G. E. Gigante, E. Alesse, M. G. Cifone, C. Fieschi, M. Reale, and P. U. Angeletti, "A role for Ca^{2+} in the effect of very low frequency electromagnetic field on the blastogenesis of human lymphocytes," *FEBS Lett.* 181: 28-32, 1985.

8. C. F. Blackman, S. G. Benane, D. E. House, and D. J. Elliott, "Importance of alignment between local DC magnetic field and an oscillating magnetic field in responses of brain tissue in vitro and in vivo," *Bioelectromagnetics* 11: 159-167, 1990.

9. M. G. Yost and R. P. Liburdy, "Time-varying and static magnetic fields act in combination to alter calcium signal transduction in the lymphocyte," *FEBS Lett.* 296: 117-122, 1992.

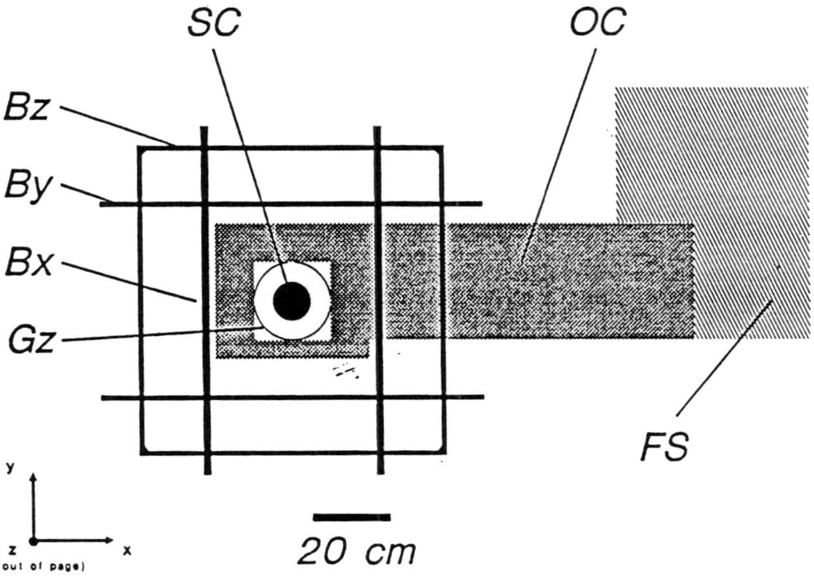

FIG. 1.—Top view of exposure/fluorescence system.

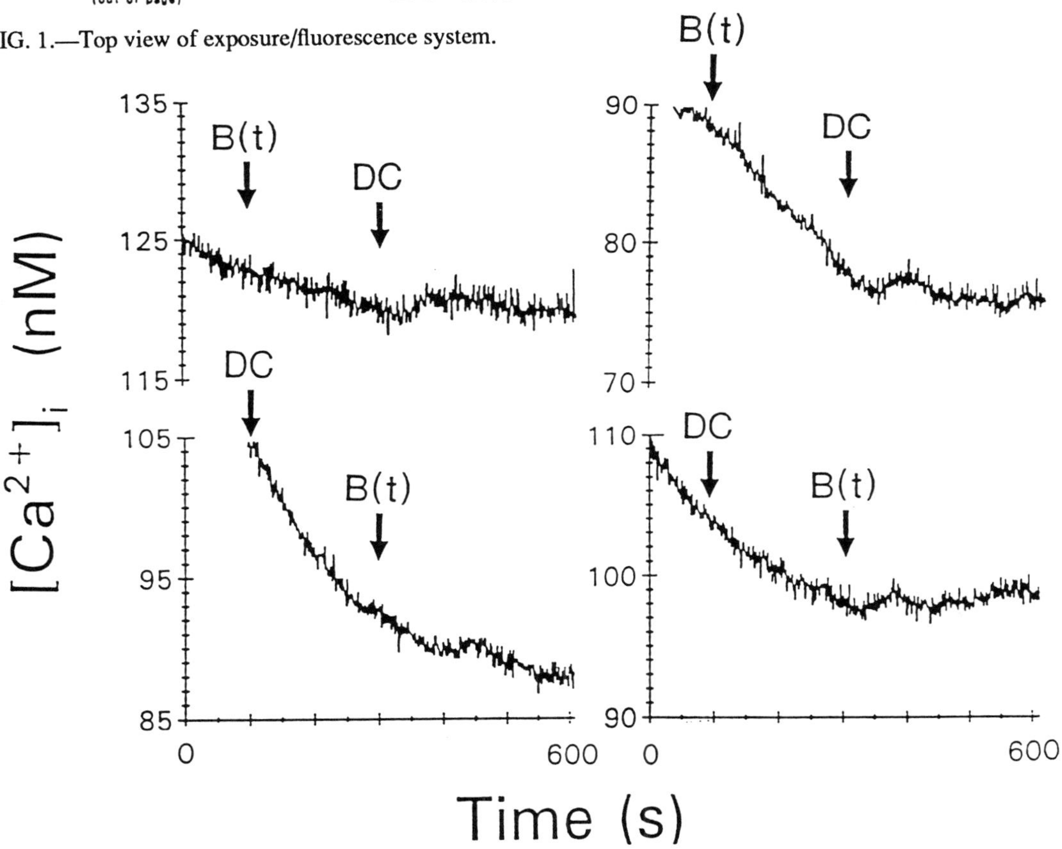

FIG. 2.—Preliminary results showing high-resolution $[Ca^{2+}]_i$ measurements in HL60 cells during on-line exposure to ELF magnetic field $B(t)$ (2.5 mT rectangular pulsed, rise time approximately 0.25 ms, 47ms repetition rate) and/or a collinear static magnetic field (DC, 0.14 mT). Exposure to $B(t)$ alone (top tracings, 100 s) or DC alone (bottom tracings, 100 s) did not increase $[Ca^{2+}]_i$. However, when both $B(t)$ and DC were applied in combination (all tracings, 300 s), a small increase in $[Ca^{2+}]_i$ resulted.

10. A. R. Liboff, R. J. Rozek, M. L. Sherman, B. R. McLeod, and S. D. Smith, "Ca^{2+}-45 cyclotron resonance in human lymphocytes," *J. Bioelectricity* 6: 13-22, 1987.

11. R. Dixey and G. Rein, "^{3}H-noradrenaline release potentiated in a clonal nerve cell line by low-intensity pulsed magnetic fields," *Nature* 296: 253-156, 1982.

12. A. R. Liboff, T. Williams Jr., D. M. Strong, and R. Wister Jr., "Time-varying magnetic fields: Effect on DNA synthesis," *Science* 223: 818-820, 1984.

13. G. P. A. Yen-Patton, W. F. Patton, D. M. Beer, and B. S. Jacobson, "Endothelial cell response to pulsed electromagnetic fields: Stimulation of growth rate and angiogenesis in vitro," *J. Cell. Physiol.* 134: 37-46, 1988.

14. W. C. Parkinson and C. T. Hanks, "Search for cyclotron resonance in vitro," *Bioelectromagnetics* 10: 129-145, 1989.

15. J. J. Carson, F. S. Prato, D. J. Drost, L. D. Diesbourg, and S. J. Dixon, "Time-varying magnetic fields increase cytosolic free Ca^{2-} in HL-60 cells," *Am. J. Physiol.* 259 (Cell Physiol. 28): C687-692, 1990.

16. E. W. Gelfand, R. K. Cheung, and S. Grinstein, "Mitogen-induced changes in Ca^{2+} permeability are not mediated by voltage-gated K^+ channels," *J. Biol. Chem.* 261: 11520-11523, 1986.

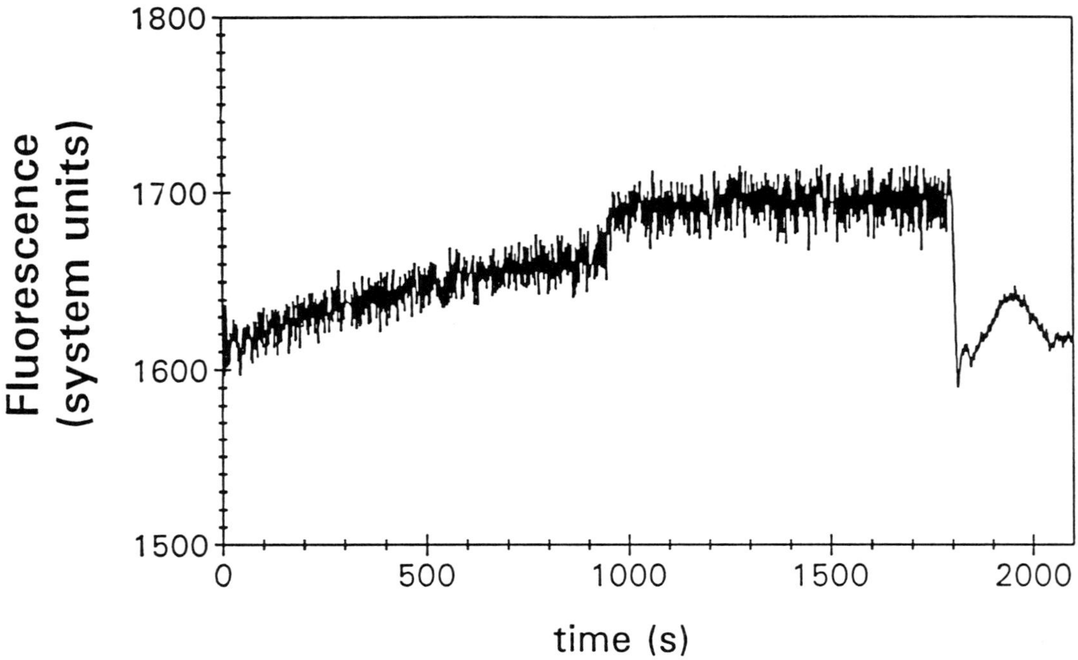

FIG. 3.—Time course of fluorescence from freshly isolated indo-1-loaded human peripheral blood lymphocytes. Cells were maintained in suspension by repeated movement of 7mm disk vertically through the suspension. Time for downward movement of disk was varied in the order 0.33 s (0-300 s), 0.29 s (300-600 s), 0.25 s (600-900 s), 0.22 s (900-1200 s), 0.2 s (1200-1500 s), and 0.18 s (1500-1800 s). Values between 1800 s and 2100 s represent fluorescence from unmixed cells.

Martin Blank, Ed., *Electricity and Magnetism in Biology and Medicine*

WEAK MAGNETOSTATIC FIELDS ALTER CALMODULIN-DEPENDENT CYCLIC NUCLEOTIDE PHOSPHODIESTERASE ACTIVITY

A. W. Bull, S. Cherng, K. A. Jenrow, and A. R. Liboff*

Oakland University, Rochester, Mich.

To probe the effect of weak magnetic fields on the interaction between calcium and calmodulin (CaM), we have measured the activity of CaM-dependent cyclic nucleotide phosphodiesterase (PDE), which catalyzes the hydrolysis of cyclic AMP. Commercially available activator-free PDE and CaM were incubated in the presence of magnetostatic fields ranging from 169 to 237 mG. Control samples were maintained at zero field. A 13% stimulation of PDE activity, over controls, was observed in samples exposed to fields in the range 192 to 204 mG. Fields outside this range did not stimulate PDE activity. These results suggest that such field-modulated PDE activity provides a physiological pathway for interactions between magnetic fields and cells.

The nature of the weak ELF magnetic interaction with the cell remains elusive. Following the original suggestion[1] that the unique physical properties of ion channels might lend themselves to an ion cyclotron resonance magnetic interaction, a number of Lorentz force models have been advanced, focusing not only on the ion channel,[2-4] but also elsewhere in the cell.[5,6] Lednev[7] suggested that the interaction site may lie in calcium-binding proteins such as calmodulin (CaM). This model, treating bound calcium ions as oscillating charges in a spherically symmetric potential, predicted altered bioresponses following cellular exposure to combinations of time-varying and static magnetic fields under cyclotron-resonance conditions. This model has been criticized,[8] among other reasons, for having the same intrinsic difficulty as previous cyclotron-resonance models, namely that, in order for a "resonance transition" to occur, the ion must enjoy mean free times between collisions that are unreasonably long. In addition, Bruckner-Lea et al.[9] found that under saturating conditions, calcium binding to calmodulin was not affected in the presence of combined AC and DC magnetic fields. Nevertheless, Shuvalova et al.[10] reported changes in CaM-dependent phosphorylation of myosin light chains for magnetic-field exposures at or near the cyclotron resonance condition for Ca^{2+}. This reaction has also been studied by Markov et al.,[11] who observed both increases and decreases in phosphorylation under a variety of weak magnetostatic and time-varying field conditions.

We have developed a second, independent experimental model to study the effects of weak magnetic fields on CaM, examining the activity of CaM-dependent cyclic nucleotide phosphodiesterase (PDE). The dependence of PDE activity on the state of CaM allows one to probe the activation of CaM by determination of PDE activity. In turn, the regulation of cyclic AMP (cAMP) levels by alterations in PDE provides a potential mechanism for alteration of cellular function.

MATERIALS AND METHODS

All solutions were prepared in plastic tubes to minimize contamination with adventitious calcium. The incubation, in a total volume of 0.5 ml, contained 40 mM Tris-HCl, CaM (0.02 units/ml), activator-free PDE (0.03 units/ml), $CaCl_2$ (1.0 μM), cAMP (1.8 mM), and varying concentrations of EGTA. The EGTA concentration was adjusted for each run to control the level of free calcium and maximize the sensitivity of the system to any changes in the interaction between Ca^{2+}, CaM, and PDE. Immediately after the addition of the cAMP to initiate the reaction, the samples were exposed for 30 min to various combinations of single-axis oriented DC and AC magnetic fields within a nonferrous, motorless incubator maintained at 37 C. Fields were applied by means of the multiple-exposure solenoid array,[12] the current version of which allows one to expose 9 samples to various intensities while maintaining three controls at zero field. Two different exposure regimens are reported here. In one, DC fields in the range 169 to 237 mG (16.9 to 23.7 μT) were applied to

*Correspondence to: A. R. Liboff (Physics), Oakland University, Rochester MI 48309.

the treated samples, while the three control specimens were kept at 0.0 mG (±0.5 mG). In the other, a DC field of 209 mG was applied together with a 209mG peak, 16.0Hz field, with controls exposed to the same AC field, but a zero DC field. The subsequent assay followed the method of Butcher and Sutherland.[13] Immediately following each exposure run, the reactions in all samples were quenched by boiling for 2 min. After cooling, the tubes were incubated in snake venom phosphodiesterase (0.1 mg/ml) for 10 min to release free phosphate from AMP. The second reaction was terminated by addition of trichloracetic acid (10% final concentration). Following the addition of ammonium molybdate, the denatured protein was removed by centrifugation and the clear supernatent was decanted to a clean tube containing Fiske-Subbarow reagent. After an additional 10 min, the absorbance at 660 nm was measured for each sample. Data are presented below in terms of the ratio of absorbance units (E/C) comparing the absorbance of magnetically treated samples to the absorbance of samples held at zero field. All samples, treated and control, were prepared together, exposed simultaneously, and assayed together. Values of E/C > 1 indicate increased calcium-CaM-dependent activation of cyclic nucleotide PDE; values of E/C < 1 indicate reduced activation.

RESULTS

Figure 1 shows the change in activation, in terms of E/C, for 40 independent trials (over 3 months) in which specimens were exposed to a DC magnetic field of 200 mG and compared with specimens simultaneously maintained at zero DC magnetic field. The data are plotted as a function of the control (i.e., zero-field) absorbance because we observed that the response of the system to the magnetic field was in part determined by the zero-field absorbance. At low enzyme activities (low absorbance), the 30min DC magnetic field tends to reduce PDE activity. However, at high PDE activities, magnetic field exposures enhance activity. There is an apparent crossover in E/C at about 0.1 absorbance unit, and some indication of a maximum magnetic effect between 0.1 and 02. It is reasonable to suggest that this absorbance region corresponds to the threshold calcium concentration for binding to calmodulin. Evidence for this conclusion is given in Fig. 2, in which zero-field absorbance is plotted against EGTA concentration. At high EGTA concentration (low calcium) the enzyme activity is minimal, as indicated by an absorbance below 0.1. At lower EGTA (and thus increasing free calcium) there is a sharp increase in enzyme activity, as shown by a rapid change in absorbance between 0.1 and 0.2. It is noteworthy that this is the same region where the effect of the magnetic field changes most rapidly. The shape of the curve in Fig. 2 illustrates the utility of measuring PDE activity as a means of assessing the Ca^{2+}/CaM interaction.

The effects of DC fields other than 200 mG were also determined. Because of the reversal in magnetically dependent enzyme activity shown in Fig. 1, we concentrated our efforts on studying magnetic-field effects at enzyme activities leading to absorbances between 0.1 and 0.5. With this restriction, 33 remaining independent determinations of E/C were found for the 200mG DC exposures shown in Fig. 1. They have a mean value of 1.133 ($p < 0.0001$). In assessing the relative effects of DC fields, we discovered that the field exposures that were attempted, ranging from 169 to 237 mG, were not all equally effective in enhancing enzyme activity. The results indicate that field intensities in the 190 to 220mG region enhance PDE activity, whereas fields outside this region are ineffective (Fig. 3).

Additional trials examined the effect on enzyme activity when combined AC and DC magnetic fields tuned to the cyclotron resonance condition for Ca^{2+} were applied instead of DC fields alone. In these experiments, the treated specimens were exposed to parallel vertical AC (200mG peak at 15.3 Hz) and DC (200mG) fields, while the controls were exposed to the same AC field but zero DC field. The relative effects on enzyme activity, again as determined in terms of absorbance ratios, were statistically insignificant, with the relative effect of AC/DC fields to DC fields alone determined as 1.09 ± 0.14 ($N = 3$).

DISCUSSION

This work provides support for the observation by Markov et al.[11] that very weak DC magnetic fields alter Ca^{2+}/CaM-dependent reactions. The fact that in the present model phosphodiesterase activation was studied (as opposed to myosin phosphorylation) lends credence to the likelihood that the magnetic field acts on the Ca^{2+}/CaM complex, which is common to both experiments. We did not observe any significant effect on enzyme activation attached to the use of cyclotron resonance AC and DC fields, over and above that which resulted from DC fields alone. The DC fields in the present work include the 209mG level used by Shuvalova et al.[10] as the DC component of their cyclotron-resonance field combination. The possibility exists that their results might have been observed even if the AC component had not been applied. Thus, although we are led to the conclusion that calcium-binding proteins can indeed serve as an interaction target for weak magnetic fields, as hypothesized by Lednev,[7] there is considerable uncertainty about the nature of this interaction.

Apart from the difficult question of explaining the interaction mechanism, important implications are attached to the possibility that weak DC magnetic fields are able to influence the Ca^{2+}/CaM interaction, ubiquitous to all eukaryotic systems. In the model that we have studied, CaM directly activates cyclic nucleotide phosphodiesterase, thus increasing the hydrolysis of cAMP. The change in activation of PDE by DC magnetic fields thereby provides a mechanism for the modulation of cellular function via alteration in the cAMP second-messenger system. It is not unreasonable to speculate that this could provide the basis for a sensitive biological control mechanism that makes use of the local geomagnetic field in a manner that has remained unsuspected until now.

The varied response shown in Fig. 3 covers a rather narrow range of DC field intensities. To probe the nature of this effect properly, it will be necessary to explore a much wider range of fields, from zero through at least 1 G. Note that the experimental procedure we employed was to maintain our controls at zero DC field. It is conceivable that some other choice, such as the local ambient field, might be a more suitable reference field at which to hold controls.

REFERENCES

1. A. R. Liboff, "Geomagnetic cyclotron resonance in living cells," *J. Biol. Phys.* 13: 99-102, 1985.
2. B. R. McLeod and A. R. Liboff, "Dynamic characteristics of membrane ions in multifield configurations of low-frequency electromagnetic radiation," *Bioelectromagnetics* 7: 177-189, 1986.
3. A. R. Liboff and B. R. McLeod, "Kinetics of channelized membrane ions in magnetic fields," *ibid.*, 9: 39-51, 1988.
4. B. R. McLeod, A. R. Liboff, and S. D. Smith, "Electromagnetic gating in ion channels," *J. Theoretical Biology* 158: 15-31, 1992.
5. A. Chiabrera, B. Bianco, F. Caratozzolo, G. Gianetti, M. Grattarola, and R. Viviani, "Electric and magnetic field effects on ligand binding to the cell membrane," in A. Chiabrera, C. Nicolini, and H. P. Schwan, Eds., *Interactions Between Electromagnetic Fields and Cells*, London: Plenum Press, 1985.
6. C.H. Durney, C. K. Rushforth, and A. A. Anderson, "Resonant AC-DC magnetic fields: Calculated response," *Bioelectromagnetics* 9: 315-336, 1988.
7. V. V. Lednev, "Possible mechanism for the influence of weak magnetic fields on biological systems," *ibid.*, 12: 71-75, 1991.
8. R. K. Adair, "Criticism of Lednev's mechanism for the influence of weak magnetic fields on biological systems," *ibid.*, 13: 231-235, 1992.
9. C. Bruckner-Lea, C. H. Durney, J. Janata, C. Rappaport, and M. Kaminski, "Calcium binding to metallochromic dyes and calmodulin in the presence of combined AC-DC magnetic fields," *ibid.*, 13: 147-162, 1992.
10. L. A. Shuvalova, M. V. Ostrovskaya, E. A. Sosunov, and V. V. Lednev, "Weak magnetic fields tuned to the paramagnetic resonance condition changes the rate of calcium-calmodulin-dependent myosin phosphorylation," *Reports of the Academy of Sciences USSR* 317: 227-229, 1991.
11. M. S. Markov, J. T. Ryaby, J. J. Kaufman, and A. A. Pilla, "Extremely weak AC and DC magnetic fields significantly affect myosin phosphorylation," in M. J. Allen, S. F. Cleary, A. A. Sawers, and D. D. Shilady, Eds., *Charge and Field Effects in Biosystems* III, Boston: Birkhauser, 1992.
12. A. R. Liboff and K. A. Jenrow, "Eight-fold multiple magnetic exposure system for resonance studies in cell culture," paper presented at 13th Ann. Mtg. Bioelectromagnetics Society, 1991.
13. R. W. Butcher and E. W. Sutherland, "Adenosine 3′-5′ phosphate in biological materials," *J. Biol. Chem.* 237: 1244, 1962.

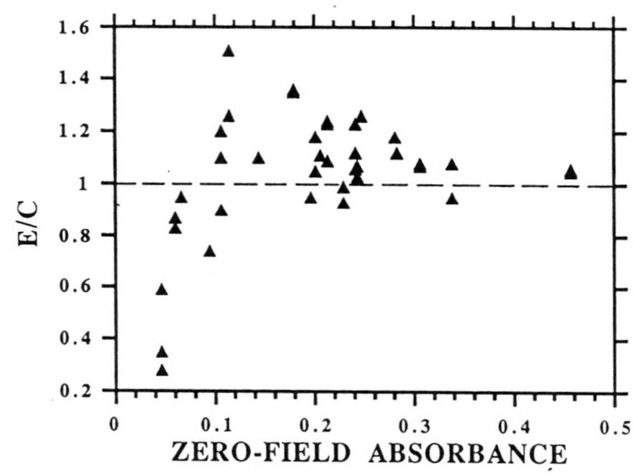

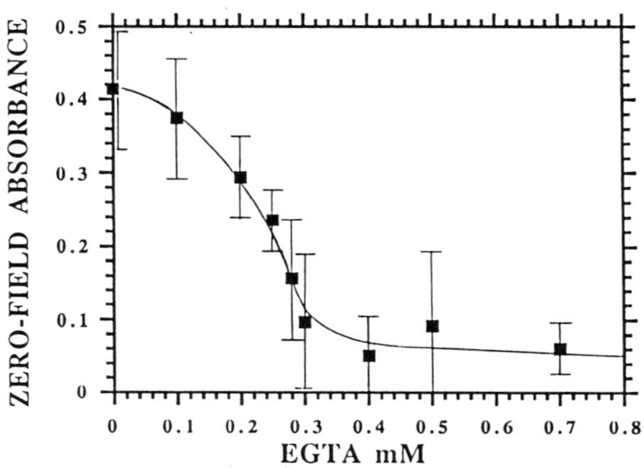

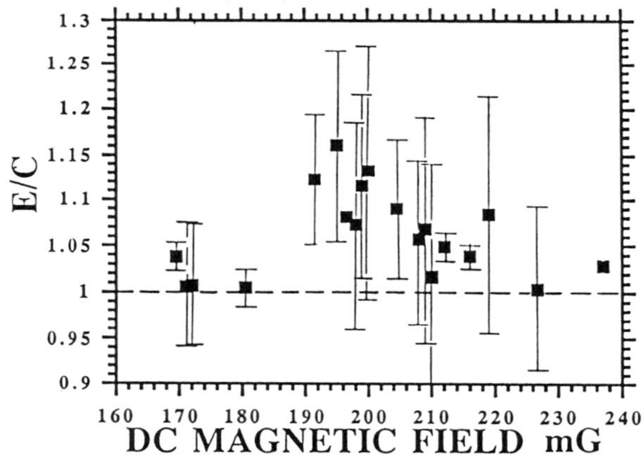

FIG. 1.—Effect of 30min uniaxial 200mG magnetostatic field exposures on Ca^{2+}/CaM-dependent activation of cyclic nucleotide PDE at 37 C, as a function of control absorbance. Controls were simultaneously exposed to zero DC field. There is enhanced enzyme activity (increased absorbance) when E/C > 1, and decreased activity when E/C < 1.

FIG. 2.—Zero-field absorbance following 30min incubations as a function of EGTA concentration. Data are presented as mean ± SD for $N \geq 5$. High EGTA concentration corresponds to low free Ca^{2+} concentration. Sharpest change in enzyme activity occurs in absorbance range 0.1 to 0.2, which is presumably where Ca^{2+} threshold for CaM activation is attained.

FIG. 3.—Effect on cyclic nucleotide activation of various DC magnetic fields, from 169 to 237 mG. Data restricted to zero-field absorbances in excess of 0.1. Data are presented as mean ± SD for $N \geq 5$, except for value at 237 mG ($N = 1$). N for 200mG point is 33, corresponding to values in Fig. 1 with absorbance greater than 0.1.

Martin Blank, Ed., *Electricity and Magnetism in Biology and Medicine*

AMBIENT RANGE SINUSOIDAL AND DC MAGNETIC FIELDS AFFECT MYOSIN PHOSPHORYLATION IN A CELL-FREE PREPARATION

M. S. Markov and A. A. Pilla

Mount Sinai School of Medicine, New York

This study evaluates the effects of sinusoidal alternating current (AC) and static (DC) magnetic fields on Ca^{2+}-calmodulin dependent myosin light chain phosphorylation in a cell-free preparation. The results show that *all* of the AC, DC, and DC/AC combination magnetic fields significantly affect myosin phosphorylation. Even exposure to only DC alterations of the ambient magnetic field was sufficient to modify myosin phosphorylation. These changes were somewhat stronger than those for exposure to all other magnetic fields.

Electromagnetic field (EMF) bioeffects have been most often related to the induced electric field, while the role of the magnetic component for other than electric-field coupling remains ambiguous. Numerous experiments performed in the past decade have demonstrated that weak electromagnetic fields elicit bioeffects in various systems.[1-6] However, for some reported EMF bioeffects, remarkably low-level DC magnetic fields can modulate the biological response. Membrane ion transport and subcellular systems[7-9] are frequently studied, but experiments on cell-free systems have also been reported.[10-16] Several theories have been recently proposed to explain the effects of extremely weak magnetic fields on biosystems. The classical approach considers the effect of the Lorentz force on a moving charged particle either in terms of cyclotron resonance of ion movement[17] or of ion binding kinetics.[18] Quantum parametric resonance theory[19] considers bound ions as harmonic oscillators whose energy levels can be split with AC/DC magnetic fields. Larmor precession as a possible mechanism for detecting static and alternating magnetic fields has also been suggested.[20] However, none of these models adequately explains the observed bioeffects, particularly for DC magnetic-field experiments.

The search for experimental evidence for the effect of magnetic fields on phosphorylation in specific enzyme systems could be quite a promising approach for screening various theoretical mechanisms. This study was therefore designed to evaluate the effect of weak sinusoidal (AC) and DC magnetic field on Ca^{2+}-calmodulin dependent myosin light chain phosphorylation in a cell-free preparation.

MATERIALS AND METHODS

Myosin light chains, myosin light chain kinase, and calmodulin were isolated from turkey gizzard.[21] The reaction mixture, adapted from Adelstein,[21] was prepared with 6 μM myosin light chains, 2 μM myosin light chain kinase, and 2 μM calmodulin in 40 mM Hepes buffer, pH 7.0; 0.5 mM magnesium acetate; 1 mg/ml bovine serum albumin; and 0.1% (w/v) Tween 80. Ca^{2+} concentration was optimized to 30 μM $CaCl_2$ for maximum phosphorylation by calibration experiments. The reaction was initiated by addition of 50 μM gamma^{32}P ATP (100-2000 cpm/pmol) to the reaction medium, and was stopped with Laemmli Sample Buffer (LSB) stopping solution containing 30 μM EDTA. Experimental and control samples were placed in 1.5ml centrifuge tubes (Eppendorf) at 37 C.

The EMF exposure system consisted of two orthogonal pairs of 60-turn 8in square coils by which both the vertical and horizontal DC magnetic field could be adjusted. A Helmholtz pair of circular (8in diameter) coils mounted horizontally was used to apply any other signal for which the parameters vary in time, in particular AC (8-20 Hz, 20.9 μT). DC magnetic fields were measured to ±0.1 μT. A specially designed mumetal box (Ammunel) permitted shielding of the exposure system and samples from any natural and extraneous AC and DC magnetic fields to ±0.1 μT. Several experiments were performed by active (blocking the

---

*Correspondence to: M. S. Markov (Bioelectrochemistry), Dept. of Orthopedics, Mt. Sinai School of Medicine, New York NY 10029.

ambient vertical and/or horizontal magnetic fields with an equal counter field) or by passive (mu-metal-box) compensations. Ambient magnetic fields were carefully monitored in all experiments. Daily (24h period) DC magnetic field variations at the experimental site were as follows: 65 ± 5 μT vertical and 15 ± 5 μT horizontal, which resulted in a magnitude of 72 ± 5 μT, 77° from horizontal. During any given experiment (2-15 min) changes in the ambient magnetic field were within ±0.1 μT. To test whether the DC magnetic field effects depended on scalar or vectorial properties of magnetic fields, the magnitude of the ambient field at the experimental site was applied alternately in the vertical and horizontal directions inside the shield.

Experiments were performed at 37 C and the reaction was allowed to run for 2-15 min. The extent of phosphorylation was determined by counting $\gamma\,^{32}P$ incorporated into myosin light chains. The mean value at each reported time point is calculated from at least 30 independent measurements. A Student's paired *t*-test was performed at each time and exposure condition.

RESULTS AND DISCUSSION

Experimental data reported elsewhere[12,13] suggest that practically *all* the applied weak magnetic fields (DC, AC, and AC/DC combination) significantly modify myosin phosphorylation. That work showed 20.9μT AC magnetic fields alter phosphorylation, as predicted theoretically[9,17,18,19] and demonstrated elsewhere experimentally.[11] Nearly equivalent effects were observed at other frequencies that did not follow prediction, although the largest effect was for a 16Hz field. A further surprising observation was made: when a 20.9μT vertical DC magnetic field was applied alone, the phosphorylation change was stronger than either AC alone or AC/DC combinations.

These results suggested the design of a specific experiment to evaluate the effects of vertical DC magnetic fields on myosin phosphorylation in the absence of AC and horizontal DC magnetic fields. The experimental procedure includes passive (by shield) or active (by application of an opposite current) compensation of horizontal DC magnetic fields and shielding of any AC magnetic fields generated in the laboratory. When the amplitude of the vertical DC magnetic field was reduced below ambient to the value of 0-50 μT, a significant decrease of phosphorylation rate to 35-50% compared to control was observed.

In contrast, phosphorylation increases to approximately 125% of the mean control value when a DC magnetic field of 110-140 μT was applied vertically. When the applied field amplitude was 200 μT, the effect reaches values of more than 180%. Figure 1 shows these results at 6 min exposure time.

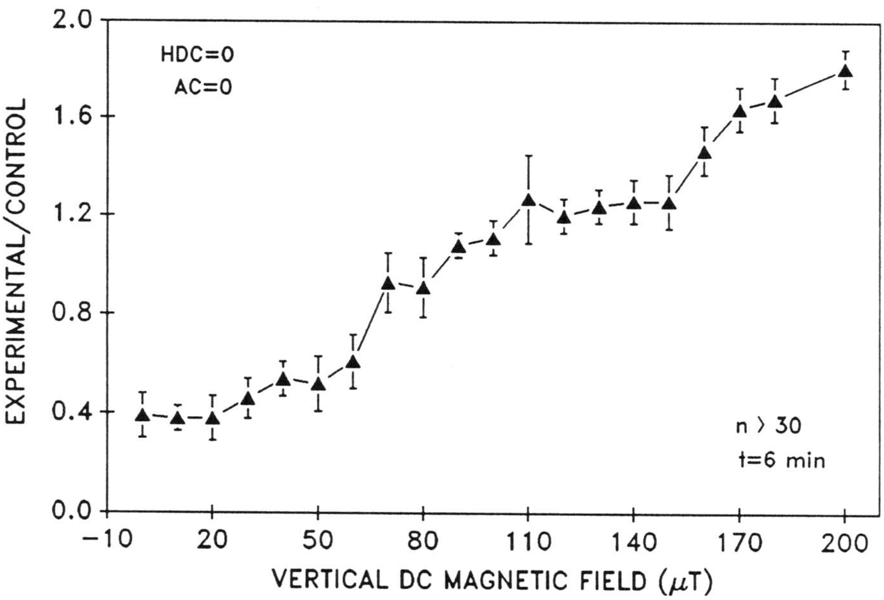

FIG. 1.—Effects of 6min exposure to vertical DC magnetic fields in absence of AC or horizontal DC fields on myosin phosphorylation.

The efficiency of shielding was tested mainly in the range of applied DC magnetic fields close to the value of the vertical component of the ambient field, and phosphorylation was estimated in detail for the range 50-100 μT. These experiments were performed either in a mu-metal box (which shielded all DC and AC fields to approximately ±0.1 μT) or by active compensation of the horizontal DC component to ±0.1 μT. These results (Fig. 2) clearly demonstrate a monotonical increase of phosphorylation rate starting at a value significantly below and ending above the values obtained at ambient magnetic field exposure. There is no significant difference between active and passive compensation techniques. At least two conclusions can be drawn: (1) extraneous AC fields played an insignificant role in these effects, and (2) alterations in the phosphorylation reaction appear to depend on the magnitude (and not on the direction) of the applied magnetic field. An experiment was designed to test the hypothesis that DC magnetic-field effects should be independent of direction. The magnitude of the DC magnetic field as evaluated at the experimental site (approximately 72 μT) was applied alternately in a vertical or horizontal direction. Essentially identical results were obtained for both.

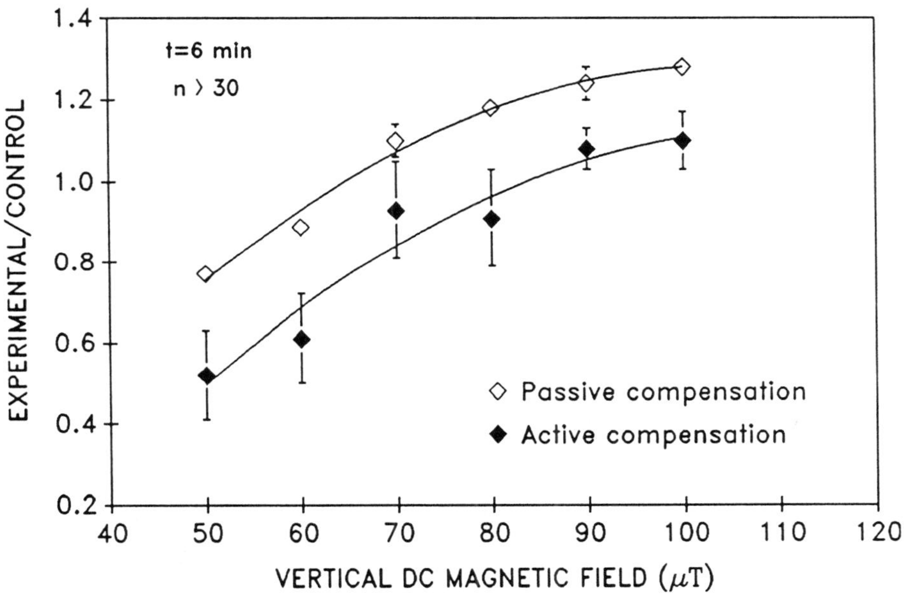

FIG. 2.—Vertical DC magnetic field effects on myosin phosphorylation. Horizontal DC field was actively or passively compensated.

In one series of experiments, the kinetics of myosin phosphorylation were evaluated for the time interval 2-15 min. The results (Fig. 3) appear to show that there is an overall shift of the kinetics curves depending on the presence or absence (shield) of magnetic field, as well as on its amplitude. For instance, a 200μT DC magnetic field significantly increases phosphorylation above ambient values. All curves follow a typical enzyme-substrate time dependence. These results suggest that the magnetic field effect occurred at short times (<2 min), i.e., during the earliest steps of phosphorylation. These changes, which may not involve a direct effect on Ca^{2+}, could then trigger phosphorylation, the measured endpoint.

Consideration of the possible targets for weak magnetic fields in this cell-free system suggests, as potential candidates, the calcium ion, the calmodulin molecule, the substrate (myosin light chain), the enzyme (myosin light chain kinase), and ATP. In contrast, most theoretical approaches consider Ca^{2+} binding to calmodulin binding sites as the most probable target(s) for magnetic-field action. However, the amplitude dependence of DC magnetic fields effects (Fig. 1) suggests that the reactivity of this cell-free enzyme system, as a whole, is affected in both negative *and* positive directions. Further, the overall shift of the reaction kinetics with DC magnetic field amplitude, as opposed to a direct effect on the time course itself (Fig. 3), indicates that Ca^{2+} binding is not a simple target pathway for the observed effects. This conclusion is reinforced by recent studies that show that equilibrium Ca^{2+} binding to calmodulin is not altered by cyclotron or parametric-resonance magnetic-field conditions.[15] That does not necessarily mean that binding kinetics are unaffected; nor does it imply that the reactivity of calmodulin itself is not altered.

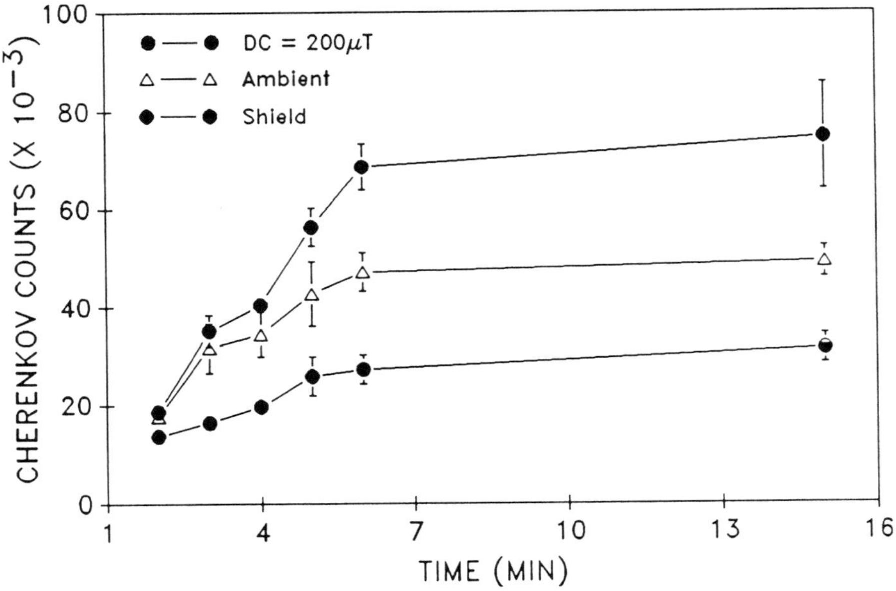

FIG. 3.—Effect of vertical DC magnetic-field amplitude on the enzyme-substrate kinetics of myosin phosphorylation.

This study clearly shows that even small alterations in ambient DC magnetic fields are sufficient to have a profound effect on myosin phosphorylation as measured in a cell-free system. Biological systems could therefore have exquisite sensitivity to weak magnetic fields. Certainly compensation and regulation mechanisms at the organism level to external electromagnetic conditions need to be considered. It is of interest to compare these results to recent in vitro and in vivo data. Weak DC magnetic field alterations of calmodulin-dependent cyclic nucleotide phosphodiesterase activity has recently been reported.[17] These data confirm the possibility that weak DC magnetic fields affect cell-free and physiologically important enzyme systems. When the static magnetic field was lowered (via shield) to 5 nT, compared to ambient levels of 30-60 μT, developing larvae of the Japanese newt exhibited developmental abnormalities.[22] In contrast, exposure of thymic lymphocytes to AC/DC magnetic field combinations, corresponding to quantum-parametric resonance conditions, altered Ca^{2+} signal transduction, whereas exposure to the DC or AC field components alone had no effects.[23] Reduction of the vertical component of the ambient DC magnetic field by 15 μT caused a 50% increase in IGF-II growth factor release from mouse calvarial cultures.[24] The results of the present study, and those mentioned above, as well as many others[2,3,16] strongly point to the necessity for careful measurement and strict control of all components of exogenous electric or magnetic fields, especially when the effects of weak fields are to be studied.[25]

REFERENCES

1. S. M. Bawin, L. K. Kaczmarek, and W. R. Adey, *Annals N.Y. Acad. Sci.* 247: 74, 1975.

2. C. F. Blackman, S. G. Benane, L. S. Kinney, W. T. Joines, and D. E. House, *Radiat. Res.* 92: 510, 1982.

3. S. K. Dutta, A. Subramonian, B. Ghosh, and R. Prashad, *Bioelectromagnetics* 5: 71, 1984.

4. C. F. Blackman, S. G. Benane, J. R. Rabinowitz, D. E. House, and W. T. Joines, *Bioelectromagnetic* 6: 327, 1985.

5. M. Grattarola, A. Chiabrera, G. Bonanno, R. Viviani, and A. Raveane, in A. Chiabrera, C. Nicolini, and H. P. Schwan, Eds., *Electromagnetic Fields and Cells,* New York: Plenum Press, 1985, 401.

6. A. A. Pilla, J. J. Kaufman, and J. T. Ryaby, in M. Blank and E. Findl, Eds., *Mechanistic Approaches to Interactions of Electric and Electromagnetic Fields with Living Systems,* New York: Plenum Press, 1987, 39.

7. S. L. Arber, *Physiol. Chem. and Physics and Medical NMR* 17: 227, 1985.

8. M. Markov and M. Blank, Eds., *Electromagnetic Fields and Biomembranes,* New York: Plenum Press, 1988.

9. A. R. Liboff, B. R. McLeod, and S. Smith, in C. T. Brighton and S. R. Pollack, Eds., *Electromagnetics in Biology and Medicine,* San Francisco: San Francisco Press, 1991, 67.

10. L. A. Shuvalova, M. V. Ostrovskaja, E. A. Sosunov, and V. V. Lednev, *Doklady Akad. Nauk USSR* 217: 227, 1991.

11. E. M. Goodman, B. Greenbaum, and M. T. Marron, 13th Ann. BEMS, Salt Lake City, 1991, 26.

12. M. S. Markov, J. T. Ryaby, J. J. Kaufman, and A. A. Pilla, in M. J. Allen, S. F. Cleary, A. E. Sowers, and D. D. Shillady, Eds., *Charge and Field Effects in Biosystems-3,* Boston: Birkhauser, 1992, 225.

13. M. S. Markov, J. T. Ryaby, S. Wang, and A. A. Pilla, in W. J. Ohley, Ed., *Proc. IEEE 18th Ann. Northeast Bioengineering Conf.* Kingston, R.I., 1992, 63.

14. M. S. Markov, S. Wang, and A. A. Pilla, *Bioelectrochem. and Bioenerg.* 30: 1992.

15. C. Bruckner-Lee, C. H. Durney, J. Janate, C. Rappoport, and M. Kaminski, *Bioelectromagnetics* 13: 147-162, 1992.

16. A. W. Bull, S. Cherng, K. A. Jernow, and A. R. Liboff, this volume.

17. A. R. Liboff, in A. Chiabrera, C. Nicolini, and H. P. Schwan, Eds., *Interactions Between Electromagnetic Fields and Cells,* New York: Plenum Press, 1985, 281.

18. A. A. Chiabrera, B. Bianco, J. J. Kaufman, and A. A. Pilla, Ref. 9, p. 27.

19. V. V. Lednev, *Bioelectromagnetics* 12: 71, 1991.

20. D. T. Edmonds, *Bioelectrochem. and Bioenerg.* (in press).

21. R. S. Adelstein and C. B. Klee, in W. Y. Cheung, Ed., *Calcium and Cell Function,* vol. 1, 1980.

22. M. Aashima, K. Shimada, and C. J. Pfeiffer, *Bioelectromagnetics* 12: 215, 1991.

23. M. G. Yost and R. P. Liburdy, *FEBS Letters* 296: 117, 1992.

24. R. J. Fitzsimmons, J. T. Ryaby, A. Weinstein, and D. J. Baylink, this volume.

25. The authors give their sincerest thanks to Dr. J. T. Ryaby for his aid in designing the reaction procedure and to Dr. R. S. Adelstein and his associates at the National Heart, Lung, and Blood Institute, NIH, for graciously supplying the enzyme, substrate, and calmodulin, and to American Medical Electronics, Richardson, TX for their substantial technical and financial support.

Martin Blank, Ed., *Electricity and Magnetism in Biology and Medicine*

SINGLE-CELL TECHNIQUES FOR MONITORING THE EFFECTS OF ELF FIELDS: FLUORESCENCE MICROSCOPY APPROACHES FOR REAL-TIME ANALYSIS OF EARLY SIGNAL TRANSDUCTION EVENTS IN LIVING CELLS

D. E. Callahan,* R. P. Liburdy, and M. F. Maestre

Lawrence Berkeley Laboratory, Berkeley, Calif.

Cell perfusion chambers containing electrodes that can generate 60Hz electric fields have been developed for fluorescence microscopy studies of immobilized living cells exposed to ELF fields. Two microscope systems were used to monitor real time changes in $[Ca^{2+}]_i$ after injection of Con-A. Preliminary cooled CCD imaging studies have been performed at 25 and 37 C, and preliminary photometry studies have been performed in the presence or absence of a 60Hz electric field (6 μA/cm^2, 0.38 mV/cm). In both studies, single-cell responses were observed that were consistent with responses observed for cell suspensions in cuvette-based fluorescence studies.

Recent cuvette-based fluorescence studies of FURA-2 loaded rat thymocyte cell suspensions indicate that concanavalin-A (Con-A) stimulated increases in intracellular free calcium ($[Ca^{2+}]_i$) are altered within several hundred seconds of exposure to 60Hz electric fields (6 μA/cm^2, 0.38 mV/cm, 37 C).[1] Although the initial rapid rise in $[Ca^{2+}]_i$ following Con-A stimulation is not affected by the presence of an electric field, the subsequent plateau phase, where $[Ca^{2+}]_i$ normally stabilizes at approximately threefold above the original baseline, is sensitive to the presence and the magnitude of a 60Hz electric field.[1,2] We now report preliminary attempts to demonstrate similar effects at the level of individual cells.

EXPERIMENTAL PROCEDURES

Preparation of FURA-2 loaded rat thymic lymphocytes has been described.[1,2] Digital imaging and photometric studies of FURA-2 loaded cells have been described by several investigators.[3,4] Thymocytes were immobilized on poly-L-lysine coated coverslips. Two fluorescence microscope systems, a photometry system (Deltascan, PTI Inc.) and a cooled CCD (Photometrics Ltd.) imaging system were used in separate studies. In photometry studies, cells were exposed at room temperature to a 60Hz field in a chamber that had ports for Con-A injection and two grooved channels containing electrodes embedded in agarose. In CCD imaging studies, a thermoregulated perfusion chamber was used to compare Con-A stimulated increases in $[Ca^{2+}]_i$ at two different temperatures (25 and 37 C).

RESULTS AND DISCUSSION

In preliminary photometry studies (Fig. 1), exposure to a 60Hz electric field (6 μA/cm^2, 0.38 mV/cm) during Con-A stimulation depressed plateau phase $[Ca^{2+}]_i$ in a manner similar to that observed in fluorescence studies of cell suspensions (cf. Fig. 12 of Ref. 1). Single-cell Con-A responses in the absence of an electric field (data not shown) were similar in appearance to the 37 C Con-A response shown in Fig. 2. Fluorescence studies of cell suspensions have demonstrated that rat thymocytes respond relatively weakly to Con-A at 25 C, compared to 37 C (Fig. 2). This is consistent with results obtained in preliminary CCD imaging studies, where few strong Con-A responses were observed when cells were stimulated with Con-A at 25 C (Figs. 3 and 4). Although the photometry system provides much greater time resolution than the CCD imaging system (Fig. 1), the present imaging system allows one to observe multiple (15-30 cells) Con-A responses in a single experiment (Figs. 3 and 4).[5]

---

*Correspondence to: D. E. Callahan, LBL Bioelectromagnetics Research Facility, University of California, Berkeley CA 94720.

REFERENCES

1. R. P. Liburdy, *NY Acad. Sci.* 649: 74-95, 1992.
2. R. P. Liburdy, *FEBS Letters* 301: 53-59, 1992.
3. R. Y. Tsien et al., *Cell Calcium* 6: 145-157, 1985.
4. J. A. Conner, *Proc. Natl. Acad. Sci. USA* 83: 6179-6183, 1986.

5. Support provided in part by the Office of Energy Management, Utilities Systems Division, the Office of Health and Environmental Research, U.S. Department of Energy under contract DE-AC-3-76SF00098, and by the National Institutes of Health under grants CA53711 (RPL) and AI08427 (MFM). Acquisition of the photometry microscope system was generously provided by B. C. Hydro and Power Authority.

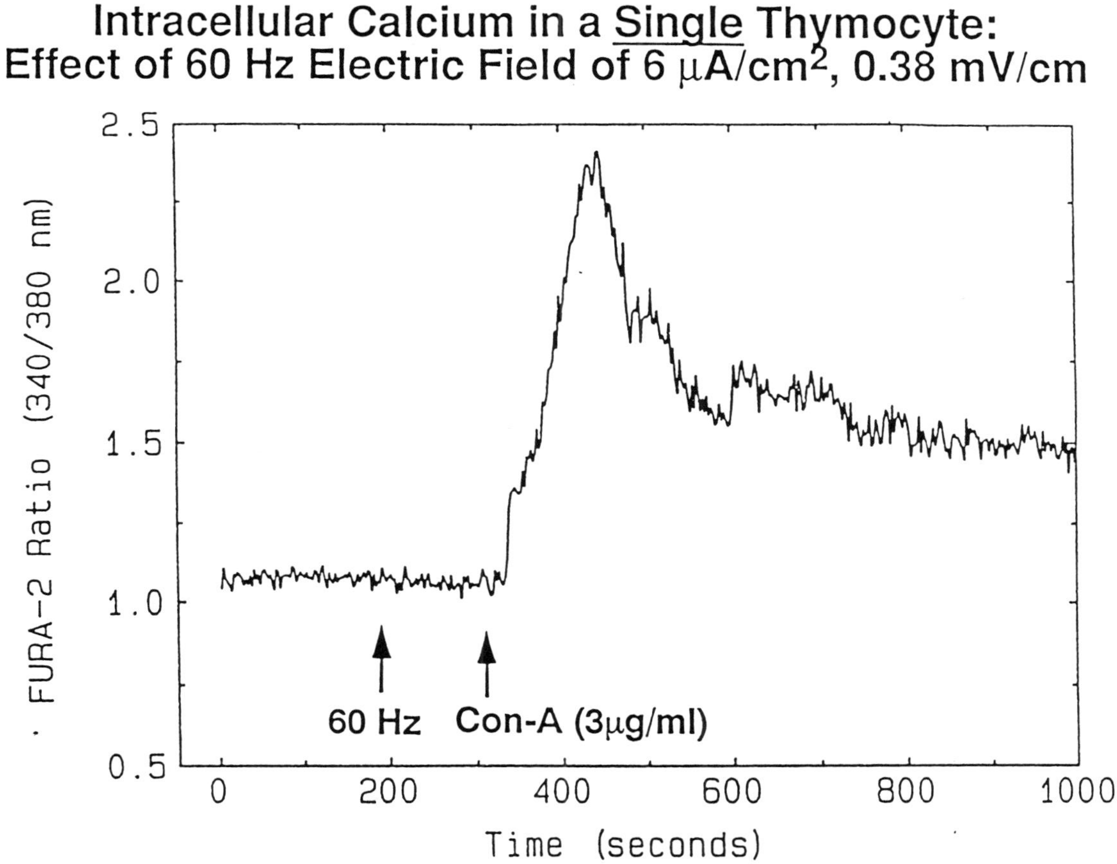

FIG. 1.—Photometry system. Two fluorescence emission (510nm) intensities were obtained at each time point with 340 and 380nm excitation. Ratio of these two emission intensities (340/380) is proportional to $[Ca^{2+}]_i$.[1-4]

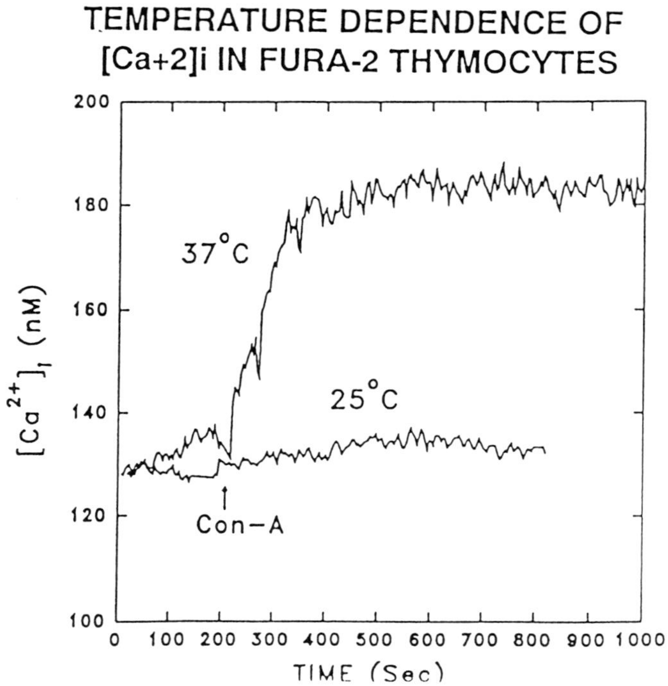

FIG. 2.—Cuvette-based study of thymocyte cell suspension performed as previously described.[2] Fluorescence ratios have been converted to $[Ca^{2+}]_i$ (nM), and Con-A responses at 25 and 37 C are demonstrated.

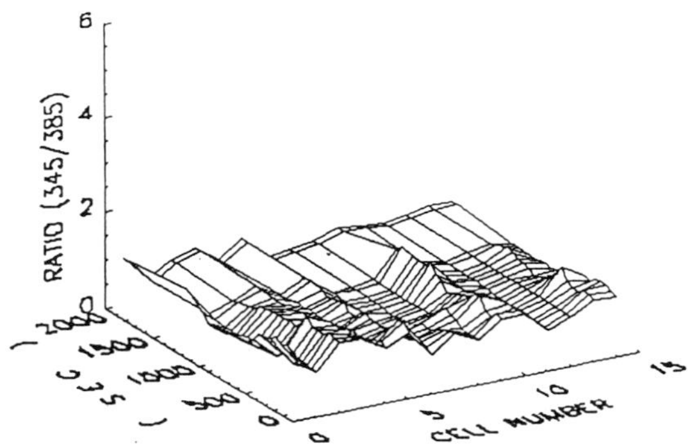

FIG. 3.—Imaging system. Two fluorescence emission images (510 nm) were obtained at 60s intervals (0-2000 s) with 345 and 385nm excitation (25 C). The mean (345/385) ratio of 15 cells was monitored as a function of time. Con-A injected at 360 s. (One cell with a ratio higher than 8.0 at time zero is not shown.)

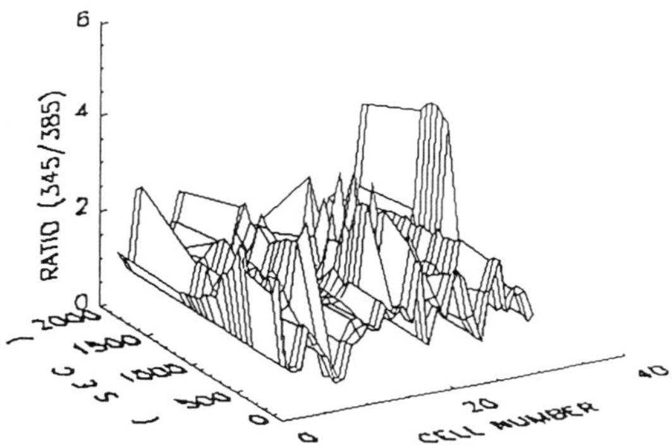

FIG. 4.—Same as Fig. 2, except that temperature was 37 C and 30 cells were monitored. Two cells (not shown) had ratios greater than 8.0 at time zero. Con-A injection at 240 s.

Martin Blank, Ed., *Electricity and Magnetism in Biology and Medicine*

TIME-VARYING AND STATIC MAGNETIC FIELDS ACT IN COMBINATION TO ALTER CALCIUM SIGNAL TRANSDUCTION IN THE LYMPHOCYTE

R. P. Liburdy*

Lawrence Berkeley Laboratory, Calif.

M. G. Yost

University of California, Berkeley

We tested the hypothesis that ELF time-varying magnetic fields act in combination with static magnetic fields to alter calcium signaling in the lymphocyte. A 60min exposure of thymic lymphocytes at 37 ± 0.05 C to a 16Hz, 421mG (42.1μT) magnetic field simultaneously with a colinear static field of 234 mG (23.4 μT) [AD/DC field intensity ratio = 1.8] inhibits calcium influx triggered by the mitogen Concanavalin-A.[1] Resting lymphocytes do not respond to the fields, which shows that mitogen activation is required. In parallel studies, we show that the nominal rise in $[Ca^{2+}]_i$ induced by Con-A is inhibited during field exposure. Split-field exposures show that the presence of the AC field or the DC field alone does not produce an effect. Exposures at AC/DC ratios of 3.8 (Bessel function first zero) and 5.3 (Bessel function first minimum) yield results "consistent" with a proposed parametric resonance theory of interaction of low-intensity magnetic fields with biological systems.[2]

A central question to interaction mechanisms of magnetic fields with biological systems is whether a time-varying (AC) magnetic field interacts in a fundamentally different manner at the cellular level when a static (DC) magnetic field is present.[1] This question is of interest since both ELF magnetic fields and geomagnetic static magnetic fields are present in the environment and workplace, and since it is critical to identify field metrics that have biological relevance.

Recently a report has specified field parameters in sufficient detail for testing the hypothesis that AC/DC field combinations interact with biological systems.[2] The parametric resonance theory states that protein-bound ions can behave as a spatial oscillator with a set of vibrational frequencies that depend on the bond energy, charge q, and mass m of the ligand bound ion. The bound ion continuously oscillates about an energy level as coordination bonds are broken and re-formed due to random thermal motion. The presence of a static magnetic field splits the energy level of the bound ion into two sublevels corresponding to electromagnetic frequencies ω_1 and ω_2 in the infrared band. The difference between these two energy levels equals the cyclotron frequency, $\omega_c = \omega_1 - \omega_2 = (q/m)B_{DC}$.

Application of a colinear AC field at the cyclotron frequency modifies the probability P of ion transitions between energy states that depends on a Bessel function argument defined by the intensity ratios of the AC and DC fields.[2] The Bessel function curve looks like a damped sinusoidal curve with a first maximum at $B_{AC}/B_{DC} = 1.8$, a first cross-over point or zero at 3.8, and a first minimum at 5.3. We interpret it to mean that at these field intensity ratios a change in biological response (positive or negative) will be observed at the maximum and minimum ratios, with no change observed at the zero ratio.

EXPERIMENTAL PROCEDURES

Preparation of rat thymocytes, description of the $^{45}Ca^{2+}$ transport assay, and description of the temperature-regulated, mu-metal shielded Helmholtz multiple-coil exposure system and dosimetry procedures were recently described,[1] as were FURA-2 loading and real-time measurement of intracellular

---

*Correspondence to: R. P. Liburdy, LBL Bioelectromagnetics Research Facility, University of California, Berkeley CA 94720. (M. G. Yost is now at the University of Washington in Seattle.)

calcium during field exposures.[3] A special Helmholtz exposure coil was fabricated and positioned around a fluorescence cuvette to permit AC/DC field exposures during on-line fluorescence measurements.

RESULTS AND DISCUSSION

Figures 1-3 present the results of AC/DC field exposures on mitogen activated calcium influx in rat thymocytes (60 min), 37 C). Figure 1 shows data for the AC/DC ratio of 1.8 (234 mG DC; 421 mG AC, 16 Hz; resonance for $^{45}Ca^{2+}$), corresponding to the first Bessel function maximum. The two left bar graphs reveal that calcium influx increased for cells treated with Con-A (1-3 µg/ml). The two right bar graphs show the effect AC/DC fields have on calcium influx in resting and activated cells. Only activated cells responded to the fields; AC/DC fields significantly inhibited calcium influx triggered by Con-A. Experiments employing separate AC and DC magnetic fields were not effective.[1]

We next performed exposures at AC/DC magnetic field ratios of 5.3 and 3.8, data that are presented in Figs. 2 and 3, respectively. At the first Bessel zero-crossing, we observed no change in calcium influx (Fig. 3); at the first Bessel minimum we observed an enhancement of calcium influx (Fig. 2). These preliminary data suggest that AC/DC field combinations have the potential to influence calcium influx in mitogen-activated lymphocytes.

The above findings also raise the important question of whether intracellular calcium is altered. To address this question we monitored FURA-2 fluorescence in thymocytes in real time during AC/DC field exposures. The significance of this approach is two-fold: it permits an instantaneous evaluation of an important signal transduction parameter, $[CA^{2+}]_i$, during field exposure; and it makes it possible to obtain information about field interaction kinetics.

Figure 4 shows the time course of $[Ca^{2+}]_i$ following AC/DC field exposure and mitogen activation. Resting thymocytes display a stable level of $[Ca^{2+}]_i$ of approximately 135 nM, unaltered by exposure at 180 s to AC/DC fields (208 mG DC; 375 mG AC, 16 Hz; AC/DC = 1.8; resonance conditions at 16 Hz for the $^{40}Ca^{2+}$ ion). This result is consistent with the calcium influx findings for resting cells in Fig. 1. When Con-A is added at approximately 260 s, unexposed cells show a characteristic rise in $[Ca^{2+}]_i$ to a level of 210 nM at 370 s. This nominal rise in $[Ca^{2+}]_i$ is significantly blunted by the AC/DC fields, although the early kinetics appear to be unaffected through the first 100 s. After 100 s the two traces diverge as time increases. A reduction in $[Ca^{2+}]_i$ levels during AC/DC field exposure is consistent with the reduced calcium influx shown in Fig. 1.

The above findings, taken together, indicate that AC and DC magnetic-field combinations can act to influence calcium signaling in mitogen-activated lymphocytes. Two *key factors* in these studies are the concepts of signal transduction and cellular activation. Several groups have reported studies using AC/DC fields but have not employed a model system with these key factors.[1] The results presented here are "consistent" with the parametric ion theory. However, further research is needed to test predictions adequately, which will involve varying the AC frequency to "detune" the interaction and systematically varying the AC/DC field ratios to cover the Bessel function curve.[4]

REFERENCES

1. M. G. Yost and R. P. Liburdy, "Time-varying and static magnetic fields act in combination to alter calcium signal transduction in the lymphocyte," *FEBS Letters* 296: 117-122, 1992.

2. L. L. Lednev, "Possible mechanism for the influence of weak magnetic fields on biological systems," *Bioelectromagnetics* 12: 71-75, 1991.

3. R. P. Liburdy, "Calcium signaling in lymphocytes and ELF fields: Evidence for an electric field metric and a site of interaction involving the calcium ion channel," *FEBS Letters* 301: 53-59, 1992.

4. This research was supported in part by the Office of Energy Management, Utilities System Division, U.S. Department of Energy, under contract DE-AC03-76SF0009 (RPL), and by the California Department of Health Services under contract 206854 (MGY).

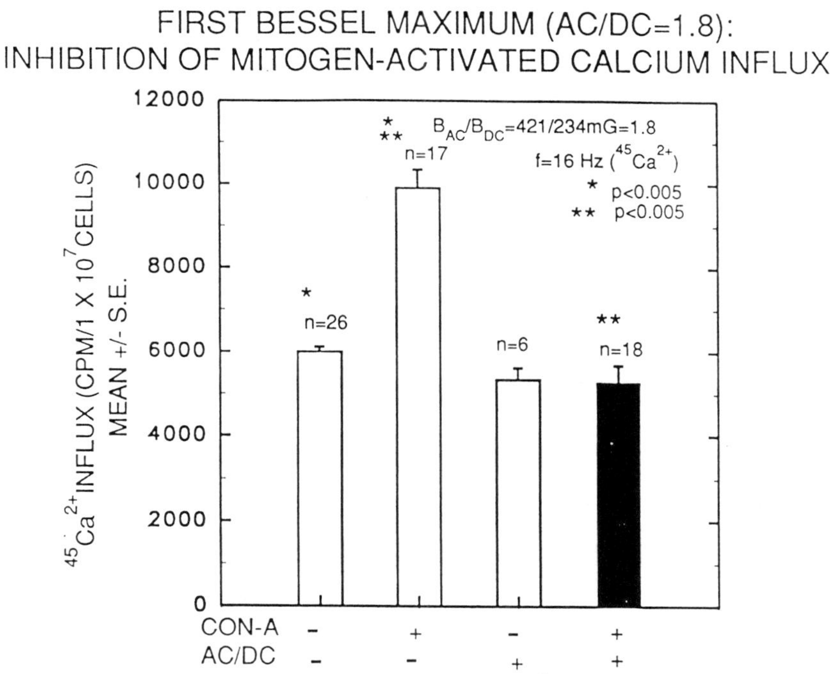

FIG. 1.—Effect of magnetic field combination AC/DC = 1.8: the first Bessel function maximum. Resonance for the $^{45}Ca^{2+}$ion: frequency, 16 Hz; AC, 421 mT, DC, 234 mG. Temperature, 37 ± 0.05 C, 60 min.

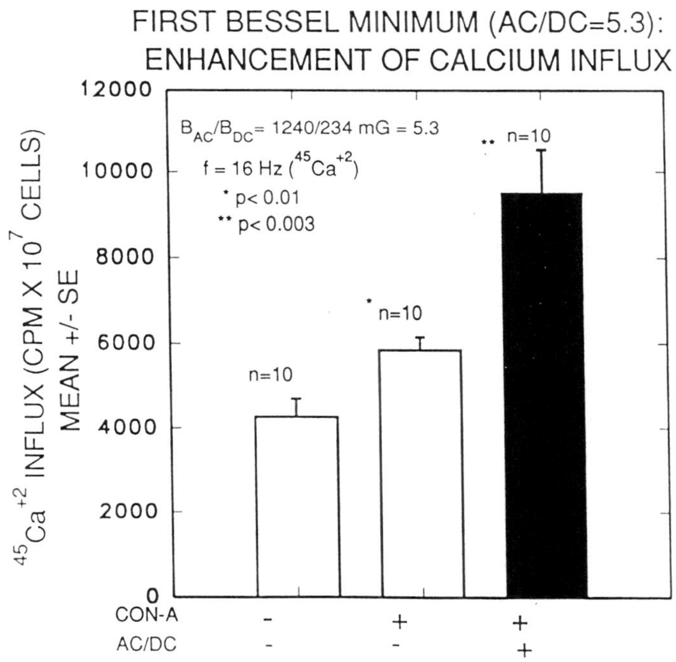

FIG. 2.—Effect of AC/DC = 3.8: the first Bessel function zero. Conditions as in Fig. 1, except AC = 901 mG.

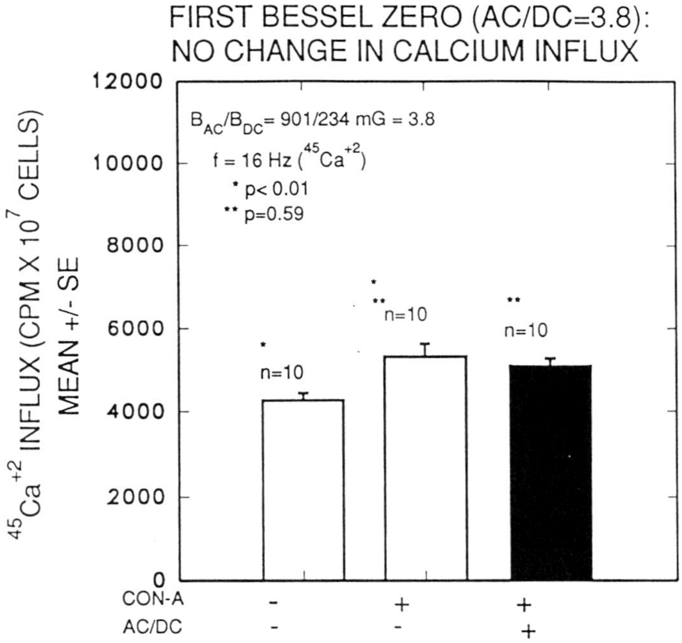

FIG. 3.—Effect of AC/DC = 5.3: the first Bessel function minimum. Conditions as in Fig. 1 except AC = 1240 mG.

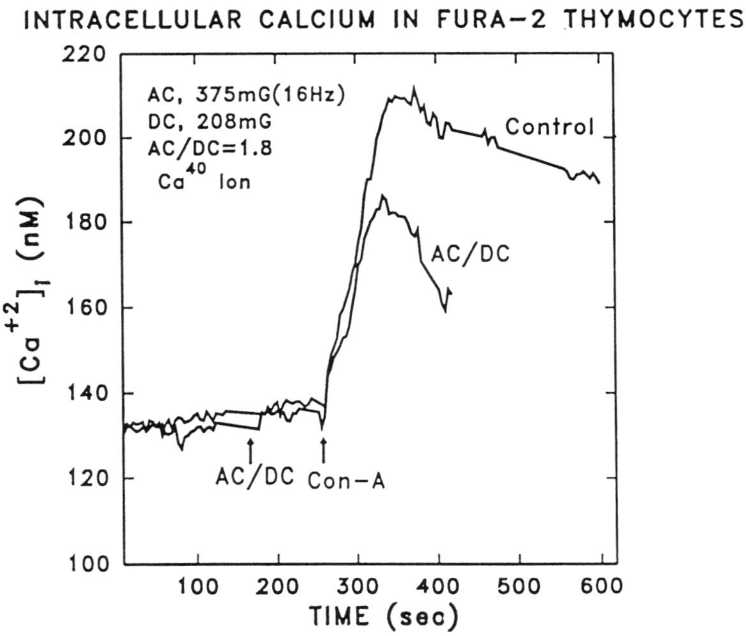

FIG. 4.—Real-time measurement of intracellular calcium during exposure to AC/DC = 1.8 (first Bessel maximum). Resonance for the $^{40}Ca^{2+}$ ion: frequency, 16 Hz; AC, 375 mG, DC, 208 mG.

Martin Blank, Ed., *Electricity and Magnetism in Biology and Medicine*

LOW-FREQUENCY ELECTROMAGNETIC FIELDS INHIBIT CELL PROLIFERATION IN CHICK EMBRYO COCHLEOVESTIBULAR GANGLION

Y. Leon, C. Valero, C. Miner, and F. Giraldez, and J. Represa*

University of Valladolid, Spain

Cochleovestibular ganglia (CVG) from 72h chick embryos were exposed to low-frequency electromagnetic fields (EMFs) ranging from 0.01 to 5 Gauss (G). CVG-explants were cultured in serum-free medium alone, and with serum (10% fetal calf serum) or NGF (100 ng/ml). EMFs inhibited CVG growth induced by serum or NGF. Growth inhibition was intensity dependent and was accompanied by a reduction in the rate of DNA synthesis as measured by [^{3}H]thymidine incorporation. These effects were associated with inhibition of NGF-induced Fos oncoprotein expression. Morphometrical analysis indicated that EMF effects are related to a decrease in the number of glial cells.

There is general agreement, although the mechanisms are still unknown, that low-frequency electromagnetic fields (EMFs) affect biological systems, and various experiments on chicken embryos indicate a significant increase in the number of malformations following exposure to EMFs.[1,2]

The development of the inner ear is an interesting example of organogenesis in the nervous system. At early developmental stages it involves the formation of the otic vesicle, which becomes the membranous labyrinth and the cochleo-vestibular ganglion (CVG), which gives rise to the sensory neurons of the ear. The CVG is of a mixed origin, deriving from the otic epithelium and the neural crest. It undergoes a distinct period of cell proliferation that precedes the differentiation of neurons and supporting cells.[3] Nerve Growth Factor (NGF) differentially regulates cell proliferation in the early developing CVG and the otic placode, probably acting on the crest-derived cell population.[4,5] Recent experiments have shown an association between cell proliferation and the expression of the nuclear proto-oncogene c-Fos in the CVG, which suggests a regulatory role for c-Fos in normal growth (unpublished observation).

The results reported here show that low-frequency EMFs ranging from 0.01 to 5 G inhibited CVG growth induced either by serum or NGF. Growth inhibition was intensity dependent, paralleled by a reduction in the rate of DNA synthesis and associated with an inhibition of NGF-induced Fos oncoprotein. The results suggest a role for EMFs in controlling cell cycle, as well as a potential teratological capacity for EMFs.

MATERIALS AND METHODS

Preparation of Explant Cultures and Exposure to EMF

Cochleo-vestibular ganglia (CVG) were aseptically isolated from 72h chick embryos and cultured as previously described.[4] Ganglia were cultured for 24 h in the presence or absence of low-frequency electromagnetic fields (50 Hz), ranging from 0.01 to 5 G in intensity. CVG explants were set up as follows: (1) in serum-free medium alone (M199), (2) medium containing serum (10% fetal calf serum), and (3) medium containing NGF (100 ng/ml).

Determination of DNA Synthesis. DNA-synthesis was measured as acid-precipitable [^{3}H]thymidine incorporation.[4,5]

Histology. CVG were fixed in 4% para-formaldehyde and processed for histology. Tissue volume was determined by morphometrical methods on serial sections as described by Represa et al.[6]

Western Blotting. Lysates from cultured CVG were analyzed by SDS-PAGE electrophoresis in a 10% gel and blotted on nitrocellulose filters. Immunodetection was carried out by use of a polyclonal immune serum (ICI-CRB). Blots were processed with a HRP-coupled secondary antibody and developed with enhanced chemiluminescence (Amersham).

---

*Correspondence to: J. Repressa (Fisiologia), Facultad de Medicina, C/ Ramon y Cajal sn., ES-47005 Valladolid, Spain.

RESULTS AND DISCUSSION

CVG explants were cultured for 24 h in either serum-free medium (OS) or with NGF (100 ng/ml), and in the presence or absence of EMFs (Fig. 1). Incubation with EMFs produced an intensity-dependent inhibition of NGF-induced growth. The numbers in Fig. 1 indicate intensities in Gauss; it can be seen that the inhibitory effect was detectable at 0.01 G and increased up to 5 G. By contrast, no significant effects were detected on serum-induced growth at intensities below 5 G.

The incorporation of ^3H thymidine into acid-precipitable material in cultures corresponding to the above experiments are illustrated in Fig. 2. DNA synthesis was reduced in parallel with the inhibition of growth. Inhibition of DNA synthesis caused by EMF was detectable at as low as 0.01 G; maximal effect occurred between 0.5 and 5 G, in CVG cultured with NGF. CVGs cultured with serum showed no significant reduction of DNA synthesis below 5 G.

The effect of EMF on cell proliferation and growth was also estimated by a measurement of the volume of tissue and cell density, obtained morphometrically from serial reconstructions of cultured ganglia. Tissue volume values paralleled those of DNA synthesis. Intensities of 0.01, 0.5, and 5 G reduced CVG growth with NGF by 29, 48, and 71%, respectively. Serum-dependent growth was inhibited by 45% with 5 G. CVG cultured with NGF at various EMF intensities showed a different balance between the two cell populations of the CVG, i.e., neurones and glial cells. Cell density values (-10^{-8} cells cm^{-3}) were 112 ± 16 glial cells and 77 ± 10 neuronal cells for NGF in absence of EMF. Cultures with NGF in presence of low-frequency EMF gave cell density values of 79 ± 10 glial cells and 84 ± 5 neuronal cells, respectively, which suggests that the inhibition of NGF-induced cell proliferation by EMF is effected mainly through a decrease of the number of glial cells. The ratio of glial cells to neurones in NGF cultures decreased from 1.4 in the absence of EMFs to 0.94 and 0.86 with 0.01 and 0.05 G, respectively. By contrast, serum-treated cultures showed an opposite effect: glial cells relatively increased (1.33 in the absence of EMFs and 2.1 at 5 G).

The possibility that the inhibitory effects of EMFs were related to c-Fos expression was examined. Figure 3 shows a Western blot analysis of CVGs cultured in presence of 50Hz EMFs at the intensities indicated by the numbers. The dominant bands had a Mr of about 56-62 kD and corresponded closely to the products of c-Fos reported by others.[7] Incubation of growth-arrested CVGs with NGF at concentration that induced cell proliferation and growth produced significant induction of Fos oncoprotein (Fig. 3, lane NGF). This growth-factor-induced increases in Fos level was partially suppressed by EMF in a dose-dependent manner.[8]

REFERENCES

1. J. M. R. Delgado, J. Leal, J. L. Monteagudo, and M. G. Gracia, "Embryological changes induced by weak extremely low frequency electromagnetic fields," *J. Anat.* 134: 533-551, 1982.

2. A. H. Martin, "Magnetic fields and time dependent effects on development," *Bioelectromagnetics* 9: 393-396, 1988; "Development of chicken embryos following exposure to 60Hz magnetic fields with differing waveforms," *ibid.*, 13: 223-230, 1992.

3. D. M. Noden and T. R. Van de Water, "The developing ear: Tissue origins and interactions," in Ruben et al., Eds., *The Biology of Change in Otolaryngology*, Amsterdam: Elsevier, 1986.

4. J. J. Represa and P. Bernd, "Nerve growth factor and serum differentially regulate development of the embryonic otic vesicle and cochleovestibular ganglion in vitro," *Dev. Biol.* 134: 21-29, 1989.

5. P. Bernd and J. Represa, "Characterization and localization of nerve growth factor receptors in the embryonic otic vesicle and cochleovestibular ganglion," *Dev. Biol.* 134: 11-20, 1989.

6. J. J. Represa, et al., "Bombesin and other growth factors activate cell proliferation in chick embryo otic vesicles in culture," *Development* 102: 87-96, 1988; "Retinoic acid modulation of the early development of the inner ear is associated with the control of *c-fos* expression," *ibid.*, 110: 1081-1090, 1990.

7. L. C. Sambucetti, M. Schaber, R. Kramer, R. Crowl, and T. Curran, "The *fos* gene product undergoes extensive post-translational modification in eukaryotic but not in prokaryotic cells," *Gene* 43: 69-77, 1986.

8. This work was coordinated by AMYS-UNESA within the Programa de Investigacion y Desarrollo Electrotécnico PIE.

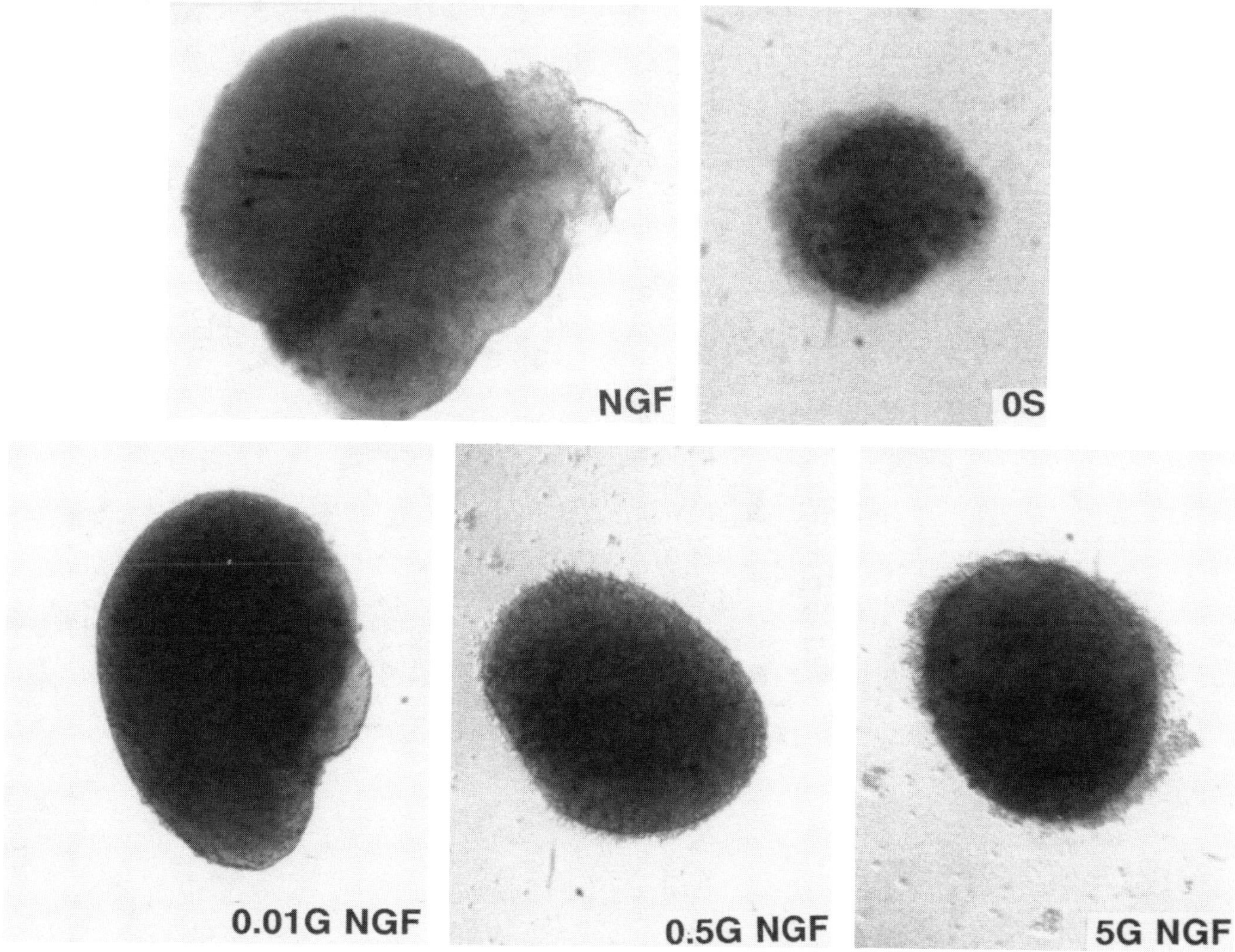

FIG. 1.—Blockade of mitogenic effect of NGF by EMF. Appearance of isolated CVG obtained from 72h chick embryos following 24 h in culture with M199 alone (OS), and with M199 containing 100 ng/ml NGF (NGF) in presence of 50 Hz. EMF ranged from 0.01 to 5 G.

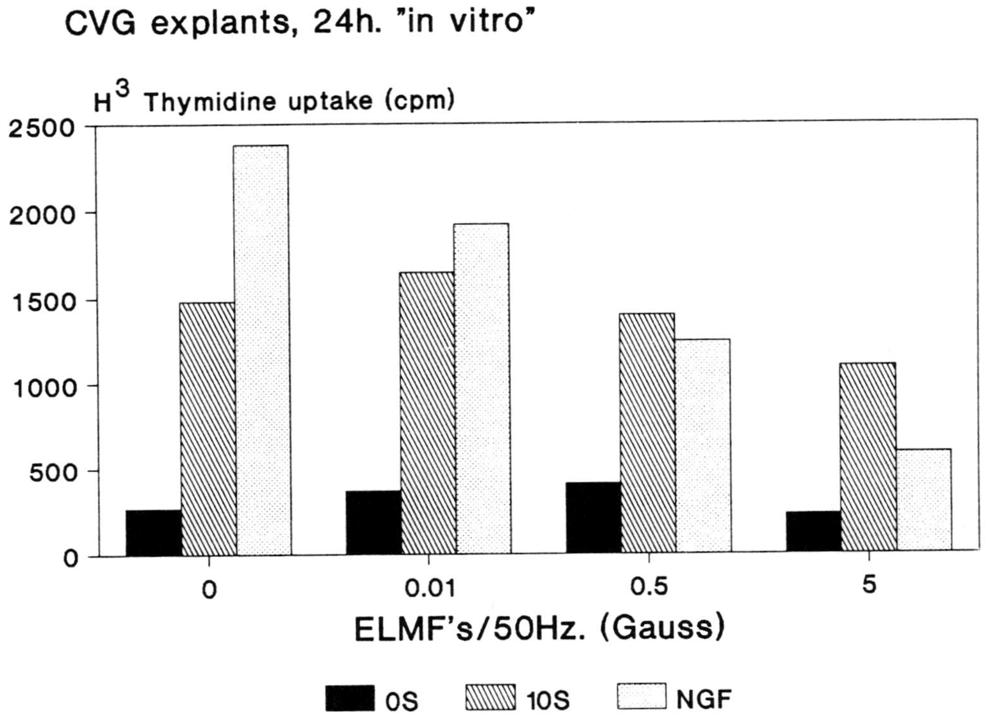

FIG. 2.—Effect of EMFs on DNA synthesis in CVG. Results were normalized with respect to the ^{3}H thymidine uptake obtained with 100 ng/ml NGF, which was 2410 ± 180 cpm. CVG explants were set up (1) in serum-free medium alone, (2) in medium containing serum (10% fetal calf serum), (3) in medium containing NGF (100 ng/ml). Values are means ±SE of five experiments.

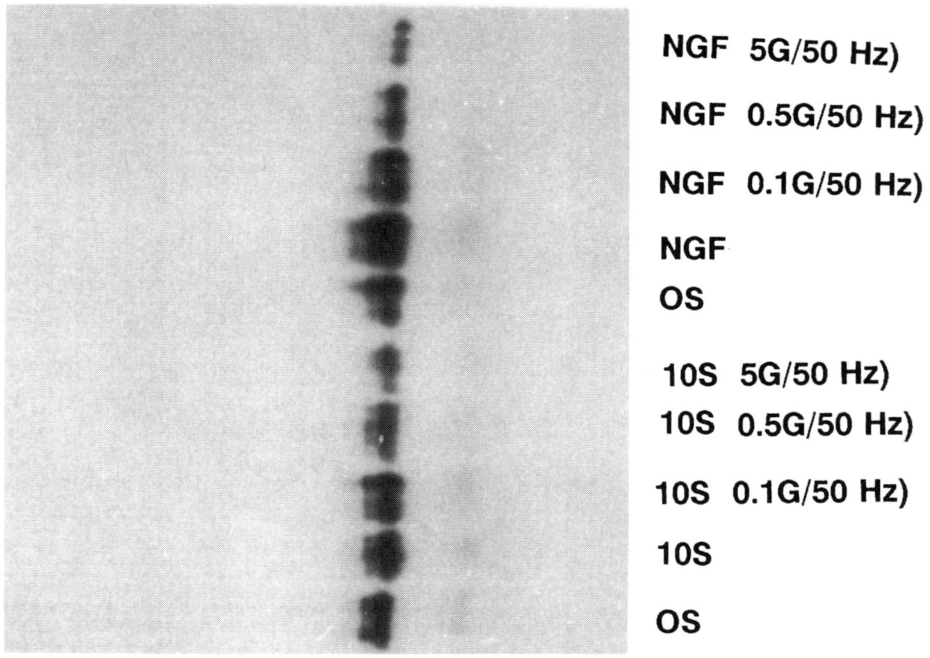

FIG. 3.—Inhibition of 55-62 kD: Fos oncoprotein by EMFs. Western blot of lysates from CVGs cultured under experimental conditions shown. Each lane was loaded with an amount of protein equivalent to that of 4 CVG of stage at 72 h. Numbers indicate relative molecular mass (Mr).

Martin Blank, Ed., *Electricity and Magnetism in Biology and Medicine*

SPATIAL AND TEMPORAL COHERENCE AFFECTS THE RESPONSE OF BIOLOGICAL SYSTEMS TO ELECTROMAGNETIC FIELDS

T. A. Litovitz,* C. J. Montrose, and P. Doinov

Catholic University of America, Washington, D.C.

We propose that living cells are affected only by electromagnetic fields that are spatially coherent over their surface. This proposition provides a solution to the so-called signal-to-noise dilemma, according to which cells respond to exogenous fields that are several orders of magnitude weaker than local endogenous fields associated with thermal fluctuations. Experiments on developing chick embryos are reported that are consistent with this data.

The association of biological effects at the cellular level with exposure to weak ELF electromagnetic fields (EMFs) has remained a controversial subject despite nearly a decade of such reports. Theoretical arguments based on signal-to-noise considerations are in large measure responsible for the skepticism.[1,2] The signal-to-noise dilemma arises because cells, existing in an electrically noisy environment, have been reported to respond to external EMFs some 100 to 1000 times weaker than local, thermally generated noise fields. To account for this Weaver and Astumian have proposed that cells integrate the electromagnetic signals, effectively narrowing their acceptance bandwidth, and thus averaging out the thermal noise.[1] However, their calculations yield averaging times much longer than exposure intervals observed to produce bioeffects. We propose that living cells discriminate against thermal-noise fields by recognizing them as *spatially incoherent*, i.e., uncorrelated at different receptor locations on the cell membrane. We suggest that biological cooperativity is an essential feature of cellular response to EMFs; a significant number of receptors at the cell membrane must be simultaneously and coherently activated (coincidence detection) to produce an effect on the biochemical functioning of the cell. Electromagnetic fields must be spatially coherent if they are to affect cell functioning. We have already demonstrated that *temporal* coherence is necessary if bioeffects are to be observed,[3] which suggests that the exquisite sensitivity to EMFs results from the cells' ability to discriminate against fields which are *both* spatially and temporally incoherent; i.e., against thermal noise. Therefore, we hypothesize that if a spatially coherent but temporally random noise field were superimposed on the coherent signal, then at some value of the signal/noise amplitude ratio, the discrimination mechanism would be defeated, and the observed field-induced bioeffect would be suppressed.

We have recently concluded a series of studies of EMF-induced avian teratogenesis (for a reasonably comprehensive view of this situation see Ref. 4 and references contained therein), using pulsed EMFs (100 Hz, 500μs duration, 1μT peak strength). These studies offer persuasive evidence ($p \ll 0.01$) that EMFs can indeed increase the rate of abnormal development in incubating chick embryos. If a spatially coherent noise field comparable in strength to the impressed EMF—a "confusion" field—were to be also applied, we would expect a reduction in the field-induced abnormalities. We have tested this hypothesis in the same White Leghorn chick embryo model with which our previous studies were carried out. The apparatus and techniques followed the "Project Henhouse" protocols.[4] In addition to pulsed EMFs, 60Hz sinusoidal (4 μT peak) and 60Hz sawtooth wave (6.4 μT peak-to-peak) fields were used. For each case three samples were examined: (1) the sham-exposed (control) sample where no field was imposed; (2) the field-exposed sample; and (3) a sample exposed to a field in which a spatially coherent noise signal was superimposed on the pulsed, sinusoidal, or sawtooth EMF. The noise spectra were flat over a bandwidth of 30-100 Hz for the pulsed field and 30-90 Hz for the sinusoidal and sawtooth wave fields; the rms noise amplitudes were 1 μT, 4 μT, and 6.4 μT, respectively. The results are presented in Table 1.

From the data it is clear that for each waveform, there is an increase in the fraction of abnormally developed embryos when the eggs are subjected to coherent external EMFs. It is also clear that this increase is greatly reduced when a confusion field is present. The statistical significance of these observations was

---

*Correspondence to: T. A. Litovitz (Vitreous State Lab.), Catholic University of America, Washington DC 20064.

TABLE 1.—Results for embryos subjected to square-pulse, sinusoidal, and sawtooth EMFs.

Exposure conditions	Live	Abnormal	Percent
100-Hz 1-μT (peak) Pulse EMF			
Sham exposed (control)	255	16	6.3
EMF exposed	152	29	19.1
EMF + noise exposed	120	9	7.5
60-Hz 4-μT (peak) Sinusoidal EMF			
Sham exposed (control)	222	19	8.6
EMF exposed	163	22	13.5
EMF + noise exposed	113	10	8.9
60-Hz 6.4-μT (peak-peak) Sawtooth EMF			
Sham exposed (control)	290	20	6.9
EMF exposed	253	34	13.4
EMF + noise exposed	157	13	8.3

tested by an examination of the appropriate null hypotheses: (1) the imposition of an EMF has no effect on developing embryos, since the difference between the EMF-exposed and control sample results are due to fluctuations in sampling; (2) the superposition of a confusion field on an EMF has no effect on developing embryos; and (3) the abnormality rate for the EMF-plus-noise-exposed and sham-exposed samples do not differ, i.e., the confusion field completely masks the effect of the imposed EMF. These hypotheses were tested by a χ^2 test. The p values (confidence levels) for rejecting these hypotheses are given in Table 2.

TABLE 2.—Statistical analysis of exposure data fom Table 1.

Exposure field	P[a] Control- EMF exposed	P[b] EMF exposed- Noise + EMF exposed	P[c] Control- Noise + EMF exposed
100-Hz 1-μT Pulse EMF	$< 10^{-4}$	0.007	0.65
60-Hz 4-μT (peak) Sine EMF	0.13	0.22	0.85
60-Hz 6.4-μT (p-p) Sawtooth EMF	0.01	0.10	0.50

[a] Probability that the difference in the control and EMF-exposed abnormality rates is due to chance

[b] Probability that the difference in the EMF-exposed and Noise-plus-EMF-exposed abnormality rates is due to chance

[c] Probability that the difference in the control and Noise-plus-EMF-exposed abnormality rates is due to chance

Several conclusions can be drawn from the results in this table. (1) Impressing ELF pulsed or sawtooth wave electromagnetic fields with amplitudes of the order of a microtesla induces significant teratogenesis in developing chick embryos; the data are less definitive for sinusoidal fields. (2) Superimposing a confusion field (spatially coherent but temporally incoherent noise) of comparable rms amplitude largely suppresses the teratogenic effect of these fields. (3) For the noise fields considered here, the abnormality rates with no field imposed and with both the temporally coherent and noise fields are nearly the same. In all cases the abnormality rate with the confusion field is slightly larger than the control sample rate; however, for none of the three waveforms is the difference statistically significant. These findings support our proposal that it is their sensitivity only to *spatially coherent* EMFs that enables living cells to respond to weak exogenous fields while remaining unaffected by the relatively large (but spatially incoherent) endogenous noise fields that are always present. The idea that cells distinguish between exogenous and endogenous fields by recognizing the latter's spatial incoherence offers a viable explanation of the signal-to-noise dilemma. Far from being "magically" exempt from signal-to-noise considerations, cellular detection mechanisms are subject to the same laws of nature as all physical systems. The masking that is provided by a spatially coherent confusion field could prove to be a basis for protecting humans against possible adverse health risks associated with environmental electromagnetic fields.[5]

REFERENCES

1. J. C. Weaver and R. D. Astumian, "The response of living cells to very weak electric fields: The thermal noise limit," *Science* 247: 459-462, 1990.

2. R. K. Adair, "Constraints on biological effects of weak extremely-low-frequency electromagnetic fields," *Phys. Rev.* A43: 1039-1048, 1991.

3. T. A. Litovitz, J. M. Mullins, and D. Krause, "Effect of coherence time of the applied magnetic field on ornithine decarbyoxylase activity," *Biochem. Biophys. Res. Comm.* 178: 862-865, 1991.

4. E. Berman, L. Chacon, D. House, B. A. Koch, W. E. Koch, J. Leal, S. Løvtrup, E. Mantiply, A. H. Martin, G. I. Martucci, K. H. Mild, J. C. Monahan, M. Sandström, K. Shamsaifar, R. Tell, M. A. Trillo, A. Ubeda, and P. Wagner, "The effects of a pulsed magnetic field on chick embryos," *Bioelectromagnetics* 11: 169-187, 1990.

5. The authors acknowledge the Walter Reed Army Institute of Research Contract for support of this work under Contract DAMD17-86-C-6260.

Martin Blank, Ed., *Electricity and Magnetism in Biology and Medicine*

SUPERPOSITION OF A TEMPORALLY INCOHERENT MAGNETIC FIELD SUPPRESSES THE CHANGE IN ORNITHINE DECARBOXYLASE ACTIVITY IN DEVELOPING CHICK EMBRYOS INDUCED BY A 60Hz SINUSOIDAL FIELD

J. M. Farrell,* M. Barber, P. Doinov, D. Krause, and T. A. Litovitz

Catholic University of America, Washington

Developing chick embryos were exposed to 60Hz sinusoidal magnetic fields under the environmental conditions specified in the "Henhouse" experiment.[1] The activity of ornithine decarboxylase (ODC) was compared with that in sham-exposed embryos for an incubation time of 15 h. A ratio of 1.74 ± 0.36 ($N = 270$) was obtained. In other experiments, ODC activity was analyzed in embryos exposed to 60Hz sinusoidal fields on which a spatially coherent, but temporally incoherent "noise" field was superimposed. This randomly generated field contained frequencies ranging from 30 to 90 Hz and an rms value comparable to the sinusoidal field's peak amplitude. Embryos so exposed had an ODC activity that was statistically indistinguishable from the controls. Thus, embryo biochemistry is shown to be modified by the imposition of a sinusoidal magnetic field. However, this modification is shown to be blocked by the superposition of temporally incoherent EM fields of comparable strengths.

Earlier investigations (such as Henhouse) have examined the possibility that extremely low frequency (ELF) EM fields induce teratogenic effects in developing organisms. Those studies have mainly used the observation of morphological abnormality to indicate a correlation between field presence and biological effect. However, since the analysis of morphological abnormality introduces a potentially non-negligible degree of subjectivity and does not easily lend itself to quantitative studies, it was considered desirable to find a quantitative measure to mark the modification of the biochemistry involved in embryonic development. Reported in this paper are the results of two related studies exploring the manner in which the activity of ornithine decarboxylase (ODC) in chicken embryos is affected by the imposition of ELF EM fields.

ODC is a key enzyme in the biochemical pathways which are responsible for polyamine biosynthesis in mammalian cells. In the chick embryo, ODC has been reported to peak twice within the first 53 h of development.[2] The first peak is observed at 15 h of development (gastrulation), the second at 23 h (the onset of neurulation and early organogenesis).

In the first of two related studies, developing chick embryos were exposed to 60Hz sinusoidal magnetic fields. This study focused on a stage slightly before the first reported peak of ODC activity (i.e., 15h incubation). ODC activity of field-exposed embryos was compared to that in sham-exposed embryos.

In the second set of experiments, chick embryos were exposed to the sum of (1) a sinusoidal magnetic field identical to that used in the first study, and (2) a temporally incoherent (but spatially coherent) EM "noise" field. In the first study, we have shown that 60Hz EM fields of amplitude weaker by 100 to 1000 times than thermal EM noise fields can affect the biochemistry of developing embryos. Litovitz et al. have speculated that cells discriminate against thermal noise by recognizing its lack of spatial coherence. They have demonstrated that the superposition of a "noise" field that is temporally incoherent but spatially coherent prevents morphological abnormalities induced in chick embryos by sinusoidal fields (which are spatially and temporally coherent).[2] In the present work, ODC activity in embryos exposed to the sum of the sinusoidal and "noise" fields was analyzed and compared with that in sinusoidal field-exposed and sham-exposed embryos.

---

*Correspondence to: J. M. Farrell (Physics), Catholic University of America, Washington DC 20064.

MATERIALS AND METHOD

Fertilized White Leghorn eggs were used and were incubated under essentially the same environmental conditions as used in the "Henhouse" experiment, with one noteworthy exception. In order to minimize nonintentional fields, the incubators were heated via separate thermal baths rather than by their own heating elements, since these elements were found to produce significant EM fields.

In both studies, six incubators were used simultaneously with ten eggs in each. Each experiment was repeated several times to assure statistical validity (Table 1). In the first study exposure was to a sinusoidal magnetic field (60 Hz, 4μT peak) for 15 h of incubation. Three incubators were used in the sinusoidal field-exposed configuration, and three in the sham configuration (no field). In the second study, six incubators were used simultaneously in three groups. Two incubators were used in the sinusoidal field-exposed configuration and two were used in the sham configuration (no field applied). In the remaining pair of incubators, two EM fields were superimposed: a sinusoidal magnetic field (60 Hz, 4μT peak amplitude) colinear with a temporally incoherent EM "noise" field (containing frequencies components from 30 to 90 Hz, 4μT rms).

After incubation, the embryos were separated according to the standard Hamburger and Hamilton stages.[4] To eliminate unintentional variability due to possible incubation before arrival at the laboratory, only embryos of the proper stage (H&H stage 3) were kept for analysis. A scalpel was then used to trim the area opaqua from the blastoderm, so that only the area pellucida remained. This procedure insured that only the embryo area proper was analyzed. The embryos were pooled according to incubator (with 4 to 6 embryos per tube), centrifuged, and frozen. A total of 270 embryos were used for the first study, and a total of 60 in the second. The assay for ODC is standard and is described elsewhere.[5]

RESULTS

The results of the first study (Table 1) support the conclusion that ELF EM fields do induce a bioeffect. In stage 3 embryos (15h incubation), ODC activity was significantly increased in field-exposed embryos compared with that in the controls.

TABLE 1.—ODC activity in embryos exposed to sinusoidal fields.

Stage	# Embryos analyzed	OCD activity exposed/control
3	270	1.74 ± 0.36

The results of the second study (Table 2) support the hypothesis that a superimposed EM noise field could mask the sinusoidal field used in the first study, thus blocking the bioeffects of the sinusoidal field. For stage 3 embryos, the ODC activity of sinusoidal field-exposed embryos was increased by field exposure. This is consistent with the result obtained in the study above. On the other hand, the ratio of activity in embryos exposed to the sum of the sinusoidal and noise fields to the controls was nearly unity. In other words, the ODC activity in embryos exposed to the sum of the coherent and noise fields was virtually indistinguishable from that in the controls.

TABLE 2.—ODC activity in embryos exposed to sinusoidal field plus noise field.

Stage	# Embryos	ODC activity coherent exposed/control*	ODC activity (coherent exp + noise)/control
3	60	1.98 ± 0.57	0.98 ± 0.06

*The ODC activity ratio for the embryos exposed to sinusoidal fields was computed only for the days on which study 2 was performed. Thus, the exact numbers differ slightly from those in Table 1, which includes all experiments in which embryos were exposed to sinusoidal fields.

CONCLUSION

These studies indicate two significant conclusions:

1. The imposition of a sinusoidal magnetic field modifies the biochemistry of developing chick embryos.
2. The modification of embryo biochemistry induced by coherent fields can be blocked by the superposition of spatially coherent, temporally incoherent EM fields with comparable strengths.[6]

REFERENCES

1. E. Berman et al., "Development of chicken embryos in a pulsed magnetic field," *Bioelectromagnetics* 11: 169-187, 1990.
2. B. Lowkvist, O. Heby, and H. Emmanuelsson, "Essential role of the polyamines in early chick embryo development," *J. Embryology and Experimental Morphology* 60: 83-92, 1980.
3. T. A. Litovitz, C. Montrose, and P. Doinov, "Spatial and temporal coherence affects the response of biological systems to electromagnetic fields," this volume.
4. V. Hamburger and H. L. Hamilton, "A series of normal stages in the development of the chick embryo," *J. Morphology* 88: 49-92, 1951.
5. M. Mullins, D. Krause, and T. A. Litovitz, "Simultaneous application of a spatially coherent noise field blocks the response of cell cultures to a 60 Hz electromagnetic field," this volume.

6. The authors acknowledge the Walter Reed Army Institute of Research Contract for support of this work under Contract DAMD17-86-C-6260.

Martin Blank, Ed., *Electricity and Magnetism in Biology and Medicine*

SIMULTANEOUS APPLICATION OF A SPATIALLY COHERENT NOISE FIELD BLOCKS RESPONSE OF CELL CULTURES TO A 60Hz ELECTROMAGNETIC FIELD

J. M. Mullins,* D. Krause, and T. A. Litovitz

Catholic University of America, Washington, D.C.

L929 murine fibroblasts display a doubling of ornithine decarboxylase activity following 4h exposures to a 60Hz sinusoidal magnetic field of 10 μT. Simultaneous application of a spatially coherent noise field of comparable amplitude eliminated enhancement of ornithine decarbyoxylase activity by the stimulating field. Lowering of the amplitude of the applied noise field restored field-induced enhancement of enzyme activity. These results are discussed in terms of the thermal-noise limit for weak-field detection.

Members of our group have examined the question of the thermal noise limit for detection of weak electromagnetic fields (EMF) by cells (see Litovitz et al., this volume). They reasoned that the spatially random nature of a thermal-noise field is fundamentally different from that of spatially coherent, applied electromagnetic fields, and that this difference could account for the ability of cells to detect weak applied fields without thermal-noise interference. According to this hypothesis, biological response to a stimulating field should be eliminated if a spatially coherent noise field of comparable amplitude is simultaneously applied.

To examine the effects of spatially coherent noise on cellular response to EMF, we have assayed enhancement of ornithine decarboxylase (ODC) activity. Logarithmically growing cultures of L929 cells display a doubling of ODC activity after exposure for 4 h to a 60Hz 10μT magnetic field.[1] ODC response was assessed under conditions in which this stimulating field was applied simultaneously with noise fields of amplitudes equal to or less than that of the stimulating field.

METHODS

Cell cultures were maintained as previously reported.[1] Experiments used three 75cm^2 flasks of logarithmically growing cells each for control, exposed, and exposed + noise conditions. Stimulating fields (60 Hz, sine wave, 10 μT, induced electric field of 0.04 μV/cm) were applied simultaneously with random-noise fields (spectral range 30-90 Hz; rms amplitude, 0.5-10 μT) via vertically oriented Helmholtz coils. The results are expressed as a ratio of enzyme activities of exposed cultures to those of matched controls, or as the percent of ODC enhancement measured for exposed + noise conditions, relative to that of matched exposed cultures.

RESULTS

Cultures of L929 cells exposed to the 60Hz stimulating field alone displayed an approximately two-fold enhancement of ODC activity (2.10 ± 0.32 times control values). However, simultaneous application of the stimulating field and a random-noise field altered this response (Fig. 1). ODC enhancement was eliminated when the stimulating field was combined with a random-noise field of 10 μT rms amplitude, but as the amplitude of the noise field was diminished to 2 and 0.5 μT, ODC enhancement was restored. ODC enhancement in response to the stimulating field thus depends on the amplitude of the applied noise field.

The ability of spatially coherent noise to interfere with ODC enhancement was examined also for Daudi human lymphoma cells. Preliminary data show that, for this cell line, a maximal enhancement (2.23 ± 0.31) of ODC activity occurred after a 6h exposure to the stimulating field. Addition of a 10 μT noise field to the stimulating field eliminated the enhancement of ODC activity.

---

*Correspondence to: J. M. Mullins (Biology), Catholic University of America, Washington DC 20064.

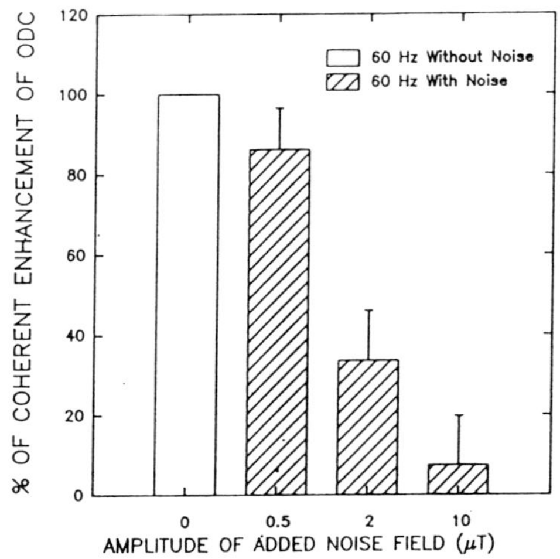

FIG. 1.—Ornithine decarboxylase response of L929 fibroblasts to EMF exposure. Cultures were exposed to 60Hz stimulating field for 4 h with simultaneous application of a spatially coherent noise field of rms amplitude ranging from 0 to 10 µT. Data are expressed as the percent enhancement of ODC activity achieved for exposed + noise samples relative to matched exposed cultures.

DISCUSSION

A spatially coherent noise field, with rms amplitude approximately equal to that of a simultaneously applied stimulating field, inhibited field-induced enhancements of ODC activity in L929 and Daudi cells. The extent of this inhibition depended on the amplitude of the noise field. These data are consistent with the inhibitory effects of spatially coherent noise on field-induced morphological abnormalities in chick embryos (Litovitz et al., this volume). Such similarity in results from work with whole embryos and with cultured cells suggests that the inhibition of field-induced effects by spatially coherent noise represents a general biological response.

Enhancements of ODC activity are thus induced by stimulating fields of microvolts/cm in the presence of random thermal noise fields, but are eliminated by spatially coherent noise fields whose amplitudes are two to three orders of magnitude lower than those of the thermal noise.[2] It is not surprising that cells might ignore the presence of thermal noise fields. Since cells experience the constant presence of random EMF produced by thermal noise, they surely have evolved in ways that render such stimulation inconsequential. The dilemma to be resolved, then, is why applied EMF far weaker than those originating from thermal noise are of consequence in altering cellular response? The concept of spatial coherence elaborated by Litovitz et al. (this volume) provides a hypothesis that answers this question.

It is a reasonable conjecture that many, if not all, effects of EMF on cells occur through interaction of the field with membrane receptor proteins. If that is so, the concept of temporal coherence previously reported,[1] and the requirement for spatial coherence demonstrated here, should provide clues to how fields and receptors may actually interact. In fact, the temporal coherence times we have noted are similar to the times required for aggregation of ligand-bound tyrosine kinase receptors,[3] and the idea of spatial coherence fits with the general concept[4] of receptor cooperativity.[5]

REFERENCES

1. T. A. Litovitz, D. Krause, and J. M. Mullins, *Biochem. Biophys. Res. Comm.* 178: 862-865, 1991.
2. R. K. Adair, *Phys. Rev.* A43: 1039-1048, 1991.
3. M. D. Hollenberg, *FASEB J.* 5: 178-186, 1991.
4. N. G. Morgan, *Cell Signaling,* Guilford: New York: Guilford Press, 1989.
5. This work was supported by the Walter Reed Institute of Research, U.S. Department of the Army, under contract DAMD 17-86-C-6260.

Martin Blank, Ed., *Electricity and Magnetism in Biology and Medicine*

AN INVESTIGATION OF THE EFFECT OF ELF ELECTROMAGNETIC RADIATION ON EMBRYOLOGICAL DEVELOPMENT IN FREE-LIVING TREE SWALLOWS, *Tachycineta bicolor*

S. D. Hill,* D. L. Beaver, P. E. Lederle, and D. L. Herman

Michigan State University, East Lansing

The effects of ELF electromagnetic fields on the embryological development of tree swallows are being investigated. The source of the fields is a frequency-modulated communications system with a mid-frequency of 76 Hz established by the U.S. Navy in the Upper Peninsula of Michigan. Development of embryos developing in nests beneath the antenna and in matched control plots distant from it is being investigated during pre-operational and operational years. Current evidence indicates that tree swallow embryology is unaffected by the communications system.

Tree swallows, *Tachycineta bicolor*, are common passerines nesting in Michigan. Their embryology is being investigated as part of a long-range environmental impact study of an ELF communications system installed by the U.S. Navy in Michigan's Upper Peninsula. Embryological development is being investigated because developing systems are known to be sensitive to insults and because factors that affect development of offspring have direct affects on stability of populations.[1] Tree swallows provide suitable experimental material because parents nest in boxes provided (which allows eggs to be readily monitored and collected) and the species is extremely plentiful in the area, so that sampling will have little effect on population size.

The ELF communications system emits a modulated signal with a mid-frequency of 76 Hz. Test and control plots in which nest boxes have been erected were established beneath the antenna and approximately 20 km distant. We are monitoring embryos from test and control plots during both pre-operational (1989 and earlier) and operational (1990 and following) years.

As specified by the original Illinois Institute of Technology and Research Institute (IITRI) Request for Proposals, test plots were established close enough to the communications system so that electric and magnetic fields measured in the soil near the earth's surface approximate 0.7 V/m and 0.03 G, respectively. Electric and magnetic fields attributable to ELF sources other than the communications system are lower by at least an order of magnitude. Control plots were selected sufficiently distant that electrical and magnetic fields attributable to the system are at least an order of magnitude below those on the test plots. Electric and magnetic fields in the air and earth attributable to ELF sources other than the system (especially 60Hz sources) were matched to within one order of magnitude between test and control plots. Electric and magnetic fields at 76 Hz and 60 Hz are independently measured by IITRI each year. Measurements of 60Hz fields have been obtained since 1983; measurements of 76Hz fields began in 1986 when testing of the system began. Mean values for 76Hz electric and magnetic fields measured in 1990, when the antenna became fully operational, are shown in Table 1. Test and control plots were matched as closely as possible with regard to physical features such as vegetation, soil type, drainage, etc.

Each year, approximately 30 clutches (about 150 eggs) are collected from nest boxes in test and control plots. Active nests are checked daily and eggs numbered sequentially. Four days after the appearance of the last egg, the entire clutch is removed and coded. Determination of abnormalities is performed as a blind study.

Developmental abnormalities related to ELF electric and magnetic fields might be caused by the direct action of radiation on developing tissues or indirectly, as a result of altered parental behavior or physiology. Both possibilities are being investigated. Since avian embryos develop in a closed system (the egg), the resources allocated to each offspring during oogenesis could have a marked influence in determining chick survival. If females forage less effectively in some situations than in others, eggs may be of lower nutrient value and chick survival compromised. To determine whether the communications system affects the

---

*Correspondence to: S. D. Hill (Zoology), Michigan State University, East Lansing MI 48824.

TABLE 1.—Mean values for 76Hz fields on test and control plots (1990) with the communications system fully operational (150 amp).

	Longitudinal electric field (mV/m)	Transverse electric field (V/m)	Magnetic flux density (mG)
Control plots			
1C4	0.044	0.001	0.002
1C6	0.074	0.001	0.004
Test plots			
1T5	62.333	11.002	13.256
1T6	181.714	7.070	10.114

quantity of nutrient deposited in eggs, each egg is weighed and measured at the time of collection. Weights of eggs from test and control plots have been compared for the years 1985-1991 by use of nested ANOVA. Although nest and year effects are present ($p < 0.01$), no difference in weights was found between test and control plots ($p = 0.25$). Since 1990 egg volumes have been determined by water displacement. Again nest and year effects are present ($p < 0.01$), but egg volumes have not differed significantly between test and control plots ($p = 0.7$).

Embryos were dissected off the yolk into Howard chick Ringer's solution, analyzed by microscopic examination and fixed in either Bouin's solution for light microscopy or 2.5% glutaraldehyde for scanning electron microscopy. Embryos were stained with either alcoholic eosin or alcoholic carmine, dehydrated, and cleared in methyl salicylate. They were then observed as stained whole mounts, photographed, then sectioned and studied by routine histological methods, if desired.

By the fourth day after the appearance of the last egg, most embryos have reached stage 20-25,[2] with the major organ systems laid down. The most commonly observed abnormalities involve the spine, brain, a failure to develop, and chaotic development. Abnormalities of the heart, branchial region, and direction of rotation also occur. Routinely 10-20% of the embryos show abnormal development. This frequency is in keeping with failure to hatch data from our study[3] and data reported by other investigators.[4] To date, no significant differences in the number of observable abnormalities have been found between test and control plots [χ^2 analysis, $p > 0.1$; $\chi^2 = 0.05$ (1991), 0.17 (1990), 0.63 (1989), 0.95 (1988)].

Adverse effects of ELF electromagnetic fields on chicken embryology have been reported by several investigators.[5-7] Chickens and tree swallows may have different susceptibilities to ELF radiation, but it seems likely that much of the difference in our results is attributable to our different experimental designs. Field strengths and waveforms in the egg/embryo microhabitat probably differ. Tree swallow embryos in the field may be partially shielded, at least from electrical fields, by the body of the incubating parent. Our study is a continuing one; however, present evidence suggests that ELF fields at our study site are not affecting tree swallow embryology.[8]

REFERENCES

1. D. Lack, *Population Studies of Birds*, Oxford: Clarendon Press, 1966.

2. V. Hamburger and H. L. Hamilton, *J. Morphol.* 88: 49-92, 1951.

3. D. L. Beaver, R. W. Hill, and S. D. Hill, *Navy ELF Communications System Ecological Monitoring Program: Small Mammals and Nesting Birds, Annual Report*, 1991 (submitted to IITRI and the U.S. Navy).

4. L. B. Chapman, *Bird-Banding* 26: 45-70, 1988.

5. A. Ubeda, J. Leal, M. A. Trillo, M. A. Jiminez, and J. M. R. Delgado, *J. Anat.* 137: 513-536, 1983.

6. J. Juutilainen and K. Saali, *Radiat. Env. Biophys.* 25: 135-140, 1986; J. Juutilainen, E. Laala, and K. Saali, *Int. J. Radiat. Biol.* 52: 135-140, 1987.

7. H. Martin, *Bioelectromagnetics* 9: 393-396, 1988.

8. This research was funded by the Submarine Communications Project Office, Space and Naval Warfare Systems Command, U.S. Navy, contracts N00039-81-C-0357, N00039-84-C-0070, and N00039-88-C-0065.

Martin Blank, Ed., *Electricity and Magnetism in Biology and Medicine*

THE EFFECTS OF INTRAUTERINE EXPOSURE OF RATS TO LOW-INTENSITY AC EM FIELDS

B. Lazetic,* B. Nikin, and N. Pekaric-Nadj

University of Novi Sad, Yugoslavia

A study on two generations of Millhood rats probed the possible influence of industrial (50 Hz, 10-30 μT) magnetic fields on intrauterine development of rats. Sexually mature animals were housed together and produced offspring during exposure to magnetic fields. The effect of the magnetic fields was judged postnatally. A significantly larger number of newborn animals per mother, significantly increased number of females, and significantly slower increase in body weight were found in the exposed group compared to the sham-exposed. When these animals became sexually mature, the same experiment was repeated with them. This second experiment uncovered a significant number of sterile animals from the first experiment and a very low survival ratio in the second generation of the exposed animals.

Low-intensity, low-frequency magnetic fields are suspected of causing many health problems, particularly miscarriages in female factory workers. Magnetic fields of 10-30 μT, 50Hz frequency are found in factories near induction motors and heaters; therefore, we performed a two-generation study on rats at these intensities to get more information on the possible effects of the magnetic fields on their intrauterine development.

MATERIALS AND METHODS

The study consisted of two experiments. In the first, sexually mature Millhood rats were housed in plastic cages, two females and one male in each, and were allowed to produce offspring. There were 10 cages in the field-exposed group and 7 cages in the sham exposed. Coils generated 50Hz, 10-30μT fields across the cages, and exposure lasted 6 h/day for 2 weeks. By the day of birth each young animal had had at least 10 days of exposure during the intrauterine development.

The second experiment was performed when the animals from the first experiment became sexually mature. Three groups of the animals were organized: first (7 cages), completely intact group *I*, grown in the environmental 50Hz fields of density less than 0.5 μT; second (10 cages), the group with exposed animals during their intrauterine development from the first experiment, *E1;* and the third group, *E2* (12 cages), the same as group *E1*, except that the animals from this group went through the same field exposure protocol again. The animals from all three groups were allowed to produce offspring.

RESULTS

The effects of the magnetic fields on the intrauterine development of rats were judged by the monitoring of postnatal development of young animals. In the first experiment we found three significant ($p < 0.01$) effects:

1. The number of newborn animals per mother was 9.2 in the field-exposed group compared with 6.7 in the sham exposed.
2. The number of females was 57% of the total number of young animals in the exposed group and 40% in the sham exposed.
3. Body weight of young animals at birth were lower in the exposed group (average 18 g) than in the sham exposed (average 25 g). The difference remained and became even more significant during the next 2 months of postnatal development (Fig. 1).

---

*Correspondence to: B. Lazetic (Physiology), Medical Faculty of University of Novi Sad, Sq. of D. Obradovic 6, 21000 Novi Sad Yugoslavia EU.

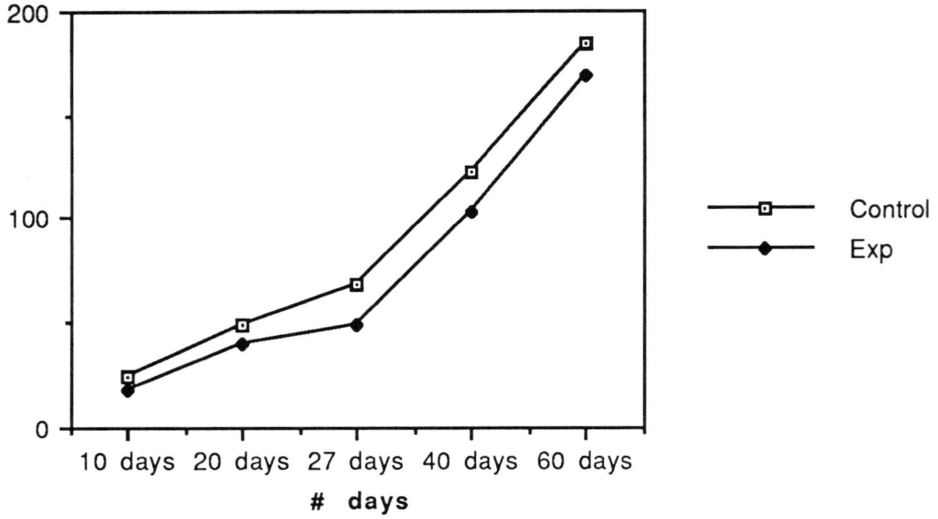

FIG. 1.—Averaged weight (in grams) of young rats from the first experiment taken at various times during first 2 months of their postnatal development. The difference was significant at the $p < 0.05$ level after 10 and 20 days of postnatal development, and $p < 0.01$ after 27, 40, and 60 days of postnatal development.

In the first experiment we found no significant effects on the fertility of parents or on the survival ratio of young animals (average 91%). But in the second experiment we found significant effects ($p < 0.007$) on two parameters:

(1) There were 8% sterile animals in *I* group (all males), 27% sterile animals in *E1* group (30% females, 70% males), and 17% sterile animals in *E2* group (all females).

(2) The survival ratio was calculated after 10 days of postnatal development. It was found to be 67% for group *I*, 32% for group *E1*, and 37% for group *E2*, and did not differ significantly in time.

At the same time we found no significant effect on the number of sibs (average 7.8) in the second experiment. The number of female animals did not differ significantly between groups (average 43%). Body weight of young animals at birth in the several groups did not differ significantly either (average 21 g).

DISCUSSION AND CONCLUSIONS

The two-generation study was carried out in order to determine the effects of 50Hz, 10-30μT magnetic fields on intrauterine development of rats. Although many other data might be of interest, in this study we monitored only the postnatal development of rats and judged the effects by that. Findings from our study seem to be in good agreement with previous findings on mice.[1]

In the first experiment we found a significantly increased number of young animals with predominantly female population born per mother in the exposed group compared with the sham exposed. We also found that body weight at birth was significantly lower in the exposed group. This effect might well not be due to magnetic fields, but might be the consequence of the greater number of newborn animals in the experimental group.

The second experiment showed that the increased female population from the first experiment was significantly sterile. The most significant effect observed in the second experiment was the effect on survival. The survival of the young animals from *I* group was unexpectedly low, possibly due to some factor we had not controlled. Even so, the exposed animals seem to have been more sensitive to this unknown factor.

REFERENCES

1. A. A. Marino, "Meta analysis of multigenerational studies of mice exposed to power-frequency electrical fields," *J. Bioelectricity* 9: 213-231, 1990.

Martin Blank, Ed., *Electricity and Magnetism in Biology and Medicine*

RANDOM DIRECTIONAL DISTRIBUTION OF ASYMMETRIC NEURITE GROWTH IN CHICK DORSAL ROOT GANGLIA EXPOSED TO PULSED ELECTROMAGNETIC FIELDS AND NERVE GROWTH FACTOR

C. H. Sutton* and M. Subramanian Vadula

Medical College of Wisconsin and VA Medical Center, Milwaukee

B. Greenebaum

University of Wisconsin, Kenosha

B. F. Sisken

University of Kentucky

Previous studies carried out in our laboratories have demonstrated that a pulse-burst pulsed electromagnetic field (PEMF) with a peak intensity of 40 G† and a repetition rate of 15 Hz produced directional outgrowth, either unidirectional or bidirectional, when the field was applied to cultured dorsal root ganglia (DRG) in the presence of nerve growth factor (NGF).[1] NGF was required for vigorous neurite outgrowth. The same 40G PEMF produced the greatest length and directionality of neurite outgrowth when combined with NGF. Under the same conditions, a 20G† field did not produce directional outgrowth, which suggested a possible threshold. The present report describes recent studies carried out in our laboratories in which DRG from our earlier experiments were reanalyzed to determine whether there was any correlation between the direction of the bundles of asymmetric neurite outgrowth and the direction of the induced electric fields in the culture medium. Our analysis does not allow us to reject the null hypothesis that the outgrowth directions were uncorrelated with the electric-field direction. Preliminary EM observations of cellular fine structure in 1μm sections suggest that neurons in ganglia exposed to a 40G PEMF remain healthy, and can demonstrate abundant neurite outgrowth if grown in the presence of 50 or 100 ng/ml NGF. Without NGF, neurons remain healthy, but display only limited outgrowth. Dense-core granules of neurotransmitter remained intact in sympathetic ganglion cells in the 40G field in vitro, even without NGF.

It is well known that when specimens in a circular culture dish are exposed to a vertical time-varying magnetic field, the induced electric field in the medium is in a circular pattern and increases linearly in strength from zero at the center to a value that depends on both the magnetic-field strength and the radius at the outer edge.[2] We have previously reported that pulsed electromagnetic fields (PEMF) and nerve growth factor (NGF) act synergistically on cultured chick embryo dorsal root ganglia (DRG) to produce vigorous neurite outgrowth.[1] We have further reported at a 40G† peak pulsed electromagnetic field (PEMF) generated by the application of bursts of a sawtooth-shaped PEMF to the ambient culture medium can produce asymmetric neurite growth patterns, whereas the same waveform with a 20G† peak intensity produces symmetric patterns.[3] In the present study, our previous dishes of DRG were re-evaluated to determine any correlation between the direction of the asymmetric neurite outgrowths and the direction of the induced electric fields in the culture medium.

MATERIAL AND METHODS

The data obtained here were taken from a reanalysis of the DRG studied in our earlier work. The methods employed were described in detail in our original report.[1] Briefly, dorsal root ganglia from 6-day-old chick embryos were dissected in Puck's saline solution containing 1.3 mM $CaCl_2$ and 119 mM NaCl. At

---

*Correspondence to: C. H. Sutton (Neurosurgery), Medical College of Wisconsin, Milwaukee WI 53295.

†The peak field strengths in Ref. 1 were reported erroneously as 80 and 40 G for exposed cells. These values were actually 40 and 20 G, respectively, upon recalculation.

least 200 dorsal root ganglia were obtained from 24 embryos for each set of runs. Ten to twelve DRG were plated in each 60×15mm tissue culture dish (Falcon), and 3 ml of Dulbecco's modified Eagle's medium (DMEM), containing 10% fetal bovine serum, 1% penicillin, and 1% streptomycin, were added. Nerve growth factor (7S NGF, Collaborative Research) was added to some of the dishes in a concentration of either 50 or 100 ng/ml. Dishes were incubated in a 5% CO_2, 95% O_2 atmosphere at 98% relative humidity at 37 C in a water-jacketed incubator (Forma Scientific). DRG were allowed to attach for 24 h before being exposed to the PEMF for 18 h. After exposure, the dishes remained undisturbed for an additional 6 h in the incubator prior to being fixed, stained, and assessed microscopically.

Helmholtz coils 10 cm in diameter were wound with 40 turns each of #16 copper magnet wire and mounted 5 cm apart on a plastic support to produce a vertical magnetic field. When energized by a signal generator (Bio-Osteogen Model 204, EBI, Fairfield, N.J.) they produced a series of bursts of 22 magnetic field pulses; the bursts were repeated at a 15Hz rate. Individual bursts consisted of 22 sawtooth-shaped pulses, each with a 200μs rise and a 20μs fall. The peak intensity of the magnetic field was 40 G, as measured by a pickup coil and an oscilloscope. Sham-exposed cultures were placed in an identical apparatus, except that the coils were not energized. The peak magnetic field in the sham apparatus was 0.04 G when it was placed in the same incubator as the exposure coils, but separated from them by a copper-shielded shelf. In all cases the field was applied vertically, producing a circumferential pattern of induced electrical currents in the 60mm-diameter culture dishes. The induced electrical field strength was zero in the center of the dish and rose linearly to a maximum value at the perimeter.

For analysis of directional growth, the measured position of each dish was recorded as outlined in Fig. 1. Each dish was oriented by a mark on the spot on its rim that pointed toward "geographic north" when in the incubator. For each ganglion, its distance from the center of the dish (*R* in Fig. 1) was measured and its angular position θ in the dish with respect to the orientation spot was calculated. Under higher-power microscopy, the neurites from each ganglion were observed and scored. As indicated in Fig. 1, the visual field was divided into quadrants—top, bottom, left, and right—by means of an ocular micrometer scale. In a manner similar to that described in our previous work,[1] the number of quadrants occupied by neurites and the average length of these neurites were recorded. The average length was determined by an eyepiece reticle with 100 divisions/mm. If the neurites were not uniformly distributed about the ganglion's circumference, the median angle of each grouping was measured with respect to the top of the field of view, corresponding to the direction of the edge of the dish (Fig. 1). These angles were recorded as α_1 for the first bundle of neurites, α_2 if there was a second bundle, and so forth. The average length of each bundle was also recorded.

RESULTS

The collected data for the directions in which bundles of neurites grew were binned in 30° groups. Groups clustered about 90° and 270° represent growth parallel to the induced electric fields; groups clustered about 180° and 360° represent growth perpendicular to these fields. As shown in Fig. 2, the direction of clustered bunches showed no preferred direction; the clustering about the four cardinal points of the circle is probably an artifact of scoring. A scatter plot of neurite length against distance of the neurite from the center of the dish showed no correlation for cultures grown with 50 ng/ml NGF and exposed to 40G fields, consistent with no correlation to the induced electrical field's direction. In sections prepared from Epon-embedded DRGs and cut at 1 μm, neuronal fine structure was preserved in specimens exposed to a 40G PEMF without NGF (Fig. 3). Dense-core granules of neurotransmitter in sympathetic ganglion cells and processes also remained intact in the 40G field while these cells were in culture, even without NGF (Fig. 4). Therefore, this relatively intense PEMF does not appear to exert a toxic or adverse effect on DRG neurons in culture.

DISCUSSION AND CONCLUSIONS

Pulsed electromagnetic fields (PEMFs) with amplitudes of 3 to 15 G have been demonstrated in several laboratories to stimulate the regrowth of neuronal processes in peripheral nerve.[4] Sisken et al.[5] have reported enhancement of neurite outgrowth from chick dorsal root ganglia (DRG) using single pulses of PEMFs at 2 Hz and at 72 Hz.[6] In our previous studies with explanted DRGs, a pulse-burst PEMF with a 40G peak intensity appeared to induce directional growth and, with the addition of NGF, also enhanced the length of outgrowth of neuritic processes. The unanticipated directional growth that resulted from exposure to the pulse-burst PEMF signal has no known parallel in the literature, but no other studies have been reported that utilized a PEMF as intense as 40 G to facilitate the growth of neural tissue in vitro.

A PEMF was used to stimulate neural regeneration in vivo by Raji and Bowden,[7] who employed a signal with a high peak power in order to stimulate regeneration of the transected common peroneal nerve in rats. This PEMF was administered for 15 min daily for periods of 3 days to 8 weeks. There was acceleration of nerve-fiber regeneration as well as an increase in the size of intraneural blood vessels and a reduction in the amount of collagenous tissue formation after treatment. The results of the present study have not demonstrated any preferential directionality of growth related to the induced electric field. Our analysis does not allow us to reject the null hypothesis that the outgrowth directions are uncorrelated with the electric-field direction. This result implies that the magnetic field itself is in all probability interacting with the neurites, a finding that is consistent with the work of Sisken.[4] Electron micrographs showed healthy neurons and nerve bundles, even on exposure to strong PEMF (40G peak intensity), suggesting that fields of this type and intensity are not detrimental to neuronal viability, although vigorous neurite outgrowth appears to require the presence of NGF. Therefore, it can be concluded that PEMFs of this intensity do not interfere with neuronal viability, and do enhance neurite outgrowth with NGF is present.[8]

REFERENCES

1. M. Subramanian Vadula, C. H. Sutton, B. Greenebaum, and B. F. Sisken, "Interaction of electromagnetic fields and nerve growth factors on nerve regeneration in vitro," in C. T. Brighton and S. R. Pollack, Eds., *Electromagnetics in Medicine and Biology,* San Francisco: San Francisco Press, 1991, 145-151.

2. H. Bassen, T. Litovitz, M. Penafiel, and R. Meister, "ELF in vitro exposure systems for inducing uniform electric and magnetic fields in cell culture media," *Bioelectromagnetics* 13: 183-198, 1992.

3. C. H. Sutton, M. Subramanian Vadula, B. Greenebaum, and B. F. Sisken, "Comparison of 40 and 80 Gauss pulsed EMF and nerve growth factor on nerve regeneration in vitro," *Proc. BEMS,* Salt Lake City, 1991, 4.

4. B. F. Sisken, M. Kanje, G. Lundborg, E. Herbst, R. Zienowicz, M. Orgel, and W. Kurtz, "Stimulation of nerve regeneration in vitro and in vivo using pulsed electromagnetic fields," *Soc. for Neuroscience* 14: 497, 1988.

5. B. F. Sisken, M. Kanje, G. Lundborg, and W. Kurtz, "Pulsed electromagnetic fields stimulate nerve regeneration in vitro and in vivo," *Restor. Neuro. Neurosci.* 1: 303-309, 1990.

6. B. F. Sisken, B. McLeod, and A. A. Pilla, "PEMF, direct current and neuronal regeneration: Effect of field geometry and current density," *J. Bioelectricity* 3: 81-101, 1984.

7. A. R. Raji and R. E. Bowden, "Effects of high-peak pulsed electromagnetic field on the degeneration and regeneration of the common peroneal nerve in rats," *J. Bone Joint Surg.* 65: 478-492, 1983.

8. This work was funded in part by the Department of Neurosurgery, Medical College of Wisconsin, and by the Research Service, Clement J. Zablocki Medical Center, Department of Veterans Affairs, Milwaukee, Wis.

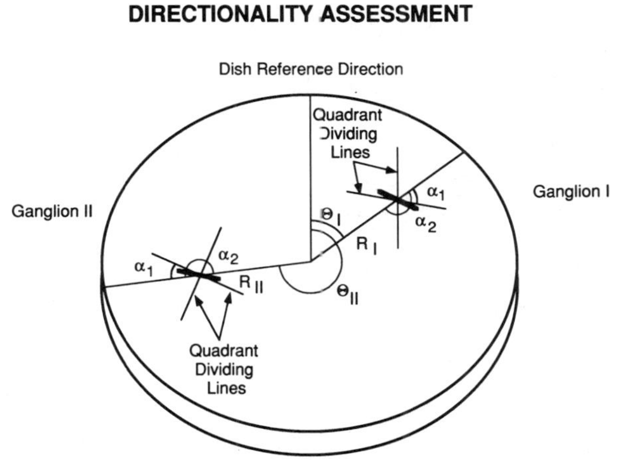

FIG. 1.—Dish containing dorsal root ganglia, indicating parameters and angles measured.

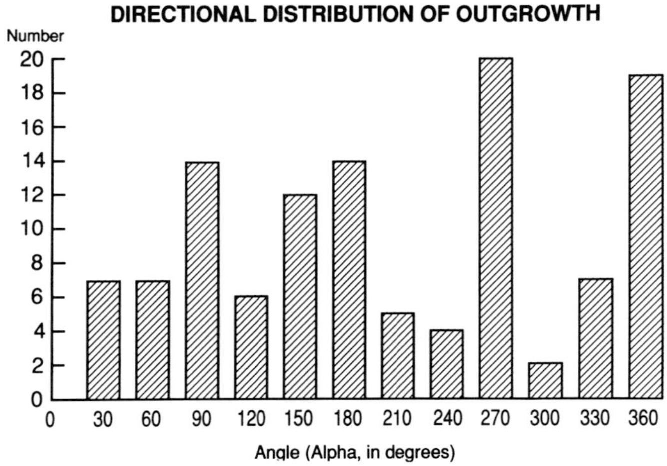

FIG. 2.—Histogram of distribution of angles labeled α in Fig. 1 formed by each bundle of directional neurite outgrowth with respect to dish rim. No asymmetry was found.

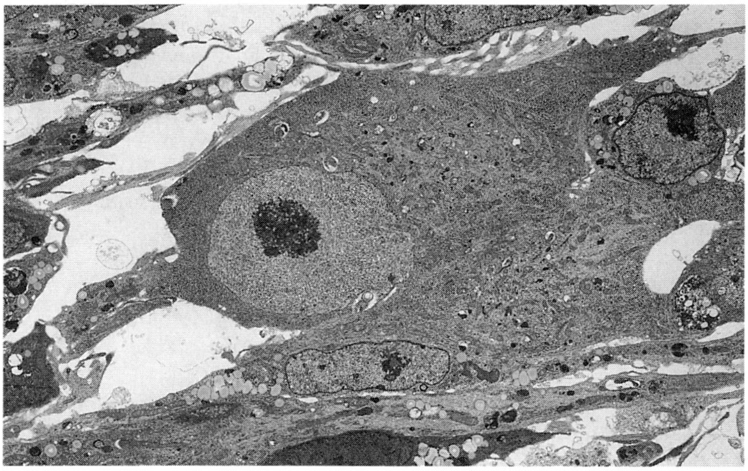

FIG. 3.—Chick dorsal root ganglion treated with 40G PEMF showing healthy neurons that display mitochondria, ribosomes, endoplasmic reticulum, and neuritic processes. (2000×.)

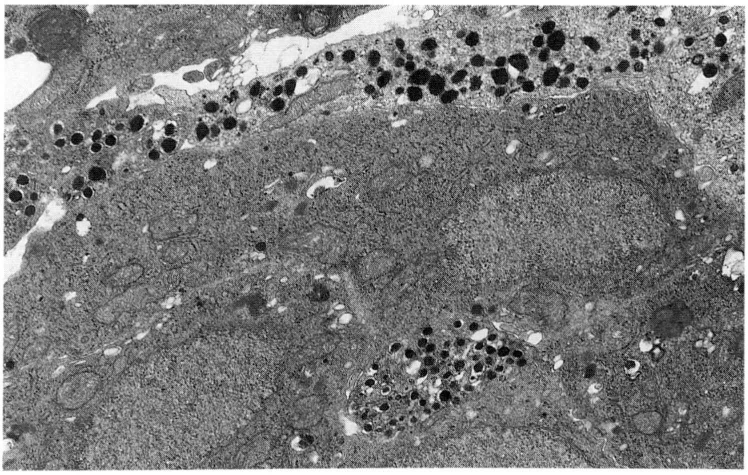

FIG. 4.—Chick sympathetic ganglion treated with 40G PEMF showing neuritic processes filled with dense-core vesicles containing adrenergic neurotransmitter. (9000×.)

Martin Blank, Ed., *Electricity and Magnetism in Biology and Medicine*

EFFECT OF PULSED MAGNETIC FIELDS ON BONE APPOSITIONAL GROWTH RATE DURING THE REPAIR OF TRANSCORTICAL HOLES: PRELIMINARY DATA

V. Canè*

University of Modena, Italy

P. Botti and S. Soana

University of Parma, Italy

The effect of pulsed low-frequency electromagnetic fields (PEMFs) on bone formation was studied via the healing process of transcortical holes bored at the diaphyseal region of the metacarpal bones of two horses exposed for 30 days to PEMFs. The mineral apposition rate (MAR), measured by means of the double tetracycline labeling technique, proved to be higher in PEMF-treated holes than in untreated ones, which suggests that PEMFs exert a positive effect on osteoblast repair activity.

Previous investigations carried out on the rate of repair of transcortical holes in shaft bones of adult horses showed that PEMFs exert a constant positive effect at diaphyseal regions, whereas at metaphyseal regions they improve or reduce the healing process as compared to controls.[1,2] Since these data were confined to a quantitative assessment of the final stage of bone repair, i.e., bone deposition, they cannot indicate whether the modulating effect of PEMFs involved the reparative osteogenesis itself or perhaps the stages preceding it.

In the attempt to solve this problem we carried out a new series of experiments, using the tetracycline labeling technique to detect whether PEMFs modulate the *mineral apposition rate* (MAR)[3] during the repair of transcortical holes at diaphyseal level.

MATERIAL AND METHODS

Two male horses 8-10 years of age were used. According to the methodology used in our previous investigations,[2] two transcortical holes were made in the lateral margin of the right and left metacarpal, at the same diaphyseal level. The metacarpals of both sides were surrounded by a pair of Helmholtz coils and the coils of the left side only were powered for 30 days by a pulse generator. The magnetic field had a peak amplitude of about 28 Gauss, a rise time of 1.3 ms, and a repetition rate of 75 Hz. As a kinetic marker of bone mineralization, tetracycline chloride was injected intravenously (25-30 mg/kg) in both horses on the 15th and 25th days. The horses were sacrificed 5 days later.

The following parameters were measured by a computer image analyzer (TESAK) on the microradiographs of the undecalcified midlongitudinal sections of the holes: (a) total area of the hole; (b) area NB occupied by the new bone, i.e., the bone laid down during 30 days; and (c) area P of the porosity (i.e., the space not occupied by bone tissue). The relative amount of bone laid down during 30 days was expressed as percent ratio of $NB/(NB+P)$. Statistical analyses were performed by two-tailed paired Student's t tests.

The mineral apposition rate (MAR), expressed in μm/day, was evaluated by measurement of the distance between the two tetracycline labels, every 140 μm; and orthogonally with respect to the labels (Fig. 2). Statistical analyses were performed by Student's t tests.

---

*Correspondence to: V. Canè (Anatomia Umana Normale), Policlinico-Via del Pozzo 71, I-41100 Modena (MO) Italy.

RESULTS AND DISCUSSION

In both PEMF-treated and untreated sides the newly formed trabeculae originate from a thin layer of new bone lining the hole surface. Comparative light-microscope observation between PEMF-treated and untreated holes shows that in the former, the stage of repair is always more advanced than in the latter. In fact, the values of $\% NB/(NB+P)$ are significantly higher ($p < 0.01$) in PEMF-treated holes (Fig. 1).

Microscopic observation under UV light of the transverse sections of the holes shows, in both PEMF-treated and untreated sides, two concentric and complete rings of tetracycline inside the above-mentioned layer of newly formed bone lining the hole surface (Fig. 2). The measurement of the distance between these two tetracycline rings shows that MAR is significantly higher ($p < 0.0001$) in PEMF-treated holes than in untreated ones (Fig. 3). Also, the fluorescence of the newly formed trabeculae is more intense in PEMF-treated holes than in untreated ones.

Tetracyclines do not label the osteoidogenetic (or osteoblast secreting) surface but the mineralization front.[4-6] What is thus being measured is not the bone appositional growth rate (BAR), but the mineral apposition rate (MAR). However, under normal conditions, when the rate of osteoid mineralization is not impaired,[7-9] as is the case in our horses, the linear rate of osteoid matrix deposition does not differ significantly in a unit of time from the linear rate of bone mineralization, and thus MAR may be considered a reliable index of BAR.

In sum, the data of MAR here reported indicate that PEMFs stimulate osteoblast activity during the healing of transcortical holes.

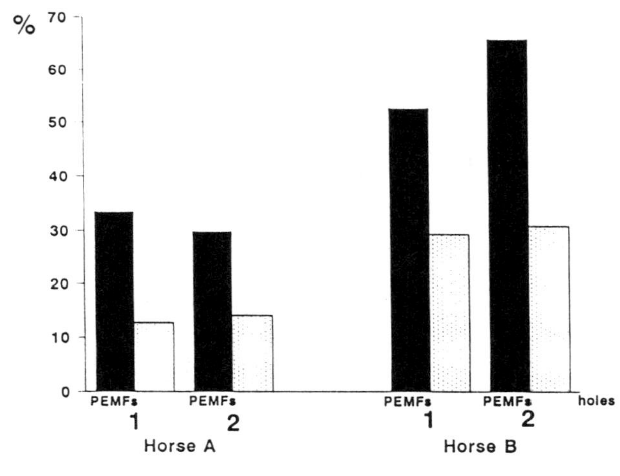

FIG. 1.—Bone deposition during 30 days in two transcortical holes made at the same diaphyseal levels of homotypical metacarpals of two horses: x axis, the two holes (1, 2) in left PEMF-treated (black columns) and right untreated (dotted columns) metacarpals; y axis, values of $\% NB/(NB+P)$.

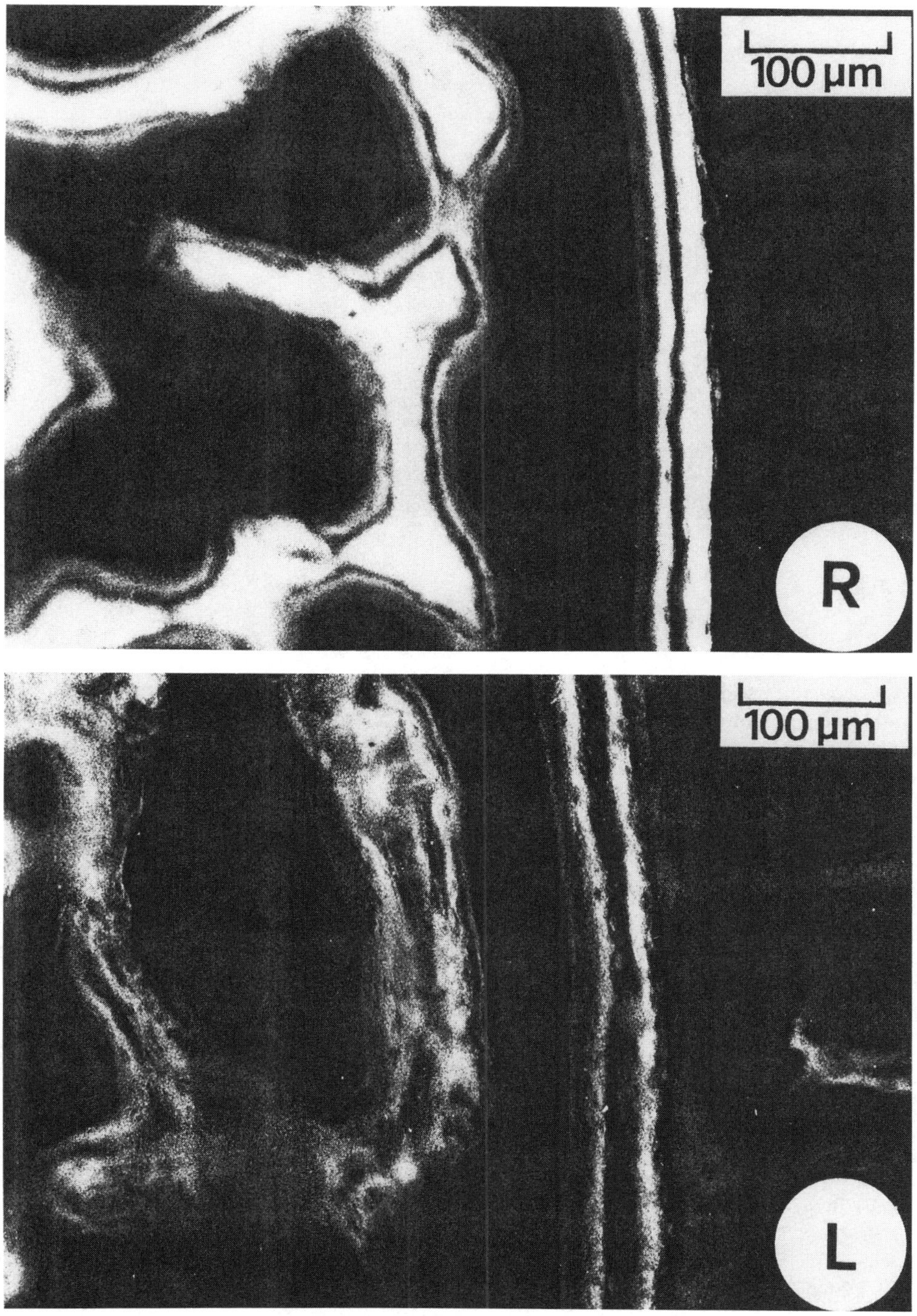

FIG. 2.—UV photomicrographs (200×) showing double tetracycline labeling in layer of newly formed bone lining the hole surface. Distance between the two labels is greater in the left PEMF-treated hole (L) than in contralateral untreated one (R). Fluorescence of newly formed trabeculae is on the whole more intense in PEMF-treated hole.

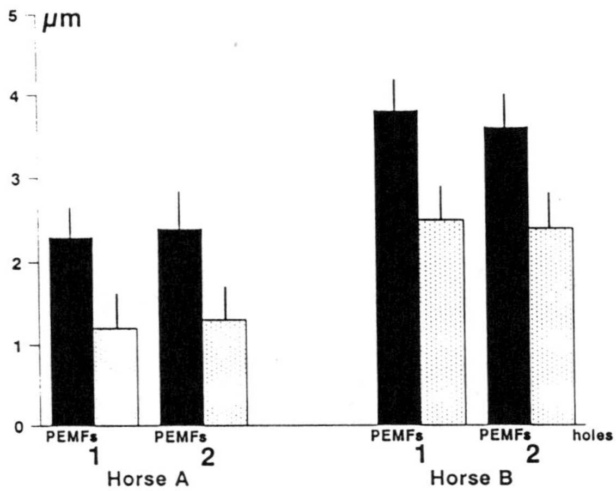

FIG. 3.—Mean values ±S.E. of mineral apposition rate (MAR) in two transcortical holes made at the same diaphyseal level of homotypical metacarpals of two horses (A, B): x axis, the two holes (1, 2) in left PEMF-treated (black columns) and right untreated (dotted columns) metacarpals of each animal; y axis, values of MAR expressed in µm/day. Each bar represents mean value calculated from 50 measurements.

REFERENCES

1. V. Canè, P. Botti, D. Farneti, and S. Soana, "Histomorphometric study of low frequency pulsing electromagnetic fields (PEMFs) on bone repair of transcortical holes," *Trans. BRAGS* 9: 19, 1989.
2. V. Canè, P. Botti, D. Farneti, and S. Soana, "Electromagnetic stimulation of bone repair: A histomorphometric study," *J. Orthop. Res.* 9: 908-917, 1991.
3. A. M. Parfitt, "Bone histomorphometry: Standardization of nomenclature, symbols and units," *Bone and Mineral* 4: 1-5, 1988.
4. A. H. Caswell and J. D. Hutchison, "Selectivity of cation chelation to tetracyclines: Evidence for special conformation of calcium chelate," *Biochem. and Biophys. Res. Comm.* 43: 625-630, 1971.
5. K. H. Ibsen and M. R. Urist, "The biochemistry and the physiology of the tetracyclines, with special reference to mineralized tissues," *Clin. Orthop.* 32: 143-169, 1964.
6. M. R. Urist and K. H. Ibsen, "Chemical reactivity of mineralized tissue with oxytetracycline," *Arch. of Pathology* 76: 484-496, 1963.
7. J. Bordier, P. Marie, L. Miravet, A. Ryckewaert, and H. Rasmussen, "Morphological and morphometrical characteristics of the mineralization front: A vitamin D regulated sequence of the bone remodeling," in P. J. Meunier, Ed., *Bone Histomorphometry*, Paris: Armour Montagu, 1977, 335-354.
8. A. J. Olah, "Quantitative relations between osteoblasts and osteoid in primary hyperparathyroidism, intestinal malabsorption and renal osteodistrophy," *Virchows Arch.* **A358**: 301-308, 1972.
9. H. Takahashi, H. Norimatsu, T. Konno, J. Inoue, and K. Yanagy, "Different patterns of tetracycline uptake in varying forms of rickets and osteomalacia," in W. S. S. Jee and A. M. Parfitt, Eds., *Bone Histomorphometry*, Levallois: Armour Montagu, 1981, 87-93.

Martin Blank, Ed., *Electricity and Magnetism in Biology and Medicine*

HISTOLOGICAL EVALUATION OF SPINE FUSION AFTER STIMULATION WITH PULSED ELECTROMAGNETIC FIELDS (PEMF)

S. Giannini and M. Di Silvestre

Istituto Ortopedico Rizzoli, Bologna, Italy

S. Guizzardi,* P. Govoni, R. Scandroglio, and R. Cadossi

Institute of Histology and Embryology, University of Parma, Italy

This study reports the histological data relative to the effect of pulsed electromagnetic fields (PEMFs) on the evolution of posterior arthrodesis induced in the lumbar vertebrae of 18 adult male Sprague-Dawley rats. After the operation, one group of six rats was stimulated with PEMFs for 18 h/day by means of a pair of coils fixed to the outside of the cage. A control group of 9 rats was given no stimulation after surgery. In the groups stimulated with PEMFs, an acceleration of the process of bone callus organization was observed as early as after four weeks, and even more so after eight; an early replacement was in fact observed of the newly formed cartilage tissue with primary bone (at 4 weeks) and subsequently with secondary bone (after 8 weeks).

The use of pulsed electromagnetic fields (PEMFs) in spinal fusions began in the past 10 years.[1,2] Their use was favored by the fact that PEMFs were simpler and more practical as well as noninvasive, and above all by the histological results obtained by Kahanowitz et al.[3] These authors used PEMFs to stimulate posterior lumbar fusions with instrumentation in dogs, with an electromagnetic coil externally applied to the spine. The animals stimulated with PEMFs in the early postoperative period showed an earlier and marked osteogenic response than the unstimulated control group.

Apart from this contribution, however, the other studies on the use of PEMFs in spinal fusions are clinical and the fusion pattern was assessed mainly with x rays.[1,2] The aim of this study is to assess callus development in rats, following posterior lumbar fusion and postoperative PEMF treatment, using histological and histochemical techniques.

MATERIALS AND METHODS

Eighteen adult male Sprague-Dawley rats weighing 350-400 g were anesthetized with an intraperitoneal injection of 2 mg/kg of body weight of sodium pentobarbital. The posterior vertebral arthrodesis of the last three lumbar vertebrae was performed according to a well-tested technique.[4,5] The spinal area was decorticated on both sides with a 0.2mm power burr. The fusion area was prepared by use of chips obtained from the spinous processes previously removed as bone grafts. No postoperative immobilization was used.

Nine out of the twelve rats underwent PEMF treatment during the postoperative period until sacrificed; two electromagnetic coils were fixed on the outside of each cage. Coils were inductively coupled and induced in a standard coil probe with an electric field varying from 2.1 mV at the center of the cage to 3 mV at the edge of the cage. Rats experienced an average induced electric field of 2.5 mV. The other nine rats, used as controls, were maintained in the same room. The rats were sacrificed 4 or 8 weeks after surgery, which made it possible to divide the animals into four groups. At the time of sacrifice, the lumbar section of the spine was removed, decalcified in EDTA, divided into two segments, and embedded in paraffin according to normal histological techniques. Our observations were focused principally on two areas: the decorticated laminas and the spaces between the articular facets. In these areas we investigated the presence of regenerating cartilage and primary or remodeling bone.

---

*Correspondence to: S. Guizzardi, Institute of Histology and Embriology, Via Gramsci 14, I-43100 Parma, Italy.

RESULTS

The animals belonging to the first group (control unstimulated animals, sacrificed after 4 weeks) showed growing cartilage tissue at the laminar level, starting from the decorticated area. This tissue was characterized by the presence of large globular cells and a scant intercellular matrix. In the deepest zones there were areas of bone tissue with interwoven fibers, indicating an initial ossification proceeding along the underlying surface, or an early osteoid transformation of the newly formed tissue. The space between the articular facets also appeared to be occupied by cartilage tissue starting from the cruentated bone surfaces. However, this tissue did not form a complete homogeneous bridge between the two bone heads; in fact, more often, in the center of the articular space only, fibrillar connective tissue was observed, rich in blood vessels.

The animals stimulated with PEMFs sacrificed after 4 weeks also showed growing cartilage tissue (Fig. 1): the newly formed tissue appeared to be organized in large trabeculae bordering on thin lacunar spaces containing a slack connective tissue rich in blood vessels. The cellular component, unlike what was observed in the control animals, appeared to consist of small elements surrounded by an abundant, intensely stained intercellular matrix. This was therefore a transition stage between cartilage and bone, earlier in these bone-stimulated animals. This tissue was also arranged to encompass and closely adhere to the bone fragments used as autologous grafts. At the level of the cruentated facets, the newly formed cartilage was marked by the presence of an intercellular matrix, in parts organized in large interwoven bands, more thickly packed than in the unstimulated control animals.

In the animals sacrificed after 8 weeks, there was no substantial difference when compared to the previous unstimulated group. There was merely a further development in the newly formed bone callus at the level of the decorticated laminae, shown by the presence of cartilage tissue with an intensely stained intercellular matrix and organized in large bands surrounding isogenic groups, and by numerous areas of tissue with morphological features more similar to primary bone than to cartilage. At the level of the cruentated articulations, the differences were even less marked: the tissue occupying the primary articular space, now forming a complete bridge between the facets, still showed morphological aspects characteristic of hyaline cartilage; however, here the intercellular matrix appeared to be more organized, and in some areas more intensely stained (Fig. 2).

In the animals stimulated with PEMFs and sacrificed after 8 weeks, newly formed cartilage tissue was almost never seen at the laminar level since it had been completely replaced by primary bone, organized to form thin trabeculae that surround large lacunae containing bone marrow. However, cartilage tissue was still present in the articular spaces, with practically the same morphological features as those observed in the animals stimulated with PEMFs after 4 weeks, with the exception of a greater presence of vessels and a more marked organization of the matrix in interwoven bands with intensely stained fibrils, attributable to calcified cartilage features (Fig. 3).

DISCUSSION

This study, carried out in rats, supplied histological results that confirm those emerging from the use of PEMFs in patients undergoing spinal fusion.[1,2] It thus appears that PEMFs have a positive effect on fusion development. In our experience this effect is particularly marked in the laminar areas: only 4 weeks after surgery, newly formed bone tissue of the woven fiber type is already present, whereas in the unstimulated control animals there is only cartilage tissue. This initial acceleration produced by PEMFs then slows down over time and after 8 weeks the pattern observed in the stimulated animals is similar to that seen in the control animals.

The positive effect of PEMFs was also observed at the level of the articular facets when compared with the control animals. However, here the differences after 4 weeks between the stimulated animals and the controls were minimal: in both cases the newly formed tissue consisted entirely of hyaline cartilage. Greater differences were observed after 8 weeks: in the stimulated animals the fusion appeared more stable, since it consisted of cartilage tissue with calcified areas, whereas in the untreated animals the spaces between the articular facets were occupied by clearly hyaline cartilage tissue.

These results lead to two possible conclusions. The first relates to the positive effect of PEMFs on spinal fusion development, but only during the initial stages: PEMFs seem to make little difference after a longer postoperative period. This result may be explained on the basis of the data provided by some in vitro studies,[6-10] in which the observation of an increase in collagen and proteoglycan synthesis has been reported as well as DNA in chondroblasts and fibroblasts in cultures subjected to PEMFs. However, no such pattern has been observed in other types of cells. It can therefore be suggested that PEMFs are effective in

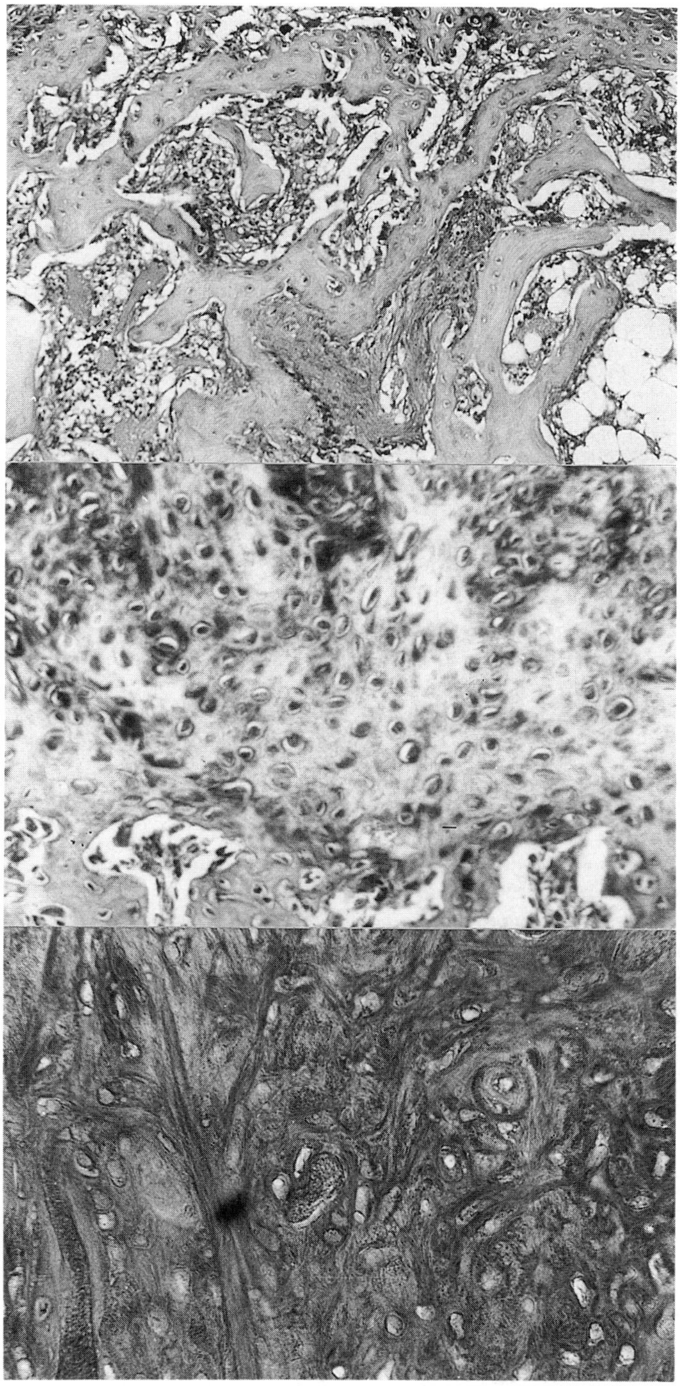

FIG. 1.—Stimulated animals sacrificed after 4 weeks. Cartilage provided with small elements surrounded by intensely stained intercellular matrix appears organized to form large trabeculae; spaces between them are occupied by slack connective tissue. Mallory's-Azan, 300×.
FIG. 2.—Control unstimulated animals sacrificed after 8 weeks. Spaces between cruentated articular facets are occupied by hyaline cartilage, whose intercellular matrix appears irregularly stained. Alcian Hematoxylin, 250×.
FIG. 3.—Stimulated animals sacrificed after 8 weeks. Marked organization in interwoven bands with intensely stained fibril bundles and numerous blood vessels is detectable in cartilage between cruentated articular facets. Mallory's-Azan, 700×.

stimulating cell lines that intervene in the early stages of bone callus formation, but they have no such effect on other cells (e.g., osteoclasts) involved in the subsequent reorganization.

The second conclusion concerns the different histological patterns observed in fusions at a laminar and an articular level. The influence of PEMFs on the latter, although present, is less marked, After 8 weeks there is still no sign of bone tissue, since the interapophyseal space is still occupied by hyaline cartilage that is only partially calcified. This result may be explained by the absence of instrumentation (internal or external) of the spinal fusion; there is therefore a delay in the reorganization of bone callus, especially at the points of greatest mobility, i.e., the articular processes.

In conclusion, it may be said that PEMFs appear, at least during the early stages, to accelerate the growth of bone callus, favoring transformation from cartilage to bone. Internal or external instrumentation of the fusion does not appear to be essential in all cases to obtain the maximum effect from PEMFs.

REFERENCES

1. V. Mooney, ''A randomized double-blind prospective study of the efficacy of pulsed electromagnetic fields for interbody lumbar fusion,'' *Spine* 15: 708-712, 1990.
2. J. W. Simmonds, ''Treatment of failed posterior lumbar interbody fusion (PLIF) of the spine with pulsing electromagnetic fields,'' *Clin. Orthop. Rel. Res.* 193: 127-132, 1985.
3. N. Kahanowitz, S. P. Arnoczky, D. Hulse, and P. K. Shires, ''The effect of postoperative electromagnetic pulsing on canine posterior spinal fusion,'' *Spine* 9: 273-279, 1984.
4. M. Di Silvestre, S. Guizzardi, S. Ruggeri, and R. Savini, ''Implants of demineralized bone matrix for preparation of posterior spine fusion in rats,'' *Abstracts*, ESDS-SRS Combined Meeting, Amsterdam, 1989, 292-293.
5. M. Di Silvestre, S. Guizzardi, A. Ruggeri, and R. Savini, ''Impianto di matrice ossea eterologa demineralizzata: Risultati preliminari di uno studio sperimentale nel ratto,'' *Chir. Org. Mov.* 73(III): 249-254, 1988.
6. R. W. Farndale and C. Murray, ''Pulsed electromagnetic fields promote collagen production in bone marrow fibroblasts via athermal mechanism,'' *Calcif. Tissue Int.* 37: 178-182, 1985.
7. R. J. Fitzsimmons, J. R. Farley, W. R. Adey, and D. J. Baylink, ''Frequency dependence of increased cell proliferation, in vitro, in exposure to a low-amplitude, low-frequency electric field: Evidence for dependence on increased mitogen activity released on culture medium,'' *J. Cell Physiol.* 139: 586, 1989.
8. J. C. Murray, M. Lacy, and S. Fitton-Jackson, ''Degradative pathways in cultured synovial fibroblasts: Selective effects of pulsed electromagnetic fields,'' *J. Orthop. Res.* 6: 24-31, 1988.
9. L. A. Norton, ''Pulsed electromagnetic field effects on chondroblast culture,'' *Reconstr. Surg. Traumat.* 19: 70-86, 1985.
10. G. A. Rodan, L. A. Bourett, and L. A. Norton, ''DNA synthesis in cartilage cells is stimulated by oscillating electric fields,'' *Science* 199: 690, 1978.

Martin Blank, Ed., *Electricity and Magnetism in Biology and Medicine*

BROADBAND EMF ACCELERATION OF BONE REPAIR IN A RABBIT MODEL IS INDEPENDENT OF MAGNETIC COMPONENT

A. A. Pilla,* M. Figueiredo, P. R. Nasser, J. J. Kaufman, and R. S. Siffert

Mount Sinai School of Medicine, New York

J. Erickson

American Medical Electronics, Richardson, Tex.

One recurrent discussion surrounding EMF bioeffects is whether the induced electric field is the primary stimulus, or whether the magnetic component is necessary. This study addresses this problem by delivery of a clinically effective EMF waveform (AME Spinal Stim®) to a fibular osteotomy site in the rabbit, either with external coils or with implanted electrodes far from the treatment site. The EMF waveform consisted of asymmetrical pulses (65 μs/195 μs) applied in a 99-pulse burst, repeating at 1.5 Hz. A new analysis of the dielectric properties of bone tissue, considering cells in gap junction contact, shows this EMF waveform to be particularly well matched to this target tissue. For both treated groups the mean maximum torque is not significantly different from the mean intact value at POD 9 vs POD 13 for controls, which shows that bone repair is significantly accelerated by a clinically effective broadband EMF waveform. These results suggest that the magnetic component of this EMF signal plays only a coupling role, which tends to support the proposal by one of us (AAP) that electrical dosimetry at the cell membrane is a primary physical factor in EMF bioeffects.

Modulation of the rate of in vitro and in vivo biological growth and repair processes using EMF signals has progressed since the first reports of the use of weak DC currents for bone repair.[1] The most common application of electrical signals at present is via pulsed electromagnetic fields (PEMF). This modality was originally created to provide a noninvasive, no-touch means of applying a specific electrical current waveform at the cell/tissue level.[2] The EMF waveform in most common use was originally created on the basis of an electrochemical model of the cell membrane.[3] It is now generally accepted that the initial target for EMF bioeffects is the cell membrane. As knowledge concerning this physical mechanism progresses, it has become clear that the waveform parameters of the induced electric field should be chosen for adequate-tissue (as opposed to single-cell) response.[4,5] The dielectric properties of the membrane target are substantially influenced by cell-cell communication. The EMF target is then a system represented by cell arrays, with significantly higher sensitivity and lower frequency response than an isolated single cell.

CELL-ARRAY BONE MODEL

Organized tissue is developed and maintained by an ensemble of complex geometry cells that have coordinated activity via gap junctions.[6] These junctions provide pathways for ionic and molecular intercellular communication.[7] The role of cooperative organization in the EMF sensitivity of biological systems has been qualitatively considered.[8] Functional modification of gap junctions by modulated microwave fields has been reported.[9] The problem of signal-to-thermal noise ratio (SNR) is central for the design of EMF waveforms capable of delivering a detectable signal to the membrane target. It has been suggested that the thermal threshold of cell sensitivity to external electric fields can be as low as 10^{-3} mV/cm only by postulating a time averaging process for the biochemical sensor.[10] However, a system of cells in ionic (electrical) contact can exhibit orders of magnitude more EMF sensitivity than for the corresponding single cell, obviating the need for complicated mechanisms.

For a cell array of length L, a useful model is a distributed-parameter linear electrical analog that permits a description of impedance per unit length for cells electrically connected in series (transmission line). Such

---

*Correspondence to: A. A. Pilla (Bioelectrochemistry), Dept. of Orthopaedics, Mt. Sinai School of Medicine, New York NY 10029.

a model allows the induced transmembrane voltage V_M to be evaluated as a function of frequency ω and position x along the array. The details of this approach can be found elsewhere.[4,5] However, a brief description of its application to bone tissue will be offered here.

The propagation of V_M in the cell array is given by

$$V_M(x,\omega) = -\frac{E(\omega)\sinh(\gamma x)}{\gamma\cosh(\gamma L)} \tag{1}$$

where $\gamma = [(R_e + R_i)Y_M(\omega)]^{1/2}$, $Y_M(\omega) = 1/Z_M(\omega)$, and R_e and R_i are, respectively, the extra- and intracellular electrolyte resistances, Z_M is the membrane impedance, all per unit array length; and $E(\omega)$ is the applied (extracellular) EMF. The exact form of Z_M depends on the model assumed. The simplest and most classic case is the membrane capacitance C_d in parallel with an ionic leak pathway R_M. The impedance of the cell array $Z(x,\omega)$ is

$$Z(x,\omega) = \frac{(R_e + R_i)}{\gamma}\tanh(\gamma x) \tag{2}$$

Equation (1) allows spatial amplification (V_M/EL) to be evaluated for an array varying from $L = 10\,\mu m$ (single cell) to $L = 1$ mm.[4] The results show that V_M can be orders of magnitude higher for a 1mm array vs that for a single cell, but only at much lower frequencies than that for a single cell target. This allows SNR to be evaluated for thermal noise which is present in all voltage dependent membrane processes. The power spectral density $S_n(\omega)$ of thermal noise is given by[11]

$$S_n(\omega) = 4kT\,\mathrm{Re}[Z(x,\omega)] \tag{3}$$

where Re signifies the real part of $Z(x,\omega)$.

Since both V_M (Eq. 1) and S_n (Eq. 3) depend on frequency, it is most revealing to use a comparison of their spectral densities to assess overall signal-to-noise ratio. Figure 1 shows the ratio of the spectrum of V_M to that of S_n for a generic asymmetric PEMF pulse (inset); $V_M(L,\omega)/[S_n(\omega)]^{1/2}$ increases with decreasing ω, which is characteristic of array response. For a single pulse (curve 1, Fig. 1), which has relatively low spectral amplitudes at low frequencies, there is a peak in the ratio about 1 kHz. This peak is higher at lower frequencies and eventually disappears as the burst count increases (20, 200 in Fig. 1).

The above analysis suggests that pulse EMF waveforms, with their sufficient pulse width *and* high burst count can easily satisfy signal-to-noise limitations in bone-tissue cell arrays by inducing a 1-10mV/cm peak electric field at the fracture site. (The effect of array length on SNR[4] reveals that adequate SNR is not obtained for a single cell at these amplitudes.) The PEMF signal chosen for this study (AME Spinal Stim®) appears to satisfy these criteria.

MATERIALS AND METHODS

We have developed a controlled bilateral fibula osteotomy in mature rabbits to test the effect of PEMF on bone repair.[12,13] Previous studies have shown that a reliable mechanical test (torsion to failure) can begin on post-operative day (POD) 8 in this model. In addition, the rate of maturation of callus formation is most rapid from day 8 through day 12.

External Coil Group

Bilateral fibula osteotomies were performed on 13 mature (3.5-5.0kg) white female New Zealand rabbits. The fibula was osteotomized by a stainless-steel circular blade 0.1 mm wide and 12.5 mm in diameter, attached to a dental handpiece rotating at 1000 rpm. An identical procedure was performed on the contralateral side. No fixation is required, since the fibula is a relatively nonweight-bearing bone, synostosed to the tibia proximally and distally, and receives only indirect mechanical stimuli.

Starting on POD 1, the active limb was treated 4 h/day by use of one small 4×6cm coil, up to sacrifice at POD 9. The fibula is approximately 2 mm in diameter and is at a depth of 3 mm from the skin. This configuration permits the use of a single coil small enough to maintain stray fields approximately 100 times lower at the level of the contralateral control osteotomy. During treatment, the rabbit was immobilized in a standard well-tolerated sling that extends the limbs and permits easy application of the coil. The limbs were maintained 6 in. apart by a 5.5in.-wide foam-rubber support. No magnetic shielding, anesthesia, or more severe restrains were required during the treatment session.[14] The induced current signal from the coil consisted of asymmetrical pulses (65/195 μs) applied in a 99-pulse burst, repeated at 1.5 Hz. The electric field at

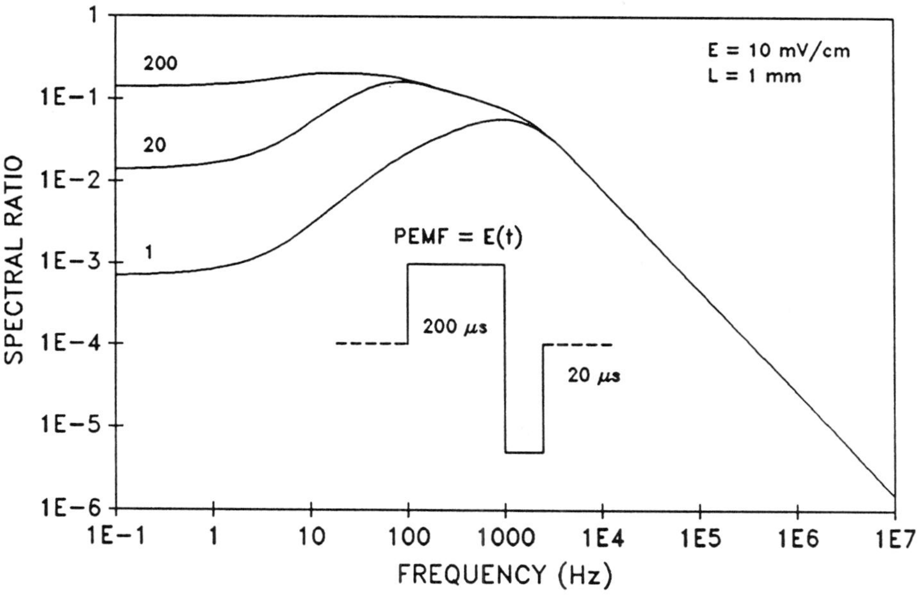

FIG. 1.—Number of pulses in a burst (burst width) affects amplitude ratio of induced transmembrane voltage to thermal noise for a physiologically relevant 1mm cell array (fracture site). This ratio approaches unity in the low-frequency range only for large burst counts. All curves are based on the generic asymmetrical PEMF signal with 200 and 20μs pulse widths in each polarity (inset). Peak electric field is 10 mV/cm for the 20μs pulse portion.

the osteotomy site was maintained at 3/1 mV/cm, corresponding to each pulse width, respectively. One active and one dummy saddle coil was anteriorly positioned over both the treated and contralateral control osteotomies. The rabbits were allowed unrestricted activity in U.S. Department of Agriculture approved cages between treatment sessions.

Implanted Electrode Group

Each of the 12 animals in this group had two stainless-steel mesh implanted electrodes (2×4 cm) flanking the osteotomy (medially and laterally) in each limb. After the completion of the osteotomy, which was performed as described above, the medial and lateral electrodes were sutured subcutaneously. Animals in this group were treated as described above. The amplitude of the input signal to the implanted electrodes was selected so that peak current levels in the tissue were approximately equivalent to that induced by the external coils, about 75/25 A, corresponding to the pulse width in each polarity, respectively.

Bilateral Control Group

Bilateral fibular osteotomies were performed on the 13 animals in this group as described for the external coil group. The 4h treatment was carried out in a standard laboratory sling, as for the external coil and implanted electrode groups. Dummy coils, identical in size and weight to those used in the active group, were placed anteriorly over each osteotomy.

Mechanical Testing Procedure

All fibulae were tested in torsion to failure at room temperature. The actuator was a DC servomotor (PMI JR16M4CH-1). Destructive testing was performed at a constant speed of 13 rad/s. Paired difference and paired percentage difference were evaluated for each treated fibula and its contralateral control. A paired one-tailed Student's t test was used to test the statistical significance of differences from intact strength for treated contralateral control and bilateral control osteotomies. For both the above cases, significance was accepted at $p \leq 0.05$.

RESULTS AND DISCUSSION

A previous study established the biomechanical healing curve for this osteotomy model.[13] Healing osteotomies were as strong as intact fibulae by POD 13. Maximal strength differences between treated and control limbs were apparent between POD 8 to 12. The results at POD 9 for both treatment modalities are shown in Fig. 2. The mean maximum torque for the coil treated limbs was 827 ± 75 cm, not significantly different from that for the mean intact value (924 ± 52 cm). The mean maximum torque for the coil group contralateral controls was 618 ± 32 cm, significantly different from both the treated and intact fibulae ($p < 0.01$). Four-hour bilateral control strengths had a mean average value of 623 ± 85 cm, significantly different from intact values ($p < 0.001$), but not significantly different from the maximum torque for the contralateral controls. In direct comparison with the above results, the implanted electrode group had mean values of 820 ± 56 cm for the treated and 639 ± 63 cm for the contralateral control fibulae. These values are significantly different from each other ($p < 0.01$). As in the coil group, the contralateral controls were not significantly different from the bilateral controls.

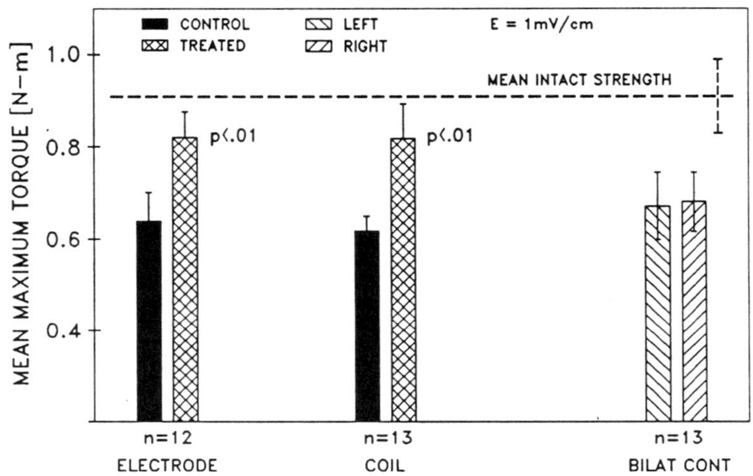

FIG. 2.—Comparison of the effect of broadband electrical signal used in this study to accelerate bone repair when applied via an external coil or an implanted electrode. Mean maximum torque of treated rabbit fibulae achieves intact value for both treatment modalities by POD 9. Contralateral controls are significantly weaker than treated side and not significantly different from bilateral controls at POD 9.

Comparison of bilateral controls with contralateral controls suggests that stray EMF, surgical bias, and systemic effects seem to play no role in the observed results. The base healing curves previously obtained for this circular-saw osteotomy model showed that mean intact strength is achieved at POD 13 for control fibulae. The results from both electrode- and coil-treated groups in this study appear to be identical, which shows that intact strength was achieved by POD 9, which represents approximately a 1.5× acceleration of acute (fresh) bone repair. Comparison of strength measurements with the actual biomechanical healing stage reveals that treated bones were in stage III or IV (i.e., callus as strong as or stronger than the bone shaft). In contrast, the majority of contralateral control osteotomies were in late stage 1 or early stage II (i.e., low stiffness, failure through the callus).

Although the use of PEMF to modulate the repair of delayed and nonunion fractures is extensive, the lack of substantial effects on fresh fracture healing has made it difficult to obtain quantitative assessments of PEMF bioeffects. The EMF sensitivity analysis presented here considers repairing bone tissue as a system of cells in gap-junction (short-circuit) contact. The transmembrane voltage to thermal noise threshold is defined on the basis of this model, which shows that it is significantly higher for cell arrays. It is suggested that the burst count of a pulsed electromagnetic field is one of the determining waveform parameters in the achievement of the induced transmembrane voltage at or above the thermal noise threshold. The Spinal Stim® waveform, used for lumbar spine fusions, was chosen for this study because it has a burst count at least 5 times that of commonly available delayed and nonunion fracture signals, at similar induced electric field levels. The result is further reinforced by a recent study on a similar rabbit model.[14] In that study, the generic PEMF waveform shown in Fig. 1 was employed in its asymmetrical and (by use of each polarity portion) symmetric forms. The authors concluded that 20μs symmetrical pulses with 250 burst count, vs the 200μs waveform with approximately 20 burst count, was most effective in causing an increase of three-point bending stiffness vs control. The authors also showed that the narrow (20μs) portion of the asymmetrical pulse was responsible for the observed bioeffects. This result follows the same theoretical predictions[3] and in vivo and in vitro experimental verifications by this Laboratory several years ago[15,16] for this induced electric-field signal.

The EMF sensitivity analysis presented here describes the manner by which a simple tissue structure increases the ratio of transmembrane voltage to applied *electric* field compared to that for an isolated single cell. This analysis was based on the assumption that the observed bioeffect is caused by an initial and perceptible change in transmembrane voltage. The presence of the magnetic field in the PEMF (coil) signal was considered to play only a carrier function, i.e., to obtain a tissue level electrical waveform having parameters designed to satisfy both electrochemical kinetics at the cell membrane (pulse widths), as well as the electrical requirements of basic cell-array tissue structure (burst count). Both coil and implanted-electrode treatment modalities caused a substantial (≈ 1.5×) acceleration of bone repair in this rabbit model. Care was taken to insure that both the wave shape and peak current values at the osteotomy site were as identical as possible for each modality. The lack of dependence of this bioeffect on the magnetic components of the coil signal ($dB/dt \approx 0.1$ G/μs, $B_{max} \approx 7G$, and $B_{avg} \approx 0.04G$) strongly suggests that the induced electric field is the bioeffective parameter.

The results of this study are somewhat better than those obtained with weak ultrasound,[12,13] which accelerated healing in this model to intact strength by POD 10, vs POD 9 in the present study. Comparison of these results with the healing curve described elsewhere[13] appears to show that the overall biomechanical healing curve has been accelerated. This is the first time, to the authors' knowledge, that an electrical signal appears to have caused an *overall* acceleration of bone repair (to torque endpoint) in a model relevant to the fresh-fracture situation. It is possible that this biophysical approach to dosimetry in a real tissue-repair situation can lead to significantly improved electrical bone-healing efficiency. Furthermore, the results to date strongly reinforce the original proposal by one of us[3] that proper *electrical* dosimetry at the cell membrane is the primary factor for EMF bioeffects.[17]

REFERENCES

1. Z. B. Friedenberg, M. C. Harlow, and C. T. Brighton, *J. Trauma* 11: 8831, 1971.
2. C. A. L. Bassett, A. A. Pilla, and R. J. Pawluk, *Clin. Orthop.* 124: 117, 1977.
3. A. A. Pilla, *N.Y. Acad. Sci.* 238: 149, 1974.
4. A. A. Pilla, P. R. Nasser, and J. J. Kaufman, in M. J. Allen, S. F. Cleary, A. E. Sowers, and D. D. Shillady, Eds., *Charge and Field Effects in Biosystems—3,* Boston: Birkhauser, 1992, 231-241.
5. A. A. Pilla, P. R. Nasser, and J. J. Kaufman, *Bioelectrochemistry and Bioenergetics* 30, 1992.
6. S. Caveney, *Ann. Rev. Physiol.* 47: 319-335, 1985.
7. S. B. Doty, *Calcif. Tissue Int.* 33: 509-512, 1981.
8. W. R. Adey, *Neurochem. Res.* 13: 671, 1988.
9. W. H. Fletcher, W. W. Shiu, D. A. Haviland, C. F. Ware, and W. R. Adey, *Proc. BEMS* (8th Annual Mtg., Madison, Wis.), 1986, p. 12.
10. J. C. Weaver and R. D. Astumian, *Science* 247: 459, 1990.
11. L. J. DeFelice, *Introduction to Membrane Noise,* New York: Plenum Press, 1981, 243-245.
12. A. A. Pilla, M. A. Mont, P. R. Nasser, S. A. Khan, M. Figueiredo, J. J. Kaufman, and R. S. Siffert, *J. Orthop. Trauma* 4: 246-253, 1990.
13. A. A. Pilla, M. Figueiredo, P. R. Nasser, J. M. Alves, J. T. Ryaby, M. Klein, J. J. Kaufman, and R. S. Siffert, in C. T. Brighton and S. R. Pollack, Eds., *Electromagnetics in Biology and Medicine,* San Francisco: San Francisco Press, 1991, 331-341.
14. D. Pienkowski, S. R. Pollack, C. T. Brighton, N. J. Griffith, *J. Orthopaedic Res.* 10: 247-255, 1992.
15. J. Assailly, J.-D. Monet, Y. Goureau, P. Cristel, and A. A. Pilla, *Bioelectrochemistry and Bioenergetics* 8: 515, 1981.
16. G. Colacicco and A. A. Pilla, *Bioelectrochemistry and Bioenergetics* 10: 119, 1983.

17. The authors are deeply grateful to Kenneth Hess, Osei Bediako, and Kingsley Gyebi for their untiring dedication to this work, and to American Medical Electronics, Richardson, Tex., for its substantial technical and financial support.

Martin Blank, Ed., *Electricity and Magnetism in Biology and Medicine*

ZETA POTENTIAL OF HUMAN BONE AND ITS RELATIONSHIP TO AGE

B. A. Monaghan,* S. R. Pollack, T. A. Corcoran, and M. E. Steinberg

University of Pennsylvania, Philadelphia

We examined the zeta potential in 73 human femoral neck specimens from 66 patients undergoing total joint arthroplasty. We found a highly statistically significant correlation between increasing magnitude of zeta potential and increasing age ($r = 0.58; p = 0.0001$). Our patient population was normal with respect to bone mineral density compared to age matched controls. We conclude that the zeta potential of bone increases during the period of adulthood and may have an adaptive role in age related changes of bone.

Electrokinetic theory states that stress generated potentials are related to the zeta potential present at the bone-fluid interface. Bassett has hypothesized that these stress generated potentials may have a role in bone remodeling.[1,2] The zeta potential of bone has been previously shown to be a dynamic property which varies with age and several models of osteoporosis in animals.[3,4] Although studies have been performed to assess the zeta potential in human bone, they have primarily focused on manipulation of the chemical composition, conductivity, and viscosity of the fluid to examine their effect on the zeta potential.[5] A preliminary study of seventeen human femoral necks by Corcoran et al. found trends of increasing magnitude of zeta potential with decreasing Singh index and thinning femoral cortices, although neither reached statistical significance.[6] The purpose of this study, then, is to augment the existing human database in an effort to determine whether statistically significant changes occur in the zeta potential with respect to patient characteristics and radiographic indices of osteoporosis. In addition, a subset of patients will also undergo dual-energy x-ray absorptiometry (DEXA) scans to assess bone mineral density. Exploring the zeta potential in human bone may provide evidence in support of Bassett's hypothesis.

MATERIALS AND METHODS

Eighty femoral neck specimens from 73 patients undergoing primary total hip arthroplasty for a variety of diagnoses were harvested at the time of surgery and frozen at −20 C. Within 24 h they were placed in Neuman's fluid, a solution which mimics the ionic composition and specific conductivity of bone interstitial fluid, and were transferred to a −80 C freezer. Within a month of surgery, each specimen was thawed and a $1 cm^3$ segment of cortical bone from the medial femoral neck (calcar femorale) was processed for electrophoretic mobility determination.[7] The specimen was loaded into a Penn Kem S3000 electrokinetic analyzer for determination of pH, specific conductivity, electrophoretic mobility, and a calculated zeta potential. Each specimen was run in triplicate.

The trabecular bone pattern of the proximal femur on preoperative anteroposterior radiograph of the hips and pelvis was assessed by assignment of a Singh index. By a variation of the method of Garn et al.,[9,10] the cortical thickness index (CTI) was calculated at a point 5 cm below the lesser trochanter after the subcortical (C) and medullary (M) diameters were measured in duplicate with Vernier calipers:

$$\mathrm{CTI} = (\mathrm{C} - \mathrm{M})/\mathrm{C}$$

A clinical impression of the bone stock as normal for age or osteoporotic was provided by the operating surgeon (MES). After approval by the Institutional Review Board and informed consent, 15 patients underwent preoperative DEXA scans in which bone mineral density of the affected femoral neck, Ward's triangle, and the greater trochanter was measured and a comparison was made to age-matched controls.

---

*Correspondence to: B. A. Monaghan (Orthopaedic Surgery), University of Pennsylvania, Philadelphia PA 19104.

Statistical analysis, consisting of ANOVA with a Scheffe F-test and simple and multiple linear regressions, was performed by means of a statistical software package (Statview II®). All values are expressed as mean ± standard deviation.

RESULTS

Of the eighty specimens processed for zeta potential determination, seven were excluded due to contamination of the specimens with marrow elements. The remaining 73 specimens of 66 patients were included in this analysis. Preoperative radiographs were suitable for evaluation of the Singh index in 53 instances. A cortical thickness index was calculated in 67 cases. Seventeen specimens (15 patients) had DEXA scan information.

The average age of the patients at the time of operation was 58 ± 16 years with a range from 25 to 94 years. Twenty-eight specimens were from males and 45 from females. Thirty-three specimens were from the left hip and 40 were from the right hip. Four specimens were from patients with a femoral neck fracture, 30 were from patients with a diagnosis of avascular necrosis (AVN), and 39 specimens came from patients with degenerative joint disease (DJD).

The average zeta potential for the entire group was −5.81 ± 2.10 mV (range −2.09 to −10.38). There was no difference in zeta potential of the population with respect to gender ($p = 0.43$) or side of the specimen ($p = 0.23$). When the specimens were grouped by decade there was a trend toward increasing zeta potential magnitude (more negative) with increasing age (Fig. 1). This result was confirmed by linear regression, which demonstrated a statistically highly significant positive correlation between increasing zeta potential magnitude and advancing age ($r = 0.58, p = 0.0001$). There was also a correlation between decreasing cortical thickness index (cortical thinning) and increasing zeta potential magnitude ($r = 0.363, p = 0.0023$). No statistically significant difference could be found when the specimens were segregated according to clinical assessment ($p = 0.91$) or Singh index. The Singh index was significantly lower in patients who were judged to be osteoporotic than in those felt to have normal bone mass ($p = 0.011$ by Mann-Whitney test). In addition, fracture patients also had a lower Singh index than DJD or AVN patients when compared by a Kruskal-Wallis test.

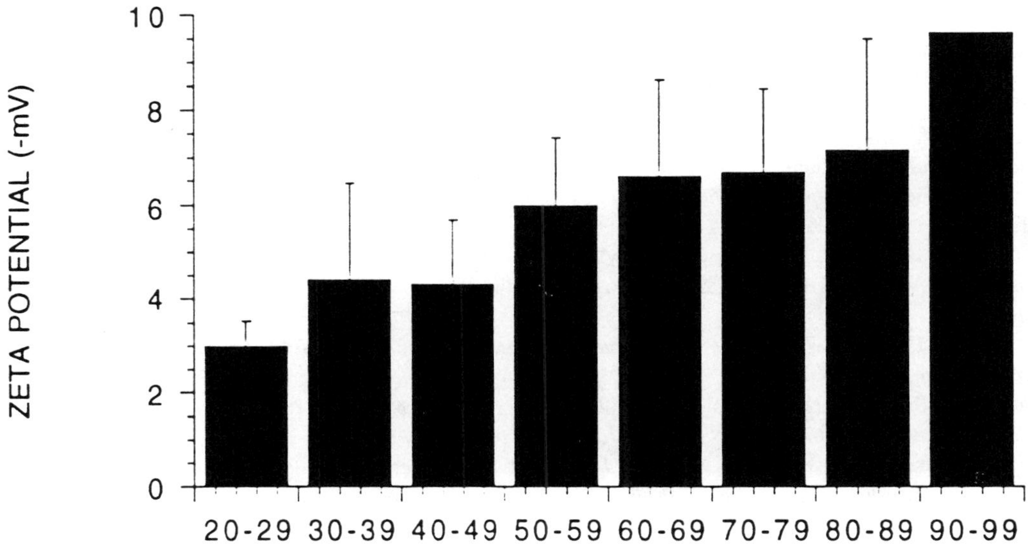

FIG. 1.—Zeta potential of human cortical bone by age in decades (mean ± S.D.).

Analysis of variance revealed that the zeta potential of bone from the avascular necrosis group was significantly smaller in magnitude than that of patients with either degenerative joint disease or femoral neck fracture (AVN: −4.75 ± 1.88 mV; DJD: −6.436 ± 1.88 mV; fracture: −7.70 ± 2.40 mV [$p < 0.05$]). However, the avascular necrosis patients were significantly younger than the other two populations (AVN =

48 ± 16 y; DJD = 63 ± 11 y; fracture 78 ± 11 y [$p < 0.05$]).

Simple linear regression between age and cortical thickness index further revealed a positive correlation, which may have been reflected in the correlation found between CTI and the zeta potential ($r = 0.407, p = 0.0006$). Fracture patients had a lower CTI (0.313 ± 0.113) than either the DJD (0.445 ± 0.070) or AVN (0.447 ± 0.077) patients ($p < 0.05$). The cortical thickness index was also significantly different when the patients were segregated by clinical assessment (CTI of normal group = 0.475 ± 0.068 vs CTI of osteoporotic group = 0.380± 0.072; $p = 0.0001$).

Linear regression was also performed between the bone mineral density as determined by dual energy x-ray absorptiometry of the three points in the affected proximal femur and the zeta potential. Bone mineral density (BMD) of the femoral neck revealed a trend toward a statistically significant correlation with increasing magnitude of zeta potential and decreasing BMD ($r = 0.463, p = 0.0613$). Decreasing bone mineral density of the greater trochanter yielded a statistically significant correlation with the increasing magnitude of the zeta potential ($r = 0.64, p = 0.0056$). Bone mineral density in Ward's triangle did not correlate with zeta potential ($r = 0.293, p = 0.2374$). The bone mineral densities in all three regions of the proximal femur was not significantly different from that of age matched controls.

DISCUSSION

The aging process in bone leads to a myriad of changes that serve to alter mineral composition and distribution, protein, composition, and charge distribution.[11] These parameters are all determinants in the magnitude of the zeta potential. Our work demonstrates that the zeta potential of human cortical bone is a property that increases by as much as a factor of 3 during adult aging. This result was in fact predictive, given the findings in two animal models of aging and strengthens their use for studying this phenomenon.[3] This increase with age occurred in a normal population because there was no evidence of osteoporosis when compared to age matched controls. The physiologic significance of these age related changes remains uncertain. Given the fact that the zeta potential represents the stimulus for the response of an endogenous electrical potential in bone, this enhanced sensitivity of the system may represent an adaptive mechanism to the process of senescence in human bone.

REFERENCES

1. C. Bassett, "Electrical effects in bone," *Scientific American* 213: 18-25, 1965.
2. C. Bassett, "Biological significance of piezoelectricity," *Calcif. Tissue Res.* 1: 252-272, 1968.
3. C. B. Craythorne, S. R. Pollack, and R. M. Kowalchuk, "The zeta potential in several models of osteoporosis of the rat," *Trans. BRAGS* 10: 53-59, 1990.
4. A. A. Spirt, "Age related changes in the zeta potential of bone and its influence upon endogenous field strength," this volume.
5. D. Pienkowski and S. R. Pollack, "The origin of stress generated potentials in fluid saturated bone," *J. Ortho. Res.* 30-41, 1983.
6. T. A. Corcoran, R. M. Kowalchuk, M. E. Steinberg, and S. R. Pollack, "The zeta potential of human femoral neck and its relationship to osteoporosis: A preliminary study of seventeen patients," Trans. BRAGS 11: 58, 1991.
7. R. M. Kowalchuk, *The Zeta Potential of Normal and Osteoporotic Bone, Ph.D. thesis, University of Pennsylvania, 1991.*
8. M. Singh, A. Nagrath, and P. Maini, "Changes in the trabecular pattern of the upper end of the femur as an index of osteoporosis," *J. Bone Joint Surg.* 52A: 457-467, 1970.
9. S. Gam, *The Earlier Gain and Later Loss of Cortical Bone,* Springfield, Ill.: C. C. Thomas, 1970.
10. S. Garn, A. Poznanski, and J. Nagy, "Bone measurement in the differential diagnosis of osteopenia and osteoporosis," *Radiology* 100: 509-518, 1971.
11. M. D. Grynpas and D. Holmyard, "Changes in the quality of bone and mineral in aging and disease," *Scan. Microscopy* 2: 1045-1054, 1988.

Martin Blank, Ed., *Electricity and Magnetism in Biology and Medicine*

THE EFFECT OF COMBINED AC/DC MAGNETIC FIELDS ON RESTING ARTICULAR CARTILAGE METABOLISM

J. T. Ryaby,* F. P. Magee, and A. M. Weinstein

OrthoLogic Corp., Phoenix, Ariz.

D. A. Grande

North Shore University Hospital/Cornell Medical College, Manhasset, N.Y.

Destruction of human articular cartilage, caused by trauma, disease, or wear, is the cause of significant human suffering, disability, and healthcare expenditures. As the average age of the American population continues to increase, the costs arising from degeneration of human articular joints will also continue to increase. Present therapies for damaged or diseased cartilage involve relief of symptoms (pain and inflammation) via pharmaceuticals or operative procedures for joint repair or replacement.[1] Growth factor therapies are at present being investigated but will require years of additional study before they may become a viable treatment option.

Therapeutic electromagnetic stimulation was first applied to un-united fractures to imitate the endogenous stress-generated potentials exhibited by living bone.[2] Articular cartilage is a hydrated load-bearing tissue that also exhibits endogenous electrical potentials upon mechanical loading. Accordingly, application of electrical fields has been demonstrated to stimulate growth of both articular and growth plate cartilage.[3,4] Theoretical mechanisms have been proposed for interaction of electromagnetic fields on cells, with the cell membrane proposed as the site of interaction. Research investigating the effect of low-frequency, low-energy electromagnetic stimulation on biological systems indicates an interaction between alternating and static magnetic fields. The theory described by Liboff and McLeod portrays this interaction as modulating ion transport across cell membranes to achieve its biological effect.[5,6] Liburdy et al. reported effects of combined AC and DC fields to stimulate 45 Ca incorporation in Con A-stimulated rat thymocytes, demonstrating an inhibition of Con A stimulation back to control values.[7] This inhibition was noted at an angular frequency of 16 Hz, which supports the paramagnetic resonance theory of Lednev.[8] Fitzsimmons et al. reported that combined AC/DC fields could induce cell proliferation and increase insulin-like growth factor II receptor expression in a clonal osteoblast cell line.[9,10] In cartilage cultures, Grande et al. demonstrated effects on cell proliferation and matrix biosynthesis in articular cartilage with AC/DC magnetic fields.[11] These results indicate biological effects of combined AC/DC magnetic fields in thymocyte, bone cell, and cartilage model systems, but further work is necessary to refine theoretical models that accurately predict calcium-dependent biological effects.

The present study is an investigation of the effect of specific combinations of AC/DC magnetic fields to stimulate extracellular matrix biosynthesis by articular cartilage and compare these responses to the effect of chondrotropic growth factors. These combined AC/DC magnetic fields are shown to increase matrix biosynthesis in a manner similar to that observed with chondrotropic growth factors.

CELLULAR MODEL SYSTEM

Full-thickness plugs of articular cartilage were harvested from the knee joints of young adult bovine calves. The samples were trephined into 4.0mm diameter cylindrical plugs (0.5 mm thick) and placed one plug/well in 24-well culture plates (Nunc) in HF12/DMEM medium, a low serum formulation designed to maintain the steady-state metabolic activity of the cartilage tissue without inducing mitotic activity.[12] It consists of a 1:1 mixture of Hams F-12 and DMEM supplemented with 0.3% calf serum, 40 μg/ml ascorbate, 4 mM glutamine, 5 μg/ml transferrin, 60 units/ml penicillin, 60 units/ml streptomycin, and 30 units/ml

*Correspondence to: J. T. Ryaby, OrthoLogic Corp., 2850 S. 36th St., Phoenix AZ 85034.

mycostatin. Cartilage was allowed to equilibrate 48 h in vitro prior to initiation of the experiments. The remaining cartilage samples were enzymatically digested with DNAse/collagenase, filtered through Nitex (120 μm), and plated at 50000 cells/cm^2. All plates were incubated at 37 C in an atmosphere of 5% CO_2 and 100% humidity.

MAGNETIC FIELD EXPOSURE APPARATUS AND TECHNIQUE

The multiwell plates were placed in the center of a round (30cm diameter) Helmholtz coil pair made from 400 turns of #28 AWG magnet wire. The coils were oriented horizontally (i.e., coplanar to the tissue culture plate) and produced a vertically oriented magnetic field. The combined AD/DC magnetic fields were generated by a function generator (BK Precision model 3011A); magnetic field measurements were made with a Schonstaedt Model 2220 digital magnetometer. AC and DC magnetic fields were created by the same pair of Helmholtz coils and were therefore parallel. The combined AC/DC magnetic field conditions utilized a DC magnetic field of 0.2 G with a superimposed sinusoidal 0.4G AC magnetic field at 15.3, 25.4, or 76.6 Hz.

The cultures were exposed to combined AC/DC magnetic fields, recombinant growth factors, or combinations of both for the indicated times. Magnetic field exposures were performed for 30 min/day for 24, 48, or 72 h.

PROTEOGLYCAN BIOSYNTHESIS

Chondrocyte cultures were labeled with $Na^{35}SO_4$ in serum-free medium for the last 4 h of culture. The cultures were washed twice with ice-cold 10mM EDTA/0.1 M sodium phosphate, pH 6.5, and the labeled proteoglycans were extracted in 4 M Guanidine-HCl/2% Triton X-100 in 50 mM NaAc, pH 5.6, overnight at 4 C. Labeled proteoglycans were separated from 35-S labeled proteins and unincorporated label by gel filtration on Sephadex G-25 (PD10) columns.[13,14] Samples of the fractions were quantitated by liquid scintillation counting; the remaining aliquots were used for DNA analysis. DNA determination was performed by the bis-benzimidazole (Hoechst 33528) assay.[15] Data are expressed as cpm/microgram DNA.

Fischer's F-test was used to test the homogeneity of the variances of all plates. Means were compared by ANOVA and Student's *t*-test, with the Bonferroni criteria used for multiple group comparisons.

RESULTS

Experiments were performed to compare the effect of short-term AC/DC magnetic-field exposure and recombinant growth factor stimulation on proteoglycan (PG) biosynthesis in both explant and monolayer cell cultures (Fig. 1). In explant cultures, exposure to the 15.3Hz combined AC/DC magnetic fields demonstrated no increase in proteoglycan biosynthesis at 24 h, but an increase of 42% ($p < 0.05$) was observed at 72 h. This same time lag was observed with insulin-like growth factor 1 (IGF1), which increased PG biosynthesis slightly (not statistically significant). In contrast, both basic fibroblast growth factor (bFGF) and transforming growth factor-beta 1 (TGF-B1) increased PG biosynthesis at 24 h by 30% and 55%, respectively, with the effect returning to control levels at 72 h. In isolated chondrocytes, TGF-B1 decreased PG biosynthesis in subconfluent cultures by 41% at 48 h; exposure to the combined AC/DC magnetic fields reversed this inhibition. Total reversal of the inhibition was observed with the 76.6Hz combined AC/DC signal.

DISCUSSION

These results indicate that combined AC/DC magnetic fields have the ability to stimulate matrix production in resting articular cartilage (as assessed by quantification of proteoglycan biosynthesis). These results are qualitatively similar to those observed with the chondrotropic agents TGF-B1 and bFGF, with the growth-factor response occurring more rapidly. The results with TGF-B1 are similar to that reported in the literature, which demonstrated that the effect of TGF-B1 shifts the balance between matrix anabolism/catabolism in favor of anabolism.[16]

Current effort is focused on long-term studies comparing the response of explant cultures to IGF1 and combined AC/DC magnetic fields. The role of IGF in the response to combined AC/DC magnetic fields has been proposed as the mechanistic mediator of tissue level response by Fitzsimmons et al.[9,10] These results indicate that modulation of growth factor response may be modulated by combined AC/DC magnetic field exposure and suggest potential therapeutic application in cartilage disease states.

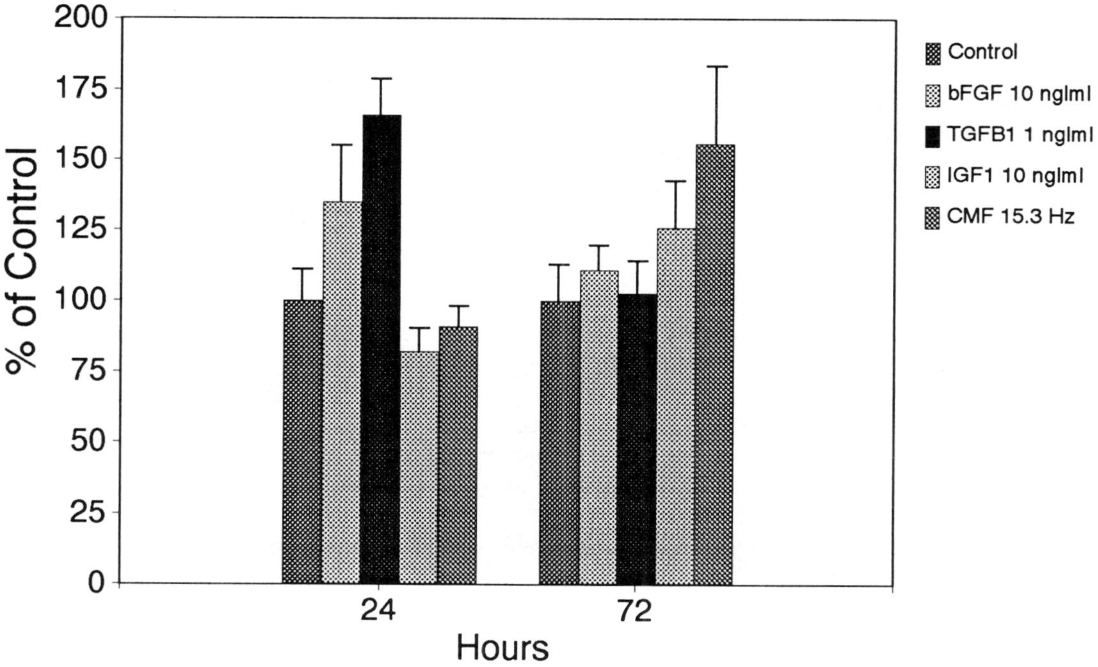

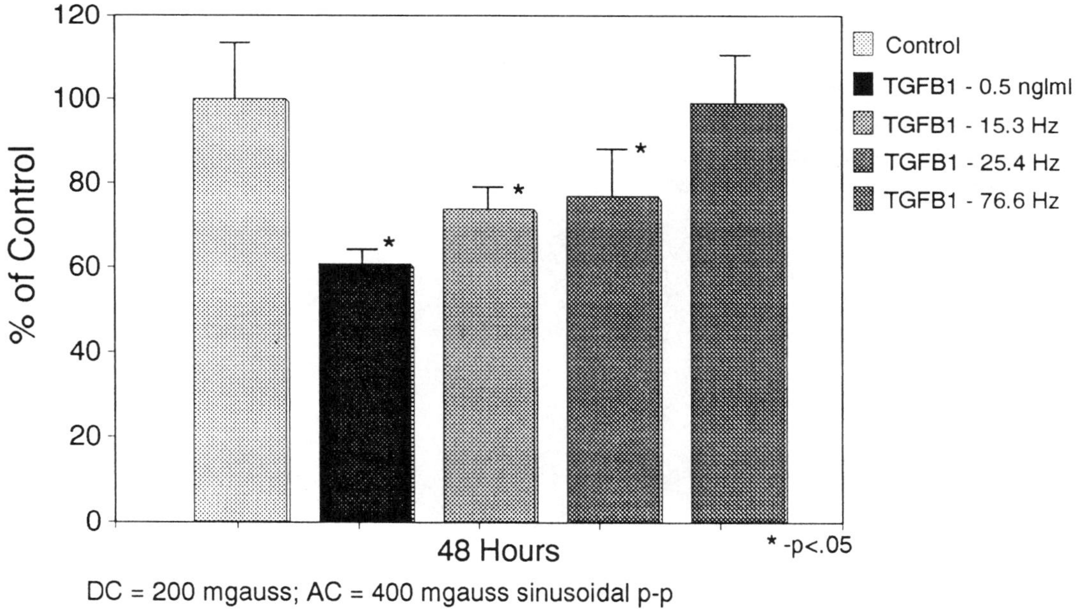

FIG. 1.—(a) Effect of growth factors and combined AC/DC magnetic fields on proteoglycan biosynthesis in explant cultures. (b) Modulation of transforming growth factor beta-1 mediated inhibition of proteoglycan biosynthesis by combined AC/DC magnetic fields in chondrocyte cultures. Groups are identified in legends and in text. Data are expressed as percentage above control.

REFERENCES

1. P. Dieppe, "Osteoarthritis: Clinical and research perspective," *Brit. J. Rheum.* 30(Suppl. 1): 1, 1991.

2. C. A. L. Bassett, "Fundamental and practical aspects of therapeutic uses of pulsed electromagnetic fields (PEMFs)," *CRC Crit. Rev. Biomed. Eng.* 17: 451, 1989.

3. J. P. Elliot, R. L. Smith, and C. Block, "Time varying magnetic fields: Effects of orientation on chondrocyte proliferation," *J. Orthop. Res.* 6: 2159, 1988.

4. C. T. Brighton, G. B. Pfeffer, and S. R. Pollack, "In vivo growth plate stimulation in capacitively coupled electric fields," *J. Orthop. Res.* 1: 42, 1983.

5. A. R. Liboff, "Cyclotron resonance in membrane transport," in A. Chiabrera, C. Nicolini, and H. P. Schwan, Eds., *Interactions Between Electromagnetic Fields and Cells,* New York: Plenum Press, 1985, 281-296.

6. B. R. McLeod and A. R. Liboff, "Cyclotron resonance in cell membranes: The theory of the mechanism," in M. Blank and E. Findl, Eds., *Mechanistic Approaches to Interaction of Electric and Electromagnetic Fields with Living Systems,* New York: Plenum Press, 1987, 89-108.

7. R. P. Liburdy, M. G. Yost, and D. J. de Manincor, "Signal transduction in the lymphocyte: Comparison of 60 Hz magnetic fields and combinations of AC/DC magnetic fields on mitogen-activated calcium influx," *Proc. BRAGS* 11: 35, 1991.

8. V. V. Lednev, "Possible mechanisms for the influence of weak magnetic fields on biological systems," *Bioelectromagnetics* 12: 71, 1991.

9. R. J. Fitzsimmons, J. T. Ryaby, F. P. Magee, and D. J. Baylink, "IGFII receptor number is increased in TE-85 cells by low amplitude, low frequency electromagnetic field exposure," *Bone Mineral Res.* (submitted for publication).

10. R. J. Fitzsimmons, J. T. Ryaby, A. M. Weinstein, and D. J. Baylink, "EMF stimulation increased IGF-II receptors in bone cells," this volume.

11. D. A. Grande, F. P. Magee, A. M. Weinstein, and B. R. McLeod, "The effect of low energy combined AC and DC magnetic fields on articular cartilage metabolism," *Proc. N.Y. Acad. Sci.* (in press).

12. K. D. Osborn, S. B. Trippel, and H. J. Mankin, "Growth factor stimulation of adult articular cartilage," *J. Orthop. Res.* 7: 35, 1989.

13. S. Vukicevic, F. P. Luyten, and A. H. Reddi, Stimulation of the expression of osteogenic and chondrogenic phenotypes in vitro by osteogenin," *Proc. Natl. Acad. Sci. USA* 86: 8793, 1989.

14. M. Yangishita, R. J. Midura, and V. C. Hascall, "Proteoglycans: Isolation and purification from tissue cultures," *Meth. Enzymol.* 138: 279, 1987.

15. C. Labarca and K. Paigen, "A simple, rapid, and sensitive DNA assay procedure," *Anal. Biochem.* 102: 344, 1980.

16. T. I. Morales, "Transforming growth factor beta-1 stimulates synthesis of proteoglycan aggregates in calf articular cartilage organ cultures," *Arch. Biochem. Biophys.* 286: 99, 1991.

Martin Blank, Ed., *Electricity and Magnetism in Biology and Medicine*

EFFECT OF ELECTROMAGNETIC-FIELD STIMULATION ON AVASCULAR NECROSIS OF THE FEMORAL HEAD

M. Bagnacani, G. Borsalino,* and F. Fornaciari

Ospedale Civile, Montecchio Emilia, Italy

We have investigated the effect of pulsed electromagnetic field (PEMF) stimulation in patients suffering from avascular necrosis (AVN) of the femoral head. Only Ficat stages I, II, and III were admitted to the study. The presence of AVN was diagnosed by x rays, scintigraphy, and CT. Patients were instructed to use the bone growth stimulator (Igeastimulator, Howmedica) 8 h/day for 6 months. Since 1988 we have treated 26 hips (22 patients). Considering the positive results in Ficat I and II stages, the patients' acceptance, and the possibility of a later surgical approach, PEMF stimulation should always be considered in the treatment of these patients.

We have investigated the effect of low-frequency, low-energy pulsing electromagnetic fields since 1987. We demonstrated, in a double-blind study, the positive effect of PEMF on the healing of femoral intertrochanteric osteotomies.[1] Bassett[2] and Aaron[3] have reported that electromagnetic stimulation can be successfully used in patients suffering from avascular necrosis (AVN) of the femoral head. AVN is a pathological condition often associated with a number of etiological factors,[4] which lead to a vascular deficit of the femoral head resulting in bone necrosis. The natural course of the disease is progressive: usually the bone collapses and an osteoarthrosis develops.[5] Several surgical treatments have been proposed for AVN patients; the most used were core decompression[6] and intertrochanteric osteotomy. Our experience relates to the intertrochanteric osteotomy, whose results are satisfactory.[7] However, as the disease often occurs in young adults, a noninvasive effective treatment, if available, is to be preferred. This opportunity has become even more important since new diagnostic techniques, such as nuclear magnetic resonance, have been adopted, which makes it very difficult to decide whether to operate on patients diagnosed at very early stages of AVN. Here we report our experience with the use of PEMF for the treatment of patients suffering from AVN.

MATERIALS AND METHODS

There were 22 patients for a total of 26 hips, 18 males and 4 females. Their average age was 42, range 30-57. The diagnosis of AVN was based on x-ray image and computed tomography (CT) scan, as well as scintigraphic examination. All x rays were rated according to Ficat staging.[8] Ficat 0 refers to very early stages of AVN, Ficat 4 to the typical osteoarthrosis roentgenographic image. Patients with Ficat I, II, and III were admitted to the protocol study. Patients were put in a nonweight-bearing regimen; however, if the patient was pain free, full or partial weight-bearing was allowed. Electromagnetic stimulation was performed by means of the stimulator supplied by Igea. The magnetic field was generated by a single coil. The stimulation parameters are: frequency 75 Hz, pulse width 1.3 ms, amplitude of electric field induced in a coil probe 2 mV.[1] Patients were instructed to use the stimulator for 8 h every day, usually during the night. A clock inside the device allowed us to check patient compliance. Two patients were considered to be noncompliant, as their average daily stimulation time was less than 2 h. Patients who discontinued the stimulation within 30 days are not included in this analysis. Patients were seen every 6 months when x rays and CT scans were taken. From the second year after the start of the stimulation, patients were seen once a year. During the study, some patients underwent hip surgery. The decision to operate was based on roentgenographic evidence of AVN progression and on clinical evaluation, presence of pain, and inadequate patient performance (based on Harris's hip score).

---

*Correspondence to: G. Borsalino (Orthopaedics), Ospedale Civile, Montecchio Emilia (RE) Italy.

RESULTS

Twenty-six hips have been stimulated; according to the Ficat rating we treated 7 patients with Ficat I, 10 with Ficat II, and 9 with Ficat III (Table 1).

TABLE 1.—Initial roentgenographic data.

	Ficat I	Ficat II	Ficat III
No hips	7	10	9

Patients have been followed for an average period of 2.6 years (Table 2).

TABLE 2.—Roentgenographic data at follow-up.

	Ficat I	Ficat II	Ficat III	Ficat Surgery
No hips	7	7	2	10
Months of follow-up	43	34	36	21

Average follow-up: 2.6 years (1-4)

At follow-up PEMF stimulation was able to prevent AVN progression in most of Ficat I and II patients. In 2 Ficat II patients we observed almost complete bone recovery, and at follow-up the AVN was rated I. Almost all patients in the Ficat III rate worsened (Table 3).

TABLE 3.—Hip exhibiting roentgenographic progression.

	Ficat I	Ficat II	Ficat III
No hips	2	3	4
	28%	30%	47%

Ten patients underwent surgery within 21 months from the start of the stimulation; 7 of them were in the Ficat III group since the start of the stimulation (Table 4).

TABLE 4.—Hip survival according to Ficat stage.

	Ficat I	Ficat II	Ficat III
No hips	7	10	9
No hips operated	0	3*	7

*Includes 2 noncompliant patients.

The decision to operate on these patients was based in most of the cases on the poor clinical results.

DISCUSSION

The results we have obtained are good for Ficat I and Ficat II patients. For Ficat III patients the results are not satisfactory, especially in consideration of the poor clinical results requiring hip surgery. Our data confirm those reported in the literature.[2,3]

We think that PEMF stimulation should always be preferred for the treatment of early Ficat stages. This therapy is extremely important considering that the recent introduction of nuclear magnetic resonance imaging permits diagnosis of AVN at Ficat 0 stage, the treatment of which by surgery is controversial to say the least.

REFERENCES

1. G. Borsalino, M. Bagnacani, E. Bettati, G. Fornaciari, R. Rocchi, S. Uluhogian, G. Ceccherelli, R. Cadossi, and G. Traina, "Electrical stimulation of human femoral intertrochanteric osteotomies: Double blind study," *Clin. Orthop. Rel. Res.* 237: 256-263, 1988.

2. C. A. L. Bassett, M. Schink-Ascani, and S. M. Lewis, "Effects of pulsed electromagnetic fields on Steinberg ratings of femoral head osteonecrosis," *Clin. Orthop. Rel. Res.* 246: 172-185, 1989.

3. R. K. Aaron, D. Lennox, G. E. Bunce, and T. Ebert, "The conservative treatment of osteonecrosis of the femoral head. A comparison of core decompression and pulsing electromagnetic fields," *Clin. Orthop. Rel. Res.* 249: 209-218, 1989.

4. H. J. Griffiths, "Etiology, pathogenesis, and early diagnosis of ischemic necrosis of the hip," *JAMA* 246: 2615, 1981.

5. M. E. Steinberg, C. T. Brighton, A. Corces, G. D. Hayken, D. N. Steinberg, B. Strafford, S. E. Tooze, and M. Fallon, "Osteonecrosis of the femoral head: Results of core decompression and grafting with and without electrical stimulation," *Clin. Orthop. Rel. Res.* 249: 199-208, 1989.

6. D. S. Hungerford, "Bone marrow pressure, venography, and core decompression in ischemic necrosis of the femoral head," *The Hip: Proc. 7th OPNE Sci. Meeting of the Hip Society*, St. Louis: Mosby, 1979, p. 218.

7. G. Borsalino, E. Bettati, F. Fornaciari, R. Rocchi, and S. Uluhogian, "Il nostro indirizzo nella esecuzione della osteotomia nella coxartrosi," *Clinica Ortopedica* 27: 103-118, 1988.

8. R. D. Ficat, "Idiopathic bone necrosis of the femoral head: Early diagnosis and treatment," *J. Bone Joint Surg.* 67(B): 3, 1985.

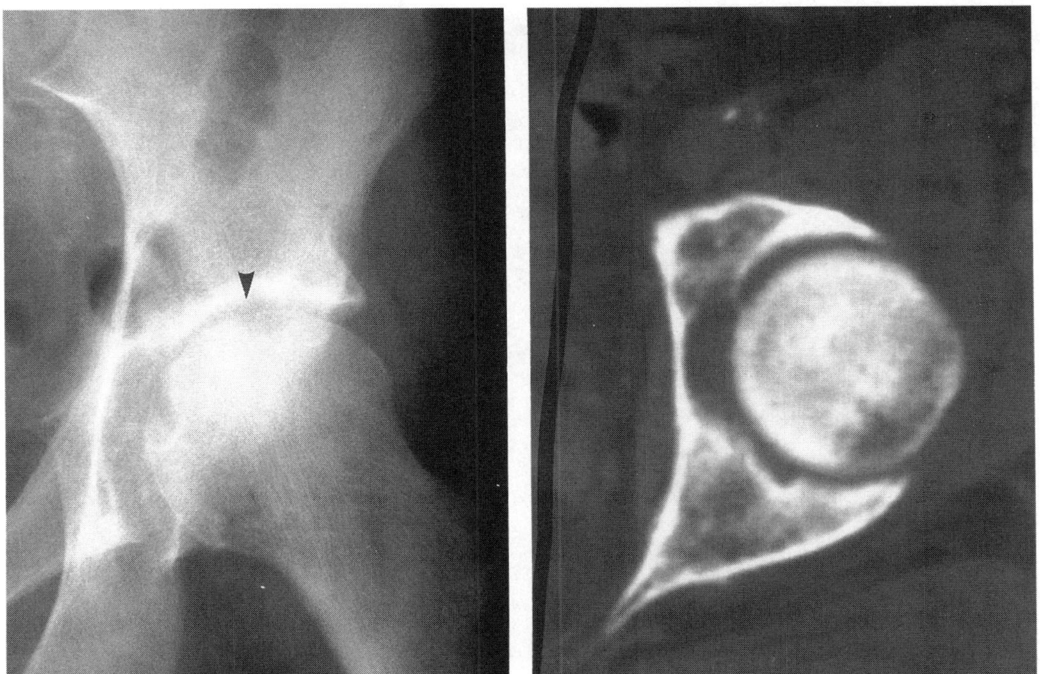

FIG. 1.—B.L. male aged 55. Left roentgenography, right CT scan of AVN. Rated Ficat II.

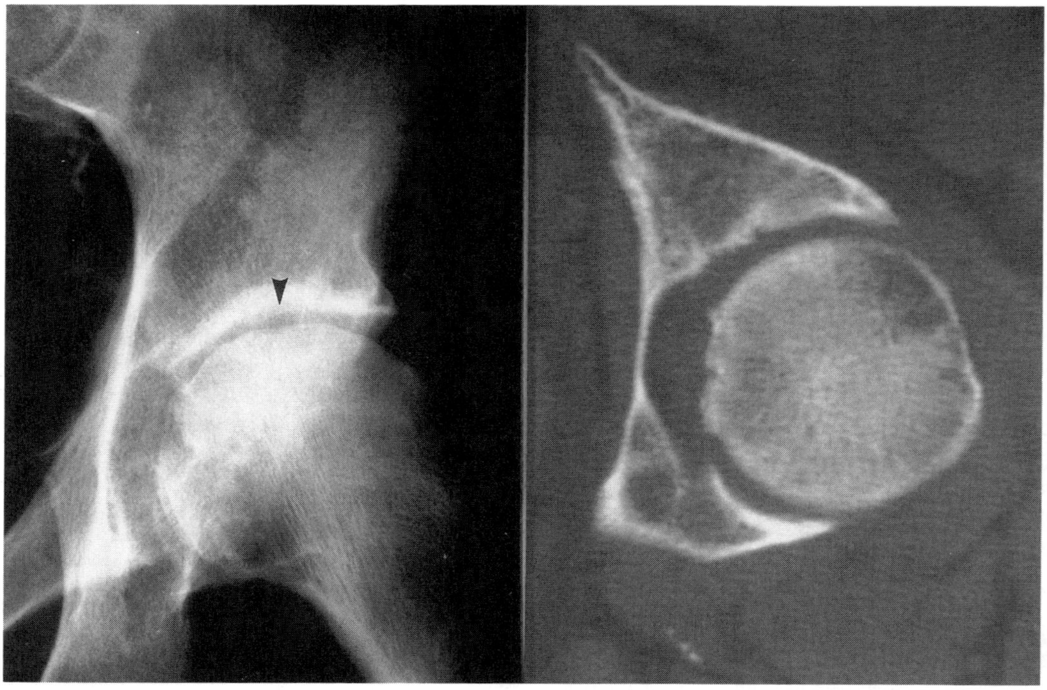

FIG. 2.—The same patient after 6 months of PEMF stimulation and 18 months of follow-up. No signs of progression of the disease, neither at the roentgenography (left) nor at the CT scan (right). The patient also had an excellent clinical result.

Martin Blank, Ed., *Electricity and Magnetism in Biology and Medicine*

EXPERIMENTAL STUDIES OF BRAIN TUMOR DEVELOPMENT DURING EXPOSURE TO CONTINUOUS AND PULSED 915MHz RADIOFREQUENCY RADIATION

L. G. Salford,* A. Brun, and J. Eberhardt

Lund University, Sweden

The effects of 915MHz microwaves, both as continuous waves (1 W) and modulated with 4, 8, 16, and 200 Hz in a rat brain glioma model were investigated. By stereotaxic technique 5000 RG2 cells were injected into the head of the right caudate nucleus in 62 Fischer 344 rats and 62 matched controls. The animals were kept unanesthetized in TEM cells and exposed for 7 h/day for 5 days a week during 2-3 weeks. Our study does not show any significant difference in tumor size between animals exposed to 915MHz microwaves and controls.

Several recent publications report an increased risk of leukemia and brain tumors among people living close to high power electrical wirings or in electrical occupations.[1-6] It has been suggested that EMFs act as a promoter late in the carcinogenesis process.[7] Adey points out that "there is evidence that intercellular communication plays an essential role in regulation of cell growth. Functional isolation of a cell from its neighbors by the separate or joint actions of EM fields and chemical cancer promoters, both acting at cell membranes, may lead to unregulated growth with tumor formation."[8] To date, no convincing laboratory evidence has been obtained indicating that EMFs cause damage to DNA at nonthermal exposure levels. Therefore, EMF exposure might play only a minor or no role in cancer induction.

In spite of the epidemiological findings, we have not found any previous work done in vivo in the laboratory to study brain tumor development during EMF exposure. For several years we have been using a rat brain glioma model for therapeutic studies in our laboratory.

We have used this model to examine the effect of 915MHz microwaves, both as continuous waves and modulated with 4, 8, 16 Hz and 217 Hz in 0.57ms pulses and 50 Hz in 6ms pulses.[9]

MATERIALS AND METHODS

The rat glioma cell line RG2, originally an ethylnitrosurea-induced rat tumor, has grown very well in infinite cell culture cycles over two decades;[10] it gives rise to glioma-like tumors when inoculated in the brain of Fischer 344 rats. By stereotaxic technique, 5000 RG2 cells in 5μl nutrient solution were injected with a Hamilton syringe into the head of the right caudate nucleus in a total of 124 rats. To avoid tumor growth extracranially, the injection site was cleaned with 70% ethanol after injection and the burr hole was sealed with wax. Groups of 2-4 animals were inoculated with matched controls at each instance with cells harvested immediately prior to the procedure.

The exposed animals were kept unanesthetized in TEM cells producing 915MHz continuous or modulated microwaves (Table 1). Exposure was started on day 5 after inoculation. The animals were exposed for 7 h/day for 5 days/week. All animals were given a 30min break after 4 h of exposure for feeding. Five coronal slices from each animal were studied microscopically in cresyl violet staining. In this way the entire telencephalon was covered except the frontal and occipital poles.

The area of the tumor in the slice where the tumor was most extended was measured by the pathologist (who was not told whether the animals had been exposed to RF or were controls), from the formula $(a \times b)\cdot(\pi/4)$, where a and b are the maximum extensions in two dimensions on the slice.

---

*Correspondence to: L. G. Salford (Neurosurgery), Lund University Hospital, S-221 85 Lund, Sweden.

TABLE 1.—Experimental parameters and exposure scheme.

Modulation frequency (Hz)	Number of treatments	Pulse length (ms)	Peak power in pulse (W)	Duty cycle	SAR (W/kg)
4	10	0.57	2	0.002	0.0077
8.33	10	0.57	2	0.005	0.016
16	13	0.57	2	0.009	0.030
50	9-13	6	2	0.3	1.00
217	9	0.57	2	0.12	0.4
CW	10-15	---	1	1	1.67

TABLE 2.—Brain-tumor area of EMF exposed rats and controls.

Modulation frequency (Hz)	Number of pairs	Tumor area controls (mm^2)	Tumor area exposed (mm^2)	Student's paired t-test p
4	12	6.6	8.4	0.48
8.33	12	9.2	8.9	0.84
16	9	4.7	6.4	0.42
217	11	7.0	5.1	0.14
50	11	11.7	14.6	0.45
CW	7	12.3	13.8	0.77
All	62	8.4	9.3	0.40

RESULTS

The rectal temperature of the animals recorded before, at 4 h, and directly after exposure did not fall or rise. All treated animals received between 9 and 15 exposures (Table 1). There were no signs of brain damage outside the tumor areas: no necrosis, gliosis, or inflammatory changes that could be ascribed to the RF exposure.

Table 2 shows results for the several modulation frequencies. The tumor area of the exposed animal and its matched control for each exposure condition are shown. Our study shows no significant difference between animals exposed for 9 to 15 eight-hour periods of 915MHz microwaves, and those not exposed. The standard deviation is large, but this is the result of the great individual variation in the model, where the status of the inoculated cells as well as that of the recipient animal is highly influential. No obvious differences are seen between the animals exposed to CW and the animals exposed to modulated fields.

DISCUSSION

Our results to date show that not even an extensive daily exposure to EMF causes tumor promotion when given from the fifth day after the start of tumor growth in the rat brain until the death of the animal, which by then has a large brain tumor. The tumor areas given in Table 2 vary greatly both between the pairs of animals and also within the pairs (as is revealed by the SD). This is an inherent problem of the animal model used, one that implies that to increase the method's sensitivity, extended groups of animals must be used. The control of body temperature during EMF exposures is important for the exclusion of thermal effects. We found no thermal effects at the EMFs used; and in separate experiments we have not seen thermal effects until the power given was five times higher than the one used here (unpublished results).

We think that our findings merit continued studies, possibly on other experimental models. One reason for the absence of significant differences between EMF-treated and control animals might be that our rat glioma tumor cell strain (RG2) is very aggressive. The cells are possibly growing at their maximal speed, producing a tumor in three weeks, and the addition of external stimuli may therefore not increase the tumor growth rate.

The proliferative activity of several of the established experimental animal brain tumors is high, with labeling indices such as Bromodeoxyuridine (BrdU) higher than for human malignant gliomas. In collaboration with the Department of Tumor Immunology at Lund University, we have recently developed a new ethylnitrosurea-induced rat glioma cell line that produces gliomas at only half the growing rate of the RG2 cells,[11] and that may be an appropriate model for continued experiments.

Another possibility is to use a xenotransplatation model we developed previously. The model is based on the implantation of a cell suspension of human gliomas into different regions of the brains of the immunocompetent Wistar rats.[12] These transplants result in tumor takes in 50% of the cases and maximally grow to diameters of 3 mm during 3 months. Such slow restrictive tumor growth may be another appropriate model for continued attempts to verify the promoting effects of electromagnetic fields.

CONCLUSIONS

In the present elaborate material no effects either of CW or modulated RF exposure could be demonstrated. These conclusions are based on one single model; other models should be applied in the continued research. Furthermore, our model only reveals tumor promotion and not tumor initiation.

REFERENCES

1. N. Wertheimer and E. Leeper, *Am. J. Epidemiol.* 109: 273, 1979.
2. L. Tomenius, *Bioelectromagnetics* 7: 191, 1986.
3. R. S. Lin, P. C. Dischinger, J. Conde, and K. P. Farrell, *J. Occup. Med.* 27: 413, 1985.
4. S. Milham Jr., *Environ. Health Perspect.* 62: 297, 1985.
5. M. A. Speers, J. G. Dobbins, and V. S. Miller, *Am. J. Ind. Med.* 13: 629, 1988.
6. T. L. Thomas, P. D. Stolley, A. Stemhagen, E. T. H. Fontham, M. L. Bleecker, P. A. Stewart, and R. N. Hoover, *J. Natl. Cancer Inst.* 79: 233, 1987.
7. C. V. Byus, S. E. Pieper, and W. R. Adey, *Carcinogenesis* 8: 1385, 1987.
8. W. R. Adey, in B. W. Wilson, R. G. Stevens, and L. E. Anderson, Eds., *Extremeley Low Frequency Electromagnetic Fields: The Question of Cancer*, Columbus, Ohio: Battelle Press, 1990.
9. L. G. Salford, A. Brun, B. Persson, and J. E. Eberhardt, in *Proc. First Congress Europ. Bioelectromagnetic Assoc.*, Brussels, 1992 (in press).
10. W. Wechsler, P. Kleihues, S. Matsumoto, K. I. Zulch, S. Ivankovic, R. Preussman, and H. Druckrey, *Ann. NY Acad. Sci.* 159: 360, 1969.
11. P. Siesjö, E. Visser, M. Lindvall, H. O. Sjögren, and L. G. Salford, in *Proc. Intern. Symp. Advances in Neuro-oncology*, San Remo, Italy, 1990.
12. L. G. Salford, A. Brun, L.-G. Strömblad, Sixth Intern. Conf. on Brain Tumor Research and Therapy, Asheville, N.C., 1985 (abstract).

Martin Blank, Ed., *Electricity and Magnetism in Biology and Medicine*

EFFECT OF LOW-FREQUENCY ELECTROMAGNETIC FIELDS ON A BREAST-CANCER CELL LINE INCUBATED WITH VARIOUS CYTOSTATIC AGENTS

S. Mikorey-Lechner*

1st University Women's Clinic, Munich, Germany

S. Johann and G. Blümel

Experimental Surgery Clinic, Technical University of Munich

W. Kraus

Institute of Medical Physics, Munich

Human breast-cancer cells (MCF7) were seeded in 96-well plates in thin layers (10^4 cells) and incubated with seven different cytostatic agents under standard conditions (methotrexate, cis-platin, cyclophosphamid, doxorubicin, mitoxantron, mitomycin, and fluorouracil). A 20Hz, 5.3mT magnetic field was applied in the incubator at 37 C, 5% CO_2. Control cells were kept at identical conditions, with the omission of the EMF. Mitochondrial activity was measured by the MTT test. After 8 days of incubation, an increase by a factor of 1.8 in mitochondrial activity was found in cell populations subjected to EMF without cytostatics. Cells incubated with cytostatic agents showed a decrease in mitochondrial activity (15-90%). The decrease of activity can be enhanced by the simultaneous application of EMF and cytostatic agents to an established breast-cancer cell line.

In 90% of all patients with a breast cancer the disease will apparently be limited to mammarian gland and the axillary lymph nodes. After the local treatment (surgery and/or radiation), more than one third of these patients will develop systemic disease and ultimately die. In metastatic disease, chemo- and hormone therapy are palliative. New drugs to improve survival of breast cancer patients continue to be developed. However, new approaches are needed to overcome resistance to chemotherapy. In 1991 Petrini et al. showed that resistance of a multidrug-resistant cell line can be overcome by exposure to a magnetic field.[1] We have used this interesting approach to study the effect of various cytostatic agents on the breast-cancer cell line MCF7. We therefore determined the mitochondrial activity of MCF7 cells by MTT assay,[2] which is a criterion for cell proliferation.

MATERIAL AND METHODS

Cell Preparation

MCF7 (ATCC HTB 22) cells are grown from the pleural effusion of a human breast adenocarcinoma. The cells were maintained in RPMI 1640 medium supplemented with 10% heat-inactivated fetal calf serum, 1% L-glutamine, sodium pyruvate, streptomycin, and penicillin. The cells were incubated at 37 C in a humified atmosphere of 5% CO_2. Cells were passaged twice a week. Cell viability was routinely determined by the trypan blue dye exclusion. Prior to the first application of the electromagnetic field, each sample was incubated for 6 h under the abovementioned conditions. The experiments were performed 4-fold and were replicated 3 times.

---

*Correspondence to: S. Mikorey-Lechner, LMU Frauenklinik I, Maistr. 11, D-8000 München 2, Germany.

Magnetic-field Exposure

A magnetodyne function generator (M70S)[3] was used to drive to solenoid coils, each of which could take up three 96-well plates. A continuous sinusoidal signal was generated at a frequency of 20 Hz and a maximum magnetic induction of 5.3 mT. The distortion factor was less than 1% as measured with a gaussmeter (Bell 640) and a spectrum analyzer. The coils were installed in an incubator at 37 C and were prevented from increasing the incubation temperature by a Haake cooling system with continuous temperature monitoring. Temperature differences between the control and experimental set-up were less than 0.15 C during application of the electromagnetic field. We used an intermittent electromagnetic field (6 h on, 6 h off) and measured the mitochondrial activity of the MCF7 cells by the MTT test.

MTT Test

The mitochondrial activity of MCF7 cells were measured by the ability of viable cells to transform tetrazolium salts into insoluble formazan. Cells were trypsinized, washed with phosphate buffered saline (PBS), and adjusted to 1000 cells/ml. Cells were incubated for 8 days under standard conditions. After this time 10 µl of yellow (dimethyl-thiazolyl-diphenyl-tetrazoliumbromide) MTT was added to each well and cells were incubated for another 4 h. The amount of MTT-formazan was quantified in an ELISA reader by absorbance at 550 nm.

Cytostatic Drugs

Seven cytostatic drugs used in clinical treatment of human breast cancer were chosen (doxorubicin, methotrexate, mitomycin, mitoxantron, cyclophosphamide, carboplatin, 5-fluorouracil). The effect of EMF is more pronounced with drugs that interfere with DNA intercalation. So called "antimetabolites" show only a minor enhancement by EMF (Table 1).

RESULTS

MCF7 cells incubated for 14 days showed a significant increase in mitochondrial activity after exposure to EMF. This increase is visible over a broad range of cell counts. The optimal cell density was found to be 500 cells/well (Fig. 1). At a lower concentration, cell growth is slower; at a concentration of 500 cells/well, cell death occurred after 6 days.

Exposure of cells (without drugs) to EMF showed a morphologic change of cell surface. Membrane ruffling and formation of microvilli were observed under scanning electron microscopy. For further investigations, we used a concentration of 10000 cells/ml. After 5 days a difference in mitochondrial activity could be observed; after 8 days a significant increase occurred (Fig. 2). The mitochondrial activity as a parameter of cell metabolism is activated by low-frequency magnetic fields. Cells incubated with various chemotherapeutic agents and simultaneously exposed to MF showed a significantly higher decrease in mitochondrial activity than control groups without MF. This effect was observed regardless of the drug used (Fig. 3).

DISCUSSION

This presentation demonstrates that MF exposure to a cell line (MCF7) leads to an increase of mitochondrial activity and morphologic changes of cell surface. Similar observations were made in nerve cells, where exposure to ELMF led to early differentiation of cells.[3] It is not known whether morphological changes are connected to proliferative activity of cells. One might speculate that MF exposure leads to cell differentiation and has an antiproliferative effect. Further investigations, especially with a clonogenic cell assay, are needed to investigate this hypothesis.

Incubation of MCF cells with 7 cytostatic agents under exposure to MF leads to a statistically significant decrease in mitochondrial activity. It is possible that decrease in mitochondrial activity leads to early death of tumor cells. According to Mosmann et al.,[2] mitochondrial activity is a useful parameter for cell viability. However, it is not known whether this test corresponds to cell differentiation or proliferation. In our study we found the MTT test useful to determine MF-induced cellular changes. Further investigations to expand these first findings to other cell lines and to prove the antiproliferative activity of MF are planned. MF increases the effect of cytostatic agents on mitochondrial activity. It will be of interest to use this method in clinical settings, with a view to a possible increase in the antitumor effect of these drugs.

TABLE 1.—Drugs and site of action.

Drug	Main effective mechanism	Effect	Reversible inhibition	Lethal effect
Methotrexate	Inhibition of dihydrofolatreductase	Inhibition of thymidin and purin synthesis: inhibition of DNA/RNA and protein synthesis	Late G1	S DNA synthesis phase
Cis-platin	Crosslinking among guanin molecules of DNA	Inhibition of RNA synthesis; DNA-network formation	S/G2 (not phase specific)	S
Cyclophosphamide	DNA alkylation, crosslinking	DNA strand breaks	G2 (not phase specific)	S
Doxorubicin	DNA-intercalation; inhibition of DNA replication and of DNA repair	Inhibition of DNA and RNA synthesis	S/G2 (not phase specific)	S/G2
Mitoxantron	Intercalation of DNA; strand breaks	Inhibiition of RNA synthesis	G2/M	S/G2
Mitomycin	Alkylating DNA; crosslinking DNA; strand breaks		G2 (not phase specific)	S/G2
5-Fluorouracil	Inhibition of thymidin; phosphorylase inhibition of DNA synthesis	Inhibition of RNA synthesis by construction of ''wrong steps''		S

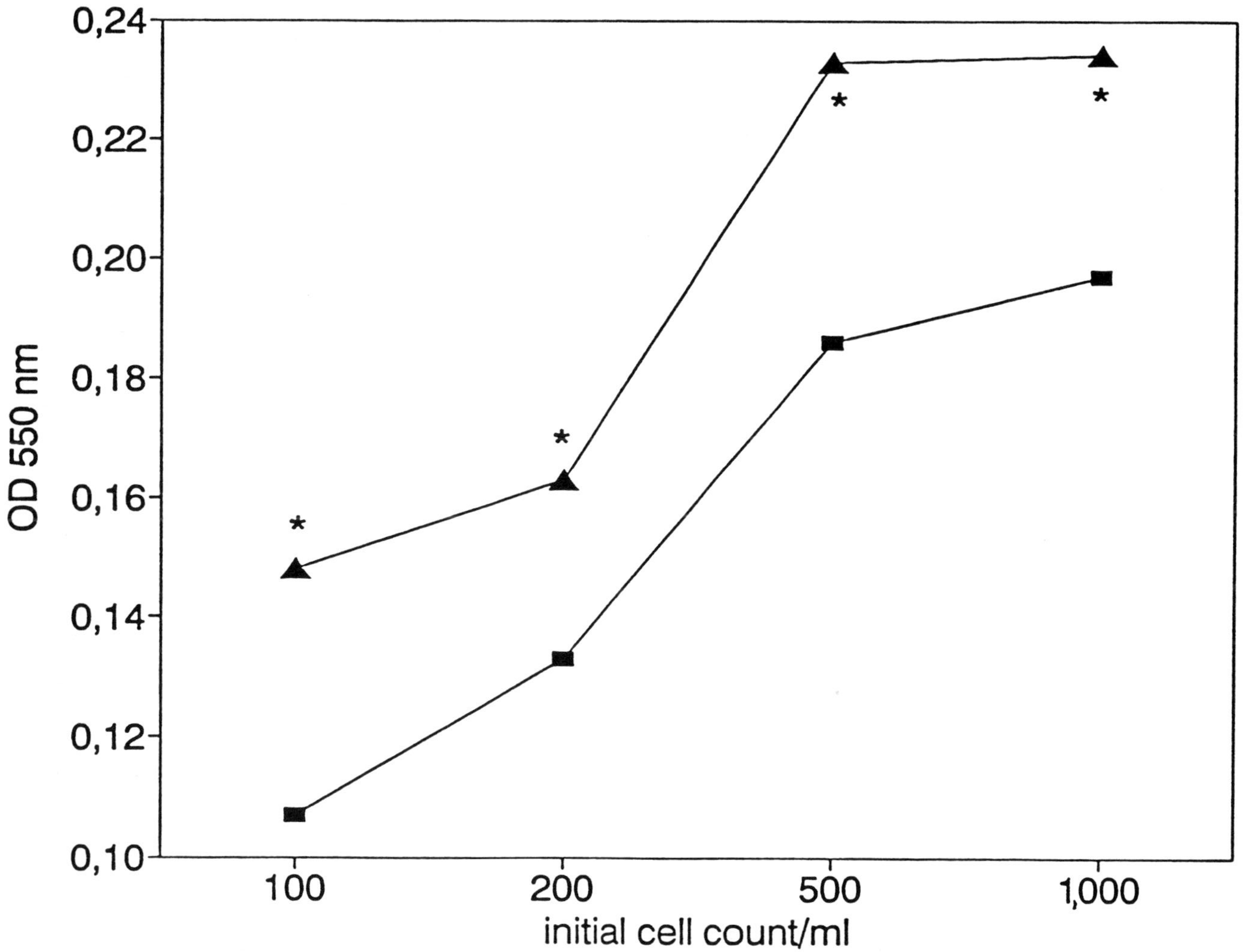

FIG. 1.—Reduction of MTT in MCF7 cells incubated for 14 days and exposed to EMF (darkened triangles) and omitting EMF (darkened bars). Cell counts as seeded at beginning of experiment. Shown are means of 36 measure points from three independent experiments; $*p < 0.05$.

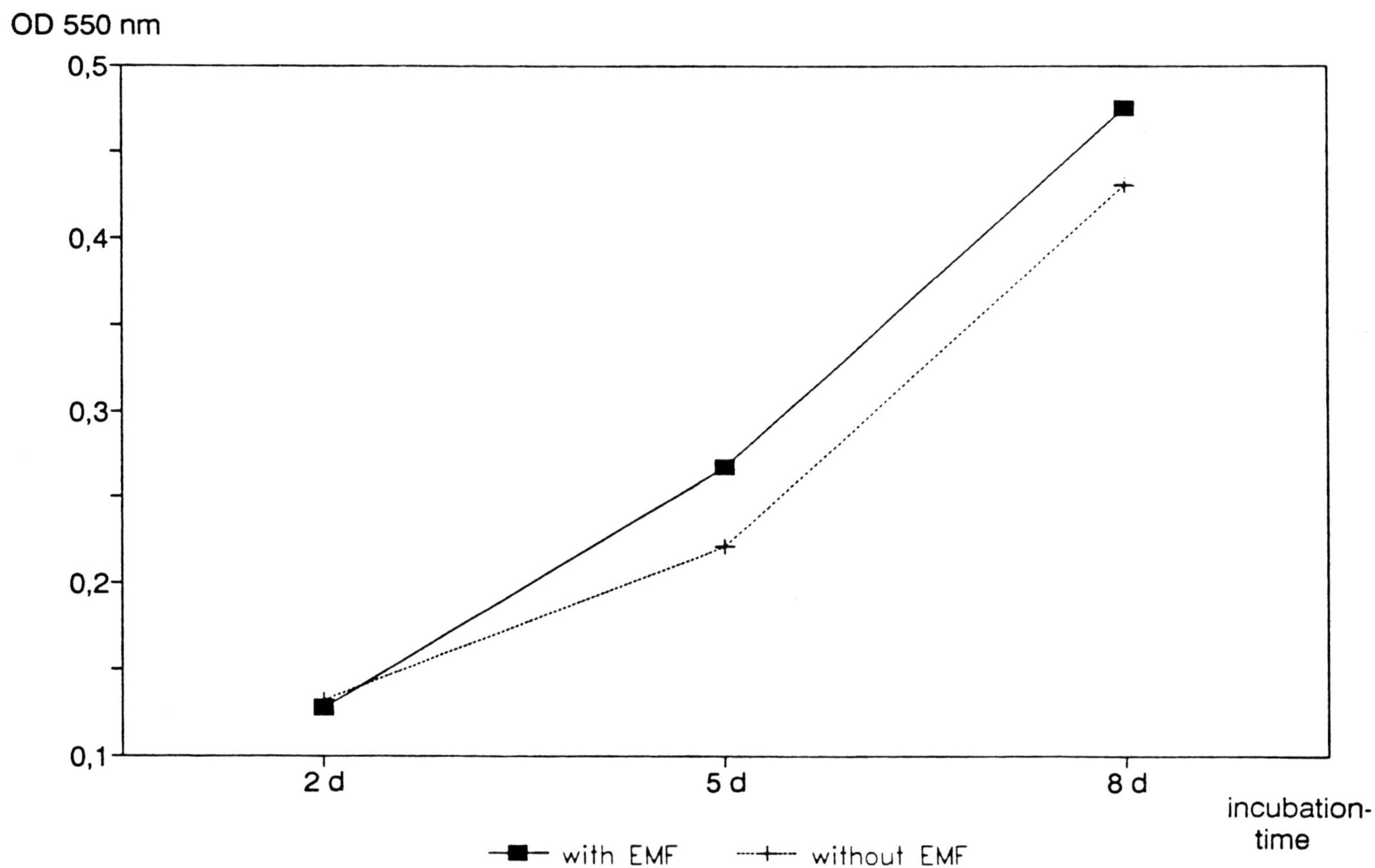

FIG. 2.—Increase in mitochondrial activity of MCF7 cells under exposure to magnetic field after 2, 5, and 8 days. The experiments were performed 4-fold and were replicated 3 times; ;$*s < 0.05$; initial cell count, 10^4 cells/ml.

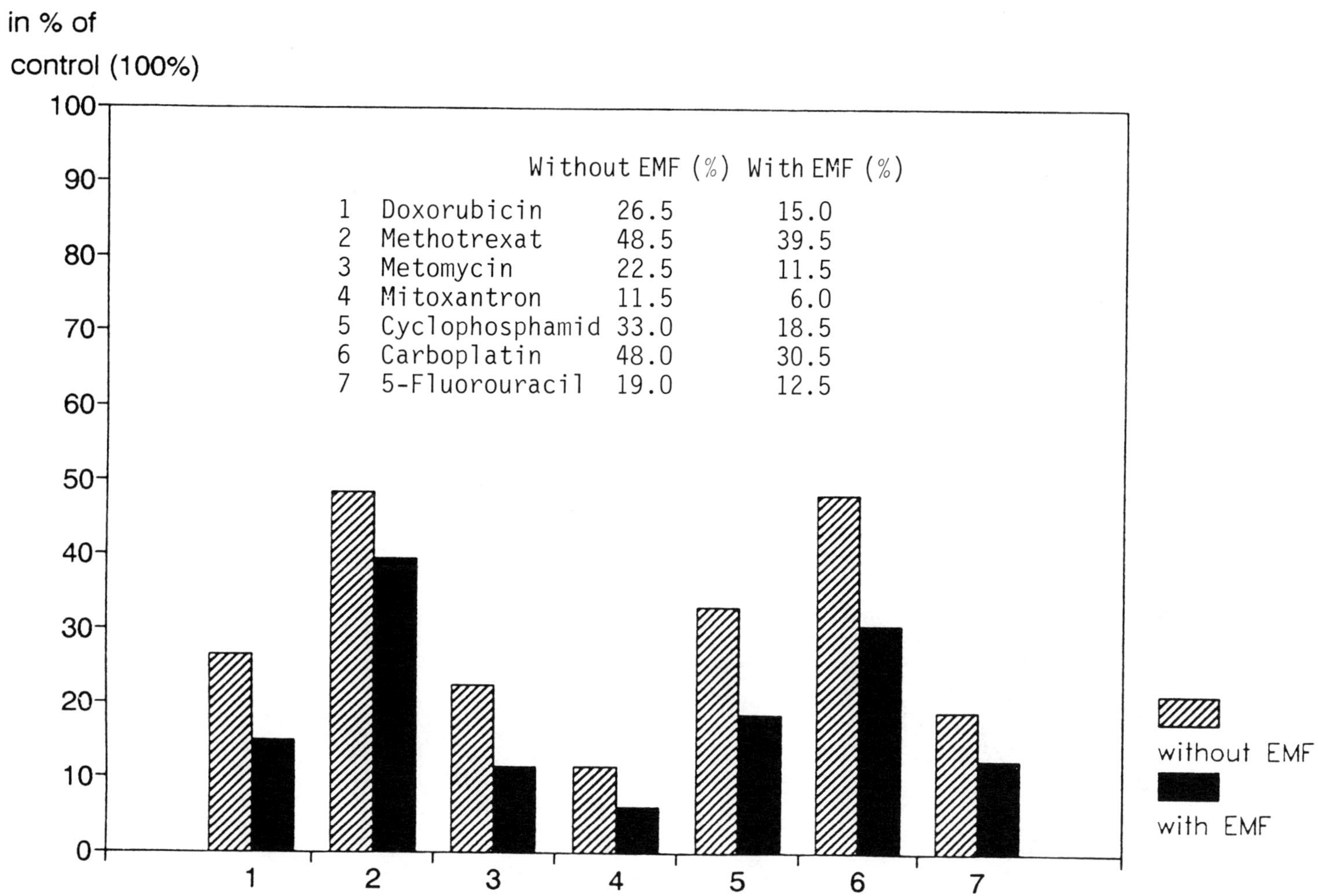

FIG. 3.—Mitochondrial activity (control 100%) after application of cytostatic agents without EMF and with EMF. Graph shows 12 measure points in 3 independent experiments; $*s < 0.05$; initial cell count, 10^4 cells/ml.

REFERENCES

1. M. Petrini, R. Cadossi, L. Mattii, A. Conte, G. Carulli, B. Grassi, and G. Ronca, *Proc. 11th BRAGS Mtg.,* Scottsdale, USA, 29: 48, 1991.
2. T. Mosmann, *J. Immunol. Meth.* 65: 55, 1983.
3. W. Kraus and F. Lechner, *Münchner Med. Wochenschrift* 42: 1814, 1972.

BIBLIOGRAPHY

T. Akamine, H. Muramatsu, H. Hamada, and T. Sakou, *J. Cell. Physiol.* 124: 247, 1985.

R. Ascherl, G. Blümel, F. Lechner, and W. Kraus, *Biochem. Bioenerg.* 14: 161, 1985.

E. Azavedo, *Clin. Radiol.* 43: 84, 1991.

W. Kraus, *Orthopäde* 13: 78, 1984.

J. C. Murray and R. W. Farndale, *Biochim. Biophys. Acta* 838: 98, 1985.

W. Mühlbauer, *Langenbecks Arch. Chir.* 337: 637, 1974.

Y. Omote, M. Hosokawa, M. Komatsumoto, T. Namieno, S. Nakajima, Y. Kubo, and H. Kobayashi, *Jpn. J. Cancer Res.* 81: 956, 1990.

F. Tabrah, M. Hoffmeister, F. Gilbert, S. Batkin, and C. Bassett, *J. Bone Mineral Res.* 5: 437, 1990.

K. Takahashi, I. Kaneko, M. Date, and E. Fukada, *Experientia* 42: 185, 1986.

T. Träger, R. Gradinger, and E. Hipp, SICOT, Montreal, 1990.

Martin Blank, Ed., *Electricity and Magnetism in Biology and Medicine*

INCIDENCE OF LYMPHOMA IN CFW MICE EXPOSED TO LOW-FREQUENCY ELECTROMAGNETIC FIELDS

E. L. Mikhail

Dalhousie University and IWK Children's Hospital, Halifax, Nova Scotia

W. Z. Fam*

Technical University of Nova Scotia, Halifax

A group of 41 mice of the CFW strain was continuously exposed to a 25mT, 60Hz electromagnetic field from conception until they were sacrificed for tests at the age of 363-418 days. Another group of 36 mice was used as controls and was kept under the same conditions, but the magnetic field was no higher than 0.05 mT. Gross and microscopic histopathological examination revealed that 78% of the exposed group of animals developed premalignant changes or malignant lymphoma, compared to only 6% in the control group which showed similar changes. These results confirm our previous report of lymphoma in another group of animals chronically exposed to this relatively strong electromagnetic field. Furthermore, statistical analysis of the results from the two studies shows a direct correlation between the duration of exposure and the incidence of lymphoma in the exposed animals.

This paper is one in a series of reports by the authors on an extensive multigeneration animal experiment carried out over the past 3 years. The objective of the study was to determine whether prolonged exposure to a relatively strong electromagnetic field can result in pathological effects. Two years ago, we reported on a group of 24 second-generation mice which were exposed to a 25mT, 60Hz magnetic field continuously for an average of 120 days.[1,2] Examination of tissue sections from the animals revealed that 58% of the males and 92% of the females had enlarged lymph nodes and hyperplasia. There were also focal lymphoid infiltration of the kidney, lung, liver, and spleen in a number of the exposed males and females. Last year, we reported on the results from a group of 55 third-generation animals continuously exposed to the field from conception until they were sacrificed at an age of 133-257 days.[3,4] Examination of tissue sections from these animals showed that 33% had developed malignant lymphoma, 13% had pre-malignant changes, and 31% had lymphoid hyperplasia. In the control group, only 5% of the animals showed lymphoid hyperplasia; the remaining 95% had no pathological findings.

This paper deals with the results of another group of 41 third-generation CFW mice, which has been continuously exposed to a 25mT, 60Hz magnetic field from conception until they were sacrificed for tests at the age of 363-418 days.

EXPOSURE SYSTEM

A general view of the exposure system[5] we used in our investigation is shown in Fig. 1. It consisted of three rectangular coils mounted on a wooden frame and separated from each other by 10 cm. Each coil measured 55×35 cm, had 260 turns of AWG #10 copper wire, and weighed 16.7 kg. Three plastic shelves divided the wooden frame into four compartments, each capable of accommodating one large cage for housing 15 animals. The cages had plastic fenestrated covers. To produce the required magnetic field, a 13 A alternating current was passed through each coil. Three cooling fans were used to blow air on the coils to maintain their temperature below 45 C. The coils were carefully constructed to minimize vibrations and the level was measured by a vibration meter and an accelerometer. The displacements at each of the shelves on which the animal cages were placed were 0.018, 0.016, 0.018, and 0.20 mm, respectively. The temperature inside the cages was measured with an alcohol thermometer while the system was energized and with all the animals

*Correspondence to: W. Z. Fam, 15214 Wetherburn Dr., Centreville VA 22020.

inside the cages. It was found to be about 1 C above the ambient room temperature, which was maintained between 22 and 24 C by two air-conditioning units. Lighting conditions in the room were such that the animals were given approximately 12 h of daylight from a large window augmented by fluorescent lights, and 12 h of darkness.

ANIMAL PROTOCOL

The animals used in this study were CFW mice obtained originally from Charles River Laboratory in Quebec, and were inbred in our laboratory by strict brother × sister mating for three successive generations. Two groups of animals were used. The experimental group consisted of 20 males and 21 females which were conceived, born, and exposed to the 25mT, 60Hz electromagnetic field continuously for 23 h/day. The control group consisted of 16 males and 20 females. They were located in the same room as the exposed group but sufficiently far from the exposure system to insure that they were not exposed to magnetic fields higher than 0.05 mT. Sterilized sawdust was used as bedding in all cages; Purina chow and water were freely available to all the animals.

After 363 days of exposure, we began to sacrifice animals from both the exposed and the control groups for tests, until by the end of 418 days all animals had been sacrificed. The animals were anesthetized with 6% pentobarbital solution for intraperitoneal injection of 90 mg/kg. A bone-marrow smear and a total blood count including differential and blood chemistry were done for each animal. A complete autopsy was then performed on all animals. Tissue sections were taken from thymus, spleen, lymph nodes, liver, kidney, lung, and bowel. Four-micrometer paraffin sections were stained with H&E for histopathological examination.

RESULTS

The pathological changes induced by exposure to this relatively strong electromagnetic field were divided into the following stages: (a) normal, (b) lymphoid hyperplasia, (c) pre-malignant changes, (d) early thymic lymphoma, and (e) advanced thymic lymphoma. This division conforms with the classification of *Tumors of the Hematopoietic System* by the International Agency for Research on Cancer–tumors of the mouse.[6] The micrographs of Figs. 2-6 illustrate some of these pathological findings as seen during gross and microscopic examination of tissue from the exposed animals. Table 1 is a summary of the results from both the control and exposed groups. It shows that 78% of the animals exposed to the electromagnetic field have developed pre-malignant changes or malignant lymphoma, whereas only 6% of the control group developed similar effects.

TABLE 1.—Comparison between pathological findings in exposed and control animals.

	Males				Females				Total			
	Exposed		Control		Exposed		Control		Exposed		Control	
Diagnosis	#	%	#	%	#	%	#	%	#	%	#	%
Normal	2	10	16	100	0	0	15	75	2	5	31	86
Lymphoid hyperplasia	4	20	0	0	3	14	3	15	7	17	3	8
Pre-malignant change	9	45	0	0	0	0	0	0	9	22	0	0
Early thymic lymphoma	4	20	0	0	14	67	2	10	18	44	2	6
Advanced thymic lymphoma	1	5	0	0	4	19	0	0	5	12	0	0

1

2

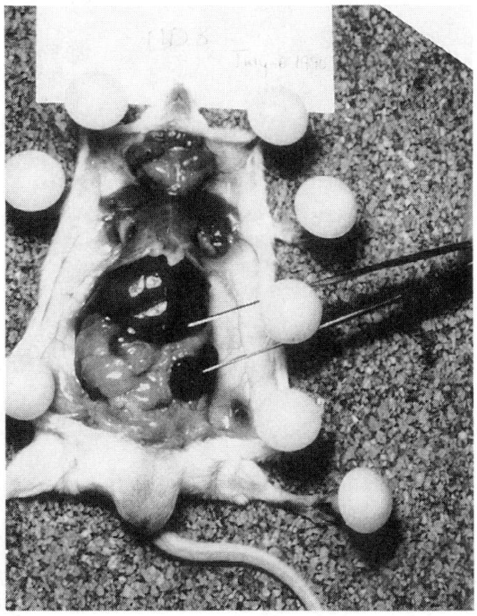

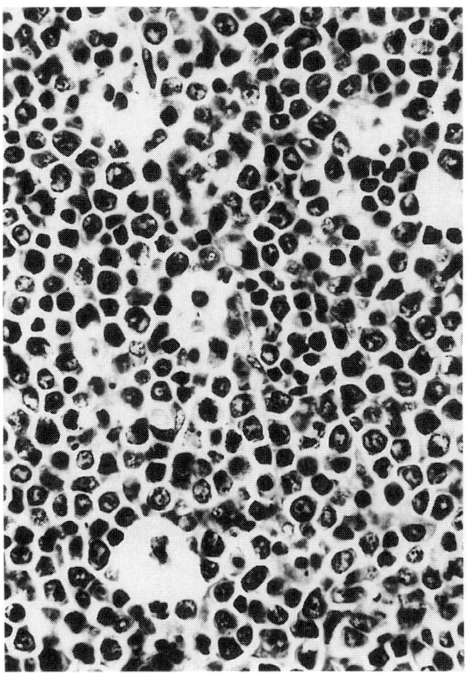

3

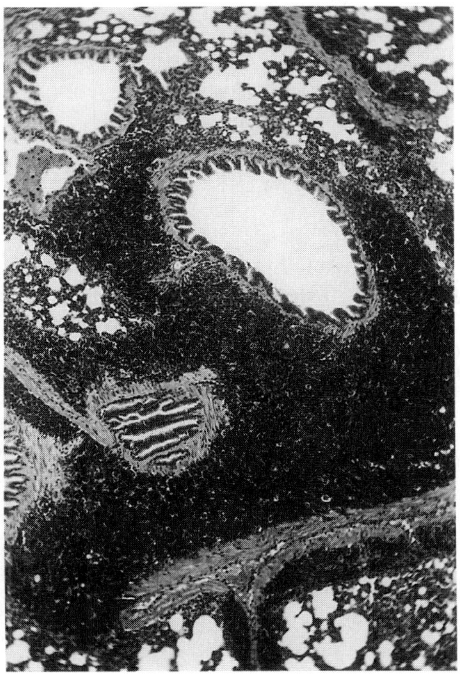

4

FIG. 1.—Exposure System.

FIG. 2.—Gross autopsy of advanced thymic lymphoma: large thymic mass, hepato-splenomegaly, lymphadenopathy.

FIG. 3.—Micrograph of lymphoid tumor.

FIG. 4.—Section of lung involved with tumor.

5

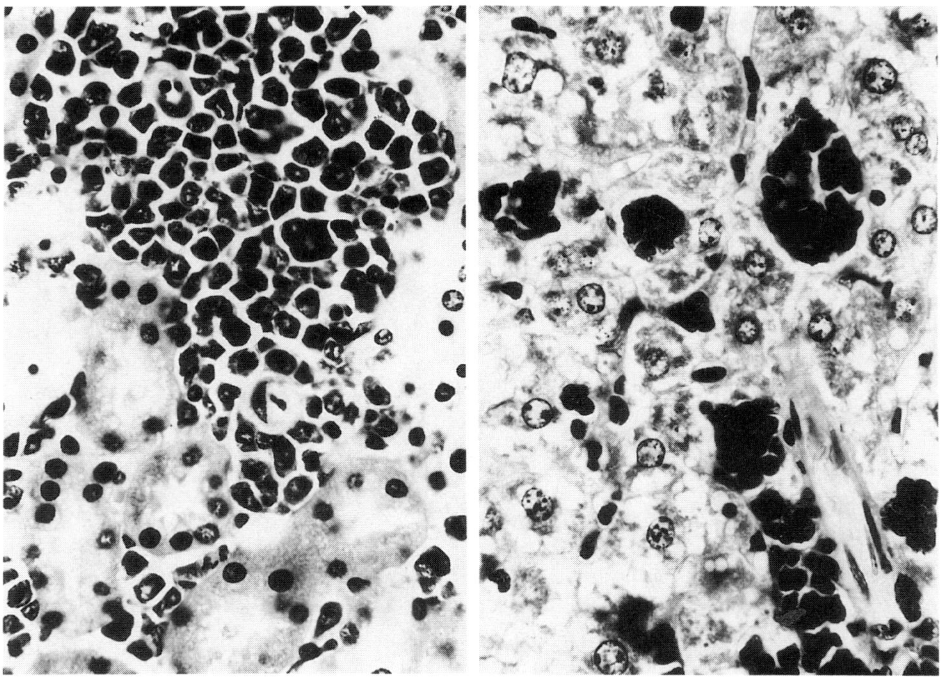

6

FIG. 5.—Section of kidney involved with tumor.
FIG. 6.—Section of liver involved with tumor.

REFERENCES

1. W. Z. Fam and E. L. Mikhail, "Biological effects in mice exposed to a 25-mT, 60-Hz magnetic field," *12th Ann. BEMS Meeting*, San Antonio, Tex., 1990, B-2-3.

2. E. L. Mikhail and W. Z. Fam, "Hematological effects of long-term exposure to electromagnetic field," Ann. Meeting, Am. Clinical Pathologists, 1990.

3. E. L. Mikhail and W. Z. Fam, "Development of lymphoma in third-generation mice exposed to 60-Hz magnetic field," *13th Ann. BEMS Meeting*, Salt Lake City, Utah, 1991, B-1-4.

4. E. L. Mikhail and W. Z. Fam, "Development of lymphoma in laboratory mice exposed to low-frequency electromagnetic fields," *60th Ann. Meeting of Royal College of Physicians and Surgeons of Canada*, Quebec city, 1991, A58.

5. W. Z. Fam and E. L. Mikhail, "A system for the exposure of small laboratory animals to a 25-mT, 60-Hz alternating or traveling magnetic field," Ref. 1, p. 62.

6. G. Della Porta, L. Chieco-Bianchi, and N. Pennelli, "Pathology of tumors in laboratory animals," *IARC Sci. Public.* 23: 527-575, 1979.

Martin Blank, Ed., *Electricity and Magnetism in Biology and Medicine*

60Hz ELECTRIC AND MAGNETIC FIELDS AND PRIMATE MELATONIN

W. R. Rogers* and J. L. Orr

Southwest Research Institute, San Antonio, Tex.

R. J. Reiter

University of Texas Health Science Center, San Antonio

In three experiments, each employing three field-exposed and three sham-exposed nonhuman primates implanted with an indwelling venous cannula to allow repeated blood sampling, we studied the effects of either 6 kV/m and 0.5 G or 30 kV/m and 1.0 G on blood melatonin patterns. Extensive quality control for the melatonin assay, computerized control and monitoring of field intensities, and consistent exposure protocols were used. No changes in nocturnal serum melatonin concentration resulted from 6 weeks of day-time exposure with "slow" field onset/offset, but a pilot experiment, without concurrent control subjects or extensive assay quality-assurance procedures, suggested that 30kV/m and 1.0G fields, with "rapid" field onset/offset, might suppress melatonin.

Separate rodent experiments conducted by Kato,[1] Reiter,[2] Wilson,[3-5] Yellon,[6] and others show that exposure to electric field (E) or magnetic field (B) can reduce nocturnal melatonin concentrations in pineal gland and/or blood. If E/B exposure reduces melatonin in humans, it could have diverse important consequences. However, no convincing direct evidence has been presented to indicate that power-frequency E/B exposure does or does not affect blood melatonin in primates, human or nonhuman. Well-known differences between rodents and primates in pineal response to light[7,8] suggest that laboratory rodents (particularly the albino strains) might not be appropriate models to predict E/B effects on human melatonin.

METHODS

Melatonin Experiment I

Subjects. Six male baboons *(Papio cynocephalus anubis)* 7-8 years old and weighing 25-28 kg were used. They were housed in fiberglass cages in an environmentally controlled facility with a 14.5:9.5 L:D cycle, with lights on between 0700 and 2230. Subjects were randomly assigned as three experimental (field-exposed) and three control (sham-exposed) subjects.

Exposure. This experiment was conducted with 60Hz fields of 6 kV/m and 0.5 G, presented concurrently by two transformers. Homogeneous, vertical E fields were generated between an overhead bus suspended 2.5 m above a large, grounded, nonferrous metal grate; homogeneous, horizontal B fields were generated by a series of conductors beneath the grate. The fields were turned on and off and monitored by computer; it took 4 s for E/B strengths to rise to, or fall from, the target value. The fields within the cages were mapped before and after the experiment and were checked every 3 weeks during the experiment. The facility operated 7 days/week, providing 12 h of E/B exposure, in three 4h blocks, during "lights on."

Blood Sampling and Assay. A chronically indwelling cannula system designed for use in E/B experiments was used to collect blood samples automatically without the confounding effects of anesthesia or stress. Following 3 weeks of recovery from surgery, three complete diurnal samplings were completed on all subjects at 1-week intervals during a 3-week pre-exposure period. A diurnal consisted of 12 blood samples taken at 2h intervals. Subjects were also sampled in the 2nd, 4th, and 6th weeks of exposure and in the 1st and 3rd weeks of post-exposure

The coded samples were divided into nine batches and delivered to UTHSCSA for analysis. laboratory. Log-logit standard curves were done in duplicate. Serum melatonin concentrations of pairs of independently

---

*Correspondence to: W. R. Rogers, SwRI Dept. of Biosci. & Bioeng., San Antonio TX 78228-0510.

coded tubes were measured in duplicate by radioimmunoassay (RIA) via the Guildhay antibody. In the analysis of each batch, formed by day of sampling, eight samples from each of three reference serum pools, in the physiologically "low," "medium," and "high" range of melatonin concentration for baboons were also analyzed in duplicate. Unless noted, all methods in Experiments I, II, and III were identical.

Melatonin Experiment II

Following the end of the 6-week post-exposure period of Experiment I, the same animals were used, without changing treatment and control groups, in a second experiment, this time involving 30 kV/m and 1.0 G. Subjects were sampled twice during a 2-week pre-exposure period and weekly during a 3-week exposure period. It took 20 s for the E/B to be turned fully on or off; field strengths changed at a linear rate.

Melatonin Experiment III

A second set of six subjects was used in an independent experiment involving 30 kV/m and 1.0 G; the design was the same as in Experiment I. Melatonin standard curves were done in triplicate, and 12 samples were tested from each of three serum pools. A set of 12 batches was constructed by subject; independently coded aliquots from each subject were in separate batches.

RESULTS

Experiment I. During exposure, as in pre- and post-exposure, the diurnal patterns of melatonin secretion of sham-exposed and field-exposed subjects were similar (Fig. 1). Analyses of variance (ANOVA) indicated that the day-vs-night melatonin concentrations differed significantly ($p < 0.05$); no important interactions were detected.

Experiment II. During both pre-exposure and exposure, the diurnal patterns of melatonin secretion of sham-exposed and field-exposed subjects were similar (Fig. 2). ANOVA confirmed presence of day-night differences and absence of differences between experimental and control groups.

Experiment III. During exposure, as well as during pre- and post-exposure, the diurnal patterns of melatonin secretion of sham-exposed and field-exposed subjects were very similar (Fig. 3). ANOVA results were the same as for Experiments I and II.

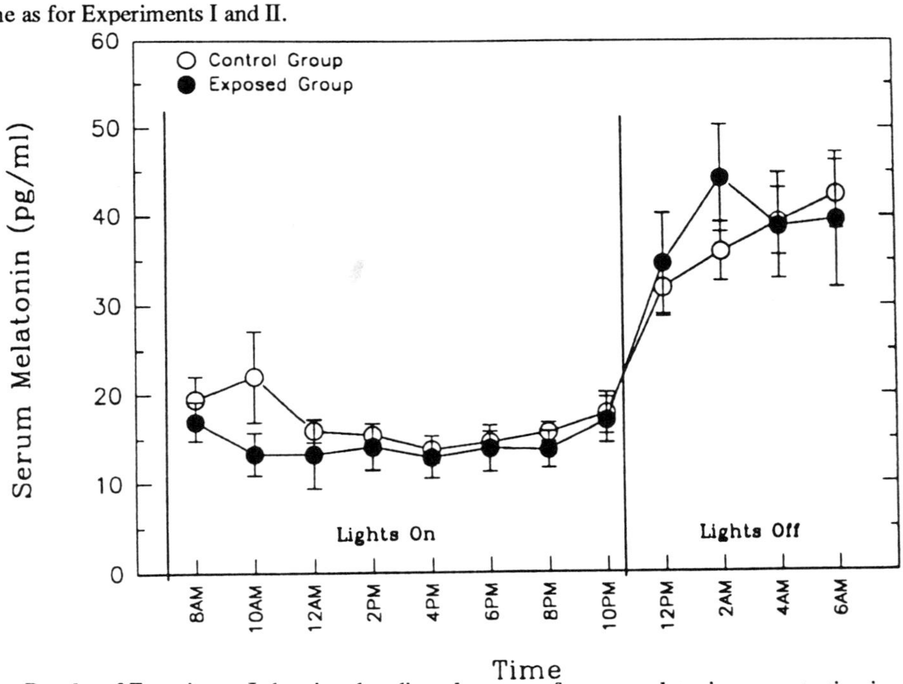

FIG. 1.—Results of Experiment I showing that diurnal pattern of serum melatonin concentration is not affected by regular day-time exposure to 6 kV/m and 0.5 G.

DISCUSSION

Experiments I, II, and III indicate that regular exposure to combined E/B for 12 h during daylight hours for 6 weeks does not appreciably affect the 24h melatonin pattern of baboons. Conclusions reached with nonhuman primates generally extrapolate well to humans.

Transients. However, it might be premature to generalize from these results and conclude that 60Hz E/B, under any exposure conditions, do not affect primate melatonin. For example, the limited results from a preliminary pilot experiment, described below, suggest that under other exposure conditions (involving fast rise-time signals, intermittency, and relatively more frequent onsets), E/B exposure might suppress circulating melatonin in primates.

Using static magnetic fields switched on and off at 5min intervals, Lerchl et al.[9] have shown that fast rise-time MF onset/offset might be a critical component affecting pineal response of rodents. Also, Wilson et al.[10] have results suggesting that urinary melatonin excretion of humans can be affected by use of electric blankets, especially those involving fast rise-time signals.

Differing Exposure Conditions. The absence of melatonin reduction in Experiments I-III and the apparent presence of melatonin suppression in the pilot experiment (below) are not necessarily contradictory, because different exposure paradigms were used. The melatonin-suppression effect in the pilot experiment could be due to speed, irregularity, or frequency of field onset/offset, or it could be due to an unrecognized sources of variation. Furthermore, although the results from Experiments I-III collectively appear quite conclusive, the results from the pilot experiment clearly are not.

PILOT EXPERIMENT

Exposure. In this engineering feasibility experiment, B onset, achieved by relay closure, was "fast;" 2 min later E was brought rapidly to 14 kV/m by relay closure and then ramped "slowly," over an 11s period, to 30 kV/m at the fixed rate of 1.5 kV/m/s. Field offset occurred in the reverse manner.

Only two subjects were used, and the within-subject design used 3-week pre-exposure and exposure periods. Through the 10th exposure day, the gradually increasing, irregular, and intermittently scheduled exposure, often with several onsets/offsets each day, occurred only during "lights-on" hours, but by the 17th exposure day the duration of daily exposure had been increased to 20 h, which means that exposure also occurred during "lights off." Sets of blood samples were collected overnight on the 10th and 20th days of exposure.

Blood Sampling and Assay. Blind duplicate analyses of four temporally ordered batches of frozen, coded serum samples were done, following extraction, via RIA with the Rollag antibody. Serum pools were not used.

Results. When sampled on the 10th exposure day, nocturnal serum melatonin concentrations of the two subjects were about 15% of the levels observed previously during pre-exposure. The same was true on the 20th exposure day (Fig. 4).

SUMMARY

A series of three experiments completed with sets of three field-exposed and three sham-exposed nonhuman primates, each fitted with chronically indwelling cannula to permit multiple blood sampling, clearly show that 6 weeks of daytime exposure to electric and magnetic fields does not appear to affect serum melatonin concentrations.[11]

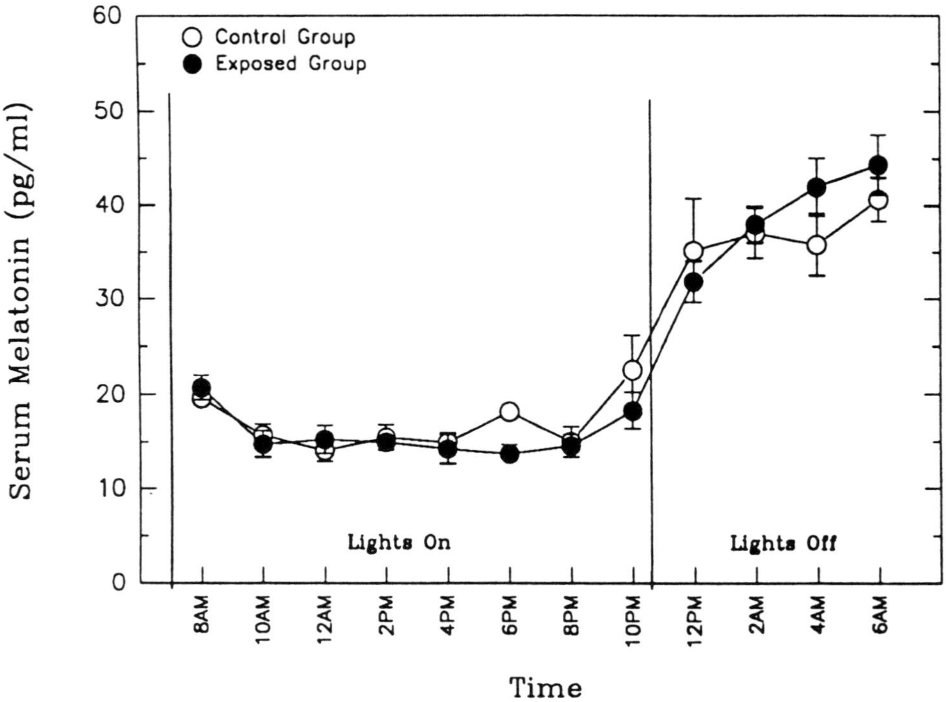

FIG. 2.—Results of Experiment II showing that diurnal pattern of serum melatonin concentration is not affected by regular day-time exposure to 30 kV/m and 1.0 G.

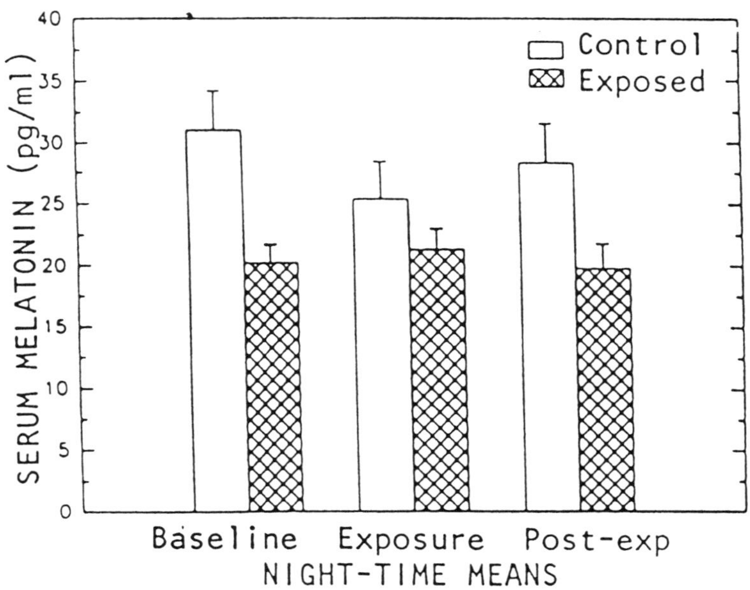

FIG. 3.—Results of Experiment III showing that nocturnal melatonin concentration is not affected by regular day-time exposure to 30 kV/m and 1.0 G.

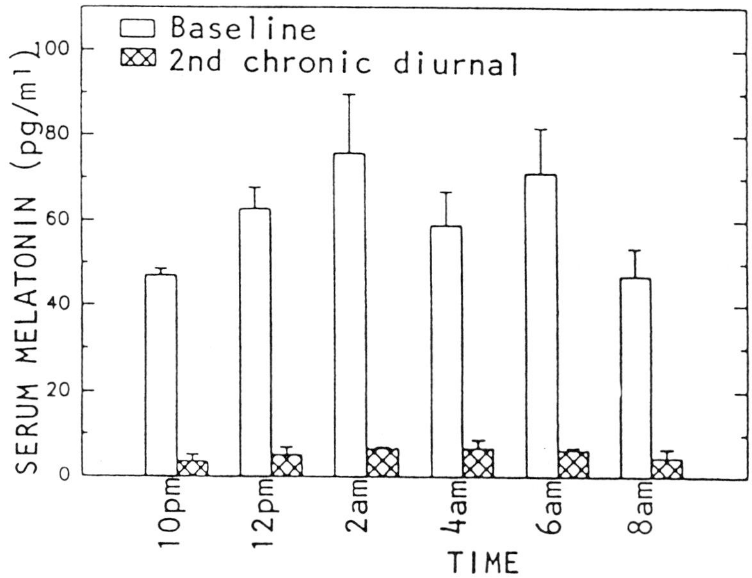

FIG. 4.—Result of pilot experiment suggesting that exposure to rapid-onset, irregularly scheduled field of 30 kV/m and 1.0 G can suppress nocturnal serum melatonin concentration.

REFERENCES

1. M. Kato, K-I. Honma, T. Shigemitsu, and Y. Shiga, "Effects of circularly polarized sine wave 50-Hz magnetic field exposure on plasma and pineal melatonin levels of rats," *Bioelectromagnetics* (in press).

2. R. J. Reiter, L. E. Anderson, R. L. Buschbom, and B. W. Wilson, "Reduction of the nocturnal rise in pineal melatonin levels in rats exposed to 60-Hz electric fields in utero and for 23 days after birth," *Life Sci.* 42: 2203-2206, 1988.

3. B. W. Wilson, L. E. Anderson, D. I. Hilton, and R. D. Phillips, "Chronic exposure to 60 Hz electric fields: Effects on pineal function in the rat," *Bioelectromagnetics* 2: 371-380, 1981.

4. B. W. Wilson, L. E. Anderson, D. I. Hilton, and R. D. Phillips, Erratum, *Bioelectromagnetics* 4: 293, 1983.

5. B. W. Wilson, E. K. Chess, and L. E. Anderson, "60 Hz electric field effects on pineal melatonin rhythms: Time course of onset and recovery," *Bioelectromagnetics* 7: 239-242, 1986.

6. S. M. Yellon, "An acute 60 Hz magnetic field exposure suppresses the nighttime melatonin rise in the pineal and circulation of the adult Djungarian hamster," presented at USDoE *Annual Review of Research on Biological Effects of 50 and 60 Hz Electric and Magnetic Fields,* Milwaukee, Wis., 1991.

7. R. J. Reiter, "Action spectra, dose-response relationships, and temporal aspects of light's effect on the pineal gland," *Ann. N.Y. Acad. Sci.* 453: 215-229, 1985.

8. M. Nozaki, M. Tsushima, and Y. Mori, "Diurnal changes in serum melatonin concentration under indoor and outdoor environments and light suppression of nighttime melatonin secretion in the female Japanese monkey," *J. Pineal Res.* 9: 211-230, 1990.

9. A. Lerchl, K. O. Nonaka, and R. J. Reiter, "Pineal gland 'magnetosensitivity' to static magnetic fields is a consequence of induced electric currents (eddy currents)," *J. Pineal Res.* 10: 109-116, 1991.

10. B. W. Wilson, C. W. Wright, J. E. Morris, R. L. Buschbom, D. P. Brown, D. L. Miller, R. Sommers-Flannigan, and L. E. Anderson, "Evidence for an effect of ELF electromagnetic fields on human pineal gland function," *J. Pineal Res.* 9: 259-269, 1990.

11. This research is sponsored jointly by the U.S. Department of Energy and the Central Research Institute of the Electric Power Industry of Japan, under contract DE-AC02-80-RA50219 to the Southwest Research Institute. The technical contributions of Jeffrey H. Lucas, H. Dwaine Smith, Lornell Barlow-Walden, Gary T. Moore, C. William Hall, Sherene S. Singh, and David H. Wood are gratefully acknowledged.

Martin Blank, Ed., *Electricity and Magnetism in Biology and Medicine*

ELF MAGNETIC FIELDS AND MELATONIN-INDUCED GROWTH INHIBITION OF ER+ BREAST CANCER CELLS

R. P. Liburdy,* R. Sokolic, and P. Yaswen

Lawrence Berkeley Laboratory, Calif.

Exposure to ELF fields is reported to depress or time-shift melatonin secretions in animals[1-4] and may be a risk factor in human breast cancer epidemiology.[5-8] In this study we asked whether an ELF magnetic field acts at the cellular level to influence growth of the breast cancer cell MCF-7 in the presence of melatonin, a known oncostatic agent. Continuous exposure to a sinusoidal 9.5-15.0 mG 60Hz magnetic field over a 7 day growth period resulted in blocking of melatonin's oncostatic effect on MCF-7 proliferation; this blocking was not observed in a background field of 1.6-3.5 mG. These findings link ELF fields, melatonin, and in vitro growth of a breast cancer cell line. The mechanism(s) of interaction is unknown and may involve modulation of signal transduction events associated with melatonin binding and MCF-7 cell growth.

Since ELF field exposure has been associated with mammary cancer risk and incidence, and ELF fields influence melatonin secretion in animals, we decided to examine whether ELF fields could directly influence melatonin-induced growth inhibition of a breast cancer cell line, MCF-7, in vitro. Melatonin, the pineal gland's major hormone, exhibits oncostatic activity on human estrogen-receptor-positive (ER^+) breast cancer cells such as MCF-7.[9,10] MCF-7 cells are derived from the pleural effusion of a mammary adenocarcinoma; when grown as a monolayer, these cells possess an epithelial-like morphology [ATTC HTB-22].

MATERIALS AND METHODS

Cells, Hormones, and Drugs. MCF-7 cells at passage 315 were a gift of Dr. David Blask of the Mary Imogene Bassett Hospital Research Institute, Cooperstown, N.Y. Cells were maintained in Dulbecco's Modified Eagle's Medium (DMEM H-21), supplemented with 10% Fetal Bovine Serum, penicillin (100 units/ml). streptomycin (100 mg/ml), and L-Glutamine (2 mm) (UCSF Cell Culture Facility, San Francisco), at 37 C in a humid atmosphere containing 5% CO_2. Melatonin was purchased from Sigma Chemical Co. (St. Louis, Mo.).

Exposure Systems. Two double-wound Merritt four-coil exposure systems were employed in matched Queue incubators (37 ± 0.1 C); parallel (active) and antiparallel (control) current flow generated the 60Hz ELF magnetic field environments of 9.5-15.0 mG (active) and 1.6-3.5 mG (background), respectively, in the central volume of the coil systems. DC field values were equivalent (130 ± 12 mG). Plate position in the central volume was systematically randomized during exposures, with matched active vs. control exposures conducted simultaneously.

Culture Techniques. MCF-7 cells were grown out in 60mm plates or T-75 flasks prior to the experiment. Cells were harvested in a 0.2% EDTA phosphate buffer from the seeding vessel and diluted in DMEM. Cells were employed at a density of 0.1×10^5 cells/35×15mm culture dish. After cells became firmly attached, 4 h after seeding, the plating medium was replaced with melatonin-enriched medium 10^{-11} M, 10^{-9} M, 10^{-7} M, 10^{-5} M) or DMEM (zero melatonin). Triplicate plates were used for each melatonin concentration. On alternate days, cells were harvested from plates in each treatment group by incubation with trypsin at 37 C for 2 min, and counted with the hemocytometer. Fresh medium, with or without melatonin as required, was fed to the remaining plates twice weekly.

---

*Correspondence to: R. P. Liburdy, LBL Bioelectromagnetic Research Facility (Bldg 74), University of California, Berkeley CA 94720.

RESULTS

The growth of MCF-7 cells in the presence of 10^{-11} M, or with no melatonin in a background 1.6-3.5mG magnetic field, is shown in Fig. 1(a). Typical exponential growth was observed over the 7-day period. When melatonin was present, an approximately 40% inhibition of growth on days 5 and 7 was detected ($p = 0.002$), consistent with reports from Blask and colleagues.[9,10,12] Figure 1(b) presents data for cells continuously exposed to a 9.5-15.0mG magnetic field. In the absence of melatonin we see an exponential growth curve for the MCF-7 cells that is superimposable with cells maintained in the background field (cf. Fig. 1a). However, when melatonin is present, the ELF field is observed to block the oncostatic growth inhibition of melatonin.

We also investigated whether this field effect is melatonin-dose dependent. Figure 2(a) depicts data (% of control) corresponding to growth of MCF-7 cells on day 5 at various melatonin concentrations in the presence of background 1.6-3.5mG field (solid curve) or the exposure 9.5-15.0mG field (dashed curve). In the absence of melatonin, cell growth for both fields was indistinguishable. As melatonin concentration was increased for the background field, maximal growth inhibition occurred at 10^{-11} M melatonin. In contrast, melatonin's growth inhibition was significantly blocked by the 9.5-15.0mG field at all melatonin concentrations. Figure 2(b) shows that a similar pattern was observed for cell growth on day 7. ANOVA across all data yields $p = 0.0014$.

In these studies no gross morphological differences were noticed between background and exposed MCF-7 cells, whether in the presence or absence of melatonin.

DISCUSSION

Our findings indicate that ELF magnetic fields at 9.5-15.0 mG act to block melatonin's growth inhibition of the breast cancer cell line MCF-7. Magnetic fields had no effect on MCF-7 cells in the absence of melatonin. Thus, although the mechanism of ELF field action is unknown, it must be mediated at the cellular level through melatonin and its action on the MCF-7 cell.

Several models have been proposed to elucidate the link between ELF field exposure and breast cancer incidence.[1] For example, one important aspect is an ELF-induced decrease in melatonin secretion in vivo and a concomitant enhancement in the production of prolactin and estrogen., which in turn may increase the growth of susceptible breast epithelial cells.[11] Our in vitro study suggests a direct interaction between ELF fields, the MCF-7 cell, and melatonin at the cellular level. This interaction most likely involves receptor activation and signal transduction (ST) pathways.

End-stage ST events include cellular proliferation and protein secretion. In in vitro studies with transformed HL60 cells, translation and transcription are altered in ELF magnetic fields.[13] Experimental evidence for a 60Hz field effect on an early-stage ST marker, $[Ca^{2+}]_i$ in mitogen-activated lymphocytes was first reported by us. Alterations in $[Ca^{2+}]_i$ were not observed during 60Hz field exposure of resting cells, but approximately 100 s after addition of mitogen, changes in $[Ca^{2+}]_i$ were detected. We postulate that alterations of an early ST marker such as $[Ca^{2+}]_i$ will propagate along the ST cascade to influence transcription and expression of such proto-oncogenes as c-MYC; we have recently provided evidence for this interaction.[15] In this regard, we hypothesize that a possible mechanism for ELF field action on MCF-7 growth may involve (a) an interference with melatonin's activity and subsequent ST sequelae, and/or (b) an activation of estrogen receptors to overcome melatonin's antimitogenic action. Future studies should address such issues.[16]

REFERENCES

1. B. W. Wilson et al., *J. Pineal Res.* 9: 259-269, 1990.
2. R. Reiter et al., DOE Contractor's Review, Abst. A-23, Milwaukee, Wisc., Nov. 1991.
3. A. Lerchl, R. Reiter, et al., *J. Pineal Res.* 10: 108-116, 1991.
4. S. M. Yellon, in Ref. 2, Abst. A-25.
5. T. Tynes and A. Andersen, *Lancet* 336: 1596, 1990.
6. G. Matanoski et al., *Lancet* 33: 737, 1991.
7. P. A. Demers et al., *Am. J. Epidemiol.* 132: 775-776, 1990.
8. R. Stevens, *Am. J. Epidemiol.* 125: 556-561, 1987.
9. S. Cos and D. E. Blask, *Cancer Letters* 50: 115-119, 1990.
10. S. M. Hill and D. E. Blask, *Cancer Res.* 48: 6121-6126, 1988.
11. M. Cohen et al., *Lancet* 2: 814-816, 1978.
12. S. Cos, D. E. Blask, et al., *J. Pineal Res.* 10: 36-42, 1991.

13. R. Goodman, *Bioelectrochem. and Bioenerg.* 25: 335-355, 1991.
14. R. P. Liburdy, *FEBS Letters* 301: 53-59, 1992.
15. R. P. Liburdy, D. E. Callahan, T. R. Sloma, and P. Yaswen, this vol., pp. ???.
16. Research supported in part by the U.S. Department of Energy under contract DE-AC03-76SF00098.

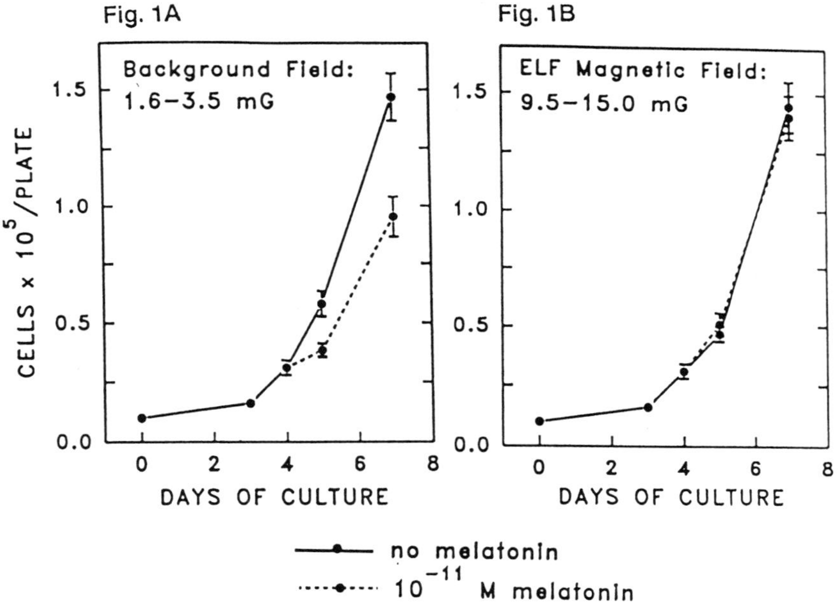

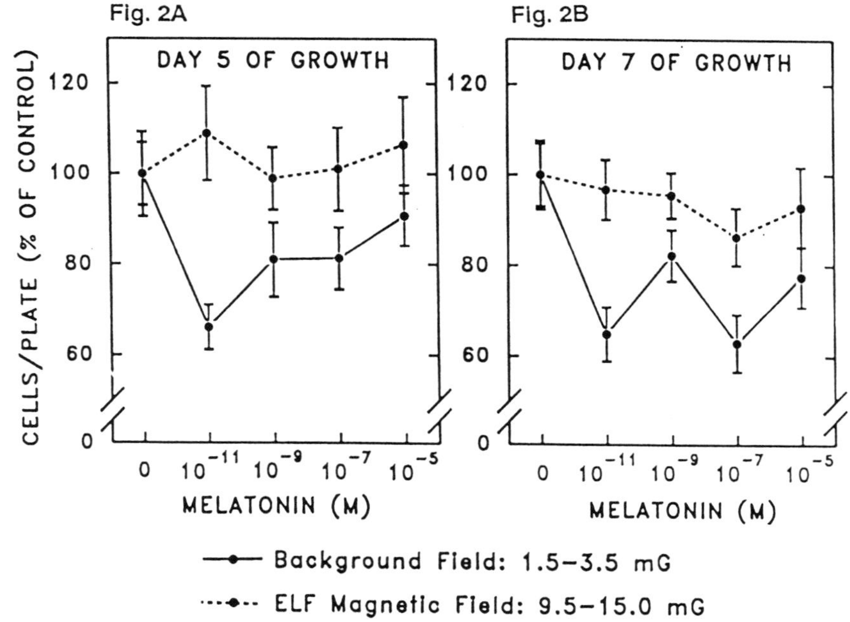

FIG. 1.—Effect of melatonin (10^{-11} M) on MCF-7 cell growth in the presence of an ELF magnetic field. (a) Melatonin exhibits typical inhibition of MCF-7 cell growth in background field of 1.6-3.5 mG (p = 0.002); (b) melatonin fails to inhibit growth of MCF-7 cells in 9.5-15.0mG field. Mean ± S.E.

FIG. 2.—Effect of melatonin dose (10^{-5}-10^{-11} M) on MCF-7 cell growth at days 5 and 7 in presence of ELF magnetic field. (a) Day 5: melatonin growth inhibition at 10^{-7}-10^{-11} M is blocked by 9.5-15.0mG ELF magnetic field; (b) day 7: melatonin growth inhibition at 10^{-7}-10^{-11} M is blocked by 9.5-15.0mG ELF magnetic field. Mean ± S.E. Multivariate ANOVA across all data yields p = 0.0014 for blocking effect of the 9.5-12.0mG ELF magnetic field.

Martin Blank, Ed., *Electricity and Magnetism in Biology and Medicine*

ENDOCRINE RESPONSES OF EWE LAMBS EXPOSED TO 60Hz ELECTRIC AND MAGNETIC FIELDS FROM A 500kV TRANSMISSION LINE

J. Lee, Jr.*

Bonneville Power Administration, Portland, Ore.

F. Stormshak, J. Thompson, P. Thinesen, L. Painter, and B. Olenchek

Oregon State University, Corvallis

D. Hess

Oregon Regional Primate Research Center, Beaverton

R. Forbes

Portland State University, Portland, Ore.

Studies of laboratory animals demonstrate that exposure to ELF electric and magnetic fields can significantly depress nocturnal levels of the pineal hormone melatonin. Other studies report that these fields may also affect levels of stress hormones. This environmental study was conducted to determine whether such effects occur in ewe lambs exposed to 60Hz fields from a 500kV transmission line. Exposure of lambs for 10 months had no apparent effects on serum concentrations of melatonin, cortisol, and progesterone. Preliminary results of a follow-up study also fail to indicate that 60Hz fields depress melatonin levels in ewe lambs.

Melatonin is produced by the pineal gland, and both pineal and serum melatonin concentrations are low during daylight hours and are elevated during darkness. Wilson et al. were the first to report that exposing rats to 60Hz electric fields of 1.7 or 60 kV/m significantly depressed the nocturnal concentration of pineal melatonin.[1] Nocturnal pineal melatonin was also reduced in young rats exposed to 60Hz electric fields in utero and for the first 23 days after birth.[2] However, Sasser et al. recently reported that 30 days of exposure to 39kV/m 60Hz electric fields did not reduce pineal melatonin levels in rats.[3]

Results of two studies suggest that power-frequency magnetic fields (as low as 10 mG) can affect melatonin levels in laboratory animals.[4,5] Other studies report that changes in dc magnetic fields can significantly affect nocturnal melatonin levels in rodents.[6]

Results of studies of stress-related hormones in laboratory rats exposed to 60Hz fields are conflicting. Wilson and Anderson suggested that effects of electric fields on pineal melatonin in laboratory rats are manifestations of stress.[7]

One function of melatonin is to provide information about daylength for regulating reproduction in species like sheep, which are seasonal breeders. Treatments that affect melatonin production in sheep can have large effects on the timing of reproduction, including the attainment of puberty in female lambs.[8]

METHODS

For experiment 1, 10 female Suffolk lambs were penned from April 1990 to February 1991 beneath a 500kV transmission line near Portland, Oregon.[9,10] The mean 60Hz electric field in the line pen ranged from 4.1 to 7.5 kV/m. The magnetic field ranged from 15 to 50 mG. Ten other lambs of the same type were

*Correspondence to: J. Lee, Jr., Bonneville Power Administration, Portland OR 97208.

penned in a control area 229 m from the line where the electric field was less than 0.01 kV/m and the magnetic field was less than 0.4 mG.

Serum melatonin and cortisol were measured by radioimmunoassay (RIA) from 6618 blood samples collected at 0.5 to 3h intervals over eight 48h periods. Serum progesterone was measured by RIA from blood samples collected twice weekly beginning when the lambs were 23 weeks of age. This hormone was used to measure the onset of puberty.

Melatonin and cortisol data were analyzed by repeated measures analysis of variance. Age at puberty was analyzed by an independent-samples t-test.

For experiment 2, 30 lambs (15 line, 15 control) were studies from April to February 1992.

RESULTS

In experiment 1, both line and control lambs showed the typical pattern of melatonin secretion consisting of low daytime and high nighttime serum levels of the indoleamine. There were no significant differences between groups in mean nocturnal melatonin concentration or in the phase or duration of the nocturnal melatonin secretion ($p > 0.05$). Mean nocturnal melatonin levels for the eight 48h samples are presented in Fig. 1.

Serum cortisol showed a circadian rhythm with highest levels during the day. There were no differences in cortisol levels between line and control groups. Mean daytime cortisol levels for the eight 48h samples are shown in Fig. 2. Age at puberty did not differ between the two groups (controls, 241.2 ± 6.8 days vs line, 236.2 ± 10.5 days).

For experiment 2, only a portion of the data has been analyzed. Preliminary data on mean nocturnal melatonin concentration for the first three of eight 48h samples are shown in Fig. 3. There is no indication of a decrease in melatonin associated with exposure.

Effects of electric and magnetic fields on the endocrine system reported in laboratory studies were not observed in this environmental study of sheep. This finding may be due to inherent species differences in response to 60Hz fields. As a possible analogy, there is a wide range in response of species to the effect of light on depressing nocturnal melatonin levels. Laboratory rats appear to be much more sensitive to light than sheep and humans. Results of this environmental study point out limitations of extrapolating from data on laboratory studies of rodents. In addition to species differences, field exposures in the laboratory may not duplicate environmental exposures.[11]

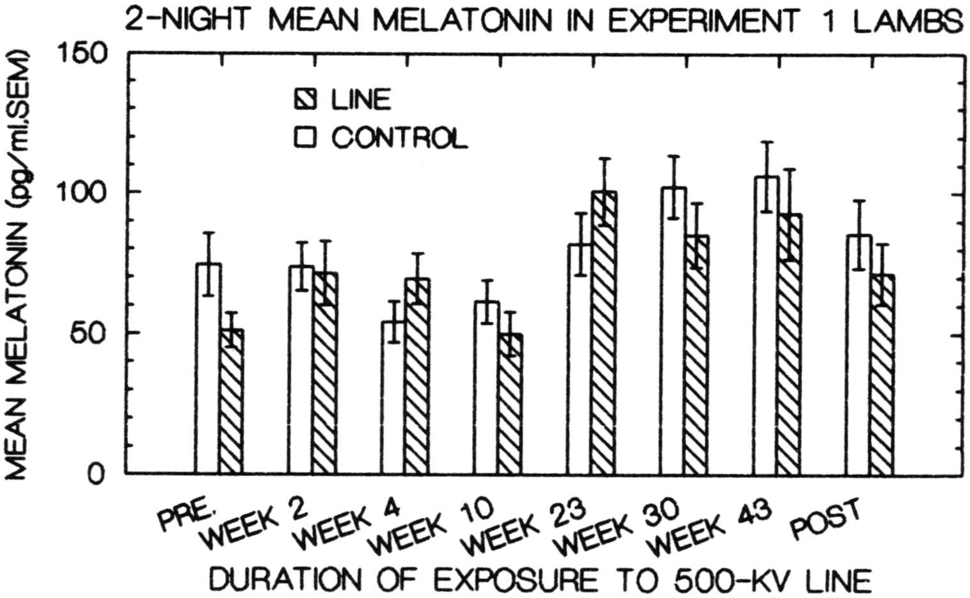

FIG. 1.—Nocturnal melatonin serum concentration for experiment 1 lambs averaged over each of two nights in eight 48h samples.

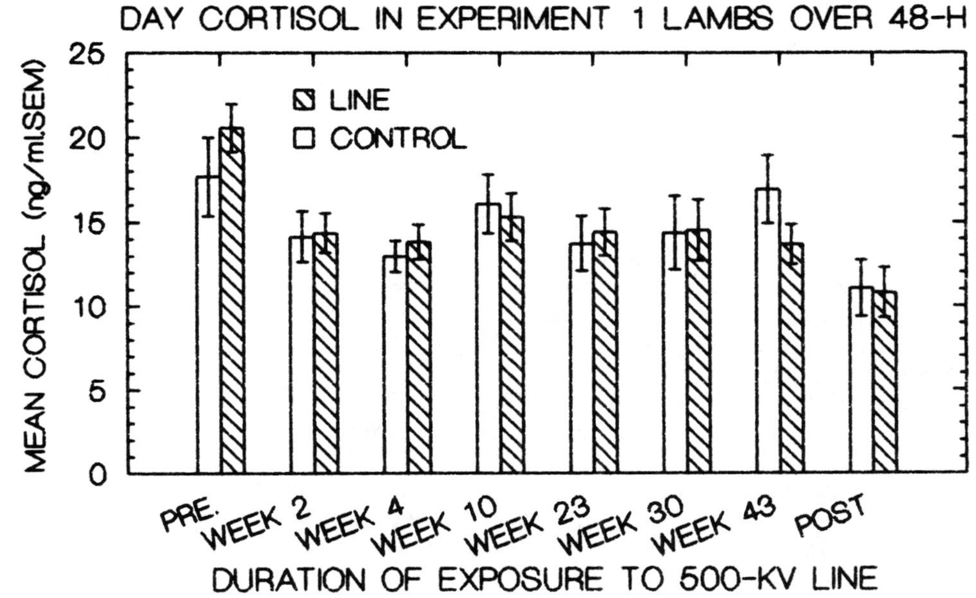

FIG. 2.—Mean serum cortisol concentrations for experiment 1 lambs for eight 48h samples.

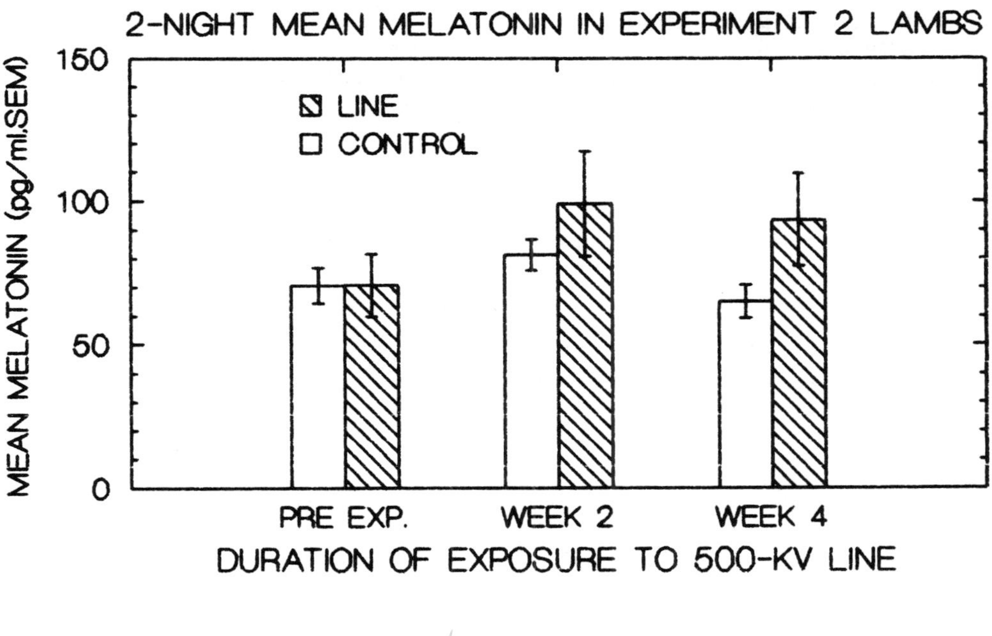

FIG. 3.—Nocturnal melatonin serum concentrations for experiment 2 lambs averaged over each of two nights in three 48h samples.

REFERENCES

1. B. W. Wilson, L. E. Anderson, D. I. Hilton, and R. D. Phillips, "Chronic exposure to 60-Hz electric fields: Effects on pineal function in the rat," *Bioelectromagnetics* 2: 371-380, 1981.

2. R. J. Reiter, L. E. Anderson, R. L. Buschbom, and B. W. Wilson, "Reduction of the nocturnal rise in pineal melatonin levels in rats exposed to 60-Hz electric fields in utero and for 23 days after birth," *Life Sciences* 42: 2203-2206, 1988.

3. L. B. Sasser, J. E. Morris, R. L. Buschbom, D. L. Miller, and L. E. Anderson, "Effect of 60-Hz electric fields on pineal melatonin during various times of the dark period," *Annual Review of Research on Biological Effects of 50- and 60-Hz Electric and Magnetic Fields,* Washington, D.C.: U.S. Department of Energy (Office of Energy Management), 1991, p. A-24.

4. S. M. Yellon, "An acute 60-Hz magnetic field exposure suppresses the nighttime melatonin rise in the pineal and circulation of the adult Djungarian hamster," *Annual Review of Research on Biological Effects of 50- and 60-Hz Electric and Magnetic Fields,* Washington, D.C.: U.S. Department of Energy (Office of Energy Management), 1991, p. A-25.

5. M. Kato, K. Honma, and T. Shigemitsu, "Effects of 50-Hz rotating magnetic fields on melatonin secretion in rats," *Abstracts, 12th Ann. Meet. Bioelectromagnetics Society* 12: 90, 1991.

6. A. Lerchl, K. O. Nonaka, and R. J. Reiter, "Pineal gland 'magnetosensitivity' to static magnetic fields is a consequence of induced electric currents (eddy currents)," *J. Pineal Res.* 10: 109-116, 1991.

7. B. W. Wilson and L. E. Anderson, "ELF electromagnetic field effects on the pineal gland," in B. W. Wilson et al., Eds., *Extremely Low Frequency Fields: The Question of Cancer,* Richland, Wash.: Battelle Press, 1990, 159-186.

8. S. M. Yellon and D. L. Foster, "Melatonin rhythms time photoperiod induced puberty in the female lamb," *Endocrinology* 119: 44-49, 1986.

9. F. Stormshak et al., *Joint HVAC Transmission EMF Environmental Study, Experiment 1,* Report prepared for Bonneville Power Administration, Portland, Oregon.

10. J. M. Lee Jr., *Seasonal Patterns of Melatonin, Cortisol, and Progesterone Secretion in Female Lambs Raised Beneath a 500-kV Transmission Line,* Doctoral dissertation, Portland State University, 1992.

11. This research was sponsored by Bonneville Power Administration, American Electric Power Service Corp., Houston Lighting and Power Co., Hydro-Québec, Salt River Project, and Western Area Power Administration.

Martin Blank, Ed., *Electricity and Magnetism in Biology and Medicine*

CHARACTERIZATION OF IMMORTALIZED CELL LINES ISOLATED FROM A SOFT-TISSUE TUMOR DERIVED FROM A RABBIT SUBCHRONICALLY EXPOSED TO 60Hz MAGNETIC FIELDS AT 1 G

J. T. Ning,* G. Bushar, L. Li, S. Toyokuni, D. Thomas, L. Cress, E. M. Czerska, and M. L. Swicord

Center for Devices and Radiological Health, Rockville, Md.

E. Elson

Walter Reed Army Institute of Research, Washington, D.C.

C. C. Davis

University of Maryland, College Park

Several immortalized cell lines have been isolated from a tumor growth of a rabbit that was undergoing ELF exposure. One of these cell lines became anchorage-independent in growth. Characterization of this cell line revealed significant responses to ELF exposure. Further study of this cell line may lead to better understanding of the mechanism of ELF interaction with the cell.

A New Zealand White male rabbit which underwent whole-body exposure to 60 Hz CW at 1 G for 6 h/day, 5 days/week developed a tumor mass approximately 1 month after initiation of the exposure regimen. At the end of the 3-month exposure period, the mass grew to approximately 10 cm in diameter. In order to characterize and elucidate any relationship between 60Hz magnetic field exposure and the development of this tumor growth, efforts were made to establish cell lines from explants of the tumor.

MATERIALS AND METHODS

At necropsy, the tumor was harvested immediately. Within 5 min of sacrifice, the tumor was minced in Hank's balanced salt solution without calcium or magnesium which had been prewarmed to 37 C in a sterile glass Erlenmeyer flask supplemented by 1mM EDTA. After mincing, the tissues were transferred to several glass Erlenmeyer flasks containing prewarmed RPMI-1640 supplemented by 15% FBS, 25 mM HEPES, 1.25 U/ml Dispase, 0.5 mg/ml collagenase, and 1 mg/ml hyaluronidase, and were incubated overnight at 37 C in a 5% CO_2 incubator. The next day, the media were removed from the Erlenmeyer flasks and replaced with RPMI-1640 without serum but supplemented by 1.25 U/ml Dispase and 0.5 mg/ml of collagenase. After 30min incubation at 37 C, the suspension was pipetted vigorously and then centrifuged at low speed. The pellets were washed three times with RPMI-1640 $\pm$ 20% FBS and then seeded into several small tissue culture flasks. Cells were incubated in a CO_2 incubator at 37 C for 24 h. After the cells appeared to have attached well to the substratum, the media were carefully pipetted off and replaced with fresh RPMI-1640 $\pm$ 20% FBS; cells were then routinely split every 3 to 4 days to isolate immortalized cell lines. Eight of the 32 flasks inoculated with minced tumor explant proliferated and exhibited foci formation without reaching confluence on day 4 of culture.

The following analyses were used to characterize the immortalized cell lines: flow cytometry, electron microscopy, two-dimensional protein gel electrophoresis, and nuclear run-off assay.

---

*Correspondence to: J. T. Ning, FDA-CDRH, Rockville MD 20857.

RESULTS

The foci of these immortalized cell lines (Fig. 1a) were selectively passaged in a cloning cylinder to transfer focus-forming cells. One of the resulting cell lines eventually exhibited anchorage-independent growth. That cell line, R3, was further characterized by flow cytometry, electron microscopy, two-dimensional protein electrophoresis, and nuclear run-off assay. Flow cytometric analysis of RNA fluorescence with acridine orange exhibited significant increase in RNA fluorescence when exposed to ELF fields (Fig. 2). Another cell line (R4), which retained anchorage-dependent growth, did not exhibit any increase in RNA fluorescence with ELF exposure (Fig. 3). Electron microscopy study of R3 did not demonstrate any virus particle harbored by this cell line (Fig. 4). Daudi, a human Burkitt lymphoma cell line which harbors the EB virus, served as the positive control and showed viral particle (not shown). Two-dimensional protein gel electrophoresis revealed apparent induction of several new proteins in exposed cells (Fig. 5) as compared to control (Fig. 6). Nuclear run-off assay demonstrated over eight to eighteen-fold increase in c-Ha-ras, N-ras, and c-sis oncogene expression (Fig. 7).

DISCUSSION

The immortalized cell lines isolated from the rabbit undergoing ELF exposure have now been continuously carried in culture for over 12 months and continue to exhibit foci formation (Fig. 1) and continue to display uncontrolled proliferation potential. This result is in contrast to primary cultures (Fig. 8) isolated by the same procedure from normal soft tissues that were immediately adjacent to the tumor. In addition, these immortalized cell lines have been tested positive for anchorage-independent growth. With 60Hz, 1G exposure, these cells showed changes in 2-D protein gel pattern and significant increase in RNA fluorescence as analyzed by flow cytometry, characteristics that are not exhibited by other immortalized cell lines.

There are three possible causes for the origin of the rabbit tumor. One is spontaneous formation. Another possible cause may be due to endogenous oncogenic virus, which is unlikely since electron-microscope study did not identify any viral particles in this cell line. The third possible reason infers that the origin of this tumor may be somehow linked to ELF exposure. To test this possibility, time and effort were spent in trying to isolate immortalized cell lines from the tumor. The observation that these rabbit tumor cell lines respond significantly to ELF fields when analyzed by multiple analytical techniques supports this possibility as the cause for the origin of the parent tumor. These anchorage-independent cells have now been inoculated into the flanks of nude mice in an attempt to assess their tumor-forming potential with respect to ELF exposure.

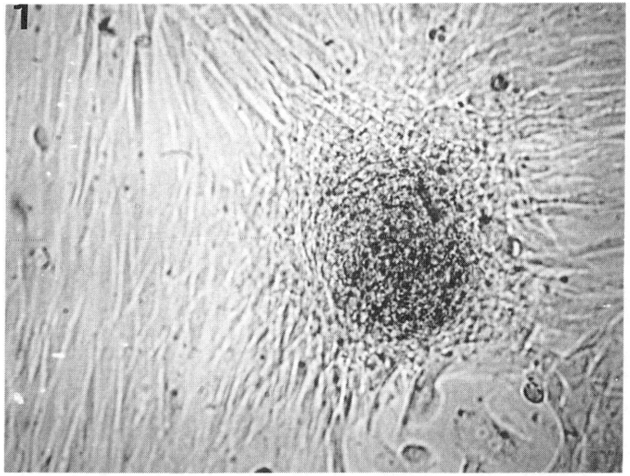

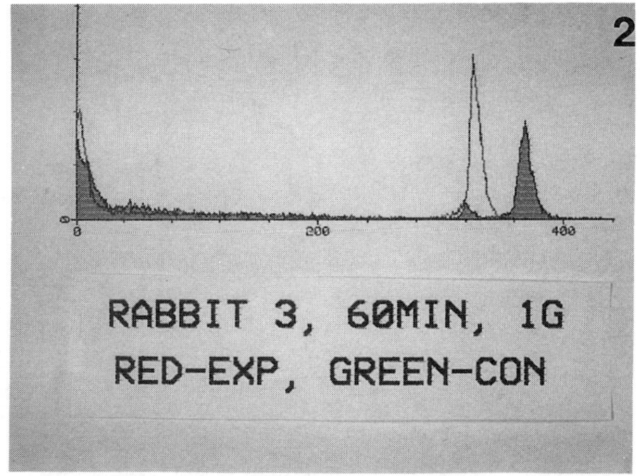

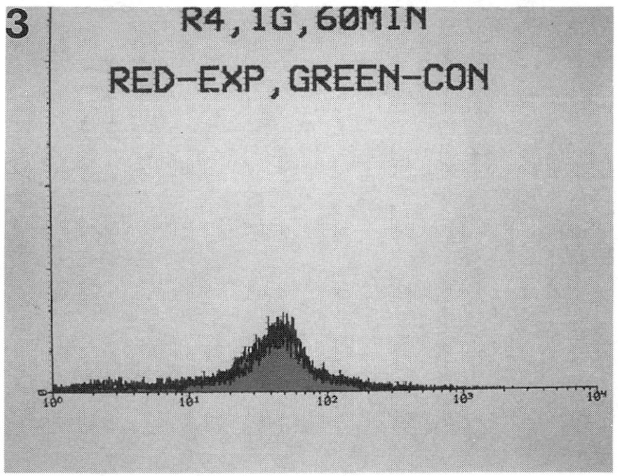

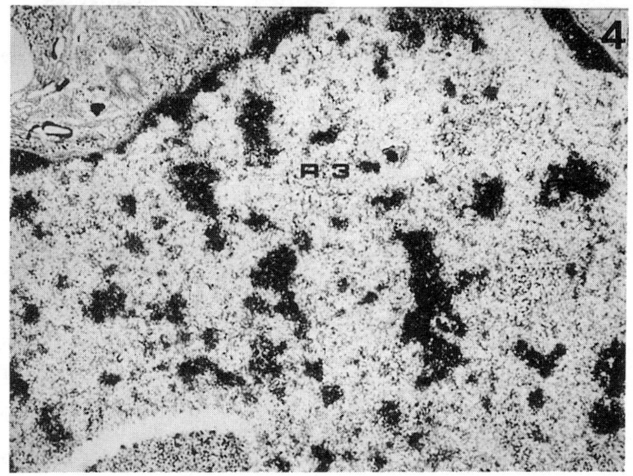

FIG. 1.—R3, focus-forming cells.

FIG. 2.—R3, RNA fluorescence histograms of exposed and control cell populations. (Increased RNA results in fluorescence shift to the right.)

FIG. 3.—R4, RNA fluorescence histograms of exposed and control cell population.

FIG. 4.—Electron micrograph of R3 cell.

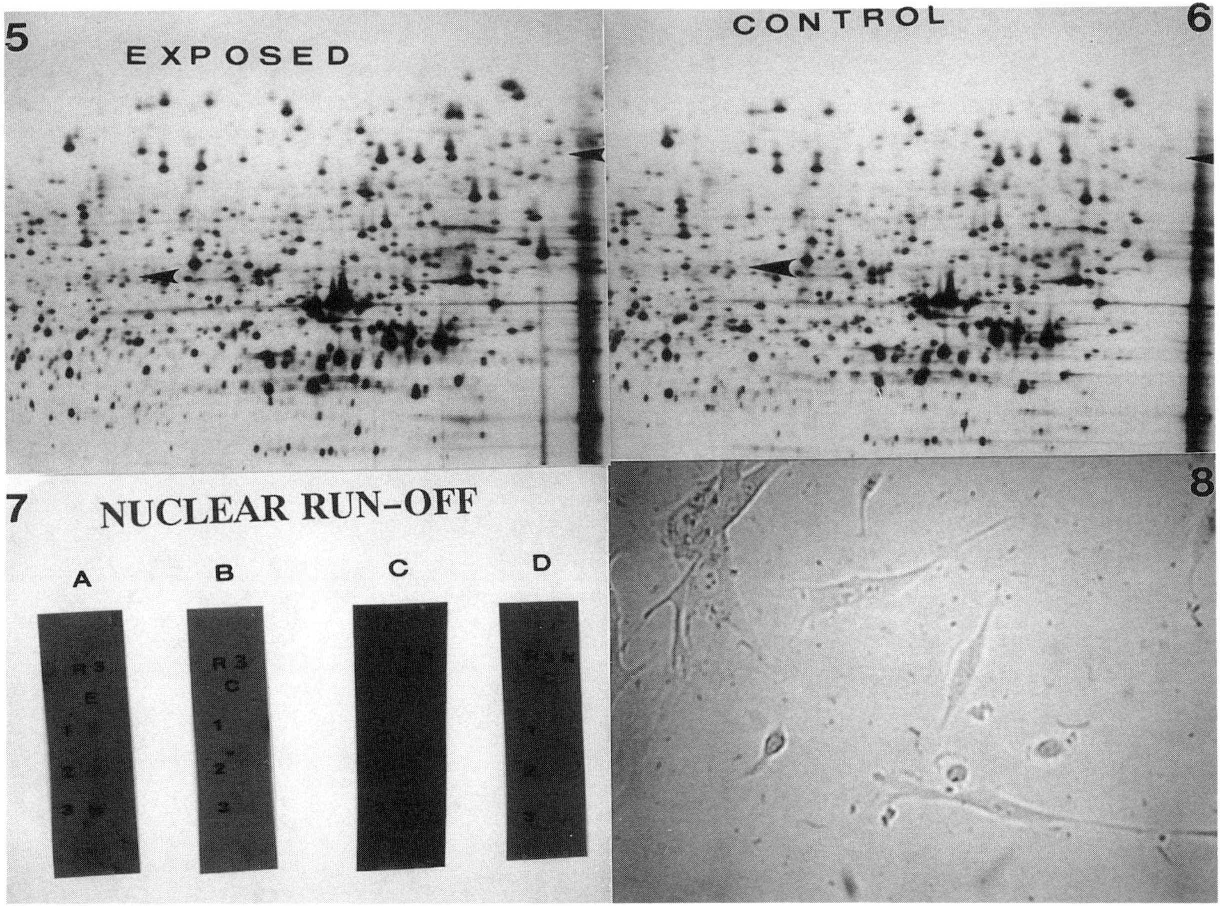

FIG. 5.—Two-dimensional protein gel of R3 exposed to ELF. (Arrows indicate new protein spots.)
FIG. 6.—Two-dimensional protein gel of R3 control.
FIG. 7.—Nuclear run-off assay: (A) R3 exposed, (B) R3 control, (C) R3 nuclei exposed, (D) R3 nuclei control.
FIG. 8.—Primary culture isolated from surrounding normal tissue.

Martin Blank, Ed., *Electricity and Magnetism in Biology and Medicine*

PROTOCOL DEVELOPMENT FOR A STUDY OF MAGNETIC-FIELD EFFECTS ON HUMAN MELATONIN

C. Graham,* H. D. Cohen, M. R. Cook, M. M. Gerkovich, and D. W. Riffle

Midwest Research Institute, Kansas City, Mo.

Public concern has increased about the possible health risks of exposure to electric and magnetic fields (EMF) generated by electric power distribution systems. Much of this concern has focused on the relationship between chronic exposure and cancer. Numerous epidemiological reports suggest a link between occupational or residential exposure to power-frequency magnetic fields and increased cancer risk.[1] Given the extensive use of electric power, even the small increases in risk reported to date could have substantial consequences for public health. Various federal agencies[2] and others (e.g., the World Health Organization) have called for research to identify a biological mechanism that could account for the observed statistical associations.

Melatonin (MLT), a hormone released by the pineal gland during the dark period of the circadian cycle, is thought to stimulate immune function and has been implicated in the control of cell proliferation and the growth of transplanted tumors.[3] In addition to its immunomodulatory effects, MLT plays a key role in the regulation of reproductive hormones implicated in a number of carcinogenic processes. Studies with rodents suggest that nocturnal MLT levels can be altered by EMF exposure,[4,5] and this relationship has been proposed as a plausible biological mechanism to account for the epidemiological findings.[6] Research is needed to determine whether exposure to magnetic fields at night suppresses nocturnal MLT levels in people.

The goal of this project was to develop and test the methods and procedures required to conduct such research under controlled experimental conditions. We first modified the 60Hz Human Exposure Test Facility at MRI[7] to perform all-night sleep studies. Computer programs were developed to control exposure conditions, record brain and heart activity during sleep, present performance tasks, and collect self-report measures. Protocols were prepared for the safe collection and processing of human blood, saliva, and urine samples for assay of MLT and its metabolites. We evaluated the feasibility of collecting and assaying nocturnal samples of sublingual and parotid saliva. Unlike urine, which yields data on metabolite levels pooled over time, salivary measures have the potential to track MLT release patterns in future studies of workplace or residential exposure. A preliminary study of MLT levels in human volunteers exposed at night to dim and bright light was performed. A future experiment to evaluate effects of intermittent exposure to 60Hz magnetic fields on nocturnal MLT levels in humans was designed.

METHODS

Five healthy young men who met selection criteria (age 21 to 35, no chronic disease or disability, nonsmoker, regular sleep habits, no night work) each participated in one training session and three all-night (2130-0800) test sessions. On the day of each test session, subjects refrained from consuming alcohol, had no caffeine after 1700, and no food after 1900. In the first session, subjects were familiarized with the laboratory and the data collection procedures. The second session was a baseline night during which subjects were exposed only to dim light (<10 lux). On the third night subjects were awakened and exposed for 1 h between 0200 and 0300 to bright light at an intensity (>8000 lux) reported to reduce serum MLT concentration. The magnetic fields were not activated in this study.

Blood samples were collected hourly via a catheter. Saliva samples were obtained every 2 h by cotton dental rolls. Urine samples were collected at the start and end of a session, and at noon the next day. Biological samples were sent to Dr. Russel Reiter for analysis of MLT and 6-OH MLT sulfate. Sleep EEG and heart rate were recorded continuously through the night. EEG data were analyzed for sleep stage and major

*Correspondence to: C. Graham, MRI, Kansas City MO 64110.

sleep parameters. Performance tests (simple and choice reaction time, time perception, speeded addition) and self-rating scales (mood, sleepiness, alertness) were administered at the start and end of a session.

RESULTS

Given the preliminary nature of this study and its purpose, this section will focus on findings relevant to MLT levels in serum and saliva, and on data illustrating physiological adaptation to the laboratory environment.

Melatonin in Serum and Saliva. Figure 1(a) shows the concentration of MLT in hourly serum samples over the test night (dashed line) and the dim-light control night (solid line) for the first three subjects. Figure 1(b) presents the same data for salivary MLT levels. The expected circadian curve with clear peaks during the control night was found. Exposure to bright light resulted in suppression of MLT. There was also evidence of a rebound effect, with MLT levels approaching control levels within 2 h. For each individual, there was good agreement in the patterns obtained during the two nights. MLT levels were variable at 2300, probably due to uncontrolled light history effects. The midnight sample obtained after 1 h in dim light was very consistent across test nights for a given individual. Data from the last two subjects could not be interpreted, as most of the data points fell within the coefficient of variation of the assay.

Relationship Between Serum and Saliva Levels. For a given subject, saliva levels as a percent of serum levels were consistent over sessions. However, the serum-saliva ratios were not constant. Figure 2 shows the data averaged over the three subjects for whom valid assay results were obtained. Ratios were higher at the beginning and end of the session than in the middle. Saliva sample volume was also higher at these points, indicating that the ratio may, at least in part, depend on flow rate. We also observed that saliva sample volume for all five subjects was increased at the 0300 collection after exposure to bright light as compared with the control night. The reasons for this finding are not clear; it could be a function of light exposure, of increased physical activity, or of changes in central nervous system arousal levels.

We do not recommend including saliva measures in studies of the effects of magnetic fields on MLT until issues about variability in serum/saliva ratios and issues about the effects of low rate on MLT concentration have been resolved. Because salivary measures could be of great value in field studies, we believe these issues should receive further investigation.

Sleep EEG. Traditional EEG scoring criteria were used to determine how long it took for subjects to fall asleep after each hourly body fluid collection (Multiple Sleep Latency Test). These data are summarized in Fig. 3. They show a clear pattern of physiological adaptation to the laboratory environment and the data collection procedures, both within and across test sessions. In session 1, sleep latency decreased in an orderly fashion over sequential collections from midnight to 0500. By session 2, the collection procedures caused little disruption of sleep latency. Data for session 3 show the clear effect of introducing the light challenge (blank box at 0200). Mean sleep latency after the 0300 blood and saliva collection was markedly increased; however, subjects quickly re-adapted and latency returned to previous levels by 0500. These patterns illustrate the need to include an initial adaptation night as a control procedure in future electrophysiological studies of night sleep in magnetic fields.

FUTURE RESEARCH DIRECTIONS

A laboratory investigation has been proposed to evaluate the effects of intermittent exposure to magnetic fields at two levels of field strength on nocturnal MLT release patterns in the human. Because MLT levels vary widely across people, each volunteer will first participate in two screening sessions to determine individual nocturnal MLT levels and the percent suppression of MLT produced by bright light. Screening will continue until 12 sets of three subjects each, matched for MLT level, have been identified. The subjects in each set will be randomly assigned to three groups: a sham-exposure control group, a group exposed at 10 mG, and a group exposed to 200 mG. A 60Hz circularly polarized magnetic field will be presented, and exposure will be intermittent (1 h on/1 h off through the night, with the field switched on and off every 15 s during "on" periods). Endpoints will include MLT concentration, heart rate, temperature, and performance measures.[8]

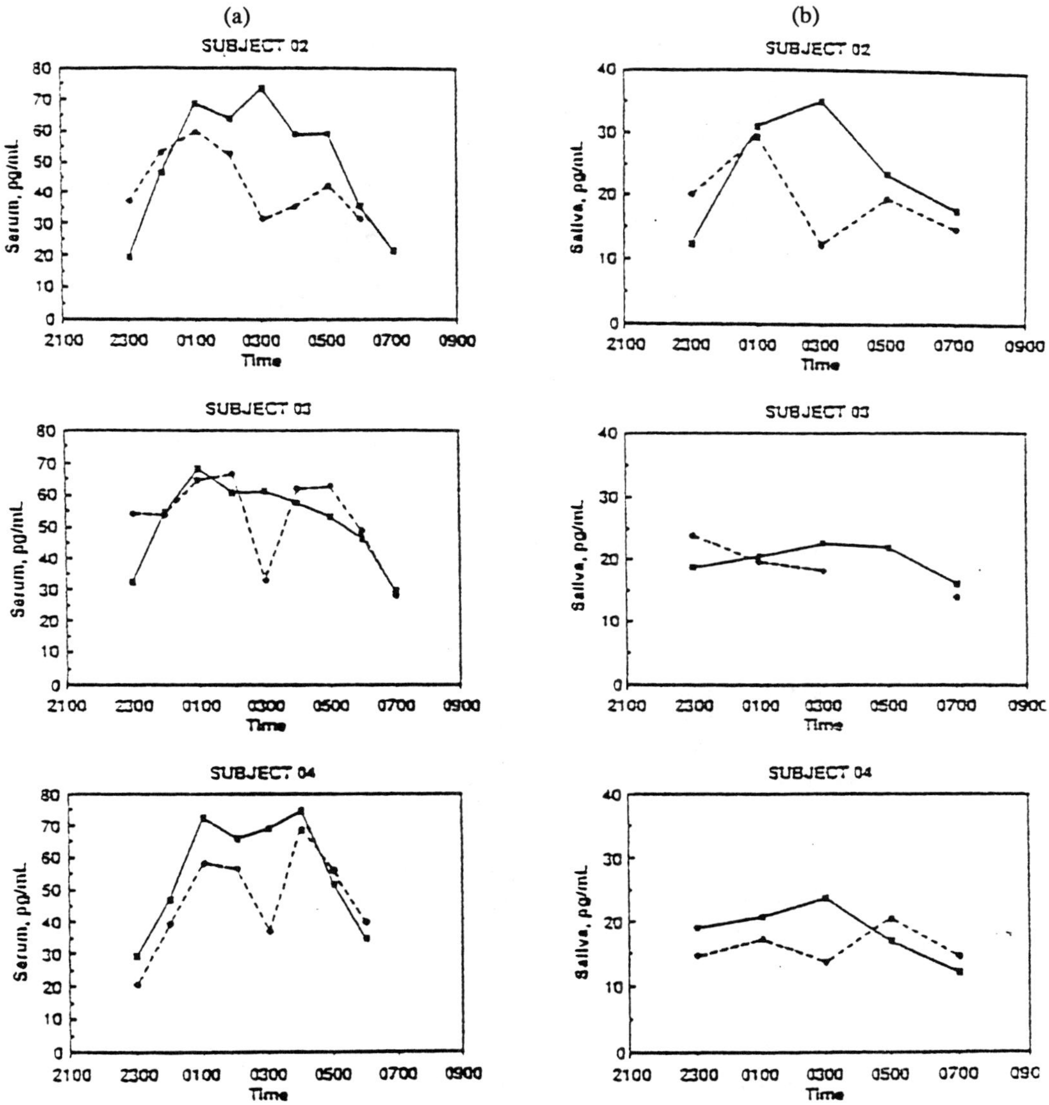

FIG. 1.—Melatonin levels in (a) human serum and (b) saliva were markedly reduced by light exposure (0200-0300) on test night (dashed line) compared with dim-light control night (solid line).

REFERENCES

1. G. Theriault, "Health effects of electromagnetic radiation on workers: Epidemiologic studies," *Scientific Workshop on the Health Effects of Electromagnetic Radiation on Workers* (NIOSH 91-37020), 1991.

2. National Institute for Occupational Safety and Health, Abstracts, Scientific Workshop on the Health Effects of Electromagnetic Radiation on Workers, Cincinnati, Ohio (NIOSH 91-37), 1991.

3. R. J. Reiter, "Pineal gland, cellular proliferation, and neoplastic growth: An historical account," in D. Gupta et al., Eds., *The Pineal Gland and Cancer,* Tübingen (Germany): Brain Research Promotion, 1988, 41-64.

4. A. Lerchl, K. O. Nonaka, K. A. Stokkan, and R. J. Reiter, "Marked rapid alterations in nocturnal pineal serotonin metabolism in mice and rats exposed to weak intermittent magnetic fields," *Biochem. and Biophys. Res. Comm.* 169: 102-108, 1990.

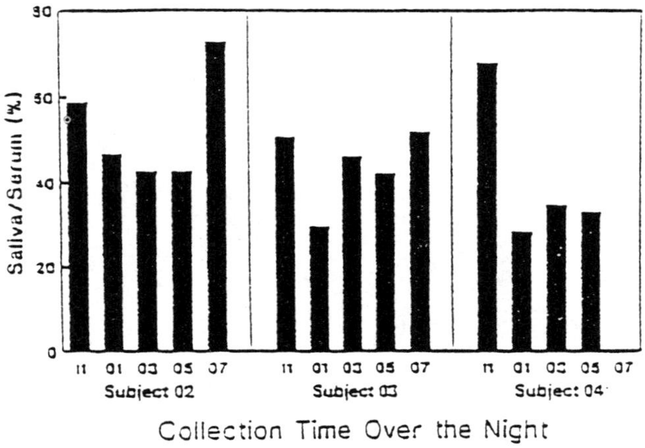

FIG. 2.—Variability observed in serum/saliva ratios over session may be related to differences in salivary flow rate.

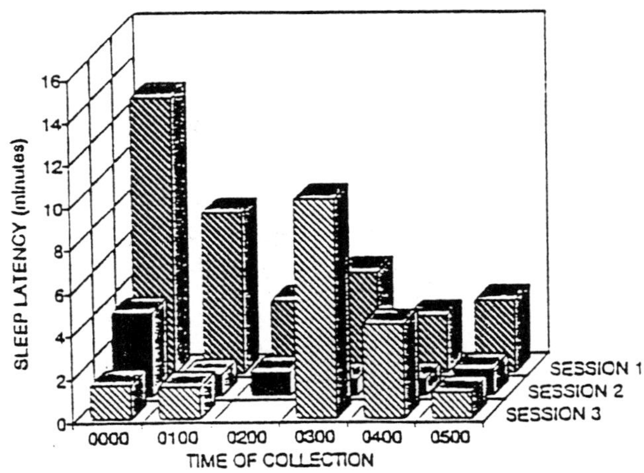

FIG. 3.—Physiological adaptation to laboratory environment occurred within and across sessions. Sleep latency after sequential body fluid collections decreased from midnight to 0500. Light exposure in session 3 (blank box, 0200) increased latency.

5. B. W. Wilson, E. K. Chess, and L. E. Anderson, "60-Hz electric-field effects on pineal melatonin rhythms: Time course for onset and recovery," *Bioelectromagnetics* 7: 239-242, 1986.

6. R. G. Stevens, "Electric power use and breast cancer: A hypothesis," *Am. J. Epidemiol.* 125: 556-561, 1987.

7. H. D. Cohen, C. Graham, M. R. Cook, and J. W. Phelps, "ELF exposure facility for human testing," *Bioelectromagnetics* 13: 169-182, 1992.

8. This work was funded by the Electric Power Research Institute under Contract RP2964-15. We thank Sherri Price, Research Nurse, and Steve Hoffman, Associate Psychologist, for their contributions to the research.

Martin Blank, Ed., *Electricity and Magnetism in Biology and Medicine*

EFFECTS OF AC MAGNETIC FIELDS ON DMBA-INDUCED MAMMARY CARCINOGENESIS IN SPRAGUE-DAWLEY RATS

M. Mevissen, U. Wahnschaffe, and W. Löscher*

School of Veterinary Medicine, Hanover, Germany

A. Stamm

Technical University of Braunschweig, Germany

A. Lerchl

University of Münster, Germany

A possible carcinogenic effect of extremely low frequency (50 or 60 Hz) magnetic fields (MF) recently suggested by epidemiological studies has engendered much concern. The aim of the present study was to determine whether low-frequency MFs at the low flux densities induce co-promoting effects in a DMBA model of breast cancer in rats. Furthermore, determinations of melatonin levels in serum were performed during MF exposure. Mammary tumors were induced by intragastric administration of 20 mg (5 mg/week) 7,12-dimethylbenz(a)anthracene in female Sprague-Dawley rats. Groups of 36 animals were either sham exposed or exposed to a 50Hz gradient MF of 0.3-1 μT for 91 days. Histopathological assessment of mammary lesions did not disclose any significant difference between MF-exposed rats and sham-exposed animals. Incidence of tumors was 61% in controls vs 67% in MF-exposed rats. Nocturnal serum melatonin levels were significantly reduced by MF exposure ($p \leq 0.05$). Thus, the data of this experiment do not indicate that a 0.3-1 μT MF induces marked tumor promoting effects in a DMBA model of breast cancer in rats.

There is accumulating evidence from epidemiological studies that exposure to MFs may increase the incidence of various types of cancer, particularly leukemia, brain tumors, and breast cancer.[1,2] However, there is little understanding of the nature of the interaction between MF and biological systems.[3] In vitro studies indicate that MFs are unlikely to act as a primary carcinogen. Animal studies provide an opportunity to discover a causal relationship between MFs and tumorigenesis. Reduction of pineal melatonin production has been implicated as a cause of tumor production.[4] The importance of reduced melatonin levels in the etiology of cancers of hormone-dependent tissues (e.g., breast) was established in experimental animals.[5] Most experimental studies on the effects of MF exposure on carcinogenesis or melatonin production used MFs at flux densities in the mT range. The relevance of such findings is questionable if field strengths and conditions (e.g., pulsing fields) are different from the time-averaged ambient levels experienced by human populations.

In the present study we present the findings on assays of promotion or co-promotion and melatonin production in rats exposed to a gradient MF of 0.3-1 μT.

MATERIALS AND METHODS

Treatment of Animals. All rats were housed in plastic standard cages in groups of 3 and received water and a standard rat diet ad libitum. Temperature was maintained at 23 ± 2 C. After 10 days acclimatization, rats were randomly subdivided into two groups (36 animals/group). Mammary tumors were induced by intragastric administration of 20 mg (5 mg/week) DMBA. At the onset of the experiment, all rats received the first administration of 7,12-dimethylbenz(a)anthracene (DMBA) at the age of 52 days and were either exposed to a 50Hz gradient MF of 0.3-1 μT or sham-exposed for a period of 91 days continuously except for

*Correspondence to: W. Löscher, Tierärtzliche Hochschule, Büntenweg 17, D-3000 Hannover, Germany.

a break for cage cleaning twice a week. Rats were palpated weekly to assess the development of mammary tumors. At termination, all rats were sacrificed and the number and size of neoplasias were determined after necropsy.

Exposure System. The exposure system consisted of six identical coils and six sham-coils. The MF in the exposure volume was not homogeneous, but ranged from 0.3 to 1 μT. The stray field for sham-exposed rats in the same room varied between 20 and 40 nT.

Melatonin Determinations. For determinations of serum levels of melatonin, blood was collected (through puncture of the retroorbital plexus) either at 11:00 a.m. or 11:00 p.m. in both sham-exposed and MF-exposed rats 8-9 weeks after the first DMBA application. Melatonin levels were determined by radioimmunoassay.[6]

Histopathology. Neoplasias and serial sectioned mammary glands were examined microscopically (H.E.-staining).

Statistical Evaluation. The incidence of tumors or hyperplasias was analyzed by the χ^2 test. Significance of differences in tumor latency, the number of tumors, and the total tumor burden was calculated by the Mann Whitney U-test and the melatonin values (means ± S.E.) were analyzed by ANOVA followed by the Newman-Keuls test.

RESULTS

There were no significant differences among the groups with regard to the body weight at the end of the experiment. Three rats of the control group died before the end of exposure period because of large bleeding mammary tumors or weight loss and vaginal bleeding. A slightly earlier development of tumors was observed in MF-exposed group, but the difference was not significant. Nor were there significant differences between MF-exposed animals and controls in the number of rats that developed tumors or in the number of tumor or tumor burden (Table 1). The incidence of tumors and hyperplasias was 72% in controls and 89% in MF-exposed rats, which was not significant in the χ^2 test. Almost all neoplasias induced by DMBA were malignant epithelial neoplasms.

TABLE 1.—Effect of MF-exposure on development and growth of mammary tumors. Average data are shown as means ± S.E.

	Sham-exposed rats (n = 36)	MF-exposed rats (n = 36)
Latency (weeks)	10.62 ± 0.44	10.09 ± 0.35
Animals with hyperplasias and/or tumors	26	32
Tumor incidence (%)	61	67
Incidence of hyperplasias and tumors	72	89
Number of tumors per tumor-bearing animal	4.3 ± 0.83	3.2 ± 0.45
Mean diameter of tumors (cm)	1.02 ± 0.061	0.98 ± 0.065

Night-time melatonin levels in serum were significantly lower in MF-exposed (288 ± 21.5 pg) compared to sham-exposed animals (350 ± 39.1 pg), whereas there was no difference between the day-time levels (sham-exposed rats: 52 ± 3.9 pg; MF-exposed rats: 46 ± 4.2 pg).

DISCUSSION

Our data indicate that subchronic exposure to low-energy MF does not induce marked tumor promoting effects in a DMBA model of breast cancer in rats. Similar to recent data from experiments with DMBA-induced skin tumors in mice,[7] the onset of palpable tumors was somewhat shorter in the rats exposed to MF. Moreover, the incidence of mammary tumors or hyperplasias tended to be higher in the MF-exposed animals, again without significance. The data on histopathological classification of mammary lesions and tumor size does not indicate that MF exposure at low flux density altered the spectrum, incidence, or size of neoplasias or hyperplasias. In contrast to our data, Beniashvili et al.[8] reported that low-frequency experiments with nitrosomethyl urea (NMU) used as chemical carcinogen enhance the induction of mammary tumors in rats.

A possible link between MF-induced alteration in melatonin synthesis and cancer has been hypothesized. Reduction of melatonin by MF may possibly increase the potential for cancer because melatonin is known to inhibit the growth of some cancers (e.g., breast). The present data reveal a significant decrease in night-time melatonin levels in MF-exposed rats compared to sham-exposed rats.

The serum control levels of melatonin were higher than those reported by other groups.[9] This result might be due to stressful stimuli (e.g., retroorbital puncture for blood sampling), because it has been recently found that stress (e.g., immobilization) applied to animals led to an increase in pineal melatonin production.[10] The reduced melatonin levels in the MF-exposed rats in our study were not associated with marked alterations in DMBA induced carcinogenesis.

REFERENCES

1. D. A. Savitz, N. E. Pearce, and C. Poole, "Methodological issues in epidemiology of electromagnetic fields and cancer," *Epidemiol. Rev.* 210: 329, 1989.

2. P. A. Demers, et al., "Occupational exposure to electromagnetic fields and breast cancer in men," *Am. J. Epidemiol.* 134: 340-347, 1991.

3. M. Villa, P. Mustarelli, and M. Caprotti, "Biological wiring configurations and childhood cancer," *Am. J. Epidemiol.* 109: 273-284, 1991.

4. B. W. Wilson, R. G. Stevens, and L. E. Anderson, *Extremely Low Frequency Electromagnetic Fields: The Question of Cancer*, Columbus, Ohio: Battelle Press, 1990.

5. L. Tamarkin, M. Cohen, D. Roselle, C. Reichert, M. Lippmann, and B. Chabner, "Melatonin inhibition and pinealectomy enhancement of 7,12-dimethylbenz(a)anthracene-induced mammary tumors in the rat," *Cancer Res.* 41: 4432-4436, 1981.

6. A. Lerchl, K. O. Honaka, and R. J. Reiter, "Pineal gland 'magnetosensitivity' to static magnetic fields is a consequence of induced electrical currents (eddy currents)," *J. Pineal Res.* 10: 109-116, 1991.

7. J. R. N. McLean, et al., "Cancer promotion in a mouse-skin model by a 60-Hz magnetic field: II. Tumor development and immune response," *Bioelectromagnetics* 12: 273-287, 1991.

8. D. S. Beniashvili, V. G. Beniashvili, and M. Z. Menabde, "Low-frequency electromagnetic radiation enhances the induction of rat mammary tumors by nitrosomethyl urea," *Cancer Lett.* 61: 75-79, 1991.

9. C. Bartsch, H. Bartsch, T. H. Lippert, and D. Gupta, "Effect of the mammary carcinogen 7,12-dimethylbenz(a)anthracene on pineal melatonin biosynthesis, secretion and peripheral metabolism," *Neuroendocrinology* 52: 538-544, 1990.

10. J. Seggie, L. Campbell, G. M. Brown, and L. J. Grota, "Melatonin and N-acetylserotonin stress responses: Effects of type of stimulation and housing conditions," *J. Pineal Res.* 2: 39-49, 1985.

STUDIES OF ELF MAGNETIC FIELD EFFECTS ON 3T3 CELL MOTILITY UNDER VARIOUS CONDITIONS

I. T. Spadinger* and B. Palcic

B.C. Cancer Research Centre, Vancouver, B.C., Canada

D. A. Agnew

Ontario Hydro, Pickering, Ont., Canada

The effect of extremely low frequency (ELF) oscillating magnetic fields on mammalian cell motility and gross morphology was examined in vitro. NIH 3T3 cells were plated at low densities onto a growth surface and their shapes and movements were tracked as a function of time by automated image cytometry techniques. Various ELF exposure parameters in the 10-800μT, 10-72Hz range, as well as various cell culture conditions (e.g., TPA exposure, deprivation of extracellular calcium) were tested. No ELF field-induced effects were observed in any of the experiments, in contrast to the effects of TPA exposure or calcium deprivation alone.

Extremely low frequency (ELF) electromagnetic fields have been reported to affect cells and tissues in a number of ways. One of the most frequently observed phenomena is the alteration of calcium homeostasis. Since calcium ions play an important regulatory role in living cells, various cellular properties may be affected by such an event. In diatoms, for example, extracellular calcium concentrations are known to affect cell mobility, and ELF field-induced changes in mobility have also be observed.[1] Similar effects could occur in mammalian cells, since the cytoskeleton is sensitive to changes in cytosolic calcium concentrations. The cytoskeleton determines gross cellular morphology and plays an active role in governing motile behavior. External factors influencing cytoskeletal structure and function can therefore lead to profound changes in cellular properties such as shape and motility.

This paper describes an extensive study undertaken to determine whether mammalian fibroblast motility and gross morphology are altered by exposure to ELF magnetic fields. ELF field-induced effects on these cellular endpoints would provide indirect evidence for disruption of calcium homeostasis in these cells, and would have implications for the possible role of ELF electromagnetic fields in the induction of cancer and fetal abnormalities, since cell migration plays a role in both these processes. Cell motility is also an important component of the wound healing process, which has been reported to be affected by ELF electromagnetic fields.[2]

MATERIALS AND METHODS

NIH 3T3 mouse fibroblasts were used in all our experiments. NIH 3T3 is an established in vitro cell line, and can be grown at low plating densities, so that the behavior of individual cells can be monitored. The cells adhere to the surface of the tissue culture vessel, and move by sending out cytoplasmic extensions and using them to pull the rest of the cytoplasm into a new position.

In each experiment, cell shape and motile behavior were measured for a selected set of cells by an automated image cytometry device which collected and analyzed images at regular intervals (i.e., every 10 min) over a period of several hours. In each experiment, approximately 100 cells were monitored before, during, and sometimes after the ELF exposure period, which lasted anywhere from 2 to 8 h. The ELF magnetic fields applied were in the 10-800μT, 10-72Hz range, and were vertical with respect to the cell sample. The geomagnetic field components were 28.3 μT (vertical) and 6.4 μT (horizontal). A great deal of attention was paid to measuring and minimizing background ELF fields, which are consistently below 0.1 μT at the site of

*Correspondence to: I. T. Spadinger (Cancer Imaging), B.C. Cancer Res. Ctr., 601 W 10th Ave., Vancouver BC Canada V5Z 1L3.

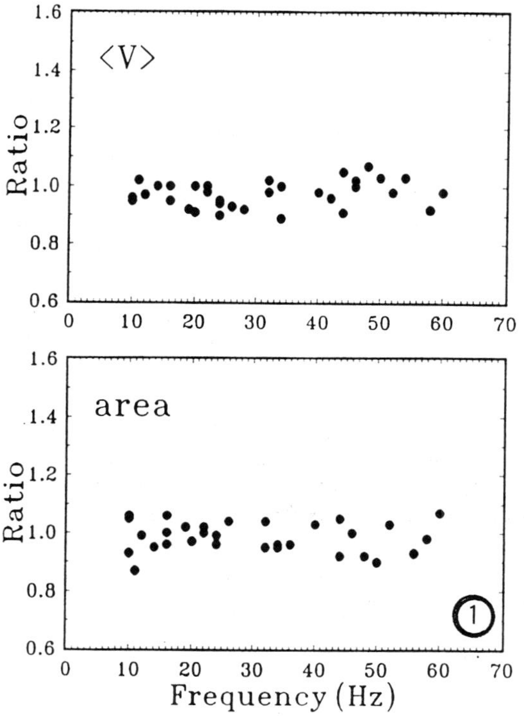

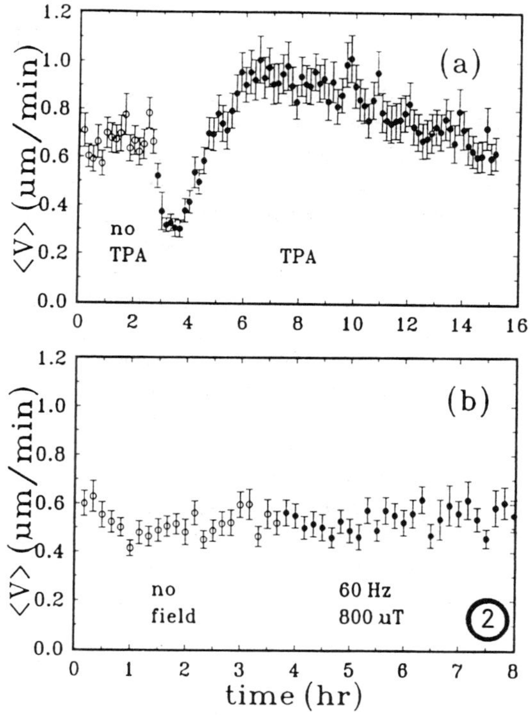

FIG. 1.—The response of cell speed (<V>) and cell area parameters to 80 μT ELF magnetic fields in the 10-60Hz range. Each point represents the ratio of averaged measurements made during a 2h exposure period to averaged data from a 2h control period immediately preceding the exposure period. Approximately 100 cells were measured for each data point.

FIG. 2.—Sample experimental records of cells exposed to TPA or ELF magnetic fields: (a) mean cell speed, <V>, prior to and after the addition of 0.1 μg/ml TPA to the growth media; (b) mean cell speed prior to and during exposure to a 60Hz, 800μT magnetic field.

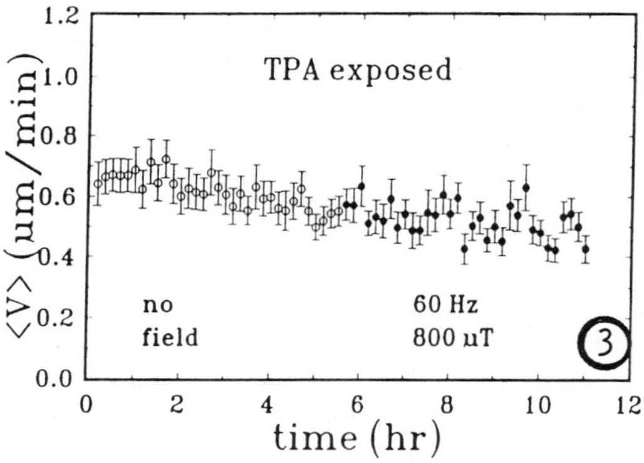

FIG. 3.—Sample motility response of a TPA-exposed cell population to a 60Hz 800μT magnetic field. The cells had been chronically exposed to TPA for several days prior to the experiment.

the experimental apparatus.

In the majority of the experiments, cells were exposed to the ELF fields under normal culture conditions (complete nutrient growth media, controlled temperature and pH). However, in some experiments TPA (a known tumor promoter) was added to the culture medium to investigate the possibility of co-promoter activity by ELF fields. In other experiments, the cells were exposed to the ELF fields while growing in calcium-reduced media. These exposures were motivated by the role of extracellular calcium concentrations in modulating the response of diatoms to ELF fields.[1]

Data collected during a given ELF field-exposure period were compared to the data from the control period immediately preceding it. Various cell shape and motility descriptors were obtained from the recorded data, which included measures of cell area (i.e., the area of the culture surface covered by the cell), circularity (a mathematical feature describing the degree of elongation of the cell), as well as the speed, direction, and directional persistence of the cell movement.

RESULTS

No reproducible effects of ELF magnetic fields on either motility or morphology were detected under any of the exposure conditions tested. The most extensive data obtained were for 3T3 cells grown under standard culture conditions and exposed for 2h intervals to 20 or 80μT ELF magnetic fields in the 10-60Hz range. Ratios of averaged results obtained during the ELF exposure periods and the control periods immediately preceding them are shown for the 80μT data in Fig. 1. Results of cell speed and cell area are shown. Within the random noise of the data, no significant individual effects and no frequency-dependent trends could be seen for these or any of the other shape or motility parameters tested.

Less extensive experiments were carried out for fields of 400 and 800 μT, including experiments in which the response of TPA-exposed cells (0.1 μg/ml, chronic exposure) and cells growing in calcium-reduced media was measured. The primary frequencies tested in these studies were 11, 22, 44, and 60 Hz. Again, no significant alterations in cell shape or motility could be attributed to the ELF fields. This result was in contrast to the effect of TPA exposure or calcium deprivation alone. TPA exposure, especially, caused profound acute and long-term changes in both shape and motility parameters. In Fig. 2 an experimental record for a cell population exposed to TPA part-way through a tracking session is compared with a similar record gathered for a cell population exposed to an ELF magnetic field. Figure 3 shows the ELF field response of a cell population chronically exposed to TPA. These cells did not appear to be affected in any way by the ELF magnetic field applied midway through the tracking session. Cell populations subjected to extracellular calcium deprivation exhibited a lower-than-normal cell speed, but also did not respond to the ELF field.

CONCLUSION

ELF magnetic field-induced effects on 3T3 cell motility or gross morphology were not observed under any of the exposure conditions tested in our experiments. No indirect evidence of ELF field effects on cytosolic calcium concentrations was therefore obtained for this cell type. These data support the findings of Parkinson and Hanks,[3] which were reported after our studies had been initiated. Using fluorescent calcium indicators to measure intracellular calcium concentrations, Parkinson and Hanks were unable to detect any ELF field-induced effects in a number of different cell lines, including one (BALB 3T3) that is very similar in origin and properties to the cells used in our studies.[4]

REFERENCES

1. S. D. Smith, B. R. MacLeod, A. R. Liboff, and K. Cooksey, *Bioelectromagnetics* 8: 215-227, 1987.
2. B. F. Sisken and E. Herbst, *Trans. IEEE* EMB-12: 1533-1534, 1990.
3. W. C. Parkinson and C. T. Hanks, *Bioelectromagnetics* 10: 129-145, 1989.
4. This work was supported by Ontario Hydro and B.C. Hydro.

Martin Blank, Ed., *Electricity and Magnetism in Biology and Medicine*

IAF CHRONIC CARCINOGENICITY EVALUATION OF 60Hz LINEAR SINUSOIDAL CONTINUOUS-WAVE MAGNETIC FIELDS IN RATS: PROTOCOL OUTLINE AND QUALITY ASSURANCE PROGRAM DEVELOPMENT

R. Mandeville,* G. Mercier, D. Oth, J.-P. Descôteaux, M. Lis, E. Franco, M. Houde, and L. Tremblay

University of Québec, Laval, Canada

The objective of this research program is to determine whether chronic exposures to 60Hz linear (single axis) sinusoidal continuous wave magnetic fields of various intensities (2, 20, 200, and 2000 μT) might increase the risk of cancer development in Fischer rats. In the past two years we have carefully designed a study protocol and intensively undertaken the construction of a dedicated animal facility to accommodate this large study. In the first two months and before starting exposure, we undertook a shakedown experiment to evaluate and map electric and magnetic field strengths in the "Exposure room," as well as the direction and strength of the earth's static magnetic field. We have also evaluated magnetic-field strength, direction, and homogeneity within each cage, as well as cage vibration, temperature, and light intensity, induced currents, and microshocks (especially under conditions of wet bedding). Ammonia level was evaluated in individual units after two adult female rats were housed for one week. In building this animal facility, we took great care to minimize the following potential confounders: electric field, noise, cage vibration, ambient temperature increase, light, and air-flow obstruction. Moreover, we have developed adequate crisis management (fire, sabotage, power outage, emergency, vandalism, accidents, etc.). Our team has put prime priority and attention on developing the protocol and the quality assurance program of this high-priority study.

Concern over the issue of an association between extremely low frequency (ELF) electromagnetic-field (EMF) exposure and increased cancer risk has led to the development of over 50 major epidemiological studies around the world. We have recently reported a comprehensive review of the literature that assessed such risk in children and adults exposed in their homes or at work[1,2] to magnetic fields generated by distribution lines. Very briefly, data derived from the large number of epidemiological studies have given rise to no firm reproducible evidence for a causal association between exposures to such fields and increased risk of certain types of cancer. The overall risk ratios tend to cluster around 1.2 and 1.5, although risk ratios of 6 or higher have been reported. Nor is there any absolute specificity with respect to cancer type or site. Positive risk associations have been reported for leukemia (mainly ALL), non-Hodgkin lymphoma, CNS tumors, melanoma, and prostate and breast cancers in males. However, no definitive conclusion can be drawn from these studies because each study has inherent weaknesses. We believe that whole-animal studies to evaluate potential hazards of magnetic-field exposures is the best choice for providing a direct answer to questions related to potential carcinogenic effects of magnetic fields. Unfortunately, until now no reports of lifetime animal exposures to magnetic fields have been published.

In the design of our chronic study various considerations have been taken into account. First, the typical values of the maximal magnetic-field intensities at 1.2 m above ground level under 345-765kV transmission lines are 0.03 mT, for a current of 2000 A per phase.[3] The second consideration was based on induced currents. Since the interaction of alternating magnetic fields with tissues is believed to be through induced currents,[4-6] scaling of the exposure conditions has been considered. The maximum induced current has been shown to be proportional to the radius of the largest loop inside the body. To scale from man to mouse, it is assumed that the maximum radius is proportional to the cube root of the body mass (volume). From a 70kg man to a 200g rat the scaling factor is about 15. It was therefore decided early in the design stage that we would design a system in which 2000μT magnetic fields would be the highest possible fields intensity. This long-term experiment will address the putative question of cancer induction by 60Hz magnetic fields and

---

*Correspondence to: R. Mandeville, Institut Armand-Frappier, University of Québec, 531 boul. des Prairies, Laval QU Canada H7N 4Z3.

provide the opportunity to investigate possible dose-response effects, in a completely blinded fashion.

STUDY DESIGN

Five groups of 50 female rats will be exposed to magnetic fields of intensities ranging from 0 (sham exposure) to 2, 20, 200, 2000 μT from birth until 2 years (104 weeks) of age. Daily exposure to the various fields will be performed for 20 consecutive hours (10 h in the dark and 10 h in the light). The number of animals needed to reach a power of 90% and a doubling in malignant tumor incidence at any of the four magnetic-field intensities has been derived by three different statistical methods. Calculations were based on data derived from the NTP (National Toxicology Program) historical database.[7-9]

Accordingly, 160 timed-pregnant female rats (after passing the weight test for pregnancy) will be randomized into five groups and immediately exposed to the corresponding magnetic field. At weaning, female offsprings will be separated from males, have an ID number tattooed on the tail, and will be distributed and housed two per cage for the duration of the study. All clinical observations and husbandry procedures will take place once per day between 9 a.m. and 1 p. m. (i.e., when the fields are off). Individual animal body weights for exposed and control animals will be recorded weekly in the first 15 weeks of the study and monthly thereafter. Mean body weights will be calculated for each group and compared to data derived from our own control colony of Fischer rats. Using these data, we expect to find a maximum mean body weight of about 330 g and mortality rates around 10% in this 2-year carcinogenicity study.[7-9] A necropsy will be performed on all exposed and control groups which die, are found moribund, or at terminal sacrifice. During necropsy, all organs and tissues will be examined for grossly visible lesions. Tissues will be preserved in 10% neutral buffered formalin. Histopathologic examination of tissues will be performed on all visible lesions in all dose groups, and on 42 different organs in all exposed animals, as specified by NTP protocols.[7-9]

TABLE 1.—Design of carcinogenicity evaluation in Fischer rats.

	Level of exposure (μT)				
Experimental condition	0	2	20	200	2000
Animals in study	50	50	50	50	50
Cage controls	50				
Sentinels (5×7)	35				

FACILITY DESIGN AND CONSTRUCTION

The magnetic-field exposure room (154 m^2) is situated in a dedicated animal facility on the IAF campus designated Building 25. The exposure room itself consists of 20 identical exposure units arranged in five groups of four units each. A separate control power panel is located in an adjacent control room. Designed by Dr. S. Harvey (Ontario Hydro) and executed by Westinghouse Canada, each unit is capable of continuous exposure of 16 rats to a horizontal continuous-wave magnetic field of specified intensity. Each group of exposure units has connections for a supply of 60Hz current for the field-generating coils and a supply of 60Hz current for separate heater windings. One group is a sham exposure group (zero current), so that only four independently adjustable field-generating coil supplies are actually required. The coil supplies are designed to provide currents in the ratio of 1000:100:10:1, yielding exposure fields of intensities of 2, 20, 200, and 2000 μT, respectively.

Four heater supplies were built in to bring the heat generated by the lower-field (2, 20, 200 μT) and sham groups up to the same level as that generated by the highest field group (2000 μT). As the heat required for each of these groups is the same to within about 1%, the heater supplies need not be independently adjusted. The line voltage is regulated and is cycled on and off through a motor-driven variable transformer. The voltage and the current on each of the four coil-supply circuits, and the currents on each of the four heater circuits, are monitored and displayed on the control panel situated in the adjacent room designated "Control

Room.'' Alarms are generated when the regulated voltage or any of the currents move outside a specified range that will nominally be ±3 to 4% around the given value of such unit. The entire system is operated under the control of a programmable timer, and will turn itself on and off according to a preset daily program. The duty cycle will be 20 h on (from 1 p.m. to 9 a.m.) and 4 h off (from 9 a.m. to 1 p.m.). A manual override is provided for testing or other purposes.

Exposure Units

Animals will be kept in custom-made polycarbonate cages with flat stainless-steel cage cover tops that hold a 500ml glass bottle with a No. 6½ rubber stopper with a glass sipper and a suspended ceramic feeder. The cage cover has been provided with a small teflon tubing so as to interrupt the full vertical loop in the stainless-steel cover. The cages are placed on a polycarbonate tray (59×59 cm). Each polycarbonate tray contains 4 cages. Two cage trays are placed inside each of the 20 electromagnetic field-generating units. Each cage is equipped with a polycarbonate card holder with a special identification color tag. Each color corresponds to a field intensity and the light indicator above each row of exposure units. This exposure system is thus completely blinded to the investigators.

Field Specifications

The specifications of the fields produced by the exposure apparatus are as follows. The figure-eight configuration of the coil wiring will give fields of high uniformity inside the volume enclosed by the coils, and a rapid decrease in field strength outside. The magnetic field within the space occupied by animals will be uniform to within ±9%. The field will be 60 Hz, sinusoidal with no special harmonic control, linear (single axis) in the horizontal plane and in a south-east orientation. The voltage will be ramped to provide smooth turn-on and turn-off of the fields. The following potential confounders have been minimized: electric field, sound, vibrations, ambient temperature increase, light, and air-flow obstructions.

Exposure conditions (magnetic-flux density) have been closely selected to simulate human exposures at home and at occupational settings. The exposure system will be operated and maintained by personnel from IREQ (Institut de recherches d'Electricité d'Hydro Québec) under the supervision of Dr. S. Maruvada. None of the personnel involved in system operation is associated in any way with the animal care. In order to limit the effect of any potential environmental confounder in this study, we have elected to develop a rotation strategy that is completely randomized. This rotation strategy is based on the colored-light field coding provided in the exposure room to designate the exposure level assigned to each of the five groups. The lights are red, green yellow, blue, and white. Permutations in cages inside each unit will be performed weekly during the entire 2 years of the experiment. Every month, units will be rotated inside the room.

ENVIRONMENTAL MONITORING SYSTEM

Both the ''Exposure'' and ''Transition'' rooms are provided with temperature, humidity, light, and noise sensors. In addition, an electric interlock system prevents entry to the exposure room when the coils are energized. An integrated system SEMS 600 (Stand Alone Energy Management System from Landis and Gyr Powers Ltd.) is provided to monitor automatically the following environmental parameters: ambient temperature, humidity, air flow, room access, light, and sound. Any special events (irregular duty cycle, abnormal conditions, outages, repairs, etc.) will also be monitored. During the course of the experiment any one of the monitored parameters can acquire values outside the acceptable range. Such nonconformance will activate alarms, and the computer will automatically activate a page telephone system that notifies the technician in charge of the alarm condition. The technician will communicate by modem to inquire about the nonconformance status and will take the appropriate corrective action.

CONCLUSION

The potential public-health impact of this ubiquitous environmental agent argues strongly for continuing and expanding research efforts to confirm or refute the reported associations with cancers in adults and children. Our effort in this area will help evaluate the potential hazards of EMF and provide a direct answer to the putative carcinogenic effect of 60Hz linear sinusoidal continuous-wave magnetic fields in normal animals chronically exposed to such fields.[10]

REFERENCES

1. R. Mandeville, "Potential cancer induction by ELF EMF: Epidemiological evidence and research protocol for animal studies," *Annual Meeting and Exhibition of the Air and Waste Management Association,* (91-133.5) 84: 1-22, 1991.

2. R. Mandeville, "IAF study on the evaluation of the potential carcinogenity of 60-Hz magnetic fields in F334 rats," Ref. 1 (92-147.01) 85: 1-13, 1992.

3. J. W. Bridges, *Biological Effects of High-voltage Electric Fields: State-of-the-art Review and Program Plan* (EPRI-381-1), Palo Alto, Calif.: EPRI, 1975. (NTIS No. PB-247-454.)

4. W. T. Kaune, "Physical aspects of ELF electric and magnetic fields: Measurements and dosimetry," in B. W. Wilson et al., Eds., *ELF EMF: The Question of Cancer,* Columbus, Ohio: Battelle Press, 1990, 159-186.

5. M. A. Stuchly and K. H. Mild, "Environmental and occupational exposure to EMF," *Trans. IEEE* EMB-6: 15-17, 1987.

6. T. S. Tenforde, "Interaction of ELF magnetic fields with living matter," in C. Polk and E. Postow, Eds., *Handbook of Biological Effects of Electromagnetic Fields,* Boca Raton, Fla.: CRC Press, 1986, 197-225.

7. J. K. Haseman, G. N. Huff, and G. A. Boorman, "Use of historical control data in carcinogenicity studies in rodents," *Toxicol. Pathol.* 12: 126-135, 1984.

8. J. K. Haseman, D. D. Crawford, J. E. Huff, et al., "Results from 86 two-year carcinogenicity studies conducted by the National Toxicology Program," *Environ. Health Report* 14: 621-639, 1984.

9. H. A. Solleveld, J. K. Haseman, and E. E. McConnell, "Natural history of body weight gain, survival, and neoplasia in the F344 rat," *J. Natl. Cancer Inst.* 72: 929-940, 1984.

10. This research program is co-sponsored by Health and Welfare Canada, Hydro Québec, and Ontario Hydro (Grant No. 6605-3493-E from the National Research and Development Program). A special grant from the Quebec Government and Institut Armand Frappier has permitted the restructuring and complete renovation of the dedicated animal facility.

Martin Blank, Ed., *Electricity and Magnetism in Biology and Medicine*

EXPOSURE ASSESSMENT IN THE DANISH EPIDEMIOLOGICAL STUDY OF CANCER IN CHILDREN LIVING CLOSE TO HV INSTALLATIONS

J. Kaad Jensen*

ELSAM, Fredericia, Denmark

E. Folkersen

Skærbækværket, Fredericia, Denmark

In Denmark it is permitted to carry overhead lines directly across residential areas. This means that several houses are placed under or very close to overhead lines. The Danish national study involves all 1720 cases of childhood cancer in the period 1968-86. The survey of the children's residential record has resulted in a classification of the addresses in an "exposed group" and an "unexposed reference group." The exposed group contains all addresses exposed to more than 0.25 μT. The exposure assessment is based on knowledge of the construction of the actual high-voltage (HV) installation, the network configuration, and an estimated load in the actual period of time. An atlas has been worked out for all addresses. For each of the addresses in the exposed group, the exposure values "extreme," "max.," "middle," and "min." are estimated. An μT-hour value can be estimated from μT-middle.

The aim of the Danish epidemiological study is to find out whether children living close to HV installations are more likely to develop cancer. Exposures down to approximately 0.25 μT are studied. The study includes all 1720 children (cases) from 0-14 years who during the period 1968-86 have been diagnosed to have had leukemia, brain tumor, or lymphoma. A number of healthy children are selected as controls. The total number of children is 6420. The information obtained from the Danish Cancer Registry (CRG) by the utilities contains an address record for each child, but no information that reveals whether an address belongs to a case or a control child.

DISTANCE CRITERIA[3]

The distance from 400 50kV substations, overhead lines, and cables to all the addresses is investigated. The *distance criteria* were determined on the basis of current levels and the wire configuration that *insures* that all addresses exposed to more than 0.25 μT are found and assessed. The background field in city areas is often 0.1-0.2 μT.

Exposed Group/Unexposed Reference Group

All addresses within the distances mentioned in Table 1, columns 2 and 4 are classified as potentially *exposed.* For each exposed address an appendix is worked out describing the substation, overhead line, or cable in question and the *location* of the home in relation to the HV installation.

An *unexposed* reference group with all addresses situated within a distance 1-3 times the exposed distance is found (Table 1, columns 3 and 5). Addresses in the unexposed reference group are considered as unexposed and their socio-economic conditions are similar to the addresses of the exposed group. As far as the 50-60kV and 132-150kV cables are concerned, no reference group has been defined.

The addresses of the unexposed reference group can be wholly or partly, and without further sorting, included in a subsequent exposure assessment if it is desired to extend the analysis to a lower exposure level than the chosen 0.25 μT.

---

*Correspondence to: J. Kaad Jensen, ELSAM, Fjordvejen 1-11, DK-7000 Fredericia, Denmark.

TABLE 1.—Exposed group and unexposed reference group.

	Overhead Lines and Substations		Cables	
	Exposed	Unexposed reference	Exposed	Unexposed reference
220-400 kV	100 m	100-300 m	20 m	20-60 m
132-150 kV	50 m	50-150 m	5 m	—
50-60 kV	25 m	25-75 m	2.5 m	—

Addresses

A total of 16 082 addresses have been studied. For each address, the study number of the child, the date of moving in, and the date of moving out are stated. A total of 30 exposed and 68 reference group addresses have been found in Denmark. For these addresses an atlas is made as a source of "inspiration." The atlas shows the addresses in relation to the transmission network, city areas, highways, etc., to make it possible to consider other factors than magnetic fields.

VALIDATION OF THE UTILITIES' ESTIMATION OF MAGNETIC FIELDS

Prior to the exposure assessment, a check of the utilities' ability to estimate the strength of a specific magnetic field has been carried out. At random, CRG chose six addresses close to HV overhead lines where there were children between 0 and 14 years old. The Danish National Institute of Occupational Health measured the magnetic field strength in each home.[1] The actual current and a cross section of the overhead line and home form the basis of the utilities' calculation of the strength of the magnetic field. The results of the calculations and measurements from the 18 measurement points are set out in Ref. 2. In the estimate of the magnetic fields, the background fields from the homes have not been considered, so that the values are typically 0.1 μT lower than the measured values. The estimates for multicircuit overhead lines have been carried out both with random phase sequence and with known phase sequence. A comparison of the estimated and the measured magnetic fields shows that they correspond to a satisfactory degree only when the actual phase sequence is applied.[3]

EXPOSURE ASSESSMENT

It is not possible to obtain exact information about the historical loads of substations, overhead lines, and cables. Measurement of magnetic fields is not sufficient for a historical exposure evaluation and is only applied in connection with a validation of the calculation method and a mapping of the magnetic fields surrounding children in their daily lives.[1] That is why the exposure assessment[3] is based on detailed historical knowledge of the actual network configuration and the operational principles.

The Danish transmission system can be divided into categories A-D:

TABLE 2.—Categories A-D for Danish transmission system.

	Characteristics	Normal max. value of full-load current	Typical utilization time (hours)
Category A	Lines issuing from power stations	75%	6000 h
Category B	Unevenly loaded lines in the transmission network	75%	5000 h
Category C	Evenly loaded lines in the meshed transmission network	50%	4500 h
Category D	Lines for distributional purposes	25%	3500 h

For each category, stylized duration curves are made. As a measure of the exposure, the following magnetic-field values are estimated for each address:

$\mu T_{extreme}$: Absolutely highest magnetic-field strength which may occur, though only for very few hours a year.

$\mu T_{max.}$: Normal maximum value of the magnetic-field strength.

μT_{middle}: Magnetic-field value that results in the actual annual μT-hour exposure. Middle value on the stylized curve.

$\mu T_{min.}$: Normal minimum value of the magnetic-field strength.

For each exposed address, $\mu T_{extreme}$, $\mu T_{max.}$, μT_{middle}, $\mu T_{min.}$ that are valid for the actual period are stated. The value is based on the distance to the home and the data for the HV installation. If the operational pattern of the HV installation has changed, exposure values are stated for each subperiod in question. Table 3 shows and example of the assessment for three selected addresses.

TABLE 3.—Examples of exposure at three addresses.

Study no.	Address year	Classification	$\mu T_{extreme}$	$\mu T_{max.}$	μT_{middle}	$\mu T_{min.}$
11460014	1976-79	Cat. B	2.0	1.31	1.0	0.30
51723012	1961-64	Cat. D	0	0	0	0
"	1964-70	Cat. D	0.27	0.13	0.05	0.04
9902022	1983-83	Cat. D	4.66	1.17	0.47	0.33

On the basis of the stated utilization time of each category, the annual dose μT-hour = 8760 μT_{middle} is calculated. At certain places in Denmark, magnetic fields from imbalanced currents occur. The effects of these currents is ignored.

With knowledge of the residential conditions/life pattern, the estimated value can be increased by the size of the background fields in the case/control in question.

WIRE CODES

Wertheimer and Leeper's (WL) method in the USA has been compared with Danish conditions. In Denmark the voltage levels 230 V and 10 kV correspond to the American 240V and 13kV levels. All Danish systems have a symmetrical three-phase construction. The magnetic field surrounding a three-phase construction is up to 3 times lower than the magnetic field surroundig a single-phase construction (common in many American states).

The exposure from 50 to 400kV overhead lines in Denmark indicates a distance criterion of 25-100 m instead of the 20 m laid down in Ref. 4. The exposure from a Danish 60/10kV substation indicates a distance criterion of 10 m, which is much smaller than the 150 m suggested by WL. The construction principles in respect to HV transmission installations are the same in the USA and Denmark.

CONCLUSION

Of the 16082 addresses that have been studied, 30 addresses belong to the exposed group containing addresses with an exposure of ≥ 0.25 μT, and 68 addresses belong to the unexposed reference group with an exposure of < 0.25 μT. The socio-economic conditions are comparable. The exposure assessment shows that with the established distance criteria, even addresses with an exposure of < 0.25 μT are included.

In connection with exposure assessment, it is crucial to know the actual phase sequence on the multicircuit overhead lines. The Danish exposure assessment is based on a classification system, the actual construction of each single installation, the network configuration, and the operational principles. The Wertheimer and Leeper wire-code classification system cannot be applied in Denmark.

REFERENCES

1. J. Skotte (National Institute of Occupational Health), "Exposure to power frequency electro-magnetic fields in occupational and residential environments in Denmark" (article in preparation by the NIOH, Copenhagen, Denmark).

2. B. C. Gellert (ELSAM, Fjordvejen 1-11, DK-7000 Fredericia), "Calculation of magnetic fields for the validation of exposure assessments."

3. J. Kaad Jensen (ELSAM) and E. Folkersen (Skæbækværket), "Address sorting and exposure assessment of the danish epidemiological study of childhood cancer."

4. N. Wertheimer (University of Colorado Medical Center, Department of Preventive Medicine, Box C-245, Denver, CO 80262) and E. Leeper, "Electrical wiring configurations and childhood cancer."

Martin Blank, Ed., *Electricity and Magnetism in Biology and Medicine*

THE EXPOSURE ESTIMATION METHOD OF THE FINNISH COHORT STUDY OF CANCER AND EXPOSURE TO MAGNETIC FIELDS

J. Valjus,* M. Hongisto, and P. Järvinen

Imatran Voima Oy, Research and Development Division, Vantaa, Finland

P. Verkasalo, K. Heikkilä, and M. Koskenvuo

Institute of Public Health, University of Helsinki

The long-term magnetic field exposure estimation method described in this report is based on geographic information systems. The routes of the 400, 220, and 110kV transmission lines were digitized and linked with the building coordinates, tower type library, and load history files (1970-1989). One of the main aims of the study was to create a more quantitative long-term exposure estimate than the well-known wire coding.[1]

This report also presents, as an interim result, information both on building density and average annual magnetic flux density in the buildings near 110-400kV lines in Finland, as well as on individual cumulative magnetic field exposure obtained during the exposure estimation process. Though the risk of cancer cannot yet be expressed, the present results have been already useful to a major power company like Imatran Voima Oy because they have given quite a good picture of the number of buildings and of the extent of magnetic field exposure near 110-400kV lines in Finland.

MATERIALS

Finland has nationwide public health service data registers and extensive person registers based on computers, which makes it possible, at rather moderate costs, to carry out an epidemiological study on the risk of cancer for people living in the vicinity of powerlines. Data collection can be carried out almost totally by means of register linkings at the National Board of Survey, the Population Register Center, and the Finnish Cancer Registry, starting from the powerline data generated by a Finnish study group that consists of both epidemiologists and engineers.

Power Lines. In Finland the 110, 220, and 400kV powerlines constitute the national grid used for electrical energy transmission from jointly operated power plants to industry and to local electrical utilities for distribution. The 400 and 220kV lines are pure transmission lines and run mostly in sparsely populated rural areas. The 110kV lines also have features of a distribution network and are common in urban areas.

The 110-400kV powerline network of 1989 was the basis of the retrospective network examination for the Finnish epidemiological study of cancer and exposure to magnetic fields.[2] The network studied consisted of 17980 km, which was 91% of the total line length of the 110-400kV national grid in 1989. The network structure of the years 1988-1970 resulted from a simplification of the network of the year 1989 on the basis of the time of the line completions. The missing powerline data are mainly associated with the 110kV network which has, in addition to a few major owners, about one hundred minor owners with varying level of available powerline data.

Buildings. The coordinates and the main structural features of every building are recorded on the land register files at the Population Register Center.

Persons. The persons living or having lived in these buildings can be identified by their social security number. The connection between the buildings and the persons comes through notices of removal, which must be filled out when moving to a new residence. The population register from 1989 to 1970 is extensive and contains data on more than six million individuals. The register of residence periods created for the Finnish cohort study includes 13.5 million residence periods for 5.8 million people. The population of Finland is 5.0 million.

*Correspondence to: J. Valjus (R&D Division), Imatran Voima Oy, P.O. Box 112, SF-01601 Vantaa, Finland.

METHODS

The 110-400kV line routes were digitized at the National Board of Survey from basic maps (1:20000) supplied by Imatran Voima Oy. The lines were identified by Imatran Voima Oy and equipped with names and line codes for line route digitization. The digitization was made by the angle coordinates of the lines. At this stage, the buildings located in 1989 nearer than 500 m from the digitized powerlines were identified and the distance of their centerpoints from the powerline(s) were calculated at the National Board of Survey on the basis of the building coordinates of the land registry supplied by the Population Register Center.

Historical data of *load currents* 1989-1970 was created at Imatran Voima Oy. Stored loadflows generated by a power system simulator were used to create the average annual load history for 1989-1984. The load history of the earlier years 1983-1970 was created on the basis of written documents and approximative procedures. The consumption of electrical energy in Finland has large seasonal variations. Data of several thousand measured load-current values of about 250 lines in both 1988 and 1989 were used to determine the time of year which corresponds to the situation of the average annual electrical energy consumption. The apparent power value at 8:30 a.m. on the second Wednesday in October was found to be an appropriate estimate for the annual average electrical power. The load currents of various powerlines were found to be very individual and no typical current estimate could be found.

Estimation of Magnetic Fields. The evaluation of the average annual magnetic flux density in the selected buildings from 1989 to 1970 was carried out at Imatran Voima Oy. The computer program used for this task is based on the theory of elliptic magnetic fields near long parallel lines. In the case of nonparallel lines (e.g., at line branching points) the transformation to a parallel-line system for every building was made.

Evaluation of Accuracy. The accuracy of the exposure estimation in buildings was studied by Monte Carlo simulation. The following error distributions could be taken into account:

- error of building location
- error of powerline location
- error of load current estimation
- error of conductor heights
- error of spacings between conductors

The Finnish building coordinate data are now being revised by the National Board of Survey in order to improve their accuracy. So far the data of 80 Finnish communities of the total number 450 have been revised. The National Board of Survey has compared the older building coordinate data with the revised data and found that 90.1% of the revised building coordinates differed by 0-20 m, 4.8% by 20-50 m, and 1.5% by 50-100 m. For 3.6% of the buildings the location error was more than 100 m, or the building no longer existed. All 269255 buildings in these 80 communities at the time of revision were included. It was assumed that the distribution of the building location error in the other communities and near powerlines is the same as in the 80 communities studied because of the large sample size (about 15% of the total number of buildings in Finland).

The powerline routes were digitized from basic maps (1:20000) prepared from aerial photographs. Though at the moment only a small number of measured tower footing coordinates are on data files, and therefore the accuracy of the powerline location could be checked by computer comparison of the digitized and the measured powerline coordinates only for a small number of towers, the accuracy of the line route digitization can be considered to be fairly good because of the reliability of the aerial photographic survey. The average error of powerline coordinates was, according to the test sample, evaluated to be 17.7 m with a standard deviation of 9.5 m.

The errors in magnetic-field calculation also depend on the errors in load currents, deviations from typical tower dimensions (e.g., special towers in angle points), and variations in conductor heights (e.g., near towers or at the centerpoint of a span). Their error distributions could be estimated on the basis of load current data and the tower library.

The reference magnetic flux densities in 1989 were calculated for 110, 220, and 400kV lines from average load currents and typical power dimensions. The estimated error distributions were used to generate random Monte Carlo error combinations to study the effects of errors in geographic data, annual average load currents, and powerline data on magnetic fields.

The accuracy of the evaluation of the annual magnetic flux density very close to a line was found to be rather poor, but beyond some 10 meters from the line, the average magnetic flux density is no longer so sensitive to the data shortages. Figure 1 presents as an example the convergence of the error as a function of distance in the case of 400kV portal tower line. Very close to the lines, even very slight changes in the exposure estimation parameters may cause large changes in the flux density and the efforts to obtain more accurate data do not lead to great improvements. Fortunately, in Finland there is a so-called building preclusion zone in the vicinity of 110-400kV lines (e.g., ±18 m from 110kV and ±31 m from 400kV portal tower lines), where permission to build is granted only in very rare cases. Outside this zone the exposure estimation is far more accurate.

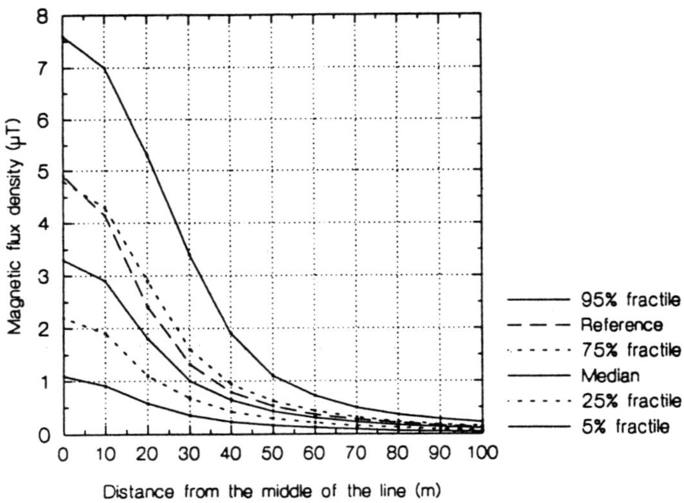

FIG. 1.—Convergence of error of magnetic flux density as a function of distance from typical 400kV portal tower line. Monte Carlo simulation is based on 22000 error combinations.

Other errors (e.g., internal magnetic fields of the buildings) could not be taken into account. However, the validity of the parallel-line estimation was analyzed and found to be satisfactory outside the building preclusion zone.

Identification of the Exposed People. All individuals with a residence period in the buildings located less than ±500m from the 110-400kV lines in the study period (1970-1989) were identified in the Population Register Center. Of those, the persons who were living or had lived in a building where a magnetic flux density over 0.01 μT was estimated even once were identified for analysis. This very low level at the start was chosen in order to enable subsequent examination of groups with different exposure. The start and end points of the residence periods and main properties of the buildings/residences were determined for epidemiological purposes.

RESULTS (Examples)

Buildings. The total number of buildings within ±500 m of 110-400kV lines in 1989 was 174770 which was 9.7% of all buildings in Finland. The building distribution near 110-400kV powerlines in 1989 (Fig. 2) is mainly made up by the 159409 buildings near 110kV lines. The corresponding number of buildings near 220kV lines was 6769, and near 400kV lines, 8592. In 1989, within the more interesting ±200 m from the lines, there were 56163 buildings of which 2675 were near 400kV, 2437 near 220 kV, and 51051 near 110kV lines.

Cumulative Exposure. Cumulative exposure of 385000 persons from 1970 to 1989 is presented in Fig. 3. The maximum exposure was 81 μT-years. Other exposure metrics like distance, average magnetic flux densities during the residence periods, and time lived in exposed residences will also be studied.

Magnetic Fields. Magnetic flux densities in the buildings within this ±200m zone in 1989 are indicated in Fig. 4a (low-field buildings) and 4b (high-field buildings). Of these, 5600 buildings with flux densities over 0.1 μT, which is the expected background exposure in Finnish residences,[3] were identified. This was 0.3% of all buildings and 10.0% of the buildings located closer than ±200 m to 110-400kV powerlines in 1989.

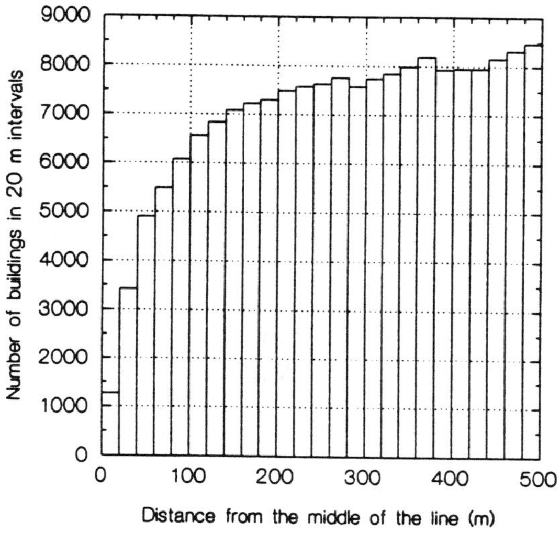

FIG. 2.—Number distribution of buildings near 110-400kV lines in 1989.

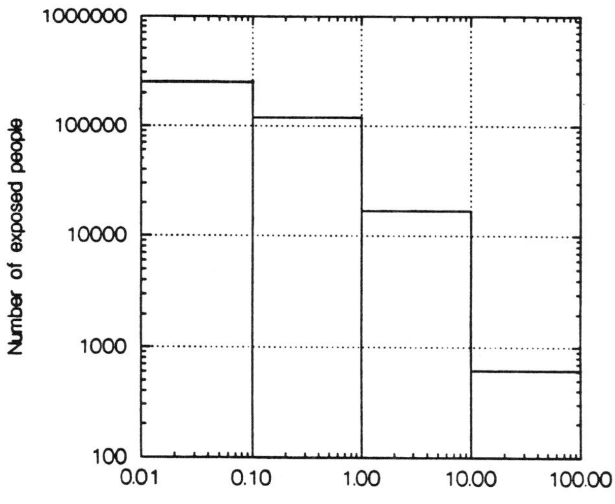

FIG. 3.—Distribution of personal exposure in microtesla-years (μTa) from the digitized 110-400kV powerlines in Finland during 1970-1989.

DISCUSSION

Very close to 110-400kV lines, where the practical limits of the exposure calculation are highest, the building density is low in accordance with Finnish practice. The average magnetic flux densities in most buildings seem to be moderate compared with the background level, and the number of buildings with high flux density is low. The exposure of people living near powerlines, of course, reflects the building distribution and average annual magnetic fields in the buildings. The number of highly exposed people is thus very low. A 110-400kV line is therefore not invariably a considerable source of magnetic field in near-by buildings. However, this may not help if correlation between even low magnetic-field exposure and cancer is to be verified.

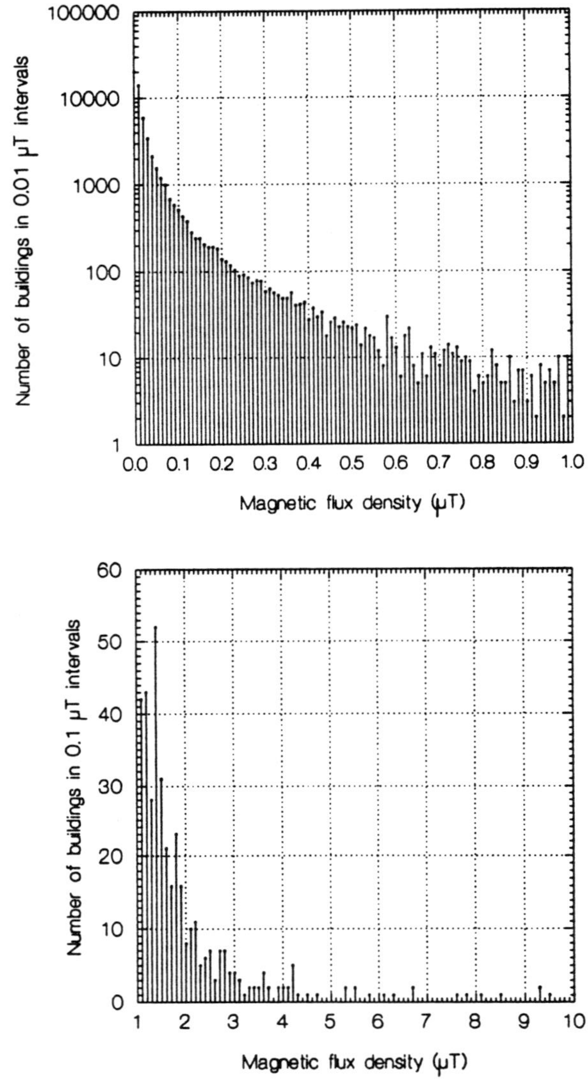

FIG. 4.—Number distribution of buildings within 200 m from 110-400kV lines according to annual average magnetic flux density in 1989. (a) Low-field buildings, (b) high-field buildings.

REFERENCES

1. N. Wertheimer and E. Leeper, "Electrical wiring configurations and childhood cancer," *Am. J. Epidemiology* 109: 273-284, 1979.

2. M. Hongisto, *Magnetic Field Exposure in the Vicinity of the Finnish National Grid* (Master's Thesis), Technical University of Helsinki, 1992, 157 pp. (In Finnish.)

3. J. Juutilainen, K. Saali, J. Eskelinen, P. Matilainen, and A. L. Leinonen, *Measurements of 50 Hz Magnetic Fields in Finnish Homes.* Research Report IVO-A-04/89, Imatran Voima Oy, Helsinki, 1989, 31 pp.

Martin Blank, Ed., *Electricity and Magnetism in Biology and Medicine*

POWER-FREQUENCY ELECTRIC AND MAGNETIC FIELD EXPOSURE: DOSIMETRIC ASSESSMENT

T. Lindh* and L.-I. Andersson

National Institute of Occupational Health, Solna, Sweden

A survey of 50Hz electric and magnetic fields in the electric power industry has distinguished three different exposure categories. They consist of the same occupational subgroups independent of whether the full shift exposure is summarized in terms of average value, median value, upper quartile value, or number of minutes per day in a magnetic field in excess of 1μT. The different indices describe different aspects of exposure. Typically, for the most exposed occupational subgroups, dosimetry measurements show an average flux density of 1 to 2.5μT for the magnetic field. The electric field strength exceeds 30 V/m for about 100 min per day.

There has been a great interest in the possible adverse health effects of exposure to magnetic fields during the past decade. In human studies most of the attention has been focused on children living near powerlines (leukemia and brain tumors), VDT operators (pregnancy outcome and skin problems), and occupational exposure in a variety of occupations (leukemia and brain tumors). Very little has been published about utility workers.[1-3] A prospective longitudinal study following a group of 529 workers employed in the electric power industry has been in progress at the National Institute of Occupational Health in Sweden since 1982. (The present study is part of a longitudinal study on health risks during work with production and distribution of electricity; principal investigators are Francesco Gamberale and Bengt Knave of the National Institute of Occupational Health, Solna, Sweden.) The health parameters studied include fertility and reproduction, blood pattern, and CNS symptoms. In addition to electric and magnetic fields, other exposure factors have been identified in the work environment. These include solvents, arsenic, noise, and strenuous physical work. The final health examination will be in 1995. This paper describes the exposure to 50Hz electric and magnetic fields.

MATERIALS AND METHODS

To estimate the exposure to 50Hz electric and magnetic fields, the study population was classified into various occupational subgroups (Table 1). Examples of work tasks performed by line and substation workers were inspection, repair, and construction work. From each subgroup a number of individuals were selected to wear a specially designed instrument that registered the strength of the 50Hz electric and magnetic field every 15 s. In a few cases a Positron dosimeter was used. It was emphasized that any prior expectation about high or low exposure should not govern the selection process. All measurements were carried out for at least 6 h, preferably during the full work shift, at 25 utility companies.

For each of the 15s readings from the instrument, the magnetic field flux density was calculated according to the equation $B = \sqrt{X^2+Y^2+Z^2}$, where X, Y, Z were the rms (root mean square) values of the magnetic field flux density in three mutually orthogonal directions of incidence. Instrument calibrations were checked regularly.

Because the detection limit for the electric field was 30 V/m, mean and median values could not always be calculated. The results are therefore given as number of minutes/day in excess of 30 V/m. The electric field is converted to "undisturbed field condition." Tests at various workplaces in the electric-power industry have shown that the converted readings are highly correlated with the (short-circuit) current that flows between a person and ground.[4] In order not to mix up measurements of the electric field taken by instruments placed at different sensor locations, results from the Positron dosimeter are not given. Including or excluding *magnetic* field data from the Positron dosimeter does not alter the overall picture in Tables 1 and 2.

---

*Correspondence to: T.Lindh, National Institute of Occupational Health, S-17184 Solna, Sweden.

Since the number of measurements within a subgroup was relatively low (generally around 10 full-shift measurements), individual extreme daily averages were excluded from the calculation of the subgroup mean exposure. A daily average was taken to be extreme if the result was more than 10 times higher than any other daily average within the subgroup.

For the purpose of the exposure health calculations, three exposure categories were formed among the occupational subgroups, based on inspection of the results, as follows. Magnetic field (B:I) < 0.15 μT; (B:II) 0.15-0.6μT; (B:III) > 0.6 μT. Electric field, (E:I) < 1 min/day with field strength >30 V/m; (E:II) 1-30 min/day; (E:III) >30 min/day.

RESULTS AND DISCUSSION

As expected, exposure to the electric field depended on line voltage (Table 1). Testing personnel worked partly in switchyards and were therefore exposed to electric and magnetic fields. Combined heat and power plant (CHP) personnel, who were working indoors, had a low to moderate exposure to the electric field. For all subgroups, field strength seldom exceeded 1.5 kV/m, except for workers in 200-400kV substations (data not shown).

The occupational subgroup least exposed to the magnetic field was distribution linemen working in the 0.4-20kV part of the network, probably because the distribution network carries low current and the lines run separately, with no other powerline in the vicinity. Transmission lines, on the other hand, carry high current and are often built in parallel with other transmission lines. That means that even if one line is de-energized, the other is presumably not. Transmission linemen were found to be more exposed to both electric and magnetic fields than distribution linemen.

Substation workers belonged to the most exposed category. Due to the complex and compact wiring configuration inside a substation/switchyard, work was often done in the vicinity of energized equipment, thereby exposing employees to electric and magnetic fields. The exposure to electric field in 0.4-20kV stations was, of course, low.

Metropolitan power plant personnel had a comparatively moderate exposure. They carry out a wide range of work tasks, including work on distribution lines and street lighting systems as well as on substations and occasionally, on transmission lines.

TABLE 1.—Exposure to 50Hz electric and magnetic fields in the utility industry.

Occupational Subgroup	n	Magnetic field (μT)	Electric field minutes/day above 30 V/m (minutes)
0.4-20kV line	11	<0.092	3.3*
Metropolitan power plant	7	0.295	16.2
Mechanical repair in CHP plant†	15	0.285	0.9
Operator in CHP plant†	16	0.232	0.4
Electrical repair in CHP plant†	11	0.76	29.4
Testing	8	0.9	86.0
20-130kV line	8	1.07	104.2
200-400kV line	6	1.27	130.6
0.4-20kV substation	10	1.32	8.0
20-130kV substation	11	1.27	103.8
Hydro power plant (electric power >10 MW)	21	1.85	22.1††
200-400kV substation	5	2.76	125.0

* n = 7 (positron dosimeter excluded); † CHP stands for combined heat and power;
†† n = 10 (positron dosimeter excluded).

It is not known today how to characterize exposure in order to have a measure that is correlated with the health effect studied, or even if such a relationship exists. A factor sometimes mentioned in discussions is that heavy exposures—peak exposures—should perhaps be given greater weight. One method to do this is to form exposure categories based on mean- and median-value. The mean value is inflated by peak values, but the median value is not. However, the grouping of data into three categories will consist of the same occupational subgroups whether the exposure is presented in terms of mean, median, upper quartile, or number of minutes/day in excess of 1 μT (Table 2).

TABLE 2.—Various indices of exposure to magnetic fields.*

Exposure category	n	Mean (μT)	Median (μT)	Upper-quartile (μT)	Number of minutes/day with levels > 1 μT (minutes per day)
Exposure category B:I	11	<0.092**	<0.046	0.081	8.2
Exposure category B:II	38	0.265 *0.21-0.32*	0.126	0.233	20.4
Exposure category B:III	83	1.406 *1.09-1.73*	0.752	1.529	117.1

* The figure in each cell is an average value of n full-shift measurement. Each of the full-shift measurements is calculated in accordance with the column heading. A 95% confidence interval for the mean value is shown in italics.

**The following full-shift mean values were measured: <0.3, <0.05, <0.07, <0.03, <0.17, <0.07, <0.03, 0.16, 0.19, 0.13, 0.08 μT.

In Table 1 extreme one-day mean values for the magnetic field were excluded in the calculation of occupational subgroup mean values. A value was taken as being extreme if it was more than ten times higher than other values. At all stages, it was only among occupations with low average exposure that extreme one-day mean values were recorded. If extreme values are included and the material is divided into exposure categories, the overall picture is not altered. However, the difference between groups will be somewhat reduced. Since extreme values are rare, more data would be necessary to obtain their true magnitude.

No answer is available today to the question of whether it is the electric field or the magnetic field, or a combination of the two, that is the factor which affects health (if effects on health can be demonstrated). As can be extracted from Table 1, classification into low (I), medium (II), and high (III) exposure will include somewhat different occupations, depending on whether the classification is based on exposure to electric or magnetic field. (Table 1 is sorted according to exposure to magnetic field).

From the epidemiological aspect, it would be desirable to determine the exposure of each of the individuals in the groups studied. However, it is reasonable to assume that the average exposure calculated over several workplaces fits relatively well for individuals among, for example, linemen and substation personnel. In these groups the individuals often work at many different workplaces within the utility. Control-room personnel can be set at the opposite end of the scale, since they have a given workplace. An average value for several control rooms then does not fit in quite so well with specific individuals, particularly since the flux density varies widely among different control rooms. The relationship between the highest and the lowest one-day mean value was greater than a factor of 100 (data not shown). Control room personnel therefore cannot be treated as a homogeneously exposed group.

The study undertaken has clarified that exposure to 50Hz electrical and magnetic fields differs among different occupational groups in the electric power industry. Whether there is any relationship between the pattern found and the health of the workers will emerge from reports to be presented later.

REFERENCES

1. M. R. Flynn, S. West, W. T. Kaune, D. A. Savitz, C. C. Chen, and D. P. Loomis, "Validation of expert judgment in assessing occupational exposure to magnetic fields in the utility industry," *Appl. Occup. Environ. Hyg.* 6(2): 141-145, 1991.

2. J. E. Deadman, M. Camus, B. G. Armstrong, et al., "Occupational and residential 60-Hz electromagnetic fields and high-frequency electric transients: Exposure assessment using a new dosimeter," *Am. Ind. Hyg. Assoc. J.* 49: 409-419, 1988.

3. T. D. Bracken, *The EMDEX Project: Technology Transfer and Occupational Measurements*, Vol. 1-3, Electric Power Research Institute Report no. EPRI EN-7048, Palo Alto, Calif.

4. T. Lindh and L.-I. Andersson, "Comparison between two power frequency electric field dosimeters," *Scand. J. Work Environ. & Health* 14 (supplement 1): 43-45, 1988.

Martin Blank, Ed., *Electricity and Magnetism in Biology and Medicine*

MEASUREMENTS OF TRANSIENT ELECTRIC AND MAGNETIC FIELDS FROM INDUSTRIAL AND HOME DEVICES

A. Thansandote* and D. W. Lecuyer

Bureau of Radiation and Medical Devices, Health and Welfare Canada, Ottawa

M. A. Stuchly and S. S. Stuchly

University of Victoria, Canada

A system for measurement and characterization of transient fields is described. It consists of active electric and magnetic field sensors, a high-speed digitizing oscilloscope, and a portable microcomputer. Measurements of transient fields were carried out for a few industrial and domestic devices. It was found that most measured electrical appliances produce high-intensity transient magnetic fields and relatively low-intensity transient electric fields as compared to the ambient fields.

Environmental transient electric and magnetic fields are generated by switching or intermittent operation of high-power industrial devices and domestic electrical appliances. Some industrial workers are chronically exposed to transient fields in the course of their jobs, whereas the general public may be exposed only occasionally. In both cases, the health risk is still unknown. The interaction of transient fields with biological systems is very complex and relatively little investigated.[1] Considerable concern among industrial workers about the possible health effects of transient fields is reflected in a recent research recommendation to develop instruments for measuring these fields.[2] In this paper, we describe a system for measurement and characterization of transient fields, and present the results of transient field measurements for some industrial and home devices.

MEASUREMENT SYSTEM

Previously developed active electric and magnetic field sensors[3,4] were used. Both sensors are free-field configuration devices that employ electrically small antennas and electronic circuitry. The electric-field sensor operates from 32 to 100 MHz (3dB roll-off frequencies) with constant sensitivity of 0.4 mV/Vm^{-1} and a rise time of 3.5 ns. The magnetic field sensor has a flat response from 17 Hz to 17 MHz with sensitivity of 0.5 V/Am^{-1} and a rise time of 13 ns. A block diagram of the measurement system is shown in Fig. 1. Measurements are made by placement of the sensors close to the transient-field sources in the locations of human exposure, or at the nearest locations of the sources where the fields do not cause an output-signal distortion. The sensors are also oriented in various directions with respect to the source in order to obtain maximum output. Transient fields detected by the sensors are simultaneously displayed on a digitizing oscilloscope and transferred to a microcomputer for recording. The recorded data are subsequently analyzed to yield the waveform of the measured field and its spectrum.

RESULTS AND DISCUSSION

The seven industrial and home devices measured were: a spot welder, an electric eraser, a hair dryer, a hair clipper, a microwave oven, a stove, and a kettle. Figure 2 shows the transient magnetic field generated by a hair clipper (120 V, 35 W) at a distance of 30 cm. The peak of the negative quasirectangular portion (13μs duration) of the transient waveform is 6 μT. The transient nature of the field is reflected by its broadband spectral density over approximately 1 MHz. It has been found that 90% of the total energy of this transient field is contained within a 44kHz bandwidth. A similar waveform was observed in the second measure-

---

*Correspondence to: A. Thansandote, Bureau of Radiation and Medical Devices, Health and Welfare Canada, Ottawa ON K1A 0L2.

ment, but the peak value and the 90% energy bandwidth were found to be 6.87 μT and 19.5 kHz, respectively. The transient electric field from the hair clipper had a peak value of 101 V/m, which occurred at the same time as the negative quasirectangular portion of the transient magnetic field. The transient electric field was found to have a 90% energy bandwidth of 4 MHz. The results of transient-field measurements for other devices are summarized in Tables 1 and 2. The transients from the sources other than hair dryers, hair clippers, and erasers were found to have predominant 60Hz magnetic field components.

TABLE 1.—Transient E-field data from some industrial and home devices.

Device	Device - Sensor Distance (cm)	Maximum Amplitude of		Maximum Amplitude of Spectral Density (Vm^{-1}/Hz)	90 % Energy Bandwidth
		Field (Vm^{-1})	Time - Derivative (MVm^{-1}/s)		
Hair Dryer 120 V, 1.5 kW	8	123 123	0.32 0.32	1.575 at ≈ 60 Hz 1.501 at ≈ 60 Hz	122 Hz 122 Hz
Hair Clipper 120 V, 35 W	30	101	389	8.34×10^{-4} at ≈ 9.8 kHz	4.09 MHz
Electric Eraser 120 V, 25 W	40	98	309	5.35×10^{-4} at ≈ 720 kHz	2.27 MHz

TABLE 2.—Transient H-field data from some industrial and home devices.

Device	Device - Sensor Distance	Maximum Amplitude of		Maximum Amplitude of Spectral Density (nT / Hz)	90 % Energy˙ Bandwidth
		Field (μT)	Time - Derivative (mT/S)		
Hair Dryer 120 V, 1.5 kW	8 cm	2.78 3.24 3.24	29.87 28.64 29.19	11.14 at ≈180 Hz 7.83 at ≈120 Hz 6.77 at ≈360 Hz	488 Hz 1025 Hz 977 Hz
Hair Clipper 120 V, 35 W	30 cm	6.87 6.02	23392 31319	0.27 at ≈4.9 kHz 0.12 at ≈9.8 kHz	19.5 kHz 43.9 kHz
Electric Eraser 120 V, 25 W	40 cm	6.40	11517	0.32 at ≈2.4 kHz	24.4 kHz
Microwave Oven 120 V, 1.35 kW	30 cm	1.53 1.52 1.87	1.72 1.37 2.01	33.28 at ≈ 60 Hz 33.54 at ≈ 60 Hz 38.37 at ≈ 60 Hz	73 Hz 73 Hz 73 Hz
Stove 120 V	30 cm	0.78	2.31	25.36 at ≈ 60 Hz	73 Hz
Kettle 120 V, 1.5 kW	10 cm	0.31	0.44	9.04 at ≈ 60 Hz	73 Hz
Spot Welder 575 V, 18 A	2.16 m	4.45 4.59	4.40 5.16	120.42 at ≈ 60 Hz 143.17 at ≈ 60 Hz	73 Hz 73 Hz

CONCLUSIONS

A system for the measurement of transient electric and magnetic fields was developed. Fields from a few industrial and domestic devices were measured. It has been found that some electrical appliances produce high-intensity transient magnetic fields and relatively low-intensity transient electric fields as compared to the ambient fields. The frequency spectra of the measured magnetic fields are in the range of 60 Hz to 44 kHz, but the electric fields of some devices have frequency spectra up to 4 MHz.

REFERENCES

1. O. P. Gandhi, *Biological Effects and Medical Applications of Electromagnetic Energy,* Englewood Cliffs, N.J.: Prentice Hall, 1990.

2. *Proc. Scientific Workshop on the Health Effects of Electric and Magnetic Fields on Workers,* Cincinnati: National Institute for Occupational Safety and Health, 1991, 205-212.

3. A. Thansandote, S. S. Stuchly, M. A. Stuchly, and M. Barski, "Broadband active E-field sensors for measurement of transients," *IEEE Trans.* IM-40: 465-468, 1991.

4. M. A. Stuchly, A. Thansandote, M. Frika, and D. T. Gibbons, "Low-frequency broadband electric and magnetic field sensors," 13th BEMS Meeting, Salt Lake City, 1991.

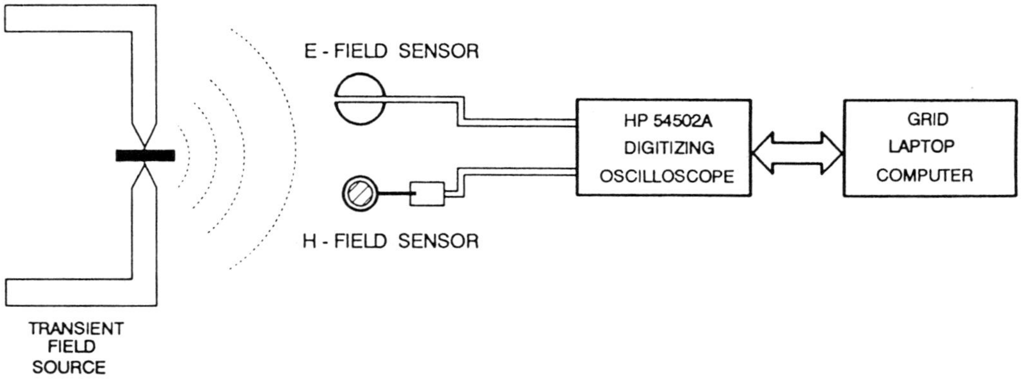

FIG. 1.—Instrumentation set-up for transient-field measurements.

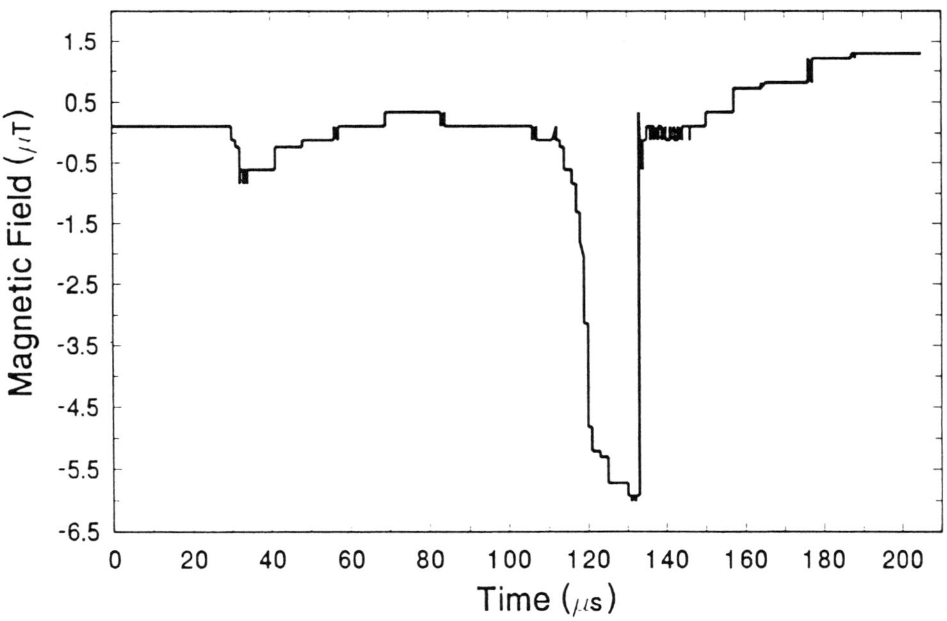

FIG. 2.—Transient magnetic field generated by hair clipper (120 V, 35 W).

Martin Blank, Ed., *Electricity and Magnetism in Biology and Medicine*

RESIDENTIAL WIRE CODE ASSOCIATIONS AND IMPLICATIONS FOR CASE-CONTROL STUDY BIAS

T. L. Jones,* C. H. Shih, D. H. Thurston, and B. J. Ware

American Electric Power, Columbus, Ohio

P. Cole

University of Alabama, Birmingham

Several studies of childhood cancer and leukemia have reported an association with certain powerline configurations. These "wire codes" were suggested as a measure of a child's residential magnetic-field exposure. A study was conducted in 1991 of 5721 residences in Columbus, Ohio, to determine residential factors, other than magnetic fields, that are associated with wire codes. Factors found to be associated include: the area in which the residence was located, the age of the transformer that served the residence, the length of time that the current occupant had lived in the residence, the household income, the type of dwelling, and the average monthly energy use. This information might be useful for further interpretation of past study results and in the design of future studies.

Several case-control studies of childhood cancer and leukemia have reported an association with certain powerline wire configurations.[1-3] These "wire codes" were suggested as a measure of a child's residential exposure to magnetic fields. However, in two of these studies, magnetic-field measurements were also taken and the reported associations were weaker than with those reported for wire codes,[2,3] which suggests that wire codes could be associated with childhood cancer due to a bias and not because they are an indicator for a true association with disease. The present study was conducted to determine other factors that might be associated with wire codes that could bias study results. Factors suggested in previous studies to be associated with high wire codes include: the inner city area,[4] rental properties,[4] residential mobility,[4,5] multiple family homes,[5] high traffic density,[5] low-income families,[5] less educated fathers,[5] and maternal smoking.[5]

METHODS

The study population consisted of 5721 residences in three adjacent telephone exchange areas in Columbus, Ohio. Residential information obtained from computer records included the age of the transformer that served each residence, the time elapsed since the start of service for the current occupant as of 26 July 1991, a marker for low household income ($6540 per household plus $3390 per person maximum), and the monthly electrical energy use. Information obtained at each residence included type of dwelling, powerline configuration, and distance from the powerline to the residence. The powerline configuration and distance information established the wire code for each residence in accordance with the Denver method.[1,2] The "very high" category is assumed to produce the highest magnetic field levels; the "very low" category the lowest levels; "ordinary high" and "ordinary low" correspond to in-between exposures. Residences served by underground ("buried") lines only are included in the very-low wire code category.

RESULTS

The results of our study show that each of the residential factors studied is associated with wire codes in some way. A summary of our results is presented in Table 1. The inner-city region has the highest proportion of homes classified as very high (17%) and ordinary high (33%). The urban area is characterized by ordinary low homes (70%), and the suburban area has a high proportion of homes classified as very low (63%).

*Correspondence to: T. L. Jones, American Electric Power, 1 Riverside Plaza, Columbus OH 43215.

Newer transformers (which could be considered indicative of newer homes) are strongly associated with lower wire codes, with 94% of the homes served by transformers less than 44 months old classified in the low-wire-code categories. A decreasing trend in the proportion of very low wire codes and an increasing trend in the proportion of very high wire codes is observed with increasing transformer age. Shorter durations of occupancy (higher residential mobility) are associated with higher wire codes. The following trends in wire code proportions with increasing durations of occupancy are noted: decrease in ordinary high, increase in ordinary low, decrease in very low. The decrease in the proportion of very-low-wire-code homes with increasing occupancy duration is a result of mobility associated with newly built homes. Low-income households are associated with high wire codes, with the greatest difference (in comparison with the non-low-income homes) observed in the very high category. However, only a small proportion (8%) of the residences were identified as "low income." Single-family homes are associated with low wire codes. Two-family homes, which are mostly found in the inner city, are associated with high wire codes. Other multiple family homes (e.g., apartments, condominiums) have a much higher proportion of high wire code homes than do single family homes (42% vs 20%). Regarding energy use, homes that use less than 300 kW-h per month have a higher proportion of high wire codes and lower proportion of low wire codes than have homes that use more energy. However, only 15% of the study homes were included in this lower-energy-use category. Table 2 provides the median values for transformer age, duration of occupancy, and energy use for each wire code classification. A trend is absent for duration of occupancy, which is at least partially due to the mobility associated with newly built homes.

DISCUSSION

The associations between wire codes and other factors observed in the present study are consistent with those observed in previous studies and have implications for case-control study bias. Residential studies have reported significant wire code associations in several ways: (1) odds ratios for low/high two-level wire codes;[1,2] (2) trends for buried/very-low/ordinary-low/ordinary-high/very-high five-level wire codes;[3] and (3) odds ratios for very-low/very-high wire codes.[2,3] Table 1 shows considerable differences in wire-code proportions across regions, which demonstrates the importance of matching cases and controls in the same neighborhood. For example, if cases were selected from the inner city and controls were selected from the urban area, a false very-low/very-high odds ratio would be observed: (0.17/0.07)/0.03/0.08) = 6.5. Studies that include children in newer homes as controls, but not as cases, would be biased due to the association between newer transformers (i.e., newer homes) and very-low and ordinary-low wire codes. A study that includes residentially mobile cases, but only stable controls, would be biased due to the observed association between residential mobility and high wire codes.[6] If controls are selected only from a moderate- to high-income population, while cases are not so restricted, the case group having the lower incomes would tend to live in more very-high and ordinary-high homes. If controls are selected only from single-family homes, while cases are selected from all homes, the case group with the multiple family homes would tend to have more very-high and ordinary-high wire codes. With respect to energy use, if the control population did not include low-energy-use homes, but the case group did, then the case group would also tend to have more very-high and ordinary-high wire code homes. (The association between low energy use and high wire codes also indicates that magnetic-field levels from in-home sources may be inversely related to magnetic-field levels from powerlines, which could be a partial explanation for the reportedly poor correlation between wire codes and measured magnetic fields.[7]) One final observation is that several factors, including income and energy use, might produce false trends across the four (or five) wire-code levels that could be incorrectly interpreted as a dose response. For example, if cases are chosen from low-income households, while low-income homes are not included in the control group, the following odds ratios would be observed across the wire code categories from very low to very high: 1.0, 1.4, 2.0, 3.4. This observation would be the result of study bias and not as a result of any dose response between exposure and disease.

In conclusion, several residential factors have been found to be associated with wire codes and have the potential to bias the results of a case-control study. Understanding what these factors are and what effect they might have on study results should be helpful. This finding is especially important in the design of studies or interpretation of associations between wire codes and disease when the cases and controls are not the same with respect to one or more of these factors.

REFERENCES

1. N. Wertheimer and E. Leeper, "Electrical wiring configurations and childhood cancer," *Am. J. Epidemiol.* 109: 273-284, 1979.

2. D. A. Savitz et al., "Case-control study of childhood cancer and exposure to 60-Hz magnetic fields," *ibid.*, 128: 21-38, 1988.

3. S. J. London et al., "Exposure to residential electric and magnetic fields and risk of childhood leukemia," *ibid.*, 134: 923-937, 1991.

4. N. W. Wertheimer and E. Leeper, "Re: 'Electrical wiring configurations and childhood leukemia in Rhode Island,'" *ibid.*, 111: 461-462, 1980.

5. D. A. Savitz, "Case-control study of childhood cancer and residential exposure to electric and magnetic fields," Supplement to contractor's final report to New York State Power Lines Project, Contract #218217, 1988.

6. T. L. Jones et al., "Selection bias from different residential mobility as an explanation for associations of wire codes with childhood cancer," *J. Clin. Epidemiology* (in press).

7. W. T. Kaune et al., "Residential magnetic and electric fields," *Bioelectromagnetics* 8: 315-335, 1987.

TABLE 1.—Proportion of residence in each wire-code classification by residential factor, Columbus, Ohio, 1991.

	Very High	Ord. High	Ord. Low	Very Low	Total Number
Total	.07	.24	.42	.27	5721
Region					
Inner City	.17	.33	.43	.07	1824
Urban	.03	.19	.70	.08	1890
Suburban	.01	.21	.15	.63	2007
Transformer Age					
<44 months	.01	.05	.38	.56	492
44-187 months	.05	.25	.33	.37	657
>187 months	.08	.26	.44	.22	4572
Duration of Occupancy					
<20 months	.08	.28	.35	.29	2039
20-91 months	.06	.24	.42	.28	2140
>91 months	.07	.18	.53	.22	1542
Income					
Low	.13	.30	.39	.18	472
Not Low	.06	.23	.43	.28	5249
Dwelling Type					
Single Family	.05	.15	.48	.32	3029
Two Family	.17	.34	.44	.05	502
Other	.08	.34	.33	.25	2190
Energy Use					
<300 kWh/mo.	.13	.31	.40	.16	861
≥300 kWh/mo.	.06	.22	.43	.29	4581
Missing data	.11	.34	.31	.24	279

TABLE 2.—Central tendency (median) of continuous residential variables for each wire-code classification, Columbus, Ohio, 1991.

	Transformer Age (months)	Duration of Occupancy (months)	Energy Use (kWh/mo.)
Very High	344	33	420
Ordinary High	252	29	520
Ordinary Low	261	51	510
Very Low	220	32	575

Martin Blank, Ed., *Electricity and Magnetism in Biology and Medicine*

RESIDENTIAL EXPOSURES TO INDOOR AIR POLLUTANTS COULD YIELD CHILDHOOD LEUKEMIA RISK LEVELS SIMILAR TO THOSE ASSOCIATED WITH 60Hz MAGNETIC FIELDS

C. E. Easterly*

Oak Ridge National Laboratory, Oak Ridge, Tenn.

Over a decade ago Easterly suggested that electromagnetic fields may be able to participate in a cooperative process leading to the expression of cancer.[1] Evidence derived from the literature is presented to support the suggestion that potentially cooperative factors other than electromagnetic fields are present in homes in sufficient quantities to result in approximately the same risk levels as are being measured in epidemiology studies of childhood leukemia and electromagnetic fields. Generally these odds ratios vary from 1.5 to 2.5.[2]

BACKGROUND AND RATIONALE

Over the past decade, the scientific community has been unable to reject the null hypothesis that 60Hz electromagnetic fields are not human carcinogens. When coupled with the inability to prove the alternative hypothesis, much concern is raised about our lack of knowledge associated with potentially harmful bioeffects of electromagnetic fields. Concurrent with the emergence of scientific interest in bioeffects of electromagnetic fields in the mid-1970s was the growing interest in potential effects of indoor air pollution, as aggravated by attempts to tighten the thermal envelope of homes. Measurements of typical chemical agents (many of which are leukemogenic or suspected leukemogens) by other investigators have resulted in an increased understanding of the likely range of concentrations that could be found within the home environment. In addition, the recent rethinking of the factors leading to bone-marrow dose to infants and children from exposure to radon and its daughters adds to the concern for the potential leukemogenic burden found in typical residences. It is the contention of the present author that residential exposures to chemical and radiochemical agents can be large enough to overshadow the effects of electromagnetic fields and that, unless epidemiologic investigations include careful total-exposure assessment, incorrect inferences could be made. Even well-designed case control studies may be susceptible to misclassification because of our incomplete understanding of how to distinguish or identify residences or neighborhoods that may contain high levels of potential leukemogens. It is even possible that there is some geologic/geographical feature that facilitates the convergence of wires carrying high currents and, for example, elevated radon levels.

ANALYSIS

Radon

Of potential childhood leukemogens, radon has not been considered previously owing to the focus on its role in lung cancer. Consideration of dose from radon dissolved in fat, from inhaled radon and thoron daughters, and accumulated ^{210}Po, suggests that the radon- and thoron-derived dose to active marrow may be greater by a factor of 10 than previously thought.[3] Henshaw suggests[4] that 13-25% of myeloid leukemia worldwide may be caused by world average exposures to radon of 50 Bq/m^3. Examination of a cumulative frequency distribution of indoor ^{222}Rn measurements[5] clearly demonstrates that a substantial number of homes experience radon levels well above 50 Bq/m^3. Nazaroff and Nero note that indoor concentrations in individual homes are sometimes above the average by an order of magnitude or more, and that ^{222}Rn concentrations in the range of 200 to 2000 Bq/m^3 occur with surprising frequency.

*Correspondence to: C. E. Easterly, ORNL Health and Safety Research Division, Box 2008, Oak Ridge TN 37919-6101.

The developing perspective of radon and its daughters, as they affect dose to the active bone marrow, provides the basis for concern about radon's potential etiological role in childhood leukemia because of (1) concentrations found by the extensive surveys performed on indoor radon, (2) calculated doses from the bone-marrow modeling of radon and its daughter products within the age-dependent human, and (3) the role of high LET dose in cell killing and cell transformation. Thus, there is a conceptual linkage between agent and disease. Given that the high-LET decay products are known cancer initiators, exposure to radon and its daughters can result in a situation that provides the first step in a sequence of events resulting in a clinical diagnosis of leukemia.[7] That is, once initiation occurs, other agents, which may not be potent initiators, can assist in the promotion or progression of the disease.

Residential Chemical Agents

Potential exposures of children to residentially based chemicals also raise the question about competing positive and negative synergisms for effects with electromagnetic fields. Any modern setting, whether residential or commercial, is associated with a complex mixture of chemical agents. Two examples of information that focus on this problem are presented below.

In the first example, an epidemiologic investigation, Lowengart et al. examined home exposures of mothers and fathers during pregnancy and nursing of children who suffer from leukemia.[8] They found odds ratios of approximately 2 to 6 ($p < 0.01$) for such seemingly innocuous residential exposures to parents as weekly exposures to household cleaning chemicals, monthly exposures to garden pesticides or herbicides, and weekly exposure to burning incense. A smaller sample of fathers exposed to chemicals at work after the birth of the child yielded similar results. These results are consistent with such previous investigations as a demonstration of elevated risks for leukemia in children whose fathers are employed in occupations where exposures to certain chemicals may occur.[9]

The second example derives from measurement studies. Quantitative studies of gas-phase organic compounds over the past decade have been conducted in more than 1500 homes in Europe and the USA. To provide an illustration of the potential role of indoor pollutants in leukemia rates, we use data from Krause et al.[10] One class of compounds they measured, the aromatics, are represented by 13 chemicals. These 13 aromatic compounds result in a mean concentration of 166 $\mu g/m^3$, but except for benzene, they have not been evaluated for carcinogenic potency. However, their abilities to potentiate carcinogenesis can be estimated according to the methods of Jones et al., which yield a weighted potency of the total aromatics of half that of benzene.[11] From the EPA slope factor for benzene, the risk is approximately 1.5 per 100000/yr.[12] This estimated effect of the single class of compounds amounts to a large fraction of the spontaneous rate of approximately 2.5 per 100000/yr. Household exposures are not limited to one class of materials within the category of VOCs; rather, the VOCs represent only a small proportion of chemicals to which children are exposed. It is inevitable that some of these chemicals will be found to be leukemogens. Thus, as with radon, the potential effects of chemical-agent exposures at home can obscure data taken to study effects of residential electromagnetic fields.

DISCUSSION

Humans do not live in a world in which interactions are closely partitioned. Rather, we live in a milieu of interactions with physical, chemical, and biological insults, and with the complex social and genetic forces that apply differently for each person. These forces and agents cannot be thought of as independent with respect to health issues; rather, they operate on each one of us in unison. Thus, having seen the single-agent influences in the examples above, and knowing that they represent only a small part of the leukemogenic burden within any child's realm, we must realize that the relationship between childhood leukemia and electromagnetic-field exposure, if present, cannot be properly understood without a serious reassessment of epidemiologic data needs and experimental paradigms.[13]

REFERENCES

1. C. E. Easterly, "Cancer link to magnetic field exposure: A hypothesis," *Am. J. Epidemiol.* 114: 169, 1981.

2. T. E. Aldrich and C. E. Easterly, "Electromagnetic fields and public health," *Environmental Health Perspectives* 75: 159-171, 1987.

3. R. B. Richardson, J. P. Eatough, and D. L. Henshaw, "Dose to red bone marrow from natural radon and thoron exposure," *Brit. J. Radiol.* 64: 608-624, 1991.

4. D. L. Henshaw, J. P. Eatough, and R. B. Richardson, "Radon as a causative factor in induction of myeloid leukaemia and other cancers," *Lancet* 335: 1008-1012, 1990.

5. International Commission on Radiological Protection, *Lung Cancer Risk from Indoor Exposures to Radon Daughters* (Annals ICRP), Oxford: Pergamon, 1987 (ICRP publication no. 50).

6. W. W. Nazaroff and A. V. Nero Jr., *Radon and Its Decay Products in Indoor Air*, New York: Wiley, 1988.

7. T. D. Jones and L. R. Glass, "Leukemia from Ra: 'Data gaps' or 'scientific evidence'?" *Health Physics* 57: 491-492, 1989.

8. R. A. Lowengart, J. M. Peters, C. Cicioni, J. Buckley, L. Bernstein, S. Preston-Martin, and E. Rappaport, "Childhood leukemia and parents' occupational and home exposures," *J. Natl. Cancer Inst.* 79: 39-46, 1987.

9. N. J. Vianna, B. Kovasznay, A. Polan, et al., "Infant leukemia and parental exposure to motor vehicle exhaust fumes," *J. Occup. Med.* 26: 679-682, 1984.

10. C. Krause, W. Mailahn, R. Nagel, C. Schulz, C. B. Seifert, and D. Ullrich, "Occurrence of volatile organic compounds in the air of 500 homes in the Federal Republic of Germany," in B. Seifert, H. Esdorn, M. Fischer, H. Rueden, and J. Wegner, Eds., *Indoor Air '87* (Proc. 45h Intern. Conf. on Indoor Air Quality and Climate, Vol. 1: Volatile organic compounds, combustion gases, particles and fibres, microbiological agents), Berlin, Federal Republic of Germany: Institute of Water, Soil, and Air Hygiene, 1987, 102-106.

11. T. D. Jones, P. J. Walsh, A. P. Watson, B. A. Owen, L. W. Barnthouse, and D. A. Sanders, "Chemical scoring by a *RA*pid *S*creening of *H*azard (RASH) method," *Risk Anal.* 8: 99-118, 1988.

12. L. A. Wallace, *The Total Exposure Assessment Methodology (TEAM) Study: Summary and Analysis* (Vol. I), Washington, D.C.: U.S. Environmental Protection Agency, EPA Report EPA-600/6-87-002a.

13. Oak Ridge National Laboratory is managed for the U.S. Department of Energy by Martin Marietta Energy Systems, Inc., under contract DE-AC05-87OR21400.

Martin Blank, Ed., *Electricity and Magnetism in Biology and Medicine*

CHILDHOOD CANCER AND RESIDENTIAL MAGNETIC FIELDS: STUDY DESIGN AND EXPOSURE ASSESSMENTS

M. Feychting* and A. Ahlbom

Karolinska Institute, Stockholm

The purpose of this study is to determine whether there is a link between ELF magnetic fields in the residential environment and the occurrence of childhood cancer.

Study Design

The study population, including the exposed as well as the unexposed, consists of those under the age of 16 who have lived within 300 m from any of Sweden's 220 and 400kV powerlines at any time during the period 1960-1985, approximately 100000 persons. The study is designed as a case-control study, as it is impossible to determine exposure for all 100000 individuals in the study population. Cases are children up to 16 years, diagnosed with cancer, who entered the study population before the time of diagnosis. Cases are identified by matching of the study population to the Swedish National Cancer Registry. Controls are a random sample of individuals from the study population, who are matched to the case on age, sex, time period, parish, and powerline. Four controls have been chosen for each of the 144 cases. Analyses will be conducted for all types of cancer together as well as for various subgroups of cancers. We plan to complete separate analyses on leukemia (39 cases) and on central nervous system tumors (33 cases). Confounding due to socioeconomic status, parental occupation at the time of conception and during pregnancy, car exhaust, parish, age, sex, and time period will be controlled statistically.

Exposure Assessment

Theoretical calculations of the magnetic field from the powerline in each home are used to determine exposure. The calculations are based on the distance to the powerline and on its configuration (i.e., type and height of the towers, distance between towers, and distance between the phases). If there is more than one powerline near the house, the ordering of the phases is also included. The mean load of the powerline for the year of interest is used in the calculations. The mean load is based on actual records of the load of each powerline made by the superintendent for the operation of the powerline each hour. We have determined exposure to 50Hz magnetic fields for the year of diagnosis and also at 1, 5, and 10 years before diagnosis, at time of birth, during pregnancy, and at the time of conception. We shall also determine the duration of the exposure.

The method of exposure assessment is validated against spot measurements taken in the home in which the case or corresponding control lived closest in time before diagnosis. Measurements were taken with an instrument specifically designed for this purpose. Measurements were taken in three different rooms, for 5 min in each room. The measurements were taken with the house current switched off. Information about the load on the powerline was obtained from the power station at the time of the spot measurement and was used for theoretical calculations of the magnetic field. The value for the calculated field was compared with the mean value of the measured field. If there were no other sources of magnetic fields in the house, the measured and calculated fields ought to correspond well with each other. On the other hand, if there were other sources of magnetic fields that we failed to discover, for example stray currents, then the measured field would be larger than the calculated field. In addition, 24h dosimetry readings were obtained for a sample of subjects in order to estimate the sensitivity and specificity of the measurements, both for schoolchildren aged 7 to 16 and for adults, by a Positron instrument. The Positron instrument was given to the subject when the spot measurements were taken. The subject then wore the instrument for 24 h. During the measurement

*Correspondence to: M. Feychting (Environmental Medicine), Karolinska Institutet, Box 60208, S-10401 Stockholm, Sweden.

period the subjects kept a diary indicating whether they were at home inside the house or out in the garden, at work, or some other place.

TABLE 1.—Distribution of magnetic field exposure from spot measurements and historical calculations (%).

	Low-power measurements			Historical calculations		
μT	All	Stockholm	Rest of Sweden	All	Stockholm	Rest of Sweden
0-0.09	60	48	76	85	84	85
0.1-0.19	19	24	13	7	7	9
0.2-	21	29	11	8	9	7
N	1329	775	554	2112	1410	702
Mean	0.18	0.24	0.10	0.07	0.08	0.06

Table 1 shows the distribution of the exposure from the spot measurements and the historical calculations. The measurements have been done for both adults and children. The results do not change when children and adults are treated separately. Looking at the spot measurements we see that the exposure is much higher in the Stockholm area than in the rest of the country. More than 50% have an exposure of 0.1 μT or more in Stockholm compared to 24% in the rest of the country. This is also illustrated by the differences in means. The mean is 0.24 μT for Stockholm and 0.10 μT for the rest of the country. The historical exposure is more evenly distributed throughout the country and lower than the spot measurements. About 15% have an exposure of 0.1 μT or more. The total mean is 0.07 μT. This difference between the historical fields and the measured fields is probably due to the change in the consumption of electrical power over the study period.

Figure 1 is a plot of the calculated magnetic fields based on the load of the powerline at the time of spot measurement vs the spot measurements made with the house current switched off, the low-power measurements. The correlation is quite good. A more thorough examination of the measurements for which the correlation is poor will be made. Figure 2 shows a plot of the calculated historical magnetic fields based on the mean load of the powerline for the year of diagnosis vs the spot measurements. Here the correlation is not so good, as is to be expected because the powerline loads have increased over time. It illustrates the importance of using historical exposures for such studies. The measurements made today are of little value in determining whether a subject was exposed at the time of diagnosis in this study.

Figure 3 shows a plot of the spot measurements vs the average of the 24h measurements. It is based on 94 subjects. The correlation between the two types of measurements is fairly good. The correlation between measurements is better when the 24h measurements are restricted to the time spent at home (Fig. 4).

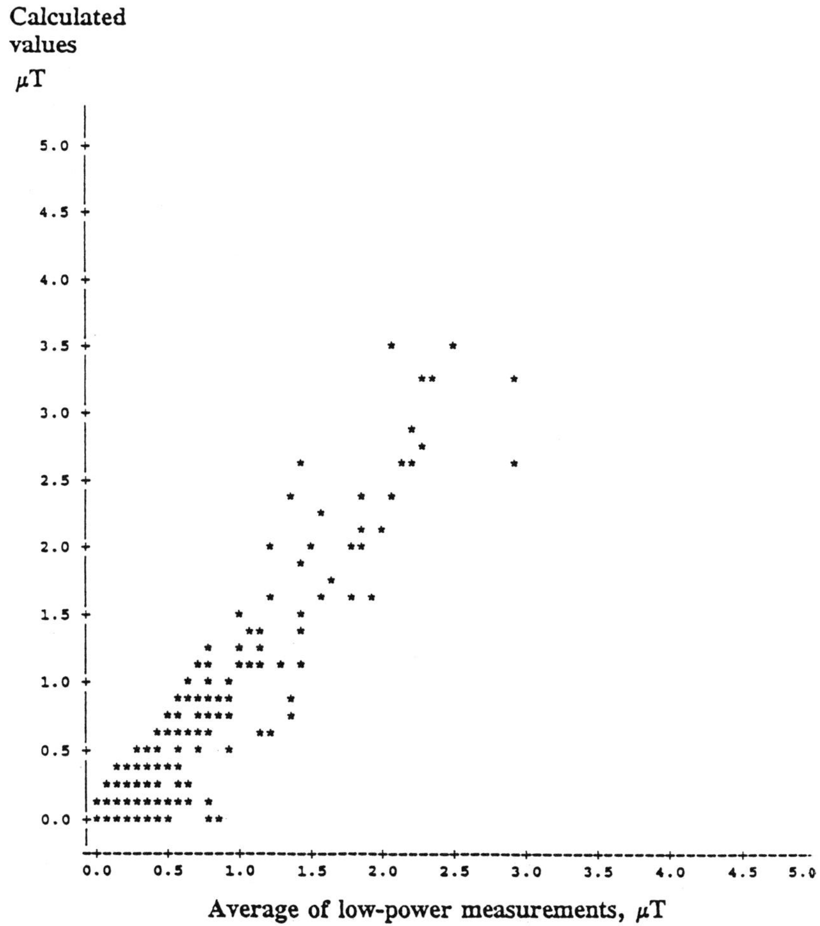

FIG. 1.—Plot of calculated magnetic fields vs low-power measurements.

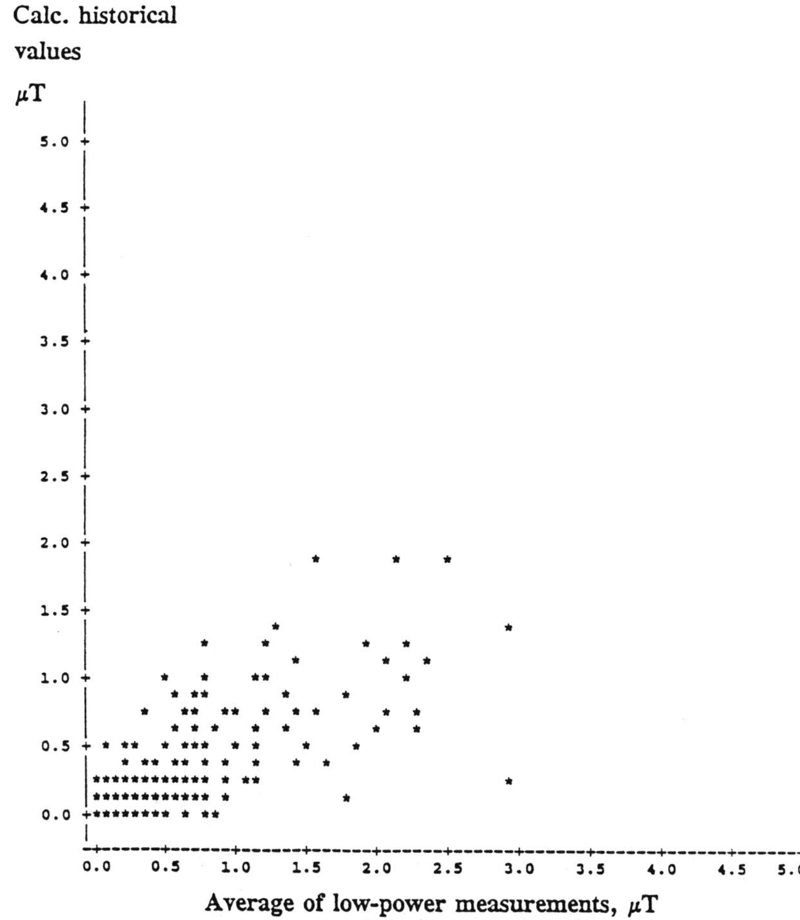

FIG. 2.—Plot of calculated historical fields by low-power measurements.

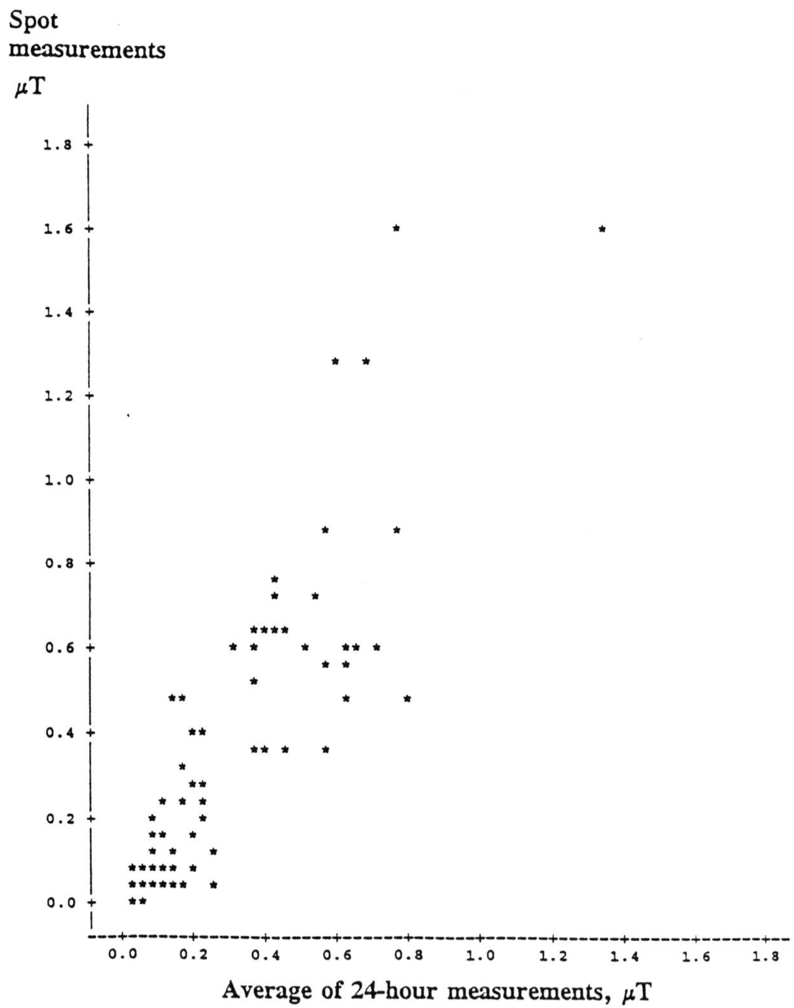

FIG. 3.—Plot of spot measurements by 24h measurements.

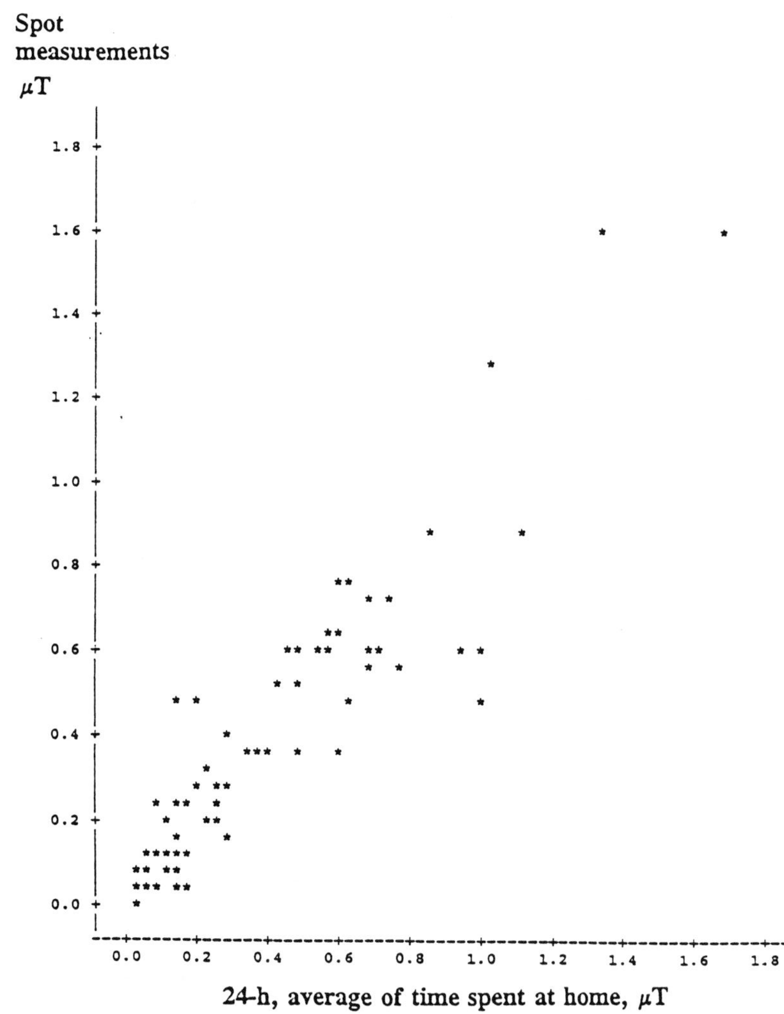

FIG. 4.—Plot of spot measurements by 24h measurements, time spent at home.

Martin Blank, Ed., *Electricity and Magnetism in Biology and Medicine*

HEALTH EFFECTS OF OCCUPATIONAL EXPOSURE TO ELF EM ENVIRONMENT

K. Ohno* and N. Miyata

Aichi Medical University, Aichi-Gun, Japan

A. Ohno

Nagoya University, Japan

K. Nakamura

Nagoya Institute of Technology

M. Nakagawa

Railway Technical Research Institute, Japan

Recently much attention is being paid to the biological effects of magnetic fields. But in Japan, the effect of magnetic fields on the actual workplace remains unclear. In this study we surveyed nine typical workplaces in two factories in which the intensity of magnetic fields was measured. In addition, 10 workers who had been continuously exposed only to magnetic-field sites were given medical check-ups and tested for sister chromatid exchange (SCE) of lymphocyte. The results may be summarized as follows: Workplace magnetic-field intensity ranged between 0.01 and 1.8 mT; no great differences between exposed and unexposed workers were observed on the basis of the medical check-up; however, the exposed workers showed a lower SCE level.

METHODS

The workplaces investigated were factories conducting performance tests of reactors and induction motors for electric-power companies, and automated battery factories. All measuring points were 1 m above the floor. A study of the magnetic-field conditions in a conventional automated factory producing cosmetic containers was made for comparison. Three-dimensional gaussmeters were used (Table 1).

TABLE 1.—Major gaussmeter specifications.

Target magnetic field	DC and AC (50 Hz to several hundred Hz)
Measuring sensor	Hall elements (3D layout)
Stored data	Vector-integrated value and integrating time of 3D

The medical examination was conducted on 10 workers consistently exposed to magnetic fields in the workplace and 10 unexposed workers. The exposed and unexposed workers were matched as to number of years employed at the factory, their age (within 5 years), and working hours. A preliminary survey was run by a questionnaire dealing with factors presumed to affect SCE (coffee, smoking, alcohol intake) and those with matching results were selected (Table 2).

*Correspondence to: K. Ohno, Aichi Medical University, Nagakute-cho, Aichi-Gun 480-11, Japan.

TABLE 2.—Age and length of work of subjects.

Exposed (age)	22	24	28	33	37	39	40	41	45	49
Unexposed (age)	22	24	27	34	37	43	39	40	40	48
Length of employment (years)	5	7	10	5	7	6	19	16	25	22

RESULTS

The level of magnetic field intensity of the workplaces ranged from 0.02 to 1.8 mT; of the control, the factory making cosmetic containers, from 0.3 to 0.5 mT. The level was almost the same in spaces occupied by multiple small electric motors and by a large single motor.

The results of general check-up showed levels within the normal ranges. Only GOT was slightly elevated among the exposed workers (Table 3), and the exposed workers showed a significantly lower SCE (Table 4). It is difficult to draw conclusions from this result, but further detailed investigation is called for.

TABLE 3.—Results of general check-up.

Item	Exposed		Unexposed
RBC ($10/mm^3$)	497.0 ± 30.6		497.6 ± 42.1
Ht (%)	46.0 ± 2.5		45.4 ± 4.0
Hb	15.7 ± 1.0		15.4 ± 1.0
GOT (U)	22.3 ± 4.1	($p < 0.01$)	17.6 ± 2.4
GPI (U)	17.4 ± 1.8		14.6 ± 4.8
r-GTP (U/I)	20.8 ± 16.6		17.0 ± 9.9
T-CHO (mg/dl)	198.2 ± 31.1		191.2 ± 23.4
TG (mg/dl))	85.8 ± 21.1		90.0 ± 31.2
B.P. (mm Hg)	118.0 ± 5.7		120.0 ± 9.6
	68.0 ± 8.8		70.0 ± 8.5
ECG	Within normal range		
Hearing	Within normal range		
Uric test	Within normal range		

TABLE 4.—SCE measurement results.

	1	2	3	4	5	6	7	8	9	10	Number/cell (mean ± S.D.)
Exposed	6.11	6.44	7.33	6.55	8.78	6.65	7.07	6.55	6.11	6.69	6.83 ± 0.78*
Unexposed	13.18	14.42	11.27	8.85	10.47	8.18	11.52	10.88	8.42	8.95	10.73 ± 1.81

*$p < 0.001$.

Martin Blank, Ed., *Electricity and Magnetism in Biology and Medicine*

ESTIMATION OF STANDARDIZED MORTALITY RATIOS IN JNR COHORT BY OCCUPATION

M. Nakagawa* and T. Koana

Railway Technical Research Institute, Tokyo

M. Tomita and M. Odaka

Central Health Institute, East Japan Railway Company

K. Ohno

Aichi Medical University, Japan

Electromagnetic field effects on workers in electric railway transportation were estimated via a Japan National Railways cohort by specific occupations. A JNR cohort of nonexecutive males over 40 years old from 1968 to 1985 was analyzed to get cumulative person-years with mortality records ascertained by death certificates. The workers were divided into six groups, one as controls and five as cases. Men in groups 2 and 3 were relatively close to ELF electromagnetic fields, and their standardized mortality ratios were lower than those of the controls, both in all deaths and in cancer deaths.

METHOD

There were 32545 JNR workers over 40 years old in metropolitan areas in 1968. The lives of JNR workers after retirement can be traced over the long term because the health-care system is firmly established and most of them become pensioners of JNR mutual benefits. Executives, public-security officers, instructors, researchers, construction workers, typographers, medical staff, and students were excluded from this cohort. The 23594 cases comprise controls, workers in occupations related to electromagnetic fields, and those working under complex environmental conditions. They could be traced up to 1985.

The entire cohort was classified into six groups: (1) clerical workers and station workers taken as controls, (2) drivers of electric locomotives and train crews, (3) workers in electric power facilities, (4) workers in station yards, (5) workers engaged in railroad maintenance and factory hands overhauling trains, and (6) workers of JNR motorbusses. Men in groups 2 and 3 were workers relatively close to ELF-electromagnetic fields; those in groups 4, 5, and 6 were exposed to more complex environmental factors. Causes of deaths were affirmed by the death certificates registered in the office of the mutual-benefit society. Standardized mortality ratios (SMRs) were calculated from cumulative person-year for 18 years and expected numbers based on 5-year age mortality rates as of 1977 in Japan, the middle year between 1968 and 1985.

RESULTS

Time-varying magnetic fields and static magnetic fields related to groups 2 and 3 were measured. Values on routine paths of facilities or in coasting train vehicles are as follows: power plant 0.1-50 μT, substation 0.1-3 μT, railway station 0.2-10 μT, DC train 0.5-5 μT (0.05-0.2 mT static), AC train 0.2-150 μT (0.1 to 4 mT static), AC/DC trains 0.5-75 μT (0.2 to 1 mT static). Maximum at power facilities or in running vehicles turned out to be several times greater values than these numbers, but the existing maximum rates are small and uncertain and the total individual exposure values cannot be readily established.

---

*Correspondence to: M. Nakagawa, Railway Technical Research Institute, Hikaricho, Kokubunji, Tokyo 185 Japan.

Table 1 shows age average and distribution as of 1968 and deaths up to 1985 in each group. Main and so-called electromagnetic field-related malignant neoplasms of each group are shown in Table 2. On average about one-third of workers died by malignant neoplasms, but of those in group 3 (electric power facilities) one-fourth died by carcinomas and without any leukemia[1,2] or lymphoma. The numbers of brain cancers[3,4] and breast cancers in men[5,6] were no larger than those expected in group 2 (drivers and train crews) and 3.

TABLE 1.—Classification of JNR workers over 40.

Group	Number in 1968	Average age	5-year age group 40-44	45-49	50-54	55-	Number of deaths	Number alive
1. Officeworker, station worker	9400	44.5	5583	2430	1331	56	842	8558
2. Train crew	2304	43.8	1539	525	217	23	175	2129
3. Electric power facilities	1094	45.2	582	273	237	2	119	975
4. Yard worker	4205	44.8	2424	1011	685	85	431	3774
5. Railroad/vehicle maintenance	4881	46.5	1955	1409	1452	65	543	4838
6. Motorbus	1714	45.3	888	447	354	25	215	1499
Total	23598		12971	6095	4276	256	2325	21273

Number of deaths were counted by the end of 1985.

TABLE 3.—SMRs of main death causes.

Group	All death	All carcinoma	Ischemic heart diseases	Cerebrovascular diseases
1. Officeworker, station worker	0.61 (−14.34**)	0.69 (−6.39**)	0.71 (−2.80**)	0.52 (−7.86**)
2. Train crew	0.55 (−7.94**)	0.60 (−3.94**)	0.71 (−1.32)	0.55 (−3.53**)
3. Electric power facilities	0.70 (−3.96**)	0.54 (−3.42**)	0.85 (−0.52)	0.53 (−2.78**)
4. Yard worker	0.67 (−8.26**)	0.85 (−2.18**)	0.89 (−0.73)	0.52 (−5.41**)
5. Railroad/vehicle maintenance	0.63 (−10.78**)	0.69 (−5.13**)	0.46 (−4.20**)	0.60 (−5.32**)
6. Motorbus	0.80 (−3.21**)	0.80 (−1.82)	0.81 (−0.81)	0.63 (−2.72**)
Total	0.64 (−21.61**)	0.71 (−9.75**)	0.69 −4.82**)	0.55 (−14.9**)

1. Z values in parentheses: *$p < 0.05$; **$p \ll 0.01$.
2. Denominators of SMR are expected deaths calculated from Japanese men aged 40-75 in 1977.

TABLE 4.—Ratios of SMR for group I as control.

Group	All death	All carcinoma	Ischemic heart diseases	Cerebrovascular diseases
2. Train crew	0.90 (−1.44)	0.87 (−1.06)	1.00 (0.02)	1.05 (0.30)
3. Electric power facilities	1.15 (1.54)	0.77 (−1.40)	1.20 (0.57)	1.00 (0.02)
4. Yard worker	1.12 (2.29)	1.22 (2.62**)	1.25 (1.42)	1.00 (−0.01)
5. Railroad/vehicle maintenance	1.07 (1.52)	1.00 (−0.06)	0.65 (−2.30*)	1.15 (1.41)
6. Motorbus	1.33 (4.25**)	1.16 (1.20)	1.14 (0.52)	1.20 (1.09)

Z values in parentheses: *$p < 0.5$; **$p < 0.01$.

SMRs of main death causes are given in Table 3. Denominators for these SMRs are expected deaths calculated from Japanese men aged 40 to 75 in 1977, the middle year between 1968 and 1985. All these SMRs are below the national average, which indicates the well-known "healthy worker effect."[7] Table 4 gives the ratios of SMR for group 1 taken as control; that is, a table of comparison among JNR occupational groups.

TABLE 2.—Number of deaths by malignant neoplasms.

		Group I Office/station worker	Group II Train crew	Group III Electric power facilities	Group IV Yard worker	Group V Railroad/vehicle maintenance	Group V Motorbus	Sum
	ICD 9th revision codes	No. of deaths (%)	No. of deaths (%)	No. of deaths (%)	No. of deaths (%)	No. of deaths (%)	No. of deaths (%)	No. of deaths (%)
All death		842 (100)	175 (100)	119 (100)	431 (100)	543 (100)	215 (100)	2325 (100)
All carcinoma	1400-2089	301 (35.7)	60 (34.3)	29 (24.4)	171 (39.7)	189 (34.8)	68 (31.6)	818 (35.2)
Esophagus	1500-1509	10 (1.2)	2 (1.1)	3 (2.5)	9 (2.1)	10 (1.8)	5 (2.3)	39 (1.7)
Stomach	1510-1519	94 (11.2)	23 (13.1)	12 (10.1)	65 (15.1)	61 (11.2)	16 (7.4)	271 (11.7)
Colon	1530-1539	12 (1.4)	2 (1.1)	0	4 (0.9)	5 (0.9)	2 (0.9)	25 (1.1)
Rectum	1540-1549	14 (1.7)	3 (1.7)	0	11 (2.6)	7 (1.3)	1 (0.5)	36 (1.5)
Liver, gallbladder	1550-1569	51 (6.1)	6 (3.4)	2 (1.7)	19 (4.4)	22 (4.1)	10 (4.7)	110 (4.7)
Pancreas	1570-1579	15 (1.8)	2 (1.1)	0	8 (1.9)	11 (2.0)	5 (2.3)	41 (1.8)
Bronchus, lung	1620-1629	44 (5.2)	9 (5.1)	6 (5.0)	19 (4.4)	29 (5.3)	10 (4.7)	117 (5.0)
Bone, soft tissue	1700-1719	4 (0.5)	2 (1.1)	0	1 (0.2)	4 (0.7)	1 (0.5)	12 (0.5)
Breast cancer of man	1750	1 (0.1)	0	0	0	0	0	1 (0.04)
Kidney	1890-1899	6 (0.7)	1 (0.6)	1 (0.8)	2 (0.5)	3 (0.6)	0	13 (0.5)
Brain, nerve	1910-1929	7 (0.8)	2 (1.1)	1 (0.8)	1 (0.2)	4 (0.7)	1 (0.5)	16 (0.6)
Lymphoma, leukemia	2000-2089	14 (1.7)	3 (1.7)	0	8 (1.9)	8 (1.5)	5 (2.3)	38 (1.6)
Other cancer		29 (3.4)	5 (2.8)	4 (3.3)	24 (5.5)	25 (4.6)	12 (5.5)	99 (4.2)
Ischemic heart diseases	4100-4149	66 (7.8)	15 (8.6)	10 (8.4)	39 (9.0)	28 (5.2)	15 (7.0)	173 (7.4)
Cerebrovascular diseases	4300-4379	142 (16.9)	34 (19.4)	18 (15.1)	67 (15.5)	106 (19.5)	34 (15.8)	401 (17.2)
Liver diseases	5700-5739	47 (5.6)	8 (4.6)	8 (6.7)	22 (5.1)	33 (6.1)	14 (6.5)	132 (5.7)
Other death		286	58	54	132	187	84	728

Significantly increased deaths are found in groups 4 to 6, which are considered to be affected by other or added complex factors. The train crews and workers in electric power facilities were not affected in their death and/or wellness by electric and electromagnetic fields of the electrified railroad system.

REFERENCES

1. S. Milham Jr., "Mortality from leukemia in workers exposed to electrical and magnetic fields," *Engl. J. Med.* 307: 249, 1982.
2. S. Tornqvist, B. Knave, A. Ahlbom, et al., "Incidence of leukemia and brain tumors in some 'electrical occupations,'" *Br. J. Ind. Med.* 48: 597-603, 1991.
3. T. Thomas, P. D. Stolley, A. Stemhagen, et al., "Brain tumor mortality risk among men with electrical and electronic jobs: A case-control study," *J. Natl. Cancer Inst.* 79: 233-238, 1987.
4. R. P. Gallagher, M. L. McBride, P. R. Band, et al., "Brain cancer and exposure to electromagnetic fields," *J. Occup. Med.* 33: 944-945, 1991.
5. T. Yynes and A. Andersen, "Electromagnetic fields and male breast cancer," *Lancet* 336: 1596, 1990.
6. P. A. Demers, D. B. Thomas, K. A. Rosenblatt, et al., "Occupational exposure to electromagnetic fields and breast cancer in men," *Am. J. Epidemiol.* 48: 597-603, 1991.
7. G. M. Kendall, C. R. Muirhead, B. H. MacGibbon, et al., "Mortality and occupational exposure to radiation: First analysis of the National Registry for Radiation Workers." *Br. Med. J.* 304: 220-225, 1992.

Martin Blank, Ed., *Electricity and Magnetism in Biology and Medicine*

RESIDENTIAL PROXIMITY TO HIGH-VOLTAGE TRANSMISSION LINES AND DEPRESSIVE SYMPTOMATOLOGY

S. McMahan,* J. Ericson, and J. Meyer

University of California, Irvine

A number of epidemiological studies indicate an association between depression and proximity to high-voltage transmission lines. However, these studies have been criticized for using surrogate measures of electromagnetic fields (EMFs) and nonstandardized measures of depression. In an effort to overcome these limitations, the Center for Epidemiological Studies Depression (CES-D) scale was administered to 152 women living either adjacent to a transmission line or one block away. Results indicate that homes adjacent to the transmission lines have an average EMF level of 4.86 mG at their front door and those one block away have an average of 0.68 mG. There was no significant difference in CES-D scores between the groups, after control for demographic variables. Possible explanations for this finding are discussed.

Electromagnetic fields (EMFs) are produced as a result of the generation, transmission, and use of electric power. Various concerns have been raised due to the reported neuroendocrine effects and circadian rhythm disruptions seen in animal models and human studies.[1] Human studies indicate that there may be an association between EMF exposure and incidence of depression.[2] Epidemiological studies indicate a correlation between EMF exposure and psychiatric illness. However, these studies have been identified as having various methodological problems. First, most research has focused on health effects resulting from or attributed to (in the case of suicide) depression. None of these studies has used any validated measures/indices for depression. (The closest was hospital admission cases.[3]) Second, most studies have been conducted in Great Britain, where the electrical distribution system runs 10Hz lower than the U.S. 60Hz frequency. Further research needs to be conducted to determine whether these findings are replicable in the USA. Another issue, addressed by Bonnell, is the lack of statistical control for demographic variables.[4] For example, although greater rates of depression occur in women, gender differences have not been addressed. Finally, surrogate measures such as proximity to source have been used in place of actual EMFs, partly because of difficulties associated with collecting such data in the past. (Only one study actually measured magnetic fields.[5])

The current study assesses depressive symptomatology in female residents living adjacent to overhead transmission lines and a less-exposed (control) group living one block away. The current study takes into account previously addressed methodological weaknesses by control for demographic factors; use of a validated objective measure to identify depressive symptomatology;[6] and use of instrumentation that can accurately assess strength of EMF exposure. Higher depressive symptomatology is hypothesized for those living on the easement when age, income, education, and length of residence in the home are controlled.

METHODS

The study was conducted adjacent to an easement containing two 4-circuit 220kV and two 66kV high-voltage transmission lines. A simple random sample of homes on the easement and homes one block away was selected. The Center for Epidemiological Studies Depression (CES-D) scale was administered to female residents during an interview conducted in their home. The interviews were conducted with 152 women during a 3-month period in 1992.

Interviews were conducted by trained research assistants and administered during evening and weekends to prevent a conflict with most work schedules. The interviews lasted approximately 20 min.

If the subject agreed to participate, the interviewer asked questions pertaining to the subject's health, life events, family history, health habits, occupation, and home life. The interviewer wore a portable electric and

---

*Correspondence to: S. McMahan (Environmental Health), UCI School of Social Ecology, Irvine CA 92717.

magnetic digital exposure (EMDEX-C) meter and recorded a sample frame number at the front door of each home. At the end of each day, the data were downloaded from the EMDEX-C to a computer file.

RESULTS

T-tests revealed no significant differences between subjects living on the easement and those living one block away for age ($M = 47$), income ($M \approx \$60000$), length of residence in the home ($M = 14$ years), education (M, some college), or ethnicity (white vs nonwhite). *T*-tests indicate that the mean electromagnetic field (EMF) at the front door of homes along the easement ($M = 4.86$ mG) was significantly higher than for homes one block away ($M = 0.86$ mG, $p < 0.001$). Although the difference was not significant, the means CES-D score for control subjects ($M = 5.63$, range = 0-28) was slightly higher than for subjects living on the easement ($M = 4.95$, range = 0-34). There was no significant difference between control subjects and those living on the easement for CES-D scores controlling for age, income, education, and length of residence in the home when entered into a logistic regression equation. (The independent variable, CES-D score, was dichotomized; values greater than eight were coded as representing higher levels of depression than expected in a normal population.) The distribution of the CES-D scores is highly skewed (Fig. 1).

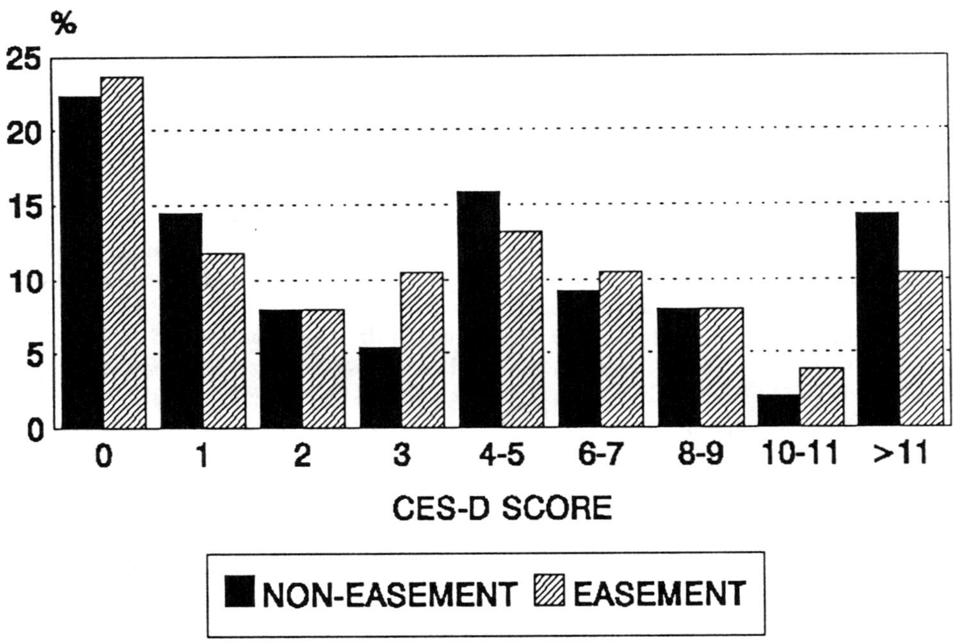

FIG. 1.—Depressive symptom scores by proximity to high-voltage transmission lines.

DISCUSSION

The hypothesis of the study was not supported; there was no association between depressive symptomatology and electromagnetic fields (EMFs) when the demographic variables were controlled. Several explanations may assist in interpreting this finding. First, Wilson speculates that some populations may be at greater risk to the effects of EMF (e.g., those with a history of bipolar depression and alcohol abuse), and the present study may have too few cases to examine these associations.[1] In addition, the high income and education levels of our subjects may act as confounding variables when one attempts to examine differences in CES-D scores. It is also possible that people may acclimate to the fluctuations in their circadian cycle caused by exposure to EMF; further research could examine this hypothesis. Indeed, subjects who moved into their homes more recently reported more depressive symptoms, but this finding was not significant. Future studies should carefully identify potential risk factors that may make an individual more susceptible to the effects of extremely low frequency EMF exposure on health.

REFERENCES

1. B. W. Wilson, "Chronic exposure to ELF fields may induce depression," *Bioelectromagnetics* 9: 195-205, 1988.
2. B. W. Wilson, C. W. Wright, J. Morris, R. S. Flanagan, R. Buschbom, M. M. McCulloch, and L. E. Anderson, "Effects of electric blanket use on human pineal gland function," *J. Pineal Res.* 9: 259-269, 1990.
3. S. Perry and L. Pearl, "Power frequency magnetic field illness in multi-story blocks," *Public Health* 102: 11-18, 1988.
4. J. A. Bonnell, W. T. Norris, J. H. Pickles, and C. Male, "Comments on environmental power frequency magnetic fields and suicide," *Health Physics* 44: 698-699, 1983.
5. F. S. Perry, M. Reichmanis, A. A. Marino, and R. O. Becker, "Environmental power-frequency magnetic fields and suicides," *Health Physics* 41: 267-277, 1981.
6. L. S. Radloff, "The CES-D scale: A self report depression scale for research in the general population," *Applied Psychological Measurement* 1: 385-401, 1977.

Martin Blank, Ed., *Electricity and Magnetism in Biology and Medicine*

JOB STRESS AND "HYPERSENSITIVITY TO ELECTRICITY"

B. B. Arnetz*

Karolinska Institutet, Stockholm, Sweden

M. Berg and S. Lidén

Karolinska Sjukhuset, Stockholm, Sweden

We followed 47 video-display unit (VDU) workers with and without work-associated skin complaints during one regular workday and a day of leisure. Employees with skin symptoms had a selective neuroendocrine stress reaction during work, but not during leisure time. They experienced more extreme job stress at the same time as they had higher control over work processes. We postulate that employees suffering from cognitive failure during VDU work develop "techno-stress." A model is proposed in which physiological signals act as unconditioned stimulus and the VDU environment as the conditioned stimuli.

One of the more common complaints among VDU users concerns the skin.[1-3] Despite much research little is known about the etiology behind VDU-associated health problems.[3,4] The purpose of this study was to assess the possible etiological importance of psychophysiological and environmental factors in the precipitation of VDU-associated skin problems.

METHODS

The study group consisted of a randomized selection from a larger cohort[2,5] of those reporting skin symptoms associated with their work ($n = 19$) and those who had no such symptoms ($n = 28$). The subjects were studied during a typical workday as well as on a day off (i.e., a day of leisure) in the employees' regular work environment. Psychosocial and psychophysiological measurements were made during a regular workday, as well as on the day of leisure. Itching behavior was registered by specially trained nurses. Blood samples were analyzed for serum cortisol, prolactin, testosterone, thyroxin, and growth hormone. Urinary concentrations of adrenalin and noradrenalin were also determined. The physical work environment was also characterized in detail, as was exposure to passive and active smoking. Subjects with and without VDU-associated skin symptoms were compared by the nonpaired t-test and two-way analysis of variance. Finally, stepwise logistic regression was applied based on the two skin groups as dependent factors. Results were age- and sex-adjusted and significance levels were set at $p < 0.05$.

RESULTS

Persons with VDU-associated symptoms had more authority over work content and work organization than those without symptoms (7.9 vs 6.3 on a scale from 2 to 10, sex-adjusted, $p < 0.05$), and experienced more occupational stress ($p = 0.01$). Furthermore, during a workday those with symptoms showed a tendency for higher global occupational stress perception, with no differences during leisure time (Fig. 1). There were no differences between groups in intellectual discretion, i.e., the development of new skills in their jobs. In contrast, compared to the control group, they exercised significantly less ($p < 0.05$). In the stepwise logistic regression analyses the following variables were significant independent predictors: itching behavior ($p < 0.01$) and authority ($p < 0.01$). In the logistic regression analysis it was found that serum testosterone, prolactin (Fig. 2), and thyroxin were significant predictors of skin symptoms during work, but not during days off. The anabolic hormone testosterone was decreased, while the stress hormones prolactin and

*Correspondence to: B. B. Arnetz (Stress Research), Karolinska Institutet, P.O. Box 60205, S-10401 Stockholm, Sweden.

thyroxin were elevated. There was nearly a three-fold increase in urinary excretion of noradrenalin during work (Fig. 3) compared to a day of leisure, without any significant group differences. There were no significant group differences in serum levels of growth hormone or cortisol. Even though there was no clinical difference in skin condition between the groups, those with symptoms had a more frequent itching behavior (Anova, $p < 0.05$, mean value 14 vs 0.1 arbitrary visual analog units). Those with skin symptoms also complained significantly more often of eye strain (Anova, $p < 0.05$). There were no systematic differences between groups with and without skin symptoms, respectively, with regard to age, gender, job functions, alcohol and coffee consumption, or smoking habits. On the average, subjects with skin symptoms had been involved in computer work 7.7 years compared to 8.6 for the reference group (n.s.). The groups did not differ with regard to neuroticism, extrovertness, psychiatric morbidity, pigmentation ability, prior and ongoing allergies or skin diseases, moodiness, or occurrence of headache. The physical work environment index was similar, as was the average distance between the employee and the VDU screen (63.5 cm ± 3.5 SEM).

DISCUSSION

Our findings show that employees with skin complaints suffer from more occupational stress compared to healthy controls working under similar objective physical indoor environmental and EMF conditions. This stress is reflected in higher stress hormone levels as well as increased subjective feelings of occupational stress, and decreased levels of the stress-protective anabolic hormone testosterone. Subjects with skin complaints also suffered more frequently from eyestrain and they scratched themselves more frequently. These findings may help explain VDU-associated facial skin sensations.

Our study indicates that there are occupational psychophysiological factors at play which may help to explain the development of VDU-associated dermatological and neurological problems. This study has shown that there are physiological differences in subjects with and without VDU-associated skin complaints. Since both groups were using VDUs to the same extent, our findings imply that factors other than electromagnetic fields must be contributing to the development of VDU-associated skin symptoms. Based on our results to date, we *postulate* the following hypothesis: Many employees working with computers suffer from occupational stress. This stress results in physiological changes characterized by elevated metabolism and increased dermal bloodflow. This response may act as an unconditioned stimulus. The conditioned stimulus in this case is the VDU environment. Once the conditioned response has been learned, the psychophysiological responses are elicited purely by the conditioned stimulus (i.e., the VDU environment).

The introduction of new technologies has resulted in major changes in work content and work organization. These changes have too often resulted in suboptimal organization, discrepancy between mental capacity and goals of the employee, and what is achievable by the software programs. The result of this discrepancy is a psychophysiological "techno-stress," as employees actively try to cope with feelings of cognitive failure resulting from a poor man-machine interface.[6,8] Occupational health aspects should be an integral part of all systems implementations.[9-11]

In conclusion, our study has found psychophysiological differences between person with and without VDU-associated skin complaints. These differences were limited to workdays only. We believe that the entire implementation strategy related to information technology must be reassessed. Specification requirements for new software systems should take human issues into consideration. Corrective measures must entail many more factors than merely installing "low-emitting" VDUs. We have now launched a state-of-the-art research and treatment program offering psychophysiological treatment and testing to sensitivity for electric and magnetic fields.[12]

REFERENCES

1. M. Berg, "Skin problems in workers using visual display terminals: A study of 201 patients," *Contact Dermatitis* 19: 335-341, 1988.
2. M. Berg, S. Lidén, and O. Axelson, "Skin complaints and work at visual display units: An epidemiological study of office employees," *J. Am. Acad. Dermatol.* 22: 621-625, 1990.
3. M. Berg, "Facial skin complaints and work at visual display units: Epidemiological, clinical and histopathological studies," *Acta Dermatovenereol. (Stockh.)*, Suppl. 150.
4. G. Swanbeck and T. Bleeker, "Skin problems from visual display units," *ibid.* 69: 46-51, 1989.
5. M. Berg, B. B. Arntez, S. Lidén, P. Eneroth, and A. Kallner, "A psychophysiological study of employees with VDU-associated skin complaints," *J. Occup. Med.*, 34: 698-701, 1992.

6. B. B. Arnetz and J. E. Arntez, "Hypersensitivity to electricity and video display units: A controlled occupational health survey in a hightech work environment" (in press).

7. A. Smith, D. Peck, and T. Clatworthy, "After-effects of working at visual display units," *Work & Stress* 3: 195-201, 1989.

8. D. E. Broadbent, P. F. Cooper, P. FitzGerald, and K. R. Parkes, "The cognitive failures questionnaire and its correlates," *Br. J. Clin. Psychol.* 21: 1-16, 1982.

9. "Work with visual display terminals: Psychosocial aspects and health (Report on a World Health Organization meeting)," *J. Occup. Med.* 31: 957-969, 1989.

10. G. Johansson, "Growth and challenge vs wear and tear of humans in computer mediated work," in B. Knave and P.-G. Widbeck, Eds., *Work with Display Units*, Amsterdam: Elsevier, 1987, 725-731.

11. G. Johansson and G. Aronsson, "Stress reactions in computerized administrative work," *J. Occup. Behav.* 4: 159-181, 1984.

12. This work was supported by the Swedish Work Environment Fund, Stockholm, Sweden.

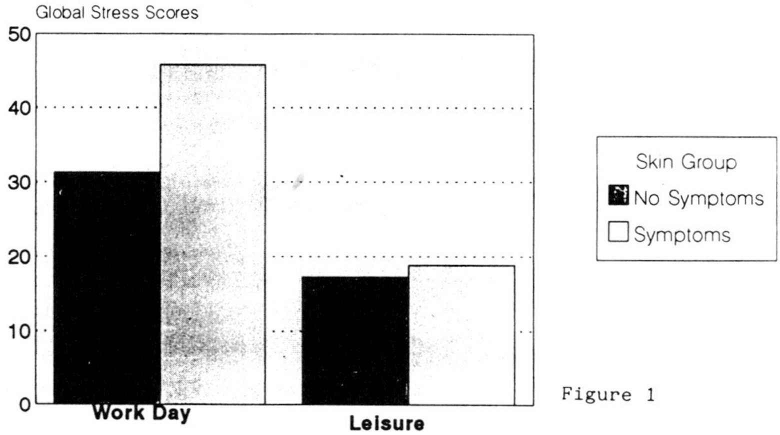

Figure 1

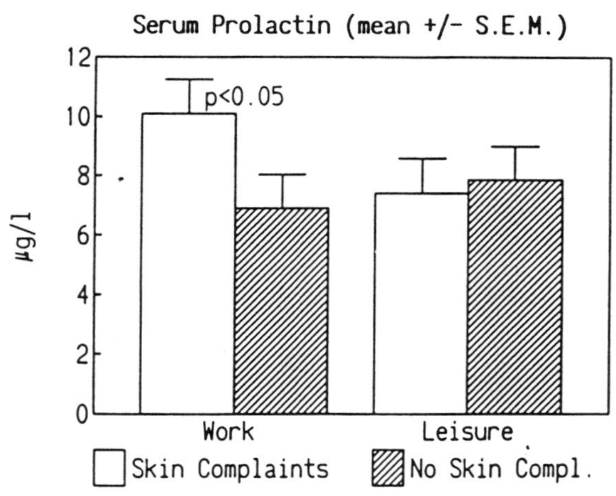

Figure 2

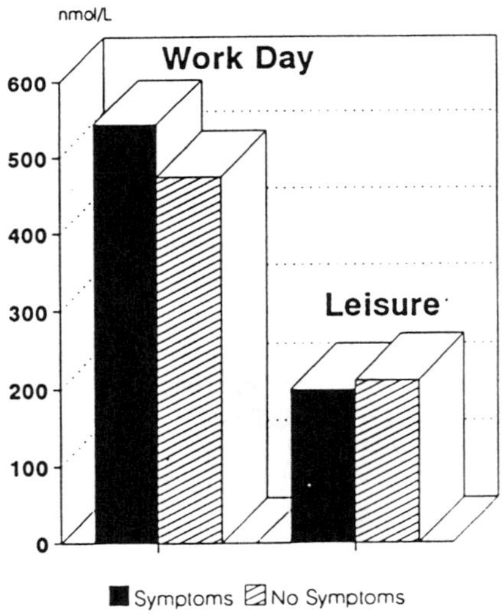

Figure 3

FIG. 1.—Global stress scores, by skin symptoms grouping, throughout workday and day of leisure.

FIG. 2.—Serum prolactin (mean ± SEM) in subjects with and without skin complaints, during workday and day of leisure.

FIG. 3.—Urinary excretion of noradrenalin in subjects with and without skin symptoms, during work and leisure. Urinary excretion was significantly higher during work, with no significant group differences.

Martin Blank, Ed., *Electricity and Magnetism in Biology and Medicine*

GEOGRAPHIC STUDIES OF CANCER RISK POSED BY EMF FROM ELECTRIC POWER FACILITIES

T. E. Aldrich*

University of North Carolina

J. W. Drane

University of South Carolina

Epidemiologic studies to evaluate the association of residential exposure to EMF and cancer risk have been inconsistent and disappointing. A persistent finding has been increased cancer risk associated with external wiring configuration. New geographic methods offer a means to study these issues by nonparametric statistical methods and Geographic Information Systems (GIS). These new methods can employ exposure scoring systems developed to assess external wiring configurations.

New technology offers a means to conduct studies of potential health effects associated with residential proximity to EMF facilities.[1] This technology uses nonparametric, space-time statistical methods and enhanced computer capabilities, Geographic Information Systems (GIS).[2] Exposure scoring systems like those developed by others may be used to assess wiring configuration codings with population-based disease registry data.[3] From GIS, the number of cases as well as the population within a designated area or transect (corridor) is determined to take into account population density. These two elements permit the calculation of adjusted disease rates,[4] or a comparison set of transects may be developed to evaluate an observed pattern of residences. An example of the spatial detail possible by use of GIS techniques is shown in Fig. 1.

EXAMPLES OF GIS STUDIES

Geographic Information System (GIS) technology has been employed to evaluate cancer cluster reports. Residences were assigned geographic coordinates for 130 malignant brain cancers and 159 controls (septicemia deaths) from 1980 to 1989. The GIS database contained information on high-voltage electrical power lines (> 100 kV) and electrical power substations (provided by Duke Power Co.). The evaluation did not evince an increased risk of brain cancer, yet these results indicate the great capability for geographic proximity studies, even to distances of hundreds of feet. A particular interest now is the use of these techniques with breast cancer in light of a recent suggestion of this potential association with EMF exposures.[5]

CONCLUSION

Geographic studies are a great asset to the burgeoning new methods for studying environmental disease. These methods may be applied within population-based registries and based on classic epidemiologic strategies.[2] Geographic files from the 1990 U.S. census will further augment this promising research tool for evaluating cancer risk from residential EMF exposures.

REFERENCES

1. T. E. Aldrich and C. E. Easterly, "Electromagnetic fields and public health," *Env. Health Perspect.* 75: 159-171, 1987.

---

*Correspondence to: T. E. Aldrich (Epidemiology), University of North Carolina, Chapel Hill NC 27599-7400.

2. T. E. Aldrich and W. D. Drane, *CLUSTER: Software to Assist Investigations of Rare Health Events,* (Diskettes and user's manual), 1990, available from the Agency for Toxic Substances, Health Studies Division, MS-31, Atlanta, GA. 30333.

3. S. J. London, D. C. Thomas, J. D. Bowman, E. Sobel, T. C. Cheng, and J. M. Peters, "Exposure to residential electric and magnetic fields and risk of childhood leukemia," *Am. J. Epid.* 134: 923-937, 1991.

4. J. W. Drane, J. M. Rodermund, and T. E. Aldrich, "Surveillance from a central cancer registry: I. A denominator-free strategy," *NCHS Publ. Health Conf. Records Stat.* 1991, Washington, D.C.

5. P. A. Demers, D. B. Thomas, K. A. Rosenblat, et al., "Occupational exposure to electromagnetic fields and breast cancer in men," *Am. J. Epid.* 134: 340-347, 1991.

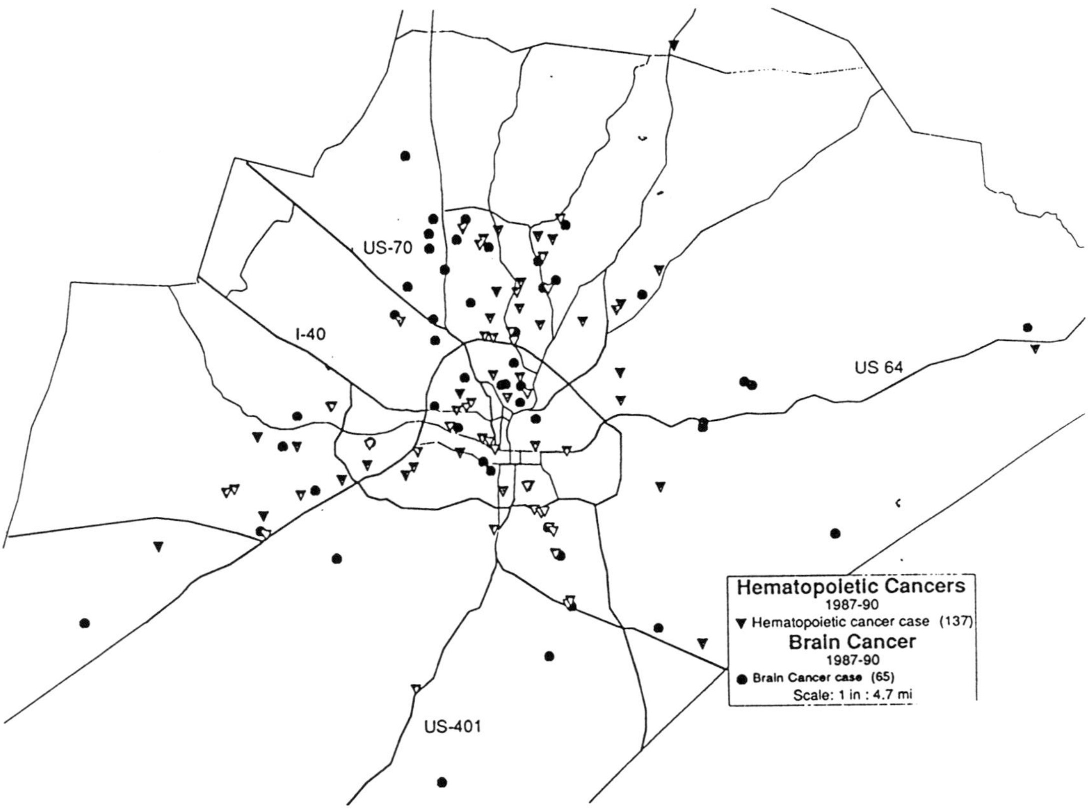

FIG. 1.—Sample GIS map showing the detail that may be achieved for studies of EMF point and line sources. Cancer cases exemplify data points and streets show line patterns; hazardous-waste points or power-lines may be represented in a similar fashion.

Martin Blank, Ed., *Electricity and Magnetism in Biology and Medicine*

NUMERICAL EVALUATION OF THE EFFECTS OF MICLASSIFICATION OF ELF MAGNETIC-FIELD EXPOSURE ON THE ESTIMATES OF RISK IN CASE-CONTROL STUDIES

V. Delpizzo*

Australian Radiation Laboratory, Yallambie

S. J. Farish and J. L. Borghesi

University of Melbourne, Australia

We investigate how nondifferential errors in assessing magnetic-field exposure may result in differential misclassification of exposure which may in turn lead to an artificial increase in the estimate of risk. We explore how this very undesirable effect depends on sample size and on the direction of the error.

Exposure assessment in epidemiological studies of the health effects of ELF magnetic field exposure has been and continues to be poor. Epidemiologists have taken comfort in the commonly accepted belief that, provided measurement errors affect equally cases and controls, misclassification of exposure can only bias the estimate of risk toward the no-effect value. However, nondifferential errors in exposure assessment very often result in differential misclassification of exposure.[1,2] For a given measurement uncertainty, misclassification is more likely to occur for subjects in a "misclassification zone" straddling the boundary between exposure classes. The mean rate of misclassification depends on the number of such subjects and on the total number of subjects in the corresponding exposure classes. Since these numbers are generally different for cases from those for controls, misclassification will usually be differential.

METHOD

We simulated case-control studies by generating a control population having a lognormal distribution of exposure with a modal value of approximately 0.4 mG and a median of 0.76 mG; and several case populations, also having a lognormal exposure distribution, but with parameters chosen to reflect different degrees of association (including zero). In this paper, we report only results obtained on the assumption of a rather strong association. Error distributions were also generated and random-matched to the "true" exposure values, which yielded an "empirical" exposure assessment. For example, we generated random numbers with distributions adjusted to reproduce the empirical distributions of the relative errors measured in a previous study,[3] in which the exposures measured by a hip-worn or a chest-worn dosimeter were compared to the whole-body average exposure, with the latter assumed to be the "true" exposure. After classifying the "subjects" according to their "correct" and "empirical" classifications, we calculated the odds ratios (ORs) in the two cases and compared them.

RESULTS

Dichotomous Classification

If there is a true association between magnetic-field exposure and the end point, the exposure distributions of cases and controls are different. This difference may result in cases being misclassified at a different rate from controls. To investigate this effect we used a very large sample size (16000 cases and 16000 controls) to minimize the effects due to irregularities in the exposure distribution curves. We regard the ORs calculated from this large sample classified on the basis of the "true" exposure as the 'true relative risk.'

---

*Correspondence to: V. Delpizzo, Australian Radiation Laboratory, Lower Plenty Rd., Yallambie VIC 3085 Australia.

A hip-worn detector tends to underestimate whole-body exposure; i.e., more exposed than nonexposed subjects are misclassified. For high cutoff values, controls are classified less reliably than cases, because there are fewer exposed controls than cases. Dichotomization at 2 mG yielded the following 2×2 tables.

	"True" classification			
	Cases		Controls	
Empirical classification	Exposed	Nonexposed	Exposed	Nonexposed
Exposed	4502	133	1985	101
Nonexposed	1007	10358	710	13204
Fraction misclassified	0.18	0.01	0.26	0.01

Accordingly, we found a small artificial increase of the ORs (2.72 rather than 2.59). The assumption of nondifferential misclassification[4] predicted an OR of 2.33.

A chest-worn dosimeter has an almost equal chance of over- and underestimating exposure. Although misclassification is still generally differential, it is less so than for the hip dosimeter. The observed odds ratio is lower than that calculated in absence of misclassification (2.51 rather than 2.59), but again not as low as predicted by the nondifferential misclassification assumption (2.30).

If there is no association between magnetic-field exposure and cancer and the sample is very large, the exposure distribution curves are virtually identical for cases and for controls. In this case, the patterns of misclassification are the same for cases and controls. However, in a smaller sample, misclassification depends also on random irregularities in the exposure distributions. For example, a clustering of cases near a cutoff value will result in a misclassification rate higher than for controls. The consequences of this type of misclassification on an individual study cannot be predicted. In the absence of a true association, an over- and underestimate of risk are equally likely. This effect adds to the usual sample-to-sample variation and may result in an increased spread in the distribution of possible ORs (in contrast to the accepted view that misclassification reduces the spread of possible ORs, by biasing the ORs toward unity).

When a true association exists, the effects of measurement errors and random variations combine. We investigated this phenomenon by repeatedly selecting samples of 250 cases and 250 controls. The results are displayed in Table 1.

TABLE 1.—Summary of results obtained from repeated simulations of case-control studies with 250 cases and 250 controls (cutoff = 2 mG; "true" relative risk = 2.6).

	Exposure assessment				
Probability of one case-control study:	"Truth"	Hip dosimeter		Chest dosimeter	
Underestimating the risk by a factor of 1.25 or more	13%	13%	(24%)	14%	(32%)
Deviating from the "truth" by a factor <1.25	72%	64%	(70%)	74%	(64%)
Overestimating the risk by a factor of 1.25 or more	15%	23%	(6%)	12%	(4%)

Note: Numbers in parentheses are the probabilities calculated under the assumption of nondifferential misclassification.[4]

Ordinal Scale Analysis

Classifying subjects into several exposure levels increases the number of subjects likely to be misclassified. However, misclassification is less differential (because the ratio of subjects in the "misclassification zone" to total subjects in the corresponding category is higher). To examine the effect of misclassification, we use as a measure of association, the gradient of the relative risk as a function of exposure, calculated by weighted linear regression.[5] Using a very large sample and assuming no misclassification, we obtained a "true" relative risk gradient of 0.91 per mG of exposure. Again, we repeatedly selected samples of 250 cases and 250 controls to investigate the combined effects of misclassification and random sampling variations on the estimate of risk. The results are displayed in Table 2.

TABLE 2.—Summary of results obtained from repeated simulations of a case-control study with 250 cases and 250 controls, classified according to a 3-level ordinal scale ("medium" is defined as an exposure ≥ 1 mG but < 2 mG; "true" relative risk gradient = 0.9/mG).

		Exposure assessment			
Probability of one case-control study:	"Truth"	Hip dosimeter		Chest dosimeter	
Underestimating the risk gradient by a factor of 1.25 or more	29%	27%	(30%)	30%	(42%)
Deviating from the true risk gradient by a factor < 1.25	51%	50%	(56%)	51%	(49%)
Overestimating the risk gradient by factor of 1.25 or more	20%	23%	(14%)	19%	(9%)

Note: Numbers in parentheses are the probabilities calculated under the assumption of nondifferential misclassification.[6]

Other Error Distributions

The examples discussed in this first report were chosen for their simplicity. In most cases of greater practical importance, the relationship between true exposure and exposure measurement error is much more complex than those we have used here. Nevertheless, a satisfactory approximation can be obtained if sufficient empirical data are available.

For example, we simulated the use of a single spot measurement taken between 9:00 a.m. and 1:00 p.m. as a surrogate for 24h monitoring of the ambient magnetic field. Errors were generated by superimposition of 3 random distributions available in the SAS language: (1) an attenuated Gaussian error; (2) a random triangular distribution, symmetrical for 1.5 mG, otherwise progressively skewed toward this value; and (3) a reduction factor of 0.8, accompanied by a random Gaussian variation around this value, to account for the fact that morning magnetic field levels are generally lower than the corresponding 24h average. Using this distribution, we found that, because of misclassification, a true OR of 2.3 is reduced to 2.2. The nondifferential misclassification assumption[4] predicted an OR of 2.0

CONCLUSIONS

The net effect of misclassification on the estimate of risk may be much lower than it had been previously estimated. However, the direction of the bias is unpredictable; therefore, it is important that measurement errors be minimized. The direction of the error is as important as its magnitude. If a true association between magnetic-field exposure and cancer exists, the true relative risk need not be much larger than that observed and may conceivably be lower.

REFERENCES

1. S. Wacholder, M. Dosemici, and J. H. Lubin, "Blind assignment of exposure does not always prevent differential misclassification," *Am. J. Epidemiol.* 134: 433-437, 1991.

2. K. M. Flegal, P. M. Keyl, and F. J. Nieto, "Differential misclassification arising from non-differential errors in exposure measurement," *ibid.*, pp. 1233-1234.

3. V. Delpizzo, "Misclassification of ELF occupational exposure resulting from spatial variation of the magnetic field," *Bioelectromagnetics* (in press).

4. K. M. Flegal, C. Brownie, J. D. Hass, "The effects of exposure misclassification on estimates of relative risk," *Am. J. Epidemiol.* 123: 736-751, 1986.

5. K. J. Rothman, *Modern Epidemiology*, Boston: Little, Brown, 1986, 341.

6. V. Delpizzo and M. R. Salzberg, "Relative risk estimate bias and loss of power in the Mantel test for trend resulting from the use of magnetic field 'spot' measurements in epidemiological studies using an ordinal exposure scale," *Bioelectromagnetics* 13: 363-378, 1992.

Martin Blank, Ed., *Electricity and Magnetism in Biology and Medicine*

MEMBRANE DYNAMICS CONTROLLED BY TRANSMEMBRANE POTENTIAL: POTENTIAL DEPENDENCE OF ERYTHROCYTE SHAPE TRANSFORMATION

R. Glaser*

Humboldt University, Berlin

Washed erythrocytes become stomatocytes if their transmembrane potential exceeds 15-27 mV; they become echinocytes at lower or negative potentials. This correlation could be found for various experimental conditions to generate the membrane potentials. It occurs at conditions of various values of internal or external pH. Drug effects can be overcome by variations of the membrane electric field. Near the threshold potential of the shape transformations, the shapes become temperature dependent (echinocytes at lower temperature, stomatocytes at higher temperature). This mechanism of field-induced shape change seems to consist of a sequence of several flip-flop reactions of membrane molecules. The initial step could be a field-induced flip-flop of small charged molecules, as well as a conformational change of a membrane constituent.

In the discussion of possible mechanisms of the effects of electric and electromagnetic fields on biological systems, one first has to consider their entire electrostatic structure in vivo and the biological function of this bioelectric field. We must learn how the intrinsic electric fields of the biological system are organized and what their biological functions are. All cells have stored electrochemical energy and can build up strong electric fields in the membrane, depending on their stimulation. Externally applied fields can modify these fields and probably trigger corresponding mechanisms. In this context one must consider not only the reactions of specific field-sensitive molecules in the membrane, but also nonspecific reactions that influence the lateral and transversal organization of the cell membrane, with various biological consequences.[1]

Human red cells are valuable model objects on which to study this problem. The transmembrane potential of these cells can be easily modified by various kinds of experimental conditions, and its change is predictable and calculable by computer models.[2] Shape changes, on the other hand, ar good indicators for collective molecular and intermolecular reactions. They are examples of mechanochemical or electromechanical processes such as vesiculation, spike and cup formation, etc., that occur in other cells.

All experiments with erythrocytes must be performed with washed erythrocytes, because the serum proteins stabilize the normocyte shape by antagonistic effects of globulins and albumins.

The results show that washed erythrocytes transform into echinocytes when the transmembrane potential decreases below a threshold value between 27 and 15 mV, or even becomes negative. However, stronger positive potentials lead to stomatocytes.[3,4] It can be shown that this transformation occurs via biconcave normocyte shapes.

These findings raise the question whether this shape transformation is really the result of the electric membrane potential or of pH differences.[7] This argument can be countered by the results given in Fig. 1. Detailed analysis of the conditions under which shape changes occur shows that no correlation exists between internal or external pH and the observed shape transformations.[8] Furthermore, we show that the correlation between transmembrane potential and shape also occurs in resealed ghosts, whereas the transmembrane potential is governed by the chloride gradient, independent of the pH conditions.[6]

Shape changes with the same potential correlation can also be generated by the application of a K potential by valinomycin, use of various potassium loaded cells,[9] and in various other ways.[5] Therefore, this phenomenon depends neither on the pH gradient nor on how the transmembrane potential is generated (by ionophores, by variations of external pH or external ionic strength, or by electric breakdown).[10] Variations of the surface potential of the cells by treatment with neuraminidase seems to have no significant effect on the shapes.[4] (Experiments with a directly applied external field are not possible, since they would require a large field strength and thus lead to a strong temperature increase in the solution.)

---

*Correspondence to: R. Glaser, Institute of Biophysics, Humboldt University, Invalidenstr. 42, D(O)-1040 Berlin, Germany.

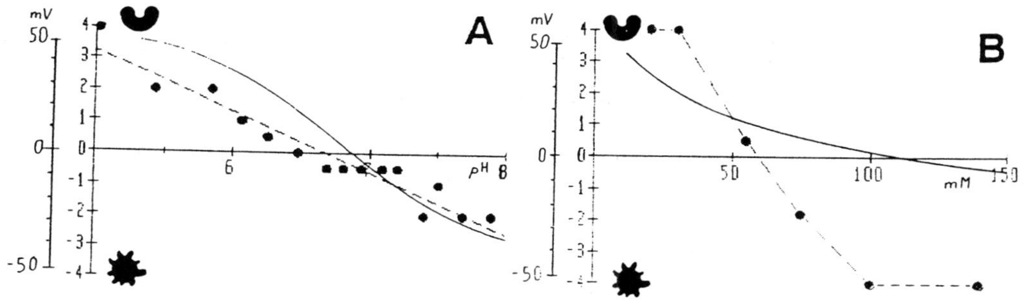

FIG. 1.—Correlation between shape of washed erythrocytes and its transmembrane potential in isotonic NaCl-sucrose solutions. Shapes are characterized by arbitrary shape factor (+4 = exclusively stomatocytes, −4 = exclusively echinocytes; for details see Ref. 4). Full lines indicate transmembrane potential, calculated for domain state ("D-state," in A); and for Cl^- pH equilibrium ("C-state," where no sodium-potassium exchange has taken place, in B). For equations, see Refs. 2-4. A: membrane potential and shapes of erythrocytes, treated with 50 μg/ml nystatin in 30 mM NaCl-sucrose solution, as a function of pH; for details, see Refs. 4 and 5. B: membrane potential and shapes of untreated erythrocytes in isotonic sucrose solutions as a function of external NaCl concentration (pH = 7.4); for details, see Ref. 6.

The threshold membrane potential (i.e., the potential at which the mean number of cells transform from one shape to another) is not constant but shifts slowly with the ionic strength of the external solution. It increases from +17 mV at 100 mM NaCl to +27 mV at 30 mM NaCl. Near these critical potentials, the shapes become temperature dependent in the range 4-37 C. Echinocytes occur at low temperatures and transform into stomatocytes if the temperature increases to a critical level.[11] If the membrane potential is several millivolts away from this threshold potential, the shapes are temperature insensitive at this interval. The electric field therefore seems to shift the phase transition point of the membrane constituents (probably lipids).

It has been known for years that anionic drugs are echinocytogenic, whereas cationic substances entering the membrane induce stomatocytes.[12-14] These influences can be overcome by modifications of the transmembrane field. For example, at a transmembrane potential of +35 mV, echinocytes generated by sodium-dodecyl-sulfate or fluvenamic acid are transformed into stomatocytes. Stomatocytes produced with propranolol, on the other hand, can be transformed into echinocytes, lowering the membrane potential to −17 mV.[5]

The kinetics of the shape transformation processes as well as its reversibility differ according to the conditions applied. Shape changes induced by transmembrane potential modifications can be converted several times,[4] by changing the membrane potential to and fro (for example, by pH titration). However, echinocyte-stomatocyte transformations that occur as a result of temperature increase are mostly irreversible.[11] Experiments using combinations with drugs and membrane potential at least indicate when a sequence of various flip-flop processes is initiated. Apparently, quick processes are performed by a direct flip-flop of small molecules, leading to a much slower exchange of lipids. A slow process with a half-time of the order of 1 h can be inhibited by vanadate, which indicates the incorporation of active flipases in this process.[15]

What is the mechanism governing this process? To answer this question, one must know what molecular changes really occur during the echinocyte-normocyte-stomatocyte transformations. Elgsaeter and Mikkelsen pointed out long ago that the bilayer-couples hypothesis[16] alone is not sufficient to explain this phenomenon.[17] Years ago we proposed another model that took account of the dynamic distribution of the pattern of membrane domains with various elastic properties.[18] Now we must conclude that the results of these experiments can be explained only by a complex model, including flip-flop reactions of molecules as well as local lateral inhomogeneities.[19]

REFERENCES

1. R. Glaser, "The influence of membrane electric field on cellular functions," in R. Glaser and D. Gingell, Eds., *Biophysics of the Cell Surface*, Springer Series in Biophysics 5: 175-195, 1989.

2. R. Glaser and J. Donath, ''Stationary ionic states in human red blood cells,'' *Bioelectrochem. Bioenerg.* 13: 71-83, 1984.

3. R. Glaser, ''The shape of red blood cells as a function of membrane potential and temperature,'' *J. Membrane Biol.* 51: 217-228, 1979.

4. R. Glaser, ''Echinocyte formation induced by potential changes of human red blood cells,'' *J. Membrane Biol.* 66: 79-85, 1982.

5.R. Glaser, T. Fujii, P. Müller, E. Tamura, and A. Herrmann, ''Erythrocyte shape dynamics: Influence of electrolyte conditions and membrane potential,'' *Biomed. Biochim. Acta* 46: 327-333, 1986.

6. P. Müller, A. Herrmann, and R. Glaser, ''Further evidence for a membrane potential shape transformation of the human erythrocyte membrane,'' *Bioscience Reports* 6: 999-1006, 1986.

7. E. M. Bifano, T. S. Novak, and J. C. Freedman, ''Relationship between the shape and the membrane potential of human red blood cells,'' *J. Membrane Biol.* 82: 1-13, 1984.

8. R. Glaser, ''Mechanisms of electro-mechanical coupling in membranes demonstrated by transmembrane potential dependent shape transformations of human erythrocytes,'' *Bioelectrochem. Bioenerg.* (in press).

9. R. Glaser, C. Gengnagel, and J. Donath, ''The influence of valinomycin induced membrane potential on erythrocyte shape,'' *Biomedica Biochimica Acta* 50: 869-877, 1991.

10. D. C. Chang and T. S. Reese, ''Changes in membrane structure induced by electroporation as revealed by rapid-freezing electron microscopy,'' *Biophys. J.* 58: 1-12, 1990.

11. R. Glaser and J. Donath, ''Temperature and transmembrane potential dependence of shape transformations of human erythrocytes,'' *Bioelectrochem. Bioenerg.* (in press).

12. B. Deuticke, ''Transformation and restoration of biconcave shape of human erythrocytes induced by amphiphilic agents and changes of ionic environment,'' *Biochim. Biophys. Acta* 163: 494-500, 1968.

13. T. Fujii, T. Sato, A. Tamura, A. Wakatsuki, and Y. Kanaho, ''Shape changes of human erythrocytes induced by various amphipathic drugs acting on the membrane of the intact cells,'' *Biochem. Pharmacology* 28: 613-620, 1979.

14. B. Isomaa, H. Hägerstrand, and G. Paatero, ''Shape transformations induced by amphiphiles in erythrocytes,'' *Biochim. Biophys. Acta* 899: 93-103, 1987.

15. J. Hartmann and R. Glaser, ''The influence of chlorpromazine on the potential-induced shape change of human erythrocyte,'' *Bioscience Reports* 11:213-331, 1991.

16. M. P. Sheetz and S. J. Singer, ''Biological membranes as bilayer couples: A molecular mechanism of drug-erythrocyte interaction,'' *Proc. Nat. Acad. Sci. USA* 71: 4457-4461, 1974.

17. A. Elgsaeter and A. Mikkelsen, ''Shapes and shape changes in vitro in normal red blood cells,'' *Biochim. Biophys. Acta* 1071: 273-290, 1991.

18. R. Glaser and A. Leitmanova, ''Transformation of human red cell shape with regard to fluid mosaic structure of the membrane,'' *Studia Biophysica* 48: 219-229, 1975.

19. This work was supported by the Volkswagen-Stiftung.

Martin Blank, Ed., *Electricity and Magnetism in Biology and Medicine*

EFFECT OF 2.45GHz MICROWAVE RADIATION ON PERMEABILITY OF UNILAMELLAR LIPOSOMES TO 5(6)-CARBOXYFLUORESCEIN: DEPENDENCE ON SOLUTES IN SOLUTION SURROUNDING THE LIPOSOMES

E. Saalman*

Chalmers University of Technology and The Nordic School of Public Health, Göteborg, Sweden

The membrane permeability of unilamellar phosphatidylcholine liposomes with trapped fluorescent marker 5(6)-carboxyfluorescein was studied under various conditions. Membrane leakage of the liposomes to 5(6)-carboxyfluorescein in the presence of various buffers, inorganic salts, or surfactants was compared on exposure to 2.45GHz microwaves and to normal heating in a waterbath. Liposomes in three buffer solutions showed increased permeability of 5(6)-carboxyfluorescein through the liposome membrane after microwave exposure compared to heating to the same temperature in a waterbath. Liposomes in inorganic salt solutions showed poor stability in membrane permeability, both with and without exposure to either microwaves or normal heat, which suggests a nonelectrostatic mechanism for membrane stabilization.

Concern that exposure to electromagnetic fields may affect biological systems has occasioned this study and also a previous paper.[1] The biological effects of microwaves have been studied extensively.[2,3] Liposomes were chosen as a model system for membranes because it has been suggested that biological membranes and, more specifically, phospholipids in natural membranes may represent the major sites of interaction with microwave radiation.[4] It is well known that leakage from the interior of liposomes and natural membranes of entrapped substances may take place.[5-11] An increase in membrane leakage under microwave exposure of liposomes and erythrocytes has also been reported.[4,12-16] In a previous paper we reported an increase in liposome membrane leakage on microwave exposure of liposomes, compared to exposure to normal heat in a waterbath, also above the phase-transition temperature.[1] To elucidate the observed effect (an increased membrane leakage after microwave exposure compared to normal heating) we have here varied the environment, i.e., the solutes in solution surrounding the liposomes.

MATERIALS AND METHODS

L-α-phosphatidylcholine (PC) unilamellar spherical liposomes were prepared and characterized as described.[1] Absorbance and fluorescence of the solutions were measured as described.[1] For the fluorescence measurements a small amount of the concentrated liposome solutions was added to various buffer solutions, inorganic salt solutions, or surfactant solutions. The lipid concentration was approximately 0.1 mg/ml (0.01 % wt) in the several solutions. Microwave exposure experiments were carried out as described.[1] The biological buffers Tris(hydroxymethyl)aminomethane (Tris), 1-2-hydroxyethylpiperazine-N-2-ethanesulfonic acid (Hepes) and 3-[N-Morpholino]propane sulfonic acid (Mops) were used. Three inorganic salt solutions were used: KCl, K_2SO_4, and $CaCl_2$. Surfactants used were sodiumdodecylsulfate (SDS), N-tetradecyl-N,N,N-trimethyl-ammonium bromide (TTAB), 3-[(3-cholamidopropyl)-dimethyl-ammonio]- 1-propane-sulfonate (Chaps) and β-octylglucoside. Liposome samples were removed from ice, left at room temperature for 1 min, and then placed in a waterbath of 38.8 C for another minute. One liposome sample (1 cm^3) was exposed to microwaves for 10 min. Another identical sample was placed in a waterbath of 38.8 C for 11 min. Other conditions concerning the experimental procedure have been described previously.[1]

---

*Correspondence to: E. Saalman (Physical Chemistry), Chalmers University of Technology, S-412 96 Göteborg, Sweden.

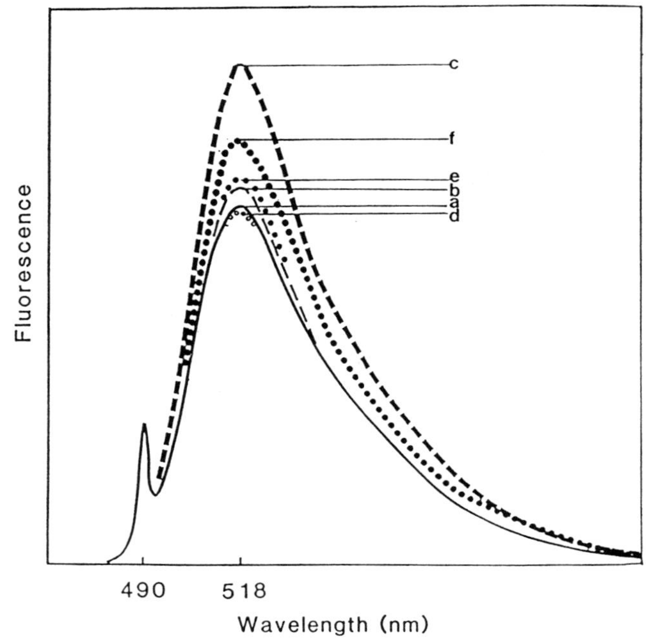

FIG. 1.—Representative fluorescence emission spectra obtained on liposome samples in 0.5cm quartz cuvette. Liposomes in 50mM Tris buffer (a) before exposure, (b) after heat exposure, (c) after microwave (2.45GHz) exposure. Liposomes in a 50mM Hepes buffer (d) before exposure, (e) after heat exposure, (f) after microwave (2.45GHz) exposure.

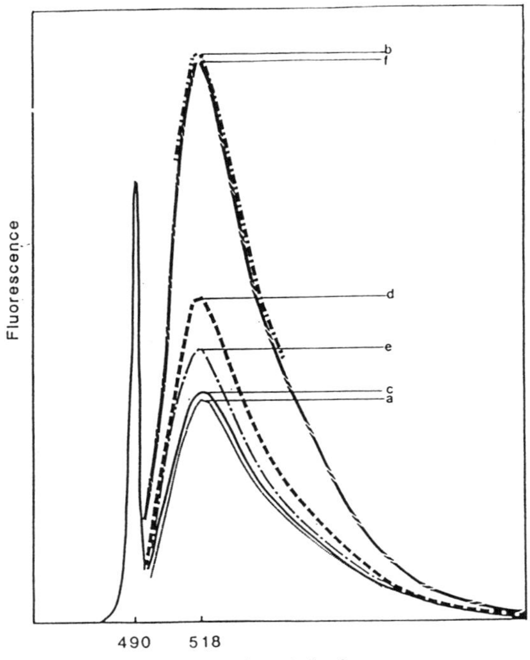

FIG. 2.—Representative fluorescence emission spectra obtained on liposome samples in 0.5cm quartz cuvette. Liposomes in 50mM KCl solution (a) before exposure, (b) after microwave (2.45GHz) exposure. Liposomes in a 50mM Tris buffer (c) before exposure, (d) after microwave (2.45GHz) exposure. Liposomes in water (e) before exposure, (f) after microwave (2.45GHz) exposure.

RESULTS

Dynamic light scattering measurements showed that the average diameter of the liposomes in the samples was 200 ± 20 nm. Table 1 shows the fluorescence intensity increase observed after 10 min heating of the liposome samples in 50 mM Tris buffer. Maximal membrane leakage was obtained after heating the liposomes to 90 C.

TABLE 1.—Spontaneous liposome membrane leakage to 5(6)-carboxyfluorescein at various temperatures, measured as fluorescence increase after 10 min at the given temperature.

Temp (° C)	% of maximal fluorescence
20	0
36	6.6
41	8.4
43	12.0
53	26.5
65	42.2
90	100

For liposomes in the three buffers there is only a small increase in fluorescence after normal heating in a waterbath (Fig. 1). Fluorescence after microwave exposure at the same temperature shows significantly increased liposome leakage in all three buffers. However, microwave exposure of liposomes in Hepes and Mops results in less pronounced leakage compared to liposomes in Tris buffer. In order to investigate how neutral salt influences the membrane leakage, liposomes in a KCl salt solution were studied (Fig. 2). Liposomes in 50 mM KCl or in water show highly increased fluorescence intensity after microwave exposure, whereas liposomes in 50 mM Tris show a much lower fluorescence increase. However, the fluorescence intensity increases when the concentration of Tris is decreased to 25 mM. We first interpreted this finding as an involvement of the electric repulsive forces at the liposome membrane surface. The high buffer concentration, 50 mM, seemed to stabilize the liposomes. We tried to isolate the effect with KCl; however, neither 25 nor 50 mM of KCl stabilized the liposomes. The stabilizing effect was no better for K_2SO_4 and $CaCl_2$. The experimental results for liposomes in the presence of surfactants were difficult to interpret owing to poor reproducibility. A high polydispersity of liposomes indicating large variations in liposome diameter was observed. Also, marked variations in leakage were observed after exposure. CF in the absence of liposomes (lipid) showed no significant dependence of fluorescence intensity on the several solutes. The wavelength of the emission maximum was the same as in the liposome solutions (518 nm); pH was carefully controlled during the experiments.

DISCUSSION

In the previous paper, we reported an increased membrane leakage after microwave exposure compared to normal heating to the same temperature in a waterbath.[1] We discussed the possibility of thermal motion of the lipid molecules in the membrane causing transient pores, which would allow leakage from the inside of the liposomes without rupture. In this paper the increase in the leakage of dye from the liposomes at increasing temperature (Table 1) implies that the thermal motion of the lipid molecules within the membrane results in the formation of such transient pores through which entrapped CF is released from the interior of the liposomes. The reason the liposomes do not disintegrate is most likely a very fast heat transport from membrane to surrounding solution.[1] However, that would not explain why 2.45GHz microwave exposure should give a higher increase in membrane leakage than normal heating. Probably effects of thermal inhomogeneity deserve most serious consideration; however, not on a macroscopic but on a microscopic level. A higher absorption of microwave energy in the proximity of the liposomes may lead to local heat effects on the membranes. For example, there is a possibility that different electric conditions in the dilute system, such as relatively higher ion concentration at the liposome surface compared to the bulk solution, may result in local heating of the liposomes. This local heating may be due to dielectric energy absorption in water bound to the membrane or to molecules adsorbed on the membrane surface.[17,18] Kaatze et al.[17] suggest the presence of water with unusual relaxation time in aqueous solutions of phospholipid micelles and vesicles. In the ionic atmosphere around a macromolecule, there is hydration water with restricted motion due to the bonding to the macromolecule. One possibility is the formation of transient pores, in addition to those formed due to thermal fluctuations, which then result in a more pronounced release of the entrapped dye compared with the normally heated control samples. If one accepts this explanation, the next problem to be solved is the determination of the particular structure of structural elements of the liposomes that are responsible for the formation of such transient pores. We first observed that increased membrane leakage after microwave exposure, compared to normal heating, had also occurred for liposomes in two other buffers, Hepes and Mops, and not only in the Tris buffer. However, the effect is less pronounced with these buffers. When normal heating is used, the membrane leakage increase is almost the same for all three buffers. Phosphatidylcholine is neutral at pH 7.4; the positive and negative charges balance each other. However, the positive charges of PC probably have a larger influence than the negative ones on molecules in the bulk solution close to the liposome membrane surface. One would expect the bulk molecules with negative charge to be preferentially attracted by the surface of the liposome membrane. Positively charged bulk molecules may also interact with the membrane if they can penetrate through the outermost (positively charged) layer and approach the negative PC charges. It is reasonable to believe that all three buffers, Tris, Hepes, and Mops, are adsorbed in the proximity of the liposome membrane. The observation of a more pronounced increase in liposome leakage, after microwave exposure, for liposomes in Tris than in Hepes or Mops may be due to the chemical structures of the buffers and their interaction with the liposome membrane. It is reasonable to believe that Tris can pass the bulky PC head-groups of the liposome membrane and thus reach the closest proximity of the liposome surface. An accumulation of Tris molecules at the membrane may be responsible for an extra heating effect (from the microwave exposure of the liposome membrane) compared to liposomes in Hepes or Mops. Water bound to Tris molecules probably shows a dielectric energy absorption. Since Tris is expected to be located very close to the liposome surface, the transfer of absorbed microwave energy, in the bound water, to the membrane may be very efficient. As for the bigger and more bulky molecules Hepes and Mops, their sulfonic acid groups are negatively charged and it is likely that they are attracted to the positively charged quaternary amine of the PC head groups pointing out from the liposome membrane surface.

REFERENCES

1. E. Saalman, B. Nordén, L. Arvidsson, Y. Hamnerius, P. Höjevik, K. E. Connell, and T. Kurucsev, *Biochim. Biophys. Acta* 1064: 124-130, 1991.

2. E. Postow and M. L. Swicord, in C. Polk and E. Postow, Eds., *Handbook of Biological Effects of Electromagnetic Fields,* Boca Raton, Fla.: CRC Press, 1986, 425-460.

3. S. M. Michaelson and J. C. Lin, *Biological Effects and Health: Implications of Radiofrequency Radiation,* New York: Plenum Press, 1987.

4. R. P. Liburdy and R. L. Magin, *Radiat. Res.* 103: 266-275, 1985.

5. C. W. M. Haest, J. De Gier, G. A. Van Es, A. J. Verkleij, and L. L. M. Van Deenen, *Biochim. Biophys. Acta* 288: 43-53, 1972.

6. D. Paphadjopoulos, K. Jacobson, S. Nir, and T. Isac, *Biochim. Biophys. Acta* 311: 330-348, 1973.

7. K. Inoue, *Biochim. Biophys. Acta* 339: 390-402, 1974.

8. M. C. Blok, E. C. M. Van der Neut-Kok, L. L. M. Van Deenen, and J. De Gier, *Biochim. Biophys. Acta* 406: 187-196, 1975.

9. M. C. Blok, L. L. M. Van Deenen, and J. De Gier, *Biochim. Biophys. Acta* 433: 1-12, 1976.

10. R. L. Magin and M. R. Niesman, *Chem. Physics Lipids* 34: 245-256, 1984.

11. R. L. Magin and J. N. Weinstein, in G. Gregoriadis, Ed., *Liposome Technology,* Boca Raton, Fla.: CRC Press, 1984, vol. 3, pp. 137-155.

12. R. B. Olcerst, S. Belman, M. Eisenbud, W. W. Mumford, and J. R. Rabinowitz, *Radiat. Res.* 82: 244-256, 1980.

13. P. D. Fisher, M. J. Poznansky, and W. A. G. Voss, *Radiat. Res.* 92: 411-422, 1982.

14. S. F. Cleary, F. Garber, and L. M. Liu, *Bioelectromagnetics* 3: 453-466, 1982.

15. R. P. Liburdy and A. Penn, *Bioelectromagnetics* 5: 283-291, 1984.

16. R. P. Liburdy and P. F. Vanek Jr., *Radiat. Res.* 102: 190-205, 1985.

17. U. Kaatze, K.-D. Göpel, and R. Pottel, *J. Phys. Chem.* 89: 2565-2571, 1985.

18. U. Kaatze, *Phys. Med. Biol.* 35: 1663-1681, 1990.

Martin Blank, Ed., *Electricity and Magnetism in Biology and Medicine*

MEMBRANE EFFECTS OF EXTREMELY LOW FREQUENCY MAGNETIC FIELDS ON K562 CELLS

W. Malorni, M. T. Santini,* S. Paradisi, E. Straface, M. Grandolfo, C. Cametti, and G. Donelli

University of Rome "La Sapienza," Italy

To investigate the possible effects of electromagnetic fields on cell integrity and function, biophysical and structural analyses were conducted on cultured cells (K562). In particular, the distribution of membrane proteins on the plasma membrane (as detected by the freeze-fracturing technique) as well as membrane dielectric properties were determined. Results obtained seem to confirm the plasma membrane as a probable target of ELF fields.

It is hypothesized that extremely low frequency (ELF) electromagnetic fields, principally originating from the production and transmission of electricity, can cause diverse effects on organisms and cells. In fact, several studies suggest the possibility that ELF field exposure can be related to human pathologies including cancer. It has become increasingly apparent by experimental studies at the cellular level that the cell membrane can play an important role in the interaction of ELF fields with living systems.[1] Several studies have shown that these fields can affect a variety of ion pumps and channels,[2-4] as well as membrane electrical properties.[5] The present investigation seeks to determine whether the exposure of K562 cultured cells to a sinusoidal 50Hz magnetic field (2.5 mT) can cause specific membrane effects. The cellular response was studied by means of electron microscopical and biophysical approaches.

MATERIALS AND METHODS

Cell Exposure

K562 cells (an erythroleukemic cell line) were grown in RPMI 1640 supplemented with 10% fetal calf serum, nonessential amino acids, 5mM L-glutamine, 100 I.U./ml penicillin, and 100 μg/ml streptomycin; 2×10^7 cells were seeded in flasks at a concentration of 5×10^5 cells/ml. Homogeneous sinusoidal 50Hz magnetic fields were generated inside two cylindrical coils; the exposure system used is described in a previous study.[5] Flasks were inserted in coils (exposed and sham-exposed) and control samples were kept in a CO_2 incubator. Exposure was continuous for 48 h at 2.5 mT.

Freeze-Fracturing

Immediately after exposure, experimental and control cells were fixed with 2.5% glutaraldehyde in phosphate buffer (0.13 M, pH 7.2) at 37 C. After fixation, cells were resuspended in the same buffer containing 25% glycerol. The cell suspensions, placed on carriers, were frozen in Freon 22. Replicas were obtained with a Balzers BAF 300 freeze-etch unit and examined with a Zeiss EM 10C electron microscope.

Dielectric Properties

The electrical conductivity of the cell suspension was measured by means of an impedance technique in the frequency range from 10 kHz to 100 MHz at a temperature of 37.0 ± 0.1 C. Due to the presence of cellular membranes, the conductivity spectra display a marked dispersion characterized by a relaxation time of about $3-5 \times 10^{-8}$ s, depending on the electrical parameters of the different components of the cell suspension. The membrane conductivity, the membrane permittivity, and the conductivity of the cytosol can be extracted from the measured conductivity of the whole suspension by means of an appropriate fitting procedure that takes into account the shape and the concentration of the cells.

---

*Correspondence to: M. T. Santini (Ultrastructures), Istituto Superiore di Sanità, University of Rome "La Sapienza," Viale Regina Elena 299, I-00161 Rome, Italy.

RESULTS

Freeze-Fracturing

The observation by transmission electron microscopy of several replicas obtained from cells exposed to a 2.5mT magnetic field revealed a random distribution of intramembrane particles in both the protoplasmic (Fig. 1) and the exoplasmic fracture faces of the plasma membrane in sham-exposed (Fig. 1a) and experimental (Fig. 1b) cells. In addition, cell cross-fractures did not display evident signs of intracellular changes, since organelles appeared morphologically preserved (data not shown).

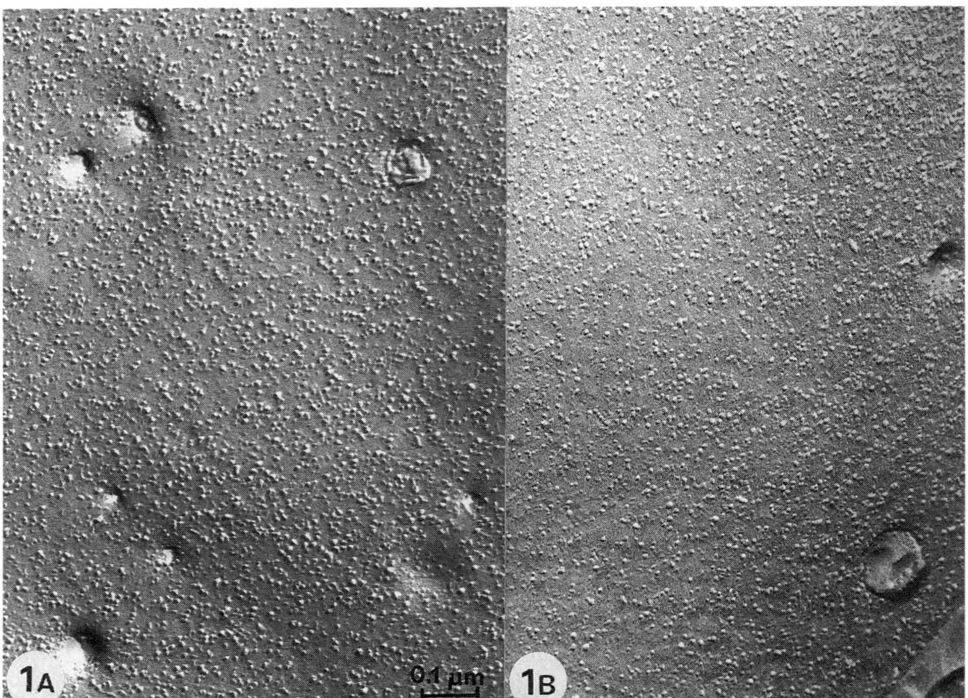

FIG. 1.—Distribution of intramembrane particles in the protoplasmic and the exoplasmic fracture faces of plasma membrane: (A) Sham-exposed; (B) experimental.

Dielectric Relaxation

The analysis of the conductivity data, on the basis of a dielectric model which considers the cells as a suspension of spherical noninteracting particles covered by a surface shell, yields the results on membrane conductivity summarized in Fig. 2. Exposure to the magnetic field significantly reduces the membrane conductivity (by a factor of about 3), whereas the membrane permittivity and the conductivity of the cytosol do not change within experimental errors (data not shown).

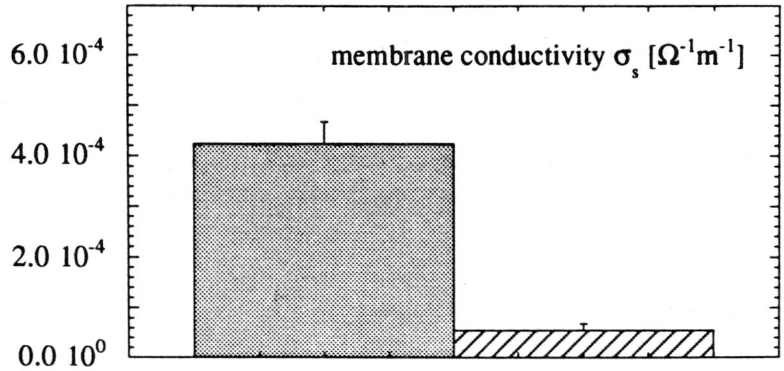

FIG. 2.—Analysis of conductivity data.

DISCUSSION

The results reported here seem to indicate that the membrane properties of in vitro cultured cells can be affected by exposure to ELF fields. In fact, as suggested by our previous papers,[5,6] membrane properties of cells as well as specific morphological features can be affected by their exposure to ELF fields in various in vitro systems, e.g., primary cultures of myoblasts[5] or the leukemic cell line K562.[6] However, these effects appear to be sublethal[6] and could also involve certain cytoskeletal elements.[7] Since the membrane conductivity is related to the ionic transport across the membrane and/or to the ionic permeation, our results also seem to indicate that magnetic field interactions can alter the overall transport mechanisms rather than the molecular arrangement of the membrane. The latter contention is supported by the freeze-fracturing analyses reported here, demonstrating that the membrane protein pattern remains unmodified after ELF field exposure.[8]

REFERENCES

1. M. Blank, in A. Marino, Ed., *Modern Bioelectricity*, New York: Marcel Dekker, 1988, 345-364.

2. T. Y. Tsong, F. Chauvrin, and R. D. Astumian, in M. Blank and E. Findl, Eds., *Mechanistic Approaches to Interactions of Electric and Electromagnetic Fields with Living Systems*, New York: Plenum Press, 1987, 187-202.

3. H. V. Westerhoff, F. Kamp, T. Y. Tsong, and R. D. Astumian, Ref. 2, 203-215.

4. F. J. Papatheofanis, *Rad. Res.* 122: 24-28, 1990.

5. M. Grandolfo, M. T. Santini, P. Vecchia, A. Bonincontro, C. Cametti, and P. Indovina, *Int. J. Radiat. Biol.* 60: 877-890, 1991.

6. S. Paradisi, G. Donelli, M. T. Santini, E. Straface, and W. Malorni, *Bioelectromagnetics*, 1992 (in press).

7. W. Malorni, S. Paradisi, E. Straface, M. T. Santini, and G. Donelli, *Proc. IRPA* 8: 1109-1112, 1992.

8. This study was supported by the CNR-ENEL Project "Interactions of energy systems with human health and environment," Rome, Italy.

Martin Blank, Ed., *Electricity and Magnetism in Biology and Medicine*

Na,K-ATPase ACTIVITY AS A MODEL FOR EM FIELD EFFECTS ON CELLS

M. Blank* and L. Soo

Columbia University, New York, NY

ATP-splitting activity of the membrane Na,K-ATPase changes when alternating currents (AC) are imposed with electrodes or induced from AC magnetic fields. The effect depends on the basal level of enzyme activity. Under optimal conditions AC decreases the activity; when the enzyme activity is lowered by the inhibitor ouabain or by lowered temperature, AC increases the activity. Both effects are frequency dependent over a broad range, with maxima at about 100 Hz. The threshold for inhibition by induced AC is higher by about an order of magnitude than for currents imposed through electrodes.

The activity of the Na,K-ATPase, the "ion pump" enzyme, is affected by AC imposed with electrodes or induced from AC electromagnetic (EM) fields.[1-3] Ion transport mediated by the enzyme is also affected by AC. AC-enhanced cation accumulation in red cells is optimal at 10^3 Hz;[4] efflux is optimal at 10^6 Hz.[5] The two processes are physiologically linked, so that it is difficult to explain the three orders of magnitude separating the optimal frequencies. One proposed mechanism suggests that the enzyme in the membrane changes conformation in the electric field.[6] Our recent measurements suggest that the field changes ion activation at the enzyme surfaces due to greater ion binding.[4,7] The two proposals are models for signal transduction in membranes.

EXPERIMENTAL METHODS

We prepare the enzyme from rabbit kidney[8] and apply sine waves from a function generator to Na,K-ATPase suspensions in a small U-tube using platinum electrodes (sheathed in plastic tubing containing salt-impregnated 1.5% Agar gels). For the induced currents, we use an apparatus designed by Kaune et al.[9] The enzyme is studied in a medium containing optimal concentrations of activating cations (Na^+ 81 mM, K^+ 16 mM, Mg^{++} 5 mM). We obtain a range of enzyme activity by changing the temperature, the ratio of activating ions, or using the specific inhibitor ouabain, with similar results for all methods. We report the effect of AC as the enzyme activity E divided by the normal activity at the time of the experiment A, i.e., the ratio E/A.

RESULTS

Figure 1 shows that the effect of applied AC on enzyme activity is a function of the enzyme activity, and can be either inhibitory or stimulatory.[1,7] In experiments changing the Na/K ratio, the effect of AC correlates directly with the enzyme activity at any particular ratio.[2] Since the activity of the enzyme varies with the ion concentrations, increasing at low activity and decreasing (or reaching a plateau) at high activity, the effect of AC could be due to changes in the binding of activating ions. Both inhibition and stimulation by AC can be explained by increased binding of activating cations.[10]

Inhibition and stimulation have similar frequency dependence, with maximal effects about 100 Hz (Fig. 2).[11] A detailed study in the 30-300Hz range shows a gradual variation with no peaks. The frequency dependence is similar to the calculated ion concentration changes at membrane surfaces due to AC;[12] increases in ion concentration would lead to greater binding.

The threshold for inhibition with an imposed electric field at 100 Hz is 5 μV/cm.[11] For an induced AC (at 60 Hz), the threshold is 52 μV/cm.[3] Both values are in the range of the experiments reporting changes in protein synthesis.[13] The presence of an additional stray magnetic field in the induced AC study may explain the higher threshold if its effect on Na,K-ATPase function is opposite to that of the E field.

---

*Correspondence to: M. Blank (Physiology and Cellular Biophysics), Columbia University, 630 W 168th St., New York NY 10032.

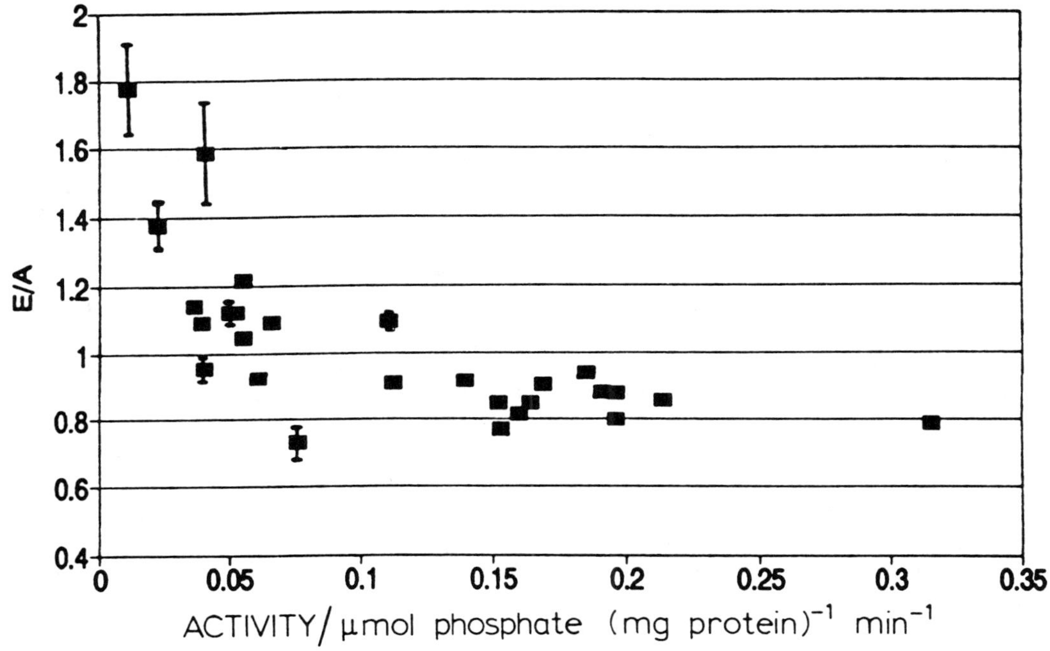

FIG. 1.—*E/A*, the ratio of enzyme activity with AC/without AC vs enzyme activity in µmoles phosphate/mg protein/min. AC frequency was 100 Hz and applied voltage was 0.01 V. Error bars are smaller than most data points. Above enzyme activity 0.12, *E/A* < 1; below it, *E/A* tends to be > 1. Low values of enzyme activity were obtained by use of 10^{-4} to 10^{-5} ouabain, and similar effects were obtained by use of low temperatures. (Reproduced, with permission, from *Bioelectrochem. Bioenerg.* 24: 51, 1990.)

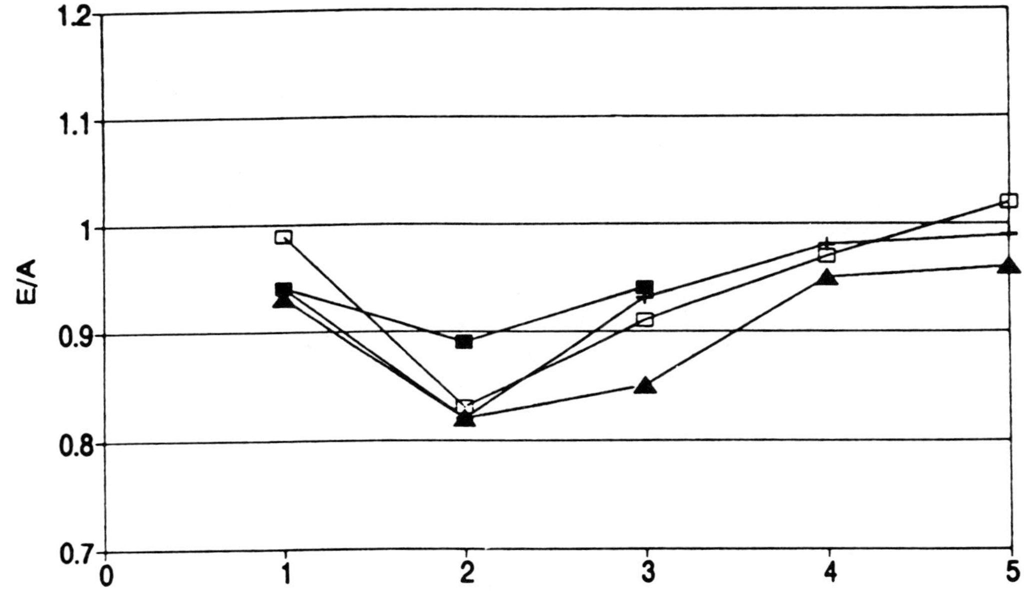

FIG. 2.—*E/A*, ratio of enzyme activity with AC/without AC vs logarithm of frequency in Hz. Results at different applied voltages are plotted separately: □ 1 mV, Δ 10 mV, □ 100 mV, +1 V. (Reproduced, with permission, from *Bioelectrochem. Bioenerg.* 24: 51, 1990.)

DISCUSSION

From our studies, it appears that a mechanism for AC effects must account for the observed dependences on the electric field and on the basal level of enzyme activity. The electroconformational coupling model[6] accounts for the effect of an electric field on the enzyme, but does not consider the effect of the enzyme activity. The ion activation model[2,7] accounts for both effects through changes in ion binding at the activation sites. In addition, the ion activation model accounts for the difference in optimal frequency between ion influx and efflux in cells. In erythrocytes, K^+ binds to the outside surface for transport inward, and is affected at low frequency. To affect Na^+ efflux from a cell, the AC signal must penetrate to the binding site inside the membrane, an effect that occurs only at much higher AC frequencies.

A mechanism must also explain how AC causes both increases and decreases in enzyme activity. Increased binding of activating ions in the field can achieve both because the activation curves show an increase in activity with increasing ion concentration, followed by a decrease in activity. When the enzyme is at optimal activity, an increase in ion concentration lowers activity. Below optimal activity, an increase in ion concentration increases the enzyme activity.

The observed frequency dependence is very close to the calculated ion concentration changes at membrane surfaces.[11] Calculations with the SCM model[14] show that ion concentrations across the cell membrane vary with the frequency, depending on ion mobilities, rate constants, and concentrations. Optimal frequencies are in the same range as observed with the ATPase.

These studies show that signal transduction can occur at the membrane Na,K-ATPase. When AC changes the ion activation sites on the outside surface, the enzyme will affect the inner composition of the cell. The results also relate to the therapeutic effects of AC signals on damaged tissues that represent far from optimal conditions.[15] We have found that when the enzyme is far from its optimum due to low temperature or an inhibitor, AC enhances the activity.[16]

REFERENCES

1. M. Blank and L. Soo, *Bioelectrochem. Bioenerg.* 22: 313-322, 1989.
2. M. Blank and L. Soo, *Bioelectrochem. Bioenerg.* 25: 51-61, 1990.
3. M. Blank and L. Soo, *Bioelectrochem. Bioenerg.* 28: 291-299, 1992.
4. E. H. Serpersu and T. Y. Tsong, *J. Biol. Chem.* 259: 7155-7162, 1984.
5. D. S. Liu, R. D. Astumian, and T. Y. Tsong, *J. Biol. Chem.* 265: 7260-7267, 1990.
6. T. Y. Tsong, D. S. Liu, F. Chauvin, A. Gaigalas, and R. D. Astumian, *Bioelectrochem. Bioenerg.* 21: 319-321, 1989.
7. M. Blank, *FASEB J.* 6: 2434-2438, 1992.
8. J. S. Britten and M. Blank, *Biochim. Biophys. Acta* 159: 160-166, 1968.
9. W. T. Kaune, M. E. Frazier, A. J. King, J. E. Samuel, F. P. Hungate, and S. C. Casey, *Bioelectromagnetics* 5: 117-129, 1984.
10. J. C. Skou, *Biochim. Biophys. Acta* 23: 394-401, 1957.
11. M. Blank and L. Soo, *Bioelectromagnetics* 13: 329-332, 1992.
12. M. Blank, *J. Electrochem. Soc.* 134: 1112-1117, 1987.
13. R. Goodman and A. S. Henderson, *Proc. Natl. Acad. Sci. USA* 85: 3928-3932, 1988.
14. M. Blank, *Biochim. Biophys. Acta* 906: 277-294, 1987.
15. C. A. L. Bassett, *Clin. Plast. Surg.* 12: 259-277, 1985.

16. We thank ONR and EPRI for their support.

Martin Blank, Ed., *Electricity and Magnetism in Biology and Medicine*

EFFECTS OF EXTREMELY LOW FREQUENCY ELECTROMAGNETIC FIELDS ON ION TRANSPORT IN SEVERAL MAMMALIAN CELLS

J. García-Sancho,* M. Montero, J. Alvarez, R. Fonteriz, and A. Sanchez

University of Valladolid, Spain

We have studied the effects of sinusoidal EMF on ion transport (Ca, Na, K) in several cell kinds (red cells, neutrophils, thymocytes, Ehrlich cells, and HL60 and U937 leukemia cells). Treatment with 50Hz and 1-20G EMF had no significant effects. Treatment with 0.2-12G EMF at the cyclotron magnetic resonance frequencies for the respective ions had no significant effects except for a 16-24% increase of the uptake of ^{42}K within a window at 14.5-15.5 Hz and 1-2 G, which was found in U937 and Ehrlich cells, but not in the other kinds of cells.

Although the mechanisms for interactions of EMF with living cells are not known, it has been proposed that they may include modifications of ion transport.[1] The physical mechanism underlying these effects has been suggested to be ion cyclotron resonance (ICR).[2] In this process, transfer of energy from a time-varying magnetic field occurs when its frequency matches the cyclotron resonance frequency (CRF) of an ion moving within a static magnetic field, such as the geomagnetic one.

In the present work we have studied the effects of sinusoidal EMF on ion transport in several cell kinds. Frequencies tested were either 50 Hz (as in EMF produced by electric appliances) or those calculated for the cyclotron resonance of ions of interest.

METHOD

The sinusoidal field was generated by a pair of Helmholtz coils spaced 17 cm apart and oriented north-south. The measured geomagnetic field in our laboratory was 0.42 G, oriented 59° below the horizontal. The several cell kinds (human red cell, HL60 and U 937 human leukemia cells, Ehrlich cells, rat thymocytes) were prepared and handled as previously described.[3-6] The cells were suspended at 1-2% in a solution containing (in mM): NaCl, 140; $MgCl_2$, 1; $CaCl_2$, 1; potassium-HEPES, 10, pH 7.4; glucose, 10; 1 ml aliquots of the cell suspensions were placed in Eppendorf tubes and incubated for 5 min at 37 C. Experimental samples were placed between the Helmholtz coils; controls were placed outside. The experiment was started by the addition of 20 μl of incubation medium containing tracer amounts of ^{45}Ca, ^{22}Na, or ^{42}K. After 60 min at 37 C the incubation period was terminated by centrifugation, followed by two washes with 1 ml aliquots of ice-cold medium. The cell pellets were extracted and counted for radioactivity. Data were compared by the two-tailed Student's *t*-test.

RESULTS AND DISCUSSION

We find no effect when 50Hz EMF (1, 12, or 20 G) were applied. No effects were found on the uptake of either ^{22} Na or ^{45}Ca at CRF (29 Hz for both ions) at 0.25, 1, or 10 G for any of the cells studied. At the cyclotron resonance frequency estimated for ^{42}K (15.5 Hz) and 1G amplitude we found a significant increase of the uptake for Ehrlich ascites tumor cells (65% increase, $p < 0.01$) and for U937 cells (28% increase, $p < 0.05$) and no significant action for the other three cell kinds studied. Treatments at 15.5 Hz and either 0.25 or 10 G had no significant effect.

Figure 1 illustrates the effects obtained in U937 cells at several frequencies and amplitudes. There was a significant increase of ^{42}K uptake at 14.5 and 15.5 Hz and 1 and 2 G, but not at the other frequencies (13.5 and 16.5 Hz) and amplitudes (0.5 G). Similar experiments (13.5-16.5 Hz) were performed with the other cell kinds at a fixed amplitude of 1 G. No effects were found for human red cells, HL60 cells, and thymocytes. For Ehrlich cells a significant increase of the uptake was found at 14.5 and 15.5 G.

---

*Correspondence to: J. García-Sancho (Fisiología y Bioquímica), Universidad de Valladolid, ES-47005 Valladolid, Spain.

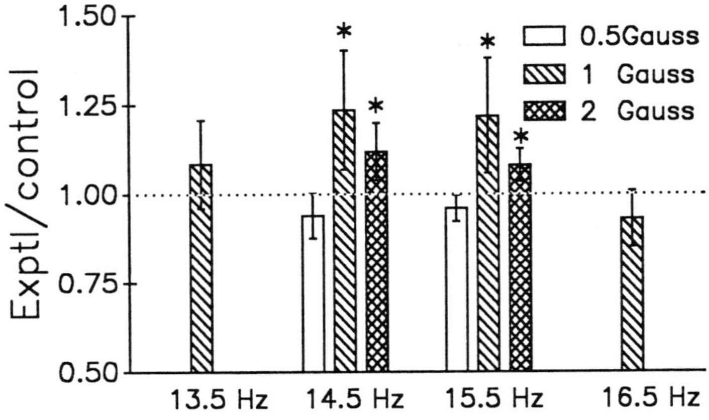

FIG. 1.—Effects of EMF on the uptake of ^{42}K by U937 cells. Values are mean ± SE of 4-8 data. *: $p < 0.05$.

The above results do not support a general role of the ICR mechanism for the interactions between EMF and ion transport. However, we found effects on ^{42}K uptake in two out of five cell kinds tested at about the ICR frequency, although at AC field intensities larger than expected. (Maximum energy transfer is predicted for EMF fields of the same intensity as the geomagnetic field.[2]) In any case, our results suggest that the effects of EMF are highly specific with regard to both the cell kind and the frequency and amplitude windows.[7]

REFERENCES

1. T. S. Tenforde, *Bioelectrochemistry and Bioenergetics* 25: 1-17, 1991.

2. A. R. Liboff, in A. Chiabrera, C. Nicolini, and H. P. Schwan, Eds., *Interactions Between Electromagnetic Fields and Cells*, New York: Plenum, 1985, 281-296.

3. J. Alvarez, J. Garcia-Sancho, F. Mollinedo, and A. Sanchez, *Eur. J. Biochem.* 183: 709-714, 1989.

4. J. Garcia-Sancho and V. L. Lew, *J. Physiol.* (London), 407: 505-522, 1988.

5. J. Garcia-Sancho, A. Sanchez, and H. N. Christensen, *Biochim. Biophys. Acta* 464: 295-312, 1977.

6. M. Montero, J. Alvarez, and J. Garcia-Sancho, *Biochem. J.* 271: 535-540, 1990.

7. This work was coordinated by AMYS-UNESA within the Programa de Investigación y Desarrollo Tecnológgico (PIE).

Martin Blank, Ed., *Electricity and Magnetism in Biology and Medicine*

ELECTRON TRANSFER IN THE UBIQUINONE REGION OF THE MITOCHONDRIAL RESPIRATORY CHAIN

G. Lenaz,* R. Fato, C. Castelluccio, M. Cavazzoni, and G. Parenti Castelli

University of Bologna, Italy

Two mechanisms by which ubiquinone may control respiration, namely diffusion and concentration limitations, were investigated. Diffusion control was excluded as the result of kinetic experiments both in the isolated bc_1 complex and in integrated electron transfer. On the other hand, it was found that ubiquinone concentration is not saturating for NAD-linked electron transfer. Both kinetic simulations and experimental studies in quinone-depleted mitochondria exhibited quinone concentrations giving half-maximum velocity of NADH cytochrome c reductase in the range of the physiological ubiquinone content.

Ubiquinone (Q) is a lipophilic redox compound acting as a mobile electron carrier between flavoprotein dehydrogenases and the bc_1 complex in the mitochondrial respiratory chain.[1] The suggested presence of a control step of respiration in the Q region[2] raises the question whether Q is directly involved in the rate control of electron transfer. Two possible mechanisms by which Q may control respiration were investigated: (a) the collisional interactions of Q with its partner enzymes could be rate-limiting, as in diffusion control;[3] (b) the Q concentration in the membrane would be non-saturating for maximum velocity of the respiratory chain (see Ref. 4 for preliminary evidence).

Lack of Diffusion Control

Diffusion-limited reactions are characterized by the low activation energy and the high viscosity dependence of k_{cat}/K_m (= k_{min}, the minimum value of the second-order association rate constant of enzyme and substrate).[5]

Ubiquinol cytochrome c reductase has a two-site ping-pong kinetic mechanism.[6] Its substrates, ubiquinol and cytochrome c (cyt.c), are lipid- and water-soluble, respectively. By temperature dependence studies we found low activation energy of $k_{min(cyt.c)}$ but high activation energy of $k_{min(ubiquinol)}$ (Table 1). Moreover, $k_{min(ubiquinol)}$ was insensitive to changes of membrane viscosity elicited either by cholesterol incorporation or fatty acid composition changes (Table 2), whereas $k_{min(cyt.c)}$ was strongly affected by aqueous viscosogens (not shown). These features suggest that the enzyme reaction is limited by cyt.c diffusion and not by ubiquinol diffusion.

TABLE 1.—Kinetic parameters of ubiquinol-2 cyt.c reductase in beef heart submitochondrial particles.

k_{cat} (s^{-1})	370
$k_{min(ubiquinol)}$ ($mM^{-1} s^{-1}$)*	13
$k_{min(cyt.c)}$ ($\mu M^{-1} s^{-1}$)	77
E_A ($k_{min(ubiquinol)}$ (kcal mol^{-1})	14
E_A ($k_{min(cyt.c)}$ (kcal mol^{-1})	1.4

* Ubiquinol concentration expressed as M in lipid phase.[8]

*Correspondence to: G. Lenaz (Biochimica), Università di Bologna, Via Imerio 48, I-40126 Bologna, Italy.

TABLE 2.—Viscosity dependence of collisional frequencies ($k_{(coll)}$) and of $k_{min(ubiquinol)}$ in ubiquinol-2 cyt.c reductase.*

	$K_{coll} \times 10^{-9}$	$k_{min(ubiquinol)} \times 10^{-3}$ ($M^{-1}s^{-1}$)
Beef heart SMP		
control	5	8
cholesterol	3.5	8
S. cerevisiae mito.		
grown at 36° → 10°	4.5	9
grown at 36°	2	10

*Collisional frequencies between ubiquinol-2 and bc_1 complex in the membrane lipids were approximated by the (diffusion-limited) quenching constants of the probe 12-(9-anthroyl) stearate by Q_3.[9] All concentration units of k_{coll} and k_{min} are expressed as molarity in the lipid phase. Viscosity in submitochondrial particles (SMP) was changed by cholesterol incorporation (from 35 nmol/mg of control to 138 nmol/mg) and in *S. cerevisiae* mitochondria by changing the growth temperature: cells grown at 36 C were less fluid than cells brought to 10 C at the end of the exponential phase.

Evidence was also provided that integrated electron transfer through the Q pool (exemplified by NADH cyt.c reductase) is not affected by an increase in the average distance between complexes I and III obtained by reconstitution of these isolated complexes at increasing phospholipid-to-protein ratios.[1] This finding is evidence that ubiquinone lateral diffusion between complexes is not limiting for electron transfer in the respiratory chain.

Saturation Kinetics

The dependence of integrated electron transfer through the Q pool on Q concentration is a complex hyperbolic function including a "poised" K_m for total Q, which reflects the V_{max} and the K_m of the two enzymes connected by the pool.[4]

Beef heart mitochondria were extracted by pentane after lyophilization;[7] varying Q levels were achieved either by partial extractions or by reconstitution. Kinetic analysis of respiratory activities mediated by the Q pool in partly depleted mitochondria exhibited saturation kinetics; the Q concentration yielding half-maximum velocity was 2.6 mM for NADH cyt.c reductase, and 0.3 mM for succinate cyt.c reductase (Fig. 1). The apparent "K_m" were larger in Q-reconstituted mitochondria, which shows that part of Q was associated to the membranes as a nonreactive pool.

The results indicate that the physiological Q concentration in the mitochondrial membrane (3-8 mM)[8] is kinetically nonsaturating for NADH oxidation (unlike succinate oxidation). Attaining the theoretical V_{max} values for NADH oxidation would require very high Q to phospholipid molar ratios, high enough to prevent molecular mixing.[4]

The finding that Q concentration is not saturating for NADH oxidation in a region of the respiratory chain where strong control may be present provides a rationale for pathological changes associated with decrease of Q content and for the beneficial effect of Q administration.[4] We can also predict from the kinetic basis of this study that exogenous Q administration might ameliorate electron transfer even when a mitochondrial respiration defect does not involve Q (e.g., cytochrome oxidase).

REFERENCES

1. G. Lenaz, *J. Membrane Biol.* 104: 193-209, 1988.
2. R. Moreno-Sanchez, T. Espinosa-Garcia, and J. C. Raja, in J. J. Lemasters et al., Eds., *Integration of Mitochondrial Function,* New York: Plenum, 1988, 297-304.
3. C. R. Hackenbrock, B. Chazotte, and S. S. Gupte, *J. Bioenerg. Biomembr.* 18: 331-368, 1986.

4. G. Lenaz, R. Fato, C. Castelluccio, M. Battino, M. Cavazzoni, H. Rauchova, and G. Parenti Castelli, in K. Folkers et al., Eds., *Biomedical and Clinical Aspects of Coenzyme Q*, Amsterdam: Elsevier, 1991, 11-18.

5. B. B. Hasinoff, *Biochim. Biophys. Acta* 704: 52-58, 1982.

6. M. Degli Esposti and G. Lenaz, *Arch. Biochem. Biophys.* 289: 303-312, 1991.

7. L. Szarkowska, *Arch. Biochem. Biophys.* 113: 519-525, 1966.

8. G. Lenaz, M. Battino, C. Castelluccio, R. Fato, M. Cavazzoni, H. Rauchova, C. Bovina, G. Formiggini, and G. Parenti Castelli, *Free Rad. Res. Commun.* 8: 317-327, 1990.

9. R. Fato, M. Battino, M. Degli Esposti, G. Parenti Castelli, and G. Lenaz, *Biochemistry* 25: 3378-3390, 1986.

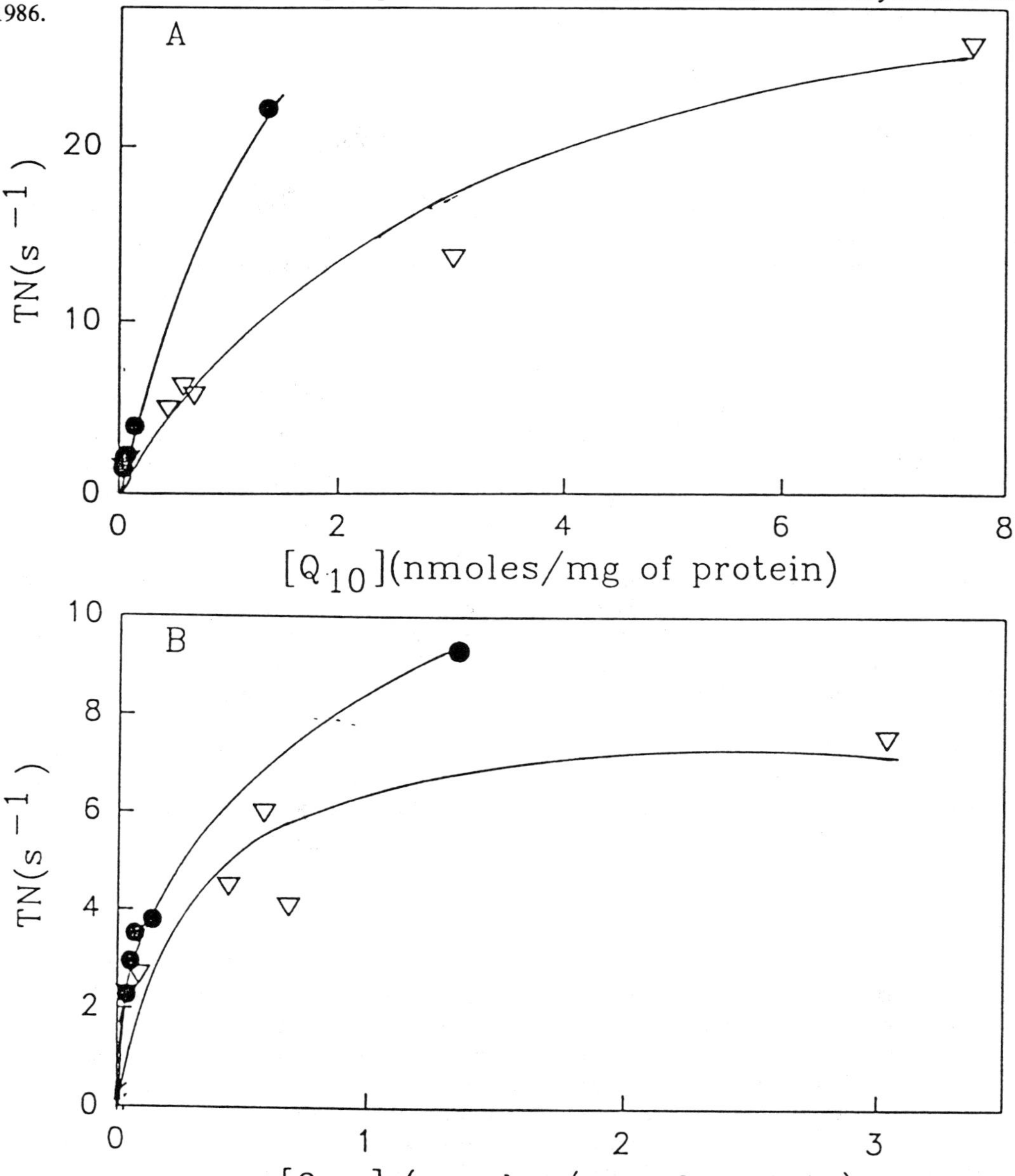

FIG. 1.—Saturation kinetics of NADH- and succinate-cyt.c reductase in pentane-extracted and reconstituted mitochondria: (•) partial pentane extraction; (▵) reconstitution with Q_{10}. (A) NADH-cyt.c reductase; (B) succinate cyt.c reductase.

Martin Blank, Ed., *Electricity and Magnetism in Biology and Medicine*

CALCIUM EFFLUX FROM TOAD HEART: A REPLICATION STUDY

A. W. Wood*

Swinburne University of Technology, Hawthorn, Australia

V. Lubinas, K. H. Joyner, and B. A. Hocking

Australian & Overseas Telecommunications Corporation, Clayton, Australia

This study is a replication of a recent study reporting enhanced calcium efflux in frog heart exposed to modulated RF radiation. An initial experimental group of 80 isolated hearts were sham-exposed or exposed at SARs of 0.24, 0.30, and 0.36 mW/kg. Results showed an enhanced efflux of 21% at 0.24 mW/kg, but in view of significant intragroup variation, a second group of 80 hearts were studied, including an SAR of 0.18 mW/kg. Here, efflux at 0.24 mW/kg was the same as control, but an enhancement at 0.30 mW/kg was observed ($p < 0.05$). A pooling of both groups showed a reduced efflux at 0.18 mW/kg.

Cardiac muscle tissue is particularly relevant to the study of EMF effects because of the syncytial arrangement of cells and the prevalence of gap junctions (that is, the internal medium is continuous between cells). These low-resistance pathways between muscle cells imply that for varying magnetic fields, the induced currents will be much larger than for electrically isolated cells, and will be greatest toward the outer surface of the heart.

Calcium plays a critical role in the normal physiology of the heart: it is responsible for excitation-contraction coupling. Calcium conductance in cardiac fibers is modified by autonomic factors. There is evidence that ELF electric and magnetic fields modify the heart rate in human volunteers, particularly if the fields are presented in an intermittent manner.[1] RF radiation modulated at ELF (16 Hz) was used in the early experiments of Adey[2] and of Blackman[3] on calcium efflux from avian brain hemispheres. Both groups have presented evidence for related effects at ELF alone, implying an uncoupling of the ELF signal from the RF carrier.

The present study has as its initial aim the replication of that of Schwartz et al.,[4] whose principal finding was an enhancement of ^{45}Ca-traced calcium efflux in frog heart exposed to 240 MHz RF amplitude-modulated by 16 Hz at 90% modulation and at an SAR of 0.3 mW/kg. We have reproduced these exposure conditions using a similar transverse electric and magnetic (TEM) system and the locally available cane toad *Bufo marinus*.

METHODS

Following the procedure of Schwartz et al.,[4] we excised the toad hearts, weighed them, attached them to pacing electrodes, and placed them in glass vials in groups of eight. The hearts were soaked in a ^{45}Ca-Ringer's solution (0.2 μCi/ml) for 30 min, rinsed twice, and then placed in fresh Ringer's for the exposure or sham-exposure period of 30 min. Hearts were removed from the bathing (or washout) solution and solubilized in separate vials. After correction for quenching, the percentage efflux was estimated from $100C_W/(C_W + C_T)$, where C_W and C_T are the total radioactivities in the washout and tissue, respectively.

Apart from the species difference, the present series of experiments differed from those of Schwartz et al. in three respects. First, since many of the hearts weighed over 1 g, the washout volume was increased from 4 to 10 ml. Second, since toad heart was not completely digested by 10ml solubilizer solution (NCS, Amersham), the remaining solid mass was digested in 2 ml of 3M NaOH and then neutralized by 2 ml of acetic acid; 1 ml aliquots of the washout, solubilizer, and caustic digest solutions were added to 10 ml of liquid scintillant (BCS, Amersham). Third, the stimulating electrodes consisted of thin wires insulated to within a

---

*Correspondence to: A. W. Wood, Swinburne University of Technology, Hawthorne VIC 3122, Australia.

few millimeters of attachment to the heart. Schwartz et al. had used carbon/Teflon electrodes, but it was considered that the amount of power absorption by these thin wires would be negligible.

Radioactive samples were assayed to 1% counting accuracy and each batch was counted twice. Quench curves were obtain by "spiking" scintillant cocktails prepared in the manner just described. Quench corrections were applied both by external standard ratio and single-channel ratio methods. In most cases quench corrections were of the order of 0-5%.

The exposure system consisted of a TEM (or Crawford) cell of dimensions 1.4×0.6×0.4 m. Standing-wave ratios were measured in the range 0.3-400 MHz. At 240 MHz the return loss was 23.1 dB. Characteristics of the TEM cell were also checked by time-domain spectroscopy.

Power levels were monitored throughout the experiment. Researchers carrying out the analyses on the heart tissue were unaware of the exposure status of the experiment. Hearts were paced at a rate of 0.5 Hz with monopolar pulses of 4ms duration and of sufficient amplitude to cause visible contraction.

RESULTS AND DISCUSSION

The experiments were conducted in two runs of 80 toads each, separated by several months. In the first run, six hearts were used for purposes other than ^{45}Ca efflux. Results from the remaining 74 hearts are shown in Table 1. In the first experiment (group of 8 hearts) the hearts were not paced and it was not certain that all counts were recovered in initial attempts to solubilize the tissue. If the results of the first experiment (which were controls) are eliminated, the mean control efflux is consistent with the means for SARs of 0.30 and 0.36 mW/kg, but a 21% enhancement ($p < 0.05$) is apparent at 0.24 mW/kg.

TABLE 1.—Percentage calcium efflux from isolated heart (240 MHz, 16Hz modulation).

Experimental group	Control	SAR (mW/kg)				
		0.15	0.18	0.24	0.30	0.36
1st run	33.2 ± 2.5 (A) (16) 36.9 ± 2.2 (B) (23)			40.0 ± 2.2 (13)*	32.2 ± 2.5 (22)	31.9 ± 1.6 (16)
2nd run	29.3 ± 1.8 (24)		26.6 ± 2.3 (16)	27.1 ± 1.0 (24)	36.8 ± 2.6 (C) (16)* 28.7 ± 1.4 (D) (8)	
Pooled 1st and 2nd runs	33.0 ± 1.5 (E) (47)		26.6 ± 2.3 (16)*	31.7 ± 1.4 (37)	34.1 ± 1.9 (38)	31.9 ± 1.6 (16)
Schwartz et al.	42.4 ± 1.4 (20)	51.3 ± 3.5 (8)*		42.0 ± 1.3 (20)	50.0 ± 1.2 (23)**	41.7 ± 1.6 (F) (20)

*,**Different from controls on same row: $p < 0.05, 0.01$, respectively. Number of hearts given in brackets.

Notes: (A) Includes all (control) results, run 1; (B) excludes experiment 1, run 1; (C) includes all (0.3 mW/kg) results, run 2; (D) excludes experiment 1, run 2; (E) if experiment 1, run 1 excluded, falls to 30.8 ± 1.4; (F) Schwartz et al.[4] also included measurements at SARs of 1.5 and 3.0 mW/kg; also measurements at AM of 0.5 Hz: none significantly different from control.

In the second run of 80 hearts, an SAR of 0.18 mW/kg was included. In this run the 0.24mW/kg value was the same as control, but a 28% enhancement ($p < 0.05$) was obtained at 0.30 mW/kg. However, of the two groups of 8 hearts at this SAR, the first is significantly higher than the second. There was also some evidence of polarization of the stimulating electrodes in the first few minutes of the experiment (before ^{45}Ca loading) due to the pacing stimulus being set too high. If the results of this experiment are eliminated, again none of the group means are different from control. If the two sets of data are compared, the mean of the controls in the second run is lower than that in the first. Pooling the data suggests, if anything, a *reduction* in efflux at 0.18 mW/kg.

When compared with Schwartz et al.'s results, our mean percentage efflux of the controls was significantly lower, but the SEM is about the same. The hearts used in the second run were notably smaller than those in the first. There was also a general downward trend in average efflux in run 1 as the experiment progressed.

CONCLUSIONS

We conclude that there is no effect at the power levels and frequency chosen; or if there is an effect, it is masked by experimental factors that remain to be identified.[5]

REFERENCES

1. C. Graham, H. D. Cohen, M. R. Cook, M. M. Gerkovitch, D. R. Riffle, and S. J. Hoffman, "Human cardiac activity in 60Hz magnetic fields," *13th Ann. Meeting Bioelectromagnetics Soc.*, 1991.

2. W. R. Adey, "Tissue interactions with nonionizing electromagnetic fields," *Physiol. Revs.* 61: 435-514, 1981.

3. C. F. Blackman, S. G. Benane, L. S. Kinney, W. T. Joines, and D. E. House, "Effects of ELF fields on calcium-ion efflux from brain tissue in vitro," *Radiation Res.* 92: 510-520, 1982.

4. J.-L. Schwartz, D. E. House, and G. A. R. Mealing, "Exposure of frog hearts to CW or amplitude-modulated VHF fields: Selective efflux of calcium ions at 16 Hz," *Bioelectromagnetics* 11: 349-358, 1990.

5. We are grateful for technical assistance from Pam Liakakos, Lisa Harris, and Zinta Bruns, and to the Australian Radiation Laboratory for providing liquid scintillation counting facilities.

Martin Blank, Ed., *Electricity and Magnetism in Biology and Medicine*

K^+ UPTAKE OF HeLa CELLS EXPOSED TO STEADY AND TIME-VARYING STRONG MAGNETIC FIELDS

H. Yamaguchi,* T. Ikehara, H. Miyamoto, Y. Kinouchi, and T. Tasaka

University of Tokushima, Japan

Rb^+ influx into HeLa cells was unaffected by exposures to DC magnetic fields of 0.5-1.7 T at a constant temperature (31.0, 37.2, or 42.5 C) and to that of 2 T at various temperatures from 10 to 45 C. In contrast, exposures to magnetic fields changing between 0.07 and 1.77 T inhibited active Rb^+ influx, provided that durations of exposures to the two magnetic flux densities were less than 10 s. The inhibition was not related to cellular energy metabolism, but possibly to an increase in negative charge on the cell surface caused by the exposures, as demonstrated by the fluorescent probes.

Membrane macromolecules are separated and orient themselves when exposed to strong DC magnetic fields, due to the anisotropy of their diamagnetic susceptibilities.[1] The exposure might influence membrane transport. However, the effect of the exposure on cation transport has not been studied as deeply.[2,3] In particular, the effects of strong time-varying magnetic fields are not as well known as the effects of ELF electromagnetic fields.[4-6] In this paper, we report on the effects of strong DC and time-varying magnetic fields on $K^+(Rb^+)$ influx into HeLa cells, which may prove useful for assessing the effects on the functions of human cells of frequent exposures to a strong magnetic field.

MATERIALS AND METHODS

HeLa S3 cells were cultured in a modified minimum essential medium[7] supplemented by 10 (v/v)% calf serum in plastic culture dishes 3.5 cm in diameter. When the cultures entered the growth phase, the culture medium was replaced with a similar medium containing 25 mM Hepes buffer (pH 7.2) and 5 mM Rb^+ instead of 5 mM K^+. The culture dishes were then placed in special copper incubators to keep the temperature of the cultures constant during the experiments on Rb^+ influx. One of the incubators was set in the gap between two polepieces (polar faces 5 cm in diameter) of an electromagnet; the other was placed outside the magnet as a control. Magnetic fields changing between 1.77 and 0.07 T were produced by switching the coil current on and off. Durations of the switching on and off time were kept equal for simplicity. Rb^+ influx was assayed for 15 min except where noted, since the influx takes place at the initial rate for at least 20 min.[8] Ouabain-sensitive (active) Rb^+ influx was estimated by subtraction of the influx in the presence of 10 μM ouabain (i.e., ouabain-insensitive, passive influx) from total Rb^+ influx. ATP was assayed by a luciferin-luciferase system, and emissions of cells loaded with fluorescent probes were measured by microfluorometry.

RESULTS AND DISCUSSION

There were no significant effects on Rb^+ influx of exposures to DC magnetic fields of 0.5-1.7 T at a constant temperature of 31.0, 37.2, or 42.5 C and to that of 2 T at various temperatures between 10 and 45 C. Arrhenius plots of these data revealed that the activation energies of control and exposed cells were 60.2 and 52.7 for ouabain-sensitive influx, and 55.7 and 51.4 kJ/mol for insensitive influx, respectively. These differences were insignificant. Therefore, DC magnetic fields up to 2 T do not significantly influence Rb^+ influx and any critical point, which suggests that the phase transition of the cell membrane is not found below 37 C.

Next, we tested the effects of exposure to time-varying magnetic fields. When the on- and off-times were longer than 30 s, there were no effects of exposures to the fields on both ouabain-sensitive and insensitive Rb^+ influxes. When the durations were less than 10 s, ouabain-sensitive Rb^+ influx was significantly inhibited, but insensitive influx was unaffected (Fig. 1). Only when the durations were 5 s was the effect

*Correspondence to: H. Yamaguchi (Physiology), School of Medicine, University of Tokushima, Tokushima 770 Japan.

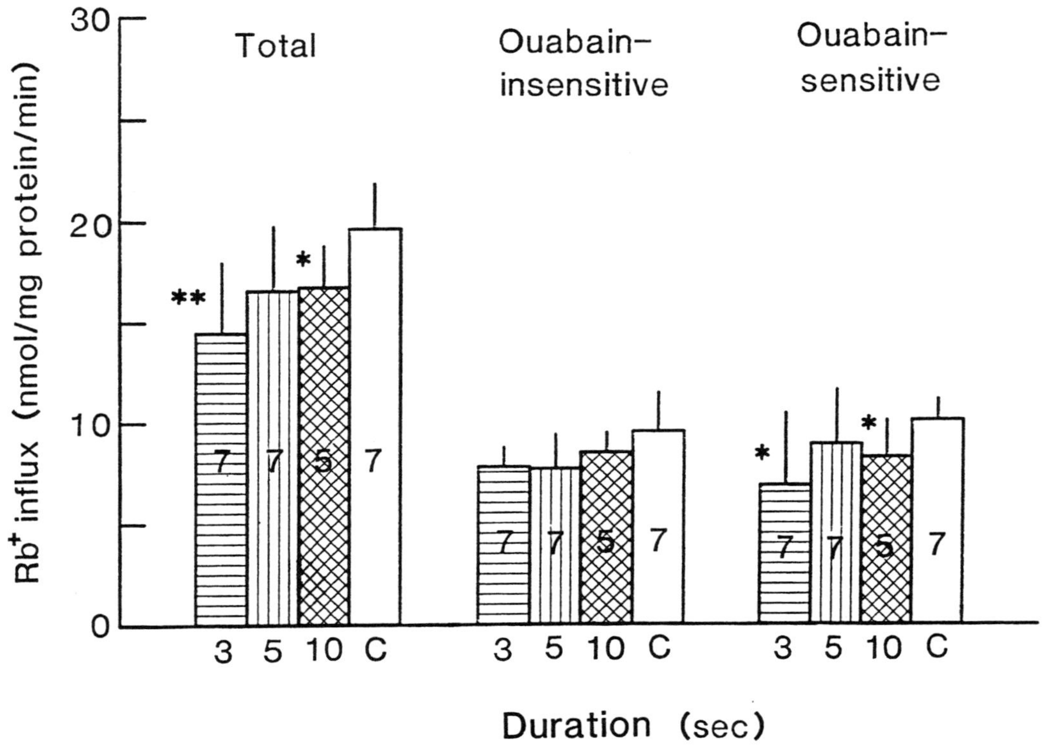

FIG. 1.—Effects of Rb^+ influx of intermittent magnetic fields with on and off time of 3, 5, and 10 s. Magnetic flux densities were 1.77 T at 'on' and 0.07 T at 'off' time. Numerals are the numbers of cases; asterisk and double asterisk represent values significantly different from controls at $p < 0.05$ and 0.01 (Dunnett *t*-test), respectively.

insignificant due to the large SD of the means. The inhibitory effect was enhanced with increase in the time of exposure (Table 1).

TABLE 1.—Effects on Rb^+ accumulation and the ATP contents in HeLa cells of exposure for various periods to intermittent magnetic field with on and off time of 3 s.

Time (min)	Ouabain (μM)	Rb^+ content (mmol/mg prot.) Unexposed	Rb^+ content (mmol/mg prot.) Exposed	ATP conent (nmol/mg prot.) Unexposed	ATP conent (nmol/mg prot.) Exposed
0	0	0		18.6 ± 1.9	
30	0	0.62 ± 0.08	0.50 ± 0.04*	24.8 ± 1.6	24.2 ± 4.0
30	10	0.33 ± 0.04	0.33 ± 0.01	23.8 ± 2.0	23.5 ± 1.4
120	0	1.09 ± 0.12	0.76 ± 0.06**	28.4 ± 2.7	26.4 ± 0.3
120	10	0.38 ± 0.00	0.32 ± 0.01	27.5 ± 3.0	23.4 ± 2.6

* and ** represent values significantly different from unexposed cells at $p < 0.05$ and 0.01 (Dunnett *t*-test), respectively.

The cellular ATP content was not changed by exposure to the intermittent magnetic field (Table 1), which shows that the inhibition of the Na^+ pump is irrelevant to any effect of the exposure on cellular energy metabolism.

The effects of the intermittent magnetic field were also tested by measurement of the emissions of cells loaded with the fluorescent probes of membrane potential, diS-C_3-(5)[8] and of pH, 4-heptadecyl-7-hydroxycoumarin.[9] Results clearly demonstrated an increase in negative charge on the cell surface.

The Na^+,K^+-ATPase activity has been reported to decrease under AC current,[10] and exposure to a pulsed magnetic field increases negative charge on the cell surface.[11] Therefore, the change in charge on the surface of HeLa cells may be closely related to the inhibition of the Na^+ pump. However, further studies are needed to confirm whether eddy current directly or increase in the surface negativity inhibits the Na^+ pump.[12]

REFERENCES

1. G. Maret and K. Dransfeld, *Top. Appl. Phys.* 57: 143-204, 1985.
2. T. Gualtierotti, *Physiologist* 7: 150, 1964.
3. T. Gualtierotti and V. Capraro, *Life Sci. Space Res.* 2: 311-316, 1963.
4. C. F. Blackman et al., *Bioelectromagnetics* 10: 35-43, 1989.
5. C. S. Collis and M. B. Segal, *J. Appl. Physiol.* 65: 124-130, 1988.
6. A. P. Stevenson and R. A. Tobey, *Bioelectromagnetics* 6: 189-198, 1985.
7. H. Miyamoto et al., in D. M. Prescott, Ed., *Methods in Cell Biology*, New York: Academic Press, 1976, vol. 13, pp. 15-27.
8. S. H. Wright et al., *Biochim. Biophys. Acta* 640: 767-778, 1981.
9. R. Pal et al., *Biochim. Biophys. Acta* 729: 185-192, 1983.
10. M. Blank and L. Soo, *Bioelectrochem. Bioenerg.* 22: 313-322, 1989.
11. O. M. Smith et al., *Bioelectromagnetics* 12: 197-202, 1991.

12. Study supported in part by grant-in-aid 03202316 for Scientific Research on Priority Areas from the Japanese Ministry of Education, Science, and Culture.

Martin Blank, Ed., *Electricity and Magnetism in Biology and Medicine*

FREQUENCY RESPONSE AND RESONANCE BEHAVIOR OF ION CHANNEL CURRENTS MODULATED BY AC FIELDS

J. Sandblom*

University of Göteborg, Sweden

P. George

University of New South Wales, Australia

The frequency responses of ion channel currents have been calculated from the equations describing the rates of transport of ions through channels represented by a pair of entrance barriers in series with an interior diffusion regime. The responses to combinations of applied DC and AC electric fields have been calculated and are represented in terms of complex admittances, from which the system is shown to behave as a double reactance network, leading in some cases to resonance effects. A further interesting property is the limiting slope of the current amplitude at high frequencies, which is found to exhibit a $f^{-3/2}$ dependence for small entrance barriers and a f^{-1} dependence for large entrance barriers.

Experimental evidence suggests that electric and magnetic fields interact with cellular transport mechanisms; particular attention has been focused on the effects on calcium exchange. Since ion channels form the primary site of ion transport across cell membranes, it is of interest to examine how fields may interact with either the channel structure or the ion transport mechanism. The response of ion diffusion to alternating fields has been previously characterized theoretically by Sandblom et al. for a homogeneous regime with fixed[1] or mobile[2] boundary concentrations and for a multicompartment system by Blank et al.[3] Sigworth has measured the frequency response of transport through channels by means of fluctuation analysis of voltage-clamped membranes and found deviations from the shot noise expected for a single-barrier crossing.[4] In order to incorporate these results in a general treatment of time-dependent currents in ion channels, we have introduced a pair of entrance barriers, one at each end of the channel, and a series of smaller barriers spanning the interior, which we treat as a homogeneous medium.

THEORY

The transport of ions across the interior set of small barriers is approximated by the Nernst-Planck (NP) electrodiffusion equations, similar to the steady-state treatment developed by Levitt.[5] In applying such an approach to time-dependent currents, a term for the displacement current is added. The barrier crossings are described by Eyring rate equations; the time-dependent NP equations are used for the diffusion in the channel interior.

This set of equations is solved for the case of a small AC field superimposed on the applied DC field (small-signal analysis). It is assumed that the DC field is uniform in the membrane and that a negligible part of the field lies across the entrance barriers. For the sake of simplicity it is also assumed that the density of ions is very small, which makes the admittance proportional to the concentrations.

ADMITTANCES

The solution to the equations can be written on the form

$$Y(j\omega) = j\omega C + C(j\omega)$$

---

*Correspondence to: J. Sandblom (Medical Physics), University of Göteborg, Medicinargatan 11, S-41390 Göteborg, Sweden.

where $Y(j\omega)$ is the complex admittance, C the total channel capacitance, ω the angular frequency, and $G(j\omega)$ the nonconservative part of the admittance. The last term is sometimes referred to as a dissipative admittance, to emphasize that the origin of the frequency dependence is a dissipative diffusion process. Figure 1 shows an example of the behavior of $G(j\omega)$ for three different barrier heights, where the heights are expressed by the term $K = kd^2/D$; k is the rate constant for leaving the channel and is normalized with respect to $\omega_D = D/d^2$, where D is the diffusion coefficient in the channel and d the channel length. It is seen that the channel admittance consists of two components. One component turns out to be determined by the time constant of the internal profile rearrangement; the other, by the time constant of loading and unloading ions in and out of the channel.

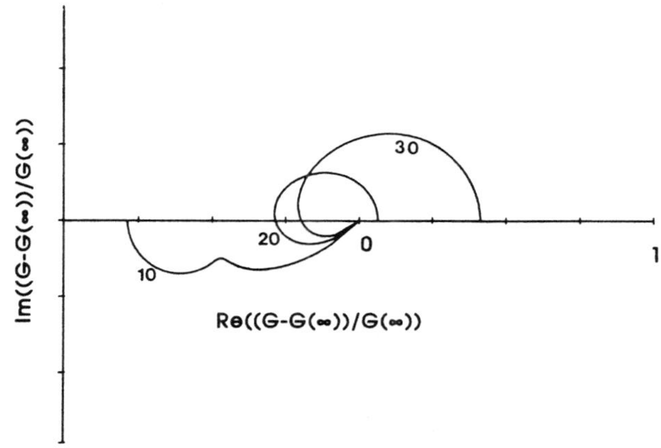

FIG. 1.—Admittance loci for three different values of K. Applied membrane potential and the equilibrium potential U_0 (in units of RT/F) are 4 and −4, respectively.

The presence of two separate time processes can give rise to resonance behavior (i.e., when one of the time processes behaves like a capacitance and the other like an inductance). Figure 2 shows an example of such a case, where the real part of $G(j\omega)$ is plotted as a function of frequency. The resonance frequency in this case is found to be roughly proportional to the square of the DC field strength.

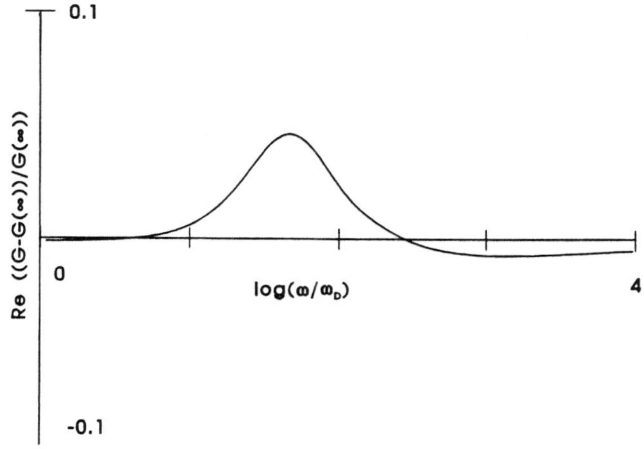

FIG. 2.—Real part of admittance plotted against log of angular frequency (normalized with respect to $\omega_D = D/d^2$) for K = 130, U = 12, and $U_0 = -8$.

FREQUENCY SPECTRUM

A further interesting finding is the behavior of the response at high frequencies. Figure 3 shows the behavior for two extreme cases of barrier heights. The limiting slope for the high-barrier (HB) case is seen to follow a f^{-1} dependence, which is typical of that for a pure depletion layer (i.e., mobile boundary concentrations).[2] The limiting slope for the low-barrier (LB) case is seen to follow a $f^{-3/2}$ dependence, which is therefore characteristic of fixed boundary concentrations. Figure 3 also includes the limiting slope expected for a purely conservative admittance with a f^{-2} dependence. The limiting slope dependence is important since, according to the Nyquist theorem, it predicts the power spectrum derived from noise measurements of single-channel currents at equilibrium.[6]

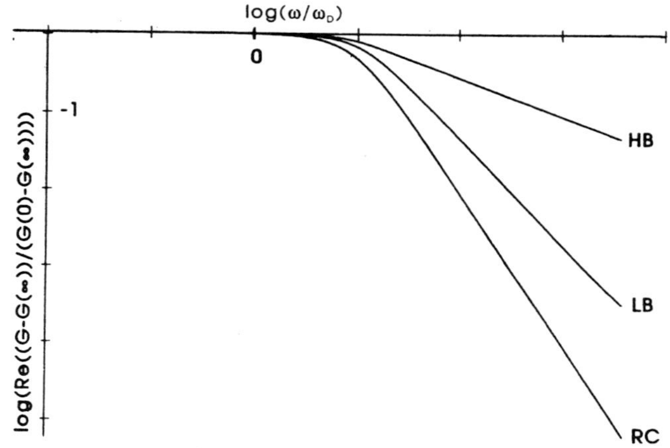

FIG. 3.—Log of real part of admittance plotted against log of normalized angular frequency for an equilibrium case ($U = U_0 = 2$), with curves calculated for high barriers ($K << 1$) indicated by HB and for low barriers ($K >> 1$) by LB. The curve labeled RC has been calculated from pure resistance capacitance network.

REFERENCES

1. J. Hägglund and J. Sandblom, "The kinetic behavior of the potassium channel in nerve membrane: A single ion electrodiffusion process," *T.I.T. J. Life Sciences* 2: 107-119, 1972.
2. J. Sandblom, J. Walker, and G. Eisenman, "The transient response and impedance locus of a mobile site membrane," *Biophys. J.* 12: 587-596, 1972.
3. M. Blank, "Theory of frequency-dependent ion concentration changes in oscillating electric fields," *J. Electrochem. Soc.* 134: 1112-1117, 1987.
4. F. Sigworth, D. Urry, and K. Prasad, "Open-channel noise: III. High-resolution recordings show rapid current fluctuations in Gramicidin A and four chemical analogues," *Biophys. J.* 52: 1055-1065, 1987.
5. D. Levitt, "Exact continuum solution for a channel that can be occupied by two ions," *Biophys. J.* 52: 455-467, 1987.

6. This work was supported by Grant 4138 from the Swedish Medical Research Council.

Martin Blank, Ed., *Electricity and Magnetism in Biology and Medicine*

MAGNETIC FIELD EFFECTS OF LECTIN BINDING TO ERYTHROCYTE MEMBRANES

L. L. Traikov, M. S. Markov,* M. A. Kuzmanova, and S. P. Ivanov

Sofia University, Bulgaria

The experiments reported in this paper concern changes in the impedance of the human erythrocyte membrane (isolated from blood prepared for transfusion) when a constant magnetic field in the range 0-90 mT was applied. The results show that phytochemagglutinin binding to the red blood cell membrane is significantly modulated by the magnetic field. The observed changes appear to depend on the period of blood conservation.

The effects of electromagnetic fields (EMF) on lectin binding to biological membranes have been studied intensively in the past decade. It has been suggested that the cell membrane is the primary target of the magnetic field.[1] Lectins, as external "messengers," are assumed to interact with membrane receptors via reversible binding processes. The magnetic field is thus capable of shifting the probability for binding to phytohemagglutinin and other lectins to various membrane components. A theoretical model, which originally focused on the interactions between lectin, receptors, and magnetic fields, has been proposed.[2] Several studies were done on phytohemagglutinin binding to lymphocytes. Immediately after lectin binds the membrane, calcium ions enter the cell, so that lymphocyte and erythrocyte stimulation by lectins appears to be a good experimental model by which to test the hypothesis for calcium involvement in electromagnetic-field effects on living systems.

The experiments reported in this paper are based on theoretical considerations, as well as on the results obtained in our previous studies of the changes in impedance parameters of human erythrocyte membranes under constant magnetic-field exposure.[3-5] The investigation of membrane electrical parameters may elucidate changes in protein and lipid domain distribution, as well as conformational changes in protein-protein and lipid-protein associates. There is no detailed information about temporal changes in the protein composition of peripheral membrane proteins located at the external cell surface of erythrocytes. However, studies of electromagnetic-field effects were focused on identifying and characterizing membrane transport alterations. The major question—which physical or physico-chemical paramaters of the membrane are modified by the electromagnetic field—remain unanswered.

Bioeffects caused by EMF have been observed to vary according to blood type, a result that suggests that the magnetic field probably affects specific antigen determinants situated on the membrane surface. Various lectins have been used in this project to permit evaluation of the specificity of their binding, as well as their ability to modify the activity of antigen determinants.

MATERIALS AND METHODS

Blood taken from healthy volunteers and prepared for blood transfusion was used in the experiments. In each container 300 ml of blood was kept in presence of 75 ml of anticoagulant solution L-12. For each experiment 50 ml of this solution was taken out of the container. After the buffy coat was removed, the blood cells were washed twice in buffered saline solution by centrifugation at 2000 G for 3 min. The relative cell concentration under this preparation was 0.70 ± 0.02. To get a different cell concentration, the preparation was diluted with buffered saline solution (Fig. 1). Phytohemagglutinin from two companies (DIFCO and SIGMA) was used in concentrations of 2-15 μl lectin (SIGMA) in 1ml cell suspension and 5-90 μl lectin (DIFCO) in 1ml cell suspension. It was found that the PHA supplied by DIFCO is less than half as effective as the PHA from SIGMA. Both lectins are blood unspecific. The lectin-binding assay employed in this work was based on incubation of PHA with erythrocyte suspension (up to 1 h) and subsequent impedance

*Correspondence to: M. S. Markov (Box 1188), Mt. Sinai Medical Center, 1 G. L. Levy Place, New York NY 10029-6574.

measurements. Magnetic fields in the 5-90mT range were applied by the MS-1 magnetic-field source described previously.[5] The pulse impedance method proposed by Pliquett and further developed in our laboratory was used as a screening method.[4-6] Changes of impedance parameters related to membrane capacitance and conductance were studied. In each run four different exposure conditions were under investigation: control, PHA-incubated only, constant magnetic field (CMF)-exposed only, and PHA and CMF exposed group.

RESULTS AND DISCUSSION

The thorough investigation of the efficiency of the applied constant magnetic field in the range of 0-90 mT on lectin binding shows that a 45mT field at 30min incubation time leads to better expressed effects, especially with regard to preventing agglutination.

It was found that the impedance parameters change significantly during the entire period of blood conservation. This outcome has been reported previously with regard to the protecting role of CMF against blood hemolysis[5,6] and was confirmed in the present study. In the first week of blood conservation, a large variation in the impedance parameters was found and led to serious difficulties in the comparison of results obtained at various terms of conservation. The observed effects of both PHA and PHA+CMF treatment are lower at days 8 and 20 than on day 14.

An experiment was designed to compare PHA effects on membrane impedance with the effects of another factor modifying surface charge, protamine. It was found that both chemicals affect membrane impedance, and that the constant magnetic field significantly decreases the effect of the chemical treatment. Moreover, it was shown in both experiments that the protecting role of the magnetic field is stronger if the physical factor is applied prior to the chemical one. The reason should be sought in a possible stabilization of both membrane structure and double electrical layer under CMF exposure.

The results also show that the agglutination of erythrocytes depends strongly on lectin concentration, as well as on the presence or absence of the magnetic field. The most important observation was that in the absence of a magnetic field, the erythrocyte agglutination increases by 43% compared to controls, whereas at any one of the applied magnetic fields, a decrease of agglutination was observed (up to 12%). The inhibition of agglutination was strongest when a magnetic field of 45 mT was applied for 30 min (Fig. 2).

It has been reported that microwaves cause an approximately twofold protein release compared with untreated erythrocytes. The amount of released proteins is only about 1% of the total protein content in erythrocytes. This small (in comparison with the total protein content) change may be a significant part of the membrane proteins, which is estimated to be about 850 fg. An ideal approach is to expose erythrocyte ghosts, but the ghost preparation creates a completely different composition inside ghost cells, which should be reflected in membrane composition and architecture. Evidence for structural alteration involving protein release from erythrocyte and lymphocyte membranes has been reported.[7,8]

To search for the reason for the observed effects, interactions of PHA with lipid-protein domains in the erythrocyte membrane should be considered. The structure of these domains is based on glycoprotein complexes, mainly of glycophorin and band-3 protein. These domains are relevant to the coupling of membrane to electromagnetic field. Even though the glycophorin and band-3 protein are structurally quite different species, both are discussed with respect to their ability to form membrane domains. Moreover, glycophorin in the human red blood cells is recognized to be a binding site for PHA[9,2,1]). Glycophorin is dispersed fairly diffusely in lipid bilayer membranes and shows little tendency for bilayer disruption.

There are several possible mechanisms by which lectins with such a modest affinity for monosaccharides might bind with very high probability to surface structure. We must take into account that a given lectin might have a high affinity for a chain of specific sugars in an appropriate linkage. It appears that these membrane domains may extrude membrane proteins, so that polar amino-acid side chains may increase when exposed to a magnetic field. At a membrane surface, all sugar-binding sites of lectin should tend to be occupied where structurally possible. Lectin binding via all four sites may often be structurally feasible, since glycoprotein headgroups are lengthy, close together, and very flexible. It appears that receptor rearrangement is an obligatory first step in the binding process.[9] This feature may significantly alter the cell surface characteristics, especially general and regional dielectric properties. It has been suggested that under microwave action proteins that were more hydrophobic and nonpolar and that carried a net cation charge were released to a greater extent than sham-treated cells. Whatever the relative importance of these various mechanisms is for any given lectin-cell interaction, it seems very likely that inter- and intramolecular headgroup rearrangements will play a key role.

It appears that electromagnetic fields can affect membrane proteins by disruption of cation (calcium and magnesium) bridges that serve to stabilize membrane structure.[8] During a phase/structural transition, striking physical changes occur in the erythrocyte membrane that involve the formation of protein/lipid conformational domains,[7,10,11] and these changes may be considered as a response of membrane structure to the applied physical factor.

The main goal of the project is to investigate whether constant magnetic fields or other ranges of electromagnetic fields can somehow modify the antigen properties of the glycoprotein complex. It thus appears that antigen determinants play an important role in the specific reaction of an erythrocyte to applied magnetic fields. Further tests of this hypothesis are going on, with the use of different bloodgroup-specific lectins.

REFERENCES

1. M. Grattarola, A. Chiabrera, G. Bonanno, R. Viviani, and A. Raveane, in A. Chiabrera, C. Nicolini, and H. P. Schwan, Eds., *Interactions Between Electromagnetic Fields and Cells*, New York: Plenum Press, 1985, 401-422.
2. A. Chiabrera, M. Grattarola, R. Viviani, and C. Braccini, *Studia Biophysica* 91: 125-131, 1982.
3. M. S. Markov, in M. Bender, Ed., *Interfacial Phenomena in Biological Systems*, New York: Marcel Dekker, 1991, 171-191.
4. M. S. Markov and F. Pliquett, *Bioelectrochemistry and Bioenergetics* 14: 495-502, 1985.
5. M. S. Markov, F. Pliquett, M. R. Kantcheva, and D. L. Abadjieva, *International Agrophysics* 1: 345-355, 1985.
6. M. S. Markov and M. R. Kantcheva, *Studia Biophysica* 90: 55-56, 1982.
7. R. P. Liburdy and P. F. Vanek, *Radiat. Res.* 109: 382-395, 1987.
8. R. P. Liburdy, A. W. Rowe, and P. F. Vanek, *Radiat. Res.* 114: 500-514, 1988.
9. N. V. Ketis and C. W. M. Grant, *BBA* 730: 359-368, 1983.
10. N. B. He and S. W. Hui, *Proc. Natl. Acad. Sci. USA* 82: 7304-7308, 1985.
11. R. P. Liburdy and P. F. Vanek, *Radiat. Res.* 102: 190-205, 1985.

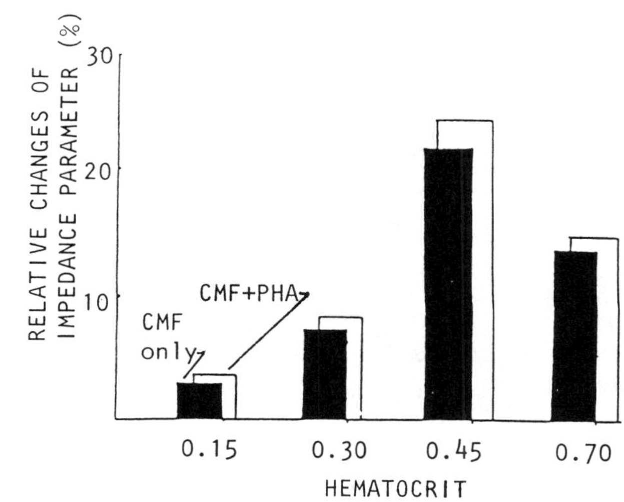

FIG. 1.—Changes in impedance parameter Δh under constant magnetic field action only, andunder combined action of magnetic field and phytohemagglutinin.

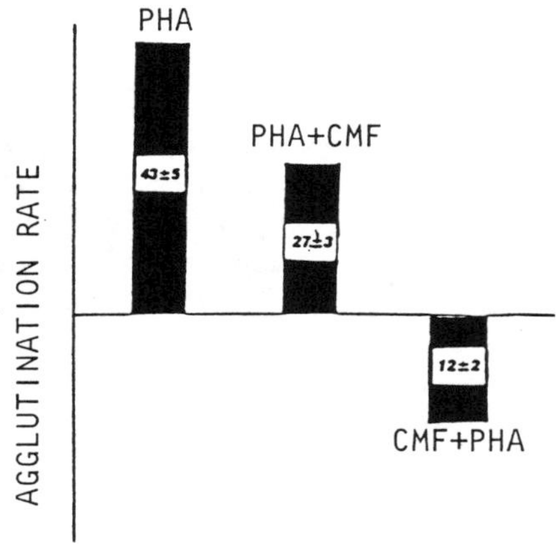

FIG. 2.—Effects of physical and chemical factors on agglutination rate. CMF significantly reduces effect of PHA, and efficiency is stronger when CMF is applied prior to PHA incubation. B = 75 mT, t = 30 min.

Martin Blank, Ed., *Electricity and Magnetism in Biology and Medicine*

ENHANCED TRANSCRIPTION OF NEUROSPORA RIBOSOMAL RNA GENES INDUCED BY 915MHz CW MICROWAVE RADIATION

M. Verma

Georgetown University, Washington, D.C.

S. K. Dutta*

Howard University, Washington, D.C.

Exposure of *Neurospora crassa* mycelial cells to 915MHz CW radiation resulted in enhanced transcription. The maximum enhancement was observed at 0.1 W/kg SAR (specific absorption rate). The RNA synthesis was determined by (i) measurement of the incorporation of ^{3}H-uridine as tricholoacetic acid precipitable material, and (ii) agarose gel electrophoresis followed by determination of the incorporated radioactivity in the specific gene transcript (17S and 26S rRNA genes) by cutting the gel and counting the slices in a liquid scintillation counter. These observations show enhancement of rRNA genes coded by RNA polymerase I as a result of exposure to low-level microwave radiation.

Several workers have shown that electromagnetic radiation (EMR) exposure of a biological system elicits a signal that leads to the activation of a variety of cellular responses,[1,2] including transcription of genes. If a step in the biochemical cycle is perturbed, it results in a variety of effects, including enhancement of gene transcripts and their products. We were interested in studying the interaction of 915MHz EMR with eukaryotic genes at the transcription level. For our studies we selected the vital process of transcription because it is the first step in the regulation of genes. The 915MHz frequency was chosen because it is widely used for clinical and general applications, e.g., warming refrigerated whole blood, thawing frozen human organs, producing differential hyperthermia in connection with cancer treatment, and rapidly reversing a patient's hyperthermic state in open-heart surgery. Furthermore, in our previous studies we have demonstrated enhanced Ca^{++} efflux in neuroblastoma IMR32 cells as a result of exposure to 915MHz EMR;[3] other investigators have found similar EMR effects in the membrane-dependent signal transduction cascade.[4]

MATERIALS AND METHODS

Culture: Neurospora crassa wild-type strain 74A was obtained from Fungal Genetics Stock Center. *N. crassa* is a filamentous fungus that completes its life cycle in three stages: conidia, sprout, and mycelia. Our experiments were performed with cells in mycelial stage of growth. The conditions for the growth and maintenance of the organism have been described previously.[5]

Chemicals. ^{3}H-Uridine (specific activity 49.7 Ci/mmole) was obtained from New England Nuclear, Inc.; agarose, Tris, EDTA, magnesium sulfate, diethyl pyrocarbonate, protease, DNase, bromophenol blue, glycerol, biotin, and components of the minimal medium were the products of Sigma Chemicals, St. Louis.

Methods. For each experiment, about 1 million cells were used. For growth and exposure conditions, 30 C temperature was maintained because the organism grows maximally at this temperature. After the exposure to 915MHz CW for 30 min, cell extract was made for the assay of RNA polymerase I (the enzyme that initiates transcription of rRNA genes) by grinding with acid-washed sea sand in Tris-HCl (50 mM), at pH 7.5, with $MgCl_2$ (10 mM) and EDTA (0.1 mM) at 4 C. The suspension was centrifuged (at 10000 g for 10 min at 4 C) and the supernatant was used as the source of the enzyme.

RNA Polymerase I Assay. The method was adopted from Grummt et al.[6] About 100 ng of template DNA (nuclear DNA of *N. crassa*) was incubated in a 50μl assay mixture containing Tris-HCl (15 mM), pH 7.5, KCl (90 mM), EDTA (0.15 mM), $MgCl_2$ (7 mM), adenosine triphosphate, cytosine triphosphate, and

*Correspondence to: S. K. Dutta (Botany), Howard University, Washington DC 20059.

guanidine triphosphate (0.5 mM each), and 1 μCi of ^{3}H-labeled uridine triphosphate. The mixture was incubated at 30 C for 30 min and 50 μl of the reaction mixture was spotted on Whatman DEAE paper. The paper was washed three times with cold 95% ethanol (10 ml each time per filter piece) and once with 80% ethanol. These steps eliminate any unincorporated radioactivity. Filters were dried and counted as described.[6] One unit of RNA polymerase I is defined as the amount that catalyzes a rate of labeled nucleotide triphosphate of 1 nmole/h in a 30min assay.

^{3}H-Uridine Uptake. Trichloroacetic acid precipitated radioactivity from the control and exposed cells was determined according to Russell et al.[7]

The Exposure System. N. crassa cells were exposed in a Crawford cell (Instruments of Industry, Model BC-110) with 915MHz CW EMR produced by a Model HP 8604B Hewlett-Packard Signal Generator as described in our previous report.[3] No temperature increase was detected during the exposure, so that the effect studied was caused by the EMR exposure per se and was not due to heat.

RESULTS AND DISCUSSION

In order to determine the RNA polymerase I activity in exposed and control cells of *N. crassa,* the following experiments were conducted.

Batches of *N. crassa* cells containing equal numbers of cells were exposed to 915MHz CW radiation at SAR 0.1 W/kg for varying dose rates. To determine the does rate dependency, we selected a range from 0.0001 to 4 W/kg. Results exhibited a linear relationship between the dose rate and the activity of the enzyme (Fig. 1A) except at higher dose rates (1 and 4 W/kg), at which the enzyme activity exposed to control ratio was lower. The explanation may be the inactivation of the enzyme at higher dose rate and the linear relationship between the activity of the enzyme and the dose rate exists in a short range. The peak value was obtained at 0.1 W/kg, which is the rationale for using 0.1 W/kg SAR for the time course experiment described above.

Because the enhancement of RNA polymerase I activity should accompany the incorporation of the precursor (uridine for RNA), the next step was to determine the uptake of ^{3}H-uridine in cells during exposure to various power levels (Fig. 1B). Enhanced uptake of ^{3}H-uridine was observed in exposed samples compared with control samples. Initially, the increased uptake of ^{3}H-uridine was in linear relationship with the dose rate, and again the peak value of the incorporation of ^{3}H-uridine was attained at 0.1 W/kg, followed by decreased incorporation (at 1 and 4 W/kg). Experiments were also conducted to insure that the incorporation of ^{3}H-uridine was not caused by transport but by incorporation per se.

As the products of RNA polymerase I are 17S and 26S rRNA, we measured the levels of these two RNA species in the exposed and control samples. First, ^{3}H-labeled RNA was isolated from the exposed and control samples and then separated on agarose gel[8] by electrophoresis. The gel was sliced (1mm thickness) with a gel slicer (Fisher Chemicals) and counted in liquid scintillation counter as described.[7] As shown in Fig. 2, enhanced levels of 17S and 26S rRNA were observed in exposed samples. The identification of the 17S and 18S rRNA was done by hybridization with rDNA probes.

The main outcome of the present report is that low-level 915MHz CW EMR induces transcription at various SAR values.[9]

REFERENCES

1. R. Goodman, C. A. L. Bassett, and A. Henderson, "Pulsing electromagnetic fields induce cellular transcription," *Science* 22: 1283-1285, 1983.

2. S. K. Dutta, W. H. Nelson, C. F. Blackman, and D. J. Bruisck, "Cellular effects in microbial tester strains caused by exposure to microwaves or elevated temperatures," *J. Environment. Pathol. Toxicol.* 3: 195-206, 1980.

3. S. K. Dutta, A. Subramanian, B. Ghosh, and R. Prashad, "Microwave radiation-induced calcium-ion efflux from human neuroblastoma cells in culture," *Bioelectromagnetics* 5: 71-78, 1984.

4. C. A. L. Bassett, "Fundamental and practical aspects of therapeutic uses of pulsed electromagnetic fields (PEMFs)," *Clin. Rev. Biomed. Engg.* 17: 451-529, 1989.

5. S. K. Dutta, "DNA homologies among heterothallic species of *Neurospora,*" *Mycologica* 68: 388-401, 1976.

6. I. Grummt, A. Kuhn, I. Bartsch, and H. Rosenbauer, "A transcription terminator located upstream of the mouse rDNA initiation site affects rRNA synthesis," *Cell* 47: 901-911, 1986.

7. P. J. Russell, J. R. Hammett, and E. U. Selker, "*Neurospora crassa* cytoplasmic ribosome: Ribosomal ribonucleic acid synthesis in wild type," *J. Bacteriol.* 127: 785-793, 1976.

8. J. Sambrook, E. Fritsch, and T. Maniatis, *Molecular Cloning: A Laboratory Manual,* Cold Spring Harbor Laboratory, N.Y., 1989.

9. This work was partly supported by a grant from NIGMS-NIH (8S06GM08016) and by EPA (R814126-01-0), and by a core grant from the Graduate School of Howard University.

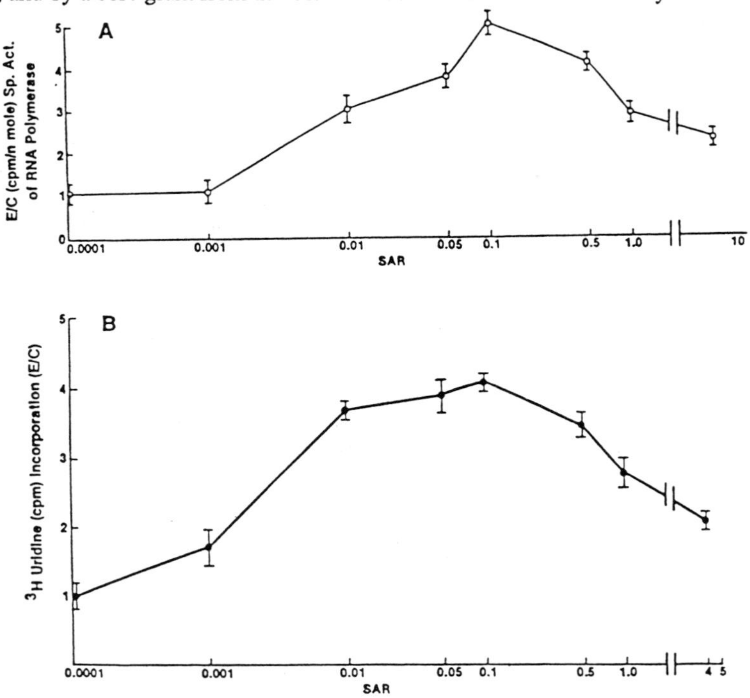

FIG. 1.—A. Dose-rate dependent induction of RNA polymerase I activity in *N. crassa;* peak of RNA polymerase I activity occurs at SAR of 0.1 W/kg. B. Uptake of ^{3}H-uridine (in counts per minute per nmole) in *N. crassa* cells under exposure to microwave radiation; peak of ^{3}H-uridine uptake occurs at SAR 0.1 of W/kg.

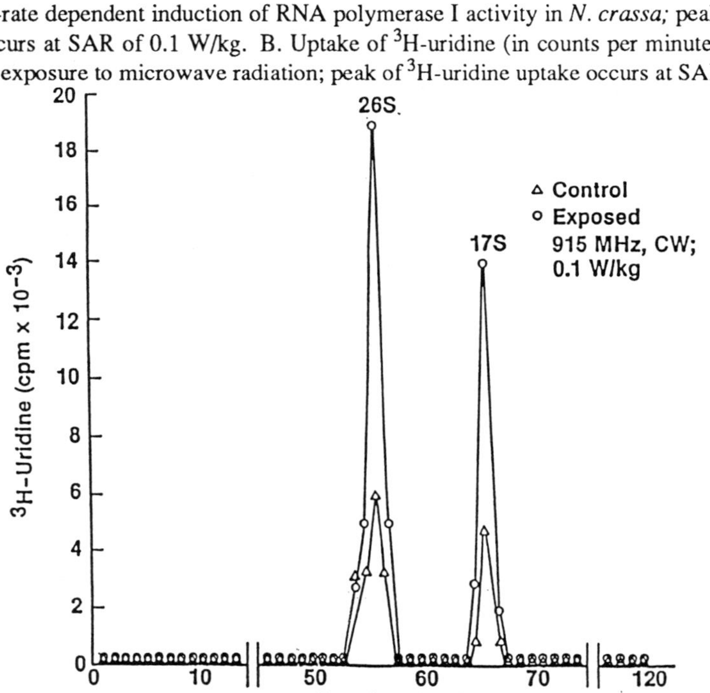

FIG. 2.—Agarose gel electrophoresis of RNA synthesized after exposure of *N. crassa* mycelial cells to microwave radiation at 915 MHz, CW, 0.1 W/kg. Gel was sliced and counts per minute were measured in each fraction (slice). Note that exposure enhanced amount of rRNA synthesis.

Martin Blank, Ed., *Electricity and Magnetism in Biology and Medicine*

EFFECTS OF 60Hz MAGNETIC FIELD EXPOSURE ON c-*fos* TRANSCRIPTION IN CCRF-CEM HUMAN T-LYMPHOBLASTOID CELLS

J. L. Phillips,* W. Haggren, W. J. Thomas, T. Ishida-Jones, and W. R. Adey

Pettis Memorial VA Medical Center, Loma Linda, Calif.

CCRF-CEM cells have been transfected with a series of *fos* promoter-CAT (chloramphenicol acetyl transferase) constructs and then exposed to a 1G, 60Hz sinusoidal magnetic field (MF). Data obtained have not indicated which promoter response elements are required for the MF-induced increase in c-*fos* transcription observed previously. Using an alternate approach, we have employed the mobility shift assay to study MF-induced effects on DNA-binding activity of specific transcription factors. Statistically significant decreases in AP-1 levels occurred in cells after 30 min of MF exposure; no changes were observed in CREB levels. In additional studies, either unchanged or increased SRF levels have been observed. These data are consistent with previously observed increased c-*fos* transcription and decreased c-*jun* transcription.

Among the many effects of MF exposure on biological systems is an effect on gene transcription. Goodman and colleagues reported in 1983 that exposure of dipteran salivary gland cells to either of two electromagnetic signals (72Hz single pulse and 15Hz pulse train) increased gene transcription as assessed by transcription autoradiography.[1] Phillips and McChesney[2] reported increased incorporation of ^{3}H-uridine into both total cellular RNA and mRNA in CCRF-CEM cells exposed to a 72Hz pulsed MF. These studies, while pointing to *general* changes in gene transcription in MF-exposed cells, do not address the important issue of which *specific* genes were affected. The first report of MF-induced changes in specific gene transcription has come from Phillips et al. who, using the nuclear run-off assay, observed a time-dependent change in the transcription of four genes (c-*fos*, c-*jun*, c-*myc*, and protein kinase C) in CCRF-CEM cells exposed to a 1G 60Hz sinusoidal MF.[3] We report here the results of studies formulated to determine the mechanism by which MF exposure increased c-*fos* transcription while decreasing c-*jun* transcription. Our goals have been two-fold: (1) to identify which response elements on the *fos* promoter are required for MF-induced changes in c-*fos* transcription, and (2) to identify changes in the levels of active transcription factors associated with key response elements in the c-*fos* promoter.

MATERIALS AND METHODS

CCRF-CEM cells, a T-lymphoblastoid cell line, were cultured in RPMI-1640 tissue culture medium containing 25 mM HEPES, 10% heat-inactivated fetal calf serum, and 6 mM glutamine. Cell cultures were set up at 5×10^5 cell/ml the night before experimentation. Cultures were always ≥ 95% viable, and were maintained in 3% CO_2/97% air at 37 C.

Cells (10^6 in 5 ml complete medium) were transfected via a liposome-mediated method by use of a commercially available reagent (DOTAP, Boehringer Mannheim) for 6 h according to manufacturer's instructions. Cells were then used for experimentation after a further 64 h incubation at 37 C. The *fos*CAT constructs used have been described by Fisch et al.[4] They were pFC700, containing a 710 bp sequence from the 5′-untranslated region of the *fos* gene (representing the complete *fos* promoter) linked to the reporter gene CAT; pFC225, similar to pFC700, but missing the sequence from −226 to −710 and thus lacking a serum response element (SRE) or AP-1 binding element; pFC60pm, missing only the calcium/cyclic AMP response element (CRE); and pFCΔ73/94, missing only the direct repeat (DR) region. In addition, all cell cultures were co-transfected with pSV2CAT, a constitutively expressed plasmid, to control for transfection efficiency.

Cells were exposed continuously to a 1G (rms) sinusoidal MF at 60 HZ as described previously[3] for times indicated in the text. Unexposed control cells were maintained in a similar incubator and treated

*Correspondence to: J. L. Phillips (Research Serv. 151), Pettis Mem. VA Med. Ctr., Loma Linda CA 92357.

identically to the exposed populations. Background 60Hz MFs in both incubators were ≤5 mG. After exposure, total cellular RNA was isolated as described.[3] S1 nuclease analysis was used to assess transcription of the *fos*CAT constructs and was performed according to Fisch et al.[5] Alternatively, nuclear extracts were prepared from exposed and control cells and mobility shift assays performed according to Fisch et al.[4] Oligonucleotides containing binding sites for AP-1, CREB (calcium/cyclic AMP response element binding protein), and SRF (serum response factor) transcription factors were obtained from Promega (Madison, Wis.), and were end-labeled with $\gamma[^{32}P]$-ATP and polynucleotide kinase. Results from S1 nuclease analyses and mobility shift assays were evaluated by image analysis of autoradiograms.

RESULTS

When CCRF-CEM cells were transfected with *fos*CAT constructs, exposed to the 1G MF for up to 120 min, and then total-cellular-RNA analyzed for expression of message containing *fos*CAT sequences (from the *fos* promoter/CAT constructs) or CAT sequences (from the control pSV2CAT plasmid), variable results (not shown) were obtained. No consistent pattern emerged that would allow us to conclude which of the response elements on the *fos* promoter were necessary for the MF-induced increase in *fos* gene transcription. (Possible reasons are discussed below.) Consequently, as an alternate approach, we have employed the mobility shift assay to look for changes in the DNA-binding activity of specific transcription factors. Figure 1 summarizes the results of nine separate experiments that measured the relative levels of two specific AP-1 complexes, as well as the CRE-binding protein CREB, in exposed cells as compared to unexposed control cells. It is seen that MF exposure induced a significant decrease in the levels of both AP-1 complexes, reaching a nadir after 30 min exposure. In contrast, MF-exposure produced no change in the level of CREB capable of binding to its recognition sequence. Figure 2 presents the results of four separate experiments in which SRF binding to the SRE was measured. In two experiments, SRF binding was unaffected by MF exposure; in two other experiments, SRF binding was increased substantially, peaking after 30 min of exposure.

DISCUSSION

Other investigators have transfected various cells with *fos* promoter/CAT constructs to investigate the essential nature of response elements in the pathways leading from chemical stimulation, but this approach has not worked well for us, for several reasons. First, our transfection efficiencies were low, perhaps not allowing us to measure changes in *fos*CAT transcription with sufficient sensitivity. Second, it has been shown that, since the *fos* gene contains 3′-intragenic regulatory sequences in addition to the 5′-promoter sequences, it is not always possible to use promoter-fusion genes to study rapid and dramatic changes in *fos* expression that occur in response to various signals.[6] Also, Fisch et al. have reported that the calcium ionophore A23187 may be able to increase *fos* expression only if the gene resides within a chromosomal context.[4]

However, our results from mobility shift assays do indicate changes in the levels of DNA-binding activity of specific transcription factors. The levels of two specific AP-1 complexes decreased significantly to 50% and 28% of control levels, respectively, after 30 min of MF exposure. These data are consistent with increased c-*fos* transcription and decreased c-*jun* transcription, as reported previously.[3] Since transcription of the *fos* gene is down-regulated by the *fos* protein, and since AP-1 consists of either *fos-jun* heterodimers or *jun-jun* homodimers, decreased AP-1 may indicate decreased *fos* protein and thus a removal of repression to *fos* gene transcription. At the same time, decreased AP-1 signifies decreased levels of *jun* protein; both are consistent with decreased c-*jun* transcription.

The results of our studies of SRF binding to the SRE are also consistent with other reports that detail the mechanism of mitogen-induced changes in c-*fos* transcription, even though they might appear to be another example of an MF-induced effect that sometimes occurs and at other times does not. It has been shown that the SRF has two distinct activities: (1) a DNA-binding activity and (2) a transcriptional-initiation stimulatory activity.[7] Consequently, depending on the cell type or the state of a given cell, the DNA-binding activity may be constitutively activated while the protein, already bound to the SRE, awaits activation of its transcriptional-initiation stimulatory function. Conversely, treatment of a cell with some agent may, through phosphorylation of the SRF, increase both its DNA-binding and transcriptional-initiation stimulatory functions. A similar situation has been reported for the CRE-binding protein CREB. Fisch et al. found no change in the amount of CREB in nuclear extracts from both HeLa and A431 cells that had been induced with epidermal growth factor, the tumor promoter TPA, or calcium ionophore A23187, although the activity of CREB to activate transcription was markedly increased.[4]

In summary, exposure of CCRF-CEM cells to a 1G sinusoidal MF at 60 Hz has been found to alter the DNA-binding activity of specific transcription factors recognized by response elements of the c-*fos* promoter. Moreover, the changes observed are consistent with our previous observation of MF-induced alterations in the transcription of both c-*fos* and c-*jun* genes. Additional studies to determine MF-induced effects on transcriptional activation functions of SRF and CREB are under way.[8]

REFERENCES

1. R. Goodman, C. A. L. Bassett, and A. S. Henderson, "Pulsing electromagnetic fields induce cellular transcription," *Science* 220: 128, 1983.

2. J. L. Phillips and L. McChesney, "Effect of 72 Hz pulsed magnetic field exposure on macromolecular synthesis in CCRF-CEM cells," *Cancer Biochem. Biophys.* 12: 1, 1991.

3. J. L. Phillips, W. Haggren, W. J. Thomas, T. Ishida-Jones, and W. R. Adey, "Magnetic field-induced changes in specific gene transcription," *Biochim. Biophys. Acta* 1132: 140, 1992.

4. T. M. Fisch, R. Prywes, and R. G. Roeder, "c-*fos* Sequences necessary for basal expression and induction by epidermal growth factor, 12-O-tetradecanoyl phorbol-13-acetate, and the calcium ionophore," *Mol. Cell. Biol.* 7: 3490, 1987.

5. T. M. Fisch, R. Prywes, M. C. Simon, and R. G. Roeder, "Multiple sequence elements in the c-*fos* promoter mediate induction by cAMP," *Genes Dev.* 3: 198, 1989.

6. K. Schilling, D. Luk, J. I. Morgan, and T. Curran, "Regulation of a *fos-lacZ* fusion gene: A paradigm for quantitative analysis of stimulus-transcription coupling," *Proc. Natl. Acad. Sci. USA* 88: 5665, 1991.

7. R. Prywes, A. Dutta, J. A. Cromlish, and R. G. Roeder, "Phosphorylation of serum response factor, a factor that binds to the serum response element of the c-*fos* enhancer," *Proc. Natl. Acad. Sci. USA* 85: 7206, 1988.

8. This work was supported by U.S.Department of Energy contract DE-AI01-90CE35035.

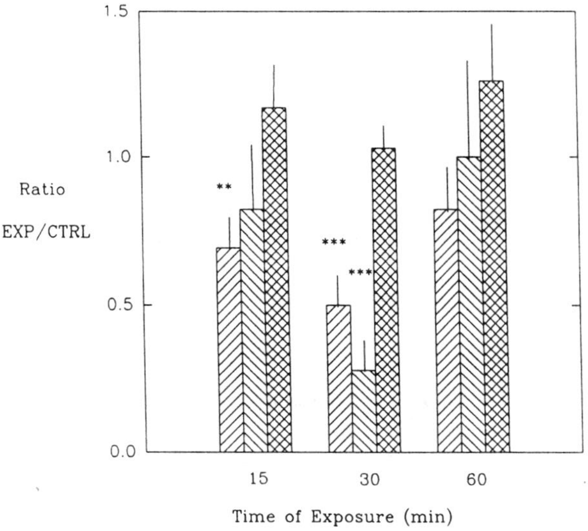

FIG. 1.—Effect of MF exposure on AP-1 and CREB DNA-binding activity. Results are expressed as ratio of activity in exposed to activity in control cell cultures, and are presented as mean value of 9 separate experiments. Vertical bars represent S.E.M. ☐, AP-1 binding protein #1; ☐, AP-1 binding protein #2; ☐, CREB. $^{**}p < 0.01$; $^{***}p < 0.001$.

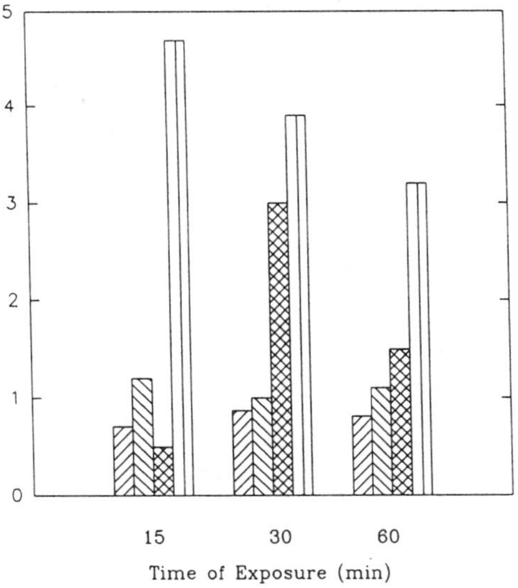

FIG. 2.—Effect of MF exposure on SRF DNA-binding activity. Results are expressed as ratio of activity in exposed to activity in control cell cultures, and are presented for 4 individual experiments.

Martin Blank, Ed., *Electricity and Magnetism in Biology and Medicine*

GENE EXPRESSION IN TUMOR CELL LINES AFTER EXPOSURE TO A 50Hz SINUSOIDAL MAGNETIC FIELD

M.-O. Mattsson* and U. Rehnholm

University of Umeå, Sweden

K. H. Mild

National Institute of Occupational Health, Umeå, Sweden

Three mammalian tumor cell lines (HL60, ELD, F9) have been used to assess the effects of a 30μT, 50Hz sinusoidal magnetic field on ornithine decarboxylase (ODC) gene expression. All these cells responded to exposure with elevated ODC enzyme activity. The increase in activity differed among cell types, ranging from about 20% in HL60 up to 5-6 times the control values in ELD-cells. Furthermore, the importance of factors as growth periods and media additives for effects obtained was corroborated.

Several recent reports have shown that weak ELF magnetic fields influence numerous cell biological processes. The observed effects include altered cell proliferation rates, changes in enzyme activity, and altered protein and mRNA synthesis. Epidemiological data, as well as some of the above-mentioned studies on the cellular level, indicate a possible role for magnetic-field exposure in promotion of certain forms of cancer, possibly by affecting regulation of proliferation (cf. Adey's recent review.[1]).

Previous studies by Byus et al.[2,3] concerning the activities of the proliferation-regulating enzyme ornithine decarboxylase (ODC) in mammalian cells during magnetic-field exposure have elicited considerable interest. The reason is that there seems to be a correlation between elevated ODC activity and cancer, as reported by Heby and Persson[4] and others. An effect on ODC activity from electromagnetic fields may thus be a mechanism behind a possible relationship between these fields and cancer. In the present study, we have set out to study the expression of the ODC gene on the transcriptional, translational, and post-translational levels. Various tumor cell lines have been used as models for exposure to a 50Hz sinusoidal magnetic field with a flux density of 30μT. The part of the study completed to date reveals an effect on the enzyme activity on magnetic-field exposure.

MATERIAL AND METHODS

Cell Culture

The following tumor cell lines have been used in the present study: human promyelocytic leukemia cell line HL60, mouse ascites tumor ELD cells, and mouse teratocarcinoma cell line F9.

HL60 cells were cultured as a suspension culture in RPMI 1640 medium containing 10% fetal calf serum (FCS) and antibiotics. ELD cells were cultured in a 1:1 mixture of DMEM and F-12 medium with addition of FCS and/or bovine serum albumin, as described in the Results section. The anchorage-dependent F9 cells were cultured in DMEM with 20% FCS added. All cells were incubated in a Forma Scientific model 3039 CO_2 incubator under controlled conditions with reference to temperature (37 C), CO_2 (5%), and relative humidity (95%).

Exposure

Cells were continuously exposed to a vertical 50Hz sinusoidal magnetic field with a calculated and measured flux density of 30μT. The exposure took place in three identical sets of serially connected Helmholtz coils (diameter 15 cm). The signal originated from a Wavetek signal generator and a broadband amplifier.

---

*Correspondence to: M.-O. Mattsson (Cellular and Developmental Biology), University of Umeå, S-901 87 Umeå, Sweden.

Continuous readings of the current in the coils (and thus the magnetic field inside the coils) was performed in all experiments by means of an oscilloscope and a printer. The time-span for exposure varied between a few hours to 3 days as outlined in Results.

Controls and exposed cells were placed in the same incubator on different shelves. The background 50Hz field in the control position during exposure was repeatedly measured and never did exceed 0.8μT. In the absence of exposure, the background field was 0.08μT or lower. The geomagnetic field was measured with a 3-D probe, and the field vector was determined to $56\pm1\,\mu$T at the locations in the incubator where the cells were placed.

Analysis of ODC-Activity

Enzyme activity in sonicated cell pellets was determined by measurement of liberated $^{14}CO_2$ in 100 μl supernatant incubated in the presence of ^{14}C-labeled ornithine and saturating levels of pyridoxal 5′-phosphate and L-ornithine.

RESULTS AND DISCUSSION

The studies performed so far indicate that all three cell lines investigated respond to exposure to a weak 50Hz field with an increase in ODC activity. The activity changes differ among cell types. The magnitude of elevation ranged from 20% up to 5-6 times the values in control cells. The effects are more prominent during earlier phases of growth in HL60 cells, whereas the largest observed effects in ELD cells were seen when cells were cultured for longer periods, e.g., when the inherent ODC activity is comparatively low.

When HL60 cells were cultured for 24, 48, or 72 h and subsequently exposed for 0.5-4 h, an increase in ODC activity of approximately 20% was seen in exposed cells as compared to controls. The most pronounced effect was recorded in cells cultured for 24 h before exposure.

ELD cells cultured for 24, 48, or 72 h and continuously exposed to the field exhibited elevated enzyme activity (Fig. 1). This effect was more pronounced after 72 h in culture. The importance of media composition in bringing these effects about can also be deduced from this figure. The data shown in Fig. 2 can be interpreted as showing that longer exposure periods caused more distinct effects, and also that exposure had more impact on the ODC activity in later periods of growth, when the inherent activity is lower.

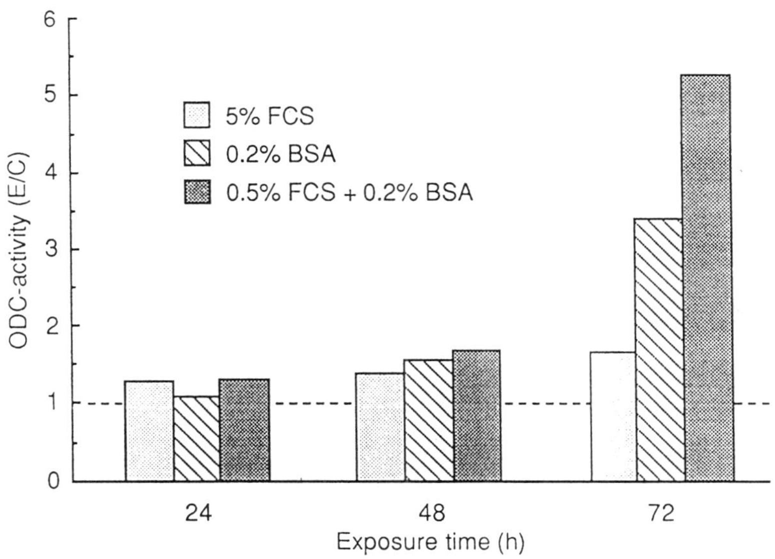

FIG. 1.—ODC activity (nmole CO_2/h/10^6 cells) in ELD cells expressed as the quotient between exposed cells and controls (E/C value). Culture of cells was done with various medium additives (FCS = fetal calf serum; BSA = bovine serum albumin). Values are means from three separate experiments.

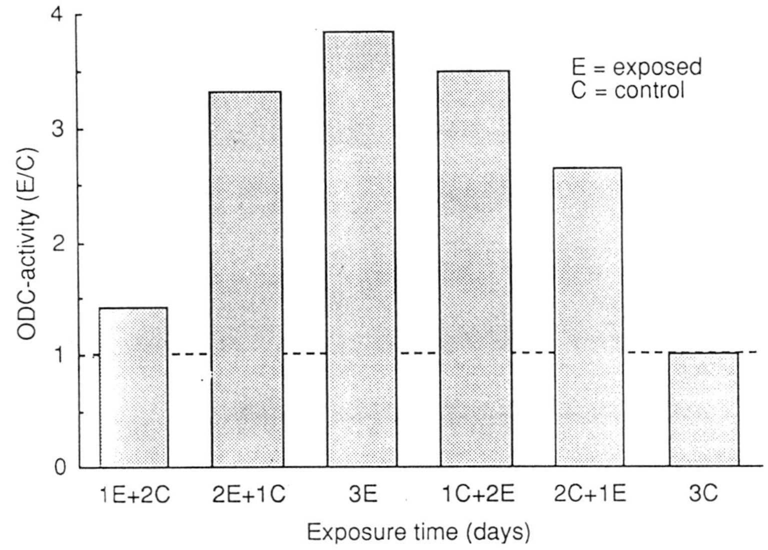

FIG. 2.—ODC activity (nmole CO_2/h/10^6 cells) in ELD cells cultured for 3 days and exposed (E) for one, two, or three of these days in different order. For instance, 2E + 1C means two days of exposure followed by one day of incubation without field (= cultured in control position). Values are means from two separate experiments.

The mouse teratocarcinoma cell line F9 has also been used for comparative purposes. The same tendency as in ELD cells can be seen (Table 1), e.g., the longer exposure (48 h) may bring about increased enzyme activity.

TABLE 1.—ODC activity in F9 cells exposed to a 50Hz, 30μT sinusoidal magnetic field expressed as E/C values. Values are given as mean ± standard deviation from three experiments.

Exposure	Mean	S.D.
24 h	0.9	0.1
48 h	2.6	0.8

These results strengthen earlier observations that ODC activity is sensitive to magnetic-field exposure.[3,5] Work is in progress to determine whether the altered enzyme activity is due to transcriptional or translational regulation. Although ODC activity is coupled to cell growth, we have not been able to see consistently any significant differences in growth between controls and exposed cells. Whether that refutes the notion of possible promoting effects of magnetic-field exposure or reflects effects on another cellular phenomenon remains to be seen.

REFERENCES

1. W. R. Adey, "Joint actions of environmental nonionizing electromagnetic fields and chemical pollution in cancer promotion," *Environm. Health Perspect.* 86: 297-305, 1990.

2. C. V. Byus, S. E. Pieper, and W. R. Adey, "The effects of low-energy 60-Hz environmental electromagnetic fields upon the growth-related enzyme ornithine decarboxylase," *Carcinogenesis* 8: 1385-1389, 1987.

3. C. V. Byus, K. Kartun, S. E. Pieper, and W. R. Adey, "Increased ornithine decarboxylase activity in cultured cells exposed to low energy modulated microwave fields and phorbol-ester tumor promoters," *Cancer Res.* 48: 4222-4226, 1988.

4. O. Heby and L. Persson, "Molecular genetics of polyamine synthesis in eukaryotic cells," *TIBS* 15: 153-158, 1990.

5. T. A. Litovitz, D. Krause, and J. M. Mullins, "Effect of coherence time of the applied magnetic field on ornithine decarboxylase activity," *Biochem. Biophys. Research Commun.* 178: 862-865, 1991.

Martin Blank, Ed., *Electricity and Magnetism in Biology and Medicine*

THE EFFECTS OF VARYING THE FREQUENCY AND MAGNETIC FIELD STRENGTH ON MORPHOLOGIC PHENOTYPE OF ELF-RESPONSIVE MODEL CELL LINE

J. T. Ning,* E. M. Czerska, J. Casamento, and M. L. Swicord

Center for Devices and Radiological Health, Rockville, Md.

H. Al-Barazi and C. C. Davis

University of Maryland, College Park

E. Elson

Walter Reed Army Institute of Research, Washington, D.C.

A model eukaryotic cell line with a marker gene under the control of a tentative ELF-responsive DNA element was constructed in order to facilitate the investigations of various exposure parameters (e.g., dose response and frequency response) that may be important in global characterization of the biological effects of ELF electromagnetic field interactions with living systems in vitro. Experiments using this model cell line which changes morphology as a result of exposure to ELF fields have yielded useful information on possible mechanisms of ELF interaction with living systems.

MATERIALS AND METHODS

The ELF responsive model cell line was constructed via transfection of primary adult rat Schwann cell isolated from sciatic nerve sheath and transfected with a plasmid containing SV-40 T antigen under the control of a metal regulatory element. An immortalized cell line was isolated. This cell line, which is driven by the SV-40 T antigen oncogene, expressed a round phenotype distinct from normal Schwann cell bipolar morphology with processes.

RESULTS

When this cell line is exposed to ELF in vitro, the phenotype changes after 48 h from the round cell shape (Fig. 1) with maximum cell density to bipolar morphology with the appearance of processes (Fig. 2) and significant decrease in cell density. This response is observed when this cell line is exposed to 60 Hz at 1 G, at 15Hz pulse trains and 10Hz bipolar pulse trains (Table 1). The onset of morphologic response is dose dependent. With increasing field strength at 60 Hz, the onset of morphologic change decreased from 6 days at 0.8 G to 1 day at 10 G (Table 2). Below 0.8 G, no morphologic response was observed after 12 days. When this ELF-responsive cell line is plated in circular tissue culture dishes and exposed to ELF, there is a gradient of morphologic changes from center to the periphery of the dish. There is no change at the center of the dish and maximal change at the periphery (Fig. 3), correlative with the gradient of the induced electric field. When the exposure system is changed so that there is a uniform magnetic field as well as an induced electric field, the pattern of morphologic change becomes randomly distributed throughout the flask (Fig. 4).

*Correspondence to: J. T. Ning, FDA-CDRH, Rockville MD 20857.

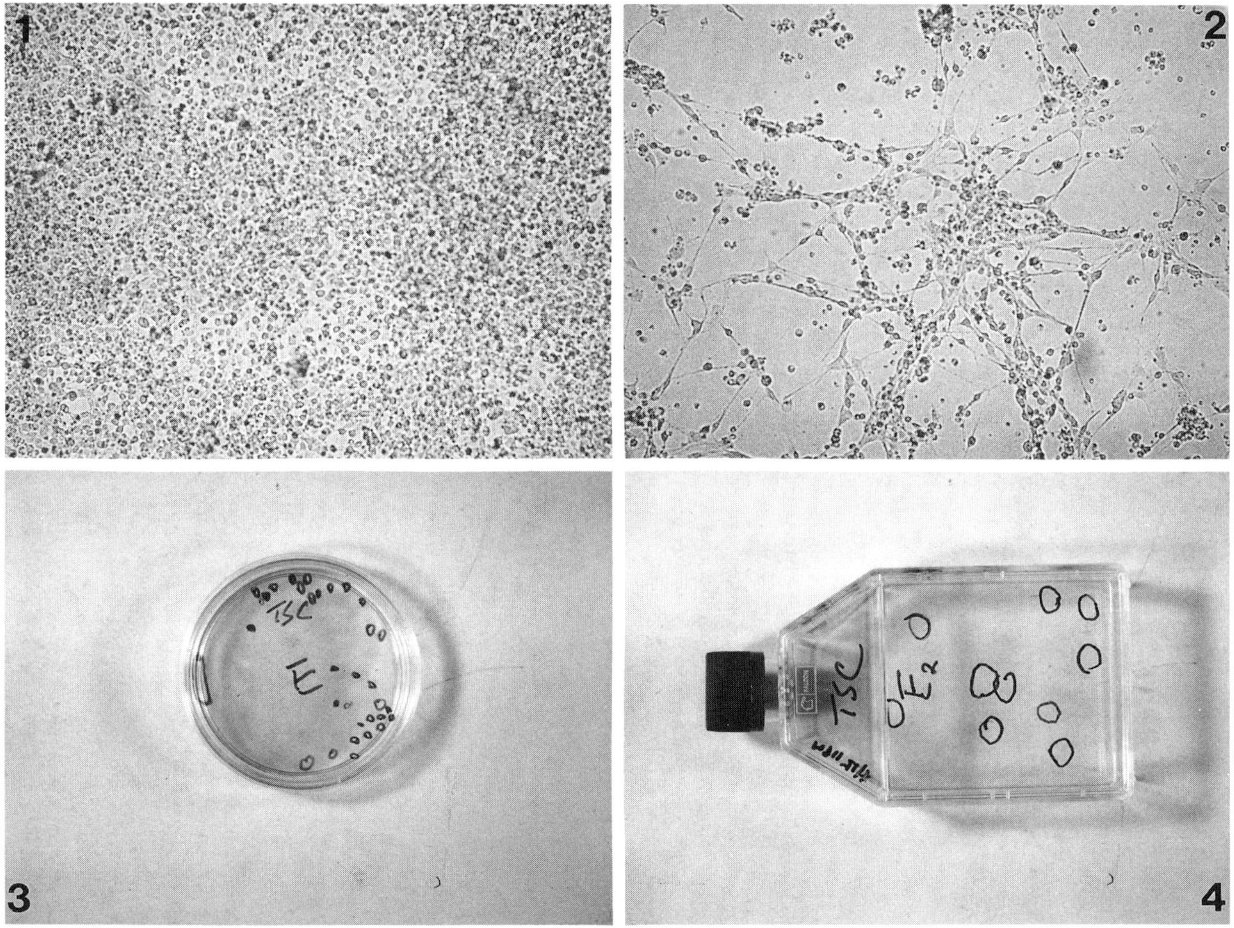

FIG. 1.—ELFR1 exhibit round morphology without ELF exposure
FIG. 2.—ELFR1 exhibit bipolar morphology after ELF exposure.
FIG. 3.—Morphologic change exhibited by ELFR1 increases from center to periphery of culture dish, proportional to magnitude of induced electric field.
FIG. 4.—Morphologic change exhibited by ELFR1 is randomly distributed in uniform magnetic and induced electric field exposure.

TABLE 1.—Effect of varying the frequency on ELFR1.

Signal	Exposure day on which response was observed
60 Hz CW, 1 G	4
15Hz pulse trains	4
10Hz bipolar pulse trains	3
Sham exposure	no response after 12 days

TABLE 2.—Effect of varying field strength on ELFR1.

Field strength (G)	Exposure day on which response was observed
0.5	No response after 12 days
0.75	No response after 12 days
0.8	6
0.9	5
1.0	4
5.0	2
10.0	1
11.0	less than 1

DISCUSSION

The geomagnetic field has been shown to be necessary for the fundamental biological activity of axonal ensheathment and myelination by the Schwann cell.[1] Bioelectric perturbation of tissue has been shown to alter phenotypic expression of chondroblasts by promotion of alkaline phosphatase activity and alteration of the structure of proteoglycans in specific cartilaginous tissues.[2] With that in mind, we assessed whether an ELF magnetic field could induce morphologic change in a Schwann cell line that can be reliably observed. The morphologic change observed was dramatic and reproducible. The cell line was ELF Responder 1 (ELFR1). The response was observed with several waveforms and the onset of morphologic change varied systematically with ELF field strength. Observation of the morphologic changes induced by ELF exposure suggests that the field effect is likely to be mediated through the induced electric field, with an apparent threshold between 0.75 and 0.8 G at 60 60 Hz.

REFERENCES

1. K. Shibib et al., "The geomagnetic field: A factor in cellular interactions? I. Magnetism and Schwann cell-axon interaction in the peripheral nerves of the newborn rat," *Neurol. Res.* 9: 225-235, 1987.

2. L. A. Norton et al., "Pulsed electromagnetic fields alter phenotypic expression in chondroblast in tissue culture," 1988.

Martin Blank, Ed., *Electricity and Magnetism in Biology and Medicine*

ELECTRIC AND MAGNETIC FIELDS: GENETIC TOXICITY DATA

J. C. Murphy,* D. A. Kaden, J. Warren, and A. Sivak

Health Effects Institute, Cambridge, Mass.

Reports of increased risk of childhood cancer associated with residential wiring configurations for electric power distribution have raised concern about the potential carcinogenicity of magnetic fields.[1-3] One of the key early events in the sequence of steps leading to cancer is likely to involve DNA damage. We have therefore reviewed the scientific literature discussing both electric and magnetic fields at 50 and 60 Hz frequencies and some miscellaneous pulsed exposures with regard to genetic toxicity. We have also reviewed the literature describing endpoints that might indirectly affect genetic toxicity.

Electric and/or magnetic fields (E/MF) have been used in a variety of biological test systems to evaluate their genetic toxicity, with most of these studies reporting no effect. A limited number of mutation studies with E/MF, performed in bacterial systems in magnetic fields,[4,5] and mammalian cells and rodents using electric and/or combined fields,[6,7] provide no evidence for a direct mutagenetic effect of E/MF.

Although most studies of chromosome abnormalities reported no effect, there were some positive findings. For magnetic fields, no chromosome aberration effects were observed in human lymphocytes exposed in vitro to fields up to 5 mT.[8,9a] One study with a pulsed magnetic-field exposure in human lymphocytes reported increases in the frequency of chromosome aberrations.[10]

Three studies examined chromosome aberrations in lymphocytes of workers exposed to both electric and magnetic fields. Two studies[11,12] reported increases in chromosome aberrations; one study[13] reported no effects of E/MF exposure. The specific exposures in two of these studies[11,13] are not accurately known. The finding in Ref. 12 is potentially important since these exposures represent actual workplace situations. Therefore, careful investigation of genetic effects under these exposure conditions is needed. (The authors suggested that this finding may be due to spark discharge exposures rather than the background electric- and magnetic-field exposures, but did not provide convincing evidence.) More information is needed about actual internal field strengths induced by spark discharge conditions. The effects in switchyard workers could also be due to chemical exposures to leakage from transformers or to spark discharges, which vaporize metals and ionize air.

Electric fields have been reported to induce 2- to 3-fold increases in chromosome aberrations in bovine lymphocytes at a current density of 2.36 $\mu A/cm^2$,[14] but not at 30 $\mu A/cm^2$.[9a] Nordenson and colleagues performed an in vitro component of their 1984 study that showed no effect of simulated background switchyard levels of electric fields.[11] However, increased frequencies of chromosome abnormalities were seen with high pulsed electric field exposures intended to simulate spark discharge.

Studies of sister chromatid exchange have been performed with electric, magnetic, and combined fields. Seven of nine studies reported no effect of exposure on SCE, both in vitro and in vivo.[8,9b,10,12-17]

An increase in micronucleus formation was reported in lymphocytes from workers exposed to E/MF[12] and in mice exposed to extremely high electric fields,[18] but not in mammalian cells exposed in vitro to E/MF.[15,19]

In addition to examining studies that investigate end points directly related to genetic damage, we also reviewed studies using end points that might be indirectly related to genetic changes, including changes in the cell cycle, mitotic index, and DNA synthesis. There were slight increases in the length of the cell cycle in two of three studies[15,20,21] and decreases in the fraction of cells in mitosis in three of eight studies.[9a,10,20,22-26] Slight increases in DNA synthesis were reported in two studies and both increases and decreases reported for a third.[17,27,28] It is not clear how such slight changes in nonmitotic portions of the cell cycle could directly influence DNA fidelity. However, such changes may indirectly affect the amount of time available for secondary processes, such as DNA repair. No such effects were seen in studies examining repair of single strand

*Correspondence to: J. C. Murphy, Health Effects Institute, 141 Portland St. (Suite 7300), Cambridge MA 02139.

breaks[29] or thymine dimers.[30]

There are serious deficiencies in the set of studies described here that limit the conclusions that can be drawn. One of them is the great variation among end points and test systems with respect to the nature of the exposures investigated. The magnetic fields investigated in these studies range over six orders of magnitude (10^{-4} to 100 mT); the electric fields range over seven orders of magnitude (10^{0} to 10^{7} V/m). However, individual exposure combinations across the range of electric and magnetic fields are only sparsely sampled. There is also extremely poor representation of conditions for combined exposure to both magnetic and electric fields. In addition, many of the exposures are unusual pulsed fields that most people would never experience.

Moreover, studies investigating the genetic toxicity of electric and magnetic fields are of variable quality in terms of methods used and of the presentation of results. Many papers report only summary data, which makes interpretation of the strength of the results impossible.

Finally, the relevance of most of these studies is questionable because most of the exposures are much higher than actual residential exposures. Kaune et al. have reported measurements made in family rooms of homes.[31] Magnetic field time-averaged 24h means were $1\times10^{-4} \pm 1.2\times10^{-4}$ mT and average (over 24 h) electric fields of 33 ± 32 V/m.

Given some positive reports and the limitations in the set of studies, an organized effort to examine both electric and magnetic fields may be warranted, with careful attention paid to the relevance of the exposure used to human exposures, including those represented in epidemiologic studies showing an elevated incidence of cancer and exposure conditions in switchyards.[32]

REFERENCES

1. S. J. London, D. C. Thomas, J. D. Bowman, E. Sobel, T. C. Cheng, and J. M. Peters, "Exposure to residential electric and magnetic fields and risk of childhood leukemia," *Am. J. Epidemiol.* 134: 923-937, 1991.

2. D. Savitz, H. Watchtel, F. Barnes, E. John, and J. Tvrdik, "Case-control study of childhood cancer and exposure to 60-Hertz magnetic fields," *Am. J. Epidemiol.* 128: 21-38, 1988.

3. N. Wertheimer and E. Leeper, "Electrical wiring configurations and childhood cancer," *Am. J. Epidemiology* 109: 273-284, 1979.

4. J. Juutilainen and A. Liimatainen, "Mutation frequency in Salmonella exposed to weak 100-Hz magnetic fields," *Hereditas* 104: 145-147, 1986.

5. R. L. Moore, "Biological effects of magnetic fields: Studies with microorganisms," *Can. J. Microbiol.* 25: 1145-1151, 1979.

6. C. I. Kowalczuk and R. D. Saunders, "Dominant lethal studies in male mice after exposure to a 50-Hz electric field," *Bioelectromagnetics* 11: 129-137, 1990.

7. M. E. Frazier, J. E. Samuel, and W. T. Kaune, "Viabilities and mutation frequencies of CHO-K1 cells following exposure to 60-Hz electric fields," in L. E. Anderson, B. J. Kelman, and R. J. Weigel, Eds., *Proc. 23rd Hanford Life Sciences Symposium: Interaction of Biological Systems with Static and ELF Electric and Magnetic Fields,* Pacific Northwest Laboratory, Richland, Washington, 1987, 255-267.

8. M. Rosenthal and G. Obe, "Effects of 50-Hertz electromagnetic fields on proliferation and on chromosomal alterations in human peripheral lymphocytes untreated or pretreated with chemical mutagens," *Mutation Res.* 210: 329-335, 1989.

9. M. M. Cohen, A. Kunska, J. A. Astemborski, D. McCulloch, and D. A. Paskewitz, (a) "Effect of low-level, 60-Hz electromagnetic fields on human lymphoid cells: I. Mitotic rate and chromosome breakage in human peripheral lymphocytes," *Bioelectromagnetics* 7: 415-423, 1986; (b) M. M. Cohen, A. Kunska, J. A. Astemborski, and D. McCulloch, "II. Sister-chromatid exchanges in peripheral lymphocytes and lymphoblastoid cell lines," *Mutation Res.* 172: 177-184, 1986.

10. A. M. Khalil and W. Qassem, "Cytogenetic effects of pulsing electromagnetic field on human lymphocytes in vitro: Chromosome aberrations, sister-chromatid exchanges and cell kinetics," *Mutation Res.* 247: 141-146, 1991.

11. I. Nordenson, K. H. Mild, S. Nordstrom, A. Sweins, and E. Birke, "Clastogenic effects in human lymphocytes of power frequency electric fields: In vivo and in vitro studies," *Radiat. Environ. Biophys.* 23: 191-201, 1984.

12. I. Nordenson, K. H. Mild, U. Ostman, and H. Ljungberg, "Chromosomal effects in lymphocytes of 400kV-substation workers," *Radiat. Environ. Biophys.* 27: 39-47, 1988.

13. M. Bauchinger, R. Hauf, E. Schmid, and J. Dresp, "Analysis of structural chromosome changes and SCE after occupational long-term exposure to electric and magnetic fields from 380 kV-systems," *Radiat. Environ. Biophys.* 19: 235-238, 1981.

14. G. d'Ambrosio, A. Scaglione, D. Di Serardino, M. B. Lioi, L. Iannuzzi, E. Mostacciuolo, and M. R. Scarfi, "Chromosomal aberrations induced by ELF electric fields," *J. Bioelectricity* 4: 279-284, 1985.

15. G. K. Livingston, K. L. Witt, O. P. Gandhi, I. Chatterjee, and J. L. Roti Roti, "Reproductive integrity of mammalian cells exposed to power frequency electromagnetic fields," *Environ. Mol. Mutagen* 17: 49-58, 1991.

16. J. M. Garcia-Sagredo, L. A. Parada, and J. L. Monteagudo, "Effect of SCE in human chromosomes in vitro of low-level pulsed magnetic field," *Environ. Mol. Mutagen.* 16: 185-188, 1990.

17. K. Takahashi, I. Kaneko, M. Date, and E. Fukado, "Effect of pulsing electromagnetic fields on DNA synthesis in mammalian cells in cultures," *Experientia* 42: 185-186, 1986.

18. S. M. El Nahas and H. A. Oraby, "Micronuclei formation in somatic cells of mice exposed to 50-Hz electric fields," *Environ. Mol. Mutagen.* 13: 107-111, 1989.

19. M. R. Scarfi, F. Bersani, A. Cossarizza, D. Monti, G. Castellani, R. Cadossi, G. Franceschetti, and C. Franceschi, "Spontaneous and mitomycin-C-induced micronuclei in human lymphocytes exposed to extremely low frequency magnetic fields," *Biochem. Biophys. Res. Commun.* 176: 194-200, 1991.

20. A. Brulfert, M. W. Miller, D. Robertson, D. A. Dooley, and P. Economou, "A cytohistological analysis of roots whose growth is affected by a 60-Hz electric field," *Bioelectromagnetics* 6: 283-291, 1985.

21. E. M. Goodman, B. Greenebaum, and M. T. Marron, "Bioeffects of extremely low-frequency electromagnetic fields—Variation with intensity, waveform, and individual or combined electric and magnetic fields," *Radiat. Res.* 78: 485-501, 1979.

22. M. Inoue, M. W. Miller, C. Cox, and E. L. Carstesen, "Growth rate and mitotic index analysis of *Vicia faba L.* roots exposed to 60-Hz electric fields," *Bioelectromagnetics* 6: 293-303, 1985.

23. R. W. Greene, "60-Hz Ultra-high voltage electric field effect on onion root mitotic index: A lack of impact," *Environ. Exp. Botany* 23: 355-360, 1983.

24. V. D. Dyshlovoi, A. S. Panchuk, and V. S. Kachura, "Effect of electromagnetic field of industrial frequency on the growth pattern and mitotic activity of cultured human fibroblastoid cells," *Cytology and Genetics* 15: 6-9, 1981.

25. D. Robertson, M. W. Miller, C. Cox, and H. T. Davis, "Inhibition and recovery of growth processes in roots of ***Pisum sativum L.*** exposed to 60-Hz electric fields," *Bioelectromagnetics* 2: 329-340, 1981.

26. M. W. Miller, M. M. Reddy, G. R. Yettewich, and G. E. Kaufman, "Lack of effect of extremely low frequency electric and magnetic fields on roots of *Vicia faba,*" *Environ. Exp. Botany* 16: 83-88, 1976.

27. A. R. Liboff, T. Williams Jr., D. M. Strong, and R. Wistar, "Time-varying magnetic fields: Effect of DNA synthesis," *Science* 223: 818-820, 1984.

28. G. A. Rodan, L. A. Bourret, and L. A. Norton, "DNA synthesis in cartilage cells is stimulated by oscillating electric fields," *Science* 199: 690-692, 1978.

29. M. E. Frazier, J. A. Reese, J. E. Morris, R. F. Jostes, and D. L. Miller, "Exposure of mammalian cells to 60-Hz magnetic or electric fields: Analysis of DNA repair of induced single-strand breaks," *Bioelectromagnetics* 11: 229-234, 1990.

30. G. L. Whitson, W. L. Carrier, A. A. Francis, C. C. Shih, S. Georghiou, and J. D. Regan, "Effects of ELF electric fields on cell growth and DNA repair in human skin fibroblasts," *Cell Tissue Kinet.* 19: 39-47, 1986.

31. W.T. Kaune, R. G. Stevens, N. J. Callahan, R. K.Severson, and D. B. Thomas, "Residential magnetic and electric fields," *Bioelectromagnetics* 8: 315-335, 1987.

32. The authors are grateful for the technical review of the manuscript by Martin Misakian and Steven Bloom. Although the development of the article was in part supported by Health Effects Institute sponsors (U.S. Environmental Protection Agency grant X-818312 and electric power companies), the views expressed in this article of course do not necessarily reflect the views or policies of HEI or its sponsors.

Martin Blank, Ed., *Electricity and Magnetism in Biology and Medicine*

A CRITICAL EVALUATION OF THE GENOTOXIC POTENTIAL OF ELECTRIC AND MAGNETIC FIELDS

J. McCann*

ICF Kaiser Engineers, Inc., Oakland, Calif.

F. Dietrich

Electric Research & Management, Inc., Pittsburgh, Penn.

A. O. Martin

Foley & Lardner, Chicago

C. Rafferty

Electric Power Research Institute, Palo Alto, Calif.

About 60 published articles were identified that tested static or ELF (extremely low frequency) electric or magnetic fields for genotoxic effects. The biological assays used spanned a wide range, from microbial systems and plants to studies in humans. Experimental results were grouped into four exposure categories: static electric, static magnetic, ELF electric, and ELF magnetic. All experiments were critically analyzed with respect to basic data quality criteria. Experiments within each exposure category were then compared to determine whether results reinforced or contradicted one another. The results of the analysis indicated that the preponderance of evidence strongly suggests that neither static nor ELF electric or magnetic fields have a demonstrated potential to cause genotoxic effects. However, there is unconfirmed evidence for weak genotoxic activity from (a) exposure to spark discharge accompanied with exposure to high ELF electromagnetic fields, and (b) co-exposures to certain mutagens or to ionizing radiation. In addition, certain exposure categories are not represented or are under-represented by tests in some genotoxicity test systems that are usually included in minimal test batteries as specified by EPA for chemicals. It is suggested that consideration be given to whether additional genotoxicity testing is warranted to fill these gaps.

A number of investigations have been carried out to test the potential genotoxicity of electric and magnetic fields. Since initiation of carcinogenesis is believed to involve DNA damage, assays for genotoxicity are considered to supply evidence relevant to carcinogenic potential. Most genotoxicity studies have been negative suggesting that EMF may not have potential as an initiator of carcinogenesis. In part because of these negative results, most recent research on the question of a causal role of EMF in the development of cancer has focused on the later stages of carcinogenesis and cancer development, namely promotion and progression. Several reviews including partial analysis of genotoxicity studies have appeared, most recently in a draft report from the U.S. Environmental Protection Agency (EPA).[1]

We conducted a systematic critical review of the published literature with the aim of assembling all known genotoxicity research involving static or ELF electric or magnetic fields. (We also identified studies involving radiofrequency and microwave radiation, though these studies were not our major focus.) The results of this study will be presented in detail elsewhere.[2] What follows is a brief summary of the preliminary results.

---

*Correspondence to: J. McCann (Environment Div.), 1800 Harrison St. (7th fl.), Oakland CA 94612.

RESULTS

Table 1 illustrates the types of genotoxicity experiments identified, by exposure category. As shown, experiments spanned a wide range of genotoxicity assays, particularly for ELF electric and ELF magnetic field exposure categories. A total of 59 experiments are represented in the table. Exposures to static electric fields are the least represented exposure category, with only 7 experiments identified.

TABLE 1.—Genotoxicity assays, by exposure category. (The numbers in this table represent a preliminary distribution of experiments into exposure categories. They may change slightly in the final report. Abstracts are not included in the table. Developmental toxicity studies in *Drosophila* have also not been included. Eight experiments involved both ELF electric and ELF magnetic field exposures; they are included in both categories.)

	Number of assays, by exposure category			
Assay type	Static electric	Static magnetic	ELF electric	ELF magnetic
Microbial systems	2	2	0	1
Drosophila	3	5	1	1
Plants	0	1	0	0
In vitro chromosome	0	5	7	10
In vivo chromosome	2	0	6	3
In vitro DNA repair	0	0	3	3
In vivo DNA repair	0	0	0	1
Dominant lethal assay	0	0	2	1
Total	7	13	19	20

All experiments were analyzed to determine whether data quality objectives were satisfied. These objectives included: use of acceptable experimental protocols, presentation of sufficient experimental details to permit a critical analysis, inclusion of critical control experiments, reproducibility and/or dose-response analysis, and use of appropriate statistical methods. Experiments within each exposure category were also compared to determine whether results reinforced or contradicted one another.

The results of analysis of these assays are briefly summarized below:

Static Electric Fields. Based on the information analyzed, it is not possible to draw a definitive conclusion regarding the possible genotoxic potential of static electric fields. Information from tests in *Drosophila* suggests that intensities up to 30000 V/m have no genotoxic effect in that test system.[3] In other test systems or at higher field intensities, positive results are reported,[4,5] but have not been replicated.

Static Magnetic Fields. All studies satisfying minimal data quality objectives which examined genotoxic endpoints reported the absence of any genotoxic effects from exposures to static magnetic fields ranging from 0.0001 to 12 T. These studies employed microbial systems,[6] *Drosophila,*[3,7] and in vitro chromosome assays.[8,9] One report[8] presents suggestive evidence of a weak genotoxic effect (chromosome aberrations) after exposure at 1.1 T with co-exposures to alpha or proton particle radiation.

ELF Electric Fields. Although results are not definitive, spark discharge accompanying exposure to high intensity ELF electric fields may have some genotoxic effects after in vivo exposure. The studies suggesting this are the reports of Nordensen et al., particularly the 1988 study,[10] and the studies of El Nahas and colleagues, particularly that of El Nahas and Oraby.[11] There is no convincing evidence of in vitro genotoxic effects in mammalian cells, though there have been several positive reports. The preponderance of studies satisfying minimal data quality objectives have not detected any genotoxic effects from ELF electric fields.

ELF Magnetic Fields. There is no convincing evidence that ELF magnetic fields produce genotoxic effects. The overwhelming preponderance of experiments analyzed are negative. The majority of studies available in this exposure category employed human peripheral lymphocytes exposed in vitro. All but two of these studies reported the absence of any genotoxic effects. The positive genotoxic effects reported by Khalil and Qassem[12] are not consistent with several other similar studies and should be viewed with caution unless independently replicated. The experiments of Rosenthal and Obe reporting a weak genotoxic effect in mammalian cells co-exposed in vitro to two known mutagens are of interest.[13] As the authors point out, this result requires further investigation to determine whether observed effects were due to the field exposure conditions or the experimental protocol.

DISCUSSION

The preponderance of available evidence strongly suggests that neither static nor ELF electric or magnetic fields have a demonstrated potential to cause genotoxic effects. There are two possible exceptions to this conclusion, both of which require further confirmation: (a) exposure to spark discharge accompanying high ELF electric or magnetic fields; and (b) co-exposure to certain mutagens or to ionizing radiation.

Three types of data gaps are relevant to considerations of the need for further genotoxicity testing. The first are suggestive positive results requiring independent replication. The second are tests or exposure conditions that are under-represented in a minimal battery of tests such as that represented in Table 2. As is evident in Table 2, the complement of reported genotoxicity tests varies considerably for the different categories of exposure. Certain exposure categories have not been tested in some genotoxicity test systems usually included in minimal test batteries, as specified by EPA for chemicals. For example, no tests were identified of static electric fields in in vitro chromosome or DNA repair systems, and no tests of static magnetic fields in in vitro DNA repair mammalian systems. Third, in some cases experimental results are available, but results are judged to be inadequate, incomplete, or conflicting.[14]

TABLE 2.—Distribution of test results into a genotoxicity test battery.*

	Static electric	Static magnetic	ELF electric	ELF magnetic
Microbial	?	–	NA	?
Drosophila SLRL	?	–	?	?
In vitro chromosome	NA	–	?	?
In vitro repair	NA	NA	–	–
In vivo chromosome	?	NA	?	?

*NA, no genotoxicity test results were identified; ?, test results were uncertain or conflicting; –, test results were negative.

REFERENCES

1. EPA (U.S. Environmental Protection Agency), "Evaluation of the potential carcinogenicity of electromagnetic fields," Office of Research and Development, Washington, D.C., EPA/600/6-90/005B, October, 1990, review draft.

2. J. M. McCann, F. Dietrich, A. O. Martin, and C. Rafferty, "A critical review of the genotoxic potential of electric and magnetic fields," *Mutat. Res.* (in press).

3. J. R. Diebolt, "The influence of electrostatic and magnetic fields on mutation in *Drosophila melanogaster* spermatozoa," *Mutat. Res.* 57: 169-174, 1978.

4. J. T. Mitchell, A. A. Marino, T. J. Berger, and R. O. Becker, "Effect of electrostatic fields on the chromosomes of Ehrlich ascites tumor cells exposed in vivo," *Physiol. Chem. Phys.* 10: 79-85, 1978.

5. A. A. Marino, T. J. Berger, J. T. Mitchell, B. A. Duhacek, and R. O. Becker, "Electric field effects in selected biologic systems," *Ann. N.Y. Acad. Sci.* 238: 436-443, 1974.

6. A. Thomas and P. G. Morris, "The effects of NMR exposure in living organisms. I. A microbial assay," *British J. Radiology* 54: 615-621, 1981.

7. P. G. Kale and J. W. Baum, "Genetic effects of strong magnetic fields in *Drosophila melanogaster:* I. Homogeneous fields ranging from 13,000 to 37,000 Gauss," *Environ. Mutagenesis* 1: 371-374, 1979.

8. T. Takatsuji, M. S. Sasaki, and H. Takekoshi, "Effect of static magnetic field on the induction of chromosome aberrations by 4.9 meV protons and 23 MeV alpha particles," *J. Radiat. Res.* 30: 238-246, 1989.

9. F. J. Peteiro-Cartelle and J. Cabezas-Cerrato, ''Absence of kinetic and cytogenetic effects on human lymphocytes exposed to static magnetic fields,'' *J. Bioelectricity* 8: 11-19, 1989.

10. I. Nordensen, K. Hansson Mild, U. Ostman, and H. Ljungberg, ''Chromosomal effects in lymphocytes of 400 kV-substation workers,'' *Radiat. Environ. Biophys.* 27: 39-47, 1988.

11. S. M. El Nahas and H. A. Oraby, ''Micronuclei formation in somatic cells of mice exposed to 50-Hz electric fields,'' *Environ. Molec. Mutagenesis* 13: 107-111, 1989.

12. A. M. Khalil and W. Qassem, ''Cytogenetic effects of pulsing electromagnetic field on human lymphocytes in-vitro: Chromosome aberrations sister-chromatid exchanges and cell kinetics,'' *Mutat. Res.* 247: 141-146, 1991.

13. M. Rosenthal and G. Obe, ''Effects of 50-Hertz electromagnetic fields on proliferation and on chromosomal alterations in human peripheral lymphocytes untreated or pretreated with chemical mutagens,'' *Mutat. Res.* 210: 329-335, 1989.

14. This work was sponsored by the Electric Power Research Institute, Palo Alto, Calif., under contracts RP2965-22 (ICF Kaiser Engineers, Inc.) and RP2965-03 (Electric Research & Management, Inc.).

Martin Blank, Ed., *Electricity and Magnetism in Biology and Medicine*

EFFECT OF AC MAGNETIC FIELDS ON ULTRAVIOLET LIGHT-INDUCED MUTATION AND MITOTIC RECOMBINATION IN *SACCHAROMYCES CEREVISIAE*

D. D. Ager* and J. A. Radul

Ontario Hydro, Toronto

In this study, we have examined whether 60Hz magnetic fields are capable of inducing mutation or mitotic recombination in the yeast *Saccharomyces cerevisiae*. In addition, we determined whether magnetic fields were capable of altering the genetic response of *S. cerevisiae* to ultraviolet light (UV). We measured the frequencies of induced mutation, gene conversion, and reciprocal mitotic crossing over for exposures to magnetic fields alone or in combination with various preliminary UV exposures. These experiments were performed on both repair proficient and deficient stains. Magnetic-field exposures did not induce mutation, gene conversion, or reciprocal mitotic crossing over in either of these stains, nor did the fields influence the frequencies of UV-induced genetic events.

The genetic effects of extremely low frequency (ELF) electric and magnetic fields have been studied for a wide range of biological systems. In general, mutational studies involving ELF magnetic fields have proved to be negative.[1-3] However, both positive and negative effects have been reported concerning the effect of ELF magnetic fields on genetic recombination.[4-6] In addition, both positive and negative effects have been observed for combined exposures of various genotoxic agents with ELF magnetic fields.[5,7] In this study, we have examined the ability of 60Hz magnetic fields to induce genetic damage in the yeast *S. cerevisiae*. To determine whether magnetic fields could enhance the mutagenic or recombinagenic effectiveness of a known carcinogen, genetic studies were carried out by combined treatments of UV and magnetic fields. For the yeast strains used in this study, it was possible to measure both mutational and recombinatorial types of DNA damage simultaneously.

MATERIALS AND METHODS

Strains and Media. *S. cerevisiae* strains BKO (*RAD*$^+$, a derivative of strain *D7*) and MLD39/1 ((*rad3-2*) have been described previously.[8,9] SD and YPD media[9] were used for routine growth.

UV. Mid-log phase cell suspensions were irradiated in open Petri dishes with 254nm UV from a General Electric germicidal lamp (0.4 $J/m^2/s$). UV was monitored with a Latarjet dosimeter. All experiments were carried out under yellow light to avoid photoreactivation.

Magnetic Field Exposure. Cells were plated onto agar Petri plates and stacked vertically within a 65cm solenoid (14 cm ID). The solenoid produced a 60Hz sinusoidal signal with a flux density of 1.0 mT. The solenoid was calibrated with a one-axis high-magnetic-field probe.[10] Cells were exposed to this field for a total of 3 days (MLD39/1; 6 days) at room temperature. The first 3 h of the exposure consisted of a 15min on-15min off treatment regime.

Incubation and Scoring. Survival was measured by the colony forming ability of cells incubated on SD at room temperature for 6 days. Gene conversion and mutation were measured by incubating the cells on medium lacking the appropriate supplement.[8,9] Each experiment was repeated at least twice from independent cultures.

---

*Correspondence to: D. D. Ager (Environmental Sciences), Ontario Hydro, 800 Kipling Ave., KD118 Toronto ON Canada M8Z 5S4.

RESULTS AND DISCUSSION

The data from studies on repair-proficient yeast are outlined in Tables 1 and 2. In these tables, "cross-overs" represent reciprocal mitotic crossover events between *ADE2* and the centromere, and "aberrant colonies" represent the sum total of all aberrant segregation patterns arising from such genetic events as monosomy, deletion, or forward mutation.[8] From these tables, it is evident that magnetic-field exposure does not significantly alter the spontaneous or UV-induced frequencies of reciprocal mitotic recombination, aberrant colony formation, gene conversion, or mutation.

TABLE 1.—Mitotic recombination in repair-proficient yeast.

UV Dose (J/m^2)	EMF Exposure (mT)	Surviving Fraction	Crossovers per 10^4 Survivors	Aberrant colonies per 10^4 Survivors
0	0	1.0 (19,185)[a]	nil	7 (14)
0	1	1.0 (20,670)	nil	4 (9)
12	0	0.97 (16,881)	17 (29)	97 (163)
12	1	0.98 (13,770)	10 (14)	94 (129)
25	0	0.92 (19,714)	28 (55)	188 (370)
25	1	0.93 (16,872)	27 (46)	183 (309)
50	0	0.70 (9,050)	81 (73)	418 (378)
50	1	0.70 (7,768)	93 (72)	417 (324)

[a] Numbers in parentheses represent the actual numbers of colonies scores.

TABLE 2.—Mutation and gene conversion in repair-proficient yeast.

UV Dose (J/m^2)	EMF Exposure (mT)	Surviving Fraction	Gene conversion (Trp^+ per 10^4 Survivors)	Mutation (Ilv^+ per 10^5 Survivors)
0	0	1.0	0.20 ± 0.08	0.05 ± 0.01
0	1	1.00 ± 0.03	0.19 ± 0.07	0.05 ± 0.01
12	0	0.99 ± 0.01	1.35 ± 0.04	1.20 ± 0.03
12	1	0.99 ± 0.02	1.39 ± 0.08	1.14 ± 0.03
25	0	0.95 ± 0.06	2.06 ± 0.64	2.10 ± 0.13
25	1	0.92 ± 0.07	2.03 ± 0.44	1.68 ± 0.14
50	0	0.72 ± 0.07	11.1 ± 1.5	12.9 ± 3.5
50	1	0.74 ± 0.10	11.3 ± 2.7	11.4 ± 1.5

The data from studies on repair deficient (*rad3-2*) yeast are given in Tables 3 and 4. In Table 3, "variant" colonies represent the sum total of all aberrant segregation patterns, including reciprocal mitotic crossovers, monosomy, deletion, and forward mutation.[9] Magnetic-field exposures do not significantly alter the spontaneous frequencies of mutation or recombination, nor did exposures modify the response of MLD39/1 to a preliminary treatment with UV.

TABLE 3.—Mitotic recombination in repair-deficient yeast.

UV Dose (J/m^2)	EMF Exposure (mT)	Surviving Fraction		Variant Colonies per 10^3 Survivors	
0	0	1.0	(11,828)	4.4	(52)
0	1	0.99	(8,509)	3.6	(31)
2	0	0.98	(5,862)	58	(338)
2	1	0.96	(5,898)	52	(307)
4	0	0.46	(7,156)	71	(509)
4	1	0.45	(7,834)	76	(595)

TABLE 4.—Mutation and gene conversion in repair-deficient yeast.

UV Dose (J/m^2)	EMF Exposure (mT)	Surviving Fraction	Gene conversion (Arg^+ per 10^5 Survivors)	Mutation (Lys^+ per 10^7 Survivors)
0	0	1.0	4.4 ± 0.6	1.0 ± 0.1
0	1	1.0 ± 0.05	4.8 ± 1.0	1.1 ± 0.1
2	0	0.98 ± 0.03	57 ± 7	23 ± 3
2	0	0.96 ± 0.05	58 ± 9	27 ± 3
4	0	0.45 ± 0.07	59 ± 9	27 ± 4
4	0	0.45 ± 0.06	60 ± 4	26 ± 9

In summary, AC magnetic fields (60 Hz) with a field intensity of 1 mT did not induce mutation, gene conversion, or reciprocal mitotic crossing over in either wild type or repair-deficient (*rad3* strains of *S. cerevisiae*). Nor did the magnetic field modify in any way the mutagenic or recombinagenic effectiveness of UV.

REFERENCES

1. J. Juutilainen and A. Liimatainen, *Hereditas* 104:145-147, 1986.
2. P. G. Kale and J. W. Baum, *Environ. Mutagen.* 1: 371-374, 1979.
3. R. D. Benz and A. L. Carsten, *Environ. Mutagen.* 8 (Suppl. 6): 10-11, 1986.
4. V. M. Mastryukova and S. V. Rudneva, *Biol. Bull. Acad. Sci. USSR* 5(3): 371-374, 1978.
5. M. Rosenthal and G. Obe, *Mutat. Res.* 210, 1989.
6. M. Bauchinger, R. Hauf, E. Schmid, and J. Dresp, *Radiat. Environ. Biophys.* 19: 235-238, 1981.
7. T. Koana, J. Wang, and M. Nakagawa, Abstract B-1-6, *BEMS 13th Annual Meeting*, 1991.
8. F. K. Zimmerman, R. Kern, and H. Rasenberger, *Mutat. Res.* 28: 381-388, 1975.
9. M. L. Dixon and R. K. Mortimer, *Mutat. Res.* 161: 49-64, 1986.
10. R. Scheer, Ontario Hydro Research Division Report 89-158-K, 1989.

Martin Blank, Ed., *Electricity and Magnetism in Biology and Medicine*

STIMULATION OF TRANSCRIPTION IN HL-60 CELLS BY ALTERNATING CURRENTS FROM ELECTRIC FIELDS

M. Blank,* L. Soo, H. Lin, and R. Goodman

Columbia University, New York, N.Y.

A. S. Henderson

Hunter College-CUNY, New York, N.Y.

Sixty-Hz *electric* fields (sine waves), at several field strengths and exposure durations, affect histone H2B and c-myc transcript levels in HL-60 cells. The observed increase in the basal levels of these normally expressed transcripts depends on both field strength and time of exposure. β2-microglobulin, a transcript known to be uninducible, is unaffected by cellular exposure to the field strengths used in these experiments. These results are qualitatively similar to the effects of *electromagnetic* fields.

Electromagnetic (EM) fields stimulate transcription in many cells.[1] In HL-60 cells, several transcripts have been studied, including histone H2B and c-myc, and various waveforms have been used. With 60Hz sine waves, the most effective EM stimulus, 8 μT peak magnetic field, gives an estimated 10 μV/m induced electric field. In the frequency range between 15 and 150 Hz, exposed human HL-60 cells show an increase in c-myc and histone H2B transcripts at each frequency, but the most pronounced effect is a fourfold increase in transcript levels at 45 Hz. In this paper, we report on the stimulation of HL-60 cells with directly applied electric fields, employing electrode systems that have been used to study the effects of alternating currents (AC) on the Na,K-ATPase.[2,3]

METHODS

HL-60 cells (5×10^5/ml) grown in a T-175cm^2 flask were used along with sham controls, and total RNA was determined by Agarose gel electrophoresis. Quantitative dot blot hybridizations were used to measure endogenous quantities of histone H2B (a "housekeeping" gene), c-myc (a nuclear proto-oncogene amplified in this cell line), and β2-microglobulin (noninducible) transcripts. Northern blots were used to substantiate the hybridizations relative to the size of the RNA for histone H2B and c-*myc*. Results are reported as the ratio E/C, experimental divided by the control.

RESULTS

Figure 1 presents the time dependence of c-myc and histone H2B at the (optimal) field intensity of 3 mV/m, which corresponds to an output voltage setting of 10 mV. Changes in transcript levels (E/C > 1.0) are significant after 5 min of exposure and appear to plateau in the range of 10 to 20 min of exposure. These experiments demonstrate that 60Hz electric fields alone stimulate transcription in HL-60 cells. Transcript levels for β2-microglobulin, normally unaffected by exposure to EM fields, were not altered by exposure to any of the electric fields used in these experiments.

Figure 2 shows the data for the dependence of the level of transcripts of both c-myc and histone H2B on the applied electric field at 20 min. The levels were approximately the same at 5 and 20 min. Transcripts of both c-myc and histone H2B are stimulated at the same magnitude of applied voltage, which corresponds to an electric field of 3 mV/m, the only applied voltage for which E/C was significantly greater than 1.0 (10 mV).

*Correspondence to: M. Blank (Physiology and Cellular Biophysics), Columbia University, 630 W 168th St., New York NY 10032.

DISCUSSION

The experiments reported in this paper have demonstrated an electric field strength for which there are increases in the level of transcripts. As in the case of stimulation by EM fields, the increase in transcript levels is not proportional to increasing time of exposure or to field strength.[1] A field strength ''window'' around 3 mV/m and an exposure duration time threshold between 2 and 5 min were observed, with a similar pattern for the two increased transcripts. This observation agrees with that of Greene et al.,[4] who found that transcription in HL-60 cells was stimulated at this magnitude of field. The field is higher by two to three orders of magnitude than the electric fields calculated for the experiments with EM fields, but comparable to threshold fields in experiments with the Na,K-ATPase.[2,3]

The course of time dependence of both transcripts (Fig. 1) suggests that stimulation of transcription at 10 mV leads to a relatively stable level for at least 20 min. If the theory of Montrose et al.[5] of stimulation described this system, one would expect increased transcription at shorter times at the higher intensities of stimulation (e.g., 0.1 or 1 V), which was not observed. These results indicate that direct stimulation by electric fields differs from EM field stimulation, where earlier studies showed that the level of transcripts behaved according to the theory.

It appears that very weak AC currents affect transcription, regardless of whether they originate from electric fields directly imposed by electrode systems or are induced by time-varying EM fields. The differences between the current (and electric-field) magnitudes in the two types of stimulation require further study. However, these results on the stimulation of transcription, coupled with the observations of Greene et al.,[4] have suggested the magnitude and duration of AC stimulation needed to affect biosynthesis. These ionic currents are of the type that flow in the body, and they may be related to those that stimulate bone reforming and healing,[6] as well as the changes in muscle phenotypic expression[7] on electrical stimulation.[8]

REFERENCES

1. R. Goodman and A. S. Henderson, *Bioelectrochem. Bioenerg.* 25: 335-355, 1991.
2. M. Blank and L. Soo, *Bioelectrochem. Bioenerg.* 22: 313-322, 1989.
3. M. Blank, *FASEB J.* 6: 2434-2438, 1992.
4. J. J. Greene, W. J. Skowronski, J. M. Mullins, and R. M. Nardone, *Biochem. Biophys. Res. Com.* 174: 742-749, 1991.
5. C. J. Montrose, T. A. Litovitz, and E. C. Elson, *Bioelectrochem. Bioenerg.* 25: 383, 1991.
6. C. A. L. Bassett, *Clin. Plast. Surg.* 12: 259, 1985.
7. S. Ausoni, L. Gorza, S. Schiaffino, K. Gundersen, and T. Lomo, *J. Neuroscience* 10: 153-160, 1990.
8. We thank EPRI, ONR, and DOE for their support.

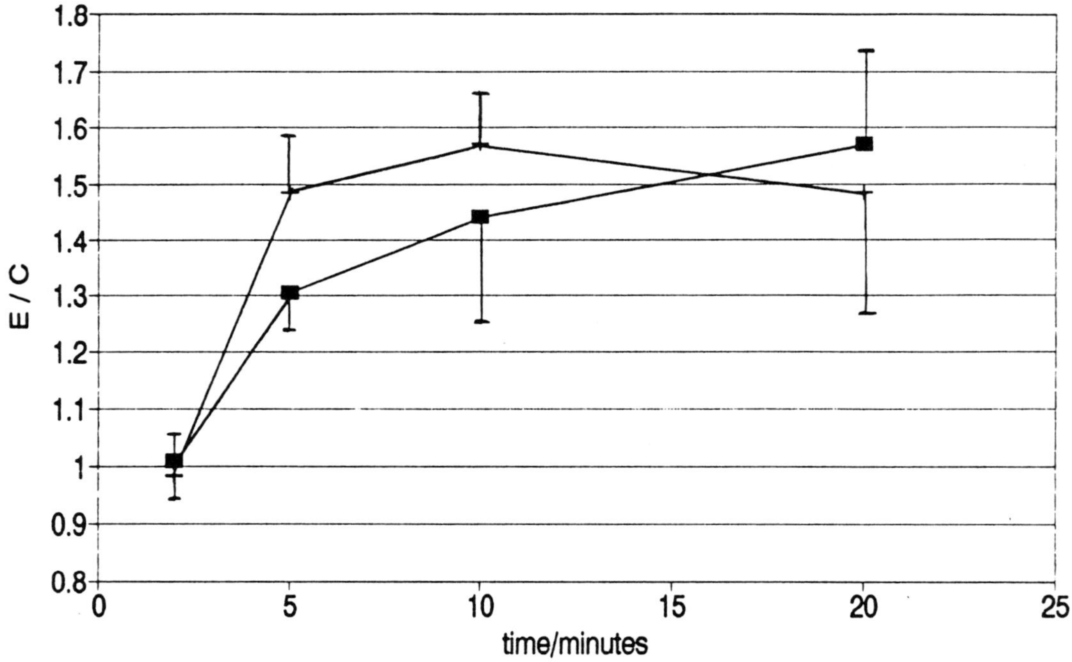

FIG. 1.—Effect of time of exposure on c-myc (□) and histone H2B (+) transcripts in HL-60 cells exposed to 60Hz fields. Data are given in E/C, counts per minute in exposed/control. Applied voltage was 10 mV, a field of 3 mV/m.

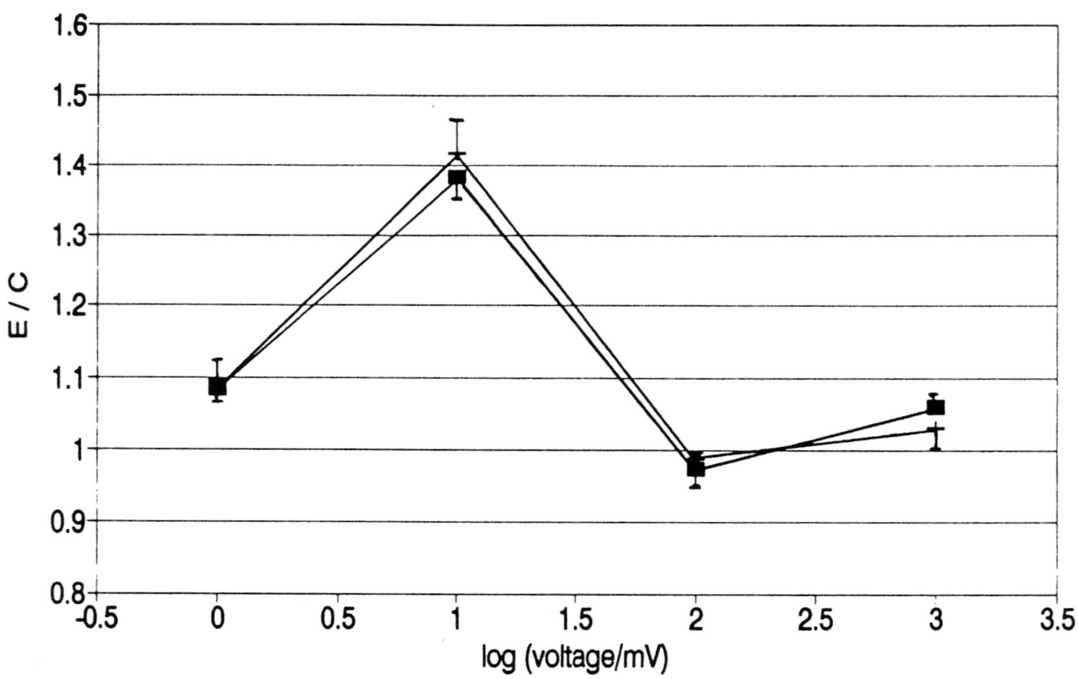

FIG. 2.—Results of measurement of c-myc (□) and histone H2B (+) transcripts in HL-60 cells following exposure for 20 min to 60Hz electric fields of varying applied voltages, plotted on a logarithmic scale. Data are given in E/C, counts per minute in exposed/control. Where data points for the two transcripts overlap, standard errors of means are pointed up for c-myc and down for histone H2B.

Martin Blank, Ed., *Electricity and Magnetism in Biology and Medicine*

60Hz MAGNETIC FIELDS DO NOT CAUSE MAJOR ALTERATIONS IN ERYTHROCYTE PLASMA MEMBRANE INTEGRITY

D. E. Godar,* W. Lee, D. B. Lyle, and M. L. Swicord

Center for Devices and Radiological Health, FDA, Rockville, Md.

We investigated whether extremely low frequency (ELF) magnetic-field exposure could result in a loss of plasma membrane integrity in erythrocytes, as occurs with x-ray exposure. Erythrocytes were irradiated with increasing doses of either x rays (0-4000 cGy), or ELF magnetic fields (0-20 G) generated by solenoids. After irradiation, samples were collected, propidium iodide was added, and a sample of this erythrocyte population was analyzed by use of a FACSCAN flow cytometer with excitation at 488 nm. Using this procedure, we demonstrate dose-dependent loss of plasma membrane integrity that occurs from x irradiation of erythrocytes but not from exposure to 60Hz magnetic fields.

We describe a slightly modified dye exclusion flow cytometry procedure,[1] which allows determination of major changes in erythrocyte plasma membrane integrity.

MATERIALS AND METHODS

Cells

Sheep erythrocytes (Advanced Biotechnologies Inc., Columbia, Md.) were stored in Alsever's solution at 4 C and diluted to about 1×10^7 cells/mL in Dulbecco's phosphate-buffered saline containing Ca^{++} and Mg^{++}. Cell densities were determined by means of a hemocytometer.

Radiation Sources, Procedures, and Post-exposure Treatment

The x-ray source was a Pantak MK.II machine operated at 250 kVcp, 15 mA, HVT = 0.4 mm Cu, with a dose rate of 2.08 Gy/min. Doses were 0, 250, 500, 1000, 2000, and 4000 cGy; 5ml samples of erythrocytes were irradiated at room temperature in 60mm culture dishes (Costar Corp., Cambridge, Mass.).

Solenoid coils were used to generate 60Hz magnetic fields of varying field strengths: 0, 0.02, 0.2, 2, and 20 G; 5ml samples of erythrocytes were placed in 50ml conical centrifuge tubes (Corning Glass Works, Corning, N.Y.) and placed in solenoid coils immersed in a water bath regulated at 37 C.

Sample temperatures were monitored by a digital multimeter and universal temperature probe. Samples of 0.5-1 ml were collected and propidium iodide (PI, MW 668 Daltons) was added to a final concentration of 5 μg/ml following all exposures.

Flow Cytometry

A Becton Dickinson FACSCAN flow cytometer was used in conjunction with the Lysis II program, version 1.0. Samples were briefly mixed prior to data collection and 10000 cells were counted from the total remaining population. Between post-exposure time points, the samples were kept in the dark. Data analysis began with the setting of a gate on the whole-cell population as determined by the forward scattering profile, followed by an analysis of the whole-cell population for fluorescent intensity in the FL2 channel where the dye (PI) emits maximally (564-606 nm). To obtain the percentage of cells with intact membranes, a histogram marker is set on the x-axis log scale as determined by the control population (>95% exclude PI), and the percentages and statistics are calculated by the program and displayed in the histogram statistics. The data are expressed in terms of relative membrane integrity; i.e., the percentage of cells in the exposed population with intact membranes (capable of excluding PI) normalized by the percentage of cells in the sham-exposed population with intact membranes.

---

*Correspondence to: D. E. Godar, CDRH-FDA (HFZ-114), 5600 Fishers Lane, Rockville MD 20857.

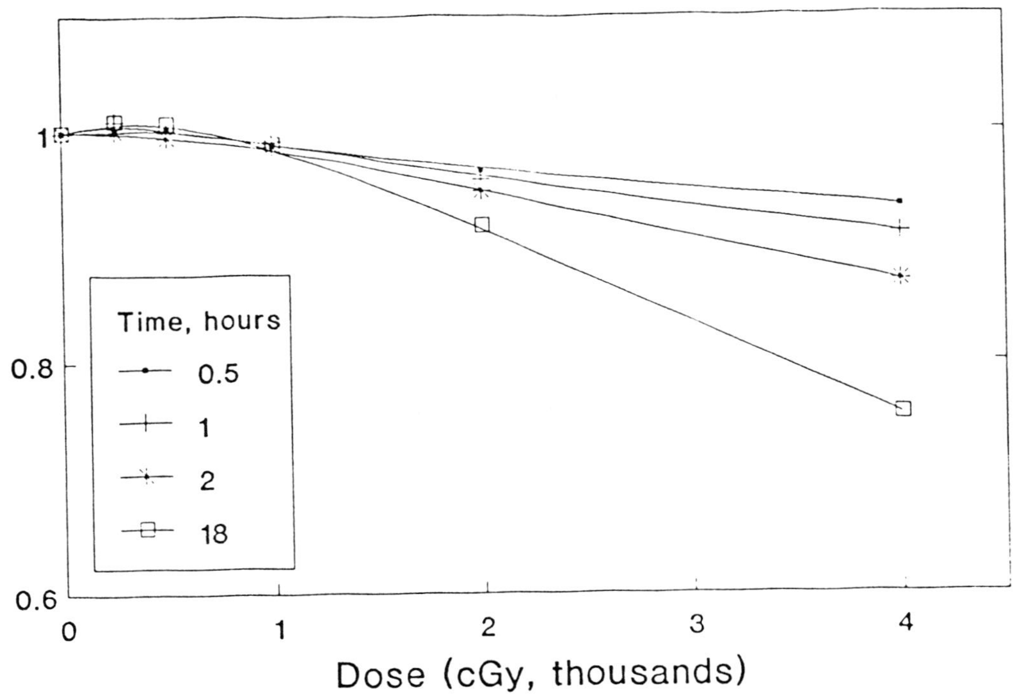

FIG. 1.—Relative membrane integrity vs x-ray dose.

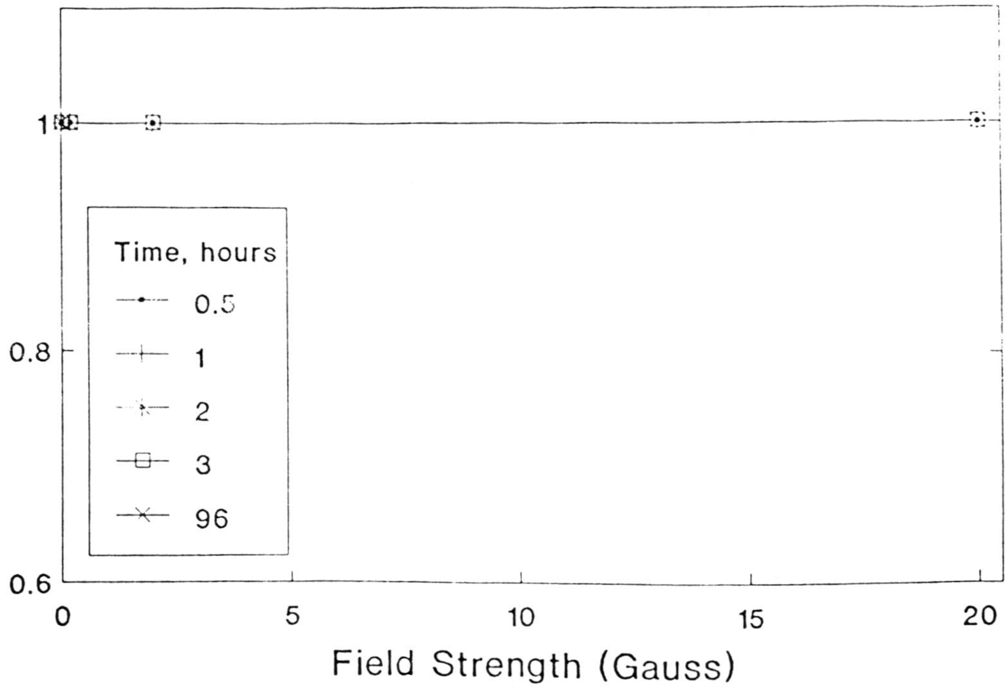

FIG. 2.—Relative membrane integrity vs 60Hz magnetic field strength.

RESULTS AND DISCUSSION

Major loss of plasma membrane integrity occurs in x-irradiated erythrocytes in a time- and dose-dependent manner (Fig. 1), but not to erythrocytes exposed to 60Hz magnetic fields (Fig. 2).

The method described in this paper, PI dye exclusion flow cytometry, measures loss of plasma membrane integrity that can occur in irradiated erythrocytes. This is a slightly modified method that is routinely employed to assess plasma membrane integrity loss in white cells.[1] The method is quantitative, reproducible, and very accurate. Propidium iodide (5 µg/ml) can only be excluded from erythrocytes with functionally intact membranes. Thus, these results do not demonstrate that absolutely no loss in membrane integrity of any kind occurs to erythrocytes exposed to 60Hz magnetic fields, but rather shows that there are no major membrane permeability changes as a result of ELF exposure. By using x irradiation as the positive control, we demonstrate that this method is valid for determining major changes in erythrocyte plasma membrane integrity.

The experimental design of the ELF solenoid exposure system permits simultaneous exposure of multiple samples under exactly the same conditions, including static magnetic fields. The erythrocyte samples were exactly the same, since they were aliquoted from the same stock solution. Sample temperatures were also the same, since the exposure temperature was regulated at 37 C and constantly monitored during the 3h ELF exposure; for the 96h exposure, the temperature was periodically checked.

We thus established that exposure of erythrocytes to 60Hz magnetic fields, unlike x irradiation, does not result in a significant loss of plasma membrane integrity.[2]

REFERENCES

1. N. P. Carter, "Measurement of cellular subsets using antibodies," in D. Rickwood and B. D. Hames, Eds., *Flow Cytometry: A Practical Approach*, New York: Oxford, 1990, 45-67.

2. We thank Dr. Christopher Davis for designing and constructing the solenoid coils and Jon Casamento for engineering support.

Martin Blank, Ed., *Electricity and Magnetism in Biology and Medicine*

EXTREMELY LOW FREQUENCY ELECTROMAGNETIC FIELDS STIMULATE ACTIVATION OF PROTEIN KINASE C IN HL-60 LEUKEMIA CELLS BY INCREASING MEMBRANE TRANSPORT OF Ca^{2+}

M. G. Monti,* L. Pernecco, M. S. Moruzzi, R. Battini, and P. Zaniol

University of Modena, Italy

B. Barbiroli

University of Bologna, Italy

We have previously shown that ELF pulsed electromagnetic fields (PEMFs) influence the activation process of protein kinase C (PKC) in HL-60 cells. We now show that increased Ca^{2+} influx is involved in the reported effect of ELF-PEMF. PKC is activated by association with membranes. Two populations of membrane-bound PKC have been described: one population is dissociated by calcium chelation and the other appears to be inserted into the membrane. Here we report data which show that an enhanced Ca^{2+}-dependent, reversible association of PKC to membranes is a result of ELF-PEMF treatment.

Ca^{2+} is an important second messenger that triggers a variety of intracellular processes in many animal cells.[1] There is evidence suggesting that changes in cytosolic free Ca^{2+} concentration during electromagnetic field exposure are involved in the mechanism responsible for many of the observed biological effects of electromagnetic fields.[2-5]

It is known that an increase in intracellular [Ca^{2+}] induces rapid and reversible association of Ca^{2+}/phospholipid dependent protein kinase C (PKC) to plasma membranes.[6,7] This enzyme has been involved in the control of cell proliferation and differentiation[8] and it appears to be the cellular phorbol ester receptor.[7-9] Association of the soluble enzyme with the cell membrane results in the activation of PKC. This step enables it to phosphorylate specific proteins and ultimately to yield the desired physiological response.[10]

In previous studies recently published we presented evidence that the activation of PKC is an early effect of exposure of HL-60 promyelocytic leukemia cells to ELF-PEMFs.[11] The present study aimed at an examination of the involvement of calcium influx in the reported action of ELF-PEMF on the PKC activation process in HL-60 cells.

MATERIALS AND METHODS

The characteristics of ELF pulsed electromagnetic field used, the ELF-PEMF treatment, and the assays of [^{3}H]4-β-phorbol-12,13-dibutyrate (PDBu) binding to whole HL-60 cells were as previously described.[11] Briefly, the shape of the pulse used was a positive triangle (50 Hz, 8 mT peak) followed by a pause of 2 ms. Cells were exposed to PEMF for 5, 10, 15, or 20 min as specified in the figures. The amount of both total membrane-bound PKC and membrane-inserted (chelator-stable) PKC was assayed by measurement of its [^{3}H]PDBu binding activity. HL-60 promyelocytic leukemia cells (2×10^6 cells/ml) were incubated in the presence of 20 nM [^{3}H]PDBu (New England Nuclear Chemicals, Germany; sp. act. 18.6 Ci/mmol) for 20 min. The total [^{3}H]PDBu binding activity was determined from the radioactivity incorporated in the whole cells collected over Whatman glass fiber filters. To measure the amount of membrane-inserted form of PKC, cells were disrupted in homogenizing buffer containing 5 mM EGTA. The cell homogenates were centrifuged at 20000 g for 15 min to remove loosely associated components from the membranes. The washed pellets were then thoroughly disaggregated in 0.8 ml of Soluene-350 (Packard, The Netherlands) and radioactivity was determined by liquid scintillation spectroscopy.

---

*Correspondence to: M. G. Monti (Chimica Biologica), Università di Modena, Via Campi 287, I-41100 Modena, Italy.

Ca^{2+} transport across the cell membrane was assayed by use of $^{45}Ca^{2+}$; HL-60 cells (2×10^6 cells/ml) were incubated with $^{45}CaCl_2$ (Amersham, U.K.) at 2 µCi/ml (total $CaCl_2$ in culture medium, 0.4 mM). Samples were removed at various time points and the cells were colleted by centrifugation at 1000 g for 4 min. The pelletted radioactivity was determined as described above.

The viability of the cultures was not altered by exposure to ELF-PEMFs used. Sample heating due to ELF-PEMF exposure was effectively removed by a water-cooling system.

RESULTS AND DISCUSSION

When HL-60 cells are assayed for the ability to take up Ca^{2+}, the results shown in Fig. 1 are obtained. Treatment with the pulsed electromagnetic field resulted in a striking increase in total calcium accumulated in the cells compared to untreated ones.

As already emphasized, PKC is physiologically active only when associated with cell membranes. The membrane-associated enzyme binds phorbol esters with a high affinity, so that the amount of active enzyme can be conveniently determined by its ability to bind [^{3}H]PDBu with a precise 1:1 stoichiometry.[12]

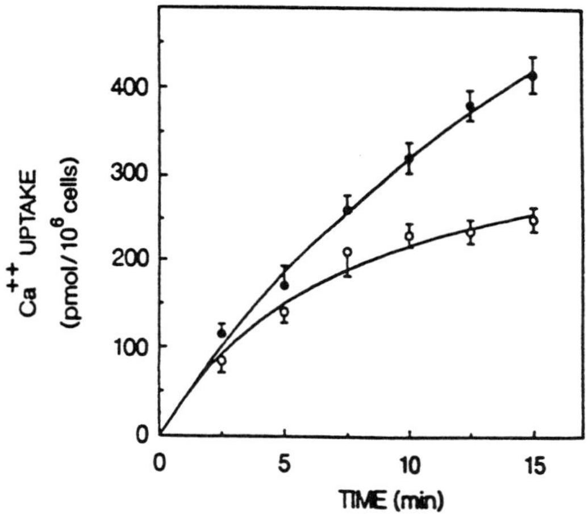

FIG. 1.—Effect of ELF-PEMF on Ca^{2+} uptake in HL-60 cells. Results are means of four separate experiments. O control cells, • treated cells.

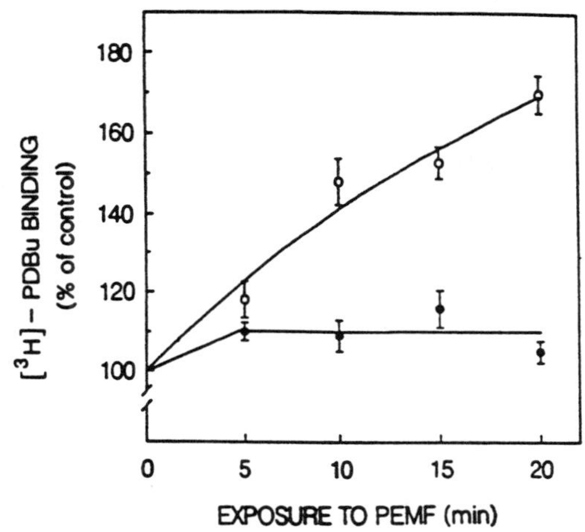

FIG. 2.—Effect of ELF-PEMF on [^{3}H]PDBu binding to intact HL-60 cells in the absence (O) or presence (•) of 1 µM verapamil. Cells were incubated with 20nM [^{3}H]PDBu for 20 min and exposed to PEMF for durations reported. Abscissa is time of exposure to PEMF during last portion of 20min incubation with label. Verapamil was added 2 min before addition of labeled PDBu. Results are expressed as percentage of [^{3}H]PDBu bound to untreated cells (65 ± 3.3 pmol/10^9 cells) and reported as average of four separate experiments.

As shown in Fig. 2, treatment of HL-60 cells with the pulsed electromagnetic field markedly increased the extent of [^{3}H]PDBu binding, the extent being dependent on the time of exposure to PEMF (open circles).

To assess whether an increased influx of Ca^{2+} could be involved in the observed effect of the EMF, experiments were performed in the presence of verapamil, a Ca^{2+} channel blocker. Results reported in Fig. 2 (closed circles) show that the presence of 1µM verapamil in the culture medium abolished the stimulatory effect of ELF pulsed electromagnetic fields. These data support the hypothesis that increasing Ca^{2+} influx is a necessary prerequisite of the observed effect of ELF-PEMFs on PDBu binding activity in HL-60 cells.

A characteristic feature of PKC in living cells is its distribution between a cytoplasmic inactive form and a membrane-bound active form. An additional intriguing feature of this enzyme is its ability to form two membrane-associated states; one is calcium-dependent and reversible, the other is an irreversible complex with the characteristics of an intrinsic membrane protein. It has been suggested that the formation of irreversible membrane-bound PKC complex is important for long-term effects in the cell.[13]

That poses the question whether the PEMFs used in our experiments could exert their stimulatory effect on PKC activation by increasing the reversible and/or irreversible membrane-bound complex. To answer this question, cells were disrupted in the presence of 5 mM EGTA, and the [^{3}H]PDBu binding activity of the particulate fraction was determined. The presence of EGTA in the homogenizing buffer insures that only the tightly bound, chelator-stable form of PKC is determined. As shown in Table 1, the [^{3}H]PDBu binding to the inserted form (chelator-stable form) of PKC was not influenced by the exposure of HL-60 cells to ELF-PEMF, which suggests that the observed effect on PKC activation in ELF-PEMF treated cells was due to an increase in reversible Ca^{2+}-dependent association of the enzyme to membranes.

TABLE 1.—Effect of ELF-PEMF and Ca^{2+} ionophore A23187 treatment on the total and chelator-stable binding of [^{3}H]PDBu in HL-60 cells.

Treatment	[^{3}H]PDBu binding to whole cells	Chelator-stable [^{3}H]PDBu binding to membrane fraction
	pmol/10^9 cells	
Control	65 ± 3.3	38 ± 1.9
ELF-PEMF	111 ± 5.8	37 ± 1.8
A23187	126 ± 6.7	44 ± 2.1

Cells were incubated with 20 nM [^{3}H]PDBu for 20 min with or without PEMF treatment. 1 μM A23187 was added 2 min before the addition of labeled PDBu. Results are means ±S.D. of three separate experiments.

To test whether an increase in Ca^{2+} influx enhances the reversible association of PKC to membranes, we studied the effect of the Ca^{2+} ionophore A23187 on the [^{3}H]PDBu binding activity of both total and chelator-stable membrane-bound enzyme. As shown in Table 1, the total extent of [^{3}H]PDBu binding to whole HL-60 cells was markedly stimulated by the addition of A23187 and this effect appeared to be due to Ca^{2+}-induced reversible association of PKC to membrane.

Our present data show that treatment of HL-60 cells with ELF pulsed EMFs promotes the reversible association of PKC to membranes, and that this process depends on an increase of Ca^{2+} transport across membranes.

We propose that the ELF pulsed EMFs used in these experiments enhance the intracellular Ca^{2+} by specific effects on plasma membrane permeability. This effect would in turn support the PKC association with membranes. We also propose that the subsequent transient activation of PKC represents the critical element in regulating the biological response of the cells to ELF-PEMF treatment.[14]

REFERENCES

1. M. J. Berridge, *Ann. Rev. Biochem.* 56: 159-193, 1987.
2. J. Walleczek and R. P. Liburdy, *FEBS Lett.* 271: 157-160, 1990.
3. J. J. L. Carson, F. S. Prato, D. J. Drost, L. D. Diesbourg, and S. J. Dixon, *Am. J. Physiol.* 259 (Cell Physiol. 28): C687-692, 1990.
4. D. B. Lyle, X. Wang, R. D. Ayotte, A. R. Sheppard, and W. R. Adey, *Bioelectromagnetics* 12: 145-156, 1991.
5. R. Cadossi, G. Emilia, C. Ceccherelli, and G. Torelli, in A. A. Marino, Ed., *Modern Bioelectricity*, New York: Marcel Dekker, 1988, pp. 451-496.
6. M. Wolf, H. LeVine III, W. S. May Jr., P. Cuatrecasas, and N. Sahyoun, *Nature* 317: 546-549, 1985.
7. M. D. Bazzi and G. L. Nelsestuen, *Biochemistry* 28: 3577-3585, 1989.
8. N. Schwantke, F. Le Bouffant, M. Dorée and C. J. Le Peuch, *Biochimie* 67: 1103-1110, 1985.
9. Y. Nishizuka, *Science* 233: 305-312, 1986.
10. R. M. Bell, *Cell* 45: 631-632, 1986.
11. M. G. Monti, L. Pernecco, M. S. Moruzzi, R. Battini, P. Zaniol, and B. Barbiroli, *J. Bioelectricity* 10: 119-130, 1991.
12. U. Kikkawa, Y. Takai, Y. Tanaka, R. Miyake, and Y. Nishizuka, *J. Biol. Chem.* 258: 11442-11445, 1983.
13. M. D. Bazzi and G. L. Nelsestuen, *Biochemistry* 27: 7589-7593, 1988.

14. This work was supported by a grant from MURST (quota 60%).

Martin Blank, Ed., *Electricity and Magnetism in Biology and Medicine*

IN VIVO DNA SYNTHESIS OF SPECIFIC GENES ENHANCED BY LOW-DOSE-RATE, 915MHz CW MICROWAVE RADIATION

S. K. Dutta*

Howard University, Washington, D.C.

M. Verma

Georgetown University, Washington, D.C.

Continuous-wave (CW) 915MHz microwave radiation is used for clinical and therapeutic purposes. In order to understand the effects of this radiation on DNA synthesis in vivo we determined the levels of DNA synthesis in *Neurospora crassa* under exposure conditions and compared them with those without any exposure. Enhanced in vivo DNA synthesis was observed when exponentially growing mycelial cells of *N. crassa* were exposed to 915MHz CW radiation at SAR = 0.1 W/kg. The enhanced DNA synthesis was determined by the ^{3}H-thymidine incorporation.

Reports on the interaction of microwaves (MW) with DNA in vitro are available from various groups and have been reviewed with descriptions of their beneficial and harmful effects.[1,2] In their studies, Sagripanti and Swicord determined the ability of MW to cause single-strand breaks as well as potentially more harmful double-stranded breaks in plasmid DNA that are conventionally related to mutagenesis and carcinogenesis.[3] However, they cautioned that because the work was conducted in vitro with pure DNA molecules, the conclusions should not be extrapolated to more complex living systems. Most of the results on the mutagenesis of MW were found to be negative or could be explained by an increase in temperature. The field of the interaction of athermal microwaves in various in vivo systems involving the studies of DNA synthesis in general and specific genes in particular is apparently still open for investigation. We undertook this challenge and studied the effect of athermal microwaves on DNA synthesis. Cleary et al. reported stimulation of cell proliferations and DNA synthesis using radiofrequency radiation under isothermal conditions.[4] Further evidence of enhanced in vivo DNA synthesis at SAR = 0.05 W/kg for the fungal cells of *N. crassa* is reported here. In this study we were interested to know whether microwave exposure (915MHz CW at SAR = 0.1 W/kg) causes induction of DNA synthesis, possibly by making them cancerous.

MATERIALS AND METHODS

Culture Conditions and DNA Probes

The conditions of the growth and maintenance of the fungus *N. crassa* have been published previously.[5]

Exposure System

The Crawford exposure cell (Instruments for Industry, Model BC110) consists of a rectangular transmission line tapered at each end to a transection that mates with a standard coaxial cable. The Crawford cell was housed within an incubator at 30 C. A signal generator (Hewlett-Packard, Model 8640) produced unmodulated 915MHz CW sinusoidal radiation, and a linear amplifier (Amplifier Research, Model 1W1000) amplified the 915MHz signal to 1 W to provide adequate input power to the transmission line. The details of the system and estimation of SAR have been described previously.[6]

Exposure of *N. crassa* Cells and Isolation of DNA

A known number of conidia (2×10^6 cells/ml) were used for each of two different experiments, and conidia were allowed to grow in minimal medium at 30 C under shaking conditions.[5] After 18 h, fully grown

*Correspondence to: S. K. Dutta (Botany), Howard University, Washington DC 20059.

mycelial cells were in the exponential phase. These cells were transferred to various flasks of surface area 25 cm^2. Two of these flasks were used for exposure and two were used as controls. The exposure was done at various power levels with 915MHz CW radiation for 30 min at 30 C. The short duration of 30 min was chosen because in an organism such as *N. crassa* the rate of DNA synthesis is high—about five nucleotides per second.[7] Just before exposure, ^{3}H-thymidine (2.5 μCi) was included in each flask. Immediately following exposure (15000 g for 20 min at 40 C), cells were harvested and washed twice with 50 mM Tris-HCl, at pH 7.5, and with 1 mM magnesium chloride (20 ml each time). Thus ^{3}H-thymidine was present only at the time of exposure. The amount of DNA synthesis was then measured by determination of the trichloroacetic acid precipitable radioactivity as described by Russell et al.[8]

Monitoring Temperature Increases

In our exposure conditions there was no increase in the temperature during the exposure; the effects that we observed were thus caused by exposure only and were not due to heat, as described also in our previous contribution.[6]

RESULTS AND DISCUSSION

When *N. crassa* cells were exposed to 915MHz radiation at SAR = 0.1 W/kg, an increased uptake (about three-fold) of ^{3}H-thymidine was observed in exposed samples compared to their control counterparts. A dose-response curve for ^{3}H-thymidine uptake was also determined with 915MHz CW radiation at various SAR values (Table 1). The results suggest that the uptake of thymidine initially increases with the increase in the dose rate of radiation. The peak value is obtained at SAR = 0.1 W/kg; a decline in the rate of uptake of thymidine is observed at higher SAR values. The uptake of ^{3}H-thymidine into DNA was determined by the trichloroacetic acid precipitation method, which inherently excluded free thymidine from the washing step. As further proof that the increase in the uptake of thymidine was caused by DNA synthesis *per se*, the activity of DNA polymerase I was determined (data not shown) and similar results were obtained. Figure 1 confirms enhanced DNA synthesis when measured quantitatively.

TABLE 1.—Uptake of ^{3}H-thymidine in *N. crassa* cells with and without exposure to 915MHz CW radiation at various SAR values.

SAR (W/kg)	Control (C) samples CPM±SE	Exposed (E) samples CPM±SE	E/C
0.0001	3155 ± 85	3312 ± 80	1.05
0.001	3265 ± 78	6399 ± 83	1.96
0.01	3506 ± 205	7292 ± 210	2.08
0.05	3136 ± 204	12199 ± 202	3.89
0.1	3264 ± 185	13872 ± 176	4.25
0.5	3365 ± 76	12585 ± 70	3.74
1.0	3166 ± 180	10922 ± 165	3.45
4.0	3256 ± 201	7977 ± 196	2.47

These values were derived from three experiments run independently. Each experiment was done with the same number of *N. crassa* cells under identical conditions. Plateau is reached when cells are exposed at SAR = 0.1 W/kg; further increase in the SAR value yields a lower E/C ratio (albeit more than the control values). Exposure time for each experiment was 30 min.

Findings regarding the potential of microwaves for affecting biological systems are conflicting. Some reports suggest that microwaves have too little energy to cause a mutation or break a covalent bond and that DNA in solution does not absorb microwaves.[9,10] Sagripanti and Swicord found that microwaves interact with specific conformation of DNA in solution (that is, linear or covalently closed circular plasmid DNA);[3] however, the presence of cuprous ions may have contributed to their results. Our studies provide the groundwork for the extension of studies related to the possible interaction of microwaves and enhanced DNA synthesis. It may be premature to pinpoint that cells become cancerous as a result of exposure to microwaves, but it is possible that microwave radiation affects the S phase of the cell cycle,[11] and thus the properties of the cell might be changed.[12]

REFERENCES

1. C. Polk and E. Postow, *CRC Handbook of Biological Effects of Electromagnetic Fields*, Boca Raton, Fla.: CRC Press, 1986.

2. S. K. Dutta and R. M. Millis, *Biological Effects of Electropollution: Brain Tumors and Experimental Models*, Philadelphia: Information Ventures, Inc., 1986.

3. J. L. Sagripanti and M. L. Swicord, "DNA structural changes caused by microwave radiation," *Int. J. Rad. Biol.* 50: 47-50, 1986.

4. S. F. Cleary, L.-M. Liu, and R. E. Merchant, "In vitro lymphocyte proliferation induced by radiofrequency electromagnetic radiation under isothermal conditions," *Bioelectromagnetics* 11: 47-56, 1990.

5. S. K. Dutta, "DNA homologies among heterothallic species of *Neurospora*," *Mycologia* 68: 388-401, 1976.

6. S. K. Dutta, A. Subramanian, B. Ghosh, and R. Prashad, "Microwave radiation induced calcium-ion efflux from human neuroblastoma cells in culture," *Bioelectromagnetics* 5: 71-78, 1984.

7. W. F. Marzluff and R. C. C. Huang, "Transcription of RNA in isolated nuclei," in B. D. Hames and S. J. Higgins, Eds., *Transcription and Translation*, Oxford: IRL Press, 1986, 89-129.

8. P. J. Russell, J. R. Hammett, and E. U. Selker, "*Neurospora crassa* cytoplasmic ribosomes: Ribosomal ribonucleic acid synthesis in the wild type," *J. Bacteriol.* 127: 785-793, 1976.

9. C. Gabriel, E. H. Grant, R. Tata, P. R. Brown, B. Gestblom, and E. Noreland, "Dielectric behavior of aqueous solutions of plasmid DNA at microwave frequencies," *Biophys. J.* 55: 29-34, 1989.

10. A. Leonard, A. J. Berteaud, and A. Bruyere, "An evaluation of the mutagenic, carcinogenic and teratogenic potential of microwaves," *Mut. Res.* 123: 31-46, 1983.

11. B. Lewin, *Genes IV*, Oxford: Oxford Publishing Co., 1990.

12. This work was partly supported by grant 8SO6GM08016 from the National Institute of General Medical Sciences and grant R814126-01 from the U.S. Environmental Protection Agency, as well as by a core grant from the Graduate School of Howard University.

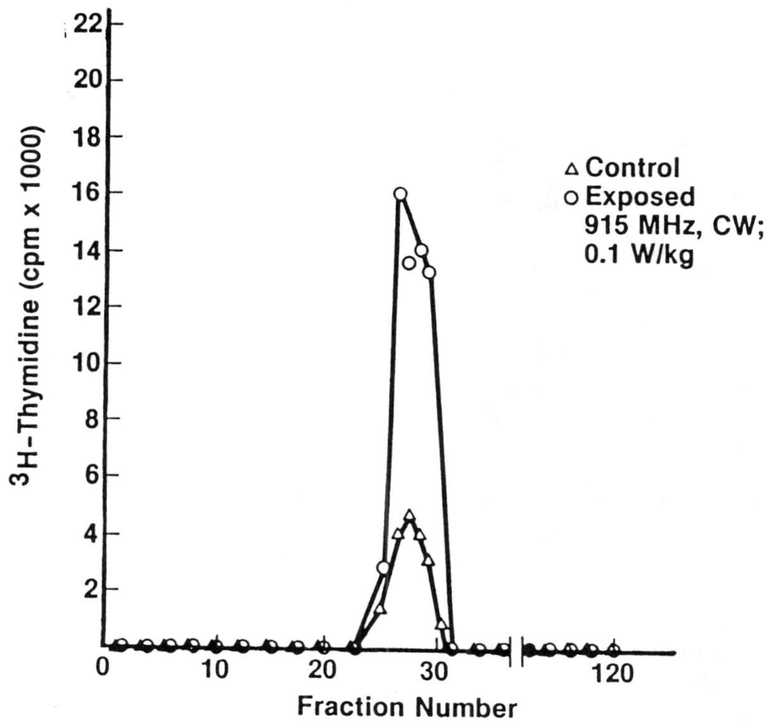

FIG. 1.—Induction of DNA synthesis by 915MHza CW microwave radiation. ^{3}H-DNA isolated from exposed and unexposed samples was electrophoresed on 0.5% agarose gel and stained with ethidium bromide (0.5 μg/ml) for 30 min. Gel was sliced and slices were counted in liquid scintillation counter.

Martin Blank, Ed., *Electricity and Magnetism in Biology and Medicine*

CHANGES IN THE DISTRIBUTION OF PROTEINS FOLLOWING ELECTROMAGNETIC STIMULATION OF *SCIARA* SALIVARY GLANDS

M. Blank,* O. Khorkova, and R. Goodman

Columbia University, New York, NY

The biosynthetic responses of *Sciara coprophila* salivary gland cells to electromagnetic (EM) fields are compared to those following heat shock (HS). Data from 2D gels, converted into protein distribution curves (% mass vs MW, % mass vs pI), show that the changes due to EM and HS are quite similar. There are new peaks at MW = 30 kD and pI = 7.1 for both EM and HS. There is also a shift of the peak in pI distribution from 5.8 to 6.3 for both EM and HS. The changes in the two pI peaks as a function of EM field strength define a dose-response curve, as well as the range of the effect. These changes may be characteristic of the cellular response to stress.

Changes in biosynthesis are stimulated by electromagnetic (EM) waves in the frequency range below 300 Hz.[1-4] The changes in the proteins are rather complicated, but there appear to be similarities in the patterns of the responses of cells to two physical stresses, EM fields, and heat shock (HS). The patterns found in the distribution of the major proteins synthesized by the salivary glands cells of *Sciara*[5-7] suggest that EM fields stimulate a pathway that is similar to that used by cells in response to HS and other kinds of physical stress. There may also be specific responses to each stimulus, as suggested by the observation that new proteins have a distribution of charges (i.e., smaller proteins are more highly charged) that would be expected if they arose from an interruption of biosynthesis. In this paper, we present evidence for clear patterns in the biosynthetic response to EM fields.

EXPERIMENTAL RESULTS

Sciara coprophila salivary gland cells maintained at normal growth temperature of 20 C and isolated from known electromagnetic field sources served as controls. Some cells were exposed to heat shock at 37 C; others, to 60Hz sinusoidal magnetic fields that varied by factor of 10 increments from 0.8 μT to 800 μT, generated by 13 × 14cm Helmholtz coils with an 8cm spacing.

Two-dimensional electrophoresis was performed by Kendrick Labs, Inc. (Madison, Wisc.). The gels were scanned and the images (along with the calibration strips) were stored as .IMG files on diskettes and analyzed by use of the software QGEL (Kendrick Labs) and QuattroPro (Borland).

Changes in the distribution of proteins in the different molecular weight ranges are shown in Fig. 1(a) for control conditions, for EM stimulation at 0.8 μT, and for HS. The main differences, due to both EM and HS, are the reduction of % mass around 20 kD, the peak at 30 kD, and the depressions around 35 kD and 70 kD. Increase in the strength of the magnetic field maintains the charges, but at 800 μT, the highest field strength studied, the effects are much smaller.

The changes in the pI distribution are more pronounced. Figure 1(b) shows a comparison of the distribution under control conditions, with the lowest field tested (0.8 μT), and with HS. Both stresses show the same features—a shift of the peak in the distribution to more neutral pI (from 5.8 to 6.3) and the development of a peak at pI = 7.1. These features appear to all levels of EM stimulation, and were weakest at 800 μT, the highest field intensity studied. If the magnitudes of the two peaks in the pI distribution are plotted as a function of increasing EM field intensity, the graph defines a dose-response curve with an approximate range between 0.8 and 800 μT.

---

*Correspondence to: M. Blank (Physiology and Cellular Biophysics), Columbia University, 630 W 168th St., New York NY 10032.

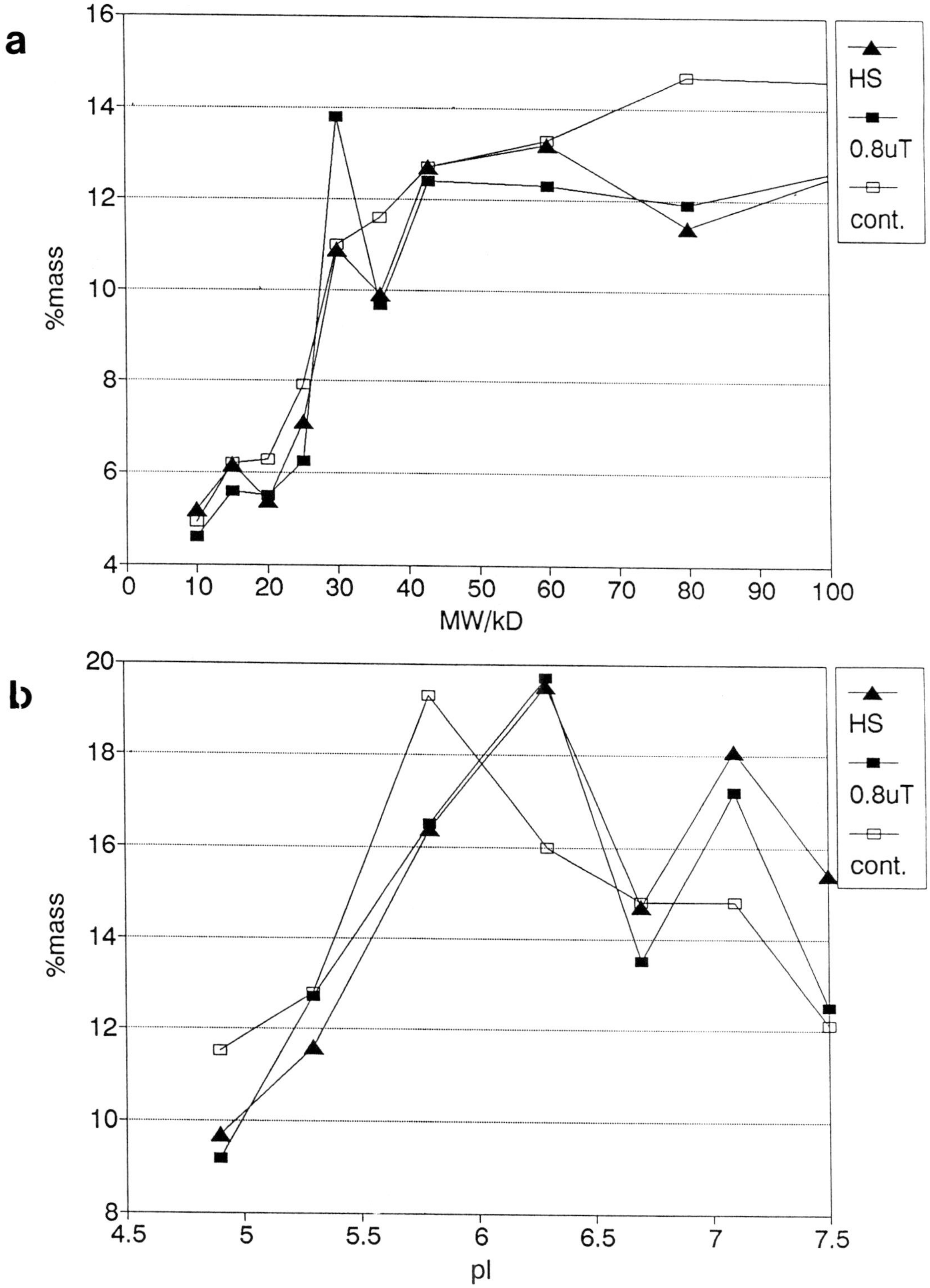

FIG. 1.—Distribution of mass as a function of MW for EM-stimulated, HS-stimulated, and control protein synthesis. % mass is plotted vs (a) MW and (b) pI for control (□), EM-exposed (0.8 μT for 45 min) (□), and HS-exposed (at 37 C for 45 min) (Δ). Control is average of 9 gels, EM is average of 3, and HS is average of 4. Standard errors are less than 1%.

DISCUSSION

Our results indicate that the patterns of proteins synthesized as a result of EM stimulation differ from the control pattern and are similar to the pattern seen in response to heat shock. The main features of the response are:

- New peak at 30 kD for both EM and HS.
- New peak at pI = 7.1 for both EM and HS.
- Shift of peak in pI distribution from 5.8 to 6.3 for both EM and HS.

Examination of both EM and HS gels, especially the prominent spot at 30 kD and pI = 7.1, suggests that phosphorylation or glycosylation, post-translational modifications, contribute to the response.

The response at the lowest intensity studied, 0.8 μT, is well above the field under ambient conditions (about 0.1 μT in the mu-metal box). The decrease in the response at 800 μT, the highest field intensity studied, suggests that higher field strengths could be ineffective or inhibitory.

The EM and HS stimuli are very different, so that the cells appear to be responding in a largely undifferentiated way to stresses of different modalities and intensities. The small differences between the two responses may indicate specific effects, but overall, the cells appear to be using the same stress-response pathway to both stimuli.

REFERENCES

1. R. Goodman and A. S. Henderson, ''Exposure of salivary gland cells to low frequency electromagnetic fields alters polypeptide synthesis,'' *Proc. Nat. Acad. Sci. USA* 85: 3928-3932, 1988.

2. R. Goodman and A. S. Henderson, ''Transcription and translation in cells exposed to extremely low frequency electromagnetic fields,'' *Bioelectrochem. Bioenerg.* 25: 335-355, 1991.

3. K. J. McLeod, R. C. Lee, and H. P. Ehrlich, ''Frequency dependence of electric field modulation of fibroblast protein synthesis,'' *Science* 236: 1465-1469, 1987.

4. R. A. Luben, C. D. Cain, M. Y. Chen, D. M. Rosen, and W. R. Adey, ''Effects of electromagnetic stimuli on bone and bone cells, in vitro: Inhibition of responses to parathyroid hormone by low-energy, low-frequency fields,'' *Proc. Natl. Acad. Sci. USA* 79: 4180-4184, 1982.

5. M. Blank and R. Goodman, ''An electrochemical model for the stimulation of biosynthesis by external electric fields,'' *Bioelectrochem. Bioenerg.* 19: 569-580, 1988.

6. M. Blank and R. Goodman, ''New and missing proteins in the electromagnetic and thermal stimulation of biosynthesis,'' *Bioelectrochem. Bioenerg.* 21: 307-317, 1989.

7. M. Blank and R. Goodman, ''Charge effects in electromagnetic stimulation of biosynthesis,'' in G. Milazzo and M. Blank, Eds., *Bioelectrochemistry III: Charge Separation Across Membranes*, New York: Plenum Press, 1990, 311-324.

Martin Blank, Ed., *Electricity and Magnetism in Biology and Medicine*

MULTIGENERATION EXPOSURE TEST OF *DROSOPHILA MELANOGASTER* TO ELF ALTERNATING MAGNETIC FIELDS

T. Kikuchi and M. Ogawa

Tokyo Electric Power Co., Japan

Y. Otaka* and M. Kato

Mitsubishi Kasei Institute of Toxicological and Environmental Sciences, Ibaraki, Japan

In previous work, we obtained negative results in a sex-linked recessive lethal mutation test using fruit flies exposed to 0.5 and 5mT 50Hz sinusoidal alternating magnetic fields. In the present work, we investigated the effects of the fields to nonlethal mutation by the Curly/Plum accumulation method and also obtained negative results. Nondominant lethal mutations were accumulated on the wild type second chromosomes for 40 generations under continuous exposure and the mutations were tested on homozygotes by sib-mating at every 10 generations. Viability distributions of the progeny were compared among the exposed and the control groups.

Two investigations on *Drosophila melanogaster* (D.m.) were performed to determine whether the power-frequency magnetic field is a mutagen. Ramirez et al. obtained a positive result in his dominant lethal test using D.m. exposed to a 1mT 50Hz sinusoidal alternating magnetic field,[1] but Walter et al. obtained a negative result in the duplicate experiment.[2] We also carried out a sex-linked recessive lethal test of the fly exposed to 50Hz sinusoidal alternating magnetic field and reported a negative result.[3]

Nonlethal spontaneous mutation affecting viability of the fly is known to occur about 20 times more frequently than lethal.[4] In the present work, we investigated whether the field affects the frequency of nonlethal mutation. The mutations were accumulated on the second chromosomes of wild-type flies under continuous exposure to the fields for 40 generations and compared with the mutations of the control according to the same method as adopted by Mukai et al.[4]

METHOD

A second chromosome, on which recessive lethal and nonlethal mutations were accumulated, was maintained for 40 generations as a heterozygote with that from Curly/Plum (Cy/Pm) fly. Cy/Pm has dominant phenotypic characteristics of curly wings and plum-colored eyes, which serve as markers. As mutations are accumulated, the viability decreases. In order to estimate the viability, homozygotes (+/+) were made by sib-mating of Cy/+ male and Cy/+ female from a culture and the number of +/+ flies and that of Cy/+ flies were counted in generations 20, 30, and 40. (Cy/Cy is lethal.) The mating scheme is shown in Fig. 1.

Culture vials were continuously exposed to 0.5 and 5mT 50Hz sinusoidal alternating magnetic fields generated by Helmholtz coils shown in our previous paper.[3] The culture room was controlled at 25±1 C, 60±5% RH. Exposure levels were same as in the previous paper. Controls were cultured in a corner of the room, where the background field did not exceed 1 μT.

RESULTS AND DISCUSSION

Lines with less than 1% of the expected number of +/+ flies in the progeny of sib-mating were regarded as lethal. Viability γ is defined as (2 × number of +/+ flies)/(number of Cy/+ flies + 1). The expected value of γ is 1.00 for the line without deleterious mutation. Figure 2 shows the viability distributions. By the F-test

---

*Correspondence to: Y. Otaka, Mitsubi Kasei Institute of Toxicological and Environmental Sciences, 14 Sunayama Hasaki-Machi, Kashima-Gun, Ibaraki 314-04 Japan.

and the Bartlett's test prior to it, no significant difference was detected between the distributions at each generation. Figure 3 shows the decreases of average viabilities with generation, where lines with less than 0.3 are defined as markedly detrimental. The rates of decrease are a little larger than that reported by Mukai et al.[4]

CONCLUSION

No significant difference was detected between the viability distributions at each generation.

REFERENCES

1. E. Ramirez, J. L. Monteagudo, M. Garcia-Gracia, and J. M. R. Delgado, "Oviposition and development of *Drosophila* modified by magnetic fields," *Bioelectromagnetics* 4: 315-326, 1983.

2. E. Walter and E. L. Carstensen, "Test for the effects of 60Hz magnetic fields on fecundity and development in *Drosophila*," *Bioelectromagnetics* 8: 351-354, 1987.

3. Y. Otaka, S. Kitamura, M. Furuta, and A. Shinohara, "Sex-linked recessive lethal test of *Drosophila melanogaster* after exposure to 50-Hz magnetic fields," *Bioelectromagnetics* 13: 67-74, 1992.

4. T. Mukai, S. I. Chigusa, L. E. Mettler, and J. F. Crow, "Mutation rate and dominance of genes affecting viability in *Drosophila melanogaster*," *Genetics* 72: 335-355, 1972.

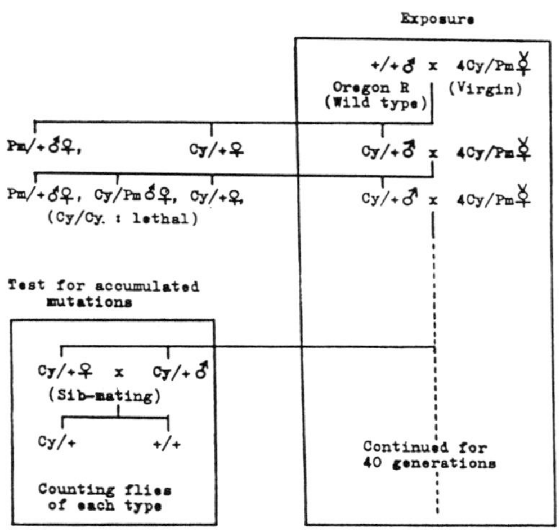

FIG. 1.—Mating scheme.

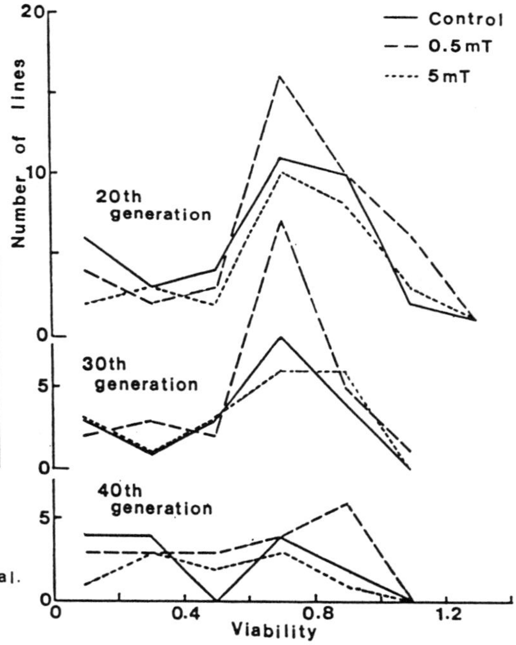

FIG. 2.—Viability distribution at each generation.

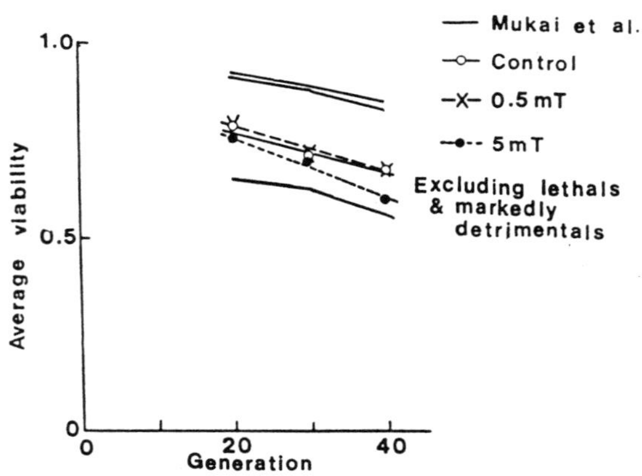

FIG. 3.—Decrease in average viability.

Martin Blank, Ed., *Electricity and Magnetism in Biology and Medicine*

MICROWAVE-INDUCED HYPERTHERMIA DISRUPTS WORKING MEMORY AND EVOKES SELECTIVE EXPRESSION OF BRAIN C-FOS

G. A. Mickley* and B. L. Cobb

Brooks Air Force Base, Tex.

P. Mason

Operational Technologies Corp., San Antonio, Tex.

L. K. Weigel

Trinity University, San Antonio, Tex.

We tested remembering of recently explored objects to determine changes in working memory following exposure to 600Mhz microwave radiation. Brain/rectal temperatures and expression of the early proto-oncogene c-fos were also measured. Rats exposed to 10, 9.3, 8.5, or 5 W/kg microwaves exhibited a reliable brain hyperthermia compared with nonirradiated control rats. Only the 9.3 and 10W/kg treatments produced a significant memory disruption. There was a noticeable difference in the distribution of c-fos between the normothermic and hyperthermic groups. These data suggest that working memory may be disrupted by microwave-induced hyperthermia. However, brain temperatures must be raised approximately 2.0 C before this amnesia is observed. The pattern of expression of c-fos may suggest candidate brain nuclei that mediate the behavioral changes we observed following microwave-induced hyperthermia.

Memory tests are among the most sensitive behavioral assessments of toxicological insult.[1] Here, we sought to determine the threshold for microwave-induced hyperthermia's disruption of a working memory task. Further, we measured the expression of the early proto-oncogene c-fos as an indicator of neural activation produced by radiation exposure.

METHODS

Animals. Subjects were male rats (325-400 g) of the Sprague-Dawley CD-VAF/Plus strain obtained from Charles River (Portage, Mich.). The animals were individually housed (within an AAALAC-accredited vivarium). Rodent chow and water were available *ad libitum*. Home cage temperature was maintained at 23-26 C under a 12/12h light-dark cycle (lights on at 0600h).

Exposure System and Dose. CW microwave energy was provided by a Microwave Cavity Laboratories, Inc. (MCL) 1 kW radiofrequency radiation (RFR) source operating at 600 Mhz. The amplified RFR signal was transmitted to a 90° corner reflector via RG-318/U coaxial waveguide. Exposures were performed within an Emerson and Cuming Eccosorb anechoic chamber maintained at 24 ± 1 C.

Rats were irradiated individually in plastic restraint holders positioned so that the long axis of the rat was parallel to the electric field. Far-field whole-body specific absorption rates (SAR = time rate of total energy absorbed divided by the total mass of the body) were determined for each animal. Determination of whole-body SAR was accomplished by use of the Dewar-flask calorimetric technique. During microwave exposure, individual rats were lightly restrained in ventilated plastic holders (3 mm thick). Each day, for up to 4 weeks prior to exposure, animals were placed in holders for 30 min and rectal probes were inserted 3 cm past the rectal sphincter in order to adapt them to these procedures.

---

*Correspondence to: G. A. Mickley (ALOEDR), 8308 Hawks Rd., Brooks AFB TX 78235.

Stereotaxic Surgery. Unilateral intracranial cannula implantation was performed in accordance with standard aseptic surgical procedures, with ketamine HCl (50 mg/kg, i.p.) and acepromazine (10 mg/kg, i.p.) used as anesthetic agents and atropine sulfate (0.4 mg/kg, i.p.) as a preanesthetic. A 1mm-diameter hole was drilled in the skull at the following stereotaxic coordinates, 3 mm posterior from bregma and 3 mm lateral to the midline. A plastic cannula was placed within the hole (just touching the dura mater) and fastened to the skull with dental cement and nylon screws. The cannula was filled with physiological saline and capped. In order to provide postsurgical analgesia, Alzet (#1003D) osmotic pumps (dispensing Buprenex, 0.3 mg/ml at 1 μl/h for 3 days post-op) were implanted subcutaneously in the nape of the neck. The pumps were removed 3 days later under methoxyflurane anesthesia.

Experimental Procedures. We used a simple test of remembering recently explored objects to determine changes in working memory. This test (developed by Ennaceur and Delacour in 1988[2]) requires a rat to explore initially a novel object within a familiar test arena (66.04 cm L×45.72 cm W×45.72 cm H, made of frosted plastic) for 10 min (Trial 1 = T1). The animal was then removed from the arena and placed inside a plastic holder and exposed for 20 min of microwave radiation (1 W/kg, $n = 13$; 5 W/kg, $n = 14$; 8.5 W/kg, $n = 8$; 9.3 W/kg, $n = 10$; and 10 W/kg, $n = 14$). Other animals ($n = 19$) were sham-irradiated for 20 min or remained in their home cage ($n = 6$). (These 2 control groups produced similar results and were combined for subsequent statistical analyses; see Results.) During microwave and sham-irradiation periods, rectal temperature was monitored with a BSD (#21-10104-002) microwave-compatible probe. Brain surface temperatures were also measured immediately before and after irradiation. Following exposure, the animal was placed back into its home cage for 10 min and then placed back into the test environment where, in addition to the now-familiar object, there was a novel object (both objects constructed from Doplos™) in a different area of the test arena. The animal was allowed to explore these objects within the arena for 3 min (Trial 2 = T2). Memory changes were evaluated by measurement of the relative time during which subjects explored the familiar versus the new stimulus objects. Normal subjects spend significantly more time in contract with new environmental components and less time with familiar objects.[2] A subject that extensively re-explores a stimulus with which it has previous experience is presumed to exhibit some memory loss associated with the object. The role (novel or new object) and the position of the two objects during T2 were counterbalanced and randomly permuted across our subject pool.

C-fos. Two hours after microwave or sham exposure, four rats (two sham-irradiated, two exposed to 8.5 W/kg) were anesthetized with sodium pentobarbital and perfused intracardially with a heparinized saline, followed by 4% paraformaldehyde/phosphate buffer. Brains were removed and placed in the buffer for 48 h, followed by soaking in 30% sucrose for 48 h. These brains were then sliced into 40μm sections. Sections were stained for c-fos expression with rabbit polyclonal antifos (AB-2, Oncogene Science) and rabbit IgG (PK-6101, Vector Laboratories), followed by 3,3'-diaminobenzidine.

RESULTS

Brain Temperatures. Radiation exposure produced a dose-dependent brain hyperthermia [$F(5,69) = 43.14$, $p < 0.01$ Fig. 1]. This brain temperature rise was statistically significant after exposure to 5.0, 8.5, 9.3, and 10.0 W/kg (Tukey HSD post hoc test, $p < 0.05$).[3]

Behavioral Measures. We calculated an amnesia index (AI) by subtracting the time during T2 spent exploring the new object from the time spent exploring the familiar object and dividing this number by the total object exploration time (Fig. 1). Radiation exposure had a significant effect on this indicator of working memory [$F(5,78) = 4.21$, $p < 0.01$]. However, individual group comparisons revealed that only 9.3 and 10.0 W/kg produced a statistically significant decrease in AI, compared with the control rats (Tukey HSD post hoc test, $p < 0.05$).

Total object exploration time was used as an indicator of generalized activity/inactivity. Although a one-way ANOVA reveal no overall change in this measure during T2 [$F(5,78) = 2.15$, $p > 0.05$], rats exposed to 10 W/kg showed less total object exploration than did sham-irradiated subjects. However, when all groups were considered, disruption of working memory (AI, see above) was not significantly correlated with this activity measure ($r = -0.0175$, $n = 84$, $p = 0.11$).

C-fos Expression. There was little expression of the c-fos protein in the sham-irradiated rat brains, but exposed rats showed substantial c-fos expression in the bed nucleus of the stria terminalis, medial preoptic region, lateral hypothalamus, paraventricular thalamic nucleus, central gray, and the pyramidal tracts.

DISCUSSION

Microwave-induced brain temperature increases above 2 C (SARs $\geq$ 9.3 W/kg) produced a disruption of working memory. Apparently, extreme hyperthermia can disrupt memory consolidation. (For review of this phenomenon, see McGaugh.[4]) These experimental results are consistent with, and extend, reports of hyperthermia-induced amnesia following exposure to hot water[5,6] or short-pulse high-power microwave.[7]

We saw no obvious behavioral signs of convulsions in our animals. Some investigators have reported that convulsions are necessary for hyperthermia-induced amnesia;[8] do not believe that behavioral convulsions must precede retrograde amnesia.[9] Although we did not make brain electroencephalographic recordings and therefore cannot totally rule out brain seizure activity, our data suggest that working memory disruption may be observed in nonconvulsive animals.

Our measure of working memory depends somewhat on locomotor activity, since movement is required to make contact with the test objects. However, there is no reason to expect that radiation-induced hypoactivity (e.g., following 10 W/kg) would selectively alter exploration of either the novel or old object during T2. In fact, the amnesia index was not significantly correlated with measures of total zone exploration time. Thus, these data suggest that the amnesia effect we observed may not be activity dependent.

The expression of c-fos in a variety of peri-ventricular nuclei point to these brain areas as particular targets of microwave-induced heating. Future studies will address the role of peri-ventricular structures in the working memory deficits observed here.

REFERENCES

1. J. Elsner, B. Hodel, K. Suter, D. Oelke, B. Ulbrich, G. Schreiner, V. Cumo, R. Cagiano, L. E. Roengren, K. E. Karlsson, and K. G. Haglid, "Detection limits of different approaches in behavioral teratology, and correlation of effects with neurochemical parameters," *Neurotoxicol. Teratol.* 10: 155-167, 1988.
2. A. Ennaceur and J. Delacour, "A new one-trial test for neurobiological studies of memory in rats: 1: Behavioral data," *Behav. Brain Res.* 31: 47-59, 1988.
3. B. J. Winer, *Statistical Principles in Experimental Design," New York: McGraw Hill, 1971.*
4. J. L. McGaugh, "Time-dependent processes in memory storage," *Science* 153: 1351-1358, 1966.
5. J. R. Misanin, R. E. Vonheyn, S. W. Bartlet, W. L. Boulden, and C. F. Hinderliter, "The effects of hyperthermia on memory in rats," *Physiol. Psychol.* 7: 339-344, 1979.
6. T. M. Rauch, D. I. Welch, and L. Gallego, "Hyperthermia impairs retrieval of an overtrained spatial task in the Morris water maze," *Behavioral & Neural Biology* 52: 321-330, 1989.
7. R. H. Lovely, R. B. Johnson, C.-K. Chou, and A. W. Guy, "Induction of retrograde amnesia for classically conditioned fear by a single microwave pulse," paper presented at the XIX URSI Assembly, 1978.
8. J. Kane and M. E. Jarvik, "Amnesic effects of cooling and heating in mice," *Psychon. Sci.* 18: 7-8, 1970.
9. S. Zornetzer and J. L. McGaugh, "Retrograde amnesia and brain seizures in mice," *Physiol. Behav.* 7: 401-408, 1971.

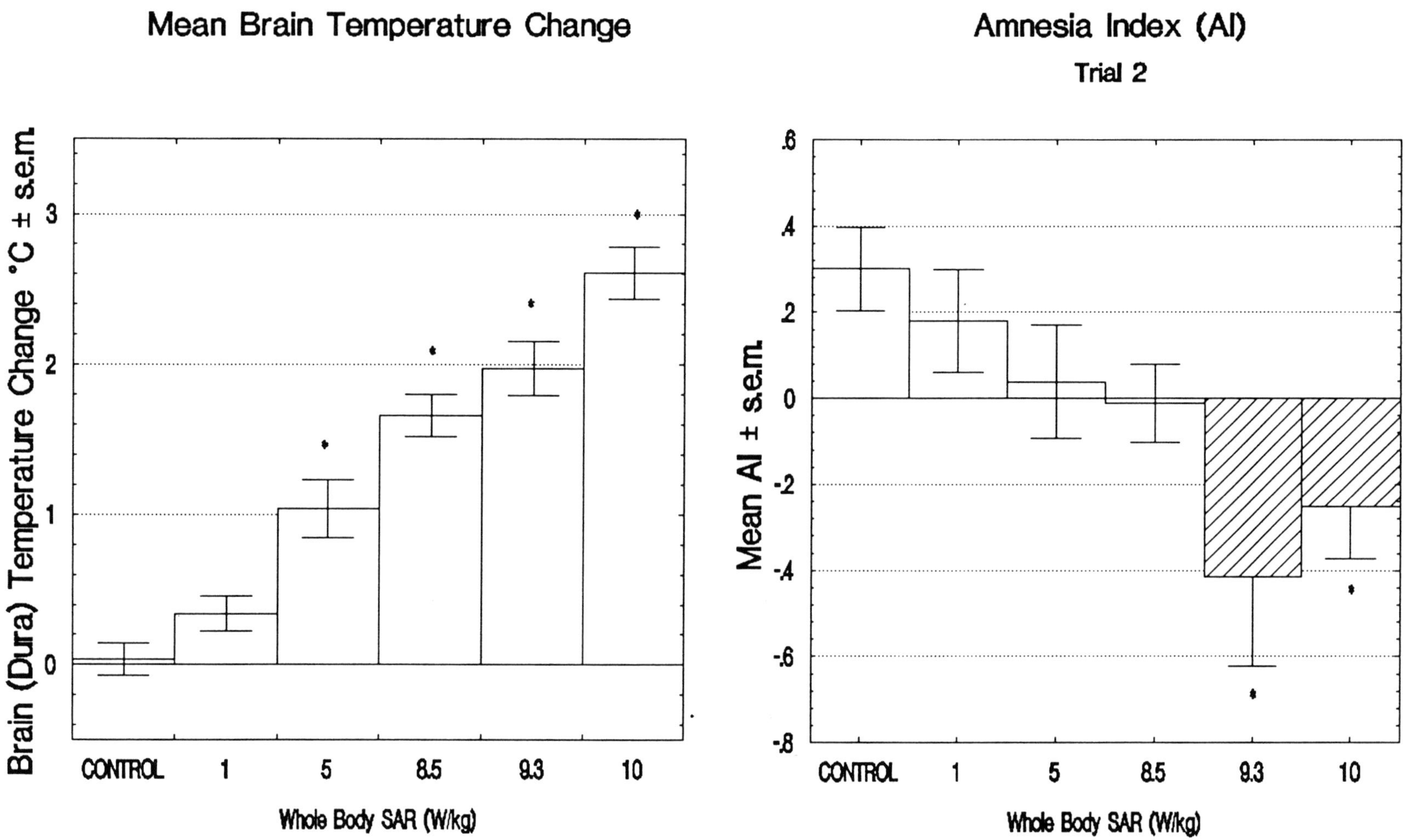

FIG. 1.—Microwave radiation exposure produces dose-dependent increase in brain temperature and (at the highest does used here) a retrograde amnesia; *p < 0.05 as compared with control.

Martin Blank, Ed., *Electricity and Magnetism in Biology and Medicine*

COMPARISON BETWEEN 'CLASSICAL' AND 'QUANTUM' MODELING OF BIOELECTROMAGNETIC INTERACTION MECHANISMS

B. Bianco and A. Chiabrera*

University of Genoa, Italy

G. D'Inzeo, A. Galli, and A. Palombo

"La Sapienza" University of Rome

In this work a compared analysis of "classical" and "quantum" modeling for the study of bioelectromagnetic nonthermal effects on cell membranes is carried out. In such models, what is to be discussed are both the hypotheses necessary to reduce the complexity of the subject to a level that can be mathematically handled, and the numerical results obtained as a function of various physical and energetic parameters.

Modeling of the mechanisms that are supposed to underlie the interaction phenomena between electromagnetic (EM) fields and biological systems is a much-discussed matter. In particular, as regards the important effects of weak ELF EM fields at the cell level,[1] theoretical investigations are typically based on microscopic physical approaches concerning the protein sites in membranes.

At first, a model has been developed by starting from a classical electrodynamic approach as an extension of the Lorentz force equation for ligands (ions, etc.) in viscous binding sites.[2] On this basis, theoretical developments[3] have demonstrated interesting agreements with some experimental data[1] and the possibility of analyzing various configurations of exposure.[4] A quantum approach has also been suggested[5] and developed,[6] in order to overcome some controversial aspects of the "classical" model, mainly the simplification of the viscosity, endogenous forces and noise.

Through a comparative analysis of both hypotheses and results of such models, some interesting considerations on the study of interaction mechanisms can be emphasized.

BASIC APPROACH

The basic point in modeling of interaction mechanisms at cell membrane level, through both a classical and a quantum approach, is represented by the hypothesis that a ligand is "free" when it is outside a "critical sphere" of radius R_c centered on the site, and is "bound" when it is inside the sphere. The probability of binding can be linked to the external electromagnetic perturbations, by evaluation of the alteration of the biological system starting from an unperturbed initial situation. The particular method of investigating this alteration is representative of the two different approaches.

A typical experimental setup for investigating bioelectromagnetic effects of weak ELF fields on cell systems is considered (Fig. 1). A Petri dish containing a biological sample is exposed to specific EM fields; in particular, it is supposed that a uniform magnetic field is given by a superposition of parallel static (dc) and time-harmonic (ac) components ($\mathbf{B}_{ext} = \mathbf{B}_{dc} + \mathbf{B}_{ac}$, $\mathbf{B}_{dc} = B_{0z}\,\mathbf{z}_0$, $\mathbf{B}_{ac} = B_{1z}\,\mathbf{z}_0 = B_1 \sin\omega t\,\mathbf{z}_0$), and the electric field is of induced type $\mathbf{E}_{ext} = \mathbf{E}_{ac} = \frac{1}{2}(\mathbf{r}+\mathbf{d})\times(\partial\mathbf{B}/\partial t)$, where $\mathbf{d}$ represents the site position referred to the center of the disk and $\mathbf{r}$ the site-ligand distance.

---

*Correspondence to: A. Chiabrera (Engg Biophysics & Electronics), University of Genoa, Via Opera Pia 11a, I-16145 Genoa, Italy.

CLASSICAL MODEL

It is possible to study from a classical point of view the dynamic effects of the EM field on the ligand (represented by the charge-to-mass ratio $\gamma = q/m$) in the protein site environment by means of the Lorentz-Langevin equation

$$\mathbf{F} = m\dot{\mathbf{v}} = q(\mathbf{E}_{ext} + \mathbf{v} \times \mathbf{B}_{ext} + \mathbf{E}_{end}) - b\mathbf{v}$$

It can be written in terms of the ion displacement $\mathbf{r}$ as

$$\ddot{\mathbf{r}} = \gamma(\mathbf{E}_{ext} + \dot{\mathbf{r}} \times \mathbf{B}_{ext} + \mathbf{E}_{end}) - \beta\dot{\mathbf{r}}$$

where the "viscosity parameter" $\beta = b/m\ [s^{-1}]$ characterizes the biological passive forces around the site, which have been chosen proportional to the charge velocity; the "endogenous parameter" χ [Vm] is linked to the restoring action of the site on the ligand, which, as a first approximation, has been chosen to be of coulombic type: $\mathbf{E}_{end} = -(\chi/r^3)\mathbf{r}$.

Even considering such a simple model, a large number of parameters, both measurable and unknown, can be varied. Several numerical solutions of the differential equations for the ion displacement $\mathbf{r}$ and velocity $\mathbf{v}$ have been obtained: a selection is presented here, by consideration of the variation of the field frequency f, of the viscosity parameter β, and of the endogenous parameter χ; the field-exposure configuration has been chosen as in some experiments on Ca^{2+}-dependent diatom motility.[1] In Fig. 2, Ca^{2+} ion displacement is plotted vs frequency and viscosity for an exposure time t_{final} starting from the initial position $\mathbf{r}_0$. In Fig. 3, Ca^{2+} ion displacement is plotted vs frequency and endogenous field.

QUANTUM MODEL

A quantum approach can be based on the equation for the density operator ρ:[6]

$$\frac{d\rho}{dt} = \frac{[H,\rho]}{jh} - \sum_e \{T_e, [T_e, (\rho - \rho_{th})]\}$$

where ρ_{th} is the density operator at thermal equilibrium. The first term on the right represents the contribution of the EM field through the Hamiltonian $H = H_0 + H_1$, which is a superposition of a Hamiltonian in absence of dynamic field $H_0 = \frac{1}{2} m\mathbf{P}\cdot\mathbf{P} + U_{end}$, and of a Hamiltonian taking into account the presence of dynamic fields, $H_1 \approx jh\,(q/m)\,\mathbf{A}\cdot\nabla + q\phi$, expressed by means of the well-known electrodynamic potentials $\mathbf{B} = \nabla \times \mathbf{A}$, $\mathbf{E} = -(\partial\mathbf{A}/\partial t) - \nabla\phi$. The second term on the right takes into account the interaction with the thermal bath.

Referring again to the same exposure conditions, through the solution of this equation we are able to evaluate directly the ligand-site binding probability alteration, which is plotted in Fig. 4 as a function of a normalized frequency. P^{av} is the probability of the Ca^{2+} ion to be inside the binding sphere when the exposure is "on," and P_{th}^{av} is the probability of the ion to be inside the binding sphere when the exposure is "off;" T is the state life-time.

CONCLUSION

The main critical aspect of the classical model can be found in a correct consideration of the biological noise, which is implicitly present in the quantum model. However, both the classical and the quantum results seem to show the need for considering different fundamental aspects which take into account the specific behavior of living matter, such as endogenous chemical reactions, molecular dynamics effects, and so on.

Correct evaluation of the conditions of exposure appears to be of primary importance in permitting a significant comparison with experimental data.

REFERENCES

1. A. R. Liboff, B. R. McLeod, and S. D. Smith, "Ion cyclotron resonance effects of ELF fields in biological systems," in Wilson, Stevens, and Anderson, Eds., *Extremely Low Frequency Electromagnetic Fields: The Question of Cancer,* Battelle Press, 1990, 251-289.

2. A. Chiabrera and B. Bianco, "The role of the magnetic field in the EM interaction with ligand binding," in Blank and Findl, Eds., *Mechanistic Approaches to Interactions of Electric and Electromagnetic*

Fields with Living Systems, Plenum Press, 1987, 79-95.

3. G. D'Inzeo, A. Galli, and A. Palombo, "Dynamic effects of EM fields on cell ions: Agreement between theoretical and experimental data," *XXIIIrd General URSI Assembly,* Prague, 1990, Abstract, 29.

4. G. D'Inzeo, A. Galli, and A. Palombo, "Effects of static and harmonic ELF fields on the cell membrane ion dynamics for different configurations of exposure," *13th Ann. BEMS Mtg.,* Salt Lake City, Utah, 1991, Abstract, 40

5. V. V. Lednev, "Possible mechanism for the influence of weak magnetic fields on biological systems," *Bioelectromagnetics* 12: 71-75, 1991.

6. A. Chiabrera, B. Bianco, J. J. Kaufmann, and A. A. Pilla, "Quantum dynamics of ions in molecular crevices under electromagnetic exposure," in C. T. Brighton and S. R. Pollack, Eds., *Electromagnetics in Medicine and Biology,* San Francisco: San Francisco Press, 1991, 21-26; and "Quantum analysis of ion binding kinetics in electromagnetic bioeffects," *ibid.,* 27-31.

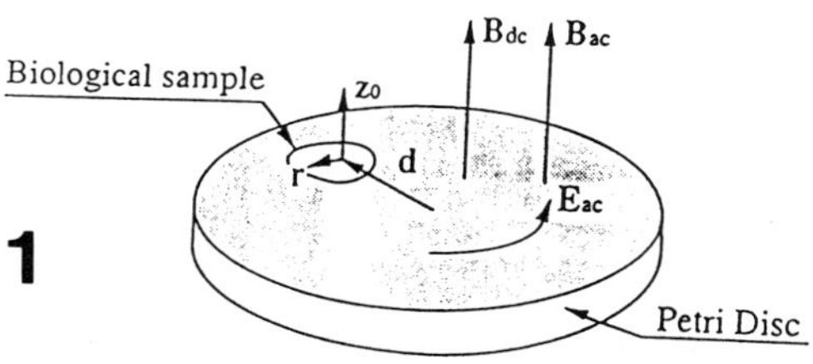

1

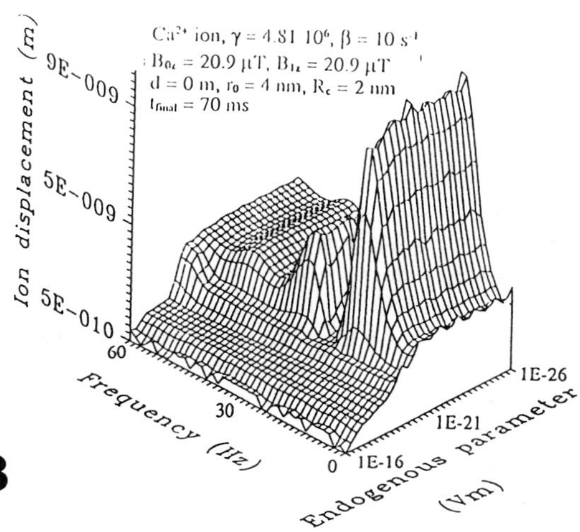

3

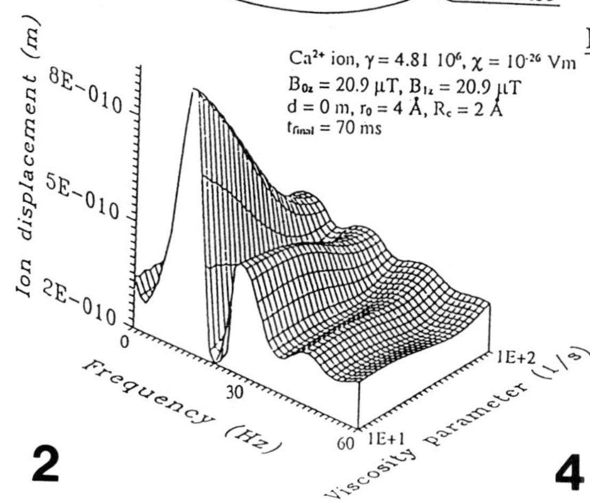

2

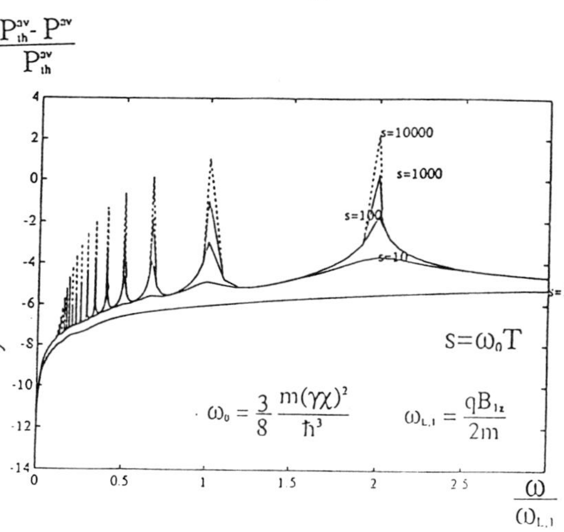

4

FIG. 1.—Typical exposure situation for ELF experiments.

FIG. 2.—Ion displacement vs frequency and viscosity parameter.

FIG. 3.—Ion displacement vs frequency and endogenous parameter.

FIG. 4.—Binding probability variation vs normalized frequency.

Martin Blank, Ed., *Electricity and Magnetism in Biology and Medicine*

ELECTRICAL MODELS FOR NERVE CELLS EXPOSED TO ELF ELECTRIC FIELDS

A. R. Sheppard* and W. R. Adey

Pettis Memorial Veterans Medical Center and Loma Linda University, Calif.

Surface potential and transmembrane potential are changed when cells are exposed to electric fields. Based on this face we present the basic concepts of established models for the electrotonic coupling of a nerve cell in an electric field and apply them to make a model for typical invertebrate pacemaker neurons. The model describes well experimental data obtained from *Aplysia* pacemaker neurons exposed to extremely low frequency electric fields.

Nerve cells in spatially uniform electric fields (EFs) have been studied experimentally in a few cases.[1-5] In each of these cases the EFs were strong enough and at low enough frequency (or static) so that membrane polarization was changed by direct electric effects. We discuss such electrotonic coupling in terms of the biophysical concepts needed to understand data, but only quote results in order to avoid repetition of well-known derivations.[6-9] The formal treatments can be distilled to a few principles by consideration of a spherical cell body and a long cylindrical process. The concepts apply to studies of cells in extremely low frequency EFs, including many cases where EFs are induced by time-varying magnetic fields, but we do not treat the time-dependent (capacitive) properties that are important with transients such as synaptic potentials and pulses, and at higher frequencies.

The potential inside a small sphere covered with a poorly conducting membrane in a uniform EF is effectively constant; equal and opposite transmembrane potentials occur at corresponding points in each hemisphere but they cancel, so there is no *net* change in transmembrane potential. The field-related change in transmembrane potential V_m is maximum at the poles of the sphere,

$$V_m \,|_{\cos\theta=1} = -(3/2)RE \qquad (1)$$

where R is the radius and E is a uniform static EF directed along the z-axis, from which θ is measured;[6] V_m is also enhanced by a factor of 3/2 elsewhere on the surface when compared to the potential difference that would occur in the medium for those points if no sphere were present. From Eq. (1), for $R = 10^{-5} m$ and $E = 1$ V/m, V_m is shifted by 0.015 mV, which may be compared to typical membrane potentials of $V_m = 75$ mV; i.e., the shift is just 0.02%.

In the same field, a long cylindrical process (representing an axon or dendritic tree) also has potential differences across the membrane at the cylinder ends, but unlike in the cell body, significant internal current flows along the axis and there are significant potential gradients inside the cylindrical process. The current that leaks into the cylinder across the surface membrane defines a characteristic cylinder length λ, for which the total membrane surface resistance equals the total axial resistance (since surface resistance falls with increasing surface area but axial resistance within the cylinder is proportional to axial length). In terms of the area-specific membrane resistivity R_m (Ω-cm^2), the internal fluid resistivity R_i (Ω-cm), and the diameter d,[8]

$$\lambda = [\tfrac{1}{4}d\,(R_m/R_i)]^{\frac{1}{2}} \qquad (2)$$

This simple but important quantity λ is known as the "length constant." However, even for a cell of fixed dimensions it is a "constant" only insofar as the membrane resistivity R_m remains constant.

One important feature of axons and dendrites in external fields is that because the membrane is leaky, there is an upper limit to the internal current and V_m at the ends. No matter how long the cable is, $V_m\,|_{max} = \lambda E$. For typical mammalian cells, $\lambda \approx 10^{-3}$ m (compare with $R = 10^{-5}$ m), so that V_m is typically 100 or more times greater at the ends of the cable-like parts of a cell than for the poles of a spherical cell

---

*Correspondence to: A. R. Sheppard, Pettis Memorial VA Medical Ctr., Loma Linda CA 92357.

body. For $E = 1$ V/m, V_m at the cylinder end cap is about λE, or of the order of 1 mV, a physiologically significant change. For much weaker fields, the potentials are still enhanced, but V_m may be far below levels significant for electrical signaling at the membrane.

Rall discovered that the complex dendritic tree of a motoneuron could be treated as if it were a single cable,[8] provided the branching pattern is such that the diameter of the parent branch d_p and the two daughter branch diameters d_1 and d_2 obey the relation $d_1^{3/2} + d_2^{3/2} = d_p^{3/2}$. Extensive morphometric work is required to determine whether an actual cell branching follows the "3/2 rule"; it is therefore common to assume that the rule applies and to test that assumption by the degree to which electrophysiological data match predictions.

MODEL

The above concepts permit assembly of a model for the electrical nature of any neuron in an EF by combination of the dendrite and axon trees into an equivalent cylinder and union of the cylinder with the sphere.[8] Thus, even anatomically complex cells can have electrically simple representations as the sum of a spherical (or punctate) cell body and a cylindrical cable. Under some conditions, the sphere may act electrically like the end cap of a long cylinder. We considered a model *Aplysia* invertebrate neuron and compared this model with experimental results.

Aplysia pacemaker neurons such as those of the right upper quadrant of the abdominal ganglion can be modeled as a spherical body ($R = 1.5 \times 10^{-2}$ cm) with a cable diameter that is effectively 2×10^{-4} cm (although the proximal axon segment is generally 10 times wider); $R_m = 10^5$ Ω-cm^2; and $R_i = 90$ Ω-cm. These values are based on anatomical and electrical studies of molluscan nerve cells.[1,10] With these values and with $E = 1$ V/m, Eq. (1) gives $V_m |_{\cos\theta=1} = 0.22$ mV and Eq. (2) gives $\lambda_{cell} = 1.8$ mm. This length constant is about twice that found experimentally for EF-exposed snail neurons.[1] For the large proximal protion of the model axon, $\lambda \approx 5.6$ mm. Since current leaks into the cell along the length of the neurite, we expect an angular dependence proportional to $\cos\theta$ in the coupling of the EF to the cell. Since the cell plus axon is at least as long as λ_{cell}, a change in membrane potential of up to $\lambda E \approx 1.8$ mV (sufficient to synchronize some cells) might appear at the soma. The model indicates there is little coupling to the cell body when the EF is not parallel to a major neurite and therefore does not carry current to the soma.

We compared this model for *Aplysia* pacemaker neurons with experiment. A uniform sinusoidal external EF was applied in the bath to test for synchronization of cell firing to the field. Custom computer routines were used to assign a phase angle of the applied waveform for each action potential and to identify synchronization when the χ^2 test showed a nonrandom distribution of phase angles. The geometric angle between the cell axis (defined by careful placement of the ganglion) and the applied EF was varied electrically by a custom-built amplifier that permitted adjustment of the vector sum of the EFs from each of two orthogonal electrode pairs while keeping total field magnitude constant. The results (Fig. 1) show greatest synchronization at 90° where the threshold for the most sensitive cells was below 0.2 V/m and 100% of the cells were synchronized at 4 V/m. In contrast, at 30° and 150° the threshold was four times higher, 0.8 V/m, and only about 60% of cells were synchronized at 4 V/m. The data clearly illustrate the cosinusoidal angular dependence and are consistent with model-related estimates of threshold field strengths from 0.36 to 7.2 mV, corresponding to 0.2-4V/m field strengths.

Others have also found results consistent with electrotonic coupling using invertebrate neurons[1,3] and turtle brain;[4,11] and Cooper suggested that gap junction-coupled tissue may also show enhanced, directional sensitivity to EFs.[12] The simple models and Fig. 1 illustrate the general pattern of electrotonic coupling to nerve cells. More refined analysis and techniques are needed to deal with realistic nerve cells that have anisotropic surface membranes, with particular attention to current flow at spike initiating zones or other specialized regions,[1] and to the modulation of synaptic properties.[13] The nonuniformity of external EFs in close-packed brain tissue also poses interesting questions. Some bioelectromagnetic research and applications involve EFs of the order of 0.2 V/m, a level at which we found that some *Aplysia* neurons could be synchronized and where we might expect similar results with vertebrate neurons after taking into account differences in morphology and membrane properties. The very much weaker EFs that occur in a body during nearly all environmental ELF exposures would not produce direct electric effects of the kind we and others observed. However, even though V_m itself may not figure importantly, the changes in surface potential along the length of the membrane follow the physical rules discussed in this paper. In this way, the concepts developed for work with nerve cells may also prove to be useful in models for other cellular effects[14,15] that depend on surface potential.[16]

FIG. 1.—Synchronization of pacemaker neuron to sinusoidal EF at pacemaker frequency was tested in eight RUQ *Aplysia* neurons. For each orientation angle we scored the percentage of cells synchronized at the indicated EF strength. Data points are joined by spline curves. Some cells were markedly more sensitive than others, accounting for the spread of responses over a 20-fold range in field strength, but as a group all exhibited similar angular dependence that fits $\cos[\theta - (\pi/2)]$, which shows that the major cell axis was perpendicular to the x-axis (0°); neurites along other directions gave weaker contributions. Sketch at right schematically illustrates pacemaker neuronal soma and neurites with major neurite structure along $\theta = \pi/2$ direction.

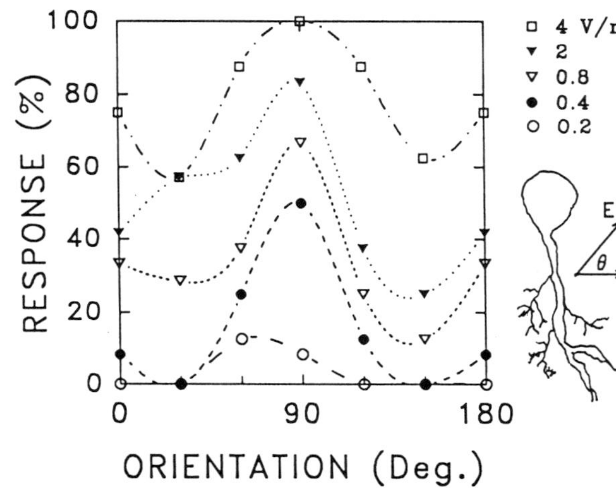

REFERENCES

1. V. N. Ierusalimsky and P. M. Balaban, "Functioning of identified snail neurones in electric fields," *J. Exp. Biol.* 131: 17-36, 1987.

2. A. R. Sheppard, S. M. Bawin, M. Burton-Sagan, and W. R. Adey, "Angular dependence of threshold for electric field effects on firing rates of *Aplysia* pacemaker cells," *Soc. Neurosci. Abstr.* 11: 506, 1985.

3. C. A. Terzuolo and T. H. Bullock, "Measurement of imposed voltage gradient adequate to modulate neuronal firing," *Proc. Natl. Acad. Sci.* 42: 687-694, 1956.

4. D. Tranchina and C. Nicholson, "Model for the polarization of neurons by intrinsically applied electric fields," *Biophysical J.* 50: 1139-1156, 1986.

5. H. Wachtel, "Firing pattern changes and transmembrane currents produced by extremely low frequency fields in pacemaker neurons," in R. D. Phillips, M. F. Gillis, W. T. Kaune, and D. D. Mahlum, Eds., *Biological Effects of Extremely Low Frequency Electromagnetic Fields,* Richland, Wash.: U.S. Dept. of Energy, 1979, 132-146.

6. K. S. Cole, *Membranes, Ions, and Impulses,* Berkeley: University of California Press, 1972, 15.

7. J. J. B. Jack, D. Noble, and R. W. Tsien, *Electric Current Flow in Excitable Cells,* Oxford: Clarendon Press, 1975.

8. W. Rall, "Core conductor theory and cable properties of neurons," in J. M. Brookhart and V. B. Mountcastle, Eds., *Handbook of Physiology,* Bethesda: American Physiological Society, 1977, vol. I, pt. 1, 39-97.

9. H. P. Schwan, "Biophysical principles of the interaction of ELF-fields with living matter: II. Coupling considerations and forces," in M. Grandolfo, S. M. Michaelson, and A. Rindi, Eds., *Biological Effects and Dosimetry of Static and ELF Electromagnetic Fields,* New York: Plenum, 1985, 243-271.

10. K. Graubard, "Voltage attenuation within *Aplysia* neurons: The effect of branching pattern," *Brain Res.* 88: 325-332, 1975.

11. C. Y. Chan and C. Nicholson, "Modulation by applied electric fields of Purkinje and stellate cell activity in the isolated turtle cerebellum," *J. Physiol.* 371: 89-114, 1986.

12. M. S. Cooper, "Gap junctions increase the sensitivity of tissue cells to exogenous electric fields," *J. Theor. Biol.* 111: 123-130, 1984.

13. S. M. Bawin, A. R. Sheppard, M. D. Mahoney, and W. R. Adey, "Influences of sinusoidal electric fields on excitability in the hippocampal slice," *Brain Res.* 323: 227-237, 1984.

14. W. R. Adey, "Electromagnetic field interactions in the brain," in E. Basar, Ed., *Brain Dynamics* 1, Heidelberg: Springer Verlag, 1988.

15. A. R. Sheppard and W. R. Adey, "The role of cell surface polarization in biological effects of extremely low frequency," in Ref. 5, 147-158.

16. This work was supported by the U.S. Department of Energy under contract DE-A101-90CE35035.

Martin Blank, Ed., *Electricity and Magnetism in Biology and Medicine*

PHYSICAL MECHANISMS FOR BIOLOGICAL EFFECTS OF ELF LOW-INTENSITY ELECTRIC AND MAGNETIC FIELDS: THERMAL NOISE LIMIT AND COUNTERION POLARIZATION

C. Polk*

University of Rhode Island, Kingston

It is shown that the "thermal noise problem" is much less severe for field effects depending on counterion polarization than for effects that require a change of transmembrane potential. Furthermore, counterion polarization is accompanied by ion-density fluctuations that extend into the intercellular medium, where phase varies rapidly with distance from the cell surface and phase and particle velocities can be of the order of 10^{-4} m/s at ELF for typical tissue properties. As a consequence, any change in electrical boundary conditions on the surface of one cell, brought about by chemical or mechanical action, very rapidly affects the magnitude of the ion current due to an applied electric field "seen" on adjacent cells. This effect on cell-to-cell communication should be particularly noticeable at the lowest frequencies, since the radial width of the charge disturbance is inversely proportional to the square root of frequency.

In recent years many reports have appeared which indicate that electric fields, or magnetic fields that induce electric fields in tissue, can influence bone development, transcription in cells, activity in lymphocytes, Ca-efflux from neural tissue, or development of avian embryos. Most of these reports involve sinusoidal fields with frequencies below 100 Hz and suggest that the intensity of the electric fields induced in the tissue, or at the cell surface, may be below 10^{-3} V/m, or in some cases, as low as 10^{-5} V/m.

Without assuming averaging over long time periods and requiring that cells be able to perform rectification as well as crosscorrelation with an endogenous signal of the correct frequency, one finds it difficult to see how "signals" of the order of 10^{-3} V/m or less could produce a potential difference across the cell membrane that would be larger than thermal noise. To show this we use the well-known expression for the thermal noise voltage V_n produced by the random motion of charges in any resistor R,

$$V_n = \sqrt{4kTR\,(\Delta f)} \tag{1}$$

where k is Boltzmann's constant and T is the absolute temperature. Assuming a bandwidth $\Delta f \sim 100$ Hz, and a typical value for the resistance of a cell membrane $R = 10^8\ \Omega$, one obtains $V_n = 13\ \mu$V. If one furthermore assumes that the membrane surrounds an idealized spherical cell of radius $a = 10\ \mu$m and uses the approximate expression relating the electric field E in the surrounding fluid to the transmembrane potential $V_t \approx 1.5aE$, one obtains a minimum value of $E = 0.86$ V/m for $V_t \geq V_n$ that is necessary to overcome thermal noise without resonance and time averaging.

This result should not be surprising, because the idealized, sharply bounded cell membrane is a system that presents a particularly unfavorable signal-to-noise situation, since the random thermal energy per degree of freedom $W = \frac{1}{2}kT$ of any closed system is independent of its volume, and the electrical input energy is equal to $\frac{1}{2}\varepsilon E^2$ (volume). Since the signal/noise ratio therefore depends on volume, it is low in a membrane with a thickness of about 6×10^{-9} m surrounding a sphere of 10^{-5}m radius. However, it is well known that the cell can sense its environment through many mechanisms other than detection of changes in transmembrane potential. Protein molecules embedded in the cell membrane, which protrude to the exterior of the cell, serve as "receptors" of chemical signals and possibly also as detectors of electrical and mechanical disturbances. Furthermore, any mechanism that affects cell-to-cell communication is likely to have substantial consequences for cell behavior.[3]

One striking property of biological tissue is its enormously high dielectric constant ε_r in comparison with most inanimate matter, and the fact that the dielectric dispersion at low frequencies does not follow the typi-

---

*Correspondence to: C. Polk (EE), University of Rhode Island, Kingston RI 02881.

cal Debye relaxation curve.[1] It is now well established that this magnitude of ε_r is due to counter-ion polarization; however, an electrochemical theory that quantitatively describes dielectric relaxation at low frequencies in agreement with experiment is only a relatively recent development.[1,2] After reference to fairly elementary concepts of macroscopic and microscopic electric polarization, we shall show that counter-ion polarization requires a considerably lower applied electric field E *at extremely low frequencies* to overcome thermal noise than does detection of an E-induced transmembrane potential. We shall then show that based on the development by Chew[2] and Chew and Sen,[1] disturbance of counter-ion polarization by an electric field may very well affect receptor behavior and cell-to-cell communication.

ELECTRIC POLARIZATION—NUMERICAL EXAMPLE FOR BIOLOGICAL CELL

In IS units, the electric displacement D is given by

$$D = \varepsilon_0 E + P = \varepsilon_0 [1 + P/(\varepsilon_0 E)] \quad \text{C/m}^2 \tag{2}$$

where ε_0 is the dielectric permittivity of free space, 8.884×10^{-12} F/m and P is the dipole moment per unit volume. The relative dielectric constant ε_r is given by

$$\varepsilon_r = 1 + [P/(\varepsilon_0 E)] \tag{3}$$

and for the large values of ε_r of interest here $\varepsilon_r \approx P/(\varepsilon_0 E)$. The dipole moment p of the average individual dipole of volume v in the dielectric material, assuming homogeneity and isotropy, is

$$p = PV = Qd \quad \text{Cm} \tag{4}$$

where Q is the magnitude of the equivalent point charges with opposite sign at each end of the dipole and d is their separation.

A typical measured value for ε_r of muscle below 20 Hz is 10^7. Fluid-saturated tissue can have even higher values such as $5 \times (10^7)$ for liver tissue and $2.5 \times (10^7)$ for lung. The electric conductivity σ of muscle in the same frequency range is about 0.5 S/m. We assume as a first approximation—on which we shall improve later—that the tissue consists of "ideal-dielectric" polarizable spheres, embedded in an ionic fluid with complex dielectric permittivity, such as salt water. We then use dielectric mixture theory with the "mixture" having conductivity $\sigma_m = 0.5$ S/m and $\varepsilon_{rm} = 10^7$, individual cells having $\sigma_c = 0$ and interstitial fluid $\sigma_i = 1$ S/m and $\varepsilon_{ri} = 80$. That gives a volume fraction (cell volume/total volume) of 0.377 and an equivalent relative dielectric constant ε_{rc} for the spherical cell of 1.33×10^7. From Eq. (3) it then follows that the equivalent polarization P_c of noninteracting dielectric spheres without the interstitial water would be equal to $1.18 E \times 10^{-4}$ C/m^2; and from Eq. (4) that the dipole moment p of the individual cells is

$$p = 4.94 E \times 10^{-19} \text{ Cm} \quad \text{and} \quad Q = 3.71 E \times 10^{-14} \text{ C}$$

if we estimate that $v = (4/3)\pi a^3$ with a cell radius $a \approx 10^{-5}$ m and use a dipole length $d = 4a/3$ (see Ref. 4).

MINIMUM ELECTRIC FIELD REQUIRED TO PRODUCE CHARGE VARIATIONS ABOVE THERMAL NOISE

The mean square value of the fluctuating charge in thermal equilibrium, within a system of electrical conductance G is given by[4]

$$<\delta q^2> = 2GKTt \quad \text{C}^2 \tag{5}$$

where t is the observation time. The conductance G in the present case is the surface conductance, which is independent of cell size:

$$G = q n_0 u \quad \text{S} \tag{6}$$

where q is the individual ionic charge, n_0 is the surface charge density, and u is the electric mobility of counter-ions.

To obtain a large, more unfavorable value (as far as noise is concerned), we assume a doubly charged ion and use $q = 3.2 \times 10^{-19}$ C. We set $n_0 \approx 10^{17}$ m^{-2} and $u = 18^{-8}$ m^2/Vs, where we assume that u at the surface is smaller than in the bulk medium by a factor of less than 10. The result is $G = 3.2 \times 10^{-10}$ S. The minimum "observation time" t in Eq. (5) is the time that lapses between charge maxima of opposite sign due to the sinusoidal time variation. For a frequency of 15 Hz it follows that $t = 1/30$ s. We must then equate

$$2Q = 7.42\times10^{-14}E_{\min} = \sqrt{2GkTt} \tag{7}$$

to evaluate the minimum electric field necessary to exceed thermal charge noise. For the selected values one obtains $E_{\min} = 4.1\times10^{-3}$ V/m, which is lower by more than two orders of magnitude than the value obtained below Eq. (1) for the electric field necessary to produce a transmembrane potential above thermal noise. The value of 4.1×10^{-3} V/m is, of course, still much higher than the 10^{-4} or 10^{-5} V/m that seem to produce biological effects in some experiments. However, two considerations must be taken into account. First, the mechanism proposed here—disturbance of surface-charge distribution—may not be the correct one to explain results at the *lowest* field intensities (10^{-5} V/m). Second, the extreme inhomogeneity of real biological systems guarantees that the actual field values in the vicinity of some cells can be higher by at least an order of magnitude than the "average" electric field in the tissue.

COUNTER-ION ASSOCIATED CELL DISTURBANCE IN INTERCELLULAR MEDIUM

It is of considerable interest to examine the electrically neutral charge cloud between adjacent cells. Chew and Sen have shown that dipole formation by counter-ion redistribution involves diffusion currents and resulting small changes of charge density beyond the double layer at the cell surface.[1,2] The perturbation density $n_\pm$, which is added to the equilibrium charge density N_0 in the intercellular medium, is given by Eq. 4.12 in Ref. 1:

$$n_+ = N_0 C\,\frac{a^3}{r^2}\,\frac{1+\lambda r}{1+\lambda a}\,\cos\theta\, e^{-\lambda(r-a)} \tag{8}$$

where we find C (given in Ref. 4) by satisfying conservation of charge at all boundaries, and the exponential decay distance λ^{-1} is

$$\frac{1}{\lambda} = \frac{1}{1-j}\sqrt{2D/\omega}\ \ \text{m} \tag{9}$$

The diffusion constant D is given by

$$D = ukT/q \quad \text{m}^2/\text{s} \tag{10}$$

Substituting the same values for u and q as before, we obtain $D = 1.34\times10^{-10}$ m^2/s and $\sqrt{2D/\omega} = 1.69\ \mu$m at 15 Hz. We note also that λ is a complex number, and since C contains the applied field $E_0 \exp(j\omega t)$ as a factor, the charge density disturbance varies as

$$n_\pm \sim e^{j\omega[t-\lambda_2(r-a)/\omega]}\, e^{-\lambda_1(r-a)} \tag{11}$$

where λ_1 and λ_2 are the real and imaginary parts of λ, respectively. From Eq. (11) it becomes apparent that the extracellular charge disturbance propagates at a phase velocity ω/λ_2, which is in the present example equal to $1.69\,\omega\,10^{-6}$ m/s. Thus the phase of the disturbance travels 5.3 μm in 1/30 s when $\omega = 2\pi(15)$. Comparing the real part of λ_1 with the distance between spherical cells of 10μm radius, which is found to be 2.3 μ from the volume fraction of 0.377 determined earlier for our example, we find that the exponential decay distance at this low frequency is larger than half the distance between adjacent cells. Thus one can expect diffusion currents not only from the charge cloud surrounding a single cell, but also propagation of ion density "waves" between adjacent cells that would be affected by any change of boundary conditions at either end.

CONCLUSIONS AND SUGGESTED EXPERIMENTS

It has been shown that in a medium, such as muscle, modification of the charge distribution on the surfaces of individual cells (or groups of cells) requires a lower value of applied electric field to exceed thermal noise than does modification of transmembrane potential. Furthermore, the radial extent of that disturbance of charge density in the electrically neutral space between cells is comparable with intercell separation. The phase of the disturbance also varies rapidly with distance. As a consequence, ion density fluctuations will be propagated between adjacent cells when a sufficiently large electric field is applied. Since the radial width of the neutral charge cloud is inversely proportional to the square root of frequency, such an effect on cell-to-cell communication would be particularly pronounced in the lower part of the ELF spectrum, perhaps below 20 Hz. At first, qualitative test of the ideas outlined here would be to perform an experiment on cells in cultures at various cell densities, where such differences give clearly different results in the *absence* of an applied electric field. The experiment might then be repeated with the application of an electric field to the culture with the lower density. If the mechanism proposed here is applicable, one should expect that the ELF field would have the same effect as an increase of cell density.

REFERENCES

1. W. C. Chew and P. N. Sen, "Dielectric enhancement due to electrochemical double layer: Thin double layer approximation," *J. Chem. Phys.* 77: 4683-4693, 1982.

2. W. C. Chew, "Dielectric enhancement and electrophoresis due to an electrochemical double layer: A uniform approximation," *J. Chem. Phys.* 80: 4541-4551, 1984.

3. W. R. Adey, "Physiological signaling across cell membranes and cooperative influences of ELF fields," in H. Fröhlich, Ed., *Biological Coherence and Response to External Stimuli,* Springer-Verlag, 1988, 148-170.

4. C. Polk, "Counter-ion polarization and low frequency, low electric field intensity biological effects," *Bioelectrochemistry and Bioenergetics* 8: 267-277, 1992.

Martin Blank, Ed., *Electricity and Magnetism in Biology and Medicine*

EFFECT OF THE ACOUSTIC EXPOSURE ON LIGAND BINDING

T. Tommasi,* E. Moggia, A. Chiabrera, and B. Bianco

University of Genoa, Italy

A quantum analysis of biological effects induced by low-intensity acoustic exposure is proposed. The ligand-binding process is theoretically investigated to evaluate the changes of the binding probability of an ion due to the acoustic field. A first-order perturbing Hamiltonian operator is evaluated, corresponding to a coulombic endogenous potential of the binding site, whose center is displaced by the acoustic field. The equation system for the density operator is found, which describes the ensemble behavior of the binding ions. A preliminary comparison between acoustic and ELF electromagnetic exposures is also made.

Ultrasound is widely used in many diagnostic and therapeutic applications in the health-related field. In the latter applications, acoustic energy can be utilized to modulate cell functions; therefore, it is important to evaluate the possible physical mechanisms involved.

Acoustic energy coupled to electrolytes generates both a mechanical and an electric field. An acoustic wave propagating through an electrolyte induces a local displacement of ions that depends on the effective masses and frictional coefficients of the solvated ions. The resulting differences in displacement of cations and anions give rise to an electrokinetic bulk effect, predicted by Debye and analyzed in Refs. 1 and 2. This effect becomes negligible at high ionic strengths (above 10 mM), so that it does not appear to be a plausible explanation for the observed bioeffects.

In this paper we consider the action of the associated mechanical field on the macromolecules of membrane sites, which are responsible for the endogenous force acting on ligands or ion messengers.

MODEL OF ACOUSTIC EXPOSURE OF A CREVIX-ION SYSTEM

We consider a charged ligand (ion) about to bind in a molecular crevix (e.g., binding site, mouth of a membrane channel, etc.). A sinusoidal acoustic exposure displaces the water molecules, and the collisions with the crevix induce a displacement of the crevix itself, which follows the same time-harmonic behavior as the acoustic wave. Possible effects of the acoustic field onto the ion are neglected; that is, the ion inside the site is thought to be "in a vacuum."

The binding site has an endogenous potential energy, $U(\mathbf{r}')$, with $\mathbf{r}' = \mathbf{r}_0$, where $\mathbf{r}_0$ is the reference center of the crevix, $\mathbf{r}$ is the vector distance of the ion from the reference frame origin, and $\mathbf{r}'$ is the actual distance between the ion and the binding site (Fig. 1). The acoustic exposure determines a time-varying $\mathbf{r}_0(t)$, which makes $U(\mathbf{r}')$ time-varying as well. To evaluate the perturbation induced, a coulombic electric potential is assumed, and a first-order perturbation is derived as follows:

$$U(\mathbf{r}') = -\frac{\zeta}{r'} \approx -\frac{\zeta}{r} - \frac{\zeta \mathbf{r}_0 \cdot \mathbf{r}}{r^3} \tag{1}$$

where ζ is a constant and the binding-site displacement $\mathbf{r}_0$ is assumed to follow the time-harmonic law $[a \cos(\omega t)\,\mathbf{i}]$, where a and ω are the displacement amplitude and the angular frequency of the acoustic wave in water, respectively, and $\mathbf{i}$ is the direction of the acoustic wave, whose direction cosines are (α, β, γ), so that $\mathbf{i} = \alpha\,\mathbf{i}_1 + \beta\,\mathbf{i}_2 + \gamma\,\mathbf{i}_3$. Equation (1) is accurate enough if the displacement a is small compared with the distance r, which permits a first-order approximation. The crevix-ion distance r can be estimated to be 1 to 10 nm; the amplitude displacement a is evaluated by means of the following expression, which gives the intensity I (power density) of a longitudinal acoustic planar wave in a fluid: $I = 1/2\,\delta v\, a^2 \omega^2$, where δ is the density of the fluid and v is the longitudinal velocity of the wave. In actual experiments, at a frequency of

---

*Correspondence to: T. Tommasi (Biophysical and Electronic Engg.), University of Genoa, Via Opera Pia 11a, I-16145 Genoa, Italy.

1 MHz, if the reference intensity used is $I = 1$ W/m^2, the amplitude displacement is as low as 0.2 nm.

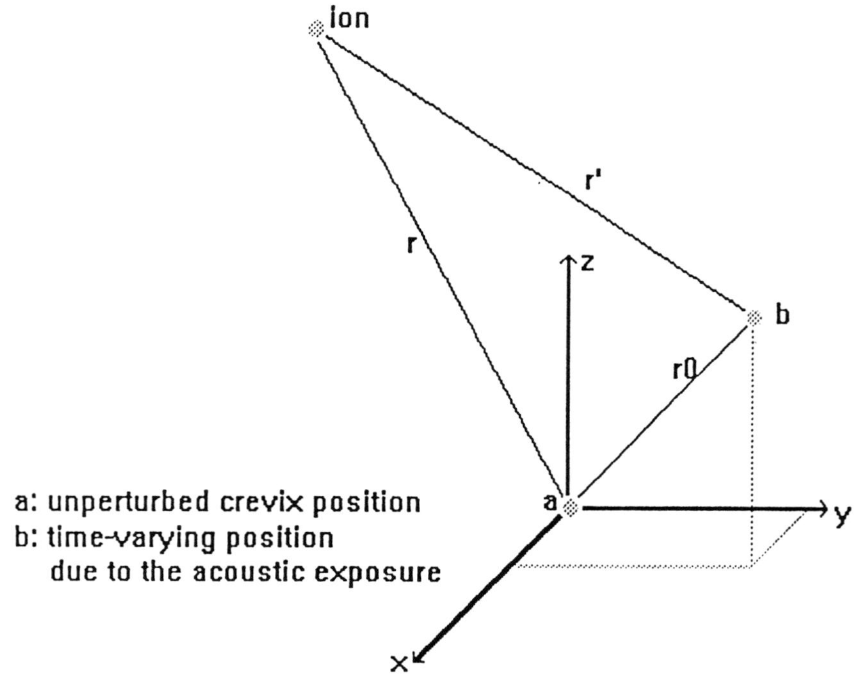

FIG. 1.—Crevix-ion system under acoustic exposure.

ION QUANTUM DYNAMICS

Equation (1) allows one to define the perturbing Hamiltonian operator H_1, as follows:

$$H_1 = -\frac{\zeta\alpha a\, \mathbf{i}_1\cdot\mathbf{r}}{r^3}\cos\omega t - \frac{\zeta\beta a\, \mathbf{i}_2\mathbf{r}}{r^3}\cos\omega t - \frac{\zeta\gamma a\, \mathbf{i}_3\cdot\mathbf{r}}{r^3}\cos\omega t \tag{2}$$

Once H_1 has been calculated, the usual methods can be applied[3,4] to evaluate the effects of the acoustic exposure on the crevix-ion system. To this end, the probability for the ion to be bound in the presence of the acoustic field should be compared with the probability calculated when no perturbation occurs. The methods for quantum analysis have been discussed elsewhere;[5-7] here we only point out the main steps to be taken that are specific for the present application.

The entries D_{nk} of a matrix $D(t)$ are computed as follows:

$$D_{nk}(t) = \frac{\mathrm{j}}{h_c}\exp[\mathrm{j}(\omega_n - \omega_k)t]\int \Phi_n{}^{*} H_1 \Phi_k\, dr \tag{3}$$

where Φ_k is the time-independent k-th eigenfunction associated with the unperturbed Hamiltonian $h_c\,\omega_k = E_k$, E_k is the eigenvalue of the respective eigenfunction, h_c is Plank's constant divided by 2π, and * stands for conjugate.

The orthonormal set of eigenfunctions is the usual hydrogen-like set. We take five eigenfunctions, defined elsewhere;[4] this choice is referred to as a "five-state" coulombic binding site. It is useful to note that the corresponding eigenvalues are $E_1 = -M\,\zeta^2/2h_c^2$, while $E_i = -M\,\zeta^2/8h_c^2$, $i = 2$ to 5, where M is the ion mass, and we let $E_2 - E_1 = 3M\,\zeta^2/8h_c^2 = h_c\,\omega_0$. Once the elements D_{nk} have been calculated, the density operator permits a statistical evaluation of the perturbation effects on the binding site.

The equation system for the elements ρ_{pq} of the density operator matrix is

$$\frac{\delta\lambda_{pq}}{dt} = \sum s\,(D_{ps}\,\lambda_{sq} - D_{sq}\,\lambda_{ps}) - C_{pq} \tag{4}$$

where $\lambda_{pq} = \rho_{pq} \exp(-j\, \omega_0 t)$ if $p = 1$ and $q \neq 1$, or $p \neq 1$ and $q = 1$; $\lambda_{pq} = \rho_{pq}$; otherwise

$$C_{pq} = \sum{}'_j \frac{\Delta_{pp} - \Delta_{ij}}{\theta_{pj}} \quad \text{if } p = q\,; \qquad C_{pq} = \frac{\Delta_{pq}}{\tau_{pq}} \quad \text{if } p = q\,; \qquad \Delta_{pq} = \rho_{pq} - \rho_{pq,th}\,,\ p,q = 1 \text{ to } 5$$

The lifetime parameters θ_{pj} and τ_{pq} are discussed in Ref. 6; $\rho_{pq,th}$ are the values assumed at thermal equilibrium by ρ_{pq} , determined by the Boltzmann distribution. The density matrix entries permit one to evaluate an average probability describing the behavior of the physical system under investigation.

RESULTS

In the present analysis, calculations of entries D, accomplished by use of Eqs. (2) and (3), give the following results:

$$D_{13} = \frac{4\gamma k'}{27\, a_0{}^2\sqrt{2}}\,; \quad D_{14} = \frac{2\beta\, k'}{27\, a_0^2} - j\, \frac{2\alpha\, k'}{27\, a_0^2}\,; \quad D_{15} = \frac{2\beta\, k'}{27\, a_0^2} + j\, \frac{2\alpha\, k'}{27\, a_0^2}$$

where $k' = [-\zeta\, a (\cos \omega t) \exp(-j\omega_0 t)]/jh_c$ and $a_0 = h_c^2/\zeta M$. $D_{j1} = D^*{}_{1j}$ ($j = 3,4,5$). The other D_{nk} entries are equal to zero.

The matrix D has also been calculated for the case of magnetic exposure (a Petri dish between Helmholtz coils), $B(t) = B_0 + B_1 \sin(\omega_1 t)$ parallel to the z-axis.[7] A comparison between the elements D of magnetic and acoustic exposures yields interesting results. The ratio between the time average of the element D_{14} for the acoustic exposure, labeled as D_{ac} , and D_{14} for the magnetic one, labeled as D_{mag}, gives the following relation: $D_{ac}/D_{mag} = a\, \omega_0^2/(d\, \omega_1\, \omega_{1,1}{}^2)$, where $\omega_{1,1} = qB_1/2M$, d is the distance between the crevix and the axis of the Petri dish, and q is the ion charge. This ratio can also be written as $D_{ac}/D_{mag} = au^2/d\Omega L^2$, where $u = h\, \omega_0/kT$, $\Omega = \omega_1/\omega_{1,1}$, $L = h_c\, \omega_{1,1}/kT$, and k is Boltzmann's constant.

Although the plausible values for the parameter u cover a wide range, for a set of reasonable values ($a = 10^{-10}$ m, $d = 10^{-2}$ m, $L \approx 10^{-12}$ m, $\Omega = 1$), D_{ac} is greater by many orders of magnitude than D_{mag}. This result is of interest, since it points out that acoustic exposure might be biologically very effective. Its implications on the binding probability have to be further investigated, and compared with the analysis carried out for electromagnetic exposure.[5]

REFERENCES

1. A. Chiabrera et al., "Therapeutic ultrasound generates electric fields in situ, with dosimetry comparable to that of PEMS," 5th BRAGS, Boston, 1985.
2. E.Yeager et al., "Ultrasonic waves and electrochemistry: I,II,III,IV," *J.A.S.A.* 25: 443-469, 1953.
3. L. Landau and E. Lifschitz, *Quantum Mechanics,* Moscow: MIR, 1966.
4. C. Cohen-Tannoudji, B. Diu, and F. Laloe, *Quantum Mechanics,* New York: Wiley, 1977.
5. A. Chiabrera et al., "Quantum dynamics of ions in molecular crevices under electromagnetic exposure" and "Quantum analysis of ion binding kinetics in electromagnetic bioeffects," in C. Brighton and S. Pollack, Eds., *Electromagnetics in Biology and Medicine,* San Francisco: San Francisco Press, 1991, 21-31.
6. B. Bianco et al., "From the Langevin-Lorentz equation to the Zeeman model of electromagnetic effects on ligand-receptor binding," *Bioelectrochemistry and Bioenergetics* 28: 355-365, 1992.
7. E. Moggia et al., "Comparison of a five-state vs a three-state coulombic Zeeman model of EMF effects on ligand binding," this volume.

Martin Blank, Ed., *Electricity and Magnetism in Biology and Medicine*

POSSIBLE MECHANISM FOR THE EFFECT OF WEAK MAGNETIC FIELDS ON BIOLOGICAL SYSTEMS: CORRECTION OF THE BASIC EXPRESSION AND ITS CONSEQUENCES

V. V. Lednev*

Institute of Theoretical and Experimental Biophysics, Pushino, Russia

The expression describing the biological effects of weak modulated magnetic fields tuned to the parametric resonance conditions is corrected. It follows from the expression in the corrected form that: (i) magnetic fields tuned to either even or odd subharmonics of the cyclotron resonance (CR) frequency for Ca^{2+} must respectively produce biological effects of opposite sign, e.g., either activation or inhibition of cell proliferation and motility; (ii) effects observed in biosystems exposed to the magnetic fields tuned to the CR conditions for K^+ may be reinterpreted as the result of action of the second subharmonic of the CR frequency for Ca^{2+}; (iii) contrary to what is often suggested there is no reason to expect that biological effects may be caused by magnetic fields tuned to the cyclotron frequencies of such ions as Na^+, Li^+, or H^+; (iv) at physiological conditions, magnetic fields may affect biosystems at the limited set of well-predicted frequencies corresponding to the harmonics and subharmonics of the CR frequencies for Ca^{2+} and Mg^{2+}, known to be chelated by the calcium-binding proteins for the relatively long time of about 1 s.

CORRECTION

In two earlier papers[1,2] I used an expression† originally derived by Podgoretzkii,[3]

$$\bar{p} \approx A_1^2 + A_2^2 + 2A_1A_2J_n(nB_1/B_0)\cos\delta \tag{1}$$

as representing the quantitative estimate of the influence of modulated, weak magnetic fields under conditions of parametric resonance on the probability of transition of Ca^{2+} from an excited vibrational level to the ground-state level in a calcium-binding protein. This expression appears to be in error. The correct expression must be written as

$$\bar{p} \approx A_1^2 + A_2^2 + 2A_1A_2J_{-n}(nB_1/B_0)\cos\delta \tag{2}$$

or, since $J_{-n}(x) = (-1)^n J_n(x)$,

$$\bar{p} \approx A_1^2 + A_2^2 + (-1)^n 2A_1A_2J_n(nB_1/B_0)\cos\delta \tag{2'}$$

which coincides with the expression obtained by Khrustalev,[4] who considered quantum-mechanically the problem of light scattering from atoms in a modulated magnetic field. The seemingly small difference between Eqs. (1) and (2) was apparently also not noticed in the review by Podgoretzkii and Khrustalev,[5] where Eq. (1) was again presented. The difference between Eqs. (1) and (2) does not depend on the particular way in which they are derived, i.e., on the use of the quantum-mechanical or classical approach; thus, Eq. (2) can be obtained by the classical electrodynamics derivation used by Podgoretzkii.[3]

Comparing Eqs. (1) and (2) one may see that Eq. (1) used in my earlier papers[1,2] is valid only for odd values of n when $\cos\delta = -1$. For even values of n the sign of the last term in Eq. (2) changes, i.e., the sign of the biological effect of the modulated magnetic field must also change.

*Correspondence to: V. V. Lednev, Institute of Theoretical and Experimental Biophysics, Academy of Sciences, Pushino 142292, Russia.

†The term $\cos\delta$ was omitted in the earlier papers.[1,2] When the electromagnetic radiation from the excited vibrational level is split into a doublet by a constant magnetic field B_0, the difference between the initial phases of the radiation from the two sublevels is $\delta = \delta_1 - \delta_2$. The other symbols in Eq. (1) are the same as in the earlier papers.[1,2] (It is a pleasure to thank O. A. Khrustalev and M. I. Podgoretzkii for helpful discussions.)

CONSEQUENCES

To illustrate this important point, let us consider some particular cases of special interest, taking into account that in a static magnetic field for the model under consideration, $\cos\delta = -1$.

1. At $n = 1$, coresponding to resonance at the cyclotron frequency, one obtains

$$\bar{p} = A_1^2 + A_2^2 + 2A_1A_2J_1(B_1/B_0) \quad (3)$$

i.e., the value of $\bar{p}$ is increased or decreased depending on whether J_1 is positive or negative. This expression was discussed earlier[1,2] and the conclusions obtained then remain valid.

2. At $n = 2$, corresponding to the resonance at the second subharmonic of the cyclotron frequency,

$$\bar{p} \approx A_1^2 + A_2^2 - 2A_1A_2J_2(2B_1/B_0) \quad (4)$$

i.e., the value of $\bar{p}$ is decreased in this case for positive values of J_2.

3. When $B_1 = 0$, $B_0 \neq 0$ and the frequency gap Ω between the sublevels is comparable to or smaller than the natural line width λ of the sublevels, one obtains[3]

$$\bar{p} = A_1^2 + A_2^2 - 2A_1A_2/[1 + (\Omega^2/\lambda^2)] \quad (5)$$

Since Ω is proportional to B_0, it follows that by changing the value of the static field one may affect $\bar{p}$ and correspondingly the functional state of the biosystem. Furthermore, by screening off or canceling of the geomagnetic field, i.e., letting $B_0 \rightarrow 0$ ($\Omega \rightarrow 0$), one obtains

$$\bar{p} \approx A_1^2 + A_2^2 - 2A_1A_2 \quad (6)$$

From the corrected Eq. (2) it also follows that results of experimental studies observed (or not observed) by exposure of biosystems to modulated magnetic fields tuned to the cyclotron frequencies of singly charged ions, such as K^+, Li^+, Na^+, or H^+, may be interpreted in a radically different way. This important point must be considered in more detail.

The lifetime of Ca^{2+}—and Mg^{2+}—ions chelated in a calcium-binding protein (i.e., in calmodulin or troponin) is about 1 s,[6] while that of K^+, Li^+, and Na^+ is only about 10^{-2}-10^{-3}.[7] In the last case the time is too short to affect ion-protein interactions by magnetic fields modulated with frequencies in the range of 10-100 Hz. Therefore, according to the model of magnetic parametric resonance, it appears to be impossible to affect biosystems by fields tuned to the CR frequencies of K^+, Na^+, and Li^+. Indeed, the attempts to observe biological effects of magnetic fields tuned to the CR frequencies of Na^+ were all negative.[8-10] The frequency of the modulated DC magnetic field of 0.261×10^{-4} T, claimed to be effective when tuned to the CR frequency of Li^+,[11] probably corresponded to the third harmonic of the CR frequency for Ca^{2+}, which appears to be biologically effective.[12]

On the other hand, a number of reports indicate that magnetic fields tuned to the CR frequency for K^+ cause substantial effects in various biological systems, e.g., inhibition of motile activity of diatomic algae[12] or inhibition of proliferation in cell culture.[13] This puzzle may be easily explained by taking into account that the CR frequency for K^+ ($q/m = 1/39$) practically coincides with the second subharmonic of the CR frequency for Ca^{2+} ($q/m = 1/20$), which according to Eq. (2) must be biologically effective and is expected to be opposite to that caused by the first harmonic. This is exactly what was observed experimentally, at least in some biosystems.[12,13]

The rather good fit between experimental data and the new predictions following from the corrected Eq. (2) seems to substantiate further the validity of the magnetic parametric resonance phenomenon as an explanation for the interaction of weak modulated magnetic fields with biosystems.

The suggested mechanism of interaction between weak magnetic fields and biosystems may function only when the excited vibrational levels of the ions chelated by the calcium-binding protein are reasonably populated. Assuming a Boltzmann distribution, it appears that this is not likely to occur in an equilibrium system at room temperature, e.g., in a solution containing only calcium ions and a calcium-binding protein. However, in a living system, when biochemical reactions are taking place, one may expect that the energy required to pump up calcium ions to an excited level may be provided by the biochemical reactions. In Ca^{2+}-calmodulin-dependent reactions this energy may be transmitted to the bound Ca^{2+} ion in a cyclical process of interaction between ATP and the Ca^{2+}-calmodulin-kinase complex. Analysis of the experimental data shows that this suggestion is not unreasonable, since binding of Ca^{2+} to the calmodulin-kinase complex seems to be affected upon binding of ATP or its unhydrolizable analogs to kinase.[14]

REFERENCES

1. V. V. Lednev, "Possible mechanism for the influence of weak magnetic fields on biological systems," Preprint, Academy of Sciences Scientific Center for Biological Research, Institute of Biophysics (in Russian), 1988.

2. V. V. Lednev, "Possible mechanism for the influence of weak magnetic fields on biological systems," *Bioelectromagnetics* 12: 71-75, 1991.

3. M. I. Podgoretzkii, "On the modulation and beats in quantum transitions," Preprint P-491, United International Institute for Nuclear Research, Dubna (in Russian), 1960.

4. O. A. Khrustalev, "On the resonant scattering of light by atoms in alternating magnetic field," Preprint P-574, United International Institute for Nuclear Research, Dubna (in Russian), 1960.

5. M. I. Podgoretzkii and O. A. Khrustalev, "Interference effects in quantum transitions," *Sov. Phys. Uspekhi* 6: 682-700, 1963 (translation from Russian).

6. J. A. Cox, "Interactive properties of calmodulin," *Biochem. J.* 249: 621-629, 1988.

7. A. Permyakov, "Parvalbumin and related calcium binding proteins," *Nauka,* 1985 (in Russian).

8. S. D. Smith, B. R. McLeod, and A. R. Liboff, "Cyclotron resonance stimulation of explanted chick femurs," 8th Ann. BRAGS Meeting.

9. A. R. Liboff, W. C. Parkinson, and D. C. Dawson, "Ion cyclotron resonance study in turtle colon," 10th Ann. BEMS Meeting, Abstract, p. 32.

10. A. R. Liboff and W. C. Parkinson, "Search for ion-cyclotron resonance in a Na^+ transport system. *Bioelectromagnetics* 12: 77-85, 1991.

11. A. R. Liboff, J. R. Thomas, and J. Schrot, "Intensity threshold for 60 Hz magnetically induced behavior changes in rats," *Bioelectromagnetics* 10: 1-3, 1988.

12. B. R. McLeod, S. D. Smith, and A. R. Liboff, "Calcium and potassium cyclotron resonance curves and harmonics in diatoms (*A. coffeaformis*)," *J. Bioelectricity* 6: 153-168, 1987.

13. Yu. A. Rochev, A. A. Narimanov, E. A. Sosunov, A. N. Kozlov, and V. V. Lednev, "Effects of weak magnetic fields on the rate of cell proliferation in culture," *Studia Biophysica* 135: 93-98, 1990.

14. A. Mamarbachi and J. A. Cox, "Quantitative analysis of the free-energy coupling in the system calmodulin calcium smooth-muscle myosin light chain kinase," *Cell Calcium* 8: 473-482, 1987.

Martin Blank, Ed., *Electricity and Magnetism in Biology and Medicine*

LARMOR PRECESSION AS A MECHANISM FOR THE BIOLOGICAL SENSING OF SMALL MAGNETIC FIELDS

D. T. Edmonds*

The Clarendon Laboratory, Oxford, England

The Larmor theorem predicts the change in motion of a charged particle subject to a strong centrally directed force when a magnetic field is applied. The field imposes and axial symmetry on the system even though the energy of interaction with the field is very small in comparison with that of its original motion. The theorem is applied to an ion such as Ca^{++} bound in a cavity of an enzyme such as calmodulin. It is also shown that small magnetic fields can cause charge-dependent coherent motion of ions trapped in the Debye layer about a charged body.

To illustrate the Larmor effect we consider a system with cylindrical symmetry and choose cylindrical coordinates R, θ, Z with a constant magnetic field applied along the Z axis. The particle of mass M and charge Q may be subject to other forces along the Z and R directions, but we assume that only the magnetic force acts around the Z axis in the θ direction. We obtain the equation of motion for θ by equating the rate of change of angular momentum to the couple acting and yields

$$\mathrm{d}(MR^2\dot{\theta})/\mathrm{d}t = -R\dot{R}QB \tag{1}$$

$$\mathrm{d}[MR^2(\dot{\theta} + W_L)]/\mathrm{d}t = 0, \quad \text{where } W_L = QB/(2M)$$

$$R^2(\dot{\theta} + W_L) \text{ is constant in time} \tag{2}$$

If at time $t = 0$ the particle passes through the origin ($R = 0$), then Eq. (2) can only be satisfied for all subsequent R if

$$\dot{\theta} = -W_L \tag{3}$$

Thus the effect of applying the magnetic field is to leave the original motion intact but to superimpose a precession about the Z axis at a rate $-W_L$. W_L is half the cyclotron frequency for that particle in that field and is called the Larmor frequency. The magnetic field also exerts a force along the R direction, but this force is assumed to be negligible in comparison with the strong central force already acting. If the original motion was harmonic motion about the Z axis along a direction perpendicular to the Z axis, the effect of the field is to cause the direction of vibration to precess at $-W_L$ about the Z axis.

Because a magnetic field exerts no force on a charged particle that moves parallel to it, more complex motion is easily dealt with by resolution of the motion parallel and perpendicular to the applied field. The motion parallel to the field is not affected and the motion perpendicular to the field precesses. Thus the Larmor precession imposes an axial symmetry on the system, even though the energy of interaction with the magnetic field is entirely negligible in comparison with the energy of the original motion. Although it appears that extra angular momentum associated with the precession has been created, in fact angular momentum is conserved when account is taken of the momentum inherent in a magnetic field. The precession rate does not depend on the frequency of the centrally directed motion, which need not be harmonic. A thermally excited vibrator which undergoes sudden random changes of frequency and amplitude still precesses at the Larmor rate.

These considerations may be applied to an ion such as calcium bound in a cavity within an enzyme such as calmodulin in the hope that they will provide an explanation of the magnetic field dependence of the catalysis rate found for the phosphorylation of light chain myosin.[1] The environment of the bound ion is remarkably similar in all four calcium binding sites of calmodulin and troponin C,[2,3] so that when bound, the

*Correspondence to: D. T. Edmonds, The Clarendon Laboratory, Parks Rd., Oxford, England OX1 3PU.

ion touches six negatively charged oxygen atoms arranged approximately octahedrally. Because of the high symmetry, the electric field at the ion is small. The enzymes are active only while calcium is bound and we may speculate that the thermally excited ion vibrating within the cavity changes the effectiveness of the enzyme depending on its direction of vibration within the cavity. The effectiveness of an applied magnetic field may then be altered by manipulation of the vibration direction through the Larmor effect if the rate of field-dependent rotation is comparable to or greater than that produced by the thermally excited shaking of the whole cavity. The vibrating ion is well shielded within the cavity and vibrations of the cavity are effective in rotating the direction of vibration only in a narrow band of frequencies close to the vibration frequency of the ion within the cavity. Since this case has been recently discussed,[4] it will not be further described here. In the same paper it was shown that manipulation of the vibration direction is also possible by oscillating magnetic fields of very small amplitude applied at an angle to the static field at particular frequencies related to the Larmor frequency in a manner similar to nuclear magnetic resonance (NMR).

As a second application of the Larmor effect, it is shown below that the thermally activated random Brownian movement of ions can lead to their coherent motion when a magnetic field is applied. Adopting the geometry described above, we consider ions trapped in a Debye layer surrounding a charged cylindrical body aligned along the Z axis. A much simplified but analytical model of this motion is obtained if a charged particle executes radial simple harmonic motion with amplitude A at angular frequency V about the surface $R = R_0$, so that $R = R_0 + A \sin(Vt)$. Substitution into Eq. (2) yields

$$\dot{\theta} = W_L[R_0^2/R^2) - 1] \tag{4}$$

where we have assumed that $\dot{\theta} = 0$ when $t = 0$. As expected from the sign of the Lorenz force, $\dot{\theta} < 0$ when $\dot{R} > 0$ and vice versa, but averaged over the complete oscillation the change in θ is not zero. In fact, during T, the period of one cycle of the oscillation, the angle θ changes by $D\theta$ in

$$D\theta = W_L T \{[1/(1 - M^2)^{3/2}] - 1\}$$

where $M = A/R_0$.

A more realistic computer simulation is shown in Fig. 1, which is for a sodium ion trapped in the Debye layer around a negatively charged cylindrical body aligned along the Z axis of radius 10 Å. The voltage in the Debye layer approximates that for a surface potential of -100 mV with a Debye length in the aqueous solution bathing the body of 10 Å, corresponding to a typical biological fluid. The ion is subject to thermally generated momentum impulses of random direction with exponentially distributed collision times with a mean collision time of 2×10^{-12} s. A magnetic field of 50 μT, typical of the earth's field, acts along the Z axis. In order to show the systematic rotation in a small time, only the momentum impulses along the R direction are applied. The impulses along the θ direction produce θ rotations that cancel to zero over time. In this simulation the rotation due to the field acting on the fluctuations in R corresponds to about 2 rad/s, or about 115°/s. The direction of rotation depends on the sign of the charge on the ion through the sign of W_L.

Thus, although ions in solution drift at low velocities which appear to rule out appreciable Lorenz forces in small magnetic fields, when the motion is decomposed into its constituent small-amplitude fast excursions, coherent motion due to the field is possible. As an application of this effect, small magnetic fields could disturb the directions of the primary ion currents that lead to differentiation[5] in a developing egg.

REFERENCES

1. M. S. Markov, J. T. Ryaby, S. Wang, and A. A. Pilla, Proc. 1st EBEA Cong., *Bioelectrochem. Bioenerg.* (in press).

2. K. A. Satyshur, S. T. Rao, D. Pyzalska, W. Drendel, M. Greaser, and M. Sundaralingam, *J. Biol. Chem.* 263: 1628, 1988.

3. Y. S. Babu, J. S. Sack, T. V. Greenhough, C. E. Bugg, A. R. Mears, and W. J. Cook, *Nature* 315: 37, 1985.

4. D. T. Edmonds, in Ref. 1.

5. L. F. Jaffe and R. Nuccitelli, *Ann. Rev. Biophys. Bioeng.* 6: 445, 1977.

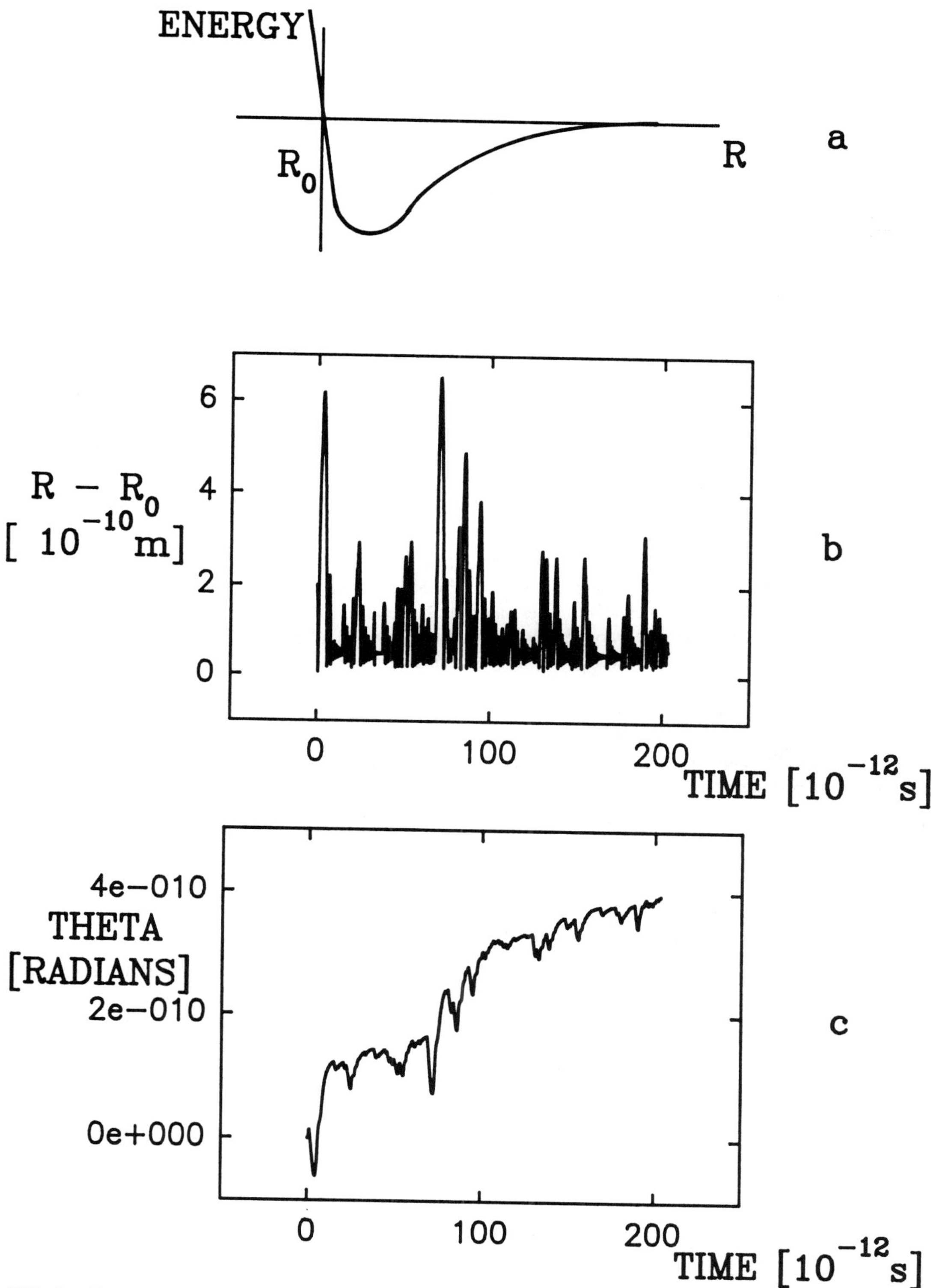

FIG. 1.—Energy profile of cation in cylindrical Debye layer; (b) computer-simulated thermally excited excursions of sodium ion along radial direction as a function of time; (c) changes in θ about the axis with time that result from action of axial magnetic field on cation moving randomly in R direction. (Parameters used in simulation are given in text.)

Martin Blank, Ed., *Electricity and Magnetism in Biology and Medicine*

COMPARISON OF A FIVE-STATE VS A THREE-STATE COULOMBIC ZEEMAN MODEL OF EMF EFFECTS ON LIGAND BINDING

E. Moggia, T. Tommasi, B. Bianco, and A. Chiabrera*

University of Genoa, Italy

A simplifying assumption about the number of states in quantum models of EMF effects on ligand binding is discussed. In particular, the three-state Coulombic model so far assumed (n = 1, l = 0; and n = 2, l = 1, m ± 1) is extended to a five-state model, by adding the two missing states corresponding to the quantum numbers n = 2, l = 0 and n = 2, l = 1, m = 0. A comparison between resulting binding probabilities for the three-state and five-state models is made. From a numerical point of view, such probabilities turn out to be similar. This result suggests that the three-state model is accurate enough for the quantum analysis proposed.

A comprehensive quantum model of the effects of low-intensity electromagnetic exposure on ion binding has been developed,[1,2] based on the density operator. For a Helmholtz-coil exposure system, the binding probability depends on the time derivative of the magnetic induction field. A steady magnetic field modulates only the effectiveness of the time-varying field which, on the other hand, is necessary to achieve any possible effect. This fact is true for both quantum-mechanics models (three-state and five-state Coulombic models) considered in this paper.

In the model, the Helmholtz coils generate a time-varying magnetic field, B_{1z}, parallel to the z axis. A constant magnetic field, along the z axis, B_{0z}, and the x-axis, B_{0x}, is added. A Coulombic endogenous potential is assumed for the unperturbed Hamiltonian, and hydrogen-like eigenfunctions are found. A perturbing Hamiltonian operator H_1 applies:

where Q and M are the charge and the mass of the ion, respectively, **B** is the magnetic induction field vector, h_c is Plank's constant divided by 2π, and d_x, d_y, 0 are the components of the vector distance between the binding site and the axis of a Petri dish filled with a conductive medium and placed between the coils.

Starting from the operator H_1, we can develop an equation system for the density operator. The density operator describes the overall behavior of the binding ions, and an average binding probability can be derived from it.

THE FIVE-STATE MODEL

The five-eigenfunction set consists of three eigenfunctions corresponding to the quantum numbers (n = 1, l = 0), (n = 2, l = 1, m = 1) and (n = 2, l = 1, m = − 1) (three-state model),[3] and of two new eigenfunctions corresponding to the quantum numbers (n = 2, l = 0) and (n = 2, l = 1, m = 0). Such functions are used as a base set for the unperturbing Hamiltonian associated to the system. This choice considers all the first and second levels of energy, which correspond to the eigenvalues E_1 and E_2, thus completing the set. The entries D_{nk} related to a magnetic field parallel to the z axis, $B_z(t) = B_{0z} + B_{1z} \sin \omega t$, are computed from

where Φ_k is the time-independent k-th eigenfunction associated with the unperturbed Hamiltonian, $h_c \omega_k = E_k$ (E_k is the eigenvalue of the related eigenfunction), and * stands for conjugate. The evaluation of the matrix

*Correspondence to: A. Chiabrera (Biophys. & Electronic Engg.), University of Genoa, Via Opera Pia 11a, I-16145 Genoa, Italy.

D allows one to compute an equation system for the density operator elements ρ_{ij}.[1]

EVALUATION OF THE BINDING PROBABILITY

The binding probability, $P^{av} = \lim_{t\to 0} \frac{1}{t}\int_0^{\infty} P(t)\,dt$, is evaluated for a sphere of radius R_c centered on the binding site, and is compared with the probability at thermal equilibrium P^{av}_{th}.

The case $B_{0x} = 0$ is briefly discussed. As a first-order approximation, the following expression holds:

$$
\begin{aligned}
P^{av}_{th}-P^{av} &= \frac{2^8}{3^8 u^3}\delta^2\Omega^2 L^4 \cdot (\rho_{11,th} - \rho_{22,th}) \\
&\cdot\Big[(\Theta_{11}-\Theta_{33})\Sigma_{l=0,\infty}[J'_l(\tfrac{1}{\Omega})]^2\Big\{\frac{(1+\eta\frac{L}{u})^2+(\frac{1}{\xi})^2+l^2\Omega^2\frac{L}{u}^2}{[-(1+\eta\frac{L}{u})^2+l^2\Omega^2\frac{L}{u}^2+(\frac{1}{\xi})^2]^2+(\frac{2}{\xi})^2(1+\eta\frac{L}{u})^2} + \\
&\frac{(1-\eta\frac{L}{u})^2+(\frac{1}{\xi})^2+l^2\Omega^2\frac{L}{u}^2}{[-(1-\eta\frac{L}{u})^2+l^2\Omega^2\frac{L}{u}^2+(\frac{1}{\xi})^2]^2+(\frac{2}{\xi})^2(1-\eta\frac{L}{u})^2}\Big\} + \\
&(\Theta_{11}-\Theta_{22})(\frac{2\sigma_1}{5}-1)\Sigma_{l=-\infty,\infty}[J'_l(\tfrac{1}{\Omega})]^2\Big\{\frac{1}{[-1+\eta\frac{L}{u}+l\Omega\frac{L}{u}]^2+(\frac{1}{\xi})^2}+\frac{1}{[-1-\eta\frac{L}{u}+l\Omega\frac{L}{u}]^2+(\frac{1}{\xi})^2}\Big\} + \\
&(\Theta_{22}-\Theta_{33})\Big[\frac{2\sigma_1\sigma_2}{4\sigma_1+\sigma_2}-1+3\cdot\frac{2\sigma_1}{5}\cdot\frac{\sigma_1-\sigma_2}{4\sigma_1+\sigma_2}\Big]\cdot \\
&\cdot\Sigma_{l=-\infty,\infty}[J'_l(\tfrac{1}{\Omega})]^2\Big\{\frac{1}{[-1+\eta\frac{L}{u}+l\Omega\frac{L}{u}]^2+(\frac{1}{\xi})^2}+\frac{1}{[-1-\eta\frac{L}{u}+l\Omega\frac{L}{u}]^2+(\frac{1}{\xi})^2}\Big\} + \\
&-2\Theta_{12}\frac{1}{(\frac{1}{\xi})^2+1}\cdot\frac{3^6}{2^6\sqrt{8}}\cdot\Sigma_{l=-\infty,\infty}[J'_l(\tfrac{1}{\Omega})]^2\Big\{\frac{[-1+\eta\frac{L}{u}-l\Omega\frac{L}{u}]+(\frac{1}{\xi})^2}{[-1+\eta\frac{L}{u}+l\Omega\frac{L}{u}]^2+(\frac{1}{\xi})^2}+\frac{[-1-\eta\frac{L}{u}-l\Omega\frac{L}{u}]+(\frac{1}{\xi})^2}{[-1-\eta\frac{L}{u}+l\Omega\frac{L}{u}]^2+(\frac{1}{\xi})^2}\Big\}\Big] \qquad (1)
\end{aligned}
$$

where $\rho_{11,th}$ and $\rho_{22,th}$ are the values of ρ_{11} and ρ_{22} at thermal equilibrium, described by the Boltzmann distribution. If one defines a linear operator Θ whose value is 1 inside the sphere of radius R_c and 0 outside, one can find the related entries, depending on R_c and $a_0 = h_c^2/\zeta M$ (ζ is a fitting parameter), so that $P^{av}_{th} = \Theta_{11}\,\rho_{11,th} + (\Theta_{22} + 3\Theta_{33})(1-\rho_{11,th})/4$.

Furthermore, the following parameters are defined:

$$
E_2\text{-}E_1 = h_c\omega_0\,;\quad \omega_{1,0z} = \frac{QB_{0z}}{2M}\,;\quad \omega_{1,1} = \frac{QB_{1z}}{2M},
$$

$$
\frac{h_c\omega_0}{kT} = u\,;\quad \frac{h_c\omega_{11}}{kT} = L\,;\quad \frac{h_c\omega_{10z}}{kT} = \eta\,;\quad \frac{\omega}{\omega_{11}} = \Omega\,;
$$

$$
\frac{d}{\dfrac{h_c\sqrt{3}}{\sqrt{8MkT}}} = \delta\,;\quad \xi = \omega_0\tau_1,\quad \frac{\theta_1}{\tau_1} = \sigma_1\,;\quad \frac{\theta_2}{\tau_1} = \sigma_2.
$$

The lifetime parameters τ_1, τ_2, θ_1 and θ_2 are due to the thermal bath interaction. They are discussed elsewhere.[2] $J_1(1/\Omega)$ represents the Bessel function of integer order 1, and k is the Boltzmann's constant.

COMPARISON BETWEEN THE THREE-STATE AND THE FIVE-STATE MODELS

The differences between P^{av}_{th} for the two models appear to be rather small. For the sake of simplicity, the case $\eta = 0$ is discussed. Some possible relations between the lifetimes were analyzed via numerical calculations. The cases considered ($\sigma_1 = \sigma_2, \sigma_1 \ll \sigma_2$, and $\sigma_1 \gg \sigma_2$) showed that the contribution due to the populations is always positive, as previously found for the three-state model.[3] We should stress that the term due to the coherences holds for the five-state model only, and that its contribution may result in a P_{av} slightly greater than P^{av}_{th}. In general, differences between probabilities found in three-state and five-state cases are not significant. In conclusion, the three-state model seems sufficient to describe the Zeeman Coulombic binding process.[4]

REFERENCES

1. B. Bianco and A. Chiabrera, "From the Langevin-Lorentz equation to the Zeeman model of electromagnetic effects on ligand-receptor binding," *Bioelectrochemistry and Bioenergetics* (in press).
2. A. Chiabrera and B. Bianco, "The effect of lifetimes on ligand modelled by the density operator," *Proc. First EBEA Congress,* Brussels, 1992.
3. A. Chiabrera et al., "Quantum dynamics of ions in molecular crevices under electromagnetic exposure," and "Quantum analysis of ion binding kinetics in electromagnetic bioeffects," in C. Brighton and S. Pollack, Eds., *Electromagnetics in Medicine and Biology,* San Francisco: San Francisco Press, 1991, 21-31.

4. This work was supported by the Italian Ministry for the University and Research (MURST 40%).

Martin Blank, Ed., *Electricity and Magnetism in Biology and Medicine*

ROLE OF POLARIZATION FORCES IN MEDIATING THE INTERACTION OF LOW-FREQUENCY ELECTRIC FIELDS WITH LIVING TISSUE

K. J. McLeod,* C. T. Rubin, H. J. Donahue, and F. Guilak

University of Pennsylvania School of Medicine

Electric field effects on cells and tissues have been shown to occur at field intensities below 1 mV/cm, precluding explanation of a cellular response through perturbations of the transmembrane potential. Moreover, cells and tissues respond to sinusoidal extracellular electric fields with a distinct frequency dependence. A physical explanation for the interaction of electric fields with living tissue must therefore identify a process of amplification, must be nonlinear for low-intensity stimuli, and must account for the frequency specificity reported for cell sensitivity. We propose that the polarization forces that arise from the electric-field gradient in the highly polarizable environment of the glycocalyx produce an interaction with cells consistent with these empirical findings.

At frequencies below 1 kHz and for cell diameters smaller than several hundred micrometers, an external electric field is essentially excluded from the interior of a cell by charge relaxation at the surface of the poorly conducting plasma membrane. This situation suggests that low-frequency electric fields can only interact with cells through the forces developed on the free charges and dipolar structures in or outside of the cell plasma membrane. Electric forces on mobile charges (Lorentz forces) give rise to the phenomenon of charge relaxation at the extracellular surface of the plasma membrane and result in the development of a transmembrane potential perturbation proportional to the extracellular field intensity and the size of the cell. The maximum amplitude of the field-induced membrane potential perturbation is extremely small, less than 1μV for a 10μm cell in a 1mV/cm electric field, substantially below the membrane noise voltage.

In addition to the Lorentz force, a polarization force distribution **F** arises due to the presence of a nonuniform electric field **E** in a polarizable medium of permittivity ε:

$$\mathbf{F} = (\mathbf{P}\cdot\nabla)\mathbf{E}$$

where, for a linear dielectric material, $\mathbf{P} = (\varepsilon - \varepsilon_0)\,\mathbf{E}$, with ε_0 the permittivity of free space.

This distributed force arises at the cell surface and is of a magnitude that depends on the gradients in the electric field surrounding the cell and the dielectric properties of the extracellular region of the cell (Fig. 1). The magnitude of this polarization force can be estimated, by use of the Maxwell stress tensor form of the force density, as

$$\mathbf{F} = \int (\varepsilon_x - \varepsilon_m)\,\varepsilon_0 E_t^2\, dA$$

where E_t is the tangential electric field component outside of the cell membrane, $\varepsilon_x - \varepsilon_m$ reflects the gradient in the relative dielectric permittivities of the extracellular space and the bulk medium, and the integral is taken over the cell surface. As ε_0 is the permittivity of free space, the magnitude of the developed force depends largely on the dielectric properties of the extracellular space, the electric field intensity, and the geometry of the cell.

CHARACTERIZATION OF ELECTRIC POLARIZATION FORCES

Four characteristics of cells and tissues have important implications for understanding the potential role of polarization force effects on living cells. They are the high effective dielectric permittivity, which provides a means for amplifying the effects of low-level fields; the ionic environment, which imposes a frequency dependence on the development of a substantial force; the permeability of the cell membrane, which per-

*Correspondence to: K. J. McLeod, SUNY Dept. of Orthopaedics, Stony Brook NY 11794-8181.

mits a cumulative response to prolonged field exposure; and the distinct morphologies of various cells, which alter the spatial integration of the force density.

Dielectric Amplification

The interface of the cell with its environment is a highly charged cell coat, or glycocalyx. In the presence of very low frequency electric fields, an ion diffusion cloud develops within the glycocalyx to compensate for the depletion of the co-ion species in this region.[1] Recent investigations on the low-frequency dielectric properties of both cell suspensions and ion-permeable membranes have demonstrated dielectric permittivities as high as $10^6 - 10^7$ in living tissue for frequencies below the alpha dispersion frequency (10-100 Hz). Although the microscopic dielectric permittivity of the glycocalyx region remains small (<100), the large dielectric enhancement at the macroscopic level produces a substantial force on the macroscopic collection of fixed charges on the cell surface and in the glycocalyx. For an extracellular field of 1 mV/cm and a relative dielectric constant of the glycocalyx of 10^7, a peak pressure develops at the surface of the cell of the order of 10^{-5} dyne/cm^2, or approximately 1 microPascal. This value is equivalent to the pressure on the tympanic membrane at the threshold of hearing.[3] However, cell and tissue responses to extracellular fields have been reported at field intensities below 100 μV/cm (Fig. 2). Because the induced force is proportional to the square of the field intensity, the cells would have to detect pressures as low as 10^{-9} P.

Frequency Dependence

Proportionality to the square of the field intensity means that the polarization forces developed at the cell surface are nonlinear. The force vector is always directed from the region of high dielectric constant to the region of low dielectric constant, or from a region of low field intensity to a region of high field intensity. The induced forces on opposite sides of any cell are in opposite directions so that the effects of the applied field are additive, independent of the phase or polarity of the electric field.

This characteristic of polarization forces would suggest that both DC and AC should produce equivalent forces at a surface for a given field intensity. That would be true for solid dielectrics, but polarization forces arising at cell surfaces demonstrate a frequency dependence because the large dielectric constant critical to the development of a significant force density depends on both diffusion and drift processes. The dielectric behavior of the tissue requires the buildup of a diffusion cloud around the glycocalyx of the cells. At sufficiently high frequencies (above 1 kHz) the diffusion cloud is restricted in growth, which results in a small dielectric enhancement. As frequency is decreased, there is increased time for the diffusion cloud to build to a size sufficient to affect current flow in the region near the cell surface. However, at progressively lower frequencies, a significant drift current, or bulk motion, of the ions results in a diminution of the dielectric constant. The ions responsible for the large polarization charge distribution at the membrane surface have electrophoretic mobilities of the order of 10^{-3}cm^2/V – s. For an applied field of 1 mV/cm at 10 Hz, the expected displacement of a typical counter ion would be less than 1 nm, but with decreasing frequency the maximum possible displacement increases proportionally. At very low frequencies (below 1 Hz) the collective motion of the extracellular ions and entrained water is sufficient to disrupt the buildup of the diffusion cloud outside of the double layer and reduce the extent of the dielectric enhancement. The net effect of the diffusional and drift processes in the ionic environment of the cell is the development of an induced force with a frequency dependence similar to that seen for the response of cells and tissues to exogenously induced electric fields in the ELF range (Fig. 3).

Temporal Integration

Proportionality to the square of the applied field intensity also means that both a steady force component and an alternating component at twice the applied field frequency is developed, and both are transmitted to the cell membrane via the glycocalyx molecules. Certain structures (such as ion channels) which may have characteristic resonances, could be directly affected by the acoustic vibrations imposed by the alternating force component.

More important, the permeability of living cell membranes permits the effect of the steady induced force to be integrated over time as well as over space by changes in cell volume. Although high field intensities are required for significant deformation under the assumption of maintaining constant cell volume, a long exposure, even at a low field intensity, affects size or shape of a permeable cell, with a subsequent effect on cell activity. The pressure on the plasma membrane surface may be capable of interacting directly with membrane protein through changes in membrane fluidity, or with nuclear activity by means of the cytoskeletal network connecting the plasma membrane with the nuclear envelope. Modulation of nuclear pore conductance or deformation of the nuclear membrane could lead to a rapid effect on gene expression.

Morphologic Parameters Affecting the Induced Force

The net force on a cell also depends on spatial parameters, so that these parameters should influence the degree to which low-frequency fields affect physiological processes. Because the buildup of the ion diffusion cloud is critical to the creation of the large dielectric enhancement, any augmentation or disruption of this cloud affects the magnitude of the induced force and the cellular response. An increase in the effective thickness of the glycocalyx, for example, would require the establishment of larger diffusion clouds and result in an increased force at the cell surface for a given field intensity. Alternately, high-velocity or turbulent fluid flow would inhibit the establishment of a large diffusion cloud. Also, the morphologic properties of the cell itself affects both the induced forces on the cell and cellular sensitivity to these induced forces. Cells with a large surface area may be more affected by an exogenous field than a smaller cell, as larger forces are obtained when the induced pressures are integrated over the larger surface. Alternately, cells with a low membrane stiffness or membrane fluidity might be expected to be particularly sensitive to extracellular electric fields.

SUMMARY

In the extremely low frequency range, electric field gradients arise in the highly polarizable region surrounding a cell, due to the low conductivity of the cell membrane, and result in a distributed volume force that acts to stretch the cell along the direction of the applied field. The magnitude and temporal characteristics of this force have been estimated; it is shown that the polarization forces at the cell surface provide qualitative, and to some degree quantitative, explanations for the major empirical observations we have obtained on cell and tissue responses to electric fields. These explanations include an intrinsic amplification process sufficient to provide a reasonable signal-to-noise ratio for detection of low-intensity (less than 1 mV/cm) electric fields; a rectification process that explains the ability of cells to respond to sinusoidal fields; appropriate frequency response characteristics; and a mechanism for temporal integration of the "signal." Therefore, we propose that it is the polarization forces arising at the cell surface that allow extremely low frequency (ELF) electric fields to interact with living cells and thereby influence the normal development and repair of tissues exposed to such fields.

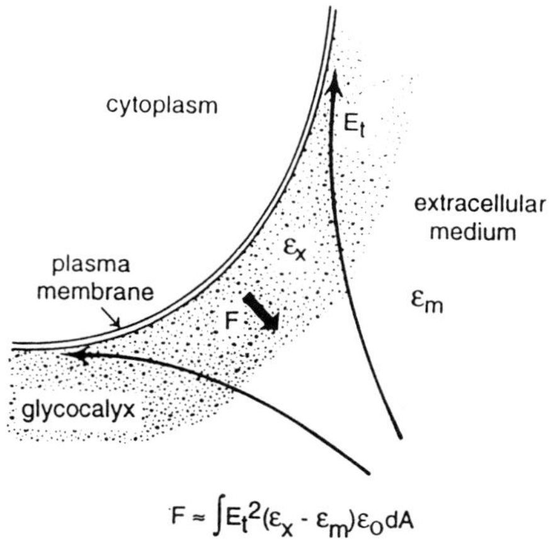

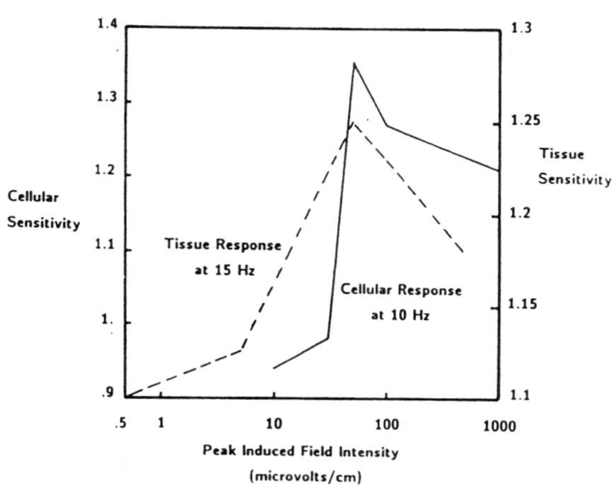

FIG. 1.—Polarization forces arise at cell surface due to electric-field gradients within glycocalyx. Factors dominating net force are field intensity, thickness and dielectric properties of glycocalyx region of the cell, and cell surface area.

FIG. 2.—Intensity response characteristics of cells and tissues to exogenously induced ELF fields. Solid line: relative sensitivity of bovine fibroblastic cells as assayed by change in protein synthesis stimulated by 10Hz electric field exposure of 14 h.[4] Dashed line: new bone growth initiated by 1 h/day of 15Hz electric field in the functionally isolated avian ulna.[5]

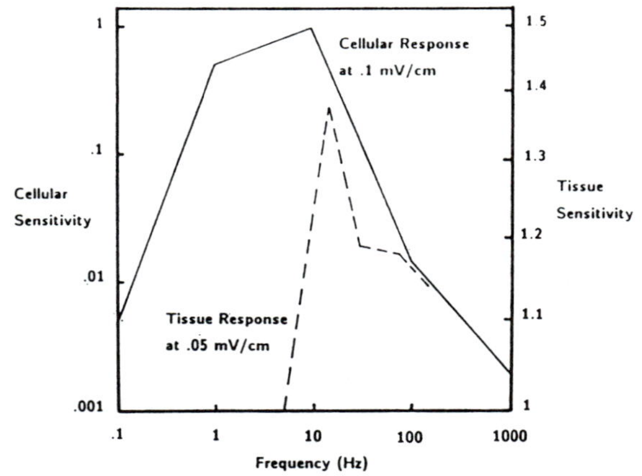

FIG. 3.—Frequency response characteristics of cells and tissues to exogenously induced ELF fields. Solid line: response of fibroblastic cells as determined by minimum field intensity required to affect protein synthesis.[4] Dashed line: response of tissue as assayed by new bone growth in avian ulna initiated bv 1 h/dav of ELF field exposure for 8 weeks.[5]

REFERENCES

1. W. C. Chew, *J. Chem. Phys.* 80: 4541, 1984.
2. K. R. Foster and H. P. Schwan, *CRC Crit. Rev. Biomed. Eng.* 17: 25, 1989.
3. J. C. R. Licklider, in S.Stevens, Ed., *Handbook of Experimental Psychology,* New York: Wiley, 1951, 985.
4. K. J. McLeod, R. C. Lee, and H. P. Ehrlich, *Science* 236: 1465, 1987.
5. K. J. McLeod and C. T. Rubin, *J. Theor. Biol.* 145: 385, 1990.

6. This work was supported by the National Aeronautics and Space Administration's Space Biology Research Associates Program (NAGW-70), The Whitaker Foundation, and National Institutes of Health Grant AR41040-01.

Martin Blank, Ed., *Electricity and Magnetism in Biology and Medicine*

MAGNETICALLY INDUCED ELECTRIC FIELDS MEASURED IN RATS AND COMPARED TO A HOMOGENEOUS RAT MODEL

D. L. Miller*

Battelle Pacific Northwest Laboratory, Richland, Wash.

A probe of approximately 2mm resolution was used to measure electric fields induced in rats by a 60Hz magnetic field at 1 mT. The induced field decreased as the probe moved into the animals, as expected, but some values differed from theoretical estimates for a homogeneous model. At the diaphragm, maximal values were 61 μV/cm (12 μV/cm SD) for axial exposure, and 66 μV/cm (7.4 μV/cm SD) for frontal exposure compared to theoretical estimates for a homogeneous model of 70.2 and 93.4 μV/cm, respectively. Further work is needed before these differences can be understood and the in vivo dosimetry of ELF magnetic fields can be fully characterized.

Extremely low frequency magnetic fields interact with biological subjects and induce small electric fields and currents,[1] which represent the actual internal exposure to an animal; their detailed characterization is essential for thorough dosimetric considerations. The induced fields can be difficult to measure or theoretically predict in detail due to the complex shape and structure of animals. One convenient simplifying approximation, which has been used in both theoretical and experimental research, is the homogeneous model. For example, we have recently made induced-field measurements in homogeneous (saline-agar) models of a rat and a human by means of a miniaturized electric-field probe.[2] This present work employed the same methodology in actual rats, in order to assess the validity of the homogeneous model and to begin the study of the detailed ELF dosimetry in vivo.

METHODS

We used a differential probe described previously[2] to obtain the internally induced electric field by measuring the voltage across the probe with a lock-in amplifier, and dividing by the calibrated probe spacing. The spacing ranged from 0.23 to 0.20 cm in this study. (A gradual size reduction results from repeated cleaning.) Tests were conducted in a 1mT field at 60 Hz, which could be oriented in either the vertical or horizontal direction. The measured field direction was perpendicular to the exposure field, and the probe was oriented perpendicular to both the exposure and the induced fields. Only the component of the induced field in phase with a reference-coil signal, which is the only component expected for a homogeneous model, was measured; a 90° out-of-phase signal was sometimes noted as well, but it was uncalibrated and is therefore not reported here. Rats weighing 611 to 738 g were sacrificed by CO_2 asphyxiation and prepared for measurements by the cutting of small holes in the body wall to insert the probe. Saline of 0.2 S/m conductivity was used sparingly as needed to irrigate the probe and maintain electrical coupling. The electric fields obtained for each animal were adjusted by a scaling factor to account for the size variation in the animals (i.e., the cube root of the ratio of 640 g to the animal's weight). The data given here represent the average of six measurements from three different animals (different animals were also used for each orientation, for a total of nine animals), with standard deviation error bars.

RESULTS

Scanning of the probe within the body was difficult in some regions and produced erratic results. The head and other areas of solid tissue could not be scanned by present methods. In addition, readings could often not be obtained within the chest cavity, since the lungs collapsed after the chest wall was penetrated, but maximum induced fields found at the body wall were measurable. In the abdominal cavity, the probe

---

*Correspondence to: D. L. Miller MS P7-53, Battelle PNL, Box 999, Richland WA 99352.

could normally be moved to scan the induced field, with the internal organs apparently sliding out of the way. As expected, the measured fields tended to decrease as the probe moved into the body of the animal, reaching zero near the center and then increasing again (with a negative sign indicating the reverse direction). However, this progression was not smooth and large variations occurred between points. These variations, possibly associated with individual organs, appeared to be reproducible in each animal, but yielded large standard deviations when results from several animals were averaged.

Results of scans with the probe moved downward into the side of animals in 0.5cm increments are shown in Fig. 1 for axial exposure and in Fig. 2 for frontal exposure. The field was horizontal, with an animal lying on its side, and the scan line across the abdominal cavity near the diaphragm. The data are compared to theoretical calculations (solid lines) based on a homogeneous model of the animal's body as an ellipsoid with approximately measured axes of 5.4 cm (thickness), 8.0 (width), and 18 cm (length excluding the head and tail). These homogeneous-ellipsoid values are a better representation of the animals used in this study than values from the previous agar-model measurements,[2] because the shape of the real rats differed considerably from the shape of the agar model. The first few points of the scan with axial exposure (Fig. 1) agree with the theoretical line. For example, 61 μV/cm (12 μV/cm SD) measured at the body wall is compared to 70 μV/cm for the homogeneous model. However, with frontal exposure (Fig. 2), the measurements are often much lower than the expected values. For example, 66 μV/cm (7.4 μV/cm SD) measured at the body wall is significantly different from 93 μV/cm for the homogeneous model ($p < 0.05$). With deeper penetration, the measurements become more uncertain and are less than expected from the simple theory.

Measurements were also made for lateral exposure with a vertical field and probing from the front, but the probe could only penetrate to the back of the abdominal cavity (about 5 cm of the 8cm thickness at the diaphragm). In addition, scans were made in the lower abdomen and the chest for all three configurations (axial, frontal, and lateral exposure). The lower-abdominal scans gave results similar to those shown in Figs. 1 and 2, but results in the chest were erratic and readings could often not be obtained. Maximal fields at the body wall are given in Table 1 for these scans.

TABLE 1.—Induced fields at body wall in μV/cm (standard deviation).

Exposure	Chest		Diaphragm		Abdomen	
Axial	55.1	(13.8)	61	(12.1)	53.1	(15.4)
Frontal	47.2	(15.3)	66	(7.4)	63.3	(40.1)
Lateral	67.8	(22.2)	102	(31.0)	101	(12.5)

DISCUSSION

Miniature-probe methodology for dosimetric characterization of ELF field exposure was applied to rats. The results presented show significant differences from theory for a homogeneous model; measured values are often less than half the expected values. The reasons for this divergence are uncertain at this time. One reason may be perturbation of the field by conductivity variations and internal membranes, similar to that observed in eggs.[3] For example, variations in tissue properties for various internal organs may be responsible for some of the variability of the measurements. In addition, the peritoneum and lungs may tend to impede an induced field oriented across the diaphragm, which might help explain the significantly reduced fields found for frontal exposure (Fig. 2) compared to simple theory. In contrast, for axial exposure, where the induced field was oriented parallel to the diaphragm, results for the same region (Fig. 1) were not significantly different from the simple theory. Further work is needed to resolve these uncertainties and to characterize fully the in vivo dosimetry of ELF magnetic fields.[4]

REFERENCES

1. T. S. Tenforde and W. T. Kaune, "Interaction of extremely low frequency electric and magnetic fields with humans," *Health Physics* 53: 585, 1987.

2. D. L. Miller, "Miniature-probe measurements of electrical fields and currents induced by a 60 Hz magnetic field in rat and human models," *Bioelectromagnetics* 12: 157, 1991.

3. D. L. Miller, "Electric fields induced in chicken eggs by 60-Hz magnetic fields and the dosimetric importance of biological membranes," *Bioelectromagnetics* 12: 349, 1991.

4. Supported by the U.S. Department of Energy (DE-AC06-76RLO-1830).

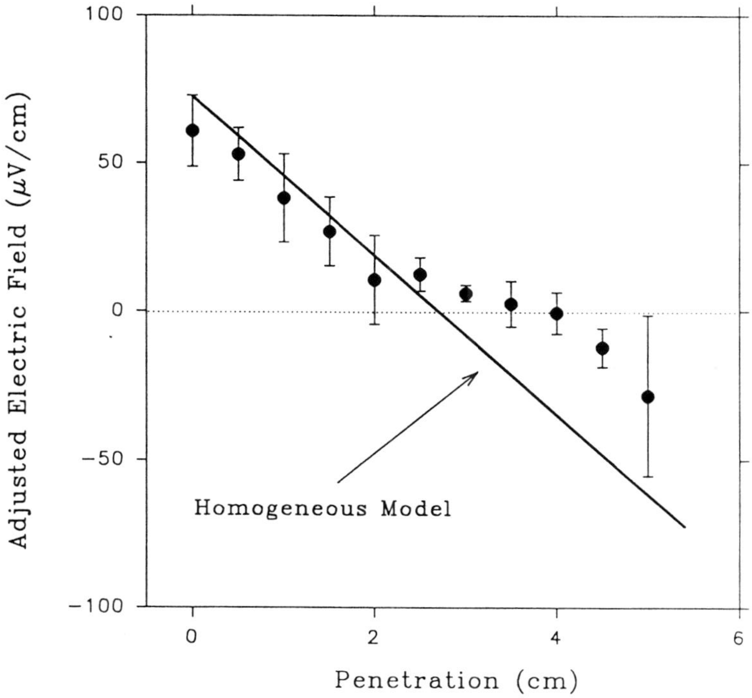

FIG. 1.—Data for axial exposure of rats with theoretical line.

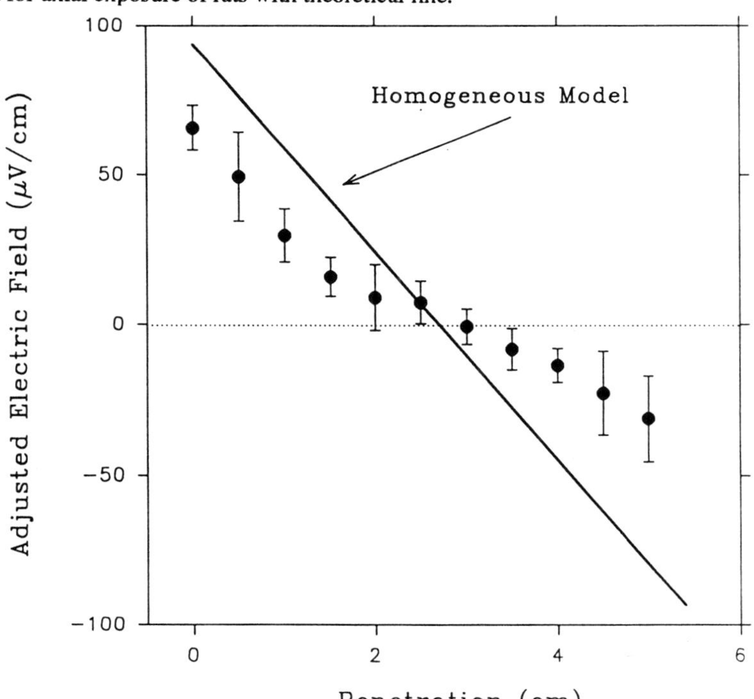

FIG. 2.—Data for frontal exposure of rats with theoretical line.

Martin Blank, Ed., *Electricity and Magnetism in Biology and Medicine*

CONDUCTING SOCKS ELIMINATE ANKLE SAR IN A FULL-SIZE MODEL EXPOSED TO NEAR-FIELD IRRADIATION AT THREE RADIOFREQUENCIES

T. A. Griner,* R. G. Olsen, and J. J. King

Naval Aerospace Medical Research Laboratory, Pensacola, Fla.

Radiofrequency (RF) currents induced in ankles and wrists near transmitting antennas can produce localized specific absorption rates (SARs) that exceed present guidelines even though whole-body average SAR is well within those guidelines. We tested the concept of conductive socks to prevent ankle SAR at three frequencies in the high-frequency (HF) band using a full-size human model near a monopole antenna. Our results showed that a conductive ripstop nylon sock effectively prevented any irradiation-induced temperature rises in the model's ankle as compared with the unprotected ankle.

Electromagnetic energy such as radiated from shipboard antennas or RF heat sealers, induces heating in personnel due to currents flowing in the body.[1] Wrist and ankle currents may generate relatively high specific absorption rates (SARs) due to the high current densities in small cross-sectional areas of the limbs.[2] The objective of this study was to determine whether a conducting sock surrounding a simulated ankle would reduce RF-induced heating. We have previously observed that conductive gloves eliminate RF burns about Navy ships, and we hypothesized that conductive socks would reduce ankle SAR by virtue of the "RF skin effect." In humans, ankle heating is a highly subjective sensation; therefore, this study was conducted to test our hypothesis quantitatively.

METHODS AND PROCEDURES

We measured local SAR in the restricted leg of a full-size muscle-equivalent human model at an outdoor groundplane irradiation facility.[3] The model was suspended from a plastic pipe stand in an upright position with the model legs touching the groundplane. We determined SAR from temperature-rise measurements obtained with nonperturbing temperature probes.[4] The model was exposed to continuous wave (CW) irradiation 1.2 m from a 10.8m whip antenna operating with a transmitter power of 1000 W at 10.2550, 16.6857, and 29.9000 MHz. (We frequently use this location relative to the antenna to provide a common reference among different experiments.) We compared an ankle covered by the conductive material with an uncovered ankle. The "anti-SAR" sock was constructed of silver/copper coated nylon cloth (Fiber Materials, Inc.). The sock covered the lower 33 cm of the model's left leg and was pinned at the top with common metal stick pins to insure conductive contact through the plastic model bag. The model leg is shown in Fig. 1. A 6cm-diameter simulated ankle was formed by constriction of the lower leg with duct tape. To measure the heating rate, the thermometer probe was inserted into the center of the constriction and allowed to stabilize for a few minutes. Then, a 15min preirradiation baseline was obtained by temperature recording at 1min intervals. Next, the model was irradiated for 15 min with the same sampling intervals. Corrections were made to the heating slope by subtraction of the baseline slope, and SAR was calculated from the resultant RF-induced slope.

RESULTS AND DISCUSSION

Table 1 shows the localized SARs along with previously measured E-field intensities at the selected location. Figure 2 shows the 29.9000MHz data and the dramatic lack of heating with use of "anti-SAR" sock. Although only three HF frequencies were used, we believe that these results are valid over a wider frequency spectrum because the phenomenon of the RF skin effect has been known to exist over many decades.

---

*Correspondence to: T. A. Griner, Naval Aerospace Medical Research Laboratory, NAS, 51 Hovey Rd., Pensacola FL 32508-1046.

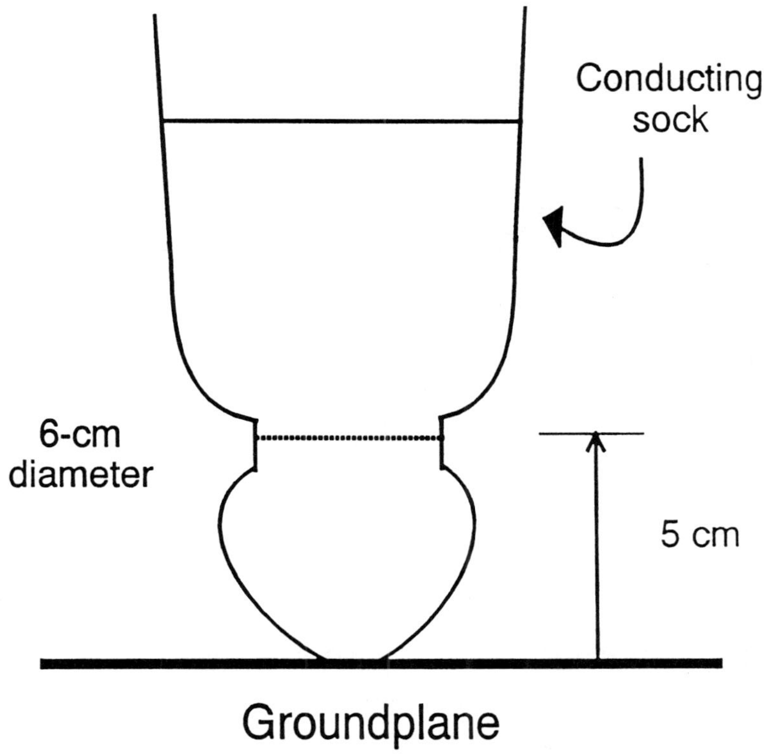

FIG. 1.—Diagram of model's simulated ankle.

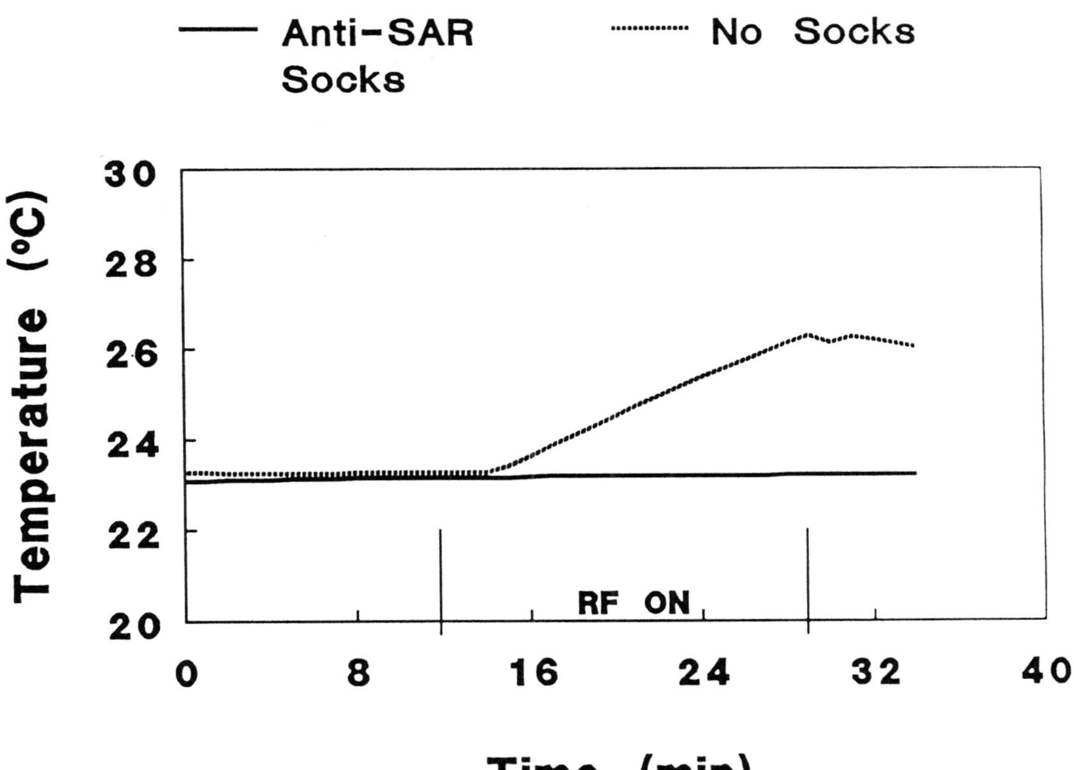

FIG. 2.—Ankle heating in model at 29.9000 MHz.

TABLE 1.—Localized SARs and E-fields.

Frequency (MHz)	Ankle SAR (W/kg) (N = 1)		E-field[a] (V/m)	
	Socks	No socks	Total	Vertical
10.2250	0	1.0	211	172
16.6857	0	3.9	332	220
29.9000	0	14.4	225	133

[a]Average of measurements at heights of 0.91, 1.37, and 1.82 m.

CONCLUSIONS

We conclude that conducting socks might be a practical method of reducing localized SAR in workers who must remain near RF emitters. The ripstop nylon used in this study would not be appropriate as a regularly worn garment; some type of conductive elastic textile is needed in order to produce a truly practical solution to the problem of excessive ankle SAR.[5]

REFERENCES

1. O. P. Gandhi, J. Y. Chen, and A. Riazi, "Currents induced in a human being for plane-wave exposure conditions 0-50 MHz and for RF sealers," *Trans. IEEE* BME-33: 757, 1986.
2. J. Y. Chen and O. P. Gandhi, "Thermal implications of high SAR's in the body extremities at the ANSI-recommended MF-VHF safety levels," *ibid.*, 35: 435, 1988.
3. R. G. Olsen and T. A. Griner, "Outdoor measurement of SAR in a full-sized human model exposed to 29.9 MHz in the near field," *Bioelectromagnetics* 10: 161, 1989.
4. R. G. Olsen and R. R. Bowman, "Simple nonperturbing temperature probe for microwave/radio frequency dosimetry," *ibid.*, 10: 209, 1989.

5. This research was sponsored by the Naval Medical Research and Development Command under work unit 63706N M000096.004-7102 DN241516. The views expressed in this article are those of the authors and do not reflect the official policy or position of the U.S. Department of the Navy, U.S. Department of Defense, or the U.S. Government.

Martin Blank, Ed., *Electricity and Magnetism in Biology and Medicine*

BRAIN ELECTROCHEMICAL POTENTIAL OSCILLATIONS: THE RESULT OF BIOCHEMICAL SYNERGISM?

A. G. Dubinin*

Russian Mendeleev Institute of Chemical Technology, Moscow

T. B. Shvets-Teneta-Gurii

Institute of High Nervous Actions and Neurophysiology, Moscow

E. L. Ivanova-Anninskaja

Russian State Medical University, Moscow

The brain electrochemical potential E of the alert rabbit, as measured by chronically implanted platinum electrodes, has been found to display oscillations (period from several seconds to 1-3 min, amplitudes from 0.1 to 20-30 mV). The probability of E oscillations has been shown to depend on various conditions: severity of the injury caused by the implantation of electrodes, local heating of the brain, administration of several pharmacological agents, level of brain-specific functional loads. E oscillations are considered to be the result of synchronization of metabolic oscillations in various biochemical systems; i.e., biochemical synergism occurs.

Brain electrochemical potential (E-rhythm) oscillations appear to have been observed by Garcia Ramos.[1] We have found that such oscillations are a rather normal phenomenon, but the probability of their appearance is defined to a great extent by the level of the brain activity.[2]

METHODS

E was recorded in chronic experiments on alert rabbits and rats by means of platinum electrodes (0.2-0.3 mm) implanted under sterile conditions and local anesthesia (2% novocaine) over the dorsal cortex up to the dura mater or pia mater to a depth of 0.5-1.5 mm. A platinum electrode implanted in the skullbone served as a reference. To record E, dc amplifiers with an input resistance of 1 MΩ were used. The system time constant (two platinum electrodes as amplifier input) was 0.3-0.5 s, which enabled us to reliably distinguish changes of E (2 s and up) from the slow E potential shifts.[3] Short-duration changes of E were summed up with the electrocorticogram; E were recorded by mechanical ink-writers. The rabbits' movements were recorded by a ballistocardiograph, placed below the chamber in which the animal was kept.

RESULTS

The quantitative estimation of readings (i.e., all fragments recorded in an experiment by an electrode) showed that the probability of E-rhythm oscillations appearing depended on the number of implanted electrodes and on the manner of their implantation. In rabbits with 8-12 electrodes implanted up to the dura mater (14 rabbits, 141 electrodes, 420 tests), this probability was lower (Fig. 1a) than in those with 22-25 electrodes, likewise implanted up to the dura mater (3 rabbits, 67 electrodes, 90 tests; Fig. 1b). The animals with 10-12 electrodes implanted all the way up to the pia mater (4 rabbits, 40-48 electrodes, 120 tests) had a higher probability of E-rhythm oscillations appearing than those with the same number of electrodes but implanted all the way up to the dura mater (Fig. 1c). And the rabbits with 10-16 electrodes implanted in the cortex had the highest probability of E-rhythm oscillations appearing (14 rabbits, 206 electrodes, 420 experi-

---

*Correspondence to: A. G. Dubinin, Russian Mendeleev Institute of Chemical Technology, Miusskaya sq. 9, 125190 Moscow, Russia.

ments; Fig. 1d). These oscillations ranged in amplitude from 0.1 mV (oscillations of lower amplitudes were neglected) up to several dozen millivolt and a duration from seconds to minutes. Normally there were a number of interimposed oscillations with various periods (Fig. 2).

E-rhythm oscillations were also found to intensify when the brain was directly injured by heating of the output of one of the electrodes located in the frontal brain areas. As for the occipital-cortex areas, in all cases of heating (3 rabbits, 16 electrodes in cortex of each one, 12 tests) E-rhythm oscillations were also formed, with a period of several seconds and amplitudes up to 1 mV (Fig. 3a).

E-oscillations also appeared (Fig. 3b-d) upon the systematic introduction of *polymyxin,* which is an inhibitor of protein kinase[4] (27 mg/kg; 3 rats, 6 electrodes in each one's cortex, 3 tests); of *phenosan-antioxidant* (30-50 mg/kg, 5 rabbits, 8-16 electrodes in each one's cortex, in 13 tests out of 20); and of *teturam,* which is the preparation inhibiting noradrenaline synthesis (100-300 mg/kg, 4 rabbits, 16 electrodes in each one's cortex, in 8 tests out of 12). The latent period before E oscillation appeared was from a few seconds to 50 min.

The probability of E-rhythm oscillations formation also depended on brain-specific functional loads. Thus, it increased in the first step of defensive orienting conditioning (Fig. 4a), a combination of light signal and electrocutaneous stimulus, during which the brain functions most intensively (29 rabbits, 390 electrodes, 870 tests; Fig. 4b). Hypnosis was brought on by stretching of the animal and with the hands, either on the back or the stomach.

DISCUSSION

According to the literature, oscillations in the oxidative metabolism of brain tissue (synchronization of periodic activity of many elements) appear under much harder conditions than such normal conditions as anoxia, edema, or drop of blood pressure,[5] i.e., when there is an acute shortage of energy carriers and a stress in oxidative metabolism arises.

Our data show, that even in the oxidative metabolism of normally functioning brain, rhythm oscillations reflected in E-oscillations can be formed. The probability of the appearance of such oscillation is well correlated with the increase of brain tissue energy requirements necessary either for compensation of various injuring actions or functional activity under conditions of stress. We assume, according to the hypothesis stated in Ref. 6, that the basis of the appearance of oscillations is a density increase of biochemical elements working in similar rhythms, which results in a summation of their activity with respect to the mechanism of biochemical energetics.

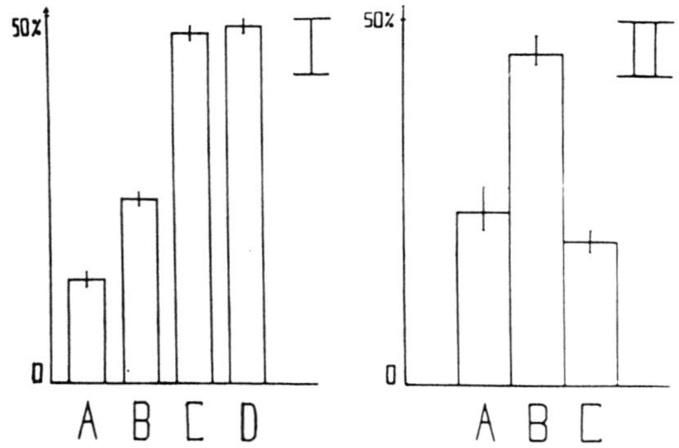

FIG. 1.—Dependence of the appearance of E-rhythm oscillations probability on degree of brain injury (I) and activity level (II). I: (a) 8-12 electrodes implanted up to pia mater, (b) 22-25 electrodes, (c) 10-12 electrodes implanted up to pia mater, (d) 10-16 electrodes implanted in cortex. II. 10-16 implanted electrodes: (a) to 40 combinations, (b) 40-120 combinations, (c) 120-220 combinations. Ordinate represents records in which E-rhythm oscillations take place ($p < 0.05$).

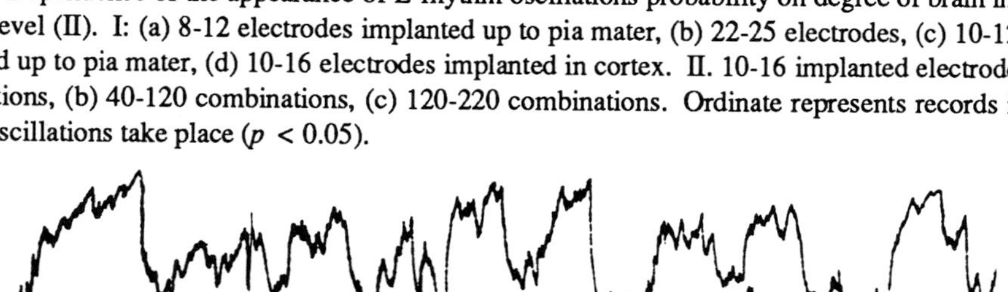

FIG. 2.—Complex record of E-oscillations being intensified in reaction to implanted electrodes. (Calibration: 1 mV, 10 s, upward deviation corresponds to positive sign of electrode.)

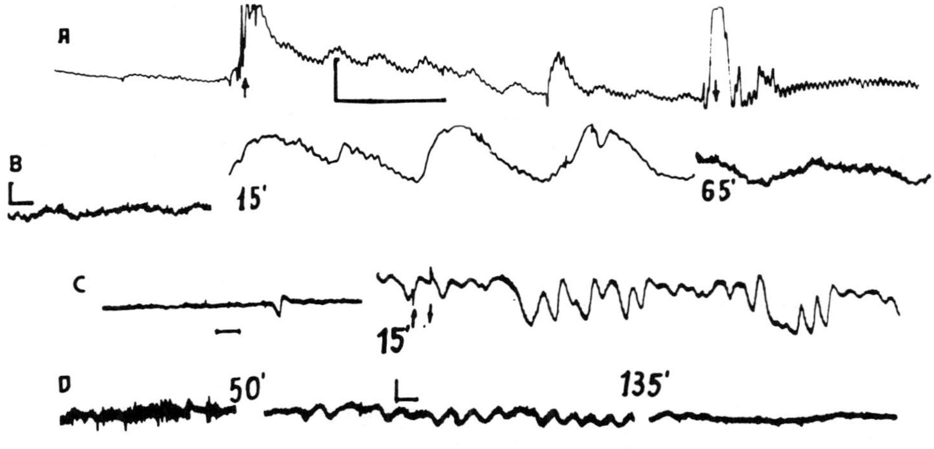

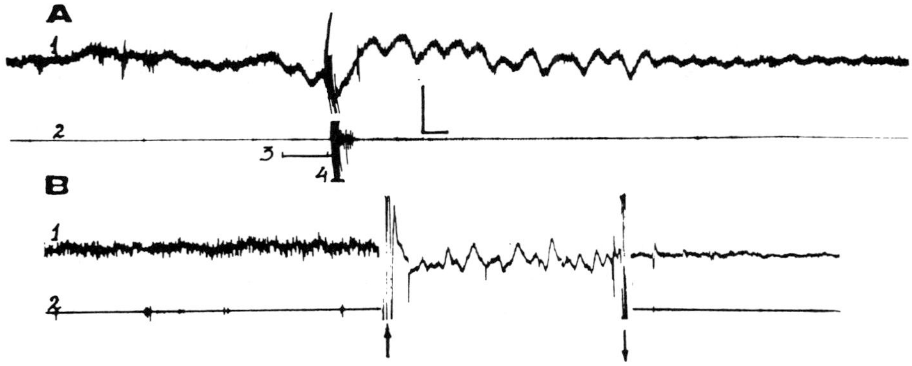

FIG. 3.—(a) Brain E-rhythm oscillations intensified by heating of one electrode; (b) after introduction of phenosane (1/p, 50 mg/kg); (c) teturam (100 mg/kg); (d) polymixin (27 mg/kg). Upward and downward arrows in (a) signify beginning and end of heating; in (c), they signify when 1mV calibration signal is switched on and off. Latent period in E oscillations and intervals in recording are marked in recording. Other marks correspond to those in Fig. 2.

FIG. 4.—E-rhythm oscillations appearing while specific activity level of brain is increased: (a) during combination of signals (light) and reinforcing (electrocutaneous-irritation) stimuli; (b) during hypnosis; 3—light mark, 4—that of electrocutaneous irritation. Upward arrow: removal of hands, rabbit enters into hypnosis; downward arrow: comes out of hypnosis. Other marks correspond to those in Fig. 3.

REFERENCES

1. F. Garcia Ramos, *Currents in Modern Biol.* 1: 125-132, 1967.

2. T. B. Shvets-Teneta-Gurii, (a) *Bioelectrochemical Activity of the Brain and Learning*, Moscow: Science, 1986, 192 pp.; (b) D. Sc. Thesis, Moscow, 1987; (c) with V. N. Maz and O. B. Kovchegova, *BAL* 108: 259-261, 1989; (d) *Proc. USSR Acad. Sciences* 315: 1014-1071, 1990.

3. T. B. Shvets-Teneta-Gurii, *Bioelectrochemical Activity of the Brain*, Moscow: Science, 1980, 205 pp.

4. K. G. Reymann et al., *Brain Res.* 440: 305-314, 1988.

5. B. Dora, K. Olaffson, B. Chance, and A. G. Kovach, *Kiserl. Orvostud.* 31: 123-129, 1979; B. Dora and A. G. Kovach, *Acta Physiol. Acad. Scient. Hungaricae* 57: 261-275, 1981; F. F. Fobis, "General discussion," in *Brainwork*, Copenhagen, 1975, 507.

6. G. G. Aldridge and E. K. Pye, *Nature* 259: 670-671, 1976.

Martin Blank, Ed., *Electricity and Magnetism in Biology and Medicine*

CURRENT-DENSITY DISTRIBUTION INDUCED IN A TWO-DIMENSIONAL INHOMOGENEOUS HUMAN MODEL BY A TIME-VARYING MAGNETIC FIELD

F. X. Hart*

University of the South, Sewanee, Tenn.

A finite-difference method is used to calculate the current density distribution produced within a two-dimensional human model by a low-frequency magnetic field. The calculation is performed by use of a commercially available spreadsheet program on a personal computer. Different values are used for the resistivities of soft tissue, bone, and brain. The electrical properties of the limbs are anisotropic; that is, different values can be used for the longitudinal and radial resistivities of both the bone and soft tissues. The current-density values are generally reduced by the introduction of inhomogeneity. The effects of anisotropy are also significant.

The calculation of the current-density distributions produced inside human and rat models by applied electric or magnetic fields is important for the establishment of the dosage experienced by a subject exposed to external sources of low-frequency fields. The results of numerical calculations performed with a commercially available spreadsheet program on a personal computer agree with measurements made on instrumented homogeneous human and rat mannequins.[1-3] Generalization of these methods to more realistic inhomogeneous anisotropic models is required to provide a better estimate of the dosage. Such a model was used to examine how changes in the electric properties of tissue can modify the three-dimensional internal field distribution in a limb to which surface electrodes have been attached.[4]

This paper describes the use of the spreadsheet method to determine the importance of including electrical anisotropy of tissue in a two-dimensional, whole-body, human model exposed to a low-frequency magnetic field. The model (Fig. 1) is inhomogeneous and partially anisotropic.

NUMERICAL METHOD

The details of this two-dimensional, finite-difference method are provided elsewhere.[2,3,,5] The calculation were performed for an applied 60Hz, 10^{-4}T magnetic field. The results can be scaled directly to other field strengths and frequencies.[2]

Three types of tissue were used in this stylized human model: bone, soft tissue, and brain. Three sets of tissue resistivities were chosen in order to examine how strongly the results of the calculation depend on the complexity of the model. In the first set (homogeneous model) the resistivities for the three tissue types were all 200 $\Omega \cdot$m. In the second set (inhomogeneous, isotropic model) the resistivities of brain, all soft tissue, and all bone were set at 167, 500, and 5000 $\Omega \cdot$m, respectively. In the third set (inhomogeneous, anisotropic model) the resistivities of the tissues in the arm and leg were made anisotropic: for bone, 2500 and 7700 $\Omega \cdot$m in the longitudinal and radial directions, respectively; for soft tissue, 167 and 1250 $\Omega \cdot$m in the longitudinal and radial directions, respectively.

RESULTS AND DISCUSSION

Table 1 illustrates the vertical (J_y) and horizontal (J_x) components of the induced current density at several selected positions, indicated in Fig. 1, for each set of tissue properties. The total current density $J_t = (J_x^2 + J_y^2)^{1/2}$ is also indicated. The current densities in the inhomogeneous models are generally much smaller than in the homogeneous model. The reduction is due to the greater resistivities in those models and to the longer path that the current takes to avoid their more resistive bone regions. Introduction of anisotropy

*Correspondence to: F. X. Hart (Physics), University of the South, Sewanee TN 37375-1000.

TABLE 1.—Current-density values in nA/cm^2 induced in a human model by 10^{-4}T, 60Hz magnetic field applied in the z direction. Positive J_y values indicate vertical current flow upward in Fig. 1; positive J_x values indicate horizontal current flow to the right in Fig. 1.

Body Position	Homogeneous Model			Inomogeneous Isotropic Model			Inhomogeneous Anisotropic Model		
	Jy	Jx	Jt	Jy	Jx	Jt	Jy	Jx	Jt
Brain	1.49	0.47	1.57	0.98	0.06	0.98	0.98	0.06	0.98
Neck	8.13	0.77	8.17	0.77	0.18	0.80	0.79	0.19	0.81
Midriff	1.80	-0.24	1.82	0.11	-0.02	0.11	0.12	-0.02	0.12
Hip	1.03	-1.18	1.56	0.04	-0.08	0.09	0.04	-0.06	0.08
Groin	-0.62	-4.60	4.65	-0.15	-1.57	1.58	-0.17	-1.69	1.70
Armpit	1.93	-0.68	2.04	0.46	-0.09	0.47	0.38	-0.07	0.39
Elbow	0.90	0.00	0.90	0.38	0.00	0.38	1.02	-0.01	1.02
Femur	-0.84	-1.12	1.40	-0.11	-0.23	0.25	-0.15	-0.21	0.26
Knee	-0.68	-0.41	0.80	-0.21	-0.13	0.25	-0.39	-0.16	0.42

in the limbs modifies the current distribution there and in neighboring regions. It is clear that inhomogeneous models must be used to provide more realistic estimates of the internal dosage.[6]

REFERENCES

1. F. X. Hart, "Use of a spread sheet to calculate the current-density distribution produced in human and rat models by low-frequency electric fields," *Bioelectromagnetics* 11: 213-228, 1990.
2. F. X. Hart, "Numerical and analytical methods to determine the current density distributions produced in human and rat models by electric and magnetic fields," *Bioelectromagnetics* 13(S1): 27-42, 1992.
3. F. X. Hart, "Electric fields induced in rat and human models by 60-Hz magnetic fields: Comparison of calculated and measured values," *Bioelectromagnetics* 13: 313-316, 1992.
4. F. X. Hart, "Changes in the electric field at an injury site during healing under electrical stimulation," *J. Bioelec.* 10: 31-51, 1991.
5. F. X. Hart and K. W. Wood, "Eddy current distributions: Their calculation with a spreadsheet and their measurement with a dual dipole antenna probe," *Amer. J. Phys.* 59: 461-467, 1991.

6. This research is supported by the Electric Power Research Institute, Palo Alto, Calif.

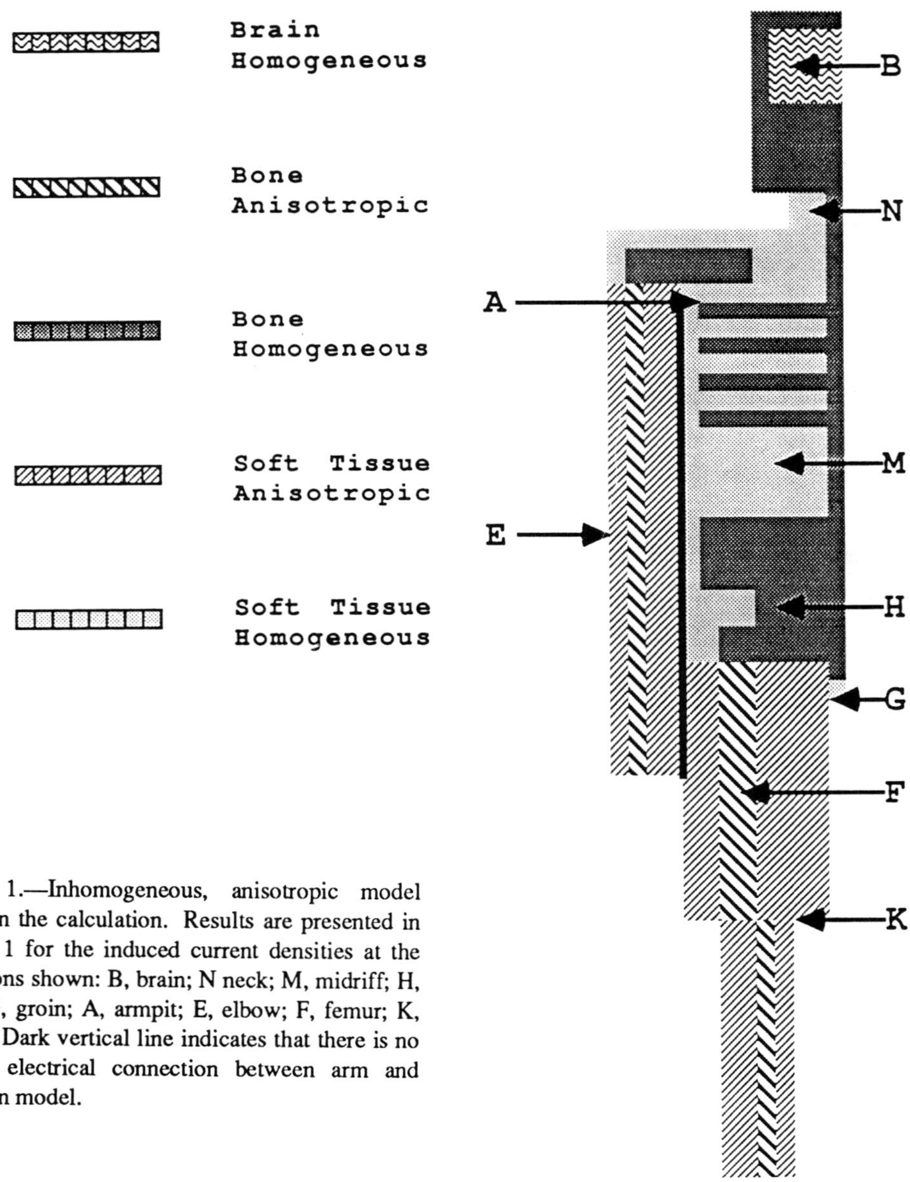

FIG. 1.—Inhomogeneous, anisotropic model used in the calculation. Results are presented in Table 1 for the induced current densities at the positions shown: B, brain; N neck; M, midriff; H, hip; G, groin; A, armpit; E, elbow; F, femur; K, knee. Dark vertical line indicates that there is no direct electrical connection between arm and body in model.

Martin Blank, Ed., *Electricity and Magnetism in Biology and Medicine*

A FREQUENCY-DEPENDENT FINITE-DIFFERENCE TIME-DOMAIN (FDTD) METHOD FOR INDUCED CURRENT AND SAR CALCULATIONS FOR A HETEROGENEOUS MODEL OF THE HUMAN BODY

C. M. Furse,* J.-Y. Chen, and O. P. Gandhi

University of Utah, Salt Lake City

This paper describes a new frequency-dependent FDTD [$(FD)^2TD$] method for ultrawide-band applications such as irradiation by extremely short pulses, where the effects of dispersive tissue properties must be considered. Based on a differential-equation approach for the time-domain representation of $\mathbf{D}(t)$, this method may be used for general dispersive media where $\varepsilon^*(\omega)$ and $\mu^*(\omega)$ can be described by rational functions in the frequency domain, or for human tissues where $\varepsilon^*(\omega)$ is described by a Debye equation with several relaxation constants. Based on a single run with a broadband pulse excitation, the $(FD)^2TD$ method is used to calculate SARs and induced currents in a heterogeneous anatomically based man model over an ultra-wide band.

A weakness of the conventional FDTD algorithm is that the dispersion of the tissue's dielectric properties is ignored and frequency-independent properties are assumed. Although this approach may be permissible for CW or narrow-band irradiation, it yields substantial errors for short pulses with ultra-wide bandwidths. We have developed a frequency-dependent finite-difference time-domain [$(FD)^2TD$] method by adding a differential equation relating to **D** and **E** to the traditional FDTD algorithm. Previously used for homogeneous models,[1] the $(FD)^2TD$ approach is now extended to the heterogeneous anatomically based model of the human body, where the tissue properties $\varepsilon^*(\omega)$ are described by second-order Debye equations. Using a single run with a broadband pulse excitation, we obtain layer-averaged SARs and induced currents at various frequencies by taking the Fourier components of the induced $\mathbf{E}(t)$ fields.

THE DIFFERENTIAL-EQUATION-BASED $(FD)^2TD$ METHOD

The procedure is similar to that used for the conventional FDTD method,[2] except that an additional differential equation relating $\mathbf{D}(t)$ and $\mathbf{E}(t)$ is used. This equation is developed from the assumption that the dispersion of the various tissues can be fitted to a Debye equation with two relaxation constants, as shown in Fig. 1 for muscle and fat:

$$\mathbf{D}(\omega) = \varepsilon^*(\omega)\,\mathbf{E}(\omega) = \varepsilon_0\left[\varepsilon_\infty + \frac{\varepsilon_{s1}-\varepsilon_\infty}{1+j\omega\tau_1} + \frac{\varepsilon_{s2}-\varepsilon_\infty}{1+j\omega\tau_2}\right]\mathbf{E}(\omega) \tag{1}$$

which may be converted to a differential equation in the time domain by assuming $e^{j\omega t}$ time dependence:

$$\tau_1\tau_2\frac{\partial^2\mathbf{D}}{\partial t^2} + (\tau_1+\tau_2)\frac{\partial\mathbf{D}}{\partial t} + \mathbf{D} = \varepsilon_0\left[(\varepsilon_{s1}+\varepsilon_{s2}-\varepsilon_\infty)\mathbf{E} + (\varepsilon_{s1}\tau_1+\varepsilon_{s2}\tau_2)\frac{\partial\mathbf{E}}{\partial t} + \varepsilon_\infty\tau_1\tau_2\frac{\partial^2\mathbf{E}}{\partial t^2}\right] \tag{2}$$

For the $(FD)^2TD$ method, we solve Ampere's law to find **D** and **E**, and Faraday's law to find **H** at each cell location. This $\mathbf{D}\rightarrow\mathbf{E}\rightarrow\mathbf{H}$ loop is then repeated for each time step.

APPLICATION OF THE $(FD)^2TD$ METHOD TO THE ANATOMICALLY BASED HETEROGENEOUS MAN MODEL

The $(FD)^2TD$ method is applied to an anatomically based heterogeneous man model previously used in Ref. 2. This model is made up of $24\times45\times135$ 1.31cm (approximately 0.5in.) cells imbedded in a total region of $42\times63\times153$ cells; 7.67 MWords of memory and approximately 165 cpuminutes are needed to run

---

*Correspondence to: C. M. Furse (El. Engg), University of Utah, Salt Lake City UT 84112.

this test case on an IBM-3090. By comparison, the traditional FDTD takes 6.37 MW words and 155 cpuminutes, but must be rerun for each frequency. Tissue properties in each cell are obtained by volume-averaging properties for all tissues in each cell. The incident plane wave is a raised cosine pulse with 3 GHz maximum frequency and $\mathbf{E}_z$ polarized (parallel to the long axis of the body), and is frontally incident on the body.

Layer-averaged current and SAR distributions computed by both FDTD and $(FD)^2TD$ methods are shown in Figs. 2 and 3, respectively, for frequencies from 20 to 915 MHz. These values differ slightly because of the slight discrepancies between measured tissue properties and their fit to the Debye equation as shown in Fig. 1.

This test case illustrates the power of the $(FD)^2TD$ method. The new method provides accurate broad-band and time-domain data for a highly dispersive model from a single run, whereas the traditional FDTD method would require a run at every frequency of interest and could not provide accurate time-domain data for this dispersive model.

CONCLUSION

This paper describes a new frequency-dependent FDTD $[(FD)^2TD]$ method for general dispersive media for which permittivity and/or permeability can be described by rational functions in the frequency domain. This approach was applied to a *heterogeneous* anatomically based model of the human body where the tissue properties $\varepsilon^*(\omega)$ are described by multiterm Debye equations. A single run with a broadband pulse excitation provides SARs and induced currents in either the time or frequency domain for this highly dispersive model.[4]

REFERENCES

1. O. P. Gandhi, J. Y. Chen, and C. M. Furse, "A frequency-dependent FDTD method for induced-current calculations for a heterogeneous model of the human body," 1992 IEEE MTT-S International Microwave Symposium, Albuquerque, N. Mex.

2. C.-Q. Wang and O. P. Gandhi, "Numerical simulation of annular phased arrays for anatomically based models using the FDTD method," *IEEE Trans* MTT-37: 118-126, 1989.

3. C. H. Durney et al., *Radiofrequency Radiation Dosimetry Handbook,* Report SAM-TR-78-22, prepared for USAF School of Aerospace Medicine, Brooks AFB, TX 78235, 1978, 2d ed.

4. The authors gratefully acknowledge the support of the National Institute of Environmental Health Sciences under grant ES03329 and the Utah Supercomputer Institute for a grant of computer time.

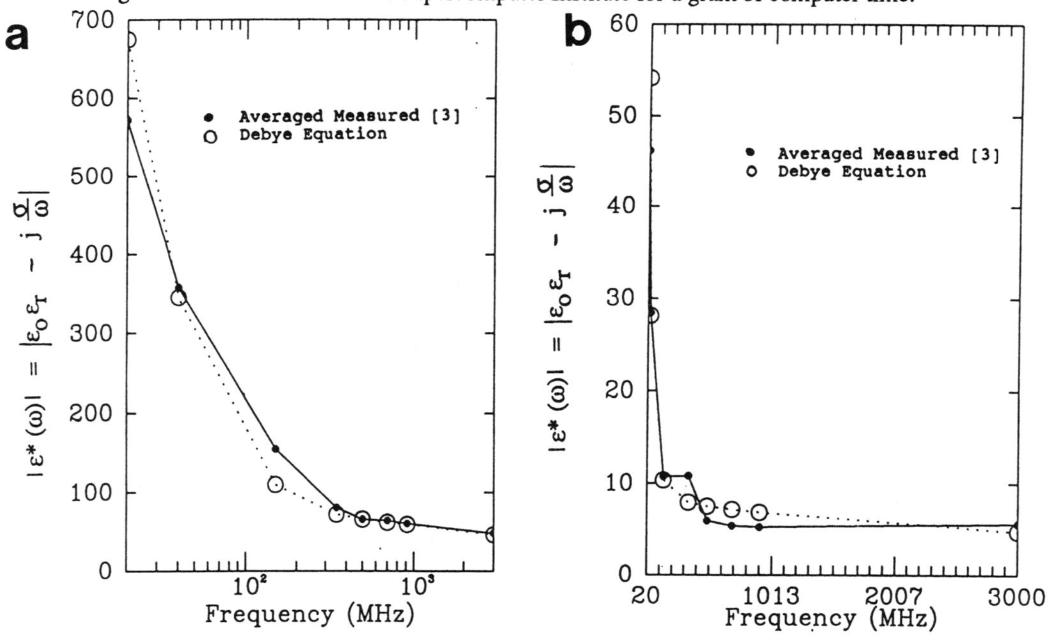

FIG. 1.—Fit of Debye equation with two relaxation constants to measured tissue properties of (a) muscle, (b) fat.

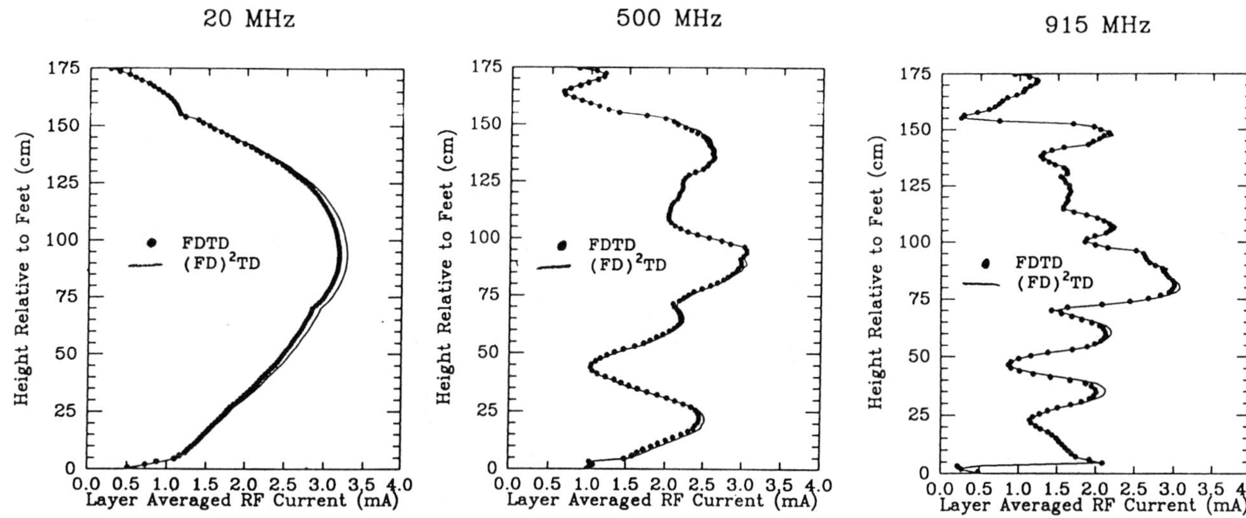

FIG. 2.—Layer-averaged RF current.

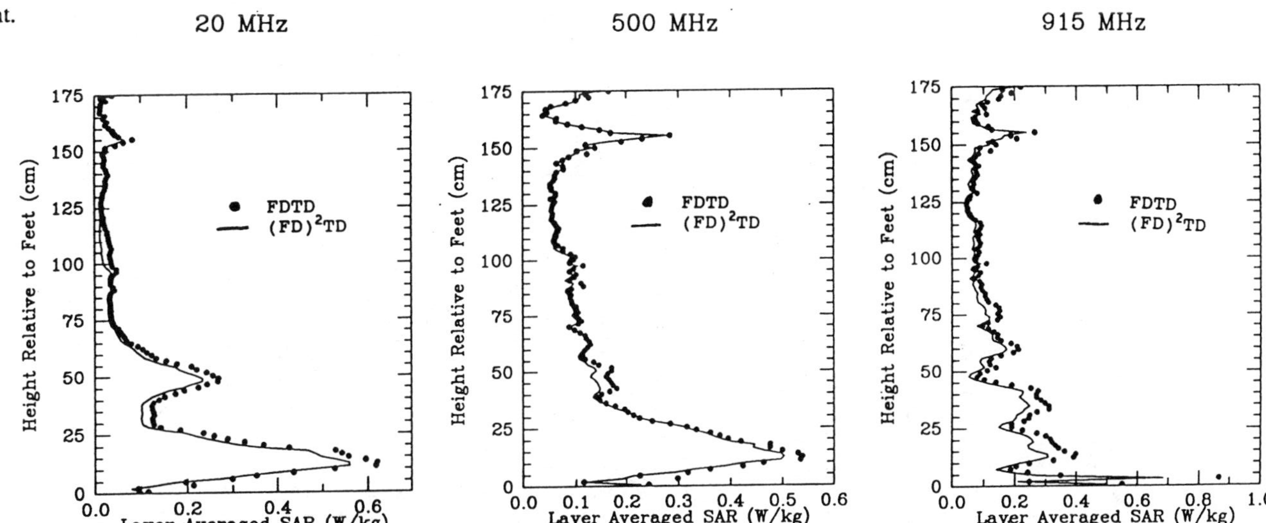

FIG. 3.—Layer-averaged SAR.

Martin Blank, Ed., *Electricity and Magnetism in Biology and Medicine*

CURRENT-DISTRIBUTION TOMOGRAPHY: A NEW TECHNIQUE FOR DETERMINATION OF INTERNAL CURRENT-DENSITY DISTRIBUTIONS

P. C. Gailey*

Oak Ridge National Laboratory, Oak Ridge, Tenn.

Measurements of the total current through various anatomical features have been performed, but no measurement technique has been available to determine the distribution of current density through the body. The new technique reported here measures the magnetic field produced by the current density at several locations, and analyzes it numerically to produce an estimate of the current density distribution inside an object. This technique offers a means of noninvasive measurement of internal current-density distribution, and will help verify mathematical modeling techniques. It may provide a method for in vivo measurement of tissue electrical properties, and be useful as a new imaging technology.

Perhaps the most challenging engineering aspect of experiments on electric- and magnetic-field bioeffects is the identification and determination of dose. There is no simple way to relate the external fields to the local fields and currents that represent the actual, albeit position-dependent, dose. The principal difficulties in relating external fields to internal exposure parameters in animal experiments are the complexity of the body shape and its highly heterogeneous internal structure. Impressive success in predicting internal fields and current densities in humans has been achieved in recent years by numerical modeling techniques.[1] These studies have employed detailed anatomical models of the body divided into small cubes with the appropriate electrical properties. Studies based on the finite-difference time-domain technique for radiofrequency (RF) fields revealed that energy absorption during field exposure is frequency dependent, with resonances that can significantly increase absorption at certain frequencies. These results have led to a frequency-dependent exposure guideline for radiofrequency fields.[2] At RF frequencies, energy absorption is considered the measure of dose. Work is now under way to determine the internal field strengths and current densities induced during exposure to extremely low frequency (ELF) fields.[3]

Because mathematical models are approximations based on various assumptions, it is desirable to verify modeling results by measurements. One method available is to measure the total current passing through the feet of an exposed animal or human, and compare this number to the total current predicted at the same location by the model. Another method is to measure heat increases or fields inside anatomically shaped human or animal models filled with one or more dielectric materials. These techniques provide valuable insights about the gross features of the models, but reveal little about the accuracy of the results within the complex heterogeneous structure inside the body. Current-distribution tomography is a new method of dealing with this shortcoming. It is a relatively simple noninvasive method for determining current density distributions inside the body by measurements of the magnetic field occurring outside the body.

THEORY

Current-distribution tomography (CDT) can be best understood by consideration of conventional current probes. Such probes are toroidal in shape, with high-permeability or air cores. In the case of an air-core current probe, wire is wrapped on an empty toroidal form. Considering each turn or wrap of the coil separately, one can imagine voltages induced in these coils by the changing flux density (produced by the current) through each turn. Because the turns are connected in series, the total circuit sums the effects of the magnetic field at each point, which yields information only about the total field or the total current through the toroid. If one now imagines the toroidal coil cut into individual turns, and these turns flattened into circles or single-turn loops, a new set of possibilities arise. Each loop will now sense only the changing flux density at its particular location. If the current through the imaginary toroid on which the loops are placed is

---

*Correspondence to: P. C. Gailey, ORNL, Oak Ridge TN 37831-6070.

distributed unevenly, each loop will, in general, sense a different magnetic field. In many cases, this information can be used to compute the current density distribution passing through the toroid.

A simple case can be used to illustrate the procedure for computing current-density distributions. Assume that several long, straight wires carrying different currents are arranged in parallel forming a bundle (e.g., a multiconductor cable). If an imaginary cylindrical shell is placed around the bundle, the position of each wire can be referenced to the center of this shell. Intersecting the cylindrical shell with a perpendicular plane, one obtains a circle along which the magnetic field will be considered. The next step is to establish a set of points on the circle equal to the number of wires in the bundle. For simplicity, assume that these points are equally spaced around the circle and that both the points and the wires are numbered arbitrarily. At point one, the magnetic field is the vector sum of the magnetic fields produced by each of the wires. In practice, a loop senses only one component of magnetic field which, for the toroid configuration, is the component tangent to the circle. The field at point p resulting from each wire is the product of the current and a geometrical factor that depends on the specific case:

$$F_p = I_1 G_{1p} + I_2 G_{2p} + I_3 G_{3p} + \cdots$$

where F_p is the magnetic field at point p, G_{mp} is the geometrical factor relating wire m to point p, and I_m is the current flowing in wire m. A similar equation can be written for the field at each point on the circle. Here, the currents I_m are unknown, and the magnetic field values F_p at each point are measured; the geometrical factors G_{mp} are constants. This system of simultaneous equations can be solved for the unknown currents I_m by any of a number of numerical techniques.

$$\bar{I} = G^{-1}\bar{F}$$

In the more general case, the measurement points need not lie on a circle or be equally spaced, but the number of measurement points must equal the number of wires. For the equally spaced circle-centered case, the geometrical factors are computed as follows:

$$G_{mp} = \frac{\cos\alpha}{2\pi R_{mp}}$$

$$\alpha = \sin^{-1}\left[\sqrt{x_m^2 + y_m^2}\,\frac{\sin(\theta_m - \theta_p)}{R_{mp}}\right] \qquad R_{mp} = \sqrt{(x_m - x_p)^2 + (y_m - y_p)^2}$$

$$\theta_p = \tan^{-1}(y_p/x_p) \qquad \theta_m = \tan^{-1}(y_m/x_m)$$

where x_p, y_p are the x and y coordinates of measurement point p and x_m, y_m are the x and y coordinates of wire m.

The technique described above assumes currents existing only along discrete lines. Most real cases of interest will involve current densities distributed over a region such as through an arm or leg. The cross section of interest must then be split into cells or regions, and the current density assumed constant over each cell. A matrix solution as described above exists for most irregular current distributions, but it is clear that the technique fails when the current distribution is cylindrically symmetric. A detailed mathematical analysis along with an evaluation of factors affecting the accuracy of the technique will be presented in a future paper.

Figure 1 shows the experimental setup used in a proof-of-concept test of CDT with the experimental parameters and results shown in the following tables. A single 100kHz signal was split among eight parallel conductors with different currents. Magnetic-field strengths were measured at various locations around the setup and then analyzed numerically to compute the currents. The computed values agreed with the known currents through each conductor with an average error of ±20%. Errors in predicting the currents can be attributed to the relatively crude arrangement for determining sensor location. A more accurate system is now being developed. The 100kHz frequency was chosen to improve sensitivity of the loop sensor. Application of CDT at lower frequencies can be accomplished by the use of more sensitive measurement techniques, including band-limited amplifiers. SQUID magnetometers may be required for determination of 60Hz body currents because of the weak fields produced by these currents and small time rat of change of magnetic field (dB/dt) at this low frequency.[4]

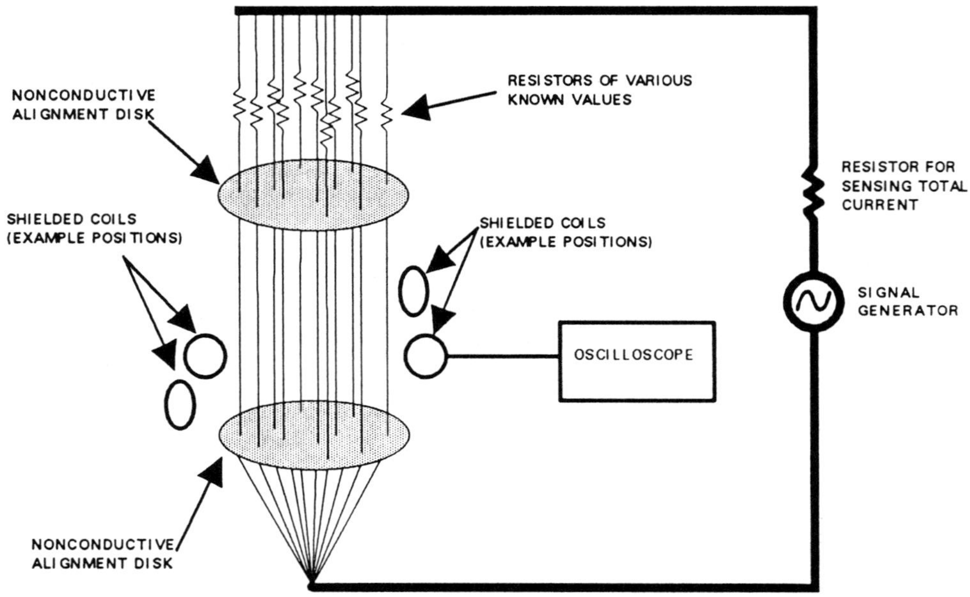

FIG. 1.—CDT concept test.

Experimental Parameters

Frequency:	100 kHz
Number of currents:	8
Loop dimensions:	6.3×0.5 cm
Number of turns:	500
Measurement circle:	6.5cm radius
Location of currents:	5cm circle
Currents:	20-113 mA

CDT Concept Test Results

Index	Actual current (mA)	CDT computed current (mA)	Percent error
0	22.7	32.0	+41.0
1	45.4	43.5	− 4.2
2	28.3	19.3	−32.0
3	75.9	62.0	−18.2
4	58.4	59.4	+ 1.8
5	45.3	48.0	+ 5.9
6	32.9	50.4	+53.3
7	112.9	120.9	+ 7.1

REFERENCES

1. O. P. Gandhi, "Electromagnetic absorption in inhomogeneous model of man for realistic exposure conditions," *Bioelectromagnetics* 3: 81-90, 1982.

2. ANSI (1982), *American National Standards Institute Safety Levels with Respect to Human Exposure to RF Electromagnetic Fields 300 kHz to 100 GHz*, New York: IEEE, Inc., C95.1-1982.

3. O. P. Gandhi and J. Y. Chen, "Numerical dosimetry at power line frequencies using anatomically based models," *Bioelectromagnetics* (in press).

4. The author gratefully acknowledges the support and encouragement of Dr. O. P. Gandhi.

Martin Blank, Ed., *Electricity and Magnetism in Biology and Medicine*

NEAR FIELD PD MEASUREMENTS

H. Trzaska*

Technical University of Wrocław, Poland

This paper deals with the power density (PD) measurement error that arises when PD is derived from E and/or H field measurements. The analysis was done for E- and H-type elemental dipoles, which gives results that maximize the errors of any other source. The analysis shows that the method may be accepted for measurements at frequencies above 300 MHz. Comparative analysis for a resonant dipole leads to a less rigorous conclusion. However, at frequencies below 100-200 MHz, in the majority of cases, the method and the instruments based on it are useless. Comments on the use of PD measurements for occupational safety and environmental protection purposes are presented.

Power density (PD) limits are accepted in many occupational safety, environmental protection, and EMC standards. The standards have created metrological needs and, as a result, appropriate meters have appeared on the market. The common denominator here is the measurement in the near field, over a wide frequency range, based on E and/or H field measurements.

There are no known methods of direct PD measurement; it is necessary to measure E- and H-field components and the phase angle between them. The method worked out by King,[1,2] based on multiple loaded loop antennas, as well as that based on Kanda's fiber optic link use,[3] for undisturbed phase information transmission through a transparent feeder, made the precise measurement of PD possible under any conditions; but the method has some limitation and requires the use of very sophisticated and expensive equipment. As a result, the most popular types are radiation hazard monitors and EMI probes in which the oldest concept of the PD measurement is applied, one that is quite reliable in the far field—the measurement of E or H and devices calibrated in PD units. This is a really good, simple, and inexpensive way, but its limitations have been nowhere discussed or even mentioned. It may cause difficulties in the interpretation of measurement results, because the readings of such a meter (and its accuracy) depend on the distance between the meter probe and the source of radiation, as well as on the structure of the source and the type (E or H) of probe applied. These problems are discussed below.

ELEMENTAL DIPOLE

Radiated power density P_r is defined as the real part of the Poynting vector P. When the electric power density P_E or the magnetic power density P_H are measured, because of standardization conditions, P_r must be equal to P_E and P_H, and

$$P_E = \frac{|E|^2}{2Z} \quad \text{and} \quad P_H = \frac{|H|^2 Z}{2} \tag{1}$$

where E and H are the electric and magnetic field intensity, respectively, and Z is the intrinsic impedance. As long as $P_r = P_E = P_H$, a PD measurement derived from E or H measurements is acceptable. However, all these quantities are functions of distance and structure of the radiation source. To investigate the equivalence of the measurements, we calculate the error η_{EE} of the PD measurement by E measurement near an elemental electric source, as well as the error η_{EH} by H measurement:

$$\eta_{EE} = \frac{|P_r - P_E|}{P_r}, \qquad \eta_{EH} = \frac{|P_r - P_H|}{P_r} \tag{2}$$

Separate calculations for the elemental magnetic dipole are not necessary because of full symmetry; i.e., $\eta_{EE} = \eta_{HH}$, $\eta_{EH} = \eta_{HE}$. Because the graphical presentation of errors given by the above formulas may be

*Correspondence to: H. Trzaska (ITA), Technical University of Wrocław, Wyspianskiego 27, 50-370 Wroclaw, Poland.

difficult for the near field, the logarithmic approach is used here:

$$\delta_{EE} = 10 \log \frac{|P_E|}{P_r}, \qquad \delta_{EH} = 10 \log \frac{|P_H|}{P_r} \tag{3}$$

For example, δ_{EE} as a function of distance kr, for several elevation angles θ, is shown in Fig. 1.

Considerations analogous to the above were examined for the measurement of the arithmetic and geometric means of P_E and P_H; the errors δ_a and δ_g (in analogy to Eq. 3) were formulated as

$$\delta_a = 10 \log \frac{|P_E + P_H|}{2P_r}, \qquad \delta_{EH} = 10 \log \frac{\sqrt{|P_E P_H|}}{P_r} \tag{4}$$

Calculation show that $\delta_{EE} > \delta_a > \delta_g$, but these are not considerable changes. For illustration, δ_g is shown in Fig. 2.

No physical meaning has been ascribed to the imaginary part of the Poynting vector, but it is worth noticing that the geometrical mean of P_E and P_H is equal to the vector modulus.

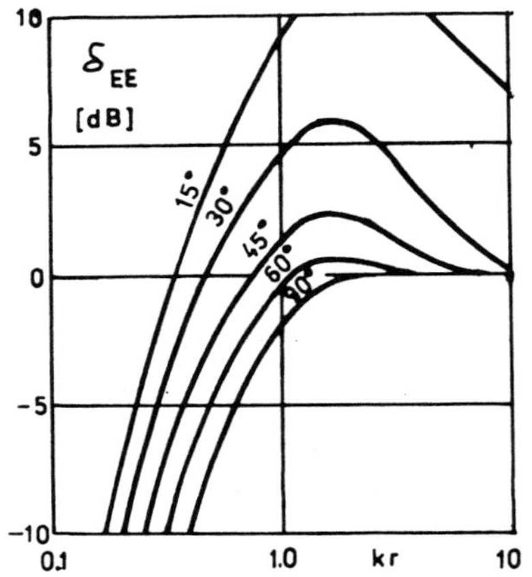

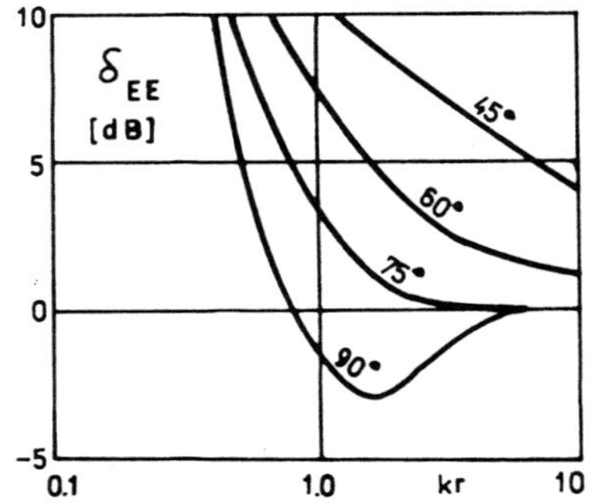

FIG. 1.—δ_{EE} vs kr (left).
FIG. 2.—δ_g vs kr (above).

RESONANT DIPOLE

It might be suggested that the above considerations should have begun at the multipole level, for which the magnitude of the errors discussed would be still larger; however, in practice even elemental sources may play only a limited role as radiators. (An example of such a radiator may be an applicator of an inductive heater.) More important in practice are resonant radiators. To check the error of PD measurements near them, considerations such as the above were repeated for symmetrical dipoles of various lengths.

Spatial E and H components of elemental sources and linear antennas are well known[4] and are not cited here because of space limitations. The results of calculations carried out for 0.1λ and 0.5λ symmetrical dipoles (where λ is the wavelength) are shown in Fig. 3. Note that all the above considerations represent only field relations: the parameters of any given meter are not taken into account (measuring antenna length, antenna-source interactions, etc.), with one exception: it is assumed that

$$|E|^2 = |E_r|^2 + |E_\theta|^2 \tag{5}$$

Because of the elliptical polarization of the EM field in the near field, the assumption is false, since E should be averaged in time; but this approach does represent available omnidirectional probes in which the spatial components of the output voltages are summed on DC.

DISCUSSION

The above estimates of PD near an elemental dipole show that, if we assume an accuracy of measurements within 3 dB and a minimum distance from a source greater than 15 cm (as given in the Polish protection standards), the method is acceptable at frequencies above 300 MHz. Similar estimates for resonant antennas show that the frequency limitation is less rigorous and the same conditions are fulfilled for

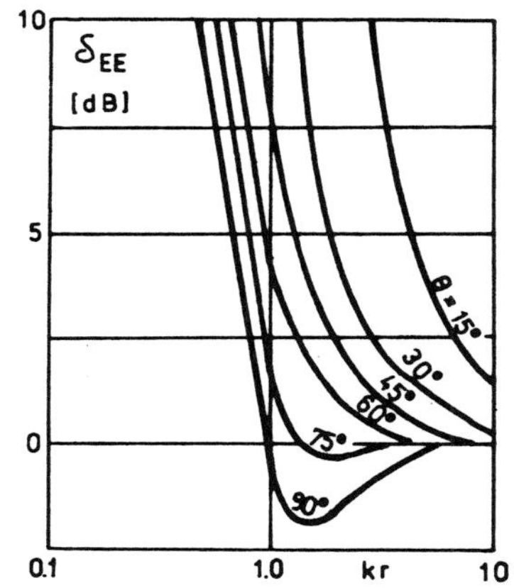

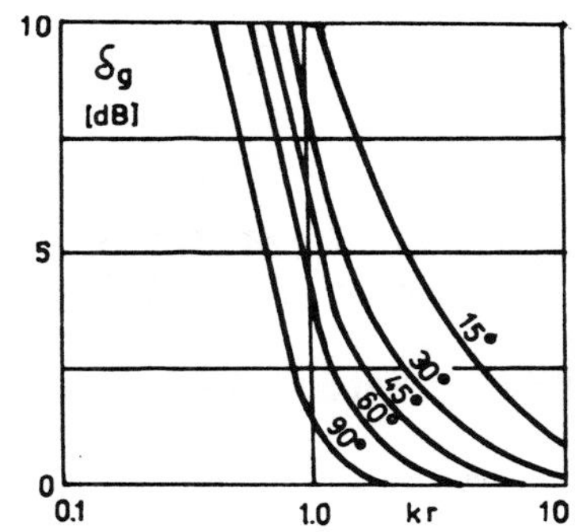

FIG. 3.—δ_{EE} of a 0.5λ (left) and 0.1λ (above) dipole.

frequencies in the range of 100-200 MHz, whereas meters for far lower frequencies are available commercially. These frequency limitations are more severe in the USA, as some American standards require measurements down to 5 cm from a source. If we take not only the error of the method into consideration (as was done in the above calculations) but also the errors of the measuring device, it will point up the limit still further. An additional disadvantage of the method is the difference in the measured data (and the accuracy of measurement) in the near field when measurements are performed with E and H probes.

All the above-mentioned limitations should eliminate the method for practical use. The question arises: what is the reason for PD measurements at all? or for that matter, why use PD in protective standards? The answer is simple. The "delinquents" here are the microwave engineers who initiated the work in the field. Multiplying PD by an effective area of an antenna gives the power absorbed by the antenna, and the first radiation hazard monitors contained an RF power meter and a horn antenna. But what works well for antennas does not necessarily work as well in the case of a living body. The relation between PD and the power absorbed by a body is not so simple. For instance, PD = 0 between capacitor plates, which does not mean that a body placed there will not be heated. The PD in free space differs considerably from that in a lossy medium; and EM energy absorption by the medium is conditioned by the E and H fields inside the medium rather than by PD measured around it. The absorption depends strongly on the medium (its shapes, electrical parameters, surroundings, homogeneity) and on the frequency. Here the temperature increase would be the most informative unless, as has been shown,[5] it is not sensitive enough. Thus, the use of PD in protection standards and measurements may lead to misinterpretations with respect to the energy transfer. Another doubt is related to the above-mentioned accuracy of measurements. To be sure, under laboratory conditions the specific absorption rate (SAR), PD, current, and any other quantity may be used, as conditions there are under strict control, and experiments are repeatable.

We are thus led to the final conclusion that E and H should be the principal quantities used in EM environment protection standards and measurements. Obviously, at microwave frequencies, because of tradition (and some support from the analysis shown), PD may be acceptable. The accuracy of PD measurement need not exceed the accuracy (or rather repeatability) of biomedical experiments; moreover, the EM field in the near field usually varies greatly, which limits the indispensable accuracy still further.

REFERENCES

1. H. Whiteside and R. W. P. King, "The loop antenna as a probe," *Trans. IEEE* AP-7: 291-297, 1964.

2. D. Hoff and G. Moenich, "Eine Sonde zur direkten Messung von Energieströmungen im Nahfeld von Sende- und Empfangsantennen," *NTZ* 27: 313-318, 1974.

3. M. Kanda, "EMC/EMI standards, measurements and calibration activities at NIST," Proc. Intern. Wrocław EMC Symp., 1992.

4. J. H. Kraus, *Electromagnetics*, New York: McGraw-Hill, 3rd ed.

5. H. Trzaska, "EM field dosimetry in biomedical investigations," *NATO ASI on Technical Advances in Biomedical Physics*, Series E, vol. 77, Amsterdam: Martinus Nijhoff, 1984, 175-186.

Martin Blank, Ed., *Electricity and Magnetism in Biology and Medicine*

AUTOMATIC MEASUREMENT OF ELECTROMAGNETIC ABSORPTION PATTERNS WITH A SCANNING THERMOCOUPLE

C. De Wagter* and L. Vercruyssen

University of Gent Hospital, Belgium

L. Martens

University of Gent, Belgium

The measuring method described permits the determination of the specific absorption rate (SAR) patterns induced by electromagnetic radiators. A 3-D scanning system is used to guide a thermocouple through a liquid phantom. At each measuring point, the SAR value is derived from the thermal history recorded during the course of a short microwave pulse. The method exploits the self-heating "artifact." An anthropomorph head phantom has been constructed to determine the absorption pattern induced by a typical telecommunication device antenna. Preliminary results are given.

There are a variety of situations in which it is important to know precisely the spatial specific absorption rate (SAR) distribution in biological tissue exposed to nonionizing radiation. In the case of intentional irradiation for therapeutic purposes in hyperthermia or diathermia, the SAR distribution in the body region treated determines the biologically effective heating pattern. Hyperthermia applicators are commonly characterized by absorption patterns measured in tissue-like phantoms under standard conditions. Moreover, in case of unintentional or accidental exposure to nonionizing electromagnetic radiation it is important to know the SAR distribution. The SAR distribution reveals where the locations of maximum electromagnetic field lie in the human body. When protective guidelines or standards are being established, the SAR distribution can be a guide for the assessment of the risks to which critical organs may be exposed.

Although computer simulations have been shown to be able to calculate SAR patterns with good accuracy in a number of realistic situations, it is generally the measuring method that forms the touchstone or that finally offers the best assessment of the risks related to a specific exposure situation.

MATERIALS AND METHODS

The Phantom and the Scanning Mechanism

The liquid muscle-like phantom consists of saline water. For each given frequency and ambient temperature, the formula in Ref. 1 relates the NaCl concentration to the relative dielectric constant ε_r and conductivity σ. We found a good agreement with the results of our dielectric measurements.

A stepper motor guides a miniature thermocouple through the phantom liquid with a step size of 1 mm. The scanning mechanism is that of an automatic 3-D water phantom commonly used for radiotherapy dosimetry. The relative SAR value is derived from the thermal transient recorded in the course of a short electromagnetic power pulse.

The Miniature Thermocouple

In each measuring point $\bar{r}$, the thermal transient is sampled during the 10s course of the microwave pulse radiated by the applicator or antenna under test. The temperature rise ΔT, mainly due to the thermocouple self-heating, can be shown to be proportional to

$$\Delta T(\bar{r},t) \propto \mathrm{SAR}(\bar{r})\tau[1-\exp(-t/\tau)] \approx \mathrm{SAR}(\bar{r})[t-(t^2/2\tau)] \qquad (1)$$

*Correspondence to: C. De Wagter (Radiophysics), University Hospital, De Pintelaan 185, B-9000 Gent, Belgium.

where τ (typically 2 s) is the time constant that characterizes the dissipation of the excess heat from the thermocouple tip along the thermocouple body and to the surrounding phantom liquid. In order to avoid uncontrollable thermally induced convection in the phantom liquid, the tip of the thermocouple was embedded in a small piece of sponge. Equation (1) shows that the overall amplitude of the self-heating transient is proportional to the local SAR. In order to enhance the signal-to-noise ratio, the recorded temperature rise is fit against the quadratic function of time that, according to Eq. (1), approximately describes the thermocouple self-heating. The combination of the short power pulses and the thermocouple self-heating causes the SAR to be measured rather than the temperature rise in the phantom itself.

Figure 1 shows a typical plot of the "temperature rise" as it is measured by the thermocouple and sampled during the microwave pulse. Only the part due to self-heating is fit against Eq. (1). The self-heating is due to the fact that the thermocouple absorbs considerably more power than the tissue-like liquid. While other methods avoid the self-heating by switching off the heating power during the temperature measurements, the current method exploits the self-heating: the "temperature" is recorded during the heating pulse and the self-heating enhances the measuring signal so that the pulse length and hence the thermal equilibrium time can be reduced considerably for a given accuracy. Considerable attention has been paid to the long-term stability of the microwave generator.

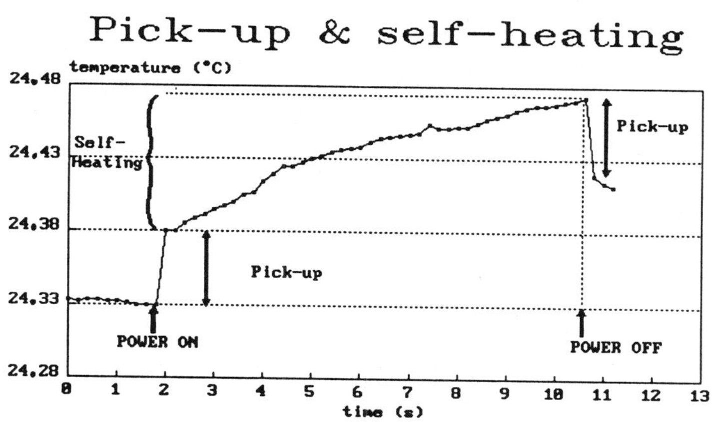

FIG. 1.—"Temperature increase" as recorded during the 10s microwave pulse.

Orienting the thermocouple perpendicularly to the applicator aperture suppresses direct electromagnetic pick-up and minimizes SAR field perturbation and polarization sensitivity. We have used a coaxially shielded Chromel-Alumel thermocouple 0.5 mm in diameter. A workstation controls the system, stores the data, and computes the iso-SAR plots or gray tone images.

The High-frequency Diode

In a second approach, the thermocouple was replaced by a high-frequency diode equipped with resistive leads. We used a Schottky barrier diode. The two out-stretched metal legs comprise the receiving microwave antenna. The legs are cut short to enhance spatial resolution and to reduce the electromagnetic field perturbation, but at the expense of the sensitivity. Therefore, the phantom structure is irradiated in a pulsed way with peak power values that may be higher than the mean power met in a realistic exposure situation. To insure reasonable accuracy, we used an excitation pulse with a peak power of 100 W. The rectified voltage is led by resistive leads to a IEEE-488 controlled digital voltmeter, which is also used for the thermocouple method. Highly conducting leads would perturb the electromagnetic field distribution too strongly. The calibration of rectified voltage versus $|E|^2$ is done at a point of high field intensity. The dipole antenna receives only the electric field component parallel to it. For a measurement of the full electric field vector, one generally needs a combination of three diodes arranged perpendicularly, or a diode with variable direction.

RESULTS

SAR Distributions of Hyperthermia Applicators

The method was initially used to characterize a 13 × 11cm TE (transverse electric) waveguide applicator developed for hyperthermia treatment application at 915 MHz. The NaCl concentration was 18.6 g/l, resulting in $\varepsilon_r = 73.2$ and $\sigma = 2.9\ \mathrm{S\ m^{-1}}$ at 20 C. Figure 2 shows the SAR distribution in the central plane that is perpendicular to the electric field, obtained by the thermocouple and the diode sensor, respectively. In the latter case, the diode was oriented parallel to the incident electric field. The contours obtained with the diode sensor are spatially averaged by the lower resolution. Figure 3 shows the results along the central axis as a function of depth. These graphs also confirm that the thermocouple method measures a signal that is proportional to $|E|^2$, where E is the electric field strength.

Verification of SAR patterns measured at 433 MHz against patterns calculated numerically using the finite-difference time-domain method is reported in Ref. 2.

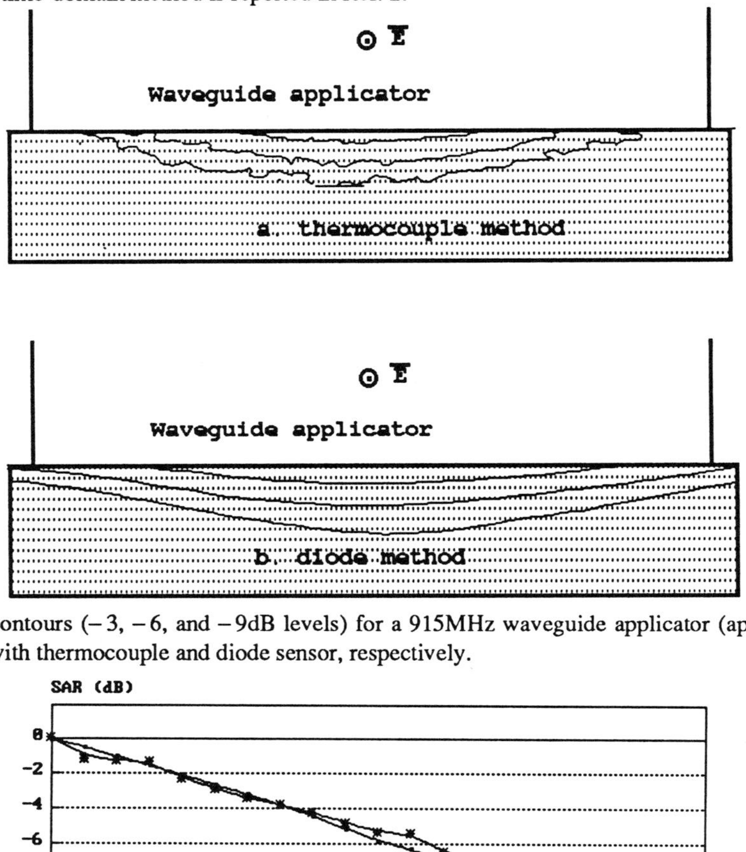

FIG. 2.—SAR contours (−3, −6, and −9dB levels) for a 915MHz waveguide applicator (aperture width 13 cm), measured with thermocouple and diode sensor, respectively.

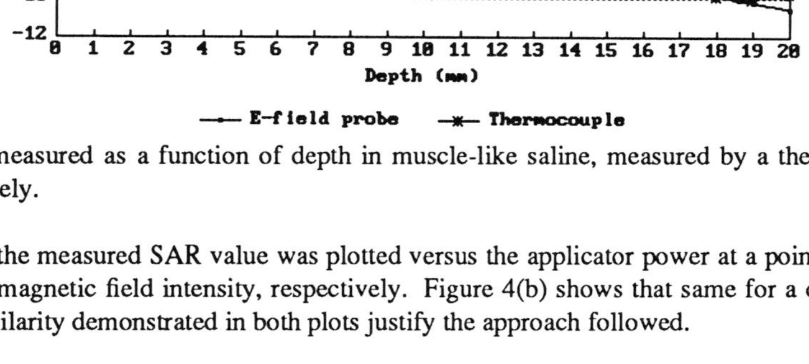

FIG. 3.—SAR measured as a function of depth in muscle-like saline, measured by a thermistor and diode sensor, respectively.

In Fig. 4(a) the measured SAR value was plotted versus the applicator power at a point of relatively low and high electromagnetic field intensity, respectively. Figure 4(b) shows that same for a diode sensor. The linearity and similarity demonstrated in both plots justify the approach followed.

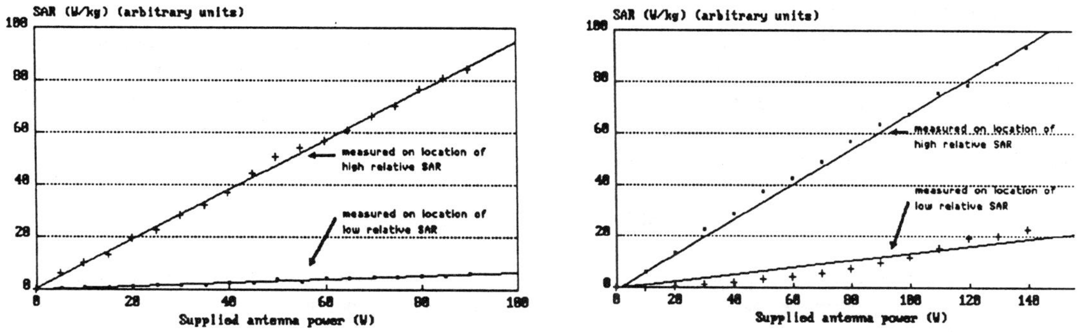

FIG. 4.—SAR vs applicator power on a location of relatively low and high field intensity (beside and in front of applicator aperture respectively) measured with (a) thermocouple sensor, (b) diode sensor.

SAR Distributions of Mobile Telephone Antenna

In a second series of experiments, SAR distributions were measured in a human head phantom irradiated by a nearby 915MHz linear antenna, which mimics the antenna of a mobile telephone device. The antenna was placed vertically in the symmetry plane of the head and in front of the nose at a distance of 15 mm. The head phantom was constructed from two perspex casts, formed by a vacuum technique using an automatic molding machine and a solid plaster model of the head of a Rando phantom, which is commonly used in radiotherapy dosimetry. With the head cast placed upside down, the sensor can freely scan the complete liquid volume.

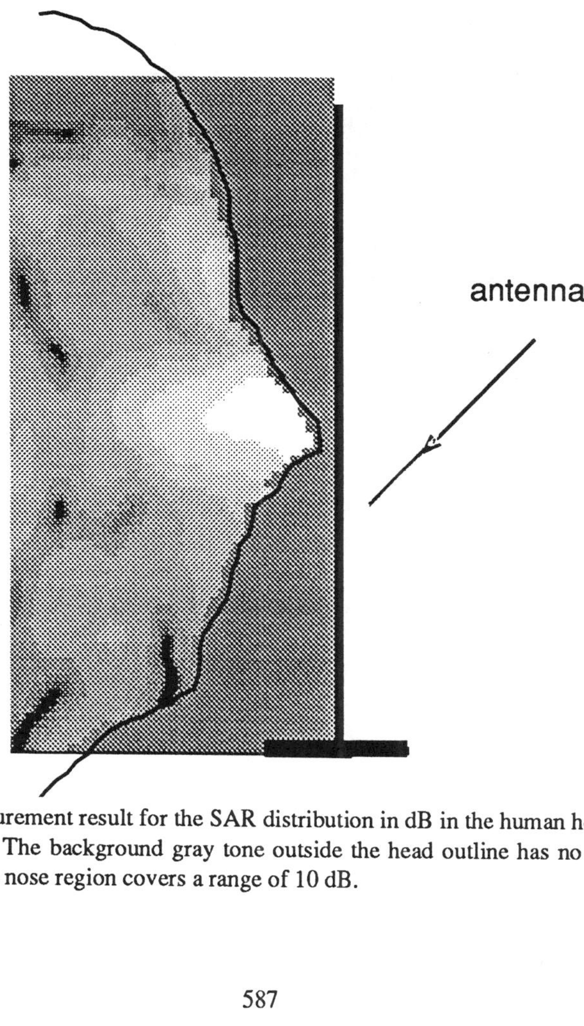

FIG. 5.—Preliminary measurement result for the SAR distribution in dB in the human head phantom irradiated by a 915MHz antenna. The background gray tone outside the head outline has no meaning in terms of SAR. The white spot in the nose region covers a range of 10 dB.

The NaCl concentration was taken to be 4.75 g/l. According to the formula given in Ref. 1, this results in $\varepsilon_r = 77.6$ and $\sigma = 0.777$ S m^{-1} at 20 C and 915 MHz.

The head outlines, which determine the scanning range, are recorded once for a specific set-up and stored in a data file. Figure 5 shows the preliminary measuring results for the SAR-distribution, measured with a diode sensor oriented parallel to the radiating antenna. To calculate the SAR from the measured electric field, we assumed that no depolarization occurs. In a forthcoming publication, the SAR distribution obtained with the thermocouple sensor will be reported.

DISCUSSION AND CONCLUSIONS

The satisfactory agreement between the electric-field measurements and the electromagnetic computations shows that the thermocouple method is reliable in the exposure situations investigated. The technique permits the measurement of absorption rate patterns in homogeneous, anatomically based phantom models under realistic exposure conditions. The method inevitably suffers from the fact that the phantom contents must be liquid and homogeneous. The method has the advantage that it supplies quantitative data in a grid that can be made compatible with that of medical imaging procedures like CT or MR. The method thus offers a valuable basis for testing computer simulations.

The thermocouple sensor makes the method rather independent of the electromagnetic frequency and polarization. The spatial resolution offered by the thermocouple is higher than that obtained with the diode sensor.

REFERENCES

1. A. Stogryn, "Equations for calculating the dielectric constant of saline water," *Trans. IEEE* MTT-19: 733-736, 1971.

2. L. Martens, J. Van Hese, D. De Zutter, and C. De Wagter, "Electromagnetic calculations using the FDTD-method for the analysis of waveguide applicators for microwave hyperthermia," *Proceedings of the 22th European Microwave Conference*, Espoo, Finland, 1992, 293-298.

Martin Blank, Ed., *Electricity and Magnetism in Biology and Medicine*

COMPUTER SIMULATIONS AND MEASUREMENTS OF ELECTROMAGNETIC FIELDS CLOSE TO PHANTOM HUMANS

K. Siwiak*

Motorola Paging and Telepoint Systems Group, Boynton Beach, Fla.

T. M. Babij, K. Zeng, and D. Yip

Florida International University

Moment method and analytical solutions to infinite cylinders were used to study the near fields of phantoms representing humans in body-mounted radio measurements. The experimentally verified results indicate that the external fields close to the body are relatively insensitive to the choice of body parameters over a wide range of permittivity and conductivity, in contrast to in vivo simulations. Pager receivers acting as sensitive miniature wireless magnetic field probes measured the magnetic fields close to humans and to phantoms representing humans. A new light-weight phantom human (SALTY-LITE) was designed by use of this analysis.

The distribution of near electric and magnetic fields close to the human body affects the performance of body-mounted radios and pager receivers. We applied an analysis using both the moment method and analytical solutions to infinitely long cylinders to model homogeneous and layered phantoms representing the human body. The results, verified by experiment, indicate that the electric and magnetic fields external to the body are relatively insensitive to the choice of body parameters over a wide range of permittivity ε and conductivity σ (in contrast to in vivo simulations, which generally require accurate representations of the various human tissues).

Pager receivers acting as miniature wireless magnetic field probes were used to measure the fields close to humans and phantom humans. We adopted a statistical technique originally developed for pager-receiver sensitivity measurements[1] for determining the levels of fields surrounding the phantoms and human test subjects.

Two phantom humans, SALTY[1] and SALTY-LITE[2] (Fig. 1), were tested and the results were analyzed. The computer simulations used NEC[3] for the low-frequency region (including whole-body resonance) and analytical infinitely long cylinder representations were implemented by use of Mathcad at frequencies above resonance. The computer-generated simulations and the results obtained from the open-field experiment are in satisfactory agreement.

This analysis was used to design the relatively small, light-weight SALTY-LITE (62 kg vs 125 kg for SALTY) phantom human. The size and weight of this new simulated body device make it appropriate for use inside small anechoic chambers at frequencies above 100 MHz.

MEASUREMENTS ON HUMANS AND ON PHANTOMS

We measured magnetic fields in the presence of the body relative to the incident fields by using pager receivers as very sensitive (in the range below 1 μA/m) wireless magnetic field sensors. All field tests were carried out between 30 and 930 MHz on the open-field antenna test site characterized in Refs. 4 and 5. The belt-level magnetic field (vertical electric-field polarization) response of a body-mounted receiver was measured for fields incident from eight equally spaced azimuth angles for an anthropometrically diverse group of human test subjects.[2] The individual human tests were averaged over the eight azimuth angles by the gain-average formula

---

*Correspondence to: K. Siwiak, Motorola Paging & Telepoint Systems Gp., 1500 Gateway Blvd., Boynton Beach FL 33426.

$$H_{avg} = \sqrt{\sum_{\phi_i} |H(\phi_i)|^2}$$

where $\phi_i = 0°, 45°, \cdots, 315°$). Those gain-averages were averaged in decibels over the individuals, as in Ref. 2 and are compared in Fig. 2 over frequency with the same eight position averages measured for the phantom humans according to

$$D = 20 \log \left[\frac{H_{avg}(\text{phantom})}{H_{avg}(\text{avg of humans})} \right]$$

The average magnetic fields around humans and around SALTY and SALTY-LITE are within 1.7 dB above 86 MHz. The SALTY response is 3 dB higher than the human response near the whole-body resonance at 66 MHz, but matches the human response at 43.5 MHz. SALTY-LITE is shorter in stature than adult humans, which is why the response relative to humans in Fig. 2 steadily decreases with frequency, all the way below the SALTY-LITE resonance at 90 MHz. An error analysis of the open-field tests is detailed in Ref. 4 and contrasted with traditional magnetic fields measurements in Ref. 5. Both SALTY and SALTY-LITE are remarkably similar to the humans with respect to external magnetic fields close to and around the body despite the radically different internal compositions of the phantoms and of humans.

ANALYSIS OF PHANTOM HUMANS

Both phantom humans were modeled at low frequencies as lossy wire antennas by the NEC with a method suggested by Guy.[6] SALTY is a saline water-filled plastic cylinder modeled as a 39-segment wire antenna with a parallel 2.29Ω resistor and a parallel 1147pF capacitor at each segment to represent a saline water column having a DC conductivity of 0.26 mho/m. SALTY-LITE is a pair of concentric plastic tubes with the intertube layer filled with saline water and is modeled by 30 wire segments, each loaded by a parallel 4.63Ω resistance and a parallel 567pF capacitance. The diameter of both SALTY and SALTY-LITE is 0.305 m (Fig. 1). An analysis was performed in free space, over an earth ground with relative permittivity $\varepsilon = 12$ and conductivity $\sigma = 0.002$ mho/m, and over an ideal perfect electric conductor (PEC) ground. The total magnetic fields relative to the incident fields for SALTY, calculated for the three ground configurations, are shown in Fig. 3, where the effects of the different ground loading are evident. The whole-body resonance near 75 MHz for the free-space case is consistent with similar calculations reported by Durney et al.[7] A secondary resonant peak evident near the third harmonic of the primary resonance is a suspected artifact of the simple model and is being investigated.

At frequencies above the whole-body resonance, the SALTY saline water column was modeled as a homogeneous lossy dielectric by the methods given by Harrington;[8] the SALTY-LITE hollow water column analysis was derived from Ref. 9. Both phantom models were validated by a generalized multilayered cylinder analysis,[10] which includes the effects of the plastic cylinder walls. Figure 4 shows the magnetic fields in front of SALTY-LITE as a function of frequency and with sodium chloride concentration as a parameter. Several resonances are evident at 1.5 g/l concentration (0.26 mho/m), which are damped out at the recommended concentration of 4 g/l (0.69 mho/m). We used the curve-fit formulas of Stogryn[11] to calculate the permittivity and conductivity of saline water. The fields for a perfect electric conductor (PEC) cylinder are shown for comparison. The calculated magnetic fields near SALTY-LITE, as could be shown for SALTY, differ by less than 3 dB above 100 MHz for the wide range of cylinder conductivities shown.

Incident plane magnetic fields in ratio to free-space fields were measured and calculated for SALTY-LITE and SALTY phantoms. Measured points, an NEC lossy wire antenna model over earth ground (solid trace), and an infinite layered cylinder calculation (dashed trace) are shown in Fig. 5 for SALTY-LITE and in Fig. 6 for SALTY. There is satisfactory agreement between experiment and analysis in both cases. The region of greatest discrepancy occurs in the transition between low-frequency and high-frequency models.

CONCLUSIONS

Lossy wire antenna models were used to predict satisfactorily the magnetic fields near a homogeneous saline water column (SALTY) and a hollow saline water column (SALTY-LITE) phantom human at frequencies that include the whole-body resonance. Homogeneous and multilayered infinitely long cylinder models yielded satisfactory results at frequencies above the resonance. Miniature pager receivers were introduced as very sensitive magnetic-field probes suitable for use on human and phantom human measurements at an open-field test site.[12]

1 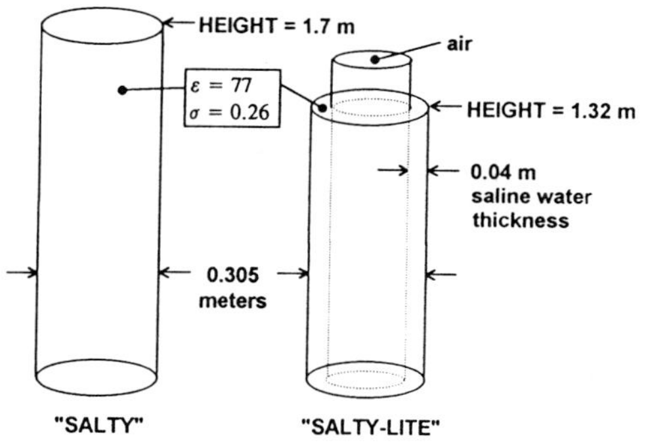

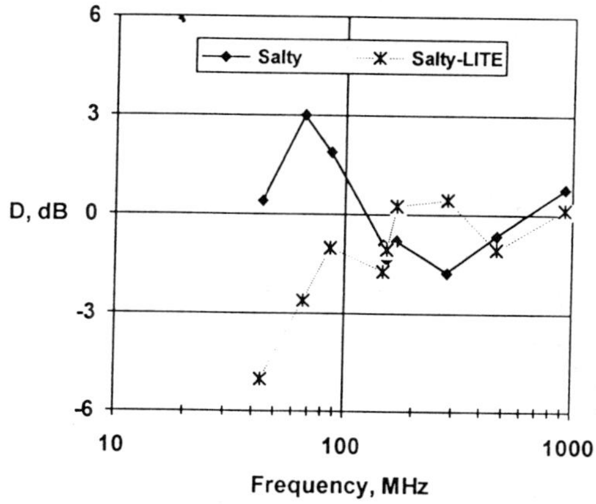 2

3 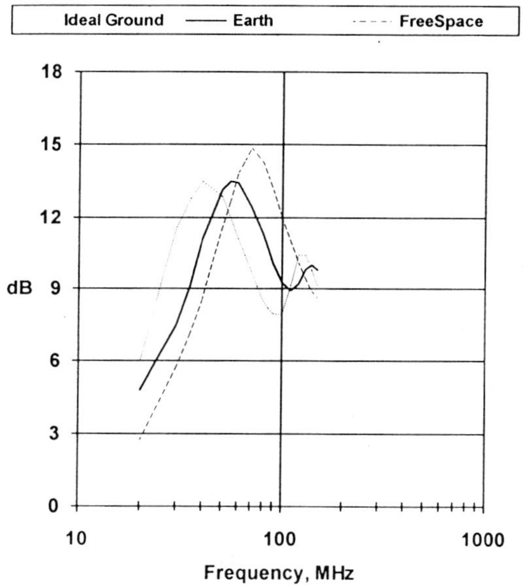

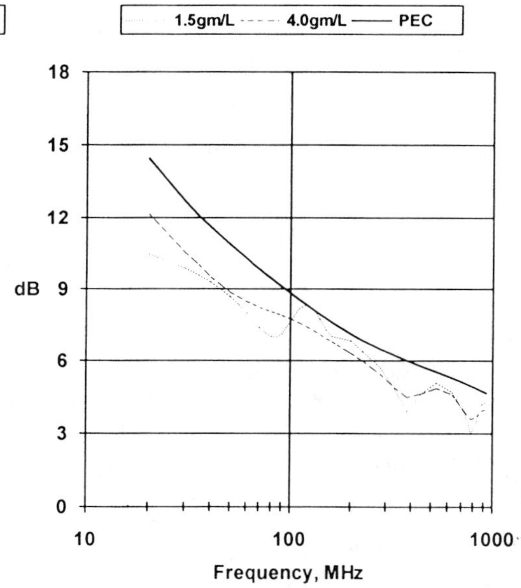 4

FIG. 1.—Magnetic fields were measured 2 cm in front of and 1 m from bottom of Standard human phantoms SALTY and SALTY-LITE.

FIG. 2.—Measured average H fields on human phantoms relative to average fields on human test subjects for SALTY and SALTY-LITE.

FIG. 3.—H_{tot}/H_{inc} 2 cm in front of and 1 m from bottom of lossy wire antenna model of SALTY.

FIG. 4.—H_{tot}/H_{inc} 2 cm in front of infinitely long SALTY-LITE cylinders for several NaCl concentrations.

5
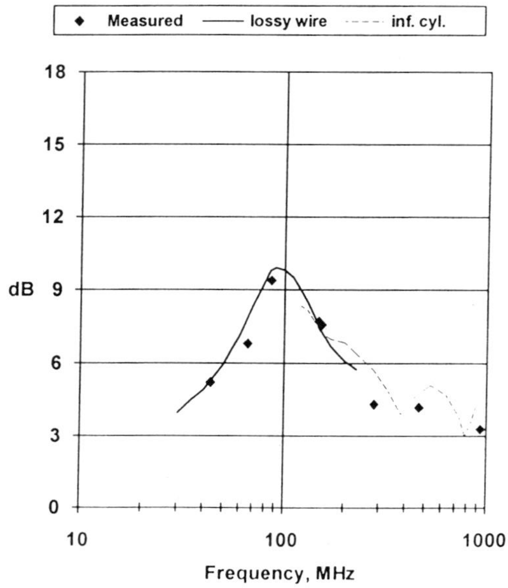

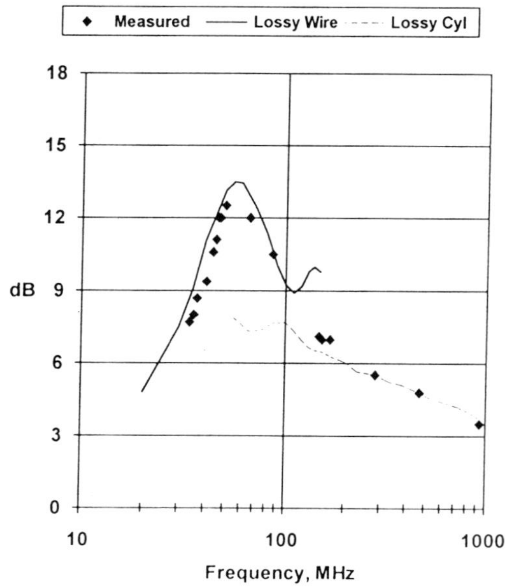

6

FIG. 5.—H_{tot}/H_{inc} 2 cm in front of and 1 m from bottom of the SALTY-LITE phantom: Analysis and measurements compared.
FIG. 6.—H_{tot}/H_{inc} 2 cm in front of and 1 m from bottom of the SALTY phantom: Analysis and measurements compared.

REFERENCES

1. "Paging Systems; European Radio Message System (ERMES), Part 5: Receiver conformance specification," prETS 300 133-5, *ETSI,* Valbonne, France, 1991.

2. K. Siwiak and W. M. Elliott III, "Use of simulated human bodies in pager receiver sensitivity measurements," *Southcon/92 Conference Record,* Orlando, Fla., 1992, 189-192.

3. G. J. Burke and A. J. Poggio, *Numerical Electromagnetics Code (NEC): Method of Moments,* NOSC TD 116, Lawrence Livermore Laboratory, Calif., 1981.

4. K. Siwiak, T. Babij, and D. Yip, "Open field antenna range analysis of measurement errors," in Ref. 2, 227-230.

5. T. M. Babij, K. Siwiak, D. Yip, and K. Zeng, "Accuracy of near field measurements involving phantom humans on open field test sites," this volume.

6. A. W. Guy, "Analysis of EMP induced currents in human body by NEC Method of Moments," *Proc. 12th Ann. Intern. Conf. IEEE Engineering in Medicine and Biology Society,* Philadelphia, 1990, 1547-1548.

7. C. H. Durney, H. Massoudi, and M. F. Iskander, *Radiofrequency Radiation Dosimetry Handbook,* USAFSAM-TR-85-73, Brooks AFB, Tex.: USAF School of Aerospace, 1986, 4th ed.

8. R. F. Harrington, *Time Harmonic Electromagnetic Fields,* New York: McGraw-Hill, 1961.

9. D. K. Misra, *Scattering of Electromagnetic Waves by Human Body and Its Applications,* Ph.D. dissertation, Michigan State University, 1984.

10. L. Ponce de Leon, *Modeling and Measurement of the Response of Small Antennas Near Multilayered Two- or Three-dimensional Dielectric Bodies,* Ph.D. dissertation, Florida Atlantic University, 1992.

11. A. Stogryn, "Equations for calculating the dielectric constant of saline water," *IEEE Trans.* MTT-19: 733-737, 1971 (corrected by K. Siwiak).

12. Kewen Zeng is a recipient of a Motorola Partnerships in Research Grant.

Martin Blank, Ed., *Electricity and Magnetism in Biology and Medicine*

ERROR ANALYSIS OF A THERMOMETRIC MICROWAVE-DOSIMETRY PROCEDURE

C. S. Gambrill,* M. L. DeAngelis, and S.-T. Lu

Ogden BioServices Corporation, Gaithersburg, Md.

A computer-aided procedure for determining microwave local specific absorption rates (SARs) from the rate of induced temperature changes was developed. Dosimetry data from rabbit ocular regions exposed to a 1.25GHz microwave source were examined with regard to the reliability of the dosimetric method. It was found that the coefficient of variance was less than 10% for measured SARs greater than 600 W/kg. A high degree of correlation among measured SARs produced by two different exposure intensities was observed ($r = 0.99$, $n = 64$, $p < 0.001$).

A critical and essential element in microwave-effects research is the determination of the dose and/or dose rate administered. The microwave dose rate is denoted by the local specific absorption rate (SAR) and the whole-body average absorption rate. In addition, the dosimetry data constitute the basis for comparisons among laboratories that employ different animal exposure methods, and the basis of current standards such as ANSI and NCRP. This paper presents results on the reliability of a current dosimetry method adapted and used in this laboratory through determinations of error analysis.

Two important aspects in determining the SAR are addressed in this paper: errors caused by changing thermal diffusion and the minimum detectable SAR as limited by the sensitivity of the measurement equipment. Changing thermal diffusion errors arise from nonhomogeneous microwave absorption. Thermal loss from the measurement site is another source of error and can be minimized by a high heat-deposition rate in a short period of time, such that the cooling becomes relatively insignificant. Analysis of tissue cooling constants indicated that the total heat loss was less than 1% of the thermal dose for exposure periods less than 10 s.

The present SAR determination method is a regimented procedure. The criteria used for data analysis are clear and concise and can be applied consistently by different analysts. The error analysis provides information on the accuracy of the method and experimental conditions required to achieve the desired accuracy.

MATERIALS AND METHODS

Five Dutch belted rabbit carcasses weighing between 1.5 and 2.0 kg were used for SAR exposures. Rabbits were euthanized with intracardiac sodium pentobarbital and pneumothorax. A Luxtron 3000 fluoroptic thermometer system was used for temperature measurements. Two Luxtron MAM-05 fiber optic probes were placed along the outer eyeball surface, one along the top and one along the bottom. Each probe contained four sensors spaced at 5mm intervals. Each rabbit was positioned with its right eye facing the transmitter antenna (W/R 650 waveguide, loaded with teflon sheets) at a distance of 3 in.

The microwave source was operated at two average power levels, 500 and 1000 W. More specifically, the transmitter parameters were 1 MW peak power, 10 μs pulse width, and either 50 or 100 pulses per second.

The Luxtron analog outputs were filtered with Krohn-Hite and Rockland filters set to 1 Hz, low pass, to reduce Luxtron output noise. The filtered outputs were recorded by means of Asystant+ Data Acquisition and Analysis software. Data acquisition started approximately 10 s before the onset of irradiation and continued until approximately 20 s after irradiation. The irradiation period was 10 s.

The acquired data were converted to the Sigma Plot data file format for analysis and divided into three sections, PRE, EXP, and POST, corresponding to pre-exposure, exposure, and post-exposure, respectively.

---

*Correspondence to: C. S. Gambrill, Ogden BioServices Corp., 6 Montgomery Village Ave. (Suite 222), Gaithersburg MD 20879.

A Sigma Plot linear regression routine was performed on each section. The dividing points between the PRE, EXP, and POST periods were adjusted until the beginning point of the EXP regression intersected the ending point of the PRE regression, while the ending point intersected the beginning point of the POST regression. An example of processed dosimetry data is shown in Fig. 1. The slopes of these three linear approximations were recorded in °C/s and used to calculate the SAR from

$$\text{SAR (W/kg)} = 3474(\text{J/kg}\cdot{}^\circ\text{C}) \cdot [\text{EXP} - (\text{PRE} + \text{POST})/2]\,({}^\circ\text{C/s})$$

The constant 3474 is the average specific heat capacity of animal tissue. The average of the PRE and POST terms was used to correct for changing thermal diffusion errors resulting from initial temperature gradients and subsequent microwave-induced temperature gradients.

RESULTS AND DISCUSSION

Local SAR determinations were replicated four times at each of the 16 measurement locations around the ocular tissue for a total of 136 measurements. The coefficient of variance (CV) for repeated measurements was plotted against the SAR (Fig. 2) to portray measurement repeatability. It was clearly seen that the CV was inversely proportional to the SAR. In fact, the CV was less than 10% when the SAR was greater than 600 W/kg (0.1 C/s).

To validate the procedure further, measurements were taken at two exposure levels and the normalized SARs were compared. Low and high exposures were obtained by application of 500 W and 1000 W of average microwave power, respectively. Normalized SAR values determined from 500 W and 1000 W exposures were plotted against each other (Fig. 3). These data points were linearly correlated. It is clearly shown in Fig. 3 that the linear correlation was nearly ideal ($r = 0.99$, $p < 0.001$, $n = 64$). Therefore, the present concise dosimetric method is deemed to be an accurate determination of the SAR.

ACKNOWLEDGMENT

Skillful technical assistance provided by David Varle and William York Jr. contributed to the success of this research. Animals used in the present study followed the guidelines set forth by the NIH and Walter Reed Army Institute of Research. This study is under contract DAMD 17-89-C-9021 with the Walter Reed Army Institute of Research, Department of Microwave Research. The views of the authors do not necessarily reflect the position of the U.S. Department of the Army or the U.S. Department of Defense (Para 4-3 AR 360-5).

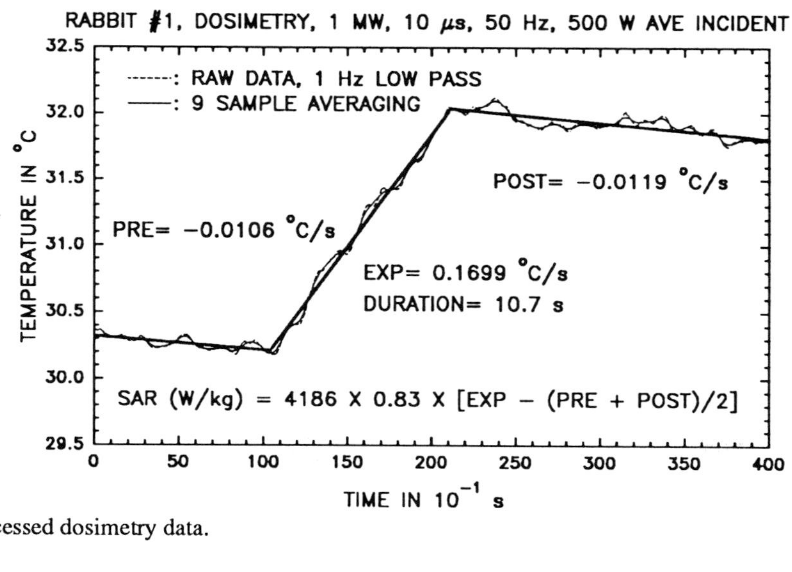

FIG. 1.—Processed dosimetry data.

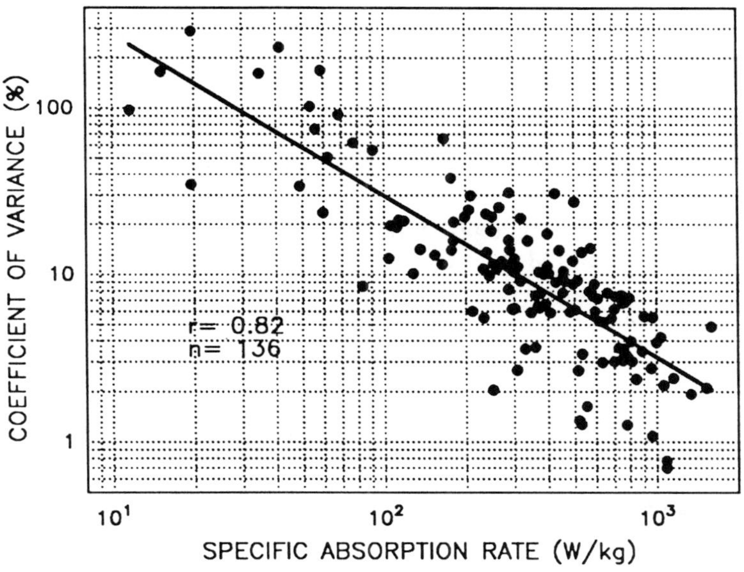

FIG. 2.—Coefficient of variance vs SAR.

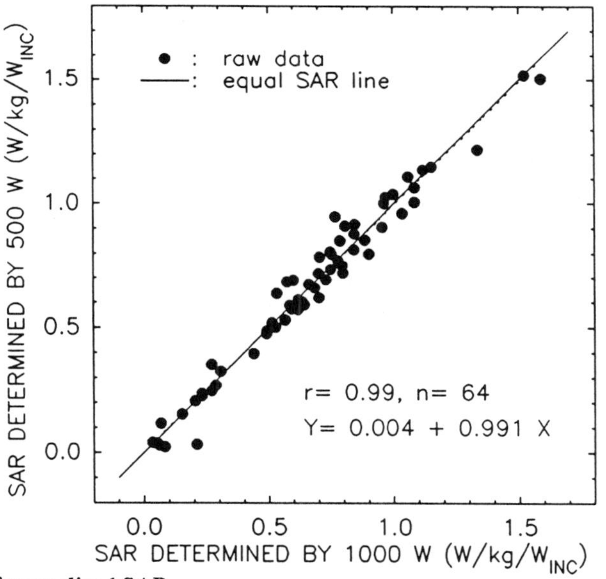

FIG. 3.—Correlation of normalized SARs.

Martin Blank, Ed., *Electricity and Magnetism in Biology and Medicine*

MILLIMETER WAVES PROTECT GAMMA-IRRADIATED RATS

M. Kuzmanova, S. Ivanova, M. Baldgiiska, B. Videlov, V. Nankova, N. Neshev, S. Alexandrov, and M. Markov*

Sofia University, Bulgaria

Millimeter waves (53.57 GHz, 1 mW/cm^2) and gamma radiation (6 Gy) were applied to white male rats (Wistar, 170 g). Nonionizing radiation was applied to the right forelimb by a Г4-142 generator. Gamma rays irradiated the whole body of the animal by use of a IGUR-1 device. The amount of ATP and electrophoretic mobility of erythrocytes, as well as spleen weight and spleen cariocyte counts, were evaluated 3, 14, and 21 days following exposure. It was found that millimeter waves substantially reduced some of the most deleterious effects of ionizing radiation, especially when they were applied prior to exposure to ionizing radiation.

Millimeter waves have been recently discussed as a factor that basically modifies the processes and systems for the generation, transmission, and reception of signals for information and regulation. Because of their unique bioelectrochemical properties, cell membranes are considered as a primary acceptor of microwave radiation, and a number of studies both of erythrocyte and lymphocyte membranes have been performed. There has been a lack of in vivo studies in comparison with cell and membrane research, partly due to the complexity of highly organized living systems such as mammals, as well as the intricate physical coupling and internal electromagnetic-field distribution. In contrast, cells provide metabolically functional entities and thus in vitro experiments permit the investigation of effects of electromagnetic fields on relevant physiological endpoints.[1,2]

Millimeter waves are increasingly applied in medicine independently or accompanied by other therapeutical factors. In particular, the combined action of x rays and millimeter waves has been reported to decrease the unfavorable effects of ionizing radiation on hematopoiesis.[3,4] One very important observation was made: if x-ray irradiation is followed by millimeter waves, a radiosensibilization is observed; whereas bone marrow is found to be defended (defence coefficient K = 2-3) if millimeter waves are applied prior to x rays.

This study was therefore designed to evaluate the effects of millimeter waves on intact and gamma-irradiated rats.

MATERIALS AND METHODS

White male rats (Wistar, 170-180 g) divided into four groups with an equal diet (a control group, only millimeter-waves-irradiated animals, only gamma-irradiated animals, and millimeter- and gamma-irradiated animals) were the subject of the experiment. The basic technique we developed for evaluation of the effects of ionizing and nonionizing radiation[5] was used for estimation of some changes in blood parameters 3, 14, and 21 days following radiation exposure. Erythrocytes were separated by three subsequent centrifugations (10 min, 2000 G) in a phosphate-buffered saline solution. After estimation of spleen weight, a 30mg piece of the spleen was homogenized in a Turk solution (5 ml of 5% acetic acid), then diluted 20 times with a Turk solution. The count of nucleus-containing cells was done in a Burker chamber by the technique of leucocyte counting. The number of cells in a whole spleen and the relative number in 1 mg was calculated.

Millimeter waves (53.57 GHz) were applied at an intensity below the thermal level at 1 mW/cm^2 by a Г4-142 generator (USSR). A digital supply unit, TEC 41 (Bulgaria) generated a constant driving voltage for the main frequency of Г4-142. Nonionizing radiation was applied to the right forelimb of the animal for 1 h daily during 5 days. Gamma radiation (6 Gy, 1 h) was applied to the whole body by an IGUR-1 system through four symmetrically mounted sources of ^{137}Cs. In that way, the equal dose allocation along the whole body of the animal was insured.

---

*Correspondence to: M. Markov (Orthopaedics), Bioelectrochemistry Laboratory, Mt. Sinai School of Medicine, 1 Gustave Levy Place, New York NY 10029-1189.

RESULTS AND DISCUSSION

ATP content analysis showed that, during the whole period of investigation, the amount of ATP in erythrocytes from nonirradiated animals remained unchanged. When 6Gy gamma irradiation was applied, a decrease of ATP content was found during the entire postradiation period. Millimeter waves only, when applied to the animals, show a decrease compared to control only 21 days after irradiation; and at day 3 and 14 a small increase was observed. Interesting results were obtained when millimeter waves were applied to animals prior to gamma irradiation: an increase of ATP content at day 3 and 14, and a decrease at day 21. The 12% decrease is about half the 26% decrease following γ-irradiation alone.

These data unambiguously show a decrease in the erythrocyte ATP content after exposure of animals to γ irradiation, a finding directly connected with disturbances of the energy balance in erythrocytes as a result of ionizing-radiation exposure. The pretreatment with millimeter waves was found to prevent the ATP decrease, especially at early postradiation periods. Moreover, a well-pronounced protection effect expressed in a significant ATP content increase was observed when millimeter waves were in use (Table 1).

TABLE 1.—Erythrocyte ATP content estimated in three groups exposed to irradiation vs control.

	Postexposure time (days)		
	3rd	14th	21st
γ irradiation	−11.68%	−22.34%	−26.11%
Millimeter waves	+ 6.88%	+ 4.03%	− 7.07%
Millimeter waves + γ irradiation	+16.86%	+ 7.84%	−12.17%

Electrophoretic mobility (EPM) changes in erythrocytes from animals treated by millimeter waves shows statistically insignificant change during the entire period of investigations. EPM of the erythrocytes from the γ-irradiated animals was found to decrease by about 10% at the 3rd day after radiation exposure and by 15% in the subsequent period. This result confirms the hypothesis that conformational changes in the components of the erythrocyte membrane occur under γ radiation. A redistribution of charged groups or domains in the membrane is the most probable way for the implementation of these changes. The treatment of animals with millimeter waves prior to γ irradiation leads to a conservation of EPM values close to control ones for the first 14 days; not until the later postradiation period does the radioprotective effect disappear. These results indicate that millimeter waves do not cause significant changes in the membrane architecture, but might stabilize membrane structure, so that in the early period after γ irradiation the membrane properties remain unchanged. Only when the stabilization potential of the cell is exhausted does radiation damage take over.

The measurements of spleen volume and weight as well as of the spleenocariocytes counts permitted the evaluation of the dynamics of ionizing and nonionizing radiation effects at the cellular level. The spleen weight (Table 2) was found to increase by approximately 30% compared to the control and remained unchanged during the entire 3-week period of investigation. Very specific spleen weight changes were observed when animals were exposed to γ radiation of 6 Gy: a 22% decrease at the 3rd day and an increase in later terms of 36% and 92%, respectively, at 14th and 21st days. The pretreatment with millimeter waves is not effective at day 3 (26% decrease of spleen weight), but significantly decreases the unfavorable effects of γ radiation later: at day 14 the spleen weight was close to normal (92% compared to control). The spleen weight at day 21 was very close to those measured when only millimeter waves had been applied.

Spleenocariocyte counts also specifically depended on the presence or absence of ionizing or nonionizing radiation (Table 3). Two numbers for each count were obtained: the total number of spleenocariocytes and the relative number (the ratio of total cell number and spleen weight). A decrease of both cell numbers was found at the 3rd day in each variant of radiation exposure, but the situation was very different in the late postradiation periods. When millimeter waves only were applied, a 59% increase of the total number was found at day 14, and the relative number returned to the control value. At day 21 the total number of cells was close to the normal, but the relative value again decreases to approximately 75% of control. The observed spleen weight increase following millimeter irradiation might be a result of an extra amount of blood kept in this organ. The decrease of spleenocariocyte number is probably a result of accelerated leucocyte excretion in the blood-vessel system. The increased cell number at day 14 may be considered to be a result of an increased proliferation as part of compensatory reactions. To test this hypothesis, a histological study of the spleen after exposure to ionizing and/or nonionizing radiation is currently in progress. Comparing data from animals exposed only to ionizing radiation with those from the group that received both millimeter waves and γ radiation, one may conclude that an atrophy of the spleen occurs under γ radiation. The extent of the atrophy of the hematopoietic organ can be used as a biological test for evaluation of postradiation changes. In addition, the combined action slowly, but monotonously normalizes the total number of spleenocariocytes toward normal. The same tendency could be seen with relative cell volume, but normalization was approached slowly because of drastic changes in the spleen volume and weight.

TABLE 2.—Spleen weight (mg) estimated after exposure of animal to ionizing and nonionizing radiation.

	Postexposure time (days)		
	3	14	21
Control	677 ± 51	677 ± 51	677 ± 51
Millimeter waves	932 ± 25	902 ± 22	902 ± 34
γ irradiation	531 ± 37	922 ± 48	1304 ± 41
Millimeter waves + γ irradiation	492 ± 30	618 ± 44	852 ± 27

TABLE 3.—Changes in spleenocariocyte counts (number of cells × 10^{-6}) as result of ionizing and nonionizing radiation.

	Postexposure time (days)		
	3	14	21
Control	710 ± 32	710 ± 32	710 ± 32
Millimeter waves	343 ± 44	1134 ± 59	690 ± 51
γ irradiation	291 ± 21	847 ± 19	1055 ± 49
Millimeter waves + γ irradiation	243 ± 33	384 ± 14	569 ± 28

The reported data are in good accordance with the four-step model of spleen radiation damage observed after 3-7 Gy whole body γ radiation. The increase of spleen weight as a response to irradiation with a sublethal dose could be connected with accumulation of the destroyed cell's elements, enlargement of the perivascular tissue, and reticulum, hydration, hyperemia, and other physiological changes. In conclusion, millimeter waves appear to reduce substantially some of the deleterious effects of ionizing radiation. It is particularly noteworthy that the protective effect of millimeter waves only occurred when they were applied prior to ionizing radiation.

REFERENCES

1. S. F. Cleary, "Biological effects of nonionizing radiation," in E. Webster, Ed., *Encyclopedia of Medical Devices and Technology*, New York: Wiley, 1988, vol. 1.

2. S. F. Cleary, L.-M. Liu, and G. Cao, in M. J. Allen, S. F. Cleary, and F. Howkridge, Eds., *Charge and Field Effects in Biosystems*, New York: Plenum Press, 1990, 221-221.

3. N. D. Devjatkov, O. V. Betzkii, A. Gelvitch, M. B. Golant, A. M. Machov, T. B. Rebrova, L. A. Sevastjanova, and A. Z. Smolenskaja, *Radiobiologia* 2: 163-171, 1981 (in Russian).

4. L. A. Sevastjanova, "Biological action of radiowaves from millimeter range on normal and cancer tissues," in *Effects of Nonthermal Millimeter Radiation on Biological Systems*, Moscow, 1983, 48-62 (in Russian).

5. M. S. Markov, S. P. Ivanova, B. P. Galutzov, L. L. Traikov, and M. A. Kuzmanova, "Biophysical test for estimation of the effects of ionizing and nonionizing radiation," *Medical & Biological Engineering & Computing* (in press).

Martin Blank, Ed., *Electricity and Magnetism in Biology and Medicine*

PERMEABILITY OF THE BLOOD/BRAIN BARRIER INDUCED BY 915MHz EM RADIATION (CW AND MODULATED) AT VARIOUS SARs

L. G. Salford,* A. Brun, J. L. Eberhardt, and B. R. R. Persson

Lund University, Sweden

Fischer-344 rats were exposed to 915MHz microwaves, both as continuous wave (CW) and modulated at 8, 16, 50, and 200 Hz. After exposure the brains were perfused with saline and albumin and fibrinogen were measured immunohistochemically in the central coronal sections. The results show albumin leakage through the blood-brain barrier (BBB) in 8% of the controls and in 35% of the animals exposed to 915MHz microwaves, with no significant difference between the results for continuous and for modulated irradiation. SAR values as low as 0.01-0.1 W/kg show significant opening of the BBB ($p = 0.015$); SAR above 3 W/kg give higher response than below ($p < 0.05$).

The normal selective permeability of the blood-brain barrier (BBB) can be altered under several neuropathological conditions such as severe trauma to the skull, in epileptic seizures, and during irradiation.[1] Several authors have reported that electromagnetic exposure and NMR examinations alter BBB permeability.[2-10] Other authors have reported difficulties in confirming these findings.[11-19] The lack of agreement among investigators might be due to the lack of an assay of BBB permeability that combines sensitivity and specificity.

In a previous investigation we have studied the permeability of the BBB after exposure to static magnetic fields, low-frequency pulsed magnetic fields and radiofrequency electromagnetic fields, and a combination of the three types of fields used for NMR imaging. We found that the exposure to these fields induced various degrees of leakage of proteins that normally do not pass the BBB.[20] In the present investigation we have studied the permeability of BBB to endogenous albumin and fibrinogen during exposure to microwaves of 915 MHz as continuous wave (CW) and modulated at various low frequencies (8-200 Hz).

MATERIALS AND METHODS

Detailed descriptions of the experimental set-up and performance are found int he previous reports.[20-23] A transverse electromagnetic transmission line cell (TEM-cell) was used for the RF exposure of rats.[23-25] The irradiation consisted of pulse-modulated 915MHz RF radiation with a pulse width of 0.57 ms at modulation frequencies of 8, 16, or 200 Hz, and 4 ms at 50 Hz. RF generator output varied between 1 and 20 W (pulsed) and between 0.2 W and 3 W (CW).

A total of 246 Fischer-344 rats, male and female, weighing from 120 to 330 g, were used in these experiments. The rats were not anesthetized during exposure. The brains were perfused with saline and then fixed with 4% formaldehyde. Central coronal sections of the brains were dehydrated, embedded in paraffin, and sectioned at 5 μm. Albumin was demonstrated with IgG fraction of rabbit antirat albumin (Cappel); fibrinogen was demonstrated with rat antihuman fibrinogen (Daco). Biotinylated swine antirabbit 1:300 was used as secondary antibody; and peroxidase conjugated avidin 1:1000 visualized with DAB (diaminobenzidine), counterstained with Meyer-HTX, was used as enzyme.

The ratios of percentage of occurrence of albumin extravasation in exposed and controls (i.e., odds ratios) were calculated and then compared with the χ^2 or Fisher's exact probability test.

$$\text{Odds ratio} = \frac{\%\ \text{BBB opening} \in \text{EXPOSED}}{\%\ \text{BBB opening} \in \text{CONTROLS}}$$

*Correspondence to: L. G. Salford (Neurosurgery), Lund University Hospital, S-221 85 Lund, Sweden.

RESULTS

Demonstration of albumin was regularly positive in normally BBB-weak areas such as tissue membranes, choroid plexus, pituitary stalk, and hypothalamus. Positivity was regarded as pathological when positive areas were located outside the BBB-weak areas, such as in the cortex, lateral thalamus, etc. The positivity of albumin was asymmetrical and spotty, often seen to be centered around small vessels. Leakage of albumin was often taken up by glial and neuronal cells in the vicinity of a capillary. These findings (particularly those which were asymmetrically positive) were regarded as carrying particular weight in this context. There was no immunopositivity for fibrinogen, neither in the exposed nor in the controls.

Two types of analyses were performed: (i) a crosstabulation of number of rats with positive findings as a function of exposure frequency, irrespective of exposure amplitude; and (ii) a crosstabulation as function of exposure amplitude, irrespective of frequency.

The results show albumin leakage in 5 out of 62 of controls and in a majority of the animals exposed to 915 MHz microwaves (Table 1). CW irradiation resulted in 14 positive findings out of 35, which differs significantly from the controls ($p = 0.0002$). With pulsed 915 MHz microwaves modulated at 200, 50, 16, and 8 Hz, 42 out of 149 were positive, which is significant at the $p = 0.001$ level. Thus, both CW and pulsed 915MHz microwaves can open the BBB to albumin passage, but there is no significant difference between CW and pulsed irradiation in this respect ($p = 0.17$). All exposed animals were grouped in various SAR intervals (Table 2). The frequency of BBB opening remains constant up to 4 W/kg with an odds ratio of about 3.3. Above 4 W/kg there is a significant further increase with an odds ratio of 5.9.

DISCUSSION

The immunohistochemical methods permit the identification of even minor leakage of protein and give an exact histological localization of this leakage. Such methods have not been previously applied to studies of EMF effects on the BBB. The leakage seems to stem from vessels around which it has often been seen to be centered. Failure to recognize this phenomenon may be due to sectioning through the periphery of areas of leakage, or through areas with albumin flow along tracts (such as in the white matter) far from the vessel of origin.

The fact that the larger fibrinogen molecules do not pass the BBB under any of the experimental conditions examined shows that the impact of EMF on the BBB is limited. Thus the combined examination for albumin and fibrinogen positivity provides a good tool for quantifying the effect of EMF on the BBB.

In sum, the effect of 915MHz microwave exposure on the BBB was shown for continuous radiation and for radiation pulsed at 8, 16, 50, or 200 Hz. Our data show that significant extravasations are induced at SAR levels as low as 0.01-0.1 W/kg, which is far below thermic levels. This result is in agreement with out previous investigation by the Evans-blue method for studying the effect of magnetic fields and RF-radiation components of NMR-imaging procedures.[20]

CONCLUSIONS

All exposed groups show a response in BBB opening compared to controls with an odds ratio of 5-3 ($0.002 < p < 0.01$).

Pulse modulation compared to CW gives an odds ratio of 0.7, which does not reveal any difference in response ($p = 0.17$).

Simultaneous comparison of all exposed groups does not reveal any difference in response at the various modulation frequencies ($p = 0.59$).

Groups exposed to SAR values as low as 0.01-0.1 W/kg show significant opening of the BBB.

The response of the BBB opening does not reveal any difference in the SAR interval 0.01-4 W/kg.

For exposed groups, the response to SAR values above 3 W/kg is significantly higher than below 3 W/kg ($0.04 < p < 0.05$).

TABLE 1.—Summary of results of postive signs of penetration of albumin through the blood-brain barrier after exposure to 915MHz electromagnetic fields, continuous and modulated at various frequencies.

Exposure condition *) Modulation frequency	Number of animals	Frequency of BBB opening (%)	Odds-ratio	Significance test *p*-value
8 Hz*)	32	25	3.1	0.02
16 Hz*)	38	26	3.3	0.01
50 Hz*)	41	34	4.2	0.001
200 Hz*)	38	26	3.3	0.01
CW (915 MHz)	35	40	5.0	0.0002
Σ Modulated	149	28	3.5	0.001
Σ CW + Modulated	184	30	3.8	0.0004
Control	62	8	1	

TABLE 2.—Summary of results for various SAR values, all modulation frequencies pooled.

Exposure condition SAR (W/kg)	Number of animals	Frequency of BBB opening (%)	Odds-ratio	Significance test *p*-value
0.01 - 0.1	34	26	3.3	0.015
0.1 - 0.4	44	32	3.9	0.002
0.4 - 2	45	22	2.8	0.04
2 - 4	38	32	3.9	0.002
4 - 5	23	48	5.9	<0.0001
Control	62	8	1	

REFERENCES

1. T.-E. O. M. Sokrab, *Neural Damage in Acute Experimental Hypertension and Epileptic Seizures: Long-term Consequences of Blood-brain Barrier Dysfunction,* thesis, Department of Neurology, University of Lund, 1989.

2. E. E. Albert, in D. G. Hazzard, Ed., *Symposium on Biological Effects and Measurements of Radiofrequency/Microwaves,* Washington, D.C.: HEW Publication FDA 77-8026, 1977, 294.

3. E. N. Albert, D. L. Brainard, J. D. Randall, and F. S. Janatta, *URSI International Symposium on Biological Effects of Electromagnetic Radiation,* Helsinki, 1978, 58.

4. K. J. Oscar and T. D. Hawkins, *Brain Res.* 126: 281, 1977.

5. Y. Ashani, F. H. Henry, and G. N. Catravas, *Radiat. Res.* 84: 496, 1980.

6. J. C. Lin and M. F. Lin, *Bioelectromagnetics* 1: 313, 1980.

7. J. C. Lin and M. F. Lin, *Radiat. Res.* 89: 77, 1982.

8. C. H. Sutton and F. B. Carrol, *Radiat. Sci.* 14: 329, 1979.

9. C. H. Sutton, R. L. Nunnally, and F. B. Carroll, *Cryobiology* 10: 513, 1973.

10. R. R. Shivers, M. Kavaliers, G. C. Teskey, F. S. Prato, and R. M. Pelletier, *Neurosci. Letters* 76: 25, 1987.

11. J. H. Merritt, A. P. Chamness, and S. J. Allen, *Rad. Environ. Biophys.* 15: 367, 1978.

12. E. Preston and G. Prefontain, *Appl. Physiol. Respir. Envionr.: Exercise Physiol.* 49: 218, 1980.

13. E. Preston, E. J. Vavasour, and H. M. Assenheim, *Brain Res.* 174: 109, 1979.

14. D. H. Spackman and V. Riley, Ref. 3, 75.

15. W. M. Williams, W. Hoss, M. Formaniak, and S. M. Michaelson, *Brain Res. Rev.* 7: 165, 1984.

16. W. M. Williams, M. del Cerro, and S. M. Michaelson, *ibid.,* p. 171.

17. W. M. Williams, J. Platnerand, and S. M. Michaelson, *ibid.,* p. 183.

18. W. M. Williams, S.-T. Lu, M. del Cerro, and S. M. Michaelson, *ibid.,* p. 191.

19. E. Moriyama, M. Salcman, and R. D. Broadwell, *Surg. Neurol.* 35: 177, 1991.

20. L. Salford, A. Brun, J. Eberhardt, L. Malmgren, and B. Persson, in C. Ramel and B. Nordén, Eds., *Interaction Mechanism of Low-level Electromagnetic Fields in Living Systems,* Oxford: Oxford University Press, 1992 (in press).

21. L. Salford, A. Brun, J. Eberhardt, and B. Persson, in *Proc. 1st Congr. Euro. Bioelectromagnetic Assoc.,* Brussels, 1992 (in press).

22. L. Salford, A. Brun, J. Eberhardt, and B. Persson, in *The blood-brain barrier: Today and tomorrow," Special issue, J. Electron Microscopy Technique* (in press).

23. J. Van Hese, L. Martens, D. De Zutter, C. De Wagter, L. Malmgren, L. G. Salford, and B. R. R. Persson, *Trans. IEEE* EMC (in press).

24. L. Martens, J. Van Hese, D. De Zutter, C. De Wagter, L. Malmgren, L. G. Salford, and B. R. R. Persson, in Ref. 21.

25. M. L. Crawford, *Trans. IEEE* EMC-16: 189, 1974.

Martin Blank, Ed., *Electricity and Magnetism in Biology and Medicine*

ACCURACY OF NEAR-FIELD MEASUREMENT INVOLVING PHANTOM HUMANS ON OPEN-FIELD TEST SITES

T. M. Babij,* K. Zeng, and D. Yip

Florida International University, Miami

K. Siwiak

Motorola Paging and Telepoint System Group, Boynton Beach, Fla.

Factors affecting the transmission between two closely spaced antennas above a perfectly conducting or real ground, as well as possible errors caused by measuring techniques are reviewed. The analysis serves as a basis for the discussion of radiation measuring methods involving radios mounted on humans and phantom humans rather than traditionally used E-H probe measurements. In this study, factors that influence the electromagnetic fields and their measurement on open-field radiation-measuring sites have been reviewed. The specialized case of radios mounted on phantom heads is detailed.

Open-field antenna test ranges are often specified for use in measurements of the sensitivity of pagers and data receivers in the VHF and UHF ranges (30 to 1000 MHz).[1,2] These receiver-sensitivity measurements are actually signal-to-noise measurements that involve statistical methods of arriving at the transmitted power required to evoke a particular receiver response (the receiver "sensitivity"). As such, errors arise from uncertainties in antenna range calibration, path loss, transmitter-generator levels, gain standards, geometric placement of the test device, ambient temperature changes, interference, and equipment malfunctions, including cables. Systematic and random errors are encountered, and absolute and relative measurements are each affected differently.

A SINGLE-AXIS E-H PROBE

Recently developed E-H probes for measuring the intensities of electric and/or magnetic fields near or within the phantom humans are presented.[3,4] The E-H probe uses dipole and loop antennas, and Schottky diode detectors that all fit in a cubic volume less than 10 mm per side. The analysis and experimental data show that the E-H probe works in the frequency range 0.5-1000 MHz. A schematic diagram of a single-axis E-H probe and its response with linear and square-law detectors are presented in Fig. 1.

3D E-H PROBE

Physical arrangements of the three orthogonal dipoles and loops can be used (Fig. 2). The coupling between the dipoles and coupling between the loops is negligible if they are thin and electrically small. The measured E-H probe receiving pattern response is shown in Fig. 3.

ANTENNA-RANGE CALIBRATION

Propagation geometry above ground on an open-field antenna site is presented in Fig. 4. This open-field range has a transmitting antenna height of 3.2 m and a range distance of 45.5 m. The electric-field strength vs height is presented in Fig. 5. The calibration factor is here defined as the field strength expressed in dB with respect to 1 μV/m at the measurement position for a transmitter generator power of 1 mW.[5,6]

Antenna-range calibration factors were calculated from the physical geometry of the range, along with a best-fit set of antenna-range ground parameters, by use of MathCAD. There is a close correspondence

---

*Correspondence to: T. M. Babij (ECE), Florida International University, Miami FL 33199.

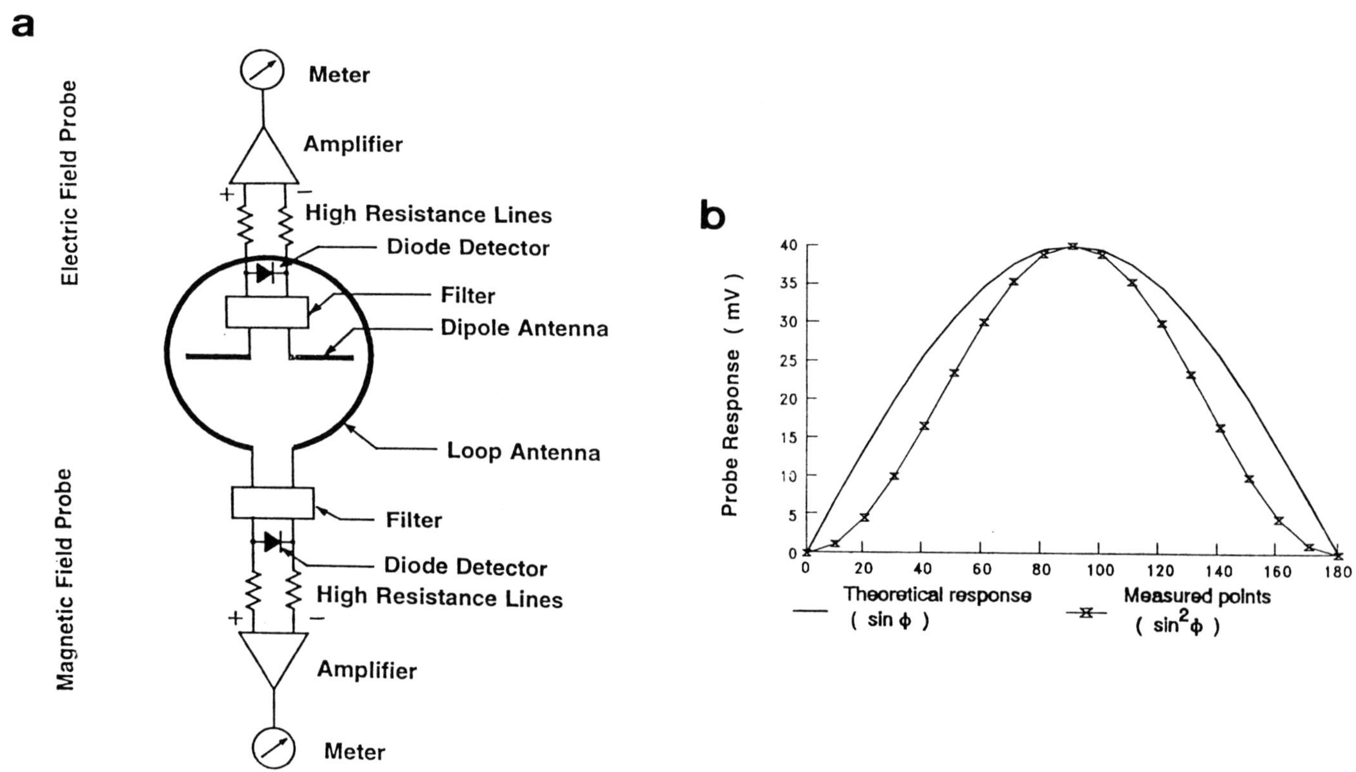

FIG. 1.—(a) Fundamental design of single-axis E-H probe; (b) antenna pattern of H probe measurement vs theoretical ideal response ($\sin \phi$ and $\sin^2 \phi$).

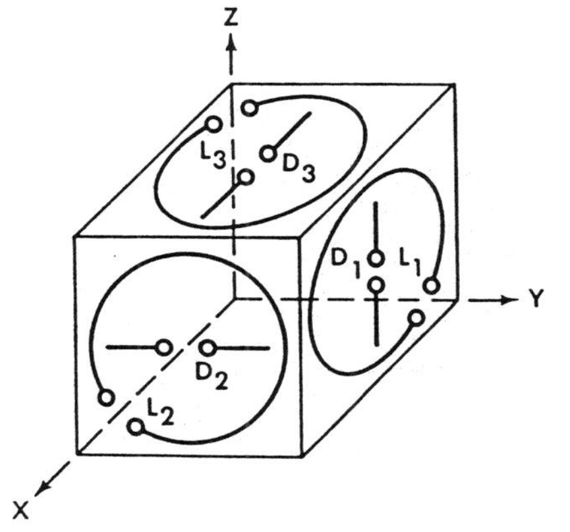

FIG. 2.—Isotropic E/H probe configuration.

between measured and calculated factors.[6,7] The standard deviation of many calibrations at several frequencies was found in Ref. 6 to be $\sigma_{cal} = 0.36$ dB.

The antenna range was modeled by a direct ray from the transmitter added to a ray reflected from a layered dielectric half-space representation of the ground. The best-fit ground parameters were a relative permittivity of $\varepsilon = 7.3$ with a conductivity of 0.002 mho/m for the upper layer, and for the lower layer 0.46 m deep with a relative permittivity of $\varepsilon = 23$ and 0.002 mho/m conductivity. Similarly good results can be obtained with a simpler model that has a homogeneous dielectric half-space representation of the ground. Plane-wave reflection coefficients are used at the ground and at the dielectric-layer interface. Another, more detailed analysis indicates that the surface-wave contribution is negligible (±0.1 dB) and it is not included here.

PAGER-SENSITIVITY MEASUREMENT METHOD

Pager sensitivity is reported as the field strength (usually expressed as the equivalent electric field in dB with respect to 1 μV/m) required for an 80% probability of responding correctly to a sent message. Statistically based methods such as the "20 call" method[1] result in the 80% calling success rate. In one experiment the sensitivity of 15 pagers was measured three times each and on three different but consecutive days. The standard deviation (σ_{all}) for the measurement was 0.45 dB. If the range-calibration errors and the pager measurement errors are treated as statistically independent, then the measurement standard deviation σ_{meas} can be found from

$$\sigma_{meas} = \sqrt{\sigma_{all}^2 - \sigma_{cal}^2} = 0.27 \text{ dB}$$

which includes transmitter-generator repeatability as well as "20 call" measurement method effects. This result compares with the maximum ±0.5 dB and "typical" ±0.2 dB quoted for the "20 call" method alone.[2]

SUMMARY OF ANTENNA-RANGE ERRORS

The calibration standard deviation, $\sigma_{all} = 0.36$ dB, takes into account the randomness inherent in geometrically locating the calibration standard with respect to the measurement point. This component is assumed inherently to include the effects due to scattering. Additional variations due to ground parameter changes during a measurement of less than ±0.1 dB can be expected during a long series of measurements. The pager measurement method standard deviation, $\sigma_{meas} = 0.27$ dB, includes transmitter-generator random and repeatability errors as well as the "20 call" measurement method random errors.

Systematic errors include the unknown actual gain of the dipole and horn gain standards. They can be estimated as ±0.4 dB in absolute gain and ±0.5 dB in mismatch and mutual coupling errors. In addition, the generator-network analyzer systematic uncertainty is ±0.5 dB.

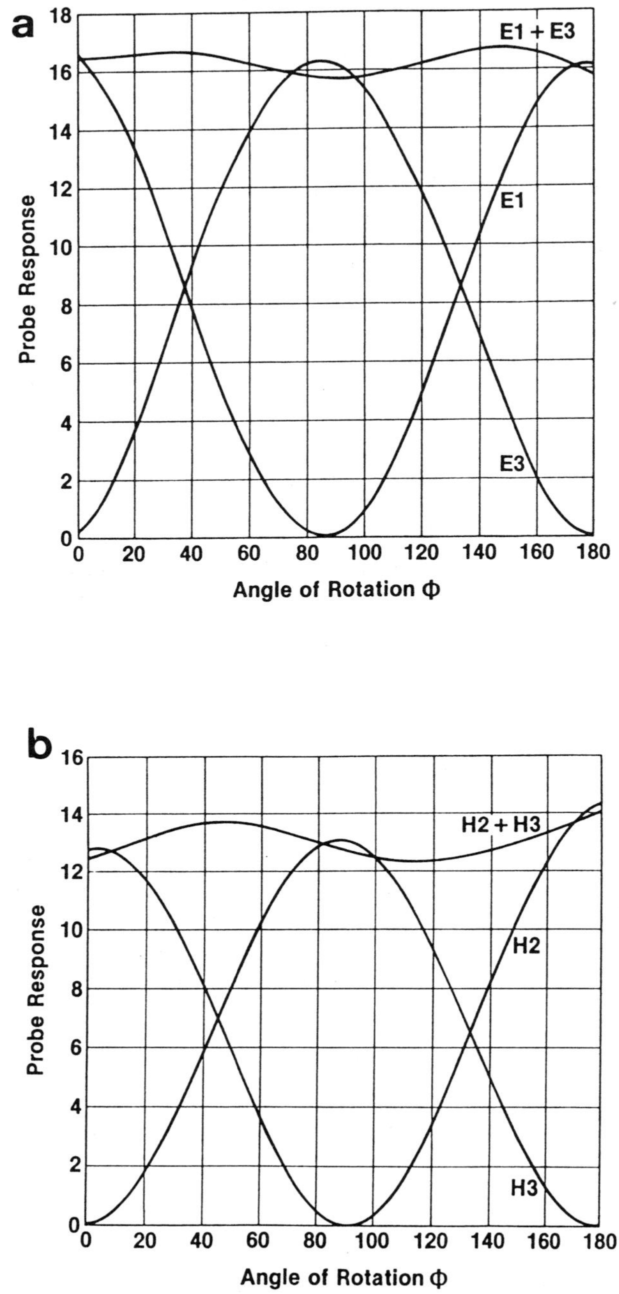

FIG. 3.—Probe response pattern of E/H probe: (a) E part, (b) H part.

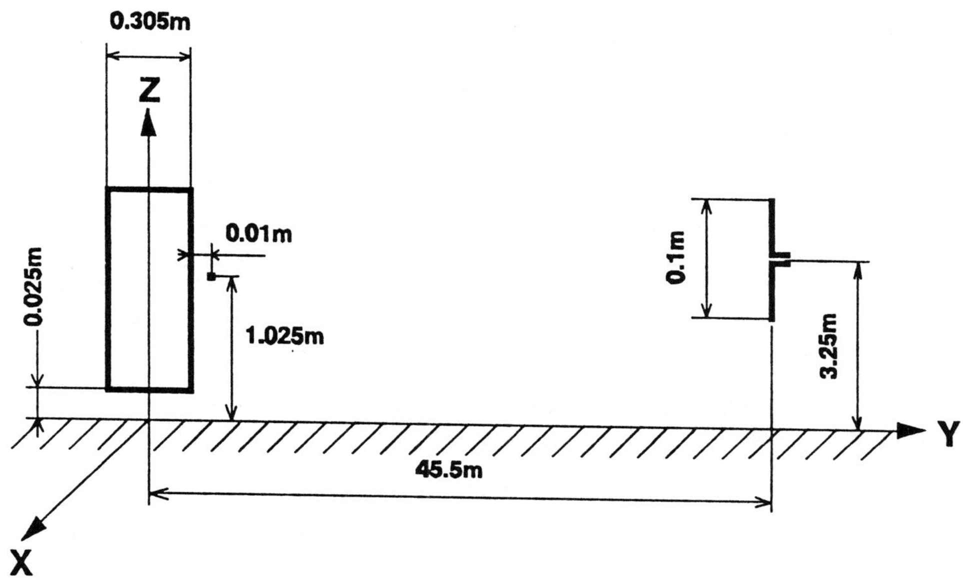

FIG. 4.—Propagation configuration above real ground ($\varepsilon_r = 12.0, \sigma = 0.002$).

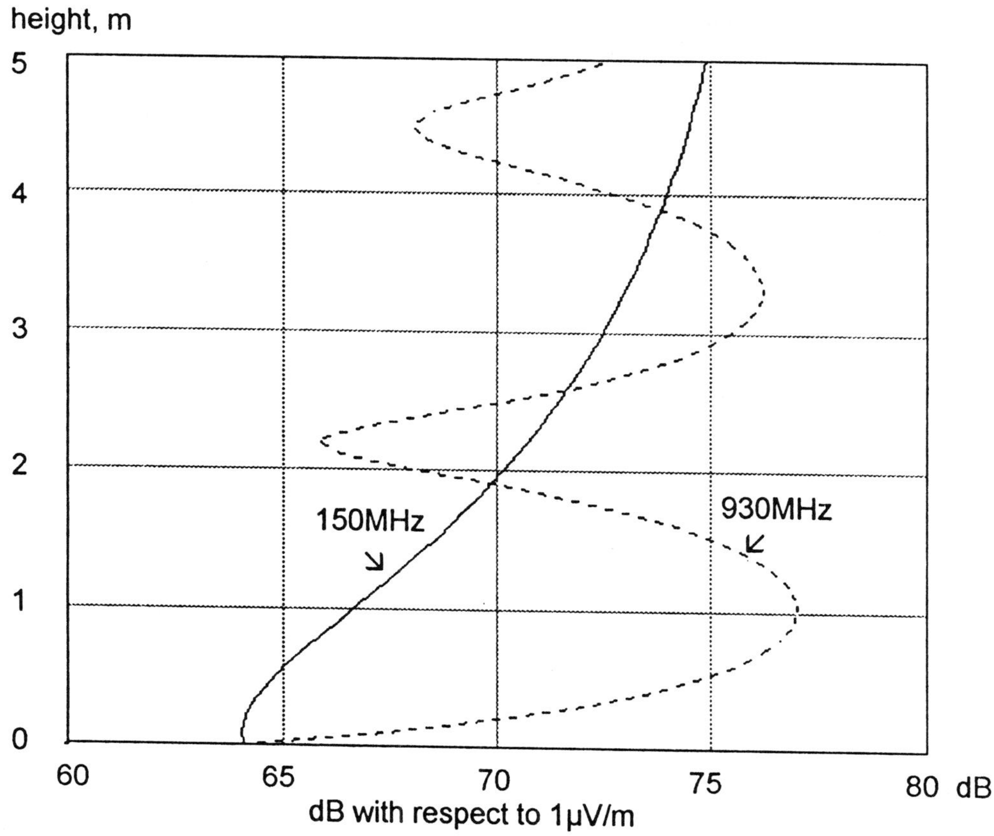

FIG. 5.—Electric-field strength vs height. Field strength is calibrated at the measurement height with standard-gain dipoles; transmitter-generator level is correlated with field strength. $\sigma_{calibrate} = 0.36$ dB due to geometric repeatability; $\sigma_{gain\text{-}std} = 0.6$ dB due to absolute gain reference uncertainty.

If all the errors are treated as statistically independent, and the error estimates are treated like standard deviations, the net standard deviation due to random errors is of the order

$$\sigma_{\text{random}} = \sqrt{\sigma_{\text{cal}}^2 + \sigma_{\text{mean}}^2 + 0.1^2} = 0.46 \text{ dB}$$

and the systematic errors total up to

$$\sigma_{\text{system}} = \sqrt{0.4^2 + 0.5^2 + 0.5^2} = 0.81 \text{ dB}$$

The systematic error represented by the systematic standard deviation will bias the sensitivity by a *fixed but unknown* amount with a standard deviation of 0.5 dB about that unknown figure. Put in another way, we can correlate to another independent sensitivity measurement (another range of similar design, a fixture, or a factory test set) to within the random spread, but we do not know the absolute figure because of the systematic errors. Any particular measurement will have an absolute accuracy to within the combined systematic and random standard deviations of about 1 dB if the measurements are carefully performed and the open-field site is carefully calibrated. Such accuracy is characteristic in electromagnetics measurements involving open-field sites.

REFERENCES

1. *Paging Systems: European Radio Message System (ERMES), Part 5: Receiver Conformance Specification*, prETS 300 133-5, 22 October 1991, ETSI, Valbonne, France.
2. *Methods of Measurement for Radio Equipment Used in the Mobile Services*, IEC 489-6, 1987, 2nd ed., Sec. 8: Reference sensitivity (selective calling).
3. T. M. Babij and H. I. Bassen, "Broadband isotropic probe system for simultaneous measurements of complex E- and H-fields," U.S. Patent 4,588,993, 13 May 1986.
4. H. I. Bassen and T. M. Babij, "Experimental techniques and instrumentation," in O. P. Gandhi, Ed., *Biological Effects and Medical Applications of Electromagnetic Fields*, New York: Prentice-Hall, 1990, 141-173.
5. A. A. Smith, "Standard-site method for determining antenna factors," *Trans. IEEE* EMC-24: 316-322, 1982.
6. K. Siwiak, T. M. Babij, and D. Yip, *Open Field Antenna Range Analysis of Measurement Errors*, Conference Record, Southcon/92, Orlando, Fla., 1992, 227-230.
7. K. Siwiak, T. M. Babij, K. Zeng, and D. Yip, "Computer simulations and measurements of electromagnetic fields close to human phantoms," this volume.

Martin Blank, Ed., *Electricity and Magnetism in Biology and Medicine*

GAP EFFECTS BETWEEN ANTENNA ELEMENTS ON THE PERFORMANCE OF AN E-H PROBE

P. Sullivan* and T. M. Babij

Florida International University, Miami

The gap effects between antenna elements of an E-H probe were evaluated by the moment-method model. The probe was designed to be electrically small and isotropic to conform to the standards for quantifying hazardous E-H fields.[1] The physical model of the probe consists of a cube structure with three dipole-loop antennas located on three of the cube faces. The length of the cube side is 50 mm, the loop diameter is 48 mm, and the dipole length is 45 mm. Orthogonal orientation of similar antenna elements (dipole or loop) minimize arbitrary polarization effects and mutual coupling. The input impedance resulted in an overall minimal mutual coupling between the dipole(active)-loop(passive) and loop(active)-dipole(passive) cases. The dipole(active)-loop(passive) case demonstrated coupling in the resistance Re(Z) when compared to the dipole self-resistance, yet in the dipole-loop input reactance Im(Z) coupling was minimal (<< 1%). Since the dipole-loop input-impedance magnitude is dominated by the reactance term, capacitive coupling is minimal.

Accurate bioelectromagnetic hazard assessment of near-zone electromagnetic fields requires the use of an electrically small, isotropic, broadband, square-law responsive E-H probe. The unknown relationship between the E and H field magnitudes complicates the determination of near electromagnetic field strength due to reactive near-field components, source-subject coupling, multipath components, and arbitrary polarization.[1] Since the E and H near field magnitudes can vary independently, they must be determined separately in the same volume at the same time to minimize measurement errors. One solution to the complex E and H field magnitudes is the use of a probe constructed to measure the E and H field components separately and within the same volume. The E-H probe consists of three orthogonal dipoles and three orthogonal loops (Fig. 1) and measures the E and H fields simultaneously and approximately within the same spatial volume.[2] The "electrically small" criterion (overall dimension $D < 0.1$ to 0.125λ, where λ is the wavelength) defines the boundary for which the probe can accurately measure the E and H fields within the same spatial volume. Arbitrary polarization effects of the E and H field components are minimized by the orthogonal orientation of the dipoles and loops with respect to each other. The objective is to measure coupling between antenna elements by use of the antenna input impedance. Single-dipole and loop-input impedances provide standard values for comparison with the various combinations of the E-H probe: dipole-dipole, loop-loop, and dipole-loop pairs. The gap between the loop elements and the gap between the dipole and loop in the dipole-loop pairs represent the location with the greatest possibility for mutual interaction due to proximity in the E-H probe construction. Reradiation of the passive conductive elements is considered to be a function of the overall impedance.

METHODS

The moment moment algorithm, as implemented in the Numerical Electromagnetic Code (NEC),[3] was used to calculate the input impedances of the E-H probe. Physical dimensions were selected to be electrically small, and to yield a broadband response for a frequency range of 10 MHz to 1 GHz. Selection of the resonance at 3 GHz, where $\lambda = 100$ mm, results in a $\lambda/2$ dipole of 50 mm; therefore a $50 \times 50 \times 50$mm cube was chosen to provide a working dimension. The probe is electrically small for $\lambda \geq 400$ to 500mm (below 600 to 750 MHz) and thus limits the broadband response to a frequency range of 10-600 MHz. The NEC moment method limits the evaluation of the low-frequency boundary to approximately 90 MHz, since the segment/wavelength ratio must be greater than 0.001.

---

*Correspondence to: P. Sullivan (Electrical & Computer Engg), Florida International University, Miami FL 33199.

The cube dimensions of $50 \times 50 \times 50$ mm constrain the loop diameter $2b$ to less than 50 mm and the dipole length $2h$ to less than the loop diameter. In order to compare the loop and dipole probe element self-impedances with published data,[4,5] the dimensional constant $\Omega = 10$, described below, was used to constrain the antenna element dimensions further. For the loop,

$$\Omega = 2 \ln (2\pi b / a) \tag{1}$$

where $2b$ is the loop diameter, and a is the wire radius. For the dipole.

$$\Omega = 2 \ln (2\pi h / a) \tag{2}$$

where $2h$ is the dipole length and a is the dipole wire radius. Choice of a wire radius of 1 mm results in a loop radius $b = 48$ mm. Dipole length $2h = 45$ mm (chosen to be smaller than the loop diameter) and $\Omega = 10$ yields a wire radius $a = 48$ mm. In addition, $\Omega \geq 9.5$ is the limit between thick- and thin-wire antenna approximations. Thick antennas are undesirable because they are subject to unpredictable and complex current modes. Input impedance was determined as the dipole length varied from 30 to 35, 40, and 45 mm. A constant dipole radius of 0.3 mm results in $\Omega = 9.11, 9.42, 9.68, 9.92(10)$, so that the 30 and 35mm dipoles can be considered thick approximations. A second series of input impedance calculations was analyzed with $\Omega = 10$ held constant, by variation in the dipole radius as the dipole length was changed. The differences were found to be negligible and are not shown.

The input impedances are calculated for the following configurations, with the notations as follows (Fig. 1): D and L represent the (D)ipole or (L)oop, respectively; the associated number 1, 2, or 3 indicates the number of the element; and the other of the letters indicates the active (first) vs passive elements. For example, D2L2 is a 1-volt applied E-field source excitation of the active element D1, with D2, L1, and L2 as the remaining passive conductive elements, when the input terminals are loaded with a 1MΩ resistive load. The loop model consisted of 16 segments and was loaded on the tenth segment or the most distal point between the loop elements. The dipole consisted of 7 segments, loaded on the fourth or central segment. The passive elements are open circuited, as in the mutual-impedance analysis of a two-port network for coupled circuits. The following list summarizes the configurations tested: D1, D2, D3, L1, L2, L3, L1D1, L2D2, L3D3, D1L1, D2L2, D3L3. The analysis of the dipole and loop elements D1 and L1 determine the self-impedance as a function of frequency over a range of 90-810 MHz. The theoretical dipole self-impedance[4] is $Z_0 = R_0 + jX_0$, where R_0 and X_0 are the self-resistance and self-reactance, respectively:

$$R_0 = 18.3\,\beta_0^2 h^2 (1 + 0.086\beta_0^2 h^2) \tag{3}$$

$$X_0 = -(396.0/\beta_0 h)(1 - 0.383\beta_0^2 h^2) \tag{4}$$

where $\beta_0 = k_0 = 2\pi/\lambda$, and the dipole length is $2h$. The dipole and loop self-impedance were validated by published results;[5,6] the dipole resistance and reactance validation are shown in Figs. 2 and 3. Published data are available only for the range 1.06-212 GHz; the above theoretical formula was used to validate the lower-frequency range from 10 MHz-1 GHz. The loop variation is shown in Figs. 4 and 5 with the loop-loop results. Validation agreement was well within 1%, for both the dipole and the loop.

RESULTS

The dipole-dipole (D2, D3) interaction was confirmed to be minimal (<<1%), which can be attributed to the wide separation (>25 mm) and the orthogonal orientation. The dipole-dipole results are not shown for D2 and D3; however, they are identical to D1 in Figs. 2 and 3. The dipole input impedance family of curves for lengths of 30-45 mm (Ω varies over 9.11-9.92) are shown in Figs. 6 and 7 with the dipole(active)-loop(passive) results. Dipole-loop results were unchanged when the antenna parameter Ω was held constant at 10 by variations in the dipole radius to accommodate the varying dipole lengths. In Figs. 6 and 7, the family of curves results from the varying effective antenna height. The loop-loop interaction (L1 vs L2, L3, Figs. 4 and 5) was less than 1% for two loops from 90 to 810 MHz. Addition of the third loop resulted in significant peaking (> 5%) approaching the first resonance (690-810 MHz). The increased coupling when the third loop is added is due primarily to the minimal gap of 1.4 mm between the loops at the proximal point, and can be optimized by reductions in the loop diameters (to be investigated as the next step).

Loop(active)-dipole(passive) mutual impedances L1 to LD1, L2 to LD2, and L3 to LD3 (Figs. 8 and 9) demonstrate the minimal (less than 1%) coupling that occurs between the excited loop and the passive dipole while the dipole length is varied from 30 to 45 mm. As the first resonance is approached, the 45mm dipole

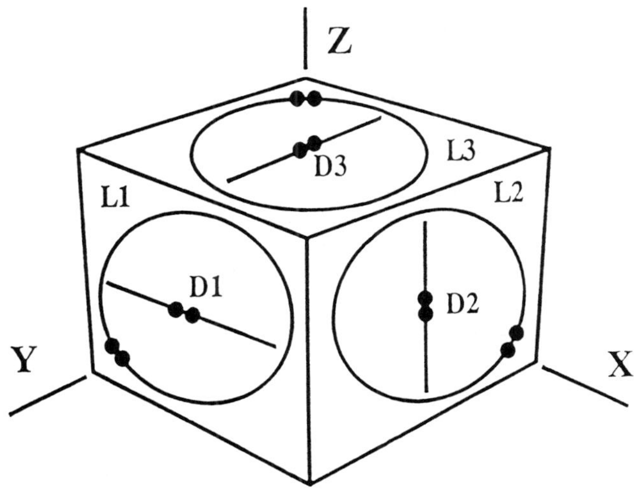

FIG. 1.—E-H probe.

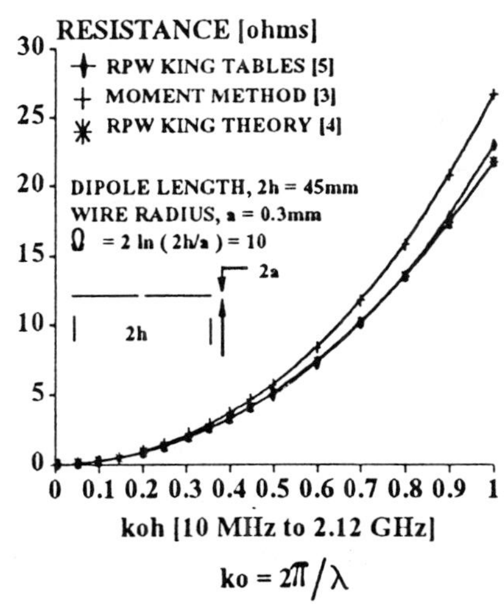

FIG. 2.—Dipole resistance validation.

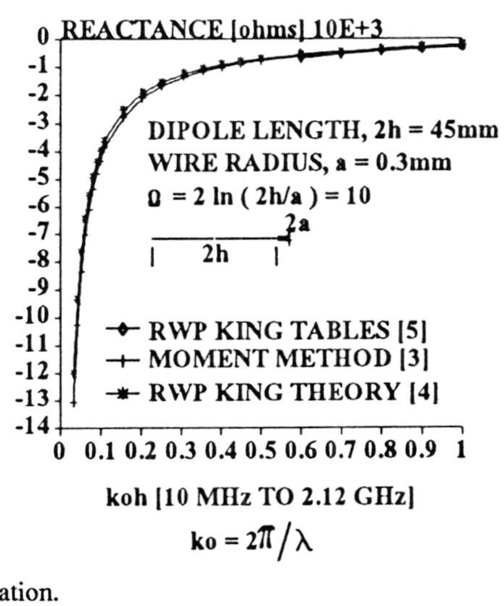

FIG. 3.—Dipole reactance validation.

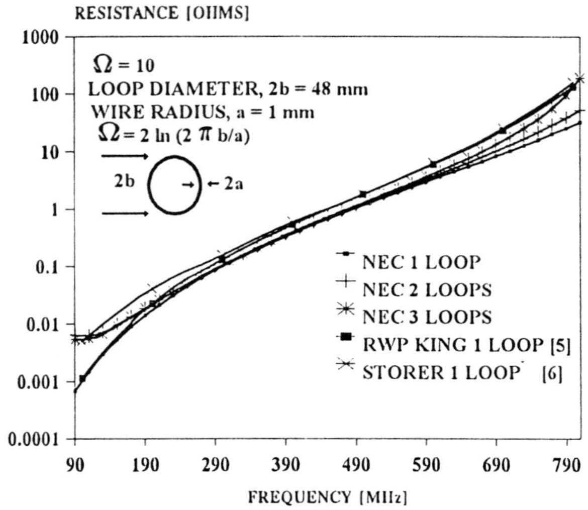

FIG. 4.—Loop resistance.

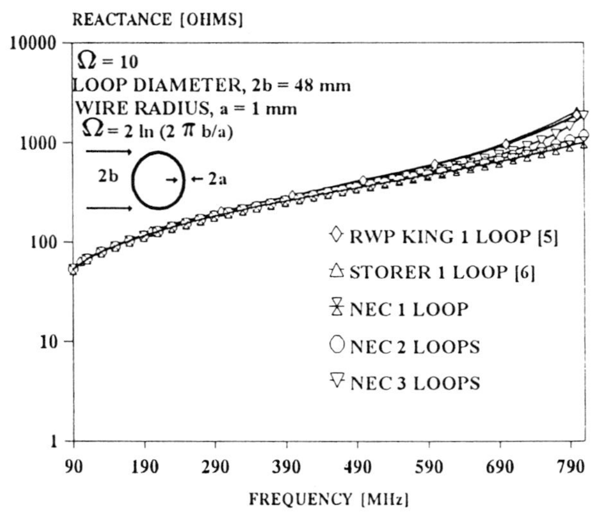

FIG. 5.—Loop reactance.

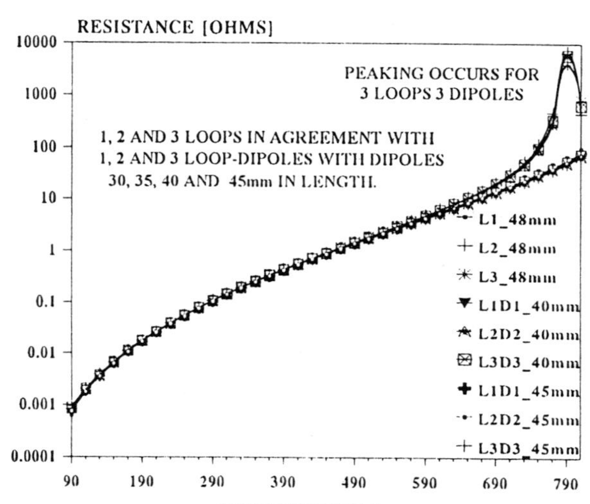

FIG. 6.—Dipole(excited)-loop resistance.

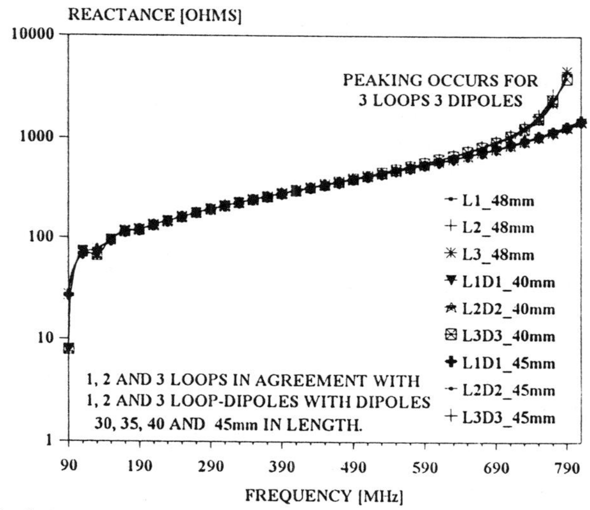

FIG. 7.—Dipole(excited)-loop reactance.

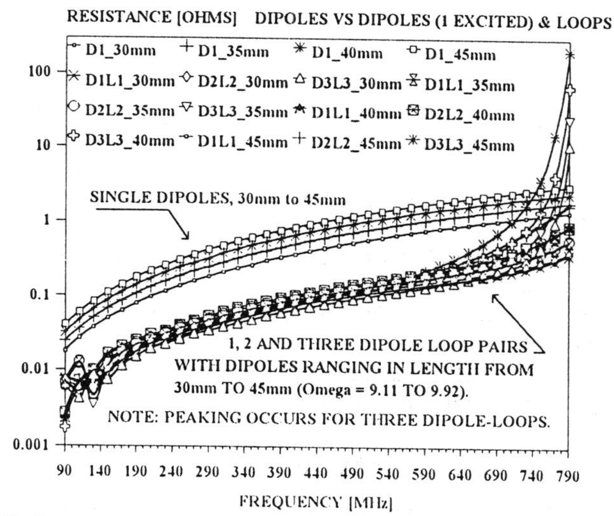

FIG. 8.—Loop(excited)-dipole resistance.

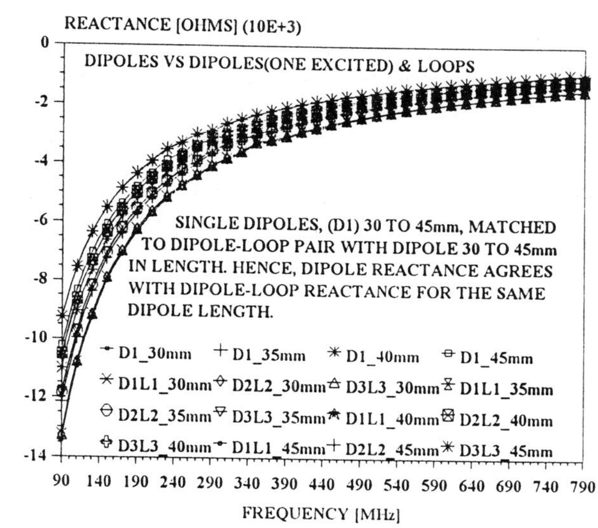

FIG. 9.—Loop-dipole reactance.

demonstrates stronger coupling; however, it remains below 5% (not shown). The optimal dipole length falls between 40 and 45 mm, but that has not yet been studied. As the dipole(passive) length is reduced from 40 to 30 mm, the active loop matches the loop self-impedance, so that coupling is minimal. Considering the minor coupling that occurs at 45 mm, the optimal dipole length is closer to 45 mm than to 40 mm. The addition of the third loop-dipole pair has the same peak as the third loop between loops only as the first resonance is approached. Therefore, the dipole has a minimal effect on the overall loop impedance, considering the proximity and the potential for re-radiated near fields.

Dipole(active)-loop(passive) (Figs. 6 and 7) has a complex interaction that has no obvious explanation. There are several possible explanations for the apparently significant coupling seen in the dipole-loop resistance (Fig. 6). One is that the order-of-magnitude difference (0.01 vs 0.1 ohms) is not really significant if we consider that the overall magnitude of the impedance is dominated by the magnitude of the reactance. The passive loop is creating an end-cap capacitance (although that seems unlikely, since it should affect the dipole input reactance more than the input resistance). Another consideration is that the passive loop circular-to-dipole mode ratio has changed.[4] In addition, the peaking that occurs as the first resonance is approached must be a byproduct of the passive loop, since the D3 testing does not result in peaking and the input resistance curves near the first resonance are remarkably similar to the L3 and LD3 curves (Figs. 4 and 8). Dipole-loop gaps ranging from 9 mm (30 mm) to 1.5 mm (45 mm) in the D3L3 configuration clearly demonstrate coupling as the first resonance is approached, with the overall increase in input impedance as the dipole length is increased.

CONCLUSIONS

The coupling between antenna elements in the E-H probe described above was shown to be minimal for all configurations tested, with the noted exception of the excited dipole and passive loops and passive dipoles. Reduction of the overall dipole length $2h$ from 45 to 30 mm to reduce coupling demonstrated a significant correlation between the dipole(active) and loop(passive) gap distance and mutual impedance as the first resonance was approached (690-810 MHz). Loop-loop coupling was also significant as the first resonance for three loops is approached. The moment method modeling of the input impedances of a complex geometrical E-H probe demonstrates that the performance can be optimized with regard to interelement coupling. The data presented have therefore shown that the loop diameters should be reduced, and have demonstrated the general effects that a passive loop has on the performance of the dipole. Further research will involve the investigation of the nature of the effect a passive loop has on the dipole input resistance, as well as the optimization of the loop-loop gap and the trade-off of useful bandwidth.

REFERENCES

1. P. F. Wacker and R. R. Bowman, "Quantifying hazardous electromagnetic fields: Scientific basis and practical considerations," *IEEE Trans.* MTT-19: 178-187, 1971.
2. T. M. Babij and H. Bassen, *Broadband Isotropic Probe System for Simultaneous Measurement of Complex E- and H-Fields,* U.S. Patent 4588993, 1986.
3. G. J. Burke and A. J. Poggio, *Numerical Electromagnetic Code (NEC)-Method of Moments,* Tech. Doc. 116, San Diego: Naval Ocean Systems Center, 1977.
4. R. W. P. King, R. B. Mack, and S. S. Sandler, *Arrays of Cylindrical Dipoles,* London: Cambridge University Press, 1968.
5. R. W. P. King, *Tables of Antenna Characteristics,* New York: IFI/Plenum, 1971.
6. J. E. Storer, "Impedance of thin wire loop antennas," *Trans. AIEE* 75(Part I): 606-619, 1956.

Martin Blank, Ed., *Electricity and Magnetism in Biology and Medicine*

OVERCONSTRAINED MOMENT-METHOD (OMoM) NUMERICAL SOLUTIONS FOR ELECTROMAGNETIC FIELD PREDICTION IN BIOLOGICAL SYSTEMS

S. Caorsi,* G. L. Gragnani, and M. Pastorino

University of Genoa, Italy

Prediction of electromagnetic-field distribution inside a complex system such as the human body is a crucial step in many medical applications (e.g., SAR-deposition control during microwave hyperthermia treatment), in evaluations of biohazards resulting from unwanted nonionizing radiation, and in other therapeutical applications of the electromagnetic field (e.g., treatment of chronic lymphedemas of extremities).

Several numerical approaches based on the method of moments (MoM) have been developed by the authors to evaluate electromagnetic absorption in simulation models of biological systems. Such approaches aimed at extending the application field of the MoM and to overcome the intrinsic limitations of the integral-equation formulation. In particular, a focusing moment method using a decomposition domain[1] was proposed, which yields an accurate SAR only in a given subdomain (obtained by a fine discretization process) without increasing the overall computational load. Recently, the authors have presented an overconstrained version of the MoM (OMoM) that enables one to consider a reduced number of discretization cells to predict a sufficiently accurate SAR distribution.[2,3] The approach is developed by use of a redundant number of testing functions (i.e., the number of testing functions is larger than that of basis functions), and a mean-square solution of the resulting rectangular system is reached by a generalization inversion algorithm. Although the method has been shown to allow a significant improvement in the SAR calculation (in terms of the cost-to-benefit ratio), a suitable choice of the redundant testing functions remains a crucial point. In the present paper, a discussion of this topic is presented. Moreover, comparisons between a classic MoM solutions and OMoM solutions are made and discussed for a test object previously considered by other authors in the literature as significant scatterer for the assessment and validation of block models.

METHOD

Let us consider the well-known electric field integral equation for the scattering by a dielectric body illuminated by a known incident field:[4]

$$\vec{E}_{tot}(\vec{r}) = \vec{E}_{inc}(\vec{r}) + j\omega\mu \int_{Body} \tau(\vec{r}\,')\vec{E}_{tot}(\vec{r}\,')\cdot\overleftrightarrow{G}(\vec{r}/\vec{r}\,')\,d\vec{r}\,' \qquad (1)$$

where τ is the complex conductivity of the body and G is the Green dyadic function for free space. It has been shown[4] that the use of this equation in the region occupied by the body introduces a singularity into the integral; one can overcome such singularity by computing the integral principal value PV.[5] According to Van Bladel's approach, one has

$$\int_{Body} \tau(\vec{r}\,')\vec{E}_{tot}(\vec{r}\,')\cdot\overleftrightarrow{G}(\vec{r}/\vec{r}\,')\,d\vec{r}\,' = PV \int_{Body} \tau(\vec{r}\,')\vec{E}_{tot}(\vec{r}\,')\cdot\overleftrightarrow{G}(\vec{r}/\vec{r}\,')\,d\vec{r}\,' - \frac{\tau(\vec{r})\vec{E}_{tot}(\vec{r})}{3k_0^2} \qquad (2)$$

Equation (1) has been reduced to an algebraic one by applying the OMoM with N hat basis functions (hence N cells B_n, $n = 1, \cdots, N$) and $M > N$ pulse testing functions, at the positions $\vec{r}_m$, $m = 1, \cdots, M$, which yields the following system of M algebraic equations with N vector unknowns:

$$\vec{E}_{tot}(\vec{r}_m) = \vec{E}_{inc}(\vec{r}_m) + j\omega\mu \sum_{n=1}^{N} \left[PV \int_{B_n} \tau(\vec{r}_n{}')\vec{E}_{tot}(\vec{r}\,')\cdot\overleftrightarrow{G}(\vec{r}_m/\vec{r}_n)\,d\vec{r}_n{}' - \frac{\tau(\vec{r}_m)\vec{E}_{tot}(\vec{r}_m)}{3k_0^2}\,\delta_{nm} \right] \qquad (3)$$

*Correspondence to: S. Caorsi (Biophys. & Electronic Engg.), University of Genoa, Via Opera Pia 11a, I-16145 Genoa, Italy.

RESULTS

The evaluation of the PV for the usual MoM with cubical cells is quite straightforward, if one applies the simple procedure suggested in Ref. 4. But great care must be exercised in using this procedure for the OMoM. In the first applications of the OMoM,[2] the *PV*s were evaluated by simple integration over a sphere centered at the testing point $\vec{r}_m$. According to the traditional MoM, a testing point is always at the center of a cell. That does not hold true for the OMoM, and the approximation of integrating over a sphere with the same volume as that of a cell and centered at a testing point may yield results that are too poor when testing points far from the cell center are used. This outcome was verified by reconsideration of the cubical scatterer previously examined in Refs. 2 and 3. The scatterer was subdivided into 27 cells, and, according to the p-th rule,[2,3] 24 different positions of the testing points were considered. Furthermore, for the evaluation of the integrals involved and in particular, for the evaluation of the *PV* contribute, each cell was further partitioned into S subcells, and $3\times3\times3$, $5\times5\times5$, and $7\times7\times7$ subcells were considered. We evaluated each *PV* by performing a usual numerical integration over $S = 1$ subcells, and by integrating over a sphere with the same volume as that of a subcell and located in the subcell containing the testing point. By doing so, we obtained a far better evaluation of the *PV* in each case in which the testing point was close to the center of a subcell, whereas results that were too rough were obtained when the testing points were far from the center of the subcells. As a consequence, an undulatory behavior of results is expected. The best results are obtained with configurations where the testing points coincide with the centers of subcells; and vice versa, the worst values are associated with configuration where the testing points are close to the boundary between two adjacent subcells. Moreover, as the behavior of the electric field is characterized by a notable gradient near the edges, accurate SAR values are obtained when one takes such variations in the electric field into account when positioning the testing points.

The results are given in Figs. 1 and 2. Figure 1 shows the behavior of the percent error in SAR values when each cell is partitioned into $3\times3\times3$ subcells. In accordance with the cell subdiscretization, we expect that there exist two possible p-configurations (including the one corresponding to the classical MoM) giving accurate *PV*s, and hence correct SAR values. In addition, there will be only one boundary (between two subcells) near which the testing points may be located. In configurations 18 and 19, the testing points almost coincide with the center of the subcells; for these configurations, minimum errors are made in the SAR evaluation, with respect to the convergence value reported in Ref. 6. Vice versa, for configurations 8 and 9, with testing points close to the boundary between two subcells, the errors on the *PV* estimation cause the SAR values to become too high (Fig. 2). The fact that such results are to be ascribed to the different approximations related to the *PV* computation is clearly pointed out in Fig. 2, which gives the results obtained with $3\times3\times3$, $5\times5\times5$, and $7\times7\times7$ subcells. This figure shows the undulatory behaviors of the values obtained for the various configurations. In the cases $5\times5\times5$ subcells and $7\times7\times7$ subcells, there are three and four possible configurations, respectively, for which one can obtain correct *PV*s. As a consequence, correct SAR computation can be performed whenever the testing points correspond to the configurations 10 and 20 in the case of $5\times5\times5$ and to the configurations 7, 14, and 21 in the case of $7\times7\times7$. By linking such points, one can plot the correct SAR behavior vs the positions of the testing points. As established in previous contributions, the testing points positioned near the edges (characterized by greater variations in the electric field) allow one to predict SAR values close to the convergence ones, yet by using very small computational resources.

CONCLUSIONS

In this paper, an overconstrained approach to the solution of dosimetry problems inside biological bodies has been reviewed, and the effect of the error made in computing the principal values of the integral for the model has been pointed out. It has been shown that such an error is responsible for the undulatory behavior (reported in previous publications) of the solutions when the positions of the testing points are varied from the center to an edge of a cell. By contrast, it has been proved that a correct evaluation of the principal values of the integrals can definitively improve the accuracy of SAR values.

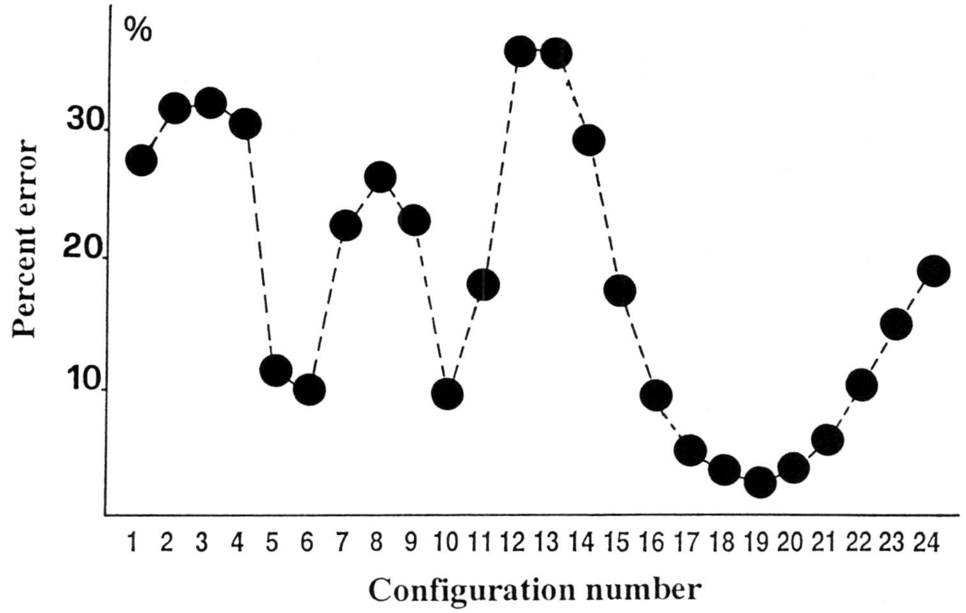

FIG. 1.—Percent error in SAR for 24 different positions of the testing points, using $3 \times 3 \times 3$ subpartitions to compute the integrals. The best approximation for the *PV*s is given by configuration No. x, while the worst approximation is given by configuration No. y. Continuous lines indicate convergence value.

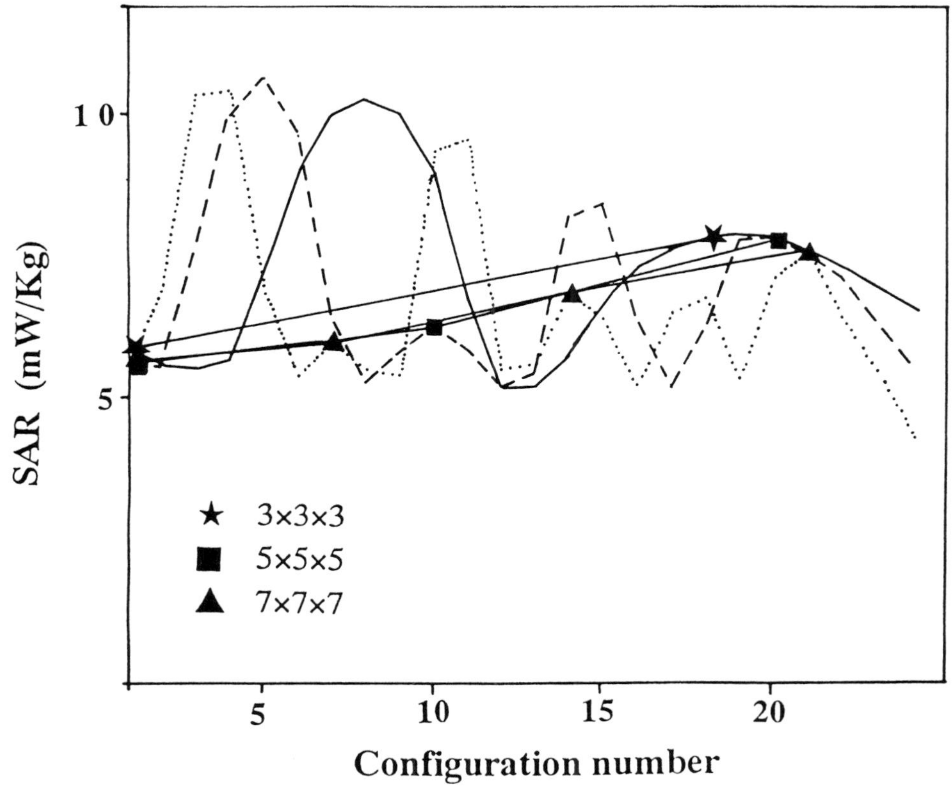

FIG. 2.—Two SAR behaviors for $3 \times 3 \times 3$ (continuous line), $5 \times 5 \times 5$ (dashed line), and $7 \times 7 \times 7$ (dotted line) subpartitions. Points at which the *PV*s are estimated to a sufficient accuracy are linked by a continuous line. Convergence value = 8.184 mW/kg.

REFERENCES

1. S. Caorsi, G. L. Gragnani, and M. Pastorino, ''A numerical approach to electromagnetic dosimetry for the human body,'' *European Transactions on Telecommunications and Related Technologies* 1: 497-503, 1990.

2. S. Caorsi, G. L. Gragnani, and M. Pastorino, ''Overconstrained version of moment method for prediction of electromagnetic energy deposition inside dielectric objects,'' *Electronics Letters* 26: 1997-1998, 1990.

3. S. Caorsi, G. L. Gragnani, and M. Pastorino, ''Use of redundant testing functions in moment-method solutions for block models'' (accepted for publication in *IEEE Trans. MTT*).

4. D. E. Livesay and K.-M. Chen, ''Electromagnetic field induced inside arbitrarily shaped biological bodies,'' *IEEE Trans.* MTT-22: 1273-1280, 1974.

5. J. Van Bladel, ''Some remarks on Green's dyadic for infinite space,'' *IEEE Trans.* AP-9: 563-566, 1961.

6. M. J. Hagmann and R. L. Levin, ''Convergence of local and average values in three-dimensional moment-method solutions,'' *IEEE Trans.* MTT-33: 649-654, 1985.

Martin Blank, Ed., *Electricity and Magnetism in Biology and Medicine*

STIMULATION OF NEURONAL DIFFERENTIATION PROTEINS IN PC12 CELLS BY COMBINED AC/DC MAGNETIC FIELDS

P. Horton

Arizona State University

J. T. Ryaby,* F. P. Magee, and A. M. Weinstein

OrthoLogic Corp., Phoenix, Ariz.

The disability associated with acute nerve trauma provides an opportunity for development of biophysical technologies to be used either as stand-alone or as adjunctive treatment devices with standard surgical techniques or medication. Spinal-cord injuries (SPI) number 10000-20000 new injuries per year in the USA, not changing over the last two decades.[1,2] Advances in medical technology have increased the life span of SPI victims, but there is still a need for research that will lead to better methods of producing increased recovery of function. Peripheral-nerve injuries account for more than 200000 patients who could benefit by improvement in treatment. Because the nervous system is the sole mechanism for receiving and transmitting signals of sensory and motor control, it is critical that a functional recovery be successful. In the case of nerve injury, the number of nerve cells does not increase, but the repair/survival of each cell takes place in an environment of proliferation of other tissue structures (fibroblasts, endothelial cells, Schwann cells).[3] Along with survival, reconnection of amputated nerve cells is of major importance in attainment of functional recovery. Successful reconnection depends on the rate of cell repair and orientation. At the cellular level, this process depends on messengers that serve as "switches" to initiate repair and regeneration. Electromagnetic fields have been demonstrated to couple to calcium-dependent cellular processes and may act as first-messenger stimuli to initiate the cellular differentiation of growth processes.[4]

The objective of this research was to determine the effect of low-frequency combined AC/DC magnetic fields on stimulation and/or induction of neuronal differentiation. Rat pheochromocytoma (PC12) cells were used as the model system.[5,6] PC12 cells respond to nerve growth factor (NGF) by slowly shifting their phenotype from an immature adrenal chromaffin-like cell to a sympathetic neuron-like cell. In the presence of NGF, they withdraw from the cell cycle, extend long neurites, and acquire the general characteristics of sympathetic neurons.[7] To determine whether exposure to combined AC/DC electromagnetic fields would stimulate the neuronal differentiation process, we attempted to make a quantitative assessment of neuronal protein synthesis after electromagnetic exposure and to correlate these changes with neurite outgrowth. In this study, we utilize a combined AC/DC magnetic field which is based on the ion resonance model[8,9] to stimulate neuronal differentiation.

EXPERIMENTAL DESIGN AND METHODS

Cell Culture

The model system employed was a clonal line of rat pheochromocytoma cells (PC12, ATCC) which was maintained in a medium consisting of 85% RPMI 1640, 10% heat-inactivated horse serum, and 5% fetal calf serum in T-75 flasks at 37 C in 6% CO_2 saturated humidity. The cells were fed three times a week and were passaged weekly by repeated forceful aspiration of medium onto the dish at split ratios of 1:3. For experiments, cells were plated in multiwell plates treated with polylysine or polyornithine[6] at a cell density of approximately 10000 cells per well in 10% serum for a minimum of 24 h and transferred to 1% serum 24 h prior to each experiment.

*Correspondence to: J. T. Ryaby, OrthoLogic Corp., 2850 S 36th St., Phoenix AZ 85034.

Magnetic-field Exposure Apparatus and Technique

The multiwell plates were placed in the center of a round (30cm diameter) Helmholtz coil pair made from 400 turns of #28 AWG magnet wire. The coils were oriented horizontally (i.e., coplanar to the tissue culture plate) to produce a vertically oriented magnetic field. The coils and tissue culture plates were supported by a nonconducting, nonmagnetic polymeric frame. The coils were energized by time-varying currents generated by a function generator (BK Precision model 3011A). The amplitude and DC offset controls of this function generator were used to control the amplitude (0.400 G peak-to-peak) of the time-varying magnetic field and the amplitude (0.200 G) of the static magnetic field, respectively. Static and dynamic magnetic fields were created by the same pair of Helmholtz coils and were therefore parallel. Magnetic field measurements were made with a Schonstedt Model 2220 digital magnetometer.

Experimental Design

There were four groups in each experiment: a control group, a positive control group [nerve growth factor (NGF) 50 ng/ml], a group exposed to continuous AC/DC combined magnetic fields, and a group exposed to AC/DC combined magnetic fields for 30 min per 24 h. The AC/DC-stimulated groups were used to test the effectiveness of the Liboff theory.[8] The time points analyzed were 24 and 72 h. These values were chosen as time points corresponding to a 25% and 50% neurite outgrowth index, respectively, for nerve growth factor.[5,6]

Enzyme-linked Immunosorbent Assay (ELISA)

A cellular ELISA was used for detection of antibodies reacting with cellular antigens.[10] Cells were fixed with 4% buffered paraformaldehyde, washed with BSA/PBS, and incubated with monoclonal antibodies specific for tubulin, MAP1, or NCAM for 2 h at 37 C. Cells were washed 3 times in PBS; then HRP conjugated secondary antibody was added and incubated for 1 h at 37 C. Cells were washed 3 times with PBS, then incubated with O-phenylene diamine HCl for 30 min at 37 C. The reaction was stopped with 3N HCl and read at 450 nm. Means were compared by Student's *t*-test with the Bonferroni criteria used for multiple-group comparisons.

Analysis of Protein Expression Utilizing Immunoprecipitation

Cells were preincubated in Methionine-free RPMI 1640 media for 1 h at 37 C. Cells were then labeled with 35S Methionine (100 μCi/ml) in Methionine-free RPMI 1640 media[11] for 2 h with either NGF, AD/DC magnetic-field exposure, or as controls at 37 C. Cells were washed 2 times with ice-cold PBS, then extracted overnight in 500 μl of ice-cold RIPA buffer (at – 20 C). Lysates were incubated with the following antibodies: NCAM, tubulin, and MAP at 4 C for 60 min with gentle agitation.[12,13] Formalin-fixed Staph A cells were used to absorb bound antibody. Immunoprecipitants were washed 3 times with 500 μl of ice-cold RIPA buffer. Samples were resuspended in 100 μl Laemmli sample buffer (LSB), subjected to SDS-PAGE, and audioradiographed on Kodak XAR5 film. A 20μl sample from each group was counted by LSC.

Immunofluorescent Labeling of Cells

Photomicrography of immunofluorescently labeled cells was utilized to visualize localization of cellular architecture and neurite outgrowth.[14,15] The two-step procedure of indirect immunofluorescence involved cell fixation with 4% buffered paraformaldehyde for 15 min; then the cells were permeabilized with 0.5% tween for 10 min. The secondary antibody used was a rhodamine conjugate. Photographs were taken with a Leitz Laborlux microscope in dark field at 40 and 100×.

RESULTS

The results from the ELISA assay based on the monoclonal antibody to NCAM are shown in Fig. 1. The 30min AC/DC exposure group at 24 h was approximately 30% above the control group ($p < 0.05$) and 35% above control at 72 h ($p < 0.01$). The NGF group was approximately 110% above control at 72 h ($p < 0.01$). The continuous AC/DC exposure group was not significantly different from the control group at either time point. ELISA results with tubulin and MAP showed no difference in synthesis for all 4 groups at 24 h and at 72 h (data not shown). The results from the immunoprecipitant assay based on the monoclonal antibody to NCAM are shown in Fig. 2. This graph shows that at all time points the amount of NCAM expression in the NGF group is significantly greater than the control group; however, between each time point the amount did not vary significantly (60-70%). The 30min AC/DC exposure group showed an increase of NCAM expression of 40% at 24 h, with no effect observed at 72 h. Immunofluorescence analysis confirmed the presence of NCAM at the cell membrane, although no differences were detectable at the microscope level. The immunoprecipitant results for MAP1 and tubulin showed no change of MAP1 synthesis with NGF treatment or AC/DC 30min exposure over time (data not shown).

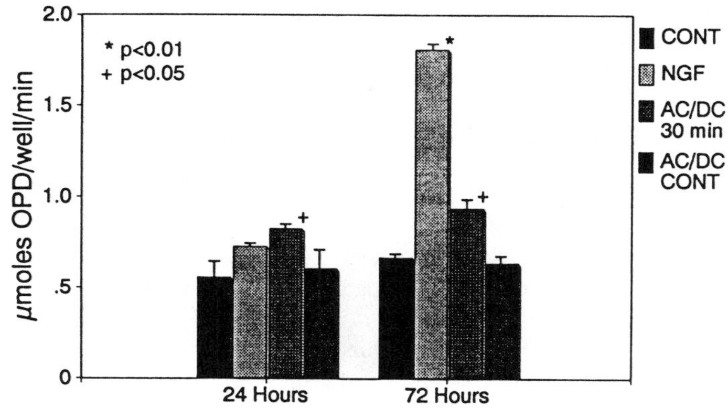

FIG. 1.—Combined effects on NCAM expression of control groups, NGF-treated groups, and groups with AC/DC combined magnetic-field exposure for 30 min as determined by immunoprecipitation as described in Methods. Data expressed as % above control. Cont = control, NGF = nerve growth factor, AC/DC 30 min = magnetic exposure.

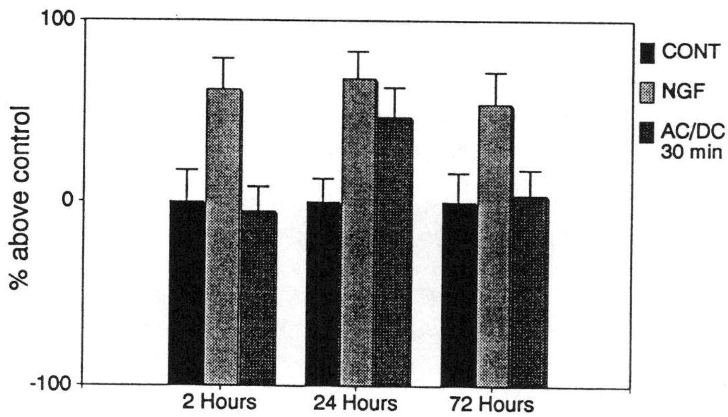

FIG. 2.—Combined effects on NCAM expression of control groups, NGF-treated groups, and AC/DC combined magnetic-field exposure groups (30 min and continuous exposure) as determined by cellular ELISA. (See Methods for description of assay.)

DISCUSSION

PC12 cells are an excellent model for studying initiation of neuronal differentiation. It has been well documented that PC12 cells will grow neurites when treated with NGF, which represents the maximum achievable differentiation rate (positive control). In the present study, the degree of expression of neuronal proteins was compared among AC/DC-exposed groups, NGF-treated groups, and control groups determined by immunoprecipitation and ELISA. Both NGF and AC/DC (30 min) increased NCAM synthesis; NGF was the more potent stimulus.

These results demonstrate that combined AC/DC magnetic fields are capable of inducing differentiation in the PC12 neuronal cell model system. Future efforts will focus on assessment of combined approaches (i.e., NGF and combined AC/DC) in order to achieve maximal induction of differentiation. These studies provide the basis for an increased effort in the development of biophysical technologies for stimulation of nerve regeneration.

REFERENCES

1. U.S. Department of Health and Human Services, Public Health Service, Center for Disease Control, Vital and Health Statistics, *Detailed Diagnoses and Procedures: National Hospital Discharge Survey,* Series 13, No. 108, 1989.

2. F. J. Seil, Ed., *Neural Regeneration and Transplantation* (vol. 6 of *Frontiers of Clinical Neuroscience,*), New York: Alan Liss, 1989.

3. G. Lundborg, *Nerve Injury and Repair,* Paris: Churchill Livingstone, 1988, ch. 5.

4. B. Sisken, "Electric and pulsed EMF effects on nerve tissue regeneration," in C. T. Brighton and S. R. Pollack, Eds., *Electromagnetics in Medicine and Biology,* San Francisco: San Francisco Press, 1991, 259-274.

5. L. A. Greene and A. S. Tischler, "PC12 pheochromocytoma cultures in neurobiological research," *Adv. Cellular Neurobiol.* 3: 373-415, 1982.

6. L. A. Greene, J. M. Aletta, A. Rukenstein, and S. H. Greene, "PC12 pheochromocytoma cells: Culture, nerve growth factor treatment, and experimental exploitation," *Methods in Enzymology* 147: 207-216, 1987.

7. L. A. Greene and A. S. Tischler, "Establishment of a noradrenergic clonal line or rat adrenal pheochromocytoma cells which respond to nerve growth factor," *Proc. Natl. Acad. Sci. USA* 73: 2424-2428, 1976.

8. B. R. McLeod and A. R. Liboff, "Cyclotron resonance in cell membranes: The theory of the mechanism," in M. Blank and E. Findel, Eds., *Mechanistic Approaches to Interaction of Electric and Electromagnetic Fields with Living Systems,* New York: Plenum Press, 1987.

9. B. R. McLeod, S. D. Smith, and K. E. Cooksey, "Ion cyclotron resonance frequencies enhance Ca^{2+}-dependent motility in diatoms," *J. Bioelectricity* 6: 1-12, 1987.

10. B. Arunachalam, G. P. Talwar, and R. Raghupathy, "A simplified cellular ELISA (CELISA) for the detection of antibodies reacting with cell-surface antigens," *J. Immun. Methods* 135: 181-189, 1990.

11. M. Nilsen-Hamilton and R. T. Hamilton, "Detection of proteins induced by growth regulators," *Methods in Enzymology* 147: 427-444, 1987.

12. A. B. Brown and G. Carpenter, "Acute regulation of the epidermal growth factor receptor in response to nerve growth factor," *J. Neurochem.* 57: 1740-1749, 1991.

13. J. M. Aletta, M. L. Shelanski, and L. A. Greene, "Phosphorylation of the peripherin 58-kDa neuronal Intermediate Filament Protein," *J. Biol. Chem.* 264: 4619-4627, 1989.

14. P. M. Conn, Ed., *Methods of Neuroscience* (vol. 2 of *Cell Culture,* San Diego: Academic Press, 1990.

15. E. Harlow and D. Lane, *Antibodies: A Laboratory Manual,* Cold Spring Harbor Laboratory, 1988.

Martin Blank, Ed., *Electricity and Magnetism in Biology and Medicine*

ELF-INDUCED PROLIFERATION AT 511 mG IN HSB-2 CELL CULTURE AS A FUNCTION OF 60Hz FIELD INTENSITY

A. R. Liboff* and K. A. Jenrow

Oakland University, Rochester, Mich.

B. R. McLeod

Montana State University

Human lymphoma cell culture exposed for 24 h to parallel AC and DC magnetic fields corresponding to two separate ion cyclotron resonance conditions for K^+ exhibited a 36% enhanced proliferation over that of controls held at zero DC field. Proliferation responses at the 30-40% level were observed over a wide range of 60Hz peak intensities, from 50 to 1500 mG. The K^+ condition of 60 Hz and 511 mG is of interest because geomagnetic intensities in the vicinity of 511 mG are common to much of the USA.

This work continues our exploration of the connection between ELF magnetic bioeffects and observations suggesting an underlying ion cyclotron-resonance signature. Even without universally accepted experimental evidence,[1,2] or more convincing theoretical arguments,[3,4] the use of ion cyclotron-resonance parameters in the design of ELF experiments makes good sense, primarily as a means of hypothesis testing, and, equally important, as an exploratory pathway in a research domain rather bare of any other roadsigns or clues.

This approach is also motivated by the fact that the charge-to-mass ratios of biologically important cations such as calcium, potassium, and magnesium in the geomagnetic field have resonance frequencies in the ELF range. Within the 250-680mG span of the geomagnetic field on the surface of the earth, one finds only two fundamental resonances at 60 Hz among the biologically interesting cations: 271 mG for Li^+ and 475 mG for Mg^{2+}. However, when one considers higher harmonics, K^+ exhibits (Fig. 1) two additional 60Hz resonance possibilities well within the geomagnetic range: 510.6 mG ($n = 3$) and 306.4 mG ($n = 5$).

The possibility of potassium resonance frequencies affecting cell response was first suggested by Liboff[5] in connection with the observation of Blackman et al.[6] that magnetostatic fields can modulate Ca efflux. Table 1 shows the frequency and field parameters that led to the results reported by Blackman et al.,[6] as well as our estimates of the resulting charge-to-mass ratios derived from these parameters under the assumption of ion cyclotron resonance. These calculated values agree rather well with the actual charge-to-mass ratio for the potassium ion, 2.465×10^6 C/kg. In addition, arguments have been raised[7] showing that the original Ca-efflux reports by Bawin et al.[8] (which involved only applied electric fields) can also be tied to K^+ resonance. Further evidence for the involvement of potassium was reported by McLeod et al.,[9] who found that diatom motility was inhibited by K^+ resonance, in contrast to the enhanced motility exhibited under Ca^{2+} resonance conditions.

It is interesting to realize that the 511mG field value mentioned above as a 60Hz resonance point for potassium is common to much of the USA. For example, the total geomagnetic field intensities in Tallahassee, Mobile, Baton Rouge, Waco, Las Vegas, and Oakland (Calif.) are all within a few milligauss of this specific potassium resonance value. Furthermore, a recent (small-sample) analysis by Bowman et al.[10] of a larger epidemiological study[11] designed to examine possible correlations between residential magnetic fields and childhood leukemia seemed to indicate an increased risk for individuals in the 60Hz environment who were also exposed to the local magnetostatic field value (506 mG), which is rather close to this K^+ 60Hz third-harmonic resonance value of 511 mG.

---

*Correspondence to: A. R. Liboff (Physics), Oakland University, Rochester MI 48309.

TABLE 1.

Fields used by Blackman et al.[6]		Charge/mass (Coul/kg) calculated via ICR	
f (Hz)	B (mG)	$n = 1$	$n = 3$
15	380	2.48×10^6	0.83×10^6
30	760	2.48×10^6	0.83×10^6
30	255	7.39×10^6	2.46×10^6

For these reasons, we have centered our attention on the possibility of magnetically induced potassium changes in cells. The present work studies the effects in human lymphoma cell culture following 24h exposure to magnetic field combinations corresponding to potassium ion cyclotron resonance conditions, as a function of both the DC field strength and the AC intensity.

MATERIALS AND METHODS

Cells of the human lymphoma line HSB-2 (ATCC CCI 120.1) were cultured in 50ml flasks in buffered, fortified RPMI 1640 medium with 10% bovine serum to concentrations between 3 and 5×10^5/ml. Nine identical one-ml aliquots were transferred to 15ml polystyrene centrifuge tubes, which were placed into the nine solenoids comprising our multiple-exposure system,[12] located within a specially designed water-jacketed incubator made of nonferrous materials, with no internal motors and remote from other magnetic and electromagnetic sources. In the operation of this system, all samples, treated and controls, are prepared together, exposed together, and assayed together. Samples are also simultaneously exposed to various preset field levels, and, if need be, to various preset field combinations. This is equivalent to performing N different experiments at the same time on identical cell samples, where N corresponds to the number of different preset fields. Each cell sample is placed in a separate vertical solenoid, eight in all; each solenoid has a different number of turns but all are connected in series, so that each has the same energizing current. Since the turns on each solenoid are different, each sample is exposed to a different DC magnetic field (or, alternatively, to a different AC intensity). The eight solenoids are arranged with their axes parallel in an octagonal array, with a ninth parallel solenoid to hold control cells in the center. This system of nine solenoids is contained in a brass housing 14 cm in diameter and 23 cm high, and is itself centered within three concentric, independent sets of Helmholtz coils, also located within the nonmagnetic incubator. These additional coils are for local-field compensation, and to apply AC and DC fields that can be superposed on the fields generated by the solenoid system. In this way, it is possible to apply a different DC field to each of the eight cell preparations, and to hold the control cells at zero DC field, while all are simultaneously exposed to a common AC field.

Eight treated cell samples and one control sample were incubated in this manner for 24 h at 36.9 ± 0.2 C, 5% CO_2, under three different exposure regimens. In one set of experiments the treated cells were exposed to a 60Hz, 511mG peak magnetic field and at the same time to DC fields varying from 470 to 540 mG (47.0 to 54.0 μT). Controls were simultaneously exposed to the same AC magnetic field, but to zero DC field (0.00+0.05 mG). In a second set of experiments, the treated cells were exposed to a 16Hz, 409mG peak magnetic field and at the same time to DC fields varying from 370 to 440 mG. The third series of experiments was similar to the first in that the eight treated cell samples were again exposed to 60Hz fields. However, in this series, the peak strength of the AC field was also changed from run to run, over the range 50-1500 mG. Controls were again exposed to the same AC field as the treated specimens, as well as to zero DC field.

Upon removal, the cells were mixed thoroughly within each tube and 0.20 ml was removed, combined with 0.1 ml of a 2% trypan blue solution, and thoroughly mixed. Cell counts were made for each of the nine specimens with a hemacytometer, where mixing was again performed immediately prior to loading the hemacytometer. Runs for which cell viability was less than 80% were excluded from further consideration. As another check on possible artifactual changes in the state of each cell batch, additional samples were incubated coincidentally with each run at ambient field conditions in a separate nearby incubator. Throughout,

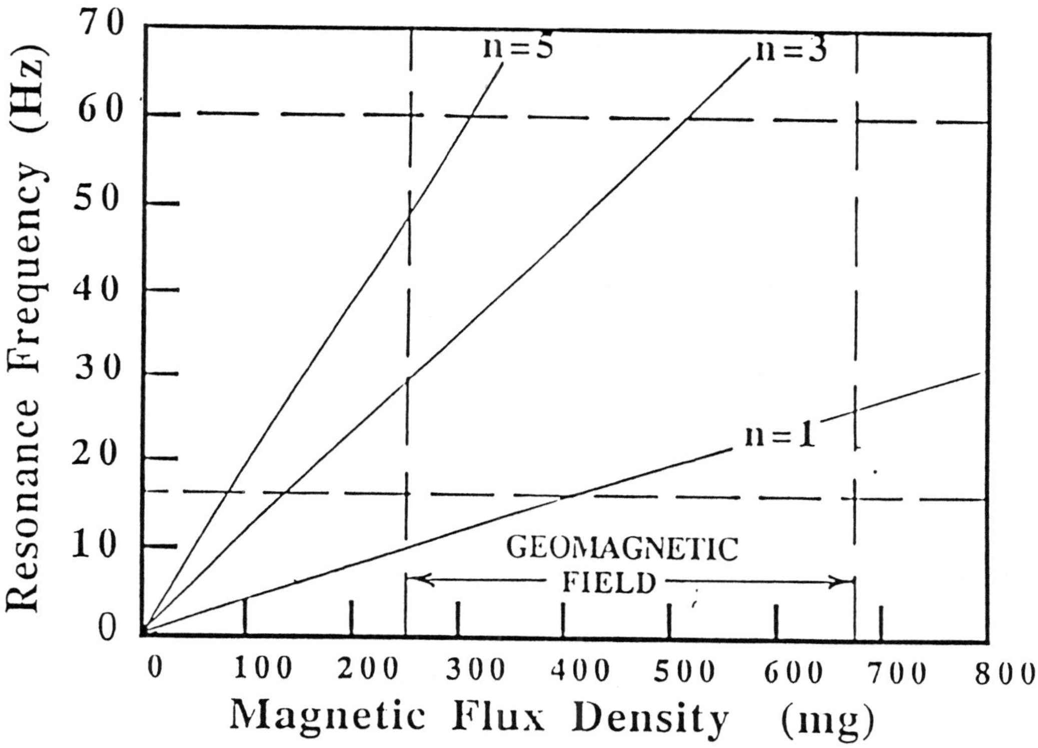

FIG. 1.—First three cyclotron resonance harmonic frequencies for K^+. Within range of geomagnetic field at earth's surface (not including magnetic anomalies), there are only two possible resonant values at 60 Hz for the DC magnetic field, one at 306.4 mG and the other at 510.6 mG. At 16 Hz one finds an additional resonance (n = 1) at 408.5 mG.

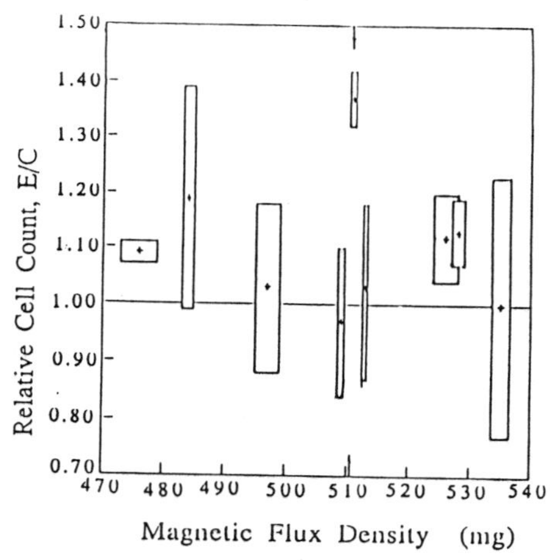

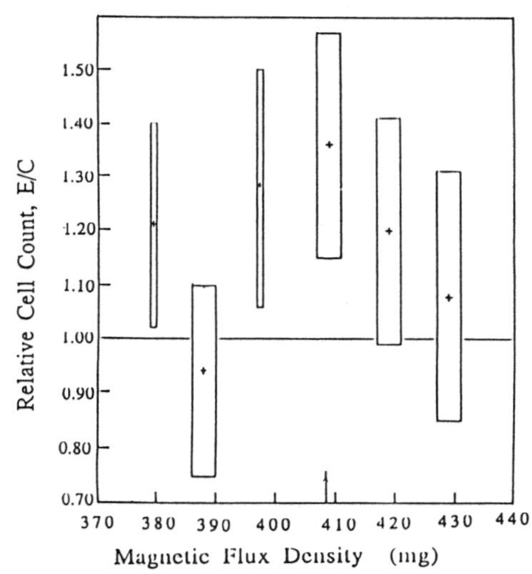

FIG. 2.—Relative proliferation of HSB-2 human lymphoma cells at 60 Hz following 24h exposures to various DC magnetic fields, based on four runs. Height of each box is twice the standard deviation. Width of each box is range of DC fields over the four runs at each point, reflecting a small variability in local field gradient from run to run. Arrows at 510.6 mG indicate the predicted DC field for K^+ cyclotron resonance expected at 60 Hz for the third harmonic. Maximum relative increase in proliferation (mean value 1.36 ± 0.05) occurs at this predicted value.

FIG. 3.—Relative proliferation of 16 Hz, based on seven runs. Maximum proliferation again occurs at predicted DC field (arrow). Mean relative increase in proliferation at this maximum (1.36 ± 0.21) is same as maximum in 60Hz results shown in Fig. 2.

extreme care was taken to insure that no additional AC or DC magnetic field components were present in the active volume of interest (14cm diameter × 23cm height). Results are expressed below in terms of E/C, the ratio of cell counts following a given 24h field exposure to control cell counts.

RESULTS

Four runs were successfully completed in which cells were simultaneously exposed for 24 h to colinear 60Hz AC (511mG peak) and DC fields, with the latter ranging from 470 to 540 mG. These data, shown in Fig. 2 in terms of the relative total count, are plotted as a function of the DC magnetic field. The vertical displacement of each point in this plot is twice the standard deviation. The horizontal width of each point is the range of DC fields at that point for all four runs. There is a sharp maximum at 511 mG (E/C = 1.36 ± 0.05).

A similar set of runs was completed at 16 Hz, with DC fields varying from 370 to 440 mG. These data are plotted in Fig. 3. The maximum is less sharp than what was observed at 60 Hz. The maximum relative increase in proliferation, compared to cells exposed to zero DC field, is E/C = 1.36 ± 0.21 (N = 7).

In a variation of the results shown in Fig. 2 for 60Hz intensity at 511mG peak, separate runs were taken in which the peak 60Hz intensity was preset to values between 50 and 1550 mG. Including the four values shown in Fig. 2, there were 17 such runs. For each of these runs, cell samples that were simultaneously exposed to 511 mG DC were counted and compared to controls at zero DC. The resulting data, for which each point is at the third harmonic cyclotron resonance condition for K^+ (511 mG at 60 Hz), are plotted in Fig. 4. The data are clearly grouped into two categories, either showing a proliferative response of about 35% or none at all.

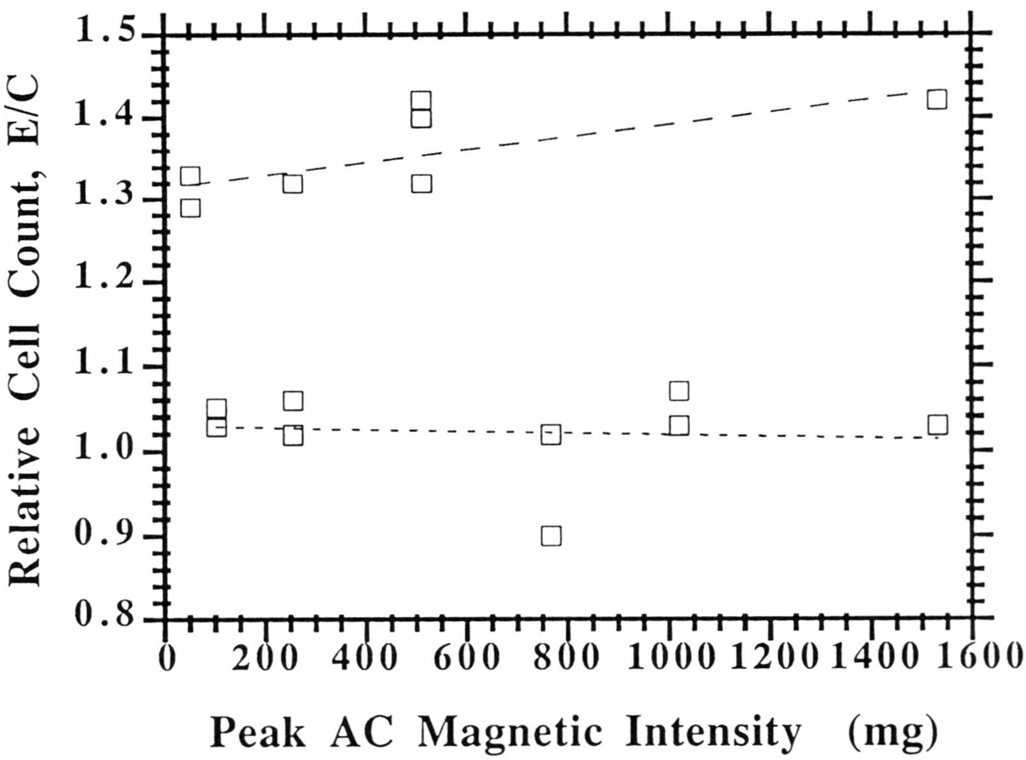

FIG. 4.—Relative proliferation as a function of 60Hz peak intensity, for seventeen separate runs. Each point is taken at the same DC magnetic field of 511 mG. The four data points boxed in at 511 mG in Fig. 2 appear here as three points, since two points (at 1.32) are superposed. Because data are apparently readily resolved into two groups, we have plotted two least-square lines.

DISCUSSION

The results shown in Figs. 2 and 3 provide evidence that there is a maximum change in proliferation in this cell line following 24 h of exposure to the two field combinations of 511 mG at 60 Hz and 409 mG at 16 Hz. As shown in Fig. 1, these two conditions correspond, respectively, to the third harmonic and to the fundamental cyclotron resonance conditions for K^+. It is interesting that the mean levels of enhanced proliferation determined in both cases were the same, namely 36%. This value is reinforced by the data presented in Fig. 4, which clearly shows that a relative enhancement of cell proliferation of between 30 and 40% can occur under potassium cyclotron resonance conditions, for AC intensities ranging from 50 to 1550mG peak. However, not all intensities resulted in enhanced proliferation. It is not clear why these data are so sharply resolved into two groups, one indicating increased enhancement with magnetic fields and the other indicating no effect whatsoever. One speculative explanation is that there is another cell-dependent variable that we have so far failed to recognize that somehow makes the cell receptive (or unreceptive) to the resonance exposure. In any event, it would be of great interest to extend these 60Hz measurements to intensities well below the 50mG lower limit reached in our studies. There are some experimental challenges still to be overcome in performing this work at AC milligauss levels.

One implication of these results, because of the way the data fit the cyclotron-resonance hypothesis, is that potassium is involved in the proliferative process. There are reports in the literature linking cell proliferation in immune cells to potassium membrane flux.[13] For example, there is a large increase in K^+ channels in murine T lymphocytes, following activation by the mitogen concanavalin A.[14] In human T-cells, a 24h exposure to Con A or to the tumor promoter TPA increases potassium conductance by a factor of two.[15]

REFERENCES

1. W. C. Parkinson and C. T. Hanks, "Search for cyclotron resonance in cells in vitro," *Bioelectromagnetics* 10: 129-145, 1989.

2. A. R. Liboff and W. C. Parkinson, "Search for ion-cyclotron resonance in an Na^+-transport system," *Bioelectromagnetics* 12: 77-83, 1991.

3. T. S. Tenforde and W. T. Kaune, "Interaction of extremely low frequency electric and magnetic fields with humans," *Health Phys.* 53: 585-606, 1987.

4. R. K. Adair, "Constraints on biological effects of weak extremely low-frequency electromagnetic fields," *Phys. Rev.* 43: 1039-1048, 1991.

5. A. R. Liboff, "Geomagnetic cyclotron resonance in membrane transport," *J. Biol. Phys.* 13: 99-102, 1985.

6. C. F. Blackman, S. G. Benane, J. R. Rabinowitz, D. E. House, and W. T. Joines, "A role for the magnetic field in the radiation-induced efflux of calcium ions from brain tissue in vitro," *Bioelectromagnetics* 6: 327-337, 1985.

7. A. R. Liboff and B. R. McLeod, "Kinetics of channelized membrane ions in magnetic fields," *Bioelectromagnetics* 9: 39-51, 1988.

8. S. M. Bawin, L. K. Kazmarek, and W. R. Adey, "Effects of modulated VHF fields on the central nervous system," *Ann. N.Y. Acad. Sci.* 247: 74-81, 1975.

9. B. R. McLeod, S. D. Smith, and A. R. Liboff, "Calcium and potassium cyclotron resonance curves and harmonics in diatoms *(A. coffeaeformis)*. *J. Bioelectr.* 6: 153-168, 1987.

10. J. D. Bowman, D. C. Thomas, and J. M. Peters, "The risk of childhood leukemia from home exposure to resonances from static and power-frequency magnetic fields," Abstr. A-40, *Ann. Rev. of Research on Biological Effects of 50 and 60 Hz Electric and Magnetic Fields,* U.S. Department of Energy, Milwaukee, 1991.

11. S. J. London, D. C. Thomas, J. D. Bowman, E. Sobel, T. C. Cheng, and J. M. Peters, "Exposure to residential electric and magnetic fields and risk of childhood leukemia," *Am. J. Epidemiol.* 134: 923-927, 1991.

12. A. R. Liboff and K. A. Jenrow, "Eight-fold multiple exposure system for resonance studies in cell culture," *Bioelectromagnetics Soc. 13th Ann. Mtg.,* Abstr. G-2-7, 1991.

13. E. K. Gallin, "Ion channels in leukocytes," *Physiol. Rev.* 71: 775-811, 1991.

14. L. S. Gray, J. D. Gnarra, J. H. Russell, and V. H. Engelhard, "The role of K^+ in the regulation of the increased intracellular Ca^{2+} mediated by the T lymphocyte antigen receptor," *Cell* 50: 119-127, 1987.

15. D. R. Matteson and C. Deutsch, "K channels in T lymphocytes: A patch clamp study using monoclonal antibody adhesion," *Nature* 307: 468-471, 1984.

Martin Blank, Ed., *Electricity and Magnetism in Biology and Medicine*

PROLIFERATIVE RESPONSES OF HUMAN FIBROBLASTS EXPOSED TO ELF MAGNETIC FIELDS

N. A. Cridland,* R. D. Saunders, and T. A. Cragg

National Radiological Protection Board, Chilton, UK

The effects of exposure to 2mT power-frequency (50Hz) magnetic fields on the proliferative responses of chromosomally normal human fibroblasts, cultured as monolayers and supplemented with suboptimal serum concentrations, were examined. Proliferative responses were monitored by following the incorporation of [^{3}H]-thymidine into acid insoluble material. Magnetic-field exposure had no effect on the proliferation of asynchronously dividing cells exposed for periods up to either 6 or 30 h. Cell cycle progression was similarly unaffected by magnetic-field exposure, although the cells were sensitive to proliferative stimuli.

It has become clear that if extremely low frequency (ELF) electromagnetic fields have any affect on human carcinogenesis at all, they are likely to do so through actions at the level of tumor promotion.[1-3] Classical tumor-promoting chemicals such as the phorbol esters typically act to increase the rate of cell proliferation; this effect must be therefore regarded as central to any investigation of the possible tumor-promoting properties of electromagnetic fields. The aim of this study was to examine the effects of exposure to power-frequency magnetic fields on proliferative responses to normal human cells. Normal human fibroblast cells were used in preference to the immortalized or even fully transformed cells that have been used in some other studies;[4-7] these latter cell types have usually escaped from the normal growth regulatory mechanisms and are therefore unlikely to provide good models for examining proliferative responses.

MATERIALS AND METHODS

Cells and Cell Culture

HF-19 normal human fibroblasts derived from fetal female lung[8] were obtained from the Medical Research Council Radiobiology Unit, Chilton, UK, and were maintained in monolayer culture in 1 × minimal essential medium (MEM), supplemented with 10% fetal calf serum (FCS) and 50 μg ml^{-1} gentamicin. The cells were cultured at 37 C under an atmosphere of 5% CO_2 in air and the medium was changed twice per week. Cells used in experiments were at a relatively low passage number (10) in order to minimize the spontaneous degenerative changes that tend to accumulate after prolonged culture. For asynchronously dividing experimental cultures, cells were released from confluent cultures by a brief digestion with trypsin.EDTA and replated in 25cm^2 (3.7×6.75cm) flasks at approximately 1×10^4 cells cm^{-2}. Cells were cultured for 2-3 days in 1 × MEM supplemented with 10% FCS, rinsed with phosphate buffered saline (PBS), and cultured overnight in 1 × MEM supplemented with FCS at the concentrations specified below. To obtain synchronous cultures, cells were seeded into 75cm^2 flasks at approximately 2×10^6 cells/flask and cultured in 1 × MEM supplemented with 10% FCS and antibiotics for at least 28 days; under these conditions the cells reach confluence after about 7 days. The cells were released by digestion with trypsin.EDTA and replated in 25cm^2 flasks at a density of 6-7×10^3 cells cm^{-2}.

Where appropriate, human recombinant basic fibroblast growth factor (bFGF) was included in culture medium at 20 ng ml^{-1}.

Quantitation of Incorporated Label

We determined the proliferative response of HF-19 cells by measuring the rate of DNA synthesis, monitored as [^{3}H]-thymidine incorporation. Labeling was terminated by rinsing of the cells twice with ice-cold PBS and lysing with 1%(w/v) sodium dodecyl sulfate (SDS). Macromolecular material was precipitated by

---

Correspondence to: N. A. Cridland NRPB, Chilton, Didcot England OX11 0RQ.

addition of an equal volume of 30%(w/v) trichloroacetic acid (TCA), 1 mg ml^{-1} thymidine. After cooling to 4 C for at least 14 h, precipitates were recovered by filtration onto GF/F filters, rinsed sequentially with 10% TCA and 70% ethanol, dried, and counted in scintillant.

Magnetic Field Exposures

Cultures were exposed to vertical 50Hz magnetic fields at a flux density of 2 mT in the NRPB magnetic field exposure facility,[9] which generates a field with a uniformity of ±5%. During exposure, sealed $25cm^2$ flasks were placed in a Perspex box, which was maintained at 37±0.5 C with hot air. Control flasks were placed in an incubator at 37±0.1 C. The temperature dependence of HF-19 cell proliferation under these conditions has been characterized, and does not vary more than ±5% between 36 and 38 C. Stray AC fields at the control flasks were ≤200 nT (measured with a Bell 9200 Gaussmeter equipped with an MOW-92-2506 magnaprobe); the static magnetic field was 45 μT. The induced current density would have been of the order of 10^{-2} A m^{-2}, assuming a conductivity of 1.5 S m^{-1}.

Asynchronous cultures subjected to 2mT, 50Hz magnetic fields for up to 6 h were incubated in the presence of [^{3}H]-thymidine (2 Ci $mmol^{-1}$; 1 μCi ml^{-1}). The concentration of FCS used to supplement the cultures was varied in order to achieve either a submaximal (0.5%) or a near-maximal (5%) proliferative stimulus. For longer exposures, cultures supplemented with 0.5% FCS were subjected to the magnetic field for 24 h and then pulse labeled with [^{3}H]-thymidine (5 Ci $mmol^{-1}$; 0.2 μCi ml^{-1}) for up to a further 6 h in the presence of the field.

Cell Cycle Progression during Magnetic Field Exposure

We produced synchronous cultures by accumulating cells in the G_0 phase of the cell cycle and then releasing them into G_1. Synchronous cultures were exposed immediately after replating at low density and [^{3}H]-thymidine (5 Ci $mmol^{-1}$; 0.5 μCi ml^{-1}) was present for the duration of the exposure. The rate of cell cycle progression was assessed by monitoring of the time taken to enter S-phase, as characterized by the onset of DNA synthesis.

RESULTS AND DISCUSSION

This study found no evidence to suggest that exposure of cultured human cells to 2mT power-frequency magnetic fields affects their proliferation. Short-term exposure did not affect the rate of DNA synthesis in cells which were submaximally stimulated and which should therefore have been capable of responding (Fig. 1a). In addition, exposure did not modulate the cellular response to serum mitogens (Fig. 1b). Similarly, longer exposures, which could have conceivably affected processes occurring early in the cell cycle, had no detectable effect on DNA synthesis (Fig. 2a), although the cells were responsive to a membrane-acting mitogen (bFGF) under the same conditions (Fig. 2b).

As an alternative approach, the effects of magnetic field exposure on the rate of cell cycle progression by submaximally stimulated cells were examined. Exposure to the magnetic field described above had no effect on the time taken to enter S-phase (Fig. 3a). By contrast, exposure to human recombinant bFGF decreased the time taken to enter S-phase, indicating that the cells were responsive to a membrane-acting proliferative stimulus (Fig. 3b).

These findings are in agreement with the results of previous studies using fibroblast,[10] melanoma,[6] lymphoblastoid,[4] and myeloid leukemia[7] cells, which found no effects on cell growth[7,9,10] or DNA synthesis.[4] However, only one of these studies employed a 50Hz field, at 200 μT;[7] the others involved a variety of pulsed waveforms at frequencies of either 15 or 72 Hz.

In general, the existing literature relating to the effects of EMF exposure on cellular proliferation reports many apparently conflicting results,[3] although that may reflect differences in exposure parameters or the responses of different cell types. For example, studies with fibroblasts exposed to a 76Hz field at 16 μT,[11] endothelial cells exposed to a 15Hz pulse burst at 0.1 mT,[12] and embryonal carcinoma cells exposed to a 100Hz pulsed field at 0.1-1 mT[5] have indicated enhanced cell growth, measured as increases in either DNA synthesis[11,12] or cell number.[5] On the other hand, experiments with fibroblasts exposed to a 21Hz, 8.4mT field[13] and melanoma cells intermittently exposed to 15Hz pulse bursts[6] have demonstrated an inhibition of cell division.

It has been suggested that magnetic fields may modulate the cellular response to mitogens; experiments involving pulsed magnetic field exposure to human lymphocytes have supported this type of mechanism.[14-19] Overall, however, the experimental evidence is again inconsistent, with reports of positive responses to 50Hz,

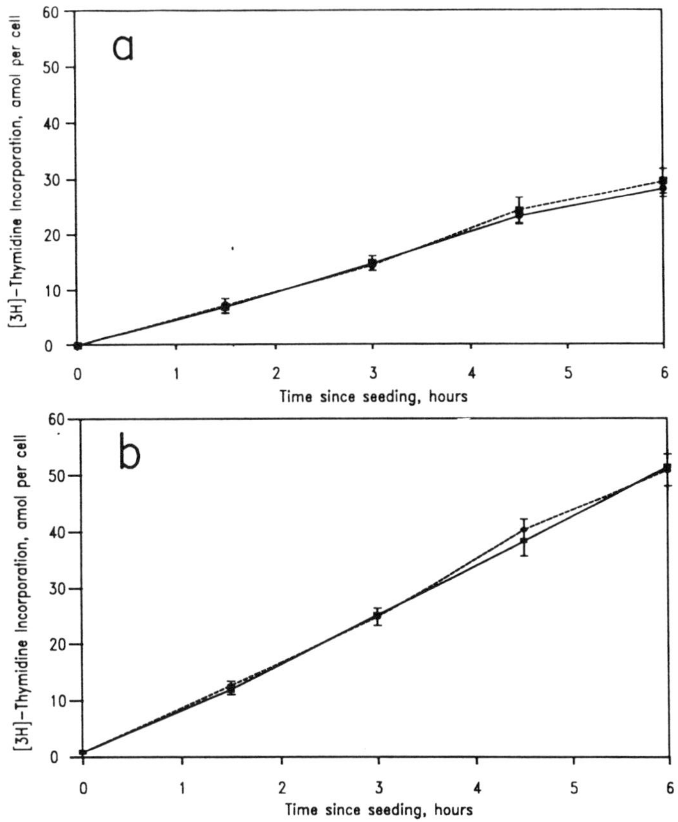

FIG. 1.—Short-term 50Hz, 2mT magnetic-field exposure has no effect on DNA synthesis by human fibroblasts. The rate of DNA synthesis by asynchronous cultures of HF-19 cells was determined by monitoring of the rate of [^{3}H]-thymidine incorporation into macromolecular material during up to 6 h of exposure. (a) Exposure did not directly stimulate proliferation. Cultures were supplemented with 0.5% FCS in order to insure that they were submaximally stimulated and therefore capable of responding. (b) Exposure did not modulate the response to mitogenic stimulation. Cultures were supplemented with 5% FCS which produces a near-maximal proliferative response. (□, control; ●, 2 mT).

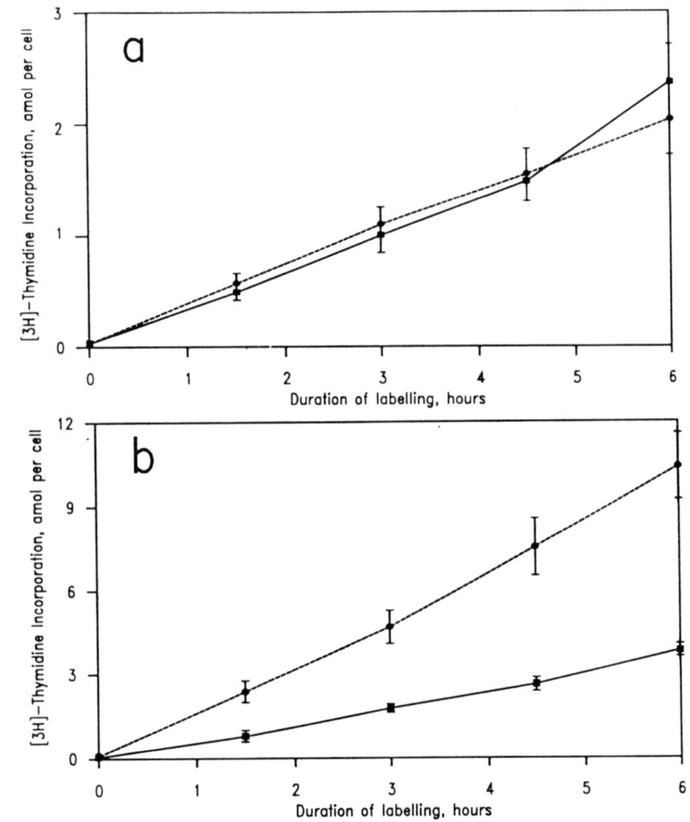

FIG. 2.—Exposure to 50Hz, 2mT magnetic fields for up to 30 h has no effect on DNA synthesis by responsive human fibroblasts. Asynchronous cultures supplemented with 0.5% FCS were exposed to the appropriate treatment for 24 h. They were then pulse labeled with [^{3}H]-thymidine for up to a further 6 h. Cultures were exposed to a 50Hz, 2mT field for up to 30 h. (□, control; ●, 2 mT). (b) Cells were responsive to a known mitogen. Cultures were incubated in the presence of a 20ng ml^{-1} human recombinant FGF for up to 30 h. (□, control; ●, 20 ng ml^{-1} bFGF).

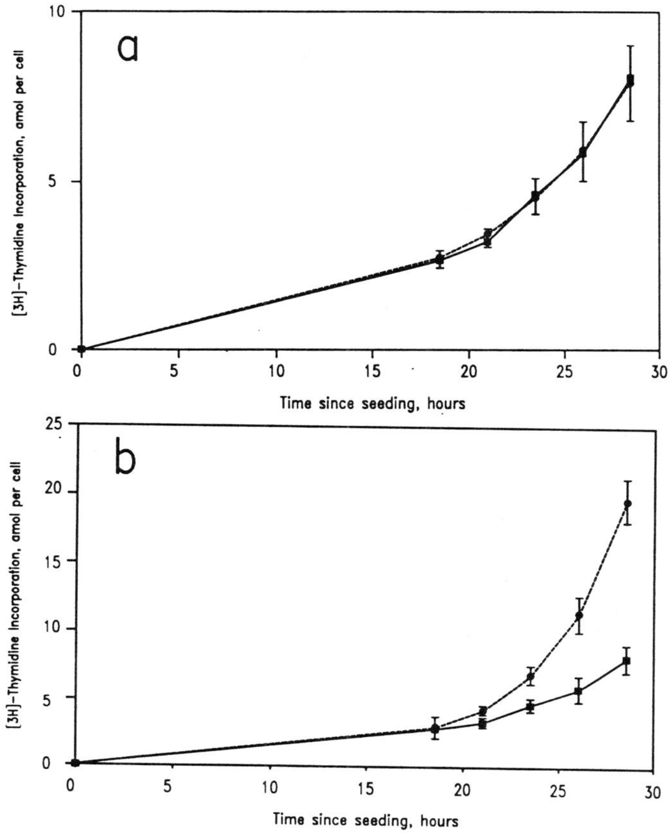

FIG. 3.—Cell cycle progression by synchronous cultures of human fibroblasts is not affected by 50Hz, 2mT magnetic-field exposure. As an alternative measure of their proliferative response, human fibroblasts were synchronized and plated in medium containing [^{3}H]-thymidine. The time taken to enter S-phase, as indicated by the onset of DNA synthesis, was taken as a measure of the rate of progression through the cell cycle. (a) Magnetic-field exposure had no effect on cell cycle progression. Cultures were exposed to a 50Hz, 2mT field from the time of synchronization. (□, control; ●, 2 mT). (b) Cells were responsive to a known mitogen. Cultures were incubated in the presence of 20 ng ml^{-1} human recombinant FGF from the time of synchronization. (□, control; ●, 20 ng ml^{-1} bFGF).

2.5mT fields[14-16] and negative responses to 1-200Hz 2.3-6.5mT fields.[17-19] No effects were found in the study described here.

It is anticipated that future work will investigate the effects of exposure to environmental levels of combined static and time varying magnetic fields.[20]

REFERENCES

1. W. R. Adey, "Nonlinear electrodynamics in cell membrane transductive coupling," *Membrane Transport and Information Storage,* New York: Alan R. Liss, 1990, 1-27.

2. NRPB, *Electromagnetic Fields and the Risk of Cancer: Report of an Advisory Group on Non-ionising Radiation,* Doc. NRPB 3, No. 1, 1992.

3. N. A. Cridland, *Electromagnetic Fields and Cancer: A Review of Relevant Cellular Studies,* Chilton, NRPB-R256 (in press).

4. J. L. Phillips and L. McChesney, "Effect of 72 Hz pulsed magnetic field exposure on macromolecular synthesis in CCRF-CEM cells," *Cancer Biochem. Biophys.* 12: 1-7, 1991.

5. T. Akamine, H. Muramatsu, H. Hamada, and T. Sakou, "Effects of pulsed electromagnetic field on growth and differentiation of embryonal carcinoma cells," *J. Cell. Physiol.* 124: 247-254, 1985.

6. D. B. Jones, R. B. Pedley, and J. T. Ryaby, "The effects of pulsating electromagnetic fields on differentiation and growth in Cloudman S91 murine melanoma cells in vitro," *J. Bioelectricity* 5: 145-169, 1986.

7. M. Fiorani, O. Cantoni, P. Sestili, R. Conti, P. Nicolini, F. Vetrano, and M. Dachà, "Electric and/or magnetic field effects on DNA structure and function in cultured human cells," *Mutat. Res.* 282: 25-29, 1992.

8. R. Cox, J. Thacker, and D. T. Goodhead, "Inactivation and mutation of cultured mammalian cells by aluminium characteristic ultrasoft x-rays: II. Dose-responses of Chinese hamster and human diploid cells to aluminium x-rays and radiation of different LET," *Int. J. Radiat. Biol.* 31: 561-576, 1977.

9. Z. J. Sienkiewicz, C. I. Kowalczuk, and R. D. Saunders, *50 Hz Magnetic Field Exposure System,* NRPB Non-ionising Radiation Department Technical Memorandum No. 4(91), 1991.

10. J. C. Murray and R. W. Farndale, "Modulation of collage production in cultured fibroblasts by a low-frequency, pulsed magnetic field," *Biochim. Biophys. Acta* 838: 98-105, 1985.

11. A. R. Liboff, T. Williams Jr., D. M. Strong, and R. Wistar Jr., "Time-varying magnetic fields: Effect on DNA synthesis," *Science* 223: 818-820, 1984.

12. G. P. A. Yen-Patton, W. F. Patton, D. M. Beer, and B. S. Jacobson, "Endothelial cell responses to pulsed electromagnetic fields: Stimulation of growth rate and angiogenesis in vitro," *J. Cell. Physiol.* 134: 37-46, 1988.

13. H. P. Rodemann, K. Bayreuther, and G. Pfleiderer, "The differentiation of normal and transformed human fibroblasts in vitro is influenced by electromagnetic fields," *Exp. Cell Res.* 182: 610-621, 1989.

14. M. Cantini, A. Cossarizza, F. Bersani, R. Cadossi, G. Ceccherelli, R. Tenconi, C. Gatti, and C. Franceschi, "Enhancing effect of low frequency pulsed electromagnetic fields on lectin-induced human lymphocyte proliferation," *J. Bioelectricity* 5: 91-104, 1986.

15. A. Cossarizza, D. Monit, F. Bersani, M. Cantini, R. Cadossi, A. Sacchi, and C. Franceschi, "Extremely low frequency pulsed electromagnetic fields increase cell proliferation in lymphocytes from young and aged subjects," *Biochem. Biophys. Res. Comm.* 160: 692-698, 1989.

16. A. Cossarizza, D. Monti, F. Bersani, R. Paganelli, G. Montagnani, R. Cadossi, M. Cantini, and C. Franceschi, "Extremely low frequency pulsed electromagnetic fields increase interleukin-2 (IL-2) utilization and IL-2 receptor expression in mitogen-stimulated human lymphocytes from old subjects," *FEBS Lett.* 248: 141-144, 1989.

17. P. Conti, G. E. Gigante, M. G. Cifone, E. Alesse, G. Ianni, M. Reale, and P. U. Angeletti, "Reduced mitogenic stimulation of human lymphocytes by extremely low frequency electromagnetic fields," *FEBS Lett.* 152: 156-160, 1983.

18. P. Conti, G. E. Gigante, E. Alesse, M. G. Cifone, C. Fieschi, M. Reale, and P. U. Angeletti, "A role for Ca^{2+} in the effect of very low frequency electromagnetic field on the blastogenesis of human lymphocytes," *FEBS Lett.* 181: 28-32, 1985.

19. N. A. Mooney, R. E. Smith, and B. W. Watson, "Effects of extremely-low-frequency pulsed magnetic fields on the mitogenic response of peripheral blood mononuclear cells," *Bioelectromagnetics* 7: 387-394, 1986.

20. The authors would like to thank the MRC Radiobiology Unit for the gift of the cells, and Dr. R. Cox for helpful advice and discussions.

Martin Blank, Ed., *Electricity and Magnetism in Biology and Medicine*

EFFECT OF 60Hz ELECTRIC AND MAGNETIC FIELDS FROM A 500kV TRANSMISSION LINE ON INTERLEUKIN 1 PRODUCTION AND ANTIBODY RESPONSE BY EWE LAMBS

S. L. McCoy* and S. H. Hefeneider

Veterans Affairs Medical Center, Portland, Ore.

B. E. Ogden and A. S. Hall

Oregon Health Sciences University, Portland

A preliminary environmental study to assess the effect of 60Hz fields from a 500kV transmission line on immune function in female sheep suggested reduced levels of interleukin-1 (IL-1) activity in four separate experiments in EMF-exposed animals ($n = 14$) as compared to controls ($n = 15$). Upon removal from EMF-exposure, two additional experiments (2 and 3 months post exposure) indicated no differences in IL-1 activity between groups. During exposure to EMF, no differences were seen in antibody development to a defined antigen between groups. Of interest, 13 of 14 EMF-exposed sheep, and none of the controls, developed a focal dermatitis, with the etiologic agent identified as *Scopulariopsis brevicaulis*. These preliminary data suggest a possible effect of EMF exposure on immune function. A follow-up study, with new animals, is currently in progress.

Several studies suggest a possible link between EMF exposure and the development of certain types of cancers. Increased incidences of lymphomas, leukemias, and brain tumors have been correlated with occupations assumed to have chronic EMF exposure.[1] Suppression of cellular immune function following in vitro exposure of cells to EMF has also been observed; however, a direct demonstration of an effect of environmental EMF exposure on immune function remains to be established.[2-4] The mechanism(s) responsible for affecting biological processes after EMF exposure are unknown. Khalil and Qassem have recently demonstrated an increased incidence of chromosomal aberrations and cell death after in vitro exposure of human lymphocytes to a pulsing EMF source.[5] Alternatively, a possible site of action by which exposure to EMF could affect immune function is at the regulation of immune responses. Cytokines are critical in the regulation and development of normal immune responses; alterations in cytokine levels have been associated with various diseases.[6-8] Diminished cytokine activity could significantly affect development of normal immune responses to vaccination and environmental antigens such as bacteria, viruses, and fungi. The current study was designed to examine whether exposure of ewe lambs to environmental EMF affected production of the immunoregulatory cytokine interleukin 1 (IL-1), which in turn could impact immune functions such as antibody development and resistance to infection.

METHODS AND RESULTS

Fourteen female sheep (8 weeks of age) were placed under constant exposure to 60Hz electric and magnetic fields from a 500kV powerline (mean electric field 5.5 kV/m, mean magnetic field 37.7 mG), with 15 age-matched female sheep in a separate pen exposed to ambient EMF (mean electric field <0.01 kV/m, mean magnetic field 0.2 mG). After 10 months of exposure, all animals were removed from the pens and pastured together under ambient EMF conditions.

To study the effect of whole-body EMF exposure on cytokine activity, we examined the production of IL-1 activity by LPS (lipopolysaccharide) stimulated leukocytes from control and line-treated animals. Whole blood was obtained from animals after continuous exposure EMF (line) or control conditions, and leukocytes were isolated by collection of buffy coats (a mixture of white blood cells), then stimulated with 20 μg/ml of the mitogen LPS per 2×10^6 cells. After incubation for 48 h at 5% CO_2, cell-free supernatants

*Correspondence to: S. L. McCoy (Immunology, 151-R), VA Medical Center, Portland OR 97201.

were collected and IL-1 activity was assessed by means of the IL-1 dependent cell line D10.S. Six IL-1 induction experiments were performed: two were done after 2 months of exposure (one drawing of blood during the day and one at night), one after 5 months exposure, and one after 6 months exposure. Each of these 4 experiments performed during exposure suggested a reduction of LPS-induced IL-1 production from EMF-exposed animals as compared to controls ($p = 0.049, 0.057, 0.062, 0.083$, Fig. 1). Two additional experiments were done 2 and 3 months after animals were removed from EMF exposure. Upon removal of animals from EMF exposure, no differences between the groups were observed ($p = 0.811, 0.571$, Fig. 2). No pre-exposure experiments were done. The differences seen between experiments in the absolute amount of IL-1 activity (μ/ml) produced over time are likely due to the efficiency of the LPS-induced activation and probably unrelated to EMF exposure. The data were analyzed by a two-tailed t-test.

To study whether this potential reduction in IL-1 activity could impact immune function, we examined the development of antibody responses to a defined antigen. If the reduction in LPS-induced IL-1 activity seen after exposure to EMF were biologically significant, that effect might impact the development of an antibody response. Animals were vaccinated with the Lepto-5 antigen (Affiliated Labs), which is routinely given to sheep to prevent Leptospira infections and is a combination of 5 serovars: ictero, canicola, hardjo, grippotyphosa, and pomona. Lepto-5 was given to animals after 5 months of either exposure to EMF or control conditions, and booster injections were administered 1 and 4 months after the initial vaccination. Antibody titers to Lepto-5 were quantified by a microagglutination assay on sera collected from animals prior to vaccination and every 2 weeks post-vaccination for a total of 12 assays. Potential differences in antibody development between the two groups were assessed separately to each serovar by a Mann-Whitney U one-tailed test for nonparametric statistics. No consistent differences were observed between control and EMF-exposed sheep in antibody development to any of the individual serovars, with the possible exception of grippotyphosa, which showed a modest reduction in titer from EMF-exposed animals as compared to controls (Fig. 3).

As an ongoing part of our study protocol, the health of all animals was monitored on a regular basis. Between 3 and 6 months of EMF exposure, 13 of 14 ewes in the line pen developed a focal dermatitis on head or neck. The lesions were observed as 2-5cm white scaly patches from which the hair/wool was easily removed. The resulting patch of alopecia did not appear markedly inflamed, crusty, or thickened. One of the lesions was biopsied and four were cultured. Three of the four cultures were identified as *Scopulariopsis brevicaulis*. (The fourth culture has not yet been characterized owing to bacterial contamination.) Histopathological examination of the biopsied lesion revealed the same fungal organism in an ectothrixic pattern. According to Dr. Arnold Kaufmann at the Centers for Disease Control (CDC), this is the first documented case of *Scopulariopsis spp.* causing an ectothrixic pattern of skin lesion. We have been unable to find any published reports of involvement of this organism in sheep dermatitis. None of the sheep in the control pen developed lesions. Following the outbreak, culture samples were collected from feeders, water bowls, and wool in both the control pen and line pen. *Scopulariopsis brevicaulis* was isolated from samples taken in both pens, which suggests that both groups of sheep had exposure to the organism. If sheep exposed to EMF were partially immune-compromised, as the IL-1 data suggests, it is possible that these animals were more susceptible to infection than the controls.

Our data suggest that environmental EMF exposure may suppress immune status by affecting interleukin-1 production. This reduction in interleukin-1 production does not appear to affect antibody development to the Lepto-5 antigen. Whether development of the fungal infection resulted from EMF-induced immune dysfunction remains to be established. A follow-up study, with new animals, is currently in progress.

REFERENCES

1. S. Milham, *Environ. Health Perspect.* 62: 297, 1985.
2. D. B. Lyle, R. D. Ayotte, A. R. Sheppard, and W. R. Adey, *Bioelectromagnetics* 9: 303, 1988.
3. R. L. Davis and S. Milham, *Am. J. Indust. Med.* 18: 79, 1990.
4. M. A. Speers, J. D. Dobbins, and V. S. Miller, *Am. J. Ind. Med.* 14: 629, 1988.
5. A. M. Khail and W. Qassem, *Mutation Res.* 274: 141, 1991.
6. M. C. Sneller and W. Strober, *J. Immunol.* 144: 3762, 1990.
7. P. Rieckmann, F. D'Alessandro, R. P. Nordan, A. S. Fauci, and J. H. Kehri, *J. Immunol.* 146: 3462, 1991.
8. M. B. Widmer and K. H. Grabstein, *Nature* 326: 795, 1987.

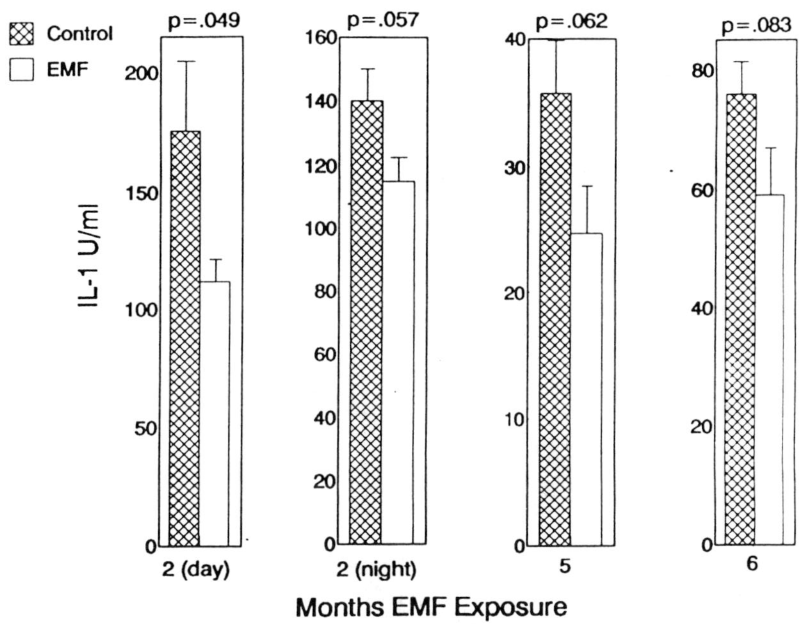

FIG. 1.—Ewe lambs were exposed to either 60Hz (from a 500kV line) or ambient EMF. Leukocytes were isolated, stimulated with 20 μg/ml LPS and cultured for 48 h. Cell-free supernatants were collected and assayed for IL-1 activity by use of the factor-dependent cell line D10.S. Data represent the μ/ml mean and standard error of the mean; differences were analyzed by a 2-tailed *t*-test.

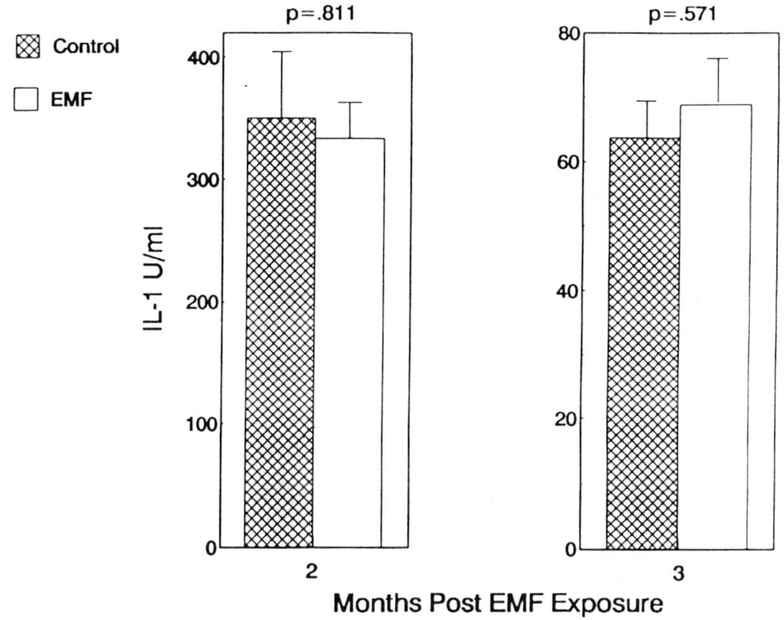

FIG. 2.—After 10 months exposure to either 60Hz (from a 500kV line) or ambient EMF, ewe lambs were removed from pens and housed together under ambient EMF conditions. IL-1 activity was assayed and data were analyzed as described in Fig. 1 legend.

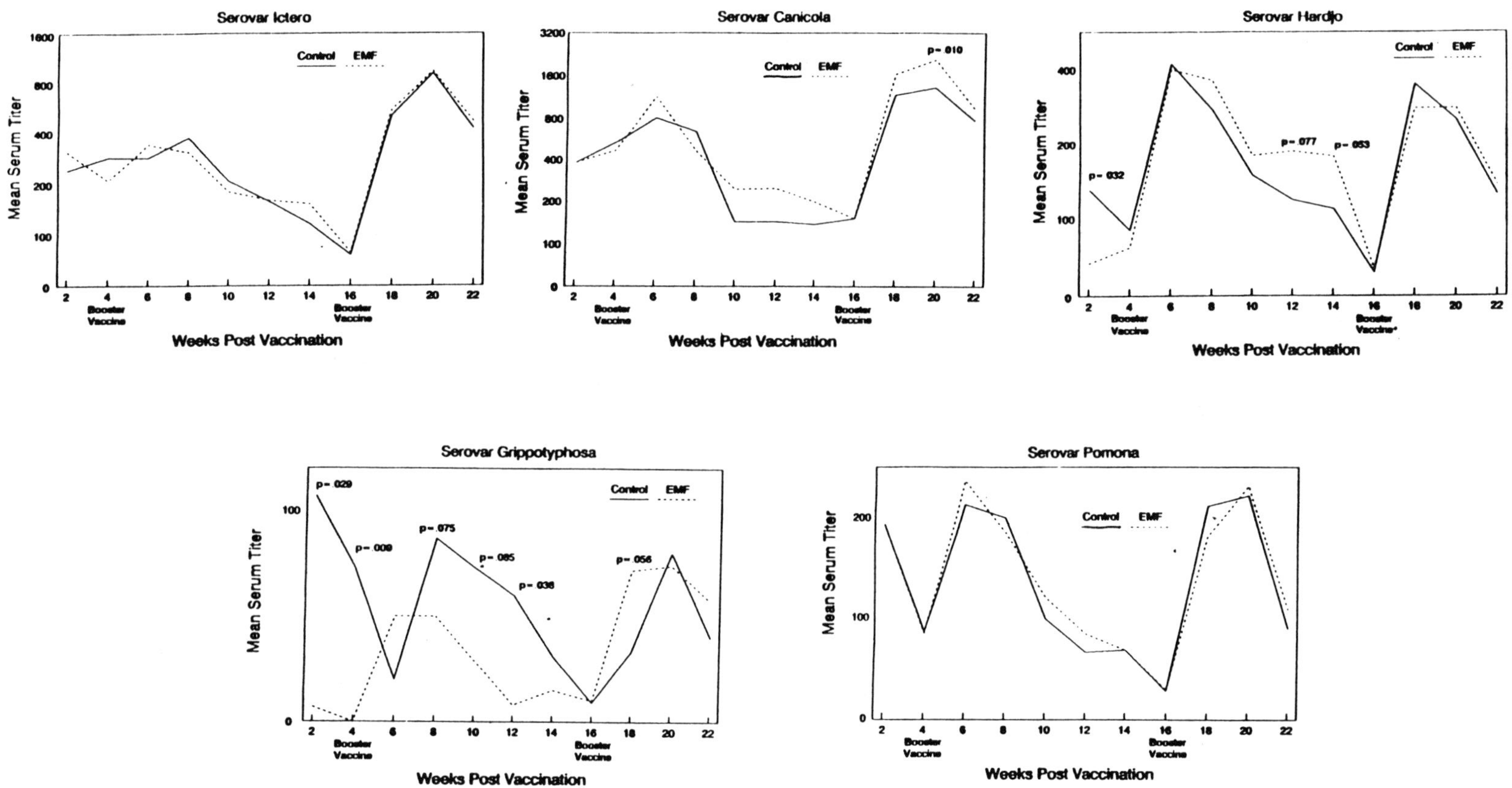

FIG. 3.—Ewe lambs were vaccinated with Lepto-5 antigen after 5 months continuous exposure to either 60Hz (from a 500kV line) or ambient EMF. Antibody titers to Lepto-5 were quantified by a microagglutination assay on sera collected prior to and every 2 weeks post-vaccination for a total of 12 assays. Data represent geometric-mean serum titers; differences in antibody development to the five individual serovars contained within the Lepto-5 antigen were assessed by a Mann-Whitney U one-tailed test.

Martin Blank, Ed., *Electricity and Magnetism in Biology and Medicine*

ACCELERATED HUMAN LYMPHOCYTE GROWTH IN A 1G, 15Hz MAGNETIC FIELD

S. Mehta,* D. Blackinton, M. Wiesendanger, and H. Wanebo

Brown University, Providence, R.I.

M. Lefebvre

Wheaton College, Norton, Mass.

D. Cherlin and C. Polk

University of Rhode Island, Kingston

Exposure to a 15Hz, 1G sinusoidal magnetic field of freshly prepared and activated human T and B-lymphocytes and EBV positive neoplastic B-cell lines (Daudi, Raji, and P3HR1) results in accelerated growth characteristics. We also observed an increase in T-cell production of the autocrine factor IL-2 in exposed cultures, while the cellular ratio of T_4/T_8 remain unchanged. The increase in T-cell proliferation may be attributed to the enhanced production of IL-2.

Epidemiological studies conducted on the incidences of cancer possibly related to electromagnetic-field (EMF) exposure in the home and work environment have designated lymphoma, nervous system tumors, and childhood leukemia as the most important types of cancer.[1-4] Low-frequency EMFs lack sufficient energy to break molecular bonds and cannot directly damage cellular DNA and lead to apoptosis.[5,6] Therefore, another hypothesis has suggested that the induced electric fields may modify electric signals that pass across cellular membrane and alter the conformation of receptors on the cell surface, resulting in modifications of biochemical processes, such as protein transcription as well as translational events.[7-9] In our studies, we have focused on the modulation of the proliferative activity of normal and neoplastic lymphocytes in vitro in the presence of a 15Hz, 1G sinusoidal magnetic field (MF). We have also examined the possible alterations in the production of a crucial cytokine, IL-2 (involved in T- and B-cell proliferative responses), and the cellular ratio of T_4 and T_8 cells during field exposure.

MATERIALS AND METHODS

Magnetic Field Exposure System

A magnetic field (MF) exposure system was located inside a 5ft^3 NAPCO (model 5100) incubator. One pair of horizontally oriented Helmholtz coils was used to produce a vertically oriented 15Hz field and also to control the vertical component of the static MF. DC and AC power supplies to this coil pair were isolated from each other by a 24mF capacitor and the DC current was adjusted exactly to cancel the vertical component of the ambient static field. A second pair of vertically oriented coils was used to cancel the horizontal component of the DC field, which resulted in the experimental exposure to a "pure" 15Hz MF of 1G (peak) amplitude.

Isolation of Lymphocytes

Human T and B-lymphocytes were obtained from heparinized peripheral blood drawn from normal healthy volunteers. Briefly, human peripheral blood mononuclear cells were isolated by use of a ficoll-hypaque gradient.[10] Monocytes were removed by adherence and T cells were isolated by negative selection as described in Ref. 11. PHA-M (GIBCO, N.Y.) was used at a concentration of 0.75% (v/v) for T-cell

*Correspondence to: S. Mehta (Surgery), Brown University Medical Ctr., Providence RI 02908.

activation, and anti-human Immunoglobulin M (15μg/ml, BIORAD, Calif.) for B-cell activation as described by S. Mehta et al.[12]

Cell Lines and Growth Factors

Epstein Barr virus (EBV) positive B lymphoblastoid cell lines, Daudi, P3HR1, and Raji were obtained from American Type Culture Collection (Bethesda, Md.). These cells were routinely maintained in RPMI containing 10% fetal bovine serum (FBS) at 37 C, 5% CO_2 at a cell density of 0.2×10^6/ml. Cultures were replenished with fresh media twice a week. Recombinant IL-2 was obtained from Genzyme Corporation (Cambridge, Mass.).

Proliferation Assays

T and B-lymphocytes at various cell densities (0.25, 0.5, and 1.0×10^6/ml) were exposed to PHA-M (0.75% v/v) and antihuman IgM (15 μg/ml) respectively in the presence and absence of an MF exposure system for a period of 3-4 days. Recombinant IL-2 was added at a concentration of 10 and 100 units/ml to assess B-cell growth. Neoplastic B-cell lines (Raji, Daudi, and P3HR1) were also cultured at cell densities 0.05-0.5×10^6/ml. Cell growth[11,12] was assessed by the addition of 1μC_i ^{3}H-thymidine (specific activity 6.7 C_i/mM) for the last 24 h of the incubation period. Cells were subsequently harvested and incorporated thymidine was assayed with a liquid scintillation counter (Model 1214, Pharmacia-LKB, Md.).

Production and Measurement of IL-2

For assaying the production of IL-2, PHA-M (0.75% v/v) was added to separated T cells (10^6/ml) and cultures were incubated in the presence and absence of MF exposure for 24 h. Cells were eliminated by centrifugation and the supernatants were stored at −70 C. Varying amounts of supernatants were added for the measurement of IL-2 to 7-14 day old PHA stimulated T-blasts (10^5/0.2ml) in a microtiter plate for 3 days at 37 C. The cell growth[13] was assessed by the addition of ^{3}H-thymidine in the manner indicated above.

Phenotypic Measurement of T-Cells

T-cells (10^6/ml) were stimulated with PHA-M (0.75% v/v) in the presence and absence of MF exposure and analyzed for T_4 and T_8 cell ratio by flow cytometry on FACScan (Becton Dickinson, Calif.).

RESULTS AND DISCUSSION

Normal Human T and B-lymphocyte proliferative responses were examined initially for the MF exposure effects in a field intensity of 1 G at 15 Hz. Freshly isolated human T-cells at various densities were continuously exposed to a suboptimal concentration of PHA-M (0.75% v/v) and rIL-2 (10 and 100 units/ml) at 37 C for a period of 3 days in the presence and absence of the MF. MF exposure increased cellular proliferation of T-cells upon PHA stimulation as compared to the unexposed control cell cultures. The increase in T-cell proliferation ranged from 1.5 to 2-fold for activated T-cells without added IL-2, 1.5 to 4-fold with IL-2 at 10 u/ml, and up to 3-fold at 100 u/ml in four independent experiments. The monocyte contamination (2-5%) of the T-cell population stimulates autocrine IL-2 production in the culture conditions where IL-2 is not added, which leads to the proliferation of PHA activated T-cells. The maximum response due to MF exposure was obtained at a cell density of 0.5×10^5 cells/well, although an increase was noted for all cell densities utilized. We then tested whether this increase in T-cell proliferation might be associated with an increase in IL-2 production. Our results indicated a 2-4 fold increase in the production of IL-2 as assayed on short term T-blast cell proliferation (7-14 day old PHA stimulated T-cells) as compared to sham control. T_4 cells are shown to produce IL-2 upon antigenic stimulation. However, the increased production did not result from a change in the T_4/T_8 cell ratio upon MF exposures.

Freshly prepared normal human B-cells were also exposed at various cell densities to various concentrations of rIL-2 in the presence and absence of 15Hz, sinusoidal MF at the 1G peak level. The proliferative stimulation index (based on anti-human IgM activated B-cell as background) was significantly greater for cells exposed to the MF, with values two to three times greater for all cell densities than those obtained for the control cell cultures. Neoplastic B-cell lines (Daudi, Raji, and P3HR1) at different cell densities also demonstrated a 1.5-3 fold increase in cell growth as measured by thymidine incorporation.

Our studies on T-cell proliferation confirmed the findings of Cossarizza et al.,[14,15] which also demonstrated an increased growth of human peripheral blood mononuclear cells during stimulation in an exposure to EMF as compared to control groups. However, previous studies[16-18] have demonstrated a decreased

Finally, preliminary data obtained by cytofluorimetric analysis of PEMF-exposed PHA-stimulated peripheral blood lymphocytes collected from aged donors and stained with propidium iodide (a classical method to assess cell cycle phases in human cells) suggested that such fields can increase the rate of cells entering the cell cycle, and particularly the G2/M transition. However, this effect was evident only in about half of the subjects studied, which suggests the presence of a high variability among human subjects.

SINUSOIDAL ELECTRIC FIELDS (EF): EFFECTS ON CELL PROLIFERATION AND CYTOTOXICITY

We have also investigated the genotoxic potential of a 50Hz ac EF with the intensity of 0.5, 2, 5, and 10 kV/m (in air), and its ability to affect important immune responses, such as the capability of human lymphocytes to respond to polyclonal stimuli and of human natural killer (NK) and lymphokine activated killer (LAK) cells to kill tumor target cell lines. A total of 75 subjects were studied. Human PBMC proliferation was assessed by stimulation of cells with various doses (0 to 10 μl/ml) of PHA or with 10 ng/ml anti-CD3 monoclonal antibody (which selectively stimulates T lymphocytes). Cell proliferation was evaluated by measurement of the ^{3}H-TdR incorporation during the last 6 h of a 72h culture. Moreover, the capability of 50 Hz ac EF of different intensities (0.5 to 10 kV/m) to affect the formation of micronuclei was also studied on cells from young donors. The ac EF was generated by a capacitive system composed of two parallel plates powered by an ac voltage generator. Such a system was placed into a CO_2 incubator and the temperature was measured with a thermoresistor. The main effects of the EF tested on the proliferation of PBMC from 21 aged subjects (mean age $\pm$ s.e. 74$\pm$1 years) and from 12 young donors (25$\pm$1 years) was a slight decrease of ^{3}H-TdR incorporation. This effect was particularly evident at 5 kV/m with optimal and supraoptimal PHA doses, and in aged subjects. The cytotoxic capability of NK cells exposed to the abovementioned EF during the assay was unaffected (Fig. 1). The same negative results were obtained on the cytotoxicity of LAK cells, generated as previously described.[10] Furthermore, the exposure did not affect the formation of micronuclei in PBMC from 33 subjects in vitro exposed to 0.5-10 kV/m EF, indicating that the fields employed were devoid of genotoxic risk.[11] In conclusion, these data show that PEMFs have strong biological effects on human lymphocytes and monocytes. The age of the donor is of critical importance in ascertaining possible biological effects of such fields on human cells. On the contrary, many immunological parameters seem unaffected by the EF at the intensity and frequency we have studied.[12]

REFERENCES

1. M. Cantini et al., *J. Bioelectricity* 5: 91-104, 1986.
2. A. Cossarizza et al., *Biochem. Biophys. Res. Commun.* 160: 692-698, 1989.
3. A. Cossarizza et al., *Aging* 3: 241-246, 1991.
4. M. R. Scarfi et al., *Biochem. Biophys. Res. Commun.* 176: 194-200, 1991.
5. A. Cossarizza et al., *Rad. Research* 118: 161-168, 1989.
6. A. Cossarizza et al., *FEBS Lett.* 248: 141-144, 1989.
7. R. Cadossi et al., *FASEB J.* 6: 2667-2674, 1992.
8. C. A. Dinarello, *Blood* 78: 1627-1652, 1991.
9. T. Kishimoto, *Blood* 74: 1-10, 1989.
10. A. Cossarzza et al., *Thymus* 14: 163-170, 1989.
11. M. R. Scarfi et al., in C. T. Brighton and S. R. Pollack, Eds., *Electromagnetics in Medicine and Biology*, San Francisco: San Francisco Press, 1991, 159-162.

12. This study was supported by the CNR-ENEL Project on Interactions of Energy Systems with Human Health and Environment, Rome.

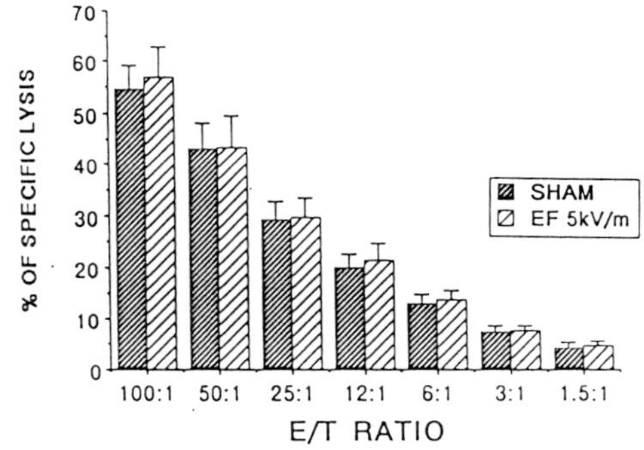

FIG. 1.—NK activity of EF-exposed peripheral blood lymphocytes from 7 young donors. Data indicate percentage of specific lysis of K562 cells in a 4h ^{51}Cr release test, expressed as mean $\pm$ s.e. Statistical analysis by two-tail paired Student's *t* test showed no significance at any point considered.

Martin Blank, Ed., *Electricity and Magnetism in Biology and Medicine*

LYMPHOMA IN MICE EXPOSED TO 60Hz MAGNETIC FIELDS: PILOT STUDIES

T. Makinodan,* W. J. Peterson, A. Kharazi, and J. M. G. Taylor

VA Medical Center, West Los Angeles, Calif. and University of California at Los Angeles

The purpose of our pilot studies is to (a) validate the experimental cancer model of Kaplan,[1] (b) establish the appropriate doses of ionizing radiation to be used in the chronic magnetic field exposure study, (c) develop tumor endpoints, and (d) enable the staff to become familiar with GLP (good laboratory practice) procedures in performing animal husbandry and pathophysiological assays. The results obtained to date suggest that the Kaplan murine lymphoma model will be an attractive model to assess the role 60Hz magnetic fields play in lymphoma formation. Several promising parameters have been identified, which should complement the classical tumor endpoint parameter, frequency of mice with lymphoma.

Recent series of epidemiologic studies[2-4] suggested that there may be an association between exposure to magnetic field and cancer. This suggestion had a disconcerting effect on the public; many feared that a causal relationship exists between exposure to magnetic field and cancer. There is thus a pressing need for more research, both epidemiologic and in animals. The urgency for animal research (the focus of this study) stems from the fact that there is no documented mechanism at present to explain how a force of very low energy can transform cells into a cancerous state.

Our objective is to determine whether a 60Hz magnetic field is a carcinogen by testing for lymphoma in mice chronically exposed to magnetic fields from less than 1 mG to 10 G. Since ionizing radiation will be used to enhance the intrinsic incidence of lymphoma, we have performed pilot studies using GLP procedures to evaluate the carcinogenicity of ionizing radiation only. The results obtained to date are presented here.

EXPERIMENTAL PROTOCOL

Female C57BL/6J mice, 4 and 6 weeks old, were subjected to either an acute exposure (radiation given in a single exposure) or chronic exposure (radiation given in four equal parts at intervals of 8 days) of ionizing radiation from a ^{60}Co source at a dose rate of 0.34-0.37 Gy/min. Mice were then inspected daily until half of the sham-irradiated mice died. Necropsies were performed of dead and moribund mice. The thymus, spleen, lymph nodes, bone marrow, liver, and other tissues with gross lesions were examined histologically for evidence of lymphoma. For histologic analysis, the tissues were frozen *en bloc* and sectioned, and sections were processed for histopathologic and immunohistologic analysis.

INCIDENCE OF LYMPHOMA

Six-week old, female C57BL/6 mice were exposed to 0, 1, 2, 4, or 6 Gy in one dose or four equal doses. The results, based on 300 days of observation, showed that (a) chronic exposure is more effective than acute exposure in lymphoma induction, (b) the lymphoma-inducing effect of ionizing radiation is dose dependent (i.e., 6 Gy > 4 Gy > 2 Gy ~ 1 Gy ~ 0 Gy, and (c) the lag time (time after radiation of appearance of mice with lymphoma) is inversely related to the dose ;of ionizing radiation.

Four-week-old, female C57BL/6 mice were then exposed to 0, 1, 2, 4, or 6 Gy in four equal doses. The results, based on 300 days of observation, showed that (a) the lymphoma-inducing effect of ionizing radiation is dose-dependent (i.e., 6 Gy > 4 Gy > 2 Gy ~ 1 Gy ~ 0 Gy), (b) the lag time is inversely related to the dose of ionizing radiation, and (c) 4-week-old mice are more susceptible to radiation-induced lymphoma than 6-week-old mice.

---

*Correspondence to: T. Makinodan (GRECC), VA Medical Ctr., Los Angeles CA 90073.

TABLE 1.—Incidence of lymphoma in 4- and 6-week-old female C57BL/6 mice 300 days after exposure to varying doses of ionizing radiation.*

Age	Total radiation dose (Gy)									
	Acute					Chronic				
(wk)	0	1	2	4	6	0	1	2	4	6
	Percent death/Lymphoma									
4	ND	ND	ND	ND	ND	0	0	0	7	69
6	0	0	1	1	3	1	0	0	9	23

*Sample size: 22-39 mice/group; ND = not done.

These results confirm those obtained initially by Kaplan and his co-workers 40 years ago.[1] Our ability to obtain data comparable to those obtained initially by Kaplan and his co-workers is remarkable, since the strain of mice we used differed and since animal husbandry has changed drastically over the past 40 years. This result underscores the attractiveness of the Kaplan murine model to assess the role 60Hz magnetic fields play in lymphoma formation.

TUMOR ENDPOINTS

The main parameter we use to assess the carcinogenicity of 60Hz magnetic fields will be *frequency of mice with lymphoma*. Therefore, one of our objectives in carrying out these pilot studies is to determine whether additional parameters can be used, which are more sensitive than or equal to the frequency of mice with lymphoma. The parameters being investigated are summarized in Table 2.

TABLE 2.—Tumor endpoints.

1. Frequency of mice with lymphoma
   A. Benign
   B. Malignant (metastatic)
2. Kinetic profile of mice with lymphoma
3. Growth potential in vitro
   A. Mortal
   B. Immortal
4. Cell surface antigens

A. Thy 1	C. CD8	E. CD44	G. TCR γ/δ
B. CD4	D. B220 (CD45R)	F. TCR α/β	H. J11D (HSA)

5. Mutations in LTR (long term repeat) of MMTV provirus
   A. Deletions
   B. Deletion + substitution

In assessing the frequency of mice with lymphoma, we found that most irradiated mice die of lymphoma which are metastatic (e.g., 32/34 in irradiated 4-week-old mice). To date, we have found only 2 unirradiated mice that died of lymphoma; both are metastatic. We anticipate that few more unirradiated mice will die of lymphoma in our ongoing pilot studies, which will enable us to determine whether the naturally occurring lymphomas are generally metastatic. If so, the frequency of mice with *benign* lymphoma may not be a good parameter in assessing the carcinogenicity of 60Hz magnetic fields.

A promising parameter under investigation is the kinetic profile of the frequency of mice with lymphoma, which is based on our preliminary findings that the time of appearance of mice with lymphoma can be shortened and the frequency of mice with lymphoma increased by exposure to irradiation. Such an outcome would suggest that 60Hz magnetic fields, acting as a promoter, can further shorten the time of appearance and increase the frequency of mice with lymphoma.

A parameter reflective of the growth potential in vitro of lymphomas has been developed. The assay is based on the ability of lymphoma cells to grow in Fisher's medium for at least 2 weeks. Those that can grow were classified as "immortal" and those that cannot as "mortal." We have found that all benign lymphoma cells are mortal, whereas only about 10% of the metastatic lymphoma cells (1/9) are mortal. This finding suggests that the frequency of immortal lymphoma cells may not be an effective parameter in assessing the carcinogenicity of 60Hz magnetic fields.

Another promising parameter is the expression of cell-surface antigens by immortal lymphoma cells. We found that immortal lymphoma cells all expressed the Thy 1 antigen, a marker for T cells, but they differed in their expression of the other test cell surface antigens. The results also suggest that immature parenchymal thymocytes are highly vulnerable to radiation-induced transformation into metastatic lymphoma cells. However, we have not found immortal lymphoma cells with cell surface antigens characteristic of progenitor parenchymal thymocytes and early postprogenitor parenchymal thymocytes. It is possible that 60Hz magnetic fields, acting as a promoter, can transform progenitor and early postprogenitor cells into metastatic lymphoma cells.

Finally, we are investigating the patterns of mutation in the LTR (long terminal repeats) region of the murine mammary tumor virus proviral gene, as a potential molecular parameter. Our preliminary results suggest that deletion in LTR is very high in chemically-induced lymphoma cells, whereas no deletion could be detected in LTR of radiation-induced lymphoma cells. It is possible that 60Hz magnetic fields, when combined with ionizing radiation, can induce deletions in LTR of a magnitude less than that induced chemically.

CONCLUSION

Our pilot studies demonstrate that the Kaplan murine lymphoma model is a very attractive model to assess the role 60Hz magnetic fields play in lymphoma formation. The results are very reproducible, as demonstrated by our ability to generate data comparable to those obtained by Kaplan and his co-workers 40 years ago. As with the C57BL/Ka strain used by Kaplan and his co-workers, the C57BL/6 strain used here as the animal model is relatively resistant to naturally occurring lymphoma (<10%), but is highly susceptible to radiation-induced lymphoma formation (e.g., >70% after chronic radiation exposure). What that means is that in our main study, in which we shall be assessing the carcinogenic potential of 60Hz magnetic fields, there will be a signal-to-noise range of 7-fold or better if we take several critical factors into account: age of mice, dose of ionizing radiation, and chronicity of radiation exposure. Finally, our search for tumor endpoints has yielded encouraging results. Thus, we have found three promising parameters to assess the carcinogenic potential of 60Hz magnetic fields, which should complement the classical tumor endpoint parameter, frequency of mice with lymphoma. The parameters are: (i) kinetic profile of the frequency of mice with lymphoma, (ii) cell surface antigens expressed by immortal lymphoma cells, and (iii) stability of the LTR proviral gene in immortal lymphoma cells.[5]

REFERENCES

1. H. S. Kaplan and M. B. Brown, "A quantitative dose-response study of lymphoid-tumor development in irradiated C57 black mice," *J. Nat. Cancer Inst.* 13: 185-208, 1952.

2. N. Wertheimer and E. Leeper, "Electrical wiring configurations and childhood cancer," *Am. J. Epid.* 109: 273-284, 1979.

3. L. Tomenius, "50 Hz electromagnetic environment and the incidence of childhood tumors in Stockholm County," *Bioelectromagnetics* 7: 191-207, 1980.

4. D. A. Savitz and E. E. Calle, "Leukemia and occupational exposure to electromagnetic fields: Review of epidemiologic surveys," *J. Occup. Med.* 29: 47-51, 1987.

5. This work was supported by U.S. Department of Veteran Affairs Medical Research Funds, Electric Power Research Institute project EPRI 2965-11, and British Columbia Hydro.

Martin Blank, Ed., *Electricity and Magnetism in Biology and Medicine*

GENOTOXICITY OF MICROWAVES

G. d'Ambrosio* and R. Massa

University of Naples, Italy

M. B. Lioi

University of Basilicata, Potenza, Italy

M. R. Scarfi and O. Zeni

CNR-IRECE, Naples, Italy

Peripheral-blood human lymphocytes cultured in vitro were exposed to continuous wave (CW) and 50Hz amplitude-modulated (AM) microwave radiation. Other exposure parameters were: frequency, 9 GHz; specific absorption rate (SAR), 90 mW/g; exposure duration, 10 min. The possible genotoxic effect was evaluated by means of cytokinesis-block micronucleus method. No genotoxic effects were induced by either exposure conditions (CW and AM) in the majority of blood samples tested. However, a micronuclei (MN) increase has been observed in blood samples taken from two subjects.

The possibility that microwave exposure induces genotoxic effects in mammalian cells has been considered by several investigators. In particular, studies on the genetic effects of microwave irradiation in vitro[1-5] and/or in vivo[6-8] have been carried out; results obtained after microwave irradiation are not univocal, but 50Hz fields demonstrated their clastogenic effectiveness rather clearly.[9,10]

Our research goal was to investigate the cytogenetic effects of microwaves, both continuous wave (CW) and 50Hz amplitude-modulated (AM) on human lymphocytes in vitro. The analysis was carried out by the cytokinesis-block micronucleus technique, which is known to be a more sensitive method than the classical scoring of aberrations.[11] Following a previous paper,[12] in which the results of CW irradiation were reported, we now report on exposures to both CW and AM. Blood samples from 21 donors were used and the relevance of individual factors was once again demonstrated.

MATERIALS AND METHODS

Microwave Exposure Set-up

The microwave exposure system was conceived so as to obtain stable and precisely known conditions in terms of both SAR and sample temperature. A waveguide system was the simplest and most reliable choice: continuous reading and control during the exposure of incident, reflected, and transmitted powers was provided. The frequency was 9 GHz (CW and 50 Hz amplitude modulated), the SAR was 90 mW/g, and the exposure time was 10 min. The apparatus assembled for our experiments is shown schematically in Fig. 1. The microwave signal (signal source HP8690B, 8-12 GHz) was amplified by a TWT amplifier (Varian VZM6991G3A, 40 dB gain) and was fed (through a 3dB attenuator, HPX382A) into a WR90 rectangular waveguide reflectometer, which allowed us to evaluate the power P_i incident on the sample and the reflected power P_r during the exposure. Two sterile test tubes (polypropylene, external diameter 12 mm, height 38 mm) containing 2 ml (1 ml each) of whole blood were longitudinally inserted into a WR112 waveguide section (transverse dimensions 28.5 × 12.6 mm). Tapered waveguide sections connected the WR112 waveguide sample holder to the WR90 reflectometer and to a power sensor measuring the power P_t transmitted beyond the sample.

---

*Correspondence to: G. d'Ambrosio (Electronic Engg), University of Naples, Via Claudio 21, I-80125 Naples, Italy.

As to thermal conditions, no attempt was made to avoid the temperature rise due to microwaves: a circulating water flow, obtained by means of a water jacket surrounding the sample holder waveguide, kept the sample at the desired initial temperature and served as a thermally stable environment. Under these conditions the controlled microwave power absorption gave a controlled temperature rise of 5 C (from 30 to 35 C). After exposure, the two exposed samples were mixed in order to obtain one sample uniformly treated.

Cell Cultures

We tested human lymphocytes from 21 healthy donors aged 21-67 years. For each subject, a control culture was set up in order to compare the exposed samples to the unexposed one. After microwave exposure, 10ml cell cultures were set up, as follows: 0.8 ml of peripheral heparinized whole blood was added to RPMI 1640 supplemented by 15% FCS, 100 μl of PHA (Gibco), and 100 μl of L-glutamine. After 44 h of incubation at 37 C, cytochalasin B (Sigma) was added to give a final concentration of 3 μg/ml according to Fenech and Morley.[11] After 72 h of growth, cultures were harvested and slides were prepared as previously described.[13] Coded slides were scored blindly by means of a Leitz microscope at 1000x magnification. The results are reported as the ratio between binucleated cells presenting MN and the total number of binucleated cells.

RESULTS AND DISCUSSION

In order to investigate the genotoxic action of microwaves, lymphocytes from 21 healthy donors were irradiated. The percentage of binucleated cells presenting MN with respect to the total number of binucleated cells is shown in Table 1. In control cultures this value ranges from 0.58 to 2.50%. This wide dispersion suggests that the comparison between untreated and microwave-treated samples must be carried out on each subject. The results obtained can be summarized as follows:

(1) in most blood samples tested (19 out of 21) the frequency of MN in cultures exposed to either CW or 50Hz AM microwaves was similar to the spontaneous frequency observed in unexposed cultures;

(2) in two subjects (number 4 and 8 in the table) a significant increase of MN in exposed samples was found.

TABLE 1.—Percentage of binucleated cells with MN. (Total number of binucleated cells scored shown in parentheses.)

Subject	Control	CW	50 Hz AM
1	1.32% (1896)	--	1.61% (4221)
2	1.75% (1774)	--	1.42% (2215)
3	1.70% (1704)	--	1.27% (3453)
4	1.18% (1702)	--	3.5 % (965)
5	2.03% (3149)	--	2.45% (2610)
6	1.19% (558)	--	1.74% (2015)
7	1.5 % (3804)	--	1.67% (2749)
8	2.16% (601)	3.22% (2171)	4.33% (3139)
9	1.65% (1213)	1.85% (1079)	1.60% (752)
10	1.5 % (1331)	1.80% (723)	1.90% (958)
11	2.5 % (1166)	3.7 % (1371)	2.7 % (2201)
12	1.74% (3039)	1.93% (1518)	2.87% (2336)
13	0.95% (3884)	1.64% (3848)	0.77% (4948)
14	1.29% (932)	1.25% (3369)	1.07% (2518)
15	0.85% (2466)	0.56% (2493)	--
16	1.24% (3056)	0.95% (3143)	0.93% (3534)
17	0.69% (727)	1.31% (2292)	0.96% (1559)
18	1.05% (2385)	0.87% (921)	1.05% (1244)
19	0.97% (1558)	1.53% (1306)	1.05% (1338)
20	0.76% (1319)	1.07% (1397)	1.46% (1029)
21	0.58% (2227)	0.63% (2836)	1.03% (970)

In conclusion, in most blood samples tested no genotoxic effects in terms of MN were induced either by CW microwave irradiations or by modulated microwaves. However, the same microwave exposure conditions gave an MN increase in two subjects.

It is well known that individual factors, such as age or health conditions, may influence the response to the MN test. Further investigation will permit correlation of the MN increase (when present) with specific individual factors.

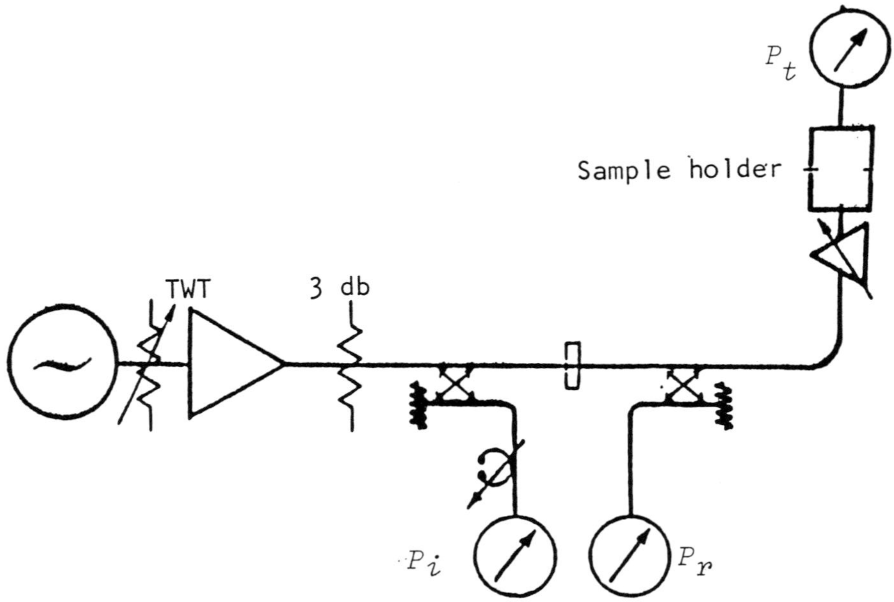

FIG. 1.—Exposure set-up: SAR, 90 mW/g; t = 10 min; T = 30 to 35 C; f = 9 GHz.

REFERENCES

1. K. T. S. Yao, *Genetics* 83: 84, 1976.
2. K. T. S. Yao, *J. Heredity* 73: 133-138, 1982.
3. D. C. Lloyd, R. D. Saunders, P. Finnon, and C. I. Kowalezuk, *Int. J. Radiat. Biol.* 46: 135-141, 1984.
4. V. Garaj-Vrhovac, D. Horvat, and Z. Koren, *Mutation Res.* 243: 87-94, 1990.
5. V. Garaj-Vrhovac, D. Horvat, and Z. Koren, *Mutation Res.* 281: 181-186, 1992.
6. P. Italiano, A. Ciaccia, P. Malacarne, P. Piffanelli, and A. Della Piccola, *Minerva Medica* 67: 1557-1560, 1976.
7. R. D. Saunders, C. I. Kowalezuk, C. V. Beechey, and R. Dunford, *Int. J. Radiat. Biol.* 53: 983-992, 1988.
8. C. V. Beechey, D. Brooker, C. I. Kowalezuk, R. D. Saunders, and A. G. Searle, *Int. J. Radiat. Biol.* 50: 909-918, 1986.
9. I. Nordenson, K. Hansson Mild, S. Nordstrom, A. Sweins, and E. Birke, *Radiat. Environ. Biophys.* 23: 191-201, 1984.
10. G. d'Ambrosio, R. Massa, D. Di Berardino, M. B. Lioi, A. Scaglione, and M. R. Scarfi, *J. Bioelectricity* 7: 239-245, 1988-89.
11. M. Fenech and A. A. Morley, *Mutation Res.* 161: 193-198, 1986.
12. G. d'Ambrosio, M. B. Lioi, R. Massa, D. Di Berardino, and M. R. Scarfi, *Recent Advances in Applied Electromagnetics,* Liguori, Naples, 1991 (in Italian).
13. M. R. Scarfi, F. Bersani, A. Cossarizza, D. Monti, G. Franceschetti, and G. Franceschi, *Biochem. Biophys. Res. Comm.* 176: 194-200, 1991.

Martin Blank, Ed., *Electricity and Magnetism in Biology and Medicine*

THE PHENOMENON OF AUTOIMMUNITY FROM THE EFFECTS OF NONIONIZING MICROWAVE RADIATION

G. Vinogradov*

A. N. Marzeev Research Institute of General and Communal Hygiene, Kiev, Ukraine

The purpose of these experiments has been the investigation of the antigen structure changes in tissue and serum protein under microwave radiation and determination of the biological role of the antibodies obtained. We used male and female white Wistar rats. The exposure to 2375MHz microwave radiation at power densities of 0.5, 0.05, and 0.01 mW/cm^2 was carried out in the special anechoic chambers for 30 days, 7 h daily. Analysis of the results of these experiments shows a possible development of autoimmunity under microwave radiation. This process is accompanied by the formation of autoantibodies that possess the damaging effect.

Microwaves are an active physical factor of the environment[1] that lead to a change in the immunological reactivity of the organism and nonspecific resistance. The data of some reports show that the processes of immunogenesis are broken by microwave radiation, more often in the direction of their suppression, and the type of infection process is changed.[1,2]

The aim of this study was the development of an experimental basis for changes in the antigen structure of tissue and serum protein under the influence of microwave radiation, and the determination of the biological role of the resulting antibodies.

White male and female Wistar rats were used in the experiment. The exposure to 2375MHz microwave radiation at power densities of 0.5, 0.05, and 0.01 mW/cm^2 was carried out in the special anechoic chambers for 30 days, 7 h daily, the power density constantly monitored. A group of sham-exposed animals was used as controls.

Antibodies to brain tissue were defined by the complement-binding reaction. Next, each serum sample was divided into two parts: one underwent immunochemical analysis, the other was used for intraperitoneal injections given to 10-days-pregnant female rats to study the effect of antibodies on fetuses according to the scheme of Ref. 3. We studied the changes in the brain antigen structure of animals exposed to microwaves by electrophoresis of the proteins of brain tissue, conducted according to the "multiphor" system in 1.0% agarose gel. In experimental samples the peaks in the γ-globulin zone (as evidenced by electrophoretic mobility and intensity of coloring) differ from those of the control samples. Of special interest are brain extracts received from animals exposed to microwaves at 0.5 mW/cm^2. In these samples the amount of γ-globulin was increased threefold compared to the controls and their electrophoretic mobility changed. The amount of β-globulins decreased threefold and the relative percentage of protein corresponding to the zone located according to their electrophoretic properties between α_1 and α_2 standard globulins decreased to half. The data obtained show that microwaves cause the appearance of a special protein with high immunogenic properties in the brain tissue of exposed animals. These results indicate a theoretical premise for the possibility of autoimmune development while the organism is under the influence of nonionizing microwave radiation.

In the same experiment, the antibrain complement-binding auto-antibodies directed against normal brain tissue were found in the blood serum. The level of antibodies ws 1.6 $\pm$ 0.19 or 2.46 $\pm$ 0.2 for exposure to 0.05 or 0.5 mW/cm^2, respectively. The exposure at 0.01 mW/cm^2 did not cause a corresponding effect.

To characterize the antibodies found, an immunochemical analysis of the serum of exposed animals was carried out. It was conducted by analytical ultracentrifuging and also by the rocket immunophoresis method with the "multiphor" system. The most interesting data were obtained for the animals exposed to 0.5 mW/cm^2.

---

*Correspondence to: G. Vinogradov (Bioscience and Biotechnology), Drexel University, Philadelphia PA 19104.

A new peak (C peak) appeared at a sedimentation constant of about 8.6. The results of the analysis in rocket immunophoresis showed its relation to globulins. Analytical ultracentrifuging data also produced evidence for the globulin nature of C peak protein. By sedimental constant and molecular weight it is closer to JgG. The rocket immune electrophoresis showed a statistically significant decrease of peak heights.

We know that not every stimulation of autoimmune reactions can be considered a pathology. Moreover, the fact of revealing antibodies cannot be the criterion of autoimmunity. The necessary condition is a proof of their damaging effect. That is why we performed the next stage of investigation of this process: revelation of the biological role of the antibodies found and study of their effect on fetuses and offspring. The data obtained prove that neither injections of the blood serum of sham-exposed rats to pregnant female animals in the absence of such antibodies, nor injections of a sodium chloride solution, change the average weight and length of the fetus, the number of offspring, and postimplantation deaths at the level of the control. No changes were observed following injections of blood serum from the rats exposed to 0.01 and 0.05 mW/cm^2, although the presence of antibodies in the serum was registered.

Analysis of the data of the possible effect of animal blood serum following exposure to microwaves at 0.5 mW/cm^2 indicates that under these conditions, one observes a significant increase in the postimplantational death rate of the fetuses: 3.04 ± 0.40, compared to 0.60 ± 0.23 in sham-exposed animal serum effects, accompanied by a significant decrease in the quantity of the offspring, as well as in weight and size. No differences were found in the effect of blood serum of control animals on the postimplantation death from the serum of animals exposed to 0.01 and 0.05mW/cm^2 microwaves, although antibrain antibodies were found in the serum of the latter.

In sum, antibodies contained in the blood serum of rats exposed to microwaves at 0.05 mW/cm^2 have no damaging effect and lead to the support of homeostasis, whereas 0.5mW/cm^2 microwaves show damaging effects on the fetus and promote an increase in postimplantation death. As an additional argument to prove the effect of autoantibodies contained in blood serum, the control investigations with blood serum of animals exposed at 0.5 mW/cm^2 exhausted brain tissue. The results of the experiment show that after the injection of white female pregnant rats with exhausted blood serum under the identical conditions caused the postimplantation death of fetuses on the level $0.70 \pm 0.37\%$, i.e., coinciding with control.

Thus the results of these experiments show that microwaves may be considered as an inductor of autoimmunity. This process is accompanied by antigen-structure changes of the tissue protein of the brain and formation of the autoantibodies that possess the damaging effect.[4]

REFERENCES (in Russian)

1. M. Shandala, *Vestnik AMS USSR* 1: 9-15, 1981.
2. H. Vinogradov, *Hygiena and Sanitaria* 3: 9-11, 1991.
3. M. Shandala and H. Vinogradov, *Vestnik AMS USSR* 10: 89-100, 1982.

4. The author thanks Prof. M. Shandala for helpful discussions and comments on the manuscript.

Martin Blank, Ed., *Electricity and Magnetism in Biology and Medicine*

PROTEIN SHEDDING AND ELF MAGNETIC FIELDS: ANTIBODY BINDING AT THE CD3 AND CD20 RECEPTOR SITES OF HUMAN LYMPHOCYTES

R. P. Liburdy,* D. E. Callahan, and J. D. Harland

Lawrence Berkeley Laboratory, Calif.

We investigated whether ELF magnetic fields influence the release of proteins from the cell surface of human lymphocytes. Protein shedding was followed by monitoring of the release of fluorescently labeled monoclonal antibodies against CD3 (pan T-cell marker) and CD20 (pan B-cell marker) cell surface determinants. Preliminary results indicate that T lymphocytes exhibit an approximate one-fold increase in anti-CD3 antibody release in the presence of a 60Hz magnetic field (220 G, 1.7 mV/cm, 60 min, 37 C) compared with isothermal control treatment. In contrast, B lymphocytes did not show a marked change in anti-CD20 antibody release in the field compared to controls. Such responses suggest that human T and B lymphocytes respond differently to ELF fields.

An important way that electromagnetic fields can alter membrane structure and function is to remove proteins from the cell surface.[1-3] Proteins at the cell surface are responsible for recognition of extracellular signals important to cellular function. Two classes of proteins, intrinsic and peripheral, constitute the membrane bilayer (Fig. 1). Peripheral proteins are loosely bound to the cell membrane, do not span the bilayer, are anchored via cationic and anionic bridges, and are readily water soluble. They are candidates for protein shedding in ELF magnetic fields.

Previously we have shown that ELF fields (220 G, 1.0 mV/cm, 60 min, 37 C) lead to enhanced protein shedding in rat thymic lymphocytes;[4] these are precursor or immature T lymphocytes. Proteins are released in the 6k-150kDa range that have a net negative surface charge. This result demonstrates that 60Hz fields release endogenous proteins from the cell surface of immature T-cells.

We now asked whether specific peripheral proteins placed on the cell surface of human T and B lymphocytes are also released by 60Hz magnetic fields. Results indicate that antibody to CD3 on T lymphocytes show an approximate one-fold increase in shedding. This increase in protein shedding is consistent with our previous findings for rat thymic lymphocytes. However, B lymphocytes did not exhibit a large change (~ 10%) in release of antibody to CD20 in 60Hz fields compared to isothermal-treated control cells.

METHODS

Monoclonal antibodies were obtained from Becton Dickinson, San Jose, Calif. The pan-T-cell monoclonal was fluorescein isothiocynate(FITC)-conjugated and directed against Leu-4 (CD3). The pan-B-cell monoclonal was phycoerythrin(PE)-conjugated and directed against Leu-16 (CD20). Human peripheral blood lymphocytes were isolated by use of Sigma Ficoll-Hypaque 1077. Monoclonal labeling procedures were as specified by the manufacturer at 4 C. Exposures to 60Hz magnetic fields were as described, with the use of special multiring Petri dishes to establish a spatially uniform induced electric field of 220 G, 1.7 mV/cm.[5-7] Control treatment of cells consisted of incubating cells in an isothermal waterbath remotely located from the solenoid, at the same time field exposures were conducted. This waterbath was completely shielded in mu-metal to eliminate endogenous 60Hz magnetic fields (<5 mG). Immediately following treatment, cells were harvested and spun in a high-speed microfuge, and the cell-free supernatants were collected for fluorescene analyses by a CytoFluor 2350 automated fluorescence plate reader (Millipore, Bedford, Mass.).

---

*Correspondence to: R. P. Liburdy, LBL Bioelectromagnetic Facility (Bldg 74), University of California, Berkeley CA 94720.

RESULTS

Previously we observed that ELF fields lead to increased protein shedding in rat thymic lymphocytes,[4] which are immature T lymphocytes. This result suggested that T lymphocytes may respond in a similar manner and release surface proteins in ELF fields. However, we wished to assess whether specific proteins would be released from the cell surface. Thus, we employed FITC-labeled anti-CD3 monoclonal antibodies to label CD3 receptor sites on mature human peripheral blood T lymphocytes.

Figure 2 depicts the results of an experiment in which FITC-labeled T lymphocytes were treated with a magnetic field or placed in an isothermal waterbath and the supernatant was analyzed for the presence of FITC-anti-CD3 antibody. We observed that CD3 antibody was released from sham treated cells, and that this release was significantly increased in the magnetic field. An approximately 73% increase in fluorescence was detected. This figure corresponds to an increase in CD3 antibody release of 19 to 41 ng/ml, which is based on a standard curve constructed by use of FITC-anti-CD3 in which fluorescence was correlated to known protein concentration of the monoclonal stock. These results are consistent with results from rat thymic lymphocytes mentioned above, in which an increase in endogenous protein release was observed.

We have not previously evaluated B lymphocytes for protein shedding. To do so we performed an experiment analogous to that described for T lymphocytes employing PE-labeled-anti-CD20 antibody. Figure 3 depicts the results of this experiment. Sham-treated B lymphocytes released anti-CD20 antibody and magnetic field treatment was associated with less protein shedding. The magnetic field resulted in approximately 11% less protein shedding. Although this difference appears statistically significant, such a slight change is probably not biologically meaningful.

DISCUSSION

We have evaluated protein shedding previously in erythrocytes exposed to electromagnetic fields.[1-3] Both gel electrophoresis and HPLC detected the enhanced release of *endogenous* proteins from the cell surface during exposures. Recently, this phenomenon was extended to detect the release of endogenous proteins from the rat thymic lymphocyte by HPLC techniques.[4] The present study reports the first evidence of changes in antibody release from human peripheral blood lymphocytes. Specifically, *T lymphocytes* respond to magnetic fields by releasing significantly more anti-CD3 antibody than do isothermal control cells. This finding is consistent with increased release of endogenous proteins from rat thymic lymphocytes.[4] *B lymphocytes* do not appear to be markedly influenced (~ 11% change) by the magnetic fields.

Protein shedding is a natural phenomenon and it is observed in the control isothermal-treated lymphocytes. Peripheral proteins (Fig. 1) represent the class of proteins loosely bound to the cell surface via noncovalent interactions such as hydrogen bonds, anionic/cationic bonds, and polarity/hydrophobic interactions. Possible mechanisms of release must involve chaotropic interference with these interactions. The mechanism of field action for protein shedding is not known, but probably involves movement of charge in the oscillating field that could disrupt noncovalent interactions. We plan further studies to characterize protein shedding using (a) cells maintained in buffers of different conductivity, which should define a dose-response correlated with induced current; and (b) cells exposed in a constant magnetic field at various induced electric fields using our multiring cell culture plate,[5-7] which should also define a correlation with induced current.[8]

REFERENCES

1. R. P. Liburdy and A. Penn, *Ann. N.Y. Acad. Sci.* 435: 302-305, 1985.
2. R. P. Liburdy and P. F. Vanek Jr., *Rad. Res.* 109: 382-395, 1987.
3. R. P. Liburdy, A. W. Rowe, and P. F. Vanek Jr., *Rad. Res.* 114: 500-514, 1988.
4. R. P. Liburdy, M. Dall'Era, and D. J. de Manicor, 13th Bioelectromagnetics Meeting, Salt Lake City, Utah, 1991, Abstr. E-1-2.
5. R. P. Liburdy, *FEBS Letters* 301: 53-59, 1992.
6. R. P. Liburdy, in C. Ramel and B. Norden, Eds., *Interaction Mechanisms of Low-Level Electromagnetic Fields in Living Systems,* London: Oxford University Press (in press).
7. R. P. Liburdy, *Ann. N.Y. Acad. Sci.* 649: 74-95, 1992.

8. Support provided in part by the U.S. Department of Energy under contract DE-AC-3-76SF00098, and NCI under grand CA53711.

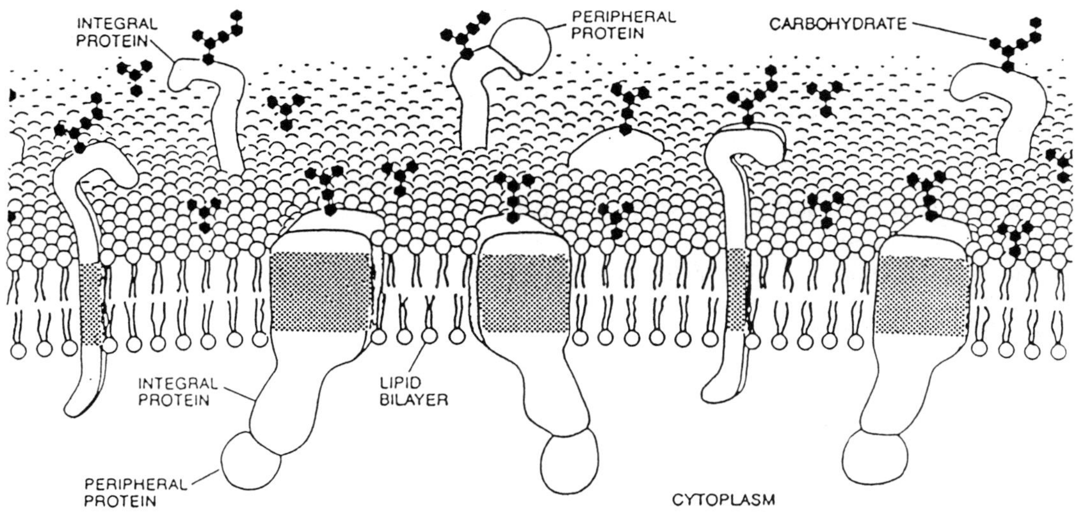

FIG. 1.—Schematic representation of peripheral and intrinsic proteins in cell membrane. Peripheral proteins are likely candidates for protein shedding; unlike intrinsic proteins, they do not span the bilayer, are bound via cationic/anionic bridges, are released by mild chaotropic agents that do not disrupt the bilayer, and are readily water soluble and not associated with membrane lipids.

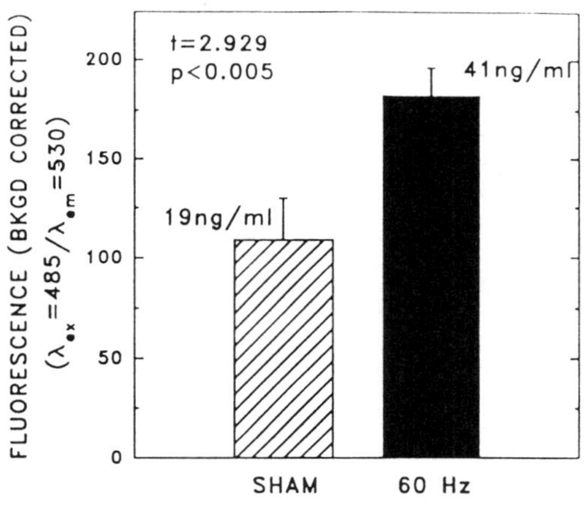

FIG. 2.—Protein shedding of anti-CD3 antibody from human T lymphocytes (mean ± S.E., n = 12).

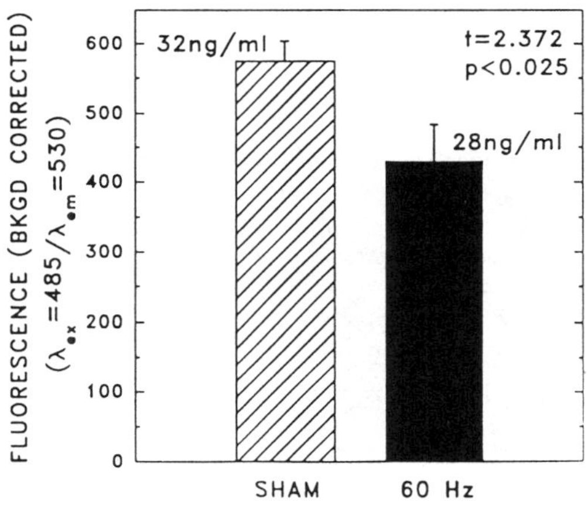

FIG. 3.—Protein shedding of anti-CD20 antibody from human B lymphocytes (mean ± S.E., n = 12).

Martin Blank, Ed., *Electricity and Magnetism in Biology and Medicine*

GENOTOXIC EFFECTS OF 50Hz ELECTROMAGNETIC FIELDS

M. R. Scarfí,* O. Zeni, and G. Franceschetti

CNR-IRECE, Naples, Italy

G. d'Ambrosio and R. Massa

University of Naples

A. Cadossi, A. Cossarizza, D. Monti, and C. Franceschi

University of Modena

M. B. Lioi

University of Basilicata, Potenza, Italy

F. Bersani

University of Bologna

Genotoxic effects of extremely low frequency (ELF) electromagnetic fields were evaluated on mammalian lymphocytes by various exposure systems. Bovine and human blood were used. The electric field (EF) was applied by means of either high-voltage plane electrodes external to the culture (capacitive coupling), or by means of current injection through the culture, via agarose bridges that avoided direct contact between electrodes and culture. The magnetic field (MF) was applied by means of Helmholtz coils, fed by a pulse generator. Chromosomal aberrations increase was found in bovine lymphocytes after heavy EF/current density stress. No micronuclei increase was found in human lymphocytes after lower electric/magnetic stress.

Recently, much attention has been paid to the possible genotoxic effects of ELF fields, in connection with environmental electromagnetic (EM) pollution.[1] In the past 10 years our group has been involved in ascertaining the genotoxic effects of 50Hz electromagnetic fields on various mammalian lymphocytes. Various biological tests and EM exposure have been carried out; the results depend on the exposure system used.

RESULTS

We set up a first group of experiments on bovine lymphocytes in order to assess the clastogenic effect of a 50Hz EF.[2] A voltage of 6.5 kV was applied to the flask containing the sample by means of external electrodes; the current density was 2.4 $\mu A/cm^2$ and the EF (evaluated into the sample) was 16 mV/m. The exposure duration was 72 h in a CO_2 incubator at 37 C. Chromosomal aberrations and sister chromatid exchanges (SCE) were detected in both control and exposed samples. A statistically significant difference was found in the frequency of aberrations, which were three times higher in exposed samples (28.27%) than in the control ones (9.75%). No difference was found in terms of SCE induction. To verify whether the observed effects were due to a culture medium alteration as a consequence of the high voltage applied, additional tests were carried out exposing the culture medium for 72 h and then culturing the lymphocytes. This treatment did not affect chromosomal aberrations induction.

---

*Correspondence to: M. R. Scarfi, CNR-IRECE, Via Diocleziano 328, I-80124 Naples, Italy.

Another experiment, also on bovine lymphocytes exposed to a 50Hz EF, was done with a different exposure set-up and electric conditions:[3] a sinusoidal electric current was injected by means of platinum electrodes and agarose bridges (Fig. 1). Four conditions were tested: (i) control cultures grown in culture flasks; (ii) sham-exposed cultures grown in exposure chambers, with no voltage applied; (iii) cultures exposed to 0.1 mA/cm^2; and (iv) cultures exposed at 1.0 mA/cm^2. The exposure duration was 45 h. No significant difference was found in the frequency of aberrations between controls (9.0%), sham-exposed (12.4%), and 0.1mA/cm^2 exposed cultures (12.35%), but a highly significant difference was found between them and cultures exposed at 1 mA/cm^2 (18.9%).

In another set of experiments[4] we exposed human peripheral blood lymphocytes from 16 healthy subjects to a 50Hz pulsed electromagnetic field for 72 h. The exposure system consisted of a pair of horizontal Helmholtz coils connected to a pulse generator; the field characteristics are shown in Fig. 2. The test used to evaluate the genotoxic effect was the cytokinesis-block micronucleus (MN) method. This is one of the most sensitive methods for assessing genotoxic effects. It is simple and more sensitive than the conventional chromosome aberration assay and is also statistically more precise because a large number of cells can be scored in a short time.[5] For 7 subjects we also exposed samples treated with mitomycin-C (MMC) in order to verify a cooperative effect between electromagnetic fields and this chemical mutagen. Table 1 lists the mean MN percentage obtained in the variously treated cultures. No statistical difference (Z test, two-tailed paired Student's *t* test, and variance analysis) has been found in the MN frequency of exposed and unexposed samples.

TABLE 1.—Mean MN percentage obtained in controls, exposed cultures (PMFs), MMC treated cultures (MMC), and exposed cultures plus MMC (PMFs + MMC).

	Control	PMFs	MMC	PMFs + MMC
Mean	1.17%	1.18%	2.67%	2.69%
S.E.	±0.07	±0.11	±0.11	±0.22

The last set of experiments was carried out by exposing human peripheral blood lymphocytes of 33 healthy donors to a sinusoidal EF for 72 h.[6] In this case the EF was applied by means of two horizontal plates powered by an AC voltage generator. Four field amplitudes were tested: 0.5, 2, 5, and 10 kV/m in air. The mean MN frequency values are listed in Table 2. Again the genotoxic effect of the field with and without MMC was evaluated by the cytokinesis-block MN method.

TABLE 2.—Mean MN frequency in control and ELF-exposed cultures.

E(kV/m)	Control (%)	EF (%)	MMC (%)	MMC + EF (%)
0.5	1.31	1.187	2.91	2.23
2.0	1.00	1.14	2.40	2.44
5.0	1.28	1.24	3.07	3.28
10.0	1.09	1.18	2.69	2.85

CONCLUSIONS

The reported experiments can be divided into two groups. In the first one we used (i) bovine blood, (ii) chromosomal aberration tests, and (iii) high electric fields; genotoxic effects were observed. In the second one we used: (i) human blood, (ii) MN test, and (iii) lower electric fields and magnetic fields; no genotoxic effects were observed. Given that human blood is more interesting than bovine, and that the MN test is a faster and more practical research tool, the rather obvious conclusion (and suggestion for future work) is that further experiments must be done on human blood and with the MN test, but with the application of the same electric-field stresses that gave chromosomal aberration increase on bovine lymphocytes.

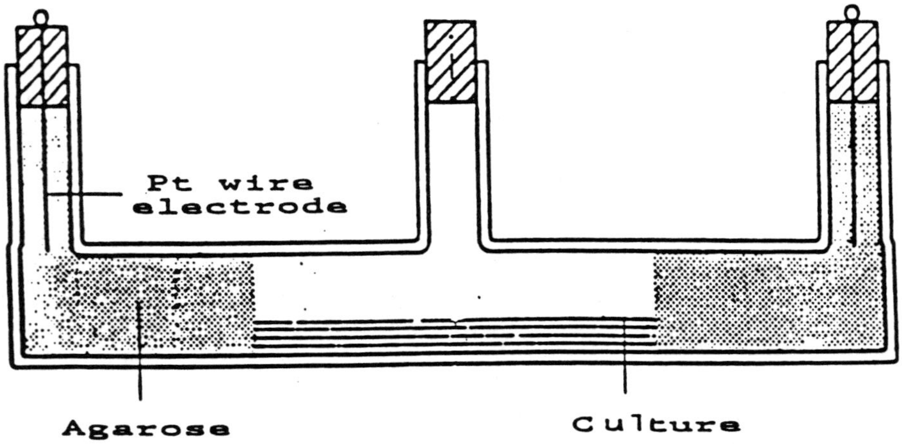

FIG. 1.—Exposure cell.

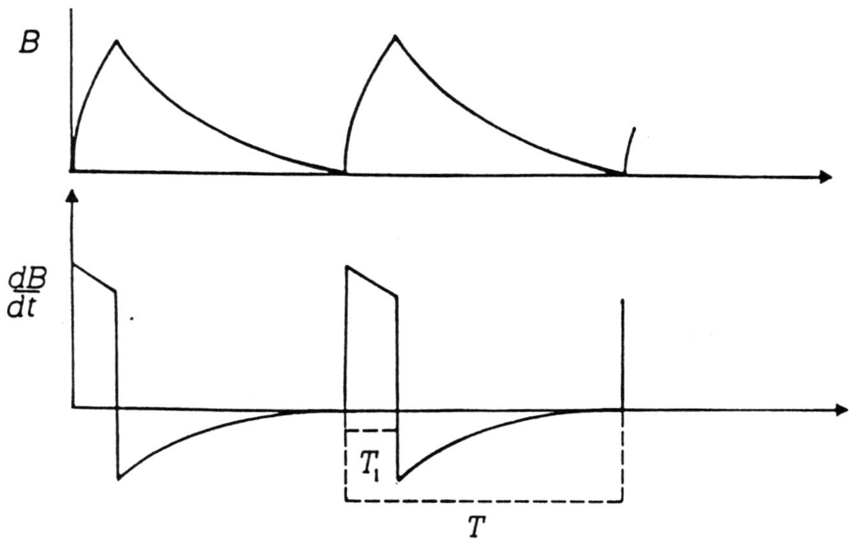

FIG. 2.—EM field characteristics. Maximum intensity of the field (upper figure), about 2.5 mT; time derivative of the magnetic field (lower figure) was proportional to induced electric field within the sample; $1/T = 50$ Hz; $T_1/T = 0.1$.

REFERENCES

1. M. R. Scarfì, G. Franceschetti, and M. B. Lioi, *Alta Frequenza* 58: 337-339, 1989.
2. G. d'Ambrosio, D. Di Berardino, L. Iannuzzi, M. B. Lioi, E. Mostacciuolo, A. Scaglione, and M. R. Scarfì, *J. Bioelectricity* 4: 279-284, 1985.
3. G. d'Ambrosio, R. Massa, D. Di Berardino, M. B. Lioi, A. Scaglione, and M. R. Scarfì, *J. Bioelectricity* 7: 239-245, 1989.
4. M. R. Scarfì, F. Bersani, A. Cossarizza, D. Monti, G. Franceschetti, C. Franceschi, *Biochem. Biophys. Res. Comm.* 176: 194-200, 1991.
5. M. Fenech and A. A. Morley, *Mutation Res.* 161: 193-198, 1986.
6. M. R. Scarfì, F. Bersani, A. Cossarizza, D. Monti, O. Zeni, M. B. Lioi, G. Franceschetti, and C. Franceschi, *Radiat. Res.* (in press).

Martin Blank, Ed., *Electricity and Magnetism in Biology and Medicine*

INCREASING ROLE OF HIGH-PEAK/LOW-AVERAGE-POWER RADIOFREQUENCY SOURCES IN TACTICAL WARFARE AND THE RADIOFREQUENCY PROTECTION GUIDELINES (RFPGs)

E. C. Elson*

Walter Reed Army Institute of Research, Washington, D.C.

In recent years, and especially since 1985, the program at Walter Reed has concentrated on biological effects of RF radiation that may be unique to or have enhanced expression resulting from high-peak/low-average-power deposition of RF energy. The interesting scientific questions are practically motivated. The employment of RF sources in both offensive and defensive applications is being explored by the U.S. Department of Defense in a variety of settings, both tactical and strategic. Therefore all three military services operate research programs with Department of Defense agencies as a cooperative enterprise. The objectives of the "directed energy" (DE) effort go beyond those of electronic warfare (EW). EW can be loosely described as a program designed to render enemy assets ineffectual by employment of electromagnetic interference (EMI) during military engagement. Employment of high-peak/low-average power can be shown to produce effects from interference to temporary electronic upset and on to irreversible damage (burnout). Hence, it is potentially far more effective than EMI in nullifying enemy objectives.

The approach to RF protection guidelines in the DE environment is that they must be consistent with existing RFPGs, in particular those of the most recent iteration of the ANSI guidelines, the SCC28 version of ANSI 1991. What that means, as a practical matter, is that the averaging of absorbed energy and the concept of limitations on the whole-body average SAR (WBA-SAR) and spatial-localized-peak SAR (SPL-SAR) will be considered *a priori* limiting, which is nothing more than a statement that the maximum thermal burden set by the community at large will not be exceeded. The strategy then is to ask whether there is reason to believe that the special conditions of high-peak/low-average power, for which there is only a meager data base, should pose any more restrictive RFPGs than those recently adopted.

The ANSI RFPGs are considered "nonthermal" in the sense that averaged over time and space, no significant departures from physiologic temperature in exposed humans are found. Such departures as could be found microscopically are not significant to organism function. Thermoregulatory mechanisms work effectively to keep temperature stable well within physiologic limits. The question is whether the averaging procedure produces a similar result in the high-peak/low-average-power domain if the RFPGs are observed. It is not *a priori* clear that microscopic fluctuations in temperature in space and time are insignificant for organism function. Until recently, the database coming from HPM research suggested no discernible effects within the time-averaged RFPGs. SCC28 had access to such an observation and used it in addressing the pulse-power case for the first time in its 1991 revision.

To what extent are these standards justified by the existing database? Much of the HPM work has been performed in military laboratories. The picture that emerges is supportive of the RFPG, with some reservation in the very high-peak/low-average-power domain. In order to bracket potential effects, exposures have exceeded the RFPGs in an experimental setting by orders of magnitude. The results of this work are considered below. An effort was made to find a threshold for a reliable and quantifiable effect whose expression could be studied at a given specific absorption and for which the specific absorption rate (SAR) and peak power could be varied.[1] These parameters would then be manipulated to be within or in excess of the ANSI RFPGs, so as to test the appropriateness of the guidelines.

The Walter Reed S-band system produces 50000 W/cm^2 on a target with an 80ns pulse width. This system is usually operated at 1 pulse every 5 s (0.8 mW/cm^2 average). A typical live-rodent experiment utilizes 200 pulses. The energy density is approximately constant through a cross section at the target about 0.5×2 m. The fluence of 1 pulse is 4 mJ/cm^2, which is equivalent over 6 min to 280 mJ/cm^2, well below the 3600

*Correspndence to: E. C. Elson (Microwave Research), Walter Reed Army Institute of Research, Washington DC 20307-5100.

mJ/cm^2 permitted by the RFPG.

The L-band system, when operated into an anechoic chamber, can produce 100 W/cm^2 peak. Pulse width can vary from 0.1 to 10 μs with repetition rates from 0 to 150 Hz. At 10 μsd, in a typical experiment, the fluence per pulse can be 1 mJ. At 4 pulses per second the system can deliver 1440 mJ in 6 min, equivalent to the RFPG-permitted fluence at 1250 MHz.

The "effect" chosen was a physiological response eventually named the "evoked body movement," so titled to convey neutrality as to the precise physiologic mechanism but to communicate that it is a reproducible, involuntary, real-time response to a microwave pulse of sufficient strength. The experimental configuration is as shown in Fig. 1. No habituation to the pulse occurs. The piezoelectric motion sensor registers the flick of the tail produced by total body motion. This motion can be reliably distinguished from random or spontaneous motions. The movement is believed to be similar to or identical with the startle response in animals. Data are displayed in Tables 1 and 2.

TABLE 1.—Comparison between frequencies of "startle" response in mice exposed to bursts of narrow pulses and to wide pulses.[a,b]

Number of Pulses Total Time Fluence	*Burst of Narrow Pulses (6 x 10^5 W/cm^2, 10 μs, 80 pps)*			*Wide Pulse (480 W/cm^2, single pulse)*			χ^2
	# of Startles	*Total Events*	*%*	*# of Startles*	*Total Events*	*%*	
2 pulses 25 msec 12 J/cm^2	*6*	*46*	*13*	-	-	-	
4 pulses 50 msec 24 J/cm^2	*14*	*46*	*30*	*14*	*47*	*30*	*0.005*[c]
8 pulses 100 msec 48 J/cm^2	*23*	*43*	*53*	*32*	*47*	*68*	*2.013*[c]
16 pulses 200 msec 96 J/cm^2	*32*	*38*	*84*	*35*	*45*	*78*	*0.548*[c]
χ^2	*75.64; df= 4; p< 0.001*			*63.91; df= 3; p< 0.001*			

[a]*adapted from Brown et al.*
[b]*All data derived from thermometric dosimetry results and "human equivalent" calculation of H. Bassen (unpublished).*
[c]*df= 1; p> 0.05.*

Recent behavioral studies have revealed "effects" at powers well within the RFPGs.[2] A number of rats were trained in an operant chamber to respond on one of two levers following presentation of a light of 5s duration or 0.5s duration. After they became well trained in the response, 5 new light pulses of intermediate duration (0.7, 1.1, 1.6, 2.4, and 3.5 s) were presented, requiring the animal to make a correct judgment of whether a given pulse was closer to 0.5 or 5 s. With repeated testing the probability that a given stimulus will produce selection of the "long" or "short" lever can be determined for all stimuli. This protocol has been used extensively in drug testing, because disturbance of the pattern of selection is considered to be a sensitive measure of disturbed cognition and the extrapolation from animal to man is considered reliable compared to many other tests. The protocol also can measure "decision confidence." If no response to a stimulus is made within 10 s, the trial is terminated and a null response is recorded. The animal subject's confidence in a judgment or decision has an inverse relationship to the decision time.

The S-band HPPM system was utilized: 3 GHz, 700MW peak power, 2W average power, 80ns pulse width, 0.2Hz repetition rate; 200 pulses were delivered over 17 min. The maximum SAR determined by whole-body calorimetry was 0.072 W/kg. Experiments were performed in an anechoic chamber and the measured maximum plane wave power density at the animal was 50 kW/cm^2 peak, 0.8 mW/cm^2 average, 4 mJ/cm^2 per pulse fluence.

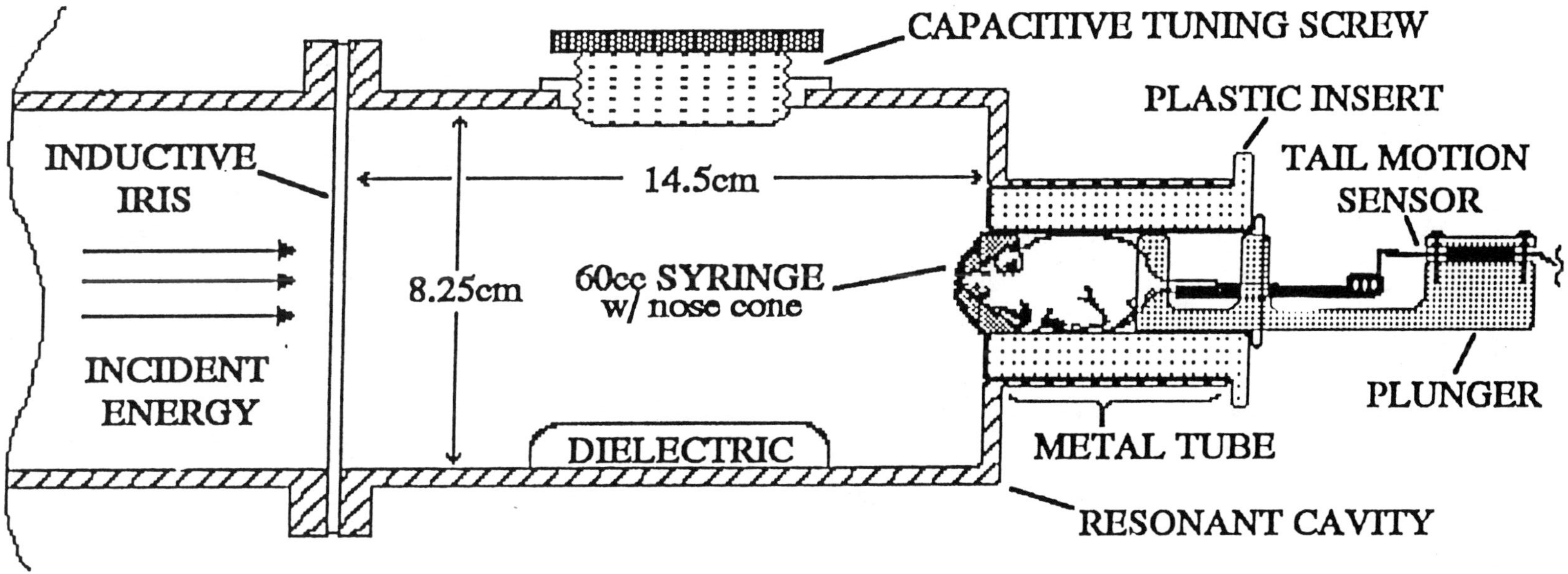

FIG. 1.—Resonant cavity (modified WR650 waveguide) and sensor.

TABLE 2.—Dependence of the "startle" response on average brain SAR at constant brain SA.[a]

Wide or Narrow Pulse Total Time, Fluence Number of Pulses Peak Power Density Average Power Density	Number of Startle Observed	Total Number of Events	Frequency of Response (%)	χ^2-Test Among Sessions[b]
Wide Pulse 0.2 sec, 59 J/cm^2 1 295 W/cm^2 295 W/cm^2	28	37	76	6.11[c]
Narrow Pulses 0.2 sec, 96 J/cm^2 16 6 x 10^5 W/cm^2 480 W/cm^2	32	38	84	0.07
Wide Pulse 0.8 sec, 59 J/cm^2 1 86 W/cm^2 86 W/cm^2	31	40	78	2.44
Narrow Pulses 2.0 sec, 96 J/cm^2 160 6 x 10^4 W/cm^2 48 W/cm^2	19	39	49	3.97
Wide Pulse 3.2 sec, 59 J/cm^2 1 22 W/cm^2 22 W/cm^2	12	41	30	1.17
χ^2-Test	χ^2= 36.72; df= 4; p< 0.001			

[a]adapted from Brown *et al* (unpublished).
[b]three test sessions separated by 24 hours between sessions.
[c]df= 2; p< 0.05.

Each rat finished all 300 trials per session, which shows that the microwave powers involved were not physically debilitating and were consistent with the measured whole-body average SAR. Nevertheless, a significant increase in null responses was observed. Figure 2 shows the number of null responses against stimulus duration for three different powers and a control. At 1.6 s, the "bisection" point, at which the animal is equally likely to select either lever, its confidence is the lowest, and the number of null responses is highest. With increasing microwaver power this lack of confidence is enhanced. Another effect found with the very high-peak power S-band system, with exposure conditions identical to the bisection experiment, was a degradation in the ability of rats to run on a treadmill.[2] Whole-body average SAR for this set of experiments was also 0.07 W/kg.

The same system and exposure parameters also demonstrated interference with consolidation of memory in rats. Rats proceed through a Y maze and learn the location of food. If exposed to the high-peak/low-average-power condition, within 30 min there is a degradation of the memory of the location of the food revealed upon repeat testing.

CONCLUSIONS

A number of investigations utilizing pulsed microwave power have sought to identify possible effects enhanced by or unique to the pulse mode. Most investigations have operated in a domain of low or moderate peak to average power ratios. The work of D'Andrea et al. showed no alteration of behavior in monkeys exposed in 60min sessions to 1.3GHz pulsed power with peak-to-average power ratios from 9 to 143 (131.8

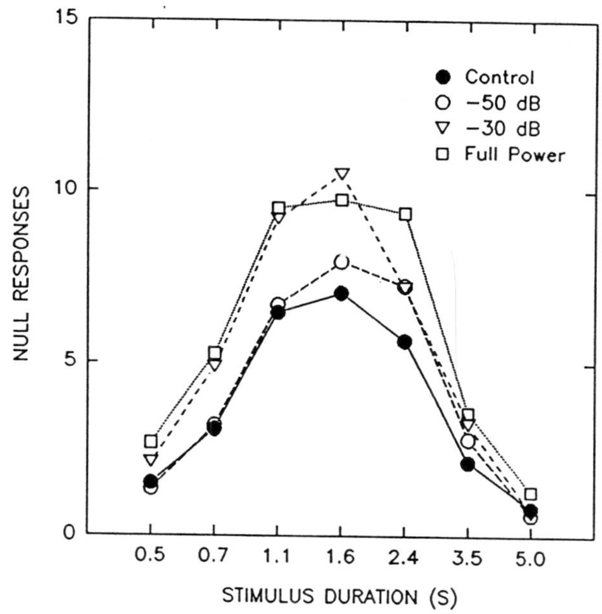

FIG. 2.—Effect of pulsed microwave energy on "decision confidence" is most pronounced at bisection point. (From Rasslear, unpublished manuscript.)

W/cm^2 peak).[3] The average power densities in this study ranged from 1 to 15 mW/cm^2, or from within to slightly above the MPE. D'Andrea cautions that "while significant effects were not observed in this study, the generalization of these findings to high-peak-power densities, other pulse rates, different carrier frequencies, as well as other behaviors is unwarranted."

The work of Raslear et al. carries the exposures to very-high-peak/low-average power within the RFPGs, 50000 W/cm^2 peak and peak/average power = 10^5 at 3 GHz. In this domain, effects are seen at 4 mJ/cm^2 per pulse. These effects may be meaningful in the world of military applications, requiring further review by cognizant scientists. At 3 GHz the permitted fluence is 3.6 J/cm^2 for pulses exceeding 100 ms. For less than 100 ms the factor of 5 reduces the fluence to 720 mJ/cm^2 per pulse. At the Raslear pulse width of 80 ns, the standard is dominated by the 100kV/m limit allowing a fluence of 0.24 mJ/cm^2, less than the Raslear fluence. It appears that in this instance the 100kV/m restriction is commendable and perhaps should be even more restrictive. On the other hand, by cutting off the peak power at 100 kV/m but holding a 6min fluence of 3.4 J/cm^2, the guideline has allowed 15000 such pulses in 6 min. This level of permissiveness will have to be addressed in view of the Raslear work. One can conclude that in this very high-peak/low-average-power domain there is a dependence on peak power per se and not merely fluence.

In a separate set of experiments at 1250 MHz, working in a range from 20 to 70 times above the RFPG permitted fluences, investigators at Walter Reed studied the "evoked body movement," which appears to be fully evolved between 12 and 100 J/cm^2 for exposure times from 25 to 200 ms. From 200 ms to 3.2 s, with total absorption or fluence held constant, the effect is seen to diminish, which reflects the decreasing rate of energy insertion.[4]

REFERENCES

1. D. O. Brown, S. Lu, S. Mathus, E. Elson, and H. Wachtel, "Reflexive responses evoked by high peak/low average power microwave pulses," unpublished.

2. T. Raslear, "Null responses, decision confidence, and temporal bisection in rats: The effect of high-power pulses microwaves," *Bioelectromagnetics* (in press).

3. J. D'Andrea, "Lack of behavioral effects in the rhesus monkey: High peak microwave pulses at 1.3 GHz," *Bioelectromagnetics* 10: 65-76, 1989.

4. The views of the author do not purport to reflect the position of the U.S. Department of the Army or the Department of Defense.

Martin Blank, Ed., *Electricity and Magnetism in Biology and Medicine*

MODIFICATION OF ACOUSTIC STARTLE AMPLITUDE BY MICROWAVE PULSES

R. L. Seaman*

Louisiana Tech University, Ruston, La.

D. A. Beblo

OGDEN BioServices Corp., Gaithersburg, Md.

T. G. Raslear

Walter Reed Army Institute of Research, Washington, D.C.

Acoustic startle with single microwave-pulse prestimuli was studied in male Long Evans rats. Heads of the animals were exposed to 1μs 1.25GHz pulses of 30 and 100 kW/kg peak SAR (30 and 100 mJ/kg) in separate experiments. Single microwave pulses were delivered at 201, 101, 51, 3, 1 ms before or 1 ms after onset of a startling noise burst. No mean startle amplitude with 30kW/kg prestimuli was significantly different from the mean startle-only amplitude. Mean startle amplitudes with 100kW/kg prestimuli at 201, 101, and 51 ms lead times were significantly smaller than the mean startle-only amplitude. Results show that a 1μs microwave pulse at 100 kW/kg but not 30 kW/kg can modify acoustic startle and that timing influences its effectiveness.

Known biological actions of pulsed microwaves include generation of acoustic and thermal energy as well as auditory perception. To test the effects of microwave pulses on a behavioral response involving sensory and motor components, acoustic startle was studied with microwave-pulse prestimuli. Acoustic startle is a well-studied behavior that is modifiable by prestimuli of a number of sensory modalities.[1]

METHODS AND MATERIALS

Subjects were male Long Evans rats (246-312 g) placed in a ventilated cylindrical plastic holder that positioned the head within WR 650 waveguide. Animal motion was detected with piezoelectric film under the holder floor. The waveguide exposure apparatus and animal holder with animal were located in a sound-isolating box with background noise of 67 dBA. The startling acoustic stimulus was a 50ms burst of broad-band noise at 106 dB peak SPL under computer control. Single 1μs microwave pulses of 1.25 GHz generated by an Epsco PH40K signal source were delivered as prestimuli, also under computer control. Forward and reflected microwave powers were detected by a bidirectional waveguide coupler and monitored on a Tektronix 2430A digital oscilloscope. The whole-body startle response was monitored on a second digital oscilloscope and recorded on a Gould TA2000 chart recorder.

Twelve animals were tested in an experiment with pulses at 30 kW/kg head SAR. Another twelve animals were tested in a separate experiment with pulses at 100 kW/kg. In each experiment, 6 experimental conditions consisted of single microwave pulses delivered at lead times of 201, 101, 51, 3, 1, and −1 ms relative to onset of the startling noise burst, with the negative lead time indicating occurrence after the onset. As control conditions, startling stimuli and prestimuli were also delivered alone to give a total of 8 conditions. Each animal experienced 10 presentations of each condition in random order over 5 testing sessions.

Startle amplitudes were measured from chart recordings for this report. The median amplitude of responses to each condition for each animal was determined. The 8 median amplitudes for each animal were normalized as follows: they were summed and each was divided by the sum.[2] If all 8 medians were the same,

---

*Correspondence to: R. L. Seaman (Biomed. Engg.), Louisiana Tech University, Box 3185, Ruston LA 71272-0001.

this procedure would give a normalized startle amplitude of 0.125 for each condition. The normalized amplitudes were then tested in one-factor repeated-measures analysis of variance (ANOVA) with a significance level of 0.05 and, as appropriate, post hoc Tukey's protected t tests.

RESULTS

Mean normalized startle amplitudes are shown in Figs. 1(A) and (B) for 30kW/kg and 100kW/kg pulses, respectively. The prestimulus-only condition (PP) gave the smallest mean amplitude for each SAR. Because no startle was seen for this condition, the mean represented typical resolution of the method. The PP condition was not included in ANOVAs and t tests.

For 30kW/kg microwave pulses, means for the startle-only (ST) condition and the startle-plus-prestimulus conditions are similar. Means of the 201, 101, and 3ms conditions are somewhat smaller than that of the ST condition. ANOVA gave $F(6,66) = 1.19, p = 0.3224$. Thus, prestimulus condition was not a significant influence on startle amplitude for 30kW/kg microwave pulses.

For 100kW/kg microwave pulses, means for all startle-plus-prestimulus conditions are smaller than the mean for the ST condition; the 51ms mean is the smallest. ANOVA gave $F(6,66) = 3.99, p = 0.0018$. Protected t tests indicated that the means of the 201, 101, and 51ms conditions were significantly different from the mean of the ST condition ($p < 0.05, p < 0.01, p < 0.01$, respectively). The t tests also indicated that each of these means was also significantly different from the mean of the -1ms condition ($p < 0.05, p < 0.01, p < 0.01$, respectively) and that the mean of the 51ms condition was significantly different from the mean of the 1ms condition ($p < 0.01$).

DISCUSSION

Based on analysis of startle amplitudes from the chart recordings, 30kW/kg 1μs pulses at the lead times tested are not effective in modifying acoustic startle in the rat. However, 100kW/kg 1μs pulses at 201, 101, and 51ms lead times, but not at 3, 1, and -1ms lead times, are effective in reducing startle amplitude. The results show that both the pulse intensity and the timing of pulses relative to the startling stimulus are relevant parameters in determining the degree of effectiveness. A more detailed analysis of startle amplitude and latency is now under way with oscilloscope waveforms stored in ASCII files.

The reductions in startle amplitude are consistent with the known inhibition of startle by sensory prestimuli.[1] Acoustic clicks of 92 dB reduce acoustic startle amplitude at lead times of 8-64 ms but increase amplitude at lead times of 1-4 ms.[3] Auditory sensation of microwave hearing[4,5] may mediate modification of the startle amplitude by microwave-pulse prestimuli.[6]

REFERENCES

1. H. S. Hoffman and J. R. Ison, "Reflex modification in the domain of startle: I. Some empirical findings and their implications for how the nervous system processes sensory input," *Psychol. Rev.* 87: 175-189, 1980.

2. J. R. Ison, G. P. Bowen, and C. Kellogg, "Potentiation of acoustic startle behavior in the rat *(Rattus-Norvegicus)* at the onset of darkness," *J. Comp. Psychol.* 105: 3-9, 1991.

3. G. R. Hammond and D. S. Leitner, "Augmentation of the rat's acoustic startle reflex by nonreflexogenic stimuli," *Behav. Neurosci.* 104: 841-848, 1990.

4. C. K. Chou, A. W. Guy, and R. Galambos, "Auditory perception of radio-frequency electromagnetic fields," *J. Acoust. Soc. Am.* 71: 1321-1334, 1982.

5. R. L. Seaman and R. M. Lebovitz, "Thresholds of cat cochlear nucleus neurons to microwave pulses," *Bioelectromagnetics* 10: 147-160, 1989.

6. This work was supported by Walter Reed Army Institute of Research contract DAMD 17-89-C-0921 to OGDEN BioServices Corp. and U.S. Army Research Office contract DAAL 03-86-D-0001 (1542 & 2776), a summer faculty program administered by Battelle. All efforts adhered to principles of the "Guide for the Care and Use of Laboratory Animals," NIH publication 85-23. The project received substantial assistance from a number of people at OGDEN BioServices Corp. and the Department of Microwave Research at WRAIR. The contributions of M. DeAngelis of OGDEN to animal-related procedures are particularly appreciated. Opinions and conclusions obtained in this work are those of the authors and do not necessarily reflect the position of the U.S. Department of the Army or the U.S. Department of Defense.

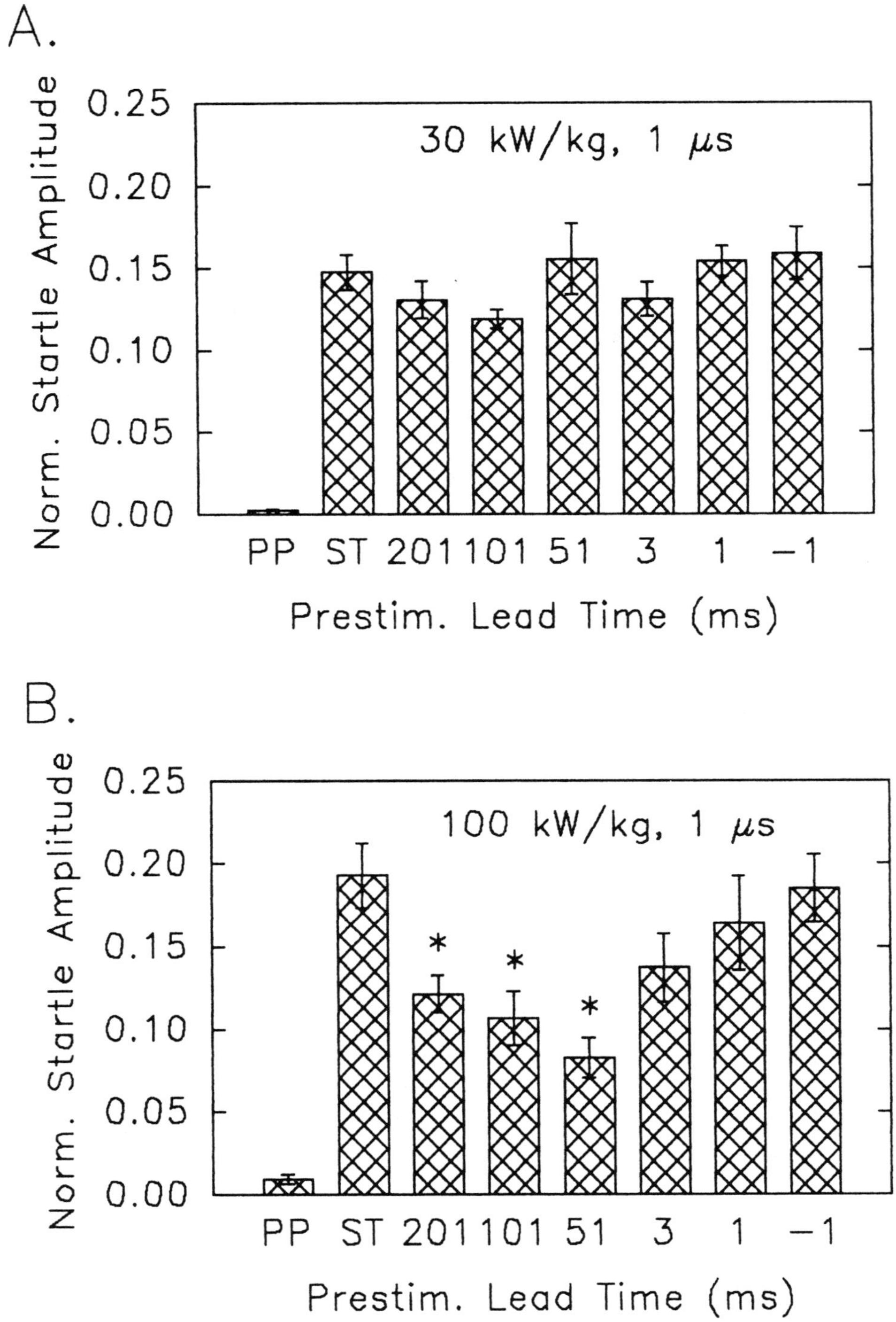

FIG. 1.—Normalized acoustic startle amplitude (mean ± s.e.m.) for microwave-pulse prestimulus lead times. Prestimulus-only condition is designated as PP; the startle-only condition, as ST. Asterisks indicate means significantly different from the mean of the respective startle-only condition. (A) 30kW/kg, 1μs, 30mJ/kg prestimuli; (B) 100kW/kg, 1μs, 100mJ/kg prestimuli.

MODIFICATION OF TACTILE STARTLE AMPLITUDE BY MICROWAVE PULSES

R. L. Seaman*

Louisana Tech University, Ruston, La.

D. A. Beblo

OGDEN BioServices Corp., Gaithersburg, Md.

T. G. Raslear

Walter Reed Army Institute of Research, Washington, D.C.

Tactile startle with single microwave-pulse prestimuli was studied in male Long Evans rats. Heads of the animals were exposed to 8μs 1.25GHz pulses of 100 kW/kg peak SAR (800 mJ/kg). Single acoustic clicks (94 dB SPL) or microwave pulses were delivered at 150, 100, 50, 0 ms before or 50 ms after onset of a startling air burst. For acoustic clicks, mean startle amplitude for the 50ms lead time was significantly smaller than the mean startle-only amplitude. For microwave pulses, mean startle amplitude for the 50ms delay was significantly larger than the mean startle-only amplitude. The results show that acoustic clicks and microwave pulses can modify tactile startle and that timing determines the effectiveness of both types of prestimuli.

Results from experiments on the modification of acoustic startle by microwave pulses indicated possible action on sensory and/or neuromuscular events. To test the effects of microwave pulses on another response involving sensory, but not necessarily auditory, and motor components, tactile startle was studied with acoustic-click and microwave-pulse prestimuli. Like acoustic startle, tactile startle is modifiable by sensory prestimuli.[1,2]

METHODS AND MATERIALS

Subjects were male Long Evans rats (246-310 g) placed in a ventilated cylindrical plastic holder that positioned the head within WR 650 waveguide. Animal motion was detected with piezoelectric film under the holder floor. The waveguide exposure apparatus and animal holder with animal were located in a sound-isolating box with background noise of 50-55 dBA. The startling tactile stimulus was a 50ms burst of compressed air delivered to the back of the neck at 48-52 psi (330-350 kPa). Either single 8μs microwave pulses of 1.25 GHz generated by an Epsco PH40K signal source or single acoustic clicks of 94 dB peak SPL were delivered as prestimuli. Forward and reflected microwave powers were detected by a bidirectional waveguide coupler and monitored on a Tektronix 2430A digital oscilloscope. Tactile, microwave, and acoustic stimuli were under computer control. The whole-body startle response was monitored on a second digital oscilloscope and recorded on a Gould TA2000 chart recorder.

Fourteen animals were tested with acoustic-click and microwave-pulse at 100 kW/kg head SAR prestimuli in separate testing sessions. During each session, 6 experimental conditions were presented in random order. Five conditions consisted of a single prestimulus delivered at lead times of 150, 100, 50, 0 and −50 ms relative to onset of the startling air burst, with the negative lead time indicating occurrence after the onset. The other condition was the startling air burst delivered alone. Each animal experienced 4 presentations of each condition in each testing session. Seven animals were tested with acoustic clicks first; the other seven, with microwave pulses first.

---

*Correspondence to: R. L. Seaman (Biomed. Engg.), Louisiana Tech University, Box 3185, Ruston LA 71272-0001.

Startle amplitudes were measured from chart recordings for this report. The median amplitude of responses to each condition for each animal was determined. The 6 median amplitudes for each animal were normalized as follows: they were summed and each was divided by the sum.[3] If all 6 medians were the same, this procedure would give a normalized startle amplitude of 0.167 for each condition. The normalized amplitudes were then tested in one-factor repeated-measures analysis of variance (ANOVA) with a significance level of 0.05 and post hoc Tukey's protected t tests.

RESULTS

Mean normalized startle amplitudes are shown in Figs. 1(A) and (B) for acoustic-click and microwave-pulse prestimuli, respectively. Experiments with acoustic startle and separate preliminary experiments had shown that neither type of prestimulus produced a startle response.

For acoustic clicks, means of the 150, 100, and 50ms conditions are smaller than the mean of the startle-only (ST) condition; the mean of the −50ms condition is larger. ANOVA gave $F(5,65) = 4.67, p = 0.0011$. Protected t tests indicated that only the mean of the 50ms condition was significantly different from the mean of the ST condition ($p < 0.01$). The t tests also indicated that the mean of the 50ms condition was significantly different from the means of the 0 and −50ms conditions ($p < 0.01$).

For 100kW/kg microwave pulses, means of the 150, 100, and 50ms conditions are somewhat smaller than the mean of the ST condition; means of the 0 and −50ms conditions are larger. ANOVA gave $F(5,65) = 3.13, p = 0.0137$. Protected t tests indicated that only the mean of the −50ms condition was significantly different from the mean of the ST condition ($p < 0.05$). The t tests also indicated that the mean of the −50ms condition was significantly different from the means of the 150, 100, and 50ms conditions ($p < 0.01, p < 0.05, p < 0.01$, respectively).

DISCUSSION

Based on analysis of startle amplitudes from chart recordings, both 94dB acoustic clicks and 100kW/kg microwave pulses of 8μs duration are capable of modifying tactile startle in the rat. The results show that the timing of either type of prestimulus relative to the startling stimulus is important in determining the degree of effectiveness. A more detailed analysis of startle amplitude and latency is now under way with oscilloscope waveforms stored in ASCII files.

The observed reduction in startle amplitude is consistent with the known inhibition of startle by sensory prestimuli.[1] Acoustic prestimuli reduce tactile startle amplitude at lead times of 50-100 ms.[2] Modification of startle amplitude by microwave-pulse prestimuli may be mediated by the auditory sensation of microwave hearing.[4,5] Although the general patterns of modification with lead time are similar for the two types of prestimulus used here, it remains to be determined why acoustic and microwave prestimuli have maximal effects at different lead times in tactile startle.[6]

REFERENCES

1. H. S. Hoffman and J. R. Ison, "Reflex modification in the domain of startle: I. Some empirical findings and their implications for how the nervous system processes sensory input," *Psychol. Rev.* 87: 175-189, 1980.

2. R. S. Mansbach and M. A. Geyer, "Effects of phencyclidine and phencyclidine biologs on sensorimotor gating in the rat," *Neuropsychopharmacol.* 2: 299-308, 1989.

3. J. R. Ison, G. P. Bowen, and C. Kellogg, "Potentiation of acoustic startle behavior in the rat *(Rattus-Norvegicus)* at the onset of darkness," *J. Comp. Psychol.* 105: 3-9, 1991.

4. C. K. Chou, A. W. Guy, and R. Galambos, "Auditory perception of radio-frequency electromagnetic fields," *J. Acoust. Soc. Am.* 71: 1321-1334, 1982.

5. R. L. Seaman and R. M. Lebovitz, "Thresholds of cat cochlear nucleus neurons to microwave pulses," *Bioelectromagnetics* 10: 147-160, 1989.

6. This work was supported by Walter Reed Army Institute of Research contract DAMD 17-89-C-0921 to OGDEN BioServices Corp. and U.S. Army Research Office contract DAAL 03-86-D-0001 (1542 & 2776), a summer faculty program administered by Battelle. All efforts adhered to principles of the "Guide for the Care and Use of Laboratory Animals," NIH publication 85-23. The project received substantial assistance from a number of people at OGDEN BioServices Corp. and the Department of Microwave Research at WRAIR. The contributions of M. DeAngelis of OGDEN to animal-related procedures are particularly

appreciated. Opinions and conclusions obtained in this work are those of the authors and do not necessarily reflect the position of the U.S. Department of the Army or the U.S. Department of Defense.

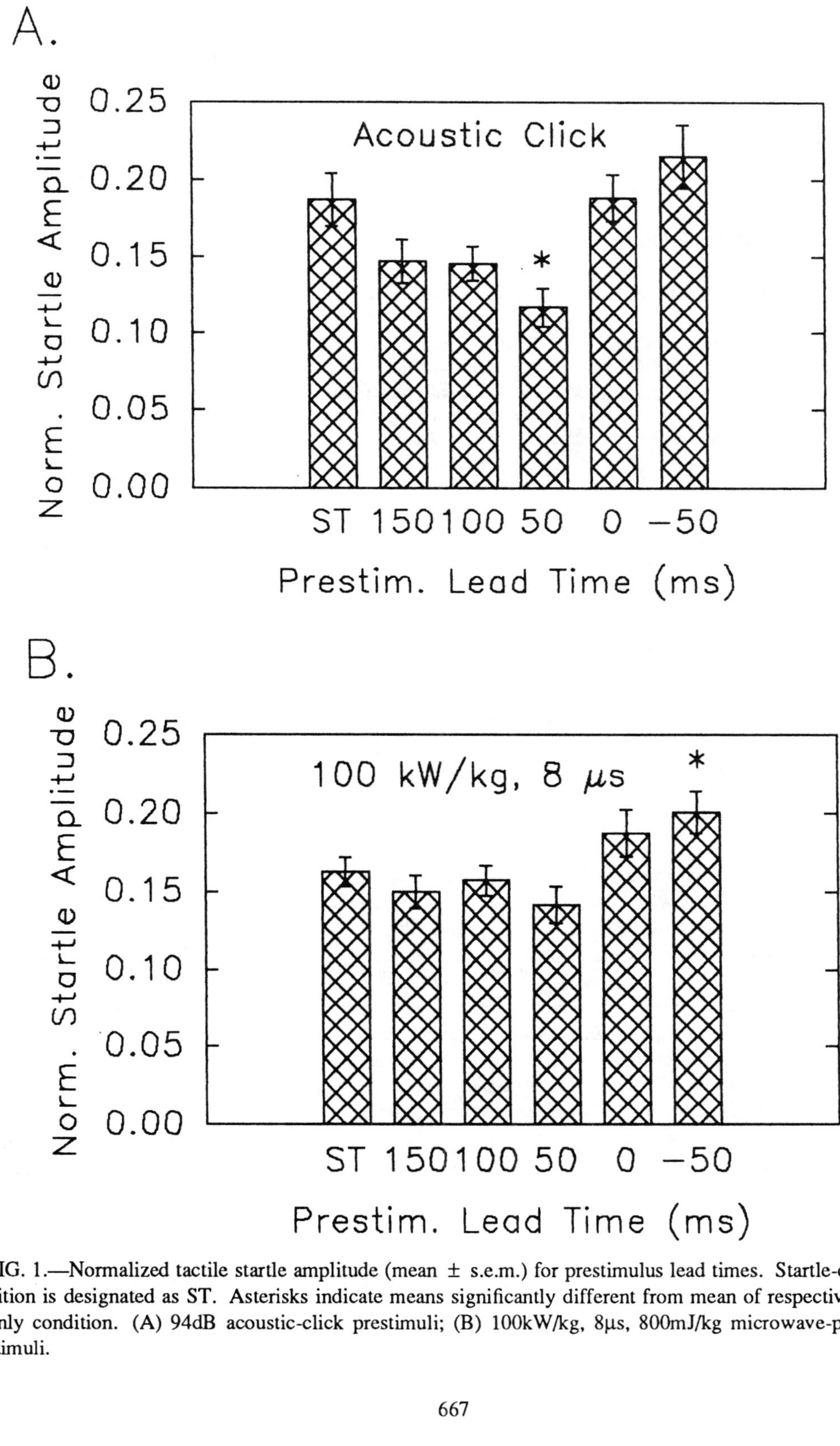

FIG. 1.—Normalized tactile startle amplitude (mean ± s.e.m.) for prestimulus lead times. Startle-only condition is designated as ST. Asterisks indicate means significantly different from mean of respective startle-only condition. (A) 94dB acoustic-click prestimuli; (B) 100kW/kg, 8μs, 800mJ/kg microwave-pulse prestimuli.

Martin Blank, Ed., *Electricity and Magnetism in Biology and Medicine*

THE EFFECTS OF HIGH-PEAK POWER PULSED MICROWAVES ON TREADMILL PERFORMANCE IN THE RAT

Y. Akyel* and M. Belt

OGDEN BioServices Corp., Gaithersburg, Md.

T. G. Raslear and R. M. Hammer

Walter Reed Army Institute of Research, Washington, D.C.

To assess the physical endurance of organisms following an exposure to 200 TEMPO high-peak-power microwave pulses (80 ns width, 1 pulse/8 s, 3 GHz, 700MW peak transmitted power), treadmill performance of rats was studied. Although the exposure system produced a time-averaged midbrain specific absorption rate (SAR) of 0.21 W/kg and time-averaged whole-body SAR of 0.07 W/kg, both well within the present safety guidelines, the results indicated that exposure to such RF fields decreased the rats' endurance, as reflected in running time on the treadmill without faltering.

Current guidelines have been developed by IEEE-ANSI, NCRP, and EPA to limit exposure to electromagnetic fields that induce the heating of tissue. Yet none of the guidelines addresses potentially adverse health effects of high-peak-power pulsed microwave irradiation that does not produce tissue heating. Since there is a report that motor function in rats is affected by pulsed microwave exposure,[1] interest should focus on the safety of the pulsed microwave fields. In this study, very short duration microwave pulses that do not violate current safety guidelines, but with peak power levels high enough to cause concern that they may have adverse effects, were used to examine the immediate effects of such microwave fields on the physical endurance of rats.

Treadmill running provides a convenient means of assessing a subject's physical endurance following various drug and environmental manipulations,[2] including ionizing radiation.[3] The length of time an animal can run on a treadmill without faltering is a measure of the animal's endurance.

METHODS AND MATERIALS

Twenty-six male albino (Sprague-Dawley) rats were trained for 5 min a day for 8 days on a treadmill (Columbus Instruments, O.) with an 8° incline and a speed of 10 m/min. To motivate running, a shock grid adjoining the moving treadmill belt was used to deliver 200ms electric pulses with a 0.5mA peak shock current (the lowest possible level on the equipment). Following the training, 9 rats were exposed to 200 microwave pulses (80 ns width, 1 pulse/8 s, 3 GHz) by use of the TEMPO exposure system with 700 MW peak transmitted power.[4] Nine of the remaining animals were sham exposed, and 8 rats served as cage controls. Sham- or microwave-exposed animals were restrained in a cylindrical plastic enclosure and positioned on an RF transparent platform inside a corner reflector. Exposures lasted approximately 25 min. Immediately following the sham or microwave exposure the animals were placed on the treadmill. Control animals were taken from their cages and placed on the treadmill. During the test session the speed of the belt was increased to 30 m/min with the same inclination. Running time was defined as the total time that each animal spent in the treadmill before receiving 3 consecutive electric shocks within 10 s without leaving the shock grid area.

---

*Correspondence to: Y. Akyel, OGDEN BioServices Corp., 6 Montgomery Village Ave. (Suite 222), Gaithersburg MD 20879.

DOSIMETRY

The microwave exposures used here were of sufficiently low average power (2 W) to avoid producing any thermal effects. To conduct whole-body and localized dosimetry, the TEMPO's vircator was replaced by a 1kW CW klystron. With the same radiating system, this produced sufficiently high average power densities to make it possible to conduct calorimetric and thermometric measurements. These measurements were used to evaluate both whole-body and spatially localized specific absorption rate (SAR) in cadavers of similar size. Local SARs were determined by means of implantable RF transparent Luxtron temperature probes. Whole-body SAR was determined using a single gradient-layer calorimeter.[5] The results of these measurements were used to extrapolate both the average and peak power SARs associated with the TEMPO exposures.

RESULTS

The endurance of the animal was defined as the total running time of the subject during the high-speed (30m/min) test trial. The mean (± SEM) running times of the cage control and sham-exposed animals were 150.13 s (± 25.65) and 152.89 s (± 35.18), respectively (Fig. 1). On the other hand, microwave-exposed animals ran for 56.44 s (± 10.69), significantly less than the two groups not exposed to microwave fields. A Kruskal-Wallis one-way ANOVA test ($\chi^2 = 6.02$, $p < 0.05$) showed that reliable differences existed among the groups.

DISCUSSION

The results indicate that exposure to 200 TEMPO pulses decreases the rats' endurance, as reflected in running time on the treadmill without faltering. Current safety guidelines suggest that whole-body SAR should be limited to 0.4 W/kg and/or a local SAR of 8 W/kg, yet the TEMPO exposure system, which was used in the present experiment, produced a time-averaged midbrain SAR of 0.21 W/kg and a time-averaged whole-body SAR of 0.07 W/kg. Thus, even though exposure to high-peak low-average-power microwave fields may be within present safety guidelines, the present data raise questions concerning the adequacy of these guidelines.[6]

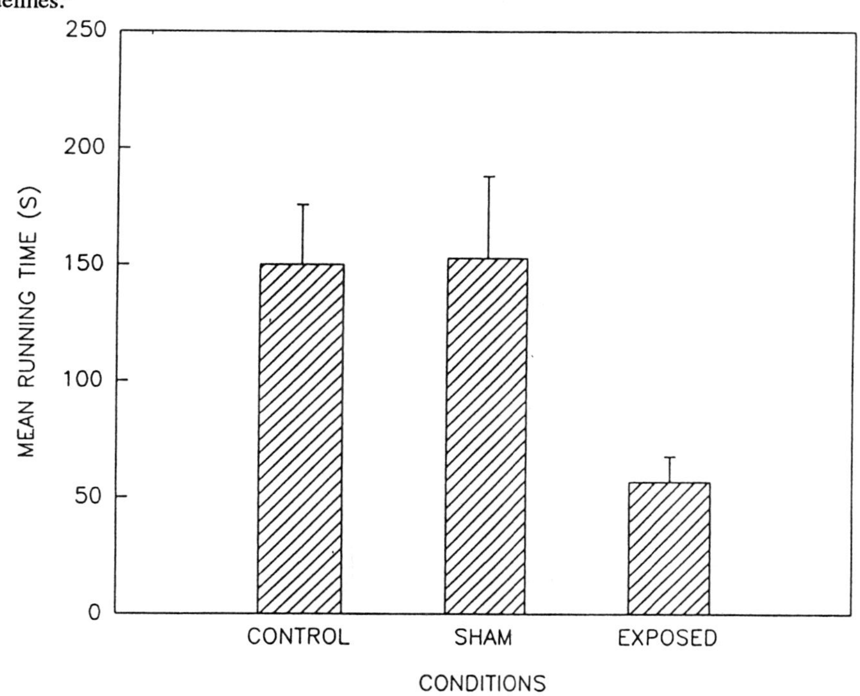

FIG. 1.—Mean running time of cage control, sham-, and microwave-exposed rats in seconds. Vertical bars depict standard errors.

REFERENCES

1. B. J. Klaunberg, J. H. Merritt, and D. N. Erwin, *Behavioral Effects of Exposure to the TEMPO High-power Microwave System,'' Report no. USAFSAM-TR-87-35, Brooks Air Force Base, TX, 1988.*

2. M. C. Gerald, ''Effects of (+)-amphetamine on the treadmill endurance performance of rats,'' *Neuropharmacology* 17: 703-704, 1978.

3. D. C. Jones, G. K. Osborn, and D. J. Kimeldorf, ''Treadmill exercise performance by the aging irradiated rat,'' *Radiation Res.* 30: 714-724, 1967.

4. F. Bates and H. Bassen, ''A biological effects exposure facility utilizing an 800 megawatt peak power pulsed microwave source,'' *Proceedings of the 11th Engineering in Medicine and Biology Society Conference,* Seattle, WA, 1989, 1134-1135.

5. S. Mathur, Y. Akyel, and S-T. Lu, ''Whole-body microwave dosimetry based on a single, gradient-layer calorimeter,'' *Bioelectromagnetics* 13: 435-438, 1992.

6. This study was supported by the U.S. Army Medical Research and Development Command under contract DAMD 17-89-C-9021 awarded to OGDEN BioServices Corp. The study was conducted in compliance with the Animal Welfare Act and other Federal regulations relating to experiments involving animals, and adheres to the principles stated in the *Guide for the Care and Use of Laboratory Animals,* NIH publication 85-23. The views of the authors do not necessarily reflect the position of the U.S. Department of the Army or the Department of Defense (Para 4-3 AR 360-5).

Martin Blank, Ed., *Electricity and Magnetism in Biology and Medicine*

THE EFFECTS OF HIGH-PEAK POWER PULSED MICROWAVE IRRADIATION ON TIME BISECTION IN RATS

T. G. Raslear*

Walter Reed Army Institute of Research, Washington, D.C.

Y. Akyel, F. Bates, and M. Belt

OGDEN BioServices Corp., Gaithersburg, Md.

The postexposure effects of pulsed microwaves of high peak power (0.07 to 700 MW) but low average power (≤2 W) on sensory, perceptual, and decision processes was studied in eight rats performing on the temporal bisection task. A maximum specific absorption rate of 0.072 W/kg, determined by whole-body calorimetry, was produced. Regardless of the power level, the rats finished all 300 trials every session, proving that there was no physically debilitating effect of exposures. Measures of sensory and perceptual function were largely unaffected. However, exposure to microwave levels lower than the current occupational safety standard produced a reliable decrease in decision-making confidence.

Because the psychophysical method of bisection provides comparable information concerning perception in rats and humans,[1] it is a useful animal model for the investigation of the toxic effects of treatments that cannot be ethically tested with humans. For instance, this method has been used to characterize the long-term effects of cholinesterase poisons.[2] The present study reports the application of that method to irradiation with high-peak power pulsed microwaves.

METHODS

Eight male albino Sprague-Dawley derived rats, 90-110 days old at the beginning of experimentation, served as subjects. The animals were maintained at 80% of their free-feeding weight. Water was freely available in their home cages.

As described previously,[1] each animal was trained in an operant chamber to respond on one of two levers following presentation of one of the two training stimulus durations (house light, 5 or 0.5 s). A correct response produced a food pellet and terminated the trial; an error merely terminated the trial. Following a 10s intertrial interval, a new trial began with presentation of one of the training stimuli. If no response was made within 10 s of stimulus termination, the trial terminated and a null response to that stimulus was recorded. During the final stage of training, the probability of each training stimulus was 0.5 and the probability of reinforcement was 0.25. Three hundred trials/day were scheduled, and null responses did not decrease the trial count.

During the test phase, five new stimulus durations, intermediate to the training stimuli, were introduced. The test stimuli were equally spaced logarithmically between the training stimuli. Each stimulus duration occurred with equal probability. Responses to the test stimuli were not reinforced. Correct responses to the training stimuli were now reinforced so as to produce the same over-all probability of reinforcement that prevailed during the training phase.

The rats performed during the test phase immediately following each of seven microwave exposure conditions: cage control (rats are taken from their home cage and placed directly into the operant test chamber), sham exposure (rats are first placed in a plexiglas cylinder in an anechoic microwave chamber for 40 min with no microwave radiation), −50, −40, −30, −20, and 0 dB attenuated power exposures [rats are first placed in a plexiglas cylinder in the same anechoic microwave chamber and are given 200 pulses (80 ns, 1 pulse/5 s) of high-power (700 MW at 0 dB attenuation) microwave irradiation (3 GHz) with the TEMPO ex-

---

*Correspondence to: T. G. Raslear (Microwave Res.), Walter Reed Army Institute of Research, Washington DC 20307-5100.

posure system.][3] All rats received three cage-control and three sham exposures, followed by an ascending and descending series of microwave exposures. All tests occurred on Tuesdays and Thursdays. With the exception of the initial sham and control exposures, rats were exposed to microwave radiation one day each week. The other test day was used to obtain additional postexposure control and sham bisection data. The remaining days were used for discrimination training.

DOSIMETRY

The microwave exposures used here were of sufficiently low average power (2 W) so as not to produce any thermal effects. To conduct whole-body and localized dosimetry, the TEMPO's vircator was replaced with a 1kW CW klystron. With the same radiating system, this arrangement produced sufficiently high average-power densities to enable calorimetric and thermometric measurements. These measurements were used to evaluate both whole-body (0.072 W/kg) and midbrain (0.21 W/kg) specific absorption rate (SAR) in cadavers of similar size. Local SARs were determined by means of implantable RF transparent Luxtron temperature probes. Whole-body SAR was determined by a single gradient-layer calorimeter.[4] The results of these measurements were used to extrapolate both the average and peak power SARs associated with the TEMPO exposures.

RESULTS

Our previous research with drugs[2] had indicated that rats fail to finish all 300 trials of a daily session at drug doses that are physically debilitating. As expected, in the absence of heating, all rats completed 300 trials in every test condition.

Figure 1 shows null responses (trials on which no response was made) as a function of stimulus duration for several of the exposure conditions. If null responses vary directly with decision confidence,[5,6] a bitonic relationship, as obtained, would be expected with a peak near the bisection point (1.6 s). By use of multiple regression, reliable quadratic ($t = -17.63, p < 0.0001$) and linear ($t = 12.82, p < 0.001$) effects of log stimulus duration were found, in addition to a reliable linear exposure condition effect ($t = 3.78, p = 0.0005$). The regression equation accounted for more than 88% of the variance in the data ($F_{45,3} = 110.75$, $p < 0.0001$).

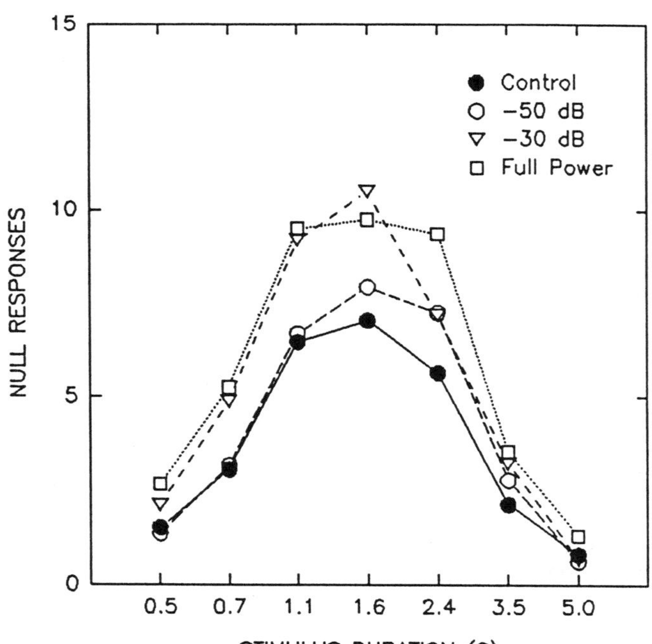

FIG. 1.—Null responses as a function of stimulus duration (s) and exposure condition. Spacing of the stimulus duration axis is logarithmic.

Figure 2 clarifies the effect of power level on null responses. Within each exposure condition, null responses are cumulated across stimulus durations. The sham-exposure condition clearly separates the three highest power levels from the lowest two. A threshold for an effect of high-power pulsed microwaves on decision processes thus occurs at or above the −30dB power level, corresponding to an SAR of approximately 7.2×10^{-5} W/kg, a value well below the currently established occupational safety standard of 0.4 W/kg.[7]

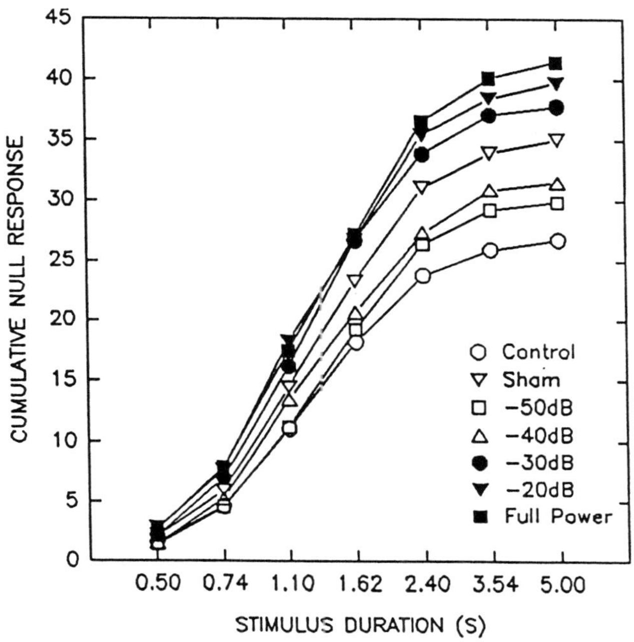

FIG. 2.—Null response cumulated as a function of stimulus duration (s) for all exposure conditions. Spacing of the stimulus duration axis is logarithmic.

The bisection point, a measure of perceptual function in this procedure, did not vary reliably as a function of exposure condition. Signal detection measures of discriminability (A′) and bias (B″) were also unaffected.

CONCLUSIONS

Taken together, the data indicate that high-power pulsed microwaves affect decision making, but not sensory or perceptual function, in rats in the temporal bisection procedure. Null responses in animal bisection procedures are similar to confidence ratings in human psychophysical procedures.[5,6] A high level of null responses indicates a low level of confidence in the decision. That is clearly seen in the present data (Fig. 1), where the highest confidence is expected at the training stimuli (0.5 and 5.0 s) and the lowest confidence is expected at the bisection point (1.6 s). A decrease in decision-making confidence was evident in subjects exposed to pulsed microwave dose levels lower than the present occupational safety standards, warranting further investigation of this effect.[8]

REFERENCES

1. T. G. Raslear, "A test of the Pfanzagl bisection model in rats," *J. Experimental Psychology: Animal Behavior Processes* 9: 49-62, 1983.

2. D. Shurtleff, T. G. Raslear, R. F. Genovese, and L. Simmons, "Perceptual bisection in rats: The effects of physostigmine, scopolamine and pirenzepine," *Physiology & Behavior* 51: 381-390, 1992.

3. F. Bates and H. Bassen, "A biological effects exposure facility utilizing an 800 megawatt peak power pulsed microwave source," *Proc. Intern. Conf. IEEE* EMB-11: 1134-1135, 1989.

4. S. Mathur, Y. Akyel, and S.-T. Lu, "Whole-body microwave dosimetry based on a single, gradient-layer calorimeter," *Bioelectromagnetics* 13: 435-438, 1992.

5. R. Pierrel and C. S. Murray, "Some relationships between comparative judgment, confidence, and decision-time in weight-lifting," *Am. J. Psychology* 76: 28-38, 1963.

6. J. P. Guilford, *Psychometric Methods,* New York: McGraw-Hill, 1954.

7. C. Polk and E. Postow, *CRC Handbook of Biological Effects of Electromagnetic Fields,* Boca Raton, Fla.: CRC Press, 1986.

8. The research described in this report was conducted in compliance with the Animal Welfare Act and other Federal statutes and regulations relating to animals and experiments involving animals and adheres to the principles stated in the *Guide for the Care and Use of Laboratory Animals,* NIH publication 85-23. The views of the authors do not purport to reflect the position of the Department of the Army or the U.S. Department of Defense (para. 4-3, AR 360-5).

Martin Blank, Ed., *Electricity and Magnetism in Biology and Medicine*

OCULAR MICROWAVE THERMOMETRIC DOSIMETRY AND THERMOMETRY IN THE RABBIT

S-T. Lu,* M. L. DeAngelis, and C. S. Gambrill

OGDEN BioServices Corp., Gaithersburg, Md.

In preparation for a study on microwave-induced retinal injury, ocular and rectal temperature changes were evaluated in rabbits exposed to a 1.25GHz pulsed microwaves with 1MW peak incident power, 10μs pulse width, and 0.86Hz pulse repetition rate for 4 h. Ocular specific absorption rates (SARs) facing or away from the waveguide applicator were 6.49 ± 0.33 and 2.11 ± 0.23 W/kg (±S.E., $n = 5$). Magnitudes of ocular temperature increments were between 2.59 ± 0.28 and 3.35 ± 0.26 C (±S.E., $n = 5$). The rectal temperature increments were 2.74 ± 0.30 C (±S.E., $n = 5$) and the rectal temperature could be used to estimate the highest ocular temperature induced by this microwave exposure.

Microwave-induced retinal neuronal degeneration was noted by Paulsson et al.[1] in rabbits; retinal detachment and pyknotic change of the pigmented epithelium in the monkey was observed by Kues et al.[2,3] Both groups observed the retinal degeneration in experimental animals subjected to repetitive pulsed microwave exposure. Kues et al. noted that the threshold specific absorption rate (SAR) for primate retinal degeneration was 3.9 W/kg for 2.45GHz pulsed microwaves (10μs pulse width, 100 pulses per second, 15 mW/cm^2 average), and 3.6 W/kg for 1.25GHz pulsed microwaves (10μs pulse width, 0.225 pulses per second, 12.5 mW/cm^2).[2,3] Kues et al. also observed abnormal electroretinographic changes and cone injury in adult Rhesus monkeys after seven 4h exposures to 1.25GHz pulsed microwaves (0.5μs pulse width, 16 pps, SAR = 3.5-4.0 W/kg, inconsistent with previous reports).[4] These threshold SARs were lower than the permissible local SAR (8 W/kg spatial average over 1 g of tissue) by ANSI[5] and the noninjurious whole-body average SAR (4 W/kg) determined by ANSI,[5] and concurred by EPA[6] and NCRP.[7] Microwave-induced retinal injuries, if proven, represent a unique injury distinguishable from other microwave-induced thermal injuries (such as skin burn, cataract, and male infertility) which required an extensive exposure (45-100 mW/cm^2 or higher) and a high local tissue temperature (42 C or higher).[6]

Due to the lateral positioning of the eyes, rabbits were chosen for study because two different SARs between two eyes can be achieved if the microwave beam is directed along the axis formed by the eyes of a rabbit. In addition, it is an important principle in toxicological study to be able to identify the same toxic effect in more than one species of experimental animals for interspecies extrapolation to human. In preparation for the evaluation of microwave-induced retinal injury, ocular microwave dosimetry was performed. Since the absorbed microwave energy is converted to heat energy, it is essential to evaluate the thermal potency of the microwave treatment. Therefore, this report includes results on ocular dosimetry and ocular thermometry in rabbits exposed to microwaves.

MATERIALS AND METHODS

Five male Dutch belted rabbits weighting 1.5-2.0 kg were used for determining the ocular specific absorption rates and temperatures changes induced by a microwave exposure. For determining temperature changes, rabbits were anesthetized with a ketamine (60 mg/h) and xylazine (8 mg/h) mixture administered as an intramuscular infusion for more than 4 h. Subsequently, the animals were euthanized with intracardiac sodium pentobarbital and pneumothorax. The carcasses were used for determining ocular SARs by a thermometric dosimetry method.[8]

For temperature changes in tissues due to microwave absorption, two Luxtron MAM-05 probes were implanted beneath each eye to 2.0 cm in depth and a third MAM-05 probe was inserted into rectum to 5.5 cm in

*Correspondence to: S-T. Lu (Microwave Research), OGDEN BioServices Corp., 6 Montgomery Village Ave. (Suite 222), Gaithersburg MD 20879.

depth. The analog outputs of Luxtron 3000 and 850 corresponding to seven ocular temperatures and one rectal temperature were smoothed with 1Hz low-pass filters (Krohn-Hite 3322 and Rockland 1022) and converted to digital signals and recorded at 0.1 Hz (Asystant+) for 4 h 16 min. The pulsed microwaves were at 1.25 GHz, 1MW peak power, 10μs pulse width, and 0.86Hz pulse repetition rate (average incident power, 8.6 W). The duration of exposure was 4 h. A Teflon-loaded WR 650 waveguide was used as an applicator. The distance between the antenna and the nearer eye was 7.6 cm.

Sixteen ocular SAR measurement points were used, four infra-ocular and four supra-ocular points for each eye. The pulse repetition rate was increased to 100 Hz (average incident power, 1 kW) for dosimetry exposures. A detailed thermometric dosimetry procedure appears elsewhere in this volume.[8] The local ocular SAR was calculated from pre-exposure (PRE), exposure (EXP), and post-exposure (POST) rates of temperature change as:

$$\text{SAR (W/kg)} = 3474\ (\text{J/kg-C°})\,[EXP - (PRE + POST)/2]\ (\text{°C/s})\ .$$

RESULTS AND DISCUSSION

The average ocular specific absorption rate of rabbits is shown in Fig. 1. The maximum absorption occurred in a short distance (0.5 cm) from the air-ocular interface. The maximum SAR was 8.75 ± 0.87 W/kg (±S.E., $n = 5$). The minimum absorption occurred at 5.0 cm from the air-ocular interface corresponding to the eyelid away from the air-ocular interface. The minimum SAR was 0.71 ± 0.11 W/kg (±S.E., $n = 5$). The ratio between maximum and minimum SARs was 12.3. The average ocular SAR were 6.49 ± 0.33 W/kg for the right eye and 2.11 ± 0.23 W/kg (±S.E., $n = 5$) for the left eye. Significant differences in SAR between eyes were noted ($t = 10.94$; $p < 0.001$).

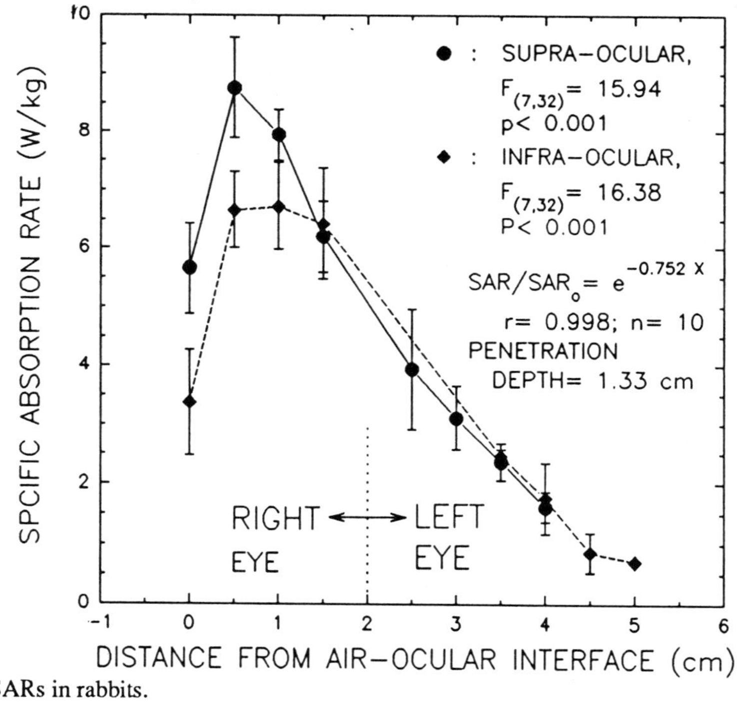

FIG. 1.—Ocular SARs in rabbits.

The pre-exposure baseline and maximum ocular temperatures for a 4h exposure are shown in Fig. 2. Apparently, this local microwave exposure could induce significant temperature changes in the eyes. The temperature increments were between 2.59 ± 0.28 ($n = 5$) and 3.35 ± 0.26 C and they were not different from each other. The ratio of the temperature increments was 1.29, a fairly uniform increase in ocular temperatures. The rectal temperature increment was within the range of microwave-induced ocular temperature increments.

The rectal-ocular temperature gradients are shown in Fig. 3. These temperature gradients decreased from positive values to negative values in 4 of 7 measurement points. The minimum temperature gradient was

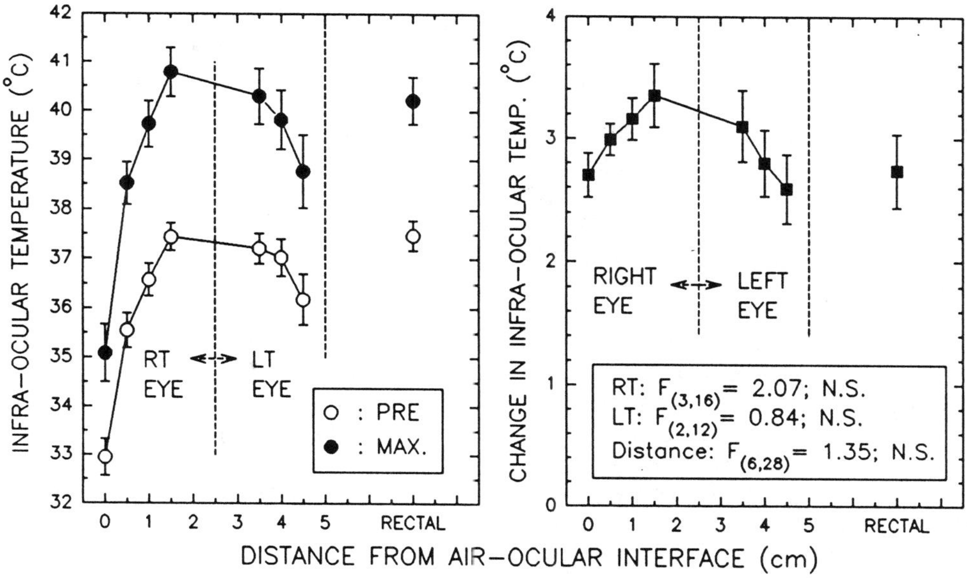

FIG. 2.—Microwave-induced ocular and rectal temperature changes.

−1.08 ± 0.19 C (n = 5) and none was lower than −1.6 C. Apparently the change in rectal temperature can be used to estimate the upper limit of ocular temperature changes in this type of local microwave exposure. Uniform ocular temperature changes from variable local SARs can assist in elucidating the planned research on mechanisms of microwave-induced ocular injury. A thermal-dependent injury may be deduced if retinal injury occurred in both eyes. An SAR-dependent injury may be deduced if ocular injury occurs in the eye with the higher SAR only.[9]

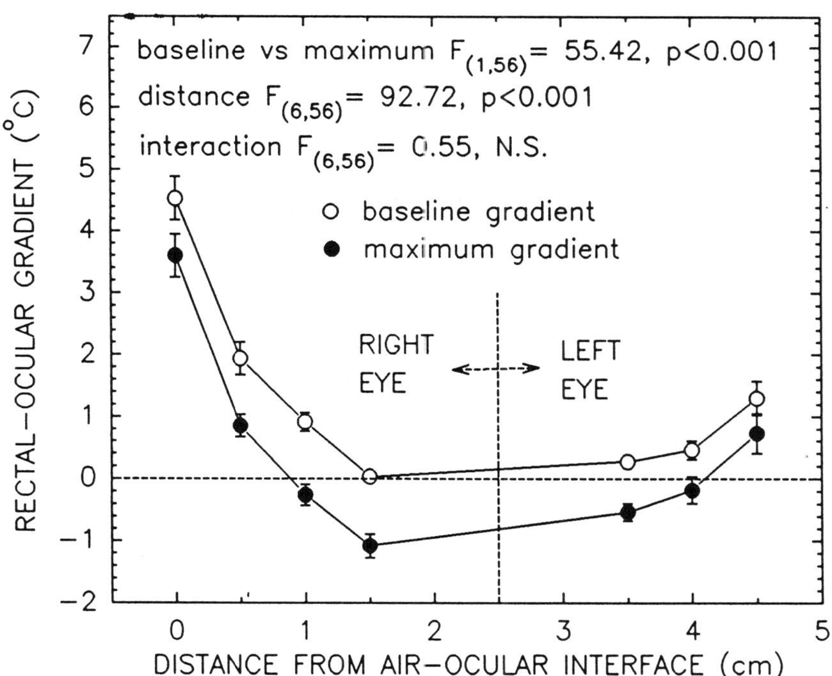

FIG. 3.—Rectal-ocular temperature gradients in rabbits.

REFERENCES

1. L. E. Paulsson, Y. Hamneius, H. A. Hasson, and J. Sjostrand, "Retinal damage experimentally induced by microwave radiation at 55 mW/cm^2," *Acta Ophthalmol.* 57: 183-197, 1979.

2. H. A. Kues, D. S. McLeod, S. A. D'Anna, G. A. Lutty, and J. C. Monahan, "Pulsed microwave-induced ocular pathology in non-human primates," Abstracts of the 11th BEMS Annual Meeting, 1989.

3. H. A. Kues, D. S. McLeod, S. A. D'Anna, G. A. Lutty, C. Gambrill, and E. C. Elson, "Ocular changes following exposure to high peak pulsed 1.25 GHz microwaves," in Ref. 2.

4. H. A. Kues, D. S. McLeod, S. A. D'Anna, M. A. Johnson, C. R. Perry, and J. C. Monahan, "Microwave-induced electroretinographic changes in the primate," Abstracts of the 13th BEMS Annual Meeting, 1991.

5. *American National Standard Safety Levels with Respect to Human Exposure to Radiofrequency Electromagnetic Fields, 300 kHz to 100 GHz,* ANSI C95.1-1982, American National Standards Institute, New York, N.Y., 1982.

6. J. A. Elder and D. F. Cahill, Eds., *Biological Effects of Radiofrequency Radiation,* EPA-600/8-83-026F, U. S. Environmental Protection Agency, Research Triangle Park, N.C., 1984.

7. *Biological Effects and Exposure Criteria for Radiofrequency Electromagnetic Fields,* NCRP Rep. 86, National Council on Radiation Protection and Measurements, Bethesda, Md., 1986.

8. C. S. Gambrill and S-T. Lu, this volume.

9. Skillful technical assistance provided by David Varle and William York Jr. contributed to the success of this research. Animal use in the study followed the guidelines set forth by NIH and Walter Reed Army Institute of Research. This study is under contract DAMD 17-89-C-9021 with the Walter Reed Army Institute of Research, Department of Microwave Research.

Martin Blank, Ed., *Electricity and Magnetism in Biology and Medicine*

EFFECT OF INTERACTION BETWEEN AMBIENT TEMPERATURE AND SAR ON MICROWAVE-INDUCED HYPERTHERMIA IN RATS

S-T. Lu*

OGDEN BioServices Corp., Gaithersburg, Md.

S. M. Michaelson

University of Rochester, N.Y. (deceased)

To assess the hyperthermic (increased core temperature) potency of microwave energy, rats were exposed to 2.45GHz microwaves at various whole-body average specific absorption rates (SARs) and at various ambient temperatures (T_a). The lowest hyperthermic SAR for each T_a was inversely proportional to T_a; the hyperthermic potency was proportional to T_a. The skin temperature increased proportionally with SAR and/or T_a. The magnitude of the microwave-induced skin temperature changes under various T_a was highly constant. Therefore, SAR could be equated to T_a elevation, i.e., 1.26 C per W/kg.

The most investigated and documented effect of radiofrequency (RF) energy on biological tissues is the transformation of energy into kinetic energy of the absorbing molecules, which produces a general heating in the absorbing medium. Michaelson noted that the magnitude of microwave-induced hyperthermia, defined as an elevation in core body temperature, was affected by ambient temperatures.[1] A homeotherm is capable of maintaining the core body temperature in a very narrow range when it is exposed to various ambient temperatures while skin temperature of the homeotherm varies with ambient temperatures. Homeotherms maintain this euthermic state by two major mechanisms, behavioral and autonomic thermoregulation. Behavioral thermoregulation during exposure to radiofrequency radiation was clearly demonstrated by Stern et al.[2] and Adair and Adams.[3] However, behavioral means of controlling or selecting a micro-environment may not always be available, or a homeotherm may intentionally override behavioral thermoregulation during engagement in other, more urgent activities. In addition, most of the studies on the biological effects of RF radiation utilized rodents exposed to such radiation under various ambient conditions. Since the majority of reliable biological effects of RF radiation are usually accompanied by an increase in core body temperature, it is important to assay the hyperthermic potency of RF radiation under various ambient conditions. The purpose of this study was to elucidate the interaction between ambient condition and RF radiation on induced hyperthermia. The tail is a major organ for heat loss in the rat.[4] Therefore, it was also included in this study.

MATERIALS AND METHODS

Sixty-four to 65-day-old male Long-Evans hooded rats from a commercial source (Charles River Breeding Laboratories, Inc., Wilmington, Mass.) were used. Their body weights were 310 ± 18 g (mean ± S.D., n = 184) ranging between 260 and 378 g. They were obtained at 35 days of age and acclimated and gentled according to procedures described previously.[5] These animals were maintained at 24 ± 1 °C on a 12h light-dark cycle (light on: 0800 to 2000 h). Feed and water were available ad libitum.

Rats were randomly subjected to 23 different treatments in foamed polystyrene cages (20×20×30 cm) with polystyrene grid bottoms and fiberglass screen tops. They were unanesthetized and unrestrained during these experiments. Rats were subjected to these experiments four at a time in an anechoic chamber (1.2×1.2 m) which was housed in a customized environmental chamber (1.8×3.0×1.8 m, Forma Scientific). The average air flow velocity inside these cages was 1.6 m/s. The treatments included various ambient temperatures

*Correspondence to: S-T. Lu (Microwave Research), OGDEN BioServices Corp., 6 Montgomery Village Ave. (Suite 222), Gaithersburg MD 20879.

(18, 24, 28, 32, and 36 °C) and various microwave exposures (1, 10, 20, 30, 40, 50, and 60 mW/cm^2) each at three different ambient temperatures (18, 24, and 28 °C). The ambient temperatures were maintained within ± 0.15 C and relative humidity between 45 and 50% at any ambient temperature. The animal was equilibrated at 24 C for 2.5 h and remained undisturbed for an additional 0.5 h during the switching to a new ambient temperature. Duration of exposure was 2 h. Colonic temperature (5 cm, YSI 423) and tail skin temperature (0.5 cm from the base of the tail, YSI 427) were taken immediately after the exposure at 1400 hours. The 2.45 GHz continuous wave microwaves were used. The average whole-body absorption rate was 0.21 W/kg per mW/cm^2.[5]

Statistical analyses included analysis of variance, a *post hoc* Dunnett's test, student *t*-test, and regression analysis. The null hypothesis was rejected at a 0.05 level.

RESULTS AND DISCUSSION

Effects of microwave exposures and ambient temperatures (T_a) on colonic temperatures (T_{co}) are shown in Fig. 1. Significant T_{co} differences were noted in rats subjected to various T_a ($F_{(4,35)} = 46.80$, $p < 0.001$)) and various microwave intensities at 18 C ($F_{(6,48)} = 42.23$, $p < 0.001$), 24 C ($F_{(6,49)} = 56.96$, $p < 0.001$), or 28 C ($F_{(6,45)} = 105.79$, $p < 0.001$). The T_{co} in rats subjected to 32 and 36 C was higher than the T_{co} in rats subjected to 18, 24, or 28 C. The minimal hyperthermic microwave intensities (threshold intensities) were 40 mW/cm^2 at 18 C, 30 mW/cm^2 at 24 C, and 10 mW/cm^2 at 28 C. The microwave hyperthermic potencies were 0.31 C/W/kg at 18 C, 0.37 C/W/kg at 24 C, and 0.46 C/W/kg at 28 C. Therefore, the threshold of microwave-induced hyperthermia were inversely proportional to T_a, but the potency of microwave-induced hyperthermia was proportional to T_a.

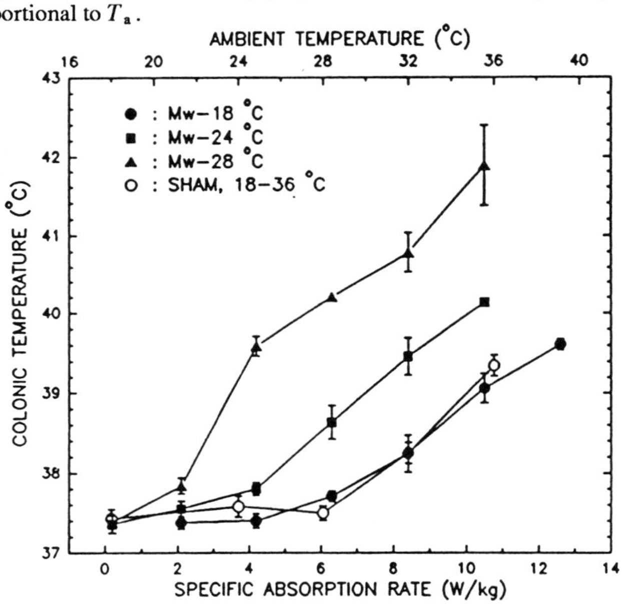

FIG. 1.—Responses of colonic temperature to changes in ambient temperatures and microwave exposures.

Changes in tail skin temperatures (T_{sk}) due to T_a or microwave exposures are shown in Fig. 2. Significant differences in T_{sk} were noted in rats subjected to various T_a ($F_{(4,35)} = 77.35$, $p < 0.001$), and to various microwave intensities at 18 C ($F_{(6,48)} = 40.99$, $p < 0.001$), 24 C ($F_{(6,39)} = 52.37$, $p < 0.001$), or 28 C ($F_{(6,45)} = 65.85$, $p < 0.001$). The minimal microwave intensities to induce increases in T_{sk} were 10, 20, and 10 mW/cm^2, respectively. Correlation between T_{sk} and T_a and between T_{sk} and SAR under various T_a were significant ($r = 0.97$ to 0.99, $p < 0.005$). The potency of T_a on T_{sk} was 0.39 C T_{sk} for every 1 C increase in T_a. The potencies of microwave radiation on T_{sk} under various T_a did not vary with T_a. They were 0.49 C/W/kg at 18 C, 0.49 C/W/kg at 24 C, and 0.48 C/W/kg at 28 C. Therefore, each 1 W/kg of the microwave absorption in rats could be equated with a 1.26 C increase in T_a.

The relation between T_{co} and T_{sk} is shown in Fig. 3. A demarcation T_{sk} (31.8 C) was clearly evident at the transition between euthermia and hyperthermia. Microwave exposures appeared to be similar to changing T_a. Hyperthermia is a result of interaction between SAR and T_a and SAR can be equated to T_a changes.[6]

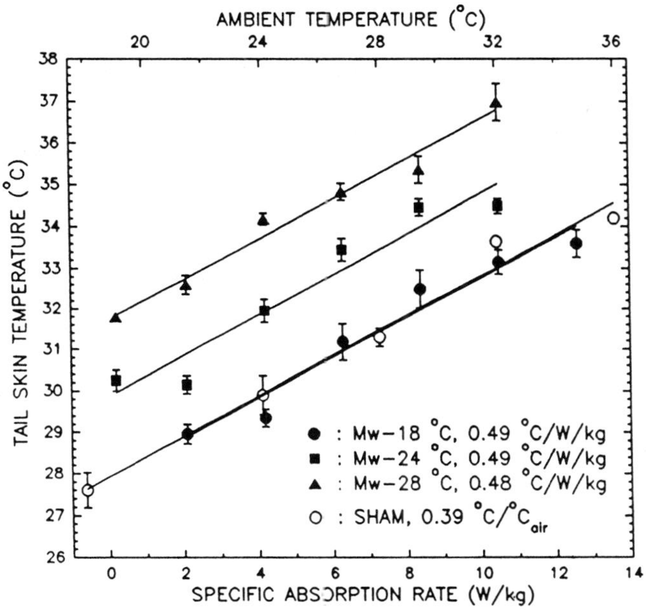

FIG. 2.—Responses of tail skin temperature to various ambient temperatures and microwave exposures.

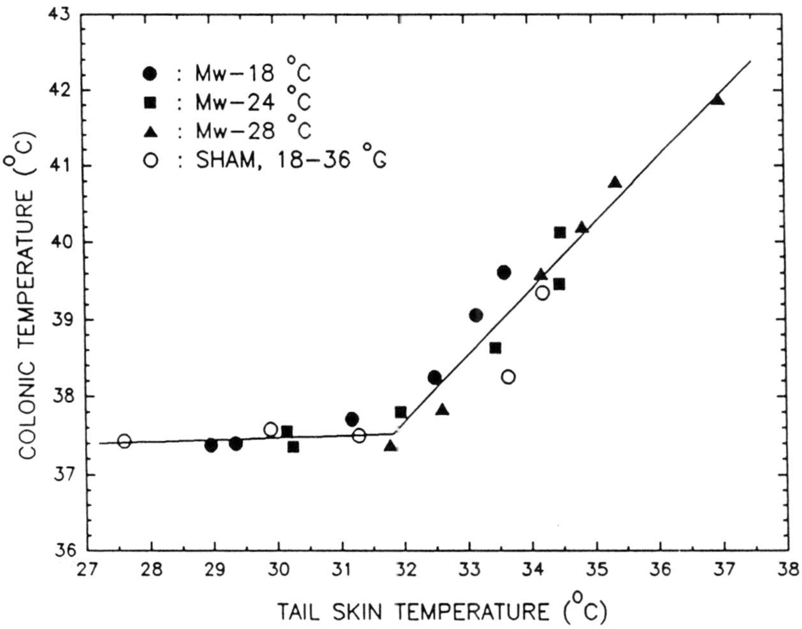

FIG. 3.—Relation between rectal and tail skin temperatures.

REFERENCES

1. S. M. Michaelson, "Thermoregulation in intense microwave fields," in E. R. Adair, Ed., *Microwaves and Thermoregulation,* New York: Academic Press, 1983.

2. S. Stern, L. Margolin, B. Weiss, S-T. Lu, and S. M. Michaelson, "Microwaves: Effect on thermoregulatory behavior in rats," *Science* 206: 1198-1201, 1979.

3. E. R. Adair and B. W. Adams, "Microwaves modify thermoregulatory behavior in squirrel monkey," *Bioelectromagnetics* 1: 1-20, 1980.

4. R. A. Little and H. B. Stoner, "The measurement of heat loss from the rat's tail," *Q. J. Exp. Physiol.* 53: 76-83, 1968.

5. S-T. Lu, N. A. Lebda, S-J. Lu, S. Pettit, and S. M. Michaelson, "Effects of microwaves on three different strains of rats," *Rad. Res.* 110: 173-191, 1987.

6. This paper was written in tribute to my advisor, the late Prof. Solomon M. Michaelson, a luminary whom we all respected and admired for his enormous contributions and advances achieved in the area of biological effects of nonionizing radiations. The work was supported by PHS grant ES 03330. The use of animals was in compliance with NIH and University of Rochester policies. The primary author is currently supported under a contract DAMD 17-89-C-9201 with the Department of Microwave Research, Walter Reed Army Institute of Research.

Martin Blank, Ed., *Electricity and Magnetism in Biology and Medicine*

ENDOGENOUS FIELDS IN INTACT, REMODELING, AND HEALING BONE

L. A. MacGinitie,* M. W. Otter, D. D. Wu, and G. V. B. Cochran

Helen Hayes Hospital, West Haverstraw, N.Y.

Streaming potential measurements in healing and remodeling canine cortical bone show that endogenous field magnitudes, frequency response, and possibly orientation depend on large bone porosity (Haversian and resorption spaces), as well as on smaller pores (matrix microporosity or canaliculi). Such fields therefore possess structural information consistent with a role as a feedback signal.

The endogenous electric fields produced by mechanical loading of hydrated bone are now thought to be due to a streaming mechanism,[1,2] manifest macroscopically by streaming potentials (SPs) and streaming currents (SCs). A convective current is produced by mechanically driven flow of the positively charged fluid past negatively charged bone surfaces. The resulting charge separation and potential difference drives a conductive "back" current either within or external to the bone. Physiologically, both SPs and SCs would be expected, since the electrical boundary condition is likely between that giving rise to SPs (open circuit) vs SCs (short circuit).[3]

A key hypothesis for this section is that during growth and remodeling, changes in bone composition, anatomy, and possibly fluid may alter the electrokinetic fields (as they alter SPs in vitro[1,2,4,5] and in vivo[6]), which could then provide feedback to the cells involved. Measurement of SPs generated by bending cortical canine bone both in vivo and in vitro (and of SCs in vitro) has demonstrated that changes in structure and composition alter field magnitudes and frequency response in healing and remodeling, compared to intact bone, for bending frequencies of 0.1-30 Hz. These frequencies include much of the range other investigators suggest may be relevant to bone remodeling.[7,8]

SPs and SCs IN HEALING AND REMODELING BONE

SPs were measured in vivo at healing, remodeling, and intact canine tibia cortical bone sites, in accordance with NYDOH and NIH guidelines. Healing sites were 4mm drill holes in the diaphysis, remodeling sites were intact bone adjacent to the drill holes, and normal sites were on the opposite unoperated limb. SPs were measured at these sites 2, 4, or 12 weeks after surgery, by use of a modification of techniques described previously.[9,10] Following in vivo tests, tibias were harvested for in vitro SP and SC measurements in 4-point bending and for histology.

SP/strain was significantly smaller for 2- and 4-week drill holes relative to remodeling or normal controls both in vivo and in vitro.[9] SPs for 12-week drill holes and remodeling controls were similar, while SPs for intact sites were 2-3 times larger. Since 12-week drill holes were healed and remodeling, the similarity of SPs between these and the remodeling controls appears reasonable. SP normalized to bone impedance (calculated from the bone + electrode system impedance) and SCs (not normalized) in vitro were similar for all sites.[11]

SP and SC kinetics were faster *on average* for healing and remodeling bone: as shown by an increase in mean SP (and SC) phase relative to strain, and an increase in mean high frequency response compared to intact values.[9] However, frequency response was variable for individual healing and remodeling bone sites in vivo.

---

*Correspondence to: L. A. MacGinitie (Engineering), Pacific Lutheran University, Tacoma WA 98447.

BONE STRUCTURE AND SP GENERATION IN BONE

Experimental studies and models to date suggest that SPs are generated by convective-current flow in small pores (radius < 200Å[4]-400Å[12]). These might be fluid spaces outside the cell processes in the canaliculi[13,14] or bone matrix microporosity.[12] This fluid apparently flows to the nearest Haversian canal.[13,15,16] SPs measured for healing and remodeling bone are consistent with this picture, and further suggest that the conductive back current flows in large (Haversian, intertrabecular, or resorption) spaces, rather than in the small pores in which the convective current flows.[11] The faster SP and SC kinetics are consistent with faster fluid relaxation due to shorter distances between the large porosity. The small SPs at 2-4 weeks appear to be due to greater bone conductance (large porosity), since SC and SP/impedance are similar for all sites, and SC does not depend on the conductive "back" current while SP does. Bone conductance had previously been attributed to Haversian canals,[17,18] an interpretation consistent with the relation between porosity and conductance in healing and remodeling bone.

Although *mean* SP magnitudes and frequency response can be explained in terms of macroscopic impedance and mean bone porosity, the magnitude and frequency response at individual sites cannot, and may depend on fine bone structure and surface characteristics (e.g., ζ potential). For example, although SP magnitude at individual sites shows a significant ($p < 0.001$) correlation with impedance, impedance alone is not a good predictor of SP magnitude.[11] In addition, some healing and remodeling sites displayed on non-monotonic (peak) frequency response or monotonic responses that were not asymptotic below 30 Hz (Fig. 1), unlike the SP of normal bone which typically increased monotonically with bending frequency to an asymptote (as observed previously in vivo[10] and in vitro[4,12]). While the increase in high-frequency (nonasymptotic) response can be explained by shorter relaxation times due to large porosity, the peak type of frequency response cannot.

The peak type of frequency response can be characterized by two relaxation times (magnitude of opposite sign) that could be due to contributions by two different pore sizes (e.g., the microporosity vs canaliculi) to the convective current. This frequency response was similar to that previously observed in normal bovine bone strips soaked in low concentrations, typically less than 0.03 M (Fig. 1). In this previous experiment, the transition from a normal, asymptotic to a peak type of response could be attributed either to a decrease in the magnitude of the convective current in smaller pores (microporosity) as the charged layer thickness increased at low concentrations, or in the effective pore radius of the larger pores (canaliculi) due to osmotic swelling of organic matter. For remodeling bone, the magnitude of convective current due to microporosity relative to canaliculi might vary with the degree of bone mineralization.

ELECTROKINETIC FIELDS AS A REMODELING FEEDBACK SIGNAL

Field magnitude, orientation, and frequency may all impact on field-modulated cell behavior. That SP and SC magnitude and frequency response depend on changes in bone structure, and possibly composition during healing and remodeling, indicates they are theoretically capable of serving as a feedback signal. Since SP magnitudes are 6-14 times smaller for healing than for normal bone and since experiments indicate that *increases* in applied field magnitude maintain or produce bone,[8] it can be argued that these SPs are unlikely to promote bone healing. However, changes in local field gradients and orientation that cannot be resolved by macroscopic measurements may also occur, and these would not be mimicked by applied fields. Microscopic measurements demonstrating cusp-like potentials around Haversian canals[16] suggest that microscopic nonuniformity in potential due to flow toward large pores may also imply differences in local potential gradients in healing or remodeling vs intact bone due to differences in large porosity. Further, a change in large pore orientation (from the transverse intertrabecular spaces of healing woven bone to the transverse Haversian and resorption spaces of remodeling bone to the longitudinal orientation of Haversian systems in normal bone) implies a change in convective current orientation. The relative contributions to convective current proposed for microporous spaces and canaliculi would imply a further change in field orientation on a smaller scale.

SCs may be as relevant as SPs as an indicator of the electrokinetic fields present during remodeling. However, SC measurements in vitro may not adequately reflect the changes in SCs during healing and remodeling that might occur under more physiological conditions, since the ratio of convective and conductive currents will depend not only on the bone structure, but on changes in fluid and electrical boundary conditions[3] at the bone surface.

The mean increase in high frequency response for healing, and particularly for remodeling bone, may relate to the observation of McLeod and Rubin that frequencies above 10 Hz are most effective in maintaining bone mass.[3,8,19]

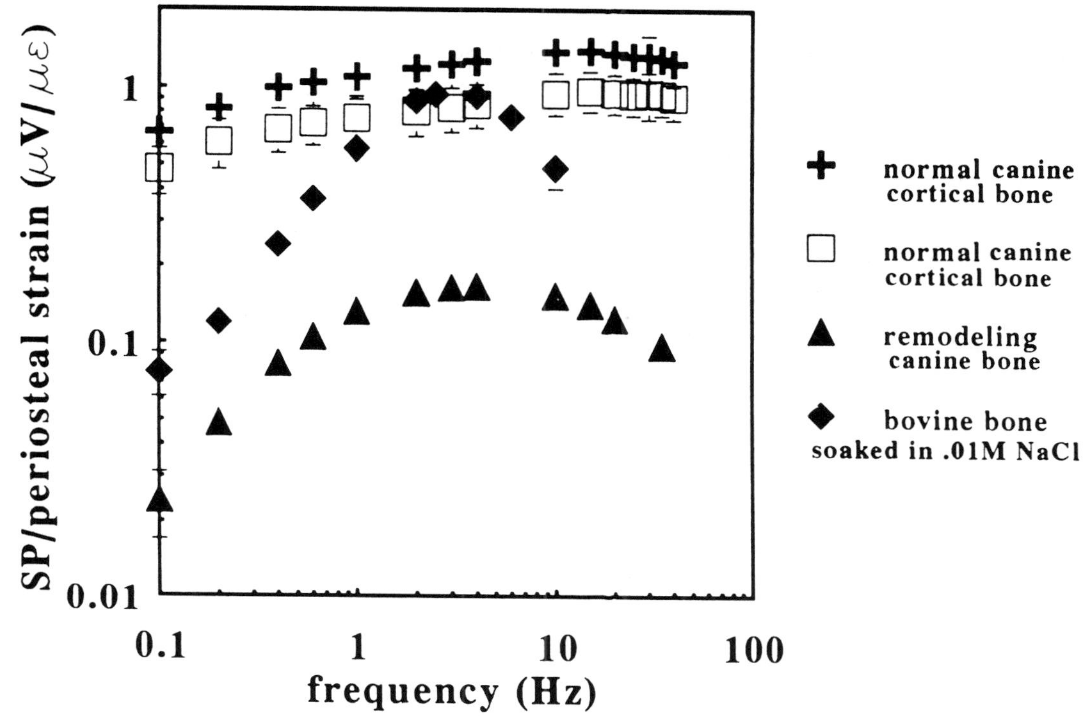

REFERENCES

1. D. Gross and W. S. Williams, *J. Biomech.* 15: 277-295, 1982.
2. D. Pienkowski and S. R. Pollack, *J. Orthop. Res.* 1: 30-41, 1983.
3. M. W. Otter et al., this volume.
4. L. A. MacGinitie et al., *J. Biomech.* (in press).
5. S. R. Pollack, *J. Biomed. Mater. Res.* 2: 677-700, 1977.
6. M. W. Otter et al., *Trans. BRAGS* 11: 61, 1991.
7. C. A. L. Bassett, this volume.
8. K. J. McLeod and C. T. Rubin, this volume.
9. L. A. MacGinitie et al., *Orthop. Trans.* 16, 1992.
10. M. W. Otter et al., *J. Orthop. Res.* 10: 710-719, 1992.
11. L. A. MacGinitie et al., (in press).
12. R. A. Salzstein and S. R. Pollack, *J. Biomech.* 20: 271-280, 1987.
13. M. W. Johnson et al., *J. Biomech.* 15: 881-885, 1982.
14. S. R. Pollack et al., *J. Biomech.* 17: 627-636, 1984.
15. L. A. MacGinitie et al., *Trans. BRAGS* 9: 15, 1989.
16. W. Starkebaum et al., *J. Biomed. Mat. Res.* 13: 729-751, 1979.
17. D. A. Chakkalakal and M. W. Johnson, *Clin. Orthop. Rel. Res.* 161: 133-145, 1981.
18. J. D. Kosterich et al., *IEEE Trans.* BME-30: 81-86, 1983.

19. Supported in part by NSF BCS-9008748, VA A-160-RA, NIH R01-AR39574, the Walter Scott Foundation, and the N.Y. State Department of Health.

Martin Blank, Ed., *Electricity and Magnetism in Biology and Medicine*

STREAMING POTENTIAL ALTERATIONS IN OSTEOARTHRITIS

W. R. Walsh* and M. G. Ehrlich

Brown University, Providence, R.I.

A shift in the mechanical loading environment results in a permutation of the external forces that influence bone development. A shift in the load and deformation patterns also alters the fluid flow and therefore the electrokinetic streaming potentials. The role of streaming potentials in cancellous bone in osteoarthritis (OA) has not been examined. Intact cancellous streaming potentials were examined to determine the electrokinetic properties of bone in the Hulth model of OA. The physical properties of bone and biochemical and histological changes of the articular cartilage were also examined. This study examines whether the electrokinetic signal in cancellous bone changes during the progression of OA.

Osteoarthritis is not only extremely frequent in aging subjects, affecting approximately 80% of individuals over the age of 70, but also has a severe impact on the working population. Experimental models of OA provide evidence of alterations in the subchondral bone and cartilage destruction. The mechanism(s) responsible for the bone changes remains unclear. Pathologic changes in subchondral bone have been implicated in the pathogenesis of OA.[1,2] There is also great discrepancy regarding the type of alterations in the subchondral cancellous bone in OA models. An increase[2] as well as a decrease[1] in the mechanical properties of bone during the progression of cartilage destruction in OA has been observed clinically and in animal models.

Cancellous bone and its surrounding bone fluid spaces represent an electrically charged porous material and an ion-containing fluid phase. In bone, an electrical double layer is established between the charged bone matrix surface and the ion-containing fluid.[3,4] Fluid flow due to external mechanical forces results in the generation of streaming potentials due to the tangential displacement of ions in the mobile portion of the electrical double layer. Streaming potentials have been suggested to be involved in the repair and remodeling of bone tissue.[5] Altered loading patterns in diarthrodial joints may alter the streaming potentials in bone and result in a modification in the biofeedback system. Altered loading and fluid flow potentials may expose bone to a new electrokinetic environment to which cells respond and bone remodels.

MATERIALS AND METHODS

Twenty-five 4-5kg New Zealand white rabbits were operated on according to the Hulth-Telhag model for OA[6] and sacrificed at 5, 7, 9, and 12 weeks following surgery. This model produces instability by transection of the medial collateral, anterior, and posterior cruciate ligaments and medial meniscetomy. The left limb was used as the experimental side. Histological-histochemical analysis was performed by Safrinin-O, iron hematoxylin and fast green, and Mankin Scale grading.[7] Electrokinetic, density, and ash-weight experiments were performed on rabbit femoral and tibial core samples. Streaming potentials from bovine tibia cancellous bone with normal-appearing articular cartilage were also performed. Uronic acid contents were determined by a modified assay of Bitter and Muir.[8]

Cancellous streaming potentials were determined in a horizontal streaming potential chamber,[4] in an aphosphate buffer.[3] Zeta potentials were calculated from the Helmholtz-Smoluchowski equation, at the slope of the pressure and streaming potential in the low-pressure (linear-flow) region. A one-way ANOVA was used to determine statistical differences.

RESULTS AND DISCUSSION

The mean grade for the severity of OA was 3.75 (1.65) for the surgical side and 2.07 (1.03) for the control side. OA in the Hulth model used does not become severe until about 6 months following surgery.[9]

---

*Correspondence to: W. R. Walsh (Orthopaedics), Brown University School of Medicine, Providence RI 02903.

Uronic acid assay results confirm no significant changes by 3 months in the glycosaminoglycans (GAG).

Zeta potentials of rabbit cancellous bone were more negative compared to control bovine cortical bone.[3,10] Samples from the lateral side had consistently more negative zeta potentials compared to the medial side from both left and right limbs after 5, 9, and 12 weeks. A reduction in the negativity of zeta potentials from the lateral and medial samples from the left (surgical) side was observed after 5, 9, and 12 weeks (Fig. 1). A slight reduction in the negativity of the zeta potential was also observed for femur samples from the control side after 5 and 12 weeks.

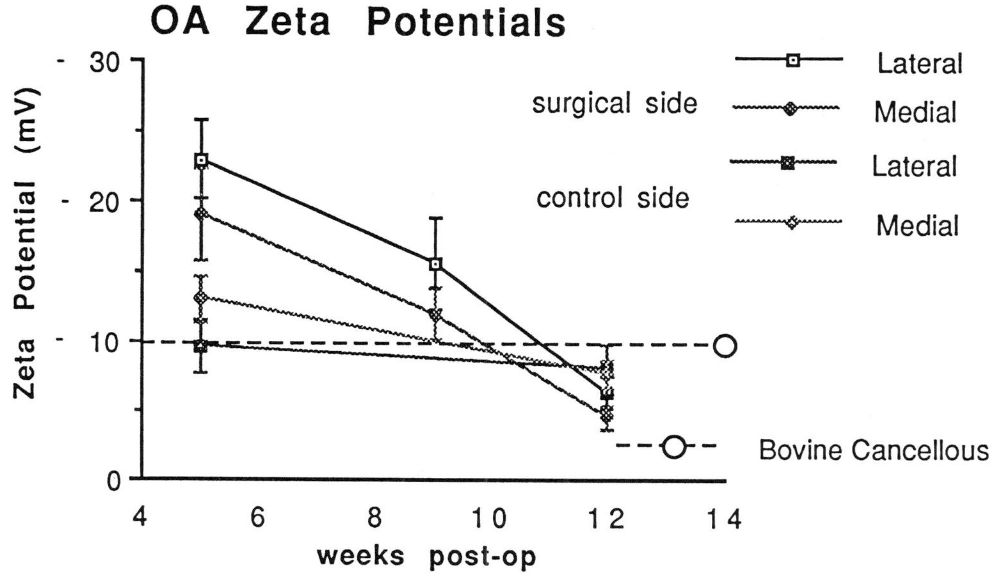

FIG. 1.—OA zeta potentials.

Typical results for streaming potentials per applied pressure for normal bovine cores with marrow are presented in Fig. 2. The relationship between pressure and potential in the pressure range examined was very linear. Streaming potentials were positive for all pressure ranges (negative zeta potentials). The mean bovine cancellous streaming potential per applied pressure was 5.24×10^{-3} mV/kPa and calculated zeta potential was -9.76 mV, which is more negative than bovine cortical bone in the same solution (-6.07 mV).[3,10] The zeta potentials of bovine cancellous bone were more negative than the 5-week OA samples in the surgical group, but similar to the 12-week OA samples, where the zeta potentials appear to have reached a steady state. This result suggests that by 12 weeks the samples were beginning to approach the zeta potential of normal bone.

The apparent densities from the 12, 9, 7, and 5-week animals were 768.0 (5.3), 771.5 (4.7), 774.0 (4.2), 773.0 (4.27) kg/m^3, respectively. The ash weight values for the 12, 9, 7, and 5 week groups were 0.603 (0.02), 0.637 (0.03), 0.624 (0.05), and 0.621 (0.04), respectively. These data do not indicate any drastic alteration in the amount of mineral or material present.

An alteration in the zeta potential suggests that a change in the electrical double layer and the fluid flow induced streaming potentials during the progression of OA. Histological grading and biochemical results confirm that significant osteoarthritic changes had not occurred. This finding suggests that the osteoarthritis in the Hulth model used here had not progressed very far. The electrokinetic changes suggest that alterations in the electrokinetic signal of bone tissue may undergo considerable change prior to cartilage changes, as measured by histological grading and biochemical assays for GAG content. Modifications in the electrical double layer, streaming potential, and zeta potential result in a new electrokinetic environment, to which cells may respond and remodel, which would lead to bony changes.

REFERENCES

1. E. L. Radin, H. G. Parker, J. W. Pugh, R. S. Steinberg, I. L. Paul, and R. M. Rose, "Response of joints to impact loading: III. Relationship between trabecular microfractures and cartilage degeneration," *J. Biomechanics* 6: 51-57, 1973.

2. M. T. Serink, A. Nachemson, and G. Hansson, "The effect of impact loading on rabbit knee joints," *Acta Orthop. Scand.* 48: 250-262, 1977.

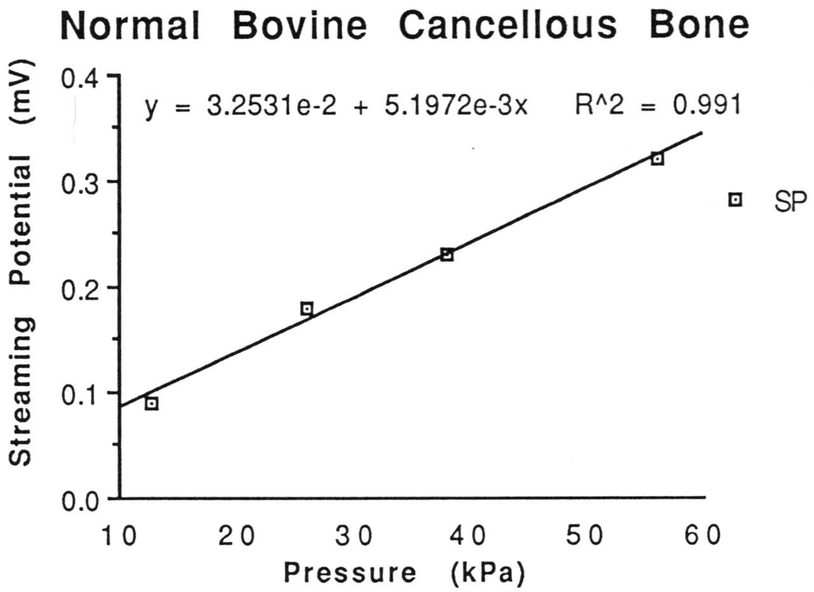

FIG. 2.—Normal bovine cancellous bone.

3. W. R. Walsh and N. Guzelsu, "Electrokinetic behavior of intact wet bone: Compartmental model," *J. Orthop. Res.* 9: 683-692, 1991; and "Streaming potentials of the vascular system of bone: Detergent treatment and phosphate ion equilibration," in C. T. Brighton and S. R. Pollack, Eds., *Electromagnetics in Biology and Medicine,* San Francisco: San Francisco Press, 1991, 37-38.

4. W. R. Walsh, T. P. Lefkoe, A. Armstrong, A. Palifox, and M. G. Ehrlich, "The role of streaming potentials in osteoarthritis," *Orthopaedic Res. Trans. 38th Ann. Meeting,* 1992, 236.

5. C. A. L. Bassett, "Biophysical principles affecting bone structure," in G. R. Bourne, Ed., *Biochemistry and Physiology of Bone,* New York: Academic Press, vol. 3, 1971, 2d ed.

6. A. Hulth, L. Lindberg, and H. Telhag, "Experimental osteoarthritis in rabbits: Preliminary report," *Acta Orthop. Scand.* 41: 5522-5580, 1970.

7. H. J. Mankin, H. Dorforman, L. Lipiello, and A. Zarins, "Biochemical and metabolic abnormalities in articular cartilage from osteo-arthritic human hips: II. Correlation of morphology with biochemical and metabolic data," *JBJS* 53A: 523-537, 1971.

8. T. Bitter and H. Muir, "A modified uronic acid carbazole reaction," *Anal. Biochem.* 4: 330-334, 1962.

9. M. G. Ehrlich, H. J. Mankin, J. Hardin, A. Grossman, C. Crispen, and D. Ancona, "Biochemical confirmation of an experimental osteoarthritis model," *J. Bone Joint Surg.* 57A: 392-396, 1975.

10. N. Guzelsu and W. R. Walsh, "Streaming potential of intact wet bone," *J. Biomech.* 23: 673-686, 1990.

Martin Blank, Ed., *Electricity and Magnetism in Biology and Medicine*

A STRUCTURAL MODEL OF THE BONE SURFACE—FLUID INTERFACE BASED ON ZETA-POTENTIAL MEASUREMENTS

R. M. Kowalchuk, S. R. Pollack,* and T. A. Corcoran

University of Pennsylvania, Philadelphia

Endogenous electrical signals in bone, thought to play a physiologically significant role in bone repair and remodeling, are electrokinetic in origin and directly proportional to the zeta potential. A summary of recent zeta potential measurements obtained from particle microelectrophoresis using bovine bone and a bone mineral in simulated bone extracellular fluid (Neuman's fluid) with calcium and phosphate excursions is presented. These results support a structural model of the bone surface/bone fluid interface which predicts that the zeta potential depends on the bone substrate proper (composed of collagen, mineral, and boundary regions), the stationery layer (wherein ions, ionic complexes, and proteins are adsorbed), and the bone extracellular fluid.

Endogenous electrical signals in bone, which are hypothesized to have a physiologically significant role in bone repair and remodeling, have been shown to be of electrokinetic origin and are directly proportional to the zeta potential.[1,2] However, bone samples used in many previous measurements of electrical potentials in bone were exposed to a variety of nonphysiologic solutions and treatments.[3-9] Consequently, our understanding of the structure of the *in situ* bone surface/bone fluid interface is still inadequate; in particular, the relative contribution of collagen, other bone mineral proteins, and mineral to the value of the zeta potential is still unclear. A summary of recent zeta-potential measurements obtained from particle microelectrophoresis of bovine bone and a bone mineral in Neuman's fluid (NF),[10] which simulates bone extracellular fluid, and in NF with calcium and phosphate concentration excursions is presented. The results are used to develop a new understanding of the structure of the bone surface/bone fluid interface.

MATERIALS AND METHODS

Cortical bone from bovine femora and a calcium-deficient hydroxyapatite (CDHA) were used as the solid substrates. After grinding with a diamond-tipped dental drill, 5 micron particles of the solid substrates were suspended in NF, or NF with controlled calcium or phosphate excursions which were physiologic and supraphysiologic. Particles were exposed only to NF during preparation and only to their testing solution thereafter.

Zeta-potential determinations for particle populations were performed in triplicate using the PenKem System 3000 Automated Electrokinetics Analyzer, as described elsewhere.[11] Briefly, charged particles suspended in a fluid medium move in response to an applied electric field with a measurable speed and direction. This electrophoretic velocity divided by the applied electric field is the electrophoretic mobility, which is directly proportional to the zeta potential according to the Smoluchowski equation for electrophoretic mobility.[12] Specific conductivity, pH, temperature, and turbidity were measured simultaneously.

Measurements were performed in NF modified by five calcium concentrations (0.2, 0.4, 0.8, 1.0, and 1.6 mM) and three phosphate concentrations (1.8, 8.3, and 14.7 mM) at 0, 2, 6, and 18 h in their respective testing solutions. Calcium and phosphate concentrations in unmodified NF are 0.48 and 1.8 mM, respectively. Multiway and one-way analysis of variance were performed, as appropriate.

RESULTS

A summary of the results is given in Table 1. The zeta potential of calcium-deficient hydroxyapatite (CDHA) in NF was negative. With increasing phosphate concentration in NF, the zeta potential became

---

*Correspondence to: S. R. Pollack (Bioengineering), 120 Hayden Hall, University of Pennsylvania, Philadelphia PA 19104.

more negative. With increasing calcium concentration in NF, the zeta potential became less negative and even positive.

TABLE 1.—Zeta potential changes in NF (mV).

	With increasing calcium (0.5× to 4×)	With increasing phosphate (1× to 7×)
CDHA	from −10 to +2	from −10 to −17
Bovine femur	from −5 to −10	from −5 to −7.5

$N = 8$ for each substrate and solution.

The zeta potential of bovine femoral cortical bone was also negative in NF and became more negative with increasing phosphate concentration in NF. However, at low calcium concentrations, the zeta potential also became more negative. With increasing calcium at *high* concentrations in NF (not shown in Table 1), the zeta potential of bone became less negative and inverted in sign.[13,14] The high-concentration result was similar to that seen for CDHA and suggests a saturation of the low-calcium concentration process.

DISCUSSION

The zeta potential of CDHA in NF, which simulates bone extracellular fluid, was negative as it was for bone.[15] The zeta potential for both CDHA and bone became more negative with the addition of phosphate. Thus, under these conditions, one could not differentiate the contributions of collagen and mineral to the zeta potential of bone. However, under conditions of increasing calcium concentration in NF, a difference between CDHA and bone was observed, which shows that the mineral phase was not the major factor in the determination of the zeta potential. This conclusion has also been drawn by others, but for different reasons.[4-6]

A proposed structural model consistent with the electrokinetic results and known histology of bone is depicted in Fig. 1. There are three major components to the model: (1) the bone substrate proper, which is composed of collagen, mineral, noncollagenous proteins, and boundary regions; (2) the stationary layer between the bone substrate proper and the slip plane, where various ions, ionic complexes, and proteins are adsorbed; and (3) the bone extracellular fluid, which can vary in composition depending on the metabolic milieu and physiologic state.

The zeta potential is defined as the electric potential at the slip surface with respect to the neutral fluid far from the interface. Figure 2 illustrates the model under conditions of a variable surface potential V_X caused by surface incorporation of ions and by the presence of adsorption within the slip surface onto a plane of adsorption. This case is similar to the Gouy-Chapman-Stern model beyond the adsorbed plane where the potential can be described by the Poisson-Boltzmann equation.[15] The combination of surface ion incorporation plus adsorbed charges within the slip plane alter the zeta potential and lead to a change in sign of the zeta potential, as indicated by ζ_2 and ζ_3. This model can explain why the change in sign of the zeta potential occurred for CDHA in NF with increasing calcium but did not occur with bone: because of the adsorption of proteins, as has been previously observed.[17] The results of the kinetic studies performed by others and supported by the results of this work suggest a time dependence explained by the kinetics of ion concentration equilibration in three interface regions: the hydration layer, the surface proper, and the bulk solid.[18]

CONCLUSIONS

A structural model was developed which predicts that the zeta potential depends on the bone substrate proper (composed of collagen, mineral, and boundary regions), the stationary layer (in which ions, ionic complexes, and proteins are adsorbed), and the bone extracellular fluid. Since the zeta potential of bone and CDHA are both negative, each may contribute to the determination of the zeta potential under certain conditions, such as increased phosphate in the fluid. However, adsorption of negatively charged proteins, ionic complexes, or protein conformational changes are needed to explain increases in zeta potential magnitude with additions of the divalent positively charged calcium ion in low concentration.[19]

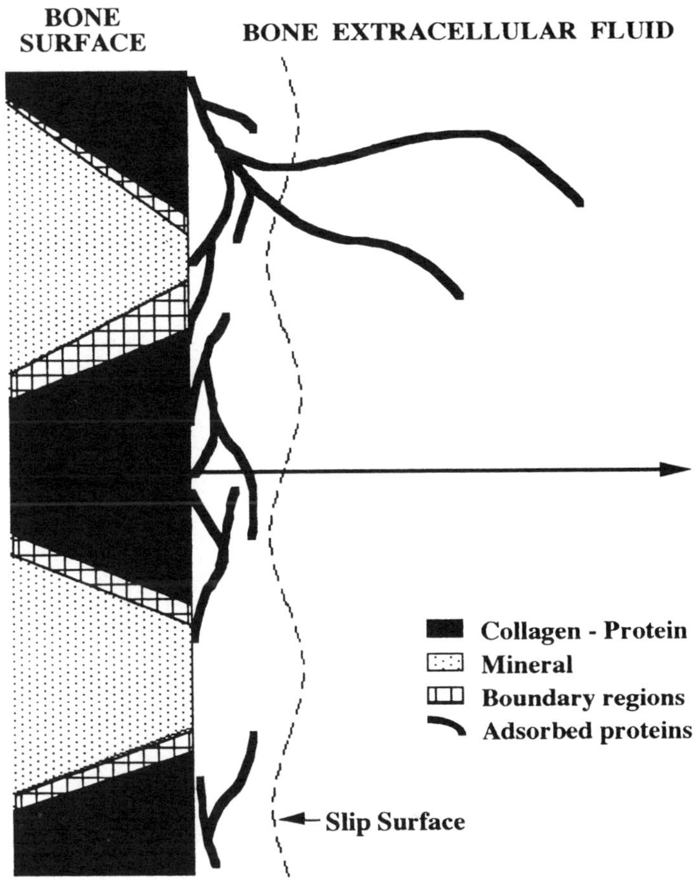

FIG. 1.—Structural model of bone surface/bone fluid interface.

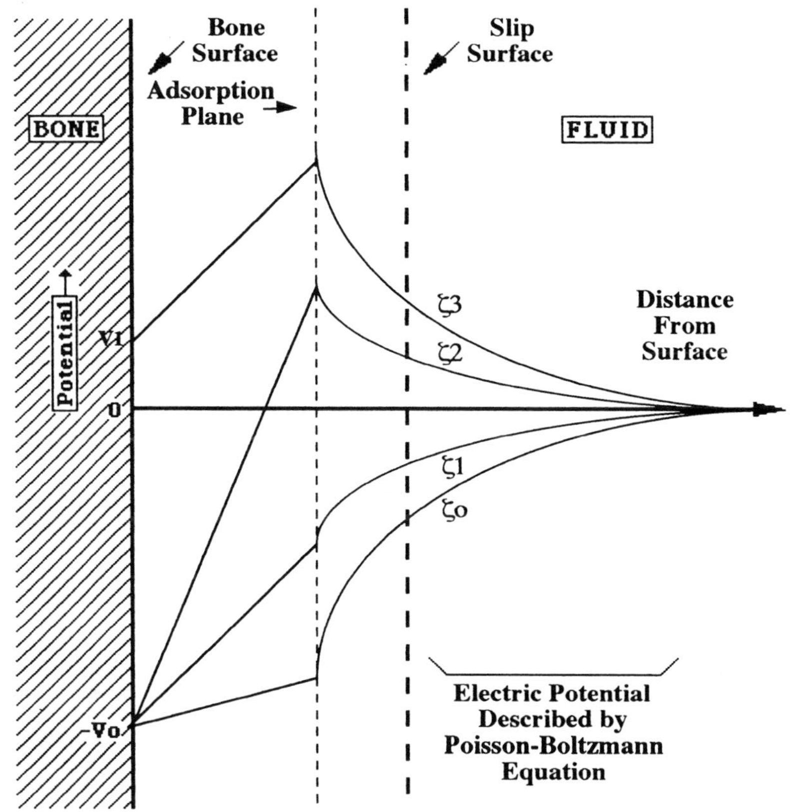

FIG. 2.—Electric potential based on model of bone surface/bone fluid interface: Variable V_X with adsorption.

REFERENCES

1. R. A. Salzstein and S. R. Pollack, "Electromechanical potentials in cortical bone: II. Experimental analysis," *J. Biomech.* 20: 271-280, 1987.

2. R. A. Salzstein, S. R. Pollack, A. F. T. Mak, and N. Petrov, "Electromechanical potentials in cortical bone: I. A continuum approach," *ibid.* 261-270, 1987.

3. D. A. Berretta and S. R. Pollack, "Ion concentration effects on the zeta potential of bone," *J. Orthop. Res.* 4: 337-345, 1986.

4. N. Guzelsu and R. Regimbal, "The origin of electrokinetic potentials in bone tissue: The organic phase," *J. Biomech.* 23: 661-672, 1990.

5. N. Guzelsu and W. R. Walsh, "Streaming potential of intact wet bone," *ibid.*, 673, 1990.

6. R. M. Kowalchuk and S. R. Pollack, "The effects of $CaCl^2$ on stress generated potentials in bovine cortical bone," *IEEE* 1987.

7. M. Otter, S. Goheen, and W. S. Williams, "Streaming potentials in chemically modified bone," *J. Orthop. Res.* 6: 346-359, 1988.

8. D. Pienkowski and S. R. Pollack, "The origin of stress generated potentials in fluid-filled bone," *ibid.*, 1: 30-41, 1983.

9. G. C. Scott and E. Korostoff, "Oscillatory and step response electromechanical phenomena in human and bovine bone," *J. Biomech.* 23: 127-143, 1990.

10. W. F. Neuman, "The milieu interior of bone: Claude Bernard revisited," *Fed. Proc.* 28: 1846-1850, 1969.

11. R. M. Kowalchuk, *The Zeta Potential of Normal and Osteoporotic Bone*, Ph.D. Thesis, University of Pennsylvania, 1991.

12. M. v. Smoluchowski, in Graetz, Ed., *Handbuch der Electrizität und des Magnetismus*, Leipzig, Barth, 1921, 366.

13. R. M. Kowalchuk and S. R. Pollack, "Stress generated potentials in bone: Effects of bone fluid steeping time and ionic composition," *J. Orthop. Res.* (in press).

14. R. M. Kowalchuk and S. R. Pollack, "The zeta potential of bone from particle electrophoresis," *ibid.*

15. R. M. Kowalchuk, S. R. Pollack, P. Ducheyne, and L. A. King, "Particle microelectrophoresis of calcium-deficient hydroxyapatite: Solution composition and kinetic effects," *J. Biomed. Mater. Res.* (in press).

16. R. J. Hunter, "Zeta potential in colloid science," in R.H. Ottewill and R. L. Rowell, Eds., *Colloid Science*, New York: Academic Press, 1981.

17. E. C. Reynolds and A. Wong, "Effect of adsorbed protein on hydroxyapatite zeta potential and *Streptococcus* mutans adherence," *Infection and Immunity* 39: 1285-1290, 1983.

18. W. F. Neuman and M. W. Neuman, *The Chemical Dynamics of Bone Mineral*, Chicago: University of Chicago Press, 1958.

19. This work was supported by National Institutes of Health SCOR Grant 5-P50-AR39226-02 and by the NIH Medical Scientist Training Program.

Martin Blank, Ed., *Electricity and Magnetism in Biology and Medicine*

AGE-RELATED CHANGE IN THE ZETA POTENTIAL OF BONE AND ITS INFLUENCE UPON ENDOGENOUS ELECTRICAL FIELD STRENGTH

A. A. Spirt* and S. R. Pollack

University of Pennsylvania, Philadelphia

A whole-bone model has been developed for calculating the magnitude of stress-generated potentials (SGPs) and the resultant electric-field magnitude in the cortex of the midshaft rat femur by use of in vivo strains. This model utilizes a 2-D biphasic analysis in conjunction with a microcontinuum model for streaming potentials in wet cortical bone. The model was used to predict the strain-dependent peak electric fields in experimental rats aged 1.5, 3, 6, 12, and 24 months at frequencies ranging from 0.5 to 30 Hz. Based on measured parameters, the strain-frequency response predicts that high-strain/high-frequency (impulse) loads produce electric fields of biologically significant magnitude, and suggests that these fields are involved in the mechanism with which mechanical forces mediate bone remodeling.

Lanyon has proposed that tissue strain level is the main controlling determinant of the remodeling response of bone to its mechanical environment.[1] However, the mechanism(s) by which tissue strain is transduced into cellular response is unknown. Rubin and Lanyon proposed that either direct cellular deformation or several strain-related phenomena may provide the cell stimuli necessary for bone remodeling.[2] These strain-related phenomena include an increase in the hydrostatic pressure of extracellular fluid within the intracortical microporous space of bone (above that of the nonstressed state); the flow of this fluid through the microporous compartment, resulting in shear stresses on cell surfaces; and the generation of endogenous electrical potentials (stress-generated potentials). This paper will focus on the endogenous electric fields that arise in the mid-shaft rat femur throughout the lifespan of the rat; results of studies on the hydrostatic pressure and flow velocity of the interstitial fluid will be presented in future publications.

It has been suggested that endogenous electric fields affect bone cell function in a dose-related manner.[3] The Zeta potential (ZP) of bone is the electrokinetic property that influences the magnitudes of SGPs.[4] In addition, the SGP magnitude depends on the cross-sectional geometry, the material properties, and the applied bending moments of the bone. We have measured, calculated, or obtained values for these parameters from the literature, and developed a whole-bone model for calculating endogenous electric fields in the midshaft rat femur throughout the lifespan of the rat. These data provide insight into the physiologic significance of electrical fields in the adaptation of bone to its mechanical environment.

THEORY

The midshaft femur was modeled as a two-phase circular cylinder (of inner and outer radii r_1 and r_2, respectively) consisting of a porous elastic matrix filled with an ionic interstitial fluid. The one-dimensional biphasic theory for bone developed by Salzstein et al.[4] was extended into a two-dimensional analysis in cylindrical coordinates (with the z-axis parallel to the long axis of the bone) in order to model whole bone in pure bending. An expression for the pressure in the fluid resulting from periodic (sinusoidal) pure bending deformation was derived for the specific bone geometry and boundary conditions (zero radial flow at the endosteal and periosteal envelopes) of the model. An electrokinetic analysis based on a microcontinuum model for streaming potentials in wet cortical bone was then applied to the whole-body model in order to describe the SGPs throughout the cortex in terms of the permeability and the driving pressure in the fluid.

The two-dimensional biphasic analysis results in the following expression for pressure in the fluid as a function of r, θ, and t

$$P(r,\theta,t) = \left\{ \left[A^* J_1 \left[i^{3/2} \sqrt{\frac{\omega}{(\mu+\lambda)k}}\, r \right] + B^* K_1 \left[i^{1/2} \sqrt{\frac{\omega}{(\mu+\lambda)k}}\, r \right] + \frac{H_A}{2} G\, r \right] \sin\theta \right\} e^{i\omega t} \quad (1)$$

*Correspondence to: A. A. Spirt (Bioengineering), 120 Hayden Hall, University of Pennsylvania, Philadelphia PA 19104.

where A^* and B^* are complex constants determined for each age group by satisfaction of the boundary conditions, J_1 is the Bessel function of the first kind (order one), K_1 is the modified Bessel function of the second kind (order one), μ and λ are the Lamé constants of the elastic matrix, H_A is the aggregate modulus of the tissue,[5] k is the permeability of the matrix,[4] G is the inverse radius of curvature of the bone, and ω is the frequency of deformation.

Following the microcontinuum model for charge distribution and the electrokinetic analysis described by Salzstein,[4] and neglecting flow in the θ-direction (the radial flow is three orders of magnitude greater), we derive the following expression for the SGP voltage:

$$V_{SGP}(r,\theta,t) = \frac{\zeta\varepsilon\phi^f}{3\eta\sigma_b T^2}\left[\frac{I_2(KR)}{I_0(KR)}\right]\cdot P(r,\theta,t) \tag{2}$$

where ζ is the Zeta potential of bone, ε is the dielectric permittivity and K^{-1} is the Debye length of the fluid, R is the mean pore radius, ϕ^f is the fluid fraction, T is the channel tortuosity, η is the viscosity of the fluid, σ_b is the conductivity of the bone, and I_0 and I_2 are modified Bessel functions of the first kind (orders zero and two, respectively). The electric field in the radial direction (the field is negligible in the θ direction) is determined from the derivative of Eq. (2) with respect to r.

MATERIALS AND METHODS

Female Wistar-derived rats were obtained from the Gerontology Research Center of the National Institute on Aging (Baltimore, Md.), and euthanized at five ages: 1.5, 3, 6, 12, and 24 months. The femora were dissected, wrapped in gauze soaked in Neuman's fluid (a fluid simulating bone extracellular fluid[6]), and frozen at −70 C until the time of testing.

Twelve femora from different animals (6 right, 6 left) from each age group were wet-ground in Neuman's fluid by means of a dental burr. Polycarbonate filters were used to exclude particles greater than 5 μm in diameter; the filtrate underwent Zeta-potential measurement in a PenKem 3000 Electrophoretic Mobility Analyzer.[7]

Eight femora (4 right, 4 left) from each age group underwent histomorphometric preparation in which thin cross sections from the femoral midshaft were embedded and mounted onto slides. A Kontron IBAS image analysis system and the Fortran program SLICE,[8] were used to compute the dimensions of a circular cylinder with equivalent medullary and cortical areas.

Another eight femora from each group underwent torsional testing to failure in a Burstein Standard Torsion Testing machine[9] in the Biomechanics Laboratory at Yale University. The shear modulus μ was determined for each specimen from the torsional rigidity at 50% failure torque divided by the mean polar moment of inertia J of the corresponding age group. The aggregate modulus H_A was computed from the shear modulus under the assumption of a mean Poisson ratio of 0.40 for all age groups.[10]

The derivative of Eq. (2) with respect to r was evaluated in order to determine the spatial and temporal variation in SGP field strength at the midshaft femur for the following values: $\sigma_b = 0.01545$ Siemens/m,[11] $K^{-1} = 7.36$ Å (calculated), η = viscosity of water, $\phi^f/T^2 = 0.261 \times 10^{-2}$, and $R = 696$ Å.[12] The following frequencies were examined: 0.5 Hz; the characteristic exercising frequencies of 2.5 Hz for 1.5-month rats, 2.1 Hz for 3-month rats, and 1.7 Hz for 6-, 12-, and 24-month rats[10] (with an interpolation of the value for the 3-month group based on femur length); 10 Hz; and 30 Hz. Equation (2) was evaluated at $r = r_2$, $\theta = +\pi/2$ radians in order to determine the maximal SGP difference per percent outer fiber strain in microstrain (OFS) from the anterolateral to postero-medial periosteal surfaces.

RESULTS

Values obtained for μ, H_A, and bone radii are listed in Table 1. Chronologic change in the Zeta potential was observed during growth, maturation, and aging in the rat femur (Fig. 1). The SGP field profile (normalized by OFS $\times \sin\theta$) across the cortex is shown in Fig. 2 for 12-month rats. The field is zero at the endosteal and periosteal surfaces (due to the zero radial fluid flow boundary condition at these borders) and is maximal at the mid-cortex. A saturation effect is seen at the higher frequencies. The other four age groups had similar profiles, but different field magnitudes.

TABLE 1.—Mean measured bone parameters (± S.D.).

Age (months)	Shear modulus (MPa)	Aggregate modulus (MPa)	Inner radius (mm)	Outer radius (mm)
1.5	1239 (124)	7436 (746)	0.923 (0.221)	1.426 (0.370)
3	3619 (576)	21713 (3458)	0.904 (0.356)	1.578 (0.502)
6	4358 (865)	26145 (5191)	0.906 (0.349)	1.679 (0.535)
12	4509 (576)	27055 (3455)	1.01 (0.384)	1.791 (0.612)
24	4598 (610)	27589 (3661)	0.959 (0.285)	1.838 (0.567)

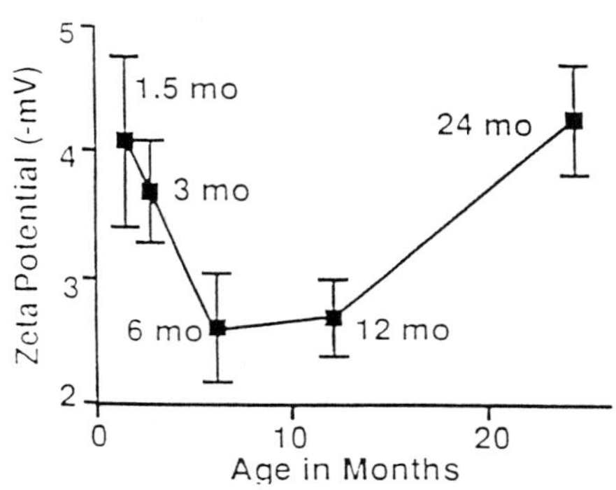

FIG. 1.—Zeta potential vs age (± S.D.).

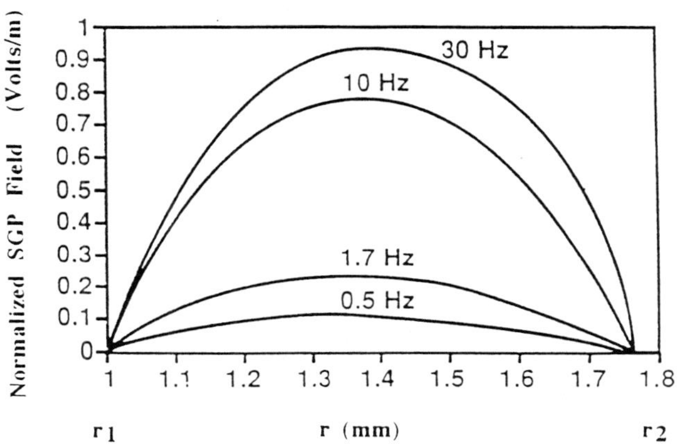

FIG. 2.—Normalized SGP field vs r at 12 months.

Figure 3 shows the predicted peak midcortical SGP field magnitude as a function of outer-fiber strain at $\theta = \pi/2$ for 12-month rats. The electric-field magnitude found to be biologically significant in most in vivo studies is in the range of 0.1 to 1 V/m.[13,14] This range has been superimposed on Fig. 3. The OFS magnitude necessary to maintain fields in the biologically significant range decreases with increased frequency.

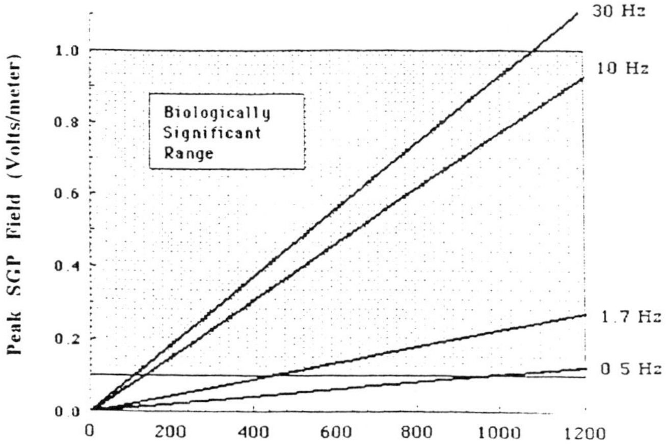

FIG. 3.—Peak midcortical SGP field vs outer-fiber strain at 12 months.

DISCUSSION

Our model predicts that higher strain levels, particularly when applied at high frequencies (i.e., impulse loads), results in biologically significant midcortical field magnitudes. The mean OFS in the midshaft femur of exercising rats (aged 6-30 weeks) was measured to be approximately 45 μstrain, and suggested to be perhaps 2-fold higher.[10] This range of strains results in field magnitudes of approximately 0.01 V/m (well below the biologically significant range) at the exercising frequency of 1.7 Hz, which suggests that higher strain, higher frequency loads are needed to generate electric fields that play a role in bone remodeling. This finding is consistent with McLeod's study, which showed a similar strain-frequency response of bone remodeling in the turkey ulna.[15] Rubin has shown that as few as 36 cycle/day will induce remodeling in the turkey ulna, with as few as 4 cycles/day needed to maintain bone mass.[16] These studies provide strong evidence that mechanical forces mediate bone remodeling through SGPs resulting from periodic high-strain/high-frequency (impulse) loads.

The normalized peak midcortical SGP field magnitude is relatively constant with age at lower frequencies, but at higher frequencies (10-30 Hz), the following pattern emerges: there is a sharp increase in field magnitude from 1.5 to 3 months of age (during the rapid growth phase of bone), a decline in the field from 3 to 6 months (during slowing of this growth), little to no change in field magnitude between 6 and 12 months (a relatively quiescent period of little bone growth or remodeling), and an increase in the field magnitude from 12 to 24 months, to a level above that of the 3-month rats. The increse in field strength from 12 to 24 months is almost entirely due to an increase in ZP, since there is little change in the mechanism that acts to increase the SGP field magnitude (via an increased ZP) in order to stimulate bone remodeling during senescence.

The SGP per % OFS across the periosteum determined from the same model exhibits a similar dependence on age as that measured by Pollack et al.[17] across the midshaft femur at comparable frequencies; however, our absolute values are 2 to 3-fold higher. Nevertheless, the similar dependence on age suggests that changes in the Zeta potential, mechanical properties, and dimensions of the bone can account for the age-related changes in the stress-generated potentials.[18]

REFERENCES

1. L. E. Lanyon, "Functional strain as a determinant for bone remodeling," *Calcif. Tissue Int.* 36: S56-S61, 1984.

2. C. T. Rubin and L. E. Lanyon, "Regulation of bone mass by mechanical strain magnitude," *Calcif. Tissue Int.* 37: 411-417, 1985.

3. S. R. Pollack, "Electrical effects in bone: Relationships to bone remodeling," in H. E. Takahashi, Ed., *Bone Morphometry,* Nishibura Co. Ltd., Japan, 1990, 170-176.

4. R. A. Salzstein, S. R. Pollack, A. F. Msk, and N. Petrov, "Electromechanical potentials in cortical bone—I: A continuum approach," *J. Biomechanics* 20: 271-280, 1987.

5. V. C. Mow and W. M. Lai, "Recent developments in synovial joint biomechanics," *SIAM Rev.* 22: 275-317, 1980.

6. W. F. Neuman, "The *milieu interieur* of bone: Claude Bernard revisited," *Federation Proceedings* 28: 1846-1850, 1969.

7. D. A. Berretta and S. R. Pollack, "Ion concentration effects on the Zeta potential of bone," *J. Orthop. Res.* 4: 337-345, 1986.

8. M. L. Nagurka and W. C. Hayes, "An interactive graphics package for calculating cross-sectional properties of complex shapes," *J. Biomechanics* 13: 59-64, 1980.

9. A. H. Burstein and V. H. Frankel, "A standard test for laboratory animal bone," *J. Biomechanics* 4: 155-158, 1971.

10. T. S. Keller and D. M. Spengler, "Regulation of bone stress and strain in the immature and mature rat femur," *J. Biomechanics* 22: 1115-1127, 1989.

11. J. D. Kosterich, K. R. Foster, and S. R. Pollack, "Dielectric properties of fluid-saturated bone: The effect of variation in conductivity of immersion fluid," *IEEE Trans.* BME-31: 369-374, 1984.

12. G. C. Scott and E. Korostoff, "Oscillatory and step response electromechanical phenomena in human and bovine bone," *J. Biomechanics* 23: 127-143, 1990.

13. C. T. Brighton, C. P. Luessenhop, S. R. Pollack, M. E. Steinberg, and F. S. Kaplan, "Treatment of castration-induced osteoporosis by a capacitively coupled electrical signal in rat vertebrae," *J. Bone Joint Surg.* 71A: 228-236, 1989.

14. E. Okereke, C. T. Brighton, and S. R. Pollack, ''Proliferative response of bone cells in vitro to capacitively coupled electrical stimulation,'' *Trans. BRAGS* 9: 56, 1989.

15. K. J. McLeod and C. T. Rubin, ''Sensitivity of the bone remodeling response to the frequency of applied strain,'' *Trans. Ortho. Res. Soc.* 17: 533, 1992.

16. C. T. Rubin and L. E. Lanyon, ''Regulation of bone formation by applied dynamic loads,'' *J. Bone Joint Surg.* 66A: 397-402, 1984.

17. S. R. Pollack, E. Korostoff, M. E. Steinberg, and J. Koh, ''Stress-generated potentials in bone: Effects of collagen modifications,'' *J. Biomed. Mater. Res.* 11: 677-700, 1977.

18. This work was supported by NIH SCOR grant 5-P50-AR-39226-02.

Martin Blank, Ed., *Electricity and Magnetism in Biology and Medicine*

OBSERVATIONS FROM MECHANICALLY AND ELECTRICALLY INDUCED BONE REMODELING

K. J. McLeod,* C. T. Rubin, and F. Guilak

State University of New York, Stony Brook

Many important aspects of the functional adaptation of bone can be explained based on the hypothesis of the mechanical stimuli being transduced to an electric field, which is then sensed by the bone cell population. However, the frequency and magnitude responses for endogenous and exogenous electric field modulation of bone remodeling do not lend strong support to this hypothesis. The data suggest that, if this hypothesis is to be retained, we must reexamine either the origin of the relevant stress generated potentials in bone, or the nature of the interaction of stress generated potentials with the bone cell population. Here, we consider several possible phenomena that could lead to apparent discrepancies in the response of bone to exogenously and endogenously induced electric fields: whether the electrokinetic fields developed in cortical bone may be excluded from the osteocytes due to fluid flow patterns; whether the relevant cell population may not be exposed to the induced fields because they are on the surfaces of the bone rather than in the bone cortex; and whether the stress generated signal may not be of electrokinetic origin.

Mechanically induced bone remodeling and the endogenous electrical currents that arise in bone tissue due to mechanical stresses have been inextricably linked since the piezoelectric properties of dry bone were first identified.[1] Subsequent experimentation has demonstrated that the recorded stress-generated potentials in moist bone are dominated by electrokinetic processes, and it is widely believed that these electrokinetic currents are connected to, if not directly responsible for, the control of bone remodeling. We have undertaken a series of investigations designed to identify the frequency and magnitude response characteristics of the bone remodeling process, as initiated by mechanical stress and exogenously induced electric fields. These studies were performed in an effort to identify the extent to which these two processes may be functionally linked.

METHODS

The experimental work in these studies has been based on the isolated adult avian ulna model of bone adaptation.[2] The preparation consists of the diaphyseal region of the adult male turkey ulna which is deprived of functional load bearing by proximal and distal metaphyseal osteotomies. The diaphyseal face of each osteotomy is covered with a stainless-steel cap to prevent reunion and fixation pins are installed transcutaneously extending through the cap and bone. These pins provide a means to apply external loading with a servo-controlled hydraulic loader and to clamp the bone during nonloading periods, which prevents any muscular strains from being imposed on the bone. Isolation surgery was performed on the left ulna of the animals and, starting 2 days post-op, the animals were exposed to either a mechanical loading protocol[3] or the exogenous induction of ELF electric fields through the use of Helmholtz coils.[4] Each in vivo preparation was either mechanically loaded for 10 min each day or underwent the electrical exposure protocol for 1 h/day. For both mechanical and electrical experiments the exposure protocol lasted for 8 weeks, after which the animal was sacrificed and the right and left ulnae were removed for morphological analysis. Changes in cross-sectional area of the treated ulnas, as compared to the intact contralateral ulnas, were obtained as functions of peak induced strain magnitude and peak induced electric field intensity, as well as exposure frequency.

*Correspondence to: K. J. McLeod, SUNY Dept. of Orthopaedics, Stony Brook NY 11794-8181.

RESULTS

Working within the frequency range of 1-200 Hz, we have found that exogenously induced electric fields have a significant effect on bone remodeling. The sensitivity of the bone tissue peaks at a frequency of approximately 15 Hz, where field intensities below 10 μV/cm are capable of maintaining bone mass under functional disuse conditions that would normally lead to a 15% loss of bone mass. Similarly, mechanical loading in this same frequency range also maintain bone mass at physiologic levels of induced strain magnitude (<1500 microstrain). However, mechanically stimulated bone remodeling demonstrates a linear increase in sensitivity as a function of frequency, up through 60 Hz. Moreover, the minimum strain magnitude necessary to maintain bone mass with a 15Hz mechanical stimulus is found to be approximately 400 microstrain. Using recent in vivo streaming potential recordings as a basis for calculating the magnitude of the endogenous electric fields induced at this strain level, we estimate that the average intracortical electric field intensity during a 400-microstrain 15Hz loading protocol would exceed 400 μV/cm,[5] with peak field intensities exceeding several millivolts/cm.[6] The optimal electric field frequencies and intensities necessary to maintain bone mass in the avian ulna model of disuse osteopenia therefore appear to be quite different under conditions of exogenously and endogenously induced fields.

DISCUSSION

The discrepancies in the field frequency and intensity levels necessary for the maintenance of bone mass raise two important issues. First, the two distinct intensity responses identify the necessity of explaining how bone tissue can respond to exogenously induced ELF electric fields at microvolt/cm levels when the endogenous mechanical environment normally induces fields of much greater intensities within a similar frequency range. Second, the discrepancies in the frequency response characteristics bring into question the relevance of mechano-electric coupling in the functional regulation of bone remodeling activity.

Although the frequency of the induced fields can be controlled in the experimental studies, there are significant differences in the spatial distribution of the electric fields when induced exogenously versus those that are induced endogenously due to a mechanical strain. These differences in spatial field patterns could be critical in determining the extent of a tissue response to the induced fields. For example, if it is assumed that the periosteal surface of bone forms a fluid-impermeable barrier, then the endogenous fields would be close to zero at the bone muscle boundary, yet this is a site where the exogenously induced fields are maximal. Because it remains unknown whether it is the osteocyte within the cortical bone or the osteoblast at the bone surface that is the primary transduction site of the physical signal that initiates bone remodeling, such macroscopic differences in the field distributions may well account for the discrepancies in the empirical data.

In addition to the macroscopic differences in the spatial distributions of the induced electric fields, there are also microscopic differences in the field patterns. An exogenously induced electric field in the extremely low frequency (ELF) range is excluded from the cell interior due to charge relaxation at the cell surface. This charge buildup provides a well-defined mechanism through which the exogenous field may interact with the cell, either through direct effects of the charge at the outer surface of the plasma membrane or through the perturbation of the membrane potential that this induced charge causes. Conversely, the electrical streaming field arising from fluid flow is excluded from the cell interior by the divergence of the fluid around the relatively fluid impermeable cell. As a result, there is little or no charge buildup on the cell surface, and correspondingly, a small or insignificant interaction of the streaming electric field with the cell.

Though differences in field distributions may explain the discrepancies in the field intensities necessary to maintain bone mass, these arguments do not explain the distinctly different frequency responses obtained for mechanical and electrical treatment protocols. Reports on in vivo measurements of streaming potentials consistently demonstrate that the peak induced fields plateau in intensity at frequency between 1 and 10 Hz. Given that the cellular sensitivity to electric fields appears to decrease at frequencies above 15 Hz, it is difficult to construct a hypothesis based on the development of streaming potentials to explain the linearly increasing response of cells to mechanical strain frequency. An alternative explanation is that the relevant mechano-electric coupling mechanism is not the electrokinetic process of streaming, but the development of piezoelectric currents in the bone.[7] Dry bone has been shown to have piezoelectric properties arising from the dipolar characteristics of the collagen matrix. Like many piezoelectric materials, dry bone demonstrates a frequency independent response for the strain induced charge in the ELF frequency range.[8] Calculation of the magnitude for the displacement current (dQ/dt) correspondingly shows an increase with frequency throughout the ELF range. If the piezoelectric coefficients remain significant in hydrated bone, and if the charge displacement takes place within a few Debye lengths of the cell surface, so that ionic relaxation does not completely shield out any effect of the induced fields, a piezoelectric transduction mechanism may be capable of explaining the characteristics of mechanical induced bone remodeling.

Alternately, transduction of the mechanical signal may not be through mechano-electric coupling, but through some direct mechanical effect. Possible processes include mechanical deformation of the cells within the bone, shear stresses on the cells arising from interstitial fluid flow, and thermal changes arising from the fluid-matrix drag occurring during dynamic loading of the bone. Each of these processes, as well as the induction of electrical currents, would be expected to be associated with different spatial patterns of remodeling activity. A more thorough understanding of the correlations between bone remodeling activity and specific loading environments could provide a decisive basis for pursuing the exact nature of the signal transduction mechanism in bone remodeling.

SUMMARY

The transduction of mechanical strain information to an electrical signal provides an elegant explanation for many of the problems of signal transduction in the process of adaptive remodeling in bone. However, our recent in vivo work has demonstrated that there is little similarity between the optimal endogenously induced and the exogenously induced electric fields necessary for the maintenance of bone mass. These results raise the important question of how a system can detect an extremely small signal in the presence of high background noise levels, and the relevance of a control parameter whose level is commonly several orders of magnitude greater than that necessary to regulate the system's response. We suggest that until more is known about the spatial distribution of remodeling activity arising from mechanical loading, one can neither confirm nor dismiss endogenous electric fields as a critical component in the normal regulation of the bone adaptation process.[9]

REFERENCES

1. E. Fukuda and I. Yasuda, "On the piezoelectric effect of bone," *J. Phys. Soc. (Jap.)* 12: 1158-1162, 1957.

2. C. T. Rubin and L. E. Lanyon, "Regulation of bone formation by applied dynamic loads," *J. Bone and Joint Surgery* 66A: 397-402, 1984.

3. K. J. McLeod and C. T. Rubin, "Sensitivity of the bone remodeling response to the frequency of applied strain," *Trans. Ortho. Res. Soc.* 17: 69, 1992.

4. K. J. McLeod and C. T. Rubin, "Frequency specific modulation of bone adaptation by induced electric fields," *J. Theor. Biol.* 145: 385-396, 1990.

5. M. W. Otter, V. T. Palmieri, L. A. MacGinitie, and G. U. Cochran, "Streaming potentials in bone are similar in vivo and in vitro," *Trans. World Congress of Biomech.* 2: 155, 1990.

6. S. R. Pollack, E. Korostoff, W. Starkebaum, and W. Iannicone, "Microelectrode studies of stress generated potentials in bone," in C. T. Brighton, J. Black, and S. R. Pollack, Eds., *Electrical Properties of Bone and Cartilage,* New York: Grune & Stratton, 1979, 69-81.

7. C. A. L. Bassett, "Electromechanical transduction and bone remodeling," this volume.

8. B. H. Pfeiffer, "Local piezoelectric polarization of human cortical bone as a function of stress frequency," *J. Biomech.* 10: 53-57, 1977.

9. This work was supported by the Whitaker Foundation.

Martin Blank, Ed., *Electricity and Magnetism in Biology and Medicine*

ELECTROMECHANICAL TRANSDUCTION IN LIVING BONE: QUESTIONS REMAIN

M. W. Otter,* L. A. MacGinitie, and G. V. B. Cochran

Helen Hayes Hospital, West Haverstraw, N.Y.

One approach to confirming the biological role (if any) of endogenous electricity within skeletal tissues is to define the characteristics of endogenous signals and compare them to exogenous signals known to affect bone. Unfortunately, numerous questions remain to be answered regarding electrical boundary conditions, the pores through which fluid flows, and the role of endogenous factors in modulating electromechanical signals before an accurate determination of the nature of endogenous fields will be possible.

Efforts to characterize electromechanical signals in bone began with in vitro tests, first in dry bone and subsequently in physiologically moist, but dead, bone. By measurements of the electrical potentials generated across mechanically deformed excised bone strips, subjected to a variety of treatments and soaked in a variety of test solutions, the electromechanical transduction mechanism in moist bone was determined to be electrokinetic in nature.[1] Fluid flow in response to mechanical deformation causes charge transport through the porous spaces of bone. In bone strips isolated from soft tissues and tested by a high-impedance voltage measuring device, charge accumulation at the bone surface creates an electric field within the bone, which drives a "back" current through the bone that is equal and opposite to the convective charge transport. The net result is a measurable voltage across the bone, known as streaming potential (SP). Recent electrode design improvements have made quantitative, repeatable measurements of SPs in vivo obtainable and thus permit new approaches.[2] However, numerous questions remain because it is difficult to detect SPs without disturbing the system.

Boundary Conditions. One question arises from the nature of the electrical boundary conditions imposed by the measuring technique. Are SPs in normal living bone smaller than measurements indicate? Depending on the electrical boundary conditions at the bone surface, electrokinetically derived phenomena may manifest themselves as streaming potentials (high-impedance surroundings resulting in charge accumulation at the bone surface), streaming currents (low-impedance pathway allowing current to flow around the bone exterior), or both (intermediate impedance, resulting in current flow with an electrokinetic potential drop across the bone and an IR potential drop across the soft tissue). In order to measure SPs in vivo, investigators have had to resort to the "sealed-boundary" electrode design,[2] which isolates a region of bone and subjects it to high-impedance boundary conditions, thereby maximizing the streaming potential. This method has great utility in that it allows the measurement of electromechanical signals in the living system and paves the way for a series of experiments in which the effects of various treatments (e.g., hormonal treatments, imposition of disuse, presence of damage and subsequent healing) on endogenous electromechanical signals may be studied. However, care must be taken in attempts to determine the magnitude of endogenous signals from these measurements, as an artificial boundary condition has been imposed. In the absence of measuring electrodes, and the surgical intervention necessary to place them on the bone surface, the surrounding soft tissue may well reduce the potential magnitude and increase streaming currents. Thus, present estimates of the magnitude of endogenous electrical fields, based on these in vivo and earlier in vitro measurements, may be too high. This possibility may account for the observation that the threshold magnitude of exogenous signals effective in altering bone architecture[3] (2 μV/cm) seems to be much lower than the local endogenous signals[4] (16 mV/cm, estimated from local SP measurements) that are thought to control bone architecture in part.[5] Unfortunately, there has been no quantification of the reduction of SPs by surrounding soft tissues to date. One solution may be to include soft tissues in the mathematical models that describe electromechanical transduction within the complex bone architecture and microstructure and are used to extrapolate from

---

*Correspondence to: M. W. Otter, Helen Hayes Hospital, West Haverstraw NY 10993.

measured streaming potentials to local electrical fields. A second approach may involve the design of an implantable biocompatible electrode around which soft tissues may heal, which would allow a more physiologically relevant measurement to be made.

Temporal Characteristics. Another set of questions arises when one considers the temporal characteristics of endogenous electrical signals. The time course of SPs is thought to be determined by the manner in which fluid flows through the complex microstructure of bone, thereby relieving the pressure gradients produced by mechanical deformation. Mathematical modeling[6] has suggested that fluid relaxation (the removal of pressure gradients by bulk transport of fluid) occurs over microseconds within large structures, such as Haversian canals (20-100 μm in diameter), so that such large canals remain at constant pressure over the much slower loading times (milliseconds) typically encountered in most experimental (and naturally occurring) situations. However, fluid cannot relax as rapidly in smaller pores. A biphasic continuum model,[7] which assumed that fluid flowed to the bone surface in order to relieve pressure gradients, was used to predict the size of pores through which fluid flows, based on the dependence of SPs on bending frequency typically observed in in vitro tests. A pore diameter of 158–702 Å was found. However, there is evidence to indicate that fluid does not flow to the bone surface, but rather flows into nearby Haversian canals.[4,8] Replacing the sample thickness by the average distance to the nearest Haversian canal would reduce the pore radius estimate, which would be consistent with the results of a study[9] in which solution concentration was used as a probe of pore diameter (40–280 Å). However, the anatomical indentity of these putative "micro"-pores is not yet clear.

It also appears that the state of the large pores into which fluid flows can affect the time course of SPs. Streaming potentials decay more rapidly in living bone than in the same samples after excision and preparation for in vitro testing,[10] which may reflect the "clogging" of Haversian canals by severed vessels in the dead bone. Furthermore, both the diameter of pores and their length (the distance to the nearest Haversian canal) can influence the relaxation times for SPs. MacGinitie et al. have shown that SPs decay more rapidly in healing and remodeling bone, perhaps due to the shorter distance over which fluid must flow to relieve pressure gradients in more porous bone.[11] In addition, SPs in healing bone appear to be smaller than in normal bone because a greater porosity leads to a reduced electrical impedance across the bone. These explanations are tentative; however, if they are correct, osteoporotic bone might be expected to produce smaller, more rapidly decaying SPs than normal bone, for the same reasons. (Greater porosity implies reduced impedance and shorter "micro"-pores.)

In order for bone suffering from disuse osteoporosis to regain a normal structure and mass upon resumption of mechanical loading, the biological transducer must provide the bone cells with an effective stimulatory signal even in the porotic state. Thus, the heightened sensitivity of bone cells to signals of higher frequency (15 vs 5 Hz) observed by McLeod and Rubin[12] may be an adaptive response that allows smaller, more rapidly decaying signals (which have a greater proportion of higher frequency components) to be an effective trigger for restoration of bone tissue.

Other Questions. Another series of questions arises from the ability of altered chemical composition of both bone fluid and matrix to alter SPs. Perfusion with simple ionic solutions alters SPs in vivo[13] and in vitro,[1] however, the fluid composition and the structure of the double layer of *living bone* has not yet been settled. In addition, the characteristics of endogenous SPs produced by various bone types (Haversian, plexiform, trabecular), by bone suffering from various pathologies (such as osteoporosis, osteoporosis, osteoarthritis, osteomalacia), and within bone being subjected to various anti-osteopetrosis regimens (such as estrogen, parathyroid hormone, calcitonin, fluoride, bisphosphonate), are yet to be fully characterized. The role, if any, of EMG fields in bone is not yet defined.

All the above questions related to boundary conditions, bone structure, and bone composition must be resolved before an accurate mathematical model of SPs in living bone can be produced, which will allow us to use the measurable quantity (the streaming potential) to calculate the local electric field to which cells are exposed. Only when such an accurate determination of the endogenous signals experienced by cells has been made will it be possible to compare endogenous fields to exogenous fields and their demonstrated effects on bone.

Finally, if control could be obtained over SP production (for example, by changing the zeta potential with a chemical or hormonal agent), the effects of experimental alterations in electromechanical transduction on the remodeling of bone may ultimately allow a direct test of the major question: Do SPs link mechanical deformation of bone remodeling?[14]

REFERENCES

1. D. A. Chakkalakal, *J. Mater. Res.* 4: 1034-1046, 1989.
2. G. V. B. Cochran et al., *J. Biomech.* 22: 745-750, 1989.
3. K. McLeod and C. Rubin, *Orthop. Trans.* 13: 353-354, 1989.
4. W. Starkebaum et al., *J. Biomed. Mat. Res.* 13: 729-751, 1979.
5. C. A. L. Bassett and R. O. Becker, *Science* 137: 1063-1064, 1962.
6. M. W. Johnson et al., *J. Biomech.* 15: 881-885, 1982.
7. R. A. Salzstein et al., *J. Biomech.* 20: 261-280, 1987.
8. L. A. MacGinitie et al., *Trans. BRAGS* 9: 15, 1989.
9. L. A. MacGinitie et al., *J. Biomech.* (in press).
10. M. W. Otter et al., *J. Orthop. Res.* 10: 710-719, 1992.
11. L. A. MacGinitie et al., *Trans. ORS* 17: 531, 1992.
12. K. McLeod and C. Rubin, *Trans. BRAGS* 9: 20, 1989.
13. M. W. Otter et al., *Trans. BRAGS* 11: 61, 1991.
14. This work was aided in part by a grant from the Orthopedic Research and Education Foundation, by NIH RO1 AR39574, VA, RR&D Grant A-160-3RA at Castle Point VAMC, the Walter Scott Foundation, and the New York State Department of Health.

Martin Blank, Ed., *Electricity and Magnetism in Biology and Medicine*

LIMITATIONS ON THE EFFECT OF APPLIED DC ELECTRICAL FIELDS ON NERVE REGENERATION

J. M. Kerns,* F. E. Levy, A. J. Fakhouri, and J. A. Gramm

Rush-Presbyterian-St. Luke's Medical Center, Chicago

J. B. Oldershaw

University of Chicago Medical Center

Many studies have shown that applied DC electrical fields may enhance nerve regeneration. We suggest that such a demonstration depends on the optimal stimulus parameters, what types of lesions are introduced, which methods of analysis are used, and when the evaluation is performed. Battery implants were used to deliver minute currents to the injured rat sciatic nerve. The vibrating probe shows a more advanced front of regenerating axons 1 week following transection. The early motor recovery measured by the twitch tension method may also be enhanced for certain lesions and treatments. In contrast, other methods (nerve, morphometry, HRP transport, SFI) show no improvement.

Many in vitro studies have demonstrated that applied DC electrical fields can enhance the growth of neurites towards the cathode.[1-5] An increase in regenerated nerve tissue was reported in adult frog limb stumps treated with DC currents,[6] which suggests that mammalian peripheral nerve regeneration might also benefit by such treatment. Various studies using the rat sciatic nerve model have described some enhancement following nerve transection and applied electric fields.[7-12] However, more recent studies have suggested that such effects may be limited.[13,14] The purpose of this report is to suggest that demonstration of a positive effect depends on what types of lesions are introduced, the optimal stimulus parameters, which methods of analysis are used, and when the evaluation is performed (Fig. 1).

Various types of lesions and repair have been studied in the adult rat sciatic nerve by use of a standard protocol approved by the institutional animal care committee. The exposed sciatic nerve or one of its branches was either crushed, transected and anastomosed, or an autograft was inserted in the nerve defect. The DC electric field was applied by implantation of small battery stimulators[15] with either wick or wire electrodes and a constant-current source in the circuit to insure a uniform current delivery. Although this mode of application has been widely used,[6,13,14,16] it is not entirely satisfactory. Electrode deposits, especially at the anode, increase the resistance and may contribute to inflammatory reactions. In our own experiments we have used current levels of 1 and 10 μA with either the cathode or anode distal to the lesion site for a duration of 1 week,[8] one month,[13] or 3 months.[17] To evaluate the progress of nerve regeneration, we have used a variety of quantitative methods (Fig. 1), which reflect functional or structural recovery at various survival periods.

The *vibrating probe* has been used to detect growth cones in the regenerating nerve.[1,8] Peaks of current density measured from the nerve surface correspond to growth cones identified by electron microscopy. The front of regenerating axons is 69% farther into the distal segment in the 1μA group than in a sham group 1 week following transection.[8] This enhancement could be due to either an increase in the rate of axonal regeneration or a decrease in the delay period.

The *twitch tension* of digital flexors can be measured to estimate the motor recovery.[18] The responses for the transection group treated with a 10μA current output and either the anode or cathode electrode located distal to the lesion were intermediate to the sham group and an untreated crush group.[17] Most previous studies conclude that an electrode configuration with the anode distal is detrimental.[11,19] Group differences were smaller at the end of 2 postoperative months, so the effects were probably temporary. The improvement of motor recovery for the 1μA treatment group occurred later and was less complete. When the crush group

---

*Correspondence to: J. M. Kerns (Anatomy), Rush-Presbyterian-St. Luke Med. Ctr, Chicago IL 60612.

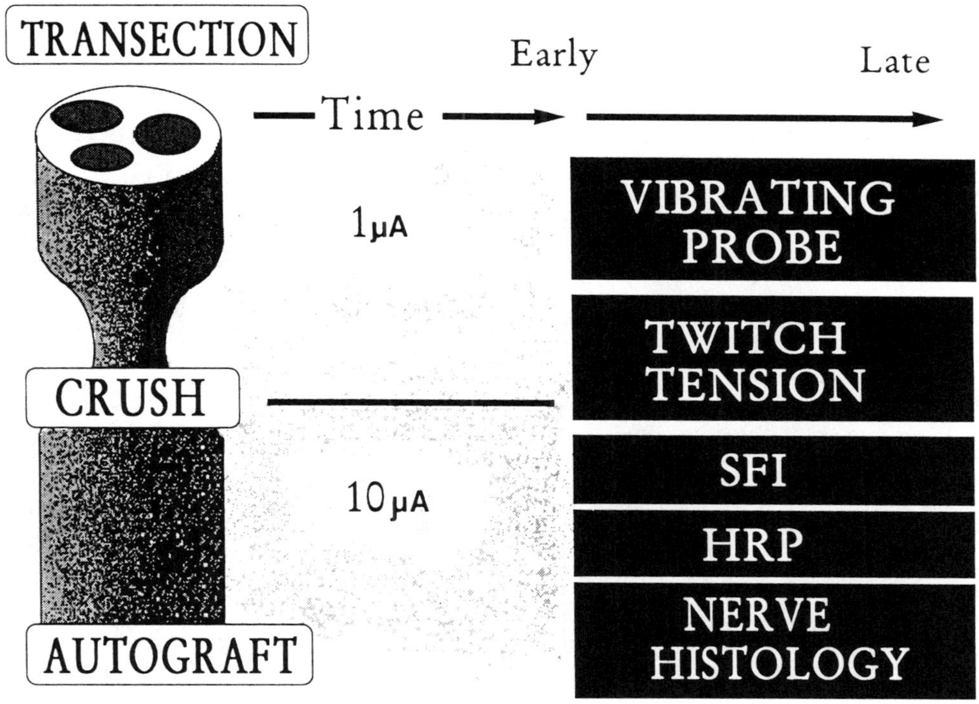

FIG. 1.—Experimental model: three lesion types treated with two current levels were evaluated at early and late postoperative times by five different methods.

was treated with 10 μA, only a window of enhancement was evident during the third postoperative week,[13] whereas no difference was observed when the same treatment was given to a transected tibial nerve. The treatment of sciatic nerve autografts showed a slight improvement compared with controls, but was not statistically significant (Fig. 2). This method shows that temporary improvements were evident among the various experimental parameters investigated.

Three other quantitative measures of nerve regeneration have shown that electric-field treatment produces little or no difference compared with controls. The *sciatic functional index* (SFI), a quantitative measure of locomotion behavior, showed no difference between the two treatment or control groups during the first 3 postoperative months (Fig. 3). Nor was there a difference between treated and untreated rats receiving a nerve crush lesion and both groups had a nearly complete recovery after 1 month.[13]

The retrograde tracer *horseradish peroxidase* (HRP) was used to determine the number of motoneurons regenerating their axons through the lesion site. At 8 days following nerve crush, the number of labeled tibial motoneurons was similar regardless of whether treated with the cathode or anode distal or if a neutral battery was used. The mean number of labeled cells at 28 dpo was 86% in the treatment group and 97% in the neutral group, but the differences were not statistically significant.[13] The results were similar in another experiment, where the tibial nerve branch was transected and anastomosed followed by treatment with a – 10μA battery implant. These data support the hypothesis that mild lesions of a peripheral nerve recover quite well without any added benefit when treated with the applied electrical fields.

Finally, *nerve morphometry* has been used to evaluate the structural recovery of regenerated axons in the distal segment. The nerve area, fiber number, and myelinated axon size were similar 28 days following nerve crush[13] or 77 days following nerve transection and repair (Fig. 4). At the earlier survival time, the fiber density in the treatment group was significantly greater than in the neutral group, which suggests that the electric field may have retarded the neural edema in the distal stump.[13]

In conclusion, these experiments suggest that some methods of evaluation show some early improvement for more severe lesions when DC electric fields are applied to injured nerves. This finding is in agreement with a recent study by McGinnis and Murphy,[14] but in contrast to most other studies. It is interesting to note that some degree of functional improvement can occur without a corresponding structural correlate. Both neural and non-neural influences may be involved, but the precise mechanisms presently remain unclear.[20]

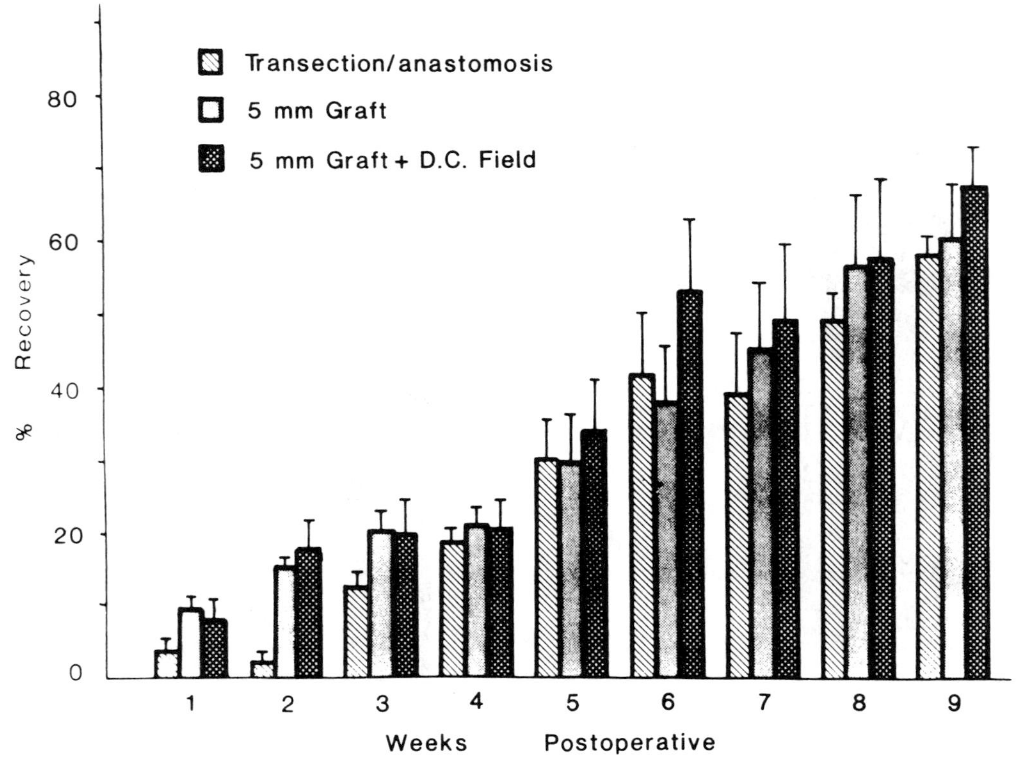

FIG. 2.—Twitch tension: recovery is given for three experimental groups (transection/anastomosis untreated, autograft untreated, and autograft – 10μA treatment); no significant group differences.

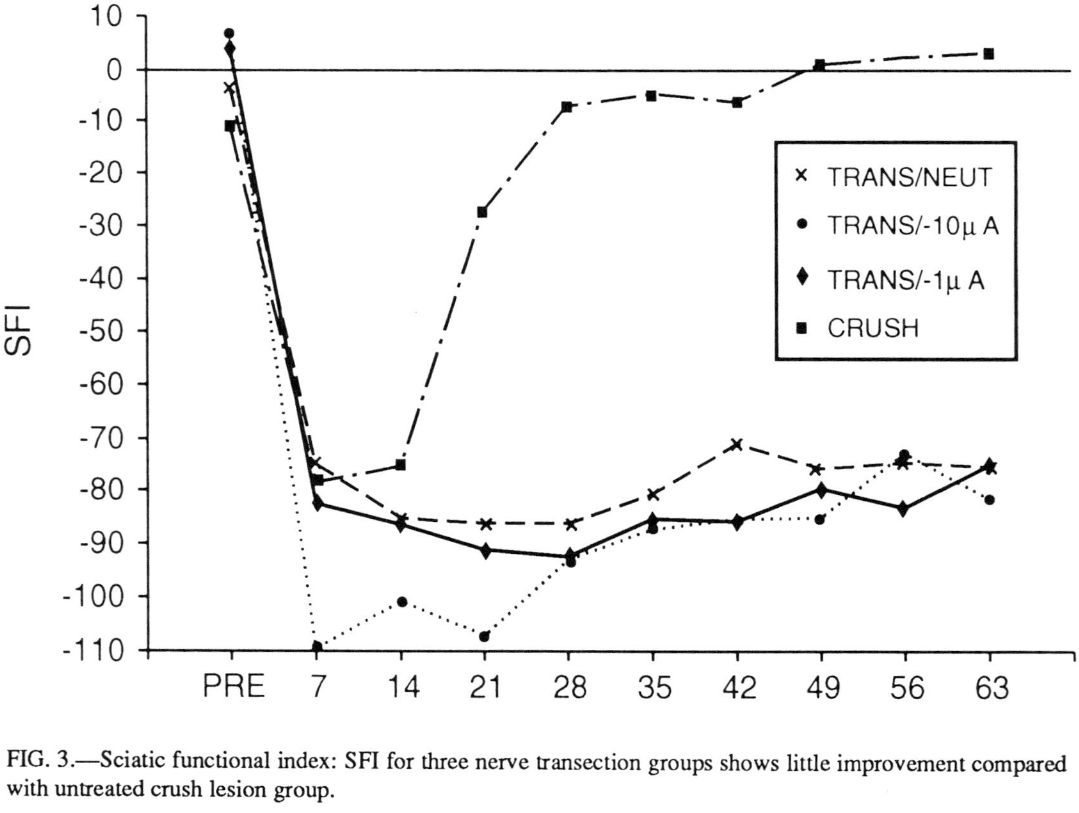

FIG. 3.—Sciatic functional index: SFI for three nerve transection groups shows little improvement compared with untreated crush lesion group.

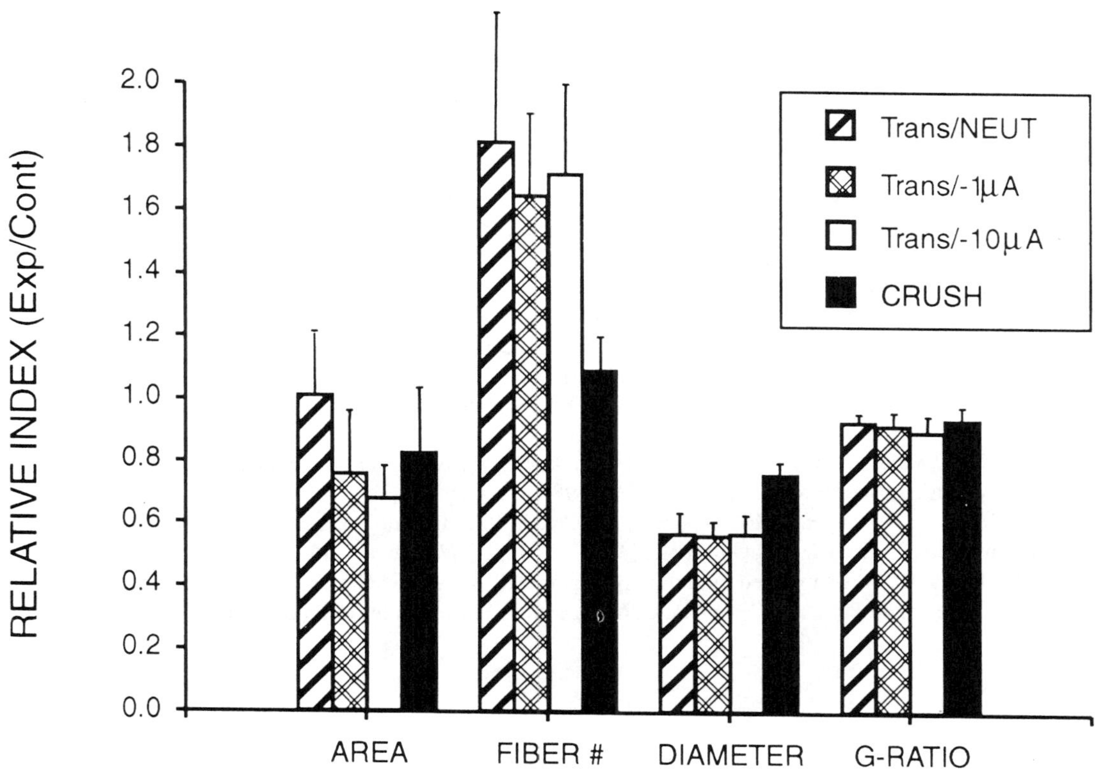

FIG. 4.—Nerve morphometry: nerve area, fiber number, fiber size, and g-ratio indices at 63 dpo are given for four experimental groups; transection groups show no significant differences.

REFERENCES

1. J. A. Freeman, P. B. Manis, G. J. Snipes, B. N. Mayes, P. C. Samson, J. P. Wikswo Jr., and D. B. Freeman, "Steady growth cone currents revealed by a novel circularly vibrating probe," *J. Neurosci. Res.* 13: 257-283, 1985.

2. L. Hinkle, C. D. McCaig, and K. R. Robinson, "The direction of growth of differentiating neurons and myoblasts from frog embryos in an applied electric field," *J. Physiol. Lond.* 314: 121-135, 1981.

3. L. F. Jaffe and M-M. Poo, "Neurites grow faster towards the cathode than the anode in a steady field," *J. Exp. Zool.* 209: 115-128, 1979.

4. C. D. McCaig, "Nerve branching is induced and oriented by a small applied electric field," *J. Cell. Sci.* 95: 605-615, 1990.

5. B. F. Sisken and S. D. Smith, "The effects of minute direct electrical currents on cultured chick embryo trigeminal ganglia," *J. Embryol. Exp. Morph.* 33: 29-41, 1975.

6. R. B. Borgens, "Artifically controlling axonal regeneration and development by applied electric fields," in R. B. Borgens, K. R. Robinson, J. W. Vanable Jr., and M. E. McGinnis, Eds., *Electric Fields in Vertebrate Repair,* New York: Alan R. Liss, 1989, 117-170.

7. J. A. Beveridge and M. J. Politis, "Use of exogenous electric current in the treatment of delayed lesions in peripheral nerves," *Plast. Reconst. Surg.* 82: 573-577, 1988.

8. J. M. Kerns, A. J. Fakhouri, H. P. Weinrib, and J. A. Freeman, "Electrical stimulation of nerve regeneration in the rat: The early effects evaluated by a circularly vibrating probe and electron microscopy," *Neurosci.* 40: 93-107, 1991.

9. L. McDevitt, P. Fortner, and B. Pomeranz, "Application of weak electric field to the hindpaw enhances sciatic motor nerve regeneration in the adult rat," *Brain Res.* 416: 308-314, 1987.

10. M. J. Politis, M. F. Zanakis, and B. J. Albala, "Facilitated regeneration in the rat peripheral nervous system using applied electric fields," *J. Trauma* 28: 1375-1381, 1988.

11. S. Ribarič, A. Stefanovska, L. Vodovnik, and P. Cvirn, "Muscle force recovery after continuous direct current stimulation of a crushed nerve," *Restor. Neurol. Neurosci.* 4: 331-338, 1992.

12. G. C. Roman, H. K. Strahlendorf, P. W. Coates, and B. A. Rowley, "Stimulation of sciatic nerve regeneration in the adult rat by low-intensity electric current," *Exp. Neurol.* 98: 222-232, 1987.

13. J. M. Kerns and C. Lucchinetti, "Effects of applied DC fields on regeneration of crushed rat sciatic nerve," *Exp. Neurol.* 117: 71-80, 1992.

14. M. E. McGinnis and D. J. Murphy, "The lack of an effect of applied DC electric fields on peripheral nerve regeneration in the guinea pig," *Neurosci.* 51: 231-244, 1992.

15. J. M. Kerns, I. M. Pavkovic, A. J. Fakhouri, K. L. Wickersham, and J. A. Freeman, "An experimental implant for applying a DC electrical field to peripheral nerve," *J. Neurosci. Meth.* 19: 217-223, 1987.

16. R. B. Borgens, J. W. Vanable Jr., and L. F. Jaffe, "Bioelectricity and regeneration: I. Initiation of frog limb regeneration by minute currents," *J. Exp. Zool.* 200: 403-416, 1977.

17. J. M. Kerns, I. M. Pavkovic, A. J. Fakhouri, and G. T. Gray, "Electrical stimulation of nerve regeneration in the rat: Functional evaluation by a twitch tension method," *Restor. Neurol. Neurosci.* 1992 (submitted).

18. J. M. Kerns, A. J. Fakhouri, and I. M. Pavkovic, "A twitch tension method to assess motor nerve function in rat," *J. Neurosci. Meth.* 19: 259-266, 1987.

19. E. Roederer, N. H. Goldberg, and M. J. Cohen, "Modification of retrograde degeneration in transected spinal axons of the lamprey by applied D.C. current," *J. Neurosci.* 3: 153-160, 1983.

20. This study was supported in part by NIH grant NS19769, BRSG grant 507RR5477, and the Enelow Fund. We are grateful to Drs. J. Freeman, G. Gray, and W. F. Hughes for technical assistance.

Martin Blank, Ed., *Electricity and Magnetism in Biology and Medicine*

AN EXAMINATION OF DIRECT-CURRENT FIELDS FOR THE TREATMENT OF SPINAL-CORD INJURY

M. G. Fehlings,* R. J. Hurlbert, and C. H. Tator

University of Toronto, Canada

Using a rat model of compressive spinal cord injury (SCI), we have examined the effects of DC fields on injured mammalian spinal-cord axons. With a 14μA DC field (epidural platinum disk electrodes with cathode caudal; longitudinal field gradient 350 μV/mm at lesion center) applied immediately after injury and maintained for 8 weeks, we observed significant recovery of clinical neurological function and improvement in hindlimb motor-evoked potentials. This outcome was associated morphologically with an increased number of surviving axons at the injury site. Reversal of the applied field (cathode rostral) did not promote recovery after acute SCI, and delayed application of a DC field (4 weeks after SCI) was ineffective. We encountered significant problems with the use of epidural metallic-disk electrodes, including inhomogeneity of the induced field and neurotoxicity. Thus, we are currently examining other techniques of DC stimulation, including use of cuff electrodes.

Much experimental evidence supports the concept that direct-current (DC) fields promote the recovery and regeneration of damaged peripheral and central nervous system axons.[1-19] In the present paper, we review our experience with the use of DC fields to treat compressive spinal-cord injury (SCI) in rats. In particular, the following points are emphasized: (i) the importance of DC field polarity in modulating recovery after acute SCI, (ii) the ineffectiveness of DC fields in promoting recovery after chronic SCI, and (iii) the problems associated with the use of epidural metallic-disk electrodes, including field inhomogeneity and neurotoxicity.

METHODS

Injury Model and Technique of DC Stimulation

Adult rats underwent a C7-T1 laminectomy and received a 1min clip compression injury of the cord at T1 with a modified aneurysm clip exerting a force of 53 g, a lesion that results in severe paraparesis or complete paraplegia.[20] After injury the rats underwent implantation of a DC stimulator. Each DC stimulator consisted of a 6.0V battery, a 3.6kΩ resistor, and a junctional field-effect transistor designed to maintain a constant output of 14 μA over a wide range of resistances.[5] The stimulators were connected to a silastic pad by 3.5 cm of insulated platinum wire. One disk-shaped electrode (surface area 2 mm^2) protruded from each end of the silastic pad; the two electrodes were separated by 10 mm. The electrode pad of the stimulator was sutured to the dura. The caudal electrode was placed distal to the injury site under the lamina of T2, with the rostral electrode positioned under the lamina of C6. The stimulator apparatus was secured to subcutaneous tissue. The current output and voltage of the DC stimulators were tested prior to insertion and following the 8 weeks of the experiment.[5]

Outcome Measures

The clinical neurological function of each rat was assessed weekly by the inclined-plane technique.[21] At the conclusion of the period of DC stimulation, electrophysiological assessment was performed by recording of motor and somatosensory evoked potentials (MEP and SSEP)[5,6] to assess the physiological integrity of the descending and ascending tracts in the cord, respectively.

The origin of intact descending axons at the injury site was determined by retrograde axonal tracing with HRP.[20] In brief, the number of HRP-labeled neurons in the brainstem and sensorimotor cortex, after introduction of the tracer into the cord at T8, was quantified. The number of intact myelinated axons at the injury site was quantified by examination of a toluidine blue-stained semithin section at 1000× by Blight's line-

---

*Correspondence to: M. G. Fehlings (Neurosurgery), Toronto Hospital, 399 Bathurst St., Toronto ON Canada M5T 2S8.

sampling technique.[22]

Field Mapping Studies

To assess accurately the electric field induced by the 14μA DC stimulators, a series of field-mapping studies was conducted in five normal rats.[23] Epidural platinum disk electrodes were positioned 10 mm apart at C6 and T2. Stimulation consisted of a balanced 20Hz sine wave, with the current limited by an on-line ammeter. Longitudinal field gradients were measured with a lock-in differential amplifier; glass microelectrodes were inserted directly into the spinal cord. Recordings were made at various positions and depths between the stimulating electrodes.

Neurotoxicity Studies

The neurotoxic effects of DC fields were examined by variation of the strength of electrical stimulation from 3 to 80 μA.[24] The morphological features of the injury site and stimulation sites were determined by examination of cross sections stained with luxol fast blue and hemotoxylin-eosin. Computer-assisted 3-D reconstructions of the lesions were generated.

RESULTS

Studies of Acute SCI

Our laboratory had previously shown that a cathode caudal DC field (14 μA) promotes recovery of clinical and electrophysiological function after acute SCI.[5,19] Furthermore, this treatment enhances the survival of axons at the injury site as determined by retrograde axonal tracing by HRP[5] and quantification of myelinated axons at the injury site.[7] To examine the effect of DC field polarity on the mechanism of these effects, we have recently completed a study of 30 rats with acute 53g injuries of the cord at T1[8]: 10 rats received a stimulator (14 μA) with the cathode oriented caudal to the lesion, 10 animals received a similar stimulator with the cathode oriented rostral to the lesion, and 10 rats received a sham stimulator.

The results of this experiment are summarized in Fig. 1. A 14μA cathode caudal DC field promoted recovery of function as assessed by behavioral testing on the inclined plane (Fig. 1a) and MEP recording (Fig. 1b). Computer-assisted counts of axons at the injury site showed that this treatment promoted increased survival of myelinated axons, including those in the ventrolateral funiculus (Fig. 1c). Studies with HRP confirmed that a cathode caudal DC field promoted survival of descending axons, particularly those in nonpyramidal tracts, including the rubrospinal and reticulospinal tracts (Fig. 1d). In contrast, a cathode rostral DC field was ineffective in promoting recovery after acute SCI. Reversal of field polarity significantly decreased the extent of retrograde labeling seen in the sensorimotor cortex and brainstem nuclei (Fig. 1d). It is thus likely that a rostrally negative DC field accentuated the extent of retrograde degeneration of descending axons in the cord.

Studies of Chronic Spinal-cord Injury

A total of 30 rats received a 53g clip compression injury of the cord at T1.[25] Four weeks after injury, the rats underwent implantation of either a functioning (14 μA, cathode caudal; $n = 15$) or a sham DC stimulator (0 μA; $n = 15$). The clinical neurological function of the rats was evaluated weekly. The animals were sacrificed after 8 weeks of DC stimulation. As shown in Fig. 2, delayed application of a DC field was ineffective in promoting recovery of gross motor function. Similarly, the latency and amplitude of MEPs and SSEPs were comparable ($p > 0.05$) in the two experimental groups. Anatomical analysis of the injury site showed a similar number of intact myelinated axons in the two groups: treatment group, 25400 ± 4800 (SEM); control group, 22400 ± 4500 (SEM); $p > 0.05$.

Field-mapping Studies

Application of 14 μA DC to the spinal cord via two epidural platinum disk electrodes (separation 10 mm) resulted in an inhomogeneous electric field. The longitudinal-field gradient in the vicinity of the disk electrodes exceeded 1.0 mm/mV and decayed exponentially with increasing distance from the electrodes. At a point midway between the two disk electrodes (corresponding approximately to where the center of a clip injury would be), the mean longitudinal-field gradient measured 350 μV/mm at the cord surface and decreased linearly with increasing depth in the cord (Fig. 3).

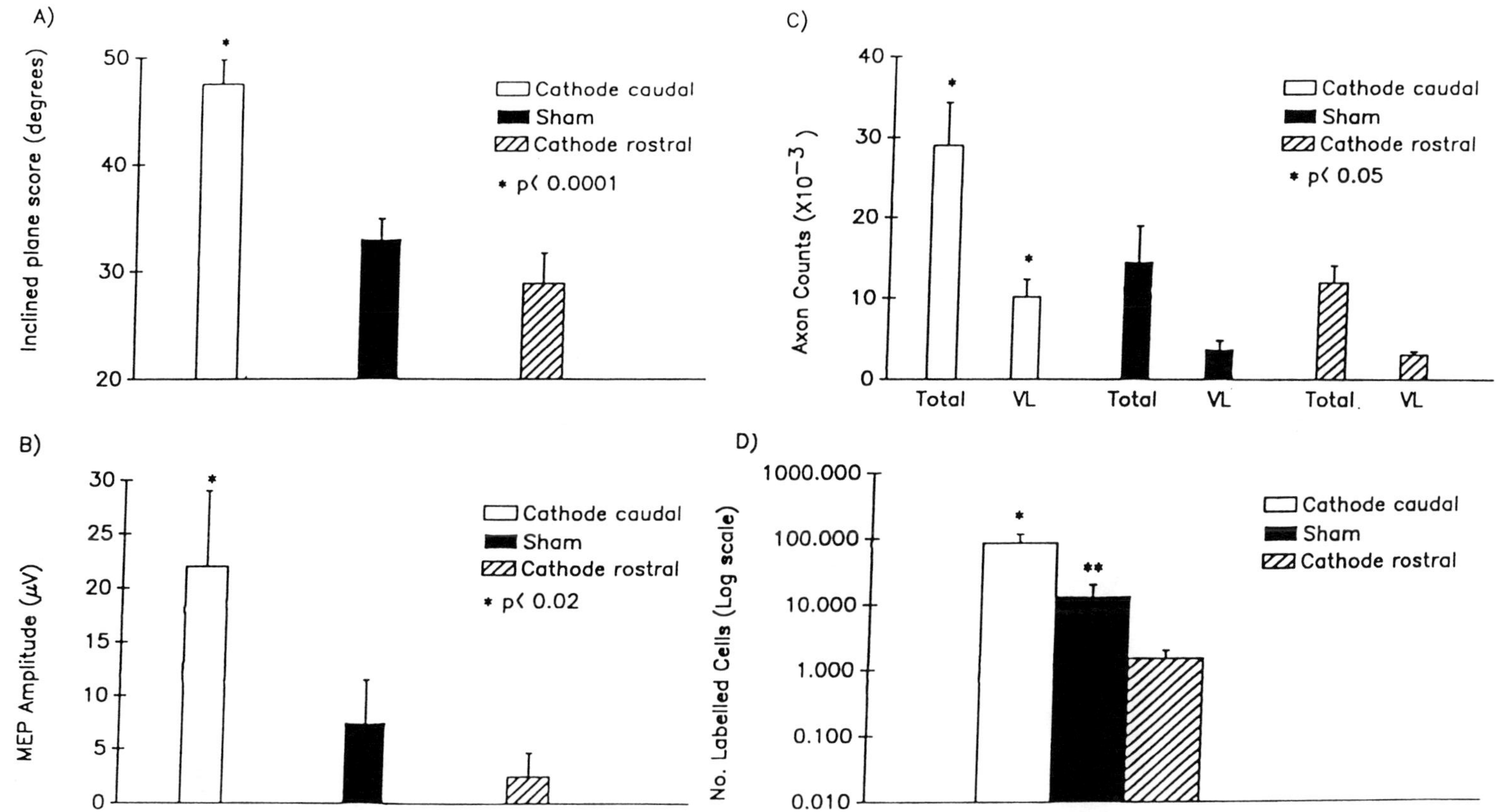

FIG. 1.—Effect of DC field polarity on recovery after compressive SCI (53g×1min clip injury at T1). Data expressed as mean ±SEM. (a) Inclined-plane performance of cord injured rats treated with caudally negative 14μA DC field significantly ($p < 0.0001$) exceeded that of rats treated with sham or cathode rostral DC stimulators; (b) amplitude of MEPs recorded from biceps femoris after cortical stimulation in rats treated with cathode caudal DC field significantly exceeded ($p < 0.02$) that of the other two experimental groups. (c) Total number of myelinated axons at injury site (Total) and number of axons in ventrolateral funiculus (VL) of rats treated with cathode caudal DC field significantly exceeded ($p < 0.05$) that of the other groups. (d) total number of neurons, retrogradedly labeled by HRP after introduction of tracer into cord at T8, were counted in the brainstem and sensorimotor cortex. Total number of labeled neurons in cathode caudal group significantly ($p < 0.05$) exceeded that of the other groups. In contrast, reversal of DC field (cathode rostral) resulted in significantly fewer ($p < 0.05$) retrogradedly labeled neurons than in the sham group. (Note logarithmic scale.)

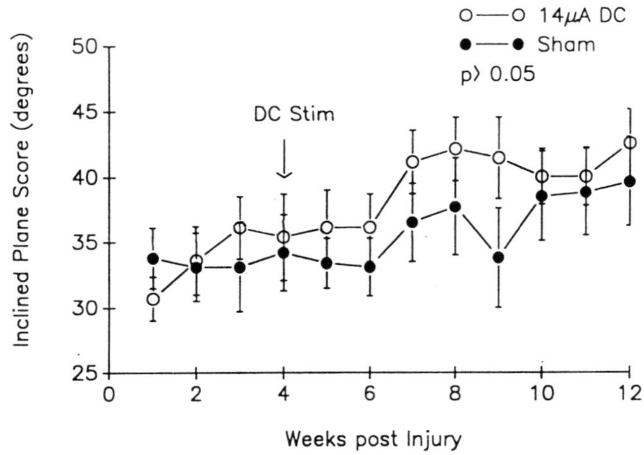

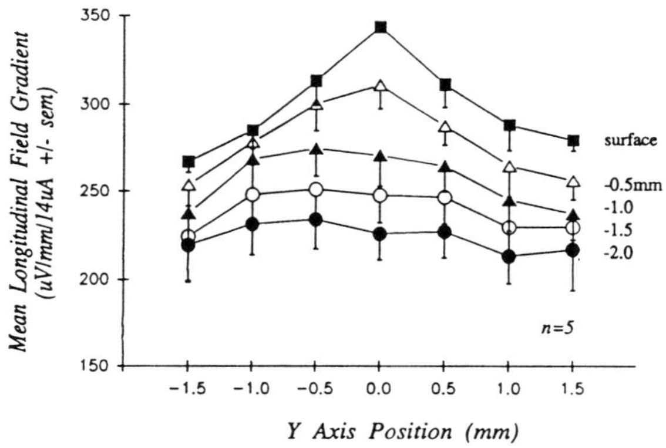

FIG. 2.—Effect of delayed application (4 weeks after SCI) of DC field (14 μA; cathode caudal) on clinical neurological function. Inclined-plane scores (mean ±SEM) of treatment and control rats were similar ($p > 0.05$).

FIG. 3.—Results of field mapping experiments in 5 normal rats (mean ±SEM). 14 μA of current (20Hz sine wave) was delivered to the cord via two epidural platinum disk electrodes (2 mm^2) separated by 10 mm. The mean longitudinal field gradient was differentially recorded at various depths in the cord by glass microelectrodes. The measurements in this figure correspond to a position midway between the two stimulating electrodes. The Y axis position corresponds to loci left (negative) or right (positive) of midline (zero point) in the transverse plane.

Neurotoxicity Studies

The use of epidural platinum disk electrodes was associated with significant neurotoxicity, particularly with applied currents exceeding 18 μA. However, some electrode-associated neural damage was occasionally seen even with currents as low as 3 μA. The damage consisted of cavitation, usually at the anode site, associated with electrochemical deposits. Microscopically, the cavitations were characterized by loss of neural tissue in the dorsal columns and dorsal gray matter, astrogliosis, and numerous chronic inflammatory cells, including macrophages.

DISCUSSION

The principal findings regarding the effect of DC fields in SCI may be summarized as follows: (i) cathode caudal DC fields (longitudinal field gradient ≈ 350 μV/mm) promote recovery after acute compressive SCI; (ii) rostrally negative DC fields are ineffective in promoting recovery of function after acute SCI, and may accentuate retrograde degeneration of descending axons; and (iii) delayed application of a DC field (4 weeks after SCI) is of no therapeutic benefit.

The neurobiological effects of DC fields include promotion of neuritic outgrowth,[9,10,12,27,28] enhancement of cell survival,[29] reduction of retrograde axonal degeneration,[26] reduction of Ca^{++} influx into transected axons,[18] reversal of conduction block at points of axonal bifurcation[30] and promotion of recovery,[5,7,11,13,16,19] and regeneration of injured axons.[1-4,11,14,15,17]

Although it is possible that DC fields may promote recovery after acute SCI by facilitating axonal sprouting or possibly regeneration (cf. Table 1), it is more attractive to postulate that the mechanism involves promoting the survival of sublethally injured fibers. Indeed, Roederer et al.[26] reported that caudally negative DC fields decreased the extent of retrograde axonal degeneration of transected lamprey spinal axons. The mechanism for this effect may involve attenuation of post-traumatic Ca^{++} gradients, as recently demonstrated by Strautman et al.[18] What is important is that the effect of DC fields on non-neuronal elements including glia, the extracellular matrix, and the microvasculature remains unknown.

Finally, we emphasize some of the technical problems inherent in the use of epidural metallic disk electrodes, including field inhomogeneity and neurotoxicity, particularly at the anode site. Thus, we are currently examining other stimulator designs, including cuff electrodes and nonmetallic conductive interfaces.

REFERENCES

1. R. B. Borgens, E. Roederer, and M. J. Cohen, "Enhanced spinal cord regeneration in lamprey by applied electrical fields," *Science* 213: 611-617, 1981.

2. R. B. Borgens, A. R. Blight, and D. J. Murphy, "Transected dorsal column axons within the guinea pig spinal cord regenerate in the presence of an applied electrical field," *J. Comp. Neurol.* 250: 168-180, 1986.

3. R. B. Borgens, A. R. Blight, and M. E. McGinnis, "Behavioral recovery induced by applied electric fields after spinal cord hemisection in guinea pig," *Science* 238: 366-369, 1987.

4. R. B. Borgens, A. R. Blight, and M. E. McGinnis, "Functional recovery after spinal cord hemisection in guinea pigs: The effects of applied electric fields," *J. Comp. Neurol.* 296: 634-653, 1990.

5. M. G. Fehlings, C. H. Tator, and R. D. Linden, "The effect of an applied direct current field on recovery from acute experimental spinal cord injury," *J. Neurosurg.* 68: 781-792, 1988.

6. M. G. Fehlings, C. H. Tator, R. D. Linden, and I. R. Piper, "Motor and somatosensory evoked potentials recorded from the rat," *EEG Clin. Neurophysiol.* 69: 65-78, 1988.

7. M. G. Fehlings, T. H. Wong, C. H. Tator, and M. Tymianski, "The effect of a direct current field on axons after experimental spinal cord injury," *Can. J. Surg.* 32: 188-191, 1989.

8. M. G. Fehlings and C. H. Tator, "The effect of direct current field polarity on recovery after acute experimental spinal cord injury," *Brain Res.* 579: 32-42, 1992.

9. C. D. McCaig, "Spinal neurite reabsorption and regrowth in vitro depend on the polarity of an applied electric field," *Development* 100: 31-41, 1987.

10. C. D. McCaig, "Nerve growth in a small applied electric field and the effects of pharmacological agents on rate and orientation," *J. Cell Sci.* 95: 617-622, 1990.

11. L. McDevitt, P. Fortner, and B. Pomeranz, "Application of a weak electric field to the hindpaw enhances sciatic motor nerve regeneration in the adult rat," *Brain Res.* 416: 308-314, 1987.

12. N. B. Patel and M. M. Poo, "Orientation of neurite outgrowth by extracellular electric fields," *J. Neurosci.* 2: 19-23, 1982.

13. M. J. Politis and M. F. Zanakis, "Treatment of the damaged rat hippocampus with a locally applied electric field," *Exp. Brain Res.* 71: 223-226, 1988.

14. M. J. Politis, M. F. Zanakis, and B. J. Albala, "Facilitated regeneration in the rat peripheral nervous system using applied electric fields," *J. Trauma* 28: 1375-1381, 1988.

15. M. J. Politis, M. F. Zanakis, and B. J. Albala, "Mammalian optic nerve regeneration following the application of electric fields," *J. Trauma* 28: 1548-1552, 1988.

16. M. J. Politis and M. F. Zanakis, "The short term effects of delayed application of electric fields in the damaged rodent spinal cord," *Neurosurgery* 25: 71-75, 1989.

17. G. C. Roman, H. K. Strahlendorf, P. W. Coates, and B. A. Rowley, "Stimulation of sciatic nerve regeneration in the adult rat by low intensity electric current," *Exp. Neurol.* 98: 222-232, 1987.

18. A. F. Strautman, R. J. Cork, and K. R. Robinson, "The distribution of free calcium in transected spinal axons and its modulation by applied electrical fields," *J. Neurosci.* 10: 3564-3575, 1990.

19. M. C. Wallace, C. H. Tator, and I. R. Piper, "Recovery of spinal cord function induced by direct current stimulation of the injured rat spinal cord," *Neurosurgery* 20: 878-884, 1987.

20. R. Midha, M. G. Fehlings, C. H. Tator, J. A. Saint-Cyr, and A. Guha, "Assessment of spinal cord injury by counting corticospinal and rubrospinal neurons," *Brain Res.* 410: 299-308, 1987.

21. A. S. Rivlin and C. H. Tator, "Objective clinical assessment of motor function after experimental spinal cord injury in the rat," *J. Neurosurg.* 47: 577-581, 1977.

22. A. R. Blight, "Cellular morphology of chronic spinal cord injury in the cat: Analysis of myelinated axons by line-sampling," *Neuroscience* 10: 521-543, 1983.

23. R. J. Hurlbert and C. H. Tator, "Electrical field mapping in the rat spinal cord," *Soc. Neurosci. Abstr.* 17(1): 723, 1991.

24. R. J. Hurlbert, E. Theriault, and C. H. Tator, "Neurotoxicity from direct current stimulation," *Soc. Neurosci. Abstr.* 16(2): 1115, 1990.

25. M. G. Fehlings and C. H. Tator, "The effect of a direct current field on chronically injured mammalian spinal cord axons," *Soc. Neurosci. Abstr.* 15(2): 1114, 1989.

26. E. Roederer, N. H. Goldberg, and M. J. Cohen, "Modification of retrograde degeneration in transected spinal axons of lamprey by DC current," *J. Neurosci.* 3: 153-160, 1983.

27. L. F. Jaffe and M. M. Poo, "Neurites grow faster towards the cathode than the anode in a steady field," *J. Exp. Zool.* 209: 115-128, 1979.

28. G. Marsh and H. W. Beams, "In vitro control of growing chick nerve fibers by applied electric currents," *J. Cell. Comp. Physiol.* 27: 139-157, 1946.

29. F. S. Kaplan, T. G. Maguire, U. M. Lee, M. E. Selzer, K. Spindler, J. Black, and C. T. Brighton, "Direct extracellular electrical stimulation influences density dependent aggregation of fetal rat cerebrocortical neurons in vitro, *Neurosci. Lett.* 94: 33-38, 1988.

30. V. Krathamer, "Modulation of conduction at points of axonal bifurcation by applied electric fields," *Trans. IEEE* BME-37: 515-519, 1990.

31. This work was supported by grants from the Medical Research Council of Canada and the University Research Incentive Fund. Drs. Fehlings and Hurlbert were Fellows of the Medical Research Council of Canada during the tenure of this investigation. Dr. Fehlings is supported by an Ontario Ministry of Health Career Scientist Award.

TABLE 1.—Neurobiological effects of DC fields.

Neurobiological effect	Authors (Ref. no.)
Promotion of neuritic outgrowth in vitro (effects polarity dependent)	Jaffe and Poo (27) Marsh and Beams (28) McCaig (9, 10) Patel and Poo (12)
Enhancement of cell survival in vitro	Kaplan et al. (29)
Reduction of retrograde axonal degeneration in vivo (lamprey)	Roederer et al. (26)
Reduction of calcium influx into transected axons	Strautman et al. (18)
Reversal of conduction block at points of axonal bifurcation	Krathamer (30)
Promotion of regeneration of peripheral axons in vivo	McDevitt et al. (11) Politis et al. (14) Roman et al. (17)
Promotion of regeneration of central axons in vivo	Borgens et al. (1-4) Politis et al. (15)
Promotion of recovery after focal brain injury	Politis and Zanakis (13)
Promotion of recovery after spinal-cord injury	Fehlings et al. (5-8) Politis and Zanakis (16) Wallace et al. (19)

Martin Blank, Ed., *Electricity and Magnetism in Biology and Medicine*

MAGNETIC FIELDS AND PERIPHERAL NERVE REGENERATION

M. Kanje* and A. Rusovan

University of Lund, Sweden

Regeneration of the rat sciatic nerve is stimulated in rats exposed to pulsed or sinusoidal magnetic fields. These findings, and some results suggesting that the effects of such fields are mediated via perturbation of Ca^{2+} fluxes in the nerve, are reviewed in this report.

We have found that the regenerative capability of the rat sciatic nerve is enhanced if the rats are subjected to magnetic-field exposure.[1-6] These studies and some recent findings suggesting that Ca^{2+} might be pertinent to the action of such fields on the regenerating nerve are reviewed here.

METHODS

We used female Sprague-Dawley rats weighing around 200 g. Regeneration of the sciatic nerve was initiated by exposing the nerve and inflicting a crush lesion on the nerve in the thigh region. The crush was labeled by an epineurial suture. Following surgery the rats were transferred to round plastic cages fitting inside a Helmholtz coil. Coils were 32 cm in diameter with 92 windings each, mounted on top of each other 16 cm apart. The coils were connected to an amplifier-function generator set-up that could deliver sinusoidal currents or current pulses sufficient to induce magnetic fields of the order of several hundred microteslas. Following exposure to the magnetic field, the sciatic nerve was re-exposed and regeneration distances were measured either by the pinch test, which measures regeneration of sensory fibers, or by immunocytochemical staining for neurofilaments in the regenerating nerve fibers. From plots of regeneration distances vs time, the regeneration rate and the initial delay can be calculated. This delay is usually, but incorrectly,[7] regarded as the time period before axons grow past the lesion site. In some experiments the regenerating segment of the nerve was perfused with drugs via a catheter connected to an implanted osmotic mini-pump.

RESULTS AND DISCUSSION

Effects of Pulsed Magnetic Fields (PEMF)

The sciatic nerves from rats exposed to a 2Hz PEMF of 300 μT during 4 h/day showed significantly higher regeneration distances than their corresponding controls.[5] This exposure increased the regeneration rate by about 20%; the initial delay remained unaffected. The stimulatory effect was independent of the orientation of the rat with respect to the magnetic field. The same degree of stimulation was observed whether the rats were restrained perpendicular or parallel to the magnetic field, or free roaming within a vertical field. This result seems to contradict the finding that stimulation of neurite outgrowth in vitro[8] depends on field orientation. However, in contrast to neurites on a flat plastic surface, the sciatic nerve in the rat has a complex trajectory. Should field orientation be important there should always, at any given moment, be some part of the nerve properly oriented with respect to the field. Moreover, both the dorsal root ganglia and its sensory cell bodies which harbor the macromolecular synthetic machinery of the nerve cells are spherical. There is no obvious reason to believe that magnetic-field orientation is of importance for such structures.

Fields Created by Sinusoidal Fields

Regeneration is also stimulated in rats exposed to sinusoidal fields.[2] Nerves from rats exposed to a 50Hz, 400μT field for 4 h/day exhibited a 20% increase in regeneration rate; the initial delay remained unaffected.

Field Strength and Exposure and Intermittent Stimulation

It is reasonable to assume that the magnetic fields act via the currents they induce in the animal. The induced currents are proportional to the frequency of the field but also to the rate of change of the field.

---

*Correspondence to: M. Kanje (Animal Physiology), University of Lund, Helgonav. 3B, S-223 62 Lund, Sweden.

Thus, by maintaining a "constant" field but changing the frequency, the currents induced in the animal can be altered. In such experiments, with sinusoidal fields, we found that the relationship between the induced currents and the regeneration effect was complex.[4] At a "constant" sinusoidal magnetic field of 50 μT, regeneration was stimulated between 250 and 1000 Hz; lower and higher frequencies were ineffective. Thus, there appears to be something resembling a frequency window for this effect.

The duration of the treatment is also important. Continuous stimulation to a sinusoidal (50Hz) field of 200 μT stimulates regeneration, whereas exposure of 4 h/day with the same field is ineffective. For a 2Hz, 400μT PEMF, exposure for 1 h/day is enough to stimulate regeneration.

Pretreatment

We have found that pretreatment with PEMF (that is, exposing the rats before regeneration is initiated and evaluated) increases the regenerative potential of the sciatic nerve.[1] Thus, if the sciatic nerve of rats that have been exposed to PEMF for, say, 4 h/day during 2 days is subjected to a crush lesion and the rats are removed from the exposure unit, regeneration distances measured 3 days later are longer than those in unexposed animals. In some unknown manner the pretreatment has conditioned the nerve. In several respects this effect resembles the increased regenerative response observed if a previously injured nerve is subjected to a second lesion—the conditioning lesion effect.[9] It is interesting that pretreatment with sinusoidal fields fail to condition the sciatic nerve.[3] The reason for this difference between PEMF and sinusoidal fields remains to be determined. However, that does not necessarily imply that the fields act via different mechanisms, since the conditioning effect could be related to the rate of change of the magnetic field. This rate is much higher for the PEMF field than for the sinusoidal field.

Mechanisms

The mechanisms by which magnetic fields affect regeneration are obscure. One possibility is that the effect is exerted directly on the nerve cell. Another possibility is that the fields act indirectly, perhaps by increasing the availability of systemic or locally produced growth-promoting substances. Thus, an increased release of growth hormone from the pituitary should result in an increased rate of regeneration.[10] The results would be similar if the fields affected the metabolism of cells surrounding the regenerating nerve fibers, including the Schwann cells which release a variety of factors that promote regeneration. Indeed, protein synthesis in such cells is affected by PEMF exposure.[6] Whatever the mechanism might be, current hypotheses suggest that the cell membrane is a prime target and that magnetic fields may alter the behavior of membrane-bound receptors and subsequently affect second-messenger systems like Ca^{2+}.

We have investigated whether Ca^{2+} could be important for the enhanced regeneration observed after magnetic-field exposure.[3] To that end the sciatic nerve was perfused with a Ca^{2+} antagonist-D600, which interferes with voltage-gated Ca^{2+} channels. Such channels are prevalent in the growth cone, whose mobility also depends on Ca^{2+}. Perfusion of the sciatic nerve proper with D600 suppressed the stimulatory effect of the magnetic-field exposure. This suppression could not be ascribed to a general toxicity of the drug. Furthermore, the effect was local, since perfusion of the ipsilateral nerve with the drug in exposed rats failed to suppress regeneration in the contralateral nerve. The possibility that outgrowth of neurites is stimulated as a consequence of perturbation of the Ca^{2+} channels by currents induced by the magnetic field must therefore be considered.[11]

REFERENCES

1. M. Kanje, A. Rusovan, B. F. Sisken, and G. Lundborg, *Bioelectromag.* (in press).
2. A. Rusovan and M. Kanje, *Exp. Neurol.* 11: 312-316, 1991.
3. A. Rusovan and M. Kanje, *Neuroreport* (in press).
4. A. Rusovan and M. Kanje, *Exp. Neurol.* (in press).
5. B. Sisken, M. Kanje, G. Lundborg, E. Herbst, and W. Kurtz, *Brain Res.* 489: 309-316, 1989.
6. B. F. Sisken, M. Kanje, G. Lundborg, and W. Kurtz, *Restor. Neurol. Neurosci.* 1: 303-309, 1990.
7. J. Sjöberg and M. Kanje, *Brain Res.* 530: 167-169, 1990.
8. B. F. Sisken, B. McLeod, and A. A. Pilla, *J. Bioelectricity* 3: 81-101, 1984.
9. I. G. McQuarrie, B. Grafstein, and M. B. Gershorn, *Brain Res.* 132: 443-453, 1977.
10. M. Kanje, G. Lundborg, and A. Edström, *Brain Res.* 439: 116-121, 1988; M. Kanje, A. Skottner, and G. Lundborg, *ibid.*, 475: 254-258.
11. This study was supported by grants from the Swedish Work Environment Fund.

Martin Blank, Ed., *Electricity and Magnetism in Biology and Medicine*

ACUTE ENDOTHELIAL PLASMINOGEN RESPONSE TO ELECTRICALLY STIMULATED CYCLING EXERCISE IN HUMAN QUADRIPLEGICS

M. S. Nash,* M. R. Ledford, J. W. Kent, and B. A. Green

University of Miami School of Medicine, Fla.

To test whether electrically stimulated cycling exercise (ESCE) stimulates fibrinolysis in persons disposed to thromboembolism by neurotrauma and chronic immobilization, four quadriplegics underwent 30 min of cycling generated by transcutaneous electrically stimulated contractions of lower extremity muscles. Blood samples drawn before, immediately after, and 30 min following ESCE were analyzed for endothelial-derived tissue Plasminogen Activator (tPA), and Plasminogen Activator Inhibitor (PAI-1). Results showed oxygen uptake increased 3.8-fold during ESCE. tPA levels increased 30% following ESCE ($p = 0.002$). PAI-1 was lowered 18.3% following ESCE ($p = 0.063$), and remained so after 30 min. We find that ESCE may activate fibrinolysis in quadriplegics via alteration of endothelial regulatory proteins.

Circulatory stasis of the lower extremities and blood hypercoagulability are known to catalyze the formation of deep venous thrombosis (DVT) in persons with chronic cervical spinal cord injury,[1] conditions that lend themselves to correction following acute exercise in persons without disability. Previous research has shown that exercise performed by nondisabled subjects acutely stimulates a hyperfibrinolytic state via release of extrinsic proteases that catalyze the conversion of plasminogen to plasmin.[2] The exercise-induced release of the tissue-type plasminogen activator (tPA) from vascular endothelium has been identified as the principal protein catalyzing this hyperfibrinolysis.[3]

The ability of persons with chronic cervical SCI to undergo cardiorespiratory exercise via arm ergometry is limited by the diminished strength and endurance of the upper extremity muscles that remain under voluntary control. Although arm ergometry in these individuals can elicit a cardiovascular response, there is no evidence that upper-extremity exercise can alter the susceptibility of the lower extremity to clotting. Moreover, the exercise itself may hasten the shoulder-joint degeneration that normally accompanies high quadriplegia. In contrast, however, electrically stimulated lower extremity cycling exercise utilizes larger muscles than arm ergometry exercise, can be more easily sustained for longer times, and elicits an acute cardiopulmonary response known to promote both elevated functional work capacity[4] and increased cardiac mass.[5] To date, the potential of electrically stimulated cycling exercise (ESCE) to promote an endothelially mediated fibrinolytic response in human quadriplegics remains unknown, yet the fibrolytic action of this exercise may have clinical significance for patients surviving SCI. The purpose of this study was to test the hypothesis that electrical stimulation cycle ergometry exercise (ESCE) stimulates the release of endothelially derived tPA and its antagonist, the fast Plasminogen Activator Inhibitor (PAI-1).

METHODS AND MATERIALS

Four quadriplegic males with complete motor lesions (Frankel Classifications A and B) of more than 1 year's duration were studied. Participants were without history of venous thrombosis and anticoagulation therapy, and were previously habituated to 30 min of steady-state ESCE twice weekly for 6 months. The treatment consisted of transcutaneous electrical stimulation (monophasic rectangular waveform, constant-current to peak output to 132 mA, 375μs pulse duration) of computer-sequenced electrically stimulated contractions of bilateral quadriceps, hamstring, and gluteus muscles. Electric current at these sites was introduced to the skin via carbon silastic electrodes coated with ECG conductor gel. On the day of testing, a venous catheter kept patent with physiologic saline was placed in the antecubital vein between 0900 and 1000

*Correspondence to: M. S. Nash (Orthopaedics & Rehab.), University of Miami School of Medicine, 5915 Ponce de Leon Blvd (5th fl.), Coral Gables FL 33146.

hours, with subjects undergoing 30 min of steady-state ESCE at the greatest resistance during which a pedal rate greater than 35 rpm could be maintained. Oxygen uptake (VO_2) during ESCE was monitored by open-circuit spirometry on a computerized metabolic measurements analyzer via a Hans-Rudolph nonrebreathing valve. Citrated blood samples drawn before exercise, immediately following 30 min of ESCE, and 30 min after completion of exercise were analyzed for tPA via an enzyme-linked immunosorbent assay (ELISA) using a peroxidase-tagged mouse monoclonal anti-tPA $F(ab')_2$ antibody (Diagnostica Stago). PAI-1 activity was quantified on acidified plasma samples by a chromogenic substrate assay (Biopool). Data were analyzed by repeated measures design analysis of variance (ANOVA), with time as the within-subjects factor. Simple-effects analyses were used to distinguish significant effects observed by the ANOVA, with 0.05 set as the level of significance both for *a priori* and *post hoc* tests.

RESULTS

ESCE stimulated a 3.8-fold increase in VO_2. The pre-exercise tPA of 4.25 ± 1.0 ng/ml (mean ± s.d.) increased by 30% immediately following ESCE ($p = 0.02$), with mean tPA following 30 min of passive recovery not differing significantly from the immediate post-ESCE sampling. An effect of ESCE on PAI-1 was also observed, with pre-exercise resting levels reduced from 10.23 ± 0.4 to 8.4 ± 1.8 IU/ml following exercise ($p = 0.063$). PAI-1 levels were 28.4% lower after exercise recovery than before (Table 1).

TABLE 1.—Effects of ESCE on tissue-type plasminogen activator and fast plasminogen activator inhibitor-1 activity (mean ± s.d, $n = 4$).

	Pre-ESCE	Post-ESCE	30 min post-ESCE	p
tPA (ng/ml)	4.25 ± 0.9	5.53 ± 0.7	5.30 ± 0.9	0.002
PAI-1 (IU/ml)	10.23 ± 0.4	8.35 ± 1.6	7.33 ± 1.8	0.063

DISCUSSION

Venous thromboembolism is a common and potentially serious clinical complication in both the acute and chronic phases of recovery from SCI.[1] Factors that contribute to elevated thrombotic tendency after SCI include immobilization-induced reduction in venous pumping of calf muscles, progressive dilation of lower extremity blood vessels, anoxia, and poor mechanical mixing of activated antithrombotic factors with their native inhibitors. These factors all encourage blood hypercoagulability, a state which is intensified in persons with cervical SCI due to circulatory stasis associated with loss of lower-extremity vasomotor tone.[6]

Cardiorespiratory exercise has been shown to foster improved circulation in healthy individuals, as well as those at risk for thromboembolism due to hypofibrinolysis. Among the circulatory changes stimulated by exercise is a 5- to 10-fold elevation of plasma fibrinolytic activity,[7] a response attributed to increased plasminogen activity in blood both during and immediately following exercise.[2,3] This exercise-induced hyperfibrinolysis has been attributed to the release of the extrinsic plasminogen activator from vascular endothelium during exercise, with the subsequent activation of plasminogen resulting in plasmin-mediated degradation of fibrin.[2] Neuromuscular stimulation has been demonstrated to mimic these effects, as electrical calf-muscle stimulation in surgical patients reduces the frequency of leg-scan detected DVT, and similar treatment in persons with acute SCI increases venous flow and fibrinolytic activity as assessed by venous plethysmography and clot lysis,[8] respectively. Nonetheless, the mechanism(s) by which this fibrinolysis is regulated and the value of treatment in DVT-prophylaxis has heretofore remained both questionable[1] and unstudied.

The present study demonstrates that ESCE promotes a significant elevation of endothelially derived tPA. The absence of a more robust response when compared to tPA release following exercise challenge in nondisabled individuals may also be explained by the limited physical work capacities of quadriplegics when they are undergoing ESCE. Previous exercise studies have shown that the rise in fibrinolytic activity in persons without disability requires an exercise intensity exceeding 50-60% of VO_{2max} representing absolute levels of oxygen consumption not attainable by quadriplegics undergoing electrically-stimulated exercise.[9] Moreover, as the exercise-induced increase in tPA is abolished following beta-adrenergic blockade[2] and is suboptimal in persons with hyperlipoproteinemia,[10] both sympathectomy and elevated blood lipid levels that typically accompany human quadriplegia may be responsible for attenuating the observed response of tPA to ESCE. However, despite the modest response of tPA to ESCE, levels of the fast antagonist of tPA, PAI-1, were reduced following ESCE, which suggests that the amount of uncomplexed tPA may have actually increased following ESCE. Thus, it is plausible that fibrinolytic activity was enhanced by an increase in "free tPA," despite the small increases in blood tPA.

We conclude that significant elevation of tPA in quadriplegics following ESCE is consistent with previous reports of endothelially regulated release of tPA and augmented fibrinolysis in persons without disability who undergo acute exercise. As lower extremity blood flow has been shown to change during ESCE via endothelially autoregulated mechanisms,[12] a larger sampling, measurement of tPA in the uncomplexed form, and direct assessment of fibrinolysis in these subjects are indicated. The physiologic significance of endothelial regulation of fibrinolysis on incidence of DVT requires additional attention.[13]

REFERENCES

1. A. G. G. Turpie, "Thrombosis prevention and treatment in spinal cord injured patients," in R. F. Bloch and M. Basbaum, Eds., *Management of Spinal Cord Injuries,* Baltimore, Md.: Williams and Wilkins, 1986.

2. M. S. El-Sayed, "Extrinsic plasminogen activator response to exercise after a single dose of propranolol," *Med. Sci. Sports Exerc.* 24: 327-332, 1992.

3. M. S. El-Sayed and B. Davies, "Effects of two formulations of a beta-blocker on fibrinolytic response to maximum exercise," *Med. Sci. Sports Exerc.* 21: 369-373, 1989.

4. K. T. Ragnarsson, S. Pollock, J. Petrofsky, W. O'Daniel, R. Edgar, and M. S. Nash, "Clinical evaluation of computerized functional stimulation after spinal cord injury: A multicenter pilot study," *Arch. Phys. Med. Rehabil.* 69: 672-677, 1988.

5. M. S. Nash, M. S. Bilsker, A. E. Marcillo, S. M. Isaac, L. A. Botelho, K. J. Klose, B. A. Green, M. T. Rountree, and J. D. Shea, "Reversal of adaptive left ventricular atrophy following electrically-stimulated exercise in human tetraplegics," *Paraplegia* 29: 590-599, 1991.

6. S. F. Figoni, "Cardiovascular and haemodynamic responses to tilting and to standing in tetraplegic patients: A review," *Paraplegia* 22: 99-109, 1984.

7. R. E. Bourey and S. A. Santoro, "Interactions of exercise, coagulation, and fibrinolysis: A brief review," *Med. Sci. Sports Exerc.* 20: 439-446, 1990.

8. R. T. Katz, D. Green, T. Sullivan, and G. Yarkony, "Functional electrical stimulation to enhance systemic fibrinolytic activity in spinal cord injury patients," *Arch. Phys. Med. Rehabil.* 68: 423-426, 1987.

9. G. L. Davis, C. F. Abilgaard, E. M. Bernauer, and M. Britten, "Fibrinolytic and hemostatic changes during and after maximal exercise in males," *J. Appl. Physiol.* 40: 287-292, 1976.

10. E. J. P. Brommer, J. H. Verheijen, G. T. G. Chang, and D. C. Rijken, "Masking ot the fibrinolytic response to stimulation by an inhibitor of tissue-type plasminogen activator in plasma," *Thromb. Haemost.* 52: 154-156, 1984.

11. D. C. Rijken, I. Juhan-Vague, and D. Collen, "Complexes between tissue-type plasminogen activator and protease inhibitors in human plasma identified with an immunoradiometric assay," *J. Clin. Lab. Med.* 101: 285-294, 1983.

12. M. S. Nash, C. M. Montefusco, and B. A. Rhodes, "Peripheral vascular response to exercise-induced and post-occlusive reactive hyperemia (PORH) in quadriplegics," *Surg. Forum* 41: 498-499, 1990.

13. This research was supported by grants from the American Biproducts Co. (#91ER002) and the Miami Project to Cure Paralysis.

Martin Blank, Ed., *Electricity and Magnetism in Biology and Medicine*

TREATMENT OF PRESSURE ULCERS BY ELECTRICAL STIMULATION OF A MONOPLEGIC PIG MODEL

A. Hyodo, S. I. Reger,* E. Reyes, and E. Z. Browne

The Cleveland Clinic Foundation, Ohio

The effectiveness of alternating-current (AC) and direct-current (DC) stimulation for treating pressure ulcers of a monoplegic pig model was tested. Twenty-nine pigs were used (11 for AC, 9 for DC, and 9 as controls). The right hindlimb was paralyzed by cutting the L2-S1 nerve roots. An 800 mm Hg pressure was applied for 48 h. Electrical stimulation was applied for 2 h/day for 4 weeks. Area and volume time constants were calculated for each wound. AC and DC stimulation appeared to enhance wound healing by reducing the wound surface area, but had little effect on wound volume.

The management of pressure ulcers imposes a formidable challenge to the attending physician who is responsible for bedridden elderly patients and patients with spinal-cord injuries or with peripheral vascular diseases. To accelerate the healing process, various pharmacological, physical, and surgical procedures have been tested.[1] Electrical stimulation has been reported to enhance the healing rate of wounds in patients and of wounds created on the normally innervated skin of animals.[1] However, the effectiveness of electrical stimulation with denervated skin in animals that simulate patients with low level spinal cord injuries has not been reported. We have developed a monoplegic pig model for grade 3 or 4 pressure ulcers (a wound extending to deep fascia or bone).[2] The goals of this study were to determine (1) whether AC and DC stimulation shorten the healing time of pressure ulcers created on denervated pig skin, (2) whether the electrical stimulation increases the vascularity of the healed granulation tissue of wounds, and (3) the relation between current density and wound healing time.

MATERIALS AND METHODS

Twenty-nine male Hanford mini pigs were used in this study. The right nerve roots of L2 through S1 were exposed through a midline incision on the pig's back, paravertebral muscle elevation from the spinous processes and laminae, and partial hemilaminectomy. Both motor and sensory components of the right L2 through S1 nerve roots were extradurally cut in the spinal canal. Pressure application began 7 days after the denervation surgery to allow for muscle atrophy. A constant pressure of 800 mm Hg was applied on the greater trochanteric area of the denervated right thigh with a spring compression indentor 3 cm in diameter. After 48 h of compression, the device was removed and followed by necrotic tissue debridement. Electrical stimulation began on the following day of the debridement procedure and continued for 2 h/day for 5 days each week until termination approximately 30 days later. Electrical current was applied through two conductive self-adhesive electrodes (Pal Plus©) placed on the normal skin 3 cm proximal and distal to wound edges, with the cathode placed distally. Eleven pigs underwent AC and 9 DC stimulation; 9 served as denervated controls. In AC stimulation, a charge-balanced tetanizing current (4 s on, 4 s off) with an amplitude of 7-10 mA and a pulse width of 300 μs was applied at 40 Hz. In DC stimulation the current flow was from proximal to distal at a constant amplitude of 0.6 mA. Daily management of the wounds consisted of a change of dressing, cleansing with saline and 0.25% acetic acid, and the administration of oral antibiotics. The volume and surface area of each wound were measured as outcome variables for the evaluation of wound closure. We measured wound volume by filling the wound cavity with sterile saline to the level of the intact skin surface while maintaining the wound surface horizontal. Wound areas were calculated by computerized planimetry of photographs of the wound with a disk of a known area for calibration. Biopsies were also obtained from wound margins at 4 weeks for the analysis of the vascularity of the granulation tissue. We measured the number and area of blood vessels in the granulation tissue by the histomorphometry of microscopic slides

---

*Correspondence to: S. I. Reger (Biomed. Engg), Cleveland Clinic Foundation, 9500 Euclid Ave., Cleveland OH 44195-6264.

stained by Hematoxylin and Eosin. We obtained current density (μA/cm^2) by dividing the amplitude of applied current (μA) by the initial area (cm^2) of each wound.

RESULTS

The volume and area of the wounds initially increased, reached a peak, and then gradually decreased. The reduction in each wound area and volume appeared to follow an exponential function.[3] Thus each variable was fitted to this function. Area and volume time constants were then calculated for each wound. Both AC and DC stimulation appeared to shorten the area time constants, but they had little effect on the volume time constants (Table 1). The number and area of blood vessels in AC- and DC-stimulated granulation tissue at the end of healing were comparable with those receiving no electrical stimulation (Table 2). In the DC-stimulation group there was a trend toward the area time constant, being shortest at a current density of 100 μA/cm^2 (Fig. 1). With AC stimulation, the correlation between the applied current density and the area time constant was not readily apparent.

TABLE 1.—Area and volume time constant (mean ± SD).

Group	AC	DC	Control
Area (days)	9.3 ± 3.7*	8.7 ± 2.8*	11.3 ± 6.7
Volume (days)	3.8 ± 0.8	4.2 ± 1.3	4.2 ± 1.5

*P = ns.

TABLE 2.—Measurements of vascularity of granulation tissue at 4 weeks (mean ± SD).

Group	AC	DC	Control
Number (mm^{-2})	20.1 ± 12.2	17.6 ± 9.2	20.5 ± 6.2
Area ($10^{-2} \times mm^2 \times mm^{-2}$)	1.6 ± 1.9	1.6 ± 0.8	1.8 ± 1.1

CONCLUSIONS

1. Surface AC and DC stimulation appeared to enhance ulcer healing in the monoplegic pig model by reducing wound surface area, but had little effect on wound volume.
2. AC and DC stimulation had no effect on the vascularity of the healed granulation tissue, as detailed by histological analysis of H and E stained specimens.
3. The optimal current density for DC stimulation occurred at 100 μA/cm^2.

REFERENCES

1. D. S. Weiss, R. Kirsner, and W. H. Eaglstein, "Electrical stimulation and wound healing," *Arch. Dermatol.* 126: 222-225, 1990.
2. S. Negami, T. F. McGovern, E. T. Reyes, E. Z. Browne, and S. I. Reger, "An experimental pressure sore model for functional electrical stimulation: Continuous pressure application on monoplegic pigs," in D. Papovic, Ed., *Advances in External Control of Human Extremities,* Belgrade: Nauka, 1990, 535-541.
3. A. Stefanovska, L. Vodovnik, et al., "Enhancement of ulcerated tissue healing by electrical stimulation," *Proc. RESNA 10th Ann. Conf.* 1987, 585-587.

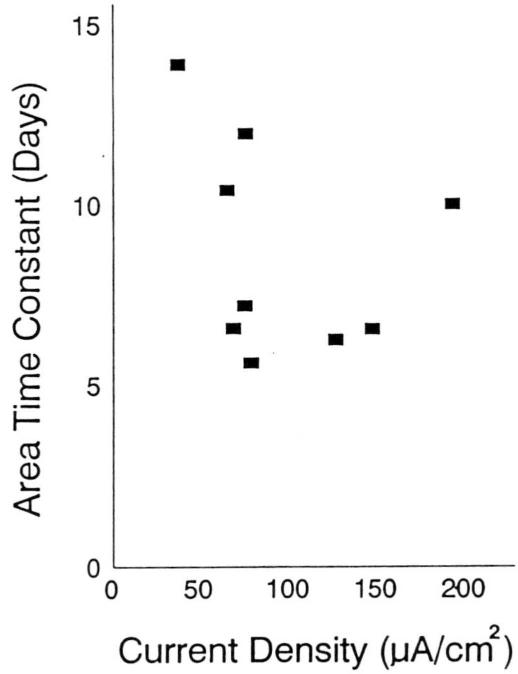

FIG. 1.—Area time constant vs current density for DC stimulation.

Martin Blank, Ed., *Electricity and Magnetism in Biology and Medicine*

PROTECTION AGAINST FOCAL CEREBRAL ISCHEMIA FOLLOWING EXPOSURE TO A PULSING ELECTROMAGNETIC FIELD

G. Grant* and G. Steinberg

Stanford University Medical Center, Calif.

R. Cadossi

Istituto Clinica Medica II, Modena, Italy

There is evidence that electromagnetic stimulation may accelerate the healing of tissue damage following ischemia. We undertook this study to investigate the effects of low-frequency PEMF exposure on cerebral injury in a rabbit model of transient focal ischemia (2 h occlusion followed by 4 h of reperfusion). PEMF exposure (280 V, 75 Hz, IGEA Stimulator) was initiated 10 min after the onset of ischemia and continued throughout reperfusion (6 exposed, 6 controls). Magnetic resonance imaging (MRI) and histology were used to measure the degree of ischemic injury. PEMF exposure attenuated cortical ischemic edema on MRI at the most anterior coronal level by 65%. On histologic examination, PEMF exposure reduced ischemic neuronal damage by 69% in this same cortical area and by 43% in the striatum. Preliminary data suggest that exposure to a PEMF of short duration may have implications for the treatment of acute stroke.

PEMFs have been shown to alter Ca^{++} efflux and decrease the release of preincubated $^{45}Ca^{++}$ from central nervous system tissues, including the cat cortex, chick embryonic brain, and in neuronal culture, although this effect varied with the frequency, amplitude, and duration of the applied field.[1,2] Calcium is implicated in many of the processes involved in neurotransmission and is a possible mediator of irreversible neuronal injury following cerebral ischemia.[3] We studied the effects of PEMFs in our rabbit model of transient focal cerebral ischemia to determine whether clinical applications of PEMF therapy could be extended to treat stroke injury.

Twelve male New Zealand white rabbits (3-3.5 kg) were anesthetized with 3% halothane delivered by mask, tracheotomized, and artificially ventilated. After catheterization of the femoral artery, systemic parameters including mean arterial pressure (MAP, end-expired CO_2 (Puritan-Bennett Capnograph), heart rate, rectal temperature, and temporalis muscle temperature were measured continuously. Arterial blood gases, hematocrit, and blood glucose were monitored intermittently. In order to control for focal tissue heating, we controlled rectal temperature at 39.0 C with a warming blanket and temporalis muscle temperature (brain temperature) at 37.0 C using a heating lamp and cooling fan.[4] Thirteen rabbits underwent transorbital clip occlusion of the left internal carotid, proximal left anterior cerebral, and proximal left middle cerebral arteries for 2 h, followed by 4 h of reperfusion. In order to confirm adequate ischemia before exposure to the PEMF, we used only animals in which the somatosensory evoked potentials were abolished completely within 10 min of arterial occlusion. Six rabbits were exposed to the PEMF at 10 min following arterial occlusion once adequate cerebral ischemia was confirmed and the field was continued throughout the time of arterial occlusion as well as during the 4 h of reperfusion. Two inductively coupled coils were used to generate the electromagnetic field directed to the head. The stimulator supplied the coil with a single pulse of electrical current (280 V) generating a magnetic field of 2.8 mT at a frequency of 75 Hz, and an impulse width of 1.3 ms (Igeastimulator, IGEA, Carpi, Italy).[5,6] The average induced electrical tension experienced by the tissue was 3.5 mV, which varied by 0.5 mV from the midline to the lateral edge of the skull. At the conclusion of the 6h experiment, rabbits were euthanized and perfused. Magnetic-resonance postfixation imaging studies were performed blind by use of a 1.5 T General Electric Sigma system with a 6in.-diameter saddle coil positioned at magnetic isocenter. We obtained 3mm-thick spin-echo T2 weighted images, using a repetition time of 2500 ms and an echo time of 25 ms. MRI is a sensitive measure of early cerebral ischemia. Five standard

---

*Correspondence to: G. Grant (Neurosurgery), Stanford University Medical Center, Stanford CA 94305.

coronal images were used to quantitate high-intensity areas representing ischemic edema. Histological examination of five standard coronal, hematoxylin-eosin stained sections (paraffin embedded, 10 μm thick) was performed blind to measure areas of ischemic neuronal damage (IND). Early ischemic neuronal damage was defined by light microscopy as moderate to severe neuronal shrinkage, increased nuclear basophilia, and a pyknotic nucleus.

PEMF exposure significantly attenuated the percentage of cortical edema assessed by MRI at the most anterior coronal level (coronal level 1) by 65%, $p < 0.001$. Furthermore, there was a trend toward attenuation of total cortical ischemic edema on MRI in the treated group by 33%, $p = 0.07$. On histopathology, the area of neocortical IND was also significantly reduced at the most anterior level in the PEMF-treated group by 69%, $p < 0.01$, and also in the striatum by 43%, $p < 0.05$ (Fig. 1). There were no significant differences between the PEMF-treated and control groups in systemic parameters.

The results of this study suggest that continued therapy with a pulsed electromagnetic field immediately following the onset of focal cerebral ischemia can significantly protect against the development of neocortical neuronal damage and ischemic edema. However, this neuroprotection achieved with PEMF exposure is not uniform throughout the area of cortical injury, but localized to an area characterized by severe ischemia in our model.

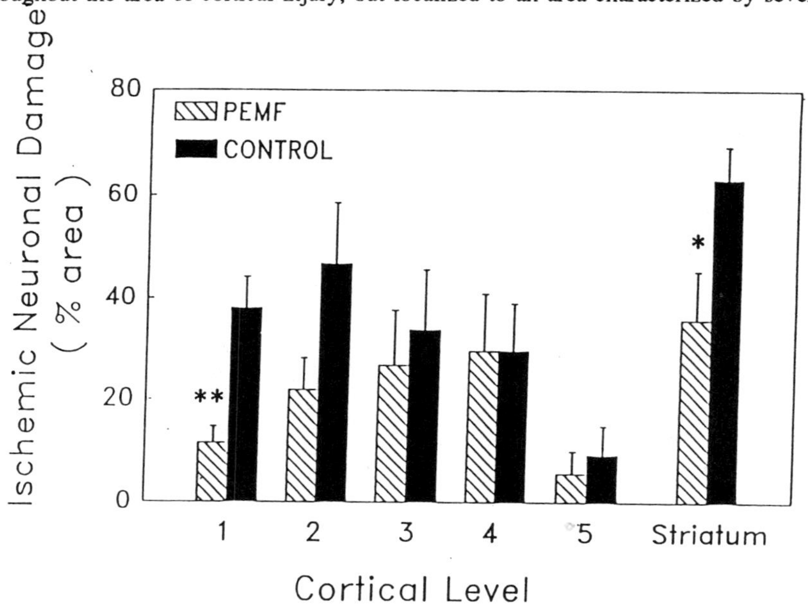

FIG. 1.—Bar graph of mean percent area of cortical and striatal neuronal damage on histologic examination at each coronal level in rabbits (1 most anterior, 5 most posterior).

REFERENCES

1. S. M. Bawin, W. R. Adey, and L. M. Sabbot, "Ionic factors in release of Ca^{++} from chicken cerebral tissue by electromagnetic fields," *Proc. Natl. Acad. Sci. USA* 75: 6314, 1978.

2. S. M. Michaelson, "Perspective on windows and calcium efflux studies," in *Interactions Between Electromagnetic Fields and Cells,* NATO ASI Series 97: 499-515, 1985.

3. D. W. Choi, "Calcium-mediated neurotoxicity: Relationship to specific channel types and role in ischemic damage," *Trends Neurosci.* 11: 465-469, 1988.

4. G. K. Steinberg, D. Kunis, J. Saleh, R. DeLaPaz, D. K. Shibata, and T. Gross, "Dextromethorphan protects against cerebral injury following transient focal ischemia in rabbits," *Stroke* 19: 1112-1118, 1988.

5. A. Albertini, G. Noera, A. Pierangeli, P. Zucchini, and R. Cadossi, "Effect of low-frequency pulsed electromagnetic fields on experimental myocardial infarcts in rats," in C. T. Brighton and S. R. Pollack, Eds., *Electromagnetics in Biology and Medicine,* San Francisco: San Francisco Press, 1991, 187-190.

6. R. Cadossi, G. Emilia, G. Torelli, G. Ceccherelli, S. Ferrari, and P. Ruggieri, "The effect of low-frequency pulsing electromagnetic fields on the response of human normal lymphocytes to phytohemagglutinin (PHA)," *Bioelectrochem. Bioenergetics* 14: 115-119, 1985.

Martin Blank, Ed., *Electricity and Magnetism in Biology and Medicine*

EFFECT OF PEMF ON HEART ISCHEMIC INJURY IN RATS

A. Albertini, G. Noera, and A. Pierangeli

University of Bologna, Italy

P. Zucchini, K. Wood, and R. Cadossi*

University of Modena, Italy

In the present study the effect of PEMF was investigated both in rats undergoing permanent coronary artery occlusion and in rats undergoing 30 min of coronary artery occlusion followed by reperfusion. Wistar rats were used; half of them were exposed to PEMF (IGEA stimulator). In rats undergoing permanent occlusion we observed a significant decrease in infarct size ($p < 0.05$). When the animals underwent reperfusion no difference in infarct size was observed between control and PEMF exposed rats. The data presented here suggest that PEMF exposure can limit the infarct size only in rats undergoing permanent occlusion of the left coronary artery.

For over 10 years, orthopedic surgeons have been using low-frequency pulsed electromagnetic fields (PEMF) for healing nonunions and delayed unions.[1-3] The utilization of PEMFs in clinics has since been applied as treatment for soft tissues, particularly venous skin ulcers.[4] PEMFs were used to treat these conditions after the results of several studies showed that PEMF stimulation promotes cell activation and particularly cell proliferation due to an effect on the cell membrane, which in turn induces an increase in the Ca^{++} influx.[5,6] PEMF exposure has been shown to have a particular effect on endothelial cell cultures that can form small vessels in vitro.[7] Several publications report that exposure of free skin flaps to PEMFs increases the amount of surviving flap, thus reducing the percentage of necrotic tissue.[8-10] In this research investigation we have studied the effect of PEMFs on ischemia as a result of the ligation of an artery under terminal circulation conditions. We elected to treat groups of rats with surgically ligated left coronary arteries with PEMF stimulation.

MATERIALS AND METHODS

Male Wistar rats weighing 350-400 g, maintained in standard stalling conditions, and fed ad libitum were used. The animals were interperitoneally anesthetized with urethane (1.2 g/kg). The animals were then intubated transtracheally and ventilated with a breathing apparatus for small animals (Harvard Apparatus, Natick, Mass.) at a rate of 54 ventilations per minute and a current volume of 1.5 ml/100 g.

The surgical procedure used was the one proposed by Vivaldi.[11] The thorax of the animals was opened at the fourth intercostal space level. After incising the pericardium and exposing the heart, we ligated the anterior descending branch of the left coronary artery at the level of its third proximal segment using a prolene 6-0 thread. In rats to be reperfused, we occluded the coronary artery by sliding a section of tubing over the two loose ends of the suture which formed a snare by clamping the tube tightly against the myocardium. After 1 h of the occlusion, the animals were reperfused by loosening of the tubing and removal of the suture.

In some rats, small (diam. 10.0 μm, 4.54×10^7 parts/ml) polystyrene fluorescent microspheres (Polyscience, Inc., Warrington, Pa.) were injected into the left ventricle after occlusion. This procedure determines the flow of blood in the myocardial region in question, and it establishes the "risk" area of myocardium.[12] The thorax of all animals was then closed with a vertical mattress suture and the pace of their assisted breathing was slowly reduced until spontaneous breathing resumed.

---

*Correspondence to: R. Cadossi, Istituto Clinica Medica II, Policlinico, Via del Pozzo 71, I-41000 Modena, Italy.

EXPOSURE

After arterial ligation, the rats were either immediately exposed to PEMF (experimental group) or isolated away from the magnetic field (control group). The rats were assigned at random to each group, but at such a rate as to insure an equal number of animals in each of the two groups.

The PEMF generator, which was supplied by Igea (Carpi, Italy), produced a magnetic field with a peak value of about 30 Gauss; the mean value of the electric field induced at the bottom of the cage on a standard probe was 2.5 mV.[5]

The animals of the experimental group were exposed to PEMFs immediately after their thorax was sutured. The animals' exposure to the magnetic field was maintained for 18 h, after which they were euthanized. After sacrifice the hearts were then removed and cut into two transverse slices: slice A below the occlusion point and slice B intermediate above the apex. Each slice was 2 mm thick.

The slices obtained in this way were then stained in a 1% triphenyltetrazolium chloride (TTC) solution at 37 C for 5 min. This procedure stains the viable myocardium red and the necrotic yellow (Fig. 1).[11] Photographs of each side of the slices were taken and the percentage of ischemic area in the left ventricle was calculated by means of an image analyzer. For the hearts in which fluorescent microspheres had been injected, an ultraviolet lamp was used in a dark room to visualize the area in which there was myocardial blood flow.[12] A photograph was taken of the fluorescent slices as well (Fig. 2).

RESULTS

In our permanent ligation study, 95 rats survived until sacrifice: 51 controls and 44 exposed to PEMF. Of these animals, 39 were injected with fluorescent microspheres as well (21 controls, 18 exposed). An additional 6 rats (3 controls, 3 exposed) were reperfused. No difference was observed in total heart weight among the two groups.

Based on measurements made, the average ischemic area of the left ventricle in the control groups was 41.3% (13.6 s.d.) and 52.1% (13.7 s.d.) for sections A and B, respectively. In the PEMF-exposed rats the ischemic area was 36.8% (12.2 s.d.) and 45.4% (16.1 s.d.) for the two slices A and B, respectively. This difference proved to be statistically significant using the Student's t test analysis ($p < .05$ and $p < .02$, respectively). When animals underwent reperfusion, no difference was observed between control and PEMF-exposed rats. Ischemic areas, section A and B, were 47.2% (18.3 s.d.) and 49.2% (22.5 s.d.), respectively, for controls and 52.1% (3.3 s.d) and 56.2% (13.9 s.d), respectively, for PEMF exposed. These data suggest that PEMF exposure can reduce the infarct size only in animals undergoing permanent occlusion of the left coronary artery.

In the rats in which microspheres were injected, photos were taken in normal light and with only ultraviolet light. These two photo slices are superposed. No difference was found for slice A. We compared the ratios of the areas of viable myocardium to the fluorescent areas. In slice B we found the ratios to be 1.05 (0.1 s.d.) in control compared to 1.14 (0.04 s.d.) for the PEMF-exposed group. This proved to be statistically significant ($p < 0.03$).

DISCUSSION

The above data show that, after ligature of the left coronary artery, the exposure of rats to an electromagnetic field can reduce the extent of the necrosis in ventricular sections by 20%. This effect is of about the same magnitude as the effect on free-flap survival. The mechanism by which PEMF decreases the amount of myocardium undergoing necrosis is not easily understandable, especially in view of the relatively short time of exposure to the electromagnetic fields. We hypothesize that PEMF may locally limit the free radicals damage and decrease the release of agents with a vasoconstrictor activity.

REFERENCES

1. C. A. L. Basset, S. N. Mitchell, L Norton, and A. A. Pilla, "A non-operative salvage of surgically resistant pseudoarthroses and nonunions by pulsing electromagnetic fields: A preliminary report," *Clin. Orthop. Rel. Res.* 124: 128, 1977.

2. G. C. Traina, R. Cadossi, G. Ceccherelli, A. Dal Monte, G. Fontanesi, F. Giancecchi, G. I. Mammi, V. Negri, F. Pisano, G. Poli, E. Rinaldi, and B. Virgili, "La modulazione elettrica della osteogenesi," *Giornale Italiano di Ortopedia* Supplement 12: 165, 1986.

3. G. Borsalino, M. Bagnacani, E. Bettati, G. Fornaciari, R. Rocchi, S. Uluhogian, G. Ceccherelli, R. Cadossi, and G. Traina, "Electrical stimulation of human femoral intertrochanteric osteotomies: Double-blind study," *Clin. Orthop. Rel. Res.* 237: 256, 1988.

4. M. Ieran, S. Zaffuto, M. Bagnacani, M. Annovi, A. Moratti, and R. Cadossi, "Effect of low-frequency pulsing electromagnetic fields on skin ulcers of venous origin in humans: A double-blind study," *J. Orthop. Res.* 8: 276, 1990.

5. R. Cadossi, F. Bersani, A. Cossarizza, P. Zucchini, G. Emilia, G. Torelli, and C. Franceschi, "Lymphocytes and low frequency electromagnetic fields," *FASEB J.* 6: 2667, 1992.

6. A. Cossarizza, D. Monti, F. Bersani, M. Cantini, R. Cadossi, A. Sacchi, and C. Franceschi, "Extremely low frequency pulsed electromagnetic fields increase cell proliferation in lymphocytes from young and aged subjects," *Biochem. Biophys. Res. Commun.* 160: 692, 1989.

7. G. P. Ameia Yen-Patton, W. F. Patton, D. M. Beer, and B. S. Jacobson, "Endothelian cell response to pulsed electromagnetic fields: Stimulation of growth rate and angiogenesis in vitro," *J. Cell. Physiol.* 134: 37, 1988.

8. D. J. Johnson and V. R. Hentz, "Electromagnetic field enhancement of rat wound tensile strength and pedicle skin flap survival," *Trans. BRAGS* 7: 78, 1987.

9. M. J. Politis, M. F. Zanakis, and J. E. Miller, "Enhanced survival of full-thickness skin grafts following the application of DC electric fields," *Plastic Reconstructive Surg.* 4: 267, 1989.

10. M. J. Im, W. P. A. Lee, and J. E. Hoopes, "Effect of electrical stimulation on survival of skin flap in pigs," *Physical Therapy* 70: 37, 1990.

11. M. T. Vivaldi, R. A. Kloner, and F. J. Schoen, "Triphenyltetrazolium staining of irreversible ischemic injury following coronary artery occlusion," *Am. J. Pathol.* 121: 522, 1985.

12. L. H. Sharon and R. A. Kloner, "Effect of early coronary artery reperfusion on infarct development in a model of low collateral flow," *Cardiov. Res.* 21: 668. 1987.

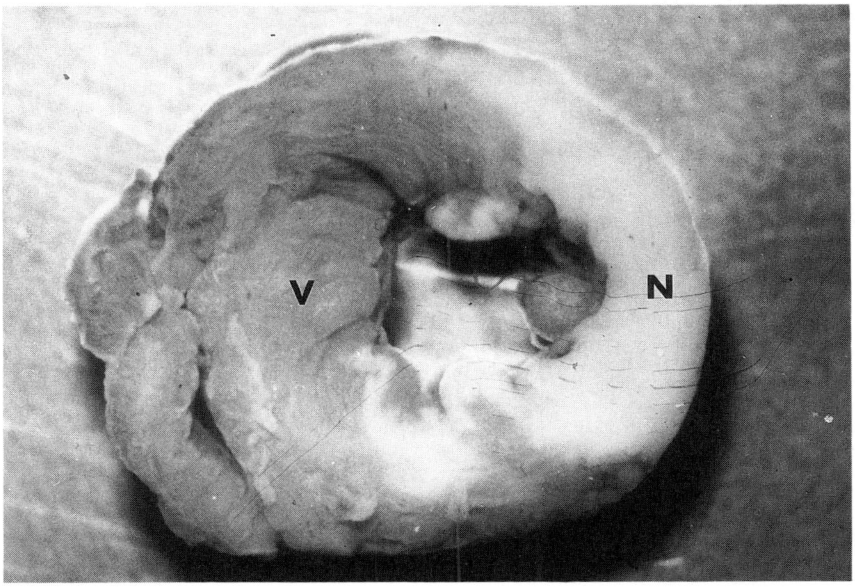

FIG. 1.—Tripheniltetrazolium chloride (TTC) staining of myocardium slice B of rat exposed to PEMF. N, necrotic area; V, viable myocardium.

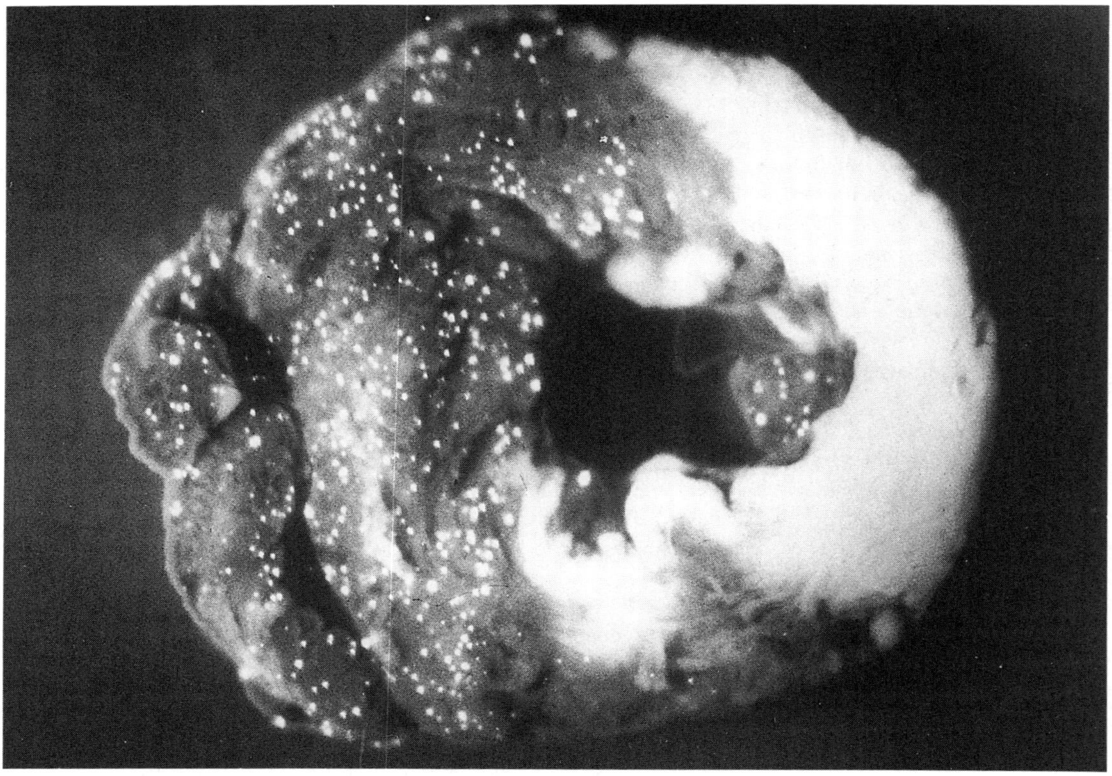

FIG. 2.—Same myocardium slice as in Fig. 1: Under ultraviolet light, area of particulate fluorescence is the perfused myocardium; area without fluorescence is the area at risk; pale area is the necrotic area.

Martin Blank, Ed., *Electricity and Magnetism in Biology and Medicine*

TRANSCRANIAL MAGNETIC STIMULATION: AN ACCURATE 3-D PREDICTIVE MODEL

G. Cerri, R. De Leo,* F. Moglie, A. Schiavoni, and M. Soverchia

University of Ancona, Italy

M. Guidi and O. Scarpino

INRCA, Ancona, Italy

A. Sarti

CINECA, Bologna, Italy

The resistive-network method is applied to the prediction of induced currents in human brain by magnetic stimulation. The model is characterized by a 3-D equivalent resistive network derived from Maxwell's equations. Time dependence of electric charge is neglected because of the low maximum frequency (10 kHz) of the pulse spectrum. The head geometry is obtained directly from a magnetic resonance image (MRI). The conductivity of each point is obtained by means of an interpolating function derived by associating known conductivity values to reference points of the MRI. This model permits an accurate prediction of current distribution.

Brain cortex magnetic stimulation is a very attractive diagnosis method from many points of view; in particular, it is painless regarding electrical stimulation, but still presents at least two major aspects to be investigated. The first concerns the characterization of induced current in the human head by magnetic stimulation; the second, the design of particular stimulators (or array of stimulators) for better focusing of the magnetic stimulus in specific areas of the brain. The model calculates the induced current in the brain by discretizing the MRI of the human head in a 3-D resistive network. Application of Maxwell's equations to each cell yields a linear matrix equation that is solved numerically.

MATHEMATICAL MODEL

The model is derived by Maxwell's equations written in integral form:

$$\oint_C \mathbf{E}\cdot d\mathbf{l} = -\frac{\partial}{\partial t}\int_S \mu_0 \mathbf{H}\cdot d\mathbf{S} \tag{1}$$

$$\oint_S \mathbf{J}\cdot d\mathbf{S} = -\frac{\partial}{\partial t}\int_V \rho\, dV \tag{2}$$

where

$$\mathbf{J} = \sigma(\mathbf{r})\mathbf{E} \tag{3}$$

The total magnetic field is given by:

$$\mathbf{H} = \mathbf{H}_0 + \mathbf{H}_s \tag{4}$$

where $\mathbf{H}_0$ is the field generated by the source and $\mathbf{H}_s$ is the field scattered by the induced currents; since $\mathbf{H}_s$ is smaller[1] by several orders of magnitude than $\mathbf{H}_0$, it can be neglected in Eq. (1). Since the magnetic pulse rise time is very low,[2] quasistatic assumptions hold and Eq. (2) becomes

---

*Correspondence to: R. De Leo (Elettronica ed Automatica), University of Ancona, Via Brecce Bianche, I-60131 Ancona, Italy.

$$\oint_S \mathbf{J} \cdot d\mathbf{S} = 0 \tag{5}$$

After discretizing the biological medium in cubic cells of size l, the Eqs. (1) through (5) assume the following expressions in a Cartesian coordinate system:

$$\sum_i E_{i_{xy}} = -\mu_0 \, l \, \frac{\partial H_{0_z}}{\partial t} \tag{6}$$

$$\sum_i E_{i_{xz}} = -\mu_0 \, l \, \frac{\partial H_{0_y}}{\partial t} \tag{7}$$

$$\sum_i E_{i_{yz}} = \mu_0 \, l \, \frac{\partial H_{0_x}}{\partial t} \tag{8}$$

$$\sum_i \sigma_i E_i = 0 \tag{9}$$

where $E_{i_{pq}}$ are the unknown electric-field components in each cell's side and H_{0_p} is the known magnetic field. The sums in Eqs. (6), (7), and (8) represent the discretized form of the closed line integral of Eq. (1), where the electric field is assumed to be constant in each side and the magnetic field is constant in the line-enclosed surface.

CONDUCTIVITY VALUE

The conductivity value of each cell is directly derived from an MRI in T2 mode. The numerical data representation of each image consists in a matrix of 256×256 integer numbers. Each number (whose value varies over $0 \leq g \leq 4095$) represents the gray level of each point; an interpolating equation associates the point conductivity to the corresponding gray level by

$$\sigma = 0.011 - 6.915 \times 10^{-5} g + 6.573 \times 10^{-8} g^2 + 1.308 \times 10^{-11} g^3 \tag{10}$$

Figure 1 shows the function and the ranges of the tissues, corresponding to the various gray levels extracted by the MRI data of normal subjects and the conductivity values inferred from Refs. 3 and 4.

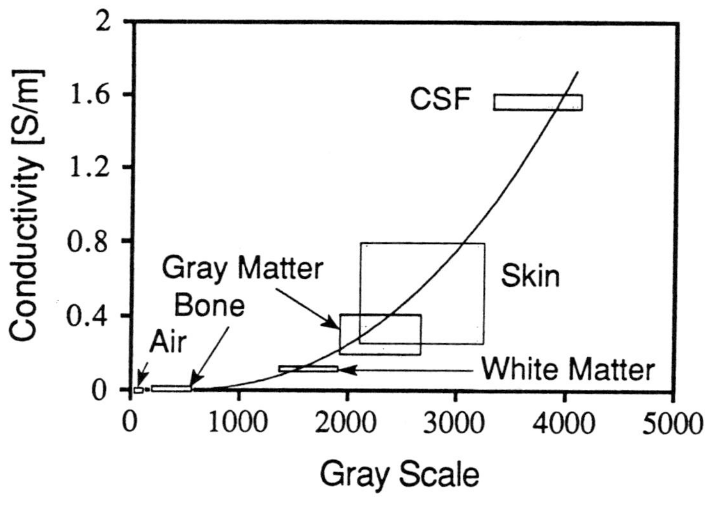

FIG. 1.—Interpolating function yields conductivity (i.e., gray level). X-edge of rectangles is range of gray levels pertaining to a tissue in the MRI; Y-edge is range of conductivity values inferred from literature.

RESULTS

The computer simulation code runs on a Cray Y-MP 4/64. NAG library functions D01ATF and F04QAF are used to evaluate the magnetic field induced by a coil and to solve the sparse linear system, respectively. The stimulator adopted for this simulation is a Novametrix Mag-Stim 200 model, provided by a coil of 19 turns with an inner diameter of 56 mm and outer diameter of 120 mm; it can deliver a peak current of 5500 A as a maximum for each pulse. In Table 1 CPU time and memory request are reported for three different discretized models.

TABLE 1.—Geometric parameters, CPU times, and memory occupation.

				Magnetic Field	System Solution	
Grid	Equations	Unknowns	Elements ≠ 0	CPU (s)	CPU (s)	Memory (Mw)
16x16x8	9,257	7,208	41,040	3.791	4.039	0.844
32x32x15	65,488	50,127	292,510	32.872	39.955	4.889
64x64x30	507,551	384,670	2,275,644	243.178	408.167	36.32

Figure 2 shows the current distribution in a horizontal section of the head, which crosses the arm motor area. Figure 3 shows the 3-D reconstruction of the head and the distribution of induced current on a single horizontal plane.

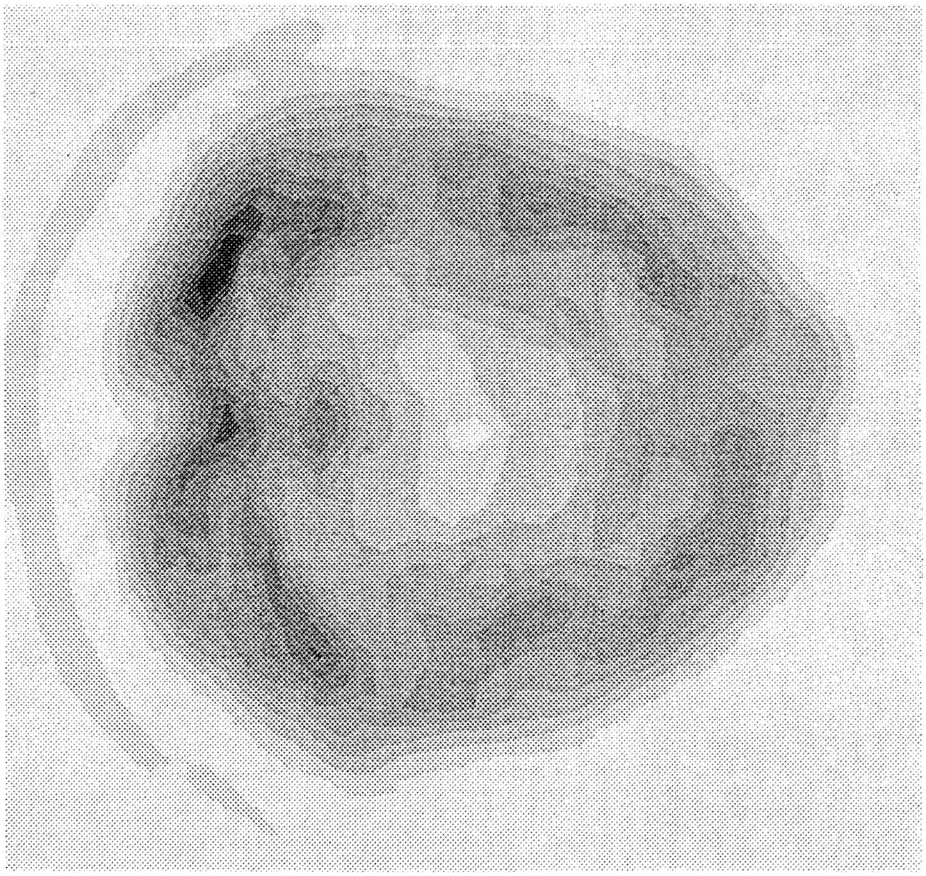

FIG. 2.—Distribution of the currents on a horizontal section of the head that crosses arm motor area.

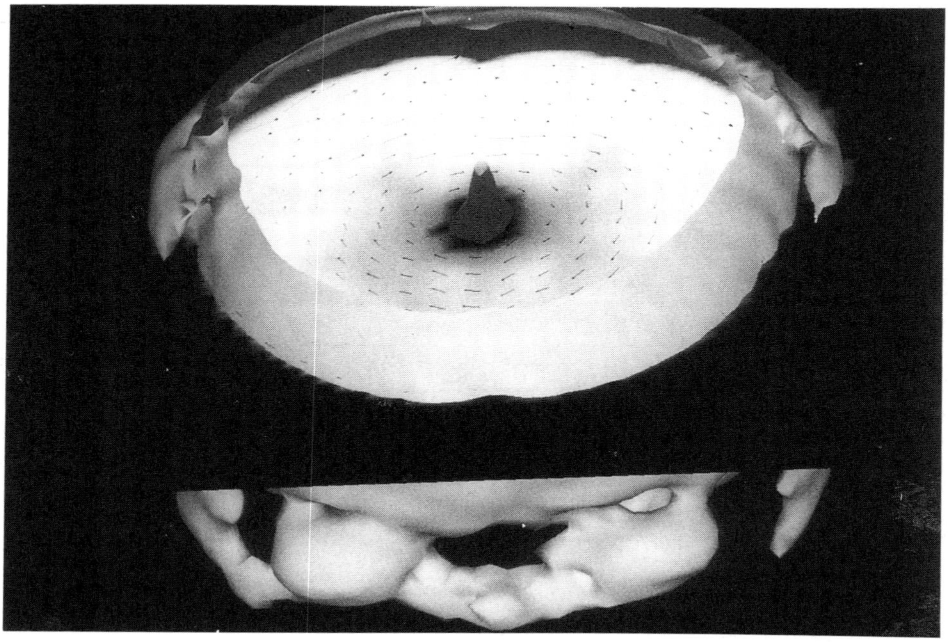

FIG. 3.—3-D reconstruction of head, and distribution of induced current on a single horizontal plane.

CONCLUSIONS

We have applied a 3-D model to the analysis of the current distribution induced by transcranial magnetic stimulation. It is the first current estimation in a realistic 3-D model derived from an MRI. Our future aim will be to optimize the parameters of stimulation to suggest alternate shapes of the coil, or array of coils, that can induce a more focused excitation.[5]

REFERENCES

1. G. Cerri, R. De Leo, O. Francescangeli, and S. Melone, "Accurate evaluation of electrical current excitation in brain cortex by magnetic field pulses," *Physica Medica* 7(3): 99-105, 1991.

2. N. Orcutt and O. M. Gandhi, "A 3-D impedance method to calculate power deposition in biological bodies subjected to time varying magnetic fields," *Trans. IEEE* BME-35(8): 577-583, 1988.

3. L. A. Geddes, "Optimal stimulation duration for extracranial cortical stimulation," *Neurosurgery* 20: 94-99, 1987.

4. K. S. Holton and C. F. Walker, "Correlation of a magnetic resonance brain image and tissue impedance," *Proc. Twelfth Ann. Intern. IEEE Conf.* EMBS-12: 92-93, 1990.

5. The Cray Y-MP 4/64 adopted for this research was supported by a grant from CINECA (Bologna, Italy). The authors are grateful to Dr. Roberto Rossi of the INRCA Radiology Unit for supplying the MRI data, and to Dr. Valeria Frontini of CINECA for assistance in computer libraries.

Martin Blank, Ed., *Electricity and Magnetism in Biology and Medicine*

VECTORIAL MAGNETIC STIMULATION OF THE HUMAN BRAIN

S. Ueno* and T. Matsuda

Kyushu University, Fukuoka, Japan

We have developed a method of vectorial magnetic stimulation of the brain using a figure-of-eight coil. We have observed that an optimal direction of stimulating currents for neural excitation exists in each functional area of the human cortex. The present study is focused on the vectorial characteristics in the magnetic stimulation of the human brain. A model is proposed to simulate nerve excitation processes elicited by magnetic stimulation. The model suggests our observations that the vectorial characteristics reflect both functional and anatomical organization of neural fibers in the brain.

Magnetic stimulation of excitable tissue has been studied by several investigators. Magnetic stimulation of the human brain was first reported by Barker et al.[1] According to their method, however, broad areas of the brain are stimulated simultaneously. We have developed a method of focal magnetic stimulation of the brain.[2] The basic idea is to concentrate induced eddy currents locally in the vicinity of a target by a pair of opposing pulsed magnetic fields that can be produced by a figure-of-eight coil. Using this method, we were able to stimulate the motor cortex of the human brain within a 5mm resolution.[3,4] Since the concentrated eddy currents beneath the intersection of the figure-of-eight coil flow in the direction parallel to the tangent of both circular coils, vectorial stimulation can be achieved. The neural fibers can be excited easily when the fibers are stimulated by the eddy currents that flow parallel to the nerve fibers. Based on this principle, we have obtained functional maps related to the hand and foot areas.[5,6] We have observed that an optimal direction of stimulating currents for neural excitation exists in each functional area of the cortex. We have also observed that the functional maps of the cortex vary with the orientation of the stimulating current. To explain the mechanism responsible for producing this anisotropic response to brain stimulation, we have developed a model of neural excitation elicited by magnetic stimulation.

In this paper, we focus on the vectorial characteristics in magnetic stimulations of the human brain and we propose a model of neural excitation elicited by magnetic brain stimulation. The model explains our observation that the optimal direction of the stimulating current vector for neural excitation reflects both functional and anatomical organization of neural fibers in the brain.

PRINCIPLE

The basic idea of focal and vectorial stimulation is to concentrate induced eddy currents in a target area by a pair of time-varying magnetic fields.[2] This method is an application of our previously proposed technique in hyperthermia to the magnetic nerve stimulation.[7,8] A figure-of-eight coil is positioned on the surface of the head so that time-varying magnetic fields pass through the head in the opposite directions around the target to be stimulated. Eddy currents can be induced in the head in such a manner that the current flows form two vortexes, which flow together at the target beneath the intersection of the figure-of-eight coil so that the eddy currents are thus concentrated at the target. A computer simulation shows that the current density at the target attains a peak which is 2-3 times higher than current densities in the surrounding regions.[2] Since the concentrated eddy currents at the target beneath the intersection of the figure-of-eight coil flow in the direction parallel to the target of both circular coils, vectorial stimulation can be achieved.

---

*Correspondence to: S. Ueno (Electronics), Kyushu University, Fukuoka 812 Japan.

FUNCTIONAL MAPPING OF THE MOTOR CORTEX

Magnetic stimulation of the human motor cortex related to the hand and foot areas was achieved. The various parts in the motor cortex were stimulated, and the corresponding muscular action potentials were recorded by electromyograms (EMGs) from the hand and foot muscles. Definite EMG responses to magnetic brain stimulation were obtained from the thenar muscles when a figure-of-eight coil was placed at a position called a "thenar point" in the motor cortex. No responses were obtained at a point 5 mm anterior to, above, or below the thenar point. In other words, the selective stimulation of the human motor cortex was realized within a 5mm resolution. When the figure-of-eight coil was rotated about the intersection of the coil, the thenar point in the motor cortex was stimulated. The maximum EMG response was obtained when the stimulating current flowed in the direction from the temporal to the frontal region; no EMG signals were obtained when the stimulating currents flowed in the opposite direction, from the frontal to the temporal region. The functional vector mapping of the human motor cortex related to the hand, arm, and foot areas was obtained. It was observed that an optimal direction of stimulating currents for neural excitation exists in each functional area in the cortex. This vector map may reflect the functional and anatomical structures of neural fibers. The localization of the target area was confirmed by MRI.

MODELING OF NEURAL EXCITATION ELICITED BY MAGNETIC STIMULATION

The existence of orientation in the brain may be paretically understood by the anatomical structure of the neural fibers in the cortex. The neural fibers can be excited easily when the fibers are stimulated by the eddy currents that flow parallel to the nerve fibers.[9] A model is introduced to describe the neural excitation by magnetically induced eddy currents. We simulate the excitation of neural fibers as a function of direction of angle between the figure-of-eight coil and a nerve axon, using the parameters of the Frankenhaeuser-Huxley model.[10] We apply Roth and Basser's model[11] to our model of neural excitation elicited by focal and vectorial stimulation with a figure-of-eight coil. When the eddy currents generated by a figure-of-eight coil flow parallel to the intersection of the spatial distribution of the eddy currents, the field makes a peak at a point beneath the intersection of the figure-of-eight coil. The sharpness of the field gradient is important for nerve excitation. The gradient of the induced eddy current reflects the source of transmembrane potential which acts to stimulate the nerve. The simulated results suggest that the direction of neural excitation is governed by the angle of inclination of nerve fibers themselves. Although the simple model for a single neuron fails to recognize the convoluted structure of the cerebral cortex, the model is valid for the explanation of the vectorial characteristics of induced eddy currents on neural excitation. In other words, the model simulates well the change of neural excitation processes with the change of angle between the figure-of-eight coil and the nerve fiber.[12,13] The existence of the vectorial characteristics of stimulating currents for neural excitation reflects both the functional and anatomical organization of neural fibers in the brain.

REFERENCES

1. A. T. Baker, R. Jalinous, and I. L. Freeston, "Non-invasive magnetic stimulation of human motor cortex," *Lancet* i: 1106-1107, 1985.

2. S. Ueno, T. Tashiro, and K. Harada, "Localized stimulation of neural tissue in the brain by means of paired configuration of time-varying magnetic fields," *J. Appl. Phys.* 64: 5862-5864, 1988.

3. S. Ueno, T. Matsuda, M. Fujiki, and S. Hori, "Localized stimulation of the human motor cortex by means of a pair of opposing magnetic fields," *Digests of Intermag. Conf.* GD-10, Washington, D.C., 1989.

4. S. Ueno, T. Matsuda, and M. Fujiki, "Localized stimulation of the human cortex by opposing magnetic fields," in S. J. Williamson et al., Eds., *Advances in Biomagnetism,* New York: Plenum Press, 1989, 529-532.

5. S. Ueno, T. Matsuda, and M. Fujiki, "Functional mapping of the human cortex obtained by focal and vectorial magnetic stimulation of the brain," *IEEE Trans.* MAG-26: 1539-1544, 1990.

6. S. Ueno, T. Matsuda, and O. Hiwaki, "Localized stimulation of the human brain and spinal cord by a pair of opposing pulsed magnetic fields," *J. Appl. Phys.* 66: 5838-5840, 1990.

7. S. Ueno, T. Tashiro, S. Kamise, T. Oosako, and K. Harada, "A paired-coil configuration for localized hyperthermia of deep tissues," *Digests of Intermag. Conf.* BD-3, St. Paul, Minn., 1985.

8. S. Ueno, T. Tashiro, S. Kamise, and K. Harada, "Localized hyperthermia by means of a paired-coil configuration: Calculation of current distributions in cubical model," *IEEE Trans.* MAG-23: 2437-2439, 1987.

9. J. P. Reilly, "Peripheral nerve stimulation by induced electric currents: Exposure to time-varying magnetic fields," *Medical & Biol. Eng. & Comput.* 27: 101-110, 1989.

10. B. Frankenhaeuser and A. L. Huxley, "The action potential in the myelinated nerve fiber of *Xenopus laevis* as computed on the basis of voltage clamp data," *J. Physiol.* 171: 302-315, 1964.

11. B. J. Roth and P. J. Basser, "A model of stimulation of a nerve fiber by electromagnetic induction," *IEEE Trans.* BME-37: 588-597, 1990.

12. S. Ueno, "Focal and vectorial magnetic stimulation of the human brain," Abstracts of the International Symposium on Biomagnetic Stimulation, Kyushu University, Fukuoka, Japan, pp. 7-11, July 1991.

13. S. Ueno, T. Matsuda, and O. Hiwaki, "Estimation of structures of neural fibers in the human brain by vectorial magnetic stimulation," *IEEE Trans.* MAG-27: 5387-5389, 1991.

Martin Blank, Ed., *Electricity and Magnetism in Biology and Medicine*

COIL OPTIMIZATION FOR NEURAL STIMULATION WITH MAGNETIC FIELD

K. P. Esselle*

Macquarie University, Australia

M. A. Stuchly

University of Victoria, Canada

Strong pulsed magnetic fields can stimulate neurons. This interaction has recently found numerous diagnostic and research applications in neurology. The objective of the present work was to analyze the induced electric fields and their spatial derivatives, as well as to evaluate the inductance of various coil configurations in order to optimize the stimulator performance. The optimization was in terms of localized depolarization of a selected neuron at a given location below the air-tissue interface, given a constant voltage of the stimulator.

Since mid-1980s magnetic stimulation of neurons has gained rapid acceptance as a diagnostic and research tool in neurology.[1] Neural stimulation occurs when an externally applied stimulus has a sufficient amplitude and duration to cause current to flow through a cellular (neural) membrane and change is potential above a threshold value. In the case of magnetic stimulation the stimulus is produced by a pulse of magnetic field from an external coil. The electric field induced in tissue is proportional to the time derivative of the coil current.

INDUCED ELECTRIC FIELD

The computation of the induced electric field in the semi-infinite tissue half-space, from a time-varying current in an external arbitrarily shaped coil usually requires several numerical steps. However, the electric field from an infinitesimal element of the coil can be evaluated analytically,[2] provided the time-variation of current is slow enough to justify quasistatic approximations. The spatial derivatives of the induced electric-field components are given by simple expressions.

Most of the best-performing stimulating coils, such as double and quadruple square coils, are composed of several linear segments. Others, such as circular coils, can be approximated by polygon-shaped coils composed of linear segments. In these cases, the fields and their spatial derivatives produced by each linear segment can be evaluated analytically, which saves a considerable amount of computer time and provides physical insight.

Several quadruple coils have been proposed for nerve stimulation.[2,3] Although there are substantial differences among them, all the quadruple coils have several common advantages over single and double coils when used for stimulation of long nerves. They produce only one stimulation spot (the primary depolarization peak) in a well-defined location (right below the coil center). Their secondary depolarization peaks and hyperpolarization peaks are relatively weak (<50% of the primary depolarization peak). If the time rate of change of current and the total number of turns are constant, properly designed quadruple coils create the same *peak* spatial derivative of the electric field as their single counterparts. However, the inductance of a quadruple coil is lower than that of a single coil with the same total number of turns, so that the quadruple coil draws a larger current and produces a stronger peak spatial derivative of electric field when used with a given stimulator (for a fixed stimulator voltage).

---

*Correspondence to: K. P. Esselle (Computing and Electronics), Macquarie University, Sydney 2109 Australia.

The shapes of quadruple coils that have been analyzed include four squares closely adjacent to each other or separated, four triangles, and a three-dimensional arrangement based on a quadruple triangular coil. The last produces the weakest hyperpolarization spot (about 20% of the peak depolarization).

OPTIMIZATION CRITERION

The voltage of the capacitors of a given magnetic stimulator is constant, and the time derivative of the coil current is proportional to this voltage and inversely proportional to the coil inductance. This dependence provides for the criterion of an optimization parameter for long nerves defined as

$$S = \frac{|\partial E_x / \partial x|_{\text{peak}}}{L\,(dI/dt)}$$

For any given coil shape, wire radius, number of coil turns, and capacitor voltage, parameter S defines how effective the coil is and permits optimization of the coil dimensions for a nerve at a certain depth below the interface. Similarly, coils of different shapes can be compared.

RESULTS

The optimization of the quadruple square coil is illustrated in Fig. 1. In addition to the parameter S, other factors should be considered in selecting a coil. The rate at which the peak value of $|\partial E_x / \partial x|$ decreases with distance from the interface is important when deep-lying nerves are excited. Figure 2 shows this parameter for various types of coils.

As far as the strength of the stimulus is concerned, unspaced quadruple triangular coils are preferred for shallow nerves, and spaced quadruple square or three-dimensional coils are preferred for deep nerves.

CONCLUSIONS

In stimulation of nerves with a pulsed magnetic field several parameters of the stimulating coil have to be considered to produce effective stimulation of a desired nerve. For long, straight nerves the activation function is the spatial derivative of the electric field along the neuronal axis. This derivative depends on the coil design, shape, and location with respect to the nerve. Ideally, the coil should produce a small region at a desired location of the nerve where the stimulus produces nerve depolarization. In practice, more than one region of depolarization as well as regions of hyperpolarization are produced. For a given stimulator (i.e., a bank of capacitors charged to a given voltage), the stimulus is inversely proportional to the coil inductance.

Simplified analyses described here, which rely heavily on analytical solutions in addition to simple numerical computations, provide tools for coil optimization and intercomparison. This approach can lead to a selection of the best coil geometry and location for selective stimulation of a desired long nerve. Furthermore, the physical insight gained from this analysis should prove useful for more complex numerical analyses of tissue models that better reflect tissue geometry and heterogeneous electrical properties. Quadruple coils of various shapes emerged from our evaluation as the optimal coils for stimulation of long straight nerves.

REFERENCES

1. L. A. Geddes, "History of magnetic stimulation of the nervous system," *J. Clin. Neurophysiol.* 8: 3-9, 1991.

2. K. P. Esselle and M. A. Stuchly, "Neural stimulation with magnetic fields: Analysis of induced electric fields," *Trans. IEEE* BME-39: 693-700, 1992.

3. B. J. Roth, R. Turner, L. G. Cohen, and M. Hallet, "New coil design for magnetic stimulation with improved focality," *Movement Disorders* 5 (Supp. 1): 32, 1990.

4. This work was performed at the Bureau of Radiation and Medical Devices, Health and Welfare Canada, Ottawa.

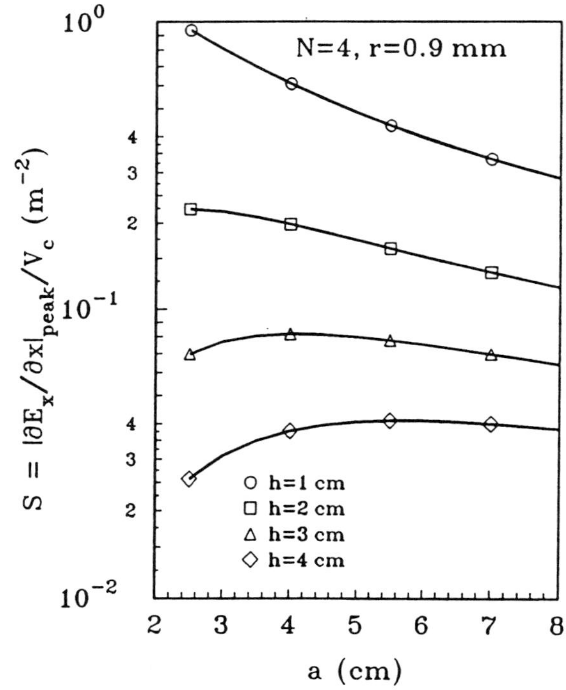

FIG. 1.—Peak electric-field spatial derivative per unit voltage S vs length of one coil section a, for 4-turn QS coils; h is depth of nerve.

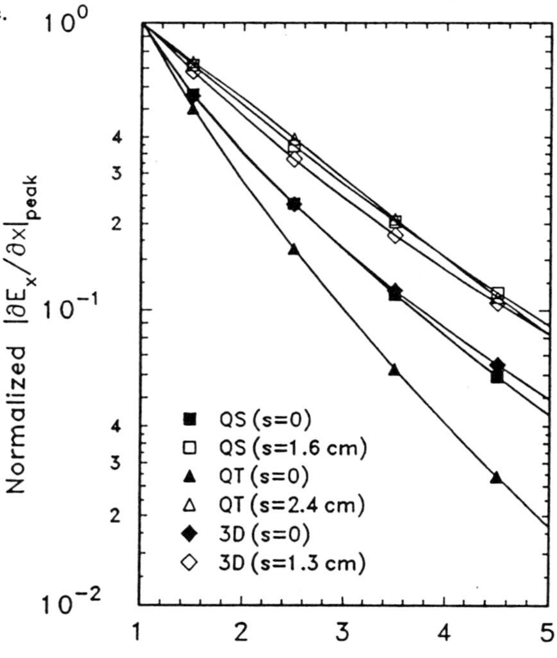

FIG. 2.—Peak electric-field spatial derivative normalized to value at 1cm depth, vs depth h measured from bottom of coil; $N = 40$.

Martin Blank, Ed., *Electricity and Magnetism in Biology and Medicine*

GENERALIZED ANALYTIC RECONSTRUCTION OF NEUROMAGNETIC IMAGING SYSTEM FROM DISCRETE SAMPLES

T. Obi,* M. Yamaguchi, N. Ohyama, and T. Honda

Tokyo Institute of Technology, Yokohama

K. Sekihara and Y. Ogura

Hitachi Ltd., Tokyo

A neuro current reconstruction method based on Generalized Analytic Reconstruction from Discrete Samples (GARDS) is proposed. GARDS is an algorithm that can perfectly describe the relation between an object in a continuous space and observed data in a discrete space. In this paper, we show first how GARDS is used in the neuromagnetic problem, and then the results of the computer simulation. From this simulation an evaluation method of the neuromagnetic imaging system is suggested.

The reconstruction of neural current distributions in a human brain from magnetic fields measured by SQUIDs has been studied by many researchers. Some researchers use a linear estimation approach based on a discrete-discrete space. This approach has the advantage that a current distribution, not several number of dipoles, can be treated. However, an image reconstructed by this method may suffer from artifacts because of the systematic error caused by making the current space discrete. Therefore, to avoid these artifacts, we have to use a continuous-discrete model that successfully describes the object in a continuous space and the measured data in a discrete space.[1,2]

In this paper, we show an approach to the estimation of the neural current distribution by the Generalized Analytic Reconstruction from Discrete Samples (GARDS),[1] a method by which objects can be estimated in continuous space from discretely measured data. The neuromagnetic imaging system can be thus modeled in a continuous object space and in a discrete measurement space. In addition, the GARDS includes explicitly a condition of an object size for the reconstruction and provides an evaluation of the neuromagnetic imaging system.

RECONSTRUCTION ALGORITHM

Let us define a continuous current distribution junction $\mathbf{j}(\mathbf{r})$ in the object space and represent magnetic data by an M-order column vector $\mathbf{B}$, where $\mathbf{B}$ is assumed to be measured by a set of M detectors. Defining a sensitivity function $\mathbf{h}i(\mathbf{r})$, which is derived from the law of Biot-Savart, we have

$$B_i = \int_v \mathbf{h}_i^T(\mathbf{r})\mathbf{j}(\mathbf{r})\,d\mathbf{r} \tag{1}$$

where v denotes the object space and the suffix shows the i-th element. Defining a linear forward mapping operator H, we may rewrite Eq. (1) as

$$\mathbf{B} = H\{\mathbf{j}(\mathbf{r})\} \tag{2}$$

Next, we define an adjoint operator $H^\dagger$ as

$$H^\dagger \mathbf{B} = \sum_{j=1}^{M} B_j\, \mathbf{h}_j(\mathbf{r}) \tag{3}$$

which means a single backward mapping. Because $HH^\dagger$ becomes a matrix, let us call it a GARDS matrix; its

*Correspondence to: Takashi Obi, Imaging Science and Engineering Laboratory, Tokyo Institute of Technology, Yokohama 227 Japan.

element is represented by

$$A_{ij} = [HH^{\dagger}]_{ij} = \int \mathbf{H}_i^T(\mathbf{r})\,\mathbf{h}_j(\mathbf{r})\,d\mathbf{r} \tag{4}$$

In order to derive an inverse operator, we apply singular value decomposition (SVD) technique to this matrix. From the eigenvalues $\{\lambda_i\}$ and eigenvectors $\{v_j\}$ $(i = 1 \sim R)$ in the measurement space, the reconstruction operator H^+ is easily derived as

$$\mathbf{H}^+ = \mathbf{H}^{\dagger} \sum_{i=1}^{R} \frac{1}{\lambda_i} \mathbf{v}_i \mathbf{v}_i^T \tag{5}$$

where the number R is a rank of the GARDS matrix. Operating H^+ to the measurement data **B**, we obtain an estimation of current distribution $\hat{\mathbf{j}}(\mathbf{r})$ as

$$\hat{\mathbf{j}}(\mathbf{r}) = H^+\{\mathbf{B}\} = H^{\dagger} \sum_{i=1}^{R} \mathbf{v}_i \frac{\mathbf{v}_i^T}{\lambda_i} \mathbf{B} \tag{6}$$

If the measured magnetic field contains noise, the reconstruction operator of Eq. (5) had better be modified to the regularized form. In this case, an optimum estimation is given by

$$\hat{\mathbf{j}}(\mathbf{r}) = H^+\{\mathbf{B}\} = H^{\dagger} \sum_{i=1}^{R} \mathbf{v}_i \frac{\mathbf{v}_i^T}{\lambda_i + c} \mathbf{B} \tag{7}$$

where c is positive and called a regularization constant. For a display of the estimation, we actually use a sampled adjoint operator $H_s^{\dagger}$ in Eq. (6).

SIMULATION

To confirm the effectiveness of our method, we carried out a computer simulation. The configuration used in the computer simulation is shown in Fig. 1. Here the object is supposed to a be circular current. Only the z component of the magnetic field is utilized in the reconstruction; 400 detectors are assumed. Figure 2 shows a comparison of estimated objects obtained by a discrete-discrete model and by a continuous-discrete model. From this comparison we see that the continuous-discrete model gives a better image than the discrete-discrete model, because the discrete-discrete model leads to artifacts caused by use of the improper model.

Next, we examine the effect on a reconstructed image caused by a change of the object space size. The results of estimation is shown in Fig. 3. In the image-reconstruction problems, the constraint of object space is generally said to improve the quality of reconstructed images. However, in the neuromagnetic imaging case, the image with the constraint may be degraded (Fig. 3b). The reason for the image degradation is as follows. Because the magnetic field decreases as the reciprocal of square of the distance between a detector point and a current point, the farther a detector is from the object, the smaller the sensitivity. As a result, the GARDS matrix elements corresponding to noneffective detectors are nearly zero and the pseudoinverse of GARDS matrix becomes more sensitive to the accuracy of calculation. In confirmation of this outcome, only half the eigenvalues and eigenvectors are utilized in the reconstruction with the same constraint. The result is shown in Fig. 4, which suggests that an optimum size of space exists.

CONCLUSIONS

We have presented a method to reconstruct current distribution in a continuos space from a discretely sampled magnetic field and confirmed the effectiveness of the continuous-discrete model. We also showed that the constraint of object space does not always improve the quality of reconstructed images.

REFERENCES

1. N. Ohyama and H. H. Barrett, *Signal Recovery and Synthesis,* New Orleans, La.: 105-107, April 1992.
2. M. S. Hämäläinen and J. Savars, *Phys. Med. Biol.* 32: 91-97, 1987.

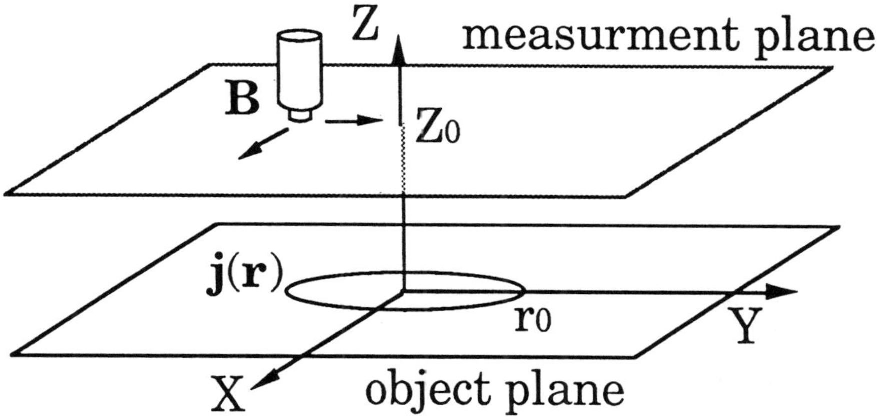

FIG. 1.—Configuration used in computer simulation (arbitrary units): measurement plane size, 20×20; original current distribution, a circle with 3 radius; distance between current plane and measurement plane, $z_0 = 3$.

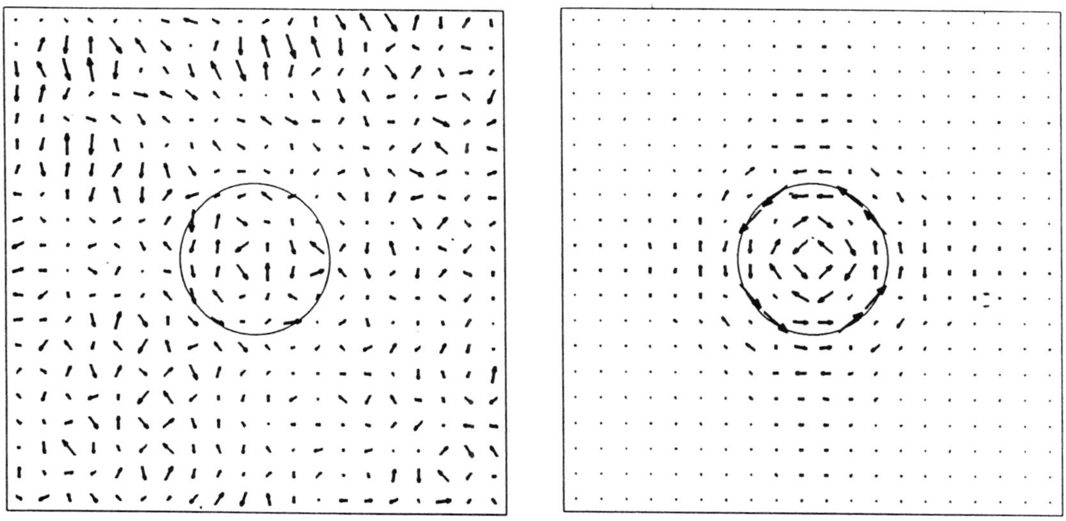

(a) (b)

FIG. 2.—Comparison of estimated object: (a) discrete-discrete model, (b) continuous-discrete model. Size of object plane, 20×20; number of sample points, 400.

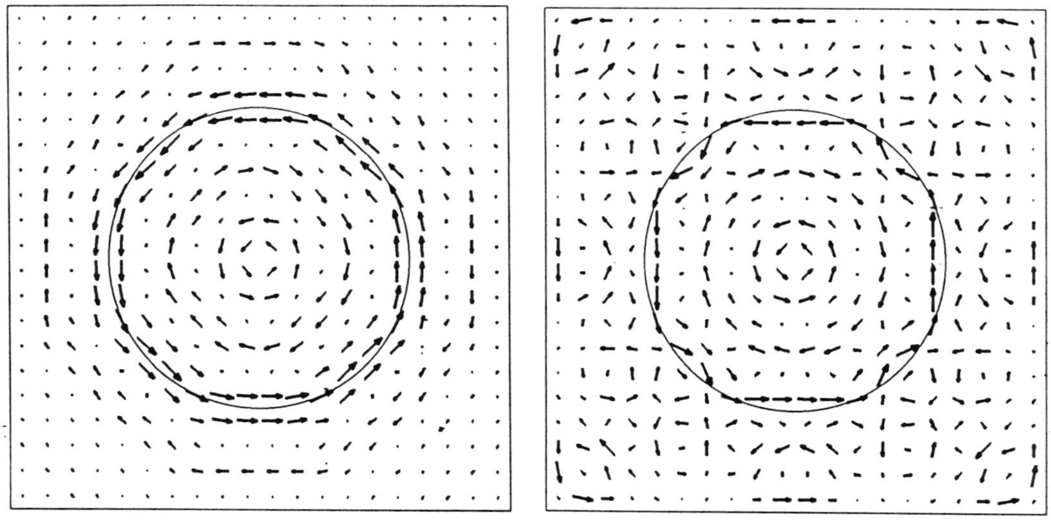

(a) (b)

FIG. 3.—Images of different object size: (a) 20×20, (b) 15×15. Region for display, 10×10.

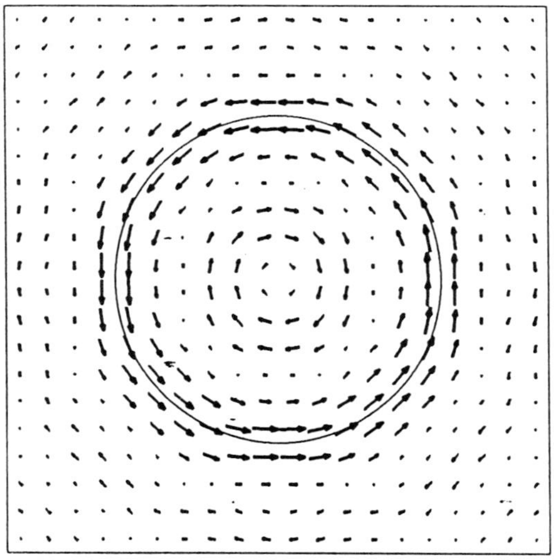

FIG. 4.—Reconstructed image based on only half the eigenvalues to reconstruct; size of object, 15×15.

Martin Blank, Ed., *Electricity and Magnetism in Biology and Medicine*

CORRELATION BETWEEN CEREBRAL CIRCULATION AND ELECTRICAL ACTIVITY (EEG) IN THE REGULATORY PHYSIOLOGICAL MECHANISM OF THE CENTRAL NERVOUS SYSTEM AFTER MICROWAVE EXPOSURE OF RATS

G. Thuróczy,* L. D. Szabo, G. Kubinyi, J. Bakos, and M. Bodo

National Research Institute for Radiobiology and Radiohygiene, Budapest

In acute experiments on rats (nembutal 40 mg/kg ip), electroencephalogram (EEG), rheoencephalogram (REG, as an index of cerebral circulation), temperature, and heart rate (HR) were measured immediately after 10 min of 2450MHz CW microwave exposure. The power densities were 10 and 30 mW/cm^2, respectively. The measured specific absorption rates (SAR) in the brain tissue of anesthetized rats were 8.3 ± 3.4 mW/g and 25 ± 5.2 mW/g, respectively. The increase in the EEG alpha (8-12 Hz) and beta (12-30 Hz) bands depended on the SAR exposure. The total power of EEG spectra increased after the 30mW/cm^2 exposure, but no changes were measurable at 10 mW/cm^2. The cerebral blood flow was activated (REG amplitude increased) after the 10mW/cm^2 exposure, which demonstrates the existence of regulation processes during microwave exposure even under anesthesia.

The correlation between the physiological modalities in microwave field-activated systemic or localized regulatory mechanisms with changes in the central nervous system (CNS) shows them not to be identical.[1,2] The quantitative evaluation of the phenomena related to CNS functions could be provided by the use of electrophysiological recording with a relevant signal-processing procedure. To clarify the mechanisms of field-activated changes in the variable physiological modalities and the fundamental mechanisms of interaction between the variables, we compared the changes of electrophysiological recordings of anesthetized and freely moving rats immediately after microwave exposure.

The psychophysiological background of the human subjective complaint has not been clarified; nor has the question, which physiological modalities are correlated with intensity and absorbed energy (SAR), carrier frequency, or modulation of microwave and RF radiation?[3,4] The compensation factors of regulation involved in the response to the radiation should play an important role in the mechanisms.

METHODOLOGY

Microwave Exposure and Dosimetry

The irradiations were performed in an anechoic room (2.55 × 1.8 × 2.9 m) with a standard horn antenna (G = 14 dB). The incident power toward the horn antenna was continuously measured by a directional coupler and a power meter. The spatial pattern of power density was measured by a radiation monitor (Narda 8616, 8623B probe). The exposure was performed in the near field of the horn antenna at 2.45 GHz. The SAR in the brain cortex was measured with a thermocouple (0.18 mm, 40 mV/°C). The average SAR in the cortex was 25 ± 5.2 mW/g at a power density of 30 mW/cm^2. The microwave frequency was 2.45 GHz CW; the duration of exposure was 10 min. The exposure levels at the rat's head were 10 and 30 mW/cm^2.

Electrophysiological Preparation and Recording

In acute experiments (5 in each group, nembutal 40 mg/kg ip), the following electrophysiological modalities were registered simultaneously: EEG (70 Hz/0.3 s), REG (time constant 3 s), brain tissue and rectal temperature, ECG (ECG-I from limbs, 150 Hz/0.3 s). The recorded signals were saved on analog memory and hard disk. The off-line signal and data processing were done on a microcomputer (IBM-AT).

---

*Correspondence to: G. Thuróczy, National Research Institute for Radiobiology and Radiohygiene, P.O. Box 101, Budapest H-1775 Hungary.

Signal and Data Processing

The outputs of the data processing are as follows:

- Heart rate (HR), beats/min from ECG R-R peak interval time
- Amplitude of rheoencephalogram (REG ampl.), minimum and maximum of 20 averaged REG pulse wave (as a measure of the CBF).[6]
- EEG frequency bands derived from the power spectrum by FFT analysis with 0.4Hz accuracy. The processing interval of the EEG signal was 10 × 2 s and the power spectra of the epochs were averaged in parallel. The separate frequency bands were delta (0.4-4 Hz), theta (4.4-8 Hz), alpha (8.4-14 Hz), and beta (14.4-30 Hz).

Before exposure (baseline) and 1, 10, 20, and 30 min after it, 20 s of registration were processed on each animal of the several groups. The results are expressed and illustrated as percentages of the *pre-exposure values*. The average and SD represent the values within the groups ($N = 5$). The variability of the pre-exposure values in the rat populations ($N = 3$, sham, CW with two doses) remained below the normal biological deviations and is without significance.

RESULTS

- The irradiation at 30 mW/cm^2 (SAR = 25 mW/g) in the brain causes an increase in EEG activity (sum of EEG frequency bands) and a slight increase in the delta band (0.4-4 Hz) (Fig. 1). We regard these changes as thermal effects.
- The 10mW/cm^2 exposure (8.3 mW/g in the brain) did not cause changes in any EEG frequency band. The 10mW/cm^2 exposure produced a slight temperature rise in the brain (Fig. 2); we regard it as a "quasi nonthermal" effect.
- The rheoencephalogram (REG) amplitude increased after both exposure levels (Fig. 3); we regard it as an increase in cerebral blood flow (CBF).
- The ECG profile and heart rate (HR) showed no changes after exposure (Fig. 4).
- The temperature increase in the cortex at 30 mW/cm^2 was definite, but at 10 mW/cm^2 the homeostasis mechanisms of the living organisms (especially in the brain) were also detectable (Fig. 2), which led to a slight temperature rise in the brain.

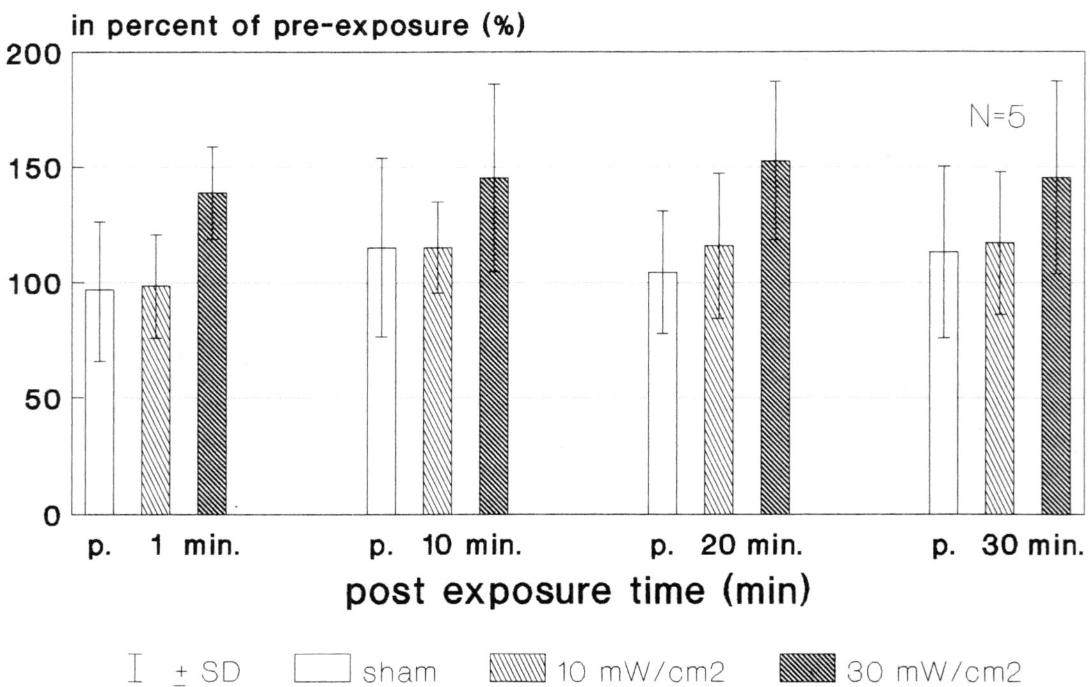

FIG. 1.—EEG activity changes after 10 min of 2.45GHz CW exposure; data are in percent of pre-exposure values.

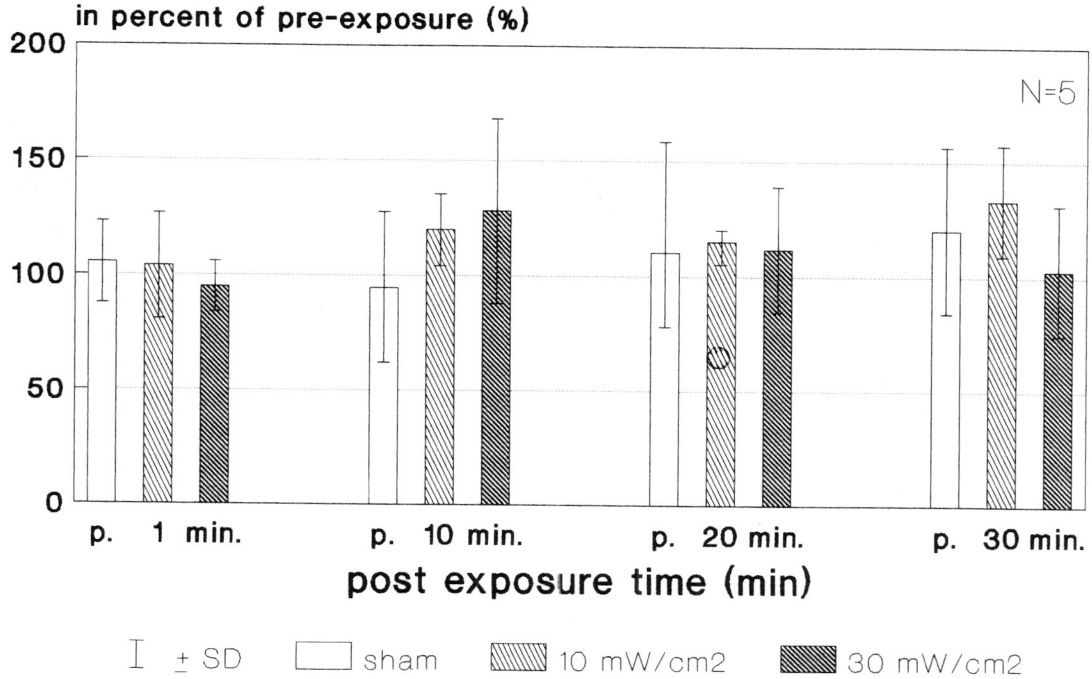

FIG. 2.—REG amplitude (as an index of cerebral blood flow) after 10 min of 2.45GHz CW exposure; data are in percent of pre-exposure values.

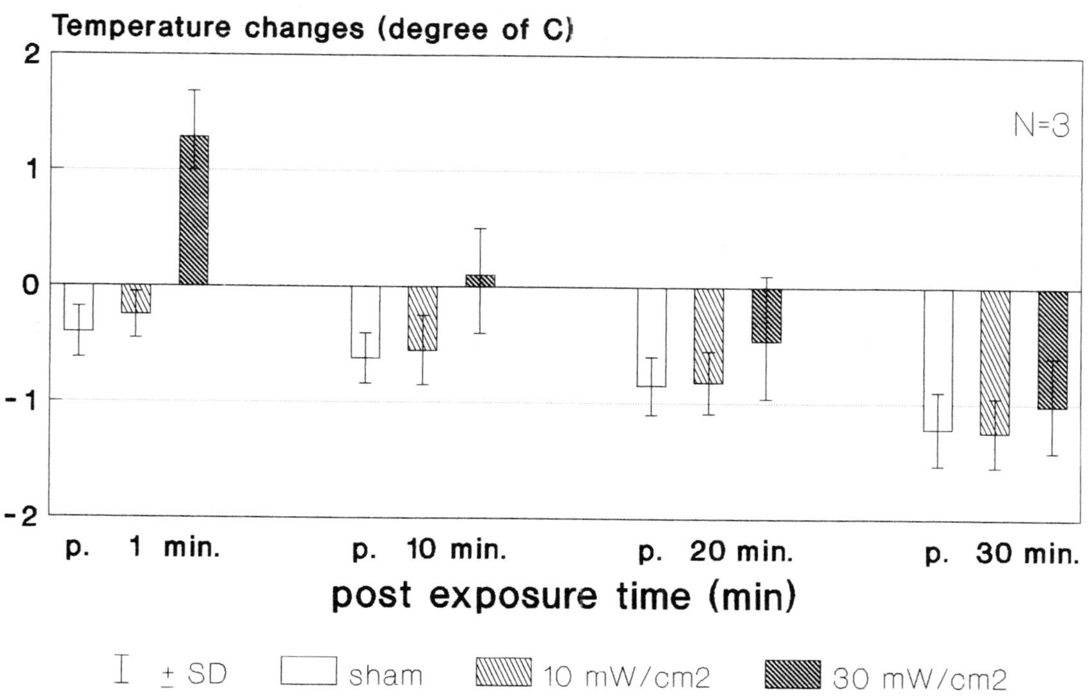

FIG. 3.—Brain temperature changes after 2.45GHz exposure to 10 and 30 mW/cm^2.

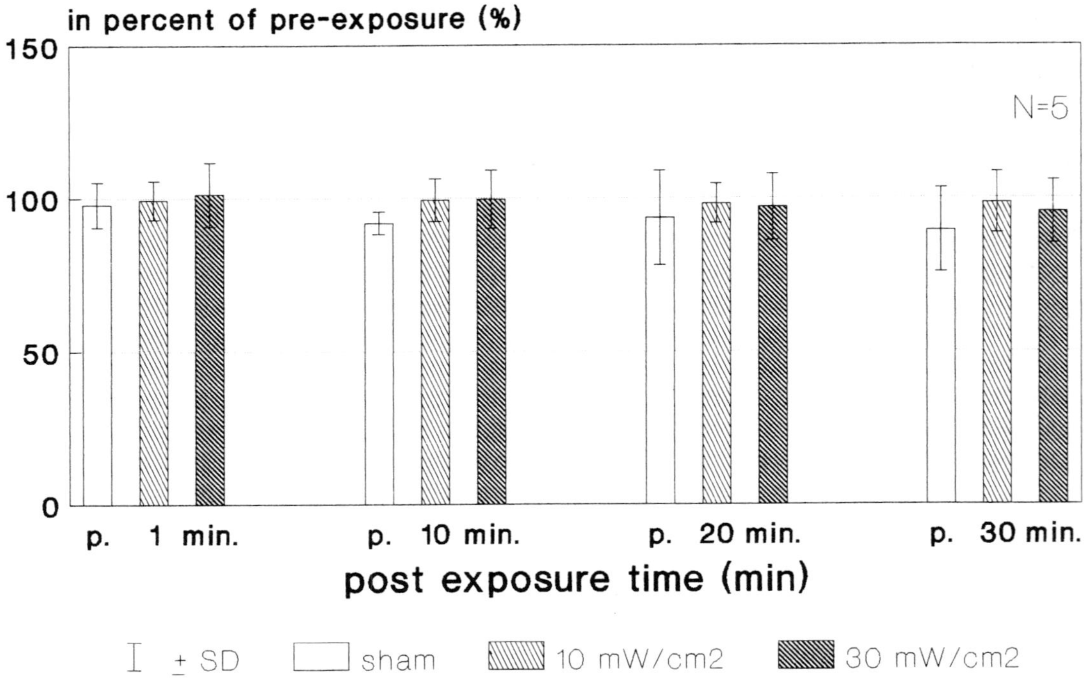

FIG. 4.—ECG heart-rate change after 10 min of 2.45GHz CW exposure; data are in percent of pre-exposure values.

DISCUSSION

At the thermal-level power density (30 mW/cm^2), the sum of EEG bands increased at the first stage of postexposure time. The CBF (REG ampl.) increased simultaneously. This regulation mechanism is probably also involved in the changes of EEG spectra during the thermal-level exposure. At an intensity of 10 mW/cm^2, the reaction of REG amplitude preceded the response of EEG spectra, which by itself demonstrates the existence of *compensation mechanisms* generated by CBF in the first stage of postexposure.

These results cannot be explained by the EEG (and the genesis of EEG) alone. The changes in the cerebral blood flow (REG ampl.) and tissue impedance showed presumable correlations between the physiological modalities measured in the regulation mechanisms. These correlations are very important with respect to the interpretation of the results in the literature.[2,4,7,8]

Because of the close correlation in the *circulation-metabolism function* system of the brain, the exposure did not cause considerable changes in the measured polygraphic modalities; below the physiological regulation in the brain stem, the compensation mechanisms can be detectable. However, at the thermal-level irradiation (30 mW/cm^2), the changes remained in the range of the physiological regulation, but the effects of metabolisms and circulation were more detectable. It is not clear whether the mechanisms and involvement of *thermoregulation system* are affected by microwaves due to the special energy absorption in the brain, as well as to the effects on the factors of the circulation-metabolism function system during microwave irradiation in correlation with the thermoregulation mechanism.

REFERENCES

1. R. P. Blackwell and R. D. Saunders, *Int. J. Radiat. Biol.* 50: 761-787, 1976.
2. R. A. Chizhenkova, *Bioelectromagnetics* 9: 337-347, 1988.
3. W. R. Adey, in J. B. Andersen, Ed., *Modern Radio Science,* Oxford: Oxford University Press, 1990, 1-36.
4. S. Takashima et al., in S. S. Stuchly, Ed., *Electromagnetic Fields in Biological Systems,* Ottawa, 1978.
5. G. Thuróczy et al., *Proc. URSI Int. Symp. on EM Theory,* Budapest: Akademiai Kiado, 1986, 241-244.
6. G. Thuróczy et al., *Proc. 6th IMEKO Conf. Measurement in Clinical Med.,* Sopron (Hungary), 1990, 307-308.
7. W. T. Joines, *Med. Res. Eng.* 12: No. 3, 1976.
8. K. J. Oscar et al., *Brain Res.* 204: 220-225, 1981.

Martin Blank, Ed., *Electricity and Magnetism in Biology and Medicine*

EXPERIMENTAL AND ANALYTICAL STUDIES ON MEMBRANE EXCITATION OF NERVE AXONS BY TRAPEZOIDAL MAGNETIC FIELD EXPOSURE

S. Ueno* and O. Hiwaki

Kyushu University, Fukoaka, Japan

This study focuses on the properties of nerve-excitation response to a series of pulsed electric fields induced by time-varying magnetic fields. Magnetic stimulation was carried out by a sequence of rectangular electric fields induced by trapezoidal magnetic fields. Nerve excitation processes were studied by means of changes in the duration of the plateau of the trapezoid. The results show that nerve excitation is suppressed when the plateau is shorter than a critical duration. An inhibitory effect was simulated well based on a neurophysiological model.

Magnetic stimulation of the human brain was first reported in 1985.[1] A focal and vectorial magnetic stimulation has been developed.[2-5] A theoretical model of nerve excitation by electromagnetic stimulation was proposed.[6] In this paper, we describe a study of the processes of nerve-excitation response to the magnetically induced rectangular electrical fields by changing the interval between two pulsed electric fields.

METHODS

A figure-of-eight coil[2-5] was used to carry out the localized stimulation and to control the direction of the stimulating currents. The basic idea is to concentrate induced eddy currents locally at a target by a pair of coils positioned outside the body, so that time-varying magnetic fields pass through the body in the opposite directions around the target (Fig. 1).[7] A coil was placed on the right clavicle to stimulate brachial plexus and the evoked electromyogram (EMG) from brachialis muscles of the right arm was recorded. The waveform of the current in the coil was trapezoidal, so that a sequence of rectangular electrical fields was induced. The pulse duration (T1, T3) was 100 μs and the interval (T2) between two pulsed electrical fields varied from 0 to 100 μs. We observed the change of the EMG amplitude with T2.

We analyzed the nerve excitation processes elicited by trapezoidal magnetic fields, using a model of magnetic stimulation of nerve fiber based on a myelinated nerve fiber model.[8,9]

RESULTS AND DISCUSSION

No obvious EMG signals were observed when T2 was shorter than 60 μs at the stimulating intensity of 10.3×10^2 T/s (Fig. 2); the T2 which gave the large EMG signals became shorter with the increase of stimulating intensity. Based on the nerve excitation model, we calculated the change in the threshold for nerve excitation related to T2; the threshold decreased with T2. According to the simulation of the change of neural membrane potential elicited by magnetically induced electric fields, the nerve was not excited when T2 was 30 μs, whereas at same intensity (current in the coil, 320 A), the nerve was excited when T2 was 40 μs. The results suggest that rapid reversals of electric fields suppress the transmembrane potentials of nerve fiber at a subthreshold level.

REFERENCES

1. A. T. Barker, R. Jalinous, and I. L. Freeston, "Non-invasive magnetic stimulation of human motor cortex," *Lancet* i: 1106-1107, 1985.

---

*Correspondence to: S. Ueno (Electronics), Kyushu University, Fukoaka 812 Japan.

2. S. Ueno, T. Tashiro, and K. Harada, "Localized stimulation of neural tissues in the brain by means of paired configuration of time-varying magnetic fields," *J. Appl. Phys.* 64: 5862-5864, 1988.

3. S. Ueno, T. Matsuda, and M. Fukiki, "Localized stimulation of the human brain by a pair of opposing pulsed magnetic fields," in S. J. Williamson et al., Eds., *Advances in Biomagnetism,* New York: Plenum, 1989, 529-532.

4. S. Ueno, T. Matsuda, and O. Hiwaki, "Localized stimulation of the human brain and spinal cord by a pair of opposing pulsed magnetic fields," *J. Appl. Phys.* 67: 5838-5840, 1990.

5. S. Ueno, T. Matsuda, and M. Fujiki, "Functional mapping of the human motor cortex obtained by focal and vectorial magnetic stimulation of the brain," *Trans. IEEE* MAG-26: 1539-1544, 1990.

6. B. J. Roth and P. J. Basser, "A model of the stimulation of a nerve fiber by electromagnetic induction," *Trans. IEEE* BME-37: 588-597, 1990.

7. S. Ueno, T. Tashiro, S. Kamise, T. Oosako, and K. Harada, "A paired coil configuration for localized hyperthermia of deep tissues," *Trans. IEEE* MAG-23: 2437-2439, 1987.

8. B. Frankenhaeuser and A. F. Huxley, "The action potential in the myelinated nerve fiber of *Xenopus laevis* as computed on the basis of voltage-clamp data," *J. Physiol.* 171: 302-315, 1964.

9. D. R. McNeal, "Analysis of a model for excitation of myelinated nerve," *Trans. IEEE* BME-33: 974-977, 1986.

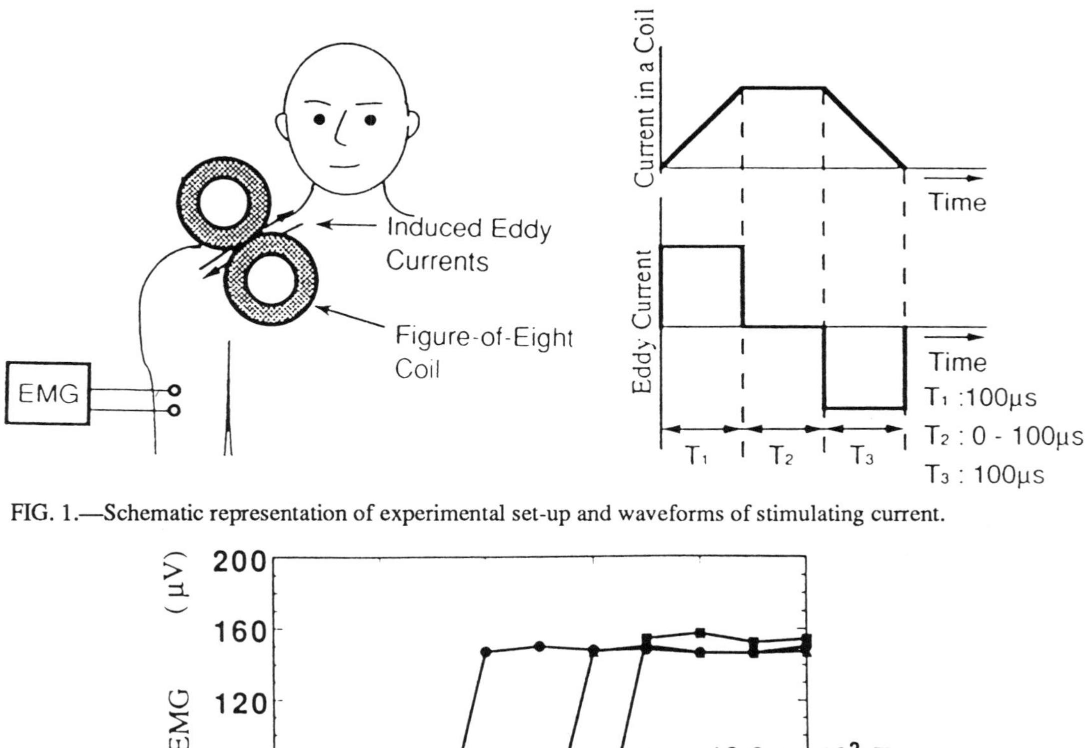

FIG. 1.—Schematic representation of experimental set-up and waveforms of stimulating current.

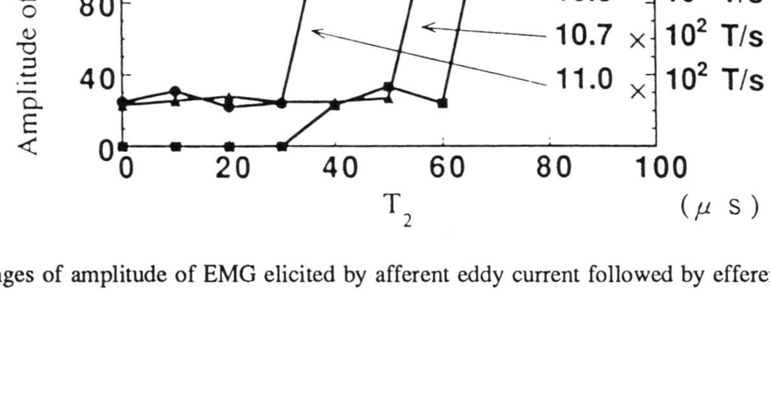

FIG. 2.—Changes of amplitude of EMG elicited by afferent eddy current followed by efferent eddy current with T2.

Martin Blank, Ed., *Electricity and Magnetism in Biology and Medicine*

ULTRAHIGH STATIC MAGNETIC FIELDS AND SIGNAL TRANSDUCTION IN THE LYMPHOCYTE

R. P. Liburdy,* M. S. Falk, and D. J. de Manincor

Lawrence Berkeley Laboratory, Calif.

Effects of ultrahigh static magnetic fields of 2.3, 7.5, and 9.0 T on calcium transport in resting and in mitogen-activated lymphocytes were evaluated. Cells were maintained in static fields at 37 ± 0.5 C for 30 min and calcium transport was followed with $^{45}Ca^{2+}$ in the absence or presence of mitogen (1 μg/ml Con-A). The 2.3T fields did not influence calcium transport; 7.5T field exposure was associated with a 45% increase in calcium transport in resting cells ($p = 0.001$) compared with control cells treated simultaneously in an isothermal, mu-metal-shielded waterbath located remote from the magnet (<30 mG static magnetic field flux density). The 9.0T fields resulted in a 26% decrease in calcium transport for resting cells and 15% for mitogen-activated cells. An apparent threshold for static magnetic field effects exists above 2.3 T.

Motivation for performing biosafety and bioeffects studies employing ultrahigh static magnetic fields derives in part from the use of NMR imaging technology in the medical setting.[1] Current FDA guidelines permit the use of static fields up to 2.0 T. There is interest in improving the spatial and temporal resolution of NMR imaging by use of static field in excess of 2.0 T.

Previously we investigated the effects of a complete NMR imaging environment on calcium transport in resting and mitogen-activated rat and in human lymphocytes.[2] Lymphocytes from rats and humans were found to be unresponsive to static fields in the resting state; they transported calcium in a manner identical to that for isothermal control cells. But when mitogen-activated rat and human lymphocytes were placed in NMR fields, they experienced a 22% and 30% increase in mitogen-activated calcium influx, respectively, compared with isothermal control cells.[2]

These findings led to studies in which two NMR imaging field components were evaluated separately. The RF (pulsed 64 μT, 100MHz carrier), and static (2.3 T) field components separately did not influence calcium transport.[3] In light of these findings we decided to investigate whether static magnetic fields greater than 2.3 T would influence calcium transport. The results of these studies are presented here.

METHODS

Preparation of rat thymocytes and procedures for performing the $^{45}Ca^{2+}$ transport assay have been described previously.[2,4] The superconducting magnet employed for 7.5 and 9.0 T field exposures is described in Ref. 5. Six Sprague-Dawley rats (250-375 g) were employed in these studies over a period of approximately 3 weeks. Three exposures to 7.5 T and three exposures to 9.0 T were performed with exposures randomly alternated on different days. Results presented here are pooled across these randomized experiments. Cells at 10^7/ml were placed in Eppendorf tubes secured on plastic rods and placed into a polyethylene bottle that served as a waterbath reservoir for distilled deionized water circulated from a remote waterbath circulator. This bottle was positioned in the middle third volume of the warm horizontal-bore magnet. Cells were maintained at 37 C for 30 min in the absence or presence of Con-A (1 μg/ml) during treatment in the magnet, or sequentially in a remote isothermal waterbath. This waterbath was shielded in mu-metal and cells inside experience a static magnetic field of <30 mG.

RESULTS

The importance of maintaining rat thymocytes at 37 ± 0.5 C is illustrated in Fig. 1. The temperature dependence of calcium transport in thymocytes at temperature between 25 and 40 C in the absence or

*Correspondence to: R. P. Liburdy, LBL Bioelectromagnetic Research Facility (Bldg 74), University of California, Berkeley CA 94720.

presence of Con-A (1 μg/ml) is different. Resting cells (no mitogen) are relatively temperature insensitive. Mitogen-activated cells are strikingly temperature dependent, with a maximal increase in calcium uptake observed at the physiologically significant temperatures of 37 and 38 C. For this reason we controlled temperature to 37 ± 0.05 C during magnetic field and control experiments.

The effect of exposure of thymocytes to a 7.5T static magnetic field is shown in Fig. 2. Resting cells in the magnetic field displayed an approximately 45% increase in basal calcium transport compared with isothermal control cells, $p = 0.001$ (left-hand side). This observation suggests that the resting state is potentially influenced by 7.5T fields. A comparison of the two open bar graphs indicates that Con-A itself resulted in a modest 16% increase in calcium influx in response to mitogen. During mitogen-activation in the absence or presence of the magnetic field, cells exhibited statistically indistinguishable levels of calcium influx (right-hand side). This finding is consistent with the null effect on mitogen-activated calcium influx we previously obtained at 2.3T.[3]

When thymocytes were exposed to 9.0 T, both resting and activated cells experienced a statistically significant reduction in calcium transport (Fig. 3). Resting cells displayed a 26% inhibition of basal calcium transport, $p = 0.000006$ (left-hand side). Activated cells showed an inhibition in calcium influx of 15%, $p = 0.0128$ (left-hand side). Comparison of the open-bar graphs indicates that Con-A itself resulted in an approximately 17% increase in calcium influx. This increase is comparable to that seen for control cells in the 7.5T experiments (Fig. 2). Comparison of the closed-bar graphs shows that cells in the 9.0T magnetic field experienced an approximately 34% increase in calcium influx due to Con-A. The levels of activation in the absence and presence of the field are modest and are not believed to be biologically significant. The above findings suggest that 9.0T magnetic fields inhibit basal calcium transport and inhibit calcium influx during mitogen activation.

DISCUSSION

Static magnetic fields can interact with bilayer membrane structures by inducing a conformational change in membrane elements that possess a significant magnetic moment.[5,6] However, biological membranes are complex, and the localized magnetic susceptibility tensors associated with the lymphocyte calcium pump or calcium ion channels in the bilayer are not yet defined. In our studies we have the added experimental consideration that the lymphocytes were not immobilized with regard to the direction of magnetic flux density during exposures; cells were free to orient in suspension. However, the transit time for a calcium ion moving through the bilayer is significantly less than the rotational diffusion constant associated with cell rotation.

Given these considerations, the most likely site of interaction remains the cell membrane, specifically the ordered structures associated with the calcium pump and calcium ion channels. Ordered repetitive domains are candidates for large summed magnetic-susceptibility tensors. Future studies are required to characterize completely the dose-response of lymphocytes to static magnetic fields at intermediate magnetic flux densities, such as 4.0 T. We plan to perform such studies.

REFERENCES

1. R. L. Magin, R. P. Liburdy, and B. Persson, Eds., *Biological Effects and Safety Aspects of Nuclear Magnetic Resonance Imaging and Spectroscopy, Annals N.Y. Acad. Sci.* 649, 1992.
2. R. P. Liburdy, in Ref. 1, 74-95.
3. *BEMS*, 1991.
4. R. P. Liburdy, *FEBS Letters* 301: 53-59, 1992.
5. R. P. Liburdy and T. S. Tenforde, *Radiation Res.* 108: 102-111, 1986.
6. T. S. Tenforde and R. P. Liburdy, *J. Theor. Biol.* 133: 385-396, 1988.

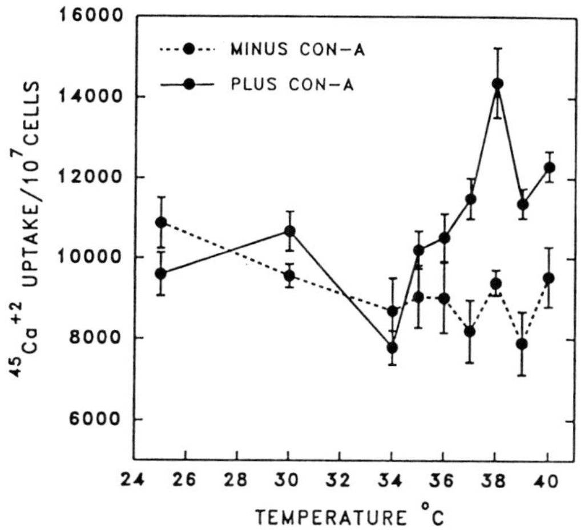

FIG. 1.—Effect of temperature on calcium transport in rat thymocytes. Mitogen-activated (1 μg/ml Con-A) calcium influx exhibits temperature sensitivity. Mean ± S.E.; n > 7.

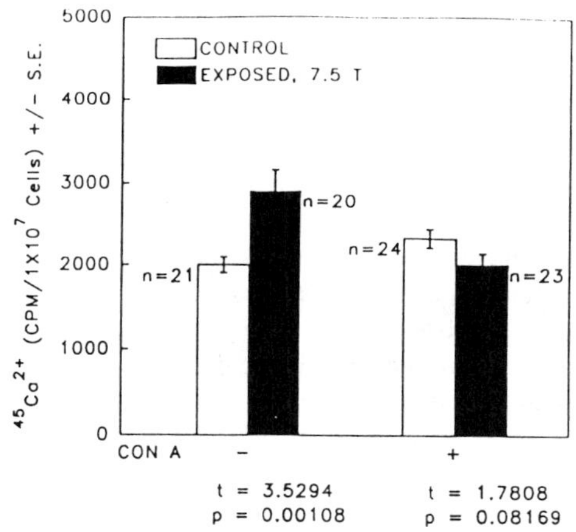

FIG. 2.—Effect of 7.5T static magnetic field on calcium transport in rat thymocytes. Mean ± S.E.

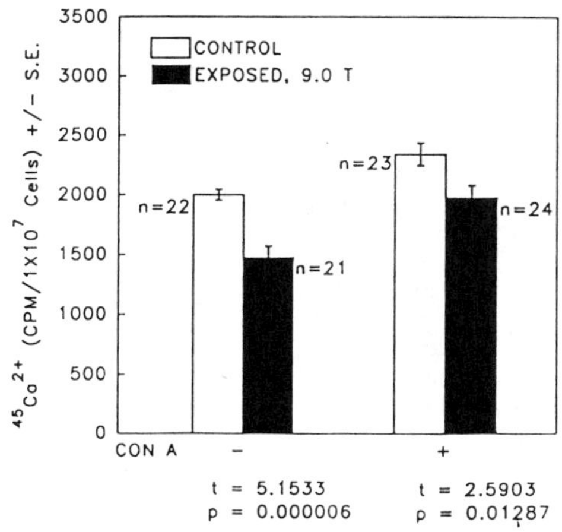

FIG. 3.—Effect of 9.0T static magnetic field on calcium transport in rat thymocytes. Mean ± S.E.

Martin Blank, Ed., *Electricity and Magnetism in Biology and Medicine*

MAGNETIC RESONANCE IMAGING AT 1.5T ATTENUATES OPIOID ANALGESIA: EVIDENCE FROM MORPHINE- AND AN ENKEPHALINASE INHIBITOR-INDUCED ANALGESIA IN SNAILS

F. S. Prato* and J. M. Wills

St. Joseph's Health Centre, London, Ont., Canada

M. Kavaliers, L. A. M. Galea, L. M. Saksida, and K.-P. Ossenkopp

University of Western Ontario

Accumulating evidence shows that magnetic fields can affect a variety of opioid-mediated behavioral and physiological functions. It has been shown that exposure to various magnetic stimuli, including those associated with magnetic resonance imaging (MRI), has significant inhibitory effects on exogenous and endogenous opioid mediated nociceptive responses of the land snail *Cepaea nemoralis* in a manner comparable to that described in vertebrates. In particular, exposure to diagnostic MRI procedures has inhibitory effects on both morphine-induced analgesia and the antinociception arising from the enhancement of endogenous enkephalin following administration of the enkephalinase inhibitor SCH 34826.

Evidence has accumulated that endogenous opioid systems, which are involved in the modulation of a broad range of basic functions, can be affected by magnetic fields. Among the more dramatic actions of magnetic stimuli in mammals are reversible modifications in the effects of exogenous opiates and endogenous opioid peptides. Natural geomagnetic disturbances arising from solar activity, earth strength, 0.5-1.5G 60Hz magnetic fields, relatively weak rotating magnetic fields, and magnetic fields associated with diagnostic magnetic resonance imaging (MRI) have all been shown to reduce the analgesic effects of morphine in mice.[1-6] These and additional findings from in vitro preparations,[7] avian orientation,[8] and spatial learning in rodents,[9] which relate the effects of magnetic fields to alterations in opioid activity, suggest that a broad range of fundamental opioid-mediated functions may be sensitive to magnetic stimuli.

Results of behavioral and physiological investigations with mollusks have shown that endogenous opioid peptides, as well as various exogenous opiate agonists and antagonists, have actions resembling those evident in mammals.[10] Morphine enhances the latency of response of the land snail *Cepaea nemoralis* to an aversive thermal stimulus (nociceptive sensitivity) in a manner analogous to that observed in rodents. The prototypic opiate antagonist naloxone suppresses and reverses these analgesic effects in morphine in *Cepaea*. Moreover, and as in mammals, exposure to stressful stimuli increases endogenous opioid activity, inducing naloxone-reversible analgesia.

As in mammals, exposure to various magnetic fields has been shown to attenuate opioid-induced analgesia in *Cepaea*.[10-12] In particular, exposure to MRI at relatively low magnetic field intensities (0.15T static field) attenuates morphine-induced analgesia similarly in both mice and snails.[6] The present report briefly considers the effects of magnetic resonance imaging at 1.5 T on morphine and endogenous opioid peptide (enkephalinergic) mediated antinociception in *Cepaea*.

PROCEDURES

Nociceptive sensitivity in hydrated snails was measured as the latency of a characteristic foot-lifting (nociceptive) response[10] to a warmed (38.5 ± 0.5 C) thermal surface. After basal nociception was determined, snails were injected into the haemocoel with either morphine sulfate (10 μg/2.0 μl saline) or the enkephalinase inhibitor SCH 34826 (SCH, 10 μg/2.0 μl saline). (SCH is an enkephalinase inhibitor that

*Correspondence to: F. Prato (Nuclear Medicine and Magnetic Resonance), St. Joseph's Health Centre, 268 Grosvenor St., London ON Canada N6A 4V2.

blocks the endopetidases that degrade enkephalins, leading to an enhancement of the functioning of the enkephalinergic system.[13,14]) The snails were then exposed to a 22min multislice MRI procedure [1.5T static field; 63MHz RF field; gradients, i.e., time-varying magnetic fields (TVMF), of 8.0 mT/m with 1ms rise times repeated at ELF frequencies], or a sham procedure designed to mimic the sound pressure and vibration produced by the imaging procedure (the main magnetic field was ramped down from 1.5 T to 0 but with a remnant field of 0.4 mT remaining), after which their nociceptive responses were again recorded.

RESULTS AND CONCLUSIONS

Sham-exposed snails treated with either morphine or SCH were analgesic, displaying significantly greater antinociceptive responses than either prior to injection or of saline-treated (not shown) animals (Fig. 1). In separate experiments not reported here, morphine-induced analgesia was reversible by naloxone, whereas the analgesia induced by the met-enkephalinase inhibitor SCH was blocked by the specific delta opioid (met-enkephalin) antagonist ICI 154, 129. This latter observation shows that the antinociception arising from SCH treatment was associated with the augmentation of endogenous enkephalin activity through the blockade of degradory endopetidases.

In contrast to the sham-exposed snails, animals exposed to the MRI procedure displayed a marked attenuation of morphine and SCH-induced analgesia (Fig. 1). This finding shows that MRI procedures can attenuate both exogenous opiate and endogenous opioid (delta opioid, enkephalin) mediated analgesia in *Cepaea* (and likely in vertebrate systems). Further experiments revealed that if exposure was confined to the TVMF, most of the MRI effect could be accounted for by the magnetic fields produced by the MRI gradients—a result similar to that we observed at 0.15 T both in mice[5] and snails.[6]

These alterations in opioid activity are likely to encompass alterations in calcium ion flux, protein kinase C activity, and other second messenger systems that are associated with opioid signal transductions and have been previously shown[11,12] to be sensitive to magnetic stimuli in *Cepaea*.[15]

REFERENCES

1. M. Kavaliers and K.-P. Ossenkopp, *Neuropharmacology* 89: 440-443, 1984.
2. M. Kavaliers and K.-P. Ossenkopp, *Naun. Schmiedbergs Arch. Pharmacol.* 336: 308-315, 1987.
3. K.-P. Ossenkopp and M. Kavaliers, *Brain Res.* 418: 356-360, 1987.
4. K.-P. Ossenkopp, M. Kavaliers, and M. Hirst, *Neurosci. Lett.* 40: 321-325, 1983.
5. F. S. Prato, K.-P. Ossenkopp, and M. Kavaliers, *Mag. Reson. Imag.* 5: 9-14, 1987.
6. F. S. Prato et al., *Ann. N.Y. Acad. Sci.* 649: 44-58, 1992.
7. G. P. Golding et al., *Neuropeptides* 5: 357-358, 1985.
8. F. Papi, P. Lusci, and P. Limota, *J. Exp. Biol.* 1666: 169-179, 1992.
9. M. Kavaliers et al., *J. Comp. Physiol.* (in press).
10. M. Kavaliers and K.-P. Ossenkopp, *Biol. Bull.* 180: 301-309, 1991.
11. M. Kavaliers and K.-P. Ossenkopp, *J. Comp. Physiol.* 162: 551-558, 1989.
12. M. Kavaliers, K.-P. Ossenkopp, and D. M. Tysdale, *Brain Res.* 554: 65-71, 1990.
13. R. E. Chipin et al., *J. Pharmacol. Exp. Therp.* 245: 829-838, 1988.
14. L. M. Saksida, L. A. M. Galea, and M. Kavaliers, *Peptides* (in press).
15. We thank Dr. A. Barnett and Schering-Plough Research for supplying the SCH 34826. The work described here was supported in part by a Medical Research Council of Canada grant to F.S.P. and Natural Science and Engineering Research Council of Canada grant to M.K. and K.-P. O.

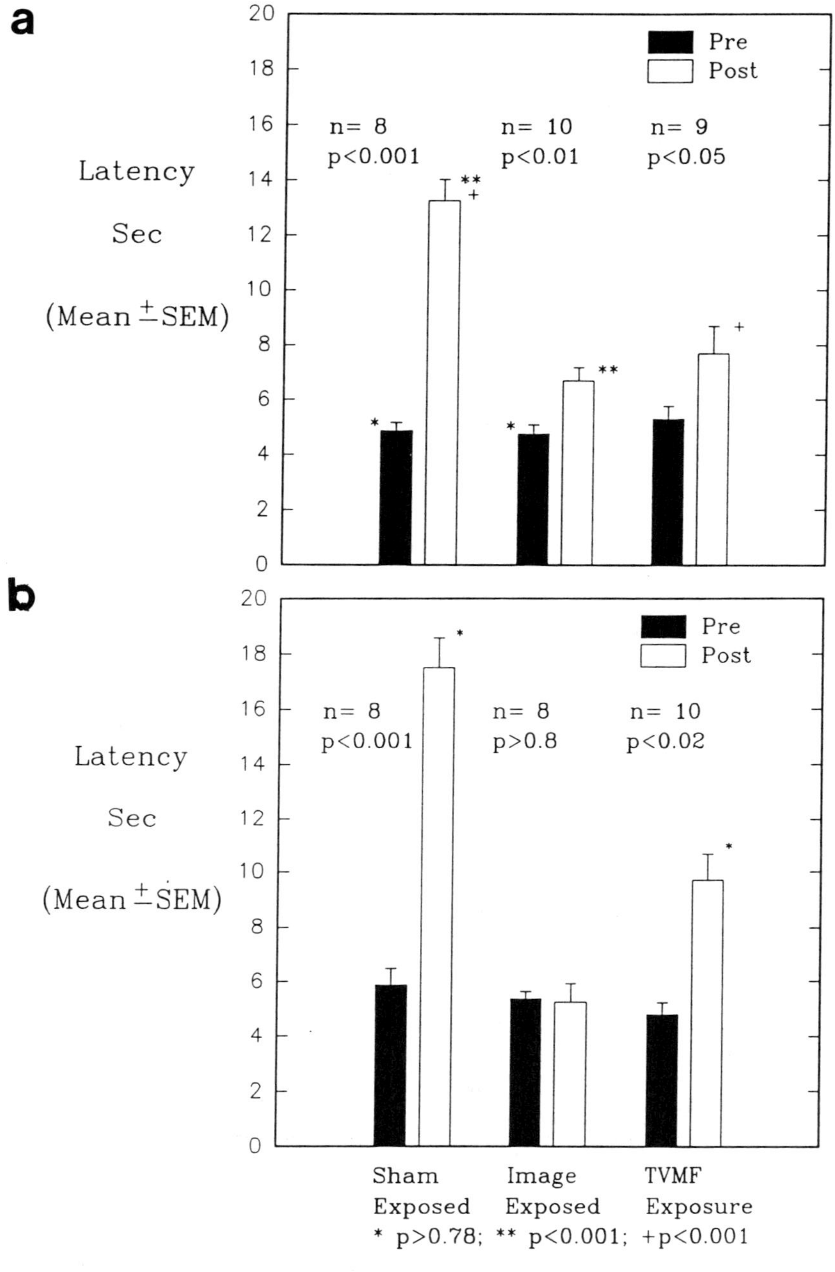

FIG. 1.—Mean time snails remained on a warm thermal surface (38.5 C) prior to foot-lifting. Snails were pre-tested (Pre) and 30 min later injected with (a) morphine (10 µg/2 µl saline) or (b) the enkephalinase inhibitor SCH 34826 (10 µg/2 µl saline), and then exposed for 22 min to a sham, full MRI, or just the TVMF. The morphine-induced analgesia, as seen in the sham-exposed snails (a), was significantly attenuated by exposure to a complete MRI procedure ($p < 0.001$) or by the TVMF ($p < 0.001$); the SCH 34826-induced analgesia, as seen in the sham-exposed snails (b), was completely eliminated by exposure to a complete MRI procedure ($p < 0.001$) and significantly attenuated by exposure to the TVMF ($p < 0.001$).

Martin Blank, Ed., *Electricity and Magnetism in Biology and Medicine*

ATTRACTION OF PARAMAGNETIC ERYTHROCYTES DURING FLOW IN AN INHOMOGENEOUS MAGNETIC FIELD

T. Shiga and A. Seiyama

Osaka University, Suita, Osaka, Japan

M. Okazaki

Government Industrial Research Institute, Nagoya, Japan

N. Maeda

Ehime University, Ehime, Japan

The effect of an inhomogeneous magnetic field on the flow of erythrocytes was systematically studied, with three experimental set-ups. The attraction of paramagnetic erythrocytes in an inhomogeneous magnetic field of 10-40 T^2/m can be detected. The degree of attraction is proportional to (a) the product of magnetic field strength and its spatial gradient, (b) the magnetic susceptibility to hemoglobin, and (c) the reciprocal of the flow velocity; it also depends on the hematocrit. Under an inhomogeneous magnetic field of above 100 T^2/m, such an asymmetric distribution of venous deoxygenated blood flow may induce some biological effect.

Three categories of the effect of magnetic fields on the blood stream may be considered. (a) The magneto-hydrodynamic effect may affect the rapid blood flow in a large vessel; e.g., a theoretical calculation predicts a 10% decrease of aortic flow velocity in a 5T magnetic field.[1,2] However, no experimental proof is available. (b) The diamagnetic interaction is known to modify the orientation of matter due to diamagnetic anisotropy in a strong uniform magnetic field; e.g., the flowing deoxygenated sickle erythrocytes orient at 0.35 T,[3,4] whereas normal discocytic erythrocytes orient at several Teslas.[5] (c) The paramagnetic interaction occurs only with paramagnetic erythrocytes. Magnetic separation of erythrocytes containing oxidized hemoglobin has been developed[6-8] by application of a very high spatial gradient of the magnetic field.

We have dealt with the paramagnetic interaction between a spatially inhomogeneous magnetic field and the blood stream[9-12] from the viewpoint of safety. Our intention was to determine, if possible, the safety limit of magnetic attractive force acting on flowing blood. The magnetic properties of erythrocytes are paramagnetic upon deoxygenation but diamagnetic upon oxygenation; the magnetic effects on the blood stream should thus depend on the spin state of hemoglobin. Here, we summarize the results of our studies, carried out with three types of in vitro experiments under a spatially inhomogeneous magnetic field: (1) the displacement of a narrow stream line of erythrocyte suspension in a wide laminar buffer flow was measured, (2) the asymmetric distribution of a flowing erythrocyte suspension in a cylindrical vessel was monitored, and (3) the acceleration of the erythrocyte sedimentation rate was detected. On the basis of these studies, we can estimate the limit of the paramagnetic interaction between the blood flow and an inhomogeneous magnetic field.

MATERIALS AND METHODS

Four types of erythrocytes, of different spin states of hemoglobin, were prepared from freshly drawn human blood after removal of plasma and buffy coat.[9-11] The erythrocytes were suspended in an isotonic buffer containing glucose; four different species were employed: (a) diamagnetic erythrocytes with oxygenated hemoglobin ($S = 0$), (b) erythrocytes with deoxygenated hemoglobin ($S = 2$), and those with high-spin methemoglobin ($S = 5/2$) and low-spin methemoglobin ($S = 1/2$).

---

*Correspondence to: ??..

An inhomogeneous magnetic field was produced with an electromagnet and a triangular iron block. Two flow systems were used (Fig. 1). One was a rectangular flat cell to establish a wide laminar buffer flow, in which a narrow stream line (ca 80 μm) of erythrocyte suspension was made, and the displacement of the stream line due to the paramagnetic attraction was measured by a microscopic photography.[9,10] The other was a glass cylinder ($\phi = 2$ mm) with a side branch from which a small portion of the erythrocyte flow was taken up to detect the change in the distribution of flowing erythrocytes, and the increase of erythrocytes due to the paramagnetic attraction was quantified by determination of hemoglobin concentration in the side branch.[11] The erythrocyte sedimentation rate was measured with a Westergren glass tube,[13] which was placed along the iron block.

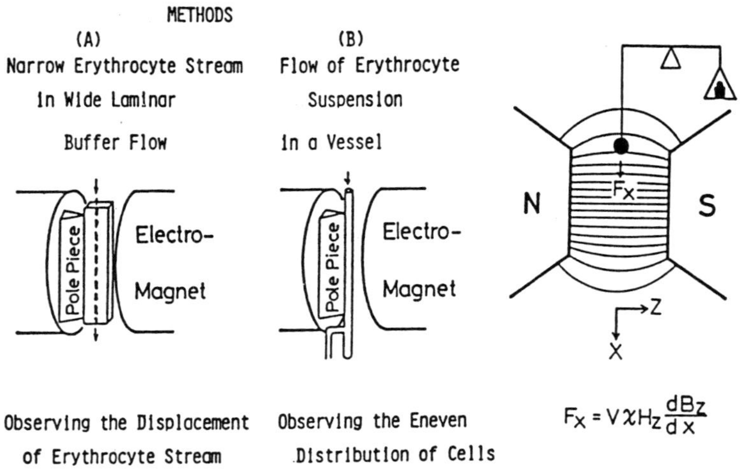

FIG. 1.—Flow systems attached to electromagnet: (A) Rectangular glass microslide (4 × 150 × 0.4 mm); (B) glass tube (OD 4.0, ID 2.0 mm) is connected to metacrylate block with main (ID 2.0) and side (ID 1.5 mm) branches. Magnetic field strength was measured with a gaussmeter whose Hall probe was moved with a micromanipulator. Right: Principle of Faraday's method for the determination of the magnetic susceptibility of material, from the magnetic attraction.

RESULTS AND DISCUSSION

The Displacement of a Paramagnetic Erythrocyte Stream Line in Laminar Flow[9,10]

Figure 2 shows the degree of displacement of the stream line (of erythrocyte suspension) in the rectangular flow cell (Fig. 1A). This displacement was proportional to (A) the product of the magnetic field strength and its spatial gradient, (B) the magnetic susceptibility of hemoglobin derivatives in erythrocytes, and (C) the reciprocal of the flow velocity. Furthermore, it was parallel to (D) hematocrit of the suspension.

Figure 3 illustrates the forces acting on a flowing paramagnetic particle. The force F_{mag} is the attractive magnetic interaction, neglecting the Lorentz force and the induced Hall effect (because they are relatively small). The frictional Stokes force F_f is equilibrated to F_{mag} within a short time. From these two equations, the velocity of displacement u is obtained, and by integrating u along the path (y axis) one can calculate an estimate of displacement L.

Although the dependencies of the displacement of the erythrocyte stream line on the parameters (Fig. 2) accord with the theory, the observed displacement was much larger than the calculated value for a single erythrocyte. Moreover, the displacement depended on the hematocrit of the erythrocyte suspension. These results can be interpreted by the assumption that a "group" of erythrocytes (in a "volume") are attracted as a whole by the magnetic force.

Asymmetrical Distribution of Flowing Paramagnetic Erythrocytes in a Vessel[11]

The attraction of erythrocytes due to a strong inhomogeneous magnetic field was observed with the set-up shown in Fig. 1(B). The quantity of erythrocytes skimmed into the side branch, which was branched rectangularly just below the magnetic field, was measured before and after exposure to the magnetic field. The change in the amount of the skimmed erythrocytes ΔV is defined as

$$\Delta V = [(\text{amount with field})/(\text{amount without field})] - 1$$

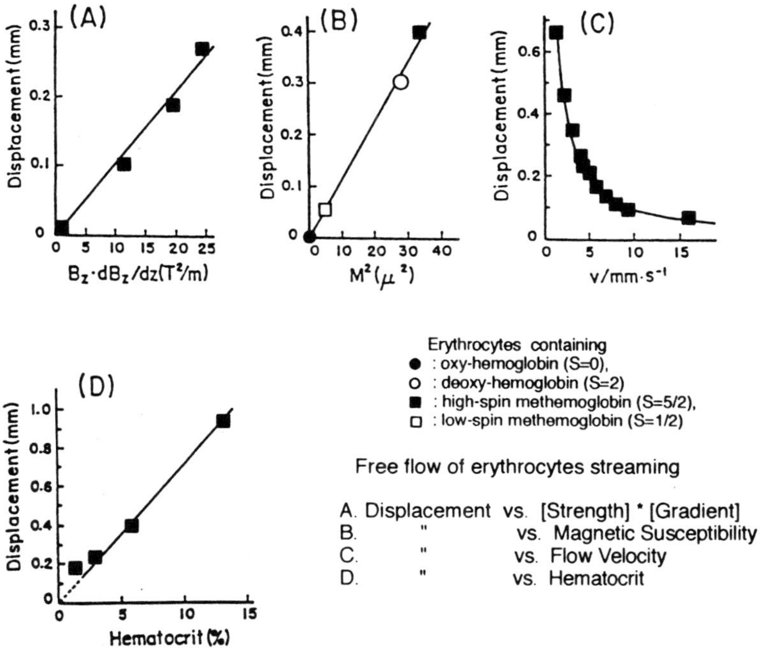

FIG. 2.—Displacement of erythrocytes stream line due to magnetic field: (A) $B_z * dB_z/dz$; (B) vs $M^2=\chi$; (C) vs flow velocity v; and (D) vs hematocrit of suspension.

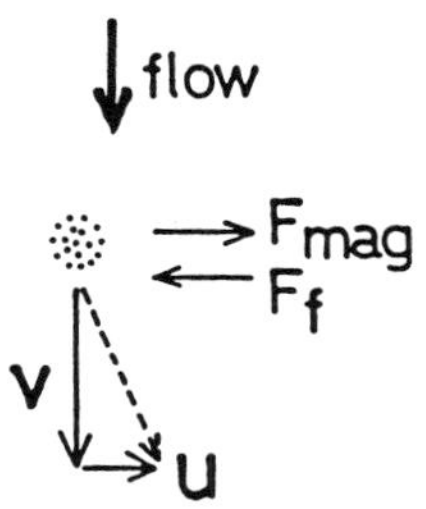

$$F_{mag} = \chi V(\mathbf{H}\cdot\nabla)\mathbf{B}$$

$$\chi_p = \frac{M^2 N \mu_w}{3kT}$$

$$F_f = 6\pi R \eta u$$

$$L = C\frac{M^2 N}{\eta T R v}\cdot\int_{y_0}^{y_1} H_z\,(dB_z/dz)dy$$

FIG. 3.—Forces acting on flowing paramagnetic particle. χ: magnetic susceptibility, V: volume, H: magnetic-field intensity, B: magnetic-flux density, M: magnetic moment, N: number of the paramagnetic center in unit volume, μ: magnetic permeability, R: hydrodynamic radius of moving particle, η: viscosity of the solvent, u, velocity of displacement due to magnetic field, v: flow velocity, L: displacement of particle due to paramagnetic attraction during flow from y_0 to y_1.

Figure 4 shows the results. ΔV increases proportionally (A) with the product of magnetic field strength and its spatial gradient, but saturates at weak magnetic field (20 T^2/m); (B) with the magnetic susceptibility of hemoglobin derivatives in erythrocytes; and (C) with the reciprocal of the flow velocity. Furthermore, the effect is maximal (D) at the optimal hematocrit of ca 5%. These phenomena are essentially explained by the paramagnetic attraction proportional to the magnetic susceptibility of erythrocytes. The saturation at high field and the decreased effect at high hematocrit may be due to collisions toward the wall and/or among erythrocytes, and to the resultant eddy current.

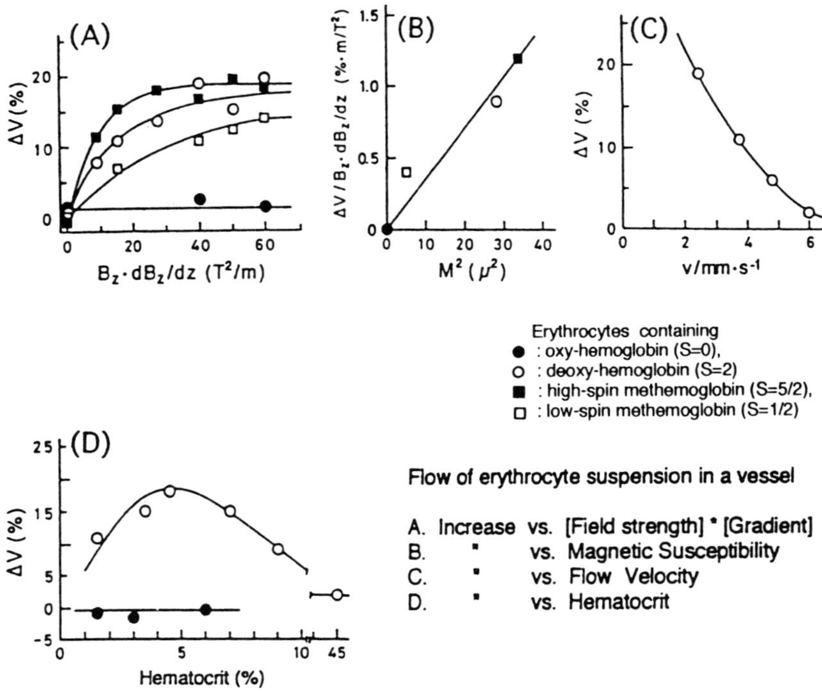

FIG. 4.—Asymmetric distribution of flowing erythrocyte suspension in a vessel due to magnetic field: (A) vs $B_z * dB_z/dz$, (B) vs M^2, (C) vs flow velocity, and (D) vs hematocrit of suspension.

Acceleration of Erythrocyte Sedimentation Rate[12]

The sedimentation rate of paramagnetic erythrocytes in a Westergren tube[13] increased with a spatially inhomogeneous magnetic field. This phenomenon is similar to so-called Boycott effect,[14] i.e., the accelerated sedimentation rate in an inclined cylinder or channel. The detailed mechanism of this effect is not fully understood.[15]

REFERENCES

1. I. I. H. Chen and S. Saha, *J. Bioelectricity* 3: 293-298, 1984.
2. V. K. Sud, P. K. Suri, and R. K. Mishra, *Studia Biophys.* 69: 175-186, 1978.
3. M. Murayama, *Nature* 206: 420-422, 1965.
4. A. Brody et al., *Invest. Radiol.* 20: 560-566, 1985.
5. Y. Yamagishi et al., *Physica B* (in press).
6. J. A. Oberteuffer, *Trans. IEEE* MAG-9: 303-306, 1973.
7. D. Melville, F. Paul, and S. Roath, *Nature* 225: 706-707, 1975.
8. E. H. Dunlop, W. A. Feiler, and M. J. Mattione, *Biotech. Advs.* 2: 63-74, 1984.
9. M. Okazaki, N. Maeda, and T. Shiga, *Experientia* 42: 842-843, 1986.
10. M. Okazaki, N. Maeda, and T. Shiga, *Eur. Biophys. J.* 14: 139-145, 1987.
11. M. Okazaki et al., *Physiol. Chem. Phys. Med. NMR* 20: 3-14, 1988.
12. M. Okazaki et al., *J. Colloid Interface Sci.* 146: 590-593, 1991.
13. Int. Committee, *Br. J. Haematol.* 24: 671-674, 1973.
14. A. E. Boycott, *Nature* 104: 532-533, 1920.
15. The work was supported in part by grants from the Ministry of Education, Science, and Culture of Japan; Nissan Science Foundation; and Ehime Health Foundation.

Martin Blank, Ed., *Electricity and Magnetism in Biology and Medicine*

DYNAMICS OF WATER UPTAKE BY TROUT (*Salmo trutta* L.) EGGS EXPOSED AFTER ACTIVATION TO MAGNETIC FIELD

K. Formicki,* A. Sobociński, and A. Winnicki

University of Agriculture, Szczecin, Poland

Fish eggs after activation absorb water and the perivitelline space is formed. The effect of a steady magnetic field (250 mT) on the course of perivitelline space formation was investigated. It was found that in the magnetic field water absorption by eggs is greatly slowed down, particularly in the preliminary phase of egg swelling (20-25 min). The magnetic field exerts no effect on the final balance of water absorbed by the egg and only slows down its absorption, thus greatly shifting in time the final formation of the perivitelline space.

Freshly shed fish eggs absorb water. The length of formation time of the perivitelline space and the amount of water absorbed by the egg depend on the specificity of fish species and on environmental conditions. In the case of salmonids, the volume of eggs after they have been shed into water increases by an average of 12-16%; the water-absorption time is synchronized with the increase in the resistance of egg membranes, caused by the hardening enzyme.[1]

Water penetration into the eggs proceeds in two stages.[2] The perivitelline gap (under the egg membrane) contains a small amount of an osmotically active substance. Osmotic pressure caused by this substance is sufficient for a small volume of water to penetrate into the perivitelline gap. The cortical alveoli then burst and hydrophilic colloids released from them[3] insure further water absorption by the egg. The hardening egg membrane loses elasticity and the water absorption process is inhibited. The egg assumes a spherical shape, and the high degree of turgor[1] insures protection of the germinal disk from mechanical damage.

Some environmental factors (e.g., temperature, salinity, pH) may affect the rate of water absorption by the egg. Thus, it was of interest to determine whether a steady magnetic field also affects this process, the more so since according to earlier results[4-8] such a field exerts an effect on many biological processes in animals, including fish.

METHODS

Trout (*Salmo trutta* L.) eggs were fertilized immediately after hatching and were divided into two equal samples (from the same females), whereupon they were activated in water. Activation and swelling of eggs of the experimental series took place in a steady magnetic field of 200 or 250 mT generated by an electromagnet; eggs in the control series were kept under identical conditions but out of the magnetic field.

At definite time intervals (from 5 to 20 min) from both samples several tens of eggs were collected, their volume was determined with measuring pipettes to within 0.01 cm^3, and the volume of one egg was calculated. All experiments were carried out at a constant temperature of 9 C.

In the statistical treatment of the results, the Student's *t*-test and a multiple-regression analysis were applied.

RESULTS

There are differences in the dynamics of water absorption between trout eggs exposed to magnetic field and control eggs (Fig. 1). In the initial period (up to 20 min) water absorption by eggs exposed to the magnetic field is slowed down. Subsequently, there appears a tendency for equalization of the water absorption rate, so that by the end of the experiment (180 min) the volumes of eggs exposed to magnetic field and control eggs are closely similar.

---

*Correspondence to: K. Formicki (Fish Anatomy and Embriology), University of Agriculture, W. K. Królewicza 3/4, 71-550 Szczecin, Poland.

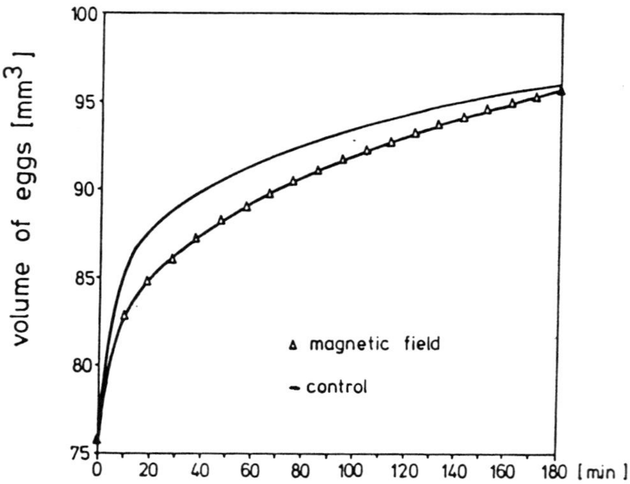

FIG. 1.—Water uptake by trout (*Salmo trutta* L.) eggs: in magnetic field, $y = 76.01889 + 3.02983\,t^{0.3628}$ ($p < 0.001$); in controls, $y = 76.35907 + 5.24454\,t^{0.2556}$ ($p < 0.001$).

DISCUSSION AND CONCLUSIONS

The present results indicate that the magnetic field influences water absorption by eggs immediately after their activation. This process does not affect the total balance of absorbed water, but consists only in slowing down the absorption during the first few minutes after activation. What causes this slowing down? It is difficult to relate this phenomenon to the effect of the magnetic field on the osmotic activity of substances in the perivitelline gap, which are decisive in the "start" of the water absorption process.[2] Moreover, earlier closing of the micropyle (main "inlet gate" immediately after activation) has to be ruled out, because it would adversely affect the fertilization process, whereas fertilization has been found to proceed more effectively in the magnetic field.[9] The slowing down of the water absorption process may be related to the effect of the magnetic field on the liquid crystals from which biological membranes (susceptible to the action of steady magnetic field) are formed.[10-13] The effect of magnetic fields of medium strength manifests itself by a reduction of the permeability of biological membranes. However, this outcome is contradicted by our recent experimental results indicating that when isolated trout egg membranes are exposed to the magnetic field, their water permeability increases.

It seems therefore that the slowing down of the formation of the perivitelline space by exposure to a magnetic field, as observed in the present experiments, arises from the action of this field on hydrophilic colloids,[3] which slows down the processes of their water saturation.

REFERENCES

1. A. Winnicki, *Bull. Acad. Pol. Sci.* (Cl II) 15: 785-787, 1967.
2. A. Winnicki and C. Cykowska, *Acta Ichthyol. Piscat.* 3: 4-9, 1973.
3. M. Bogucki, *Protoplasma* 9: 345-367, 1930.
4. A. J. Kalmijn, in K. Schmidt-Koenig and W. T. Keeton, Eds., *Animal Migration, Navigation, and Homing*, New York: Springer-Verlag, 1978, 354-355.
5. J. L. Kirschvink, M. M. Walker, S.-B. Chang, A. E. Dizon, and K. A. Peterson, *J. Comp. Physiol.* 157: 375-381, 1985.
6. M. Lindauer and H. Martin, in *Sensory Physiology*, Berlin-Heidelberg: Springer-Verlag, 1985, 123-145.
7. G. Maret, J. Kiepenheuer, and N. Boccara, Eds., *Biophysical Effects of Steady Magnetic Fields*, New York: Springer-Verlag, 1986.
8. E. L. Carstensen, *Biological Effects of Transmission Line Fields*, New York: Elsevier, 1987.
9. K. Formicki, A. Sobociński, A. Winnicki, and Z. Mongialo, *Acta Ichthyol. Piscat.* 21: 99-104, 1991.
10. J. J. Johnson and R. S. Porter, *Liquid Crystals and Ordered Fluids*, New York: Plenum Press, 1970.
11. A. Lösche, *Post. Fizyki* 24: 407-422, 1973.
12. K. Rościszewski, *Post. Fizyki*, 24: 393-406, 1973.
13. M. M. Labes, in T. S. Tenforde, Ed., *Magnetic Field Effects on Biological Systems*, New York: Plenum Press, 1979, 85-86.

Martin Blank, Ed., *Electricity and Magnetism in Biology and Medicine*

PULSED ELECTROMAGNETIC FIELD THERAPY FOR POSTMENOPAUSAL OSTEOPOROSIS

T. W. Bilotta,* A. Zati, A. Mignani, G. Mari, O. Davoli, and E. Zucchini

Istituto Ortopedico Rizzoli, Bologna, Italy

We have carried out a clinical study to evaluate pulsed electromagnetic fields efficacy on postmenopausal osteoporosis and to see it at work with a concurrent uptake of calcitonin. Our aim was to verify a possible synergic action of the two treatments, since magnetic fields are bone-healing activators and calcitonin is an inhibitor of bone tissue reabsorption. For our study we have selected 100 postmenopausal women aged from 55 to 65 years and suffering from osteoporosis, excluding patients undergoing drug treatment or suffering from chronic diseases that could interfere with bone metabolism; 54 of the hundred completed therapy in 24 months. At the beginning of treatment and every six months, the patients underwent a check-up of radius bone mineral content.

Osteoporosis is a condition characterized by a reduction of the bone mass and a deterioration of the bone tissue microarchitecture that involves an enhanced fragility leading to a higher fracture risk. One of the most common physiotherapic techniques for osteoporosis treatment is the application of extremely low frequency (ELF) pulsed electromagnetic fields. The interest in the use of electricity to influence regulation of the bone metabolism started in 1954, when Yasuda showed that mechanical deformation of the bone generated an electrical potential. In the 1960s Bassett, Becker, and others analyzed the characteristics of those micropotentials and came to the conclusion that the normal motor activity produces bioelectrical signals that are variable and related to the regulation of the specific cell activity undergoing dynamic stimulation. Since then, several authors have demonstrated the possibility of obtaining biostimulating effects through the use of electromagnetic fields that could mimic the effects of mechanical deformation.

Research carried out so far indicates that pulsed electromagnetic fields can reactivate quiescent metabolic bone cells with a considerable effect on experimental osteoporosis in animals and on fracture healing both in humans and in animals. As the pulsed electromagnetic fields efficacy on postmenopausal osteoporosis seems less evident, we have carried out a clinical study in order both to evaluate and see it at work with an accompanying uptake of calcitonin. Our aim was to verify a possible synergic action of the two treatments as the magnetic fields are bone-healing activators and calcitonin is an inhibitor of bone-tissue reabsorption.

MATERIALS AND METHODS

For our study we have selected 100 postmenopausal women aged 55 to 65 suffering from osteoporosis, excluding patients undergoing drug treatment or suffering from chronic diseases that could interfere with bone metabolism; 54 of the hundred completed therapy in 24 months.

Patients were grouped and treated as follows.

Group 1: Nasal spray synthetic salmon calcitonin 100 I.U. per day for 2 months and then withdrawn for another 2 months. This routine was repeated 6 times.

Group 2: Pulsed electromagnetic field therapy, 30 min daily for 20 days every 6 months, 4 times.

Group 3: Both the above treatments.

Group 4: No therapy (only check-up).

The pulsed electromagnetic field had the following characteristics: monodirectional sinusoidal waves at 50 Hz and 50 Gauss for 10 min; alternating 50 and 100Hz monodirectional sinusoidal waves at 50 Gauss for 4s pauses (i.e., 4 s at 50 Hz, 4 s at 100 Hz, 4 s pause) for 10 min; and 100Hz monodirectional sinusoidal waves at 50 Gauss for another 10 min.

*Correspondence to: T. W. Bilotta, Istituto Ortopedico Rizzoli, Via G. C. Pupilli 1, I-40163 Bologna, Italy.

At the beginning of treatment and every 6 months patients were tested for radius bone mineral content, alkaline phosphatase, phosphorus, calcium, and hydroxyproline urinary. Bone mineral content was determined at the distal and proximal radius (where the infra-bone distance in 5 mm) by a single photonic absorptiometry measurement (270 Norland densitometer, I 125); every measurement was determined in three scans and their mean is reported in our study.

RESULTS

Figure 1 shows the variations in bone density. The mineral content of the compact bone at the average point of the radius remained constant in the four groups. With regard to the spongiosa bone at the distal point, we have verified the following:

Group 1: 0.42% average annual increase.
Group 2: 1.44% average annual decrease, $p < 0.05$.
Group 3: 2.3% average annual increase, $p < 0.05$.
Group 4: 3.03% average annual decrease, $p < 0.05$.

As for the other parameters, we have found no significant modification except for the reduction of urinary hydroxyproline in patients under calcitonin therapy.

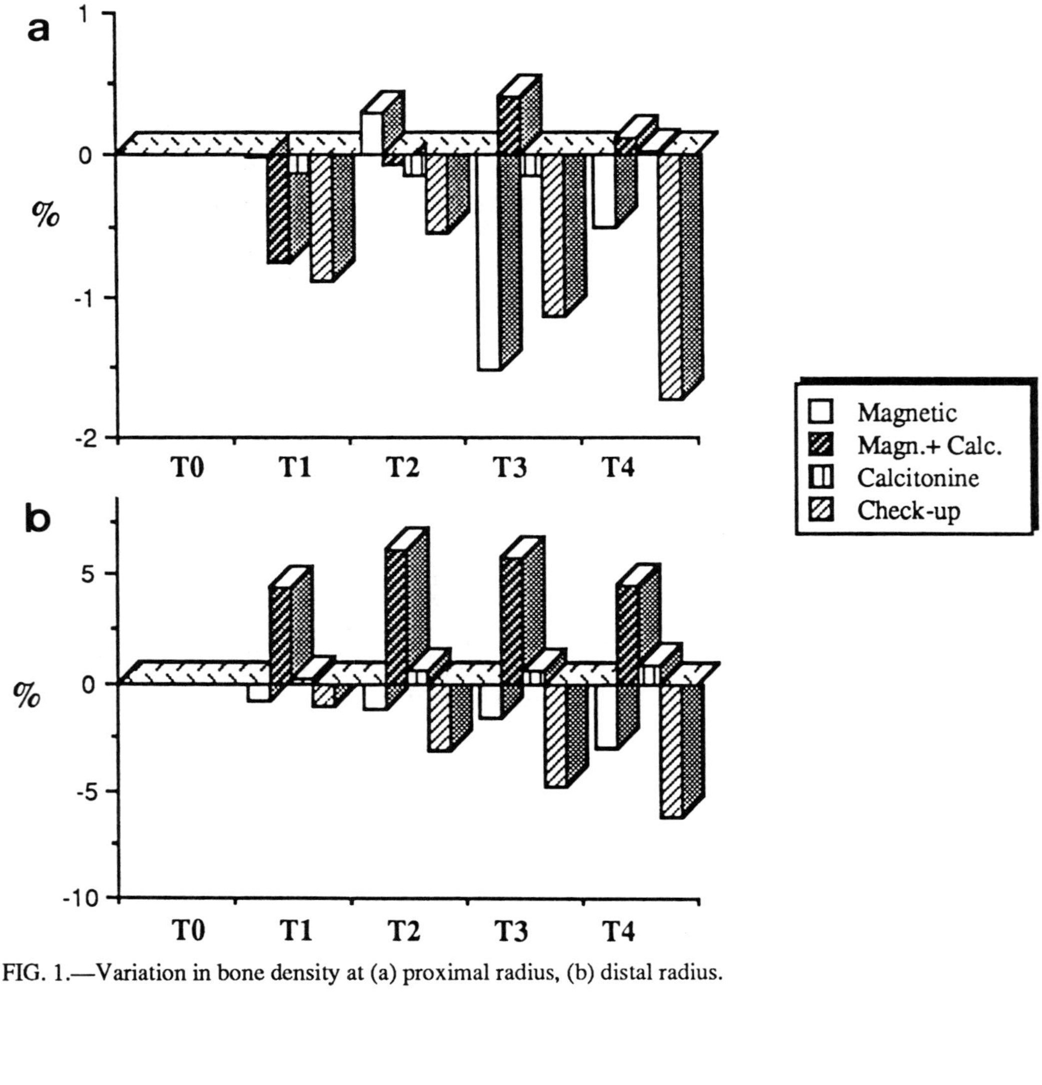

FIG. 1.—Variation in bone density at (a) proximal radius, (b) distal radius.

CONCLUSIONS

Even though pulsed electromagnetic fields cannot stop bone loss in menopausal women, they can slow it down if they are applied at the intensity and times suggested, which confirms the effects of pulsed electromagnetic fields on bone metabolism.

Our study shows that the response of bone tissue to pulsed electromagnetic fields is less intense and fast than the formation of bone callus in pseudoarthrosis. We believe that the differences in the biological effects of pulsed electromagnetic fields can be accounted for by the different situations responsible for reduced osteogenic activity. Exposure to pulsed electromagnetic fields is usually much longer for pseudoarthrosis than it is for idiopathic osteoporosis treatments.

The association of pulsed electromagnetic fields with calcitonin in Group 3 not only made the bone decrease stop but showed a significant increase in bone mass. This increase is much higher than that obtained by drug treatment alone, which proves that pulsed electromagnetic fields therapy is a useful and successful treatment in postmenopausal osteoporosis.

REFERENCES

1. C. A. L. Basset, R. J. Pawluk, and A. A. Pilla, "Augmentation of bone repair by inductively coupled electromagnetic fields," *Science* 184: 575-577, 1974.

2. T. W. Bilotta, D. Carlucci, A. Giulioni, G. Mari, and L. Prioli, "Utilizzo dei C.E.M.P. a bassa frequenza a bassa intensità nel trattamento dell'osteoporosi (risultati ad 1 anno di distanza)," Simposio internazionale nel tema Elettrobiologia oggi, in onore di L. Galvani, Bologna, 1989, 29-30 (abstract).

3. T. W. Bilotta, A. Giulioni, A. Zati, F. Fichera, and D. Carlucci, "Approccio farmacologico e fisioterapico all'osteoporosi senile," *Reumatismo* 42: 150-152, 1990.

4. I. Binderman, Z. Shimshani, and D. Somjen, "Biochemical pathways involved in the translation of physical stimulus into biological message," *Calcif. Tissue Int.* 36 (Suppl.): 82-85, 1984.

5. R. S. Feldman, N. S. Krieger, and A. H. Tashjian Jr., "Effects of parathyroid hormone and calcitonin on osteoclast formation in vitro," *Endocrinology* 107: 1137-1143, 1980.

6. J. Friedman, W. Y. Au, and L. G. Raisz, "Responses of fetal rat bone to thyrocalcitonin in tissue culture," *Endocrinology* 82: 149-156, 1968.

7. N. Haimovici and G. B. Languasco, "Campi magnetici pulsati e tessuto osseo: Considerazioni teoriche e sperimentali," in F. Bistolfi, Ed., *Campi Magnetici in Medicina*, Turin: Minerva Medica, 1986, 387-405.

8. G. A. Rodan, "Mechanical and electrical effects on bone and cartilage cells: Translation of the physical signal into a biological message," in H. G. Barrer, Ed., *Orthodontics: The State of the Art*, Philadelphia: University of Pennsylvania Press, 1981, 315-322.

9. C. T. Rubin, K. J. McLeod, L. E. Lanyon, and S. Brook, "Prevention of osteoporosis by pulsed electromagnetic fields," *J. Bone and Joint Surg.* 71A: 411-417, 1989.

Martin Blank, Ed., *Electricity and Magnetism in Biology and Medicine*

EFFECTS OF 50Hz ELECTROMAGNETIC FIELDS ON THE HUMAN HEART

L. Korpinen* and J. Partanen

Tampere University of Technology, Finland

This paper covers two studies on the effect of 50Hz electric and magnetic fields on the cardiovascular system. The field source was a 400kV transmission line. In the field exposure and also outside the field, ECG and EEG were measured by ambulatory methods. Blood pressures were also registered and the subjects executed some cardiovascular autonomic function tests: orthostatic test, Valsalva manoeuvre, and deep breathing during the test. Analysis of the ECG recordings showed that extrasystoles or arrhythmias were as frequent in the field as outside it. In the first study a small decrease in the heart rate was observed after the field exposure. However, this finding was not registered in the second study. During the field exposure the diastolic blood pressure seemed to increase slightly. The systolic blood pressure seemed to be the same in and outside the field.

FIRST STUDY

In the first study two groups were examined. Altogether, 26 voluntary men (Group 1) and 27 transmission-line workers (Group 2) were measured. Their ECG was recorded with a Holter recorder both in and outside the field. The measurement period of Group 1 was 1 h (½ h in the field and ½ h outside the field); that of Group 2 took 1-3 h. In Group 1 the fields were 1.7-4.9 kV/m and 1.1-7.1 μT; in Group 2 they were 0.2-11.2 kV/m and 1.0-16.2 μT. The ECG analysis showed that there were no extrasystoles or arrhythmias. The only significant observation was a decrease in the heart rate in Group 1 during the exercise (walking) outside the field after exposure. The drop cannot be explained with the first measuring method. Therefore we began a new study.

SECOND STUDY

Subjects and Test Environment

In the second study there were two test groups; Group 1 (26 male volunteers) were measured in real field exposure, Group 2 (15 male volunteers) in sham fields. The subjects of Group 1 were 1 h outside the field (period 1), then 1 h in field exposure under a 400kV transmission line (period 2), and then again 1 h outside the field (period 3). Under the 400kV line the field varied from 3.5 to 4.3 kV/m, and from 1.4 to 6.6 μT. Group 2 spent the whole test period (3 h) in a 33kV outdoor testing station; the sham field was switched on and off by manual opening and closing of the disconnector indicated by the subjects.

Methods

A similar test protocol was used for both groups. Their ECG and EEG were measured by ambulatory methods.[1] Their blood pressure was also measured every 10 min, except in the orthostatic tests, where additional measurements were taken. In the blood-pressure measurements the systolic and diastolic blood pressures and the average heart rate were registered by means of an automatic noninvasive cuff method. Before and after the field exposure the subjects executed some cardiovascular autonomic function tests: orthostatic test, Valsalva manoeuvre, and deep breathing.[2] During the field exposure the subjects did some Ruddel's mental arithmetic tasks. The aim of these tasks was to keep the subjects active.

Results

*Correspondence to: L. Korpinen, Tampere University of Technology, P.O. Box 6D2, SU-33101 Tampere, Finland.

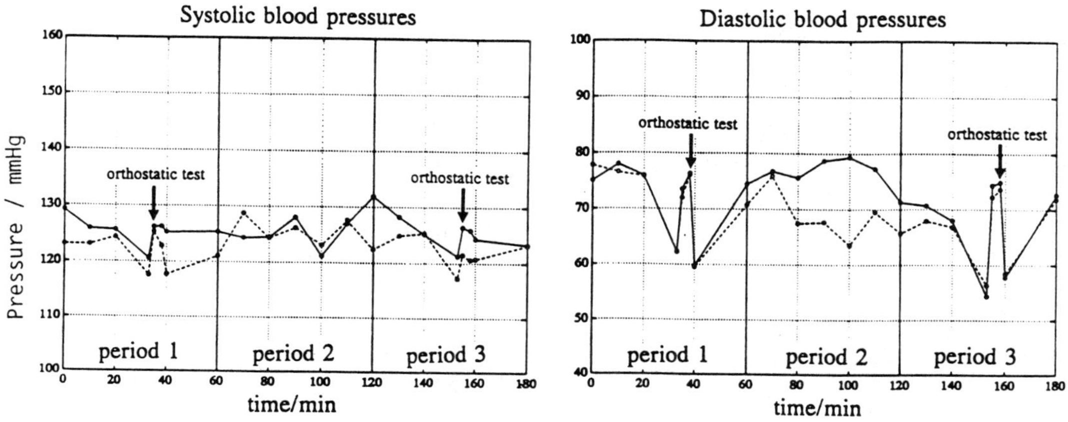

FIG. 1.—Mean values of systolic and diastolic blood pressures in real field (n = 11), solid line; in sham field (n = 13), dashed line.

So far, the ECG and blood pressure measurements have been analyzed. Figure 1 presents the mean values of the 1min pulse rates for both groups. The solid line represents the pulse rate of Group 1 (real field exposure); the dashed line, the pulse rate of Group 2 (sham field). No decrease in the pulse rate after field exposure similar to that observed in our first study was indicated by the results shown in Fig. 1. The difference between the results of studies 1 and 2 has not yet been analyzed in detail.

Figure 2 shows the mean values of the systolic and diastolic blood pressures. The solid line represents Group 1 (real field exposure); the dashed line, Group 2 (sham field). The systolic blood pressures are exactly the same in both groups during the whole test period. The diastolic blood pressure seems to be higher in Group 1 than in Group 2 during the field-exposure period.

CONCLUSION

This paper presents two studies of the effect of 50Hz electric and magnetic fields of a 400kV transmission line on the cardiovascular system. The ECG and EEG of subjects exposed to the fields were measured by ambulatory methods. The blood pressures were also registered and the subjects executed some cardiovascular autonomic function tests: orthostatic test, Valsalva manoeuvre, and deep breathing during the test.

The extrasystoles or arrhythmias were as frequent in the field as outside it. In the first study a small decrease in the heart rate was observed after the field exposure. However, this finding could not be registered in the second study. The systolic blood pressure seemed to be the same in and outside the field. During the field exposure the diastolic blood pressure seemed to increase slightly. This result cannot yet be explained but will be studied further.

REFERENCES

1. H. L. Kennedy, *Ambulatory Electrocardiography Including Holter Recording Technology*, Philadelphia: Lea and Febiger, 1981, pp. 11-14.

2. S. Piha, *Cardiovascular Autonomic Function Tests*, Publications of the Social Insurance Institution, Turku, Finland, 1988.

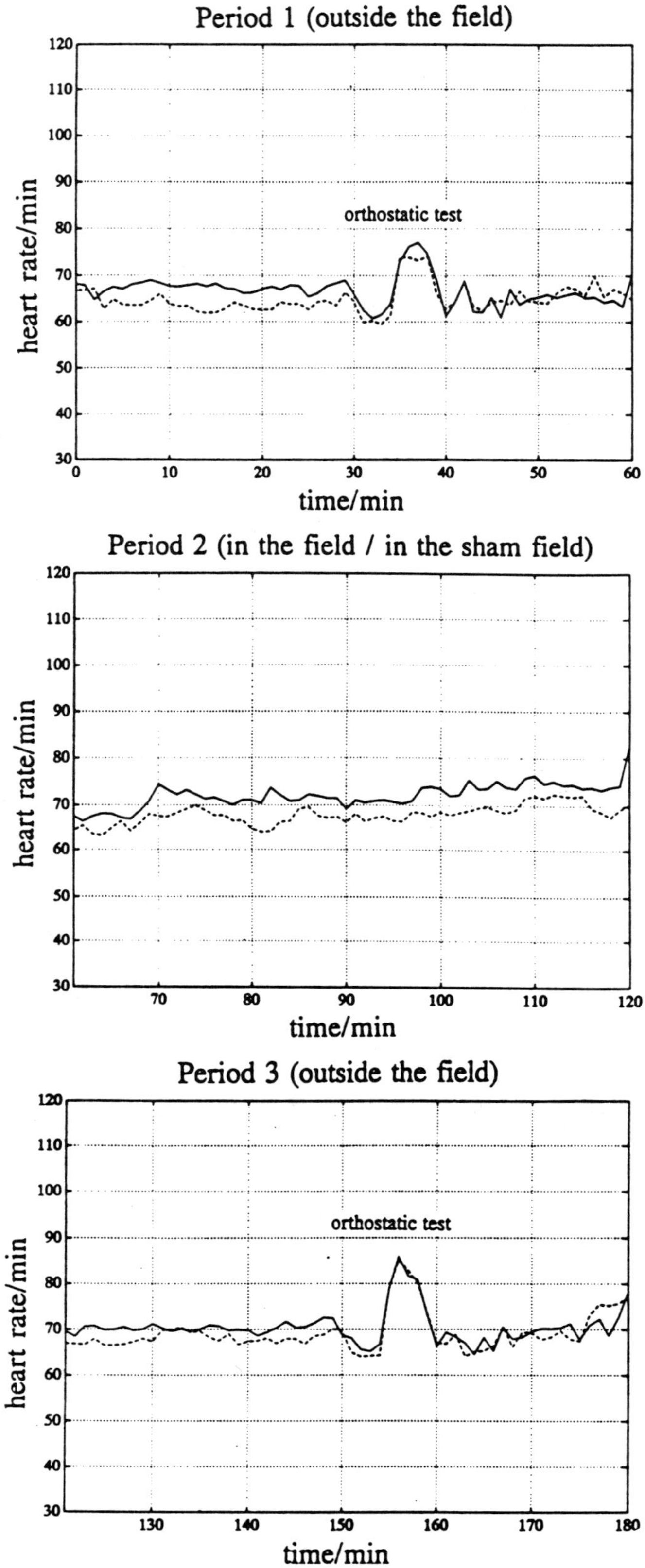

FIG. 2.—Mean values of 1min pulse rates in real field (n = 13), solid line; in sham field (n = 13), dashed line.

Martin Blank, Ed., *Electricity and Magnetism in Biology and Medicine*

HUMAN CARDIAC RESPONSE TO 60Hz FIELDS: EXPOSURE PARAMETERS AND INDIVIDUAL DIFFERENCES

C. Graham,* H. D. Cohen, M. R. Cook, M. M. Gerkovich, and D. W. Riffle

Midwest Research Institute, Kansas City, Mo.

There is growing public concern about the possible health risks associated with exposure to the electric and magnetic fields generated by power-distribution systems. Environmental risk is often a function of effects that accumulate over time and interact with dose level. Thus, the primary goal of our research program has been to determine, under controlled laboratory conditions, whether exposure to 60Hz electric and magnetic fields has effects on well-established measures of human physiology and performance. Our research strategy has been (1) to identify which human functions appear to be affected by exposure; (2) to determine whether these effects are reproducible; (3) to understand the specific circumstances under which they occur; and (4) to identify the underlying mechanisms involved.

Project studies are performed in the Human Exposure Test Facility at Midwest Research Institute.[1] This facility is designed to present uniform, corona-free, 60Hz electric and magnetic fields under controlled conditions of ambient temperature and relative humidity. Electric fields up to 16 kV/m and magnetic fields up to 400 mG can be presented together or separately. The facility has two capabilities that make it unique. The first is a double-blind control system that makes it possible for human volunteers to be exposed to real or sham fields with neither the volunteers nor experimenters knowing which condition is in effect at any given time. Thus, effects due to field exposure can be distinguished from effects due to an individual's expectations about exposure. The facility is also capable of recording human heart and brain activity during, as well as before and after, exposure to high-voltage fields. This capability aids in tracking the time course of any effects observed.

Several reports indicate that power-frequency electric and magnetic fields couple with and influence the human cardiovascular system. Since Sazanova's initial observations in 1967 of a slowing of pulse rate in Soviet switchyard workers, field-related effects on heart rate have been reported in rats, dogs, and humans.[2-6] We have now evaluated exposure effects on more than 200 healthy young men at various levels of field strength, and under both continuous and intermittent exposure conditions. As shown in Table 1, our laboratory-based studies[7] show that compared to sham exposure, field exposure has a significant effect on the human cardiovascular system. The cardiac slowing response does not appear to be a simple linear function of exposure duration or field strength. Slowing occurred after 2 h and after 6 h of exposure. It was observed in four studies of exposure to combined fields of 9 kV/m and 200 mG, but did not occur in two studies of exposure to stronger fields (12 kV/m and 300 mG), or when subjects were exposed to weaker fields (6 kV/m and 100 mG). Slowing of heart rate is observed under continuous exposure conditions; the response appears to be differentially sensitive to certain combinations or "windows" of field strength.

In contrast, periods of brief intermittent field exposure can induce both speeding and slowing of heart rate. Figure 1 shows the results of intermittent exposure to combined electric (9 kV/m) and circularly polarized magnetic (200 mG) fields. For exposed subjects, intermittent fields were presented in alternating 45m on/off periods. For sham-exposed subjects, the fields were never activated. Intermittent exposure induced both speeding and slowing of the heart. Heart rate speeding occurred shortly after the intermittent exposure periods began (A), and again shortly after they ended (C). During both field-on (B) and field-off (D) periods, the heart slowed. Sham-exposed subjects did not show this alternating pattern of heart-rate speeding and slowing. Differences between sham and exposed conditions are significant at $p < 0.01$.

Individuals within the exposed groups appeared to differ markedly in their sensitivity to field exposure. Additional analyses indicated that the resting baseline heart rate of a person prior to exposure was a significant ($p < 0.01$) predictor of such individual differences in field-related cardiac change. Subjects with

*Correspondence to: C. Graham, MRI, Kansas City MO 64110.

TABLE 1.—Heart rate slows more after continuous field exposure than after sham exposure; effect is not linear function of exposure duration nor of field strength.

Study	Number of subjects	Field strength	Exposure duration (h)	Heart rate slowing? (yes/no)	Significance level ($p \leq$)
1	18	6 kV/m, 100 mG	6	no	ns
2	12	9 kV/m, 200 mG	2	yes	0.05
3	12	"	6	yes	0.02
4	18	"	6	yes	0.025
5	30	"	6	yes	0.01
6	18	12 kV/m, 300 mG	6	no	ns
7	28	"	6	no	ns

slower baseline heart rate (mean = 60 bpm) show the full pattern of cardiac speeding and slowing during intermittent exposure (Fig. 2). In contrast, subjects with faster baseline heart rate (mean = 80 bpm) show only the slowing effect. Recent work in our laboratory also has suggested that the effects of intermittent exposure on the heart may be due more to the magnetic field than to the electric field. Exposure to the magnetic field (200 mG) alone produces the complex pattern of speeding and slowing observed previously with exposure to combined electric and magnetic fields. However, exposing subjects only to the electric field (9 kV/m) produced only a small slowing response that did not differ from the effects of sham exposure.

FUTURE DIRECTIONS

The effects on heart rate cannot be considered health risks at present. They are not outside the normal range, and they do not last very long beyond the time of actual exposure to the fields. Their significance lies in the indication they provide that field exposure at ambient levels can interact with and influence human physiology. We need to understand the underlying mechanisms involved in this interaction. Changes in heart rate may be a reflection of more important changes occurring in other areas of the cardiovascular system. Investigations are needed of field effects on physiological, neural, and biochemical cardiac control mechanisms. For example, alterations in stroke volume, blood pressure, and/or respiration can result in a slowing of heart rate, but the pathways and control systems by which such slowing is brought about are quite different. Future studies should be designed to test specific hypotheses derived from current biophysical models of cardiac control function.

There is also a need to investigate individual differences in cardiovascular response. Studies utilizing behavioral and pharmacological manipulations may be particularly valuable. We have also noted a curious phenomenon when the same individuals are repeatedly exposed to both real and sham field over a relatively short time. Individuals exposed first to real fields, and then to sham fields show greater responsivity than those exposed in the reverse order. Other laboratories have also noted such effects,[8] but as yet no explanation has been found for this puzzling phenomenon. Future research to address these issues should help clarify the mechanisms by which exposure to power-frequency fields interacts with and alters human function.

REFERENCES

1. H. D. Cohen et al., "ELF exposure facility for human testing," *Bioelectromagnetics* 13: 169-182, 1992.

2. P. Cerretelli et al., "1000-kV Project: Research on the biological effects of 50-Hz electric fields in Italy, in R. D. Phillips et al., Eds., *Biological Effects of Extremely Low-Frequency Electromagnetic Fields,* Springfield, Va.: NTIS, 1979, 241-257.

3. G. Fischer et al., "Influence of line frequency electric fields on the heart rate of rats," cited in E. L. Carstensen, *Biological Effects of Transmission Line Fields,* New York: Elsevier, 1987.

4. D. S. Gann, Final Report: Electric Power Research Institute, RP98-02, Johns Hopkins University, cited in E. L. Carstensen, Ref. 3.

5. T. V. Kalyada et al., "Biological action of a 50-Hz electric field on the human body," cited in J. M. Lee et al., *Electrical and Biological Effects of Transmission Lines,* Portland, Or.: Bonneville Power Administration, 1989.

6. R. Waibil, "Objective proof of the influence of 50-Hz electric fields on man," cited in E. L. Carstensen, Ref. 3.

7. C. Graham et al., Final Report: Immunological and Biochemical Effects of 60-Hz Electric and Magnetic Fields in Humans (Report No. DE90006671), Springfield, Va.: NTIS.

8. B. T. Stollery et al., "Physiological functioning during human exposure to 50-Hz electric currents," report to Central Electricity Generating Board, Leatherhead, Surrey, UK, 1989.

9. This research was funded by U.S. Department of Energy grant DE-FG01-89CE34025.

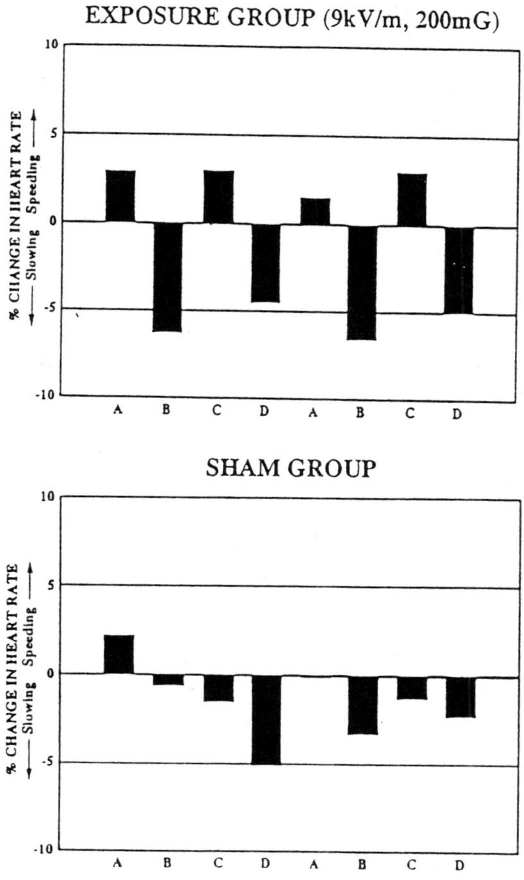

FIG. 1.—Intermittent exposure to combined electric and magnetic fields (9 kV/m, 22 mG) induces alternating pattern of speeding and slowing of human heart rate significantly ($p \leq 0.01$) different from sham exposure.

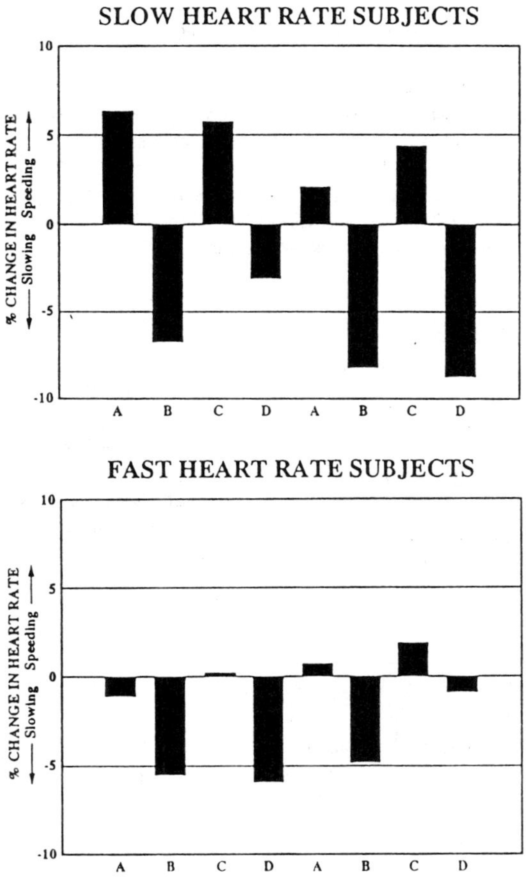

FIG. 2.—Pre-exposure heart rate (HR) is significant ($p \leq 0.01$) predictor of individual cardiac reactivity. Subjects with slower baseline HR (mean = 80 bpm) show both speeding and slowing of HR during intermittent exposure. Subjects with faster baseline HR (mean = 80 bpm) show only slowing effect.

Martin Blank, Ed., *Electricity and Magnetism in Biology and Medicine*

FIBRINOLYTIC RESPONSES TO DESMOPRESSIN ACETATE [DDAVP] INFUSION IN EXERCISE-TRAINED AND UNTRAINED QUADRIPLEGICS

M. S. Nash,* M. Multach, K. Banovac, J. A. D'Errico, S. L. Miller, M. R. Ledford, J. W. Kent, and K. J. Klose

University of Miami School of Medicine, Fla.

Desmopressin acetate [DDAVP] was administered to six exercise trained and five sedentary quadriplegics to determine whether endothelial function is altered by 6 months of electrical stimulation cycle ergometry exercise. Blood samples drawn at rest and at 20, 40, and 60 min following DDAVP (0.4 μg/kg BW) were analyzed for tissue type Plasminogen Activator (tPA) and Plasminogen Activator Inhibitor (PAI-1). tPA was higher in trained than in sedentary subjects at rest and following DDAVP. Native PAI-1 was higher in quadriplegics than normals, and higher in trained than in sedentary quadriplegics in all samples following DDAVP ($p = 0.1452$). Electrically stimulated training may alter endothelial regulation of extrinsic fibrinolytic-mediating proteins.

There is growing speculation that physical conditioning may reduce the susceptibility of healthy individuals to thromboembolism.[1] This assumption is based on evidence that acute exercise stimulates a 5- to 10-fold increase in fibrinolysis,[1,2] a process by which fibrin contained in developing thrombi is enzymatically digested. The principal stimulus for this fibrinolysis involves the release of the tissue-type plasminogen activator (tPA) from vascular endothelium during exercise.[3] There is also evidence that sustained morphological and functional changes of vascular endothelium accompany exercise training, adaptations that reduce the likelihood of thrombus formation.[4,5]

Conditions accompanying spinal cord injury (SCI) which include physical immobilization, physiological deconditioning, and hyperlipidemia have long been associated with increased incidence of deep venous thromboembolism (DVT).[6] Thrombotic tendency following SCI is also fostered by lower extremity stasis from loss of skeletal muscle pumping and vasomotor paralysis.[7] These conditions are acutely relieved following transcutaneous electrical stimulation of the calf muscles, a procedure that increases lower extremity blood flow and stimulates fibrinolysis as assessed by venous plethysmography and reduced clot lysis times, respectively.[8] It has been speculated that electrically stimulated cycling exercise might similarly alter thrombotic tendency,[9] although the sustained effects of this activity on fibrinolysis have yet to be determined.

In past studies, the effects of exercise training on vascular endothelium have been assessed by measurement of the release of tPA following occlusion-ischemia provoked by an inflation cuff.[5] This technique poses a risk for those with SCI. In the alternative, administration of a vasopressin analog desmopressin acetate [DDAVP] also stimulates the release of tPA, as well as its fast antagonist, Plasminogen Activator Inhibitor (PAI-1).[10] The purpose of this study was to determine whether the release kinetics of endothelially derived tPA and PAI-1 following DDAVP administration differ among electrically stimulated exercise-trained quadriplegics, matched sedentary cohorts, and nondisabled control subjects.

METHODS

Subjects were six exercise-trained quadriplegic males, five sedentary quadriplegic cohorts matched for gender, age, and duration of injury, and six nondisabled male controls. All quadriplegic subjects had neurologically complete motor injuries at the C5 and C6 levels (Frankel Grades A and B) for longer than one year. Training consisted of 6 months of twice weekly cycle ergometry using computer-sequenced electrical stimulation (monophasic rectangular wave form, constant current to peak output of 132 mA, 375μs pulse duration)

*Correspondence to: M. S. Nash (Orthopaedics & Rehab.), University of Miami School of Medicine, 5915 Ponce de Leon Blvd (5th fl.), Coral Gables FL 33146.

of the bilateral quadriceps, hamstring, and gluteus muscles.[9] Electrical current was introduced to these sites via carbon silastic electrodes coated with ECG conductor gel.

On the day of testing, 10ml blood samples were drawn from resting subjects via an antecubital venous catheter between 0900 and 1030 hours. DDAVP (0.4 μg/kg BW) was then given intravenously in 20 ml of 0.9% saline over a 20min period, with blood samples drawn immediately, 20, 40, and 60 min after drug administration. Citrated samples were analyzed by an enzyme-linked immunosorbent assay (ELISA) for tissue-type Plasminogen Activator (tPA) using a peroxidase-tagged mouse monoclonal anti-tPA $F(ab')_2$ (Diagnostica Stago). PAI-1 activity was determined on acidified plasma by a chromogenic substrate assay (Biopool). Data were analyzed by repeated measures design analysis of variance (ANOVA), with time as the within-subjects factor, and grouping as the between-subjects factor.

RESULT

A significant within-subjects effect both for tPA and PAI-1 was sustained by all groups following DDAVP administration. Resting tPA levels were 12% higher in exercise-trained than sedentary cohorts, and 5.3, 18.8, and 42.5% higher at 20, 40, and 60 min following DDAVP, respectively (Table 1). Nonetheless, the test for interaction of time and grouping failed to reach statistical significance. Native PAI-1 was higher in quadriplegics than in normals regardless of conditioning status. PAI-1 levels in exercise-trained subjects were higher than sedentary cohorts in all samples following DDAVP administration (Table 2). The absence of significant interaction of time and grouping may be attributed, in part, to the small sample size and the large intersubject variability.

TABLE 1.—Responses of tissue Plasminogen Activator (ng/ml) following DDAVP infusion in electrically stimulated exercise-trained and sedentary quadriplegics, and nondisabled controls (mean ± s.d., n = 6).

	Controls	Exercised	Sedentary	p
Pre-DDAVP	3.67 ± 2.0	4.68 ± 3.3	4.18 ± 2.4	ns
20 min	13.7 ± 4.5	16.6 ± 11.0	14.4 ± 8.2	ns
40 min	8.8 ± 2.3	13.9 ± 11.0	11.7 ± 6.6	ns
60 min	6.2 ± 2.1	10.4 ± 7.6	7.3 ± 5.4	ns

TABLE 2.—Responses of the fast Plasminogen Activator Inhibitor (IU/ml) following DDAVP infusion in electrically stimulated exercise-trained and sedentary quadriplegics, and nondisabled controls (mean ± s.d.; n = 6).

	Controls	Exercised	Sedentary	p
Pre-DDAVP	6.03 ± 2.2	11.5 ± 9.8	11.6 ± 7.5	ns
20 min	0.9 ± 2.0	4.6 ± 9.0	3.5 ± 4.8	ns
40 min	1.5 ± 1.7	3.3 ± 7.0	1.2 ± 2.7	ns
60 min	2.1 ± 2.1	3.5 ± 5.7	2.4 ± 3.7	ns

DISCUSSION

Previous work has shown that morphological and functional adaptation of endothelium accompany exercise training. Neovascularization of skeletal muscle characterized both by increased capillary density and capillary:fiber ratio has been shown following physical conditioning.[4] This angiogenesis has also been histologically verified following chronic electrical stimulation.[11] Moreover, the fibrinolytic response of healthy persons to venous occlusion is heightened following physical conditioning.[5] This heightened fibrinolytic activity has also been shown to be a function of level of fitness, with marathon runners having higher ^{125}I-plasma clot lysis than either recreational runners or sedentary individuals.[12]

A recent investigation in our laboratory has shown a significant acute effect of electrically stimulated exercise on the release of tPA, a response suggesting that the activity is of sufficient work intensity to stimulate endothelial response. This response is likely the result of hemodynamic changes induced by augmented lower extremity blood flow[1] experienced during cycling.[7] The present study suggests that responses of electrically stimulated exercise-trained quadriplegics differ from their sedentary counterparts. The extent to which these differences may be attributed to the angiogenesis or altered endothelial function remains to be investigated by histological techniques. Nevertheless, the results of this study suggest that the manner in which tPA and PAI-1 are either stored, released, or resynthesized may be altered by electrically stimulated exercise training. The extent to which the observed differences are meaningful in terms of reduced susceptibility to DVT awaits additional study.[13]

REFERENCES

1. J. A. Colwell, "Effects of exercise on platelet function, coagulation, and fibrinolysis," *Diabetes Metab. Rev.* 1: 501-512, 1986.

2. R. E. Bourey and S. A. Santoro, "Interactions of exercise, coagulation, platelets, and fibrinolysis: A brief review," *Med. Sci. Sports Exerc.* 20: 439-446, 1988.

3. M. S. El-Sayed, "Exercise intensity-related responses of fibrinolytic activity and vasopressin in man," *Med. Sci. Sports Exerc.* 22: 494-500, 1989.

4. P., Brodal, F. Ingjer, and L. Hermansen, "Capillary supply of skeletal muscle fibers in untrained and endurance trained men," *Am. J. Physiol.* 232: H705-712, 1977.

5. R. S. Williams, E. E. Logue, F. L. Lewis, T. Barton, N. W. Snead, A. G. Wallace, and S. V. Pizzo, "Physical conditioning augments the fibrinolytic response to venous occlusion in healthy adults," *N. Engl. J. Med.* 302: 987-991, 1980.

6. J. E. Greenleaf and S. Kozlowski, "Physiological consequences of reduced physical activity during bed rest," *Exerc. Sports Sci. Rev.* 10: 84-119, 1982.

7. M. S. Nash, C. M. Montefusco, and B. A. Rhodes, "Peripheral vascular response to exercise-induced and post-occlusive reactive hyperemia (PORH) in quadriplegics," *Surg. Forum* 41: 498-499, 1990.

8. R. T. Katz, D. Green, T. Sullivan, and G. Yarkony, "Functional electrical stimulation to enhance systemic fibrinolytic activity in spinal cord injury patients," *Arch. Phys. Med. Rehabil.* 68: 423-426, 1987.

9. K. T. Ragnarsson, S. Pollock, J. Petrofsky, W. O'Daniel, R. Edgar, and M. S. Nash, "Clinical evaluation of computerized functional stimulation after spinal cord injury: A multicenter pilot study," *Arch. Phys. Med. Rehabil.* 69: 672-677, 1988.

10. V. Vicente, I. Alberca, and P. M. Mannucci, "Reduced effects of exercise and DDAVP on factor VIII-von Willebrand factor and t-plasminogen activator after sequential activation of both stimuli," *Thromb. Heamost.* 51: 129-130, 1984.

11. O. Hudlicka, H. Hoppler, and E. Uhlmann, "Relationship between the size of the capillary bed and oxidative capacity in various cat skeletal muscles," *Pflügers Arch.* 410: 369-375, 1987.

12. E. W. Ferguson, L. L. Bernier, G. R. Banta, J. Yu-Yahiro, and E. B. Shoomaker, "Effects of exercise and conditioning on clotting and fibrinolytic activity in men," *J. Appl. Physiol.* 62: 1416-1421, 1987.

13. This research was supported by grants from the American Bioproducts Co. (#91ER002) and the Miami Project to Cure Paralysis.

Martin Blank, Ed., *Electricity and Magnetism in Biology and Medicine*

INTERACTION MECHANISM OF ELECTRIC FIELDS WITH CELLS: ALTERATION OF BINDING SITE ACCESS OF MEMBRANE PROTEINS

D. Mycue, J. Zahn, M. Zahn, and J. C. Weaver*

Massachusetts Institute of Technology, Cambridge, Mass.

A continuum model for membrane protein and a portion of a cell membrane is used to compute numerically the electrostatic energy of the protein/membrane system for various protein displacements while the protein charge and transmembrane voltage U is kept constant. Throughout, the charge on the membrane is allowed to vary so as to maintain a constant U. The resulting plots of energy vs displacement are interpreted as binding energies for association of the protein with the membrane, with emphasis on the possibility that protein equilibrium positions can change as U is varied, which in turn may alter binding access to external molecules and therefore affect signaling.

The mechanism of interaction of electric and magnetic fields with cells is of great interest at present. Not only are beneficial applications of fields increasingly recognized, there is considerable controversy regarding "weak" environmental fields. In both cases theoretical models of interaction mechanisms are valuable, because models allow the dependence on important parameters to be identified and can provide order-of-magnitude estimates of thresholds for "weak" field effects when combined with quantitative estimates of intrinsic noise.[1] A number of mechanism candidates have been identified that involve the changes in transmembrane voltage $U(t)$ caused by an external field. These include a general class for which geometric changes of the cell (whole cell deformation[2]), the cell membrane (electroporation[3]), or of membrane enzymes (electroconformational coupling[4]) occur. If the physical deformation modifies one or more biochemical pathways, altered cell behavior is possible. Here we consider another candidate mechanism, the possible alteration of protein binding site access through changes in protein protrusion.[5]

METHODS

A numerical software package (Ansoft Axisymetric Electrostatic Solver, Ansoft, Pittsburgh, Pa.) was used to solve Poisson's equation numerically for the protein/membrane systems shown in Figs. 1 and 2. For the protein and membrane having constant charge, the electrostatic energy $W_e(x)$ is computed from[6]

$$W_e = \frac{1}{2}\int_{\text{all space}} \mathbf{E}\cdot\mathbf{D}\,dV; \quad f_x = -(\partial W_e/\partial x)_{q_{p1},q_{p2},q_{m1},q_{m2}} \tag{1}$$

where f_x is the force on the protein. As the protein location is changed, the transmembrane voltage necessarily changes if the membrane charges q_{m1} and q_{m2} are to remain constant. In order to treat the case of long-term constancy of $U = U_{m1} - U_{m2}$, the hybrid energy W_h was also calculated. In this case, the charge q_{m1} and q_{m2} on each side of the membrane was allowed to vary in order to maintain constant U; W_h was therefore computed as[7]

$$W_h = q_{m1}U_{m1} + q_{m2}U_{m2} - W_e; \quad f_x = +(\partial W_e/\partial x)_{q_{p1},q_{p2},U_{m1},U_{m2}} \tag{2}$$

The present computations do not involve the Poisson-Boltzmann equation for the electrolyte regions, and therefore will be in error for charge located within the electrolyte. However, the contribution to the energy by charged regions located immediately adjacent or within the membrane should be reasonably represented.

---

*Correspondence to: J. C. Weaver (20A-128), Harvard-MIT Div. of Health Sciences and Technology, Cambridge MA 02139.

FIG. 1.—(a) First continuum model for membrane protein located in circular patch of membrane. Negative charge is assumed to be uniformly distributed within two spherical volumes of the protein, with each charge (q_{p1}, q_{p2}) and transmembrane voltage U held constant. (b) Corresponding electrostatic energy of the protein/membrane system as a function of protein displacement normal to membrane; $q_1 = q_2 = 5 \times 10^{-19}$ C, which is a tightly bound case (top); hybrid energy for the same conditions as above (bottom). (c) Three cases based on (a) that range from tight to loose binding as the charge magnitude is reduced from $q_1 = q_2 = 5 \times 10^{-19}$ C to 3×10^{-19} C; last case has well depth of about 10kT for $U = 70$ mV.

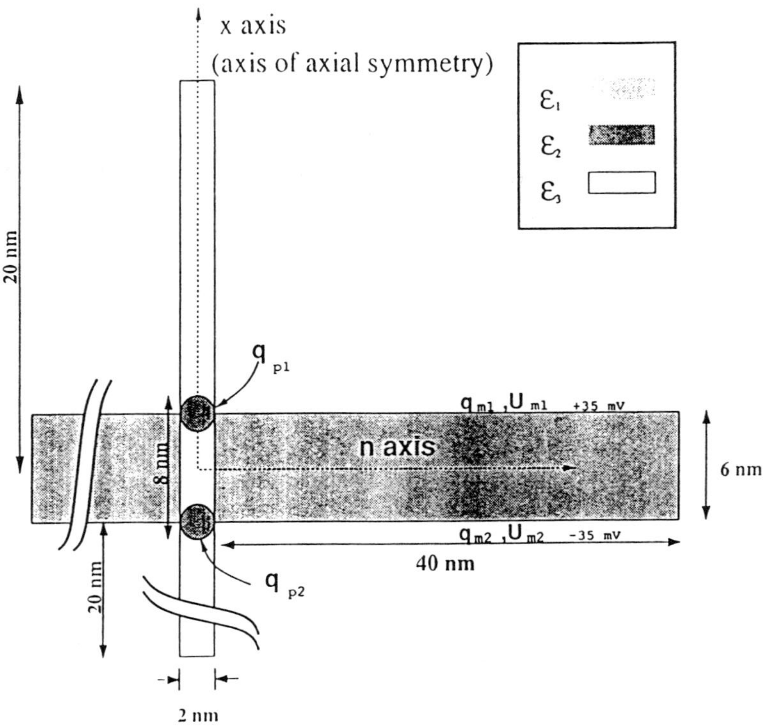

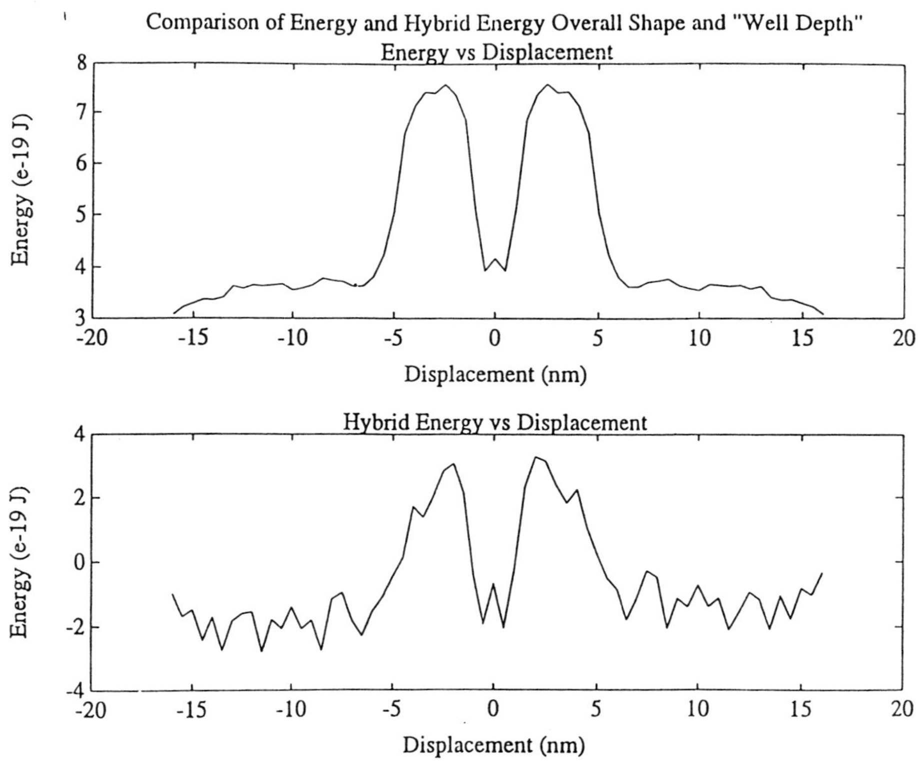

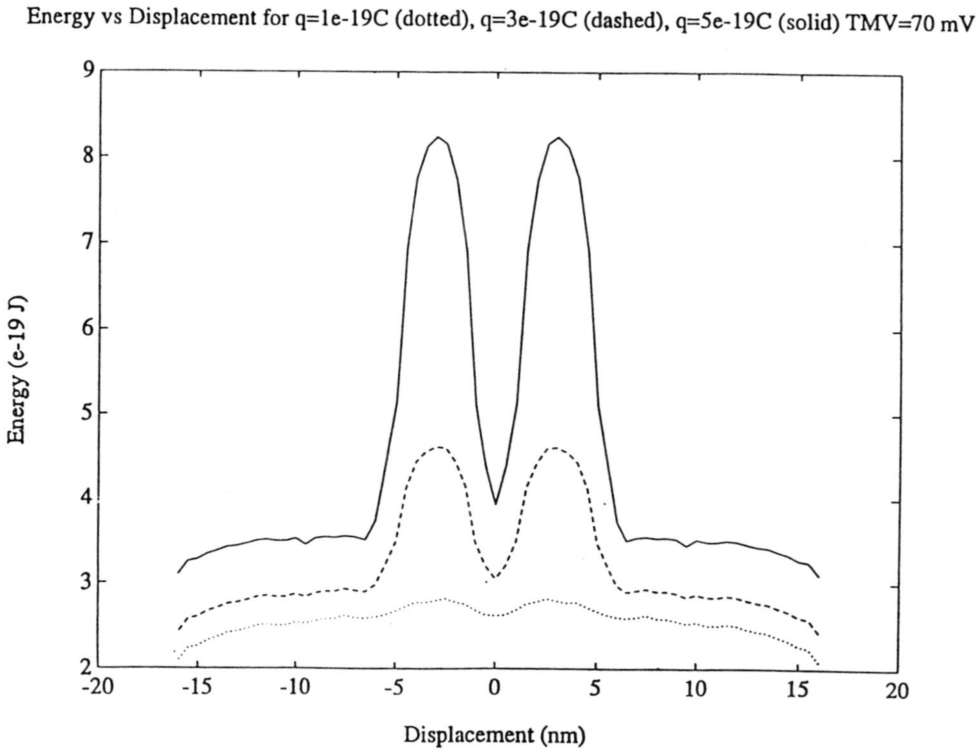

RESULTS AND DISCUSSION

As shown in Fig. 1, a simple continuum model with large volume spherical charge distributions at the two ends of the protein is stably associated with the membrane, in agreement with widely accepted qualitative views. The potential energy "well" ΔW ranges from very deep ($q = 5\times10^{-19}$ C, $\Delta W \approx 10^{-19}$, $J \approx 100kT$) to shallow ($q = 10^{-19}$ C, $\Delta W \approx 10^{-20}$, $J \approx 10kT$). This range reflects the large energy change associated with insertion of charge into the low-dielectric-constant interior of the membrane. The high Born energy cost of bringing charge near or into a low dielectric constant membrane has been appreciated since the calculations of small ion entry for various different membrane configurations.[8]

Another simple model (Fig. 2) is relevant to proteins with charged regions located within the membrane. Alternately, it is relevant to electro-insertion by electroporation.[9] In the second model ΔW is smaller because the charge is always located in a small dielectric constant medium, and this minimizes changes in the interaction with the low-dielectric-constant membrane. The second model can also be used to represent an aqueous pore, in which ΔW is even lower, because in this case the fixed region within the membrane has a large dielectric constant.

Finally, inspection of the preliminary result of Fig. 2(b) reveals a shift in the local minimum, i.e., the well moves in response to ΔU (in this case a large 30 mV). Work in progress will seek a refined computation in order better to determine the magnitude of $\partial \bar{x}/\partial U$. The preliminary result only allows a bound to be estimated, namely $\partial \bar{x}/\partial U < 0.5$ nm/30 mV $= 1.7\times10^{-2}$ nm/mV.[10]

REFERENCES

1. J. C. Weaver and R. D. Astumian, "The response of cells to very weak electric fields: The thermal noise limit," *Science* 247: 459-462, 1990.

2. M. Winterhalter and W. Helfrich, "Deformation of spherical vesicles by electric fields," *J. Colloid and Interfac. Sci.* 122: 583-586, 1988.

3. D. C. Chang, B. M. Chassy, J. A. Saunders, and A. E. Sowers, Eds., *Guide to Electroporation and Electrofusion,* New York: Academic Press, 1992.

4. R. D. Astumian, P. B. Chock, T. Y. Tsong, and H. V. Westerhoff, "Effects of oscillations and energy-driven fluctuations on the dynamics of enzyme catalysis and free-energy transduction," *Phys. Rev.* A-39: 6416-6435, 1989.

FIG. 2.—(a) Second continuum model, with three charges (here each 1.6×10^{-19} C) localized to smaller uniform spherical volumes and U = 70 mV. (b) Corresponding electrostatic energy of the protein/membrane system as a function of displacement normal to membrane. Two cases are presented: cylindrical protein within membrane has dielectric region that moves with protein charges, for which relative dielectric constant is $\varepsilon = 2$, and transmembrane voltage is U = 70 mV (top); the same, but with U = 100 mV (bottom).

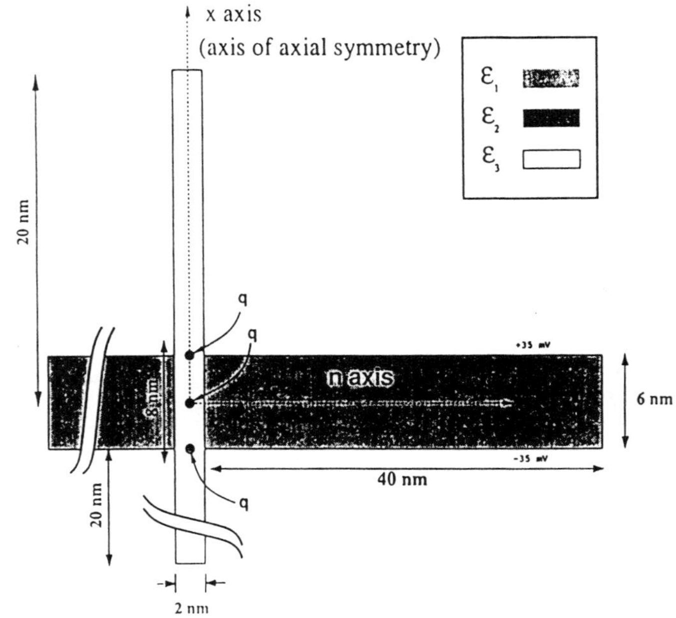

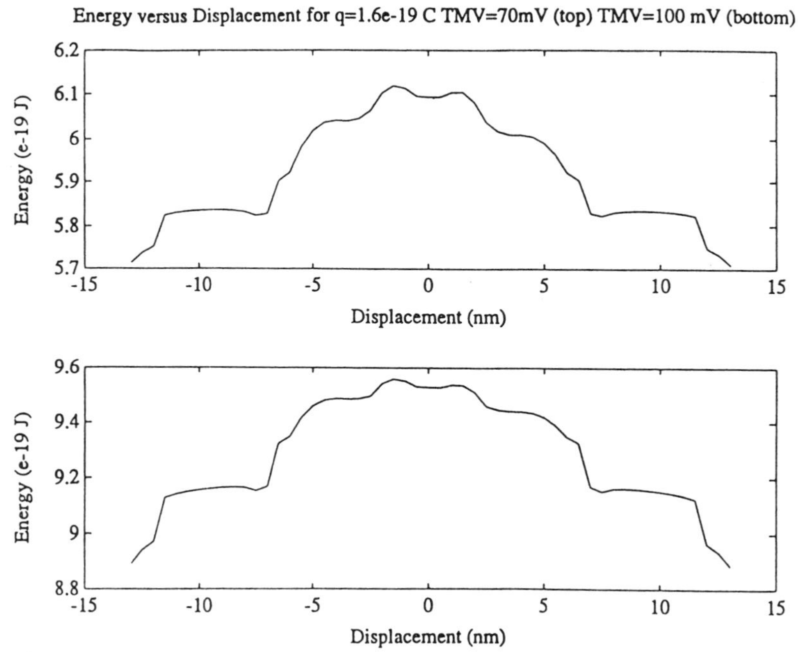

5. J. C. Weaver, "Membrane electroconformation changes: Progress in theoretical modelling of electroporation and of protein protrusion alteration," in J. Allen, S. F. Cleary, and F. M. Hawkridge, Eds., *Charge and Field Effects in Biosystems*, New York: Plenum, 1992, 3 ed., 475-496.

6. M. Zahn, *Electromagnetic Field Theory: A Problem Solving Approach*, New York: Wiley, 1979.

7. H. H. Woodson and J. R. Melcher, *Electromechanical Dynamics* (Part 1: Discrete Systems), New York: Wiley, 1968.

8. V. A. Parsegian, "Energy of an ion crossing a low dielectric membrane: Solutions to four relevant electrostatic problems," *Nature* 221: 844-846, 1969.

9. Y. Mouneimne, P-F. Tosi, Y. Gazitt, and C. Nicolau, "Electro-insertion of Xeno-glycophorin into the red blood cell membrane," *Biochem. Biophys. Res. Comm.* 159: 34-40, 1989.

10. This work was supported by NIH grant ES06010, and by a computer equipment grant from Stadtwerke Düsseldorf, Düsseldorf, Germany.

Martin Blank, Ed., *Electricity and Magnetism in Biology and Medicine*

COMPARISONS OF HAND WARMING AMONG TWO RF GLOVES AND A DC GLOVE

J. R. Lloyd,* R. G. Olsen, and B. J. Van Matre

Naval Aerospace Medical Research Laboratory, Pensacola, Fla.

Radiofrequency (RF) rewarming studies have indicated a potentially more efficient modality for active thermal protection of extremities. Two different 27.12MHz RF gloves (electrically balanced and unbalanced) and a direct-current (DC) resistive wire glove were compared for ability to warm cold-stressed hands. Each glove received a net input power of 15 W while hand surface skin temperatures (t_{sk}) reflected the progression of heating over a 5min period. Temperature data were recorded from eight locations on each hand. Our results show significantly higher average hand temperatures (t_{hand}) with the unbalanced RF glove.

Hands exposed to cold environmental conditions show reduced work performance from the fingers even if the exposure time is relatively short.[1] The need for longer personnel exposures in remote areas and harsh environments demands that active thermal protection systems be less encumbering, lighter in weight, and highly efficient in converting energy to tissue warmth. Previously investigated as an efficient core-warming method,[2] RF energy has been shown to be an effective method in rewarming cold-stressed hands and feet.[3] We show in this study a comparison among two RF glove designs and a DC-heated glove.

MATERIALS AND METHODS

Two pairs of series finger coils, designed to resonate at 27.12 MHz, were constructed in the same manner as coils used in a previously study.[4] Each of these hand coil assemblages were fitted over the outside of a glove of the same type (knitted nylon) as used in the DC test but with the resistive wire extricated. Delivery of RF energy to each hand coil of one pair was through RG-58A (50 Ω) coax cable soldered directly onto the coil, forming an electrically unbalanced system. Identically configured coax cables were used for the second pair of RF coils (balanced system) that coupled through Amidon toroidal baluns (T-68-6 wound 1:1) and 0.01μF capacitors. A pair of $28V_{dc}$ gloves (Vacuum Reflex Ltd.) functioned as the third modality of warming. RF equipment, setup, and temperature data acquisition were identical to that used in an earlier study.[4]

Each of six subjects experienced whole-body exposure to 5 C air for each type of rewarming glove. Hand-surface skin temperatures were monitored at eight locations per hand; four probes were positioned at the palmar tips of index, ring, little, and middle palm hand; the remainder four probes were at the dorsal base of the same fingers and back of the hand. Although pretreatment cooling rates for hands varied widely among subjects, rewarming was initiated only when any monitored temperature on either hand cooled below 15.0 C. Warming continued for 5 min followed by 5 min of monitored cooling before termination of the test.

RESULTS

Graphed data (Fig. 1) of t_{hand} responses to generated heat for all eight locations of the left hand show better overall warming from the electrically unbalanced RF glove. A similar response was seen in the right-hand data. Repeated measures analysis of variance (ANOVA) showed a difference during irradiation at the $p < 0.01$ level of significance for both hands between the RF unbalanced and DC gloves. The ANOVA indicated no difference between the RF balanced and DC gloves.

*Correspondence to: J. R. Lloyd (Research), Naval Aerospace Medical Res. Lab., 51 Hovey Rd. (NAS), Pensacola FL 32508-1046.

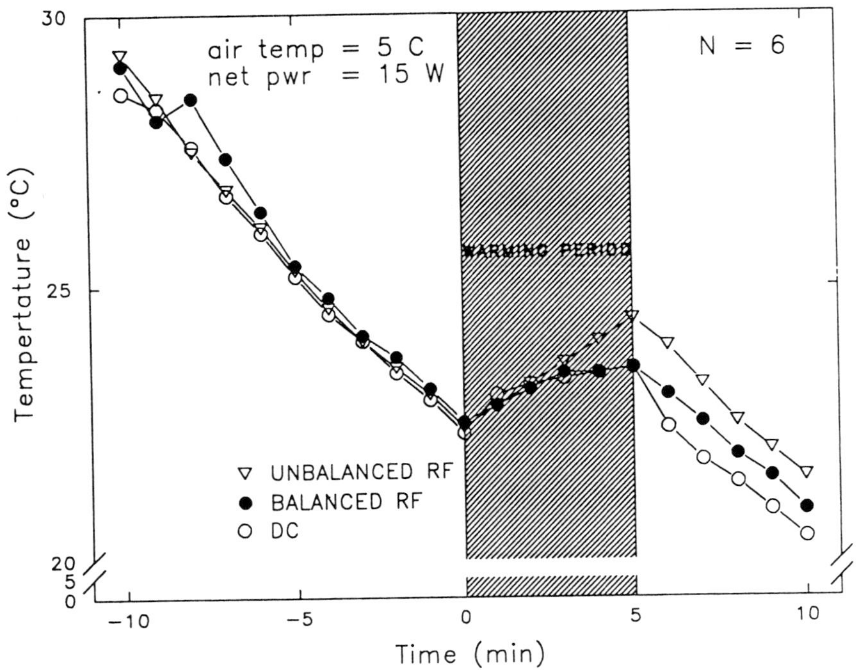

FIG. 1.—Averaged surface skin temperatures (t_{hand}) from eight locations of left hand.

DISCUSSION AND CONCLUSION

The post-irradiation cooling period was important in contrasting the methods of warming. During this period, we saw that initial responses from the RF gloves were quite different from those of the DC glove, particularly for the left hand, as shown in Fig. 1. During the first minute after irradiation, t_{hand} of the DC glove dropped twice as much as that for the RF gloves, a clear indication that the total energy transferred to the deep tissues of the hand was greater for the RF gloves than for the DC gloves. In addition, Fig. 1 shows that during irradiation, t_{hand} of the unbalanced RF glove increased nearly twice as much as that of the DC glove in spite of the fact that the temperature probes were very close to the DC heating elements.

We believe that our method of thermometry may have biased the during-irradiation data in favor of the DC gloves; nevertheless, the overall outcome has shown that RF energy provides a superior means of heat deposition in the hands.[5]

REFERENCES

1. S. M. Horvath and A. Freedman, "The influence of cold upon the efficiency of man," *J. Avia. Med.* 18: 158-164, 1947.

2. R. L. Hesslink Jr., S. Pepper, R. G. Olsen, S. B. Lewis, and L. D. Homer, "Radiofrequency (13.56 MHz) energy enhances recovery from mild hypothermia," *J. Appl. Physiol.* 67: 1208-1212, 1989.

3. J. R. Lloyd and R. G. Olsen, "Radiofrequency energy for rewarming of cold extremities," *Undersea Biomed. Res.* (in press).

4. R. G. Olsen, "RF energy for warming divers' hands and feet," in M. E. O'Connor, R. H. C. Bentall, and J. C. Monahan, Eds., *Emerging Electromagnetic Medicine,* New York: Springer-Verlag, 1990, 135-144.

5. This research was sponsored by Commander, U.S. Naval Special Warfare Command. The views expressed in this article are those of the authors and do not reflect the official policy or position of the U.S. Department of the Navy, U.S. Department of Defense, nor the U.S. Government. Volunteer subjects were recruited, evaluated, and employed in accordance with the procedures specified in U.S. Department of Defense Directive 3216.2 and Secretary of the Navy Instruction 3900.39 series. These instructions are based on voluntary informed consent and meet or exceed the provisions of prevailing national and international guidelines.

Martin Blank, Ed., *Electricity and Magnetism in Biology and Medicine*

HUMAN ENDOGENOUS 60Hz FIELDS

J. A. Bergeron*

GE Corporate R&D, Schenectady, N.Y.

Human endogenous fields are dominated, at rest, by the heartbeat. The electrocardiogram (ECG) is complex. Fourier analysis was used to assess the 60Hz content by calculation for an idealized ECG having only the triangular spike (R) and by display of the Fourier transform in real time (FFT) for lead I. The calculations showed about 90% of the amplitude between the fundamental (1.2 Hz) and the 20th harmonic. A secondary maximum (0.05 of fundamental) appeared between harmonics 42 and 75. The FFT exhibited 60 Hz about 0.001 of the R maximum whether "windowed" on the QRS complex or elsewhere.

Despite growing concern about alleged health hazards due to exposure to very weak ambient 60Hz fields, little attention has been given to the 60Hz fields produced in the body by physiological activity; i.e., the endogenous 60Hz fields. Consideration of such fields is clearly important as a reference for assessing the relevance of the fields induced in the body by exposure to weak ambient 60Hz fields. In the human at rest the dominant source of endogenous ELF is the heart. The electrocardiogram (ECG) is a complex waveform repeated about once per second. Typically, the most obvious feature (R wave) is an approximately triangular spike of about 0.04s duration with a period of about 0.82 s and about 1mV peak as sensed at the periphery. The intensity increases with proximity to the heart. The intensity at the surface of the heart is sufficient to excite a nerve, as in the original experiment used to establish the electrical activity of the heart beat.[1] Golden et al. obtained the Fourier transform between DC and 10 kHz of 203 lead averaged cycles from 10 normal men.[2] The mean intensity observed at 60 Hz relative to the fundamental at 1.2 Hz was about 0.02. The ratio of the fundamental to the R maximum (1.5 mV) was 0.056. The ratio of 60 Hz to the R maximum was 0.0011. The 60 Hz was in a secondary maximum suggestive of possible "noise" pickup despite the high quality of the experiments.

EXPERIMENTAL

Calculations. We performed Fourier harmonic analysis of the idealized R spike (triangular) and observed a weak secondary maximum between harmonic numbers 40 and 80 (Fig. 1). The maximum was at about harmonic 56 (68 Hz). The intensity relative to the fundamental (1.2 Hz) was 0.046.

Human Experiments. We obtained real-time Fourier transforms of the ECG (Lead I, arm-to-arm) with vector averaging ($n = 60$) at a resolution of 2 Hz. Dual channel operation with 100ms "windows" allowed comparison of the 60Hz component within the QRS complex and subsequent portions such as the nominally isopotential segment between the T and P "waves" (Fig. 2a and b).

A battery-powered differential amplifier based on the 284J (Analog Devices) was used to sense the potential difference (0-1 kHz) between the arms. The right leg lead was tied to the guard. The unamplified (1:1) output was applied to the input of the 35665A dynamic signal analyzer (Hewlett-Packard). The subject sat facing west on a wooden chair inside a metal mesh enclosure that also included the battery-powered differential amplifier. The electrodes were commercial adhesive "disposables." The DC resistance between leads ranged from 20 to 50 kΩ. The ELF E and B fields in the unoccupied enclosure at head height were measured on three axes (compass directions and vertical) with the 140 and 160 meters (Electric Field Measurements). The E field ranged from 0.1 to 0.3 V/m; the B field was 0.1 mG or less. Under the measurement conditions the 60Hz "noise" within the analyzer was 25-29 nV rms (25Ω termination). When the leads were attached to the vertices of a triangle of 100K resistors simulating the resistance of the body, the 60Hz component was 64.6-64.7 nV rms. The 60Hz component of the ECG was 1.1-1.2 μV when the maximum in the ECG was 1

---

*Correspondence to: J. A. Bergeron, GE Corp. R&D (Bldg K1, Rm 3B14), Box 8, Schenectady NY 12301.

mV, greater by a factor of about 20 than observed with the 100K "dummy."

DISCUSSION

The measurement of weak biopotentials is always complicated by potential pickup of 60Hz "noise." The prior observation of a secondary maximum around 60 Hz in the FFT of the ECG might have reflected such noise.[2] However, we found that just such a secondary maximum can be expected as a component of a triangular spike with the duration and repetition of the R "wave." In addition, our experimental measurements agree with the findings of Golden et al.[2] The ratio of the 60Hz component to the R maximum was 0.0011. The 60Hz intensity appeared relatively constant over the cycle, as if the R were not the only origin.[3]

REFERENCES

1. R. A. Kolliker and J. Meuller, "Nachweis der negativen Schwankung des Muskelstroms am natürlich sich contrahirenden Muskel," *Verhandl. phys. med. Gesellsch. Würzburg* 6: 528, 1856.

2. D. P. Golden Jr. et al., "A spectral analysis of the normal resting electrocardiogram," *Trans. IEEE* BME-??: 366, 1973.

3. The contributions of L. H. String in the selection, design, and construction of equipment and the experimental measurements were essential for this study. Helpful discussions and exchange of information with R. K. Adair are also acknowledged.

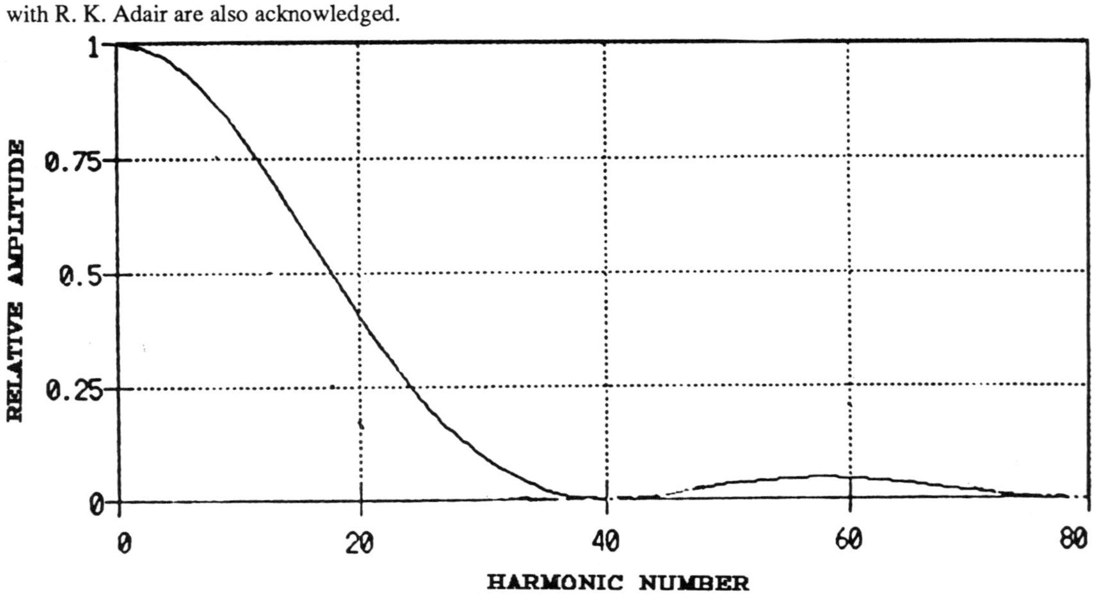

FIG. 1.—Relative amplitude vs harmonic number for triangular spike (0.04s base, 0.8s period).

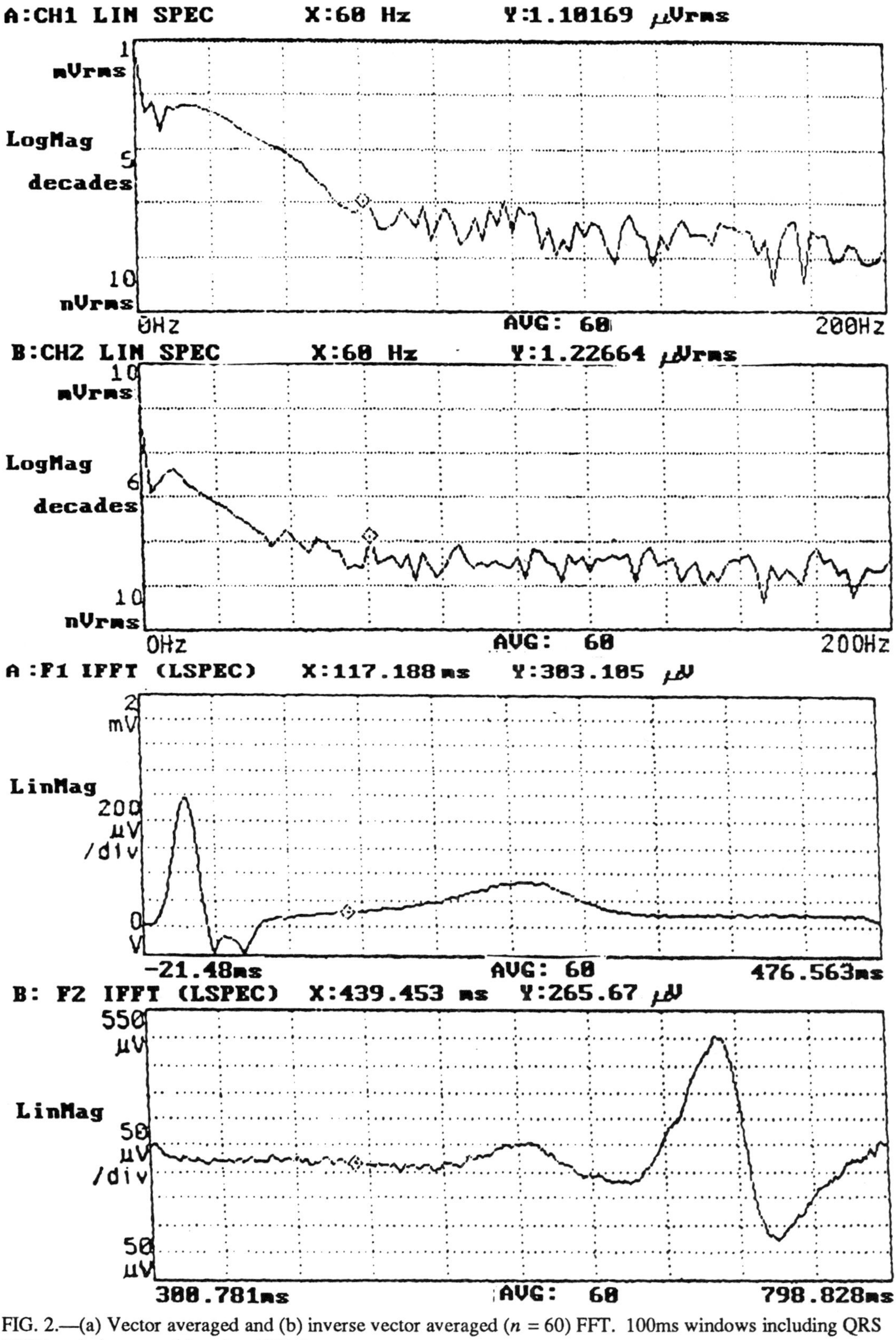

FIG. 2.—(a) Vector averaged and (b) inverse vector averaged (n = 60) FFT. 100ms windows including QRS (CH1) or in isopotential region (CH 2).

Martin Blank, Ed., *Electricity and Magnetism in Biology and Medicine*

ACUTE EFFECTS OF ELECTRICALLY STIMULATED CYCLING EXERCISE ON OXYHEMOGLOBIN AFFINITY IN QUADRIPLEGICS

M. S. Nash,* J. R. Hughes, and B. A. Green

University of Miami School of Medicine, Fla.

R. E. Isaacks

Miami Veterans Administration Medical Center, Fla.

To determine whether exercise alters blood oxygen affinity [P_{50}] in quadriplegics, four subjects underwent 30 min of cycling via electrical stimulation of quadriceps, hamstring, and gluteus muscles. Blood drawn before, after, and 30 min following cycling was analyzed for intraerythrocyte 2,3-bisphosphoglycerate (2,3-DPG) and P_{50}. Electrically stimulated cycling failed to alter the blood oxygen dissociation curve during exercise, although a slight rightward shift was noted afterward. This shift was not attributable to 2,3-DPG [4.055 ± 0.375 before, to 3.994 ± 0.492 micromoles/ml RBC (mean ± s.d.) after stimulation (p = n.s.)]. We conclude that erythrocyte regulation of O_2 delivery is unaltered by cycling exercise in quadriplegics.

Oxygen (O_2) transport and utilization during exercise are facilitated by work responses that increase cardiac output, shunt blood to areas with low resting flow, and augment the arteriovenous oxygen difference. The last mechanism is facilitated through tissue oxygen delivery regulated by the blood-tissue O_2 gradient, as well as intraerythrocyte mechanisms that facilitate O_2 transport from blood to metabolically active tissues.[1] This reaction is controlled by the well-characterized effector of the oxyhemoglobin complex, 2,3-bisphosphoglycerate (2,3-DPG), whose acute transitory elevation during cardiorespiratory exercise promotes a rightward shift in the oxyhemoglobin dissociation curve.[2] This shift favors both erythrocyte O_2 dissociation and tissue O_2 delivery, as well as elevation of the partial pressure at which hemoglobin is 50% O_2 saturated [P_{50}].[3]

Facilitation of O_2 transport and the prevention of tissue hypoxia are important for persons with spinal-cord injury (SCI), as many sustain chronic circulatory stasis and intermittent tissue hypoxia associated with immobilization-related physical deconditioning, vasomotor paralysis, and ischemic necrosis of skin and muscle. That persons with cervical SCI can now undergo extended periods of steady-state cycling exercise through activation of lower extremity muscles with computer-sequenced transcutaneous electrical current[4] suggests that these muscles experience increased tissue oxygenation during exercise. To date, however, the mechanism by which this oxygenation occurs has not been examined. The purpose of the study was to test the hypothesis that electrically stimulated cycling alters erythrocyte 2,3-DPG and P_{50} in quadriplegics.

METHODS

Subjects were four male volunteers with complete motor lesions (Frankel Grades A and B) between the C5 and C7 levels. Subjects who were previously habituated to electrically stimulated cycling exercise twice weekly for 6 months underwent 30 min of steady-state cycling by computer-sequenced transcutaneous electrically stimulated contractions of the bilateral quadriceps, hamstring, and gluteus muscle groups (monophasic rectangular waveform, constant current to peak output of 132 mA, 375μs pulse duration). Imposed resistance was greatest during which subjects could maintain a constant pedal rate greater than 35 rpm, but did not exceed 1/8 kp. Metabolic responses for oxygen uptake (VO_2) and respiratory exchange ratio (RER) were monitored by open-circuit spirometry on a metabolic measurements analyzer via a Hans-Rudolph nonre-

---

*Correspondence to: M. S. Nash (Orthopaedics & Rehab.), University of Miami School of Medicine, 5915 Ponce de Leon Blvd (5th fl.), Coral Gables FL 33146.

breathing valve. Blood was taken via an indwelling antecubital venous catheter immediately before exercise, immediately after 30 min of steady-state cycling, and 30 min following cessation of exercise. Erythrocyte 2,3-DPG was quantified by the phosphoglyceromutase method, with P_{50} measured via a modification of the Langmuir and Chow polarographic technique. Hematocrit (Hct) was determined by standard microcapillary tube centrifugation. Hemoglobin was quantified spectrophotometrically by the cyanmethemoglobin reagent method. Data were analyzed by multivariate repeated measures ANOVA, with significance accepted at the 0.05 level.

RESULTS

Results showed that ESCE stimulated a 4.5-fold increase in steady-state VO_2. Respiratory exchange ratios (RER = VCO_2/VO_2) as high as 1.15 were noted in all subjects in the early phase of cycling, and greater than unity throughout their steady-state exercise. Pre-exercise values for all dependent measures (Table 1) were within the reference range for persons without disability. No significant effect of exercise on P_{50} was sustained by the study subjects, although a slight decreasing trend immediately following ESCE and recovery may represent a mild thermic shift in the oxyhemoglobin dissociation curve. This trend is not attributable to changes in 2,3-DPG, as their concentrations decreased insignificantly from 4.055 ± 0.375 (mean ± s.d.) to 3.994 ± 0.492 micromoles/ml RBC immediately following exercise, but returned to baseline following exercise recovery (Table 1). Absence of a clinically significant change in Hct suggests that the usual hemoconcentration observed during aerobic exercise in nondisabled persons was not experienced by these subjects during ESCE. If anything, the transitory change observed in volume-dependent measures (i.e., Hct and Hb) may be due to a hemodilution effect attributable to entry of extravasated fluid into the peripheral circulation from the normally edematous lower extremities of these subjects.

TABLE 1.—Effects of electrical stimulation cycle ergometry exercise on erythrocyte 2,3-bisphosphoglycerate (2,3-DPG), P_{50}, hematocrit, and hemoglobin (mean ± s.d., n = 4).

	Pre-exercise	Post-exercise	Post-recovery	*p*
2,3-DPG (μm/ml RBC)	4.055 ± 0.376	3.994 ± 0.492	4.105 ± 0.38	ns
P_{50}	27.22 ± 0.77	26.93 ± 0.69	26.73 ± 1.1	ns
Hct (%)	42.38 ± 1.48	41.59 ± 1.50	42.03 ± 1.6	ns
Hb	134.72 ± 0.89	127.0 ±12.65	129.91 ±11.9	ns

DISCUSSION

The absence of a significant effect of electrically stimulated exercise on erythrocyte metabolism contrasts with several reports in which erythrocyte 2,3-DPG and P_{50} of nondisabled persons were acutely elevated following cardiorespiratory exercise.[5] Several of the reports suggest that elevation of 2,3-DPG most often accompanies submaximal exercise of long duration, during which the demand for skeletal muscle oxygen delivery and utilization is greatest.[6] In contrast, the rise in oxygen uptake in quadriplegics undergoing ESCE was modest when compared with those of nondisabled individuals undergoing cardiorespiratory exercise. This nominal rise in VO_2 may be attributed to (1) the effects of physical immobilization and nervous system decentralization on biochemical and contraction properties of skeletal muscles, (2) unique activation of skeletal motor units by transcutaneous electrical stimulation, and (3) atypical exercise responses of persons with autonomic nervous system decentralization secondary to cervical spinal-cord injury.

It is known that skeletal muscles undergo biochemical and functional changes favoring characteristics of Type II (fast) motor units following either physical deconditioning[7] or nervous system decentralization, both of which are experienced by persons with cervical SCI. Thus, the recruitment of these fibers during exercise favors anaerobic and not oxidative metabolism. This observation is supported by the metabolic findings of this study, in which RERs above unity were maintained even under steady-state conditions with low imposed workloads, evidence for the anaerobic character of the exercise. Additionally, activation of skeletal muscle fibers by transcutaneous electrical current is known to differ from recruitment under central command. This difference results from activation of the large-diameter peripheral nerve fibers, which have low thresholds for electrical stimulation.[8] As these large-diameter axons also preferentially innervate large motor units (Type II), the recruitment of these muscle fibers also favors glycolytic metabolism. Thus, while low intensity movements under voluntary command preferentially activate Type I (slow) motor units with the highest oxidative capacity, the reciprocal response is observed in electrically stimulated muscle. Last, it is known that autonomic dysfunction associated with cervical spinal cord injury may alter exercise responses favoring glycolytic metabolism, a response that has been experimentally mimicked in nondisabled persons undergoing exercise following administration of a beta-adrenergic blocking agent.[9] It is of interest that the response of the subjects under beta-blockade, including a reduction in steady-state VO_2 and elevation of RER, parallels those observed for quadriplegics undergoing ESCE.

The great reliance on glycolytic metabolism to fuel muscle contractions in quadriplegics undergoing ESCE may represent a limiting factor in substrate availability for 2,3-DPG synthesis. It is known that 2,3-DPG is synthesized in the Rapoport-Luebering cycle from the glycolytic triose intermediate glyceraldehyde-3-phosphate (G-3-P), which is then phosphorylated to 1,3 bisphosphoglycerate and converted to 2,3-DPG via the phosphoglycerate mutase reaction. Otherwise, 1,3-DPG may also be synthesized by phosphorylation of 3-phosphoglyceric acid. As these reactions are driven by G-3-P substrate availability from glycolysis, high glycolytic activity may limit G-3-P availability for these reactions. Moreover, it is known that high levels of glycolytic metabolism during exercise inhibit specific enzymes in the Embden-Meyerhof pathway,[7] including its principal regulatory enzyme, phosphofructokinase. Thus, supply of substrate from 2,3-DPG may be limited both by competition during glycolytic ATP synthesis and by reduction of total glycolytic metabolism from reduced enzymatic activity.

CONCLUSION

We conclude that quadriplegics undergoing ESCE experience neither erythrocyte 2,3 DPG nor P50 alterations. This outcome contrasts with results reported for nondisabled exercisers, and may be attributable to a modest elevation of VO_2 during exercise or to factors related to high glycolytic metabolism sustained during exercise. The atypical metabolic responses of these subjects require additional study.[10]

REFERENCES

1. J. P. Klein, H. V. Forster, R. D. Stewart, and A. Wu, "Hemoglobin affinity for oxygen during short-term exhaustive exerciser," *J. Appl. Physiol.* 48: 236-242, 1980.
2. P. Lunen, P. Hespel, S. Van Oppens, R. Fiocchi, W. Goosens, E. Vanden Eynde, and A. Amery, "Erythrocyte 2,3 diphosphoglycerate and serum enzyme concentrations in trained and sedentary men," *Med. Sci. Sports Exerc.* 18: 174-179, 1986.
3. J. W. Eaton, J. A. Faulkner, and G. J. Brewer, "Response of the human red cell to muscular activity," *Proc. Soc. Exptl. Biol. Med.* 132: 886-887, 1969.
4. K. T. Ragnarsson, S. Pollock, J. Petrofsky, W. O'Daniel, R. Edgar, and M. S. Nash, "Clinical evaluation of computerized functional stimulation after spinal cord injury: A pilot multicenter study," *Arch. Phys. Med. Rehabil.* 69: 672-677, 1988.
5. K. Remes, P. Vuopio, and M. Harkonen, "Effect of long-term training and acute physical exercise on red cell 2,3 diphosphoglycerate," *Eur. J. Physiol.* 42: 199-207, 1979.
6. H. D. Meen, P. H. Holter, and H. E. Refsum, "Changes in diphosphoglycerate after exercise," *Eur. J. Appl. Physiol.* 46: 177-184, 1981.
7. A. J. Sergeant, C. T. M. Davies, R. H. T. Edwards, et al., "Functional and structural changes after disuse of human muscle," *Clin. Sci. Mol. Med.* 52: 337, 1977.
8. N. I. Spielholz, in J. V. Basmajian and S. L. Wolf, Eds., *Therapeutic Exercise,* Baltimore, Md.: Williams and Wilkins, 1990, 68-69.
9. M. A. van Baak, J. A. G. Wijnen, C. deHaan, H. Struyker Boudier, and L. van Bortel, "Beta-blockade and lipolysis during endurance exercise," *Med. Sci. Sports Exerc.* 24: S176, 1992.

10. This research is supported by the Miami Project.

Martin Blank, Ed., *Electricity and Magnetism in Biology and Medicine*

INFLUENCE OF ELECTRICALLY STIMULATED CYCLING EXERCISE ON H-REFLEXES IN HUMAN QUADRIPLEGIA

M. S. Nash* and N. I. Spielholz

University of Miami School of Medicine, Fla.

To test whether decreased motor neuron excitability explains observed relaxation of spastic muscle following electrically stimulated cycling exercise (ESCE) in persons with spinal cord injury (SCI), four subjects (3 quadriplegic, 1 paraplegic) underwent 30 min of ESCE via transcutaneous activation of quadriceps, hamstring, and gluteus muscle groups. H and M onset latencies, and H_{max}/M_{max} amplitude ratios were measured before and after ESCE. Resting H/M of SCI subjects differed from nondisabled controls ($p < 0.01$), although pre-ESCE H/M (0.65 ± 0.29, mean ± s.d.) remained unchanged following treatment (0.66 ± 0.23). We conclude that decreased muscle tone following ESCE is not due to reduction of motor neuron excitability.

Skeletal muscle hypertonicity is a potentially complicating clinical outcome of spinal-cord injury (SCI). Involuntary muscle contractions and skeletal muscle rigidity following SCI are attributable to diminished or lost influence of supraspinal inhibition on spinal cord alpha-motor neurons, interneurons, and muscle spindle Ia afferents.[1] Other neural factors that contributed to spinal spasticity may include sprouting of unaffected intraspinal fibers, denervation hypersensitivity, transsynaptic interneuronal degeneration, and progressive loss of internuncial spinal cord neurons.[1] Uncontrolled spasticity represents a serious obstacle to the well-being of persons with SCI, especially when it decreases functional range of motion, promotes joint contracture, compromises optimal sitting posture, disrupts daily activities, or fosters pain.

In addition to pharmacological management[2] and surgical remedies for spasticity,[3] muscle relaxation in spastic hemiparetics has been noted following transcutaneous electrical stimulation.[4] This response is mediated via (1) stimulation of Ia fibers resulting in facilitation of target muscles and inhibition of antagonists, (2) stimulation-induced contraction of antagonist muscles that slowly stretch spastic-muscle Golgi tendon organs producing reflex inhibition, (3) inhibition via the long loop transcortical reflex mechanism following stimulation of secondary endings, or (4) other unidentified mechanism(s).[4,5]

In addition to the relaxing effect of electrical stimulation on spastic muscle, intensity-dependent relaxation of skeletal muscle has recently been shown following treadmill exercise in nondisabled (nonspastic) persons.[6] Tonic relaxation in these subjects was attributed to diminished excitability of the lower motor neuron pool determined by a reduced ratio of the maximal Hoffman's (H) to Muscle (M) waves. These findings parallel earlier work in nondisabled subjects showing acute and chronic tranquilizing effects of moderate intensity exercise on muscle action potentials[7] and H/M ratios.[8] Similar muscle relaxation has been observed following steady-state electrically simulated cycling exercise (ESCE)[9] in spastic quadri- and paraplegics, although the mechanism(s) responsible for this response remain to be investigated. The purpose of this study was to test the hypothesis that electrically stimulated cycling exercise alters the excitability of the lower motor neuron pool as assessed by the H-reflex in individuals with cervical and thoracic upper motor neuron lesions.

METHODS AND MATERIALS

Four spinal cord injured individuals (3 quadriplegic, 1 paraplegic) with stable complete motor lesions (Frankel Grades A and B) of more than 1 year's duration underwent 30 min of steady-state cycling exercise by use of transcutaneously activated computer-sequenced contractions of the bilateral quadriceps, hamstring, and gluteus muscle groups (monophasic rectangular wave form, 40 Hz, 375μs pulse duration, constant

---

*Correspondence to: M. S. Nash (Orthopaedics & Rehab.), University of Miami School of Medicine, 5915 Ponce de Leon Blvd (5th fl.), Coral Gables FL 33146.

current to 130 mA). H-reflex studies were performed immediately before and after cycling with a Cadwell 5200A EMG system. Before ESCE, subjects were prepared in the prone position, with a pair of recording electrodes placed over the soleus muscle on the dominant side, and subsequent stimulation of the tibial nerve in the popliteal fossa. The surface electrodes were secured with tape and left in place throughout the study and exercise periods. Stimuli used were 300μs square-wave pulses at a rate of 0.5/s and a sweep speed of 5 ms per division. Stimulus intensity was increased until a maximum M wave was obtained, and then decreased and adjusted to produce a maximum H wave. Responses were recorded and stored for measurement of onset latencies and peak-to-peak amplitudes. Four gender-matched individuals without SCI served as control subjects for resting H-reflexes. Data were analyzed by paired Student's *t*-tests, with significance accepted at 0.05 level.

RESULTS

Results showed cycling exercise to elevate VO_2 an average of 4.15-fold above resting levels. As anticipated, significant differences were observed between resting H/M ratios of quadriplegic and control subjects ($p < 0.01$), reflecting presence of an upper motor neuron lesion and heightened excitability of the lower motor neuron pool (Table 1). No significant effects of electrically stimulated cycling on H or M waves, or H/M ratios were observed, with resting H/M ratio of 0.64 ± 0.29 (mean ± s.d.) remaining virtually unchanged following cycling (0.66 ± 0.23) (Table 2).

TABLE 1.—Comparison of resting H and M waves and the H/M ratio in quadriplegics and nondisabled controls.

	Quadriplegics	Nondisabled controls	*p*
H (μV)	6406 ± 418	7625 ± 2021	
M (μV)	12938 ± 6699	18250 ± 6972	
H/M	0.64 ± 0.29	0.41 ± 0.08	<0.01

TABLE 2.—Effects of steady-state ESCE on H and M waves and the H/M ratio in quadriplegics.

	Pre-ESCE	Post-ESCE	*p*
H (μV)	6406 ± 418	7688 ± 2473	
M (μV)	12938 ± 6699	13562 ± 6972	
H/M	0.64 ± 0.29	0.66 ± 0.23	ns

DISCUSSION

Spasticity is a disorder accompanying upper motor neuron lesion resulting in velocity-dependent increases in tonic reflexes and hyperexcitability of spinal reflexes. Skeletal muscle hypertonia is attributable to diminished (or absent) supraspinal inhibition of descending voluntary motor pathways,[1] although there is current evidence suggesting that myofibrillar-specific "muscle stiffness" in resting muscle arising from weak crosslinking of actin and myosin is also facilitated following nervous system decentralization.[10]

The H-reflex is an electrophysiological measure that may be used to characterize spinal cord alpha motor neuron excitability. In persons without disability, H-reflex excitability has been shown to decrease in an intensity-dependent fashion following cardiorespiratory exercise.[6] The change in reflex excitability has been suggested as grounds for the "muscle-relaxing" effects of exercise, findings that are contrary to those observed in the present study of individuals with quadriplegia undergoing steady-state cycling exercise via electrically stimulated muscle contractions.

Several studies have proposed mechanisms by which voluntary and electrically stimulated exercise reduce muscle rigidity. One hypothesis suggests that electrical current may alter spasticity through proprioceptive responses from contracting muscles that alter the influence imposed by cortical efferents upon lower motor neuron pools through a long loop transcortical mechanism.[11] This effect has been demonstrated in persons with hemiplegia secondary to cerebrovascular accident and neurologically incomplete SCI, but the subjects in the present study had sustained motor complete lesions and their lower motor neuron responses are not under supraspinal control. It has also been proposed that elevation of temperature in the body or brain stem accompanying exercise might lead decreased muscle spindle activity and synchronized cortical electrical activity,[12] responses that have been associated with states of relaxation. Similarly, the descending inhibition promoted by elevated temperature would be blocked by SCI, which would explain an absence of reduction of H/M during and following ESCE.

The increased tone of spastic muscles is generally attributed to uninhibited neural activity, yet evidence suggests that a history-dependent mechanical phenomenon of skeletal muscle may also influence its tone. According to this evidence, a stiffness, or "thixotropy," is normally present between actin-myosin crossbridges, even in relaxed muscle, and the inflexibility of these bridges depends on their recent history of use.[13] Basically, muscle use reduces this mechanical stiffness and disuse increases it. Furthermore, it has been suggested that upper motor neuron injuries increase resting crossbridge stiffness above normal.[13] Thus, hypertonia (or increased muscle resistance to passive elongation) may be due in part to increased mechanical stiffness and not entirely to increased activity of neural center.[10] The reduction of muscle stiffness (tone) following electrically stimulated muscle contractions may thus reflect a transient reduction in thixotropic bonding caused by the 30 min of alternating shortening and lengthening of the muscles, a factor that could explain decreased tone despite the experienced elevations of the H/M ratios of subjects with SCI.

CONCLUSIONS

We conclude that no significant effect of steady-state ESCE on excitability of the motor neuron pool is sustained by persons with SCI, although the anticipated differences in H/M ratio between subjects with SCI and nondisabled controls were observed. The findings of this study do not dispute prior observations that muscle rigidity is reduced in persons with SCI undergoing electrically stimulated cycling exercise, but challenge the notion that muscle relaxation accompanying this treatment is related to alterations in excitability properties of peripheral nerves and spinal interneurons.

REFERENCES

1. R. R. Young and B. T. Shahani, "Spasticity in spinal cord injured patients," in R.F. Bloch and M. Bausbaum, Eds., *Management of Spinal Cord Injuries,* Baltimore, Md.: Williams and Wilkins, 1986.

2. R. R. Young and P. J. Delwaide, "Drug therapy-spasticity," *N. Engl. J. Med.* 304: 28-33, 96-99, 1981.

3. R. Abbot, M. Johann, N. I. Spielholz, and F. Epstein, "Selective posterior rhizotomy and spasticity: Relationship between interoperative abnormal muscle contractions and the tone examination," in T. S. Park, L. H. Phillips, and W. J. Peacock, Eds., *Neurosurgery State of the Art Reviews: Management of Spasticity in Cerebral Palsy and Spinal Cord Injury,* Philadelphia: Hanley and Belfus, 4: 471-475, 1989.

4. V. Alfieri, "Electrical treatment of spasticity," *Scand. J. Rehab. Med.* 14: 177-182, 1982.

5. L. Vodovnik, B. R. Bowman, and P. Hufford, "Effects of electrical stimulation on spinal spasticity," *Scand. Rehab. Med.* 16: 29-34, 1984.

6. R. Bulbulian and B. L. Drabos, "Motor neuron excitability: The Hoffman reflex following exercise of low and high intensity," *Med. Sci. Sports Exerc.* 18: 697-702, 1986.

7. H. A. DeVries and G. M. Adams, "Electromyographic comparison of single doses of exercise and meprobamate as to the effects on muscular relaxation," *Am. J. Sports Med.* 51: 134-141, 1972.

8. H. A. DeVries, R. A. Wiswell, R. Bulbulian, and T. Moritani, "Tranquilizer effect of exercise: Acute effects of moderate exercise on spinal reflex activation level," *Am. J. Phys. Med.* 60: 57-66, 1981.

9. K. T. Ragnarsson, S. Pollock, J. Petrofsky, W. O'Daniel, R. Edgar, and M. S. Nash, "Clinical evaluation of computerized functional stimulation after spinal cord injury: A multicenter pilot study," *Arch. Phys. Med. Rehabil.* 69: 672-677, 1988.

10. V. Dietz and W. Berger, "Normal and impaired regulation of muscle stiffness in gait: A new hypothesis about muscle hypertonia," *Exp. Neurol.* 79: 680-687, 1983.

11. M. Wiesendanger, D. G. Ruegg, and G. E. Lucier, "Why transcortical reflexes?" *J. Canad. Soc. Neurol.* 295, Aug. 1975.

12. H. A. DeVries, P. Beckmann, H. Huber, and P. Dieckmeir, "Electromyographic evaluation of the effects of sauna on the neuromuscular system," *J. Sports Med.* 8: 61-69, 1968.

13. M. Lakie and L. G. Robson, "Thixotropy: The effects of stretch size in relaxed frog muscle," *Q. J. Exp. Physiol.* 73: 127-129, 1988.

14. A. Hufschmidt and K.-H. Mauritz, "Chronic transformation of muscle in spasticity: A peripheral contribution to increased tone," *J. Neurol. Neurosurg. Psychiat.* 48: 676-685, 1985.

15. This research is supported by the Miami Project. The authors express thanks to Cadwell, Inc., for use of the EMG unit.

Martin Blank, Ed., *Electricity and Magnetism in Biology and Medicine*

EXTREMELY LOW FREQUENCY PULSED ELECTROMAGNETIC FIELDS INCREASE THE RATE OF COLLAGEN MATURATION DURING HEALING OF EXPERIMENTAL SKIN WOUNDS IN RATS

V. Ottani,* M. Morocutti, M. Franchi, G. Giuliani Piccari, and A. Ruggeri

University of Bologna

M. Valli and C. Balduini

University of Pavia

P. Zaniol

University of Modena

We have previously shown that ELF electromagnetic fields (PEMFs) influence the rate of experimental skin wound healing in the rat. Using the same experimental model studied under light and electron microscopy and with biochemical assays, we have now shown that treatment with ELF-PEMFs increases the rate of collagen production during the early phases of healing, accompanied by a faster decrease of acid-soluble collagen content (i.e., a faster rate of collagen maturation). Type III collagen is reduced following PEMF treatment. These effects do not interfere qualitatively nor quantitatively with the process of wound healing itself.

Although considerable information is now available on the effects of pulsed electromagnetic fields (PEMFs) on calcified tissues both in humans and in experimental models, relatively little is known about the effects of PEMFs on the regeneration of soft tissues.[1] Early experiments from our group have shown that ELF PEMFs increase the rate of rat liver regeneration following partial hepatectomy, without interfering with the control mechanisms regulating the end point of regeneration.[2-4] That ELF-PEMFs are able to increase the rate of soft-tissue regeneration has also been shown in another experimental model: the experimental skin wound in the rat.[5] The aim of this study was to investigate the mechanisms underlying this effect. Using light and electron microscopy and biochemical methods we studied the modifications of collagen and proteoglycans composition in the matrix leading to a regenerated mature skin tissue.

EXPERIMENTAL MODEL

Skin wound healing is a well-known experimental model that has long been used to study the mechanisms of soft-tissue regeneration. As a result of histological, histochemical, and electron microscopy (EM) studies, the process of wound healing has been divided into three successive phases: the inflammatory, the proliferative, and the reorganization phases of healing.[6]

During the first phase, within a few seconds after injury and for several minutes thereafter, the damaged tissue shows a vascular and hemostatic response, followed by a typical inflammatory reaction with infiltration by leukocytes, lymphocytes, macrophages, and mast cells.

The subsequent proliferative phase is characterized by repair of connective and epithelial tissues accompanied by a large ingrowth of new capillaries to restore circulation. Fibroblasts migrate from the nearby connective tissue and form syncitia. From this phase on, fibroblasts produce fibrous protein collagen and proteoglycans. Repair of the epithelium is also characterized by migration of germinative cells that multiply and give rise to differentiating cells, which in turn keratinize. The invasion of capillaries buds is responsible for the granulated appearance of wound surface.

---

*Correspondence to: V. Ottani (Anatomia Umana Normale), University of Bologna, Italy.

The final (remodeling) phase is characterized by a gradual shrinkage and maturation of connective tissue in the wound. Contraction starts several days after the initial wounding and stops after about six weeks. The remodeling of connective tissue with reorganization and orientation of collagen fibrils as a result of various forces exerted within the wound continues for many months.

MATERIALS AND METHODS

Animals, surgery, and treatment were as previously described.[5] Briefly, four-month-old male albino Wistar rats (350-380 g) were used. Under aseptic conditions a square of skin 3×3 cm was removed from the dorsal toraco-lumbar region. Immediately after surgery and every 12 h thereafter, rats were exposed to 30min ELF-PEMF treatments. The shape of the pulse was a positive triangle (50 Hz, 8mT peak) followed by a pause of 2 ms. Control rats were left in the apparatus for the same length of time after it was switched off.

Light and electron microscopy. Wound and surrounding skin samples of sham and PEMF-exposed rats were fixed in 4% formalin or in 2% glutaraldehyde in 0.1M phosphate buffer, pH 7.4. Paraffin-embedded sections were stained with hematoxylin-eosin, Azan-Mallory, picro-Sirius, and Alcian Blue.[5] For EM processing, glutaraldehyde-fixed specimens were postfixed in 1% OsO_4, dehydrated in graded ethanols, and embedded in Araldite. Thin sections were stained in uranyl acetate and analyzed by a Siemens Emilskope 101 EM operated at 80 kV.

Biochemistry. Collagens were extracted from defatted tissue and purified by repeated salt precipitations after limited pepsin digestion in acetic acid. Collagen chains were identified on sodium dodecyl sulfate electrophoresis on polyacrilamide gel (SDS/PAGE), and quantified by densitometric scanning of gels stained with Comassie Brilliant Blue.[7] Glycosaminoglycan chains were obtained by papain digestion in acetate buffer of the whole tissue. Glucidic chains were purified by ethanol precipitation in the presence of acetate, and identified by electrophoresis on cellulose acetate. Statistical significance was determined by Student's *t*-test for unpaired data.

RESULTS AND DISCUSSION

Results of biochemical assays performed during healing at 3, 6, 9, 12, 21, and 42 days after wound creation are shown in Fig. 1.

Total collagen (Fig. 1A) increased during healing in both groups of animals: treated rats reached the normal skin content of unoperated rats 12 days after surgery; untreated rats reached the same amount in 21 days. This result shows that PEMF treatment leads to a larger production of collagen during the early phases of regeneration, as confirmed by the histological and ultrastructural pictures of specimens from treated rats: they showed, 12 day postoperatively, early fibrous evolution of scar tissue with the appearance of a number of bundles of thick collagen fibrils averaging 80 nm (Fig. 2B). Control rats at the same stage exhibited extracellular matrix still including a filamentous and microfibrillar aperiodic component together with thin collagen fibrils (40-50 nm) arranged in bundles (Fig. 2A).

Three days after the operation, almost all collagen was in the acid-soluble form (Fig. 1B) and showed no significant difference between the treated and untreated groups. However, at 6 days, acid-soluble collagen in the treated groups was half that of the controls; and 12 days after the operation the treated group returned to the value of the unoperated rats, whereas the untreated animals still presented high acid-soluble collagen levels. As acid-soluble collagen represents an immature form of collagen without cross links, this result shows that maturation is faster under the action of PEMFs. These results are consistent with the early formation of thick bundles of large collagen fibrils observed in thin sections in PEMF-treated rats.

Type III collagen (Fig. 1C) increased at 3 and 6 days, and decreased thereafter (at day 9, 12, and 21 postoperatively) at a much faster rate in PEMF-treated animals than in controls. Of note is the increase of type III collagen at the 42nd postoperative day in both groups, which suggests a remodeling of the tissue and formation of papillary dermis at this state of the healing process.

Figure 1D reports the content of glycosaminoglycans (GAGs) showing no difference between the two sets of samples. The matrix produced during healing was about twice as much as in normal skin, and back to normal content 42 days after operation. In normal skin samples, the only GAG detected on acetate cellulose electrophoresis was dermatan sulfate (DS); in wound repair, as well as DS, chondroitin sulphate (ChS) and a hybrid ChS/DS form were synthesized, indicative of newly formed tissue.

Pulsed electromagnetic fields used in our experiments increase the rate of wound healing by stimulating protein synthesis and by promoting faster extracellular processing of collagen, leading to faster formation of mature tissue. Although the molecular mechanisms underlying this action of PEMFs are still unclear, it is tempting to speculate that PEMFs can increase the polarity of protein secretion in fibroblasts, and/or allow a correct and rapid alignment of collagen molecules. It follows that the assembly of collagen into fibrils, and the subsequent action of lysyl oxidase and cross link formation, show a shorter lag time leading to earlier tissue deposition and organization.

Under our experimental conditions, the electromagnetic fields used can increase the rate of collagen production and extracellular collagen maturation without interfering either qualitatively or quantitatively with the phases of wound healing. It remains to be determined whether this result is achieved through an increased production and/or activity of post-translational enzymes.

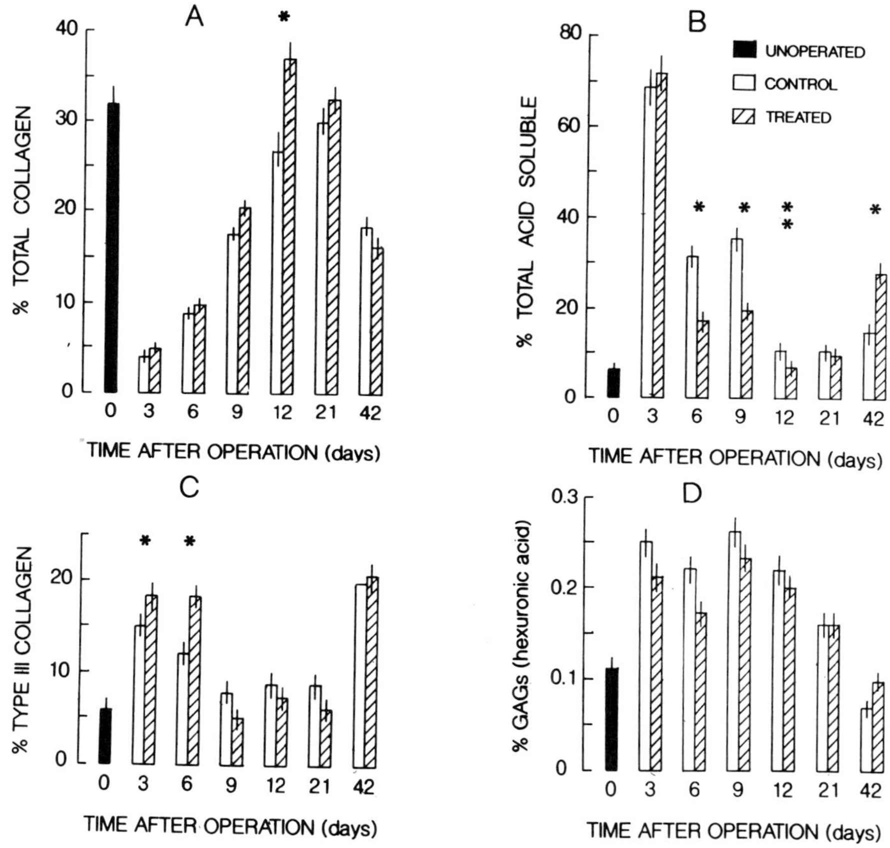

FIG. 1.—Scar tissue content of total collagen (A), acid-soluble collagen (B), type III collagen (C), and glycosaminoglycans (GAGs) (D) during healing of experimental skin wound. Data are expressed as percentage of dry weight. Dark bar, unoperated rats; open bar, sham-exposed rats; crosshatched bar, PEMF-exposed rats.

REFERENCES

1. A. A. Marino, Ed., *Modern Bioelectricity,* New York: Marcel Dekker, 1988.

2. V. Ottani, M. G. Monti, G. Piccinini, L. Pernecco, P. Zaniol, A. Ruggeri, and B. Barbiroli, "Pulsed electromagnetic fields increase the rate of liver regeneration after partial hepatectomy," *Proc. Soc. Expl. Biol. Med.* 176: 371-377, 1984.

3. V. Ottani, M. G. Monti, M. Morocutti, M. Ferri, R. Strocchi, A. Ruggeri, and B. Barbiroli, "Influence of pulsed electromagnetic fields on regenerating rat liver after partial hepatectomy," *J. Anat.* 139: 253-263, 1984.

4. R. Battini, M. G. Monti, M. S. Moruzzi, S. Ferrari, P. Zaniol, and B. Barbiroli, "ELF electromagnetic fields affect gene expression of regenerating rat liver following partial hepatectomy," *J. Bioelectricity* 10: 131-139, 1991.

5. V. Ottani, V. De Pasquale, P. Govoni, M. Franchi, P. Zaniol, and A. Ruggeri, "Effects of pulsed extremely-low-frequency electromagnetic fields on skin wounds in the rat," *Bioelectromagnetics* 9: 53-62, 1988.

6. R. Ross and G. Odland, "Human wound repair," *J. Cell. Biol.* 39: 152-168, 1968.

7. B. Sykes, B. Puddle, M. Francis, and R. Smith, "The estimation of two collagens from human dermis by interrupted gel electrophoresis," *Biochem. Biophys. Res. Commun.* 72: 1472-1480, 1976.

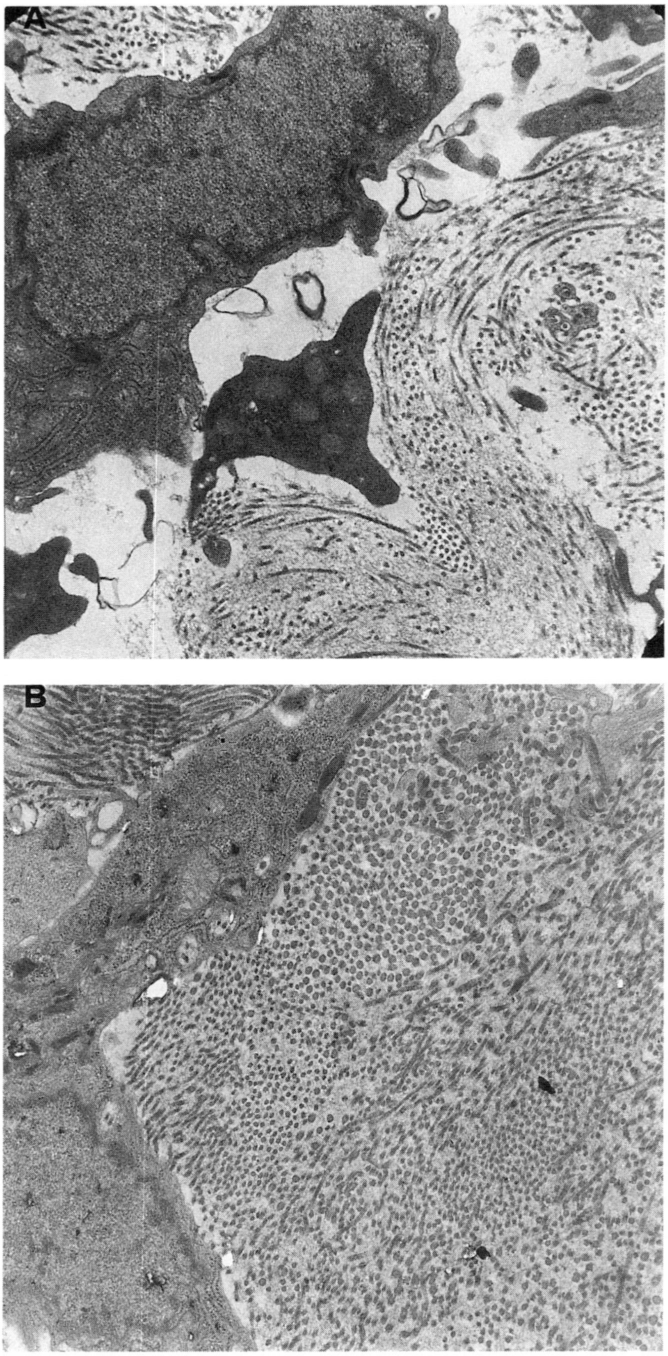

FIG. 2.—Electron micrographs of wound sections 12 days postoperatively. (A) Sham-exposed rats (original magnification 8000×). (B) PEMF-exposed rats (original magnification 8000×).

Martin Blank, Ed., *Electricity and Magnetism in Biology and Medicine*

INTRACAVITARY HYPERTHERMIA AND RADIATION OF ESOPHAGEAL CANCER

C. K. Chou,* J. A. McDougall, K. W. Chan, N. Vora, H. Howard, C. Straud, and L. Terr

City of Hope National Medical Center, Duarte, Calif.

Microwave applicators were designed for esophageal heating and radiation. Heating patterns were measured on phantom muscle with infrared thermography. The applicators have been evaluated on 6 pigs and 3 esophageal cancer patients. Pre- and post-hyperthermia were performed in conjunction with the I-192 brachytherapy. All three patients tolerated the procedure well and developed only short-term mild esophagitis. Two patients are now 3 months post treatment and are doing well with complete disappearance of dysphasia. Preliminary results indicate that this modality can be used without serious problems; continued follow-ups are being conducted to evaluate the safety and efficacy of the treatment.

Hyperthermia has been investigated mostly for treating superficial tumors and has been found to be efficacious, especially in combination with radiation or chemotherapy. Heating deep-seated tumors with external microwave or RF fields is very difficult because the fields have limited penetration. Certain tumor sites in hollow viscera or cavities can be treated with intracavitary techniques. Despite improvement in treatment protocols, current therapies involving surgery, radiation, and chemotherapy of esophageal cancer have not significantly improved the survival rate in the past several decades. Our goal is to improve the treatment results by combining intracavitary radiation and hyperthermia with external radiation. Intracavitary radiation has the advantage of localizing radiation but causes ulceration due to high radiation dosage. Hyperthermia can potentiate the effects and therefore reduce the required dose.

METHODS

Applicator. The polyurethane tube has a center lumen 2.67 mm ID to accommodate a microwave antenna, and 6 outer lumens (1.32 mm ID) for nonperturbing temperature sensors or radiation seeds. A Teflon tube was attached to the surface of the applicator in a one-turn helix (8 cm long) for additional temperature monitoring. The microwave antenna was a monopole made from flexible QMI-600 cable (2.1 mm OD) 10 cm long from the tip to the center of the 1mm gap. The center conductor was connected to the outer conductor at the tip. The operating frequency was 915 MHz.

Heating Patterns. Before and after heating, thermograms were taken of a surface of a $28 \times 28 \times 8.5$cm phantom muscle block covered with polyester screen. The polyurethane applicator with the antenna inserted was placed on the polyester screen side of the block and then covered with additional phantom muscle. The rates of temperature rise are shown in contour plots, line scans, and profile scans.

Animal Study. Three Yucatan and three domestic male pigs (43-54 kg) were anesthetized with Katemine/Xylazine (20/2 mg/kg, IM) and maintained with Isoflurane (1.5-2.5%). The chest region was exposed and Teflon tubes (22 gauge) for inserting Luxtron fiberoptic probes were attached to the esophageal muscle at various locations relative to the microwave antenna junction, 5 sensors near the aorta side and 5 on the opposite side. A Teflon tube was attached longitudinally along the outer surface of the esophagus for temperature mapping. The chest was closed after the tubes were in place. Temperatures inside the applicator and the helical tubing were also mapped. Forty watts of 915MHz power was applied. When steady state was reached, temperatures were recorded. The esophagus was removed during autopsy to determine any obvious tissue effects. Histology was done with light and electron microscopy.

---

*Correspondence to: C. K. Chou (Radiation Oncology), City of Hope National Medical Center, Duarte CA 91010.

Patient Treatments. Two male and one female patients were treated with intracavitary hyperthermia and brachytherapy. Two patients had a primary carcinoma of distal esophagus. One other patient had recurrent carcinoma of lung producing obstruction of distal esophagus. The first patient received 55.8 Gy external radiation in 7.5 weeks, followed by two intracavitary hyperthermia and brachytherapy treatments. The applicator was inserted through a nostril into the esophagus. The first hyperthermia treatment was 40 W for 60 min, followed by an intracavitary radiation with I-192 of 17.5 Gy at 0.75 cm. The second hyperthermia and 12.4Gy intracavitary radiation were delivered 2 weeks later. The second patient with recurrent carcinoma of lung and obstruction of esophagus was given external radiation 20 Gy in 2 weeks with good response, but tumor and symptoms recurred within two months. He then was given 20 Gy additional external radiation in 2 weeks followed by intracavitary hyperthermia and brachytherapy. This treatment entailed hyperthermia at 40 W for 60 min, followed by 30 Gy I-192 intracavitary radiation for two days and a second hyperthermia. The third patient with distal esophageal cancer received 45 Gy external radiation in 4.5 weeks, followed by intracavitary hyperthermia and brachytherapy. She received 30 Gy of I-192 and pre- and post-hyperthermia treatments.

RESULTS

Phantom. Figure 1 shows thermograms of the 10cm monopole. The maximum heating rate is 5 cm anterior to the junction. The heating length (>50% heating rate) is longer than 15 cm; the voltage standing-wave ratio (VSWR) is 3.5.

Animals. Figure 2(a) shows the temperature of pig No. 5 treated with 40 W forward power (11 W reflected). Curve 1 shows temperatures in the esophageal wall near the aorta; curve 6, in the esophageal wall on the opposite side. The aorta side was in general similar to the other side. Curves 2 and 3 are temperatures in the peripheral lumens inside the applicator; data of curve 4 were measured in the helical tube on the surface of the applicator. Curve 5 data were measured by a bare Luxtron 2000 sensor in the esophagus outside the applicator that measured the inner surface temperatures of the esophagus. These temperatures ranged from 43.3 to 45.3 C between 5.5 and 8 cm. Curve 7 are temperatures outside the esophageal wall. The maximum temperature inside the esophageal muscle was 44.4 C measured 6 cm distal to the gap, which was consistent with the thermogram. The temperatures in the peripheral lumens proximal to the gap were higher than the distal temperatures, which is due to the antenna self-heating from current loss and is different from the RF heating. Postmortem examination of the esophagus showed edema at 5-6 cm distal to the antenna gap, consistent with the animal temperature measurement and thermograms.

Light microscopy showed vacuolization and swollen oval cellular nuclei in the heated area. Collagen in the *lamina propria* from the heated area seem to be stretched and their bundles are also parallel to the epithelial surface. Nuclei of fibroblasts in the collagen fabric are elongated along its fibers in the heated region of the esophagus; organization of this collagen is less complicated than in the control. Electron microscopy of epithelial cells in the heated area showed the presence of numerous vacuoles in its cytoplasm and at the cell boundaries.

Patients. The three patients tolerated the treatment very well. Figure 2(b) shows temperature data for treatment 2 of patient 1 (40 W forward, 11 W reflected). The upper four curves are the temperatures within the peripheral lumens. The maximum was at the gap. The antenna self-heating is also seen here. The lower line is the temperatures on the surface of the applicator inside the spiral tubing. The mean and SD are 45.29 ± 2.07 C. Extrapolated to the pig data, the esophageal muscle temperatures are expected to be 42-44 C. Temperature data for treatment 2 of patient 3 is shown in Fig. 2(c). The first two patients showed mild esophagitis symptoms at 1-week and 3-week follow-ups, which can be attributed to the combination of hyperthermia, radiation, and tube irritation. Three months post treatment both were doing well, with complete disappearance of dysphasia. The third patient also showed mild esophagitis at 1 week examination and was able to eat well. At longer follow-ups (3, 6, 12 months) CT scans, barium x-ray, and endoscopic examinations will be conducted to evaluate the tumor response.

CONCLUSIONS

1. A 7-lumen polyurethane applicator was designed to accommodate a microwave antenna and sensors to monitor temperatures or radiation seeds to provide brachytherapy.

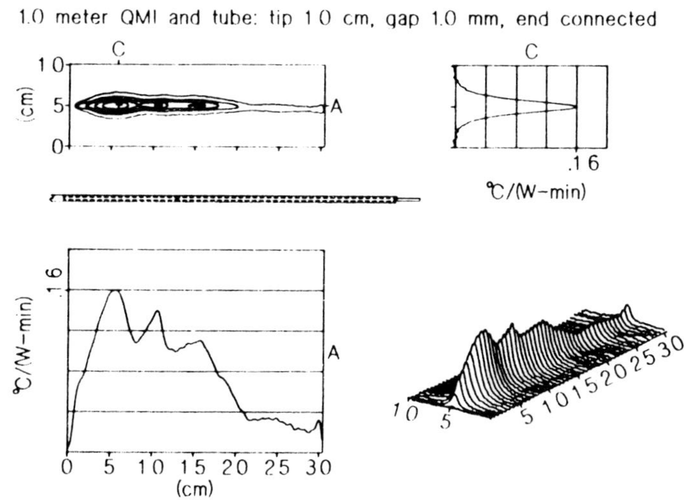

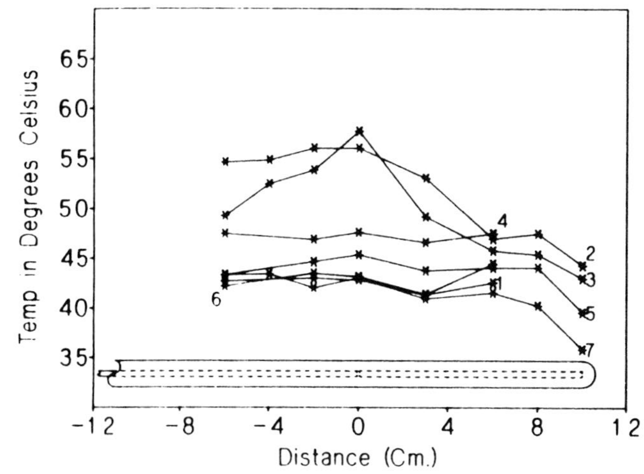

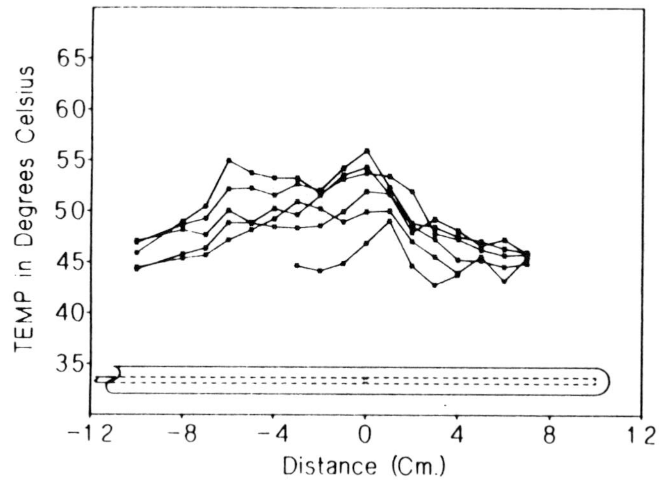

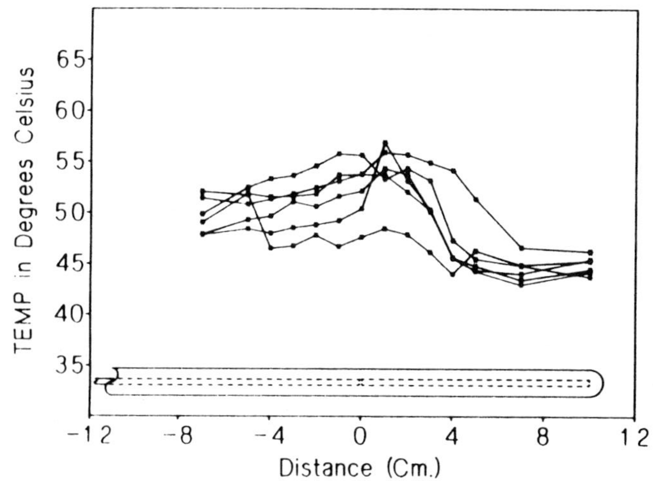

FIG. 1.—Thermograms of 10cm monopole.

FIG. 2.—EMTemperature data with esophageal applicator, 40 W forward and 11 W reflected, for three subjects: (a) pig No. 5; (b) patient No. 1, second treatment; (c) patient No. 3, first treatment.

2. The heating pattern measured in the phantom was consistent with animal temperature data in predicting hot spots.

3. Animal results provide correlation between temperatures measured inside the applicator and in the esophageal muscle wall.

4. Patients tolerated the treatment well and developed only mild esophagitis.

5. Preliminary results on three patients indicate that this modality can be used without serious problems. Continuing follow-ups are being conducted to evaluate the safety and efficacy of the treatment.[8]

REFERENCES

1. M. A. Astrahan, M. D. Sapozink, G. Luxton, T. D. Kampp, and Z. Petrovich, "A technique for combining microwave hyperthermia with intraluminal brachytherapy of the oesophagus," *Int. J. Hyper.* 5: 37-51, 1989.

2. B.-S. Hou, Q.-B. Xiong, and D.-J. Li, "Thermo-chemo-radiotherapy of esophageal cancer: A preliminary report of 34 cases," *Cancer* 64: 1777-1782, 1989.

3. H. Kai, H. Matsufuji, Y. Okudaira, and K. Sugimachi, "Heat, drugs, and radiation given in combination is palliative for unresectable esophageal cancer," *Int. J. Rad. Oncol. Biol. Phys.* 14: 1147-1152, 1988.

4. A. A. Kochegarov, N. K. Muratkhodzhaev, and S. A. Alimnazarov, "Hyperthermia in the overall treatment of esophageal cancer," *Vopr. Onkol.* 26(5): 19-24, 1980 (in Russian).

5. D. J. Li and B. S. Hou, "Preliminary report on the esophageal cancer by intraluminal microwave hyperthermia and chemotherapy," *Cancer Treatment Report* 71: 1013-1019, 1987.

6. H. Matsufuji, K. Ikejiri, H. Kai, K. Sugimachi, and K. Inokuchi, "Superficial spreading esophageal cancer controlled by hyperthermo-chemo-radiotherapy: Case report," *J. Surg. Oncol.* 31: P-243-245, 1986.

7. Z. Petrovich, M. Astrahan, K. Lam, T. Tildon, G. Luxton, and J. Jepson, "Intraluminal thermoradiotherapy for carcinoma of the esophagus," *Endocurie./Hypertherm. Oncol.* 4: 155-161, 1988.

8. This work was supported by NCI grant 1R01 CA56116-01.

Martin Blank, Ed., *Electricity and Magnetism in Biology and Medicine*

EFFECTS OF LOW-ENERGY EMISSION THERAPY (LEET) ON SLEEP STRUCTURE

R. Hajdukovic,* M. M. Mitler, and M. Erman

Scripps Clinic and Research Foundation, La Jolla, Calif.

B. Pasche

Brigham and Women's Hospital, Boston, Mass.

Low-energy emission therapy (LEET), utilizing intrabuccally emitted 27MHz amplitude-modulated electromagnetic fields generated by the Symtonic Device (Symtonic SA, Renens, Switzerland), was used in the treatment of chronic psychophysiological insomnia. The study design was controlled double-blind study with strict inclusion/exclusion criteria and screening procedures.[1,2] The effectiveness of this treatment on objective measures of sleep, such as sleep latency, total sleep time, sleep efficiency, and EEG has also been previously reported.[2-5] We now report the effects of LEET on sleep structure and sleep fragmentation in patients with chronic psychophysiological insomnia.

METHODS

The study employed a parallel-groups, placebo-controlled, repeated-measures design on a total of 30 subjects between the ages of 21 and 50. The screening procedure included review of sleep diaries, sleep interview to exclude subjects with primary sleep disorders such as sleep apnea or periodic leg movements in sleep syndrome, and psychiatric interview to exclude subjects with depression and other major psychiatric disorders. Subjects were excluded from the study if they had a positive history for significant medical illness, chronic alcoholism, drug abuse, psychotic illness, significant affective disorder, sociopathy, hypochondriasis, mental retardation, epilepsy, diagnosed CNS disease, or pregnancy. Screening polysomnography was performed and subjects who met two out of three sleep criteria were admitted into the treatment. The sleep criteria were sleep latency of more than 30 min, total sleep time of less than 6 h, and sleep efficiency of less than 85%.

The patients were randomly assigned to active or inactive (placebo) groups under double-blind conditions. Treatments were given to the subjects in the late afternoon, three times per week for a 4-week period. The duration of the treatment was 20 min. The subjects were lying comfortably on a bed, with the aluminum-coated mouthpiece of the Symtonic device placed in the mouth. The subjects kept the mouthpiece in place by holding the shielded coaxial cable that connects the mouthprobe to the device.

The second polysomnography was performed at the end of the 4-week treatment period. Standardized sleep scoring criteria were used by a scorer who was blinded to the treatment condition (active or inactive-placebo treatment). Sleep parameters assessed for change with this treatment included stage 1 and 2 NREM sleep, slow-wave sleep (SWS— stage 3+4 NREM sleep), REM sleep, wake after sleep onset (WASO), number of awakenings (# awake) and the duration of the longest awakening. The data shown below reflect the means across all subjects in the active or inactive group. All parameters are presented in absolute values (minutes or numbers).

Statistical analysis for this part of the study included a two-factor repeated-measures ANOVA to assess change scores (pre-treatment vs post-treatment) between the two treatment groups. One-way ANOVAs were performed to compare pre-treatment vs post-treatment within the active and inactive groups.

*Correspondence to: R. Hajdukovic, Scripps Clinic, 10666 N. Torrey Pines Rd., La Jolla CA 92037.

RESULTS

Data presented in Table 1 reflect the results obtained from the analysis of sleep structure in the active and inactive group, pre-treatment vs post-treatment. Stage 2 NREM sleep, stage 3+4 NREM sleep, and REM sleep were statistically significantly increased in the active-group post-treatment. Significant decrease of stage 1 NREM sleep and increase of stage 3+4 NREM sleep were noticed in the inactive-group post-treatment.

TABLE 1.—Sleep structure. (Asterisks indicate statistically significant changes.)

	Stage 1 NREM	Stage 2 NREM	Stage 3+4 NREM	REM
Active group (n = 15)				
Pretreatment	44.32	148.12	29.94	43.52
Posttreatment	37.79	199.36	44.22	74.40
p values	0.23	0.009*	0.023*	0.001*
Inactive group (n = 15)				
Pretreatment	47.03	204.26	27.34	58.34
Posttreatment	32.09	187.66	45.40	60.83
p values	0.03	0.29	0.007*	0.84

Data presented in Table 2 indicate the results of the analysis of sleep fragmentation. The active group showed a statistically significant decrease of wake time after sleep onset (WASO) and of the duration of the longest awakening in the post-treatment condition. The inactive group had significant decrease in the number of awakenings post-treatment. This group did not show changes in the wake time after sleep onset and the duration of the longest awakening.

TABLE 2.—Sleep fragmentation. (Asterisks indicate statistically significant changes.)

	WASO	# awake	Longest awakening
Active group (n = 15)			
Pretreatment	149.48	21.06	46.53
Posttreatment	75.89	20.46	21.45
p values	0.0006*	0.82	0.006*
Inactive group (n = 15)			
Pretreatment	104.58	29.40	61.26
Posttreatment	102.21	19.80	51.66
p values	0.93	0.001*	0.67

DISCUSSION

The results reported in this paper present only a small part of an extensive study on the physiologic effects of low-energy emission therapy. Statistically significant results were seen in sleep structure and sleep fragmentation. Within-group analyses showed significant differences between the pre-treatment and post-treatment conditions of the active group for stage 2 NREM sleep, REM sleep, and stage 3+4 NREM sleep, indicating significant increases in these sleep stages after treatment with LEET. WASO and the longest awakening were also significantly decreased in the active group in the post-treatment condition. The inactive group showed significant pre- vs post-treatment differences for stage 1 NREM sleep, stage 3+4 NREM sleep, and the number of awakenings. Between-group analyses revealed no significant intergroup differences. However, there were significant group-by-treatment interactions for stage 2 NREM sleep, WASO, and the number of awakenings.

CONCLUSIONS

The significant increases in stage 2 NREM sleep, stage REM sleep, and slow-wave sleep, as well as significant decreases in WASO and in the duration of the longest awakening, indicate physiological effects of LEET treatment. Although the number of awakenings did not change significantly, the subjects in the active group had significantly shorter individual periods of wakefulness during the night. These findings, coupled with previous data on sleep parameters, EEG, and psychological test measures, support the hypothesis that LEET has significant physiologic activity in humans that leads to improved sleep.[6]

REFERENCES

1. M. Erman et al., ''Insomnia: A comparison of subjective and objective sleep variables,'' *Sleep Res.* 19: 220, 1990.
2. M. Erman et al., ''Effectiveness of low energy emission therapy (LEET) in the treatment of insomnia,'' *Sleep Res.* 19: 221, 1990.
3. B. Pasche et al., ''Diagnosis and management of insomnia,'' *New England J. Med.* 323: 486-487, 1990.
4. M. Erman et al., ''Low energy emission therapy (LEET) treatment for insomnia,'' *Sleep Res.* 20A: 309, 1991.
5. R. Hajdukovic et al., ''Increase in Stage 2 NREM sleep in chronic insomniacs after 4 week treatment with low energy emission therapy, H-2-1,'' *12th Ann. Mtg. Bioelectromagnetic Soc.*, 1990.

6. This research was supported by Symtonic SA, Renens, Switzerland.

Martin Blank, Ed., *Electricity and Magnetism in Biology and Medicine*

EFFECT OF ELECTROMAGNETIC FIELDS ON SPINAL FUSION: A PROSPECTIVE STUDY WITH A CONTROL GROUP

G. I. Mammi* and R. Rocchi

C. Magati Hospital, Scandiano (RE), Italy

M. Di Silvestre

Istituto Ortopedico Rizzoli, Bologna, Italy

We have investigated 35 patients undergoing posterolateral lumbosacral arthrodesis (PLA). Patients were randomly assigned to 2 groups: one PEMF stimulated (experimental), the other not stimulated (control). All patients followed the same postoperative orthopedic protocol. Patients of the experimental group were instructed to use PEMF for 8 h/day for 3 months. Roentgenographies were taken for each patient at 2, 4, 6, and 8 months after surgery. Roentgenographies taken at 4 months were rated according to Dawson. Patients in the control group scored on average 2.3; patients in the experimental group scored 3 ($p < 0.04$). These results demonstrate the positive effect of PEMF on PLA healing.

Since 1983 we have investigated the effect of low-frequency pulsed electromagnetic fields (PEMF) on the healing of nonunions and delayed unions.[1] We also reported the positive effect of PEMF stimulation associated with autologous bone graft for nonunion healing.[2] We have demonstrated the positive effect of PEMF on bone healing in a double blind study, in a group of patients undergoing tibial valgus osteotomy.[3] Recently, PEMF have been used in patients undergoing spinal surgery.[4] Electrical stimulation has been proposed to promote the healing of failed fusions and to enhance the healing rate of recent posterolateral lumbosacral arthrodeses (PLA).[5,6]

We report here our experience with a group of patients undergoing PLA because of the painful instability of the lumbosacral spine.

MATERIAL AND METHODS

Thirty-five patients were operated on by the same senior surgeon (G. I. Mammi). In all cases cancellous autologous bone graft was used to perform the PLA. The surgical procedure was the same for all patients. In the postoperative period patients were maintained in a cast for 3 months. Within 7 days from surgery half the patients in the experimental group began the use of the Igeastimulator (Howmedica). The remaining patients did not undergo stimulation and will be used as a control group.

The average age was 45 in the control group (20 patients) and 47 in the experimental group (15 patients). Six males were included in the control group and 5 in the experimental one.

Patients in the experimental group were instructed to use PEMF for 8 h/day for 3 months. A single coil was positioned outside the cast in the lumbosacral region. The electromagnetic-field generator was set to induce a 2mV electric field at 75 Hz in a coil probe; the impulse width was 1.3 ms. These parameters proved to enhance bone healing in horses[7] and spinal fusions in rats.[8]

Patients were seen 2, 4, 6, and 8 months after surgery. At each visit roentgenographies were taken to check the PLA evolution. The PLA healing was rated according to Dawson's classification:[9] 0, absence of callus; 1, insufficient bilateral fusion; 2, insufficient monolateral fusion; 3, monolateral fusion with ipertrophy; 4, bilateral fusion with hypertrophy. Following Savini's suggestion,[6] the roentgenographic image taken 4 months after surgery was chosen to compare the PLA healing status of the control group with that of the experimental one (Fig. 1). The roentgenograms collected were rated by an orthopedic surgeon unaware of the experimental status of the patients.

*Correspondence to: G. I. Mammi (Orthopaedics), C. Magati Hospital, Scandiano (RE) Italy.

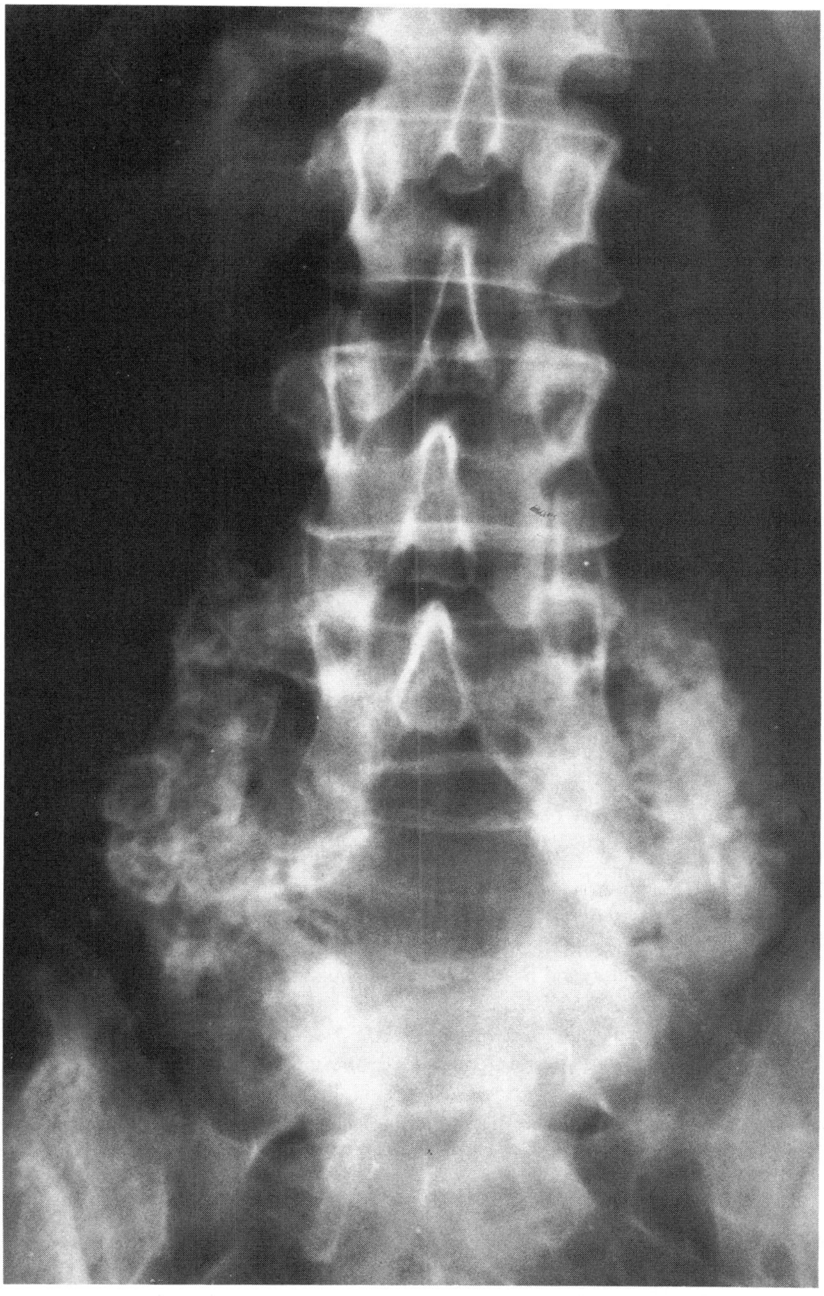

FIG. 1.—Male aged 46 who underwent PLA and PEMF stimulation. Roentgenography taken 4 months after surgery, rated 4 according to Dawson.

RESULTS

In the control group the rating of the roentgenograms taken 4 months after surgery gave an average value of 2.3; in the experimental group it was 3, $p < 0.04$. Figure 2 shows the relative distribution for the two groups of patients in the five categories of the Dawson classification, showing that there is a shift toward higher rating when PLA are stimulated with PEMF. We reported the same effect in a study involving tibial valgus osteotomies.[3]

DISCUSSION

The results reported here are particularly important in view of the ability of PEMF stimulation to reduce the number of PLA with insufficient fusion. The use of electromagnetic stimulation results in an earlier recovery for patients in the experimental group compared to control patients. These data confirm what was previously reported in the literature.[4,6] We believe that PEMF stimulation, started within 7 days after surgery, reduces the healing time and can be advantageously employed in patients undergoing spinal fusion with autologous cancellous bone graft.

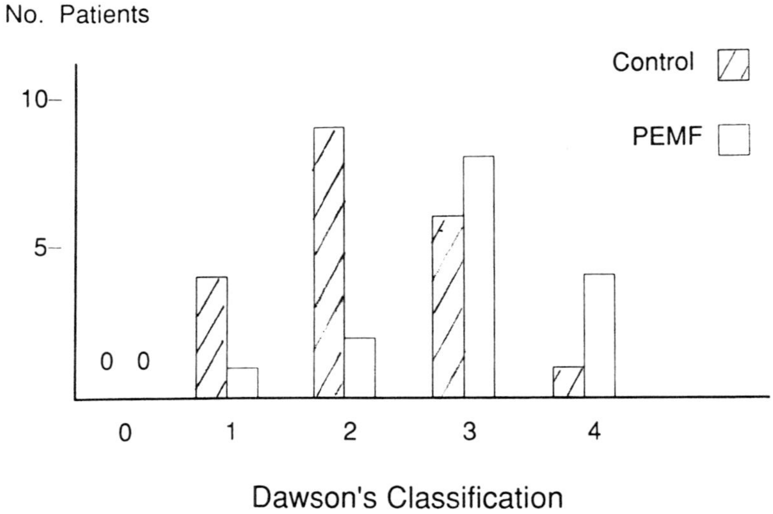

FIG. 2.—Rating of PLA healing in control and PEMF-stimulated groups of patients. Evaluation was based on roentgenograms taken 4 months after surgery.

REFERENCES

1. G. C. Traina, R. Cadossi, G. Ceccherelli, A. Dal Monte, G. Fontanesi, F. Giancecchi, G. I. Mammi, V. Negri, F. Pisano, G. Poli, E. Rinaldi, and B. Virgili, "La modulazione elettrica della osteogenesi," *Giornale Italiano di Ortopedia*, Suppl. 12(2): 165-176, 1986.

2. G. Fontanesi, F. Giancecchi, R. Cadossi, R. Rotini, G. Bugli, G. I. Mammi, L. De Sessa, and I. Tartaglia, "Campi elettromagnetici pulsati a bassa frequenza ed innesti ossei," *SERTOT* 25: 1-5, 1983.

3. G. I. Mammi, R. Rocchi, R. Cadossi, and G. C. Traina, "Effect of PEMF on the healing of human tibial osteotomies: A double blind study," in C. T. Brighton and S. R. Pollack, Eds., *Electromagnetics in Medicine and Biology*, San Francisco: San Francisco Press, 1991, 323-326.

4. R. M. Barden and G. L. Sinkora, "Bone stimulators for fusions and fractures," *Nursing Clinics of North America* 26: 89-103, 1991.

5. J. W. Simmons, "Treatment of failed posterior lumbar interbody fusion (PLIF) of the spine with pulsing electromagnetic fields," *Clin. Orthop. Rel. Res.* 193: 127-132, 1985.

6. R. Savini, M. Di Silvestre, G. Garguilo, and N. Bettinin, "The use of pulsing electromagnetic fields in posterolateral lumbosacral spinal fusion," *J. Bioelectricity* 9: 9-17, 1990.

7. V. Cané, P. Botti, D. Farneti, and S. Soana, "Electromagnetic stimulation of bone repair: A histomorphometric study," *J. Orthopaedic Res.* 9: 908-917, 1991.

8. S. Guizzardi, M. Di Silvestre, P. Govoni, R. Strocchi, and R. Scandroglio, "Effetti dei campi elettromagnetici pulsanti (CEMP) sull'evoluzione del callo di fusione vertebrale: Studio istologico," *Acta Biomedica de "L'Ateneo Parmense* 61: 227-235, 1990.

9. E. Dawson, M. Lotysch, and M. R. Urist, "Intertransverse process lumbar arthrodesis with autogenous bone graft," *Clin. Orthop. Rel. Res.* 154: 90-96, 1981.

Martin Blank, Ed., *Electricity and Magnetism in Biology and Medicine*

TRANSCEREBRAL IMPACT OF A 40.68 MHz ELECTROMAGNETIC FIELD ON THE FUNCTIONAL CONDITION OF THE HYPOTHALAMUS

V. D. Sidorov and S. B. Pershin*

National Scientific Center for Rehabilitation and Physiotherapy, Moscow

An experimental study was performed to clarify the advance of the biological effect of UHF bitemporal therapy. Realization of the effect proceeds at the expense of alterations in the potency of the neuromediatory and endogenic opioid systems of the brain. In addition, the study covered modulations of peripheral endocrine glands by hypophysis, glucocorticoids, and catecholamines by the adrenals; proliferation of the lymphoid tissue; and functional activity of humoral immunity.

Expression of neuromediators-biogenic amines and hypothalamus endogenous opioid peptides in rabbits was studied under the impact of bitemporal microwave therapy (40.68 MHz, 20 W). The investigation was carried out in 36 males weighing 2.7 to 3.3 kg. The latter were subjected to a daily 6min therapy for 12 days. As a result of the microwave treatment, the hypothalamus of the tested animals demonstrated an increase in the content of noradrenaline by 21.6%; serotonin, by 28.2%; and leucin-enkephaline by 56%, accompanied by decrease in the concentration of hemovanillic acid by 34.8% and of B-endorphine by 21.8%. No changes of dopamine and 5-oxyindo-leacetic acid were detected (Table 1). Certain changes in the hypothalamus of the laboratory animals pointed to its functional activity dynamics in the electromagnetic fields,[3] thus causing some changes in the pituitary body trophic function. The stimulation of corticotrophic function was expressed by the growth of adrenal gland mass by 28%, higher blood concentration of cumulative 11-oxycorticosteroids by 37%, and higher level of urine adrenalin by 53% (Table 2). As a consequence of the above changes[1] there was a suppression of humoral immunity, bringing about a 22% drop in the count of splenic hemolysin-forming cells.[2] Nevertheless, no regression of the proliferative processes in the lymphoid tissue was registered. The altered potency of the brain neuromediator and endogenous opioid systems seems to be one of the mechanisms of the microwave biologic effect under the bitemporal impact of the electromagnetic field in question.

TABLE 1.—Changes in concentration of biogenic amines and hypothalamus endogenous peptides in rabbits under the UHF bitemporal therapy (M+m).

Index	Groups		p
	Control	Test	
Noradrenaline (nmol/g)	3.99 ± 0.31	4.85 ± 0.22	< 0.001
Dopamine (nmol/g)	2.00 ± 0.24	2.28 ± 0.32	> 0.1
Hemovanillic acid (nmol/g)	2.30 ± 0.23	1.5 ± 0.20	< 0.01
5-oxyindoleacetic (nmol/g)	0.540 ± 0.085	0.590 ± 0.045	> 0.1
5-oxytriptamine (nmol/g)	1.25 ± 0.22	1.60 ± 0.24	< 0.05
Leucin-enkephaline (pmol/g)	168.2 ± 13.4	264.0 ± 22.1	< 0.05
B-endorphine (pmol/g)	66.6 ± 2.4	52.1 ± 5.4	< 0.002

We conclude that the application of the UHF bitemporal therapy (1) increases the content of serotonin, noradrenaline, and leucin-enkephaline in the hypothalamus; (2) stimulates the corticotrophic function of the hypophysis, increasing glucocorticoid and catecholamine synthesis and adrenal gland mass; and (3) leads to

*Correspondence to: S. B. Pershin, Novoarbatsky Prospect 50, 121099 Moscow, Russia.

TABLE 2.—Changes in indices of the adrenal, spleen, and thyroid functional condition under the UHF bi-temporal impact (M+m).

Groups	11-oxycortico-steroids	Adrenal gland mass (mg)	Adrenaline (pmol/20 ml urine)	Noradrenaline (pmol/20 ml urine)
Test (n = 24)	294 ± 19	286 ± 20	0.52 ± 0.07	3.43 ± 0.42
Control (n = 12)	215 ± 13	223 ± 24	0.34 ± 0.04	2.40 ± 0.53
p	< 0.01	< 0.05	< 0.05	> 0.1

Groups	Spleen				
	PFC in 10^6 splenocytes	PFC in all spleen	General count of splenocytes $\times 10^6$	T_3 (pmol/l)	T_4 (pmol/l)
Test (n = 24)	2.99 ± 0.23	3227 ± 325	1079 ± 79	7.58 ± 0.33	17.0 ± 2.4
Control (n = 12)	3.82 ± 0.30	3836 ± 424	951 ± 150	6.92 ± 0.35	20.7 ± 3.0
p	<0.05	>0.1	>0.1	>0.1	>0.1

PFC, plaque-forming cells = antibody-forming cells; T_3, triiodthyronine; T_4, thyroxine.

the suppression of humoral immunity, bring about a drop in the count of slenic hemolysin-forming cells.

REFERENCES

1. H. Besedovsky and E. Sorkin, "Network of immunoneuroendocrine interactions," *Clin. Exp. Immunol.* 27: 1-12, 1977.

2. N. K. Jerne and A. A. Nordin, "Plaque formation in agar by single antibody producing cells," *Science* 140: 405, 1963.

3. S. M. Michaelson, "Microwave and radiofrequency radiation," in *Nonionizing Radiation Protection*, WHO European Office, Copenhagen, 1982, pp. 97-174.

Martin Blank, Ed., *Electricity and Magnetism in Biology and Medicine*

IMMUNOREDUCING EFFECT OF INFRARED LOW-ENERGY LASER RADIATION USED LOCALLY IN PILOT STUDIES

S. B. Pershin,* T. V. Konchugova, and A. A. Minenkov

National Scientific Center for Rehabilitation and Physiotherapy, Moscow

The aim of the experiment was to study the immunomodulating potential of impulse infrared low-energy laser radiation (IRLR) applied to the zone of adrenals and head, as well as to evaluate the expediency of this factor's application in the immunodeficit condition induced by cyclophosphane (CF).

It was discovered that IRLR treatment causes immune suppression, expressed in the drop in the count of direct plaque-forming cells and hemaglutinin titer.

The increase of the immunodepression effect was registered in animals who got a 4mg/kg CF injection and were subjected to localized treatment of adrenals with the above-mentioned laser rays. It may be argued that the immunosuppression effect caused by IRLR is related to the increase of adrenal cortex glucocorticoid activity. The results obtained suggest the possibility of localized laser treatment of adrenals in the clinic to reduce the dose of injected cytostatic. The purposeful change of endogenic endocrine regulation of immune reactivity by modified physical factors has essential relevance for the therapy of the immunodeficit condition. Immunodepressants currently used are characterized by a number of adverse side effects.[2,3] Thus, research aimed at reducing the toxicity of cytostatic therapy seems to be rather urgent.

The aim of the present paper is to study the influence of IRLR on the immune system of rats in normal condition and after CF injection.

METHODS

The apparatus "Lita," which served as the source of radiation, has the following characteristics: radiation wavelength, 0.89 μm; pulse repetition frequency, 1500 Hz; average power 25 W. The daily dose of radiation was 0.9 Joule for 10 days. The animals were divided into two groups. In the first group, the adrenals were subjected to radiation; in the second, the head was treated with IRLR. Animals who underwent a "placebo" treatment served as controls. In all rats, including the control group, the primary immune response to thymus-dependent antigen was induced by immunization with sheep erythrocytes on the tenth day after the end of the treatment. The animals were sacrificed on the fourth day after the immunization. The general number of nuclear cells (splenocytes), the number of direct plaque-forming cells (PFC) according to Jerne-Nordin,[5] and serum hemaglutinin titer were all counted in the animals' spleen. On the eighth day after the end of treatment the animals underwent intraperitoneal injection of CF (dose 4 mg/kg, 2nd subgroup; 8 mg/kg, 3rd subgroup).

RESULTS

The IRLR treatment of areas of adrenals and head in normal rats leads to a marked immune suppression, expressed in the drop of the PFC count to 32% (as compared to control group data) in the 1st group and to 22% in the 2nd group, as well as to the drop of hemaglutinin titer to 77% and 63%, respectively (Table 1). There was an increase of immune suppression in the 2nd group as compared to control 2 group, which got CF injections and "placebo." This difference in the level of immunodepression was more marked when the treatment of adrenals was localized.

The data difference obtained in the third group and in control 3 proved to be irrelevant. Comparing the levels of PFC in the 2nd group with control 3, we note (Table 1) that in the 1st group the difference is not large (only 39%), whereas in the 2nd group it is considerably more marked (95%).

*Correspondence to: S. B. Pershin, Novoarbatsky Prospect 50, 121099 Moscow, Russia.

TABLE 1.—Immune data in rats, subjected to cyclophosphane injections (M+m) combined with IRLR treatment of adrenals (1st group) and head (2nd group).

Animal groups	PFC per 10^6 splenocytes	PFC per all spleen $\times 10^3$	Titer of serum hemaglutinin $\log_2$	General count of spleen cells $\times 10^6$
Control 1	836.5 ± 72.5	590.9 ± 61.2	5.4 ± 0.2	724.4 ± 34.5
Control 2 (4 mg/kg)	402.5 ± 31.9***	349.6 ± 25.3***	4.6 ± 0.2*	827.6 ± 52.2
Control 3 (8 mg/kg)	193.1 ± 14.8***	138.1 ± 12.0***	3.5 ± 0.3***	722.0 ± 46.2
Group 1				
IRLR	269.5 ± 21.3***	202.3 ± 28.8***	4.2 ± 0.3**	821.2 ± 67.1
IRLR (+4 mg/kg)	267.9 ± 23.2***	192.2 ± 17.6***	4.3 ± 0.3**	758.0 ± 34.7
IRLR (+8 mg/kg)	176.7 ± 15.9***	119.4 ± 18.3***	4.3 ± 0.2**	669.4 ± 37.3
Group 2				
IRLR	182.0 ± 34.6***	154.7 ± 11.1***	3.4 ± 0.3***	1006.3 ± 66.1***
IRLR (+4 mg/kg)	377.8 ± 54.3***	327.2 ± 43.1***	4.7 ± 0.3	892.5 ± 35.1**
IRLR (+8 mg/kg)	173.5 ± 20.3***	145.6 ± 13.6***	3.6 ± 0.4***	793.1 ± 37.3

$^*p < 0.05$; $^{**}p < 0.01$; $^{***}p < 0.001$.

Falling back on our previous research data, obtained during localized microwave radiation of the adrenals' areas, it can be argued that the immunodepression effect is correlated with the increase of adrenal cortex glucocorticoid activity.[1]

The results of the research are suggestive of the possibility of an increase in immunodepression by a combined treatment by low-energy laser radiation of adrenals and cytostatic injections.

Thus, the experiment suggests the possibility of using CF-induced immunosuppression in combination with laser radiation, applied with the aim of reducing the dose of injected cytostatic while keeping up the required level of immunodepression.

CONCLUSIONS

1. Pulsed infrared low-energy laser radiation (IRLR) of adrenals and head in normal rats brings about a marked immune suppression.
2. IRLR treatment of adrenals combined with 4mg/kg CF injection causes the increase of immunodepressive effect of the cytostatic.
3. The level of immune suppression in rats subjected to 4mg/kg CF injection under IRLR treatment of adrenals approaches that in animals with doubled CF injection dose (8 mg/kg).

REFERENCES

1. V. M. Bogolyubov, I. D. Frenkel, S. B. Pershin, et al., "Immunodepressive and immunostimulating impact of decimeter waves in primary immune response," *Problems of Curortology, Physical Therapy and Physical Training* 2: 13-17, 1982.
2. Y. K. Napolov and K. B. Borisov, "Immunodepressants applied in immune system pathology," *Pharmacol. & Toxicol.* 54: 82-84, 1991.
3. V. A. Nassonova and M. G. Astapenko, *Clinical Rheumatology: Manual for Physicians*, Moscow: AMS USSR, 1989.
4. S. B. Pershin, I. D. Frenkel, V. D. Sidorov, and E. L. Kretlova, "Immunomodulating effect of microwaves in rabbits subjected to injection of inhibitors of nuclear synthesis," *Pharmacol. & Toxicol.* 5: 124, 1987.
5. N. K. Jerne and A. A. Nordin, "Plaque formation in agar by single antibody producing cells," *Science* 140: 405, 1963.

Martin Blank, Ed., *Electricity and Magnetism in Biology and Medicine*

PREVENTION OF EXPERIMENTAL OSTEOPENIA BY USE OF COMBINED AC/DC MAGNETIC FIELDS

J. T. Ryaby,* F. P. Magee, and A. M. Weinstein

OrthoLogic Corp., Phoenix, Ariz.

R. J. Fitzsimmons, and D. J. Baylink

Pettis Memorial Veterans Hospital, Loma Linda, Calif.

Osteoporosis is defined as a decrease in amount of bone mass per unit volume of bone (mineral density), which can lead to fragility and bone fractures in the elderly, either spontaneously or after minimal trauma. Primary osteoporosis is the most prevalent form, referring principally to postmenopausal osteoporosis in women (Type I) and aging-associated osteoporosis in men and women (Type II). Secondary osteoporosis has many clearly defined causes; the most common are primary hyperparathyroidism, hyperthyroidism, and rheumatoid arthritis. Prevention or reversal of osteoporosis is one of the largest medical therapeutic markets for which no effective treatment is currently available. Current pharmacological approaches to therapy include bisphosphonates, calcitonin, and estrogen replacement therapy, all potent inhibitors of bone resorption.[1] The limitation of these therapies is the inability to stimulate new bone formation. New therapies, such as fluoride, may in fact act by stimulation of new bone formation.[2]

Electrical stimulation has been proposed as a potential treatment of osteoporosis since it is established that current electrical and magnetic stimulation paradigms are capable of stimulating new bone formation. Brighton and co-workers have shown that use of a low-voltage, high-frequency, capacitively coupled electrical signal can both prevent and reverse osteoporosis in disuse and hormonal osteopenic model systems.[3,4] Rubin et al., using an avian ulna disuse model, showed a significant increase in bone formation in the ulna when treated with pulsed electromagnetic fields.[5]

These studies clearly demonstrate the potential for utilizing electromagnetic stimulation technology in the prevention and treatment of osteoporosis. The objective of the present study was to define an appropriate therapeutic regime by combined AC/DC magnetic fields in a osteopenic animal model. Previous results have shown prevention of regional bone loss by use of combined AC/DC magnetic fields.[6] These results demonstrate that combined AC/DC magnetic fields can prevent bone loss in a site-specific manner with magnitude similar to that observed with hormonal replacement therapy.

MATERIALS AND METHODS

Animal Model

Castration-induced osteopenia (orchidectomy) of the adult male retired breeder rat was the model used in these studies. The animals were housed two per cage and were fed a commercial rat chow and deionized water ad libitum. Animals were randomly placed in their respective groups; combined AC/DC magnetic field stimulation was begun on the first postoperative day, and continued for 60 days. All animals were then sacrificed by barbiturate overdose. The L1 and L4 vertebrae were removed and ash weights determined. The right and left tibiae were immediately fixed in 10% buffered formalin for 24 h and then transferred to 70% ethanol for bone densitometry and histomorphometric analysis. The right and left femora were frozen for densitometric or biochemical analysis. Serum samples were frozen at −70 C for osteocalcin and testosterone analysis.

*Correspondence to: J. T. Ryaby, OrthoLogic Corp., 2850 S. 36th St., Phoenix AZ 85034.

Experimental Design

The combined AC/DC magnetic fields used in this study were based on the ion resonance theory.[6] Three frequencies (15.3, 25.4, and 76.6Hz sinusoidal AC magnetic fields) were tested; the exposure was four times 30 min/day. Testosterone positive controls were performed by use of time-release testosterone pellets (35mg/60day release) according to Turner et al.[7]

Bone Densitometry

For regional bone loss studies, single-photon absorptiometry (SPA) was performed according to the method of Hagaman et al.[8] Twelve density slices (three scans per site) along the length of the tibiae were taken for each bone in a water bath to eliminate soft-tissue artifacts. For site-specific bone loss studies, five equal quadrants were taken from the tibio-fibular insertion proximal to the growth plate and scanned three times per site. The bone mineral content and the bone width were measured, and the bone mineral density was calculated.

Biochemistry

Serum samples taken at sacrifice were analyzed for osteocalcin as an index of bone formation.[9] Rat osteocalcin and testosterone were determined by RIA. Biochemical markers were correlated with changes in bone density to identify bone processes affected by magnetic field exposure.

RESULTS

In the regional bone loss studies, the left tibia of the castrated rats had a significant loss in ash weight (4.7%, $p < 0.05$) when compared with the noncastrated control animals (data not shown). Densitometric analysis provided quantitative assessment of bone loss corrected for volume changes and showed that exposure to the 76.6Hz AC/DC magnetic field at both 0.4 and 1.2 G prevented the bone loss due to castration. The mineral density profiles obtained by single-photon absorptiometry were consistent with the ash weight findings (Fig. 1). Since bone loss is not evenly distributed over the volume of bone, high-resolution site-specific bone-loss studies were performed. Castration caused significant bone loss to occur in the distal and central regions of the tibia of 15.6% and 6.9%, respectively ($p < 0.05$). Combined AC/DC magnetic field (76.6 Hz) exposure prevented this bone loss, increasing bone density above noncastrated controls even at sites proximal to the central region (Fig. 1). Testosterone replacement therapy also prevented the bone loss due to castration, consistent with previous findings.[9] As can be seen in Fig. 3, osteocalcin levels in serum (markers of bone turnover) showed a significant increase in the castrated group ($p < 0.01$). Treatment with either combined magnetic fields or testosterone reduced osteocalcin levels to within 15% of control (noncastrated) values, consistent with the densitometric analysis.

DISCUSSION

This study demonstrates that combined AC/DC magnetic fields can prevent bone loss due to castration in aged breeder rats. These results are consistent with previous studies with capacitively coupled electric fields[4] and combined AC/DC magnetic fields.[6] The present study extends previous work[6] by demonstrating prevention of site-specific bone loss. The significance of these results is further enhanced by the results observed with hormonal replacement therapy. The magnitude of the response to combined AC/DC magnetic fields is similar to that observed with testosterone replacement, which demonstrates the potency of this biophysical stimulus. These results suggest that combined AC/DC magnetic fields have therapeutic potential for treatment of site-specific bone loss in clinical situations.

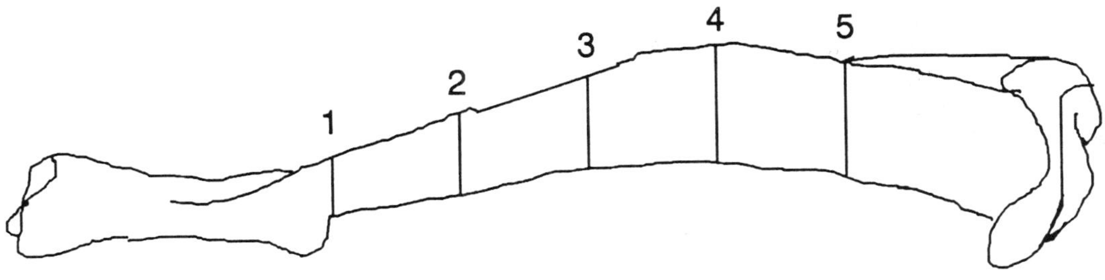

	1	2	3	4	5
CONT	0.289 ±0.018	.275 ±0.013	.266 ±0.003	.252 ±0.016	.201 ±0.013
ORX	.243* ±0.003	.255* ±0.008	.267 ±0.019	.241 ±0.017	.202 ±0.014
TEST	.272 ±0.015	.278 ±0.010	.274 ±0.012	.257 ±0.015	.202 ±0.021
MF	.290 ±0.039	.288 ±0.012	.280 ±0.022	.249 ±0.012	.233* ±0.014

FIG. 1.—Site-specific bone loss studies. CONT = Control (noncastrated), ORX = castrated, TEST = testosterone replacement, MF = combined magnetic fields (0.2 G DC, 0.4 G AC, 76.6 Hz). Data are expressed as milligrams per square centimeter. [$*p < 0.05$ (vs control)].

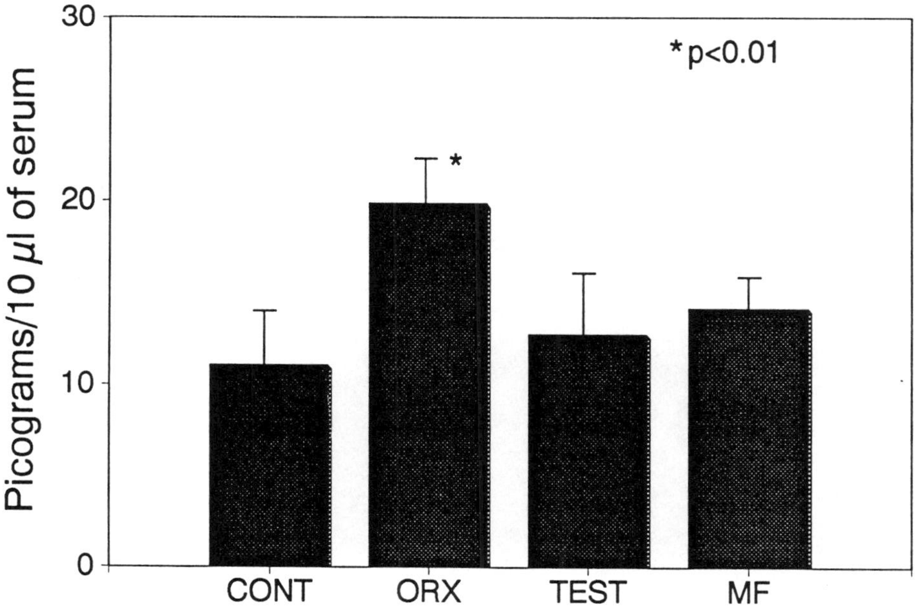

FIG. 2.—Bone turnover studies (serum osteocalcin levels). CONT = Control (noncastrated), ORX = castrated, TEST = testosterone replacement, MF = combined magnetic fields (0.2 G DC, 0.4 G AC, 76.6 Hz). Data are expressed as picograms/microliter serum.

REFERENCES

1. A. M. Parfitt, ''Bone remodeling: Relationship to the amount and structure of bone, and the pathogenesis and prevention of fractures,'' in B. L. Riggs and L. J. Melton III, Eds., *Osteoporosis: Etiology, Diagnosis, and Management,* New York: Raven Press, 1988.

2. C. Y. C. Pak, K. Sakhaee, J. E. Zerwekh, C. Parcel, R. Peterson, and K. Johnson, ''Safe and effective treatment of osteoporosis with intermittent slow-release sodium fluoride: Augmentation of vertebral bone mass and inhibition of fractures,'' *J. Clin. Endocrinol. Metab.* 68: 150-159, 1989.

3. C. T. Brighton, G. T. Tadduni, and S. R. Pollack, ''Treatment of sciatic denervation disuse osteoporosis in the rat tibia with capacitively coupled electrical stimulation,'' *J. Bone Joint Surg.* 67A: 1022, 1985.

4. C. T. Brighton, C. P. Luessenhop, S. R. Pollack, D. R. Steinberg, M. E. Petrick, and F. S. Kaplan, ''Treatment of castration induced osteoporosis by a capacitively coupled electrical signal in rat vertebrae,'' *J. Bone Joint Surg.* 71A: 228, 1989.

5. C. T. Rubin, K. J. McLeod, and L. E. Lanyon, ''Prevention of osteoporosis by pulsed electromagnetic fields,'' *J. Bone Joint Surg.* 71A: 411-417, 1989.

6. F. P. Magee, A. M. Weinstein, R. J. Fitzsimmons, D. J. Baylink, and B. R. McLeod, ''The use of low-energy combined AC and DC magnetic fields in the prevention of osteopenia,'' in C. T. Brighton and S. R. Pollack, Eds., *Electromagnetics in Biology and Medicine,* San Francisco: San Francisco Press, 1989, 171-176.

7. G. K. Wakley, H. D. Schutte, K. S. Hannon, and R. T. Turner, ''Androgen treatment prevents loss of cancellous bone in the orchidectomized rat,'' *J. Bone Mineral Res.* 6: 325, 1991.

8. J. R. Hagaman, T. V. Sanchez, and R. C. Myers, ''The effect of lactation on the mineral distribution profile of the rat femur by single photon absorptiometry,'' *Bone* 6: 301, 1985.

9. J. B. Lian and C. M. Gundberg, ''Osteocalcin: Biochemical considerations and clinical applications,'' *Clin. Orthop. and Rel. Res.* 226: 267-291, 1988.

Martin Blank, Ed., *Electricity and Magnetism in Biology and Medicine*

RECONNECTION OF SEVERED INVERTEBRATE AXONS BY LASERS AND ELECTRIC FIELDS

G. Rodziewicz* and P. Qi

SUNY-HSC College of Medicine, Syracuse, N.Y.

D. Yogev, A. Todorov, and J. Fendler

Syracuse University, Syracuse, N.Y.

After their axons have been severed, central and peripheral nerve cells can achieve functional recovery either by regrowth of the injured cells past the site of the injury or by direct repair of the injured cell at the site of injury. Direct repair of a severed nerve cell axon promises the advantages of preserving the viability and existing connections of the axon distal to the cut, as well as allowing rapid return of function distal to the injury.

We present the rationale, technical background, and anatomical results of our attempts to reconnect the medial giant axon of the earthworm *(Lumbricus terrestris)* in vitro. Directional energy (laser pulses and electric-field pulses) were applied to severed axons. Axons were considered reconnected either if Lucifer Yellow dye, iontophoretically injected into the proximal axon, passed across the reconnected segment into the distal axon; or if 1-2μm thick serial sections studied with light microscopy showed cytoplasmic continuity across the reconnected segment. Electron microscopy of selected cases confirmed the light-microscopy evidence of cytoplasmic continuity. Recovery of function in systems such as severed spinal cord and peripheral nerve requires the integrity of the axons transmitting information across the injured segment. Axon reconnection promises the rapid return of axon integrity, a necessary condition for the functional recovery of these systems.

Nervous-system injury produces its devastating effects by disrupting the functional integrity of axons in nervous-system tissue. This central neurological deficit, the axon's inability to transmit information from one side of its damaged axon segment to the other side, has been the focus of studies attempting regrowth of an intact axon across the damaged axon segment. We are developing a new approach to the central neurological deficit in spinal cord injury: direct reconnection of the damaged axon.

This approach is based on well-established work in cell fusion.[1] It is hypothesized that cell-fusion techniques allow two or more different cells to appose and connect their cell membranes, forming one large cell (Fig. 1). Cell-fusion techniques are widely used in single-cell systems to create hybridomas[2] and for the introduction of macromolecules into cells.[3] We have extended two well-known cell-fusion techniques, the application of laser pulses[4-6] and electric fields[2,3,7] to the problem of axon reconnection in neural tissue.

MATERIALS AND METHODS

The earthworm ventral nerve cord containing the MGA was dissected and prepared according to Krause & Bittner.[7] The ventral nerve cord was pinned onto a silicone rubber stage and bathed in earthworm saline[7] at 21 C. A double-edged razor blade was used to crush the ventral nerve cord, severing the medial and two lateral giant axons, while leaving the outer nerve sheath intact. This process, carried out under the dissecting microscope, allowed the operator to see the retraction of the medial and two lateral giant axons away from the knife blade, with the outer nerve sheath remaining intact. This insured apposition of the two severed MGA segments.

*Correspondence to: G. Rodziewicz (Neurosurgery), SUNY-HSC College of Medicine, 750 E. Adams St., Syracuse NY 13210.

a.

b.

c.

d.

e.

f.

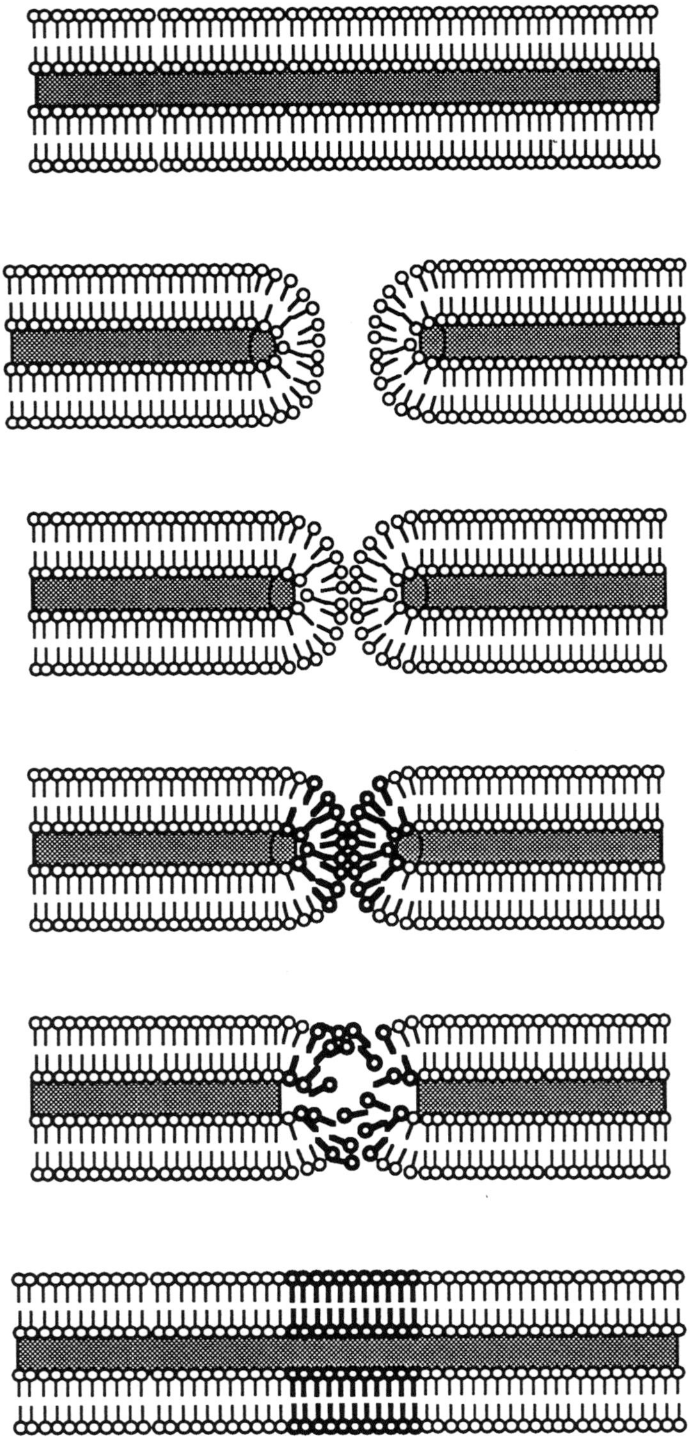

FIG. 1.—Hypothetical mechanism of cell membrane fusion in a severed axon: (a) intact axon, with lipid bilayers of the cell membrane surrounding central cytoplasm (gray hatching); (b) cut axon, with lipid bilayers of cell membrane sealing off cut ends; (c) cut ends are apposed, bringing the two cell membranes into close proximity; (d) apposed cell membranes are targeted by fusion stimulus; (e) fusion stimulus is applied, causing disruption of apposed cell membranes; (f) apposed cell membranes re-anneal into one cell membrane, re-establishing continuity of axon.

Laser Technique

A Lambda-Physik excimer EMG 101 MSC laser beam was focused through a dichroic mirror and quartz lens mounted in a Zeiss microscope to provide 308nm, 15ns laser pulses. Technical details can be found in Ref. 8. After 10-15 min of incubation postcutting, application of 2-5 pulses of this laser light, focused on the apposed proximal and distal cell membranes, resulted in the reconnection of the severed MGAs.

Electric-field Technique

The earthworm saline was replaced with mannitol buffer (0.3 M mannitol, 0.5 mM $MgCl_2$, 0.5 mM $CaCl_2$ in distilled water) prior to severing the axons. Micromanipulators were then used to place two microelectrodes (Microprobe Inc. WE30031.5A), the anode in the rostral MGA segment and the cathode in the caudal MGA segment, at a total distance of 200-300 µm. Electrical pulses with an amplitude of 80-200 V were applied through both electrodes and the tissue between the electrode tips. Technical details can be found in Ref. 9. Five minutes after severing, 3-8 electric field pulses were applied. Five minutes subsequent to electric field application with earthworm saline, and the ventral nerve cord containing the MGA was allowed to survive 0.5-2 h at 21 C.

Between 30 and 60 min after axon reconnection, Lucifer Yellow CH (Sigma), 3% w/v in water, was injected iontophoretically into either the rostral or caudal MGA segment, 1-2 mm away from the crushed area. The injection duration varied from 5 to 20 min. The diffusion of Lucifer Yellow through the crushed MGA segment was followed with real-time videotaping.

For light and electron microscopy, the ventral nerve cord containing the MGA was fixed in 2.5% glutaraldehyde in 0.1 M phosphate buffer (pH 7.4), osmicated, dehydrated, and embedded in low-viscosity Spurr's medium (Polysciences Co., Warrington, Pa.). Longitudinal horizontal serial sections were cut. For light microscopy, sections 1 µm thick were stained with 0.1% Toluidine Blue dye, and examined with an Olympus Vanox-T microscope. For electron microscopy, 90nm sections were cut, stained with uranyl acetate and lead citrate, and examined with a JEOL JEM-100CX II transmission electron microscope.

RESULTS

Criteria. Criteria for successful axon reconnection were either cytoplasmic continuity demonstrated by serial longitudinal sections or passage of Lucifer Yellow dye from one side of the reconnected segment to the other.

In both laser and electric-field reconnection, Lucifer Yellow dye, iontophoretically injected into one MGA segment, was seen to pass through the reconnected MGA segment into the other MGA segment. In contrast, the Lucifer Yellow dye did not diffuse into the other MGA segments that were not exposed to the electric field pulses or that did not fuse after exposure to fusion stimuli. Some leakage of Lucifer Yellow dye into the surrounding ventral nerve cord (but not into the caudal MGA) occurred in the controls when the nerves were injected with Lucifer Yellow dye within 15 min of severing, possibly because the severed MGA cell membranes did not completely seal within 15 min in this experimental system. We observed this leakage of dye both in mannitol and earthworm saline solutions.

Longitudinal serial sections under light microscopy demonstrated the absence of a cell membrane/glial-cell sheath boundary between the rostral and caudal segments in the successfully reconnected axons, whereas a cell membrane/glial-cell sheath boundary was found to exist between the unconnected axons.

Transmission electron microscopy of eight axons confirmed the light microscopic finding in reconnected axons: there was continuity between the axoplasm of rostral and caudal axons.

Electric-field Reconnection Results. Successful axon reconnection was achieved in 20 of 100 electric-field reconnection experiments (20%). Three reconnected cells studied with serial longitudinal sections were also processed for electron microscopy.[9]

Laser Reconnection Results. With laser reconnection attempts, we achieved a 40% success rate: in 20 experiments, 8 MGAs were successfully reconnected.[9]

DISCUSSION

Axon Reconnection

The direct repair of damaged axon segments promises advantages that may not be available with techniques aimed at nerve cell regeneration. These advantages may derive from the rapid return of axon integrity, which may prevent the "die-back" of proximal axons and death of distal axons common in regeneration models. Because the axons distal to the injury may be preserved, their distal connectivity will also be preserved and thus circumvent the problem of identifying and re-establishing the appropriate distal connections of regenerating axons. Finally, rapid return of function may occur, as the reconnected axons transmit information to their targets, circumventing end-organ disuse atrophy and secondary-neuron changes which are limitations to the re-establishment of target-organ function by regenerating axons. We postulate that functional recovery may occur after only a small percentage of axons are successfully reconnected.

Cell-fusion Techniques

The fundamental advantage shared by both laser pulses and electric fields as means of axon reconnection is that each of these techniques can be delivered in a vectorial fashion; that is, they can be directed precisely to a particular area with a particular orientation. This precise direction makes it possible to deliver laser pulses or electric fields to the closely apposed, membrane-covered ends of cut axons, in such a way that only the apposed proximal and distal axon membranes receive the energy necessary to cause axon reconnection, which would allow the reconnection of proximal to distal axons but not the connection of axons lying side by side. Specific advantages of each technique are discussed in our papers.[8,9]

Experimental Model

Attempts at the reconnection of invertebrate axons with polyethylene glycol[1] have led to the development of the in vitro earthworm MGA model which we have adopted for our experiments. This single large axon in vitro invertebrate preparation provides experimenters with a model where there is immediate, clear evidence of the success or failure of reconnection experiments. This earthworm MGA model is now being used with our two-directional energy techniques of axon reconnection, laser pulses, and electric-field application.

THE FUTURE

Future work in this axon reconnection model involves recording the passage of action potentials past the reconnected earthworm MGA segments, as well as preliminary experiments on multi-axon systems.

We are developing a new approach to the central neurological deficit in nervous system injury: direct reconnection of the damaged axon. In contrast to nerve-regeneration techniques, axon reconnection offers the promise of rapid return of axon integrity, which may greatly enhance the functional recovery of neurological systems. The data obtained from this work will form the foundation necessary to extend these techniques in a logical progression to progressively higher-order systems, perhaps even to the damaged human central nervous system.

REFERENCES

1. E. Sowers, Ed., *Cell Fusion,* New York: Plenum Press, 1987.

2. D. A. Stenger, R. T. Kubiniec, W. J. Purucker, H. Liang, and S. W. Hui, "Optimization of electrofusion parameters for efficient production of hybridomas," *Hybridoma* 7: 505-518, 1988.

3. T. Y. Tsong, "Electroporation of cell membranes," *Biophysical J.* 60: 297-306, 1991.

4. E. Schierenberg, "Laser-induced cell fusion," in Ref. 1, pp. 409-418.

5. R. Wiegand Steubing, S. Cheng, W. H. Wright, Y. Numajiri, and M. W. Berns, "Single beam optical trapping and micromanipulation of mammalian cells," *SPIE: Laser-tissue Interaction,* 1202: 272-280, 1990.

6. R. Wiegand, G. Weber, K. Zimmermann, S. Monajembashi, J. Wolfrum, and K. Greulich, "Laser-induced fusion of mammalian cells and plant protoplasts," *J. Cell Sci.* 88: 145-149, 1987.

7. T. L. Krause and G. D. Bittner, "Rapid morphological fusion of severed myelinated axons by polyethylene glycol," *Proc. Natl. Acad. Sci. USA* 87: 1471-1475, 1990.

8. D. Yogev, A. T. Todorov, P. Qi, Y. H. Fendler, and G. Rodziewicz, "Laser-induced reconnection of severed axons," *Biochem. Biophys. Res. Comm.* 180: 874-880, 1991.

9. A. T. Todorov, D. Yogev, P. Qi, Y. H. Fendler, and G. Rodziewicz, "Electric-field-induced reconnection of severed axons," *Brain Res.* 582: 329-334, 1992.

Martin Blank, Ed., *Electricity and Magnetism in Biology and Medicine*

NOVEL ELECTROMAGNETIC-IMAGING TECHNIQUE FOR MEDICAL APPLICATIONS

S. Caorsi,* M. Baracco, G. L. Gragnani, M. Pastorino, and M. Sartore

University of Genoa, Italy

A new approach to microwave imaging for biomedical applications is presented. The method is based on a modified Born approximation that allows one to obtain an *image* of the cross section of a cylindrical dielectric object by assuming an a priori knowledge of the complex permittivity distribution in a model of the object. The approach starts with an integral-equation formulation of the inverse scattering problem in terms of the Green function for 2D geometries, using TM-wave illumination. A multi-illumination multiview scheme is applied to exploit multiple information about the biological body under test.

In recent years, electromagnetic waves (at microwave frequencies) have been widely used for nondestructive evaluations, in particular, for medical diagnoses.[1] Approaches based on integral-equation formulations[2-7] have been developed in order to overcome the drawback due to the inverse-scattering solution in the Fourier transform domain (based on first-order approximations). The approach presented in this paper utilizes a previously proposed pseudoinverse-based multiview method[6] and allows one to extend the application range of the first-order Born approximation by considering a model of the unknown object as a priori knowledge. The use of a priori knowledge is very important for medical imaging, as the actual structure examined may differ only partially from the idealized one.

ALGORITHM DESCRIPTION

We consider the cross section A of a cylindrical, inhomogeneous biological body contained in a fixed investigation area S and illuminated by a TM wave (Fig. 1). The scattered electric field inside an observation domain is given by:

$$E_{tot}^{obs}(\bar{r},\Psi) = E_{inc}^{obs}(\bar{r},\Psi) + j\omega\mu_0 \int_S j\omega\varepsilon_0 [\tilde{\varepsilon}_r(\bar{r}',\Psi) - 1] E_{tot}^{inv}(\bar{r}',\Psi) G_0(\bar{r}/\bar{r}',\Psi)\, d\bar{r}' \quad (1)$$

where $\tilde{\varepsilon}_r(\bar{r},\Psi)$ is the complex dielectric permittivity inside S. We introduce an a priori knowledge by expressing $\tilde{\varepsilon}_r(\bar{r},\Psi)$ in the form $\tilde{\varepsilon}_r(\bar{r},\Psi) = \tilde{\varepsilon}_m(\bar{r},\Psi) + \tilde{\varepsilon}_v(\bar{r},\Psi)$, where $\tilde{\varepsilon}_m$ represents the estimated distribution and $\tilde{\varepsilon}_v$ the corresponding variation. By substituting in Eq. (1), we obtain

$$\begin{aligned} E_{tot}^{obs}(\bar{r},\Psi) = {} & E_{inc}^{obs}(\bar{r},\Psi) + j\omega\mu_0 \int_S j\omega\varepsilon_0 [\tilde{\varepsilon}_m(\bar{r}',\Psi) - 1] E_{tot}^{inv}(\bar{r}',\Psi) G_0(\bar{r}/\bar{r}',\Psi)\, d\bar{r}' \\ & + j\omega\mu_0 \int_S j\omega\varepsilon_0 \tilde{\varepsilon}_v(\bar{r}',\Psi) E_{tot}^{inv}(\bar{r}',\Psi) G_0(\bar{r}/\bar{r}',\Psi)\, d\bar{r}' \end{aligned} \quad (2)$$

If $\tilde{\varepsilon}_m(\bar{r},\Psi)$ is close enough to the real value $\tilde{\varepsilon}_r$, the latter term can be neglected. Then, application of the moment method (MM) to the resulting equations (for N cells, M measurement points, and P views) reduces the problem to the solution of the following linear systems:

$$\begin{bmatrix} E_{tot}^{obs}(1,p) \\ \bullet \\ E_{tot}^{obs}(M,p) \end{bmatrix} = \begin{bmatrix} E_{inc}^{obs}(1,p) \\ \bullet \\ E_{inc}^{obs}(M,p) \end{bmatrix} + [A] \begin{bmatrix} j\omega\varepsilon_0 [\tilde{\varepsilon}_m(1,p) - 1] E_{tot}^{inv}(1,p) \\ \bullet \\ j\omega\varepsilon_0 [\tilde{\varepsilon}_m(N,p) - 1] E_{tot}^{inv}(N,p) \end{bmatrix} \quad p = 1, \cdots P \quad (3)$$

These systems are ill-conditioned, hence a pseudoinversion algorithm is used, as suggested in Refs. 3 and 5. Once E_{tot} inside the body is known, we apply the MM to the EFIEs related to S and obtain:

---

*Correspondence to: S. Caorsi (Biophys. & Electronic Engg), University of Genoa, Via Opera Pia 11a, I-16145 Genoa, Italy.

$$\begin{bmatrix} E^{\text{inv}}_{\text{tot}}(1,p) \\ \bullet \\ E^{\text{inv}}_{\text{tot}}(N,p) \end{bmatrix} = \begin{bmatrix} E^{\text{inv}}_{\text{inc}}(1,p) \\ \bullet \\ E^{\text{inv}}_{\text{inc}}(N,p) \end{bmatrix} + [C] \begin{bmatrix} j\omega\varepsilon_0[\tilde{\varepsilon}_r(1,p)-1]E^{\text{inv}}_{\text{tot}}(1,p) \\ \bullet \\ j\omega\varepsilon_0[\tilde{\varepsilon}_r(N,p)-1]E^{\text{inv}}_{\text{tot}}(N,p) \end{bmatrix} \quad p=1,\cdots P \tag{4}$$

Finally, from these systems, we derive the dielectric permittivity of the body.

NUMERICAL RESULTS

The proposed approach was tested on multilayer cylinders, for which analytical solutions are available. Figure 2(a) is a pictorial view of a lossless cylinder ($\varepsilon_r = 6$) contained inside a λ_0-sided square domain S ($N = 64$, $M = 64$, $f = 10$ GHz). The testing points are located on a circumference [diameter $(22/15)\lambda_0$]. Figure 2(b) shows the dielectric reconstruction for $P = 12$ and for $\tilde{\varepsilon}_m(\bar{r}) = 6.5$ if $\bar{r} \in A$, and $\tilde{\varepsilon}_m(\bar{r}) = 1.0$ if $\bar{r} \notin A$. Figure 2(c) gives the plot of the quantity $\xi = \sum_{i=1}^{N} |\varepsilon_r(i) - \varepsilon_r^*(i)| / \sum_{i=1}^{N} |\varepsilon_r(i)|$, where $\varepsilon_r(i)$ and $\varepsilon_r^*(i)$ stand for the original and reconstructed dielectric permittivities of the i-th cell, respectively. We assumed $\tilde{\varepsilon}_m(\bar{r}) = \varepsilon_{mi}$ if $\bar{r} \in A$, and $\tilde{\varepsilon}_m(\bar{r}) = 1.0$ if $\bar{r} \notin A$.

The dielectric reconstruction of a two-layer cylinder [$\varepsilon_r^{ext} = 1.8$, $\varepsilon_r^{int} = 50$, $N = 36$, $M = 16$, diameter $= (22/15)\lambda_0$] is displayed in Fig. 3: (a) original scatterer; (b) reconstructed distribution for an exact estimate and $P = 1$; (c) exact estimate and $P = 10$. Figure 4 shows the reconstruction of a three-layer cylinder ($\varepsilon_r^1 = 20$, $\varepsilon_r^2 = 2$, $\varepsilon_r^3 = 25$, $M = 64$, $N = 64$, $P = 15$, exact estimate).

REFERENCES

1. L. E. Larsen and J. H. Jacobi, Eds., *Medical Applications of Microwave Imaging,* New York: IEEE Press, 1986.

2. D.K. Ghodgaonkar, O. P. Gandhi, and M. M. Hagmann, "Estimation of complex permittivities of three-dimensional inhomogeneous biological bodies," *IEEE Trans.* MTT-31: 442-446, 1983.

3. M. M. Ney, A. M. Smith, and S. S. Stuchly, "A solution of electromagnetic imaging using pseudoinverse transformation," *IEEE Trans.* MI-3: 155-162, 1984.

4. W. W. Guo and T. C. Guo, "Dielectric imaging by microwave inverse scattering and a technique to stabilize matrix inversion," *Proc. 2nd Int. Symp. on Recent Advances in Microwave Technology,* Beijing, China, 1989, 33-38.

5. S. Caorsi, G. L. Gragnani, and M. Pastorino, "An approach to microwave imaging using a multiview moment-method solution for a 2-D infinite cylinder," *IEEE Trans.* MTT-39: 1062-1067, 1990.

6. S. Caorsi, G. L. Gragnani, and M. Pastorino, "A multiview microwave imaging system for two-dimensional penetrable objects," *IEEE Trans.* MTT-39: 845-851, 1991.

7. S. Caorsi, G. L. Gragnani, and M. Pastorino, "Reconstruction of dielectric permittivity distributions in arbitrary 2D inhomogeneous biological bodies by a multiview microwave numerical method," *IEEE Trans. Medical Imaging* (in press).

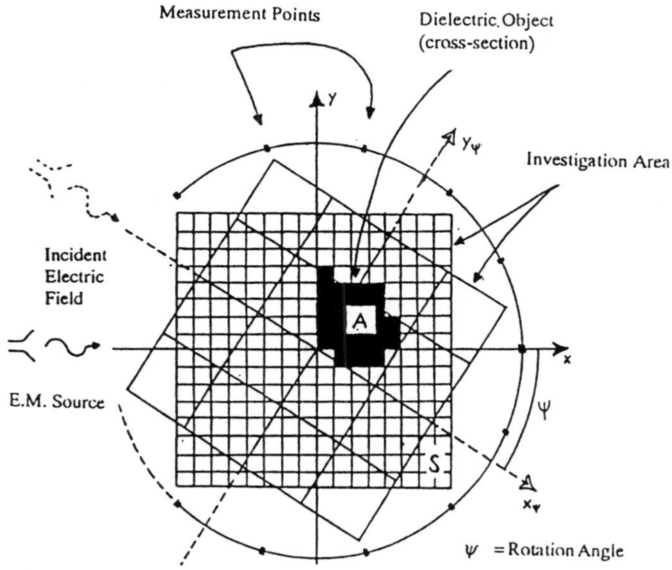

FIG. 1.—Problem geometry.

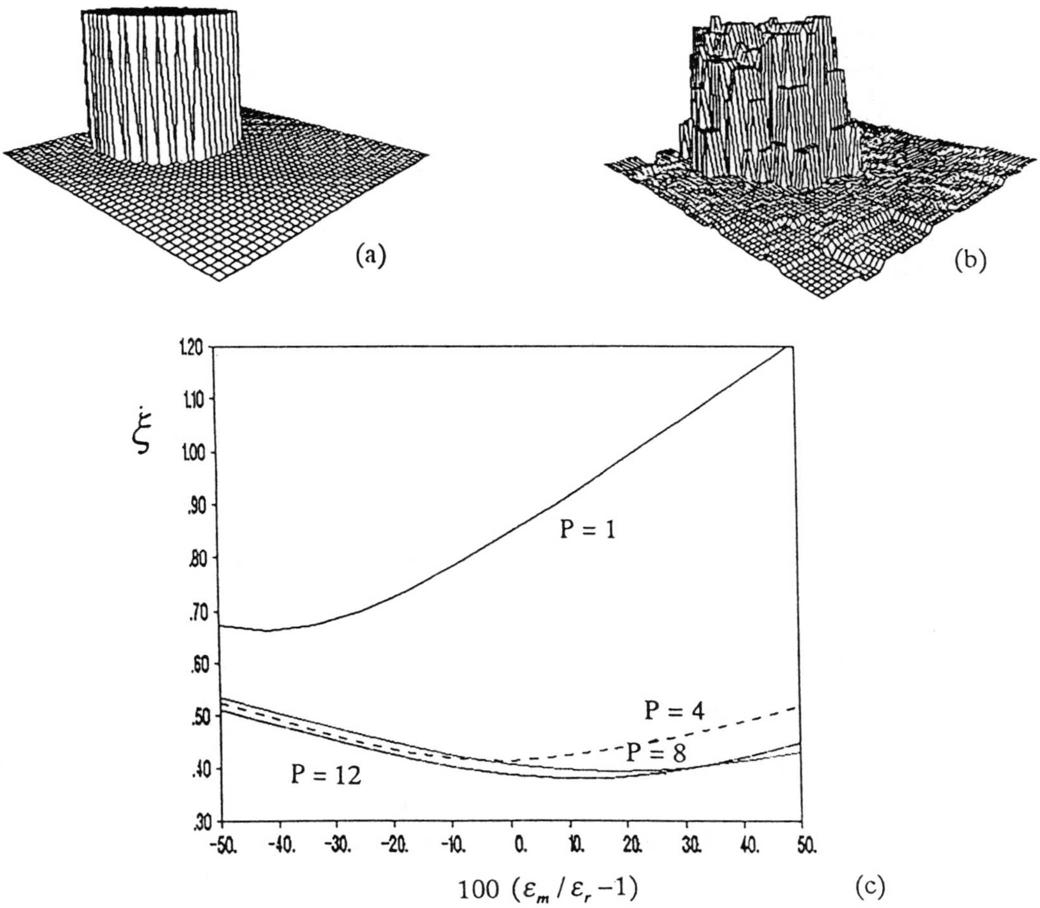

FIG. 2.—Dielectric reconstruction of a circular cylinder: (a) original, (b) reconstructed distribution, (c) plot of the ξ parameter. Unit plane-wave illumination.

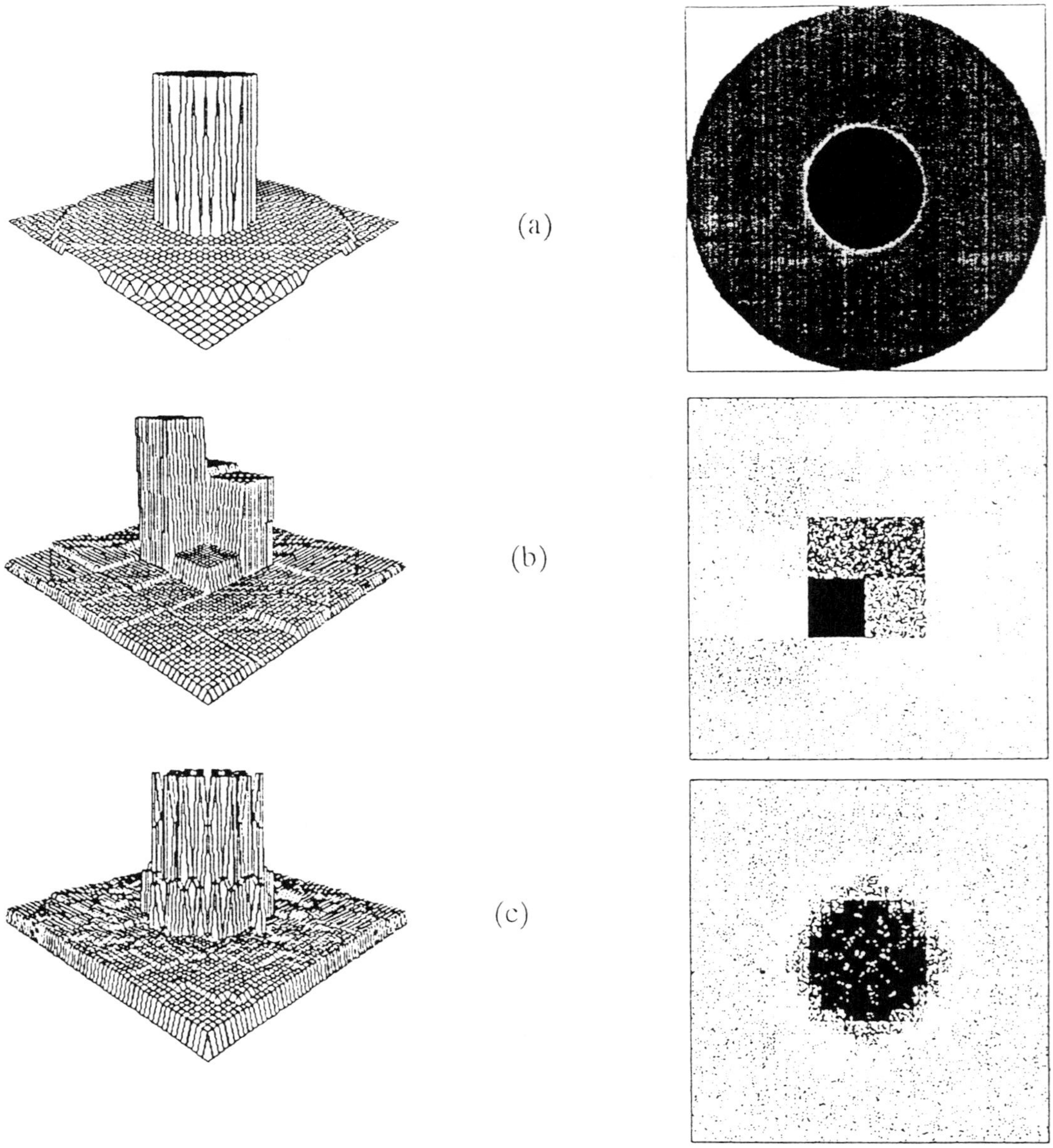

FIG. 3.—Dielectric reconstruction of a two-layered cylinder: (a) original, (b) reconstructed distribution (P = 1), (c) reconstructed distribution (P = 10).

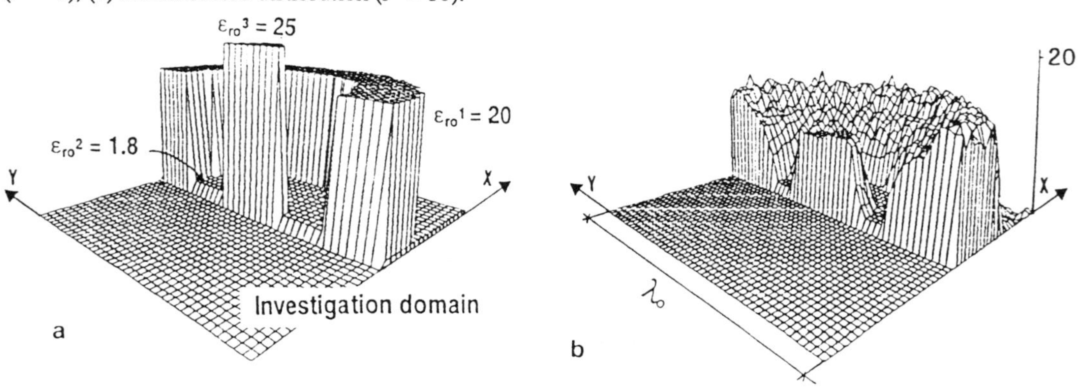

FIG. 4.—Dielectric reconstruction of a three-layered cylinder: (a) original, (b) reconstructed distribution (P = 15); only half domain is shown.

Martin Blank, Ed., *Electricity and Magnetism in Biology and Medicine*

ELECTRICAL MEASUREMENT OF EDEMA

P. Héroux*

McGill University, Montréal, Canada

M. Bourdages

Institut de Recherche d'Hydro-Québec, Varennes

Electrical impedance spectroscopy is a technique for tissue-level measurement of intra- or extracellular edema and cell-membrane capacitance by use of (1) a pair of thin, simple metallic electrodes, minimally invasive, (2) the dissipation factor at many frequencies to assess tissue (dissipation factor = R/X, the ratio of resistance to reactance), and (3) mathematical algorithms for dissipation factor spectrum interpretation. Illustrations of the method in muscle (edema), liver (toxicity), and brain cortex (reaction to anesthetic overdose) are depicted. Since the complicated aspects of the method are based on microelectronics, it has wide potential for medical diagnostics because of its speed, reliability, and simplicity for use.

Our technique of electrical impedance spectroscopy (EIS) allows in vivo and on-line measurement of cellular structures averaged over thousands of cells at a time. The following sections illustrate the principles of the method and show typical results obtained in the continuous monitoring of intra- and extracellular edema in muscle, liver, and brain cortex under conditions of thermal stress, metabolic toxicity, and interruption of blood flow. Remarkable characteristics of EIS are ease of use and high reliability.

METHOD

A simple push of the probe into the tissue is followed by activation of automated measurement sequences yielding the desired determination. Although the mathematical algorithms of the method may be complex, this complexity is entirely invisible to the end user. EIS is defined by (1) a pair of thin, simple metallic electrodes, minimally invasive; (2) use of the dissipation factor at many frequencies to assess tissue (dissipation factor = R/X, the ratio of resistance to reactance); and (3) mathematical algorithms for dissipation factor spectrum interpretation, yielding physiologically relevant variables.

EIS Probe

The probe consists of a pair of slender (0.17mm) conductive electrodes resistant to corrosion connected to two miniature coaxial cables, held by an insulating plate. Platinum is the best electrode material, but mechanically soft. The small diameter-to-spacing ratio of the electrodes insures that no dependence on cell orientation will occur (for example, transverse or parallel orientation with respect to muscle fibers).

Dissipation Factor

In typical EIS measurements, the dissipation factor is determined at selected points within a range of 1 to 500 kHz. The dissipation factor is resistant to artifacts caused by variations in implantation depth and in the effective area of the electrode because it uses the *ratio* of resistance to reactance at each frequency. It has been shown that a given tissue (e.g., muscle) has a unique signature in all animal species (interspecies similarity) and that different tissues or organs have different signatures, reflecting micro-anatomical variations between them.

Mathematical Modeling

We have developed two styles of interpretation algorithms for dissipation-factor data.

The first is a general curve-fitting procedure that allows five independent variables to be optimized from any given dissipation-factor spectrum. This curve-fitting analysis simulates the circuit shown in Fig. 1, which

---

*Correspondence to: P. Héroux (Electropathology Lab.), Royal Victoria Hospital, 1130 Pine Ave. West, Montréal QU Canada H3A 1A3.

permits discrimination between contributions from electrode polarization and those due to the tissue itself; the fit procedure takes 2 h per measurement on a 386-7 25MHz microprocessor. Results computed with this procedure are shown in Figs. 2-4.

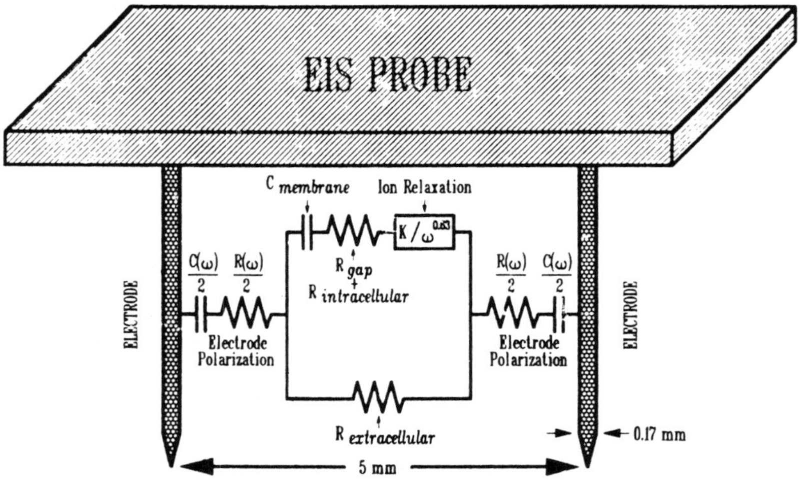

FIG. 1.—Equivalent circuit of EIS measurements.

The second is a specialized software that allows a single variable of interest to be rapidly extracted from the dissipation-factor data. For example, our algorithm for assessment of burn severity [''EIS(C/R)''] can be completed, including measurements and computations, in 10 s, so that the technique can be used in demanding theaters such as operating rooms, where speed is of importance.

EIS Readings

Figure 2 shows extracellular conductance, or $1/R_{ext}$ (indicator of extracellular edema) as a function of time after a microwave burn, as measured by an EIS probe implanted in the gluteus of a rat. The animal received a 52.7 C (lethal) burn in a limited region of the muscle and was thereafter continuously monitored. A large body of experimental data has been assembled to relate burn temperature to specific EIS indicators, which shows considerable expansion of extracellular space over time in burns. Results from basic data can be compiled by detailed modeling or by fast algorithms (10 s) designed to give surgeons indications on tissue viability. From 11 870 Ω in normal muscle, the value for R_{ext} can shrink 50-fold in burns to as little as 200 Ω. Because muscle cells have axial symmetry and because the expansion of extracellular space is very pronounced, the changes associated with burns show prominently on histological plates.

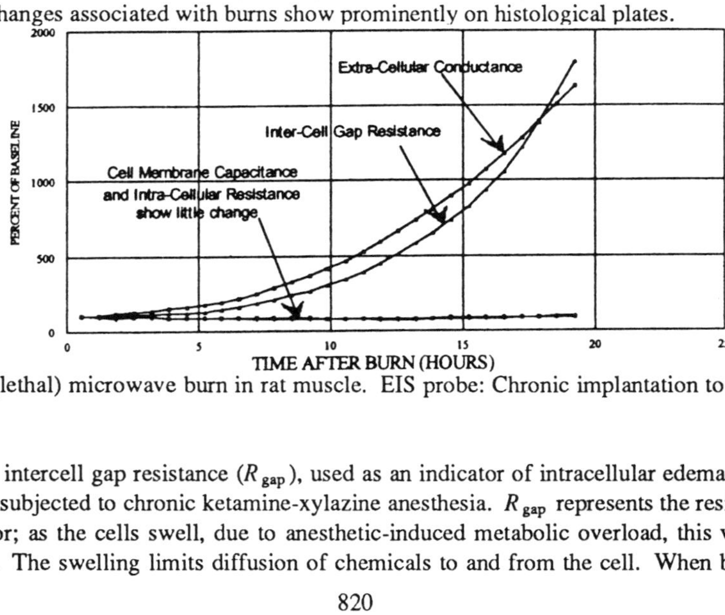

FIG. 2.—52.7 C (lethal) microwave burn in rat muscle. EIS probe: Chronic implantation to 3.5mm depth in gluteus.

Figure 3 plots intercell gap resistance (R_{gap}), used as an indicator of intracellular edema as a function of time in a rat liver subjected to chronic ketamine-xylazine anesthesia. R_{gap} represents the resistance from one cell to its neighbor; as the cells swell, due to anesthetic-induced metabolic overload, this value diminishes (from 10 to 1 Ω). The swelling limits diffusion of chemicals to and from the cell. When blood circulation stops, cell membrane capacitance falls rapidly, indicative of catastrophic events taking place in the tissue. Similar variations, but occurring over a much shorter time span, are shown in Fig. 4 for the rat brain.

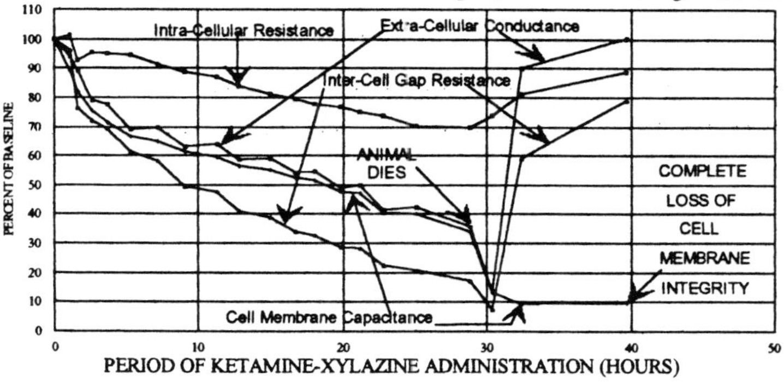

FIG. 3.—Progressive toxicity and death in rat liver. EIS probe: Chronic implantation to 3.5mm depth.

Figure 4 shows a plot of cell membrane capacitance C_m in the brain cortical tissue of a rabbit given an intraperitoneal barbiturate overdose. Lethal stress in the brain cortex can occur over minutes rather than hours as in muscle and liver. Irreversible damage is often accompanied by loss of integrity of the cell membranes. Alterations seen after 10 min are interpreted as cell swelling with reduction of extracellular space to 45% of baseline, together with a 30% reduction of cell membrane capacitance. The C_m loss of 30% is likely to be lethal.

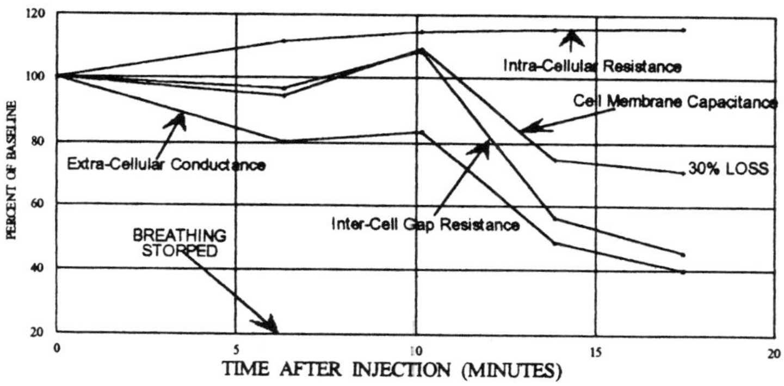

FIG. 4.—Lethal somnotol injection in rabbit brain. EIS probe: 1mm penetration in brain cortex.

CONCLUSION

EIS is a very accurate technique, particularly for following small changes over time. Its fineness and precision are related to the accuracy of electrical measurements, to the qualities of the dissipation factor as a basic variable, and to the inherent power of spectral analysis. Oddly enough, EIS can represent inexpensive high-technology in medical diagnostics.

ACKNOWLEDGMENTS

The authors gratefully acknowledge the aid of Jean Dumas and Frank Huang. Beyond the title organizations, the following supported this work: Electric Power Research Institute, Canadian Electrical Association, and Hydro-Québec-Environnement.

Martin Blank, Ed., *Electricity and Magnetism in Biology and Medicine*

ECOLOGICAL STUDIES ON EFFECTS OF ELF ELECTROMAGNETIC FIELDS: MAXIMAL COLD-INDUCED RATE OF OXYGEN CONSUMPTION IN MICE AND BIRDS

R. W. Hill,* D. L. Beaver, and P. Lederle

Michigan State University, East Lansing

Construction of a new 76Hz communications system provided opportunity to evaluate potential effects of ELF EM fields on free-living wildlife resident nearby. Deermice and black-capped chickadees inhabiting a test site near the antenna and a control site 20 km away were compared during winters of preoperational and operational years. The maximal rate of oxygen consumption was selected for study because it is a measure of overall physiological fitness and winter thermoregulatory capacity. No effects of the EM fields on the maximal rate of oxygen consumption were detected.

The recent construction of the U.S. Navy 76Hz communications system in the western Upper Peninsula of Michigan has provided a unique opportunity to assess effects of ELF EM fields on free-living animals because of the possibility of before-after study. Research on a test site, near the system antenna, and on an ecologically similar control site, 20 km away, was started in 1986, before system operation began. Study has continued during full operation of the system.

We have sought to assess effects on growth, reproduction, physiological health, and navigation capabilities in birds and mammals resident near the system antenna. We elected to study species of small body size because of their relatively limited ranges of movement. In this paper we report on the maximal rates of oxygen consumption of deermice, *Peromyscus maniculatus gracilis* (20 g adult body weight), and black-capped chickadees, *Parus atricapillus* (11 g).

An animal's maximal rate of oxygen consumption is considered a useful index of overall physiological health and physical fitness.[1,2] Brooks and Fahey state that "one definition of physical fitness is maximal oxygen consumption."[2] Astrand and Rodahl argue that "the maximal oxygen uptake is probably the best laboratory measure of. . .physical fitness," if by fitness we mean the capacity of an individual for prolonged heavy work.[1] An animal's maximal rate of oxygen consumption can vary with environmental conditions[3,4] and factors such as disease and age. Because the maximal rate of oxygen consumption may in principle be set by oxygen-uptake limits of the pulmonary system, oxygen-transport limits of the circulatory system, or metabolic oxygen-utilization limits of tissues,[5] measures of the maximal rates of oxygen consumption of animals exposed to EM fields may provide a unitary means of detecting EM-field-induced alterations in any of several key physiological systems. Maximal rates of oxygen consumption also help determine the resistance of free-living mammals and birds to winter cold because thermoregulatory heat production is proportional to oxygen consumption.

THE PHYSICAL AND OPERATIONAL PROPERTIES OF THE COMMUNICATIONS SYSTEM

The antenna of the system is a suspended wire, 90 km long, positioned in a cleared right of way. The control site has a matched right of way. Antenna current is 150 A. Communication is by frequency modulation between 72 and 80 Hz, with a center frequency of 76 Hz.

EM fields generated by the system have been extensively measured by the Illinois Institute of Technology Research Institute. The flux density of the 76Hz magnetic field is 31 mG near the antenna and falls progressively to 2 mG at a perpendicular distance of 200 m. The 76Hz electric field produced in the earth is 52 mV/m near the antenna, falling gradually to 17 mV/m at 200 m away. The 76Hz electric field produced in the air is 34 V/m near the antenna, dropping sharply to 49-150 mV/m at 25-200 m. At the control site, the 76Hz EM fields are no greater than 7% of the corresponding fields existing 200 m away from the antenna at

---

*Correspondence to: R. W. Hill (Zoology), Michigan State University, East Lansing MI 48824.

the test site. The 60Hz EM fields at both sites are of low intensity.

METHODS

Animals were captured at the test and control sites during winter (January-March) in preoperational years (1986, 1987), when the communications system was not functional; and during fully operational years (1990, 1991). At the test site, traps for mice were set 25-150 m from the antenna, and birds were netted 50 m from the antenna.

Captured animals were caged outdoors. Each individual's maximal rate of oxygen consumption was elicited by cold exposure in a helium-oxygen atmosphere.[6] Measures were made at three or more temperatures to assure that a true maximal rate was evoked and were usually completed within 5 days of capture. We have evidence that the peak rate of oxygen consumption is stable for 14 days after capture under the conditions of captivity in our work.

RESULTS AND DISCUSSION

Two-way analysis of covariance was used to test for differences between sites and years (covariate = body weight). In the deermice (Table 1), the mean maximal rate of oxygen consumption was lower in operational years than preoperational ones ($p = 0.01$). However, the test and control sites did not differ ($p = 0.14$) and there was no significant interaction between site and time period ($p = 0.98$). In the chickadees (Table 2), rates were again lower in operational years than preoperational ones ($p < 0.001$). Rates at the test site were lower than those at the control site ($p = 0.04$), but the interaction between site and time period was robustly nonsignificant ($p = 0.44$). Thus, for the chickadees there was a preexisting difference of unknown cause between the sites, and operation of the communications system did not alter its magnitude. The reason for the decline in rates between preoperational and operational years is unknown; methodological artifacts have been ruled out.

TABLE 1.—Maximal rates of oxygen consumption of deermice resident at the test and control sites during preoperational and operational years.

		Maximal rate of oxygen consumption [mL O_2/(g h)]	
Time and site	Number of animals	Mean	SD
Preoperational years			
Test site	17	19.2	1.6
Control site	18	20.2	1.6
Operational years			
Test site	23	18.6	2.0
Control site	23	18.7	1.8

TABLE 2.—Maximal rates of oxygen consumption of chickadees resident at the test and control sites during preoperational and operational years.

		Maximal rate of oxygen consumption [mL O_2/(g h)]	
Time and site	Number of animals	Mean	SD
Preoperational years			
Test site	15	24.4	2.1
Control site	20	25.2	1.8
Operational years			
Test site	26	23.3	2.0
Control site	22	23.9	1.8

We conclude that operation of this 76Hz communications system has not affected the physiological fitness or thermogenic capability of nearby resident mammals and birds, as assessed by measures of metabolic capacity.[7]

REFERENCES

1. P. O. Astrand and K. Rodahl, *Textbook of Work Physiology*, New York: McGraw-Hill, 1977, 2d ed.
2. G. A. Brooks and T. D. Fahey, *Exercise Physiology: Human Bioenergetics and Its Applications*, New York: Wiley, 1984.
3. W. R. Dawson and R. L. Marsh, in C. Bech and R. E. Reinertsen, Eds., *Physiology of Cold Adaptation in Birds*, New York: Plenum, 1989.
4. S. J. Wickler, *Physiol. Zool.* 53: 338-346, 1980.
5. D.P. M. MacLaren, H. Gibson, M. Parry-Billings, and R. H. T. Edwards, *Exerc. Sports Sci. Rev.* 17: 29-66, 1989.
6. M. Rosenmann and P. Morrison, *Am. J. Physiol.* 226: 490-495, 1974.

7. This research was funded by the Submarine Communications Project Office, Space and Naval Warfare Systems Command, U.S. Navy, Contracts N00039-81-C-0357, N00039-84-C-0070, and N00039-88-C-0065.

Martin Blank, Ed., *Electricity and Magnetism in Biology and Medicine*

INCIDENCE OF 50 AND 100Hz ELF MAGNETIC FIELD ON DIGESTIVE TRANSIT IN MICE: MODULATION OF ATROPINE EFFECT

R. Santini*

Institut National des Sciences Appliquées

The incidence of a pulsed (50 and 100Hz) South magnetic field (90 μT) on small intestine transit and atropine effect was studied in mice. Both frequencies significantly potentiated (Kruskal-Wallis nonparametric rank test) the atropine-induced delay in intestinal transit, but did not significantly affect intestinal transit per se.

Only few reports are available on the effect of electromagnetic radiation on digestive tract motility. These data report either a reduction or an acceleration of food transit in the dog, and a slackening of peristaltic movements in the guinea pig exposed to microwaves.[1] Previous studies from this laboratory have demonstrated a specific inhibitory effect of low-dose microwaves on rat digestive motility.[2] Recent studies have reported interactions between drugs and electromagnetic fields: interaction of microwave with amphetamine,[3] chlorpromazine,[4] ethanol,[5] and inhibition of morphine induced analgesia in mice exposed to a magnetic field.[6]

The purpose of the present study was to test the effect of a pulsed South magnetic field on mice transit and on atropine's inhibitory effect on the intestinal transit.

MATERIALS AND METHODS

Intestinal transit was studied in mice by a technique derived from Loewe[7] and previously described by Santini et al.[8] The transit marker used was phenolsulfone phtalein (PSP) and was given "per os" at $T = 0$ in volume of 0.3 ml. At $T = \pm 10$ min the mice were decapitated and the PSP migratory front was observed by application of 20% sodium hydroxide (NaOH) to the intestine (from the caecum toward the pylorus). A purplish color revealed the presence of PSP. The percentage of PSP migration was calculated for each mouse by measurement of the length of the small intestine and of the intestinal migration of the dye from the pylorus.

The experiments were carried out on 57-OF1 adult female mice at 100 Hz and on 60 mice at 50 Hz. Animals had fasted for 24 h and were randomly assigned to 4 groups for each experiment, as follows.

Group 1: reference group for transit, which was given 0.1 ml of distilled water intraperitoneal (I.P.) at $T = -10$ min. This group was then sham exposed to the magnetic field (just after I.P. injection) for 10 min before and 10 min after PSP administration.

Group 2: reference group for the magnetic field, which was given 0.1 ml of distilled water I.P. at $T = -10$ min and was then exposed to the magnetic field for 10 min before and 10 min after PSP administration.

Group 3: atropine group, which was given 1 mg/kg/I.P. of atropine sulfate (in distilled water volume 0.1 ml) at $T = -10$ min and was then sham exposed to magnetic field (just after atropine injection) for 10 min before and 10 min after PSP administration.

Group 4: atropine + magnetic field group, which was given 1 mg/kg/I.P. of atropine sulfate at $T = -10$ min and was then exposed to the magnetic field (just after atropine injection) between 10 min before and 10 min after PSP administration.

The magnetic field was generated by a coil connected to a MAGPULSE (Soc. SEDATELEC, F-69540 Irigny, France) apparatus: the coil was 60 mm in diameter and 15 mm thick. It produces a square unipolar magnetic field with a 75% duty cycle. The value of the magnetic field (measured with a digital electric and magnetic field meter model 130, Electric Field Measurements, Box 326, Stockbridge MA 01266) was 90 μT,

---

*Correspondence to: Roger Santini INSA, 20 ave. A. Einstein, F-69621 Villeurbanne, France.

and the frequencies were 50 and 100 Hz. The animals were exposed individually to the south side of the coil. The data are expressed as mean ± SEM of percent migration. Comparisons between groups were performed by the Kruskal-Wallis nonparametric rank test on individual percentages of PSP migration.

RESULTS

Results (Fig. 1) show that (a) 20 min exposure to the magnetic field used reduces mice digestive transit (−10.9% for 50 Hz and −15.5% for 100 Hz), but those reduction are not statistically significant (group 1 vs group 2); and (b) the atropine slackening effect of transit is significantly enhanced with 50 and 100Hz magnetic field exposure (group 3 vs group 4): +35.4% ($p < 0.02$) and +27.7% ($p < 0.001$), respectively.

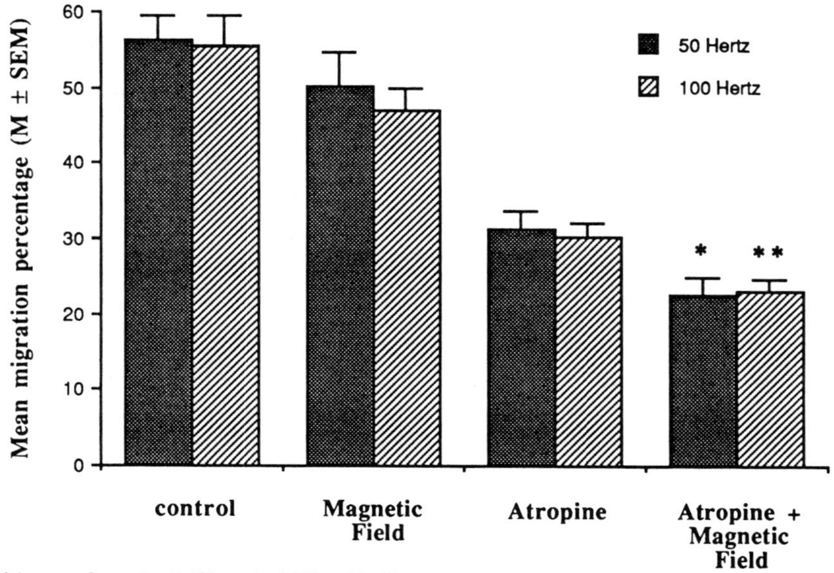

FIG. 1.—Incidence of a pulsed (50 and 100Hz) 90μT magnetic field on mice digestive transit and on atropine effect. Group 3 vs group 4: *$p < 0.02$; **$p < 0.001$.

DISCUSSION

This study shows that the inhibition of transit time by atropine is enhanced under a pulsed (50 or 100Hz) magnetic field applied for 20 min. These results are in agreement with recent studies indicating that electromagnetic field can interact with pharmacological compounds.[3-6]

The interaction between atropine and magnetic field may arise at least at two levels: modification of the mesenteric circulation, which might increase atropine intestinal availability; and action on specific muscarinic receptors' sensitivity to the drug under a magnetic environment.

An inhibiting action of magnetic field on acetylcholine level cannot be invoked because no significant effect was shown here under magnetic field alone (group 2 vs group 1).

In a view of radiation protection against nonionizing radiation and possible interactions between medications and electromagnetic fields, it is fair to conclude that people who are treated with medications have to take the electromagnetic environment into account.[9]

REFERENCES

1. S. Baranski and P. Czerski, *Biological Effects of Microwaves,* Stroudsburg, Pa.: Dowden-Hutchinson and Roos, 1976.

2. R. Santini, "Effect of low level microwave irradiation on the duodenal electrical activity of the unanesthetized rat," *J. Microwave Power* 17: 329-334, 1982.

3. H. Lai, A. Horita, C. K. Chou, and A. W. Guy, "Psychoactive drug response is affected by acute low-level microwave irradiation," *Bioelectromagnetics* 4: 205-214, 1983.

4. J. R. Thomas, J. Schrot, and R. A. Banvard, ''Behavioral effects of chlorpromazine and diazepam combined with low-level microwave,'' *Neurobehav. Toxicol.* 2: 131-135, 1980.

5. D. L. Hjeresen, A. Francendese, and J. M. O'Donnel, ''Microwave attenuation of ethanol-induced hypothermia,'' *Bioelectromagnetics* 9: 63-78, 1988.

6. M. Kavaliers and K. P. Ossenkopp, ''Magnetic fields differentially inhibit Mu, Delta, Kappa and Sigma opiate-induced analgesia in mice,'' *Peptides* 7: 449-453, 1986.

7. S. Loewe, ''Bioassay of laxatives on monkeys (Rhesus) and on lower mammalians (dyemeal methods),'' *J. Am. Pharm. Ass.* 28: 427-422, 1939.

8. R. Santini, M. Tessier, and P. Belon, ''Effects of low-doses treatment with cuprum on neostigmine digestive action in female mice,'' *Drug Development Res.* 24: 231-233, 1991.

9. Support for this research was provided by the French National Ministry of Education (M.E.N.) under Chap. 66-71, Art. 50, Para. 80, ''Soutien de programmes.''

Martin Blank, Ed., *Electricity and Magnetism in Biology and Medicine*

CAN A LOW-LEVEL ELECTRIC FIELD INDUCE STRESS IN MICE?

L. de Bruyn,* L. de Jager, and J. I. de Wet

University of the Orange Free State, Bloemfontein, South Africa

To determine whether chronic stress is induced by exposure to an electric field, the adrenal glands of mice were examined, as chronic stress is known to increase the lipid volume of all three adrenocortical zones. Experimental mice were exposed to a field of 10 kV/m for 23 h per day from birth to maturity. They were sacrificed and the lipid profile of the adrenal cortex of the experimental and control mice was assessed light-microscopically in a double-blind study. Statistical analysis showed a significant increase of lipid in the *z. glomerulosa* of exposed male mice, which suggests their exposure to a chronic stress situation.

In the controversy regarding the health effects of electric fields, chronic stress is named as a possible adverse effect and could pose as a health hazard. Chronic stress causes an increase in the lipid volume of all three zones of the adrenal cortex.[1] The present study investigated the effects of stress induced by a 10kV/m electric field on the mouse (*Mus musculus*) by a comparison of the lipid profile of the three adrenocortical zones of experimental mice with those of controls.

MATERIALS AND METHOD

Twenty-one pairs of mice were exposed to a 10kV/m electric field for 23 h per day from birth to maturity. Control mice were subjected to identical conditions barring the electric field. After the mice were sacrificed between the ages of 7 and 14 months, the adrenal glands were removed and fixed in 10% formalin, and sections were prepared and stained with haematoxylin and eosin. A double-blind light-microscopic study to assess the lipid profile of the adrenal cortex was undertaken. The lipid content of each individual cortical zone was scored as a percentage of 100. Males and females were assessed in separate groups. The results were analyzed by means of a Student t-test to evaluate the difference between the mean percentage lipid content of the corresponding zones of the experimental and control mice ($p < 0.05$ indicates a significant difference).

RESULTS

Female control mice contained statistically significant higher mean lipid values ($p < 0.05$) in all zones of the adrenal cortices compared with control males (Fig. 1).

There was a significant increase ($p = 0.004$) in the mean lipid content of the *z. glomerulosa* of the male experimental mice. This increase is demonstrated in Fig. 1, which also shows that no significant difference was found in the *z. fasciculata* of the males ($p > 0.05$). The *z. reticularis* of the males was not considered, as it contained very little lipid and was difficult to distinguish from the inner *z. fasciculata*.

Figure 3 is a light micrograph of an adrenal of an experimental male mouse showing large quantities of lipid in the cells of the *z. glomerulosa*. In contrast, Fig. 2 shows a control male adrenal where the corresponding cells contained much less lipid. In all the zones of the female mice (Fig. 1) the differences found were not significant ($p > 0.05$).

*Correspondence to: L. De Bruyn (Anatomy), University of the Orange Free State, P.O. Box 339, Bloemfontein 9300, South Africa.

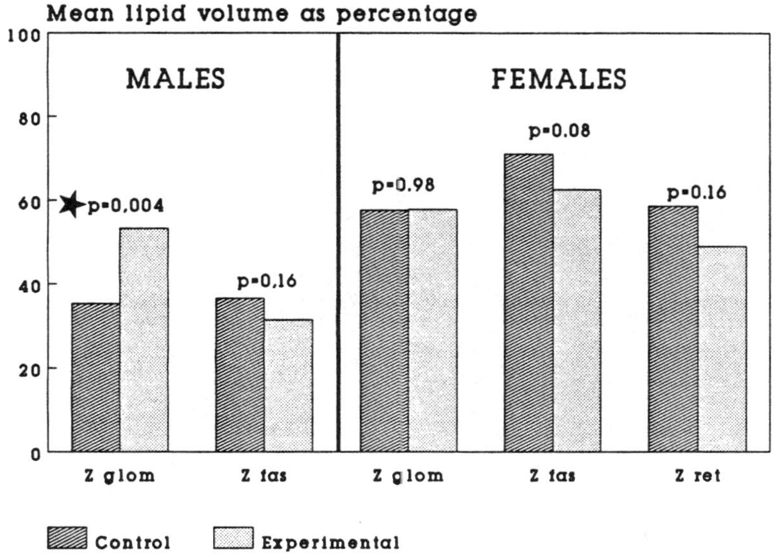

FIG. 1.—Mean percentage lipid in *z. glomerulosa*, *z. fasciculata*, and *z. reticularis* of the adrenal cortices of adult male and female mice; $p < 0.05$ indicates a statistically significant difference.

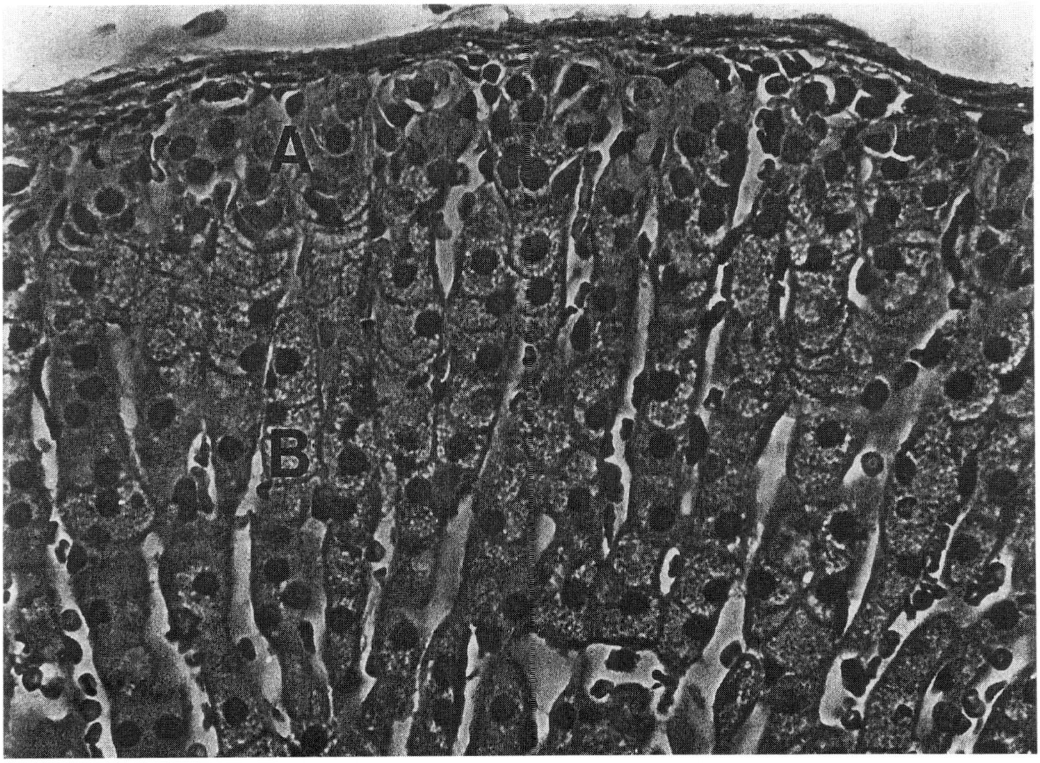

FIG. 2.—Adrenal cortex of a control male mouse showing the *z. glomerulosa* (A) and *z. fasciculata* (B). Note the small quantities of intracytoplasmic lipid in the *z. glomerulosa* (H & E×792).

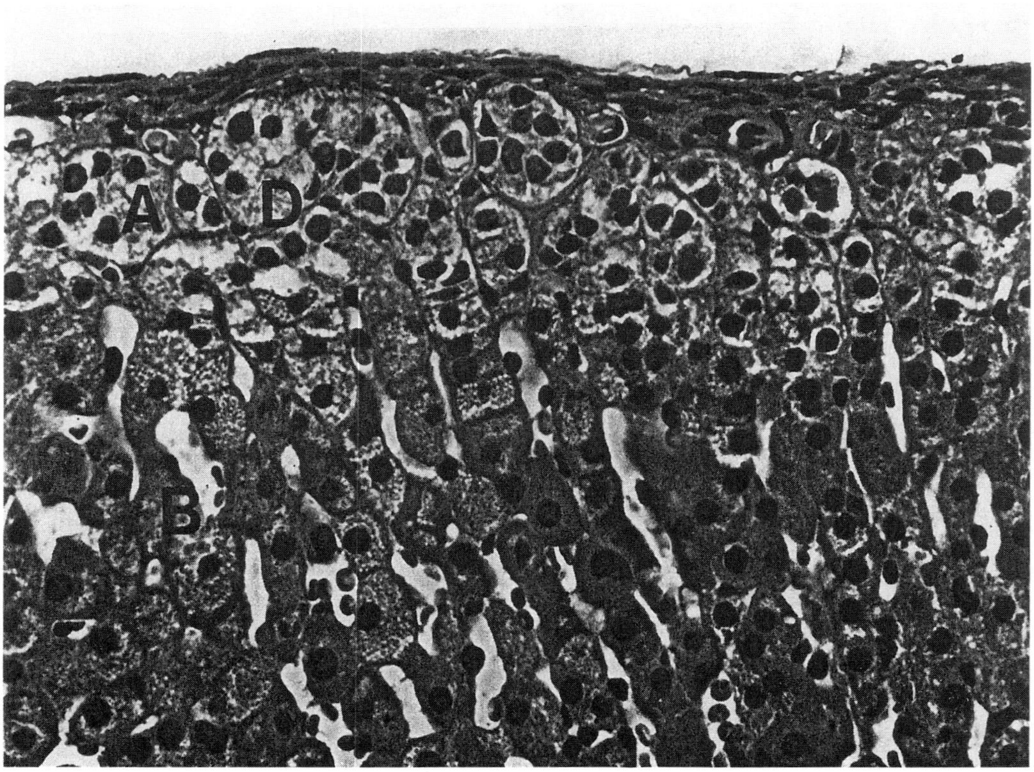

FIG. 3.—Adrenal cortex of a male experimental mouse showing the *z. glomerulosa* (A) and *z. fasciculata* (B). Note the abundance of intracytoplasmic lipid (D) in the *z. glomerulosa* (H & E×792).

DISCUSSION AND CONCLUSION

Nicander reported a relatively higher abundance of lipoid in the adrenal cortex of female mice when compared to male mice.[2] The observation of a statistically significant higher mean lipid value in the female control mice compared to the male control mice confirms his findings.

Nussdorfer reported an increase in the lipid volume of all three zones of the adrenal cortex as a result of chronic stress.[1] The increased lipid content of the experimental male group acts as evidence of these experimental mice being exposed to a chronic stressor. We thus conclude that the electric field could act as a stressor as it was the only difference in the conditions to which the experimental and control mice were subjected. The reason for the differences in response between the male and female mice is not clear. Hormonal factors may have played a part, a point that should be elucidated by further study. In living organisms, chronic stress may predispose to conditions such as decreased resistance to infections, malignant diseases, and enhanced aging. The electric field could therefore be a health hazard by acting as a stressor.[3]

REFERENCES

1. G. G. Nussdorfer, *Int. Rev. Cytol.* 98: 1, 1986.
2. L. Nicander, *Acta Anat.* 16: 1, 1952.
3. The support of the National Energy Council is gratefully acknowledged.

Martin Blank, Ed., *Electricity and Magnetism in Biology and Medicine*

EPIDIDYMAL SPERM COUNTS IN MICE EXPOSED TO A 10kV/m ELECTRIC FIELD

L. de Jager,* D. K. Hamlet, and L. de Bruyn

University of the Orange Free State, Bloemfontein, South Africa

The effect of an electric field on the reproductive system was evaluated by means of epididymal sperm counts. An experimental group of mice was exposed to a 50Hz, 10kV/m electric field from conception and a control group was sham exposed. The mice were sacrificed at the age of 60 days (sexually matured) and at age 18 months (old mice). Epididymal sperm counts were performed and the results were statistically analyzed. A statistically significant higher sperm count was found in the experimental group at both ages, which suggests that the electric field influenced spermatogenesis. Sequential studies on sperm motility and hormonal levels are indicated.

The reproductive system, a highly sensitive system that relies on precise coordination of a complex series of events, is particularly vulnerable to environmental insult and has been the subject of a large volume of research on electric and magnetic field effects.[1,2] To elucidate the possible effects of power-frequency fields on the fertility of mammals, we determined the effect of long-term exposure to an electric field of 10 kV/m on the concentration of mouse spermatozoa.

MATERIALS AND METHODS

Experimental Animals

The male offspring of 10 pairs of parent mice in both a control and experimental group, *Mus musculus*, Balb/C were used. The experimental group was exposed to a 10kV/m electric field (50 Hz) effectively in specially designed exposure cages. Both groups of mice were exposed to the same environment except for the electric field (22 C in a 12h light, 12h dark photoperiod, and 50% humidity). The mice had access to food and water *ad libitum*. The parent mice of the experimental group were placed under the electric field after being paired. On the day of birth each litter was standardized to 4 offspring and at 21 days of age they were weaned and the males housed separately. At the age of 55 days males from both the control and experimental groups were further assigned to two subgroups: one subgroup ($n = 16$ for controls and $n = 11$ for experimentals) destined to be sacrificed at age 60 days (sexually mature), the other subgroup ($n = 6$ for controls and experimentals) to be sacrificed aged 18 months (very old). The mice of the subgroup destined to live to age 18 months were paired with females.

Method for Sperm Counts

The mice were sacrificed by means of cervical dislocation. The preparation of spermatozoa for sperm counts was performed according to the method described by Searle and Beechey.[3] Four sperm counts were performed for each mouse and the mean value of the four counts was used. Data for sperm counts were first evaluated by normal probability plots of the variables in each group, after which the mean values of the control and experimental groups for each age group were compared by Student's t-test.

RESULTS

The mean epididymal sperm counts of 60-day-old mice in the control and experimental groups are shown in Fig. 1. At age 60 days the control male mice showed an average epididymal sperm count of $217 \times 10^6 \pm 97$; the experimental males showed an average sperm count of $324 \times 10^6 \pm 142$. The experimental group thus showed a higher average epididymal sperm count than the control. With statistical analysis the

---

*Correspondence to: L. de Jager (Anatomy), University of the Orange Free State, P.O. Box 339, Bloemfontain 9300, South Africa.

difference proved to be significant ($p = 0.029$).

Figure 1 shows the mean epididymal sperm concentrations at age 18 months in the control and experimental groups. At age 18 months the epididymal sperm count was considerably lower in both the control and experimental groups. The controls showed an average epididymal sperm count of $38 \times 10^6 \pm 26$ and the experimental males an average sperm count of $95 \times 10^6 \pm 25$. The epididymal sperm count in the experimental group was significantly higher than in the control group ($p = 0.002$).

DISCUSSION

The effects of occupational and environmental toxins on the reproductive potential of humans and animals are of increasing concern. Environmental exposure of humans and animals to electric and magnetic fields, both in the home and occupationally, poses the question whether fertility is affected.

Studies on variation in sperm morphology and sperm counts as a function of age demonstrated many changes at the time of puberty and during aging. The present study was therefore designed to examine the epididymal sperm concentration in sexually mature mice (60 days of age) and late in life (aged 18 months).

A statistically significant higher sperm count was found in the experimental group at age 60 days as well as at age 18 months. The electric field therefore did not negatively influence the sperm-producing function of the testes but resulted in a higher sperm concentration.

We hypothesize that the phenomenon of a higher epididymal sperm concentration recorded in the experimental group in this study, especially at old age, could be explained on the basis of altered spermatogenesis control mechanisms. Both the control and experimental groups were kept in the same room under identical conditions. As the electric field constituted the only difference in environmental conditions between controls and experimentals we hypothesize that it could have played a role in the observed differences in sperm counts (directly or indirectly, e.g., through stress-related responses).[4] The hypothesis will be further tested in follow-up studies in which hormonal levels will be monitored.

FIG. 1.—Mean epididymal sperm counts of control and experimental mice at 60 days and at 18 months ($p < 0.05$).

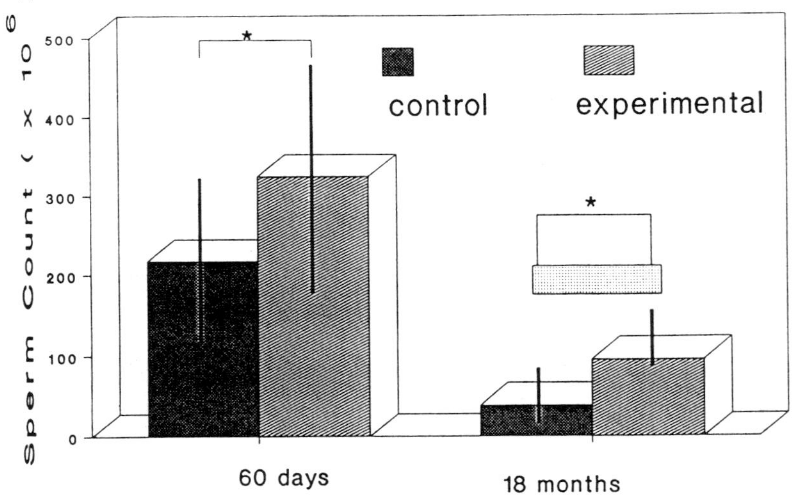

REFERENCES

1. A. A. Marino, M. Reichmanis, R. O. Becker, B. Ulrich, and J. M. Cullen, "Power frequency electric field induced biological changes in successive generations of mice," *Experientia* 35: 309-311, 1980.

2. A. L. Carsten, "Mutagenicity and toxicity of electric and magnetic fields," in report of New York State Department of Health Power Line Project, 1987.

3. A. G. Searle and C. V. Beechey, "Sperm-count, egg-fertilization and dominant lethality of x-irradiation of mice," *Mutation Res.* 22: 63-72, 1974.

4. F. C. Leung, C. A. Poindexter, D. N. Rommerein, J. A. Creim, and L. E. Anderson, "Evidence of stress in rats exposed to 60Hz electric field," in *Biological Effects from Electric and Magnetic Fields, Air Ions, and Ion Currents Associated with High-voltage Transmission Lines,* Contractors' Review, Phoenix, Arizona, 1988.

Martin Blank, Ed., *Electricity and Magnetism in Biology and Medicine*

EFFECTS OF 50Hz ELECTRIC FIELDS ON THE LIFE EXPECTANCY OF MICE

L. de Jager* and L. de Bruyn

University of the Orange Free State, Bloemfontein, South Africa

To determine the effect of an electric field on the life-span of mice, male and female mice were paired and separated into two treatment groups, E-field exposed and sham exposed. Exposed mice were subjected to a 50Hz, 10kV/m electric field 20 h/day. Sixty offspring from each treatment group (generation 1) were allowed to live for 18 months; forty offspring from generation 1 (generation 2) were allowed to die naturally. A post-mortem examination was performed on all mice. The results were statistically analyzed by the χ^2 test. A statistically significant higher incidence of deaths before 18 months was observed in the exposed group of the first generation and the second generation experienced a higher incidence of death at an earlier age. The cause of death was established where possible. There was no increase in the incidence of malignant tumors or any specific single cause of death. The results of this study indicate that continuous long-term exposure to a 10kV/m electric field may cause a shortened life expectancy.

A survey of the literature indicates that there are interactions between biological systems and electric fields. Research shows that extremely low frequency (ELF) electromagnetic fields can cause biological effects such as circadian rhythm shifts[1] and reduced function of lymphocytes.[2] Could such biological effects cause a reduced life expectancy? The objective of our study was to determine the effect of an electric field on the life span of mice.

MATERIALS AND METHOD

Male and female mice, *Mus musculus,* Balb/C were paired and separated into two treatment groups, E-field exposed and sham exposed. Exposed mice were subjected to a 50Hz, 10kV/m electric field for 20 h/day. Both groups of mice were exposed to the same environment except for the electric field (22 C in a 12h light, 12h dark photoperiod, and 50% humidity). Offspring from each treatment group were paired at sexual maturity, identified as generation 1, and allowed to live for 18 months (3/4 of the expectant life span of mice in captivity) before they were sacrificed. Offspring from generation 1 were paired, identified as generation 2, and allowed to lead a normal life (except for the electric field) and die naturally. Exposed mice of generations 1 and 2 were thus subjected to the E-field from conception until death.

The age at which death occurred was recorded in generation 1 mice dying before 18 months and in all generation 2 mice. A post-mortem examination was performed on all mice to search for evidence of disease, and in mice that had died naturally an attempt was made to establish the cause of death. The results of the age of naturally occurring death were statistically analyzed by the χ^2 test.

RESULTS

A higher incidence of deaths before 18 months was observed in the exposed group of the first generation, especially in the age group 6 to 12 months and the results proved to be statistically significant (Table 1). According to statistical prediction the possibility of dying before 18 months was 7-41% higher in the experimental group than in the control group. The results of the second generation, which was left to die naturally, indicated that the experimental group experienced a higher incidence of death at an earlier age than the control group (in the period before 18 months as well as before 24 months). The results of the second generation are considered to be clinically important; in the experimental group the possibility of dying before 18 months was between −6% and 31% higher than in the control group and between −1% and 42% higher in the experi-

*Correspondence to: L. de Jager (Anatomy), University of the Orange Free State, P.O. Box 339, Bloemfontein 9300, South Africa.

mental group before age 24 months (Table 2).

The cause of death was established where possible and no increase in the incidence of malignant tumors was found; in most cases death was a result of nonspecific causes.

DISCUSSION

The results of this study indicate that continuous long-term exposure to a 10kV/m electric field caused a shortened life expectancy. The likelihood of mice dying at an earlier age was higher for the experimental mice than for the controls. In the first generation a statistically significant difference was found in the expected outcome. The results of the second generation are considered to be of clinical importance; it was found that the possibility of dying before the age of 24 months was as high as 42% for the experimental group.

No specific single cause of death could be established and we concluded that the E-field caused a physiological effect that lowered the resistance of the mice to infections or made them succumb to minor illness and caused them to die at a younger age.

REFERENCES

1. B. W. Wilson et al., "Chronic exposure to 60-Hz electric field: Effects on pineal function in the rat," *Bioelectromagnetics* 2: 371-380, 1981.

2. D. B. Lyle et al., "Suppression of T-lymphocyte cytotoxicity following exposure to sinusoidally amplitude modulated HF fields," *Bioelectromagnetics* 4: 281-292, 1983.

TABLE 1.—Results of first-generation mice: naturally occurring deaths before 18 months ($p < 0.05$, statistically significant difference).

	Control	Experimental
Number of deaths before 18 months	10	21
Number of mice sacrificed at 18 months	50	21
Total	60	42

χ^2 analysis: $p = 0.0007$, 95% Cl: 0.073; 0.49 (7.3%; 49%)

TABLE 2.—Results of second-generation mice: life expectancy at age 18 and 24 months ($p < 0.05$, statistically significant difference).

	Control	Experimental	χ^2
Deaths before 18 months	7	12	$p = 0.2913$
Deaths after 18 months	32	27	95% Cl: −0.06; 0.31 (−6%; 31%)
Total	39	39	
Deaths before 24 months	17	25	$p = 0.1118$
Deaths after 24 months	22	14	95% Cl: −0.011; 0.42 (−1.1%; 42%)
Total	39	39	

Martin Blank, Ed., *Electricity and Magnetism in Biology and Medicine*

PATHOLOGICAL EVALUATION OF RABBITS SUBCHRONICALLY EXPOSED TO 60Hz MAGNETIC FIELDS AT 1 G AND 5 G

J. T. Ning,* S. Toyokuni, D. Thomas, E. M. Czerska, J. Casamento, and M. L. Swicord

Center for Devices and Radiological Health, Rockville, Md.

C. C. Davis

University of Maryland, College Park

E. Elson

Walter Reed Army Institute of Research, Washington, D.C.

Numerous in vitro and in vivo biological effects due to exposure to ELF fields have been reported.[1] However, there is a paucity of data relating those findings to health effects. We have initiated a pilot animal study that would address possible disease pathology. Twelve young adult 3kg New Zealand White male rabbits were divided into three groups. One group served as the control and was exposed to ambient magnetic fields ranging from 1 to 8 mG. The second and third groups were exposed to 1 and 5G 60Hz magnetic fields, respectively, for 6 h/day, 5 days/week, for the 14-week period. The animals were examined visually throughout the entire subchronic 14-week period. One rabbit exposed to the 1G magnetic field developed a tumor growth on the left posterior flank 1 month into the exposure regimen. At the end of 14 weeks, complete autopsies were performed and specimens from the heart, lungs, spleen, adrenal glands, kidneys, liver, gallbladder, pancreas, stomach, jejunum, ileum, large intestine, urinary bladder, testis, submandibular and sublingual glands, femur, quadriceps femoris muscle, cerebrum (cortex and hippocampus), cerebellum, and retina, as well as any tumor or enlarged lymph nodes, were obtained. Each specimen was fixed with 2.5% paraformaldehyde, embedded in paraffin, and sectioned at 4 µm. Hematoxylin and eosin stain, Pearls' iron stain, and reticulin stain were used. There were no positive findings in any of the organ specimens with the exception of the lung, liver, and urinary tract. In the hepatocytes of exposed animals, lipofuscin deposition was more remarkable. Slight hemosiderin deposition in hepatocytes of control animals was less prominent in the exposed animals. One of the control rabbits showed an abscess in the lung, likely to be of bacterial origin. All exposed rabbits revealed cystitis with erosion, squamous metaplasia, and lymphocyte and plasma cell infiltrations in the bladder mucosa, indicating chronic inflammation. One of the exposed rabbits showed chronic inflammation in the renal pelvis. The tumor in the flank was diagnosed as an abscess. However, this pilot animal study revealed no significant disease pathology associated with subchronic exposure of 60Hz magnetic fields at 1 and 5 G.

Extremely low frequency (ELF) electromagnetic fields (EMFs) have been reported to cause various biological effects.[1] Tests of ELF medical devices under clinical trials for potential therapeutic applications in clinical medicine are on the increase. Concurrent with the optimism as to new modalities of treatment, there is also concern for possible adverse effects on human patients, heightened by recent reports of increased risk of cancer (especially leukemia and certain central nervous-system cancers) being linked to occupational EMF exposure.[2-4] Even though ELF devices are currently being used clinically to treat nonunion fractures, only limited numbers of studies have been done on animal models with regard to health effects. This situation led us to initiate this subchronic study to assess possible disease pathology associated with 60Hz ELF magnetic field exposure at 1 and 5 G.

MATERIALS AND METHODS

Twelve young adult 3kg New Zealand White male rabbits were divided into three groups. One group served as control and was exposed to ambient EMF in the Twinbrook Research Laboratory animal-care

*Correspondence to: J. T. Ning, FDA-CDRH, Rockville MD 20857.

facility. The second and third group were exposed to 1 and 5G ELF fields at 60 Hz, respectively, for 6 h/day, 5 days/week for 14 weeks. Each day, the animals to be exposed were placed in plastic cages that were then placed in a large Helmholtz-coil exposure system. All rabbits were housed in the same room in the animal-care facility and received the same food and water ad lib. The temperature, humidity, and lighting were controlled and monitored throughout the experiment. The rabbits were examined daily for any observable gross changes. After 14 weeks of exposure, the rabbits were humanely sacrificed and necropsy was performed. Specimens from the heart, lung, spleen, adrenal gland, kidney, liver, gallbladder, pancreas, stomach, jejunum, ileum, colon, urinary bladder, testis, submandibular and sublingual gland, femur, quadriceps femoris muscle, cerebrum (cortex and hippocampus), cerebellum, and retina, as well as any tumor or enlarged lymph nodes, were obtained. Parafin-embedded histologic sections were prepared and stained with hematoxylin and eosin. For liver and kidney specimens, Pearls' iron stain and reticulin stain were also used. In addition, kidney specimens were processed for transmission electron microscopy (TEM).

RESULTS

Examination of the H&E-stained tissue sections for morphology under light microscopy revealed no abnormal findings in any of the exposed rabbits' organs with the exception of the lung, liver, and urinary tract. In the hepatocytes of the ELF-exposed rabbits, lipofuscin granules were more prominent than in control animals (Fig. 1); deposition was observed mostly in the periportal area. In the hepatocytes of control rabbits, slight deposition of hemosiderin was observed around the central vein, but in those of exposed rabbits, hemosiderin deposition was less prominent. One of the control rabbits showed abscess in the lung, likely to be of bacterial origin. All exposed rabbits had mild to severe cystitis with erosion, squamous metaplasia, and lymphocyte and plasma cell infiltrations in the bladder mucosa, indicating chronic inflammation. However, in the control group, only one rabbit showed chronic cystitis. One of the exposed rabbits showed chronic inflammation at renal pelvis. Reticulin stain revealed no abnormality in the glomeruli of the exposed animals. TEM showed normal-appearing basement membrane in the glomeruli of control and exposed kidneys, without abnormal deposition or fusion of the podocytes.

Examination of the tumor from the rabbit exposed to 60Hz ELF fields at 1 G showed a nonindurated cystic soft tissue mass in the left posterior flank. The tumor was movable by palpation. When first noticed after 1 month of exposure, the tumor measured 4 cm in diameter. The tumor steadily increased in size to approximately 10 cm in diameter at necropsy. During necropsy, a yellowish amorphous semiliquid content was observed after the tumor sac was opened. Pathological examination under light microscopy showed an abscess with chronic inflammation (Fig. 2). In some part of the abscess wall, fibroblasts with slight atypia were observed (Fig. 3). No evidence of neoplasm was found. Several portions of the abscess wall were inoculated in tissue culture media in an attempt to isolate possible transformed cell lines. Isolation and characterization of cell lines derived from the abscess wall are presented in a companion paper.

DISCUSSION

Histological examinations revealed no major changes in morphology except prominent lipofuscin deposition in the liver and urinary-tract infection in the exposed animals. Lipofuscin represents the indigestible residues of autophagic vacuoles formed during aging or atrophy, and its derivation through lipid peroxidation of polyunsaturated lipids of subcellular membranes is suggested.[5] Whether chronic exposure to ELF leads to increased cellular lipid peroxidation will be an interesting area for further study. Hemosiderin is an iron-protein complex storing excess iron in cytoplasm.[5] Iron kinetics may be changed by exposure to ELF fields to effect decreased storage of iron. Since excess iron appears to be associated with carcinogenesis,[6] further evaluation of the effect of ELF on iron metabolism is warranted.

We examined the kidneys by both light microscopy and electron microscopy in order to clarify the possible role of kidney in the decreased serum albumin and increased urinary protein described in a companion paper. These studies showed normal structure of endothelium, basement membrane, and podocytes. Thus, the kidney is not responsible for the altered protein patterns in the serum and urine, provided that protein leakage is always associated with morphological changes detectable by electron microscopy. One of the possible explanations for protein alteration would be immunosuppression (since all exposed animals showed chronic urinary-tract infection), which would explain serum albumin decrease, urinary protein increase, and high frequency of urinary tract infection. Further investigation is necessary for the evaluation of immunological factors such as T cell, B cell, polymorphonuclear leukocyte (PMN), and macrophage function.

1a

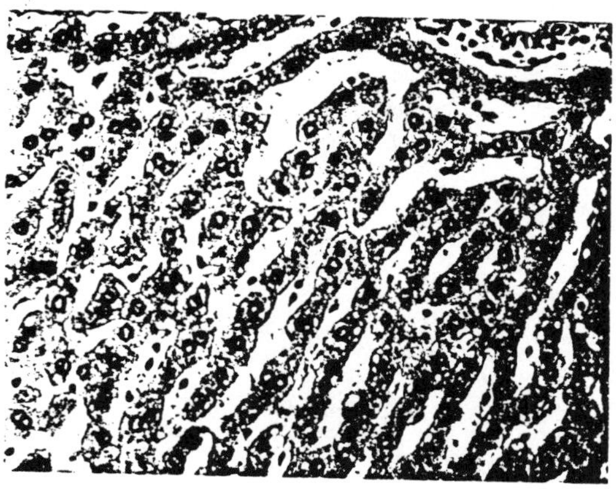

1b

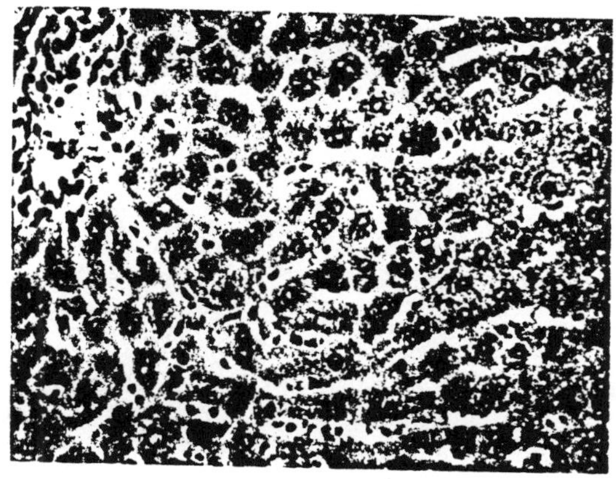

1c

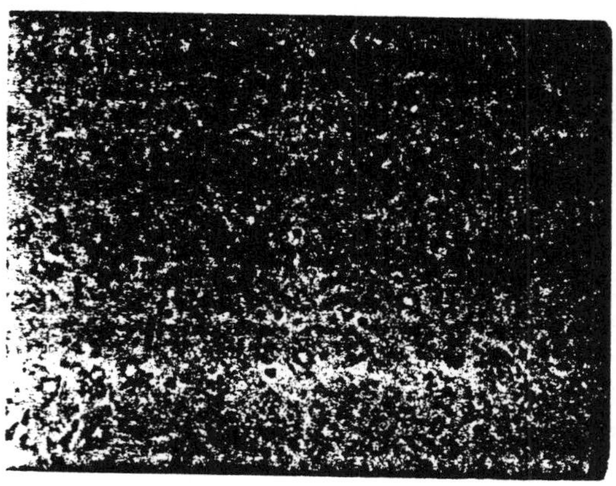

2

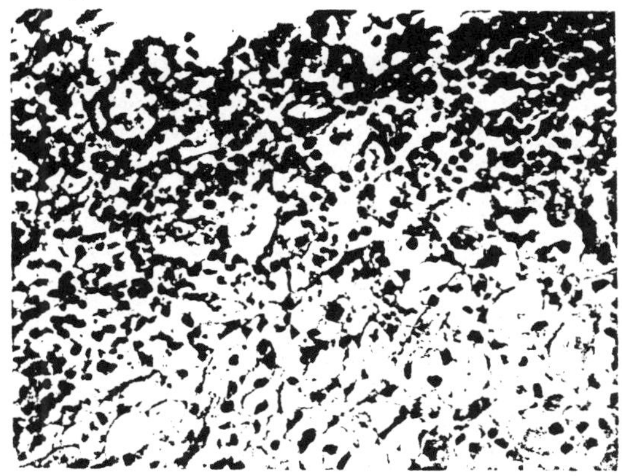

3

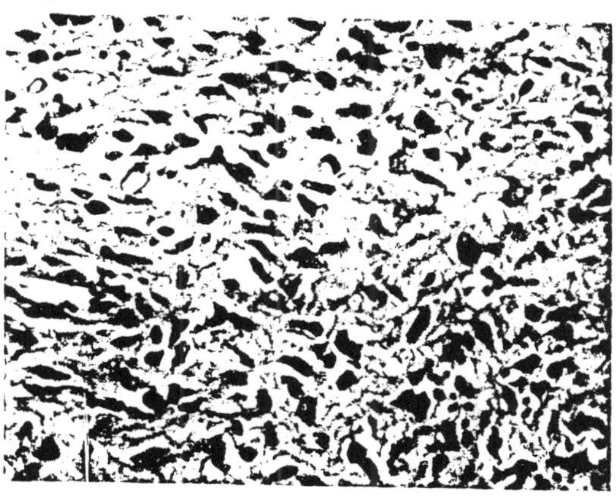

FIG. 1.—(a) Control rabbit, liver: hepatocytes show no cytoplasmic granules (H&E, 400×); (b) 60Hz 1G-exposed rabbit, liver: hepatocytes at the periportal area show cytoplasmic brown pigment deposition (H&E, 400×); (c) same as (b): negative results by Pearls' iron staining indicate that these intracytoplasmic granules are lipofuscin (Pearls' iron stain, 400×).

FIG. 2.—Wall of the tumor that appeared in 60Hz 1G-exposed rabbit. Diagnosis, abscess wall. Erosion, infiltration of polymorphonuclear leukocytes and macrophages are prominent (H&E, 400×).

FIG. 3.—Abscess wall. Fibroblasts with slight atypia were seen in a small area of abscess wall (H&E, 400×).

The tumor found on the hip of one of the exposed rabbits was identified as a non-neoplastic abscess. Several cell lines resembling fibroblast in phenotype were established from cells isolated from the abscess wall. No bacterial contamination was observed. One cell line became anchorage independent and dramatically responded to further ELF exposure. Since we found fibroblasts in a local area of the abscess wall that was slightly atypical, those atypical cells may be the origin of our cell lines. Why an abscess was formed on the hip is unknown. We observed abundant PMN in the tumor wall and cavity. The reactive oxygen species produced by PMN or macrophage[7] may be responsible for the initiation of fibroblasts.

REFERENCES

1. H. L. Konig, A. P. Krueger, L. Siegnot, and W. Sonnin, *Biological Effects of Environmental Electromagnetism,* New York: Springer-Verlag, 1981, 218-247.

2. S. M. Michaelson, "Influence of power frequency electric and magnetic fields on human health," *Ann. N.Y. Acad. Sci.* 502: 55-75, 1987.

3. D. A. Savitz and E. E. Calle, "Leukemia and occupational exposure to electromagnetic fields: Review of epidemiologic surveys," *J. Occup. Med.* 29: 47-51, 1987.

4. F. C. Garland, E. Shaw, E. D. Gorham, C. F. Garland, M. R. White, and P. J. Sinsheimer, "Incidence of leukemia in occupations with potential electromagnetic field exposure in U.S. Navy personnel," *Am. J. Epidemiol.* 132: 293-303, 1990.

5. R. S. Cotran, V. Kumar, and S. L. Robbins, *Robbins Pathologic Bases of Disease,* Philadelphia: W. B. Saunders, 1989, 4th ed., 24-26.

6. R. G. Stevens, D. Y. Jones, M. S. Micozzi, and P. R. Taylor, "Body iron stores and the risk of cancer," *New England J. Med.* 319: 1047-1052, 1988.

7. B. Halliwell and J. M. C. Gutteridge, *Free Radicals in Biology and Medicine,* Oxford: Clarendon Press, 1989, 2nd ed., 372-390.

Martin Blank, Ed., *Electricity and Magnetism in Biology and Medicine*

LIFE SPAN EXPOSURE TEST OF RATS TO 50Hz SINUSOIDAL ALTERNATING MAGNETIC FIELDS

M. Yasui and T. Kikuchi

Tokyo Electric Power Co., Japan

Y. Otaka* and M. Kato

Mitsibushi Kasei Institute of Toxicological and Environmental Sciences, Ibaraki, Japan

An exposure system for 50Hz sinusoidal alternating magnetic fields was constructed and a 104-week continuous exposure test was performed on 48 F344 rats of each sex per exposure level. The system consisted of a specific pathogen-free exposure room containing magnetic field generating coils of 0.5, 5 mT and sham, and was run without trouble throughout the experiment; 72.9-83.3% of the animals of each level had survived at the end of exposure. No significant difference was detected among the exposure levels by the Armitage's χ^2 test of the results of palpation during the exposure and macroscopic examination of almost all the organs. Microscopic pathological examination is now in progress.

In relation to the effects of power-frequency magnetic fields on human health, Walsh[1] and Makinodan[2] published plans of large-scale rodent chronic exposure tests using x-ray and γ-ray as initiators of cancer. We also constructed a magnetic field exposure system and performed a carcinogenicity test using rats.[3]

EXPOSURE SYSTEM

The system consisted of a specific pathogen-free room, furnished with attachments such as a clean-air conditioner, an autoclave sterilizer, etc., and magnetic field generators; 50Hz sinusoidal alternating magnetic fields of 0.5 and 5 mT were generated by coils of the type reported by Harvey.[4] Forty-eight rats of each sex were housed in the space of ±10% homogeneous field. Total high harmonics ranging from 2nd to 49th of the field comprised less than 0.1% of the fundamental wave. Acoustic noise was 52.6 dB(A) and no temperature rise was detected in operation. The background field strength at the sham exposure position was less than 1 μT. The system was driven without trouble throughout the experiment.

METHOD

Forty-eight F344/DuCrj rats of each sex were exposed to the fields from 5 to 109 weeks of age according to OECD test guideline 451. Animals were observed carefully almost every day, and palpated and weighed once a week. Food consumptions were measured every week in the early stage and every 13 weeks afterward. Peripheral blood smears were prepared 12 and 18 months after commencement of the exposure. The average exposure time was actually 22.6 h/day, because exposure of the investigators had to be avoided. Animals found in moribund condition through the daily observations were sacrificed and dissected, together with those found dead. All survivors were sacrificed on schedule at the end of 104-week exposure and all the organs were macroscopically inspected together with those of animals found dead or killed in moribund condition. Microscopic pathological examination of the organs is is now in progress.

RESULTS

No significant difference was observed in body weight gain and food consumption between the exposed and the sham groups. Fifteen rats with head tilt accompanied by cataracts and 19 female rats with occasional

---

*Correspondence to: Y. Otaka, Mitsibushi Kasei ITES, Sunayama 14, Hasaki-machi, Kashima-gun, Ibaraki 314-02 Japan.

epilepsy at palpation were found by daily observations, but there was also no significant difference. Figure 1 shows the survival curves. There is no significant difference among the survivals. It seems that the male 0.5mT survival curve has a little different shape from that of the control, but the causes of death are so diverse that the contributions to the statistical test are very small. Macroscopic findings on spleen, pituitary, brain, skin, subcutis, and mammary gland are shown in Table 1. In Table 1 there is no significant difference of the incidence among the exposed and the sham groups. Including 30 organs other than those shown in Table 1, whole organs will be taken into consideration in the course of microscopic examination.

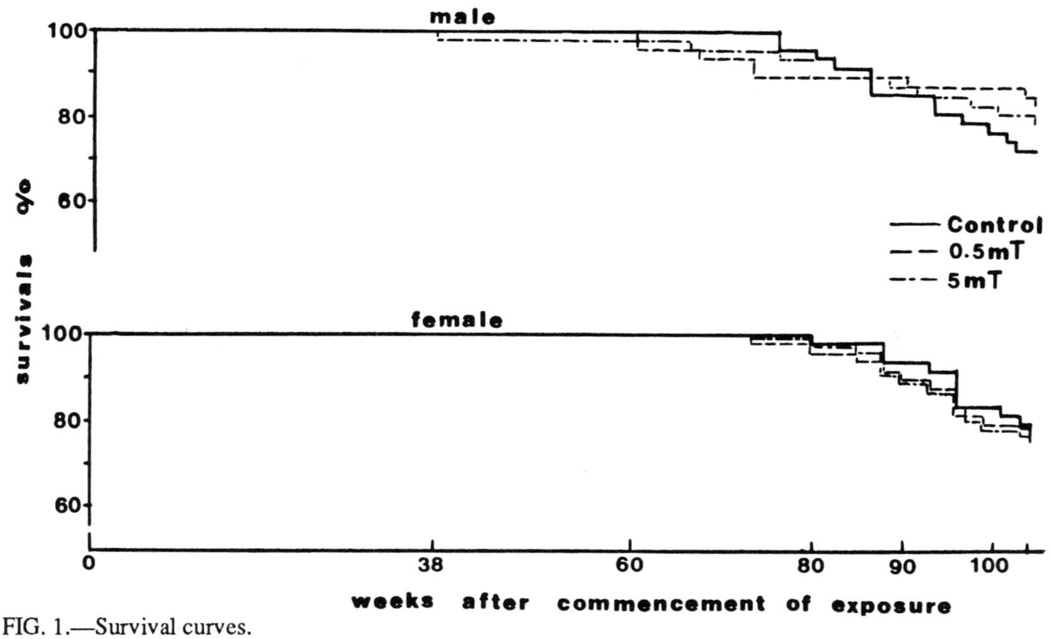

FIG. 1.—Survival curves.

CONCLUSION

No significant difference of the incidence of findings concerning carcinogenicity was obtained among the exposed and the sham-exposed groups at the stage of macroscopic examination of all organs.

REFERENCES

1. M. L. Walsh, "The Ontario Hydro 60Hz electric and magnetic field risk assessment program," *BEMS 1988 Meeting (Abstract B-04-7).*

2. T. Makinodan, *Leukemia and Brain Cancer in Animals Exposed to 60Hz Magnetic Fields*, EPRI proposal RFP 2965-11, 1989.

3. M. Yasui and Y. Otaka, *Bioelectromagnetics* (in press).

4. S. M. Harvey, *Magnetic Field Rodent Reproductive Study (MFRRS): Exposure System*, Report No. 88-125-K, Ontario Hydro, 1988.

TABLE 1.—Incidence of macroscopic findings in some organs.

Organ Findings	Male			Female		
	Control	0.5 mT	5 mT	Control	0.5 mT	5 mT
Spleen						
Atrophy	1	0	0	0	0	1
Deformation	1	0	0	0	0	0
Discoloration	2	1	1	0	1	2
Enlargement of	11	9	11	12	11	9
lymph follicle	1	0	0	0	0	0
Follicular atrophy	5	1	1	9	5	5
Fragility	1	0	0	0	0	0
Grayish dot/patch	0	0	2	0	0	0
Nodule	1	2	0	0	0	0
Roughness of surface	0	0	0	0	0	1
Rupture/hemorrhage	0	0	0	1	1	0
Pituitary						
Cyst	1	1	2	2	2	0
Discoloration	0	0	0	0	1	0
Enlargement	0	2	1	2	5	3
(red)	5	0*	0*	5	5	3
Nodule	1	2	4	10	9	8
Reddish change	0	0	0	0	1	0
Reddish dot/patch	4	5	7	15	15	11
Whitish dot/patch	0	0	1	0	0	0
Brain						
Impression	1	0	1	6	6	5
Reddish dot/patch	0	0	1	1	1	0
Skin						
Alopecia (focal)	0	3	0	0	2	1
Alopecia	2	1	0	6	5	2
Lichenification	5	10	11	21	22	24
Nodule	2	2	2	0	1	0
Subcutis						
Abscess	0	0	0	3	4	3
Focal hemorrhage	1	0	0	0	0	0
Nodule	14	16	20	15	16	12
Mammary gland						
Enlargement	0	0	0	4	1	1
Nodule	0	0	0	0	1	0

*$p < 0.05$, significant by Armitage's χ^2 test.

Martin Blank, Ed., *Electricity and Magnetism in Biology and Medicine*

CHRONIC EXPOSURE OF SHEEPSHEAD MINNOWS TO A 60Hz ELECTROMAGNETIC FIELD

J. R. Lloyd*

Naval Aerospace Medical Research Laboratory, Pensacola, Fla.

L. R. Goodman and J. A. Couch

U.S. Environmental Protection Agency, Gulf Breeze, Fla.

Sheepshead minnows *(Cyprinodon variegatus)* were reared in a reference aquarium and in an aquarium within a 60Hz electromagnetic field (EMF) in a 156-day preliminary experiment. Modified Helmholtz coils surrounding the exposure aquarium generated a horizontal 1.0 to 1.25 mT rms magnetic field throughout the water volume. Two groups of fish and samples of their embryos were exposed in overlapping exposure periods. Fish in Group I grew from juveniles to reproducing adults during 54 days of exposure. Exposure of Group II began with artificially spawned embryos that hatched and grew to reproducing adults while exposed for 137-152 days. Naturally spawned progeny from both groups were incubated to hatching in the same saltwater aquaria as parents. Although some significant effects ($\alpha = 0.05$) on test endpoints were observed between treatments, they were not consistently significantly different across exposure groups, life stages, etc. No behavioral and histopathological effects related to the EMF were observed.

A variety of biological effects resulting from experimental exposure of animals to 60Hz electromagnetic field (EMF) radiation has been reported.[1,2] Nevertheless, EMF exposure hazards, dose responses, and temporal relationships on biosystems are still largely inconclusive. Experimental exposures have typically been conducted for less than full-life cycles; few studies have examined chronic animal exposures through all major life stages. Many of the animal models have relatively long life spans, which make life-cycle tests with them very expensive. Sheepshead minnows, on the other hand, have relatively short life spans, and their biology is well documented in the scientific literature. This preliminary study was conducted to determine the possible effects of long-term exposure to a nearly constant 60Hz EMF on the sheepshead minnow.

MATERIALS AND METHODS

Equipment

Sheepshead minnows were reared in two glass aquaria (ID 89×44×26 cm high with a water depth of approximately 20 cm) receiving flowing seawater. One aquarium was within an experimentally induced EMF (EMF treatment), and the other aquarium was maintained under similar conditions except not subjected to an experimentally induced EMF (reference treatment).

A 60Hz, 12.0A current ($\approx 25\ V_{ac}$) supplied energy to the coils generating the EMF around the exposure aquarium. Four square coils (0.6×0.6 m), spaced 30.5 cm apart, vertically surrounded the aquarium/frame assemblage (Fig. 1). Each coil consisted of 26 windings of #10 gauge copper wire covered with foam pipe insulation. The four coils were connected in series to form a modified Helmholtz coil system. Electrical current in the coils was controlled by a variable autotransformer and monitored by an ammeter. A wooden frame, similar to that used for the EMF treatment, and sham coils were installed around the reference aquarium.

Magnetic and electric fields were measured with a Model MSI-20 (Magnetic Sciences International) gaussmeter and a HI-3600-02 (Holaday Industries, Inc.) electric-field meter.

*Correspondence to: J. R. Lloyd, Naval Aerospace Medical Research Lab., NAS, 51 Hovey Rd., Pensacola FL 32508-1046.

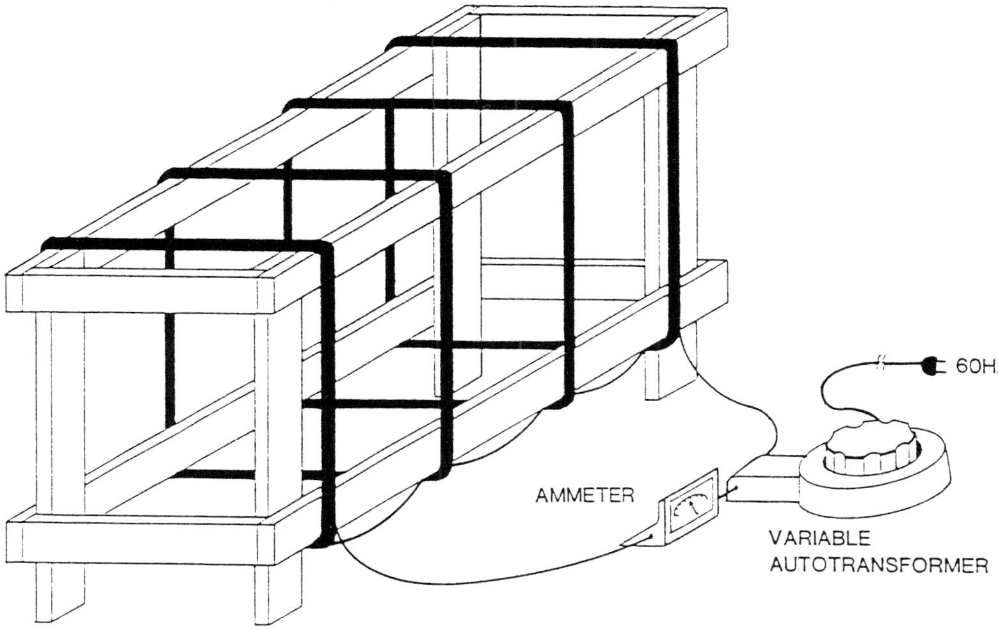

FIG. 1.—A simplified diagram of the exposure system.

Procedure

Forty juvenile fish (Group I) were stocked in each of two aquaria. The next day, 80 artificially spawned embryos (Group II) were added to 4 incubation cups (20 embryos per cup) in each aquarium. The incubation cups were placed in a removable incubation tray in each aquarium.

Beginning on day 42, after fish in Group I began reproductive behavior, three females and two males were transferred to a spawning chamber in each aquarium to collect egg production and fertility data. Samples of resulting embryos were incubated to hatching, counted for survivability, and then removed from the aquaria. On day 54, all adults from Group I were measured for standard length and weight, and preserved.

All fish in Group II were removed from the incubation cups on day 29 and photographed for subsequent length determination. Forty-nine fish per aquarium were transferred to incubation trays. On day 56, the fish were measured for standard length, and 40 per treatment were released into their respective aquaria.

Reproduction by fish in Group II was monitored, as described for Group I fish, in three spawning trials of 5-7 days, beginning on day 132. Samples of embryos collected were incubated to hatching. After each spawning trial, fish from the spawning chambers were measured, weighed, and preserved. On day 153, all remaining adult fish were also measured, weighed, and preserved. Hatching of embryos stopped on day 156, and the experiment was terminated. Preserved adult fish from Group II were examined externally, sectioned, and examined histologically.

Twenty-four magnetic-field density readings were used to map the field within the aquatic volume of each aquarium: one at each of the eight corners of the water volume, eight midway along the width and length of the aquaria, and eight midway in the water column. When a 12A current was recorded in the coils, magnetic-field densities ranged from 1.00 to 1.25 mT. Because of limited access to the aquaria and the meter's size, electric-field measurements were made only at the center of the aquarium. During the experiment, the horizontal electric-field density was recorded at 87 V/m inside the exposure aquarium. Postexperimental (aquarium empty and all laboratory building power off except overhead lights) horizontal field strength measured 13 V/m with the coil system energized. When the coil system was de-energized, the background field measured 1.7 V/m.

RESULTS

Of the endpoints measured, only the following were significantly different ($\alpha = 0.05$) between treatments: (1) mean survival of Group II embryos was lower in the EMF treatment (84% hatch) than in the reference treatment (91% hatch); (2) mean lengths of Group II fish in the EMF treatment were higher than that of reference fish on days 29 (6% greater) and 56 (4% greater); (3) mean egg production by fish exposed to the EMF was lower than that of reference fish in Spawning Trial I; and (4) mean survival of one batch of embryos produced by fish in Group II was lower in the reference aquarium than in the EMF treatment.

No histopathological or behavioral effects related to chronic exposure to EMF were observed.

DISCUSSION

Although statistically significant differences between the reference and EMF treatments were observed for some test endpoints, those endpoints were not consistently significantly different across life stages, exposure groups, etc. Survival of Group II embryos was significantly lower in the EMF treatment than in the reference treatment. This effect was not observed when progeny of Groups I and II were incubated to hatching within the EMF.

Significant effects on length observed in the EMF treatment on days 29 and 56 of Group II were not observed on days 137-153, whereas growth of Group I fish was not significantly different between treatments. The significant effects of Group II may have resulted from increased food availability due to differential fouling of incubation cup mesh by algae or to differential interruptions of water flow between treatments. A decrease in water temperature prior to Spawning Trial I may have contributed to the significant effect observed between treatments for that spawning group. Mean egg production was not significantly different between treatments in Spawning Trials II and III. A drop in water temperature to 21 C during incubation of the last batch of progeny probably contributed to the significantly reduced survival of that group of reference embryos.

Results of this experiment demonstrate that sheepshead minnows can survive, grow, and reproduce while chronically exposed to an EMF of 1.0-1.2 mT. Although some significant differences in responses were observed between treatments, no consistent adverse effects were observed that can be positively attributed to chronic exposure to the EMF.[3]

REFERENCES

1. L. E. Anderson, "Biological effects of extremely low-frequency and 60 Hz fields," in O. P. Gandhi, Ed., *Biological Effects and Medical Applications of Electromagnetic Energy,* Englewood Cliffs, N.J.: Prentice Hall, 1990, 196-235.

2. Biological Studies Task Team, J. M. Lee, Jr. (Chair): *Electrical and Biological Effects of Transmission Lines: A Review,* U.S. Department of Energy, Bonneville Power Administration, DOE/BPA—945-DE89-017650, Portland, Ore., 1989.

3. The views expressed in this article are those of the authors and do not reflect the official policy or position of the U.S. Department of the Navy, U.S. Department of Defense, or the U.S. Government.

Martin Blank, Ed., *Electricity and Magnetism in Biology and Medicine*

DOSE-DEPENDENT HEALTH EFFECTS EXHIBITED IN RABBITS EXPOSED TO 60Hz MAGNETIC FIELDS AT 1 G AND 5 G

J. T. Ning,* E. M. Czerska, J. Casamento, S. Toyokuni, and M. L. Swicord

Center for Devices and Radiological Health, Rockville, Md.

C. St. Charles, F. E. Einsmann, and B. Perry

McGuire VA Medical Center, Richmond, Va.

E. Elson

Walter Reed Army Institute of Research, Washington, D.C.

C. C. Davis

University of Maryland, College Park

Recent epidemiological studies have reported a possible association between presumed exposure to 60Hz magnetic fields and health effects of persons working with power generation and transmission equipment. (For a review, see Coleman and Beral, *Int. J. Epidemiol.* 17: 1-13, 1988.) We have initiated a pilot animal study to assess possible relationships between 60Hz magnetic field exposure and any observable health effect. Twelve 3g New Zealand White male rabbits were divided into three groups. One group served as a control and was exposed to ambient magnetic fields ranging from 1 to 8 mG. The second and the third group were exposed to 1 and 5G magnetic fields, respectively, for 6 h/day, 5 days/week, for the 14-week period. One rabbit exposed to 1G magnetic field developed a tumor growth on the left posterior flank 1 month into the exposure regimen. At the end of 14 weeks, complete autopsies were performed and specimens from all major organ systems were obtained for histopathologic studies. Blood and urine specimens were obtained by cardiac and suprapubic punctures, respectively, prior to sacrifice. Coded blood and urine specimen were analyzed by an independent clinical laboratory blinded to the study. Analysis of the data indicates that blood albumin decreased and urine proteins increased with increasing magnetic-field exposure. Urine protein electrophoresis also exhibited a dose-dependent accumulation of a single protein band (migrating in the alpha 2 region of the human serum standard) with increasing exposure. These trends suggest that 60Hz magnetic fields may somehow interact with the kidney glomerular basement membrane to produce the dose-dependent health effects observed in this study. The appearance of this single unidentified protein band with ELF exposure in the urine may serve as a convenient marker for monitoring extremely low frequency (ELF) exposure.

Many reports have been published in the literature suggesting chronic health effects with electromagnetic field (EMF) exposure.[1,2] Animal studies with mice have found increased body-weight gain, increased organ weight, and elevation of certain circulatory hormones with exposure to EMF.[2-5] In vivo studies on rabbits showed transient delay in growth when exposed outdoors, but the results were not reproduced indoors with 10kV/m electric fields.[6] A pilot study to assess 60Hz magnetic fields at 1 and 5 G was initiated to search for any health effects detectable in the rabbit that may be linked to ELF exposure.

MATERIALS AND METHODS

Twelve 3kg New Zealand White male rabbits were divided into three groups. One group served as control and was exposed to ambient EMF in the Twinbrook Research Laboratory animal-care facility. The second and third groups were exposed to 1 and 5 G ELF at 60 Hz respectively for 6 h/day, 5 days/week for 14 weeks. Each day, the animals to be exposed were placed in plastic cages, which were then placed in a

*Correspondence to: J. T. Ning, FDA-CDRH, Rockville MD 20857.

large Helmholtz-coil exposure system. All rabbits were housed in the same room in the animal-care facility and received the same food and water ad lib. The temperature, humidity, and lighting were controlled and monitored throughout the experiment. The rabbits were examined daily for any observable gross changes. After 14 weeks of exposure, the rabbits were humanely sacrificed and necropsy was performed. Histological specimens were obtained from the following organs: heart, lung, spleen, adrenal gland, kidney, liver, pancreas, stomach, jejunum, ileum, colon, urinary bladder, testis, quadricep femoris muscle, femur, and submandibular gland. Paraffin-embedded slides were prepared and stained for each specimen. Serum and urine specimen were also obtained. The following clinical laboratory tests were performed on coded specimens: complete blood count (CBC), chemistry profile, urine analysis, and serum and urine electrophoresis.

RESULTS

Table 1 shows the effect of the exposure on body weight; there are no significant differences between control and ELF-exposed rabbits. Table 2 shows a comparison of organ weights of each rabbit adjusted for total body weight. There is an apparent 20% increase in lung weight of rabbits exposed to 5G ELF. No significant differences were found in CBC or chemistry profile, except dose-dependent decreases of albumin concentration in the serum (Table 3). Concurrently, there was dose-dependent leakage of proteins into the urine (Table 4). Because of the positive correlation between both serum albumin and total protein levels in the urine and the magnetic-field strength, serum and protein electrophoresis was performed (Figs. 1 and 2). Serum protein electrophoresis pattern scans are shown for control (Fig. 3) and for 1G ELF exposure (Fig. 4). Moreover, Fig. 1 shows an additional migrating band in two of the four rabbit serum exposed to ELF as compared to control, which is illustrated in the scan profiles of Figs. 3 and 4. Figure 2 also shows a new protein band in the urine of exposed rabbits which co-migrates with the alpha 2 region of human serum standard; control rabbit urine lacks this protein band. This new band is barely visible in the urine samples from rabbits exposed to 1G ELF, but is clearly visible in the urine samples from rabbits exposed to 5G ELF.

TABLES: Effect of subchronic ELF exposure ($n = 4$) on
(1) body weight (g)
(2) specific organ weight (in % of body weight)
(3) serum albumin (g/dl)
(4) urine total protein.

1

	MEAN	S.D.
CONTROL	3452	354
ELF, 1G	2798	375
ELF, 5G	2988	306

2

ORGAN	CONTROL Mean	S.D.	ELF, 1G Mean	S.D.	ELF, 5G Mean	S.D.
BRAIN	0.30	0	0.38	0.05	0.32	0.05
HEART	0.20	0	0.22	0.05	0.22	0.05
KIDNEY	0.28	0.05	0.30	0	0.30	0
LUNG	0.35	0.06	0.38	0.05	0.48	0.05
LIVER	1.80	0.24	2.28	0.28	2.18	0.22
TESTIS	0.12	0.05	0.18	0.05	0.18	0.05

3

	Mean	SEM
CONTROL	4.0	0.3
ELF, 1G	3.4	0.2
ELF, 5G	3.1	0.3

4

	TOTAL PROTEIN
CONTROL	NEG
ELF, 1G	TRACE - 30MG
ELF, 5G	30MG - 300MG

DISCUSSION

Unlike the reports in the literature, this pilot study did not show increases in total body weight nor any organ-specific weights. The higher lung weights in rabbits exposed to 5G ELF were probably due to pooling of blood into the lung and the resulting congestion during sacrifice. Instead, this study demonstrated positive effects of subchronic exposure to 60Hz ELF on serum albumin and urine protein. It is of interest that the perturbations observed correlate with ELF field strength. We observed a dose-dependent decrease of serum albumin and corresponding leakage of protein into the urine. One possible explanation is that ELF may be proliferation of human PBL in EMF. Also, experiments performed on IL2 production and utilization showed that at 24 h, pulsed EMFs did not significantly increase IL2 production, and at 48 h there was a decrease in IL2 production. This result is in direct contrast to our results with IL2 production. We believe these inconsistencies may be due to differences in the physical characteristics of the EMFs utilized. To our knowledge no comparative studies are available on the effects of MF on human B-cell proliferation with IL-2 at present. Studies are in progress in our laboratory to determine the role of growth factor receptors during accelerated growth of both normal and neoplastic lymphocytes at a cellular and molecular level.

REFERENCES

1. N. Wertheimer and E. Leeper, *Am. J. Epidemiol.* 113: 345, 1982.
2. D. Savitz and E. Calle, *J. Occup. Med.* 29: 47, 1987.
3. R. Reiter, L. Anderson, R. Buschbom, and B. Wilson, *Life Sci.* 42: 2203, 1988.
4. G. Matanoski, in R. Pool, *Science* 249: 1097, 1990.
5. M. Bauchinger, R. Huff, E. Schmid, and J. Drisp, *Radiat. and Environ. Biophysics* 19: 235, 1981.
6. R. Pool, *Science* 249: 1379, 1990.
7. R. Liburdy, A. Rowe, and P. Vanek, *Radiat. Res.* 114: 500, 1988.
8. R. Lubin, in R. Pool, *Science* 249: 1380, 1990.
9. R. Goodman and A. Henderson, in R. Pool, *Science* 249: 1381, 1990.
10. A. Boyum, *Scand. J. Clin. Lab. Invest.* 21 (suppl. 97), 1968.
11. S. Mehta, R. Sandler, R. Ford, S. Sharma, and A. Maizel, *Lymph. Res.* 5: 49, 1986.
12. S. Mehta, D. Conrad, R. Sandler, J. Morgan, R. Montagna, and A. Maizel, *J. Immunol.* 135: 3298, 1985.
13. J. Mier and R. Gallo, *Proc. Natl. Acad. Sci. USA* 77: 6134, 1980.
14. A. Cossariza, D. Monti, F. Bersani, M. Cantini, R. Cadossi, A. Sacchi, and C. Franceschi, *Biochem. Biophys. Res. Comm.* 160: 692, 1989.
15. A. Cossariza, D. Monti, F. Bersani, R. Paganelli, G. Montagnani, R. Cadossi, M. Cantini, and C. Franceschi, *FEBS Letters* 248: 141, 1989.
16. P. Conti, G. Gigante, M. Cifone, E. Alesse, G. Janni, M. Reale, and P. Angeletti, *FEBS Letters* 162: 156, 1983.
17. H. Aro, E. Eerola, A. Aho, and R. Pentinen, *J. Orthopaedic Res.* 2: 23, 1984.
18. P. Conti, G. Gigante, M. Cifone, E. Alesse, C. Fieschi, M. Bologna, and P. Angeletti, *FEBS Letters* 199: 130, 1986.

Martin Blank, Ed., *Electricity and Magnetism in Biology and Medicine*

ELECTROMAGNETIC FIELDS AFFECT CELL PROLIFERATION AND CYTOKINE PRODUCTION IN HUMAN CELLS

A. Cossarizza,* M. Capri, S. Salvioli, D. Monti, and C. Franceschi

Institute of General Pathology, Modena, Italy

F. Bersani

University of Bologna, Italy

R. Cadossi, P. Zucchini, S. Angioni, and F. Petraglia

University of Modena, Italy

M. R. Scarfi

CNR-IRECE, Naples, Italy

EXTREMELY LOW FREQUENCY PULSED ELECTROMAGNETIC FIELDS (ELF-PEMFs)

Our previous studies have shown that the in vitro exposure of human peripheral blood mononuclear cells (PBMC) to ELF-PEMFs was able to increase in vitro phytohemagglutinin (PHA)-induced lymphocyte proliferation, measured by ^{3}H-thymidine (^{3}H-TdR) and bromo-deoxy-uridine incorporation assays.[1,2] Microtiter plates containing the cell cultures were placed between a pair of Helmholtz coils (maintained parallel to the plates) powered by a pulse generator (Igea, Carpi, Italy). The pulse duration was about 2 ms and the repetition rate 50 Hz, yielding a duty cycle of 1/10. The intensity of the magnetic field was 2.5 mT; the average time variation of the field was of the order of 1 T/s. The induced voltage was about 2 mV. The induced electric field inside each well was in the plane parallel to the microtiter plate surface, and was estimated to be 0.02 mV/cm.

Such an effect was much more evident when cells from aged donors or from subjects affected by a syndrome of precocious aging such as Down's syndrome were used.[2,3] However, these fields were devoid of genotoxic effects (as assessed by the cytokinesis block micronucleus method, a powerful cytogenetic test) and did not interfere with DNA repair capability of human lymphocytes exposed to genotoxic agents.[4,5]

New data were generated, in order to assess (i) the reproducibility of these results; (ii) the mechanism(s) by which ELF-PEMFs are able to act as co-mitogens in PHA-stimulated PBMC, and the kind of cells affected; and (iii) the phase(s) of the cell cycle most sensitive to the PEMFs.

Thus, the exposure system was given to investigators of two other laboratories (Prof. P. Conti, University of Chieti and Dr. M. Petrini, University of Pisa). They were able to reproduce the same abovementioned potentiating effects (data not shown), suggesting that this system can be assumed as a reliable biosensor to study the effects of electromagnetic fields on human cells. Incidentally, in the course of several years we have obtained the same increased cell proliferation in PEMF-exposed PHA-stimulated PBMC from aged people.

As far as the second question is concerned, it is known that the proliferation of human PBMC is regulated and modulated by several cytokines. We previously demonstrated that IL-2 is more efficiently utilized in PEMF-exposed cells. A higher percentage of such lymphocytes expressed IL-2 receptor p75 chain (CD25 antigen),[6] as shown by cytofluorimetric analysis. Here, we present data suggesting that the production of other cytokines is also affected by ELF-PEMFs. Indeed, when PBMC from healthy donors were exposed to the fields, dramatic increases in the production of IL-1β and IL-6 were observed in comparison with

*Correspondence to: A. Cossarizza, Institute of General Pathology, Via Campi 287, I-41100 Modena, Italy.

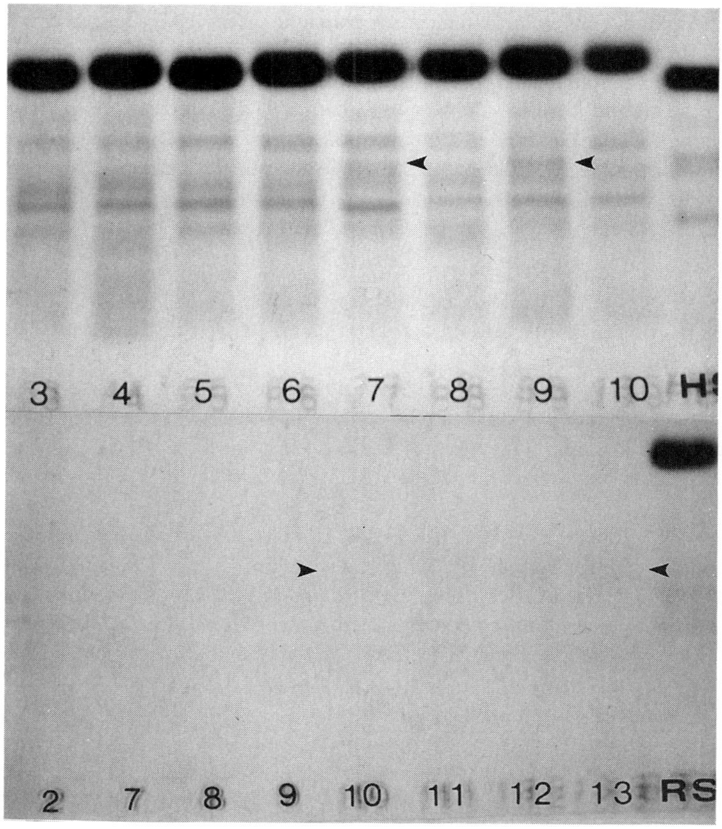

1

2

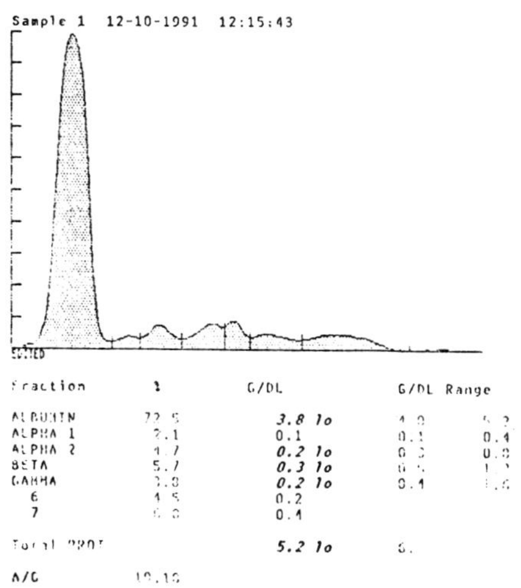

3

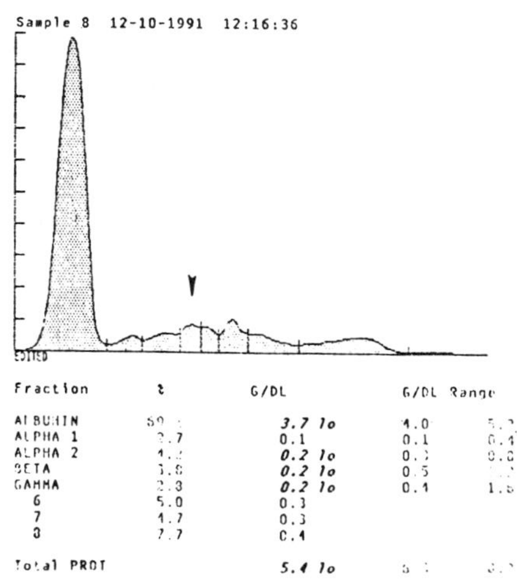

4

FIG. 1.—Serum protein electrophoresis: lanes 3-5, control; 6-9, 1G ELF; 10, 5G ELF; HS, human serum standard.

FIG. 2.—Urine protein electrophoresis: lane 2, control; 7-9, 1G ELF; 10-13, 5G ELF; RS, rabbit serum.

FIG. 3.—Protein electrophoresis, control.

FIG. 4.—Protein electrophoresis, 1G ELF.

interacting with the kidney glomerular basement membrane to effect the leakage of this specific protein into the urine with ELF exposure. It is surprising that analysis by urine protein electrophoresis produces a dose-dependent appearance of a protein band co-migrating to the alpha 2 region of the human serum standard. The indentity of this new protein band should be established; by its electrophoretic pattern of migration, it is unlikely to be albumin. The appearance of this urine protein band may serve as a convenient marker for ELF exposure if it is reproducible. It would be of interest to investigate whether this phenomenon can also be found in primates and humans.

REFERENCES

1. J. Douglas, "Electromagnetic fields and human health," *EPRI J.* 9(4): 14-21, 1984.

2. M. G. Morgan, H. K. Florig, I. Mair, and D. Lincoln, "Power line fields and human health," *IEEE Spectrum* 2: 62-68, 1985.

3. A. A. Marino, R. O. Becker, and B. Ullrich, "The effect of continuous exposure to low frequency electric fields on three generations of mice: A pilot study," *Experientia* 32: 565-566, 1976.

4. A. A. Marino, T. J. Berger, B. P. Austin, R. O. Becker, and F. X. Hart, "In vivo bioelectrochemical changes associated with exposure to extremely low frequency electric fields," *Physiol. Chem. Phys.* 9: 433-441, 1977.

5. A. A. Marino, M. Reichmanis, R. O. Becker, B. Ullrich, and J. M. Cullen, "Power frequency electric field induces biological changes in successive generations of mice," *Experientia* 36: 309-311, 1980.

6. H. A. Hansson, C. Darnfors, B. Rozell, and S. Stemme, "Effects of experimental exposure of power-frequency electromagnetic fields on the nervous system," *23rd Hanford Life Sciences Symposium: Interaction of Biological Systems with Static and ELF Electric and Magnetic Fields,* Richland, Wash., 1984.

Martin Blank, Ed., *Electricity and Magnetism in Biology and Medicine*

EFFECT OF CONSTANT MAGNETIC FIELDS ON FISH MELANOPHORES IN VIVO AND IN VITRO

P. Wannitikul

Khon Kaen University, Thailand

A. Winnicki and K. Formicki

Szczecin University, Poland

The effect of a steady magnetic field on the melanophores of *Salmo gairdneri* Rich., *Salmo trutta* L., and *Anguilla anguilla* L. was studied in vivo and in vitro. The material was exposed for 10-30 min to steady state magnetic fields (160, 250, 310, 380, and 600 mT). The melanophore index read (after completion of the experiments) from the fixed preparations obtained for the exposed and control groups was a measure of the magnetic-field effect. It was found that in all experimental variants steady magnetic fields caused an increase in the dispersion of the melanosomes in melanophores by about one order of magnitude on the 5-score scale of the Hogben and Slome melanophore index.[1]

Fish are not indifferent to the magnetic field, which causes substantial changes in the vital functions of the organism as a whole, as well as in various organs, tissues, cells, and even intracellular structures.[2-4] The magnetic effects may differ according to dependence on the physical characteristics of the field and on the structural specificity of the exposed organism.

Melanophore cells are sensitive to many internal and external stimuli and factors, such as nervous, hormonal, humoral, chemical (e.g., oxygen), physical (e.g., light), and others. Reports on the action of the magnetic field on melanophores of various animals[5,6] are quite scarce and are nonexistent in fish. We resolved to determine whether (and to what an extent) a steady magnetic field acts on the melanophore cells of fish.

MATERIALS AND METHODS

Counting of the melanophore index was performed in the tailfin of rainbow trout (*Salmo gairdneri* Rich.), in the dorsal fin of eel (*Anguilla anguilla* L.), and in trout (*Salmo trutta* L.) fry skin, all exposed to a steady magnetic field.[1] This index was compared with that obtained for parallel control samples kept under identical conditions, but outside the range of the magnetic field. Fish (in vivo) or fin sections (in vitro) were placed between the poles of an electromagnet with regulated fields (160, 250, 310, 380, and 600 mT) for 10-30 min, whereupon the fish and fin sections were fixed in 6% glutaraldehyde. The index was read from microphotographs. Unexposed control samples were treated identically as to fixation and reading. The analysis of variance (ANOVA) was applied for statistical treatment of the results.

RESULTS

In vitro and in vivo experiments showed that the magnetic field significantly influences melanosome dispersion in the melanophores of fish fins and skin (Fig. 1). Melanosome dispersion measured on a 5-score scale of the melanophore index was greater by about one order of magnitude in all experimental variants than in the control samples. The differences were significant.

DISCUSSION

The behavior of the melanosomes in melanophores (aggregation, dispersion) is regulated in vivo by the nervous and hormonal system. It can thus be assumed that the same mechanism underlies the observed effect of melanosome dispersion in our in vivo experiments as well. However, it became evident that a steady magnetic field of medium strength may also directly affect the melanosomes, as shown by in vitro experiments

---

*Correspondence to: K. Formicki (Fish Anatomy and Embryology), University of Agriculture, Ul. K. Królewicza 3/4, 71-550 Szczecin, Poland.

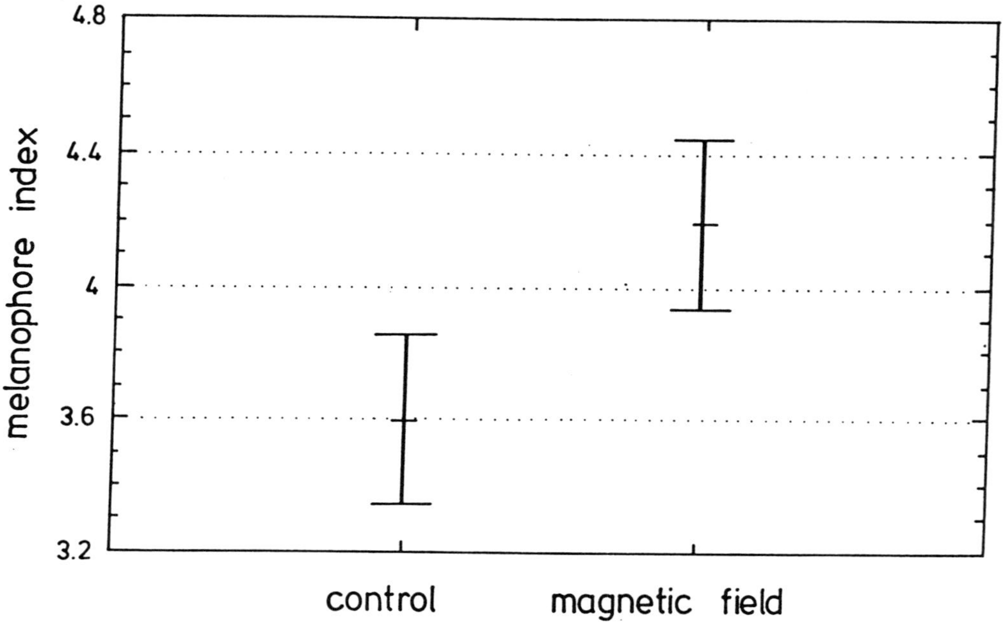

FIG. 1.—In vitro effect of steady magnetic field (310 mT) on melanophore dispersion in skin of trout (*Salmon trutta* L.).

whose final results (a significant rise of the melanophore index) is nearly identical with that of the in vivo tests. Our results contradict the findings of Leucht,[5,6] according to which a magnetic field leads to aggregation of the melanin granules in melanophores. This inconsistency may be due to our using a different experimental material and (more important) to our use of a much higher magnetic-field strength. Nevertheless, both Leucht's results[5,6] and ours indicate that the melanosomes and melanophore cells are not indifferent to the action of a magnetic field and always react to it in a specific manner.

The origin of these effects and responses is associated with the specific macromolecular structure of melanin (free radicals and paramagnetic centers),[7,8] as well as with an ion flow through sodium-calcium canals modulated by a magnetic field.[9] Further studies are required for elucidation of these phenomena.

There is an interesting similarity between the effects on melanophores of magnetic fields of medium strength, and of electromagnetic radiation (light). This similarity permits a broad interpretation in terms of the evolutional origin of melanophore cells (from the lateral carina of the primary nervous plate), as well as in terms of their biological protective role against all kinds of electromagnetic radiation. This approach may be supported by our recent observation that immature *S. gairdneri* individuals exposed to a magnetic field display a characteristic rainbow coloring, which suggests that not only the melanophores but also other fish chromatophores are sensitive to magnetic field.

REFERENCES

1. L. T. Hogben and D. Slome, *Proc. R. Soc. London* B108: 10-53, 1931.
2. T. Tenforde, Ed., *Magnetic Field Effects on Biological Systems,* New York: Plenum Press, 1979.
3. G. Maret, J. Kiepenheuer, and N. Boccara, Eds., *Biophysical Effects of Steady Magnetic Fields,* Berlin: Springer-Verlag, 1986.
4. K. Formicki, *The Effect of a Steady Magnetic Field on the Early Ontogenetic Stages of Salmonids in Light of Present Knowledge of the Influence of These Fields on Living Organisms,* dissertation, University of Agriculture, Szczecin, 1991 (in Polish).
5. T. Leucht, *Naturwissenschaften* 74: 192-193, 1987.
6. T. Leucht, *ibid.,* 441-443.
7. F. W. Cope, R. J. Sever, and D. Polis, *Archs. Biochem. Biophys.* 100: 171-177, 1963.
8. R. C. Sealy et al., in W. A. Pryor, Ed., *Free Radicals in Biology,* New York: Academic Press, 1980, vol. 4, p. 209.
9. E. Findl, in M. Blank and E. Findl, Eds., *Mechanistic Approaches to Interactions of Electric and Electromagnetic Fields with Living Systems,* New York: Plenum Press, 1987, 15-38.

Martin Blank, Ed., *Electricity and Magnetism in Biology and Medicine*

BABOON WORK-STOPPAGE: 60Hz ELECTRIC AND MAGNETIC FIELDS

J. L. Orr* and W. R. Rogers

Southwest Research Institute, San Antonio, Tex.

Transient (1-3-day) cessation of learned behavior to earn food pellets ("work stoppage") was observed in baboon subjects in a series of experiments involving exposure to vertical electric fields above the 12kV/m behavioral detection threshold previously determined in a psychophysical task with a different group of baboons. Exposure to 6kV/m electric fields in combination with 0.5G magnetic fields did not result in work stoppage, nor did exposure to 30 kV/m in combination with 1.0G magnetic fields in two experiments. The reason for the difference in outcome in the presence of magnetic fields is not understood.

Learned (operant) behavior was one of the neurobehavioral endpoints studied in a series of concurrent experiments at Southwest Research Institute.

METHOD

General Experimental Design

Main experiments consisted of 6-week periods of pre-exposure baseline, exposure, and post exposure. Main experiments are indicated by a roman numeral. Experiments I, II, and III involved different sets of male baboon subjects. The exposed and control groups in Experiment IV were the same groups used in Experiment III. During the exposure period, fields were present 12 h/day in three blocks of 4 h each during the daytime and evening "lights on" hours for 7 days/week. Every 3 weeks, there was a two-day interruption of exposure as the baboons were moved out of the facility for cleaning and maintenance, scheduled in the 2nd and 5th week of each of the three periods. Following completion of main experiments, additional exposures indicated by letter suffixes were conducted with the same baboons. Exposure IIA was a re-exposure; Experiments IA, IIB, and IVA were crossover exposures of the former control group.

Facility

The experimental facility is shown in top view in Fig. 1. The space is open, with a common air-handling system. The cages are constructed of fiberglass which, when dry, is transparent to electric and magnetic fields. Homogeneity and stability of the fields were verified by extensive measurements and on-line monitoring with an IBM PC.

In all experiments described here, the operant-behavior experiments were conducted simultaneously with experiments that studied the social behavior of groups of eight young adult male baboons in group cages. In Experiments III and IV, groups of three baboons equipped with indwelling venous catheters were also housed individually in the exposure region on each side of the facility.

Behavioral Task

Experiments I, IA, II, IIA, and IIB. The behavioral task used in these experiments was a multiple schedule of fixed-ratio 30 (FR30) and differential reinforcement of low rates 20 s (DRL20). The schedules were in effect for alternating 15min periods in six sessions of 30 min each near the start and end of each 4h exposure block. The FR30 schedule delivers a banana-flavored food pellet for every 30 depressions of the response pushbutton and typically generates high rates of responding. The DRL20 provides a food pellet for the first response 20 or more seconds since the last response and typically generates a low rate of responding.

The experimental design was intended to seek performance changes between the experimental periods of 6 weeks each. However, when the first exposure was conducted, we observed that there was a transient

---

*Correspondence to: J. L. Orr, SwRI, San Antonio TX 78228-0510.

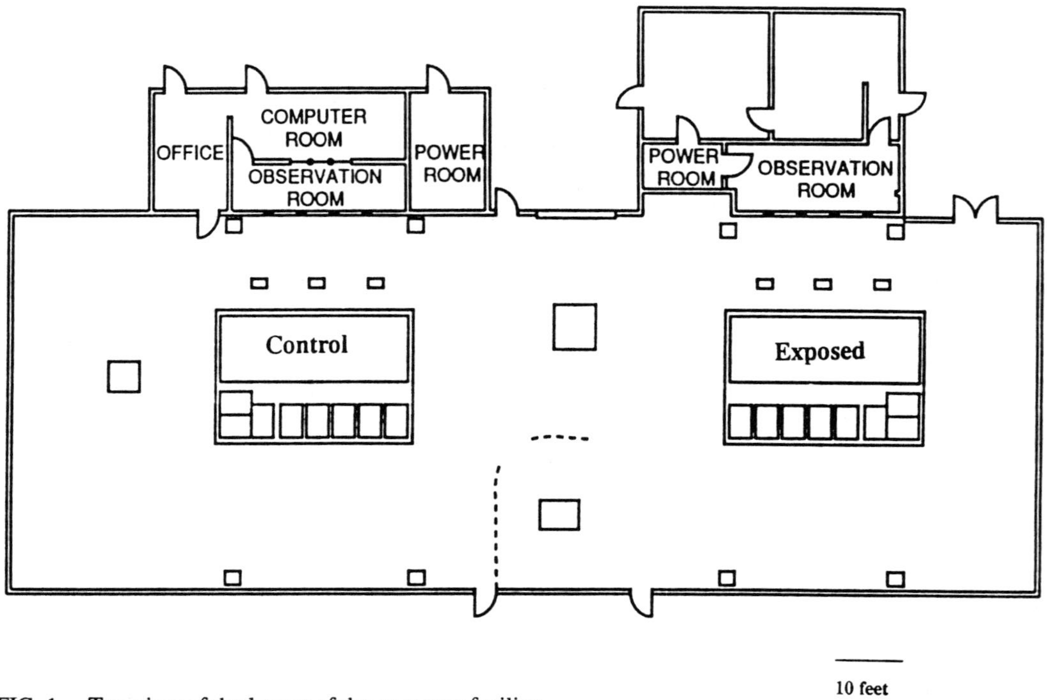

FIG. 1.—Top view of the layout of the exposure facility.

suppression of responding with a time course of a day or so. This transient was not statistically detectable when periods were compared but was statistically detectable when the data were analyzed on a daily basis for the week before and the first week of exposure.

Experiments III and IV. Experiments I and II showed transient work stoppage but not changes on performance once responding returned. To evaluate a cognitive function, the behavioral procedure was changed to a delayed match-to-sample task. In this task, the baboons had to remember the flash rate of a stimulus light across delay intervals of 0 to 16 s. The delayed-match-to-sample task performance showed a decrease in accuracy with increasing delay, which is characteristic of this type of task.

In Experiments III and IV, work performance was measured by the number of trials initiated by depression of the illuminated center pushbuttons to produce a target flash rate. This observing response is considered to be the analog of variables such as number of DRL rewards or number of ratios completed in the earlier experiments.

Comparisons Across Experiments

Because two different behavioral tasks were used, it was necessary to select an index for comparison across the studies. For each group, baseline means and standard deviations were calculated from the group means for the behavioral measure on the seven sessions preceding the first day of exposure. The group mean on the first day of exposure minus the baseline mean was divided by the baseline standard deviation to give a z-score. This index shows how far the group mean data for the first day of exposure is from the baseline mean in terms of the variability (units of standard deviation) of the baseline data.

Field Onset Observations

The general response to the initial field exposure was observed by the authors and others from observing rooms adjacent to the exposure area from where the social groups can be viewed. Animals performing the operant behavior task could be observed by closed-circuit television.

RESULTS

Figure 2 shows the entire series of experiments with the baseline/first-day-of-exposure comparison using the z-score approach described above.

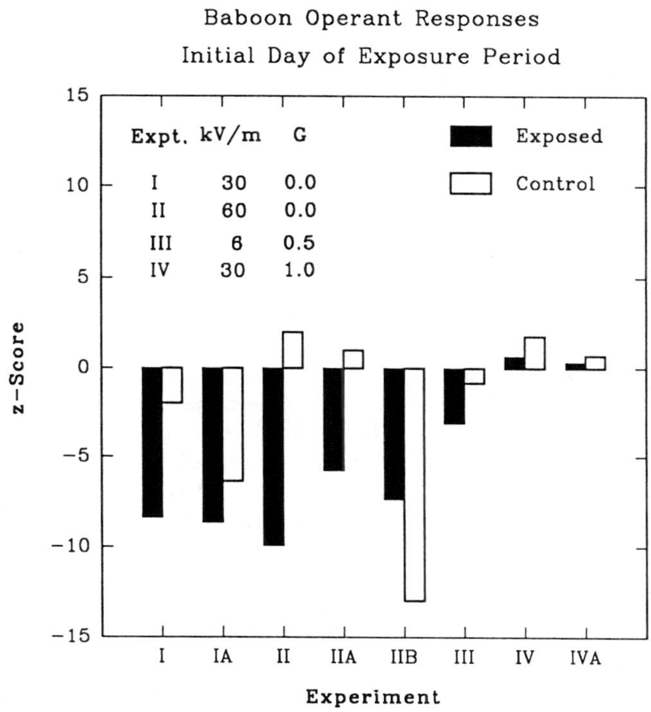

FIG. 2.—The *z*-scores for work stoppage.

Experiments with electric fields alone (Experiments I through IIB) had *z*-score decreases in the behavioral index for the exposed group. Crossover experiments (IA and IIB) also had large reductions in the responding of the previously but not currently exposed control group.

Experiments including magnetic fields (Experiments III through IVA) did not reveal suppression.

The exposed social groups of eight young adult baboons exhibited much greater than normal activity and vocalization during the initial exposures, which declined over the first few exposures during Experiments I, IA, II, and IIB. This pattern was not observed in the social groups in Experiments III, IV, or IVA, although in IV and IVA it appeared that the baboons were perceiving the field; we observed the occurrence of scratching and minor startle responses.

DISCUSSION

Several differences exist between Experiments I through IIB relative to Experiments III through IVA. In addition to the inclusion of magnetic fields, the time period of the experiments, a change from iron to aluminum floor grate for magnetic exposure, different behavioral task, different groups of baboons, different sources of the social group baboons, and differing responses of the social groups to the onset of field exposure have been identified as differences between the two series of experiments. In all but Experiment III, the electric fields were at least 30 kV/m and presumably readily detectable.

One interpretation for the lack of work stoppage in Experiment IV is that the same animals had previously been exposed to lower-strength fields in Experiment III. This interpretation does not explain Experiment IVA, in which the control and exposed groups switched roles. The baboons who served as controls in Experiment III and IV experienced only stray fields of approximately 10 mG before Experiment IVA.

The data for Experiment IV are consistent with the interpretation that magnetic-field exposure produces a change in brain function, perhaps through the release of endorphins or specific neurotransmitters, which offset or cancelled the expected excitatory effect of electric-field exposure observed in Experiments I, IA, II, and IIB. However, magnetic fields were not expected to block the expected suppression of behavior by the 30kV/m electric field in Experiments IV and IVA, and in the absence of other data supporting a neurophysiologic mechanism other, perhaps more likely, possibilities must be considered.

One parsimonious explanation of the entire set of findings is that work stoppage in the operant behavior task is a result of the combination of current or past suprathreshold electric exposure *and* the stimuli from the intense, short-term disruption in the normal social-group behavior. Work stoppage *only* occurred on occasions when there was initial disruption in the exposed social group, including greatly increased activity and vocalizations, and only in groups with current or previous field exposure. Sounds (and probably smells) were accessible to the control groups which, with a previous exposure history paired with social group disruption, exhibited work stoppage in Experiments IA and IIB in the absence of exposure but *with* a social group disruption in progress. Under this interpretation, the work stoppage is not a direct effect of exposure to electric fields; rather it is an indirect effect of the disruption in the social group and a history with a clearly perceptible novel stimulus, the electric field.

This conclusion leads to a testable hypothesis: does work stoppage occur during initial electric-field exposure in the absence of social groups? If work stoppage can be demonstrated with electric fields without the presence of social-group responses, other important questions include: (1) does concurrent magnetic field exposure during initial electric-field exposure block work stoppage; and (2) what is the relation between work stoppage and the 12kV/m detection threshold? The testing during initial exposure of individual baboons (without social groups present) on the original floor grating with electric fields only at 30 or 60 kV/m would provide important information. If work stoppage occurs, it would demonstrate that the presence of an "excited" social group is *not* necessary for work stoppage and the possibility of direct effects should be investigated further.

ACKNOWLEDGMENT

This research was sponsored jointly by the Office of Energy Management of the U.S. Department of Energy (DOE) and the Central Research Institute of The Electric Power Industry of Japan under DOE contract DE-AC02-80RA50219.

Martin Blank, Ed., *Electricity and Magnetism in Biology and Medicine*

DOSIMETRY TECHNIQUE FOR BIO-BEHAVIORAL EXPERIMENTS

C. J. Sherry* and T. J. Walters

Brooks AFB, Tex.

A dosimeter was designed that (1) was sensitive to changes in both the ''E'' and ''H'' fields; (2) could be used to collect power density information during exposures; (3) would not be damaged by high power fields; and (4) was more compact than traditional power density devices and could be used to ''map'' fields more accurately. The dosimeter consisted of 4.000 ml of deionized water in a 5ml polypropylene test tube with the probe of a Model 101 Vitek temperature sensor inserted 38 mm into the center of the tube via a 36mm-long plastic guide.

Typically, the incident power density of the microwave or radio frequency field is determined before and after each experimental exposure by a Model 8616 Loral Microwave-Narda Electromagnetic Radiation Monitor, with a Model 8623D Isotropic Probe. To confirm that the probe is isotropic in this application, we determined the incident power density placing the probe at the location of the subject and making nine separate readings, rotating the probe 45° between each reading; the first and last reading were taken at the same position. We obtained the mean and standard deviation of the individual readings and took the average incident power density as the sum of the mean and the standard deviation. We obtained the incident power at 40 W so as not to damage the probe, and extrapolated to the actual power by multiplying the value obtained with the Loral Microwave-Narda Electromagnetic Radiation Monitor by the reading of the forward-power meter at the time of exposure. In our behavioral experiments, there was a lack of relationship between the behavior of rats during exposure and the reported forward power and/or calculated power density. It was decided that the forward-power meter may not provide the best metric to use in determining power density. Therefore, an alternate metric should be developed.

INITIAL METHOD

To determine the effect of the total field, we irradiated 20 ml of simulated muscle material in a 25 × 150mm polyethylene scintillation vial each time a rat was exposed and noted the rate of temperature change of the simulated muscle material, as well as the total change in temperature. We determined the temperature of the simulated muscle material by placing the probe of a Model 101 Vitek temperature sensor 38 mm into the tube, via a plastic guide 36 mm long, and recorded the output of the Vitek on a two-channel Model Brush 220 Gould chart recorder. Since the simulated muscle material is a solid and since the exposure times are relatively brief (seconds), the rise in temperature of the simulated muscle material continued after the end of the RFR exposure.

RECOMMENDED METHOD

After this pilot work, we decided to replace the simulated muscle material with deionized water. The water was obtained from a Continental Water Systems Co. deionizer. Deionized water (4.000 ml) was pipetted into a 5ml polypropylene test tube (12 × 75 mm, round bottom) by means of a Model EDP-Plus Rainin Instrument Co. motorized microliter pipette. A 36mm guide was mounted in a plastic top designed to fit tightly over the test tube. The test tube was replaced after each exposure and the top with the guide was placed on the new tube. The tube was held in place with a jig fashioned from Model SM, 10 psi Dow Chemical Co. Styrofoam extruded polystyrene insulation foam. The orientation of the test tube top and the probe was held constant, with respect to the horn, by the alignment of a mark on the test tube top with another mark on the top of the sytrofoam jig. A Vitek probe was inserted 38 mm into the center of the tube via a 36mm plastic

---

*Correspondence to: C. J. Sherry (Systems Res. Labs), Box 35505, Brooks AFB TX 78235-5301.

guide. The output of the Vitek temperature sensor was recorded on one channel of a Model Brush 220 Gould chart recorder.

RESULTS

The relationship between recorded forward power and rate of change in water temperature (Fig. 1) shows that there is an overlap in the change in water temperature at various reported forward powers. This result strongly suggests that the reported forward powers (200, 300, and 400 W) were not significantly different from each other, and may be the reason for the lack of correlation between forward power and the rat's behavior, as well as between the animal's behavior and calculated power density, since forward power is the metric used in these calculations.

Since the test tubes containing the water are considerably smaller than the Narda probe, it is possible to "map" the power density in a confined space, such as in an operant conditioning chamber. An operant chamber for use in microwave fields was modeled after the Model E10-10 Coulbourn Instruments (Lehigh Valley, Pa.) Large Modular Test Cage. The chamber was constructed of 8mm-thick clear Plexiglas and was $32 \times 27 \times 34$ cm ($L \times W \times H$). The right and left walls consisted of two tracks designed to accept various modules, modeled after the Coulbourn Response Lever (E21-03), a Triple Cue Lamp Module (E11-02), and a Pellet Trough (E14-06), as well as a series of blank spacer modules. The floor consisted of Plexiglas rods 5 mm in diameter and spaced on 2cm centers. The rods were raised 3 cm off the base of the cage. The space between the base and the bars was designed to accommodate a drop pan to collect urine and feces.

The test tube/Vitek Probe dosimeter was used to "map" the fields at the floor level of the operant chamber, 7.6 cm above the floor, and another 7.6 cm above that level. Nine separate readings were taken at each level in a grid with 10cm centers from the front to the back of the chamber and 8cm centers from the sides of the chamber. The results obtained by this dosimetry technique are shown in Fig. 2. This method allowed us to detect "hot spots" in the operant chamber that would have been missed if we collected dosimetry information with a Narda probe, because the probe was too large to fit into the corners where the "hot spots" occurred.

When exposure durations are short and/or when low forward powers are used, water may not be the most sensitive fluid to use, as it has a relatively high specific heat. Another fluid, ethylene glycol (Texaco Anti-Freeze Coolant), may be more sensitive since it has a lower specific heat and would thus respond to changes in energy deposition more rapidly than deionized water.

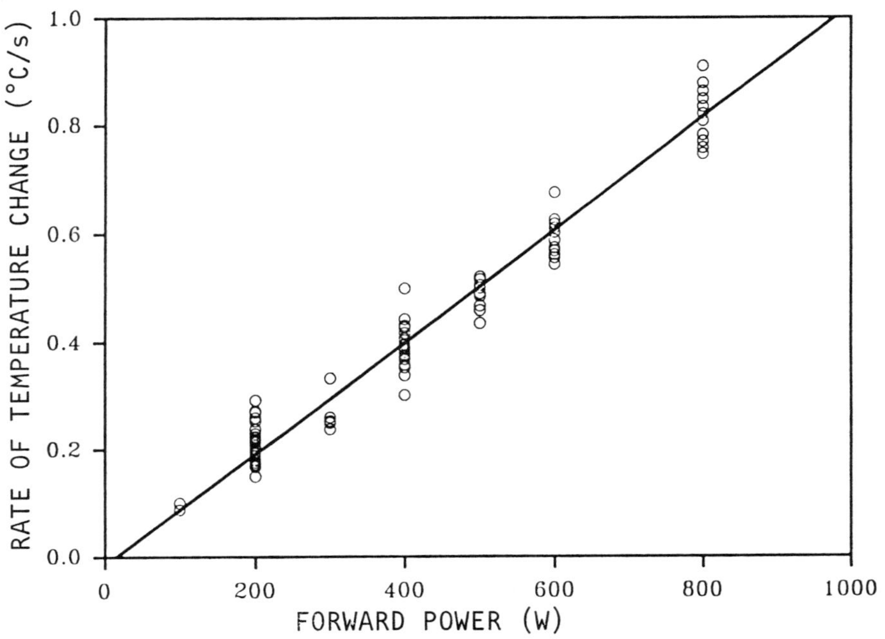

FIG. 1.—Relationship between reported forward power and rate of change in water temperature.

CONCLUSIONS

The water/Vitek or ethylene glycol/Vitek dosimeter is more sensitive to changes in average power density in the near field than the more traditional power-density device.

The water/Vitek or ethylene glycol/Vitek dosimeter can be used to collect power-density information during exposures and is independent of power density (that is, it will not be damaged by high-power fields).

The water/Vitek or ethylene glycol/Vitek dosimeter is more compact than more traditional power-density devices and can thus be used to "map" the fields more accurately.

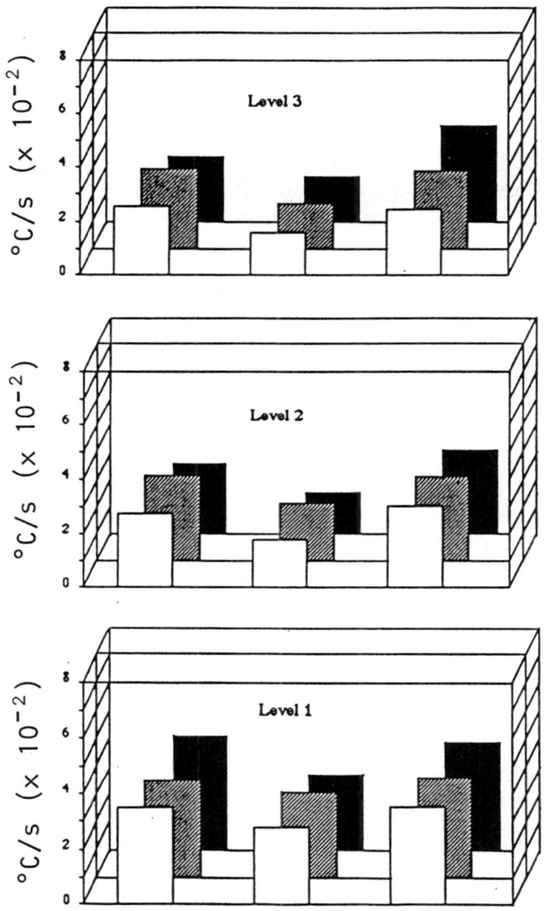

FIG. 2.—"Map" of interior of operant chamber. Heights of bars in histogram are proportional to rate of change in water temperature. Level 1 was recorded at floor of operant chamber, level 2 7.6 cm above floor, level 3 another 7.6 cm above that level. Power density was highest at floor level in the chamber corners. (These "hot" spots would have been missed with more traditional power-density devices.)

Martin Blank, Ed., *Electricity and Magnetism in Biology and Medicine*

HIGH-BANDWIDTH MEASUREMENTS OF TRANSIENT MAGNETIC FIELDS IN RESIDENTIAL ENVIRONMENTS

J. L. Guttman,* J. C. Niple, and J. M. Silva

ENERTECH Consultants, Campbell, Calif.

Preliminary high-bandwidth measurements of transient magnetic fields in residential environments have been performed; transient electric fields and currents were also measured. These measurements have revealed transients with frequency components ranging from below 60 Hz to at least 500 MHz (the limit of the recording instrumentation) from switching of common household appliances. High-frequency measurements of the time rate of change of flux density dB/dt were made by means of a multigap loop-type sensor and broadband measurements of flux density B were made with an active loop antenna. High-frequency measurements of the time rate of change of electric displacement (dD/dt) were made with a dipole-type sensor. Simultaneous measurements of transient currents in residential wiring were done by use of precision current transformers, to investigate current propagation and to validate the magnetic-field measurements.

MEASUREMENT INSTRUMENTATION

A conceptual schematic of the instrumentation system is shown in Fig. 1. The system comprises an ensemble of sensors to measure transient magnetic fields, electric fields, and currents; baluns to convert differential sensor outputs to single-ended 50 ohm outputs; signal amplifiers; and a LeCroy Model 7200 digitizing oscilloscope mainframe with a Model 7234 2-channel plugin for data acquisition. The LeCroy oscilloscope has an analog bandwidth of 400 MHz and a maximum sampling rate of 1 Gigasample/second (GSa/s).

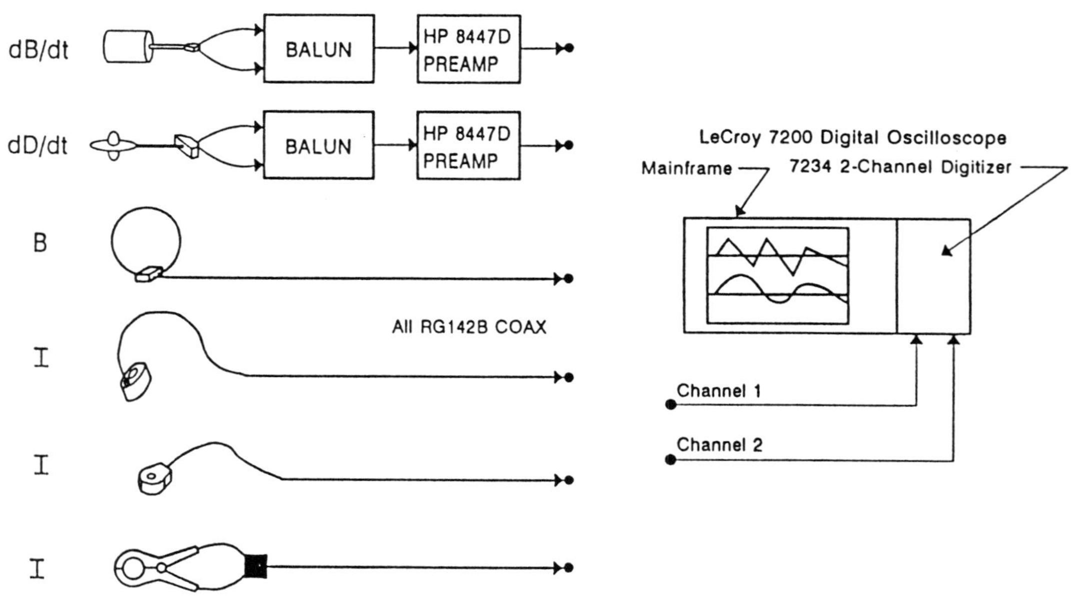

FIG. 1.—Conceptual schematic of transient measurement system.

*Correspondence to: J. L. Guttman, ENERTECH Consultants, 300 Orchard City Drive (Suite 132), Campbell CA 95008.

Table 1 lists the various probes used for the measurements, their function, bandwidth specifications, and calibration constant in terms of voltage output per unit input of the quantity sensed.

TABLE 1.—Sensors used for residential transient measurements.

Sensor type	Model	Response	Bandwidth	Calibration
Magnetic field	Prodyn B-60	dB/dt	DC — >1.8 GHz	10^{-3} V/T/s
Magnetic field	EMCO 6502	B	9 kHz — 30 MHz	$9.5{*}\times10^{7}$ V/T
Electric field	Prodyn AD-40	dD/dt	DC — >1 GHz	$4.4{*}\times10^{-12}$ V/V/m/s
Current	Prodyn I-075-1B	I	70 kHz — 100 MHz	5 V/A
Current	Pearson 2877	I	300 Hz — 200 MHz	1 V/A
Current	Fluke 80i-600	I	10 Hz — 50 kHz	$5{*}\times10^{-2}$ V/A

Only two channels were recorded simultaneously; transient data were acquired by various pairings of field and current sensors. Data sets obtained from numerous measurements of specific transient sources by means of various pairs of sensors were merged to provide a qualitative/semiquantitative picture of transient events.

EXPERIMENTAL PROCEDURE AND RESULTS

In residences, the field sensors were placed at various locations; e.g., centers of rooms and near appliances, service panels, power cords, etc. The current sensors were placed at various locations in a given household circuit: at the switch hot and neutral, at the service panel circuit breaker hot and neutral, on the service entrance hots and neutral, and plumbing pipes. Various loads/appliances were then switched on or off, or allowed to run, and the resulting transient signals were recorded. Measurements were also performed in two neighboring rural residences sharing a pole-mounted distribution transformer. In this case loads were switched in one residence and transients were readily detected in the other.

Transients were generated at practically every switching event, on or off, from every household appliance tested. These transients exhibit extremely complex signatures with frequency content covering the entire range of the measurement instrumentation.

Transients are generated from some appliances in multiple ways by various internal components. For example, with a refrigerator, a transient is generated when the door is opened due to the light turning on, and another transient is generated when the compressor motor turns on. In addition to "on-off" type switching transients, some appliances, like handheld hair dryers and heating pads, generate randomly occurring transients during normal operation due to motor brush sparking and heater circuit switching. For lamps controlled by electronic "dimmer" switches, a transient is generated when the lamp is switched on, and once on, a repetitive transient is generated due to the electronic switching that occurs during each cycle.

The measured transients are therefore categorized into two main types, depending on how they are generated: random switching transients generated when a load is switched on or off, either manually or automatically; and repetitive transients associated with electronic controllers such as dimmer switches. Measured transient waveforms depend on source, sensor, data acquisition system, and measurement position.

Current transients propagate through residential wiring and plumbing. The propagation velocity measured in ROMEX-type wiring was $0.67c$ (where c is the velocity of light), similar to coaxial cable. The highest observed frequency components are strongly attenuated within 10 m of the switch along residential wiring, but signals in the megahertz range are detected at distances of the order of 100 m.

Magnetic-field transients from random switching have two components: prompt high frequencies that radiate from the vicinity of the switch, and delayed lower frequencies due to transient currents conducted in residential wiring. These transients exhibit strong spatial dependence on amplitude and frequency content. Repetitive transients typically exhibit a single dominant frequency component with spatially dependent amplitude. The measured flux density amplitude ranged from 0.01 to 100 mG, depending on source and position.

SUMMARY OF RESULTS

- Transient fields are generated by practically every switching event.
- Measured transient waveform signatures depend on source, sensor response, and recording system bandwidth.
- Frequency content is broadband, covering a range from less than 60 Hz to at least 500 MHz.
- Narrowband repetitive transients are generated by electronic controllers.
- Magnitude of measured flux density ranged from 0.01 to 100 mG, depending on source and position.
- Transient currents propagate within residential wiring and plumbing, with frequency-dependent attenuation.
- Switching events in one residence generate transient magnetic fields in other residences with shared secondary.

ACKNOWLEDGMENT

This work was supported by the Electric Power Research Institute under research contract RP2966-07. The support and guidance of Drs. Stan Sussman and Rob Kavet is gratefully acknowledged.

Martin Blank, Ed., *Electricity and Magnetism in Biology and Medicine*

ELECTRIC FIELD AND CURRENT INDUCED IN REALISTIC MODELS OF HUMAN BODIES BY OVERHEAD POWER TRANSMISSION LINES

H. Y. Chen

Yuan-Ze Institute of Technology, Taoyuan Shian, Taiwan

M. F. Iskander*

University of Utah

We utilize the volume integral equation formulation (VIEF) and the image method to calculate the electric effects of coupling 765kV powerlines on the human body. The validity of the VIEF was checked by a comparison of current density distributions with those obtained from numerical calculations and experimental measurements reported in recent literature for homogeneous human bodies. Comparisons between the VIEF results and Deno's data for current distributions showed some differences. The results of induced electric fields were found to be very small compared with the external electric field, in the range from 0.7 to 54 mV/m. It was also found that the induced body currents are independent of the dielectric constant and the conductivity of homogeneous human bodies.

The large electric fields produced by the EHV and UHV powerlines can cause electric effects and be potentially hazardous when electrically coupled to humans. These electric effects, including induced currents and fields as well as absorbed power, may cause biological effects and potential health hazards including cellular effects, growth and development effects, genetic effects, cardiovascular effects, hematopoietic effects, immunity effects, and neurophysiological effects.

The volume integral equation formulation (VIEF) together with the image method were used to calculate the electric effects due to the coupling of high-voltage powerlines to humans at extremely low frequencies (ELF). Humans were simulated by realistic models with inhomogeneous permittivity. Considerations were given to a lineman standing on the ground under energized transmission lines and moving in free space. For the free-space case, calculations were made for a lineman from a few meters above ground and for standing near an energized conductor.

The validity of the VIEF was first checked by other numerical methods and experimental measurements.[1,2] Obtained results of internal electric field and current distributions are presented. Comparisons between the results of the VIEF and Deno's measured distributions of surface currents on a copper-foil-covered mannequin for the grounded model[3] are discussed.

THE VOLUME INTEGRAL EQUATION FORMULATION

Based on Livesay's analysis,[4] the volume integral equation may be expressed as

$$\left[1+\frac{\tau(\bar{r})}{3j\omega\varepsilon_0}\right]\bar{E}(\bar{r})-PV\int_V \tau(\bar{r}')\cdot\overleftrightarrow{G}(\bar{r},\bar{r}')\,dV' = \bar{E}^i(\bar{r}) \qquad (1)$$

where

$$\overleftrightarrow{G}(\bar{r},\bar{r}') = -j\omega\mu_0\left[\overleftrightarrow{I}+\frac{\nabla\nabla}{k_0^2}\right]\Psi(\bar{r},\bar{r}'), \quad \Psi(\bar{r},\bar{r}') = \frac{\exp(-jk_0|\bar{r}-\bar{r}'|)}{4\pi|\bar{r}-\bar{r}'|}$$

$$k_0 = \omega(\mu_0\varepsilon_0)^{1/2}, \quad \tau(\bar{r}) = \sigma(\bar{r})+j\omega[\varepsilon(\bar{r})-\varepsilon_0], \quad \overleftrightarrow{I} = \text{unit dyad}$$

The *PV* symbol is called the "principal value" of the integral, meaning that the integral is carried out by ex-

*Correspondence to: M. F. Iskander (EE), University of Utah, 3280 MEB, Salt Lake City UT 84112.

clusion of a small volume surrounding the field point first and then letting this small volume approach zero; σ is the conductivity of the lossy dielectric object, μ_0 and ε_0 are the permeability and permittivity of free space, and $\bar{r}$ and $\bar{r}'$ indicate the locations of the field and source points, respectively; $\bar{E}(\bar{r})$ is the unknown total electric field inside the dielectric objects to be determined, and $\bar{E}^i(\bar{r})$ is the known incident electric field, which may be determined by use of image theory and the quasistatic approximation, as well as the assumption that the earth is a perfect conductor.[5]

NUMERICAL RESULTS

Figure 1 shows the configuration of a 765kV powerline system used for the calculations of electric effects due to coupling to the human body. The maximum values of the unperturbed electric fields have a wide variety in the range from 9 to 22 kV/m for calculation heights changing from 0 to 10 m above ground.

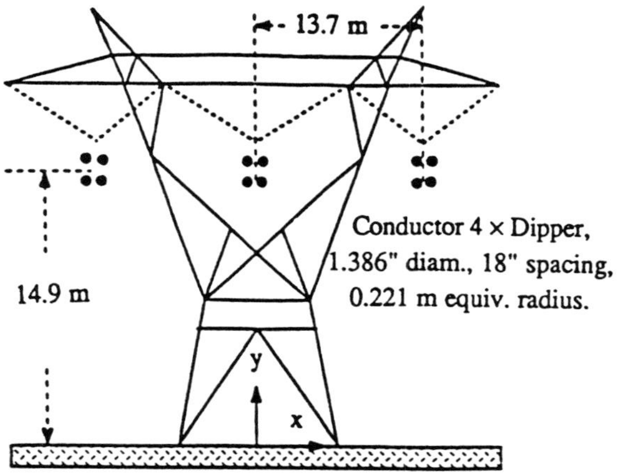

FIG. 1.—Configuration of a 765kV powerline system.

In order to check the validity of the VIEF and the image method, a homogeneously grounded human model with 152 cubic cells of various sizes ranging from 1.25 to 3 cm^3 was constructed as shown in Fig. 2, and the voltage of the power system was adjusted to minus 1212.4 kV, so that an approximately uniform electric field of 10 kV/m in the vertical direction (+y axis) was obtained at the origin near the ground plane. The human model was 45 cm high with arms at the side, and the dielectric constant and conductivity of the homogeneous human body were taken to be $\varepsilon = 1.0 \times 10^6 \varepsilon_0$ and $\sigma = 0.02$ S/m. The numerical results for the induced current densities calculated by the VIEF agree very well with those obtained by the surface-charge integral equation (SCIE) method[2] and the experimental measurements,[1] as shown in Fig. 3. The average error is within 5-10% of the amplitude.

After a check of the validity and accuracy of the solution procedure, the VIEF and the image method were then used to calculate the electric effects of coupling 765kV powerlines to the human body. The dielectric constant and conductivity of human skeletal tissues including the head, neck, shoulder, arms, and legs were similarly adopted from the canine skeletal muscle tissue. In this case, $\varepsilon = 1.32 \times 10^6 \varepsilon_0$ and $\sigma = 0.56$ S/m. The dielectric constant and conductivity of the human brain tissues were adopted by $\varepsilon = 1.18 \times 10^6 \varepsilon_0$ and $\sigma = 0.05$ S/m, and $\varepsilon = 1.47 \times 10^6 \varepsilon_0$ and $\sigma = 0.11$ S/m were simulated for high-water-content tissue such as the abdomen. It was found that the induced body currents obtained by the VIEF have a smaller value than the surface currents measured by a metallized mannequin[3] except for a small portion of the head and legs. The maximum induced body currents were compared with the short-circuit currents obtained by Deno's empirical formula by use of the equivalent-capacitance method, $I_s = 5.4 \times 10^{-9} h^2 E_g$ (A), where h (m) is the height of a man and E_g (V/m) is the electric field near the ground.[3] There was good agreement between the VIEF and Deno's results. The difference is about 5-7%, depending on the shape of the model (separation distance between the two feet).

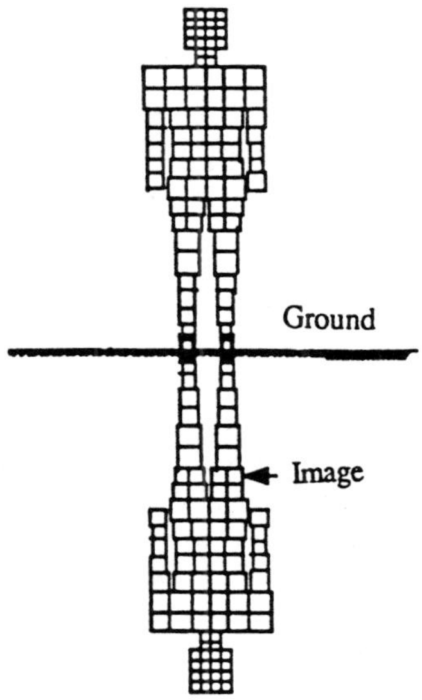

FIG. 2.—Model of man standing beneath a 765kV powerline system and its image.

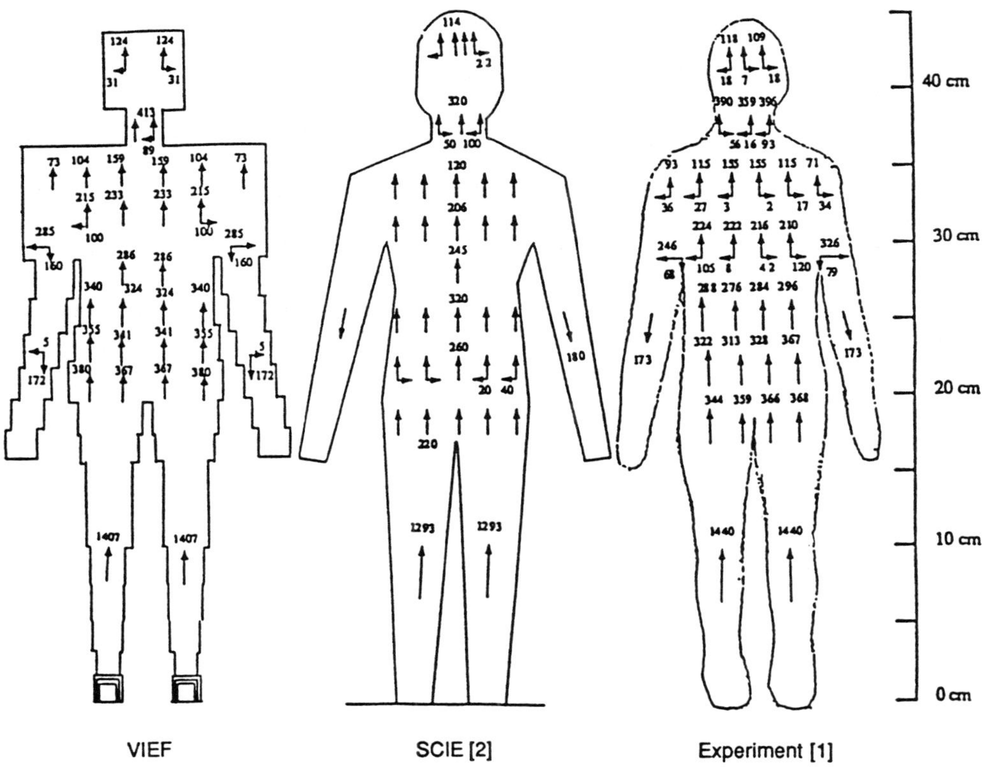

FIG. 3.—Comparisons of results obtained of induced body current densities with Chen et al.'s numerical results[2] and Kaune's experimental data[1] for grounded human exposed to 10 kV/m, 60Hz electric field in vertical direction. (Induced current densities are given in units of nA/cm^2.)

The second test example was chosen for the inhomogeneous model of a lineman moving in free space. The height between the bottom of the feet and the ground plane varies from 0 to 10 m. The maximum values of the electric field, 53.9 mV/m, and induced current, 225 μA, are obtained in the neighborhood of an energized conductor; the minimum values are found near the ground in free space. The values of induced currents and electric fields for the grounded cases are in between. It was also found that the maximum induced electric current is independent of the dielectric constant and conductivity.

CONCLUSIONS

The coupled electric fields and current distribution to the human body from 765kV powerlines were calculated by the VIEF procedure. The validity of the VIEF was checked by comparison of the current density distributions with those obtained by Kaune et al.'s experimental measurements[1] and Chen et al.'s SCIE method[2] for homogeneous human bodies. Although the basic procedures of the VIEF and the SCIE are different, there is an agreement between the results obtained. The results of induced electric fields and currents obtained by the VIEF together with the image method have been presented. The VIEF method provides an accurate description of electric effects induced in human bodies exposed to high-voltage power systems. The results obtained of current distributions inside a grounded man were different from those obtained by Deno, who measured distributions of surface currents on a copper-foil-covered mannequin. It is shown that the results of our calculations are lower than Deno's values. However, the maximum induced currents inside a grounded man are very close to the short-circuit currents obtained by Deno's empirical formula using the equivalence capacitance method. The results of induced electric fields are very small compared with the external electric field: the range of induced electric fields varies from 0.7 to 54 mV/m for a man beneath the 765kV power system. It is also shown that the induced currents are independent of the dielectric constant and the conductivity of homogeneous human bodies.

REFERENCES

1. W. T. Kaune and W. C. Forsythe, "Current densities measured in human models exposed to 60-Hz electric fields," *Bioelectromagnetics* 6: 13-32, 1985.

2. K. M. Chen, H. R. Chuang, and C. J. Lin, "Quantification of interaction between ELF-LF electric fields and human bodies," *Trans. IEEE* BME-33: 746-756, 1986.

3. D. W. Deno, "Currents induced in the human body by high voltage transmission line electric field measurement and calculation of distribution and dose," *Trans. IEEE* PAS-96: 1517-1527, 1977.

4. D. E. Livesay and K. M. Chen, "Electromagnetic fields induced inside arbitrary shaped biological bodies," *Trans. IEEE* MTT-22: 1273-1280, 1974.

5. Electric Power Research Institute, *Transmission Line Reference Book, 345 kV and Above*, Electric Power Research Institute, Palo Alto, Calif., 1982, 329-419.

Martin Blank, Ed., *Electricity and Magnetism in Biology and Medicine*

END-USE MAGNETIC-FIELD MEASUREMENTS AND CHARACTERIZATION

G. Rauch*

Electric Research & Management, Inc., Felton, Calif.

M. Samotyj

Electric Power Research Institute, Palo Alto, Calif.

D. Robertson

Electric Research & Management, Inc., State College, Pa.

Many research efforts over the past decade have focused on assessing exposure from electric-power transmission and distribution facilities. However, very few studies have attempted systematically to characterize the magnetic fields from appliances and other electrical end-use devices. The goals of the measurement program now under way are to document the temporal, spatial, and spectral characteristics of the magnetic field emitted by appliances and other electrically powered end-use equipments. A number of specific short-term goals that support the desired objectives include development of an accurate and efficient measurement protocol, and development of a measurement test fixture to facilitate efficient measurements and conduct pilot measurements.

Scientific research that addresses health impacts of magnetic-field exposure has yet to produce a reliable indication of precisely what field exposure aspect, if any, leads to adverse effects. The lack of a dose metric is less of a hindrance in characterizing fields near power-system facilities where fields are predominantly 50/60 Hz, have reasonably well understood temporal characteristics, and often have predictable attenuation patterns. But for electrical equipments that have very complex magnetic-field characteristics, the situation is different. Absent scientific guidance about dose, it is shortsighted to disregard the potentially significant frequency and temporal characteristics of appliance magnetic fields by characterization of their measured field values by a single number.

The technical challenges associated with end-use device magnetic field characterization include: temporal characteristics of end-use equipments are often nonrepetitive; spatial characteristics are highly nonuniform; and spectral content is often complex. Examples of end-use devices that create measurement challenges are small appliances with brush-type universal motors that produce high-frequency fields from commutator sparking, and electronic equipment that can produce fields with high-order harmonics and other, nonharmonic frequency components. Frequent short-term field fluctuations can also result from intermittent use, routine cycling, motor-starting currents, etc.

The measurement protocol that is required to capture appliance magnetic field characteristics in this project is based on application of the *MultiWave*™ System,[1] which was modified for this application by addition of small triaxial magnetic-field sensors with 300kHz bandwidth, increased A/D sample rate for higher-frequency response, and increased sample periods for expanded low-frequency resolution.

APPLIANCE MEASUREMENT SYSTEM

Some data regarding the magnetic fields generated by operation of specific appliances are available,[2] but they are normally limited in detail. For example, the majority of available data were obtained at the power frequency only, although many appliances generate appreciable magnetic-field levels into the hundreds of kilohertz. Based on these and other considerations, a list of specifications for the design of a measurement sys-

*Correspondence to: G. Rauch, Electric Research & Management, Inc., 6630 Hwy 9 (Suite 203), Felton CA 95018.

tem was established. The specification were as follows:

1. Accommodate all types of appliances found in households.
2. As far as possible, test fixtures must be constructed from nonmetallic materials, to prevent any distortion of the field generated by the appliance under test.
3. Determine magnetic-field levels and attentuation characteristics from the surface of the appliance to 3 m from the surface.
4. Determine the characteristics of the magnetic fields at all frequencies within the range of 60 Hz to 3 kHz. Scalar magnetic flux density measurements in the bandwidths 3-30 kHz and 30-300 kHz should also be recorded.
5. Determine phase relationships between line voltage, line current, and magnetic field.
6. System field-measurement sensitivity: 0.2 mG to 200 G from the appliance surface to 30 cm from the surface; 0.02 mG to 20 G at 100 cm from the surface; 0.01 mG to 1 G at 300 cm from the surface.

TEST FIXTURE

A prototype test fixture was developed to facilitate magnetic-field data collection. Four basic components make up the fixture, including a mounting platform, sensor spars, spar support structure, and the *MultiWave* monitoring system. Figure 1 is a sketch of the text fixture with a full complement of sensors and sensor spars.

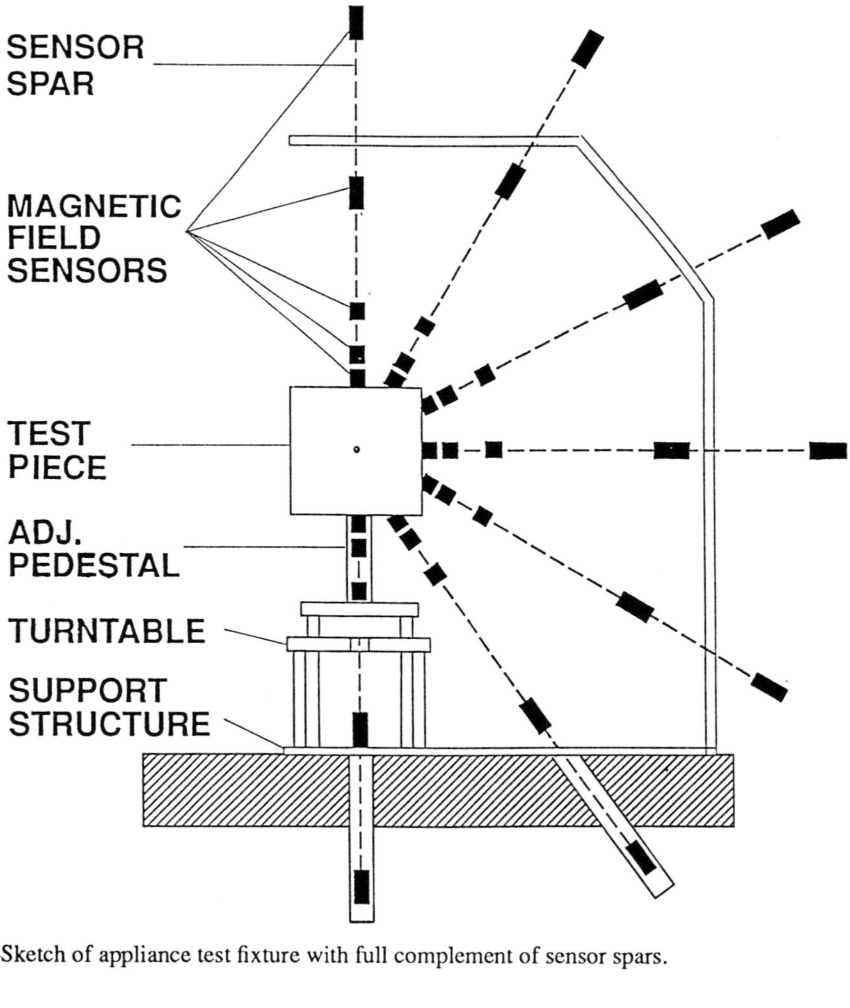

FIG. 1.—Sketch of appliance test fixture with full complement of sensor spars.

PILOT MEASUREMENTS

The first spatial magnetic-field data in which the prototype system was used were taken around a can opener. This common appliance exhibited strong ELF fields and no VLF or LF fields. The measured data exhibited a five-decade range of flux density across the five sensor positions (surface and 0.1, 0.3, 1, and 3 m) along the sensor spar. This procedure is indicative of the challenge in illustrating spatially diverse data in three-dimensional displays.

CONCLUSIONS

Several measured appliances indicate a number of considerations to be made in developing a measurement protocol. First, the protocol must be flexible to accommodate appliance byproducts such as the generation of heat, light, and motion. Proper methods for loading the appliance must be developed. Waveform changes at 300 cm could be detected, although they are small even in the case of strong-field sources such as the can opener that was tested. Hence, operating the measurement fixture in an environment with minimal background field levels will aid in the measurement of these fields. The ratios among the ELF, VLF, and LF flux densities was also broader than expected; the ability to change VLF and LF gains independent of the ELF gains must be developed. Also, the display of data needs further development. Finally, a method of consolidating the data into a single number or small set of numbers should be considered.

REFERENCES

1. *Measurement of Power System Magnetic Fields by Waveform Capture,* EPRI TR-100061, Project RP2942-1, Final Report, February 1992. 2. J. R. Gauger, "Household appliance magnetic field survey," *Trans. IEEE* PAS-104(9): September 1985.

Martin Blank, Ed., *Electricity and Magnetism in Biology and Medicine*

EXPOSURE TO POWER-FREQUENCY ELECTROMAGNETIC FIELDS IN OCCUPATIONAL AND RESIDENTIAL ENVIRONMENTS IN DENMARK

J. Skotte*

National Institute of Occupational Health, Copenhagen

The purpose of this study is to assess personal exposure to power-frequency electromagnetic fields in Denmark. Exposure to electrical and magnetic 50Hz fields is measured with personal dosimeters over 24h periods in both occupational and residential environments. The study includes electrical utility workers, office workers, industrial workers, and people living near high-voltage transmission lines. With the participation of 301 volunteers, 396 dosimeter measurements were carried out. Some preliminary results are reported and compared with results from other studies.

METHODS

Measurements were carried out with personal dosimeters collecting and storing data on 50Hz electric and magnetic fields and transient electric fields in the 5-20MHz range. The dosimeter measured magnetic fields in the range 0.012-200 μT and electric fields in the range 0.6-10000 V/m, both in 16 bins.

The dosimeters were tested for calibration, linearity, frequency response, directivity, timing, etc. The calibration was within ±15% for all dosimeters for both electric and magnetic fields. Several practical problems with the dosimeters occurred. The measurement period was 24 h and the sample rate of the dosimeters was 5 s. The volunteers wearing the dosimeters logged start and stop times of various activities in a diary during the measurement.

All volunteers noted the time spent at the workplace (W), home (H, dosimeter worn), sleeping (S, dosimeter not worn), and traveling, shopping, etc. (T). The volunteers from electric utility organizations also noted the time spent in 20 different work environments. (The volunteer normally wears the dosimeter in a belt.)

PARTICIPATING GROUPS

Main group:	Volunteers:	Measurements:
Electric utility workers	129	196
Office workers (Off)	55	83
Industrial workers	62	62
People living near transmission lines	49	49

The electrical utility workers are grouped into five subgroups according to their primary work environment roughly following the same criteria as in the American Emdex project:

Primary environment of electrical utility workers	Volunteers	Measurements
Generation (Gen)	46	65
Transmission (Tran)	13	17
Distribution (Dist)	33	51
Substation (Sub)	22	46
Other (Misc)	15	17

*Correspondence to: J. Skotte, Arbejdsmiljøinstituttet, Lersø Parkallé 105, DK-2100 Copenhagen Ø Denmark.

The industrial workers were grouped into two subgroups:

Ind1: Examples of highly exposed jobs in the industry (e.g., welders and electric furnace workers), 24 measurements.

Ind2: Examples of "normal" industrial exposure consisting of workers in an automobile repair shop, a plate shop, a machine shop, and a plant producing telecommunications equipment, 30 measurements.

The group of people living near transmission lines included residences less than 100 m from 400kV, 50 m from 132/150kV, and 25 m from 50/60kV overhead lines; and less than 5 m from 132kV buried cables. Distances were measured from the right-of-way center to the nearest part of the residential building. The current load of the lines and cables was recorded during the dosimeter measurement periods.

ANALYSIS

A dosimeter measurement is split into the activity periods (W-H-S-T) according to records in the diary. A dosimeter measurement often has two W periods and two or more H and T period. Periods matching the same activity are combined and a histogram describing the distribution of single measurements for the activity is produced. In this way a dosimeter measurement for a person results in four histograms, one for each of the combined W-H-S-T periods. Finally, the magnetic and electric field means are calculated for these periods as a measure of "dose" (excluded periods less than 4 h). Periods in which the dosimeter is not worn are excluded from analysis. The electric field measurements are useful for detecting periods during which the dosimeter is not worn.

The magnetic-field *means* for the primary work and nonwork environments are shown in Fig. 1. In the electric utilities substation, workers generally have the highest field exposures and distribution workers the lowest. A few of the generator facility workers have very high exposures. Welders and electric furnace workers (Ind1) typically have exposures greater by a factor of 10 or more than workers in the electrical utility facilities. The exposures of office workers and the general industrial workers (Ind2) seem comparable. The nonwork exposures are lower than work exposures.

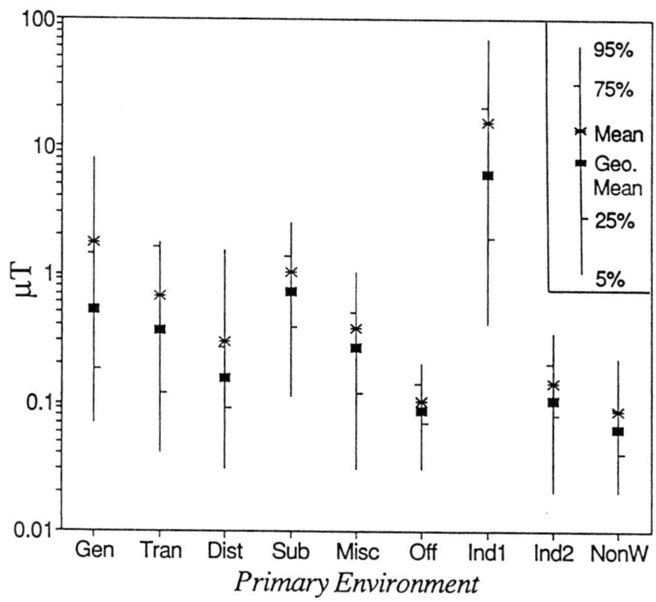

FIG. 1.—Distribution of magnetic-field workday (W) means for eight primary work environment groups and nonwork period (H+T) means for all electrical utility, office, and industrial workers. Mean, geometric mean, and 5, 25, 75, and 95 percentiles are shown for all groups. The number of measurements in the Tran and Misc groups is small (17); the nonwork period group is large (295).

The magnetic field *means* in homes near and far off transmission lines are shown in Fig. 2. The mean current during the dosimeter measurements in the home was below 500 A (typically, 100-400 A).

The Danish and the American Emdex results are compared in Fig. 3. In the Danish homes the geometric mean of all single magnetic-field measurements is 0.031 μT compared with 0.065 μT in the Emdex project. This factor-of-two difference is supposed to be caused mainly by differences in electric power consumptions in Danish and American homes and by the 110/220V difference in distribution voltage.

CONCLUSIONS

The magnetic-field exposure in Danish homes is approximately half the exposure in American homes. Exposures of welders and electrical furnace workers are typically greater by a factor of 10 or more than for th electrical utility workers. The typical magnetic field exposure in homes near transmission lines is 7 times that of magnetic-field exposure in homes distant from transmission lines. Very-low-level magnetic fields are not found in homes near transmission lines. Large variations are found in all groups.

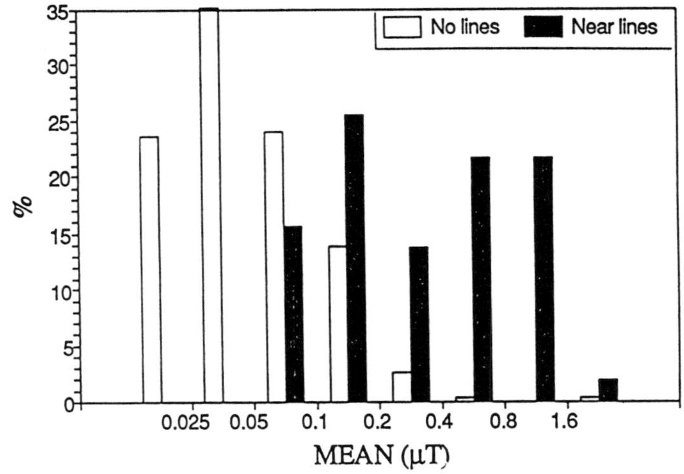

FIG. 2.—Distribution of magnetic field means in homes (H-periods) near transmission lines and cables (51 measurements) and in homes far off lines and cables (267 measurements). Geometric mean is 0.31 μT for the Near Lines group and 0.045 μT for the No Lines group; 84% in the Near Lines group have magnetic-field means in their home above 0.1 μT, but in the No Lines group 83% are below this level. Magnetic-field means below 0.05 μT are not found in the Near Lines group. The current load of the lines and cables in the Near Lines group was relatively low.

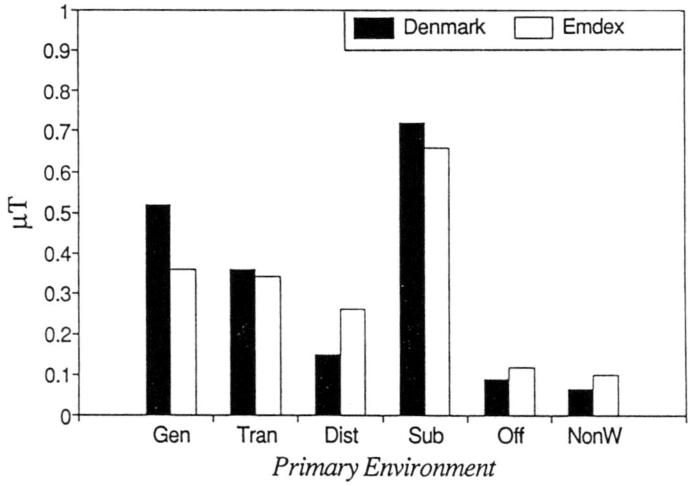

FIG. 3.—Geometric mean of magnetic field means for primary work and nonwork environments from this study compared to results from American Emdex project. Results show similar trends; largest relative differences are found for distribution workers. It is supposed that one reason for this difference is the different structure of the Danish and the American distribution systems. (Note also the relatively large difference in the nonwork environment.)

Martin Blank, Ed., *Electricity and Magnetism in Biology and Medicine*

VISUALIZATION OF WORKER EXPOSURE TO MAGNETIC FIELDS

L.-I. Andersson,* L. Rask, and T. Lindh

National Institute of Occupational Health, Solna, Sweden

A magnetic-field measurement and video-recording technique was developed to study exposure of substation workers to magnetic fields in the electric-power industry. The exposure level is observed as a constantly moving bar graph on the video film recorded during work. This technique enables a visual interpretation and analysis of the ''invisible'' magnetic field to which the worker is exposed. The method is useful to document worker exposure to physical hazards, to visualize sources of high and low exposure, to show how the exposure level varies, and for worker training designed to prevent or reduce unnecessary exposure to hazards.

When exposure measurements are made with dosimeters to determine the daily exposure and the collected data are presented as a line graph, it may be difficult to know which work tasks have determined the shape of the exposure graph. Even if the workers take notes of what they have been doing during the workday, it is often difficult to understand details of the graph. In order to clarify and visualize the cause of the momentary exposure levels, the PIMEX® method has been used in combination with a dosimeter in magnetic field measurements.[1]

METHODS AND INSTRUMENTATION

Five substation workers were randomly selected from a cohort of 500 people employed in an ongoing epidemiologic study in the Swedish electric-power industry. Their exposures to magnetic fields were recorded during one workday. The five selected work shifts represented three different types of work tasks. The first and fourth day were spent in inspection work in 400, 200, and 70kV substations. The third and fifth day were devoted to preventive maintenance in 70 and 400kV substations, and the second day to control-room work. The PIMEX® method means that the worker is video filmed and the signal from a direct reading instrument is mixed with the video signal and presented as a moving bar in the picture (Fig. 1). The magnetic-field measurements were carried out with a custom-built direct reading instrument, Radians Innova BM-2 AI. The instrument measures the magnetic-flux density simultaneously in three orthogonal directions and thus yields the rms value of the magnetic-field vector. The sensors consist of three 600-turn airwound coils with an area of 10 cm^2 each.

To achieve a wide dynamic range of three decades, the output amplifier was designed with a logarithmic gain. The instrument had four frequency ranges and five measurement ranges; 50 Hz and 0.01-20 μT was used in this case.

During the measurements the worker wore the instrument in a harness with the sensor on the chest. The output signal was transmitted by radio to a videomixer, which produced the bar graph encoded in the recorded video signal. A Positron dosimeter was used to achieve a continuous registration of the magnetic field and to determine the exposure at levels above 20 μT, which was the upper limit for the PIMEX bar. Data from the dosimeter were transferred to a PC, where a diagram for the specific workday was extracted from the Positron datafile. Selected material of the recorded material was edited to a short video film. In the film, each workday presented begins with a line graph showing the magnetic field strength vs time. Specific parts of the graph were marked and the exposure levels during various work tasks were visualized.

---

*Correspondence to: L.-I. Andersson, National Institute of Occupational Health, S-171 84 Solna, Sweden.

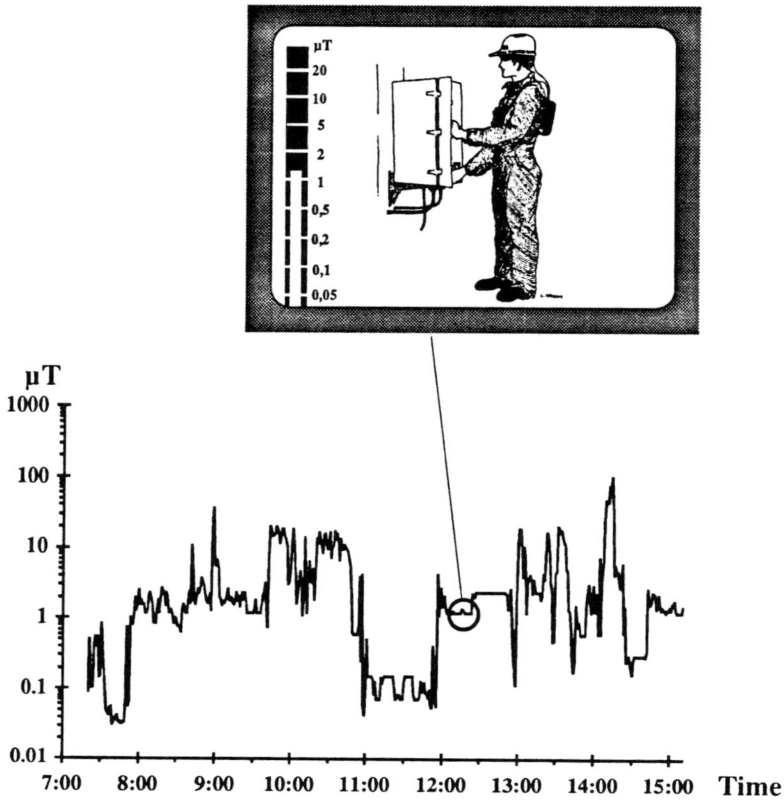

FIG. 1.—Variation of day exposure to magnetic field during inspection of 70 and 400kV substations. Video film augments exposure and work tasks.

CONCLUSION

The method permits subsequent review and determination of the cause of the variability of worker exposure and is useful for the documentation of worker exposure to physical hazards, the visualization of sources of high and low exposure, and as a record of how the exposure varies. The method is useful in guiding workers how to prevent or reduce unnecessary exposure to physical agents. The recorded material shows that the exposure level can be reduced by a factor of 10 if the work is carried out when parts of the switchyard are de-energized. Vans with mobile workshops are often parked in the switchyard, which results in unnecessary exposure that can be easily avoided if the car is parked away from live conductors.

REFERENCES

1. G. Rosén and I.-M. Andersson, "Video filming and pollution measurement as a teaching aid in reducing exposure to airborne pollutants," *Ann. Occ. Hyg.* 1: 137-144, 1989.

Martin Blank, Ed., *Electricity and Magnetism in Biology and Medicine*

INVESTIGATION OF THE RELATIONSHIP BETWEEN RESIDENTIAL WIRE CONFIGURATIONS AND MEASURED MAGNETIC-FIELD PARAMETERS

M. G. Yost*

University of Washington, Seattle

J. D. Wu

University of California, Berkeley

G. Lee, R. Neutra, and L. Hristova

California Department of Health Services, Berkeley

D. Duane and T. Tarsus

Kaiser Permanente, Oakland, Calif.

This study investigates the hypothesis that wire configuration codes are related to some measurable parameters of the alternating field in a residence. Data were collected from 220 homes randomly sampled from 800 Northern California residences for which wire configuration codes were obtained. In addition, indoor and outdoor magnetic-field measurements were performed. They included average bedtime resultant magnetic field and spot measurements under powerlines and at the front door of a residence. The results obtained from the basic statistics indicate that wiring codes do crudely classify the bedtime average magnetic field. However, only a weak correlation exists between the bedtime average magnetic field and wiring codes. When principal-component analysis was performed for six variables extracted from these measurements, the results showed that the model generated by the six variables can achieve the same classification as the wiring codes, and that about 70% variance is explained by the model. The results conclude that of the ensemble of measured factors, the main variables associated with wiring codes are bedtime time-weighted average resultant magnetic field, bedtime rate of change of magnetic field, spot measurements under powerlines, powerline types, and vertical inclination angle of resultant magnetic field.

In epidemiological studies of the health effects of electromagnetic-field (EMF) exposure, investigators have used wire-configuration codes to estimate and classify the EMF exposure that people receive in residential environments. The principle of categorizing EMF exposure by wiring codes is based only on the powerline configuration system and the proximity of lines to a residence. However, if wiring codes are indeed a surrogate for a measurable EMF exposure, it is reasonable to ask how sensitive and specific the wiring codes are for classifying the exposure; what the relevant measurement-related variables that determine relevant exposure to EMF are; and how these variables might be associated with the wiring codes. These questions are crucial to understanding the characteristics of EMF exposure. The purpose of this study was to explore the relationship between wiring codes and measured magnetic-field variables in residential exposure to EMF, by use of the well-known statistical technique of principal-component classification analysis.

A total of 220 homes were randomly selected from 800 Northern California residences recruited in an epidemiological investigation of reproductive effects and exposure to EMF. The wiring codes of the homes were determined following the scheme used by Wertheimer and Leeper in their Denver study.[1] Personal 24h exposures to EMF were measured for each participant by an EMDEX dosimeter. In addition, specific sets of outdoor spot measurements were collected around these homes when the wire configuration was coded. The data-analysis task focused mainly on the use of the field measurements to predict the wiring codes of homes, and to compare the predicted codes with the observed wiring codes.

---

*Correspondence to: M. G. Yost, School of Public Health, University of Washington, Seattle WA 98195.

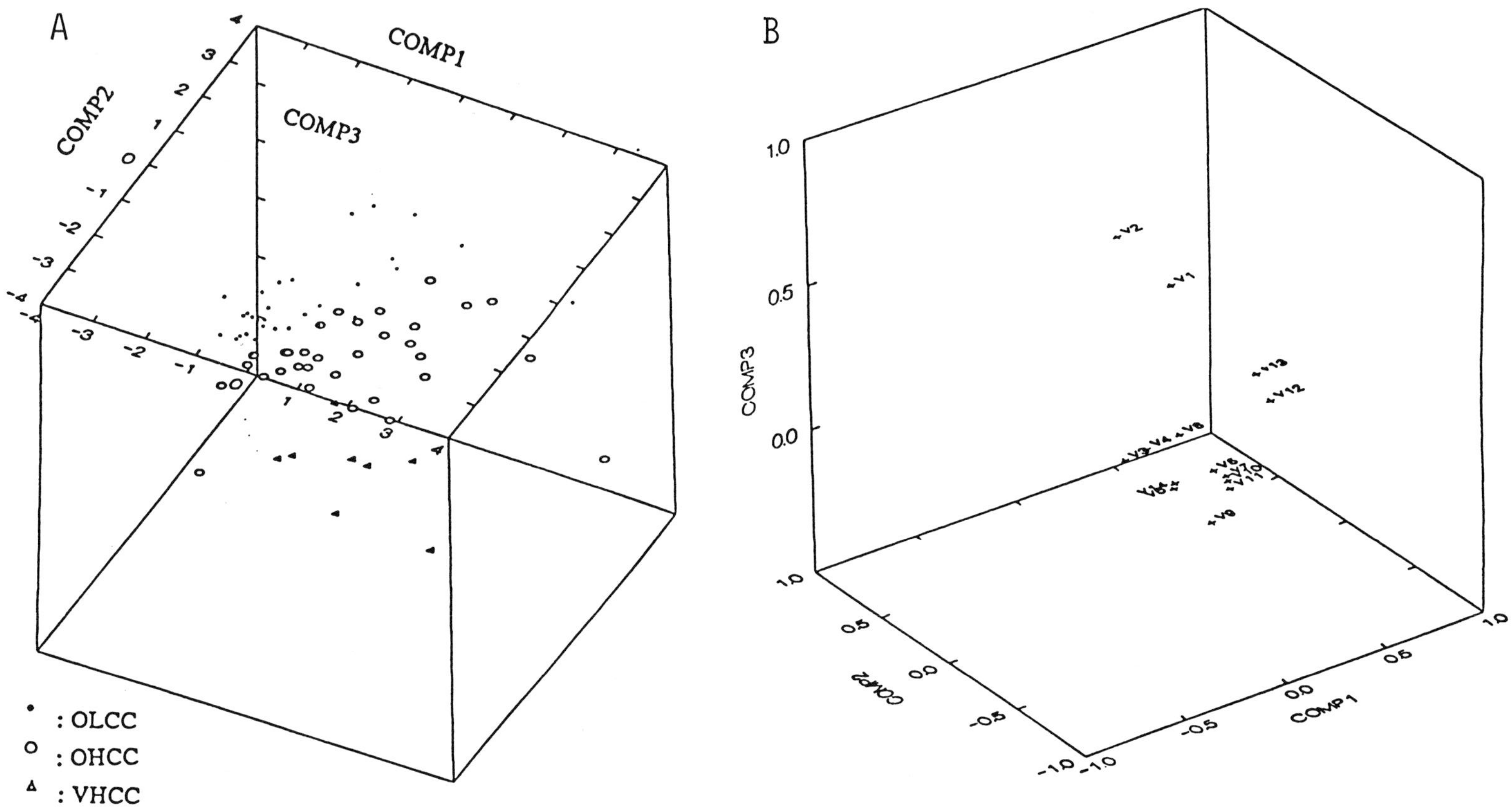

FIG. 1.—(A) Scatterplot of homes in first three principal-component space (training data set). Different symbols show the various wiring-code categories. Note that the categories cluster in the component space with little overlap between categories. (B) Relationship of variable loadings. V4 is the observed wiring-code variable. It represents the origin of the graph. The magnitude of a variable's distance from the origin indicates the contribution of a variable to the classification model. Note that similar types of measurement variables cluster in the component space.

The principal-component analysis (or modeling) was performed with the computer package called Soft Independent Modeling of Class Analogy (SIMCA). Many of the homes in the study were served by buried powerlines, so that it was impossible to obtain measurements related to the powerlines around these homes. Therefore, all these homes and a few (6) homes without complete measurements for other reasons were excluded from the model generation. Two thirds of the 128 valid homes were used as the "training" data set to build the classification models. The remainder were used as a test data set to validate the effectiveness of the classification models.

RESULTS

The average resultant magnetic field measured during bedtime is consistent with the rank-order exposure categories of wiring codes. The bedtime average resultant magnetic field for buried, ordinary low (OLCC), ordinary high (OHCC), and very high (VHCC) current configuration categories are 0.58, 0.65, 1.26, and 1.92 mG, respectively. However, a regression analysis shows that a low explained variance ($R^2 = 6.8\%$) exists between the bedtime average resultant magnetic filed and the wiring codes. Attempts to develop an alternative coding scheme, based on the powerline types and proximity of the external powerlines around these homes, achieved a modest increase in the explained variance in the bedtime average resultant magnetic field, to around 20%.

Thirteen variables extracted from the measurements were used to conduct the principal-component analysis. The first three components explained 65.3% of variance in the training data set (Table 1). In a 3-D distribution of objects in the first three components, three clusters of objects can be distinguished (Fig. 1A). This result shows that the principal-component analysis successfully classified the wiring categories based on these chosen variables. Once the model for each wiring-code category was established from the training data set, the test data set was put into the models to explore the validity (sensitivity and specificity) of these models. The results from the test data set are quite consistent with those of the training data set (Table 2).

An examination of the plots of the distribution of the 13 variable loadings for the first three components yielded two interesting features. First, the 13 variables can be roughly separated into four clusters (Fig. 1B). One cluster consists of the variables (V5, V7, V9, V10, and V11) related to the outdoor magnetic field spot measurements. Another is composed of variables (V12 and V13) described by powerline types. The third one consists of the variables related to bedtime resultant magnetic field measurements (V1 and V2). The last one includes two variables of the inclination angle between the AC magnetic filed and the geomagnetic (DC) field (V6 and V14). Second, what that implies is that it may be possible to pick only one variable from several highly associated variables to simplify the models. After repeated performance of the procedures of model building for different combinations of the 13 variables, the results show that the models could be simplified by use of only six variables with only a slight penalty in terms of the proportion of explained variance.

A total of 128 objects combined from both the training and test data sets were used to generate the final models. The percentages of variance explained by the first three components are shown in Table 1. The sensitivity and specificity of the model for each wiring code category were calculated and are shown in Table 2. The results show that the sensitivity of each wiring code model is equal to or greater than 80.0% and the specificity of the OLCC, OHCC, and VHCC models are 74.6, 63.8, and 99.1%, respectively. The variable loading for the first three components and three wiring code categories are graphed in Fig. 1(B). The magnitude of a variable's distance from the origin indicates the contribution of a variable to the classification model. Similar types of measurement variables cluster in the component space.

TABLE 1.—Percentage of variance explained by first three components.

Model	Comp. 1 (%)	Comp. 2 (%)	Comp. 3 (%)
All categories	40.5	53.8	71.0
OLCC	21.1	37.5	47.7
OHCC	27.3	49.3	63.6
VHCC	48.8	78.8	84.7

TABLE 2.—Sensitivity and specificity of three wiring code models.

	OLCC	OHCC	VHCC
Sensitivity	83.1%	83.3%	80.0%
	(54/65)	(40/48)	(12/15)
Specificity	74.6%	63.8%	99.1%
	(47/63)	(51/80)	(112/113)

DISCUSSION

Several reports have associated wiring codes with the occurrence of cancer in nearby residences.[2-4] However, as noted in the findings published by London et al.,[5] there has been little support for the hypothesis that increased exposure to AC magnetic fields, as assessed by 24h resultant measurements or spot measurements, increased the risk of childhood leukemia. They found a stronger association of elevated leukemia risk with an indirect measurement of magnetic field (wiring codes) than with a direct measurement (24h magnetic field measurements or spot measurements). This paradox underscores the importance and necessity of understanding the relationship between wiring codes and exposure to measurable aspects of the magnetic-field environment in residences.

The models built in this study can classify residences in a manner similar to those based on wiring codes, which demonstrates that these variables are associated with the wiring codes. When considering the application of the models to disease outcomes, one might also use these models to classify the outcomes of diseases. If this idea works well, we can compare the differences in variables of wiring-code models and those of disease-outcome models. Thus, the "hazardous" factors of magnetic-field exposure could be determined and the characteristics of magnetic-field exposure pertaining to the wiring codes would also be known. This approach may be helpful for resolving the controversy outlined above.

Although the models built in this study were successful in probing the important characteristics of magnetic fields described by the wiring codes, several additional factors could be explored in a future study to improve the accuracy and completeness of the models. For example, the magnetic-field polarization could be measured to assess whether linear, circular, or elliptical magnetic fields are associated with line types; and the frequency spectrum or harmonic content of the magnetic field could be analyzed and added to the models to determine whether it plays an important role in wiring-code classification. Finally, more data on wiring codes should be collected to validate the models, especially data for the VHCC category, which has been associated with disease risk most often.

CONCLUSION

Based on the models generated from the principal-component analysis, several factors appear to be associated with wiring codes. The main variables associated with wiring codes are bedtime average resultant magnetic field, bedtime rate of change of magnetic field, spot measurements under powerlines surrounding a residence, powerline types, and the vertical inclination angle of the resultant magnetic field. Other characteristics of magnetic field not measured in this study may also be important in future studies. A similar classification model-building approach could be applied to disease outcomes and may provide clues about the relevant exposure variables in EMF health effects.

REFERENCES

1. N. Wertheimer and E. Leeper, "Electrical wiring configurations and childhood cancer," *Am. J. Epidemiology* 109: 273-284, 1979.

2. N. Wertheimer and E. Leeper, "Adult cancer related to electrical wires near the home," *Intern. J. Epidemiology* 11: 345-355, 1982.

3. L. Tomenius et al., "50-Hz electromagnetic environment and the incidence of childhood tumors in Stockholm County," *Bioelectromagnetics* 7: 191-207, 1986.

4. D. A. Savitz, H. Watchtel, F. A. Barnes, E. M. John, and J. G. Tvrdik, "Case-control study of childhood cancer and exposure to 60-Hz magnetic fields," *Am. J. Epidemiology* 328: 21-38, 1988.

5. S. L. London, D. C. Thomas, J. D. Bowman, E. Sobel, T. C. Cheng, and J. M. Peter, "Exposure to residential electric and magnetic fields and risk of childhood leukemia," *Am. J. Epidemiology* 134: 923-937, 1991.

Martin Blank, Ed., *Electricity and Magnetism in Biology and Medicine*

WIRE CODES AND RESIDENTIAL EXPOSURES RELATED TO MAGNETIC FIELDS AND GROUND CURRENTS*

Thanh Dovan*

State Electricity Commission of Victoria, Australia

W. T. Kaune

EM Factors, Richland, Wash.

R. I. Kavet

EPRI, Palo Alto, Calif.

An intensive program of measurements was conducted in 81 Denver, Colorado, homes that had previously been measured and wire coded in 1985. This program included measurements of geomagnetic fields, power-frequency magnetic fields and their harmonics, ground currents and their spectral components, radon, and home water quality. In this paper, we examine the relationship between wire codes and residential exposures related to magnetic fields and ground currents. The average of all geomagnetic field measurements was 52.7 μT. Within homes and between homes standard deviations were 2.6 μT and 3.5 μT, respectively. Spot 60Hz magnetic flux density (MFD) measurements under low-power conditions in different rooms of homes were highly correlated ($r = 0.8$-0.9), and measurements made in rooms under low- and high-power conditions were also highly correlated. The 24h average ground currents of about 20% of our study homes exceeded 0.5 A, and the 24h average MFD in these homes were elevated, regardless of their wiring codes. In these homes, the 24h ground currents and home average MFD were substantially correlated ($r = 0.7$). Analysis of variance was performed on log-transformed ground currents with wire codes as the grouping variable; the only significant difference was between homes with underground wiring and all other homes. Preliminary results of home water quality assays did not show any significant correlations between ground currents and levels of contaminant.

In the two most recent epidemiological studies of childhood cancers and exposure to power-frequency electric and magnetic fields (EMF),[1,2] exposure assessment consisted principally of wire coding and EMF measurements. In both studies the authors reported an association between the incidence rates of certain types of cancers and wire codes, but found no significant association between cancer rates and measured MFD. The finding that wire codes were a better predictor of cancer incidence in children than were spot MFD measurements is somewhat puzzling, if we assume that magnetic fields are truly related to cancer incidence, because a spot measurement would seem to be a more direct measure of the exposure of interest than the use of a wire code. This result suggests the possibility that the true exposure in these studies was related to some factor associated with wire codes but not with measured values of power-frequency MFD in homes. This factor could be completely unrelated to magnetic fields, or could be conceivably some aspect of magnetic fields that is not captured by the type of measurements that have so far been made. Candidates that have been discussed for such an exposure include traffic density and/or air pollution,[3] contaminants in drinking water,[4] and aspects of magnetic fields such as abrupt changes in field strength or high-frequency transients.

The present study, informally known as the "Back-to-Denver" (BTD) study, revisited 81 of the Denver, Colorado, homes that had been originally wire coded and measured by Savitz and colleagues in 1985. The main goals of the study were to examine the temporal stability spot measurements and repeatability of wire codes,[5] and to identify possible confounders of high wire codes. The study included measurements of geomagnetic fields, 24h and spot measurements of residential power-frequency MFD and ground currents (including their spectral components), radon, and home water quality. This paper presents the results related

*This work was carried out while authors Dovan and Kaune were with ENERTECH Consultants, Campbell, Calif. 95008. Correspondence to T. Dovan, SECV, Power Grid Development Dept., GPO Box 2765Y, Melbourne 3001 Australia.

to the second goal, namely the relationship between wire codes and residential exposures related to magnetic fields and ground currents.

SAMPLE DESIGN

The population from which residences were sampled consisted of the 579 case and control homes that were wire coded by Savitz and colleagues. The BTD sample was stratified along two dimensions: wire code (LCC/HCC) and case/control status. In this way, four strata were defined: HCC cases, HCC controls, LCC cases, LCC controls. The overall goal, based on statistical and resource considerations, was to obtain 20 samples in each of the four strata. Eighty-one original homes were measured, consisting of 18 HCC cases, 20 HCC controls, 20 LCC cases, and 23 LCC control homes.

MEASUREMENT PROTOCOLS AND RESULTS

Wire Coding. The wire coding procedure used in this study followed closely that used in the Savitz study. The repeatability of wire code and the materials and methods of the BTD study including its measurement protocol has been reported elsewhere.[5]

Spot Geomagnetic-field Measurement. Measurements of magnitude and orientation of the earth magnetic field and its three orthogonal components were made at room centers in three "basic" rooms of each study home (bedroom of the youngest child, living/family room, and principal dining area) by means of a fluxgate magnetometer. The average of all geomagnetic field measurements was 52.7 μT. The within-home variations in measured geomagnetic fields of 2.6 μT were almost as large as the between-home variations of 3.5 μT.

Spot Measurement. MFD measurements were made under low- and high-power use conditions in all rooms of the study homes. Under each power-use condition, spot measurements were taken 1 m above the center of the floor. Spot measurements of ground current were also made in the home waterpipe under low- and high-power conditions. A summary of the results is presented below:

- Low-power spot MFD measurements in different rooms of homes were highly correlated (r = 0.8 to 0.9).
- Spot MFD measurements under low-power and high-power conditions were highly correlated (r = 0.6 to 0.8 for different rooms of homes).
- Linear regression analysis of the relationship between low-power and high-power MFD suggests that residential magnetic fields measured in different rooms of BTD homes increased by 30 to 50% from low-power condition to high-power condition.

Twenty-four-hour MFD and Ground-current Measurements. Twenty-four-hour MFD measurements in this study were made in the three basic rooms mentioned above. In each room, an EMDEX meter was used to record the three orthogonal components of the room ambient MFD at a sampling rate of one measurement every 30 s. Similar 24h ground-current measurements in the home waterpipes were made by a clamp-on current sensor and an EMDEX meter. The results are summarized as follows.

- Measured home-averaged MFD (average of the three basic rooms) exhibited a clear diurnal rhythm. Low-power spot MFD and 24h average MFD were well correlated (r = 0.7 to 0.8 for different rooms of homes).
- In about 20% of BTD homes, 24h average ground currents in waterpipes exceeded 0.5 A. Average MFD in these homes are elevated regardless of wire code.
- Analysis of variance was performed on log-transformed ground currents with wire codes as the grouping variable: the only significant difference was between homes with underground wiring and all other homes.

Water Quality Measurement. Two water samples were collected at each residence early in the morning. The first sample contained water that stood overnight in the pipe between home and street water main. The second water sample was collected after the home water system had been purged with "fresh" water from the main line. Water samples were then promptly analyzed for the following parameters: copper, iron, lead, zinc, cadmium, sodium, hardness (as $CaCO_3$), alkaline, pH, and chloride. Preliminary analysis of home water quality showed no significant correlations between ground currents and levels of contaminant.[6]

REFERENCES

1. D. A. Savitz, H. Wachtel, F. Barnes, E. John, and J. Tvrdik, "Case-control study of childhood cancer and exposure to 60-Hz magnetic fields," *Am. J. Epidemiology* 128: 21-38, 1988.

2. S. J. London, D. C. Thomas, J. D. Bowman, E. Sobel, T. C. Cheng, and J. M. Peters, "Exposure to residential electric and magnetic fields and risk of childhood leukemia," *Am. J. Epidemiology* 134: 923-937, 1991.

3. D. A. Savitz and L. Feingold, "Association of childhood cancer with residential traffic density," *Scand. J. Work Environmental Health* 15: 360-363, 1989.

4. R. I. Kavet and J. M. Silva, *An Alternative Hypothesis for Association Between Distribution Wiring Configurations and Cancer: Planning Phase,* Report to Electric Power Research Institute (EN-6853), Palo Alto, Calif. 94303.

5. T. Dovan, W. T. Kaune, and D. A. Savitz, "Repeatability of residential power frequency magnetic fields and wire codes," *Bioelectromagnetics* (in press).

6. Major funding for the study was provided by California Department of Health Services; additional funding was provided by Southern California Edison Company and the U.S. Environmental Protection Agency.

Martin Blank, Ed., *Electricity and Magnetism in Biology and Medicine*

GROUND CURRENTS AND MULTIGROUNDED NEUTRAL DISTRIBUTION LINES

S. Maurer*

New York Institute of Technology

A Pspice analysis of a lumped equivalent circuit for single-phase multigrounded neutral distribution (MGND) lines yields ground-current characteristics for single and multiple loads. Graphs of neutral line voltage, neutral to earth voltage, and neutral currents are used to emphasize the effects of substation and large-user ground impedance on generation of ground currents in nearby utility users. The earth conducting channel can carry a large amount of the phase current, leaving a net current on the two conductor line. This situation produces larger magnetic fields that fall with distance R at the slower rate of $1/R$.

Primary circuit ground currents are a consequence of grounding requirements of the National Electric Safety Code[1] for MGND lines. Neutral currents flow into and out of a conducting earth channel beneath MGND lines (Fig. 1). Currents between this channel and the neutral line pass through the earth via utility users' neutral and ground connections, as well as via other required grounding lines. These currents appear on grounded wires and water pipes and as a result can create substantial power-frequency magnetic fields over an extended region. The earth channel current (as much as 60% or more of the phase current) also yields a net nonzero current on the distribution line, which has the effect of introducing a $1/R$ magnetic field far from the line.

CIRCUIT MODELS

A lumped circuit model has been developed for a single-phase MGND line (Fig. 2). The elements shown for a line section are the primary (ϕ), neutral, and mutual impedances for sections between grounded poles (nodes). For the example evaluated, we assume a distribution transformer at each pole. The line impedances are calculated according to Carson's equations, derived for ground return currents.[2] There are several grounding paths at each distribution transformer. The division of neutral-to-earth current depends inversely on the impedance to remote earth of each of the paths to ground. These currents are largest and hence most noticeable in the vicinity of substations and large power users. The ground impedances (essentially resistances) at each node are the parallel combination of all grounding paths. For a node having a distribution transformer with multiple users, pole grounds, etc., the ground resistance can become quite small (10 ohms or less). These ground resistances are those measured to remote earth. Neutral and neutral-to-earth currents for single and multiple loads on MGND lines have been evaluated by use of Pspice.

ILLUSTRATIONS FOR A SUBURBAN MGND LINE

An example, a single-phase MGND line of 51 nodes and 200 ft between nodes, with a net grounding resistance of 10 ohms at each node, has been evaluated for single and multiple loads. The resistive loads are described by currents and resistances whose values are reflected into the primary circuit by transformer ratios. For a single load and long lines with many nodes, the neutral-to-earth currents (and the neutral voltage) tend to zero. As shown by Levey,[3] when this condition occurs, the following equation can be derived for the relation of the earth channel current I_E and the primary phase current I_ϕ:

$$I_E = [1 - (z_M / z_N)] I_\phi$$

---

*Correspondence to: S. Maurer (EE), N.Y. Institute of Technology, 1855 Broadway, New York NY 10023.

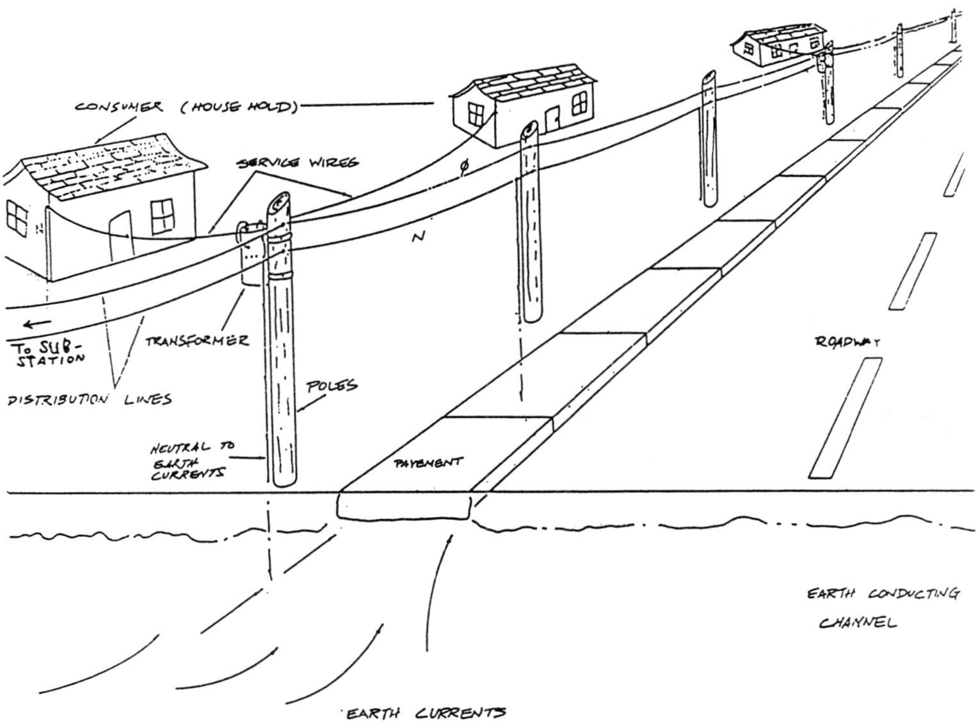

FIG. 1.—Currents through earth channels.

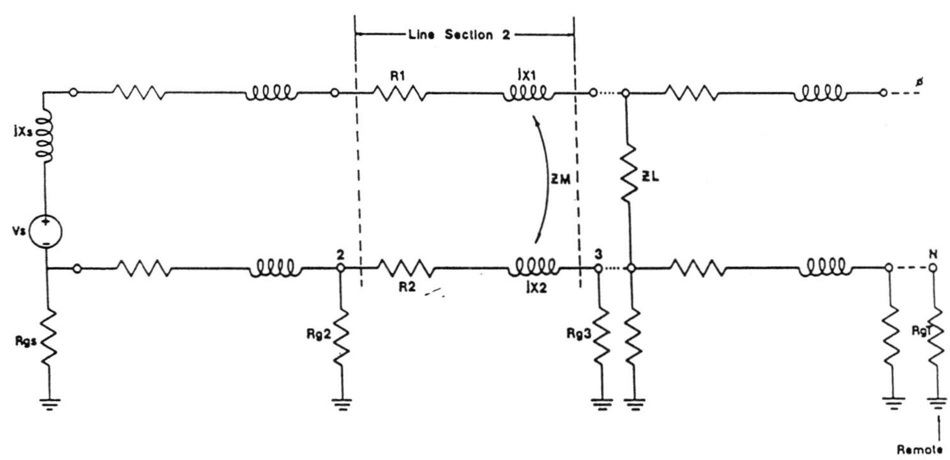

FIG. 2.—Equivalent circuit of multisection distribution line.

The magnitude of the ratio of mutual-to-neutral impedances per unit length, z_M / z_N , can be substantially less than unity, which means that a large part of the return current can flow in the earth channel. Figure 3 for a single 100A load shows this condition occurring at nodes 21 or 24. A careful look at the neutral-to-earth current phases shows that the ground currents enter the earth near the load and return to the neutral near the substation. The minimum points are where the reversal takes place. Large substation grounds allow more of the earth currents to return through users' ground circuits near the substation. They can in turn be diverted to power users' ground wires, water pipes, CATV shields, etc.; in some cases strong currents can create regions of much higher magnetic fields. A low substation ground resistance reduces this effect to a considerable extent. Similar results are found when 50 2A loads are evaluated (Fig. 4). When a large user's ground resistance is varied, the nearby utility customers ground currents are affected in much the same way as if the substation resistance were varied (Fig. 5). The ground current at each node is roughly inversely proportional to the node's ground resistance (Fig. 6); the neutral voltage remains relatively unaffected.[4]

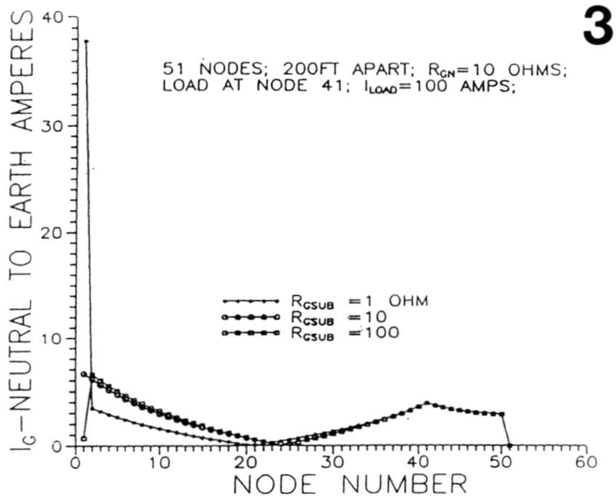

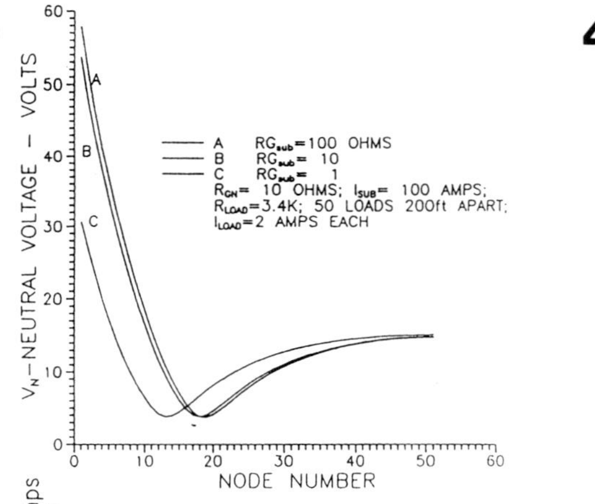

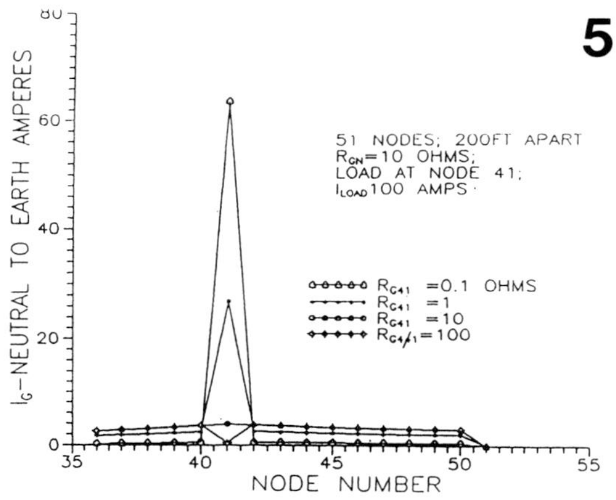

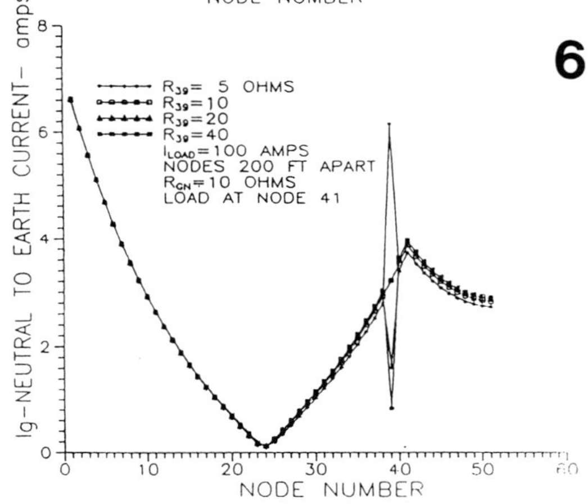

FIG. 3.—Neutral-to-earth current vs node.

FIG. 4.—Neutral-to-remote earth voltage vs node number.

FIG. 5.—Neutral-to-earth current vs node.

FIG. 6.—Neutral-to-earth current vs node.

REFERENCES

1. *National Electric Safety Code,* 1990, Sect. 9.

2. *Electric Transmission and Distribution Reference Book,* Pittsburgh: Westinghouse Electric Corp., 1964, 4th ed., p. 41.

3. L. Levey, "Calculation of ground fault currents using an equivalent circuit and a simplified ladder network," *IEEE Trans.* PAS-101: August 1982.

4. This work was supported by ESEERCO under contract EP90-57.

Martin Blank, Ed., *Electricity and Magnetism in Biology and Medicine*

A SIMPLE ELECTROMAGNETIC PULSE (EMP) INTENSITY METER FROM A PHOTOGRAPHIC FLASH METER

J. R. Forstall* and R. G. Olsen

Naval Aerospace Medical Research Laboratory, Pensacola, Fla.

Electromagnetic pulses (EMPs) strong enough to damage electronic equipment over a large geographical area are generated by nuclear exo-atmospheric explosions. The effects of EMP fields are being studied predominantly by the military services using simulators. We developed an EMP field-strength meter that is portable, hand held, and battery operated. The EMP meter indicates field strength by illuminating a bar graph of nine light-emitting diodes (LEDs). The meter was calibrated with peak E-fields in the range of 0.5 to 80 kV/m. Portability and simplicity of the device suggest the possibility of using multiple instruments for mapping all the required locations near a simulator with single pulse.

Electromagnetic pulses strong enough to damage electronic equipment over a large geographical area are generated by nuclear exo-atmospheric explosions.[1] Since the mid-1960s, defense contractors and the military have studied the effects of EMP fields using simulators.[2] These simulators generate electric fields (E-fields) as high as 100 kV/m with pulse durations of 100-300 ns. Instruments associated with the quantitation of EMP parameters are usually large computer-controlled systems with sensor probes connected via fiber-optic cable. We developed an EMP E-field meter that is relatively small, battery operated, and portable. The instrument consists of a modified and shielded commercially available photographic flash meter. The EMP meter indicates the field strength of a pulse by illuminating a bar graph of nine light-emitting diodes (LEDs). Portability and simplicity of the device suggest the use of multiple instruments for mapping intensity variations as an alternative to firing one pulse for each location.

METHOD

In preparing for a series of EMP-related dosimetric experiments, we noticed certain similarities between an EMP simulator and a photographic strobe light device. Both units charge up and terminate with a brief, high-intensity discharge and both are governed by the same physical laws of propagation of energy. The idea was conceived to adapt the simplest form of commercially available photographic flash meter to respond to the EMP simulator output. One can produce a practical field sensor by replacing the flash meter's photocell with a transistor amplifier and by adding a full-wave bridge rectifier connected to a monopole telescoping antenna (10-50 cm) for pulse detection. Sensitivity of the device is controlled by extension or lowering of the antenna. Electromagnetic shielding of the circuitry is also required to eliminate interference from strong fields. The instrument shown in Fig. 1 has a bar graph LED readout that accommodates a wide range of EMP field intensities. We retained the meter's internal circuitry that makes each succeeding light on the bar graph represent a two-fold increase in field intensity. Field strength readings were taken with the meter placed on a nonconductive stand 1.2 m high.

RESULTS

Calibration of the meter with peak E-fields of 0.5-80 kV/m was conducted at the Navy EMP simulator site near Solomons, Md. Table 1 shows the range of field-strength values for both antenna configurations (extended and lowered). With the antenna extended, the useful range of the sensor was extended below 1.4 kV/m. The table was derived by a comparison of the number of illuminated LEDs to field-strength data in kV/m obtained from precision field-monitoring equipment and from site maps of previous surveys. Simulator pulses were accurately detected 70 ft from the simulator antenna and up to distances of more than 1 mile

---

*Correspondence to: J. R. Forstall, NAMRL Code 27, 51 Hovey Rd., Pensacola FL 32508-1046.

from the site. Calibration for higher intensities could not be accomplished owing to asymmetric near-field propagation of the simulator.[3]

TABLE 1.— Calibration data obtained for antenna extended and lowered.

Number of bar graph lights illuminated	Electric field (E-field) intensity kV/m	
	Antenna extended	Antenna lowered
1	0.4 - 0.6	1.4 - 2.0
2	0.8 - 1.4	2.5 - 4.0
3	1.5 - 2.1	5.0 - 9.0
4	2.2 - 4.5	10 - 18
5	5 - 9	20 - 36
6	10 - 18	40 - 80
7	20 - 36	---

DISCUSSION AND CONCLUSIONS

When the instrument was removed from the nonconductive stand and handheld, both antenna positions yielded a consistently higher reading. We attribute this to electrical lengthening of the antenna via capacitive coupling to the body.

The EMP field intensity meter is constructed with parts and materials costing less than $300. The device is relatively small and rugged and has a dynamic range of more than two orders of magnitude. It accomplishes a task for which a $4500 instrument is normally used.[3]

REFERENCES

1. K. L. Gronhaug, *Induction in Horizontal Conductors by an EMP with Vertical Polarization*, Kjeller, Norway: Report 359-F/136, Forsvarets Forskings-Institutt, Norwegian Defence Research Establishment, 1970.

2. J. M. Muhm, "Mortality investigation of workers in an electromagnetic pulse test program," *J. Occup. Med.* 34: 287-292, 1992.

3. This research was sponsored by the Naval Sea Systems Command under work unit N000249 WR13050. The views expressed in this article are those of the authors and do not reflect the official policy or position of the U.S. Department of the Navy, U.S. Department of Defense, or the U.S. Government

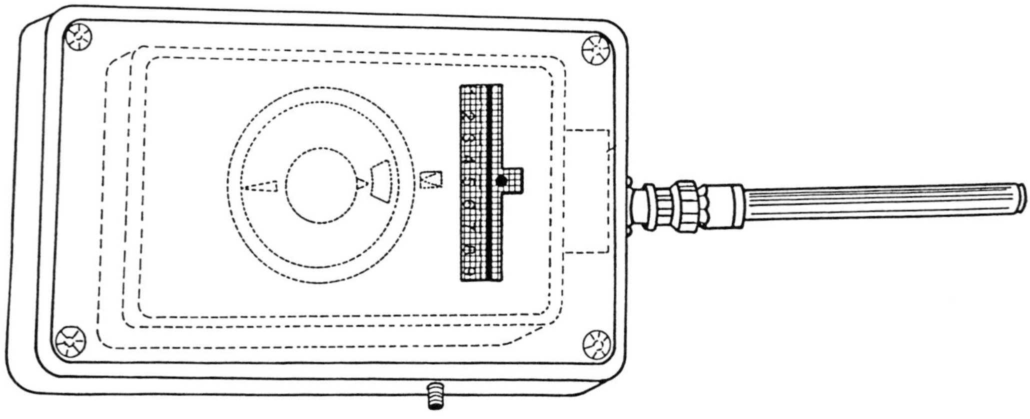

FIG. 1.—EMP field sensor shown with antenna retracted.

Martin Blank, Ed., *Electricity and Magnetism in Biology and Medicine*

COMPUTER-CONTROLLED MICROWAVE EXPOSURE SYSTEM: RESULTS OF AN IN VITRO CYTOGENETIC STUDY ON HUMAN PERIPHERAL BLOOD LYMPHOCYTES

C. De Wagter*

University of Gent Hospital, Belgium

L. Martens

University of Gent, Belgium

L. Verschaeve and A. Maes

VITO, Mol, Belgium

A system has been developed to expose cell suspensions to microwave radiation of 2450 MHz. Two modes of automatic control are provided: temperature control and E(lectric)-field control. Each mode can deliver continuous-wave or 50Hz pulsed radiation. Human peripheral blood lymphocytes were exposed for 0, 30, and 120 min. A cytogenic analysis showed a marked increase in the frequency of chromosomal aberrations. These results indicate that 2450MHz microwaves may indeed damage the genetic material under normothermal conditions.

Although there is much evidence in favor of the nongenotoxicity of microwaves, it has been reported in some papers[1] that microwaves may be mutagenic. In order to verify this assumption, we developed a rigorously controllable system to expose cell suspensions to microwave radiation of 2450 MHz, which is the frequency commonly used in domestic microwave ovens. Human peripheral blood lymphocytes were exposed while being kept at a constant temperature of 37 C. An in vitro cytogenic analysis was carried out.

MATERIALS AND METHODS

Microwave Exposure System

The exposure system developed permits irradiation of cell suspensions ranging in volume between 5 and 50 cc. Figure 1 gives a general view of the exposure system containing the controlling microcomputer, measuring equipment, microwave generator, and applicator. The applicator is based on a WR-340 waveguide (Fig. 2). The test tube containing the cell suspension is placed in the waveguide applicator. A stand-up collar choke, consisting of a cylindrical waveguide under cutoff, prevents excessive microwave leakage from the insertion hole. A rotating magnet can be used to stir the solution continuously to improve the heating uniformity in case of temperature control. Also, gas cooling can be applied in order to influence the suspension temperature further.

The microwaves are generated by a magnetron operating at 2450 MHz, which is the frequency commonly used in domestic microwave ovens. The nominal maximum output power of the magnetron is 1000 W. The magnetron generates a maximum peak power of 300 W for any duty cycle. Two magnetron power supplies are incorporated to drive the magnetron in a pulsed or a continuous-wave mode.

In the first mode, a standard LC-stabilized half-wave voltage doubler is used that makes the magnetron oscillate during one alternation of each 50Hz line cycle. This is the classic principle as it is commonly implemented in domestic microwave ovens. The resulting microwave output is 50 Hz pulsed with a duty cycle of about 10% for the power intensities considered. Electronic leveling of the microwave power output is realized by a phase-controlling triac circuit in the primary of the high-voltage transformer.[2]

*Correspondence to: C. De Wagter (Radiophysics), University Hospital, De Pintelaan 185, B-9000 Gent, Belgium.

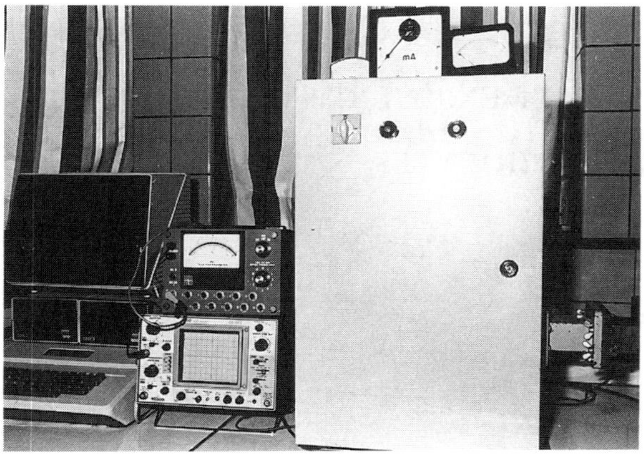

FIG. 1.—Exposure system containing the controlling microcomputer, measuring equipment, microwave generator, and applicator.

FIG. 2.—Test tube containing 5 cc of human blood to be exposed, lifted out from microwave applictor.

Second, the modality of continuous-wave radiation has been worked out. Since the dynamic impedance of magnetrons is in general low, small variations in the applied voltage cause appreciable changes in electrical current. What that means is that a magnetron is a typical current-fed device. Therefore, a highly stabilized direct-current regulator was constructed for driving the magnetron continuously. The main principles are those followed in Ref. 3 for a similar application. The current regulator is built up from a chain of 20 commercially available low-voltage regulators with a minimum amount of external components.

Two different modes of automatic control are provided: temperature and E(lectric)-field control. In the temperature-controlled mode, the microwave output is adjusted so that one can rapidly obtain and maintain a set temperature. The temperature is measured in the test tube by a commercially available thermistor microprobe. In the E-field controlled mode, the aim is to follow closely a prescribed E-field time course with minimal elementary time steps of 2 s. A self-constructed E-field probe is inserted in the cell suspension to measure the main electrical field component. Figure 3 depicts the details of the probe. It consists of a high-frequency diode whose leads form the receiving dipole antenna. This probe delivers the appropriate feedback signal to the microcomputer to control the microwave power output of the generator.

Each mode can deliver 50Hz pulsed or continuous-wave radiation. Table 1 summarizes the various modes.

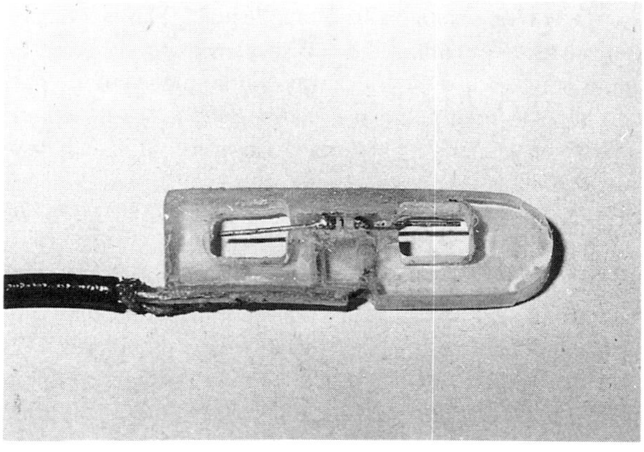

FIG. 3.—Self-constructed E-field probe to be inserted in test tube in case of E-field control.

TABLE 1.—Combinatorial diagram illustrating the various modes offered by the microwave exposure system. "X" indicates the mode currently used to expose blood lymphocytes.

	Type of control	
Type of modulation	Temperature control	E-field control
50Hz pulsed	x	•
Continuous wave	•	•

Cytogenetic Experiments

The exposure system described has been used to exposure human peripheral blood lymphocytes to 50Hz pulsed microwave radiation. The mode used is indicated in Table 1. The microwave intensities delivered were just high enough to keep the temperature constant at 37 C. The average microwave power as about 0.4 W for 5 cc of blood, and was substantially needed to compensate for the heat losses. The exposure time was 30 and 120 min, respectively. A control sample was placed in a waterbath at the same temperature. A cytogenetic analysis was performed on the peripheral blood lymphocytes from two different donors. Cells were cultivated during 48 h for analysis of structural chromosome aberrations, or 72 h for analysis of micronuclei (in cytochalasin B blocked cells).

RESULTS AND CONCLUSIONS

Results of our three independent investigations (one donor investigated once, the other twice) revealed that microwave exposure does not influence the cell cycle nor the frequency of sister chromatid exchanges (Table 3). However, as appears from Tables 2 and 4, a statistically highly significant increase was found in the frequency of chromosome aberrations and micronuclei in cultures exposed for 120 min. Of particular interest is the occurrence of a few dicentric chromosomes and acentric fragments that are usually associated with ionizing radiation exposure. The complete results have been reported in Ref. 4.

The above preliminary results were shown to be highly reproducible and indicate that 2450MHz microwaves may cause cytogenetic damage in human lymphocytes under normothermic exposure conditions.[5]

TABLE 2.—Chromosome aberrations in 2450MHz microwave-irradiated human lymphocytes at constant temperature of 37 C. Exposure time is 0 (control), 30, and 120 min, respectively.

DONOR 1 / EXPERIMENT 1

	Number of cells	gaps & isogaps	breaks	excess acentric	dicentrics + ac. fragm. +	dicentrics + ac. fragm. -	other aberrations	cells with aberrations 1	2	3	total	%cells with aberrations	aberrations per 100 cells
control	500	5	8	1	0	0	0	12	1	0	13	2.6	2.8
30 min.	200	4	1	3	1	0	0	9	0	0	9	4.5	4.5
120 min.	200	7	9	13	2	0	3	16	6	2	24	12.0	17.0

DONOR 1 / EXPERIMENT 2

	Number of cells	gaps & isogaps	breaks	excess acentric	dicentrics + ac. fragm. +	dicentrics + ac. fragm. -	other aberrations	cells with aberrations 1	2	3	total	%cells with aberrations	aberrations per 100 cells
control	200	2	0	3	0	0	0	3	1	0	4	2.0	2.5
30 min.	200	8	2	4	0	0	0	8	2	0	10	5.0	6.0
120 min.	200	8	4	4	0	1	0	15	0	0	15	7.5	7.5

DONOR 2 / EXPERIMENT 1

	Number of cells	gaps & isogaps	breaks	excess acentric	dicentrics + ac. fragm. +	dicentrics + ac. fragm. -	other aberrations	cells with aberrations 1	2	3	total	%cells with aberrations	aberrations per 100 cells
control	100	3	2	3	0	0	0	8	0	0	8	8.0	8.0
30 min.	100	2	3	2	0	1	0	6	1	0	7	7.0	8.0
120 min.	100	4	2	6	0	0	0	12	0	0	12	12.0	12.0

TABLE 3.—Sister chromatid exchanges and cell kinetics in microwave-exposed human lymphocytes. M1, M2, and M3 are first, second, and third division metaphases.

DONOR 1 / EXPERIMENT 1

	mean # if SCE's per >50 M2 cells	cell kinetics (200 metaphases) 48h cultures %M1	%M2	%M3	72h cultures %M1	%M2	%M3
control	5.88	91	9	0	14	41	45
30 min.	5.56	93	7	0	19	47	34
120 min.	7.16	88	12	0	20	40	40

DONOR 2 / EXPERIMENT 1

	mean # if SCE's per >50 M2 cells	cell kinetics (200 metaphases) 48h cultures %M1	%M2	%M3	72h cultures %M1	%M2	%M3
control	5.95	98	2	0	25	39	36
30 min.	6.94	100	0	0	25	44	31
120 min.	7.75	99	1	0	26	36	38

TABLE 4.—Micronuclei in microwave-exposed human lymphocytes.

DONOR 1 / EXPERIMENT 1

	number of cells	cells with # micronuclei 1	2	3	total number of micronuclei	micronuclei per 1000 cells
control	1292	4	3	0	10	7.7
30 min.	1000	11	0	2	17	17.0
120 min.	1672	35	2	0	39	23.3

DONOR 1 / EXPERIMENT 2

	number of cells	cells with # micronuclei 1	2	3	total number of micronuclei	micronuclei per 1000 cells
control	1000	7	0	0	7	7.0
30 min.	1000	11	0	0	11	11.0
120 min.	1000	15	3	0	21	21.0

DONOR 2 / EXPERIMENT 1

	number of cells	cells with # micronuclei 1	2	3	total number of micronuclei	micronuclei per 1000 cells
control	2000	16	0	0	16	8.0
30 min.	1000	11	0	0	11	11.0
120 min.	1000	22	0	0	22	22.0

REFERENCES

1. V. Garaj-Vrhova, D. Horvat, and Z. Koren, "The effect of microwave radiation on the cell genome," *Mutation Res.* 243: 87-93, 1990.

2. C. De Wagter, M. De Pourcq, and W. Van Loock, "Microwave apparatus for controlled heating of aqueous solutions," *Proc. 19th Ann. Microwave Power Symp.*, Minneapolis, 1984, 196-199.

3. C. De Wagter, P. Gheeraert, and W. Van Loock, "Narrowband microwave generator using a commercial magnetron," *Proc. 20th Ann. Microwave Power Symp.*, Chicago, 1985, 124-127.

4. L. Verschaeve, A. Maes, A. Arroyo, C. De Wagter, and L. Martens, "In vitro study on the genotoxicity of 2450MHz microwave irradiation in human peripheral blood lymphocytes," *Proc. 8th IRPA World Congress*, Montreal, 1992, 1144-1147.

5. The authors thank L. Vercruyssen for his help in running the experiments.

Martin Blank, Ed., *Electricity and Magnetism in Biology and Medicine*

DESIGN CONSIDERATIONS FOR AN ELECTROMAGNETIC SADDLE-COIL TRANSDUCER

A. M. J. van Amelsfort*

Eindhoven University of Technology, The Netherlands

The effects of position and dimensions of a two-saddle-coil stimulator on the electromagnetic field in an anisotropic four-layer limb model surrounded by air is investigated. Under the assumption that the electric-field amplitude forms the effective signal in the stimulation of bone, the efficiency of a saddle-coil stimulator can be improved by positioning of the coils in such a way that the fracture zone is located near one of the transverse end-planes of the coils (the area of maximum electric-field strength), and by elimination of the DC component from the coil current.

In investigations of therapeutic applications of low-power electromagnetic fields it is recognized that the local electric-field strength and the local magnetic-field strength are two basic electromagnetic signals. The reaction to low-power electromagnetic fields implicitly excludes the local density of dissipated electromagnetic power as an electromagnetic signal. Depending on the type of application, one of the two signals or a combination of these two signals can be considered to be effective. The spatial distribution of the two signals is influenced by dimensions and position of the transducer. Knowledge of that influence may help to set up new experiments and to design more effective or efficient stimulators. In this paper results are presented of a parametric study concerning the electromagnetic field in an anisotropic four-layer limb model surrounded by air and excited by a two-saddle coil transducer.

THE MODEL

In the past a procedure has been developed for the determination of the electromagnetic field distribution in a set of coaxial circularly cylindrical layers generated by two saddle-coils and a time-harmonic coil current.[1] This procedure has been applied to compute the electromagnetic field distribution in the limb model, consisting of marrow, anisotropic bone, anisotropic muscle, and fat, surrounded by air. The two-saddle-coil stimulator is modeled by two saddle-shaped loops of electric current of 1 Ampere-turn, operating at a frequency of 4.3 kHz.

EFFECT OF COIL DIMENSIONS

In order to keep the amount of data tractable, the investigation of the effect of the coil dimensions on the electromagnetic field is restricted to three different points in the bone layer, A, B, and C (Fig. 1). A and B are both located in the transverse plane $z = 0$. Because the electric-field strength reaches its maximum value just outside the transverse end plane of the coils, we have chosen C within this plane. As a consequence the axial coordinate of C changes with the axial length of the coils. We can study the effects of the saddle-coil parameters on the electromagnetic field by considering the effects on the axial electric field on B (E_z), the radial electric field in C ($E_r = E_y$), and the radial magnetic field in A (H_r). These three fields are normalized with respect to the corresponding field strengths for the case when the coil dimensions have their reference values, as listed in the Reference Table, according to $f_i = F_i / F_{i,reference}$, where (F_i, f_i) denote either (E_z, e_z), (E_r, e_r), or (H_r, h_r). It follows from the Reference Table that the electric-field strength in C is almost 70 times larger than in B. The effect of the dimensions of the saddle-coils in the field strengths at the three points are plotted in Figs. 2 through 6.

*Correspondence to: A. M. J. van Amelsfort (EH6.26), Eindhoven University of Technology, P.O. Box 513, NL-5600 MB Eindhoven, The Netherlands.

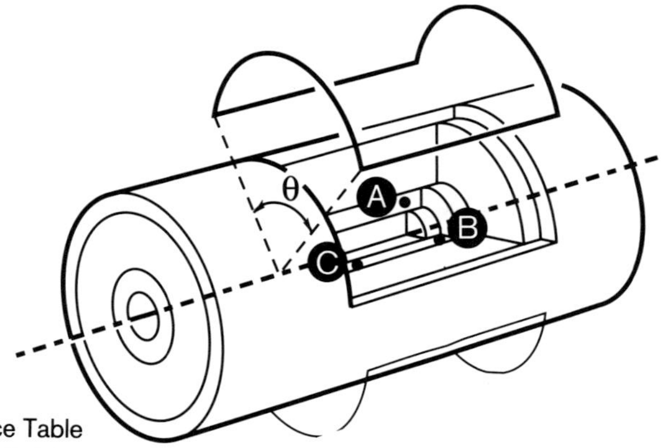

Reference Table

Reference coil		Reference field strength	
axial coil length	120 mm	H_r in A	6.34 A/m
coil flare angle	91.6 degrees	E_z in B	0.25 mV/m
coil radius	75 mm	E_r in C	17.0 mV/m

FIG. 1.—Field components and points of observation.

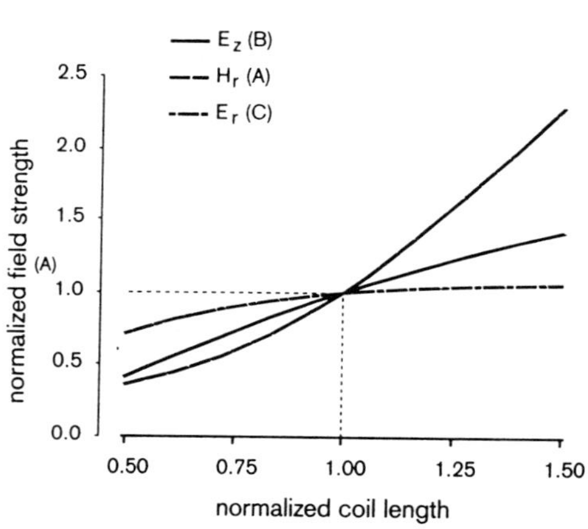

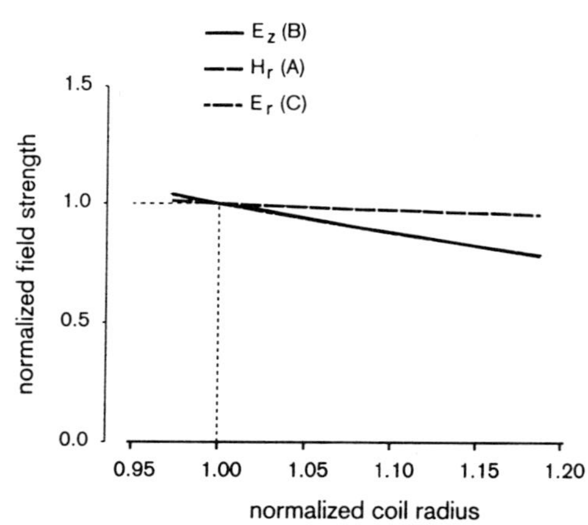

FIG. 2.—Field strength as a function of axial coil length. The three field strengths increase with increasing axial coil length, which can be understood as a result of the increase in coil circumference. Although sensitivity of radial magnetic and electric field to changes in coil length decreases with increasing axial length, sensitivity of axial electric field increases.

FIG. 3.—Field strength as a function of coil radius. All field strengths decrease with increasing coil radius. With increasing coil radius, both the coil-to-limb distance and coil circumference increase. Apparently increase in coil-to-limb distance plays a more dominant role than increase in coil circumference.

DISCUSSION AND CONCLUSIONS

It is demonstrated that the effect of the coil dimensions on the field strengths depends on the choice of the field component and the point of observation. The largest effect is found for the axial electric-field component in B when the axial coil length is changed (Fig. 2). A threefold of the axial length corresponds to a sixfold of the axial electric-field strength. A much larger effect on the electric-field strength is found by the choice of the point of observation: in C the electric field is about 70 times stronger than in B. In the case of electromagnetic stimulation of bone, these quantitative results enforce the need for a clear identification of the effective electromagnetic signal.[2] Under the assumption that the electric-field amplitude acts as the effective signal, the efficiency of a saddle-coil stimulator can be improved if we position the coils in such a way that the fracture zone is located near one of the transverse end planes of the coils, the area of maximum electric-field strength. Furthermore, as only the magnetic field of time-dynamic coil-currents induce an electric field, a possible direct current (DC) component in the coil current is redundant and may be removed without affecting the therapeutic effectiveness of the stimulator, which leads to a dramatic reduction of the power consumption. Finally, the study reveals that the field strengths are maximum if the flare angle of the saddle coils is 180° and/or the coil radius equals the limb radius.

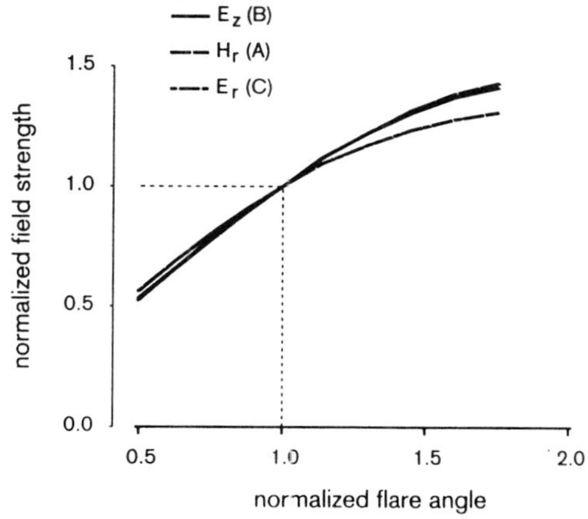

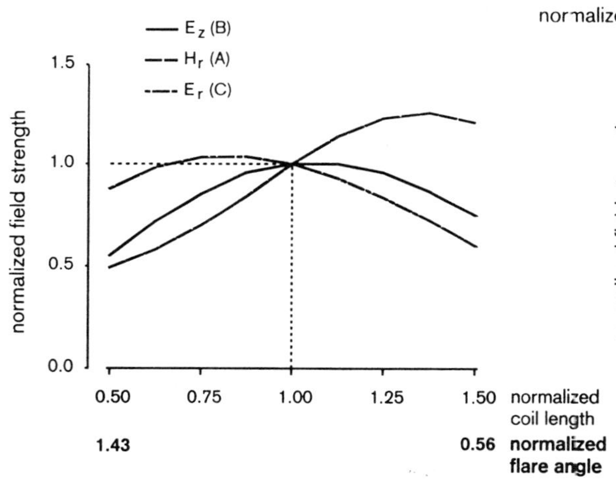

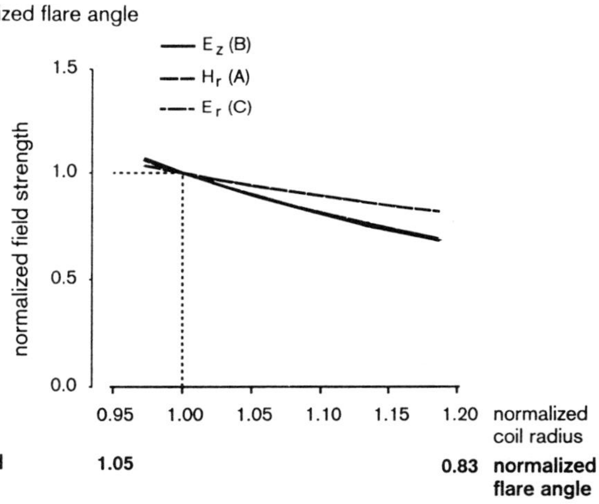

FIG. 4.—Field strength as a function of flare angle. Field strengths increase almost equally with increasing flare angle. This again can be understood as a result of lengthening of the coil circumference.

FIG. 5.—Field strength as a function of both axial coil length and flare angle, with coil radius and circumference kept constant. All field strengths show a local maximum, each for a different value of coil parameters. As an increase in axial length implies a decrease in flare angle, on can qualitatively verify this local maximum by combining the plot for axial coil length with a reverse plot for flare angle.

FIG. 6.—Field strength as a function of coil radius and flare angle, with axial length and circumference kept constant. Amplitude of the field components decrease monotonically with both increasing coil radius and decreasing flare angle. Because increase in coil-to-limb distance is not compensated by increase in coil circumference, field strengths decrease more rapidly than when flare angle is kept at its reference value.

REFERENCES

1. A. M. J. van Amelsfort, *An Analytical Algorithm for Solving Inhomogeneous Electromagnetic Boundary-value Problems for a Set of Coaxial Circular Cylinders,* Ph.D. Thesis, Eindhoven University of Technology, 1990. (Available from James Clerk Maxwell Foundation, Box 7615, NL-5601 JP Eindhoven, The Netherlands.)

2. A. M. J. van Amelsfort and T. Scharten, "On computed electromagnetic field amplitudes during therapeutic application," in C. T. Brighton and S. R. Pollack, Eds., *Electromagnetics in Medicine and Biology,* San Francisco: San Francisco Press, 1991, 237-246.

Martin Blank, Ed., *Electricity and Magnetism in Biology and Medicine*

EFFECTS OF MICROWAVE FIELDS ON AMINO ACID METABOLISM IN MOUSE MACROPHAGE CELL CULTURES

H. A. Schwertner*

Wilford Hall USAF Medical Center, Lackland AFB, Tex.

D. K. Robinson and J. L. Kiel

Armstrong Laboratory, Brooks AFB, Tex.

R. G. Dunham

Southwest Foundation for Biomedical Research, San Antonio, Tex.

In this study, we sought to determine whether microwave radiation has an effect on amino acid metabolism. For the study, mouse macrophage cells were exposed for 30 min to 2450 MHz (103.5 W/kg) and amino acids were analyzed at 24 and 48 h. Statistical differences between microwaved samples and sham controls were found for alanine and lysine at 24 h and for alanine, isoleucine, phenylalanine, and tyrosine at 48 h. Group-time interactions, i.e., differences in amino acids between 24 and 48 h for microwave vs. sham controls, were statistically significant for phenylalanine, tyrosine, leucine, isoleucine, arginine, and methionine.

The aim of this study was to determine whether microwave radiation has an effect on amino acid metabolism of mouse macrophage cells. For the study, we examined changes in amino acid concentrations in cell culture media at various times after exposure to microwave radiation. This experimental approach was chosen because it permits an accurate measurement of a wide range of amino acids under well-controlled experimental conditions. We also felt that amino acid concentration changes might be sensitive indicators of microwave exposure effects because they are intermediate metabolites in a wide range of metabolic pathways and conditions.

MATERIALS AND METHODS

For the study, RAW 264.7 mouse macrophage cells (Abelson leukemia virus-transformed monocyte/macrocytes) were grown in RPMI 1640 medium containing 10% (V/V) fetal calf serum. Five samples containing 2.0 ml of 2.54×10^6 cells/ml were exposed to microwave radiation (2450 MHz; mean SAR 103.5 W/kg) for 30 min and five samples were treated as sham controls. Temperature in the microwave-exposed and sham-control samples was continuously maintained at 37 ± 0.1 C with a computer-controlled feedback system.

After microwave or sham exposure, the samples were incubated at 37 C and 1ml aliquots were removed at 24 and 48 h. The samples were centrifuged to remove the cells and aliquots of the supernatants were transferred to separate tubes for amino acid analysis. Individual amino acids were analyzed with a Beckman Model 7300 Amino Acid Analyzer.

RESULTS

Analysis of variance was used to determine whether the individual amino acids in samples exposed to microwave radiation differed from those in sham-exposed samples. Statistically significant differences between microwave and sham-exposed samples were found for alanine and lysine at 24 h; and for alanine, isoleucine,

---

*Correspondence to: H. A. Schwertner, Wilford Hall USAF Med. Ctr. (RD), Lackland AFB TX 78236-5319.

phenylalanine, and tyrosine at 48 h ($p \leq 0.05$).

The 24h and 48h data were combined to determine whether there was a group (or treatment) effect, a time effect, or an interaction of group (treatment) with time effect. Only alanine showed a statistically significant group effect (Table 1). Most of the amino acids showed a time effect. Group by time interactions (i.e., differences in the amino acid concentrations between 24 and 48 h for microwave and sham treatments) were statistically different for phenylalanine, tyrosine, leucine, isoleucine, arginine, and methionine. The changes in phenylalanine (Fig. 1) and tyrosine between 24 and 48 h were almost identical. Likewise, the changes in isoleucine concentrations closely resembled those of leucine.

TABLE 1.—Combined analysis of 24 and 48h data showing group, time, and group by time interactions of microwave vs sham-exposed cell supernatant amino acid concentrations (probability values). *Statistically significant ($p < 0.05$; only p values below 0.2 are shown).

Amino acid	Group	Time	Group by time
Alanine	0.0057*	0.0001*	0.1886
Arginine	0.1919	0.0001*	0.0301*
Isoleucine	0.0594	0.0301*	0.0213*
Leucine	0.1375	0.0622	0.0235*
Methionine	---	0.0001*	0.0443*
Phenylalanine	0.1197	---	0.0137*
Tyrosine	0.0965	---	0.0064*

DISCUSSION

The experimental results indicate that microwave radiation has an effect on the amino acid metabolism of macrophage cells. The changes are most apparent for phenylalanine, tyrosine, leucine, isoleucine, arginine, and methionine. Even though changes in these amino acids were seen, it remains to be determined whether they are due to the thermal effects of microwave radiation or the effects of microwave radiation per se. Many biological studies involving microwave radiation have been criticized because of inadequate thermal controls.[1,2] In our studies, we controlled the temperatures to within ±0.1 C; however, the changes in amino acids could still be due to the relatively high thermal flux (SAR 103.5 W/kg) through the cell cultures.

We think that the changes seen in our studies are not due to the thermal effects of microwave radiation, for several reasons. First, if thermal heating were responsible for the amino acid changes, we would expect to see an immediate effect rather than a delayed effect. Second, the concentrations of phenylalanine, tyrosine, leucine, and isoleucine in the sham and microwave-treated samples diverged with time. If the changes were due to temperature, we would not expect the curves to diverge. Rather, we would expect the curves of sham and microwave-treated samples to be in the same direction, though with different slopes. Our results suggest that microwave radiation most likely resulted in enzyme induction that caused the delay in microwave response and the different metabolic responses for the sham and microwave-treated samples.[3]

REFERENCES

1. S. M. Michaelson and J. C. Lin, *Biological Effects and Health Implications of Radiofrequency Radiation,* New York: Plenum Press, 1987.

2. W. Grundler, U. Jentzch, F. Keilmann, and V. Putterlik, ''Resonant cellular effects of low-intensity microwaves,'' in H. Fröhlich, Ed., *Biological Coherence and Response to External Stimuli,* New York: Springer, 1988, 65-85.

3. This research was supported in part by the U.S. Air Force Office of Scientific Research. The views expressed in this paper are those of the authors and do not reflect the official policy of the U.S. Department of Defense or other departments of the U.S. Government.

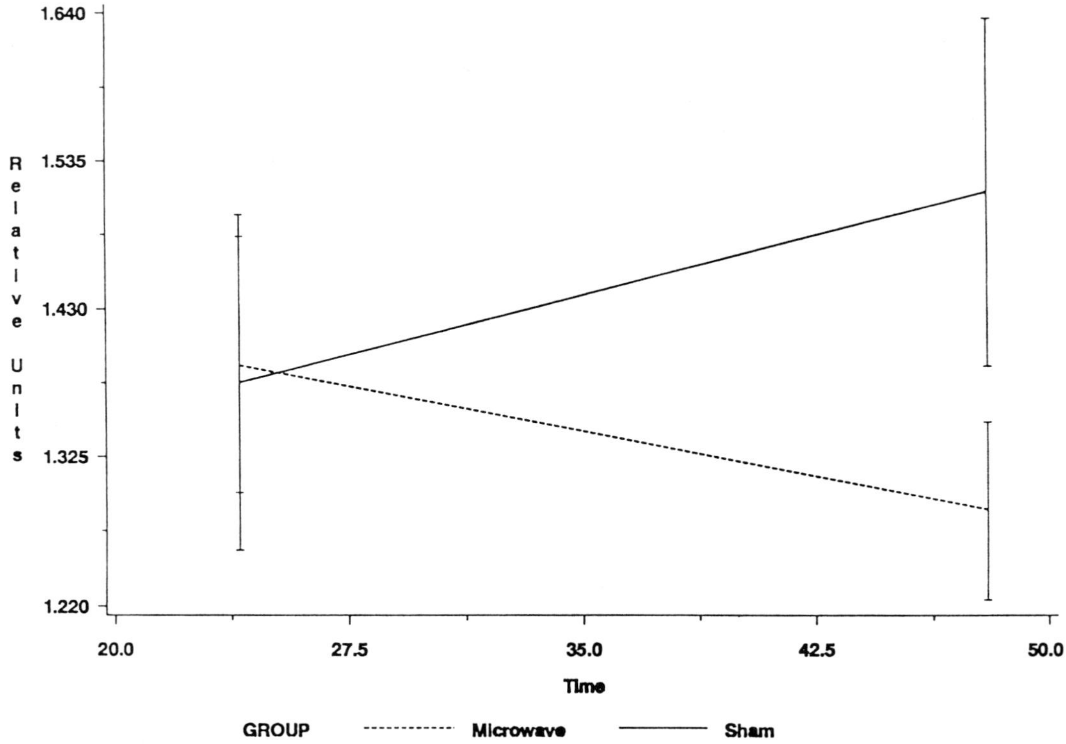

FIG. 1.—Changes in phenylalanine concentrations in sham and microwave- samples (mean ± SD).

Martin Blank, Ed., *Electricity and Magnetism in Biology and Medicine*

ELECTROSTIMULATION OF CHO CELLS: CONSEQUENCES ON CELL ADHESION

S. Bogey* and J. Teissié

CNRS-CRBGC, Toulouse, France

Most cellular metabolic activities are under the control of the transmembrane potential. This control should play a part in cell adhesion. The implication of transmembrane potential has been investigated by the application of external electric fields on Chinese hamster ovary cells suspension. Electropulsation led to a rapid increase of adherent cells in comparison with unpulsed control. Cell adhesion depends on frequency, intensity, and pulse duration. The dependence on frequency suggests that electrogenic activities are playing a key role. An effect of the electric field has been observed on cell growth and on intracellular calcium level.

Cell adhesion to extracellular matrix components and to plastic dishes is a complex process that depends on multiple interactions between matrix receptors, various ligands, and the cytoskeleton.[1,2] The various interactions give rise to stable cell-to-cell bonds, induce cellular division, and lead to cell differentiation and contact inhibition of cellular motility. A real problem in biotechnology is to control and stabilize cell adhesion for culture in serum-free medium. Most of the cell surface membrane proteins likely to be involved in cell adhesion are known, such as collagen, proteoglycan, elastin, and structural glycoproteins such as fibronectin and laminin. Several stages of cell adhesion to substrates can be distinguished: the initial attachment is followed by cell spreading and subsequent development of focal adhesion sites.[3] On the other hand, extracellular calcium, which is indispensable for the organization of the cytoskeleton, is also necessary to cell adhesion.[4] Ellem and Kay have pointed out that cell attachment and replication may be influenced by transmembrane electron flow.[5] Although the molecules playing a role in cell adhesion are known, their mode of action remains unclear. By involving charged surface interactions, cell adhesion is relevant to bioelectrochemistry.

A cell can maintain a steady-state transmembrane potential by an asymmetric distribution of ionic species in both sides of the membrane. The application of an electric field to a cell (electropulsation) induces a local and reversible change of the transmembrane potential added to the native potential (-70 mV for animal cells).[6] When a spherical cell in suspension is exposed to an electric field, this effect can be expressed as $\Delta\Psi = 1.5rE\cos\theta$, where r, E, and θ are, respectively, the outer radius of the cell, the applied field strength, and the angle between the field line and the normal to the point of interest on the membrane.[7]

Electropulsation has been used to check the influence of this potential difference on cell adhesion.

MATERIALS AND METHODS

Cell culture. Chinese hamster ovary (CHO) cells were cultured at a density of 10^5 cells per culture dish (Nunc, Denmark) in Eagle's minimum essential medium (MEM medium, Gibco) containing 5% newborn calf serum (SVF, Gibco), 5 nM methotrexate, 1% penicillin, streptomycin, and 2% glutamin.[8] The culture medium was changed every week. Cells were kept at 37 C in a 5% CO_2 incubator (Jouan, France).

Application of electric pulses to the cell. Confluent CHO cells were dissociated by use of 2 ml of trypsin/EDTA saline (0.07% p/v, Gibco, Glasgow) for 2 min. To stop the reaction, 30 ml of MEM medium supplemented with 5% newborn calf serum was added. Cells in suspension were centrifuged for 5 min at 400 G (1200 rpm, C-500 centrifuge, Jouan, France) and resuspended in a serum-free medium (Sanofi, France) at a concentration of 5×10^5 cells per milliliter. 500 µl of the suspension was poured between two thin flat stainless-steel parallel electrodes on a 1.8×1.8 cm^2 culture dish (Greiner, Germany) precoated by collagen (10 µg/cm^2). The electrodes were connected to a voltage generator which gave square-wave electric pulses

---

*Correspondence to: S. Bogey, CNRS Centre de Recherche de Biochimie, 118 rte de Narbonne, F-31062 Toulouse CEDEX, France.

(CNRS cell electropulser, Jouan, France). The direction of the field between each pulse was automatically inverted. The voltage pulses applied to the cells were monitored with an oscilloscope incorporated into the cell electropulser. The frequency f and pulse duration t were modified simultaneously so that ft = constant. Voltage pulses were then applied.

Determination of electropulsed adhesion. Cells were pulsed in the serum-free medium, incubated from 10 min to 24 h (~ 1 generation) at 37 C in a 5% CO_2 incubator (Jouan, France). The medium was discarded and cells were washed twice with phosphate saline buffer (PBS: 10 mM phosphate buffer, 104 mM NaCl, 3 mM KCl, pH 7.4); 1 ml of permeant dye, cristal violet (Gurr, 0.1% p/v) in phosphate buffer (10 mM, pH 7.2) with sucrose (250 mM) and $MgCl_2$ (1 mM) was added to the cells and incubated 15 min at room temperature. The cells were washed three times with PBS and the dye was extracted by an acetic acid solution (10%) for 15 min. We measured the optical density of each sample at 595 nm.[9] The percentage of adherent cells was the ratio of number of adherent electropulsed cells to the number of adherent nonelectropulsed cells times 100.

Determination of intracellular free calcium concentration. 1 ml of cells was suspended at a density of 3×10^6/ml in serum-free medium containing 10 µl Fluo-3/Am (500 µM in anhydrous DMSO, Molecular Probes) and 10 µl/ml pluronic acid F127 at 25% for 30 min at room temperature with gentle agitation of the cell suspension. Cells were then centrifuged for 5 min at 400 G and washed twice with serum-free medium to remove all traces of extracellular fluo-3/Am. Voltage pulses were then applied. The Ca^{2+} fluorimetric titration was measured with a JY3 spectrofluorimeter (Jobin Yvon, France) (λ_{exc} = 500 nm and λ_{em} = 525 nm) with 5×10^5 cells in a final volume of 2 ml. At the end of an experiment, fluo-3 fluorescence intensities were calibrated by $[Ca^{2+}]_i = Kd(f - f_{min})/(f_{max} - f)$ where Kd is 400 nM at vertebrate ionic strength according to Kao.[11]

RESULTS

Effect of Electric Field Parameters on Cell Adhesion

The application of square wave pulses of 50 V/cm electric field to CHO cells suspension, $ft = 25 \times 10^{-3}$, during 2.5 min, led to a rapid increase of the adherent population 10 min after the electrostimulation, when the spontaneous adhesion is weak, resulting in a 100% increase (Fig. 1a). A bell-shaped curve is obtained as a function of frequency, with an optimum value at 200 Hz. The cell adhesion speed also depends on the intensity (optimum at 50 V/cm) and on the pulse duration (optimum at 0.125 ms).

After 1h, the effect was decreased to 30% due to the spontaneous adhesion (data not shown). Moreover, we observed an effect of the electric field on the cell growth: 24 h after electropulsation, only 50% of electropulsated cells grew (Fig. 1b), which suggests that electropulsation either kills 50% of the cells or decreases the growth rate by 50%.

The application of square-wave pulses at 50 V/cm at 100 Hz for 1 min yields a strong increase of the adherent population early after the electropulsation when we increase the pulse duration (Fig. 2a). On the other hand, when we vary the pulse intensity, an optimum is obtained at 50 V/cm or less (55% more adherent cells in comparison with control) with 50µs pulse duration for 5 min (Fig. 2b). A linear decrease in viability is also observed from 0 to 150 V/cm 24 h after the electropulsation (data not shown).

A linear increase in the number of adherent cells, 10 min after pulsing, is observed after 1-5 min of electropulsation when we pulsed the cells at 50 V/cm, 100 Hz, and a pulse duration of 500 µs. The viability 24 h after the electropulsation did not change under these conditions (data not shown).

Effect of Electric Field Parameters on Intracellular Free Calcium Concentration

Because of the importance of calcium in adhesion,[11] the effect of the electric field on intracellular calcium level has been measured (50 V/cm, 2.5 min, ft = constant with frequency up to 1200 Hz). Also, after cell electrostimulation, a 50% decrease is observed as a function of frequency (Fig. 2b). The decrease of calcium is very fast. As soon as 50 Hz was reached, a 50% decrease in $[Ca^{2+}]_i$ was observed, 5 min after pulses application.

Effect of Electric Field Parameters on ATP

Unlike what has been shown by Teissié in *Escherichia coli*,[12] we could not measure ATP synthesis in CHO.

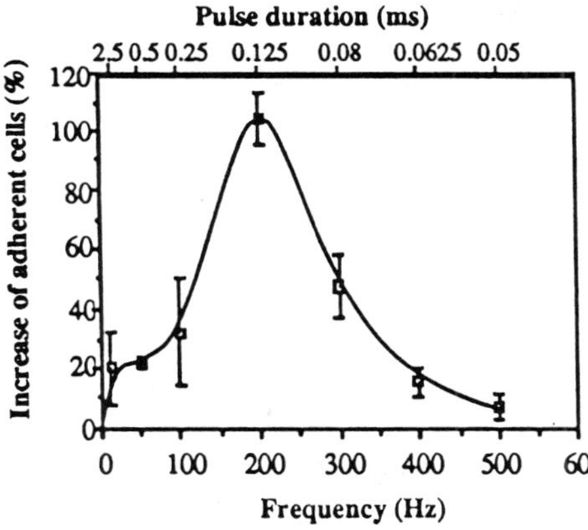

Figure 1a.

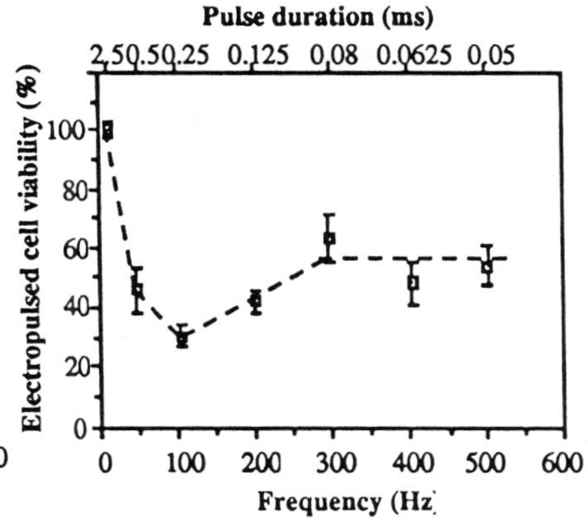

Figure 1b.

FIG. 1.—(a) Effect of frequency-pulse duration impulsions on cell adhesion; (b) electropulsed cell viability 24 h after electropulsation. 2×10^5 cells/ml were electropulsed for 2.5 min with a pulse intensity of 50 V/cm, $ft = 25 \times 10^{-3}$, with f up to 600 Hz. The percentage of adherent cells was 100 times the ratio of the number of electropulsed cells to the number of standard cells.

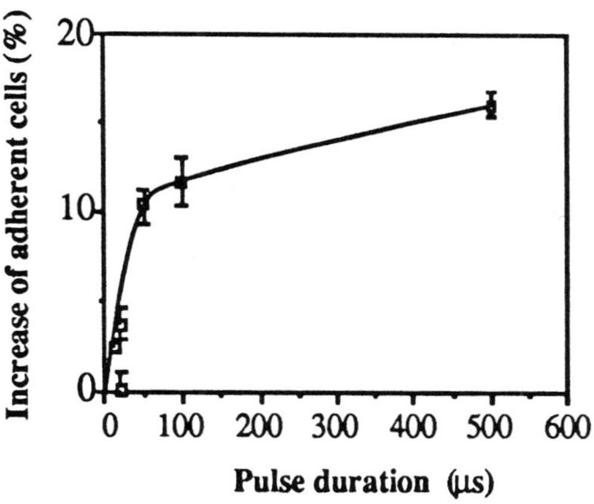

Figure 2a.

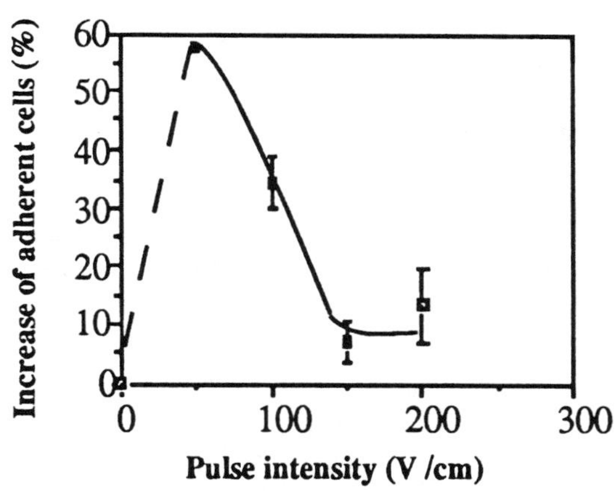

Figure 2b.

FIG. 2.—(a) Effect of pulse duration of cell adhesion. Cells were electropulsed with a pulse intensity of 50 V/cm for 1 min at 100 Hz. (b) Effect of pulse intensity on cell adhesion. Cells were electropulsed for 5 min with a pulse duration of 50 µs at 100 Hz.

DISCUSSION

Electropulsation by square-wave pulses of 50 V/cm on CHO cell suspensions led to a rapid increase of the adherent population at the early time after electrostimulation (10 min) with an optimum frequency at 200 Hz (Fig. 1). As previously described by others,[7,13] the main effects of the electropulsation are electrochemical reactions that take place in the neighborhood of the electrodes, an increase in the temperature of the suspension by the Joule effect, and an induction of a transmembrane potential difference.[14] The increase in temperature depends linearly on the pulse number. The electrochemical reactions depend on pulse duration. Both effects are independent of frequency. The bell-shaped curve as a function of frequency suggests that this stimulation is mediated by a modulation of the transmembrane potential, as predicted by the electrocoupling theory.[15] The increase of adherent population is obtained at an intensity pulse and pulse duration above the critical values and the effect is optimum at an optimal frequency (here, 200 Hz).

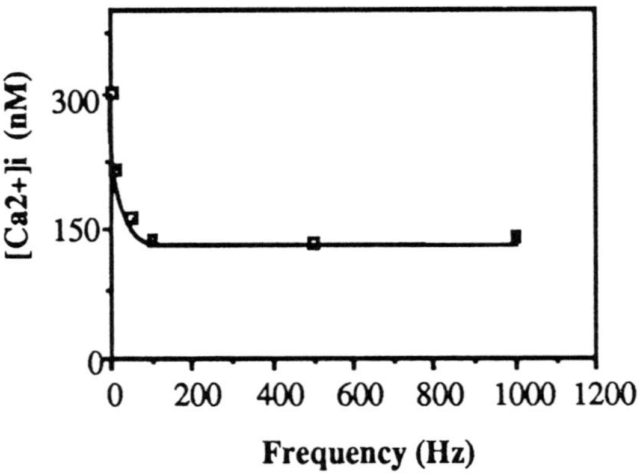

FIG. 3.—Effect of frequency-pulse duration impulsions on intracellular free calcium concentration. 3×10^6 cells/ml in serum-free medium were electropulsed with a pulse intensity of 50 V/cm during 2.5 min for calcium and during 1 min for ATP, ft = constant with f up to 1200 Hz.

On the other hand, the results presented in this study demonstrate the induction of an efflux of calcium following a stimulation of cells by electric fields. It has been shown previously that when bone cells are exposed to capacitively coupled external electric fields, calcium uptake is observed.[16] But the results given in Fig. 3(a) show that calcium efflux increases monotonosly with the frequency of the applied electric field up to a plateau. No clearcut correlation between adhesion activation and Ca^{2+} efflux was obtained. Both processes depend on the frequency of the applied pulses. But a resonant frequency (200 Hz) is clearly present only for adhesion.

Cell adhesion can be stimulated by applying external electric field whose physical parameters are well defined. The trigger of the adhesion's activation is the transmembrane potential difference.

REFERENCES

1. B. Öbrink, *Exp. Cell Res.* 163: 1-21, 1986.
2. K. Hynda, R. Kleinman, J. Klebe, and G. R. Martin, *J. Cell. Biol.* 88: 473-485, 1981.
3. J. Bereiter-Hahn, M. Luck, T. Miebach, H. K. Stelzer, and M. Voth, *J. Cell. Sci.* 96: 171-188, 1990.
4. G. P. Downey, C. K. Chan, S. Trudel, and S. Grinstein, *J. Cell. Biol.* 110: 1975-1982, 1990.
5. K. A. O. Ellem and G. F. Kay, *Biochem. Biophys. Res. Com.* 112: 183-190, 1983.
6. M. P. Rols and J. Teissié, *Biophys. J.* 58: 1089-1098, 1990.
7. T. Y. Tsong, T. T. Tsong, E. Kingsley, and R. Siciliano, *Biophys. J.* 16: 1091-1104, 1976.
8. P. Ferrara, F. Pecceu, E. Marchese, N. Vita, W. Rosham, and J. Lupker, *FEBS Lett.* 226: 47-52, 1987.
9. W. Kueng, E. Silber, and U. Eppenberger, *Anal. Biochem.* 182: 16-19, 1989.
10. J. P. Y. Kao, A. T. Harootunian, and R. Y. Tsien, *J. Biol. Chem.* 264: 8179-8184, 1989.
11. M. P. A. Davies, T. J. Hallam, and J. E. Merritt, *Biochem. J.* 267: 13-16, 1990.
12. J. Teissié, *Biochemistry* 25: 368-373, 1986.
13. T. Y. Tsong, *Bioscience Reports* 3: 487-505, 1983.
14. J. Teissié and T. Y. Tsong, *Biochemistry* 20: 1548-1554, 1981.
15. T. Y. Tsong, *T.I.B.S.* 14: 89-92, 1989.
16. A. Danon and R. Korenstein, *Bioelectrochem. Bioenerg.* 13: 49-54, 1984.

Martin Blank, Ed., *Electricity and Magnetism in Biology and Medicine*

EMF-STIMULATED BONE-CELL PROLIFERATION

R. J. Fitzsimmons* and D. J. Baylink

Pettis VA Hospital, Loma Linda, Calif.

J. T. Ryaby and F. Magee

Orthologic Inc., Phoenix, Ariz.

This study demonstrates the ability of an extremely low-amplitude EMF to increase bone cell proliferation, possibly by increasing IGF-II release.

Electromagnetic fields (EMFs) with amplitudes below the potential of the cell plasma membrane (1.66×10^{-14} J/m) have been shown to produce biological effects. Even more remarkable is that EMFs with amplitudes below the thermal noise level (6.4×10^{-21} J) have also been reported to have biological effects. However, the detection of an energy source by a biological mechanism when that source is below the noise level has been the subject of considerable debate. In these studies we have shown that EMFs produced by a coil arrangement with calculated potentials of 2.4 to 12×10^{-23} J/m can reproducibly increase bone-cell proliferation in vitro. These studies also indicate that an important regulator of bone, the IGF-II system, is affected by EMF exposure and may account for the increase in bone-cell proliferation. Furthermore, these studies show that the mechanism of action involves a frequency-dependent component.

METHODS

Cell Culture. An osteosarcoma cell line TE-85 and normal human bone cells (HBC, derived from rib) were grown in 10% calf-serum DMEM until used for an experiment. For experiments, cells were plated in 1% calf-serum DMEM for 24 h, followed by an additional 24 h serum-free before the start of an experiment. Tissue-culture plate wells were 16 mm in diameter (24/plate).

Cell Proliferation. Cell cultures were exposed to an EMF for 30 min, followed by assessment of cell proliferation 18 h later by ^{3}H-thymidine incorporation into TCA precipitated material.[1]

IGF-II Assay. Culture medium was collected and pooled from 6 wells 18 h after a 30min EMF exposure in BSA-coated polypropylene tubes. IGF-II concentration was determined by a radioreceptor assay (RRA) run in duplicate, which included IGF-I to reduce/eliminate IGF-binding protein artifacts.[2]

EMF Apparatus and Calculations. Tissue-culture plates were exposed to an EMF by a Helmholtz-coil configuration monitored by a flux-gate magnetometer. The EMF parameters were adjusted to a DC component of 200 mg (20 μT) and an AC Component of 400 mG (40 μT) p-p at the indicated frequencies. The resulting parameters were:

Background thermal noise, KE = (3/2) kT (at 37 C) = 6.4×10^{-21} J
E (plasma membrane), E_m = V/m = 10 mV/100 nM = 1×10^{5} V/m or 1.66×10^{-14} J/m
E (induced in culture), E_r = 3.14(frBo), where f = frequency and r = radius,

$$E_r = 15.3\times10^{-5}\ \text{V/m} \quad \text{or} \quad 2.4\times10^{-23}\ \text{J/m at } 15.3\ \text{Hz}$$

$$E_r = 76.6\times10^{-5}\ \text{V/m} \quad \text{or} \quad 1.2\times10^{-22}\ \text{J/m at } 76.6\ \text{Hz}$$

---

*Correspondence to: R. J. Fitzsimmons, Research Service (151), VA Medical Center, Loma Linda CA 92357.

RESULTS

Because bone formation depends primarily on osteoblast number,[3] we initially assessed the effect of EMF exposure on bone cell proliferation. An EMF with the DC component adjusted to 200 mG (normal DC background, 395 mG) had no significant effect on cell proliferation (Fig. 1). However, when a 15.3Hz AC signal was added, a significant increase in cell proliferation occurred in both TE-85 cells and normal human bone cells.

According to the theory of ion resonance, the DC and AC components must be both controlled so as to couple to a specific ion. Using the formula $f_c = (B/2\pi)(q/m)$, at a DC component of 200 mG, an AC component of 15.3 Hz couples to calcium. Furthermore, an AC field of 76.6 Hz would be close to the 5th harmonic of calcium and the 3rd harmonic of magnesium. When we exposed TE-85 cells for 30 min to an EMF at 76.6 Hz, we found an increase in cell proliferation of 28 ± 7%, $p < 0.01$, similar in magnitude to that observed with the 15.3Hz signal.

Because IGF-II is one of the most abundant growth factors in bone, we investigated the effect of EMF exposure on IGF-II concentration in culture medium. TE-85 cells were exposed for 30 min to an EMF with 200 mG DC and 400 mG AC at the indicated frequency. EMF exposure increased IGF-II levels in a frequency-dependent manner, with a peak increase near 76.6 Hz (Fig. 2).

DISCUSSION

A very interesting hypothesis is one that considers an EMF with energy characteristics below the thermal background level as biologically effective. Such a hypothesis does exist; its correctness is the subject of energetic debate. The present contribution presents evidence that EMFs below the noise level *can* increase bone-cell proliferation. Since this observation is reproducible, the hypothesis must be considered to correspond to a real phenomenon. At present there is no comprehensive theory that adequately explains and/or predicts the action of extremely low-amplitude EMFs on biological systems. To develop such a theory, the conditions for an effective EMF must first be determined: the frequency window, energy input/cycle, orientation, waveform, duration of exposure (i.e., total energy input), etc.

In these studies the results indicate that simply changing the DC component is not a significant stimulus for cell proliferation. By adding the information contained in an oscillating field we produce a significant stimulus that causes the bone cell to increase its proliferation rate. This effect may be mediated through calcium, since EMFs tuned to the 1st and 5th harmonic of calcium (according to ion-resonance theory) were effective. A possible biological/biochemical mechanism for the increase in cell proliferation is that EMF exposure increases release of IGF-II, which in turn stimulates cell proliferation. Consistent with this hypothesis is the observation that the increase in IGF-II is adequate to account for the increase in cell proliferation; i.e., addition of IGF-II in the concentration range measured increases cell proliferation.

In conclusion, these data show that an extremely low-amplitude EMF-exposure increases cell proliferation. Furthermore, EMF exposure increases IGF-II concentration in bone cells in a frequency-dependent manner. Because the correlation between energy input at the various frequencies and IGF-II concentration is not significant, the data indicate that frequency is the dominant factor in determining a response.

REFERENCES

1. D. Gospodarowicz, H. Bialecki, and G. Greenberg, "Purification of fibroblast growth factor activity from bovine brain," *J. Biol. Chem.* 253: 3736, 1978.

2. S. Mohan, C. M. Bautista, S. J. Herring, T. A. Linkhart, and D. J. Baylink, "Development of valid methods to measure insulin-like growth factors I and II in bone cell conditioned medium," *Endo.* 126: 2534, 1990.

3. H. Gruber, J. Ivey, E. Thompson, E. Chesnut, and D. J. Baylink, "Osteoblast and osteoclast number and cell activity in postmenopausal osteoporotics," *Calcif. Tiss. Int.* 34: S31, 1982.

4. S. Mohan and D. J. Baylink, "Bone growth factors," *Clin. Orthop. Res.* 263: 30, 1991.

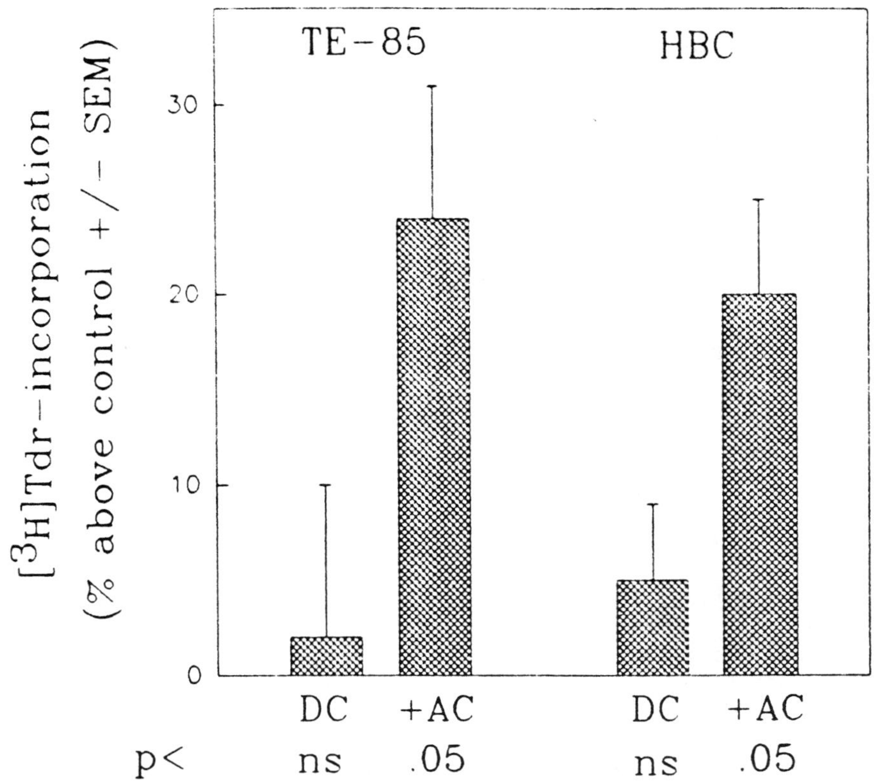

FIG. 1.—Increase in cell proliferation under EMF exposure.

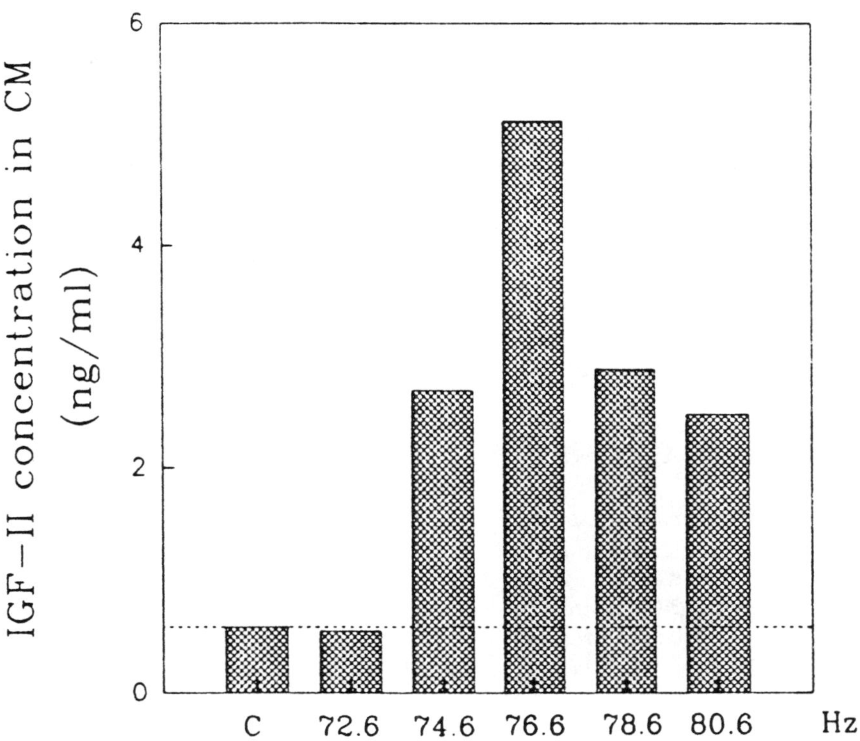

FIG. 2.—Increase in IGF-II concentration in culture medium under EMF exposure.

Martin Blank, Ed., *Electricity and Magnetism in Biology and Medicine*

EFFECT OF ELF ELECTROMAGNETIC EXPOSURE ON PRECIPITATION OF BARIUM OXALATE

R. Berton,* D. Beruto, B. Bianco, A. Chiabrera, and M. Giordani

University of Genoa, Italy

This paper deals with the effect of low-intensity ELF electromagnetic fields on the nucleation and precipitation of barium oxalate from aqueous solutions of barium nitrate. Fewer nuclei that grow faster are produced, as compared with control solutions. Electromagnetic exposure induces changes in the interfacial energy, which in turn increases both the total apparent free energy of activation for the nucleation process and the subsequent crystal growth kinetics.

The effects of low-intensity, low-frequency electromagnetic fields on bioelectrochemical reactions are considered with growing interest in the scientific literature.[1] Efforts concerning physical modeling and theoretical investigations for experiments have been directed toward explaining how an exogenous field can modify both the trajectories of messenger ions and the related biochemical phenomena.[2-7] All available models are still inadequate. Except for very few cases (e.g., Ref. 8), experimental evidence is difficult to obtain, mainly due to the extremely high complexity and variability of living matter.

It is therefore of interest to study the effects of low-intensity ELF fields on the nucleation and precipitation processes of inorganic materials, as the information that can be derived from these relatively simple systems might be useful both in investigating the fundamental mechanisms of more complex biophysical processes and in predicting new materials for a wide range of applications.

In this paper, we further investigate the precipitation of barium oxalate from an aqueous solution of barium nitrate (initially considered in Ref. 9) and from a hydroalcoholic solution (ethanol/water, 1:3 v/v) of barium nitrate. There is evidence of effects of magnetic and electric fields on the nucleation process. More specifically, some of us have shown experimentally that ELF fields affect the precipitation process in important materials, such as calcium carbonate[10] and magnesium-aluminum hydroxides.[11] This paper extends such experimental observations to other materials and allows one to gain a deeper insight into the effects of ELF fields on the precipitation process.

Experimental

Four sets of precipitation experiments were carried out under the same experimental conditions, with and without low-intensity sinusoidal ELF exposure. For the experiments under electromagnetic exposure, the reactor was aligned with the symmetry axis of two Helmholtz coils. The signal-generator frequency was set to 1300 Hz and the magnetic induction was set to an amplitude $B_M = 3$ mT, with the same protocol as described in Ref. 9.

RESULTS AND DISCUSSION

Figure 1 shows the precipitation data obtained without the ELF field vs time, and those obtained by similar runs but in the presence of a low-frequency, low-intensity sinusoidal field. An earlier set of data corresponds to experiments performed with water as a solvent (curves A and B).[9] A second set of data corresponds to new experiments performed under analogous conditions, but with a mixture of ethanol/water (1:3 v/v). Taking into account all possible sources of experimental errors, one can state that (a) the effect of the applied ELF field did not change the total time required to get a complete precipitation; and (b) the ELF electromagnetic field significantly increased the barium concentration left in the aqueous solution at the intermediate times.

---

*Correspondence to: R. Berton (DIBE Biophysics), Via Opera Pia 11A, I-16145 Genoa, Italy.

The results of an XRD analysis of the solid precipitate phases obtained at the beginning and the end of the precipitation process show that at the beginning of the process, the barium oxalate precipitates as $BaC_2O_4 \times nH_2O$, with n less than 2, and that at the end of the process, such a product is transformed into $BaC_2O_4 \times 2H_2O$.[9,12] Experimental evidence also shows that the ELF exposure increases the crystal growth kinetics, since *for the same precipitation time* the average dimensions of the precipitate particles under ELF exposure are 5 times larger than those of the controls.[9] The dominant mechanism should thus explain the decrease in the nucleation rate (i.e., the number of nuclei produced over a given time interval is smaller) and the higher rate of growth following nucleation (i.e., the aforesaid produced nuclei are *more reactive*).

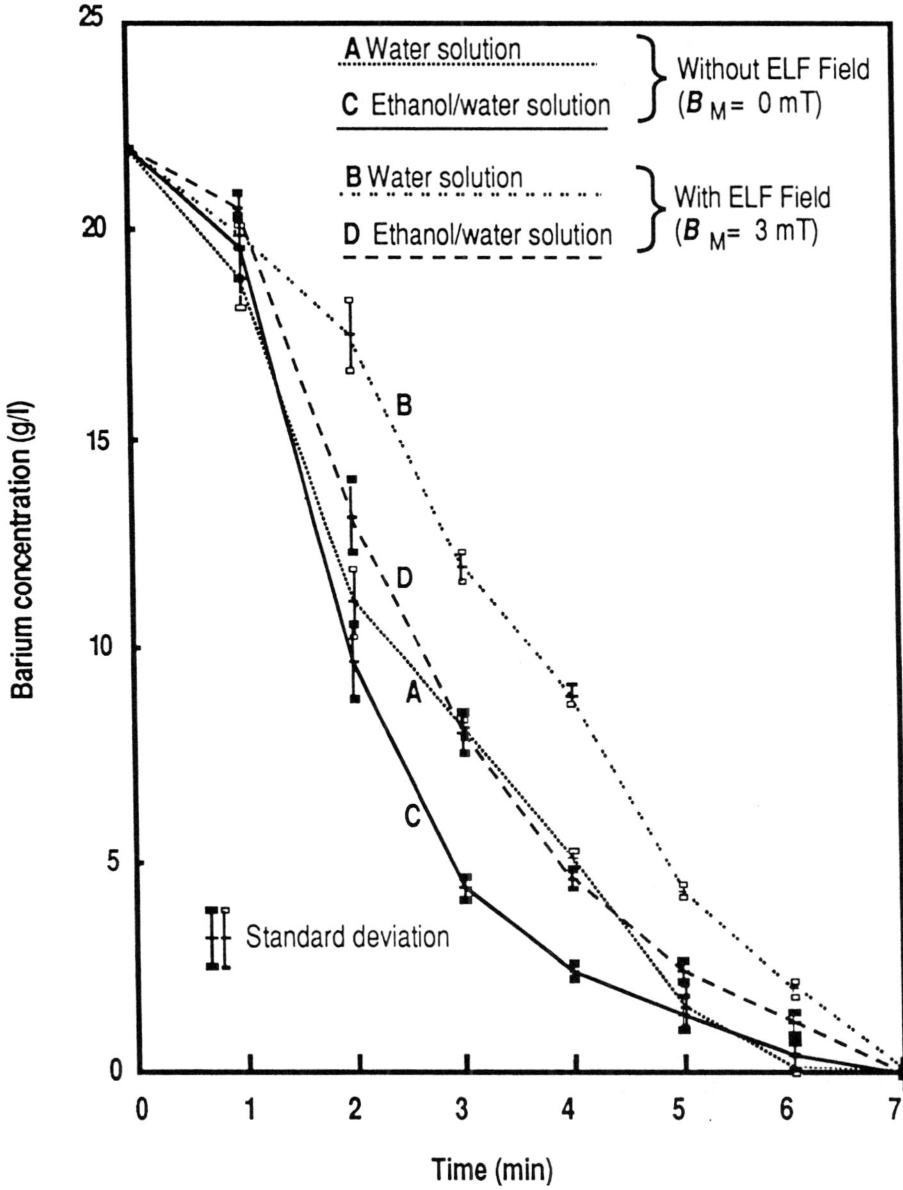

FIG. 1.—Precipitation data in presence and absence of ELF fields.

In Ref. 9, we concluded that the applied ELF field should produce an increase in the interfacial energy of the cluster of critical nuclei, and such a mechanism seems to be indeed the only one that can explain the above phenomena. We tested this preliminary conclusion by performing the same experiments in a solvent less polar than water, i.e., an ethanol aqueous solution (curves C and D). Even if the lower solubility of barium oxalate increases the kinetics of precipitation, the effect of the ELF field is similar to that obtained in aqueous solutions. Therefore, the ELF field acts on the interfacial energy of the cluster of critical nuclei.

These results could be useful in explaining physiological nucleation phenomena. For instance, it has been recognized that bone nucleation can be induced by an electromagnetic field. Our results prove that an electromagnetic field favors the growth of crystals larger than those that can be obtained without application of any exogenous field. From a mechanical point of view, this result does not necessarily constitute a benefit, despite the faster growth kinetics.

CONCLUSIONS

1. Low-intensity ELF sinusoidal fields produce fewer nuclei that grow faster. The reported experimental results are in agreement with others related to the precipitation of different inorganic systems.

2. The experimental data obtained with different solvents can only be explained if one assumes that ELF exposure increases the interfacial energy of the critical cluster of ions. As a result, the total apparent free energy of activation for the nucleation process is increased, and so is the subsequent crystal growth kinetics.

3. A small amount of energy is required to produce changes in the interfacial energy, in agreement with the low intensity of ELF exposure. These results could be useful in explaining other experimental data on different systems (e.g., fracture healing).[13]

REFERENCES

1. C. T. Brighton and S. R. Pollack, Eds., *Electromagnetics in Medicine and Biology*, San Francisco: San Francisco Press, 1991.

2. B. Bianco and A. Chiabrera, "From the Langevin-Lorentz to the Zeeman model of EM effects on ligand-receptor binding," *Bioelectrochemistry and Bioenergetics* 28: 355, 1992.

3. A. Chiabrera, B. Bianco, J. J. Kaufman, and A. A. Pilla, in C. Ramel, Ed., *Bioelectromagnetic Resonance Interactions: Endogenous Fields and Noise*; Interaction mechanisms of low-level electromagnetic fields, Oxford: Oxford University Press, 1992, 164-179.

4. V. V. Lednev, *Bioelectromagnetics* 12: 71, 1991.

5. F. X. Hart, *J. Bioelectricity* 9: 1, 1990.

6. A. Chiabrera, B. Bianco, J. J. Kaufman, and A. A. Pilla, "Quantum dynamics of ions in molecular crevices under electromagnetic exposure," in Ref. 1, 21-26.

7. A. Chiabrera, B. Bianco, J. J. Kaufman, and A. A. Pilla, "Quantum analysis of ion binding kinetics in electromagnetic bioeffects," in Ref. 1, 27-31.

8. C. F. Blackman, L. S. Kinney, D. E. House, and W. T. Joines, *Bioelectromagnetics* 10: 115, 1989.

9. R. Berton, D. Beruto, B. Bianco, A. Chiabrera, and M. Giordani, "Effect of ELF electromagnetic field on precipitation of barium oxalate," *Trans. First Congress EBEA*, Brussels, 1992, 8; *Bioelectrochemistry and Biomagnetics* (in press).

10. D. Beruto, R. Botter, and M. Giordani, "Interactions between ELF electromagnetic fields and microstructural processes: Calcite and argonite formation from aqueous solutions," *ibid*, 43.

11. D. Beruto, R. Botter, M. Giordani, and G. Giannetti, in H. Hausner, G. L. Messing, and S. Hirano, Eds., *Ceramic Powder Processing Science*, Cologne: Deutsche Keramische Gesellschaft, 1989, 77.

12. *Powder Diffraction File Search Manual*, JCPDS, Pub. International Center for Diffraction Data, Park Lane, Swarthmore PA 19081, 1988.

13. The authors thank D. Tomaso Aste for very helpful discussions, and Dr. Bruno Peripimeno and Mrs. Carla Sanna for valuable technical assistance. This work was partially supported by the National Research Council (CNR) of Italy within the framework of the Progetto Finalizzato "Special Materials for Advanced Technologies," by CSM, and by Italian MURST (40%)—Bioelectromagnetics.

Martin Blank, Ed., *Electricity and Magnetism in Biology and Medicine*

THE EFFECTS OF 1.5MHz ELECTRIC FIELD STIMULATION IN PRIMARY BONE CELL CULTURES

O. L. da Silva

University of Sao Paulo, Sao Carlos, Brazil

S. R. Pollack* and K. A. Reinbold

University of Pennsylvania, Philadelphia

The mechanism by which ultrasound increases bone healing rate and alters cellular biochemistry may involve direct mechanical interaction or interaction by piezoelectrically induced fields. We investigated the effect of electric fields of 1 V/m at 1.5 MHz on primary rat calvarial bone cells using Fura-2 fluorescence technology. Increases in cytosolic calcium concentrations of amplitude several times baseline were observed due to field exposure. Bradykinin was used as a positive agonist control. Field exposure also altered bradykinin responses, even though the bradykinin was added after the applied field was turned off.

The stimulation of bone growth using ultrasound has been investigated for clinical use for fresh fractures in the rabbit fibular osteotomy model.[1] Low-intensity (20 and 45mW/cm^2) ultrasound at 1.5 MHz has been shown to decrease adenylate cyclase activity in subconfluent and confluent MC3T3 osteoblast cultures.[2] The cellular mechanism of transduction of ultrasound is not known but may involve direct mechanical interaction at some membrane receptor or channel.[3,4] Brighton et al.[5] showed that cyclical biaxial mechanical strain on rat calvarial bone cells increased metabolites of the inositol phosphate pathway which are known to lead to increased cytosolic Ca^{2+} concentration.[5] The ultrasound transduction mechanism may also involve electric-field interactions caused by the piezoelectric effect in the case of bone or charged fluid flow over cells in the case of tissue culture.[6] In this study we investigate the effect of continuous-wave electric fields at the constant amplitude of 1 V/m p-p and at a frequency of 1.5 MHz on the real-time cytosolic Ca^{2+} concentration in primary cultures of rat calvarial bone cells. The value of 1 V/m was chosen for study based on the work of Silva and Duarte.[6]

MATERIALS AND METHODS

Calvaria were dissected from 1 to 3-day-old Sprague-Dawley rats (Charles River) and serially digested in 0.2% collagenase.[7] Cells of the first two digestions were discarded; those of the last two were pooled in tissue culture medium (37.5% MEM, 36.5% Earle's B.S.S., 10% NCTC-135, 10% newborn calf serum and antibiotics). The cells were then plated onto modified 35mm glass bottom microwell dishes (MatTek Corp., Ashland, Mass.) at a density of 20000 cells/cm^2 and subjected to incubation at 37 C in 21% oxygen, 5% carbon dioxide, and maximal humidity with the tissue culture medium (TCM) being changed on days 1 and 3. The cells were used 2-5 days after plating before the culture had reached confluence. Seventy-eight cells were examined from the isolations of approximately 80 rat puppies from 6 mothers.

Twenty µl of an 8% (w/v) pluronic: DMSO solution was added to 50 µg of Fura-2 AM (Molecular Probes, Inc., Eugene, Ore.) and allowed to sit in the dark for 5 min. This solution was then added to 20 ml Liebovitz (L-15) medium (Sigma Chem. Co., St. Louis, Mo.), vortexed for 2 min, and sonicated for 5 min, which resulted in a 2.5 µM Fura 2-AM:L-15 cell loading solution. The TCM was then aspirated from a cell culture dish and the dish was rinsed three times with L-15. One ml of the cell loading solution was then added to the dish and floated in a covered 37 C water bath for 45 min. The cells were then rinsed three times with a modified Hanks' solution (137 mM NaCl, 5.4 mM KCl, 0.44 mM KH_2PO_4, 0.34 mM Na_2HPO_4, 1.25

---

*Correspondence to: S. R. Pollack (Bioengg), University of Pennsylvania, 120 Hayden Hall, 240 S. 33rd St., Philadelphia PA 19104-6392.

mM $CaCl_2$, 0.8 mM $MgSO_4$, 5.6 mM D-glucose, 20 mM HEPES, and 4.2 mM $NaHCO_3$) that was adjusted to a pH of 7.4. The dishes each contained 2 ml Hanks' prior to experimentation. The cultures were studied at room temperature (23 ± 2 C).

The culture dishes were placed on the microscope stage of an Olympus IMT-2 inverted microscope (Tokyo, Japan). Epifluorescent excitation at wavelengths of 340 and 360 nm yielded 510nm emission images of the field of view of the 20× objective that were captured by a SIT camera with associated image processing hardware and stored on CD-ROM. Pixel-to-pixel ratios of the emitted image intensities were computed and compared to calibration curves to yield values for the cytosolic Ca^{2+} concentration on a single-cell basis.

Dipole electrodes of 0.25mm-diameter platinum wire spaced 1.0 cm apart were immersed in the Hanks' solution over the glass coverslip cemented to the based of the culture dish. A sinusoidal voltage of 0.01 V p-p was applied at a frequency of 1.5 MHz. The cells of interest for the study laid along the line joining the two electrodes where the field was 1 V/m.

After a 2min baseline, the field was turned on for 18 min and then turned off. To verify cellular integrity, 0.5 ml of 0.25μM bradykinin (BK), a known Ca^{2+} agonist, was then added to the 2 ml of solution already in the dish, which yielded a final concentration of 0.05 μM BK and images recorded for another 9.5 min. The fluorescence intensity images from the 340 and 360nm excitations were recorded every 30 s for a total of 29.5 min.

RESULTS

Figure 1 shows the typical responses for single cells from a room-temperature control experiment. The ordinate (in nM Ca^{2+}) is the ratio of the intensity of the 510nm emission image from excitation at 340 nm to the intensity of the 510nm emission image from excitation at 360 nm, which is converted to Ca^{2+} concentration by comparison to calibration curves. The abscissa is in units of time, with a total of 29.5 min shown. Note the constant baseline and the BK response used here as the positive agonist control. Also note the spontaneous oscillation of an amplitude of 218% over baseline. The characteristic shape of the BK response (a rapid rise in concentration followed by a nearly exponential decay of the baseline) is called the "normal" response.

Figure 2 shows the cytosolic Ca^{2+} concentration for two experimental cells subjected to a 1.5MHz field 1 V/m in amplitude. The oscillatory behavior of the response is evident and the width of the Ca^{2+} spiking is of the order of 1 min. The two cells in Fig. 3 are under the same conditions as those in Fig. 2, but show a drastically different response to field exposure. In these cells, increases in Ca^{2+} concentration over baseline are longer in duration (over 5 min) and there is no oscillatory behavior as compared to Fig. 2. Also note the variable latency between field application and cellular response. Alterations of the BK responses as compared to controls are also observed and are labelled as "abnormal."

Table 1 summarizes the results achieved to date for 73 cells studied. The total number of cells studied is lower than the amount stated previously because some cells (6% of the total, or 5 of 78) were unresponsive to BK stimulation and could not be counted in the total analysis.

TABLE 1.—Results of 1.5MHz electric-field study: E = 1 V/m, 23 C (room temperature).

Study type	Number of cells	Bradykinin Nrm/Abnrm	Number of cells with spikes	Number of spikes	Spike amplitude (mean of % BL)
Control	32	25/7	2	3	121
1.5 MHz	73	45/28	27	47	228

DISCUSSION

There is a pronounced real-time response to 1.5MHz, 1V/m electric fields applied to these primary bone cell cultures. This response is characterized by Ca^{2+} spiking (oscillations) in 37% of the cells studied during the 18 min of stimulation. The mean Ca^{2+} spike amplitude over baseline is 228% and is often comparable to the normal BK response amplitude. The mean amplitude of these spikes is also larger than that of spontaneous oscillations, by a factor of two. The spikes occur with variable latency ranging from approximately 30 s following the application of the field to more than 15 min. Two different classes of spikes have been observed with respect to duration. Some have half-widths of approximately 1 min (Fig. 2); others have halfwidths of 5 or more min (Fig. 3). In addition, 38% of the cells exhibit altered or abnormal BK response after the field is turned off. These findings strongly suggest that the mechanism(s) underlying the biological effects of ultrasound may involve not only direct mechanical coupling to membrane receptor systems but also electric fields that inevitably accompany ultrasound in vivo. Future work must attempt to identify specific site(s) of interaction and clarify the question of "how" ultrasound is transduced to relevant second messengers.[8]

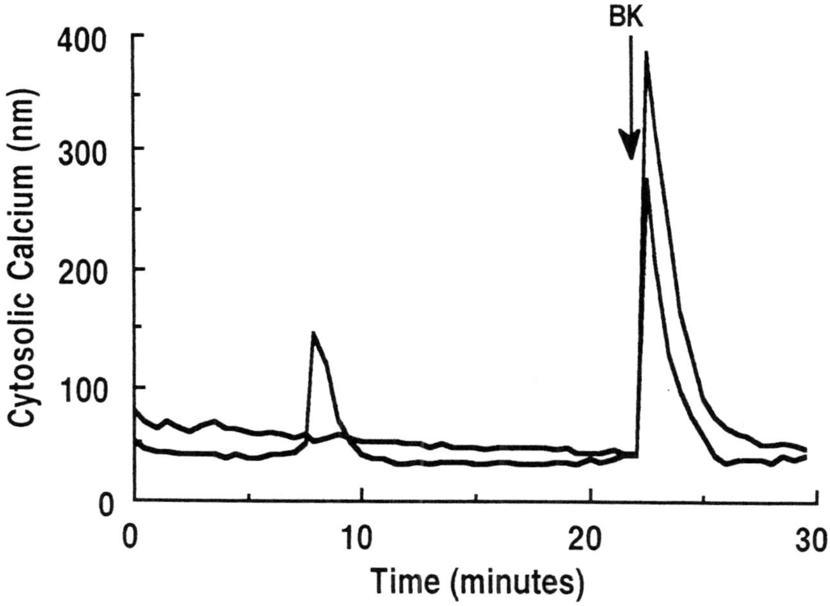

FIG. 1.—Calcium concentration profiles for two different control cells, 23 C (room temperature).

REFERENCES

1. A. A. Pilla, M. A. Mont, P. R. Nasser, S. A. Khan, M. Figueiredo, J. J. Kaufman, and R. S. Siffert, "Noninvasive low-intensity pulsed ultrasound accelerates bone healing in the rabbit," *J. Orthopaedic Trauma* 4: 246: 1990.
2. J. T. Ryaby, J. Mathew, A. A. Pilla, and P. Duarte-Alves, "Low-intensity pulsed ultrasound modulates adenylate cyclase activity and transforming growth factor beta synthesis," in C. T. Brighton and S. R. Pollack, Eds., *Electromagnetics in Biology and Medicine*, San Francisco: San Francisco Press, 1991, 95-100.
3. F. Sacks, "Mechanical transduction in biological systems," *CRC Critical Reviews in Biomed. Eng.* 16: 141, 1988.
4. D. Somjen, I. Binderman, E. Berger, and A. Harrell, "Bone remodelling induced by physical stress is prostaglandin E_2 mediated," *Biochim. Biophys. Acta* 627: 91, 1980.
5. C. T. Brighton, B. J. Sennett, J. C. Farmer, J. P. Iannotti, C. A. Hansen, J. L. Williams, and J. Williamson, "The inositol phosphate pathway as a mediator in the proliferative response of rat calvarial bone cells to cyclical biaxial mechanical strain," *J. Orthopaedic Research* 10: 385-393, 1992.
6. O. L. da Silva and L. R. Duarte, "A proposed mechanism for ultrasonic bone growth stimulation," *Trans. Bioelectrical Repair and Growth Soc.* 9: 17, 1989.
7. J. S. Brand and T. J. Hefley, "Collagenase and the isolation of cells from bone," in T. G. Pretlow II and T. P. Pretlow, Eds., *Cell Separation: Methods and Applications*, New York: Academic Press, 1984, 265-283, 1984.

8. This work was supported by National Science Foundation grant BCS89-21492.

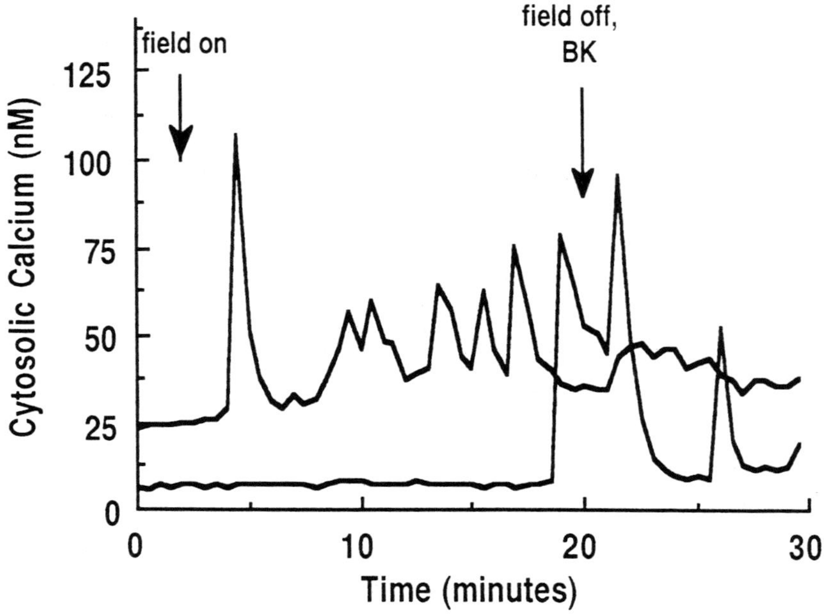

FIG. 2.—Calcium concentration profiles showing distinctive calcium spiking for two different cells during electric-field exposure. E = 1 V/m, 1.5 MHz, 23 C (room temperature).

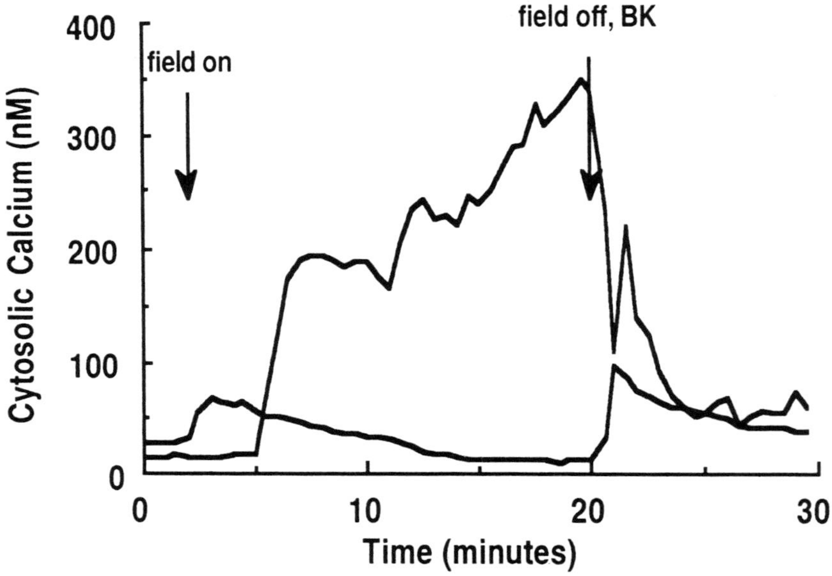

FIG. 3.—Calcium concentration profiles displaying long duration or sustained increases in calcium concentration for two different cells during electric-field exposure. E = 1 V/m, 1.5 MHz, 23 C (room temperature).

Martin Blank, Ed., *Electricity and Magnetism in Biology and Medicine*

ELECTRICAL IMPEDANCE PROTERTIES OF ERYTHROCYTES

J.-Z. Bao and C. C. Davis*

University of Maryland, College Park, Md.

R. E. Schmukler

CDRH, U.S. Food and Drug Administration, Rockville, Md.

We describe measurements of the electrical impedance of human erythrocytes in the frequency range from 1 Hz to 10 MHz, and for temperatures from 4 to 40 C. The impedance spectra of erythrocytes display constant-phase-angle (CPA) characteristics. Based on the geometry of cells embedded in a filter, a circuit model is proposed for the cell-filter-electrolyte system. A CPA element has been used to describe the ac response of the interface between cell surface and electrolyte solution. The CPA and other elements of the circuit model are determined by a complex nonlinear least-squares (CNLS) fit. The temperature dependence of the membrane capacitance shows a surprising increase at about 37 C, which suggests a phase transition.

Impedance measurements of biological cells can provide fundamental electric and dielectric information about their membranes and cytoplasm, which will be helpful in understanding the structure and function of these cells, and their interaction with the environment. The measurement method used in this study is the cell-embedding technique, in which the cells are embedded in the pores of a polycarbonate filter that greatly reduces the shunt current pathway around the cells in the pores. This technique also provides physiological conditions for the cells under measurement and forces the cells embedded in the filter into a well-defined geometric state.

We report here on impedance measurements of human erythrocytes in the frequency range from 1 Hz to 10 MHz and for temperatures from 4 to 40 C. We also describe an accurate calibration procedure for a four-electrode measurement system. A circuit model based on the geometric form of the cells is proposed and a constant-phase-angle (CPA) element, i.e.,

$$Z = A/(j\omega)^{\alpha} \tag{1}$$

where A is a constant, $j = \sqrt{-1}$, ω is the angular frequency, and $0 < \alpha < 1$ is used to describe the interface between the cell surface and the electrolyte environment. CPA and other elements are determined by a complex nonlinear least-squares (CNLS) fit, which simultaneously fits the real and imaginary parts of the experimental data to the circuit model. Many types of system exhibit a region of CPA behavior, and some recent explanations have been based on a presumed self-similarity in the system or in its relaxation dynamics. It is fairly well known that an interface CPA element can reflect the fractal nature of the interface, with the exponent α approaching unity as the interface becomes increasingly smooth.[1-4] Impedance measurements as a function of temperature strongly suggest that a phase transition occurs in the cell membrane at about 37 C: the marked increase of the membrane capacitance over a small temperature range may be caused by a structural change of the lipid molecules in the membrane.[1,2] Given the importance of the cell membrane in the functioning of the cell, it may be quite significant that a phase transition occurs so close to physiological temperature.

EXPERIMENT AND SYSTEM CALIBRATION

Impedance measurements are performed in an electrochemical conductance chamber (Fig. 1) by a four-electrode technique, which avoids the distortion of the impedance data by the electrode-electrolyte interface

*Correspondence to: C. C. Davis (EE), University of Maryland, College Park MD 20742.

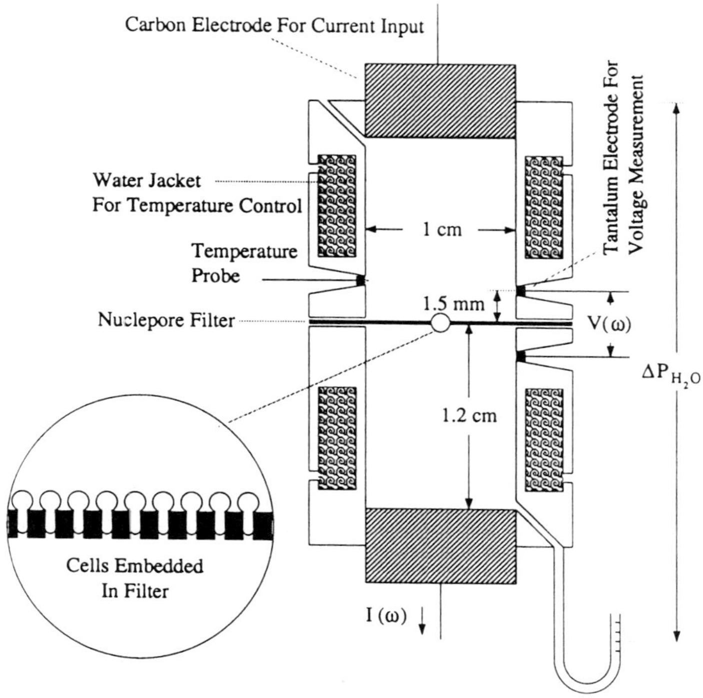

FIG. 1.—Electrical impedance measurement chamber.

polarization at the two current input electrodes (carbon). A polycarbonate Nuclepore filter is clamped between the upper and lower halves of the chamber. The core instrument in the measurement system is a Solartron Model 1260 Impedance/Gain-Phase Analyzer, which is controlled by a PC. The amplitude of the voltage source generator output of the Impedance Analyzer is chosen to be 10 mV, which gives a voltage drop across the filter of less than 3mV (a relative small perturbation on the static membrane potential of 13 mV). Care is taken to insure that the filter is fully wetted and that there are no air bubbles in the chamber. The erythrocytes are layered onto the filter uniformly with a syringe. After about 10 min, a hydrostatic pressure (about 30 cm of water) is applied. Under this hydrostatic pressure the cells already in the pores will be embedded deeper, and the empty pores will be filled with nearby cells, which must move toward the empty pores. Eventually all the pores will be filled if enough cells are present and sufficient time has elapsed.[1,2,5-7]

The impedance measurement is subject to errors. In the low-frequency range (<10 Hz) small errors result mainly from the ac coupling of voltage and current amplifiers: these effects are small and do not affect the frequency profile of the spectrum very much. But in the high-frequency range (>1 MHz) the spectrum shows a sharp roll-off, which is in a very important frequency range in a study of the spectra of erythrocytes, because there may be a dispersion in this range. Consequently, effective control and correction of systematic errors becomes necessary. There are at least two sources that contribute to high-frequency errors: (1) stray fields and parasitic elements of the measurement chamber and connecting cables, and (2) instrumentation artifacts related to the impedance analyzer, especially when it is operated at low generator levels in a four-terminal configuration. Although the sources of error at high frequencies can be classed into the above two categories, these errors can hardly be separated. In order to obtain reliable measurements over a wide frequency band, it is necessary to include an error-correction procedure.

Based on a linear assumption of measurement system and some simple circuit network derivations,[1,5,6] we have

$$Z_m(\omega) = \frac{Z_t(\omega)A_1(\omega) + A_2(\omega)}{Z_t(\omega) + A_3(\omega)} \tag{2}$$

where Z_t is the true value of the impedance under measurement; Z_m is the measured value given by the impedance analyzer; and A_1, A_2, and A_3 are complex constants for a given measurement system and frequency, and can be determined from the experiment. In order to do this, three reference impedance measurements have to be made, and for each frequency point we have to solve a 3×3 complex matrix equation from Eq. (2).

Comparisons between calibrated and measured results for erythrocytes embedded in the filter are given in Fig. 2. These results show that low-frequency errors and high-frequency roll-off errors are well corrected. Most interesting, in Fig. 2 (a) the apparent dispersion at about 8 MHz of erythrocytes embedded in a filter disappears after calibration, which confirms that there is no other dispersion in the upper frequency range of the measurement.

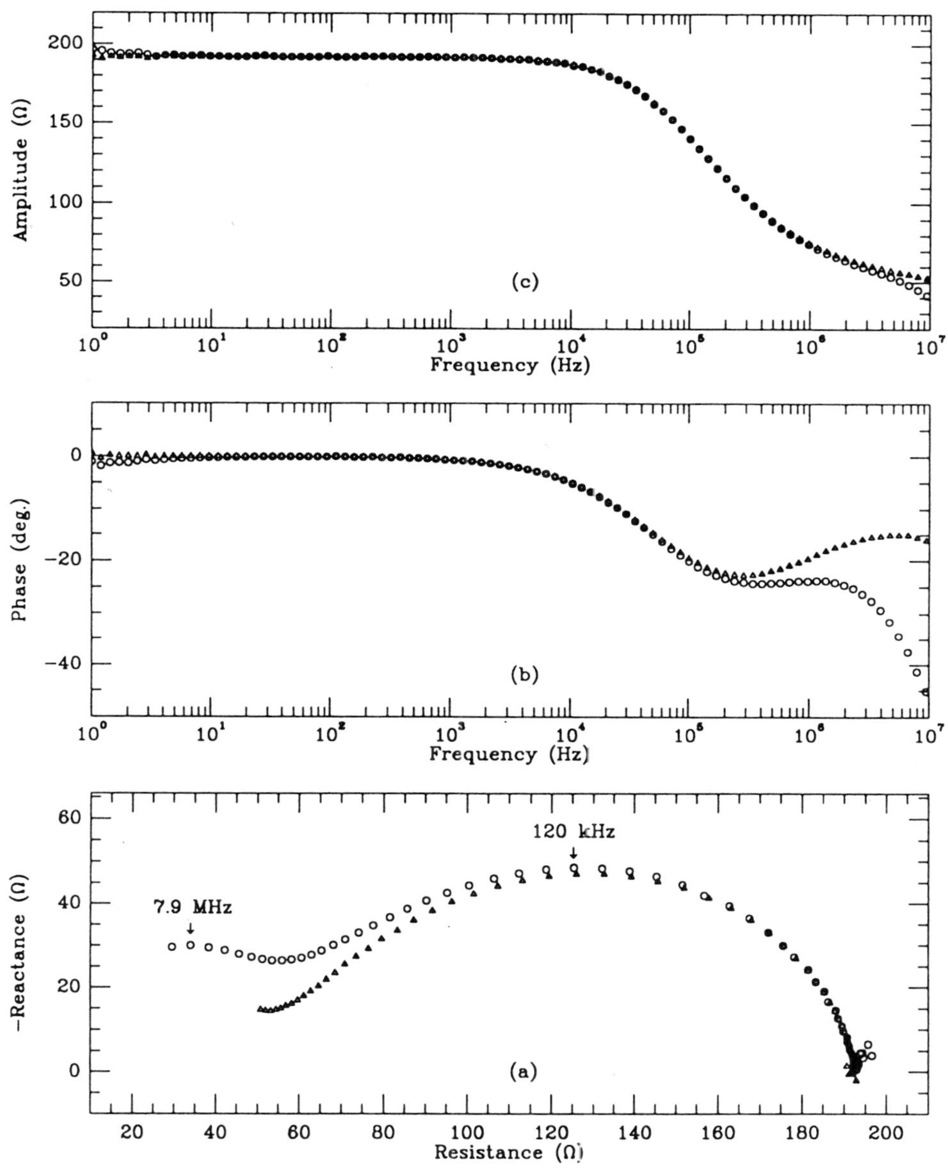

FIG. 2.—Comparison between measured (o) and calibrated (Δ) results for erythrocytes embedded in a fiber.

CIRCUIT MODEL AND COMPLEX NONLINEAR LEAST SQUARES (CNLS) FIT

In general the frequency response of a system does not uniquely determine the equivalent circuit used to model that response. However, it seems reasonable to hope that for any specific real extended physical system the actual response is best represented by a unique circuit model. The erythrocytes embedded in the pores of a filter are also depicted schematically in Fig. 1, which is based on observations made by transmission electron microscopy. Because all the cells embedded in the filter are approximately equivalent and in parallel, we can use a simple circuit based on the geometric form of the embedded cells to model the whole cell-filter-electrolyte system. We find that erythrocytes impedance spectra display CPA character from a complex impedance plane plot, where the unsymmetrical impedance plane arc exhibits a dispersion frequency at low frequencies (kHz range) and a CPA response at high frequencies (MHz range). It is well known that an impedance with a power-law frequency dependence cannot be described in terms of a finite, resolvable set of frequency-independent RC elements, no matter how they are chosen. Thus, we propose the overall circuit model for the cell-filter-electrolyte system shown in Fig. 3, where C_m and R_i are independent of frequency and Z_{int} is a CPA element.

After the equivalent circuit model is proposed, we need to fit the model to the experimental data. The estimation of parameter values of an equivalent circuit from measured impedance data is important in obtaining quantitatively accurate representations of the frequency response of the system under measurement. In a CNLS fit the objective function S is defined as

$$S(\vec{P}) = \sum_{n=1}^{N} \{W_R(\omega_n)[Z^R_{mod}(\omega_n, \vec{P}) - Z^R_{exp}(\omega_n)]^2 + W_I(\omega_n)[Z^I_{mod}(\omega_n, \vec{P}) - Z^I_{exp}(\omega_n)]^2\} \quad (3)$$

where N is the number of data points; ω_n are the angular frequency points; subscript and superscript R and I denote real and imaginary parts, respectively; W is the weighting factor; Z_{mod} is the analytical complex impedance expression of the circuit model; $\vec{P} = \{P_1, \cdots, P_M\}$ are M real parameters of the model; and Z_{exp} are the experimental data. We find the best-fit parameters $\vec{P}_{min}$ by minimizing the object function $S(\vec{P})$ using the Levenberg-Marquardt algorithm.

RESULTS AND DISCUSSION

A typical CNLS fit of the spectra of erythrocytes embedded in a filter at 26 C is shown in Fig. 4, which is a 3-D perspective plot. The circuit parameters found from this CNLS fit are $C_m = 19.18 \pm 1.81$ nf, $R_i = 25.15 \pm 3.81\Omega$, $\alpha = 0.552 \pm 0.033$, $A = 32.37 \times 10^4 \pm 1.62 \times 10^4$, $R_l = 162.47 \pm 1.47\Omega$, $R_f = 4.89 \pm 1.77\Omega$, $R_s = 24.93 \pm 3.16\,\Omega$, and $C_f = 238.81 \pm 28.18$ nf. We convert the above parameters to cell constants, shown in Table 1, by using the geometry of the embedded cells and the filter characteristics (pore density $4.22 \times 10^5/cm^2$, pore diameter 2.05 μm, filter thickness 13.15 μm).

The power-law frequency dispersion is described by a CPA element, which may provide insight into understanding cell communication with the outside medium. An assumption of molecular roughness of the cell surface provides an explanation for this CPA behavior. The temperature dependence of the spectra of cells embedded in a filter is shown in Fig. 5. CNLS fits of these data show that membrane capacitance increases markedly at 37 C in Fig. 6, which suggests a phase transition in the cell membranes near the physiological temperature. Based on this evidence for an increase in the membrane capacitance, there is a possibility that a "rippled" phase might exist in the erythrocyte membrane as it does in pure lipid bilayers,[8] or there may be a sharp increase of membrane dielectric constant at the phase transition caused possibly by an increased permeability to water molecules.[9]

TABLE 1.—Erythrocyte cell constants.

Specific membrane capacitance ($\mu F/cm^2$)	0.901 ± 0.036
Cytoplasm conductivity (S/m)	0.413 ± 0.031
Interface CPA element exponent	0.538 ± 0.026
Interface CPA element constant ($\times 10^{11}$)	1.06 ± 0.12
Leak resistance around a cell (MΩ)	58.01 ± 0.19
Resistance due to the unfilled part of a pore (MΩ)	1.59 ± 0.23

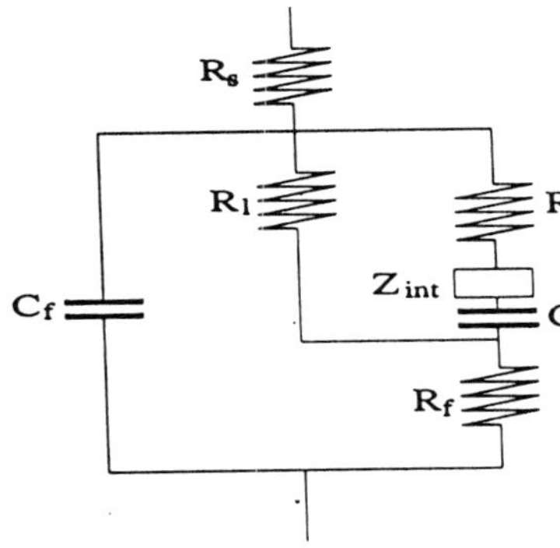

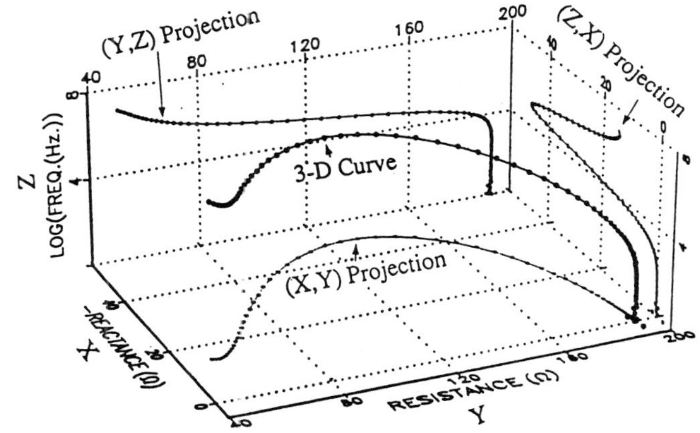

FIG. 3.—Circuit model for the whole cell-filter-electrolyte system. R_s is the bulk resistance of the electrolyte solution, R_l is the leak resistance around the cells embedded in the filter, R_i is the interior resistance of the cells, C_m is the membrane capacitance of the cells, Z_{int} is the interface impedance (CPA element) between the cell surface and the electrolyte solution, C_f is the filter capacitance, and R_f is the resistance due to the unfilled part of the pores of the filter.

FIG. 4.—A 3-D perspective plot and its projections for typical calibrated experimental data (•) and CNLS fit (solid lines) for erythrocytes embedded in a filter at T=26 C.

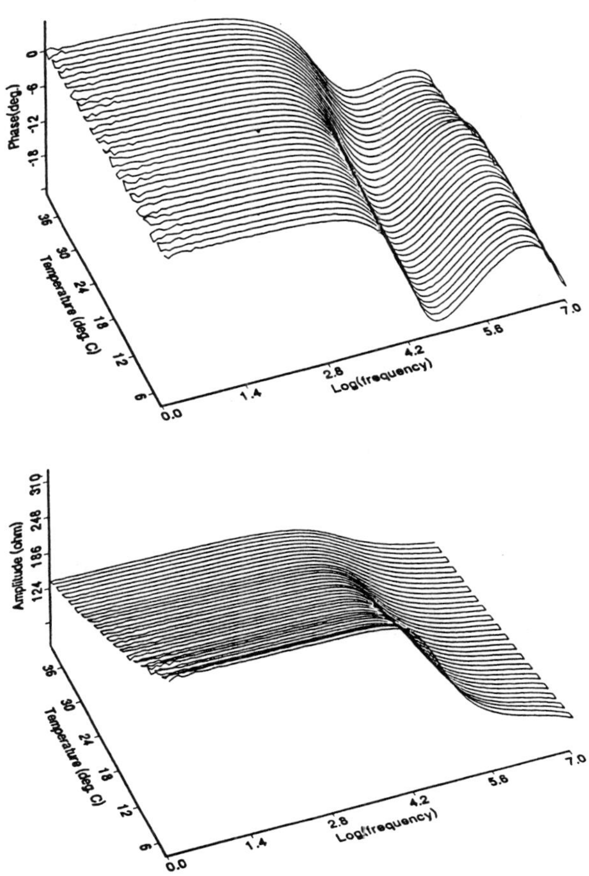

FIG. 5.—Temperature dependence of the spectra of embedded cells from 4 to 40 C.

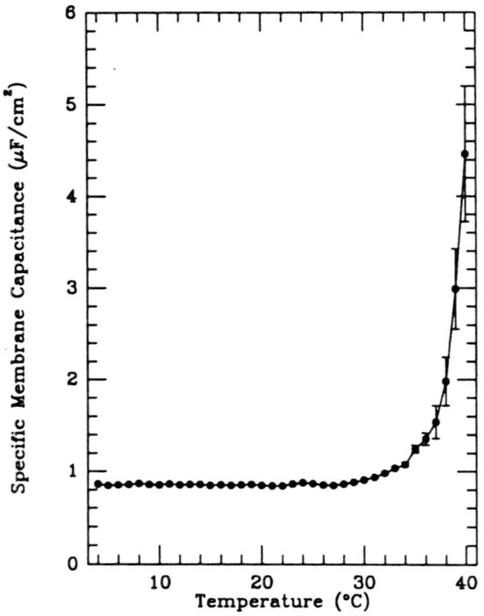

FIG. 6.—Temperature dependence of specific membrane capacitance.

REFERENCES

1. J. Z. Bao, *Impedance Spectroscopy of Human Erythrocytes: Constant Phase Angle Characteristics and a Membrane Phase Transition*, Ph.D. thesis, University of Maryland, 1991.

2. J. Z. Bao, C. C. Davis, and R. E. Schmukler, "Frequency domain impedance measurements of erythrocytes," *Biophysical J.* 61: 1427-1434, 1992.

3. L. A. Dissado and R. W. Hill, "Constant-phase-angle and power-law regimes in the frequency response of a general determinate fractal circuit," *Phys. Rev.* B37: 3434-3439, 1988.

4. S. H. Liu, "Fractal model for the ac response of a rough interface," *Phys. Rev. Letters* 55: 529-532, 1985.

5. J. Z. Bao, C. C. Davis, and R. E. Schmukler, "Frequency-domain impedance measurements of erythrocytes with error correction and complex nonlinear data fitting," paper presented at 13th Ann. Mtg. Bioelectromagnetic Society, 1991.

6. J. Z. Bao, C. C. Davis, and R. E. Schmukler, "Impedance spectroscopy of human erythrocytes: System calibration and nonlinear modeling" *Trans. IEEE* BME (in press).

7. R. Schmukler, "Measurements of the electrical impedance of living cells in frequency domain," in *Charge and Field Effects in Biosystems—2*, New York: Plenum Press, 1989, 357-372.

8. D. C. Wack and W. W. Webb, "Measurement of modulated lammellar $P_{\beta'}$ phase of interacting lipid membranes," *Phys. Rev. Letters* 61: 1210-1213, 1988.

9. This work was supported by FDA/CDRH through Contract 223-84-6038. We are grateful to Mays Swicord for his help.

Martin Blank, Ed., *Electricity and Magnetism in Biology and Medicine*

IRREVERSIBLE TRANSITION OF THE DIELECTRIC β-DISPERSION OBSERVED DURING A 45′ HYPERTHERMIC TREATMENT ON V79 FIBROBLASTS

A. Bonincontro

University of Rome "La Sapienza," Rome

C. Mari* and G. F. Mariutti

National Institute of Health, Rome

Dielectric spectroscopy can be used to investigate biological phenomena, which are characterized by structural and/or functional alterations at molecular and cellular levels. We have investigated the effects of hyperthermic treatments on a cell suspension of Chinese hamster V79 fibroblasts, by performing conductivity measurements in the frequency range 10 kHz-100 MHz. Cells were heated directly in a cylindrical waveguide partially filled with the cell suspension and connected to the measuring instruments. Under these experimental conditions, we qualitatively observed a dual phase effect on cell suspension conductivity. In the first 20 min, at low frequencies, cell suspension conductivity progressively decreases. After this time interval, the phenomenon reverses, the conductivity gradually increases, and (in the case of 44 C treatment), practically disappears.

Hyperthermia alone or associated with ionizing radiation and/or drugs is one of the clinical modalities for the treatment of malignant tumors. To obtain the maximum therapeutic benefit from such combined treatments, it is important to know the specific effect of each physical or chemical factor at the cellular, subcellular, and molecular levels.

Several pieces of experimental evidence lead one to consider the hyperthermic damage to cellular membranes in general, and to plasmatic membrane in particular, as the most important effect for heat cell killing. However, it is not yet clear which is the specific role of the most important membrane components to cause the critical alterations related to heat-induced cell death. As a general indication, temperature elevation increases membrane fluidity. Changes in cell-membrane fluidity result in a different degree of order in the phospholipid acyl chains. This effect in turn can cause conformational transitions of membrane proteins and membrane-bound proteins, thus altering the complex regulatory function of the cell membrane system such as selective permeability, active transport processes, and enzymatic activities. Hyperthermia has been shown to alter transport capabilities and permeability of the cell membrane.

In our previous work we have studied the effects of hyperthermia on ion-transport capability of the cell membrane by means of a dielectric spectroscopy technique.[1] In this work we present the behavior of the cell suspension when conductivity measurements are carried out during hyperthermic treatment. The first 20 min of hyperthermic treatment are characterized by a substantial and progressive reduction of cell suspension conductivity at low frequencies. After this time the phenomenon reverses and the conductivity of the suspension gradually increases, reaching the value measured before the treatment, and after a time interval depending on temperature the β dispersion practically disappears.

CELL CULTURE, HYPERTHERMIC TREATMENT, AND CONDUCTIVITY MEASUREMENTS

Chinese hamster lung fibroblasts V79 were grown as a monolayer in Eagle's minimum essential medium (E MEM) supplemented with 15% foetal bovine serum, 50 μg/ml gentamycin 0.48 g/l sodium bicarbonate, and 20 mM Hepes buffer. In all the experiments, the volume of each sample of cell suspension was 1.5 ml. To obtain the optimum fractional volume (i.e., volume of cell pellet over volume of suspension), 2×10^8 cells

*Correspondence to: C. Mari (Physics Lab.), National Institute of Health, Viale Regina Elena 299, I-00161, Rome, Italy.

are required. All heat treatments were carried out on cells reaching the early stage of confluency.

Immediately before hyperthermic treatment, cells are harvested by treatment of the monolayer with a trypsyn (0.01%) and EDTA (0.02%) solution obtained from Flow Labs. The sample is then centrifuged and washed twice in 15ml Dulbecco's phosphate buffered saline solution (PBS) kept to 3-4 C and the cell pellet is resuspended in PBS to obtain a 1.5ml total volume of cell suspension. The same experimental procedure is followed to prepare the control samples.

Hyperthermic treatments were performed directly in the dielectric cell connected to the measuring instruments. Under these experimental conditions, the phenomena that take place on cells are in our opinion more representative of the effects induced by hyperthermia on cells of solid tumors.

Conductivity measurements were carried out between 10 kHz and 100 MHz by means of two impedance analyzers (Hewlett Packard model 4192A and model 4193A). Both these instruments were driven and controlled by a computer. Each measurement required a few minutes. The measuring cell is described elsewhere[2] and the calibration followed that of Bottomley.[3] The error in conductivity measurements was estimated to be within 1% in the entire range of frequencies. The dielectric cell has been thermostated by means of a water jacket connected to a thermostate Haake mod. E8, so that all the measurements were carried out at the selected temperature within 0.1 C.

DATA ANALYSIS

The dielectric behavior of a cell suspension can be described by the complex dielectric constant $\varepsilon^* = \varepsilon' - j\varepsilon''$ or alternately by the complex conductivity $\sigma^* = \sigma' + j\sigma''$, where ω is the angular frequency, and ε_0 is the permittivity of free space, $j = \sqrt{-1}$, and the relations between the imaginary and real parts of the conductivity and dielectric constants are $\sigma' = \sigma_0 + \omega\varepsilon_0\varepsilon''$ and $\sigma'' = \omega\varepsilon_0\varepsilon'$. The experimental values of conductivity σ vs frequency f showed the presence of a dielectric relaxation (β-dispersion), so that the data fit the Cole-Cole equation.[4]

The free parameters in the fit procedure are the static conductivity σ_0; the conductivity increment $(\sigma_\infty - \sigma_0)$, where σ_∞ is the conductivity at frequencies well above those of the dispersion; the relaxation frequency f_c; and the Cole-Cole α parameter. The conductivity increment is substantially a measurement of the membrane conductivity (essentially ionic transport). The cell can be represented as a sphere covered by a shell of low conductivity (the cell membrane) randomly distributed in a continuous medium, under the hypothesis that no dielectric relaxation of single component of the system is evident in the experimental frequency interval (Hasted-Pethig).[5]

A complete mathematical treatment based on the original theory of Pauly and Schwan[6] was reviewed and applied to cell suspensions by Asami et al.[7-9]

RESULTS AND DISCUSSION

We have investigated the effects on cell membrane conductivity of hyperthermia during the treatment by partially filling the cylindrical waveguide that acts both as measuring cell and a sample holder. The course of the measurements therefore represents a real-time picture of the effects induced by the treatment. No controls are needed during the treatment because the measurements carried out on the same sample are considered in terms of changes due to the time of treatment.

A cell suspension maintained at 20 C does not change its conductivity after 4 h. The same effect has been observed after 50 min at 37 C (Fig. 1). When the temperature is increased to values normally used during hyperthermic treatments, we observed a phenomenon that we have never found before in our experiments: a substantial decrease of cell suspension conductivity at low frequency that shows a peak 10 min after the beginning of the treatment, after which the phenomenon reverses; the suspension gradually increases in conductivity, and after 40-60 min the dispersion practically disappears. Figure 2 illustrates this effect measured on a sample heated to 44 C.

The results are a convincing indication that the effects of hyperthermia on cells are rather complex. In considering the effects on the cell membrane, one can speculate that more determinant factors are involved in membrane damage. Figure 3 clearly illustrates the two phases of the observed phenomenon. The first phase, which is characterized by a decrease of cell-suspension conductivity, can be interpreted as a fast inactivation of the active transport processes across the cellular membrane. This effect could be produced by structural alterations of proteins and enzymes involved in the ion pumps or by reduced ATP availability. The subsequent increase of conductivity might in our opinion be interpreted as a structural damage of the membrane that leads to the disappearance of the electrochemical gradient and to the formation of holes that destroy the regulatory activity of the membrane itself. We think that both the functional and structural effects are acting and interacting to produce the damage on the membrane, but the relative intensity and speed of these two components depend strongly on the temperature value. Further experiments are in progress to elucidate these aspects of our results.[10]

REFERENCES

1. A. Bonincontro and G. F. Mariutti, *Phys. Med. Biol.* 33: 557-568, 1988.
2. C. Ballario, A. Bonincontro, and C. Cametti, *Z. Naturforsch.* 39C: 160-166, 1984.
3. P. A. Bottomley, *J. Phys.* E11: 413-414, 1988.
4. R. Pethig and D. B. Kell, *Phys. Med. Biol.* 32B: 933-970, 1987.
5. R. Pethig, *Dielectric and Electronic Properties of Biological Materials,* New York: Wiley, 1979.
6. H. Von Pauly and H. P. Schwan, *Z. Naturforsch.* 14: 125-131, 1959.
7. K. Asami, T. Hanai, and N. Koizumi, *J. Membr. Biol.* 28: 169-180, 1976.
8. K. Asami, T. Hanai, and N. Koizumi, *Biophys. J.* 31: 215-228, 1980.
9. A. Irimajiri, K. Asami, T. Ichinowatari, and Y. Kinoshita, *B.B.A.* 896: 203-213, 1987.

10. The authors thank Dr. Paola Di Ciaccio and Mr. Marco Sabatini for their technical assistance in the preparation of the manuscript.

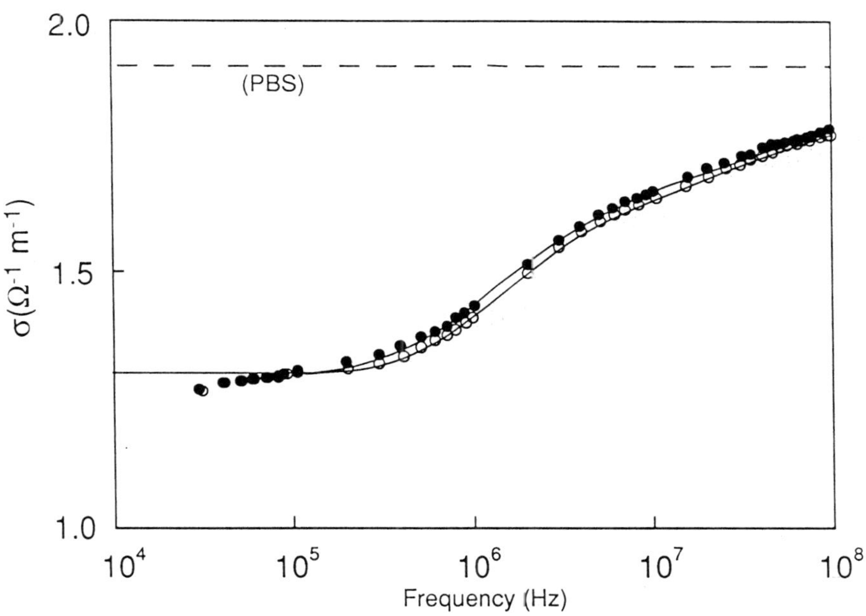

FIG. 1.—Conductivity of cell suspension of V79 fibroblasts maintained for 50 min at 37 C: (●) $t = 0$ min at 37 C, (O) $t = 50$ min at 37 C.

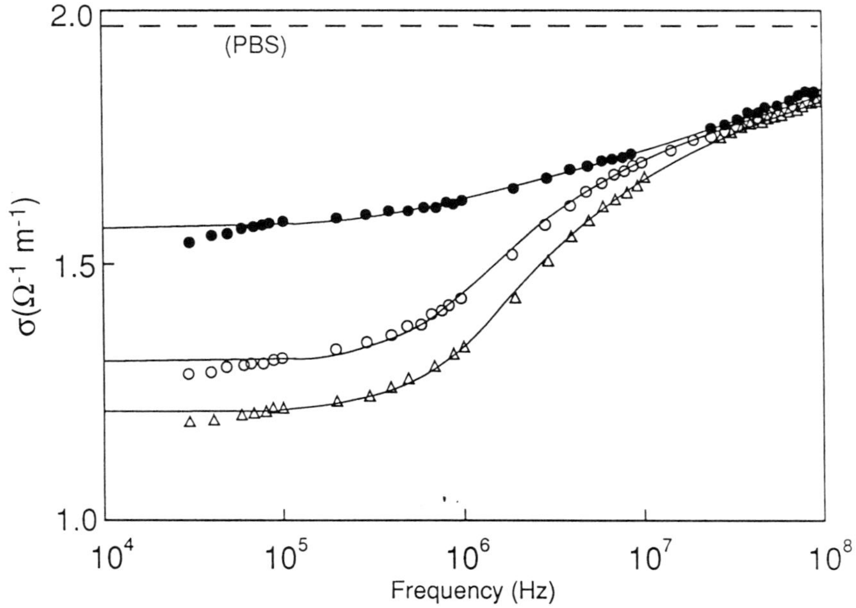

FIG. 2.—Conductivity of cell suspension of V79 fibroblasts during single hyperthermic treatment at 44 ± 0.1 C. Curves represent conductivity of suspension after: (○) t = 0 min at 44 C, (Δ) t = 15 min at 44 C, and (●) t = 45 min at 44 C.

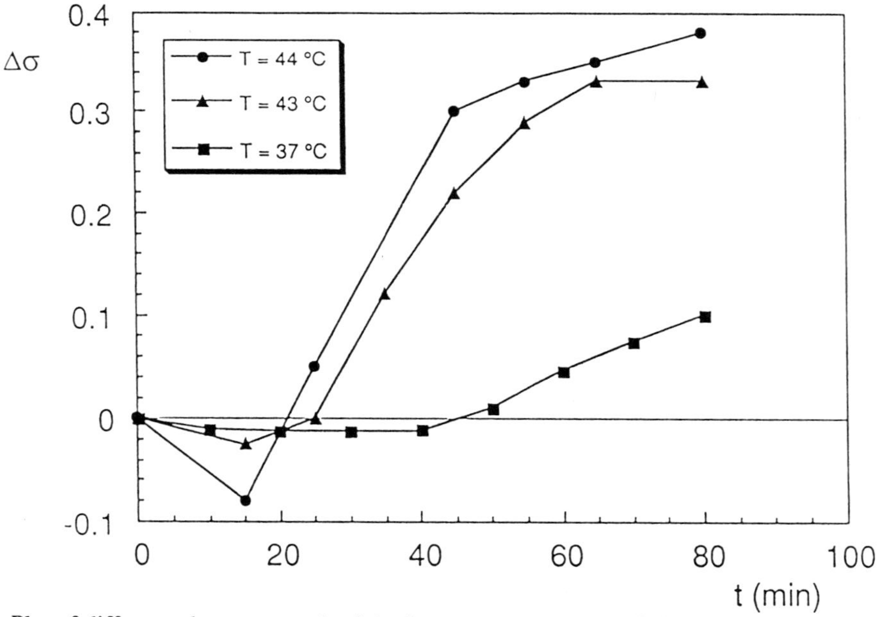

FIG. 3.—Plot of difference between conductivity increment measured at 0 time $\Delta\sigma_0$ and that measured during heating $\Delta\sigma_t$; $\Delta\sigma = \Delta\sigma_0 - \Delta\sigma_t$.

Martin Blank, Ed., *Electricity and Magnetism in Biology and Medicine*

VISCERAL VASODILATION PRECEDES CIRCULATORY SHOCK DURING MILLIMETER-WAVE IRRADIATION

M. R. Frei*

Trinity University, San Antonio, Tex.

R. E. Berger

Operational Technologies Corp., San Antonio, Tex.

J. R. Jauchem

Armstrong Laboratory, Brooks AFB, Tex.

Recent research has shown that circulatory shock [dramatic decrease in mean arterial pressure (MAP)] can be induced by 35GHz radiofrequency irradiation (RFR). It was suspected that shock was preceded by visceral vasodilation, as occurs during environmentally induced hyperthermia. To verify this effect, rats were exposed to 35GHz RFR while subcutaneous (T_s) and colonic (T_c) temperature, arterial blood pressure, respiratory rate, EKG, and visceral blood flow (mesenteric artery) were recorded. Heart rate (HR) and MAP initially increased during irradiation; however, when T_s reached a range of 43-45 C, a marked visceral vasodilation occurred. This reaction was followed by a dramatic decrease in MAP, an increase in respiratory rate, and a continued increase in HR.

In mammals, a primary mechanism of heat loss is through subcutaneous vasodilation; additional blood is routed to the surface so that heat may be dissipated. Under normal circumstances, increased peripheral blood flow is offset by decreased blood flow to the viscera, so that mean arterial pressure (MAP) is maintained. However, severe hyperthermia may result in the medical emergency called heat stroke, a condition that is characterized by a dramatic decrease in MAP (circulatory collapse or shock).

Previous studies have demonstrated relationships among environmentally induced hyperthermia (core temperature >41.5 C), visceral vasodilation, and the onset of circulatory shock.[1,2] Recently, investigations were conducted to determine the patterns of thermal distribution and the physiological effects of whole-body exposure to radiofrequency radiation (RFR) at millimeter wavelengths (MMW).[3] During the course of exposure, the subcutaneous temperature T_s increased significantly more than the colonic temperature T_c (37 to 48-50 vs 37 to 39 C, respectively). Heart rate (HR) increased significantly during irradiation, as during heat stroke. However, in heat stroke, due to environmental heating, these reactions normally occur at $T_c \geq 41.5$ C or higher, whereas in the MMW-exposed animals shock occurred at $T_c < 39$ C.

OBJECTIVES

The objectives of present study were to investigate the relationships among MMW exposure, superficial heating, and the onset of circulatory shock, and to determine whether visceral vasodilation precedes cardiovascular shock during MMW irradiation as occurs during shock induction by environmental heating.

METHODS

Ketamine-anesthetized male Sprague-Dawley rats were exposed in E orientation (long axis parallel to electric field) to MMW (35GHz RFR) at power densities that resulted in whole-body average specific absorption rates of 13 W/kg.

*Correspondence to: M. R. Frei (Biology), Trinity University, 715 Stadium Dr., San Antonio TX 78212.

Carotid arterial blood pressure, EKG, respiratory rate, and temperature change at five sites [left (T_s) and right subcutaneous (sides toward and away from the MMW wave source), colonic (T_c), tympanic, and tail] were continuously monitored and recorded. Mesenteric arterial blood flow was measured by an electromagnetic-flow probe. During experimentation, readings of temperature and physiological changes were obtained at 0.5 C increments of T_c change by use of a PC-compatible system.

RESULTS

During the course of exposure, the T_s (side toward MMW source) increase was significantly greater than the T_c increase. The HR significantly increased during irradiation; MAP increased initially and then precipitously decreased (Fig. 1). At T_s = 43-45 C (T_c < 39 C), a dramatic increase in mesenteric blood flow rate occurred that was accompanied by a marked decrease in MAP (Fig. 2). This sequence of events is consistent with the onset of circulatory shock due to environmental heating.

SUMMARY

Whole-body exposure to MMW resulted in severe peripheral heating while the core temperature remained below 39 C. The MMW exposure was accompanied by elevated HR and severely depressed MAP, which are characteristic of decreased cardiac output due to hypovolemia. These results show that elevated peripheral temperature induced visceral vasodilation, which in turn caused a decrease in functional blood volume that led to the onset of circulatory collapse. As in the case with environmental heating, the mechanism of shock induction due to peripheral heating remains unresolved.

REFERENCES

1. K. C. Kregel and C. V. Gisolfi, "Circulatory responses to vasoconstrictor agents during passive heating in the rat," *J. Appl. Physiol.* 68: 1220-1227, 1990.

2. C. V. Gisolfi, R. D. Matthes, K. C. Kregel, and R. Oppliger, "Splanchnic sympathetic nerve activity and circulating catacholamines in the hyperthermic rat," *J. Appl. Physiol.* 70: 1821-1826, 1991.

3. M. R. Frei, J. R. Jauchem, and R. E. Berger, "Circulatory shock during superficial heating with millimeter waves," *The Physiologist* 34: 246, 1991.

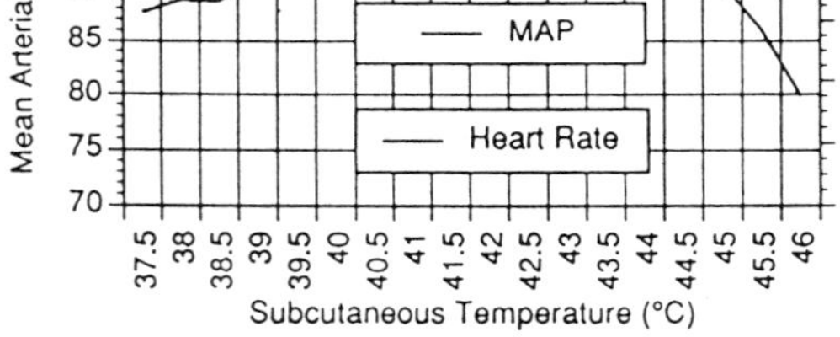

FIG. 1.—Heart rate and blood pressure changes vs subcutaneous temperature change during whole-body millimeter-wave irradiation of rats (n = 13).

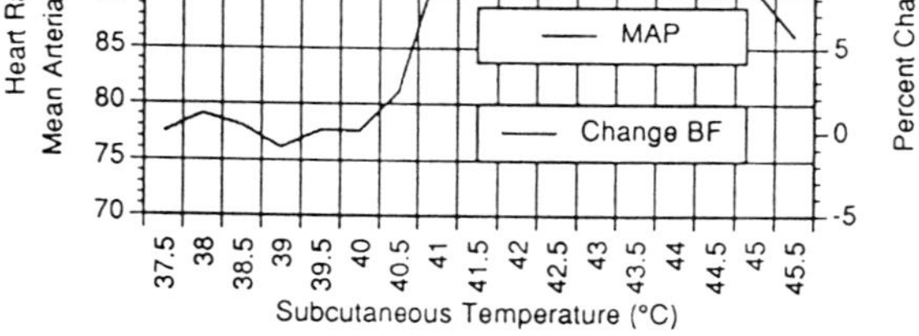

FIG. 2.—Blood pressure and percent change of blood flow vs subcutaneous temperature change during whole-body millimeter-wave irradiation of rats (n = 13).

Martin Blank, Ed., *Electricity and Magnetism in Biology and Medicine*

CALCULATION OF ELECTROMAGNETIC FIELDS INDUCED IN OPERATORS OF PORTABLE RADIO TRANSMITTERS

L. Martens,* D. De Bisschop, J. De Moerloose, and D. De Zutter

University of Gent, Belgium

J. De Poorter and C. De Wagter

University Hospital of Gent, Belgium

The Finite Difference Time Domain (FDTD) method is applied to the calculation of the fields induced in the human head of an operator using a portable radio transmitter. An accurate near-field model for the antenna is presented. Three-dimensional electromagnetic calculations have been performed on a macroscopic electromagnetic model of the human head. The electric field distribution has been determined in various tissues of the human head.

The applications of portable radio transmitters are growing very fast, not only for police communications but also in the private sector. There is a large expansion in the applications of mobile communication systems. The flexibility offered by portable transmitters introduces a lot of applications for these devices. Wireless networks are very attractive and already under development.[1] The operators that use these networks frequently will be nearly continuously irradiated by the near fields of the transmitter antennas. This development initiated our research on the safety aspects of the portable transmitter antennas.

In the past, no risk of heating the operator's head has been found,[3] but other field effects are still under study and controversial. For example, the effects of low-level electromagnetic fields on the permeability of the blood-brain barrier and brain tumor development are studied in Refs. 4 and 5. Although the effects are of a microscopic nature, it is important to know the macroscopic field distribution in the head of the operator and to determine the locations of maximal electric field to which the biological effects may be related.

The macroscopic fields can be determined from a solution of Maxwell's equations in the human head. However, that is a complex situation because of near-field effects and the presence of the highly inhomogeneous lossy head. Kuster et al. calculated the fields by the multiple multipole (MMP) method.[5] They used a $\lambda/2$ (half-wavelength) dipole and a simple homogeneous phantom model for the human head. The calculations of Fujiwara et al. are based on a closed-form formula for a homogeneous sphere irradiated by a half-wavelength dipole antenna.[6]

In this paper, calculations based on the finite difference time domain (FDTD) method are presented. This numerical method solves Maxwell's equations in the time domain. The method can be very conveniently applied to complex inhomogeneous structures such as the human head, and complex sources such as the portable transmitter antenna.

FDTD METHOD AND IMPLEMENTATION OF ANTENNAS

The FDTD method was introduced by Yee.[7] The algorithm is based on the finite difference discretization of Maxwell's equations in space and time domain.[5] Yee used a special unit cell of discretization that simplifies the solution of the discretized equations from initial to steady state in the case of sinusoidal excitation. Space and time sampling increments are chosen in such a way that no aliasing of the continuous field distribution is found and that the stability of the time-marching algorithm is guaranteed.[8]

In order to include wire antennas that are much thinner than the discretization cell unit, the FDTD equations must be changed to include the near-field behavior of the antenna. The near field is assumed to vary as $1/r$ if $r < \lambda/10$. The tangential electrical field on the wire is zero. The antenna also has a gap, across which

---

*Correspondence to: L. Martens (Electromagnetism and Acoustics), University of Gent, Sint-Pietersnieuwstraat 41, B-9000 Gent, Belgium.

the excitation voltage is applied. The gap has a length of one space increment. The improved equations are found by application of the Ampere and Faraday laws along a well-chosen contour. These equations are derived in Ref. 9.

CONFIGURATION

The calculations were done for the ISM frequency 915 MHz, which will give nearly the same results as for the 900MHz portable transmitters that will be used in future personal communication networks.

A realistic and accurate near-field model was developed for the antenna of the portable transmitter. This model consists of an infinitely thin $\lambda/4$ wire antenna placed on a perfectly rectangular box. Between the wire antenna and the box, a gap is available for voltage excitation. Figure 1 shows two examples of antennas modeled: (a) a $\lambda/2$ dipole antenna, and (b) an accurate model of the portable transmitter.

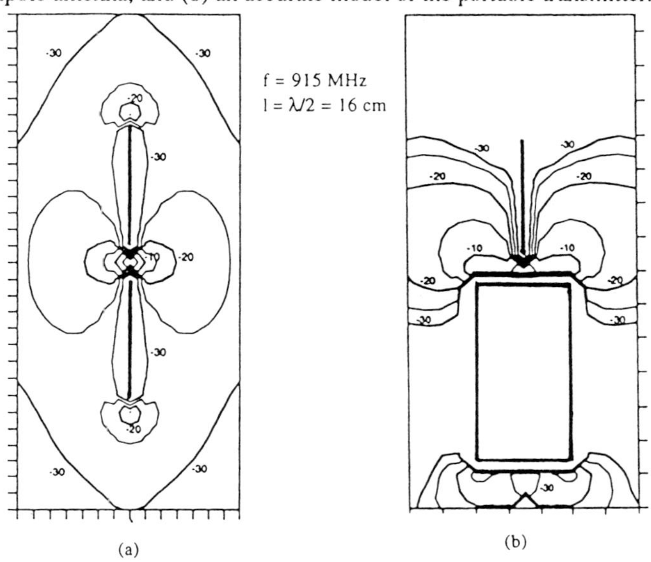

FIG. 1.—(a) Half-wavelength dipole antenna and (b) accurate model for portable transmitter: contour curves (in dB) of longitudinal electric-field component in free space at 915 MHz.

A three-dimensional electromagnetic model for the head was derived from an MR image provided by the MR Department of the University Hospital of Gent (Fig. 2a). White and gray brain matter, muscle, cerebral spinal fluid (CSF), fat, and bone were distinguished from the MR image by use of a semi-automatic computer algorithm. This procedure resulted in a 3-D macroscopic electromagnetic model for the human head (Fig. 2b) that could be directly used as input for the FDTD algorithm.

RESULTS

The longitudinal E_z component for the $\lambda/2$ dipole antenna and the portable transmitter antenna at 915 MHz in free space is plotted in Fig. 1. The near field changes strongly due to the presence of the box of the portable transmitter.

As a next step, electromagnetic fields were calculated for the case of the head irradiated by a $\lambda/2$ dipole antenna. Figure 3 shows the $\sigma|E|^2$-distribution normalized to the maximum in the head. We note a maximal absorption of the field near the gap, which is clear from the fact that the field in free space is also maximal near the gap. This is a preliminary result. We are currently doing the calculations for the head irradiated by the portable transmitter with the accurate model.

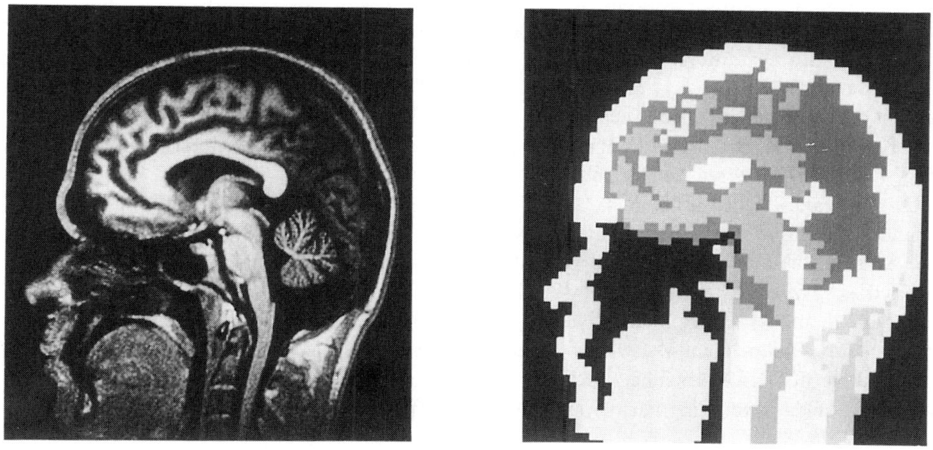

(a) (b)

FIG. 2.—(a) Original MR image of human head and (b) processed image for calculations input.

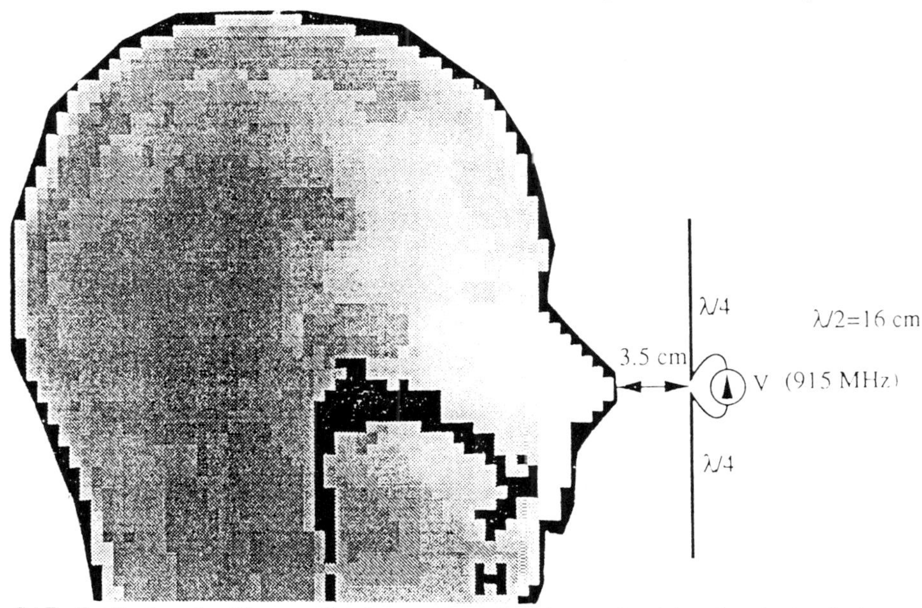

FIG. 3.—SAR distribution (in dB normalized to maximum) in human head irradiated with λ/2 antenna at 915 MHz.

CONCLUSIONS

The FDTD algorithm is very flexible for calculation of fields in a complex structure such as the human head irradiated with a portable radio transmitter antenna. To determine the accuracy of the method, measurements of the fields in a homogeneous human head phantom with a realistic shape are going on. As a next step, the locations of the maximal field need to be correlated with possible biological effects.

REFERENCES

1. T. A. Freeburg, "Enabling techniques for wireless in-building network communications: Four technical challenges, four solutions," *IEEE Communications Mag*. April 1991.

2. Q. Bolzano, O. Garry, and F. E. Steel, "Heating of biological tissue in the induction field of VHF portable radio transmitters," *Trans. IEEE* VT-27: 51-56, 1979.

3. L. G. Salford, A. Brun, J. L. Eberhardt, and B. R. R. Persson, "Permeability of the blood-brain barrier in rats induced by radiofrequency EM-radiation continuous-wave and modulated at 8, 16, 50, and 200 Mhz," *Trans. 1st Congr. European Bioelectromagnetics Assoc.*, Brussels, 1992, 62.

4. L. G. Salford, A. Brun, J. L. Eberhardt, and B. R. R. Persson, "Experimental studies of brain tumour development during exposure to continuous and pulsed 915 MHz radiofrequency radiation," *ibid.*, 65.

5. N. Kuster and R. Ballist, "MMP method simulation of antennae with scattering objects in the closer nearfield," *Trans. IEEE Magnetics* 25: 2881-2883, 1989.

6. O. Fujiwara, H. Higashihama, and T. Azakami, "Calculation of face-SAR due to portable transmitter," *Intern. Wroclaw Symp. Electromagnetic Compatibility*, 1990, 169-172.

7. K. S. Yee, "Numerical solution of initial boundary value problems involving Maxwell's equations in isotropic media," *Trans. IEEE* AP-14: 302-307, 1966.

8. A. Taflove and M. E. Brodwin, "Numerical solution of steady state electromagnetic scattering problems using the time-independent Maxwell's equations," *Trans. IEEE* MTT-23: 623-630, 1975.

9. A. Taflove, K. R. Umashankar, B. Beker, F. Harfoush, and K. S. Yee, "Detailed FDTD analysis of electromagnetic fields penetrating narrow slots and lapped joints in thick conducting screens," *Trans. IEEE* AP-36: 247-257, 1988.

Martin Blank, Ed., *Electricity and Magnetism in Biology and Medicine*

ASSESSMENT OF THE EFFECTS OF EXTREMELY LOW FREQUENCY ELECTROMAGNETIC RADIATION ON GROWTH AND MATURATION IN NESTLING TREE SWALLOWS AND DEERMICE

D. L. Beaver,* R. W. Hill, and P. E. Lederle

Michigan State University, East Lansing

The effects of ELF fields on the growth and maturation of nestlings of free-living tree swallows and deer mice were studied during 1985-1991 at the site of the U.S. Navy's submarine antenna, a 76Hz system, in the Upper Peninsula of Michigan. Both species were studied in test plots at the antenna and control plots 30 km west. Pre-operation data (1985-1989) were compared to operational data (1990-1991). No significant differences were found for four variables of nestling growth; body mass (tree swallows and deer mice), tarsus, ulna and wing length (tree swallows), and two variables of nestling maturation (eyes open, both species; feather eruption, tree swallows and incisor eruption, deer mice) of both animals.

Since 1985 the growth and maturation of nestling tree swallows (*Tachycineta bicolor*) and deermice (*peromyscus maniculatis gracilis*) have been studied at the site of the U.S. Navy's submarine communication antenna in the Upper Peninsula of Michigan.[1] The antenna was constructed in 1986 and tested at low power and for short durations until the fall of 1989. For our growth studies, the antenna became operational in the 1990 research season. The antenna produces a frequency-modulated output centered on 76 Hz. Power input is 150 A. Field strengths on the test plot adjacent to the antenna averaged 4.59 mG, 39.43 mV/m, and 2.00 V/m for the magnetic, longitudinal, and transverse fields, respectively. At the control sites 30 km west, these field strengths were 0.004 mG, 0.074 mV/m, and <0.001 V/m, or roughly 2 to 5 orders of magnitude smaller. The test and control sites were selected for equivalent and low 60Hz background fields prior to construction of the antenna.[2]

The species studied are abundant in the area of the antenna. Tree swallows winter in South America and breed in the upper latitudes of North America. They occur in open spaces in the forest, where they feed on aerial insects. They readily accept nest boxes for breeding, which makes studies of large number so of young possible. The young are altricial and require extended care in the form of regular feeding and warming by both parents after hatching. The young fledge about 18 days after hatching.

Deermice are abundant, year-around residents, in upland forests. They utilize the litter layer as well as trees and shrubs for feeding. They are very adept climbers. They have an omnivorous diet that includes seeds and insects. Females may give birth to several litters in spring and summer. The female rears the nestlings without the male's help. Nesting occurs in nest boxes, although with some reluctance.

METHODS

Tree Swallows

Studies were conducted on test plots located from 0 to 200 m from the antenna. Natural openings along the route of the antenna were used, as was the right-of-way itself. Nest boxes made of cedar lumber were placed on wooden posts 1.5 m high at intervals of 30 m from other boxes. Of 165 nest boxes placed on test plots, 36 were designated for use in growth studies. Control plots 20-30 km to the west were matched to test plots in vegetation type. Sham antenna corridors were cut on control plots to account for the effect of the antenna right-of-way. A total of 239 nest boxes were placed on control plots. Of these, 75 were designated for growth studies.

Growth variables measured were body mass (g), and length of tarsus, ulna and wing (mm) from hatching to day 15, which is near fledging age. Age at eye opening and feather eruption were recorded as major maturational events. Maximum attained size was the variable used for body mass, tarsus, and ulna.[3] The

---

*Correspondence to: D. L. Beaver (Zoology), Michigan State University, East Lansing MI 48824.

wing is still growing at fledging. An exponential model was used to estimate the growth constant for wing and this model was used as the test variable.[3,4]

Deermice

Test plots were established in upland forests along the route of the antenna. The right-of-way cut necessitated the plots at least 75 m from the antenna, which left a buffer zone of habitat. The maximum distance of the outer edge of the plot from the antenna was 200 m. Control plots were established in similar habitat 20-30 km west. Sham right-of-way cuts were made to control for the antenna right-of-way effect on test plots and a similarly sized habitat buffer was established.

Growth of nestling deermice was studied in large outdoor enclosures made of plexiglass supported by a wood frame. These enclosures measured 5.8×6.1 m and were placed in the forest next to sites where mice were captured. The enclosures were necessary to prevent females from moving their litters in response to disturbance caused by investigators taking measurements of growth. Body mass (g) was measured from day 4 to 22 after birth, which is about the time of weaning.[5] Growth data were fitted to a linear model. The slope was used as the test variable. Age at incisor eruption and eye opening were recorded as major maturational events.

RESULTS

Means of tree swallow growth variables (tarsus, ulna, body mass, and wing growth constant) and maturational events (eye opening and appearance of feathers) were examined in a nested ANOVA design. The main effects due to antenna operation (pre-operation 1985-1989, operation 1990-1991) and to test-vs-control plot location were not significantly different ($F_{operation}$ range 0.0001 to 1.94, $p > 0.24$; F_{plot} range 0.0003 to 4.446, $p > 0.10$). In the tests of growth and maturational variables, less than 9% of the variance in the model was accounted for by operation and less than 1% by plot main effects, whereas up to 23% of the variance was due to differences among nests within plots and for different years within operation periods.

Means for deer mouse growth (body mass) and maturational variables (eye open and incisor eruption) were not significantly different in a nested ANOVA design ($F_{operation}$ range 0.59 to 1.64, $p > 0.27$; F_{plot} range 0.008 to 2.34, $p > 0.10$). However, plot and operation showed a significant interaction for growth and maturational measures ($F = 8.3, p < 0.004$ for body mass; $F = 55.69, p < 0.0001$ for eye opening; and $F = 7.17, p < 0.007$ for incisor eruption). Lower growth rates on control compared to test plots for 1990 were followed by higher growth rates in 1991 with the control plot slightly higher than the test plot. The interaction effect for mean age at eye opening appears to be due to a decrease on test plots and an increase on control plots. The interaction effect for mean age at eruption of incisors appears due to a gradually decreasing age to eruption on control plots, and it increases on test plots. At this time we do not attribute these interactions to antenna operation, but we do not know what may be causing the interaction.

For growth of body mass and maturational events in the deer mouse, the main effects of operation and plot account for 2% to 5% of the variance. Interaction of operation with years and plot with years account for up to 10% of the variance. Years and nests by themselves account for 10% to 30% of the variance in the model, respectively.[6]

REFERENCES

1. D. L. Beaver, R. W. Hill, and S. D. Hill, *U.S. Navy ELF Communications System Ecological Monitoring Program: Small Mammals and Nesting Birds*, Ann. Rep. to IIT Research Institute and the U.S. Navy, 1991.

2. D. P. Haradem and J. R. Gauger, *ELF Communications System Ecological Monitoring Program: Electromagnetic Fields Measurements and Engineering Support—1990*, IIT Research Institute, Chicago, Ill., 1991.

3. R. Zach, "Growth curve analysis: A critical reevaluation," *Auk* 105: 208-209, 1988.

4. R. J. O'Connor, *The Growth and Development of Birds*, New York: Wiley, 19??.

5. R. W. Hill, "The amount of maternal care in *Peromyscus leucopus* and its thermal significance for the young," *J. Mammology* 53: 774-790, 1972.

6. This research was supported by the Department of Navy Space and Naval Warfare Systems Command, contracts N00039-81-C-0357, N00039-84-C-0070, and N00039-88-C-0065.

Martin Blank, Ed., *Electricity and Magnetism in Biology and Medicine*

MEASUREMENTS OF RF CURRENT AND LOCALIZED SAR NEAR A SHIPBOARD RF HEAT SEALER

R. G. Olsen,* T. A. Griner, and B. J. Van Matre

Naval Aerospace Medical Research Laboratory, Pensacola, Fla.

We used a human model to assess exposure of shipboard radiofrequency (RF) heat sealer operators. We measured specific absorption rate (SAR) and body currents with the model located at the operating controls. Specific absorption rates in both ankles and one wrist were measured. The highest SARs and currents occurred with the model's "hand" extended, contacting the insulated tuning knob. There is evidence that permissible exposure limits (PELs) might be overly protective and operators could remain inside PEL lines under appropriate conditions; however, shipboard heat sealers with pervasive metal surroundings pose particular problems, and operators should remain clear of the device during operation.

Industrial RF heat sealers are known to produce RF current in the operator.[1] The problem is intensified aboard ships because the decks, bulkheads, and overheads are metal and act as RF energy reflectors. The American National Standards Institute (ANSI) currently describes PELs in terms of a criterion SAR and theoretically derived limits based on far-field, uniform irradiation. Localized SAR rates as high as 182 W/kg have been calculated in various cross sections of the leg in anatomically based computer models irradiated with PEL conditions.[2] The revised standard poses a 100 mA per leg RF current-to-ground limit for frequencies above 1.0 MHz. More knowledge about RF body current is needed to relate this new parameter to existing information. The objective of this study was to determine the RF exposure parameters near heat sealers and assess methods of reducing exposure and preventing shocks and burns to operating personnel.

METHODS AND PROCEDURES

Figure 1 shows the RF heat sealer as installed in the ship. The PEL lines shown were established by previous RADHAZ surveys for half-power (4kW) operation. The metal sliding boom projected from the overhead to about 30 cm above the deck. The 70kg model[3] was maintained upright in the operator position. The left forearm and lower legs were constricted to simulate a wrist and ankles. Average (near-field) power densities were calculated from E-field probe measurements (Narda 8616/8662) taken at levels of lower, middle, and upper body heights.

Measurements of RF body current were made with prototype RF current meters procured from the University of Utah. One device (Model GC-2) was used to measure RF body-to-ground current, and another GC-2, modified to be handheld, was used to measure hand-contact current. Specific absorption rates in both ankles and one wrist were calculated from RF-induced temperature rises that were measured with nonperturbing temperature probes.[4] Localized SARs were obtained with the model "standing" (arms down) at the controls; they were repeated with the left arm extended and touching the nonmetallic "Plate Tuning" knob. Only ankle SARs were obtained in the first configuration; in the second; SARs of the left "wrist" were also acquired. The temperature probes were inserted into the midregion of the constructed portions (ankles, wrist) of the extremities. Current-to-ground and hand-contact current were also measured in the model for these configurations.

RESULTS

Results are shown in Table 1. The normal operator position showed relatively high power densities at the operating frequency of 27.12 MHz. We attribute the high SARs to high RF power and the cavity-like nature of the metal-lined room (approximately 4.5×6×2.5 m). The touching configuration resulted in significantly

*Correspondence to: R. G. Olsen, Naval Aerospace Med. Res. Lab., 51 Hovey Rd., NAS, Pensacola FL 32508-1046.

higher ankle SARs. The disparity in left/right ankle SAR was probably due to the metal bulkhead near the model's left shoulder. Localized SAR and RF current-to-ground were further enhanced in the "left-hand touch" arrangement because of the addition of RF contact current.

TABLE 1.—Localized SAR in full-sized human model at heat sealer operator position (average power density, 122 mW/cm^2).

Condition	Model site	Measured local SAR (W/kg ± SD) (N = 3)	RF current to ground (mA ± SD) (N = 2) (both ankles)	RF contact current (mA ± SD) (N = 3)
Arms down	Left ankle	126.5 ± 11.5	1402 ± 4.2	----
	Right ankle	32.6 ± 2.6		----
Left-hand touch	Left ankle	223.9 ± 25.5	1938 ± 54.4	----
	Right ankle	64.9 ± 5.5		----
	Left wrist	86.7 ± 17.9		1087 ± 54.4

Cursory RF current measurements showed that a 10cm thickness of foam insulation under the feet provided a 75% reduction in current to ground at the operator position. Insulated pads might allow operators to remain near the sealer controls without exceeding the ANSI standard.

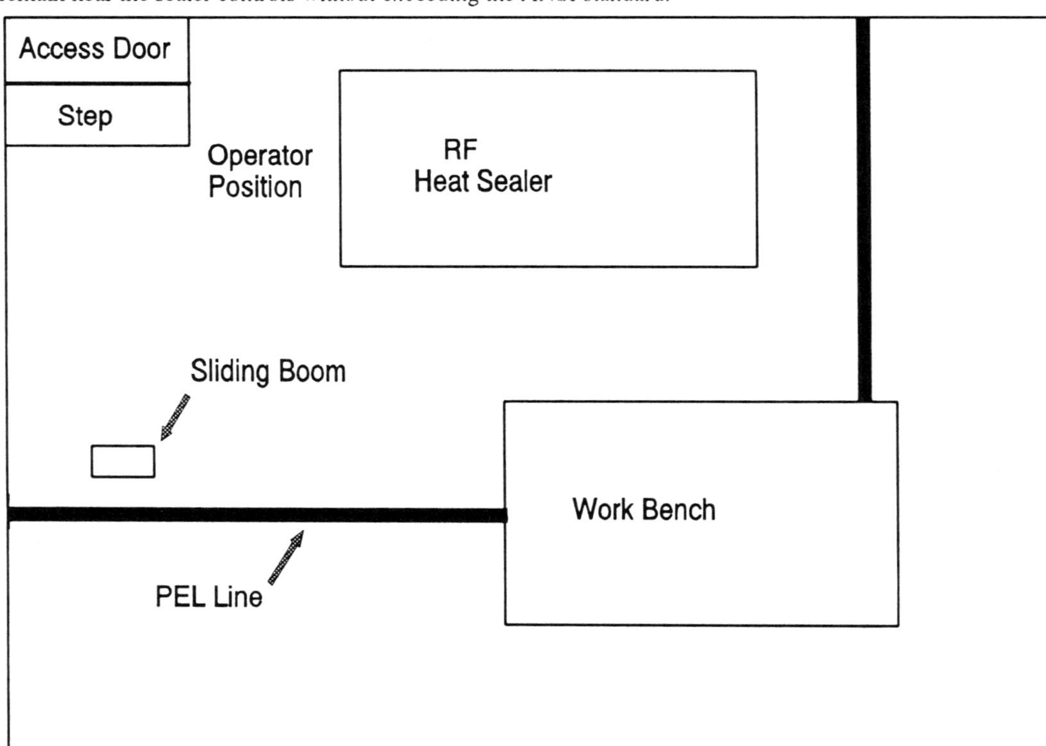

FIG. 1.—Layout of shipboard RF heat-sealer room.

DISCUSSION AND CONCLUSION

Results clearly show potential for high localized SARs in personnel near shipboard RF heat sealers. Localized SARs measured in this study are the highest we have ever recorded in a full-sized human model. Extreme precautions must be taken to prevent injury and overexposure to thermalizing levels of RF radiation. Although currents in a shoe-wearing individual would be lower than in the model, the normal operating location of shipboard heat sealers must be considered a restricted zone. We have also shown that RF current to ground can be significantly reduced with the addition of insulating materials on the deck. This method of RF current reduction should be considered when it is necessary to reduce SAR in workers who must remain close to RF heat sealers.[5]

REFERENCES

1. O. P. Gandhi, J. Y. Chen, and A. Riazi, "Currents induced in a human being for plane-wave exposure conditions 0-50 MHz and for RF sealers," *IEEE Trans.* BME-33: 757, 1986.

2. J. Y. Chen and O. P. Gandhi, "RF currents induced in an anatomically based model of a human for plane-wave exposures (20-100 MHz)," *Health Physics* 57: 89, 1989.

3. R. G. Olsen, and T. A. Griner, "Outdoor measurement of SAR in a full-sized human model exposed to 29.9 MHz in the near field," *Bioelectromagnetics* 10: 161, 1989.

4. R. G. Olsen and R. R. Bowman, "Simple nonperturbing temperature probe for microwave/radio frequency dosimetry," *Bioelectromagnetics* 10: 209, 1989.

5. This research was sponsored by the Naval Environmental Health Center under work unit N68546 WR00028. The views expressed in this article are those of the authors and do not reflect the official policy or position of the U.S. Department of the Navy, U.S. Department of Defense, or the U.S. Government.

Martin Blank, Ed., *Electricity and Magnetism in Biology and Medicine*

EFFECT OF PULSED ELECTROMAGNETIC FIELDS ON FUNCTIONAL RECOVERY AFTER RAT SCIATIC NERVE CRUSH AS ASSESSED BY VIDEO FOOTPRINT ANALYSIS: PRELIMINARY RESULTS

J. L. Walker,* S. Guarnieri, P. Meade, and B. F. Sisken

University of Kentucky, Lexington

Pulsed electromagnetic fields (PEMF) have been shown to stimulate nerve regeneration. To determine whether PEMF also enhance functional recovery, we used a crush lesion in the rat sciatic nerve model and assessed functional recovery by return of normal 1-5 toe spread on footprints analyzed by video recording. Four sham and four experimental animals were restrained 4 h/days 1-5 post injury, during which the experimental animals received total body PEMF exposure (0.3 mT, 2 Hz). No effect on functional recovery of 1-5 toe spread could be demonstrated by exposure to total body PEMF.

Pulsed electromagnetic fields (PEMF) have been shown to accelerate neurite formation in culture.[1] Zienowicz et al.[2] demonstrated improved functional recovery after nerve transection and repair with exposure to PEMF using footprint analysis in the rat sciatic nerve model. Sisken et al.[3] demonstrated accelerated nerve regeneration in the first 6 days with PEMF exposure to a crush lesion in the rat sciatic nerve using pinch test assessment. The purpose of this study was to determine the effect of PEMF on functional recovery of a crush lesion in the rat sciatic nerve model.

Footprint analysis to quantitate functional recovery in the rat sciatic nerve model was first described by DeMedinacelli et al.[4] and subsequently modified by Carlton and Goldberg[5] and Bain et al.[6] Each of these investigators proposed measuring different variables, but all agreed that the 1-5 toe spread was a valid measurement of functional recovery following sciatic nerve injury in the rat model. This variable is affected by tibial, peroneal, and combined sciatic nerve function[6] and appears to depend primarily on motor axonal contributions.[7] Because of problems with smeared prints and incomplete toe-print separation, we chose to use video recording of foot position during gait to measure this variable.

METHOD

Crush Lesion

Eight adult Sprague-Dawley rats (200-300 g) were anesthetized by use of 0.1 cc/100 g body weight pentobarbital by intraperitoneal injection. The right sciatic nerve was exposed between the *biceps femoris* and *vastus lateralis*. At the level of the *piriformis* tendon, just distal to branches to the hamstrings, a crush lesion was created with pliers modified by a small wire to prevent cutting when applied for 20 s. Two additional animals for control received no surgery and were allowed normal cage activity.

PEMF

After wound closure, operated animals were placed in plastic restrainers 4 h/day on days 1-5 post injury. During restraint, 4 experimental animals were placed between paired Helmholtz coils and exposed to total body PEMF [3 G (0.3 mT), repetition rate 2 Hz]. Four sham animals were placed between coils, but no current was applied.

Footprint Analysis

On days 5, 8, 10, 12, 15, 17, 19, 22, 24, 26, 29, 33, 36, and 47 post injury, all animals walked across a frosted plexiglass surface and the plantar aspect of the feet was recorded by videotape. Control animals were taped on the same days. The tapes were reviewed and images selected during active walking rather than standing or rearing. The 1-5 toe spread of injured and uninjured feet was measured with a Kontron Image

---

*Correspondence to: J. L. Walker (Orthopaedic Surgery), Shriners Hospital, 1900 Richmond Rd., Lexington KY 40502.

Processing System IPS X and SEM IPS software. A ratio of injured:uninjured 1-5 toe spread was calculated and differences between control, sham, and experimental animals were evaluated by a repeated measures ANOVA.

RESULTS

The results are presented in Fig. 1. A significant difference ($p < 0.001$) was found between the control animals and those receiving a crush injury. However, no statistical difference ($p > 0.05$) in functional recovery following crush injury could be demonstrated between animals exposed to 5 days PEMF (experimental) and those without PEMF exposure (sham).

DISCUSSION

Although the rate of nerve regeneration has been previously shown to increase by 22% in the first 6 days, we were unable to measure any difference in functional recovery following crush injury lesion treated with 5 days of PEMF exposure. Possible explanations for this discrepancy are as follows.

1. Our method of assessing functional recovery (1-5 toe spread) or large animal variations does not permit measurement of small differences.
2. PEMF, despite increasing regeneration rate, do not affect functional recovery in crush lesions.
3. Maintenance of the acceleration requires continued applications of PEMF throughout the period of functional recovery.
4. The early acceleration of nerve regeneration is not maintained through the period of functional recovery.[8]

REFERENCES

1. B. F. Sisken, M. Kanje, G. Lundborg, and W. Kurtz, "Pulsed electromagnetic fields stimulate nerve regeneration in vitro and in vivo," *Restor. Neurol. Neurosci.* 1: 303, 1989.
2. R. J. Zienowicz, B. A. Thomas, W. H. Kurtz, and M. G. Orgel, "A multivariate approach to the treatment of peripheral nerve transection injury: The role of electromagnetic field therapy," *Plast. Reconstr. Surg.* 87: 122, 1991.
3. B. F. Sisken, M. Kanje, G. Lundborg, E. Herbst, and W. Kurtz, "Stimulation of rat sciatic nerve regeneration with pulsed electromagnetic fields," *Brain Res.* 485: 309, 1989.
4. L. DeMedinacelli, W. Freed, and R. Wyatt, "An index of the functional condition of the rat sciatic nerve based on measurements made from walking tracks," *Exp. Neurol.* 77: 634, 1982.
5. J. M. Carlton and N. Goldberg, "Quantitating integrated muscle function following reinnervation," *Surg. Forum* 37: 661, 1986.
6. J. R. Bain, S. E. MacKinnon, and D. A. Hunter, "Functional evaluation of complete sciatic, peroneal, and posterior tibial nerve lesions in the rat," *Plast. Reconstr. Surg.* 83: 129, 1989.
7. H. Wietholter, S. Eckert, and A. Stevens, "Measurement of atactic and paretic gait in neuropathies of rats based on analysis of walking tracks," *J. Neurosci. Methods* 32: 199, 1990.

8. This work was supported in part by the Orthopaedics Research and Education Foundation and by NIH grant NS 29621-01.

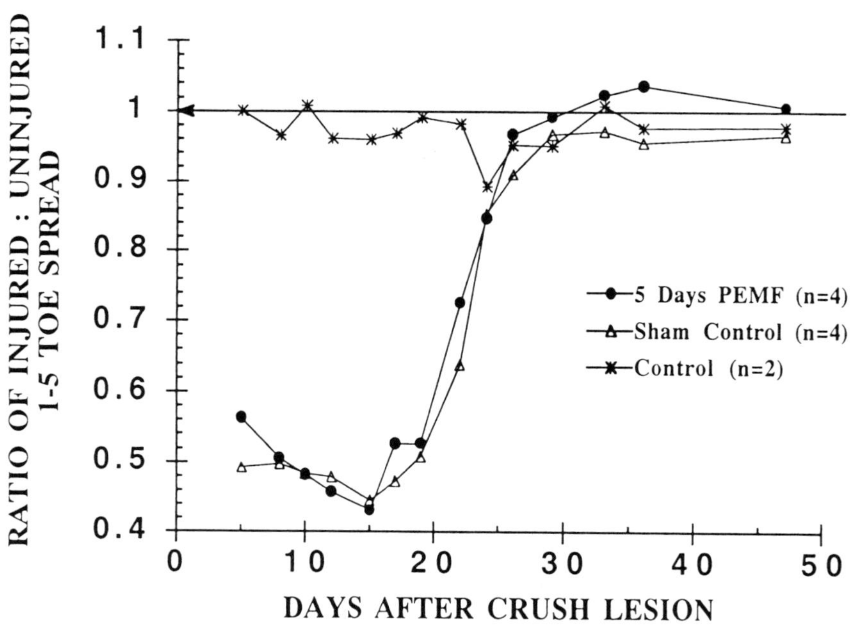

FIG. 1.—Effect of PEMF on 1-5 toe spread following sciatic crush lesion in the rat model.

AUTHOR INDEX